Introductory Algebra

Eighth Edition

Margaret L. Lial
American River College

John Hornsby
University of New Orleans

Terry McGinnis

Boston • San Francisco • New York
London • Toronto • Sydney • Tokyo • Singapore • Madrid
Mexico City • Munich • Paris • Cape Town • Hong Kong • Montreal

Publisher	Greg Tobin
Editor in Chief	Maureen O'Connor
Project Editor	Lauren Morse
Editorial Assistant	Marcia Emerson
Managing Editor/Production Supervisor	Ron Hampton
Text and Cover Design	Dennis Schaefer
Supplements Production	Jason Miranda
Production Services	Elm Street Publishing Services, Inc.
Media Producers	Sharon Smith and Sara Anderson
Marketing Manager	Jay Jenkins
Marketing Coordinator	Tracy Rabinowitz
Prepress Services Buyer	Caroline Fell
Technical Art Supervisor	Joseph K. Vetere
First Print Buyer	Hugh Crawford
Composition Services	Pre-Press Company, Inc.
Cover Photo	© Daryl Benson/Masterfile
Cover Image	Walkway through field: Kouchibouguac National Park, New Brunswick, Canada

Photo Credits All photos from PhotoDisc, except the following: Beth Anderson, pp. 23, 142 right, 157, 164, 165 left, 324, 464, 539, 543 left, 664; Mitchell Layton/Corbis, pp. 50, 543 right; John Hornsby, pp. 64 top, 570; Comstock, pp. 92, 602; Aris Messinis/AFP/Getty Images, p. 107, 144 left; Reuters/Corbis, pp. 134, 169 right, 528, 536 right; Lou Dematteis/ Corbis, p. 140; Albert Ferreira/Reuters/Corbis, p. 142 left; NASA, pp. 144 right, 389 top and bottom, 592, 652; Digital Vision, pp. 160, 208, 221, 403, 455 top, 468; SuperStock, p. 169 bottom left; Maureen O'Connor, p. 191; AP/Wide-world, p. 273, 308; PunchStock/BrandX Pictures, p. 298; Corbis RF, p. 301, 309, 310, 627; 20th Century Fox/Paramount/The Kobal Collection, p. 302; Steve Starr/Corbis, p. 328; James Leynse/Corbis, p. 331; U.S. Department of Commerce, p. 455 bottom; Tom Wagner/Corbis Saga, p. 466; The Kobal Collection, pp. 471, 554; David Bergman/Corbis, p. 536 left; Bettmann/Corbis, p. 541; Ben Wood/Corbis, p. 559; Nik Wheeler/Corbis, p. 598 top and bottom left; Associated Press/Goldsboro News—Argus, p. 624; Roger Ressmeyer/Corbis, pp. 629, 659; Guy Motil/Corbis, p. 680.

Library of Congress Cataloging-in-Publication Data

Lial, Margaret L.

Introductory algebra.—8th ed. / Margaret L. Lial, John Hornsby, Terry McGinnis.

p. cm.

Includes index.

ISBN 0-321-27921-2 (Student Edition)

ISBN 0-321-29224-3 (Hardback)

1. Algebra. I. Hornsby, E. John. II. McGinnis, Terry. III. Title.

QA152.3.L557 2005

512.9—dc22 2004052579

1 2 3 4 5 6 7 8 9 10 VHP 08 07 06 05

Contents

Index of Applications

Astronomy/Aerospace

Automotive

Biology

Business

Chemistry

Construction

Consumer

Economics

Education

Environment

Sports/Entertainment

Statistics/Demographics

Transportation

Videotape and CD Index

SECTION	EXERCISES
R.1	15, 19, 41, 45, 57
R.2	17, 25, 31, 41, 45
1.1	7, 13, 19, 21, 27, 39, 43, 47, 53, 65, 67, 77
1.2	11, 19, 25, 29, 35, 43, 51, 53, 63, 67, 69
1.3	17, 19, 25, 31, 35, 37, 39, 43
1.4	5, 7, 9, 33, 57, 59
1.5	7, 13, 15, 23, 33, 35, 39
1.6	7, 9, 13, 15, 37, 51, 73
1.7	39, 49, 59, 65
1.8	15, 19, 23, 27, 29, 31, 41, 49, 51, 55
2.1	45
2.2	49, 51, 53
2.3	5, 9, 11, 13, 19, 27, 43, 49
2.4	9, 15, 41, 49, 53
2.5	15, 19, 41, 59
2.6	5, 9, 15, 27, 29, 43, 47
2.7	23, 33, 55
3.1	19, 23, 39, 71
3.2	3, 27, 29, 33
3.3	1, 23, 25, 27, 47, 51
3.4	37
4.1	5, 7, 19, 21, 23, 25
4.2	5, 9, 11, 17, 25
4.3	5, 13, 17, 33
4.4	13, 19, 21, 35
4.5	9, 15, 17
5.1	23, 25, 27, 29, 37(a), 69, 75
5.2	7, 13, 27, 33, 35, 43, 45, 47, 49, 73
5.3	13, 19, 21, 29, 33, 37
5.4	21, 29
5.5	59, 63
5.6	7, 9, 13, 21, 23, 25
5.7	5, 17, 19, 23, 31
5.8	15, 17, 19, 23, 29, 31
6.1	3, 7, 9, 13, 15, 29, 43, 49
6.2	13, 15, 19, 51
6.3	31, 33
6.4	7, 15, 23, 43
6.5	13, 25, 35, 49
6.6	43, 45, 51
6.7	23, 27
7.1	9, 17, 31, 41, 43, 49
7.2	5, 7, 11, 13, 15, 31, 33
7.3	11, 19, 23, 27, 39
7.4	13, 19, 31, 49, 55
7.5	15, 23
7.6	35, 39, 41
7.7	5, 15, 17, 21, 27
7.8	3, 7, 13, 17
8.1	77, 93
8.2	5, 9, 15, 23, 27, 61, 77
8.3	5, 7, 11, 15, 29, 31, 33
8.4	7, 11, 15, 17, 25, 27, 43
8.5	5, 9, 11, 17, 21, 51, 55
8.6	5, 7, 11, 19, 21, 33, 39
9.1	31, 33, 35
9.2	13, 15, 29
9.3	11, 19, 39
9.4	11, 13
9.5	9, 13, 15, 25, 27, 33

Index of Focus on Real-Data Applications

Preface

The eighth edition of *Introductory Algebra* continues our ongoing commitment to provide the best possible text and supplements package to help instructors teach and students succeed. To that end, we have tried to address the diverse needs of today's students through an attractive design, updated figures and graphs, helpful features, careful explanations of topics, and a comprehensive package of supplements and study aids. We have taken special care to respond to the suggestions of users and reviewers and have added new examples and exercises based on their feedback. Students who have never studied algebra—as well as those who require further review of basic algebraic concepts before taking additional courses in mathematics, business, science, nursing, or other fields—will benefit from the text's student-oriented approach.

This text is part of a series that also includes the following books:

- *Essential Mathematics*, Second Edition, by Lial and Salzman
- *Basic College Mathematics*, Seventh Edition, by Lial, Salzman, and Hestwood
- *Prealgebra*, Third Edition, by Lial and Hestwood
- *Intermediate Algebra*, Eighth Edition, by Lial, Hornsby, and McGinnis
- *Introductory and Intermediate Algebra*, Third Edition, by Lial, Hornsby, and McGinnis.

HALLMARK FEATURES

We believe students and instructors will welcome the following helpful features.

Chapter Openers New and updated chapter openers feature real-world applications of mathematics that are relevant to students and tied to specific material within the chapters. Examples of topics include the Olympics, credit card debt, and movie revenues. (See pp. 107, 193, and 273—Chapters 2, 3, and 4.)

Real-Life Applications We are always on the lookout for interesting data to use in real-life applications. As a result, we have included new or updated examples and exercises from fields such as business, pop culture, sports, the life sciences, and technology that show the relevance of algebra to daily life. (See pp. 142, 208, and 308.) A comprehensive Index of Applications appears at the beginning of the text. (See pp. v–vii.)

Figures and Photos Today's students are more visually oriented than ever. Thus, we have made a concerted effort to include mathematical figures, diagrams, tables, and graphs whenever possible. (See pp. 196, 246, and 316.) Many of the graphs use a style similar to that seen by students in today's print and electronic media. Photos have been incorporated to enhance applications in examples and exercises. (See pp. 134, 328, and 455.)

Emphasis on Problem Solving Introduced in Chapter 2, our six-step problem-solving method is integrated throughout the text. The six steps, *Read, Assign a Variable, Write an Equation, Solve, State the Answer*, and *Check*, are emphasized in boldface type and repeated in examples and exercises to reinforce the problem-solving process for students. (See pp. 133, 301, and 445.) New **PROBLEM-SOLVING HINT** boxes provide students with helpful problem-solving tips and strategies. (See pp. 133, 135, and 529.)

Also new to this edition of the text is Appendix A Strategies for Problem Solving. (See pp. 687–700.) This appendix provides examples of additional problem-solving techniques, such as working backward, using trial and error, and looking for patterns. A wide variety of applications are included.

- ***Learning Objectives*** Each section begins with clearly stated, numbered objectives, and the included material is directly keyed to these objectives so that students know exactly what is covered in each section. (See pp. 31, 133, and 341.)

- ***Cautions and Notes*** One of the most popular features of previous editions, **CAUTION** and **NOTE** boxes warn students about common errors and emphasize important ideas throughout the exposition. (See pp. 235, 367, and 498.) The text design makes them easy to spot: Cautions are highlighted in bright yellow and Notes are highlighted in purple.

- ***Calculator Tips*** These optional tips, marked with calculator icons, offer basic information and instruction for students using calculators in the course. (See pp. 26, 160, and 387.) An Introduction to Calculators is included at the beginning of the text. (See pp. xxi-xxv.)

- ***Margin Problems*** Margin problems, with answers immediately available at the bottom of the page, are found in every section of the text. (See pp. 69, 303, and 481.) This popular feature allows students to immediately practice the material covered in the examples in preparation for the exercise sets.

- ***Ample and Varied Exercise Sets*** The text contains a wealth of exercises to provide students with opportunities to practice, apply, connect, and extend the algebraic skills they are learning. Numerous illustrations, tables, graphs, and photos have been added to the exercise sets to help students visualize the problems they are solving. Problem types include writing, estimation, and calculator exercises as well as applications and multiple-choice, matching, true/false, and fill-in-the-blank problems. In the *Annotated Instructor's Edition* of the text, writing exercises are marked with icons so that instructors may assign these problems at their discretion. Exercises suitable for calculator work are marked in both the student and instructor editions with calculator icons . (See pp. 377, 455, and 479.)

- ***Relating Concepts Exercises*** These sets of exercises help students tie together topics and develop problem-solving skills as they compare and contrast ideas, identify and describe patterns, and extend concepts to new situations. (See pp. 266, 298, and 362.) These exercises make great collaborative activities for pairs or small groups of students.

- ***Summary Exercises*** Based on user feedback, every chapter now includes at least one set of in-chapter summary exercises. These special exercise sets provide students with the all-important *mixed* review problems they need to master topics. Summaries of solution methods or additional examples are often included. (See pp. 247, 299, and 525.)

- ***Study Skills Component*** A desk-light icon at key points in the text directs students to a separate *Study Skills Workbook* containing activities correlated directly to the text. (See pp. 24, 189, and 423.) This unique workbook explains *how* the brain actually learns, so students understand *why* the study tips presented will help them succeed in the course. Students are introduced to the workbook in the To the Student section at the beginning of the text.

- ***Focus on Real-Data Applications*** These one-page activities present a relevant and in-depth look at how mathematics is used in the real world. Designed to help instructors answer the often-asked question, "When will I ever use this stuff?," these activities ask students to read and interpret data from newspaper articles, the Internet, and other familiar, real sources. (See pp. 140, 278, 410, and 486.) The activities are well-suited to collaborative work and can also be completed by individuals or used for open-ended class discussions. A comprehensive Index of Focus on Real-Data Applications appears at the beginning of the text. (See pp. ix and x.) Instructor teaching notes and extensions for the activities are provided in the *Printed Test Bank and Instructor's Resource Guide.*

- ***Test Your Word Power*** To help students understand and master mathematical vocabulary, this feature can be found in each chapter summary. Key terms from the chapter are presented along with four possible definitions in a multiple-choice format. Answers and examples illustrating each term are provided. (See pp. 258, 319, and 391.)

- ***Ample Opportunity for Review*** Each chapter concludes with a Chapter Summary that features Key Terms with definitions and helpful graphics, New Symbols, Test Your Word Power, and a Quick Review of each section's content with additional examples. A comprehensive set of Chapter Review Exercises, keyed to individual sections, is included, as are Mixed Review Exercises and a Chapter Test. Beginning with Chapter 2, each chapter concludes with a set of Cumulative Review Exercises that cover material going back to Chapters R and 1. (See pp. 257, 391, and 457.)

- ***Diagnostic Pretest*** A diagnostic pretest is included on p. xxix and covers material from the entire book, much like a sample final exam. This pretest can be used to facilitate student placement in the correct chapter according to skill level.

WHAT IS NEW IN THIS EDITION?

You will find many places in the text where we have polished individual presentations and added or updated examples, and exercises, and applications based on reviewer feedback. Specific content changes you may notice include the following:

- Section 2.1 on solving linear equations in one variable includes twice as many examples as in the previous edition. Linear equations with no solution or infinitely many solutions, formerly covered in this section, have been moved to Section 2.3.
- Section 3.4 includes new exposition, an example, and exercises on graphing linear equations using slope and y-intercept.
- Systems of linear equations and inequalities are presented in Chapter 4, earlier than in the previous edition.
- All new sets of summary exercises appear in Chapters 1–5 and 8.
- Appendix A Strategies for Problem Solving and Appendix C Mean, Median, and Mode are new to this edition.

WHAT SUPPLEMENTS ARE AVAILABLE?

For a comprehensive list of the supplements and study aids that accompany *Introductory Algebra*, Eighth Edition, see pages xv and xvi.

ACKNOWLEDGMENTS

Previous editions of this text were published after thousands of hours of work, not only by the authors, but also by reviewers, instructors, students, answer checkers, and editors. To these individuals and all those who have worked in some way on this text over the years, we are most grateful for your contributions. We could not have done it without you. We especially wish to thank the following reviewers whose valuable contributions have helped to refine this edition of this text.

Randall Allbritton, *Daytona Beach Community College*
Jannette Avery, *Monroe Community College*
Linda Beattie, *Western New Mexico University*
Jean Bolyard, *Fairmont State College*
Tim C. Caldwell, *Meridian Community College*
Russell Campbell, *Fairmont State College*
Bill Dunn, *Las Positas College*
Lucy Edwards, *Las Positas College*
J. Lloyd Harris, *Gulf Coast Community College*
Edith Hays, *Texas Woman's University*

Karen Heavin, *Morehead State University*
Christine Heinecke Lehmann, *Purdue University—North Central*
Terry Haynes, *Eastern Oklahoma State College*
Elizabeth Heston, *Monroe Community College*
Harriet Kiser, *Floyd College*
Valerie Lazzara, *Palm Beach Community College*
Valerie H. Maley, *Cape Fear Community College*
Susan McClory, *San Jose State University*
Pam Miller, *Phoenix College*
Jeffrey Mills, *Ohio State University*
Linda J. Murphy, *Northern Essex Community College*
Celia Nippert, *Western Oklahoma State College*
Elizabeth Olgilvie, *Horry-Georgetown Technical College*
Larry Pontaski, *Pueblo Community College*
Diann Robinson, *Ivy Tech State College—Lafayette*
Rachael Schettenhelm, *Southern Connecticut State University*
Lee Ann Spahr, *Durham Technical Community College*
Carol Stewart, *Fairmont State College*
Cora S. West, *Florida Community College at Jacksonville*
Johanna Windmueller, *Seminole Community College*
Gabriel Yimesghen, *Community College of Philadelphia*

Over the years, we have come to rely on an extensive team of experienced professionals. Our sincere thanks go to these dedicated individuals at Addison-Wesley, who worked long and hard to make this revision a success: Greg Tobin, Maureen O'Connor, Jay Jenkins, Lauren Morse, Marcia Emerson, Sharon Smith, Sara Anderson, Tracy Rabinowitz, Ron Hampton, and Dennis Schaefer.

Thanks are due Gina Linko, Phyllis Crittenden, and Elm Street Publishing Services for their excellent production work. Barb Brown provided invaluable assistance updating the real data used in applications throughout the text. Abby Tanenbaum and Paul Lorczak did an outstanding job accuracy checking page proofs. Special thanks to Bernice Eisen who prepared the Index and Becky Troutman who compiled the Index of Applications.

As an author team, we are committed to the goal stated earlier in this Preface—to provide the best possible text and supplements package to help instructors teach and students succeed. We are most grateful to all those over the years who have aspired to this goal with us. As we continue to work toward it, we would welcome any comments or suggestions you might have via e-mail to math@awl.com.

Margaret L. Lial
John Hornsby
Terry McGinnis

Student Supplements

Student's Solutions Manual
- By Jeffery A. Cole, *Anoka-Ramsey Community College*
- Provides detailed solutions to the odd-numbered, section-level exercises and to all margin, Relating Concepts, Summary, Chapter Review, Chapter Test, and Cumulative Review Exercises

ISBN: 0-321-28580-8

Study Skills Workbook
- By Diana Hestwood and Linda Russell
- Provides activities that teach students how to use the textbook effectively, plan their homework, take notes, make mind maps and study cards, manage study time, and prepare for and take tests
- Text desk-light icon at key points directs students to correlated activities in the workbook

ISBN: 0-321-28581-6

Videotape Series
- Features an engaging team of lecturers
- Provides comprehensive coverage of each section and topic in the text

ISBN: 0-321-28582-4

Digital Video Tutor
- Complete set of digitized videos on CD-ROM for student use at home or on campus
- Ideal for distance learning or supplemental instruction

ISBN: 0-321-28584-0

New! Additional Skill & Drill Manual
- Provides additional practice and test preparation for students

ISBN: 0-321-33167-2

Addison-Wesley Math Tutor Center
- Staffed by qualified mathematics instructors
- Provides tutoring on examples and odd-numbered exercises from the textbook through a registration number with a new textbook or purchased separately.
- Accessible via toll-free telephone, toll-free fax, e-mail or the Internet at www.aw-bc/tutorcenter

MathXL **MathXL® Tutorials on CD**
- Provides algorithmically generated practice exercises that correlate at the objective level to the content of the text
- Includes an example and a guided solution to accompany every exercise and video clips for selected exercises
- Recognizes student errors and provides feedback; Generates printed summaries of students' progress

ISBN: 0-321-28579-4

Instructor Supplements

Annotated Instructor's Edition
- Provides answers to all text exercises in color next to the corresponding problems
- Includes icons to identify writing and calculator exercises

ISBN: 0-321-28585-9

Instructor's Solutions Manual
- By Jeffery A. Cole, *Anoka-Ramsey Community College*
- Provides complete solutions to all even-numbered, section-level exercises

ISBN: 0-321-28577-8

Answer Book
- By Jeffery A. Cole, *Anoka-Ramsey Community College*
- Provides answers to all the exercises in the text

ISBN: 0-321-28576-X

Adjunct Support Manual
- Includes resources designed to help both new and adjunct faculty with course preparation and classroom management
- Offers helpful teaching tips correlated to the sections of the text

ISBN: 0-321-28586-7

Printed Test Bank and Instructor's Resource Guide
- By James J. Ball, *Indiana State University*
- Contains two diagnostic pretests, six free-response and two multiple-choice test forms per chapter, and two final exams
- Includes teaching suggestions for every chapter, additional practice exercises for each objective in every section, a correlation guide from the seventh to the eighth edition, phonetic spellings for all key terms in the text, and teaching notes and extensions for the Focus on Real-Data Applications in the text

ISBN: 0-321-28575-1

TestGen
- Enables instructors to build, edit, print, and administer tests
- Features a computerized bank of questions developed to cover all text objectives
- Available on a dual-platform Windows/Macintosh CD-ROM

ISBN: 0-321-28578-6

InterAct Math Tutorial Web site: www.interactmath.com Get practice and tutorial help online! This interactive tutorial web site provides algorithmically generated practice exercises that correlate directly to the exercises in the textbook. Students can retry an exercise multiple times with new values each time for unlimited practice and mastery. Every exercise is accompanied by an interactive guided solution that provides helpful feedback for an incorrect answer. Students can also view a worked-out sample problem that steps them through an exercise similar to the one they're working on.

MathXL **MathXL®** MathXL is a powerful online homework, tutorial, and assessment system that accompanies your Addison-Wesley textbook in mathematics or statistics. With MathXL, instructors can create, edit, and assign online homework and tests using algorithmically generated exercises correlated at the objective level to the textbook. All student work is tracked in MathXL's online gradebook. Students can take chapter tests in MathXL and receive personalized study plans based on their test results. The study plan diagnoses weaknesses and links students directly to tutorial exercises for the objectives they need to study and retest. Students can also access supplemental video clips and animations directly from selected exercises. MathXL is available to qualified adopters. For more information, visit our web site at www.mathxl.com, or contact your Addison-Wesley sales representative for more information.

MyMathLab **MyMathLab** MyMathLab is a series of text-specific, easily customizable online courses for Addison-Wesley textbooks in mathematics and statistics. MyMathLab is powered by CourseCompass—Pearson Education's online teaching and learning environment—and by MathXL—our online homework, tutorial, and assessment system. MyMathLab gives instructors the tools they need to deliver all or a portion of their course online, whether students are in a lab setting or working from home.

MyMathLab provides a rich and flexible set of course materials, featuring free-response exercises that are algorithmically generated for unlimited practice and mastery. Students can also use online tools, such as video lectures, animations, and a multimedia textbook, to independently improve their understanding and performance. Instructors can use MyMathLab's homework and test managers to select and assign online exercises correlated directly to the textbook, and they can import TestGen tests into MyMathLab for added flexibility. MyMathLab's online gradebook—designed specifically for mathematics and statistics—automatically tracks students' homework and test results and gives the instructor control over how to calculate final grades. Instructors can also add offline (paper-and-pencil) grades to the gradebook.

MyMathLab is available to qualified adopters. For more information, visit our Web site at www.mymathlab.com or contact your Addison-Wesley sales representative.

Feature Walk-Through

Chapter Openers Chapter openers feature real-world applications of mathematics that are relevant to students and tied to specific material within the chapters.

Systems of Linear Equations and Inequalities

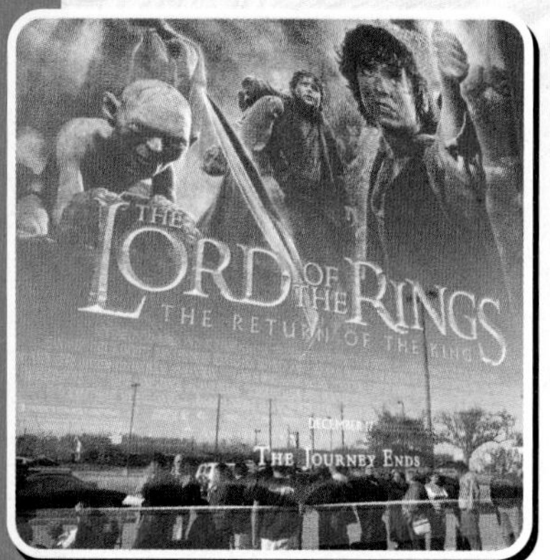

Although Americans continued their fascination with Hollywood and the movies in 2003, movie attendance and revenues dipped for the first time since 1991. Nonetheless, some 1.52 billion tickets were sold and revenues exceeded $9 billion for the second year in a row. The top box office draws of the year—*The Lord of the Rings: The Return of the King* and *Finding Nemo*—attracted scores of adults and children wishing to get away from it all for a few hours. (*Source:* Exhibitor Relations Co., Nielsen EDI.)

In Exercise 13 of Section 4.4, we use a *system of linear equations* to find out how much money these top films earned.

Section 3.2 Graphing Linear Equations in Two Variables **221**

37. Use the results of Exercises 35(b) and 36(b) to determine the target heart rate zone for age 30.

38. Should the graphs of the target heart rate zone in the **Section 3.1** exercises be used to estimate the target heart rate zone for ages below 20 or above 80? Why or why not?

39. Per capita consumption of coffee increased for the years 1995 through 2000 as shown in the graph. If $x = 0$ represents 1995, $x = 1$ represents 1996, and so on, per capita consumption y in gallons can be modeled by the linear equation

$$y = 1.13x + 20.67.$$

COFFEE CONSUMPTION

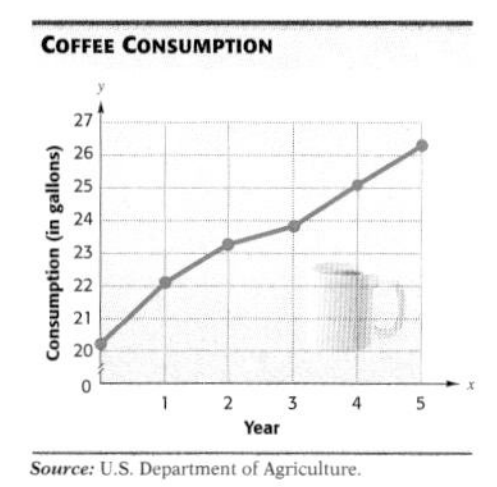

Source: U.S. Department of Agriculture.

(a) Use the equation to approximate consumption in 1995, 1996, and 1998 to the nearest tenth.

40. Sporting goods sales y (in billions of dollars) from 1995 through 2000 are modeled by the linear equation

$$y = 3.606x + 41.86,$$

where $x = 5$ corresponds to 1995, $x = 6$ corresponds to 1996, and so on.

SPORTING GOODS SALES

Sales (in billions of dollars): 0, 20, 40, 60, 80, 100; Year: 5, 6, 7, 8, 9, 10

Source: U.S. Bureau of the Census.

(a) Use the equation to approximate sporting goods sales in 1995, 1997, and 2000. Round your answers to the nearest billion dollars.

Figures and Photos Today's students are more visually oriented than ever. Thus, a concerted effort has been made to include mathematical figures, diagrams, tables, and graphs whenever possible. Many of the graphs use a style similar to that seen by students in today's print and electronic media. Photos have been incorporated to enhance applications in examples and exercises.

Relating Concepts These sets of exercises help students tie together topics and develop problem-solving skills as they compare and contrast ideas, identify and describe patterns, and extend concepts to new situations. These exercises make great collaborative activities for pairs or small groups of students.

RELATING CONCEPTS (EXERCISES 41–46) For Individual or Group Work

Attending the movies is one of America's favorite forms of entertainment. The graph shows movie attendance from 1991 to 1999. In 1991, attendance was 1141 million, as represented by the point P(1991, 1141). In 1999, attendance was 1465 million, as represented by the point Q(1999, 1465). We can find an equation of line segment PQ using a system of equations, and then we can use the equation to approximate the attendance in any of the years between 1991 and 1999. ***Work Exercises 41–46 in order.***

MOVIE BOX OFFICE ATTENDANCE/ ADMISSIONS

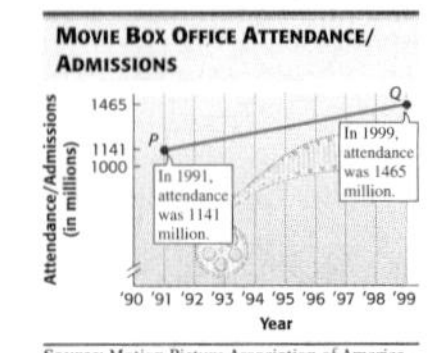

Source: Motion Picture Association of America.

41. The line segment has an equation that can be written in the form $y = ax + b$. Using the coordinates of point P with $x = 1991$ and $y = 1141$, write an equation in the variables a and b.

42. Using the coordinates of point Q with $x = 1999$ and $y = 1465$, write a second equation in the variables a and b.

43. Write the system of equations formed from the two equations in Exercises 41 and 42, and solve the system using the elimination method.

44. What is the equation of the segment PQ?

45. Let $x = 1998$ in the equation of Exercise 44, and solve for y. How does the result compare with the actual figure of 1481 million?

46. The data points for the years 1991 through 1999 do not lie in a perfectly straight line. Explain the pitfalls of relying too heavily on using the equation in Exercise 44 to predict attendance.

Focus on

Real-Data Applications

The Magic Number in Sports

The climax of any sports season is the playoffs. Baseball fans eagerly debate predictions of which team will win the pennant for their division. The *magic number* for each first-place team is often used to predict the division winner. The **magic number** is the combined number of wins by the first-place team and losses by the second-place team that would clinch the title for the first-place team.

For Group Discussion

To calculate the magic number, consider the following conditions.

The number of wins for the first-place team (W_1) plus the magic number (M) is one more than the sum of the number of wins to date (W_2) and the number of games remaining in the season (N_2) for the second-place team.

American League

East Division	W	L	Pct.	GB
New York	94	58	.618	—
Boston	89	63	.586	5
Toronto	79	73	.520	15
Baltimore	68	83	.450	25½
Tampa Bay	60	92	.395	34
Central Division	**W**	**L**	**Pct.**	**GB**
Minnesota	84	69	.549	—
Chicago	80	72	.526	3½
Kansas City	79	73	.520	4½
Cleveland	66	88	.429	18½
Detroit	38	114	.250	45½
West Division	**W**	**L**	**Pct.**	**GB**
Oakland	92	61	.601	—
Seattle	87	66	.569	5
Anaheim	72	81	.471	20
Texas	68	85	.444	24

Source: USA Today.

1. First, use the variable definitions to write an equation involving the magic number. Second, solve the equation for the magic number. Write the formula for the magic number.

2. The American League standings with about 10 games left in the 2003 baseball season are shown at the left. There were 162 regulation games in the 2003 season. Find the magic number for each team. The number of games remaining in the season for the second-place team is calculated as

$$N_2 = 162 - (W_2 + L_2).$$

(a) AL East: New York vs Boston

Magic No. ______

(b) AL Central: Minnesota vs Chicago

Magic No. ______

(c) AL West: Oakland vs Seattle

Magic No. ______

[*Note:* For the National League in the 2003 season, Atlanta and San Francisco were runaway winners of the Eastern and Western Divisions, respectively. Chicago, Houston, and St. Louis were locked in a dead heat for the Central Division lead, which was eventually won by Chicago.]

Focus on Real-Data Applications These one-page activities found throughout the text present even more relevant and in-depth looks at how mathematics is used in the real world. Designed to help instructors answer the often-asked question, "When will I ever use this stuff?," these activities ask students to read and interpret data from newspaper articles, the Internet, and other familiar, real sources. The activities are well suited to collaborative work and can also be completed by individuals or used for open-ended class discussions.

Calculator Tip Using a calculator to perform the arithmetic in Example 5 reduces the possibility of errors.

Work Problem 5 at the Side.

OBJECTIVE 4 Find percentages and percents. Percents are ratios where the second number is always 100. For example, 50% represents the ratio of 50 to 100, 27% represents the ratio of 27 to 100, and so on. We can use the techniques for solving proportions to solve percent problems. Recall from **Section R.2** that the decimal point is moved two places to the left to change a percent to a decimal number. For example, 75% can be written as the decimal .75.

Calculator Tip Many calculators have a percent key that does this automatically.

We can solve a percent problem by writing it as the proportion

$$\frac{amount}{base} = \frac{percent}{100} \quad \text{or} \quad \frac{a}{b} = \frac{p}{100}.$$

The amount, or **percentage,** is compared to the **base** (the whole amount). Since ***percent* means *per 100,*** we compare the numerical value of the percent to 100. Thus, we write 50% as

$$\frac{p}{100} = \frac{50}{100}. \quad p = 50$$

Calculator Tips These optional tips, marked with calculator icons, offer basic information and instruction for students using calculators in the course.

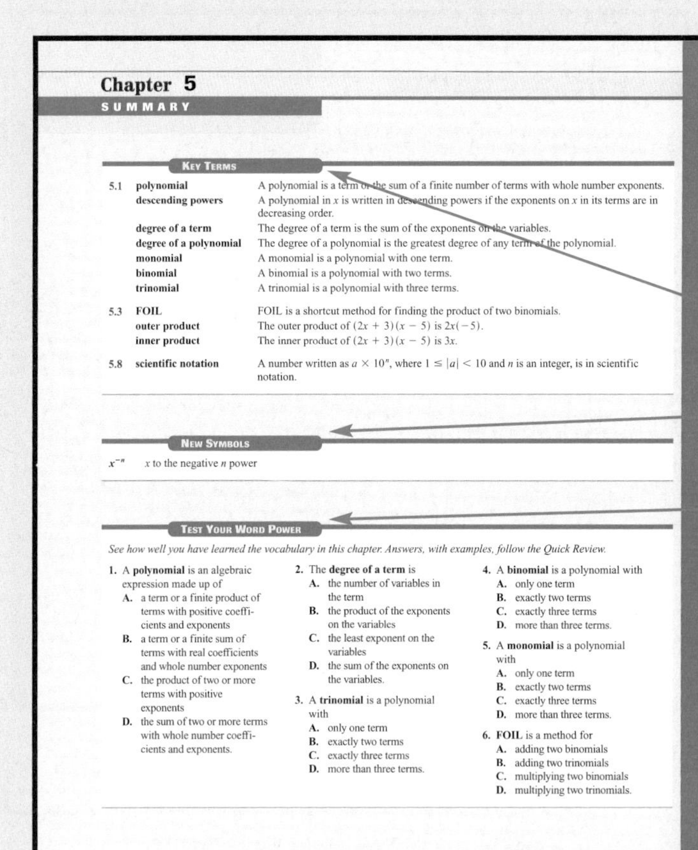

Chapter 5
SUMMARY

KEY TERMS

5.1	**polynomial**	A polynomial is a term or the sum of a finite number of terms with whole number exponents.
	descending powers	A polynomial in x is written in descending powers if the exponents on x in its terms are in decreasing order.
	degree of a term	The degree of a term is the sum of the exponents on the variables.
	degree of a polynomial	The degree of a polynomial is the greatest degree of any term of the polynomial.
	monomial	A monomial is a polynomial with one term.
	binomial	A binomial is a polynomial with two terms.
	trinomial	A trinomial is a polynomial with three terms.
5.3	**FOIL**	FOIL is a shortcut method for finding the product of two binomials.
	outer product	The outer product of $(2x + 3)(x - 5)$ is $2x(-5)$.
	inner product	The inner product of $(2x + 3)(x - 5)$ is $3x$.
5.8	**scientific notation**	A number written as $a \times 10^n$, where $1 \leq \lvert a \rvert < 10$ and n is an integer, is in scientific notation.

NEW SYMBOLS

x^{-n} x to the negative n power

TEST YOUR WORD POWER

See how well you have learned the vocabulary in this chapter. Answers, with examples, follow the Quick Review.

1. A **polynomial** is an algebraic expression made up of
   A. a term or a finite product of terms with positive coefficients and exponents
   B. a term or a finite sum of terms with real coefficients and whole number exponents
   C. the product of two or more terms with positive exponents
   D. the sum of two or more terms with whole number coefficients and exponents.
2. The **degree of a term** is
   A. the number of variables in the term
   B. the product of the exponents on the variables
   C. the least exponent on the variables
   D. the sum of the exponents on the variables.
3. A **trinomial** is a polynomial with
   A. only one term
   B. exactly two terms
   C. exactly three terms
   D. more than three terms.
4. A **binomial** is a polynomial with
   A. only one term
   B. exactly two terms
   C. exactly three terms
   D. more than three terms.
5. A **monomial** is a polynomial with
   A. only one term
   B. exactly two terms
   C. exactly three terms
   D. more than three terms.
6. **FOIL** is a method for
   A. adding two binomials
   B. adding two trinomials
   C. multiplying two binomials
   D. multiplying two trinomials.

391

End-of-Chapter Material One of the most admired features of the Lial textbooks is the extensive and well-thought-out end-of-chapter material. At the end of each chapter, students will find a Summary that includes the following:

Key Terms are listed, defined, and referenced back to the appropriate section number.

New Symbols are listed for easy reference and study.

Test Your Word Power helps students understand and master mathematical vocabulary. Students are quizzed on Key Terms from the chapter in a multiple-choice format. Answers and examples illustrating each term are provided.

A Chapter Test helps students practice for the real thing.

Chapter 5 Test 399

Chapter 5
TEST

Perform the indicated operations.

1. $(5t^4 - 3t^2 + 7t + 3) - (t^4 - t^3 + 3t^2 + 8t + 3)$ 1. ______
2. $(2y^2 - 8y + 8) + (-3y^2 + 2y + 3) - (y^2 + 3y - 6)$ 2. ______
3. Subtract. 3. ______
   $9t^3 - 4t^2 + 2t + 2$
   $9t^3 + 8t^2 - 3t - 6$

Simplify, and write each answer with only positive exponents.

4. $(-2)^3(-2)^2$ 4. ______
5. $\left(\frac{6}{m^2}\right)^3, \quad m \neq 0$ 5. ______
6. $3x^2(-9x^3 + 6x^2 - 2x + 1)$ 6. ______
7. $(2r - 3)(r^2 + 2r - 5)$ 7. ______
8. $(t - 8)(t + 3)$ 8. ______

392 Chapter 5 Exponents and Polynomials

QUICK REVIEW

Concepts	*Examples*
5.1 *Adding and Subtracting Polynomials* **Addition** Add like terms.	Add. $2x^2 + 5x - 3$; $5x^2 - 2x + 7$; $7x^2 + 3x + 4$
Subtraction Change the signs of the terms in the second polynomial and add to the first polynomial.	Subtract. $(2x^2 + 5x - 3) - (5x^2 - 2x + 7)$ $= (2x^2 + 5x - 3) + (-5x^2 + 2x - 7)$ $= -3x^2 + 7x - 10$
5.2 *The Product Rule and Power Rules for Exponents* For any integers m and n:	
Product rule $a^m \cdot a^n = a^{m+n}$	$2^4 \cdot 2^5 = 2^9$
Power rules (a) $(a^m)^n = a^{mn}$	$(3^4)^2 = 3^8$
(b) $(ab)^m = a^m b^m$	$(6a)^5 = 6^5a^5$
(c) $\left(\frac{a}{b}\right)^m = \frac{a^m}{b^m} \quad (b \neq 0)$.	$\left(\frac{2}{3}\right)^4 = \frac{2^4}{3^4}$
5.3 *Multiplying Polynomials* Multiply each term of the first polynomial by each term of the second polynomial. Then add like terms.	Multiply. $3x^3 - 4x^2 + 2x - 7$; $4x + 3$; $9x^3 - 12x^2 + 6x - 21$; $12x^4 - 16x^3 + 8x^2 - 28x$

Quick Review sections give students not only the main concepts from the chapter (referenced back to the appropriate section), but also an adjacent example of each concept.

Review Exercises are keyed to the appropriate sections so that students can refer to examples of that type of problem if they need help.

Chapter 5
REVIEW EXERCISES

[5.1] *Combine terms where possible in each polynomial. Write the answer in descending powers of the variable. Give the degree of the answer. Identify the polynomial as a* monomial, binomial, trinomial, *or* none of these.

1. $9m^2 + 11m^2 + 2m^2$

2. $-4p + p^3 - p^2 + 8p + 2$

3. $12a^5 - 9a^4 + 8a^3 + 2a^2 - a + 3$

4. $-7y^5 - 8y^4 - y^5 + y^4 + 9y$

Add or subtract as indicated.

5. Add.
$$\begin{array}{r} -2a^3 + 5a^2 \\ \underline{-3a^3 - a^2} \end{array}$$

6. Add.
$$\begin{array}{r} 4r^3 - 8r^2 + 6r \\ \underline{-2r^3 + 5r^2 + 3r} \end{array}$$

7. Subtract.
$$\begin{array}{r} 6y^2 - 8y + 2 \\ \underline{-5y^2 + 2y - 7} \end{array}$$

8. Subtract.
$$\begin{array}{r} -12k^4 - 8k^2 + 7k - 5 \\ \underline{k^4 + 7k^2 + 11k + 1} \end{array}$$

9. $(2m^3 - 8m^2 + 4) + (8m^3 + 2m^2 - 7)$

10. $(-5y^2 + 3y + 11) + (4y^2 - 7y + 15)$

11. $(6p^2 - p - 8) - (-4p^2 + 2p + 3)$

12. $(12r^4 - 7r^3 + 2r^2) - (5r^4 - 3r^3 + 2r^2 + 1)$

[5.2] *Use the product rule or power rules to simplify each expression. Write the answer in exponential form.*

13. $4^3 \cdot 4^8$

14. $(-5)^6(-5)^5$

15. $(-8x^4)(9x^3)$

16. $(2x^2)(5x^3)(x^9)$

17. $(19x)^5$

18. $(-4y)^7$

19. $5(pt)^4$

20. $\left(\frac{7}{5}\right)^6$

85. There are 13 red balls and 39 black balls in a box. Mix them up and draw 13 out one at a time without returning any ball. The probability that the 13 drawings each will produce a red ball is 1.6×10^{-12}. Write the number given in scientific notation without exponents. (*Source:* Warren Weaver, *Lady Luck*, Doubleday & Company, 1963.)

86. According to Campbell, Mitchell, and Reece in *Biology Concepts and Connections* (Benjamin Cummings, 1994, p. 230), "The amount of DNA in a human cell is about 1000 times greater than the DNA in *E. coli*. Does this mean humans have 1000 times as many genes as the 2000 in *E. coli*? The answer is probably no; the human genome is thought to carry between 50,000 and 100,000 genes, which code for various proteins (as well as for tRNA and rRNA)."
Write each number from this quote using scientific notation.

(a) 1000 **(b)** 2000

(c) 50,000 **(d)** 100,000

Mixed Review Exercises require students to solve problems without the help of section references.

MIXED REVIEW EXERCISES

Perform the indicated operations. Write with positive exponents. Assume that no denominators are equal to 0.

87. $19^0 - 3^0$

88. $(3p)^4(3p^{-7})$

89. 7^{-2}

90. $(-7 + 2k)^2$

91. $\dfrac{2y^3 + 17y^2 + 37y + 7}{2y + 7}$

92. $\left(\dfrac{6r^2s}{5}\right)^4$

93. $-m^5(8m^2 + 10m + 6)$

94. $\left(\frac{1}{2}\right)^{-5}$

95. $(25x^2y^3 - 8xy^2 + 15x^3y) \div (5x)$

96. $(6r^{-2})^{-1}$

97. $(2x + y)^3$

98. $2^{-1} + 4^{-1}$

99. $(a + 2)(a^2 - 4a + 1)$

100. $(5y^3 - 8y^2 + 7) - (-3y^3 + y^2 + 2)$

101. $(2r + 5)(5r - 2)$

102. $(12a + 1)(12a - 1)$

103. Find a polynomial that represents the area of the rectangle shown.

$2x - 3$

$x + 2$

104. If the side of a square has a measure represented by $5x^4 + 2x^2$, what polynomial represents its area?

$5x^4 + 2x^2$

Cumulative Review Exercises
CHAPTERS R–5

Work each problem.

1. $\frac{2}{3} + \frac{1}{8}$

2. $\frac{7}{4} - \frac{9}{5}$

3. $8.32 - 4.6$

4. 7.21×8.6

5. A retailer has \$34,000 invested in her business. She finds that last year she earned 5.4% on this investment. How much did she earn?

Find the value of each expression if $x = -2$ and $y = 4$.

6. $\dfrac{4x - 2y}{x + y}$

7. $x^3 - 4xy$

Perform the indicated operations.

8. $\dfrac{(-13 + 15) - (3 + 2)}{6 - 12}$

9. $-7 - 3[2 + (5 - 8)]$

Decide what property justifies each statement.

10. $(9 + 2) + 3 = 9 + (2 + 3)$

11. $-7 + 7 = 0$

12. $6(4 + 2) = 6(4) + 6(2)$

Solve each equation.

13. $2x - 7x + 8x = 30$

14. $2 - 3(t - 5) = 4 + t$

15. $2(5h + 1) = 10h + 4$

16. $d = rt$ for r

17. $\frac{x}{5} = \frac{x - 2}{7}$

18. $\frac{1}{3}p - \frac{1}{6}p = -2$

19. $.05x + .15(50 - x) = 5.50$

20. $4 - (3x + 12) = (2x - 9) - (5x - 1)$

Solve each problem.

21. A 1-oz mouse takes about 16 times as many breaths as does a 3-ton elephant. (*Source: Dinosaurs, Spitfires, and Sea Dragons*, McGowan, C., Harvard University Press...

22. If a number is subtracted from 8 and this difference is tripled, the result is three times the number. Find this number, and you will learn how many times a dolphin rests during...

Cumulative Review Exercises gather various types of exercises from preceding chapters to help students remember and retain what they are learning throughout the course.

An Introduction to Calculators

There is little doubt that the appearance of handheld calculators three decades ago and the later development of scientific and graphing calculators have changed the methods of learning and studying mathematics forever. For example, computations with tables of logarithms and slide rules made up an important part of mathematics courses prior to 1970. Today, with the widespread availability of calculators, these topics are studied only for their historical significance.

Calculators come in a large array of different types, sizes, and prices. ***For the course for which this textbook is intended, the most appropriate type is the scientific calculator,*** which costs \$10–\$20.

In this introduction, we explain some of the features of scientific and graphing calculators. However, remember that calculators vary among manufacturers and models, and that while the methods explained here apply to many of them, they may not apply to your specific calculator. ***This introduction is only a guide and is not intended to take the place of your owner's manual.*** Always refer to the manual whenever you need an explanation of how to perform a particular operation.

SCIENTIFIC CALCULATORS

Scientific calculators are capable of much more than the typical four-function calculator that you might use for balancing your checkbook. Most scientific calculators use *algebraic logic*. (Models sold by Texas Instruments, Sharp, Casio, and Radio Shack, for example, use algebraic logic.) A notable exception is Hewlett-Packard, a company whose calculators use *Reverse Polish Notation* (RPN). In this introduction, we explain the use of calculators with algebraic logic.

Arithmetic Operations To perform an operation of arithmetic, simply enter the first number, press the operation key (+, −, ×, or ÷), enter the second number, and then press the = key. For example, to add 4 and 3, use the following keystrokes.

Change Sign Key The key marked +/− allows you to change the sign of a display. This is particularly useful when you wish to enter a negative number. For example, to enter -3, use the following keystrokes.

Memory Key Scientific calculators can hold a number in memory for later use. The label of the memory key varies among models; two of these are M and STO. The M+ and M− keys allow you to add to or subtract from the value currently in memory. The memory recall key, labeled MR, RM, or RCL, allows you to retrieve the value stored in memory.

Suppose that you wish to store the number 5 in memory. Enter 5, then press the key for memory. You can then perform other calculations. When you need to retrieve the 5, press the key for memory recall.

If a calculator has a constant memory feature, the value in memory will be retained even after the power is turned off. Some advanced calculators have more than one memory. It is best to read the owner's manual for your model to see exactly how memory is activated.

Clearing/Clear Entry Keys The C or CE key allows you to clear the display or clear the last entry entered into the display. In some models, pressing the C key once will clear the last entry, while pressing it twice will clear the entire operation in progress.

Second Function Key This key, usually marked 2nd, is used in conjunction with another key to activate a function that is printed *above* an operation key (and not on the key itself). For example, suppose you wish to find the square of a number, and the squaring function (explained in more detail later) is printed above another key. You would need to press 2nd before the desired squaring function can be activated.

Square Root Key Pressing $\sqrt{\ }$ or $\sqrt{x}$ will give the square root (or an approximation of the square root) of the number in the display. On some scientific calculators, the square root key is pressed *before* entering the number, while other calculators use the opposite order. Experiment with your calculator to see which method it uses. For example, to find the square root of 36, use the following keystrokes.

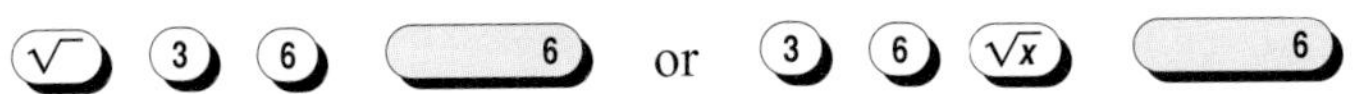

The square root of 2 is an example of an irrational number (**Chapter 8**). The calculator will give an approximation of its value, since the decimal for $\sqrt{2}$ never terminates and never repeats. The number of digits shown will vary among models. To find an approximation for $\sqrt{2}$, use the following keystrokes.

Squaring Key The x^2 key allows you to square the entry in the display. For example, to square 35.7, use the following keystrokes.

The squaring key and the square root key are often found on the same key, with one of them being a second function (that is, activated by the second function key previously described).

Reciprocal Key The key marked $1/x$ is the reciprocal key. (When two numbers have a product of 1, they are called *reciprocals*. See **Chapter R.**) Suppose that you wish to find the reciprocal of 5. Use the following keystrokes.

Inverse Key Some calculators have an inverse key, marked INV. Inverse operations are operations that "undo" each other. For example, the operations of squaring and taking the square root are inverse operations. The use of the INV key varies among different models of calculators, so read your owner's manual carefully.

Exponential Key The key marked x^y or y^x allows you to raise a number to a power. For example, if you wish to raise 4 to the fifth power (that is, find 4^5, as explained in **Chapter 1**), use the following keystrokes.

Root Key Some calculators have this key specifically marked $\sqrt[x]{\ }$ or $\sqrt[x]{y}$; with others, the operation of taking roots is accomplished by using the inverse key in conjunction with the exponential key. Suppose, for example, your calculator is of the latter type and you wish to find the fifth root of 1024. Use the following keystrokes.

Notice how this "undoes" the operation explained in the exponential key discussion.

Pi Key The number π is an important number in mathematics. It occurs, for example, in the area and circumference formulas for a circle. By pressing the (π) key, you can display the first few digits of π. (Because π is irrational, the display shows only an approximation.) One popular model gives the following display when the (π) key is pressed.

3.1415927 — An approximation for π

Methods of Display When decimal approximations are shown on scientific calculators, they are either *truncated* or *rounded*. To see how a particular model is programmed, evaluate 1/18 as an example. If the display shows .0555555 (last digit 5), it truncates the display. If it shows .0555556 (last digit 6), it rounds the display.

When very large or very small numbers are obtained as answers, scientific calculators often express these numbers in scientific notation (**Chapter 5**). For example, if you multiply 6,265,804 by 8,980,591, the display might look like this:

5.6270623 13

The 13 at the far right means that the number on the left is multiplied by 10^{13}. This means that the decimal point must be moved 13 places to the right if the answer is to be expressed in its usual form. Even then, the value obtained will only be an approximation: 56,270,623,000,000.

GRAPHING CALCULATORS

While you are not expected to have a graphing calculator to study from this book, we include the following as background information and reference should your course or future courses require the use of graphing calculators.

Basic Features

In addition to the typical keys found on scientific calculators, graphing calculators have keys that can be used to create graphs, make tables, analyze data, and change settings. One of the major differences between graphing and scientific calculators is that a graphing calculator has a larger viewing screen with graphing capabilities. The screens below illustrate the graphs of $Y = X$ and $Y = X^2$.

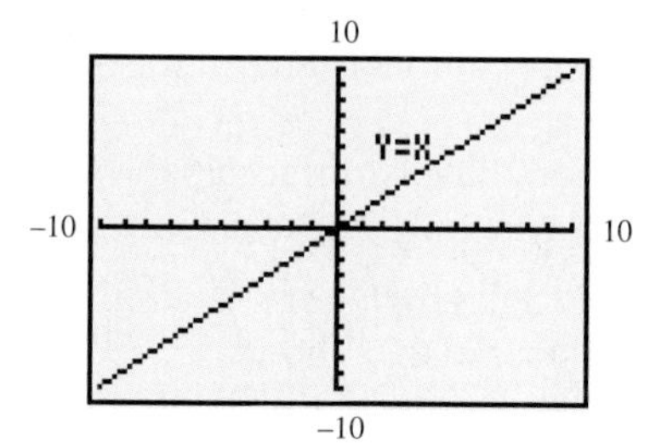

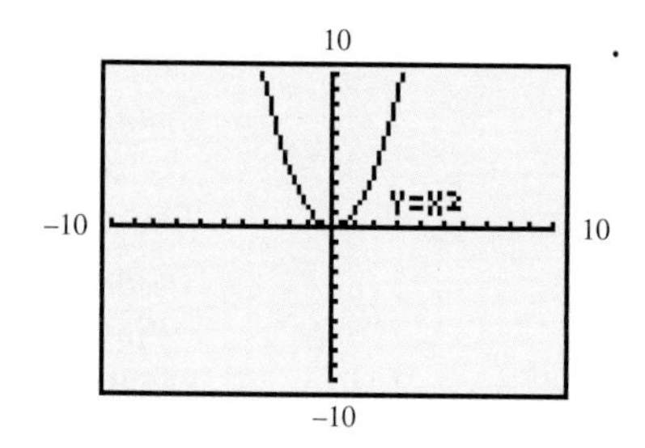

If you look closely at the screens, you will see that the graphs appear to be jagged rather than smooth, as they should be. The reason for this is that graphing calculators have much lower resolution than computer screens. Because of this, graphs generated by graphing calculators must be interpreted carefully.

Editing Input

The screen of a graphing calculator can display several lines of text at a time. This feature allows you to view both previous and current expressions. If an incorrect expression is entered, an error message is displayed. The erroneous expression can be viewed and corrected by using various editing keys, much like a word-processing program. You do not need to enter the entire expression again.

Many graphing calculators can also recall past expressions for editing or updating. The screen on the left below shows how two expressions are evaluated. The final line is entered incorrectly, and the resulting error message is shown in the screen on the right.

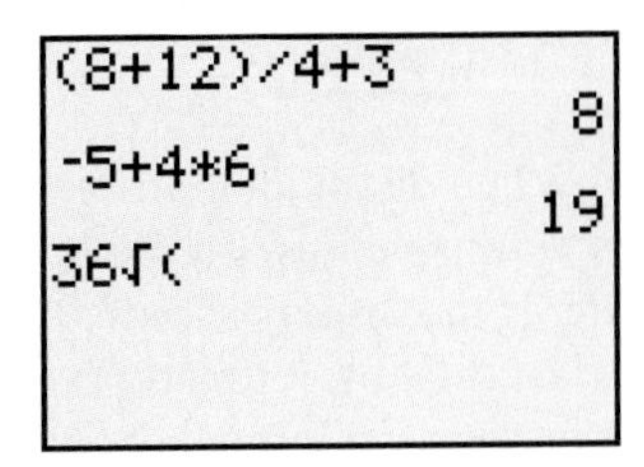

Order of Operations

Arithmetic operations on graphing calculators are usually entered as they are written in mathematical expressions. For example, to evaluate $\sqrt{36}$, you would first press the square root key, and then enter 36. See the screen on the left below. The order of operations on a graphing calculator is also important, and current models assist the user by inserting parentheses when typical errors might occur. The open parenthesis that follows the square root symbol is automatically entered by the calculator so that an expression such as $\sqrt{2 \times 8}$ will not be calculated incorrectly as $\sqrt{2} \times 8$. Compare the two entries and their results in the screen on the right.

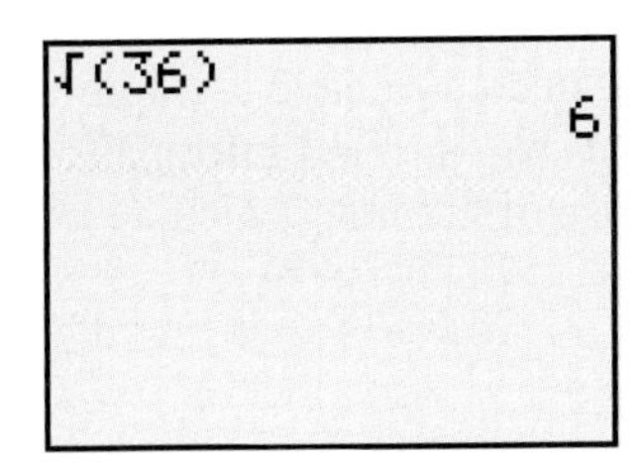

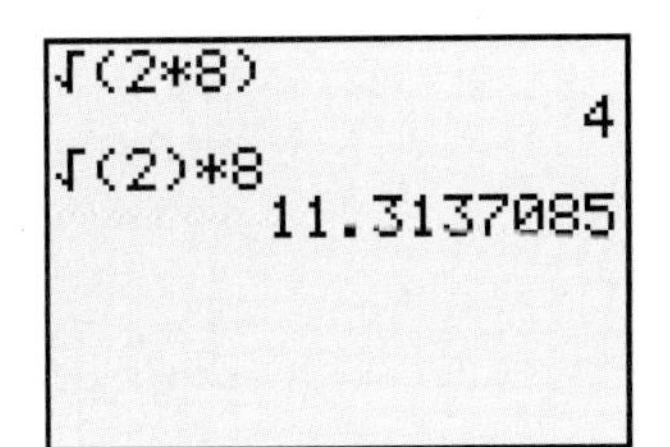

Viewing Windows

The viewing window for a graphing calculator is similar to the viewfinder in a camera. A camera usually cannot take a photograph of an entire view of a scene. The camera must be centered on some object and can capture only a portion of the available scenery. A camera with a zoom lens can photograph different views of the same scene by zooming in and out.

Graphing calculators have similar capabilities. The xy-coordinate plane is infinite. The calculator screen can only show a finite, rectangular region in the plane, and it must be specified before the graph can be drawn. This is done by setting both minimum and maximum values for the x- and y-axes. The scale (distance between tick marks) is usually specified as well. Determining an appropriate viewing window for a graph can be a challenge, and sometimes it will take a few attempts before a satisfactory window is found.

The screen on the left shows a standard viewing window, and the graph of $Y = 2X + 1$ is shown on the right. Using a different window would give a different view of the line.

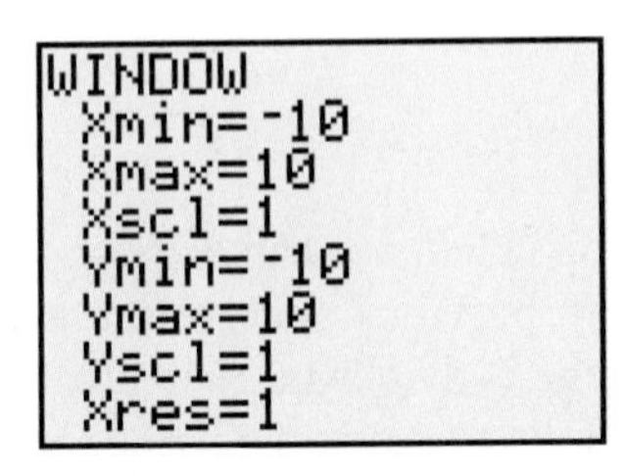

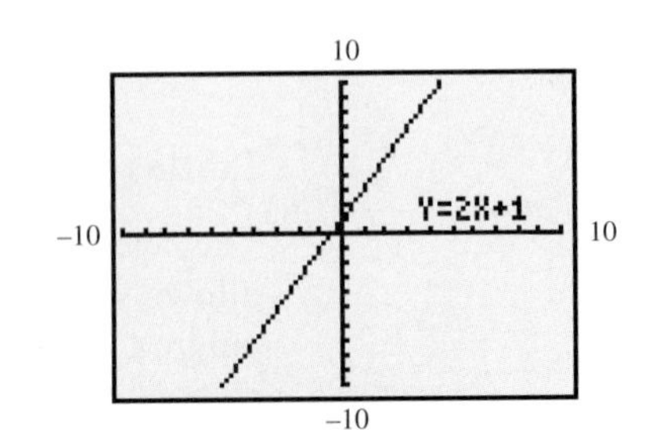

Locating Points on a Graph: Tracing and Tables

Graphing calculators allow you to trace along the graph of an equation and display the coordinates of points on the graph. See the screen on the left below, which indicates that the point (2, 5) lies on the graph of Y = 2X + 1. Tables for equations can also be displayed. The screen on the right shows a partial table for this same equation. Note the middle of the screen, which indicates that when X = 2, Y = 5.

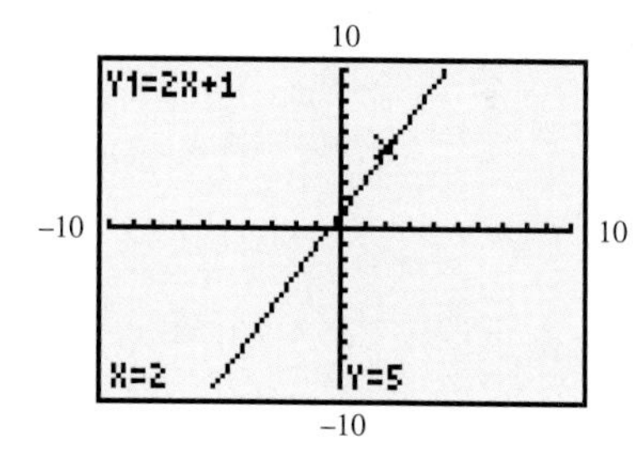

X	Y1
-1	-1
0	1
1	3
2	5
3	7
4	9
5	11

Y1 ■ 2X+1

Additional Features

There are many features of graphing calculators that go far beyond the scope of this book. These calculators can be programmed, much like computers. Many of them can solve equations at the stroke of a key, analyze statistical data, and perform symbolic algebraic manipulations. Calculators also provide the opportunity to ask "What if . . . ?" more easily. Values in algebraic expressions can be altered and conjectures tested quickly.

Final Comments

Despite the power of today's calculators, they cannot replace human thought. ***In the entire problem-solving process, your brain is the most important component.*** Calculators are only tools and, like any tool, they must be used appropriately in order to enhance our ability to understand mathematics. Mathematical insight may often be the quickest and easiest way to solve a problem; a calculator may neither be needed nor appropriate. By applying mathematical concepts, you can make the decision whether or not to use a calculator.

To the Student: Success in Algebra

There are two main reasons students have difficulty with mathematics:

- Students start in a course for which they do not have the necessary background knowledge.
- Students don't know how to study mathematics effectively.

Your instructor can help you decide whether this is the right course for you. We can give you some study tips.

Studying mathematics *is* different from studying subjects like English and history. ***The key to success is regular practice.*** This should not be surprising. After all, can you learn to play the piano or ski well without a lot of regular practice? The same is true for learning mathematics. Working problems nearly every day is the key to becoming successful. Here is a list of things that will help you succeed in studying algebra.

1. **Attend class regularly.** Pay attention in class and take careful notes. In particular, note the problems your teacher works on the board and copy the complete solutions. Keep these notes separate from your homework.
2. **Ask questions.** Don't hesitate to ask questions in class. Other students may have the same questions but be reluctant to ask them, and everyone will benefit from the answers.
3. **Read your text carefully.** Many students go directly to the exercise sets without taking time to read the text and examples. Reading the *complete* section and working the margin problems will pay off when you tackle the homework problems.
4. **Reread your class notes.** Before starting your homework, rework the problems your teacher did in class. This will reinforce what you have learned. Teachers often hear the comment, "*I understand it perfectly when you do it, but I get stuck when I try to work the problem myself.*"

5. **Practice by working problems.** Do your homework only *after* reading the text and reviewing your class notes. Check your work against the answer section or the *Student's Solutions Manual*. If you make an error and are unable to determine what went wrong, mark that problem and ask your instructor about it. Then work more problems of the same type to reinforce what you have learned.
6. **Work neatly.** Write symbols neatly. Skip lines between steps. Write large enough so that others can read your work. Use pencil. Make sure that problems are clearly separated from each other.

7. **Review the material.** After completing each section, look over the text again. Decide on the main objectives, and don't be content until you feel that you have mastered them. (In this book, objectives are clearly stated both at the beginning and within each section.) Write a summary of the section or make an outline for future reference.
8. **Prepare for tests.** The chapter summaries in the text are an excellent way for you to review key terms, new symbols, and important concepts from the chapter. After working through the chapter review exercises, use the chapter test as a practice test. Work the problems under test conditions, without looking at the text or answers until you are finished. Time yourself. When you have finished, check your answers against the answer section and rework any that you missed.

9. **Learn from your mistakes.** Keep all graded assignments, quizzes, and tests that are returned to you. Be sure to correct any errors on them and use them to study for future tests and the final exam.

10. **Be diligent and don't give up.** The authors of this text can tell you that they did not always understand a topic the first time they saw it. Don't worry if you also find this to be true. As you read more about a topic and work through the problems, you will gain understanding; the thrill of finally "getting it" is a great feeling. Listen to the words of the late Jim Valvano: ***Never give up!***

NOTE

Reading a list of study tips is a good start, but you may need some help actually *applying* the tips to your work in this mathematics course.

Watch for this icon as you work in this textbook, particularly in the first few chapters. It will direct you to one of 12 activities in the *Study Skills Workbook* that comes with this text. Each activity helps you to actually *use* a study skills technique. These techniques will greatly improve your chances for success in this course.

- Find out *how your brain learns new material*. Then use that information to set up effective ways to learn mathematics.
- Find out *why short-term memory is so short* and what you can do to help your brain remember new material weeks and months later.
- Find out *what happens when you "blank out" on a test* and simple ways to prevent it from happening.

All the activities in the *Study Skills Workbook* are practical ways to enjoy and succeed at mathematics. Whether you need help with note taking, managing homework, taking tests, or preparing for a final exam, you'll find specific, clearly explained ideas that really work because they're based on research about how the brain learns and remembers.

Diagnostic Pretest

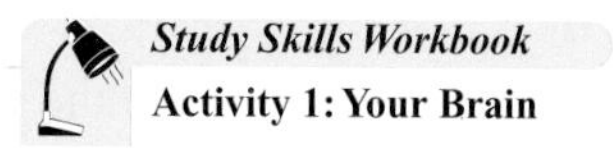

[Chapter R]

1. Find the quotient and write it in lowest terms.

$$\frac{42}{5} \div \frac{7}{15}$$

1. ______________

2. Find the sum and write it in lowest terms.

$$6\frac{7}{8} + 3\frac{2}{3}$$

2. ______________

3. Subtract $38 - 9.678$.

3. ______________

4. **(a)** Convert .99% to a decimal.

(b) Convert 4.72 to a percent.

4. (a) ______________

(b) ______________

[Chapter 1]

5. Select the lesser number from this pair: $|-35|, -|35|$.

5. ______________

Perform the indicated operations.

6. $(-3)(-8) - 4(-2)^3$

6. ______________

7. $\dfrac{-6 + |-11 + 5|}{4^2 - (-9)}$

7. ______________

8. Evaluate $\dfrac{6r - 2s^2}{-3t}$ if $r = -5$, $s = -3$, and $t = 4$.

8. ______________

[Chapter 2]

9. Solve and check $-8(x - 3) + 12x = 15 - (2x + 3)$.

9. ______________

10. The two largest cities in the United States are New York and Los Angeles. In 2002, the population of New York was 4,285,335 greater than the population of Los Angeles, and there were a total of 11,883,297 people living in these two cities. Find the population of each city, using these population figures. (*Source: World Almanac and Book of Facts,* 2004.)

10. ______________

11. Find the measure of each marked angle.

11. ______________

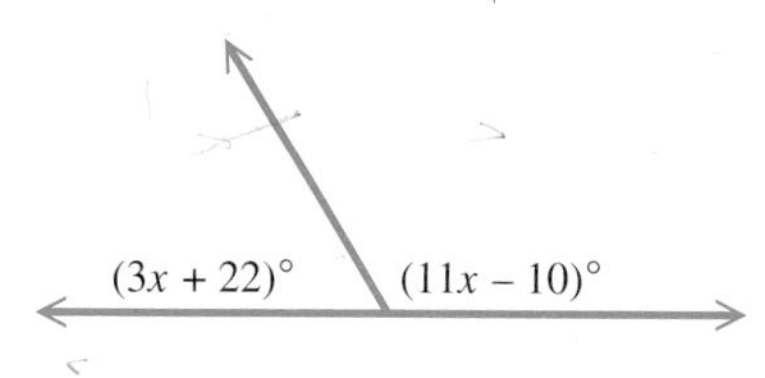

12.

12. Solve $-4x + 8 \geq -12$, and graph the solutions.

[Chapter 3]

13. x-intercept: ____________

y-intercept: ____________

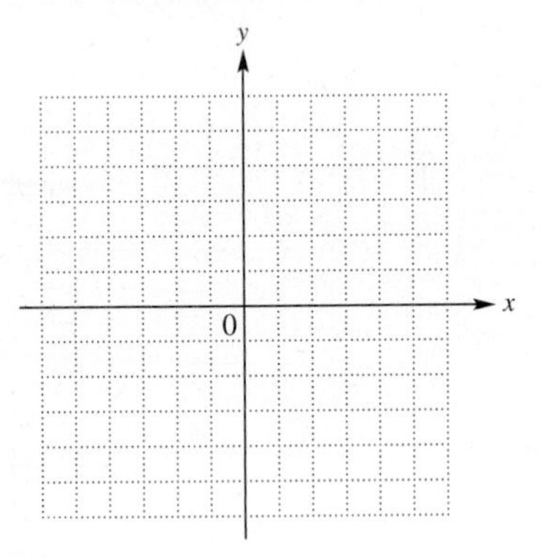

13. Graph $2x - 5y = 10$. Give the x- and y-intercepts.

14. ____________

14. Find the slope of the line through $(-2, 5)$ and $(1, -7)$.

15. ____________

15. Write an equation in slope-intercept form for the line through $(-5, 6)$ and $(1, 0)$.

16.

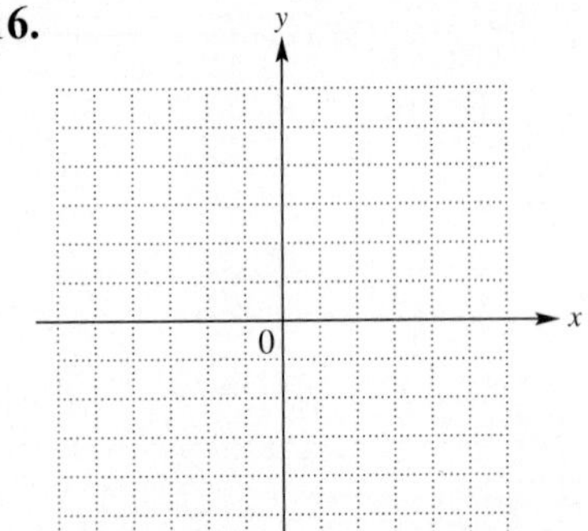

16. Graph $2x + y > 4$.

[Chapter 4]

Solve each system of equations.

17. ____________

17. $3x + y = 9$
$x - y = -1$

18. ____________

18. $-4x + 7y = 3$
$12x - 21y = 9$

19. ____________

19. Write a system of equations and use it to solve the problem.

Marla and Rick left from the same place at the same time and traveled in opposite directions. Marla drove 8 mph faster than Rick. After 2 hr, they were 228 mi apart. Find Marla's and Rick's speeds.

20.

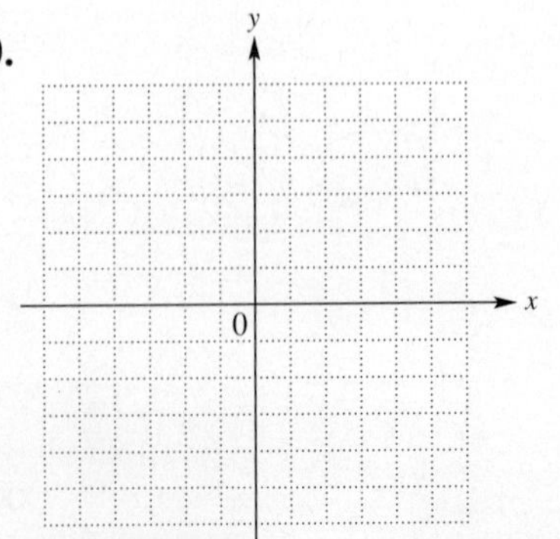

20. Graph the solution of the system of inequalities.

$$3x - 5y < 15$$
$$y \geq -2x$$

[Chapter 5]

21. Subtract $(4m^3 - 5m^2 + m - 8) - (6m^3 - 5m^2 + 10m - 3)$.

21. ______

22. Multiply $(7z + 3w)^2$.

22. ______

23. Evaluate the expression $5^{-1} + 2^{-1} - 3^0$.

23. ______

24. **(a)** Write 445,000,000 in scientific notation.

(b) Write 2.34×10^{-4} without exponents.

24. (a) ______

(b) ______

[Chapter 6]

25. Factor $3x^2 + 2x - 8$.

25. ______

26. Factor $16n^2 - 49$.

26. ______

27. Solve $t^2 - 2t = 15$.

27. ______

28. The length of the cover of a road atlas is 4 in. more than the width. The area is 165 in.^{2}. Find the dimensions of the cover.

28. ______

[Chapter 7]

29. Write $\frac{x^2 + x - 20}{x^2 - 16}$ in lowest terms.

29. ______

30. Divide. Write your answer in lowest terms.

$$\frac{2r + 1}{r - 4} \div \frac{6r^2 + 3r}{4 - r}$$

30. ______

31. Subtract. Write your answer in lowest terms.

$$\frac{z^2}{z - 3} - \frac{z}{z + 3}$$

31. ______

32. Simplify.

$$\frac{\frac{1}{y} + \frac{1}{y + 2}}{\frac{1}{y} - \frac{1}{y + 2}}$$

32. ______

[Chapter 8]

Simplify when possible.

33. ________

33. $4\sqrt{300} - 8\sqrt{75}$

34. ________

34. $(\sqrt{5} - \sqrt{7})^2$

35. ________

35. Rationalize the denominator.

$$\frac{4\sqrt{5}}{\sqrt{2}}$$

36. ________

36. Solve $\sqrt{3r} + 6 = 15$.

[Chapter 9]

37. ________

37. Solve $(x - 3)^2 = 20$.

38. ________

38. Solve $2r^2 + 3r - 1 = 0$.

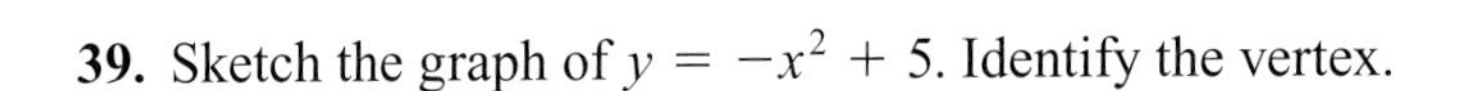

39. vertex: ________

39. Sketch the graph of $y = -x^2 + 5$. Identify the vertex.

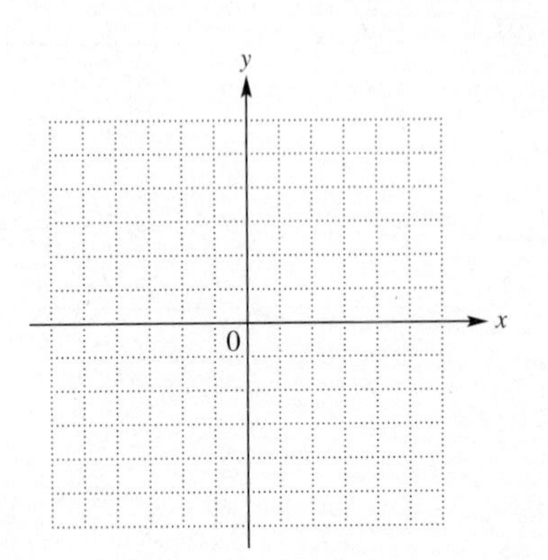

40. ________

domain: ________

range: ________

40. Decide whether the relation

$$\{(-2, 4), (-1, 1), (0, 0), (1, 1), (2, 4)\}$$

represents a function. Give the domain and range.

Prealgebra Review

R.1 Fractions

OBJECTIVES

1. Identify prime numbers.
2. Write numbers in prime factored form.
3. Write fractions in lowest terms.
4. Multiply and divide fractions.
5. Add and subtract fractions.

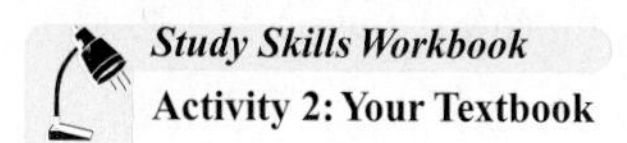
Study Skills Workbook
Activity 2: Your Textbook

The numbers used most often in everyday life are the **whole numbers,**

$$0, 1, 2, 3, 4, 5, \ldots$$

and **fractions,** such as

$$\frac{1}{3}, \quad \frac{5}{4}, \quad \text{and} \quad \frac{11}{12}.$$

The parts of a fraction are named as follows.

$$\text{Fraction bar} \longrightarrow \frac{4}{7} \begin{array}{l} \longleftarrow \text{Numerator} \\ \longleftarrow \text{Denominator} \end{array}$$

If the numerator of a fraction is smaller than the denominator, we call it a **proper fraction.** A proper fraction has a value less than 1. If the numerator is greater than or equal to the denominator, the fraction is an **improper fraction.** An improper fraction that has a value greater than 1 is often written as a **mixed number.** For example,

$$\frac{12}{5} \text{ may be written as } 2\frac{2}{5}.$$

$\uparrow$ Improper fraction — $\uparrow$ Mixed number

OBJECTIVE 1 Identify prime numbers. In work with fractions, we will need to write the numerators and denominators as products. A **product** is the answer to a multiplication problem. When 12 is written as the product $2 \cdot 6$, for example, 2 and 6 are called **factors** of 12. Other factors of 12 are 1, 3, 4, and 12. A whole number is **prime** if it has exactly two different factors (itself and 1). The first dozen primes are listed here.

$$2, 3, 5, 7, 11, 13, 17, 19, 23, 29, 31, 37 \quad \text{Prime numbers}$$

A whole number greater than 1 that is not prime is called a **composite number.** Some examples follow.

$$4, 6, 8, 9, 10, 12 \quad \text{Composite numbers}$$

The number 1 is neither prime nor composite.

1 Tell whether each number is *prime* or *composite*.

(a) 12

(b) 13

(c) 27

(d) 59

(e) 1806

2 Write each number in prime factored form.

(a) 70

(b) 72

(c) 693

(d) 97

Answers

1. **(a)** composite **(b)** prime **(c)** composite **(d)** prime **(e)** composite
2. **(a)** $2 \cdot 5 \cdot 7$ **(b)** $2 \cdot 2 \cdot 2 \cdot 3 \cdot 3$ **(c)** $3 \cdot 3 \cdot 7 \cdot 11$ **(d)** 97 is prime.

EXAMPLE 1 Distinguishing between Prime and Composite Numbers

Decide whether each number is *prime* or *composite*.

(a) 33

33 has factors of 3 and 11 as well as 1 and 33, so it is composite.

(b) 43

Since there are no numbers other than 1 and 43 itself that divide *evenly* into 43, the number 43 is prime.

(c) 9832

9832 can be divided by 2, giving $2 \cdot 4916$, so it is composite.

Work Problem 1 at the Side.

OBJECTIVE 2 Write numbers in prime factored form. To factor a number means to write it as the product of two or more numbers. Factoring is the reverse of multiplying two numbers to get the product.

Multiplication	Factoring
$6 \cdot 3 = \mathbf{18}$	$\mathbf{18} = 6 \cdot 3$
↑ ↑ Factors, ↑ Product	↑ Product, ↑ ↑ Factors

In algebra, a dot $\cdot$ is used instead of the $\times$ symbol to indicate multiplication because $\times$ may be confused with the letter x. A composite number written using factors that are all prime numbers is in **prime factored form.**

EXAMPLE 2 Writing Numbers in Prime Factored Form

Write each number in prime factored form.

(a) 35

Factor 35 as the product of the prime factors 5 and 7, or as $35 = 5 \cdot 7$.

(b) 24

We use a factor tree, as shown below. The prime factors are circled.

		24
Divide by the smallest prime, 2, to get	$24 = 2 \cdot 12.$	(2) · 12
Now divide 12 by 2 to find factors of 12.	$24 = 2 \cdot 2 \cdot 6$	(2) · 6
Since 6 can be factored as $2 \cdot 3$, $24 = 2 \cdot 2 \cdot 2 \cdot 3$, where all factors are prime.	$24 = 2 \cdot 2 \cdot 2 \cdot 3$	(2) · (3)

Work Problem 2 at the Side.

OBJECTIVE 3 Write fractions in lowest terms. A fraction is in **lowest terms** when the numerator and denominator have no factors in common (other than 1). The following properties are useful.

Properties of 1

Any nonzero number divided by itself is equal to 1; for example, $\frac{3}{3} = 1$.

Any number multiplied by 1 remains the same; for example, $7 \cdot 1 = 7$.

Writing a Fraction in Lowest Terms

Step 1 Write the numerator and denominator in prime factored form.

Step 2 Replace each pair of factors common to the numerator and denominator with 1.

Step 3 Multiply the remaining factors in the numerator and in the denominator.

(This procedure is sometimes called "simplifying the fraction.")

3 Write each fraction in lowest terms.

(a) $\frac{8}{14}$

(b) $\frac{35}{42}$

(c) $\frac{120}{72}$

EXAMPLE 3 Writing Fractions in Lowest Terms

Write each fraction in lowest terms.

(a) $\frac{10}{15} = \frac{2 \cdot 5}{3 \cdot 5} = \frac{2}{3} \cdot \frac{5}{5} = \frac{2}{3} \cdot 1 = \frac{2}{3}$

Since 5 is a common factor of 10 and 15, we use the first property of 1 to replace $\frac{5}{5}$ with 1.

(b) $\frac{15}{45} = \frac{3 \cdot 5}{3 \cdot 3 \cdot 5} = \frac{1 \cdot 3 \cdot 5}{3 \cdot 3 \cdot 5} = \frac{1}{3} \cdot \frac{3}{3} \cdot \frac{5}{5} = \frac{1}{3} \cdot 1 \cdot 1 = \frac{1}{3}$

Multiplying by 1 in the numerator does not change the value of the numerator and makes it possible to rewrite the expression as the product of three fractions in the next step.

(c) $\frac{150}{200}$

It is not always necessary to factor into *prime* factors in Step 1. Here, if you see that 50 is a common factor of the numerator and the denominator, factor as follows:

$$\frac{150}{200} = \frac{3 \cdot 50}{4 \cdot 50} = \frac{3}{4} \cdot 1 = \frac{3}{4}.$$

NOTE

When writing a fraction in lowest terms, look for the largest common factor in the numerator and the denominator. If none is obvious, factor the numerator and the denominator into prime factors. *Any* common factor can be used and the fraction can be simplified in stages.

For example, $\frac{150}{200} = \frac{15 \cdot 10}{20 \cdot 10} = \frac{3 \cdot 5 \cdot 10}{4 \cdot 5 \cdot 10} = \frac{3}{4}.$

Work Problem 3 at the Side.

OBJECTIVE 4 Multiply and divide fractions.

Multiplying Fractions

To multiply two fractions, multiply the numerators to get the numerator of the product, and multiply the denominators to get the denominator of the product. The product must be written in lowest terms.

ANSWERS

3. (a) $\frac{4}{7}$ (b) $\frac{5}{6}$ (c) $\frac{5}{3}$

4 Find each product, and write it in lowest terms.

(a) $\frac{5}{8} \cdot \frac{2}{10}$

(b) $\frac{1}{10} \cdot \frac{12}{5}$

(c) $\frac{7}{9} \cdot \frac{12}{14}$

(d) $3\frac{1}{3} \cdot 1\frac{3}{4}$

ANSWERS

4. (a) $\frac{1}{8}$ (b) $\frac{6}{25}$ (c) $\frac{2}{3}$ (d) $\frac{35}{6}$ or $5\frac{5}{6}$

EXAMPLE 4 Multiplying Fractions

Find each product, and write it in lowest terms.

(a) $\frac{3}{8} \cdot \frac{4}{9} = \frac{3 \cdot 4}{8 \cdot 9}$ Multiply numerators. Multiply denominators.

$= \frac{3 \cdot 4}{2 \cdot 4 \cdot 3 \cdot 3}$ Factor.

$= \frac{1}{2 \cdot 3} = \frac{1}{6}$ Write in lowest terms.

(b) $2\frac{1}{3} \cdot 5\frac{1}{2} = \frac{7}{3} \cdot \frac{11}{2}$ Write as improper fractions.

$= \frac{77}{6}$ or $12\frac{5}{6}$ Multiply numerators and denominators; write as a mixed number.

Work Problem 4 at the Side.

Two fractions are **reciprocals** of each other if their product is 1. For example, $\frac{3}{4}$ and $\frac{4}{3}$ are reciprocals because

$$\frac{3}{4} \cdot \frac{4}{3} = 1.$$

The numbers $\frac{7}{11}$ and $\frac{11}{7}$ are reciprocals also. Other examples are $\frac{1}{5}$ and 5, $\frac{4}{9}$ and $\frac{9}{4}$, and 16 and $\frac{1}{16}$.

Because division is the opposite or inverse of multiplication, we use reciprocals to divide fractions.

Dividing Fractions

To divide two fractions, multiply the first fraction by the reciprocal of the second. The result, called the **quotient,** must be written in lowest terms.

The reason this method works will be explained in **Section 1.6.** However, as an example, we know that $20 \div 10 = 2$, and $20 \cdot \frac{1}{10} = 2$.

EXAMPLE 5 Dividing Fractions

Find each quotient, and write it in lowest terms.

(a) $\frac{3}{4} \div \frac{8}{5} = \frac{3}{4} \cdot \frac{5}{8} = \frac{3 \cdot 5}{4 \cdot 8} = \frac{15}{32}$

Multiply by the reciprocal of the second fraction.

(b) $\frac{3}{4} \div \frac{5}{8} = \frac{3}{4} \cdot \frac{8}{5} = \frac{3 \cdot 8}{4 \cdot 5} = \frac{3 \cdot 4 \cdot 2}{4 \cdot 5} = \frac{6}{5}$ or $1\frac{1}{5}$

(c) $\frac{5}{8} \div 10 = \frac{5}{8} \div \frac{10}{1} = \frac{5}{8} \cdot \frac{1}{10} = \frac{5 \cdot 1}{8 \cdot 10} = \frac{5 \cdot 1}{8 \cdot 2 \cdot 5} = \frac{1}{16}$

Write 10 as $\frac{10}{1}$.

Continued on Next Page

(d) $1\frac{2}{3} \div 4\frac{1}{2} = \frac{5}{3} \div \frac{9}{2}$ Write as improper fractions.

$= \frac{5}{3} \cdot \frac{2}{9}$ Multiply by the reciprocal of the second fraction.

$= \frac{10}{27}$ Multiply numerators and denominators.

CAUTION

Notice that *only* the second fraction (the divisor) is replaced by its reciprocal in the multiplication.

Work Problem 5 at the Side.

5 Find each quotient, and write it in lowest terms.

(a) $\frac{3}{10} \div \frac{2}{7}$

(b) $\frac{3}{4} \div \frac{7}{16}$

(c) $\frac{4}{3} \div 6$

(d) $3\frac{1}{4} \div 1\frac{2}{5}$

OBJECTIVE 5 Add and subtract fractions. The result of adding two numbers is called the **sum** of the numbers. For example, since $2 + 3 = 5$, the sum of 2 and 3 is 5.

Adding Fractions

To find the sum of two fractions with the *same* denominator, add their numerators and keep the *same* denominator.

EXAMPLE 6 Adding Fractions with the Same Denominator

Add. Write sums in lowest terms.

(a) $\frac{3}{7} + \frac{2}{7} = \frac{3 + 2}{7} = \frac{5}{7}$ Add numerators; denominator does not change.

(b) $\frac{2}{10} + \frac{3}{10} = \frac{2 + 3}{10} = \frac{5}{10} = \frac{1}{2}$ Write in lowest terms.

Work Problem 6 at the Side.

6 Add. Write sums in lowest terms.

(a) $\frac{3}{5} + \frac{4}{5}$

(b) $\frac{5}{14} + \frac{3}{14}$

If the fractions to be added do not have the same denominator, the procedure above can still be used, but only *after* the fractions are rewritten with a common denominator. For example, to rewrite $\frac{3}{4}$ as a fraction with a denominator of 32,

$$\frac{3}{4} = \frac{?}{32},$$

we must find the number that can be multiplied by 4 to give 32. Since $4 \cdot 8 = 32$, we use the number 8. By the second property of 1, we can multiply the numerator and the denominator by 8.

$$\frac{3}{4} = \frac{3}{4} \cdot 1 = \frac{3}{4} \cdot \frac{8}{8} = \frac{3 \cdot 8}{4 \cdot 8} = \frac{24}{32}$$

ANSWERS

5. (a) $\frac{21}{20}$ or $1\frac{1}{20}$ (b) $\frac{12}{7}$ or $1\frac{5}{7}$ (c) $\frac{2}{9}$ (d) $\frac{65}{28}$ or $2\frac{9}{28}$

6. (a) $\frac{7}{5}$ or $1\frac{2}{5}$ (b) $\frac{4}{7}$

7 Add. Write sums in lowest terms.

(a) $\frac{7}{30} + \frac{2}{45}$

(b) $\frac{17}{10} + \frac{8}{27}$

(c) $2\frac{1}{8} + 1\frac{2}{3}$

(d) $132\frac{4}{5} + 28\frac{3}{4}$

Finding the Least Common Denominator (LCD)

Step 1 Factor all denominators to prime factored form.

Step 2 The LCD is the product of every (different) factor that appears in any of the factored denominators. If a factor is repeated, use the largest number of repeats as factors of the LCD.

Step 3 Write each fraction with the LCD as the denominator, using the second property of 1.

EXAMPLE 7 Adding Fractions with Different Denominators

Add. Write sums in lowest terms.

(a) $\frac{4}{15} + \frac{5}{9}$

Step 1 To find the LCD, factor the denominators to prime factored form.

$$15 = 5 \cdot 3 \quad \text{and} \quad 9 = 3 \cdot 3$$

3 is a factor of both denominators.

Step 2

$$\text{LCD} = 5 \cdot 3 \cdot 3 = 45$$

(15 = 5 · 3; 9 = 3 · 3)

In this example, the LCD needs one factor of 5 and two factors of 3 because the second denominator has two factors of 3.

Step 3 Now we can use the second property of 1 to write each fraction with 45 as the denominator.

$$\frac{4}{15} = \frac{4}{15} \cdot \frac{3}{3} = \frac{12}{45} \quad \text{and} \quad \frac{5}{9} = \frac{5}{9} \cdot \frac{5}{5} = \frac{25}{45}$$

Now add the two equivalent fractions to get the required sum.

$$\frac{4}{15} + \frac{5}{9} = \frac{12}{45} + \frac{25}{45} = \frac{37}{45}$$

(b) $3\frac{1}{2} + 2\frac{3}{4} = \frac{7}{2} + \frac{11}{4}$ Change to improper fractions.

$= \frac{14}{4} + \frac{11}{4}$ Get a common denominator.

$= \frac{25}{4}$ or $6\frac{1}{4}$ Add; write as a mixed number.

(c) $45\frac{2}{3} + 73\frac{1}{2}$

We use a vertical method here.

$$\begin{array}{r} 45\frac{2}{3} = 45\frac{4}{6} \\ +\ 73\frac{1}{2} = 73\frac{3}{6} \\ \hline \end{array}$$

Add the whole numbers and the fractions separately.

$$118\frac{7}{6} = 118 + \left(1 + \frac{1}{6}\right) = 119\frac{1}{6}$$

Work Problem 7 at the Side.

Answers

7. (a) $\frac{5}{18}$ (b) $\frac{539}{270}$ or $1\frac{269}{270}$ (c) $\frac{91}{24}$ or $3\frac{19}{24}$ (d) $161\frac{11}{20}$

The **difference** between two numbers is found by subtracting the numbers. For example, $9 - 5 = 4$, so the difference between 9 and 5 is 4. We find the difference between two fractions as follows.

Subtracting Fractions

To find the difference between two fractions with the *same* denominator, subtract their numerators and keep the *same* denominator.

If the fractions have *different* denominators, write them with a common denominator first.

EXAMPLE 8 Subtracting Fractions

Subtract. Write differences in lowest terms.

(a) $\frac{15}{8} - \frac{3}{8} = \frac{15 - 3}{8}$ Subtract numerators; denominator does not change.

$= \frac{12}{8} = \frac{3}{2}$ Lowest terms

(b) $\frac{15}{16} - \frac{4}{9}$

Since $16 = 2 \cdot 2 \cdot 2 \cdot 2$ and $9 = 3 \cdot 3$ have no common factors, the LCD is $16 \cdot 9 = 144$.

$\frac{15}{16} - \frac{4}{9} = \frac{15 \cdot 9}{16 \cdot 9} - \frac{4 \cdot 16}{9 \cdot 16}$ Get a common denominator.

$= \frac{135}{144} - \frac{64}{144}$

$= \frac{71}{144}$ Subtract numerators; keep the same denominator.

(c) $2\frac{1}{2} - 1\frac{3}{4} = \frac{5}{2} - \frac{7}{4}$ Write as improper fractions.

$= \frac{10}{4} - \frac{7}{4}$ Get a common denominator.

$= \frac{3}{4}$ Subtract.

Alternatively, we could use a vertical method.

$$\begin{array}{r} 2\frac{1}{2} = 2\frac{2}{4} = 1\frac{6}{4} \\ -\ 1\frac{3}{4} = 1\frac{3}{4} = 1\frac{3}{4} \\ \hline \frac{3}{4} \end{array}$$

Work Problem 8 at the Side.

We often see mixed numbers used in applications of mathematics, as shown in Examples 9 and 10 on the next page.

8 Subtract.

(a) $\frac{9}{11} - \frac{3}{11}$

(b) $\frac{13}{15} - \frac{5}{6}$

(c) $2\frac{3}{8} - 1\frac{1}{2}$

(d) $50\frac{1}{4} - 32\frac{2}{3}$

ANSWERS

8. (a) $\frac{6}{11}$ **(b)** $\frac{1}{30}$ **(c)** $\frac{7}{8}$ **(d)** $17\frac{7}{12}$

EXAMPLE 9 Solving an Applied Problem Requiring Addition of Fractions

The diagram in Figure 1 appears in the book *Woodworker's 39 Sure-Fire Projects.* It is a view of a bookcase/desk. Add the fractions in the diagram to find the height of the bookcase/desk to the top of the writing surface.

We must add the following measures (″ means inches):

$$\frac{3}{4},\quad 4\frac{1}{2},\quad 9\frac{1}{2},\quad \frac{3}{4},\quad 9\frac{1}{2},\quad \frac{3}{4},\quad 4\frac{1}{2}.$$

We begin by changing $4\frac{1}{2}$ to $4\frac{2}{4}$ and $9\frac{1}{2}$ to $9\frac{2}{4}$, since the common denominator is 4. Then we use the method of Example 7(c).

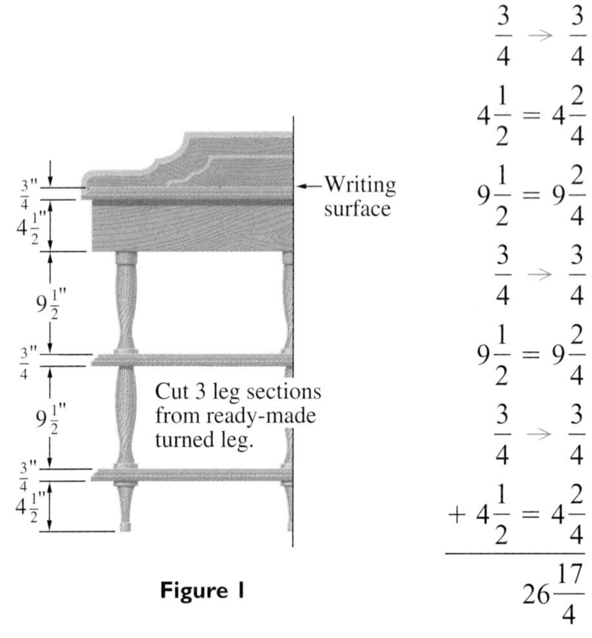

Figure 1

$$\begin{aligned}
\frac{3}{4} &\rightarrow \frac{3}{4}\\
4\frac{1}{2} &= 4\frac{2}{4}\\
9\frac{1}{2} &= 9\frac{2}{4}\\
\frac{3}{4} &\rightarrow \frac{3}{4}\\
9\frac{1}{2} &= 9\frac{2}{4}\\
\frac{3}{4} &\rightarrow \frac{3}{4}\\
+\,4\frac{1}{2} &= 4\frac{2}{4}\\
\hline
& \quad 26\frac{17}{4}
\end{aligned}$$

Since $\frac{17}{4} = 4\frac{1}{4}$, $26\frac{17}{4} = 26 + 4\frac{1}{4} = 30\frac{1}{4}$. The height is $30\frac{1}{4}$ in.

Work Problem 9 at the Side.

9 Solve the problem.

To make a three-piece outfit from the same fabric, Wei Jen needs $1\frac{1}{4}$ yd for the blouse, $1\frac{2}{3}$ yd for the skirt, and $2\frac{1}{2}$ yd for the jacket. How much fabric does she need?

EXAMPLE 10 Solving an Applied Problem Requiring Division of Fractions

An upholsterer needs $2\frac{1}{4}$ yd of fabric to cover a chair. How many chairs can be covered with $23\frac{2}{3}$ yd of fabric?

To better understand the problem, we replace the fractions with whole numbers. Suppose each chair requires 2 yd, and we have 24 yd of fabric. Dividing 24 by 2 gives the number of chairs (12) that can be covered. To solve the original problem, we must divide $23\frac{2}{3}$ by $2\frac{1}{4}$.

$$\begin{aligned}
23\frac{2}{3} \div 2\frac{1}{4} &= \frac{71}{3} \div \frac{9}{4}\\
&= \frac{71}{3} \cdot \frac{4}{9}\\
&= \frac{284}{27} \quad \text{or} \quad 10\frac{14}{27}
\end{aligned}$$

Thus, 10 chairs can be covered with some fabric left over.

Work Problem 10 at the Side.

10 Solve the problem.

A gallon of paint covers 500 ft^2. (ft^2 means square feet.) To paint his house, Tram needs enough paint to cover 4200 ft^2. How many gallons of paint should he buy?

ANSWERS

9. $5\frac{5}{12}$ yd

10. $8\frac{2}{5}$ gal are needed, so he must buy 9 gal.

R.1 Exercises

FOR EXTRA HELP

Addison-Wesley Math Tutor Center

MathXL

Digital Video Tutor CD 1 Videotape 1

Student's Solutions Manual

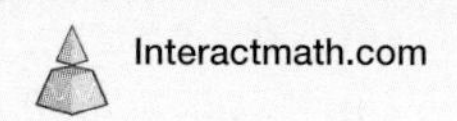

MyMathLab

Interactmath.com

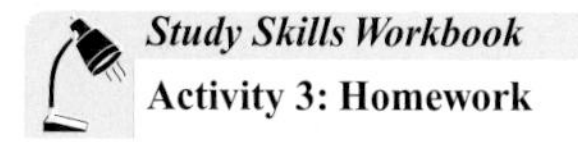

Study Skills Workbook Activity 3: Homework

Decide whether each statement is true *or* false. *If it is* false, *say why.*

1. In the fraction $\frac{3}{7}$, 3 is the numerator and 7 is the denominator.

2. The mixed number equivalent of $\frac{41}{5}$ is $8\frac{1}{5}$.

3. The fraction $\frac{17}{51}$ is in lowest terms.

4. The reciprocal of $\frac{8}{2}$ is $\frac{4}{1}$.

5. The product of 8 and 2 is 10.

6. The difference between 12 and 2 is 6.

Identify each number as prime, composite, *or* neither. *See Example 1.*

7. 19 **8.** 99 **9.** 52 **10.** 61

11. 2468 **12.** 3125 **13.** 1 **14.** 83

Write each number in prime factored form. See Example 2.

15. 30 **16.** 40 **17.** 252 **18.** 168

19. 124 **20.** 165 **21.** 29 **22.** 31

Write each fraction in lowest terms. See Example 3.

23. $\frac{8}{16}$ **24.** $\frac{4}{12}$ **25.** $\frac{15}{18}$ **26.** $\frac{16}{20}$

27. $\frac{15}{75}$ **28.** $\frac{24}{64}$ **29.** $\frac{144}{120}$ **30.** $\frac{132}{77}$

31. For the fractions $\frac{p}{q}$ and $\frac{r}{s}$, which can serve as a common denominator?

A. $q \cdot s$

B. $q + s$

C. $p \cdot r$

D. $p + r$

32. Which is the correct way to write $\frac{16}{24}$ in lowest terms?

A. $\frac{16}{24} = \frac{8 + 8}{8 + 16} = \frac{8}{16} = \frac{1}{2}$

B. $\frac{16}{24} = \frac{4 \cdot 4}{4 \cdot 6} = \frac{4}{6}$

C. $\frac{16}{24} = \frac{8 \cdot 2}{8 \cdot 3} = \frac{2}{3}$

D. $\frac{16}{24} = \frac{14 + 2}{21 + 3} = \frac{2}{3} + \frac{2}{3} = \frac{4}{3}$

Find each product or quotient, and write it in lowest terms. See Examples 4 and 5.

33. $\frac{4}{5} \cdot \frac{6}{7}$

34. $\frac{5}{9} \cdot \frac{10}{7}$

35. $\frac{1}{10} \cdot \frac{12}{5}$

36. $\frac{6}{11} \cdot \frac{2}{3}$

37. $\frac{15}{4} \cdot \frac{8}{25}$

38. $\frac{4}{7} \cdot \frac{21}{8}$

39. $2\frac{2}{3} \cdot 5\frac{4}{5}$

40. $3\frac{3}{5} \cdot 7\frac{1}{6}$

41. $\frac{5}{4} \div \frac{3}{8}$

42. $\frac{7}{6} \div \frac{9}{10}$

43. $\frac{32}{5} \div \frac{8}{15}$

44. $\frac{24}{7} \div \frac{6}{21}$

45. $\frac{3}{4} \div 12$

46. $\frac{2}{5} \div 30$

47. $2\frac{5}{8} \div 1\frac{15}{32}$

48. $2\frac{3}{10} \div 7\frac{4}{5}$

49. In your own words, explain how to divide two fractions.

50. In your own words, explain how to add two fractions that have different denominators.

Find each sum or difference, and write it in lowest terms. See Examples 6–8.

51. $\frac{7}{12} + \frac{1}{12}$

52. $\frac{3}{16} + \frac{5}{16}$

53. $\frac{5}{9} + \frac{1}{3}$

54. $\frac{4}{15} + \frac{1}{5}$

55. $3\frac{1}{8} + \frac{1}{4}$

56. $5\frac{3}{4} + \frac{2}{3}$

57. $\frac{7}{12} - \frac{1}{9}$

58. $\frac{11}{16} - \frac{1}{12}$

59. $6\frac{1}{4} - 5\frac{1}{3}$

60. $8\frac{4}{5} - 7\frac{4}{9}$

61. $\frac{5}{3} + \frac{1}{6} - \frac{1}{2}$

62. $\frac{7}{15} + \frac{1}{6} - \frac{1}{10}$

Use the chart, which appears on a package of Quaker Quick Grits, to answer the questions in Exercises 63 and 64.

63. How many cups of water would be needed for eight microwave servings?

64. How many teaspoons of salt would be needed for five stove top servings? (*Hint:* 5 is halfway between 4 and 6.)

	Microwave	Stove Top		
Servings	**1**	**1**	**4**	**6**
Water	$\frac{3}{4}$ cup	1 cup	3 cups	4 cups
Grits	3 Tbsp	3 Tbsp	$\frac{3}{4}$ cup	1 cup
Salt (optional)	Dash	Dash	$\frac{1}{4}$ tsp	$\frac{1}{2}$ tsp

Solve each applied problem. See Examples 9 and 10.

65. A motel owner has decided to expand his business by buying a piece of property next to the motel. The property has an irregular shape, with five sides as shown in the figure. Find the total distance around the piece of property. This is called the **perimeter** of the figure.

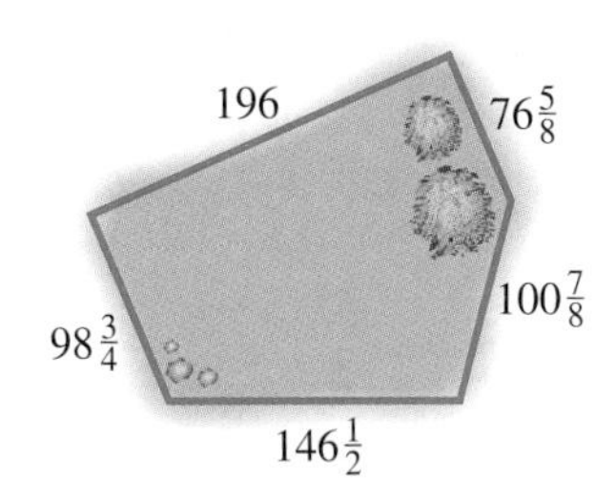

Measurements in feet

66. A triangle has sides of lengths $5\frac{1}{4}$ ft, $7\frac{1}{2}$ ft, and $10\frac{1}{8}$ ft. Find the perimeter of the triangle. See Exercise 65.

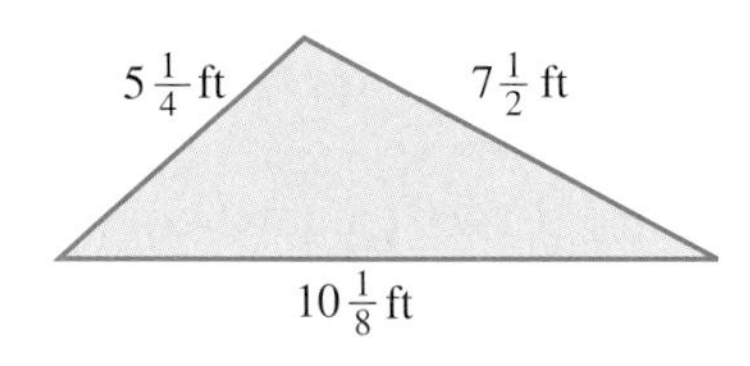

67. A hardware store sells a 40-piece socket wrench set. The measure of the largest socket is $\frac{3}{4}$ in., while the measure of the smallest socket is $\frac{3}{16}$ in. What is the difference between these measures?

68. Two sockets in a socket wrench set have measures of $\frac{9}{16}$ in. and $\frac{3}{8}$ in. What is the difference between these two measures?

69. Under existing standards, most of the holes in Swiss cheese must have diameters between $\frac{11}{16}$ and $\frac{13}{16}$ in. To accommodate new high-speed slicing machines, the USDA wants to reduce the minimum size to $\frac{3}{8}$ in. How much smaller is $\frac{3}{8}$ in. than $\frac{11}{16}$ in.? (*Source:* U.S. Department of Agriculture.)

70. Tex's favorite recipe for barbecue sauce calls for $2\frac{1}{3}$ cups of tomato sauce. The recipe makes enough barbecue sauce to serve 7 people. How much tomato sauce is needed for 1 serving?

71. It takes $2\frac{3}{8}$ yd of fabric to make a costume for a school play. How much fabric would be needed for 7 costumes?

72. A cake recipe calls for $1\frac{3}{4}$ cups of sugar. A caterer has $15\frac{1}{2}$ cups of sugar on hand. How many cakes can he make?

More than 8 million immigrants were admitted to the United States between 1990 and 1998. The pie chart gives the fractional number from each region of birth for these immigrants. Use the chart to answer the following questions.

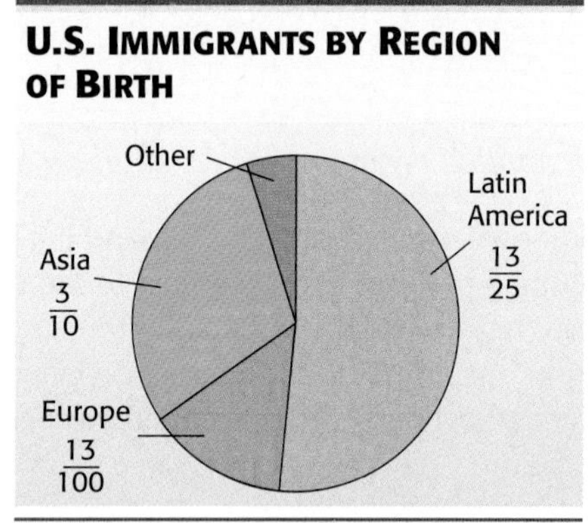

Source: U.S. Bureau of the Census.

73. What fractional part of the immigrants were from other regions?

74. What fractional part of the immigrants were from Latin America or Asia?

75. How many (in millions) were from Europe?

76. How many more immigrants were from Latin America than all of the other regions combined?

R.2 Decimals and Percents

OBJECTIVES

1. Write decimals as fractions.
2. Add and subtract decimals.
3. Multiply and divide decimals.
4. Write fractions as decimals.
5. Convert percents to decimals and decimals to percents.

Fractions are one way to represent parts of a whole. Another way is with a **decimal fraction** or **decimal,** a number written with a decimal point, such as 9.4. Each digit in a decimal number has a place value, as shown below.

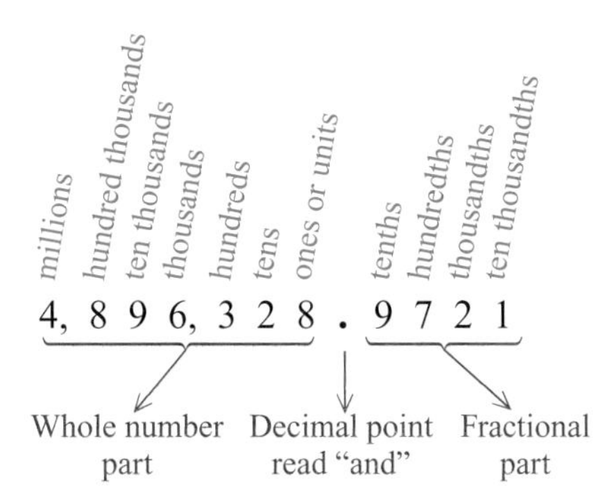

Each successive place value is ten times larger than the place value to its right and is one-tenth as large as the place value to its left.

Prices are often written as decimals. The price \$14.75 means 14 dollars and 75 cents, or 14 dollars and $\frac{75}{100}$ of a dollar.

OBJECTIVE 1 Write decimals as fractions. Place value is used to write a decimal number as a fraction. For example, since the last digit (that is, the digit farthest to the right) of .67 is in the *hundredths* place,

$$.67 = \frac{67}{100}.$$

Similarly, $.9 = \frac{9}{10}$ and $.25 = \frac{25}{100}$. Digits to the left of the decimal point indicate whole numbers, so 12.342 is the sum of 12 and .342, and

$$12.342 = 12 + .342 = 12 + \frac{342}{1000} = \frac{12{,}000}{1000} + \frac{342}{1000} = \frac{12{,}342}{1000}.$$

These examples suggest the following rule.

Converting a Decimal to a Fraction

Read the name using the correct place value. Write it in fraction form just as you read it. The denominator will be a **power of 10,** a number like 10, 100, 1000, and so on.

For example, we read .16 as "sixteen hundredths" and write it in fraction form as $\frac{16}{100}$. The same thing is accomplished by counting the number of digits to the right of the decimal point, then writing the given number without a decimal point over a denominator of 1 followed by that number of zeros.

EXAMPLE 1 Writing Decimals as Fractions

Write each decimal as a fraction. Do not write in lowest terms.

(a) .95

We read .95 as 95 hundredths, so the fraction form is $\frac{95}{100}$. Using the shortcut method, since there are two places to the right of the decimal point, there will be two zeros in the denominator.

Continued on Next Page

1 Write each decimal as a fraction. Do not write in lowest terms.

(a) .8

(b) .431

(c) 20.58

2 Add or subtract as indicated.

(a)
$$\begin{array}{r} 68.9 \\ 42.72 \\ +\ \ 8.973 \\ \hline \end{array}$$

(b)
$$\begin{array}{r} 32.5 \\ -\ 21.72 \\ \hline \end{array}$$

(c) 42.83 + 71 + 3.074

(d) 351.8 − 2.706

$$.95 = \frac{95}{100}$$
2 places 2 zeros

(b) $.056 = \dfrac{56}{1000}$
3 places 3 zeros

(c) $4.2095 = 4 + .2095 = 4 + \dfrac{2095}{10{,}000} = \dfrac{42{,}095}{10{,}000}$
4 places 4 zeros

Work Problem 1 at the Side.

OBJECTIVE 2 Add and subtract decimals.

EXAMPLE 2 Adding and Subtracting Decimals

Add or subtract as indicated.

(a) 6.92 + 14.8 + 3.217

Place the digits of the numbers in columns, with decimal points lined up so that tenths are in one column, hundredths in another column, and so on.

$$\begin{array}{r} 6.92\ \ \, \\ 14.8\ \ \ \ \\ +\ \ 3.217 \\ \hline 24.937 \end{array}$$
Decimal points lined up

A good way to avoid errors is to attach zeros to make all the numbers the same length.

$$\begin{array}{r} 6.92\ \ \, \\ 14.8\ \ \ \ \\ +\ \ 3.217 \\ \hline \end{array} \quad \text{becomes} \quad \begin{array}{r} 6.920 \\ 14.800 \\ +\ \ 3.217 \\ \hline 24.937 \end{array}$$
Attach zeros.

(b) 47.6 − 32.509

Write the numbers in columns, attaching zeros to 47.6.

$$\begin{array}{r} 47.6\ \ \ \ \\ -\ 32.509 \\ \hline \end{array} \quad \text{becomes} \quad \begin{array}{r} 47.600 \\ -\ 32.509 \\ \hline 15.091 \end{array}$$

(c) 3 − .253

A whole number is assumed to have the decimal point at the right of the number. Write 3 as 3.000; then subtract.

$$\begin{array}{r} 3.000 \\ -\ \ \ .253 \\ \hline 2.747 \end{array}$$

Work Problem 2 at the Side.

OBJECTIVE 3 Multiply and divide decimals. We multiply decimals by slightly modifying multiplication of whole numbers. We will sometimes use the times symbol, ×, instead of a dot, to avoid confusion with the decimal point.

ANSWERS

1. (a) $\frac{8}{10}$ (b) $\frac{431}{1000}$ (c) $\frac{2058}{100}$
2. (a) 120.593 (b) 10.78 (c) 116.904 (d) 349.094

Multiplying Decimals

Ignore the decimal points and multiply as if the numbers were whole numbers. Then add together the number of **decimal places** (digits to the *right* of the decimal point) in each number being multiplied. Place the decimal point in the answer that many digits from the right.

EXAMPLE 3 **Multiplying Decimals**

Multiply.

(a) 29.3 × 4.52

Multiply as if the numbers were whole numbers.

$$\begin{array}{r} 29.3 \\ \times\ \ 4.52 \\ \hline 586 \\ 1465\ \\ 1172\ \ \\ \hline 132.436 \end{array}$$

1 decimal place in first number
2 decimal places in second number
1 + 2 = 3
3 decimal places in answer

(b) 7.003 × 55.8

$$\begin{array}{r} 7.003 \\ \times\ \ 55.8 \\ \hline 56024 \\ 35015\ \\ 35015\ \ \\ \hline 390.7674 \end{array}$$

3 decimal places
1 decimal place
3 + 1 = 4
4 decimal places

(c) 31.42 × 65

$$\begin{array}{r} 31.42 \\ \times\ \ 65 \\ \hline 15710 \\ 18852\ \\ \hline 2042.30 \end{array}$$

2 decimal places
0 decimal places
2 + 0 = 2
2 decimal places

The final 0 here can be dropped and the result can be expressed as 2042.3.

Work Problem 3 at the Side.

To divide decimals, convert the divisor to a whole number.

Dividing Decimals

Change the **divisor** (the number you are dividing *by*) into a whole number by moving the decimal point as many places as necessary to the right. Move the decimal point in the **dividend** (the number you are dividing *into*) to the right by the same number of places. Move the decimal point straight up and then divide as with whole numbers.

$$\begin{array}{r} 5 \\ 25\overline{)125} \end{array}$$

Divisor → 25; Quotient ← 5; Dividend ↑ 125

3 Multiply.

(a) 2.13 × .05

(b) 9.32 × 1.4

(c) 300.2 × .052

(d) 42,001 × .012

Answers

3. **(a)** .1065 **(b)** 13.048 **(c)** 15.6104 **(d)** 504.012

4 Divide.

(a) $14.9\overline{)451.47}$

(b) $.37\overline{)5.476}$

(c) $375.1 \div 3.001$

ANSWERS

4. **(a)** 30.3 **(b)** 14.8
(c) 124.99 (rounded)

EXAMPLE 4 Dividing Decimals

Divide.

(a) $233.45 \div 11.5$

Write the problem as follows.

$$11.5\overline{)233.45}$$

To change 11.5 into a whole number, move the decimal point one place to the right. Move the decimal point in 233.45 the same number of places to the right, to get 2334.5.

$$11.5.\overline{)233.4.5} \qquad \text{Move one decimal place to the right.}$$

To see why this works, write the division in fraction form and multiply by $\frac{10}{10}$ or 1.

$$\frac{233.45}{11.5} \cdot \frac{10}{10} = \frac{2334.5}{115}$$

The result is the same as when we moved the decimal point one place to the right in the divisor and the dividend.

Move the decimal point straight up and divide as with whole numbers.

$$\begin{array}{r} 20.3 \\ 115\overline{)2334.5} \\ \underline{230} \\ 345 \\ \underline{345} \\ 0 \end{array} \qquad \text{Move the decimal point straight up.}$$

In the second step of the division, 115 does not divide into 34, so we used zero as a placeholder in the quotient.

(b) $73.85\overline{)1852.882}$ (Round the answer to two decimal places.)

Move the decimal point two places to the right in 73.85, to get 7385. Do the same thing with 1852.882, to get 185288.2.

$$73.85.\overline{)1852.88.2}$$

Move the decimal point straight up and divide as with whole numbers.

$$\begin{array}{r} 25.089 \\ 7385\overline{)185288.200} \\ \underline{14770} \\ 37588 \\ \underline{36925} \\ 66320 \\ \underline{59080} \\ 72400 \\ \underline{66465} \\ 5935 \end{array}$$

We carried out the division to three decimal places so that we could round to two decimal places, obtaining the quotient 25.09.

Work Problem 4 at the Side.

A shortcut can be used when multiplying or dividing by powers of 10.

Multiplying or Dividing by Powers of 10

To ***multiply*** by a power of 10, ***move the decimal point to the right*** as many places as the number of zeros.

To ***divide*** by a power of 10, ***move the decimal point to the left*** as many places as the number of zeros.

In both cases, insert 0s as placeholders if necessary.

EXAMPLE 5 Multiplying and Dividing by Powers of 10

Multiply or divide as indicated.

(a) $48.731 \times 100 = 48.73.1 = 4873.1$

We moved the decimal point two places to the right because 100 has two zeros.

(b) $48.7 \div 1000 = .048.7 = .0487$

We moved the decimal point three places to the left because 1000 has three zeros. We needed to insert a zero in front of the 4 to do this.

Work Problem 5 at the Side.

To avoid misplacing the decimal point, check your work by estimating the answer. ***To get a quick estimate, round the numbers so that only the first digit is not zero,*** using the rule for rounding. For more accurate estimates, the numbers could be rounded to the first two or even three nonzero digits.

Rule for Rounding

If the digit to become 0 or be dropped is 5 or more, round up by adding 1 to the final digit to be kept.

If the digit to become 0 or be dropped is 4 or less, do not round up.

For example, to estimate the answer to Example 2(a), round

6.92 to 7, 14.8 to 10, and 3.217 to 3.

(5 or more; 4 or less; 4 or less)

Since $7 + 10 + 3 = 20$, the answer of 24.937 is reasonable. In Example 4(a), round 233.45 to 200 and 11.5 to 10. Since $200 \div 10 = 20$, the answer of 20.3 is reasonable.

OBJECTIVE 4 Write fractions as decimals.

Writing a Fraction as a Decimal

Because a fraction bar indicates division, write a fraction as a decimal by dividing the denominator into the numerator.

5 Multiply or divide as indicated.

(a) 294.72×10

(b) 19.5×1000

(c) $4.793 \div 100$

(d) $960.1 \div 10$

Answers

5. (a) 2947.2 **(b)** 19,500 **(c)** .04793 **(d)** 96.01

6 Convert to decimals. For repeating decimals, write the answer two ways: using the bar notation and rounding to the nearest thousandth.

(a) $\frac{2}{9}$

(b) $\frac{17}{20}$

(c) $\frac{1}{11}$

EXAMPLE 6 Writing Fractions as Decimals

Write each fraction as a decimal.

(a) $\frac{19}{8}$

$$\begin{array}{r} 2.375 \\ 8\overline{)19.000} \\ \underline{16} \\ 3\,0 \\ \underline{2\,4} \\ 60 \\ \underline{56} \\ 40 \\ \underline{40} \\ 0 \end{array}$$

$\frac{19}{8} = 2.375$

(b) $\frac{2}{3}$

$$\begin{array}{r} .6666\ldots \\ 3\overline{)2.0000\ldots} \\ \underline{1\,8} \\ 20 \\ \underline{18} \\ 20 \\ \underline{18} \\ 20 \end{array}$$

The remainder in the division in part (b) is never 0. Because 2 is always left after the subtraction, this quotient is a **repeating decimal.** A convenient notation for a repeating decimal is a bar over the digit (or digits) that repeats. For instance, we can write $.6666\ldots$ as $.\overline{6}$. We often round repeating decimals to as many places as needed. Rounding to the *nearest thousandth,*

$$\frac{2}{3} = .667. \quad \text{An approximation}$$

CAUTION
When rounding, be careful to distinguish between *thousandths* and *thousands* or between *hundredths* and *hundreds,* and so on.

Work Problem 6 at the Side.

OBJECTIVE 5 Convert percents to decimals and decimals to percents. An important application of decimals is in work with percents. The word **percent** means "per one hundred." Percent is written with the sign %. One percent means "one per one hundred" or "one one-hundredth."

$$\mathbf{1\% = .01} \quad \text{or} \quad \mathbf{1\% = \frac{1}{100}}$$

EXAMPLE 7 Converting Percents and Decimals

(a) Write 73% as a decimal.
Since $1\% = .01$,

$$73\% = 73 \cdot \mathbf{1\%} = 73 \times \mathbf{.01} = .73.$$

Also, 73% can be written as a decimal using the fraction form $1\% = \frac{1}{100}$.

$$73\% = 73 \cdot \mathbf{1\%} = 73 \cdot \left(\frac{\mathbf{1}}{\mathbf{100}}\right) = \frac{73}{100} = .73$$

(b) Write 125% as a decimal.

$$125\% = 125 \cdot 1\% = 125 \times .01 = 1.25$$

Continued on Next Page

Answers
6. (a) $.\overline{2}$, .222 (b) .85 (c) $.\overline{09}$, .091

(c) Write $3\frac{1}{2}\%$ as a decimal.
First write the fractional part as a decimal.

$$3\frac{1}{2}\% = (3 + .5)\% = 3.5\%$$

Now change the percent to decimal form.

$$3.5\% = 3.5 \times .01 = .035 \qquad 1\% = .01$$

(d) Write .32 as a percent.
Since .32 means 32 hundredths, write .32 as $32 \times .01$. Finally, replace .01 with 1%.

$$.32 = 32 \times \mathbf{.01} = 32 \times \mathbf{1\%} = 32\%$$

(e) Write 2.63 as a percent.

$$2.63 = 263 \times .01 = 263 \times 1\% = 263\%$$

NOTE
A quick way to change from a percent to a decimal is to move the decimal point two places to the left. To change from a decimal to a percent, move the decimal point two places to the right.

Divide by 100; Move 2 places left. (Percent → Decimal)

Decimal Percent

Multiply by 100; Move 2 places right. (Decimal → Percent)

EXAMPLE 8 Converting Percents and Decimals by Moving the Decimal Point

Convert each percent to a decimal and each decimal to a percent.

(a) 45% = .45

(b) 250% = 2.50

(c) .57 = 57%

(d) 1.5 = 1.50 = 150%

(e) .327 = 32.7%

Work Problem 7 at the Side.

Calculator Tip In this book, we do not use 0 in the ones place for decimal fractions between 0 and 1. Many calculators (and other books) will show 0.45 instead of just .45 to emphasize that there is a 0 in the ones place. Graphing calculators do *not* show 0 in the ones place. Either way is correct.

7 Convert as indicated.

(a) 23% to a decimal

(b) 310% to a decimal

(c) .71 to a percent

(d) 1.32 to a percent

(e) .685 to a percent

ANSWERS
7. **(a)** .23 **(b)** 3.10 **(c)** 71% **(d)** 132% **(e)** 68.5%

Focus on

Real-Data Applications

Decimalization of Stock Prices

When the New York Stock Exchange (NYSE) was founded in 1792, Thomas Jefferson suggested that stock prices be based on a decimal system. Instead, stock prices were based on the Spanish *eight-reales* coin, or the Spanish milled dollar, which was a commonly used and legal currency in the United States until 1857.

The NYSE decision to base stock prices on an archaic coin resulted in the practice of representing a **price per share** as a mixed number with fractional parts of $\frac{1}{2}$, $\frac{1}{4}$, $\frac{1}{8}$, or $\frac{1}{16}$. After two centuries, the Securities Exchange Commission (SEC) and the NYSE proposed a change in stock pricing to a decimal system consistent with that used by foreign stock exchanges. If they had only listened to Thomas Jefferson!

On August 28, 2000, the U.S. stock markets began the 8-month process of decimalization, which was completed by April 9, 2001. The pilot program included seven stocks on the New York Stock Exchange and six stocks on the American Stock Exchange.

One of the primary issues had been whether to price stocks in one-cent increments or five-cent increments. Under the fraction-based pricing scheme, there were only 16 price changes per $1.00. At five-cent increments, there were 20 price changes per $1.00 compared to 100 price changes per dollar for one-cent increments. Trading in smaller increments increased competition because it lowered the spread between "bid" and "ask" prices—the difference between what a buyer is willing to pay for a security and what the seller is offering for a security.

Assume that stocks are priced in five-cent increments and that on April 9, 2001, you owned 50 shares of Allied Technology, priced at $\$65\frac{7}{16}$ or $65.4375 per share, resulting in an equity value of

$$\$65.4375 \times 50 = \$3271.875.$$

With decimalization, the price per share must be adjusted to the nearest nickel less than the original price, or $65.40. Since the equity is unchanged, the new number of shares would be

$$\$3271.875 \div \$65.40 = 50.02866972,$$

which is reported to the nearest thousandth as 50.029 shares. For a one-cent incremental scheme, the adjusted price would have been $65.43 and the adjusted number of shares 50.006.

For Group Discussion

For each stock given, calculate the equity, the adjusted price per share, and the number of shares (to the nearest thousandth).

Stock	Before Decimalization		Equity	After Decimalization	
	Number of Shares	Price per Share		Price per Share	Number of Shares
1. Aeroflex (five-cent pricing)	50	$\$61\frac{7}{8}$	______	______	______
2. Philadelphia Suburban (one-cent pricing)	100	$\$23\frac{1}{16}$	______	______	______

R.2 Exercises

FOR EXTRA HELP

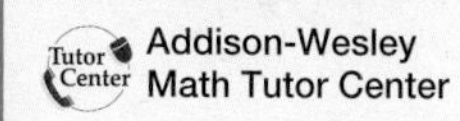
Addison-Wesley Math Tutor Center

MathXL

Digital Video Tutor CD 1 Videotape 1

Student's Solutions Manual

MyMathLab

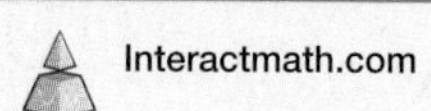
Interactmath.com

1. In the decimal 367.9412, name the digit that has each place value.

(a) tens **(b)** tenths **(c)** thousandths

(d) ones or units **(e)** hundredths

2. Write a numeral that has 5 in the thousands place, 0 in the tenths place, and 4 in the ten thousandths place.

3. For the decimal number 46.249, round to the place value indicated.

(a) hundredths **(b)** tenths

(c) ones or units **(d)** tens

4. Round each decimal to the nearest thousandth.

(a) $.\overline{8}$ **(b)** $.\overline{5}$

(c) .9762 **(d)** .8642

5. For the sum 35.89 + 24.1, which is the best estimate?

A. 40 **B.** 50 **C.** 60 **D.** 70

6. For the difference 119.83 − 52.4, which is the best estimate?

A. 40 **B.** 50 **C.** 60 **D.** 70

7. For the product 84.9 × 98.3, which is the best estimate?

A. 7000 **B.** 8000 **C.** 80,000 **D.** 70,000

8. For the quotient 9845.3 ÷ 97.2, which is the best estimate?

A. 10 **B.** 1000 **C.** 100 **D.** 10,000

Write each decimal as a fraction. Do not write in lowest terms. See Example 1.

9. .4 **10.** .6 **11.** .64 **12.** .82

13. .138 **14.** .104 **15.** 3.805 **16.** 5.166

Add or subtract as indicated. Make sure that your answer is reasonable by estimating first. Give your estimate, and then give the exact answer. See Example 2.

17. 25.32 + 109.2 + 8.574

18. 90.527 + 32.43 + 589.83 + 399.327

19. 28.73 − 3.12 **20.** 46.88 − 13.45 **21.** 43.5 − 28.17 **22.** 345.1 − 56.31

23. 32.56 + 47.356 + 1.8

24. 75.22 + 123.96 + 3.897

25. 18 − 2.789

26. 29 − 8.582

Multiply or divide as indicated. Make sure that your answer is reasonable by estimating first. Give your estimate, and then give the exact answer. See Examples 3–5.

27. .2 × .03 **28.** .07 × .004 **29.** 12.8 × 9.1 **30.** 34.04 × .56

31. 57.2 ÷ 8 **32.** 73.36 ÷ 14 **33.** 19.967 ÷ 9.74 **34.** 44.4788 ÷ 5.27

35. 57.116×100 **36.** $.094 \times 1000$ **37.** $1.62 \div 10$ **38.** $24.03 \div 100$

39. Explain in your own words how to add or subtract decimals.

40. Explain in your own words how to **(a)** multiply decimals and **(b)** divide decimals.

Write each fraction as a decimal. For repeating decimals, write the answer two ways: using the bar notation and rounding to the nearest thousandth. See Example 6.

41. $\frac{1}{8}$ **42.** $\frac{7}{8}$ **43.** $\frac{1}{4}$ **44.** $\frac{3}{4}$

45. $\frac{5}{9}$ **46.** $\frac{8}{9}$ **47.** $\frac{1}{6}$ **48.** $\frac{5}{6}$

49. In your own words, explain how to convert a decimal to a percent.

50. In your own words, explain how to convert a percent to a decimal.

Convert each percent to a decimal. See Examples 7(a)–(c), 8(a), and 8(b).

51. 54% **52.** 39% **53.** 117% **54.** 189% **55.** 2.4%

56. 3.1% **57.** $6\frac{1}{4}\%$ **58.** $5\frac{1}{2}\%$ **59.** .8% **60.** .9%

Convert each decimal to a percent. See Examples 7(d), 7(e), and 8(c)–(e).

61. .75 **62.** .83 **63.** .004 **64.** .005

65. 1.28 **66.** 2.35 **67.** .3 **68.** .6

One method of converting a fraction to a percent is to first convert the fraction to a decimal, as shown in Example 6, and then convert the decimal to a percent, as shown in Examples 7 and 8. Convert each fraction to a percent in this way.

69. $\frac{3}{4}$ **70.** $\frac{1}{4}$ **71.** $\frac{5}{6}$ **72.** $\frac{11}{16}$

73. Brand new tires have a tread of about $\frac{10}{32}$ in. By law, if 80% of the tread is worn off, the tire needs to be replaced.

(a) What fraction of an inch of tread (in $\frac{1}{32}$ of an inch) indicates that a new tire is needed?

(b) How much tread wear (in $\frac{1}{32}$ of an inch) remains (according to the legal limit) if the tread depth is $\frac{4}{32}$?

The Real Number System

A report in June 2003 indicated that of the 844,000 eateries in the United States, the number serving fast food was 177,000. Total sales of fast food that year were projected to be \$120.9 billion. Despite these numbers, a recent trend indicates that consumers are opting for more healthy food, as indicated by McDonald's first-ever quarterly loss for the last three months of 2002. During that year, its domestic sales decreased 1.5%. (*Source:* "McDonald's losses reflect nation's changing tastes," *USA Today*, Jan. 24, 2003; USDA, National Restaurant Association Center for Science in the Public Interest, "Fast Food Nation" by Felix Schlosser.)

Increases and decreases can be represented by positive and negative numbers. In this chapter we examine such numbers. In Exercises 77 and 78 of Section 1.5, we consider increases and decreases in sales in the fast food industry.

1.1 Exponents, Order of Operations, and Inequality

OBJECTIVES

1. Use exponents.
2. Use the order of operations guidelines.
3. Use more than one grouping symbol.
4. Know the meanings of $\neq$, $<$, $>$, $\leq$, and $\geq$.
5. Translate word statements to symbols.
6. Write a statement that changes the direction of an inequality symbol.

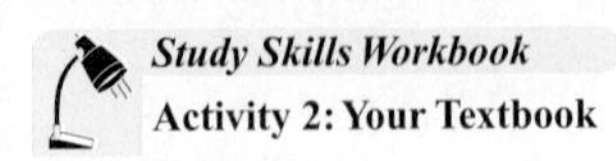

1 Find the value of each exponential expression.

(a) 6^2

(b) 3^5

(c) $\left(\frac{3}{4}\right)^2$

(d) $\left(\frac{1}{2}\right)^4$

(e) $(.4)^3$

In preparation for the study of algebra, we begin by introducing some basic ideas and vocabulary that we will be using throughout the course.

OBJECTIVE 1 Use exponents. In algebra, we use a raised dot for multiplication as shown in **Chapter R,** where we factored a number as the product of its prime factors. For example, 81 is written in prime factored form as

$$81 = 3 \cdot 3 \cdot 3 \cdot 3,$$

where the factor 3 appears four times. Repeated factors are written in an abbreviated form by using an *exponent.* The prime factored form of 81 is written with an exponent as

$$\underbrace{3 \cdot 3 \cdot 3 \cdot 3}_{\text{4 factors of 3}} = 3^4. \quad \text{(4: Exponent; 3: Base)}$$

The number 4 is the **exponent** and 3 is the **base** in the **exponential expression** 3^4. The exponent always follows the base. Exponents are also called **powers.** We read 3^4 as "3 to the fourth power" or simply "3 to the fourth."

EXAMPLE 1 Finding Values of Exponential Expressions

Find the value of each exponential expression.

(a) $5^2 = 5 \cdot 5 = 25$ 5 is used as a factor 2 times.

Read 5^2 as "5 to the second power" or, more commonly, "5 squared."

(b) $6^3 = 6 \cdot 6 \cdot 6 = 216$ 6 is used as a factor 3 times.

Read 6^3 as "6 to the third power" or, more commonly, "6 cubed."

(c) $2^5 = 2 \cdot 2 \cdot 2 \cdot 2 \cdot 2 = 32$ 2 is used as a factor 5 times.

Read 2^5 as "2 to the fifth power."

(d) $7^4 = 7 \cdot 7 \cdot 7 \cdot 7 = 2401$ 7 is used as a factor 4 times.

Read 7^4 as "7 to the fourth power."

(e) $\left(\frac{2}{3}\right)^3 = \frac{2}{3} \cdot \frac{2}{3} \cdot \frac{2}{3} = \frac{8}{27}$ $\frac{2}{3}$ is used as a factor 3 times.

Work Problem 1 at the Side.

CAUTION
Squaring, or raising a number to the second power, is *not* the same as doubling the number. For example,

$$3^2 \text{ means } 3 \cdot 3, \text{ not } 2 \cdot 3.$$

Thus $3^2 = 9$, not 6. Similarly, cubing, or raising a number to the third power, does *not* mean tripling the number.

ANSWERS

1. (a) 36 (b) 243 (c) $\frac{9}{16}$ (d) $\frac{1}{16}$ (e) .064

OBJECTIVE 2 Use the order of operations guidelines. Many problems involve more than one operation. To indicate the order in which the operations should be performed, we often use *grouping symbols.* If no grouping symbols are used, we apply the order of operations discussed below.

Consider the expression $5 + 2 \cdot 3$. To show that the multiplication should be performed before the addition, parentheses can be used to write

$$5 + (2 \cdot 3) = 5 + 6 = 11.$$

If addition is to be performed first, the parentheses should group $5 + 2$ as follows.

$$(5 + 2) \cdot 3 = 7 \cdot 3 = 21$$

Other grouping symbols used in more complicated expressions are brackets [ ], braces { }, and fraction bars. (For example, in $\frac{8-2}{3}$, the expression $8 - 2$ is considered to be grouped in the numerator.)

To work problems with more than one operation, use the following **order of operations.** This order is used by most calculators and computers.

Order of Operations

If grouping symbols are present, simplify within them, innermost first (and above and below fraction bars separately), in the following order.

Step 1 Apply all **exponents.**

Step 2 Do any **multiplications** or **divisions** in the order in which they occur, working from left to right.

Step 3 Do any **additions** or **subtractions** in the order in which they occur, working from left to right.

If no grouping symbols are present, start with Step 1.

A dot has been used to show multiplication; another way to show multiplication is with parentheses. For example, 3(7) means $3 \cdot 7$ or 21. Also, $3(4 + 5)$ means 3 times the sum of 4 and 5. By the order of operations, the sum in parentheses must be found first, then the product.

EXAMPLE 2 Using the Order of Operations

Find the value of each expression.

(a) $4 \cdot 5 - 6$

Using the order of operations given in the box, first multiply 4 and 5, then subtract 6 from the product.

$$\begin{aligned} 4 \cdot 5 - 6 &= 20 - 6 && \text{Multiply.} \\ &= 14 && \text{Subtract.} \end{aligned}$$

(b) $9(6 + 11)$

Work first inside the parentheses.

$$\begin{aligned} 9(6 + 11) &= 9(17) && \text{Add inside parentheses.} \\ &= 153 && \text{Multiply.} \end{aligned}$$

Continued on Next Page

(c) $6 \cdot 8 + 5 \cdot 2$

Perform any multiplications from left to right, then add.

$$6 \cdot 8 + 5 \cdot 2 = 48 + 10 \quad \text{Multiply.}$$
$$= 58 \quad \text{Add.}$$

(d) $2(5 + 6) + 7 \cdot 3 = 2(11) + 7 \cdot 3$ Add inside parentheses.

$$= 22 + 21 \quad \text{Multiply.}$$
$$= 43 \quad \text{Add.}$$

(e) $9 + 2^3 - 5$

Following the order of operations, calculate 2^3 first.

$$9 + 2^3 - 5 = 9 + 8 - 5 \quad \text{Use the exponent.}$$
$$= 12 \quad \text{Add, then subtract.}$$

Work Problem 2 at the Side.

2 Find the value of each expression.

(a) $7 + 3 \cdot 8$

(b) $2 \cdot 9 + 7 \cdot 3$

(c) $7 \cdot 6 - 3(8 + 1)$

(d) $2 + 3^2 - 5$

OBJECTIVE 3 **Use more than one grouping symbol.** An expression with double parentheses, such as the expression $2(8 + 3(6 + 5))$, can be confusing. We avoid confusion by using square brackets, [ ], in place of one pair of parentheses.

EXAMPLE 3 Using Brackets and Fraction Bars as Grouping Symbols

Find the value of each expression.

(a) $2[8 + 3(6 + 5)]$

Begin inside the parentheses. Then follow the order of operations.

$$2[8 + 3(6 + 5)] = 2[8 + 3(11)] \quad \text{Add.}$$
$$= 2[8 + 33] \quad \text{Multiply.}$$
$$= 2[41] \quad \text{Add.}$$
$$= 82 \quad \text{Multiply.}$$

(b) $\dfrac{4(5 + 3) + 3}{2(3) - 1}$

Simplify the numerator and denominator separately.

$$\frac{4(5 + 3) + 3}{2(3) - 1} = \frac{4(8) + 3}{2(3) - 1} \quad \text{Add inside parentheses.}$$
$$= \frac{32 + 3}{6 - 1} \quad \text{Multiply.}$$
$$= \frac{35}{5} \quad \text{Add and subtract.}$$
$$= 7 \quad \text{Divide.}$$

Work Problem 3 at the Side.

3 Find the value of each expression.

(a) $9[(4 + 8) - 3]$

(b) $\dfrac{2(7 + 8) + 2}{3 \cdot 5 + 1}$

Calculator Tip Calculators follow the order of operations given in this section. Try some of the examples to see that your calculator gives the same answers. Be sure to use the parentheses keys to insert parentheses where they are needed. To work Example 3(b) with a calculator, you must put parentheses around the numerator and the denominator.

ANSWERS

2. (a) 31 **(b)** 39 **(c)** 15 **(d)** 6

3. (a) 81 **(b)** 2

OBJECTIVE 4 Know the meanings of $\neq$, $<$, $>$, $\leq$, and $\geq$. So far we have used only the symbols of arithmetic, such as $+$, $-$, $\cdot$, and $\div$ and the equality symbol $=$. The equality symbol with a slash through it means "is *not* equal to." For example,

$$7 \neq 8$$

indicates that 7 is not equal to 8.

If two numbers are not equal, then one of the numbers must be less than the other. The symbol $<$ represents "is less than," so "7 is less than 8" is written as

$$7 < 8.$$

Also, we write "6 is less than 9" as $6 < 9$.

The symbol $>$ means "is greater than." We write "8 is greater than 2" as

$$8 > 2.$$

The statement "17 is greater than 11" becomes $17 > 11$.

Keep the meanings of the symbols $<$ and $>$ clear by remembering that ***the symbol always points to the lesser number.***

Lesser number $\longrightarrow$ **8** < 15

$15 >$ **8** $\longleftarrow$ Lesser number

Work Problem 4 at the Side.

Two other symbols, $\leq$ and $\geq$, also represent the idea of inequality. The symbol $\leq$ means "is less than or equal to," so

$$5 \leq 9$$

means "5 is less than or equal to 9." ***If either the $<$ part or the $=$ part is true, then the inequality $\leq$ is true.*** The statement $5 \leq 9$ is true because $5 < 9$ is true. Also, $8 \leq 8$ is true because $8 = 8$ is true. But $13 \leq 9$ is not true because neither $13 < 9$ nor $13 = 9$ is true.

The symbol $\geq$ means "is greater than or equal to," so

$$9 \geq 5$$

is true because $9 > 5$ is true.

EXAMPLE 4 Using the Symbols $\leq$ and $\geq$

Tell whether each statement is *true* or *false*.

(a) $15 \leq 20$ The statement $15 \leq 20$ is true because $15 < 20$.

(b) $12 \geq 12$ Since $12 = 12$, this statement is true.

(c) $\dfrac{6}{15} \geq \dfrac{2}{3}$

To compare fractions, write them with a common denominator. Here, 15 is a common denominator and $\frac{2}{3} = \frac{10}{15}$. Now decide whether $\frac{6}{15} \geq \frac{10}{15}$ is true or false. Both statements $\frac{6}{15} > \frac{10}{15}$ and $\frac{6}{15} = \frac{10}{15}$ are false; therefore, $\frac{6}{15} \geq \frac{2}{3}$ is false.

Work Problem 5 at the Side.

OBJECTIVE 5 Translate word statements to symbols. Word phrases or statements often must be converted to symbols in algebra. The next example illustrates this.

4 Write each statement in words, then decide whether it is *true* or *false*.

(a) $7 < 5$

(b) $12 > 6$

(c) $4 \neq 10$

(d) $28 \neq 4 \cdot 7$

5 Tell whether each statement is *true* or *false*.

(a) $30 \leq 40$

(b) $25 \geq 10$

(c) $40 \leq 10$

(d) $21 \leq 21$

(e) $3 \geq 3$

ANSWERS

4. **(a)** Seven is less than five. False
(b) Twelve is greater than six. True
(c) Four is not equal to ten. True
(d) Twenty-eight is not equal to four times seven. False

5. **(a)** true **(b)** true **(c)** false
(d) true **(e)** true

EXAMPLE 5 Converting Words to Symbols

Write each word statement in symbols.

(a) Twelve **is equal to** ten **plus** two. $12 = 10 + 2$

(b) Nine **is less than** ten. $9 < 10$
Compare this with 9 less than 10, which is written $10 - 9$.

(c) Fifteen **is not equal to** eighteen. $15 \neq 18$

(d) Seven **is greater than** four. $7 > 4$

(e) Thirteen **is less than or equal to** forty. $13 \leq 40$

(f) Six **is greater than or equal to** six. $6 \geq 6$

Work Problem 6 at the Side.

6 Write in symbols.

(a) Nine is equal to eleven minus two.

(b) Seventeen is less than thirty.

(c) Eight is not equal to ten.

(d) Fourteen is greater than twelve.

(e) Thirty is less than or equal to fifty.

(f) Two is greater than or equal to two.

OBJECTIVE 6 Write a statement that changes the direction of an inequality symbol. Any statement with $<$ can be converted to one with $>$, and any statement with $>$ can be converted to one with $<$. We do this by reversing both the order of the numbers and the direction of the symbol. For example, the statement $6 < 10$ can be written as $10 > 6$.

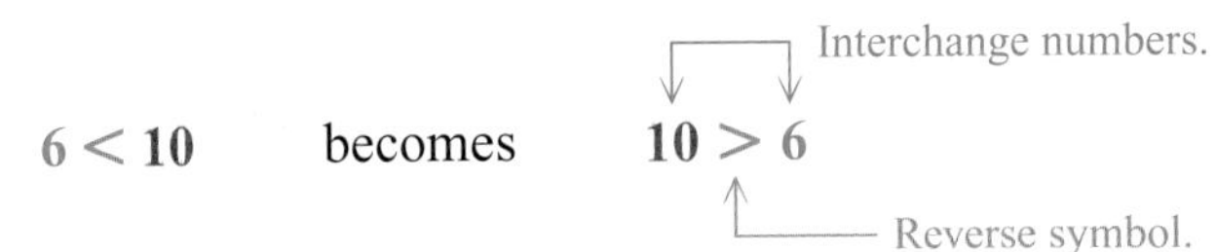

EXAMPLE 6 Converting between < and >

Parts (a)–(d) show the same statements written in two equally correct ways. In each inequality, the symbol points toward the lesser number.

(a) $9 < 16$, $16 > 9$

(b) $5 > 2$, $2 < 5$

(c) $3 \leq 8$, $8 \geq 3$

(d) $12 \geq 5$, $5 \leq 12$

Work Problem 7 at the Side.

7 Write each statement with the inequality symbol reversed.

(a) $8 < 10$

(b) $3 > 1$

(c) $9 \leq 15$

(d) $6 \geq 2$

Here is a summary of the symbols of equality and inequality.

Symbol	Meaning	Example
$=$	Is equal to	$.5 = \frac{1}{2}$ means .5 is equal to $\frac{1}{2}$.
$\neq$	Is not equal to	$3 \neq 7$ means 3 is not equal to 7.
$<$	Is less than	$6 < 10$ means 6 is less than 10.
$>$	Is greater than	$15 > 14$ means 15 is greater than 14.
$\leq$	Is less than or equal to	$4 \leq 8$ means 4 is less than or equal to 8.
$\geq$	Is greater than or equal to	$1 \geq 0$ means 1 is greater than or equal to 0.

CAUTION
The symbols of equality and inequality are used to write mathematical ***sentences.*** They differ from the symbols for operations ($+$, $-$, $\cdot$, and $\div$), discussed earlier, which are used to write mathematical ***expressions*** that represent a number. For example, compare the sentence $4 < 10$, which gives the relationship between 4 and 10, with the expression $4 + 10$, which tells how to operate on 4 and 10 to get the number 14.

ANSWERS
6. **(a)** $9 = 11 - 2$ **(b)** $17 < 30$ **(c)** $8 \neq 10$ **(d)** $14 > 12$ **(e)** $30 \leq 50$ **(f)** $2 \geq 2$
7. **(a)** $10 > 8$ **(b)** $1 < 3$ **(c)** $15 \geq 9$ **(d)** $2 \leq 6$

1.1 Exercises

FOR EXTRA HELP

Student's Solutions Manual

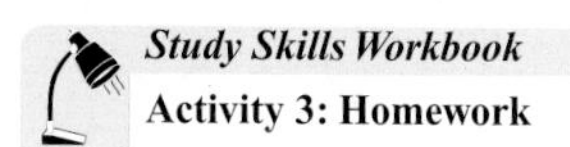

Decide whether each statement is true *or* false. *If it is false, explain why.*

1. Exponents are also called powers.

2. Some grouping symbols are $+$, $-$, $\cdot$, and $\div$.

3. When evaluated, $4 + 3(8 - 2)$ is equal to 42.

4. $3^3 = 9$

5. The statement "4 is 12 less than 16" is interpreted $4 = 12 - 16$.

6. The statement "6 is 4 less than 10" is interpreted $6 < 10 - 4$.

Find the value of each exponential expression. See Example 1.

7. 7^2 **8.** 4^2 **9.** 12^2 **10.** 14^2

11. 4^3 **12.** 5^3 **13.** 10^3 **14.** 11^3

15. 3^4 **16.** 6^4 **17.** 4^5 **18.** 3^5

19. $\left(\frac{2}{3}\right)^4$ **20.** $\left(\frac{3}{4}\right)^3$ **21.** $(.04)^3$ **22.** $(.05)^4$

23. When evaluating $(4^2 + 3^3)^4$, what is the *last* exponent that would be applied? Explain your answer.

24. Which are not grouping symbols—parentheses, brackets, fraction bars, exponents?

Find the value of each expression. See Examples 2 and 3.

25. $13 + 9 \cdot 5$ **26.** $11 + 7 \cdot 6$ **27.** $20 - 4 \cdot 3 + 5$ **28.** $18 - 7 \cdot 2 + 6$

29. $9 \cdot 5 - 13$ **30.** $7 \cdot 6 - 11$ **31.** $18 - 2 + 3$ **32.** $22 - 8 + 9$

33. $\frac{1}{4} \cdot \frac{2}{3} + \frac{2}{5} \cdot \frac{11}{3}$ **34.** $\frac{9}{4} \cdot \frac{2}{3} + \frac{4}{5} \cdot \frac{5}{3}$ **35.** $9 \cdot 4 - 8 \cdot 3$ **36.** $11 \cdot 4 + 10 \cdot 3$

37. $2.5(1.9) + 4.3(7.3)$ **38.** $4.3(1.2) + 2.1(8.5)$ **39.** $10 + 40 \div 5 \cdot 2$

40. $12 + 8^2 \div 8 - 4$ **41.** $18 - 2(3 + 4)$ **42.** $30 - 3(4 + 2)$

43. $5[3 + 4(2^2)]$ **44.** $6\left[\frac{3}{4} + 8\left(\frac{1}{2}\right)^3\right]$ **45.** $\left(\frac{3}{2}\right)^2\left[\left(11 + \frac{1}{3}\right) - 6\right]$

46. $4^2[(13 + 4) - 8]$ **47.** $\frac{8 + 6(3^2 - 1)}{3 \cdot 2 - 2}$ **48.** $\frac{8 + 2(8^2 - 4)}{4 \cdot 3 - 10}$

49. $\dfrac{4(7 + 2) + 8(8 - 3)}{6(4 - 2) - 2^2}$

50. $\dfrac{6(5 + 1) - 9(1 + 1)}{5(8 - 4) - 2^3}$

Tell whether each statement is true *or* false. *In Exercises 53–62, first simplify each expression involving an operation. See Example 4.*

51. $8 \geq 17$

52. $10 \geq 41$

53. $17 \leq 18 - 1$

54. $12 \geq 10 + 2$

55. $6 \cdot 8 + 6 \cdot 6 \geq 0$

56. $4 \cdot 20 - 16 \cdot 5 \geq 0$

57. $6[5 + 3(4 + 2)] \leq 70$

58. $6[2 + 3(2 + 5)] \leq 135$

59. $\dfrac{9(7 - 1) - 8 \cdot 2}{4(6 - 1)} > 3$

60. $\dfrac{2(5 + 3) + 2 \cdot 2}{2(4 - 1)} > 1$

61. $8 \leq 4^2 - 2^2$

62. $10^2 - 8^2 > 6^2$

Write each word statement in symbols. See Example 5.

63. Fifteen is equal to five plus ten.

64. Twelve is equal to twenty minus eight.

65. Nine is greater than five minus four.

66. Ten is greater than six plus one.

67. Sixteen is not equal to nineteen.

68. Three is not equal to four.

69. Two is less than or equal to three.

70. Five is less than or equal to nine.

Write each statement in words and decide whether it is true *or* false. *(Hint: To compare fractions, write them with the same denominator.)*

71. $7 < 19$

72. $9 < 10$

73. $\dfrac{1}{3} \neq \dfrac{3}{10}$

74. $\dfrac{10}{7} \neq \dfrac{3}{2}$

75. $8 \geq 11$

76. $4 \leq 2$

Write each statement with the inequality symbol reversed. See Example 6.

77. $5 < 30$

78. $8 > 4$

79. $12 \geq 3$

80. $25 \leq 41$

The table shows the number of pupils per teacher in selected states. Use this table to answer the questions in Exercises 81–83.

81. Which states had a figure greater than 13.9?

82. Which states had a figure that was at most 14.7?

83. For which states were scores not less than 13.9?

U.S. PUBLIC SCHOOLS, FALL 2001

State	Pupils per Teacher
Alaska	16.7
Texas	14.7
California	20.5
Wyoming	12.5
Maine	12.3
Idaho	17.8
Missouri	13.9

Source: National Center for Education Statistics.

1.2 Variables, Expressions, and Equations

To make general statements about numbers in algebra, letters called **variables** are used. Different numbers can replace the variables to form specific statements. For example, in **Section 1.7** we see that

$$a + b = b + a.$$

This statement is true for any replacements of the variables a and b, such as 2 for a and 5 for b, which gives the true statement

$$2 + 5 = 5 + 2.$$

An **algebraic expression** is a collection of numbers, variables, operation symbols, and grouping symbols, such as parentheses, square brackets, or fraction bars.

$x + 5$, $2m - 9$, and $8p^2 + 6(p - 2)$ Algebraic expressions

In $2m - 9$, the $2m$ means $2 \cdot m$, the product of 2 and m; $8p^2$ represents the product of 8 and p^2. Also, $6(p - 2)$ means the product of 6 and $p - 2$.

OBJECTIVES

1. Evaluate algebraic expressions, given values for the variables.
2. Convert phrases from words to algebraic expressions.
3. Identify solutions of equations.
4. Translate word statements to equations.
5. Distinguish between *expressions* and *equations*.

Study Skills Workbook
Activity 4: Note Taking

OBJECTIVE 1 Evaluate algebraic expressions, given values for the variables. An algebraic expression has different numerical values for different values of the variables.

EXAMPLE 1 Evaluating Expressions Given Values of the Variable

Find the value of each algebraic expression if $m = 5$ and then if $m = 9$.

(a) $8m$

$8m = 8 \cdot 5$ Let $m = 5$.
$= 40$ Multiply.

$8m = 8 \cdot 9$ Let $m = 9$.
$= 72$ Multiply.

(b) $3m^2$

$3m^2 = 3 \cdot 5^2$ Let $m = 5$.
$= 3 \cdot 25$ Square.
$= 75$ Multiply.

$3m^2 = 3 \cdot 9^2$ Let $m = 9$.
$= 3 \cdot 81$ Square.
$= 243$ Multiply.

CAUTION
In Example 1(b), notice that $3m^2$ means $3 \cdot m^2$; it ***does not*** mean $3m \cdot 3m$. ***Unless parentheses are used, the exponent refers only to the variable or number just before it.*** To write $3m \cdot 3m$ with exponents, use parentheses: $3m \cdot 3m = (3m)^2$.

Work Problem 1 at the Side.

1 Find the value of each expression if $p = 3$.

(a) $6p$

(b) $p + 12$

(c) $5p^2$

EXAMPLE 2 Evaluating Expressions with More Than One Variable

Find the value of each expression if $x = 5$ and $y = 3$.

(a) $2x + 5y = 2 \cdot 5 + 5 \cdot 3$ Replace x with 5 and y with 3.
$= 10 + 15$ Multiply.
$= 25$ Add.

Continued on Next Page

ANSWERS
1. (a) 18 (b) 15 (c) 45

2 Find the value of each expression if $x = 6$ and $y = 9$.

(a) $4x + 7y$

(b) $\dfrac{4x - 2y}{x + 1}$

(c) $2x^2 + y^2$

(b) $\dfrac{9x - 8y}{2x - y} = \dfrac{9 \cdot 5 - 8 \cdot 3}{2 \cdot 5 - 3}$ Replace x with 5 and y with 3.

$= \dfrac{45 - 24}{10 - 3}$ Multiply.

$= \dfrac{21}{7}$ Subtract.

$= 3$ Divide.

(c) $x^2 - 2y^2 = 5^2 - 2 \cdot 3^2$ Replace x with 5 and y with 3.

$= 25 - 2 \cdot 9$ Use the exponents.

$= 25 - 18$ Multiply.

$= 7$ Subtract.

Work Problem 2 at the Side.

Calculator Tip An Introduction to Calculators in the front of this book explains how to perform arithmetic operations and evaluate exponentials with a calculator.

OBJECTIVE 2 Convert phrases from words to algebraic expressions.

PROBLEM-SOLVING HINT
Sometimes variables must be used to change word phrases into algebraic expressions. This process will be important later for solving applied problems.

EXAMPLE 3 Using Variables to Change Word Phrases into Algebraic Expressions

Change each word phrase to an algebraic expression. Use x as the variable to represent the number.

(a) The **sum** of a number and 9
"Sum" is the answer to an addition problem. This phrase translates as

$$x + 9 \quad \text{or} \quad 9 + x.$$

(b) 7 **minus** a number
"Minus" indicates subtraction, so the translation is

$$7 - x.$$

Note that $x - 7$ would *not* be correct because we cannot subtract in either order and get the same results.

(c) A number **subtracted from 12**
Since a number is subtracted *from* 12, write this as

$$12 - x.$$

Compare this result with "12 subtracted from a number," which is $x - 12$.

Continued on Next Page

Answers

2. (a) 87 (b) $\frac{6}{7}$ (c) 153

(d) The product of 11 and a number

$$11 \cdot x \quad \text{or} \quad 11x$$

(e) 5 divided by a number

$$5 \div x \quad \text{or} \quad \frac{5}{x}$$ $\frac{x}{5}$ would not be correct here.

(f) The product of 2 and the difference between a number and 8

$$2(x - 8)$$

We are multiplying 2 times another number. This number is the difference between a number and 8, written $x - 8$. Using parentheses around this difference, the final expression is $2(x - 8)$.

CAUTION

Notice that in translating the words "the difference between a number and 8" the order is kept the same: $x - 8$. "The difference between 8 and a number" would be written $8 - x$.

Work Problem 3 at the Side.

OBJECTIVE 3 Identify solutions of equations. An **equation** is a statement that two expressions are equal. Examples of equations are

$$x + 4 = 11, \quad 2y = 16, \quad \text{and} \quad 4p + 1 = 25 - p. \quad \text{Equations}$$

To **solve** an equation, we must find all values of the variable that make the equation true. Such values of the variable are called the **solutions** of the equation.

EXAMPLE 4 Deciding Whether a Number Is a Solution of an Equation

Decide whether the given number is a solution of the equation.

(a) Is 7 a solution of $5p + 1 = 36$?

$$5p + 1 = 36$$
$$5 \cdot 7 + 1 = 36 \quad \text{Replace } p \text{ with 7.}$$
$$35 + 1 = 36 \quad \text{Multiply.}$$
$$36 = 36 \quad \text{True}$$

The number 7 is a solution of the equation.

(b) Is $\frac{14}{3}$ a solution of $9m - 6 = 32$?

$$9m - 6 = 32$$
$$9 \cdot \frac{14}{3} - 6 = 32 \quad \text{Replace } m \text{ with } \tfrac{14}{3}.$$
$$42 - 6 = 32 \quad \text{Multiply.}$$
$$36 = 32 \quad \text{False}$$

The number $\frac{14}{3}$ is not a solution of the equation.

Work Problem 4 at the Side.

3 Write as an algebraic expression. Let x represent the variable.

(a) The sum of 5 and a number

(b) A number minus 4

(c) A number subtracted from 48

(d) The product of 6 and a number

(e) 9 multiplied by the sum of a number and 5

4 Decide whether the given number is a solution of the equation.

(a) $p - 1 = 3$; 2

(b) $2k + 3 = 15$; 7

(c) $8p - 11 = 5$; 2

ANSWERS

3. **(a)** $5 + x$ **(b)** $x - 4$ **(c)** $48 - x$ **(d)** $6x$ **(e)** $9(x + 5)$

4. **(a)** no **(b)** no **(c)** yes

5 Change each sentence to an equation. Let x represent the number.

(a) Three times the sum of a number and 13 is 19.

(b) Five times a number is subtracted from 21, giving 15.

6 Decide whether each is an *equation* or an *expression.*

(a) $2x + 5y - 7$

(b) $\frac{3x - 1}{5}$

(c) $2x + 5 = 7$

(d) $\frac{x}{y - 3} = 4x$

ANSWERS
5. **(a)** $3(x + 13) = 19$ **(b)** $21 - 5x = 15$
6. **(a)** expression **(b)** expression **(c)** equation **(d)** equation

OBJECTIVE 4 Translate word statements to equations. We have seen how to translate phrases from words to expressions. Sentences given in words are translated as equations.

EXAMPLE 5 Translating Word Sentences to Equations

Change each word sentence to an equation. Let x represent the number.

(a) Twice the sum of a number and four is six.

"Twice" means two times. The word *is* suggests equals. With x representing the number, translate as follows.

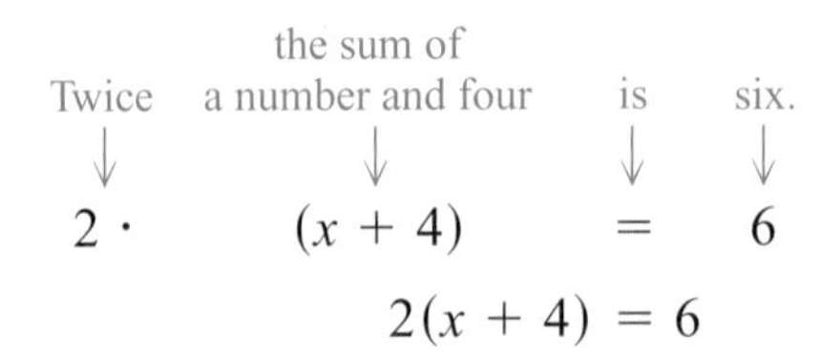

$$2(x + 4) = 6$$

(b) Nine more than five times a number is 49.

Use x to represent the unknown number. Start with $5x$ and then add 9 to it. The word *is* translates as =.

$$5x + 9 = 49$$

(c) Seven less than three times a number is eleven.

Here, 7 is *subtracted* from three times a number to get 11.

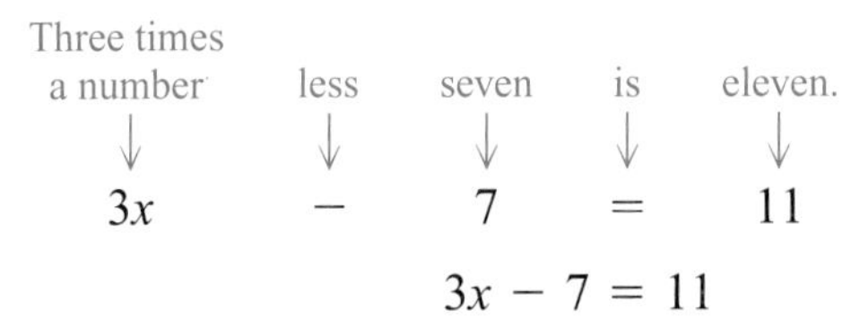

$$3x - 7 = 11$$

Work Problem 5 at the Side.

OBJECTIVE 5 Distinguish between *expressions* and *equations*. Students often have trouble distinguishing between equations and expressions. ***Remember that an equation is a sentence (with an = symbol); an expression is a phrase that represents a number.***

$$4x + 5 = 9 \qquad 4x + 5$$

Equation (to solve) — Expression (to simplify or evaluate)

EXAMPLE 6 Distinguishing between Equations and Expressions

Decide whether each is an *equation* or an *expression.*

(a) $2x - 5y$

There is no equals sign, so this is an expression.

(b) $2x = 5y$

Because of the equals sign, this is an equation.

Work Problem 6 at the Side.

1.2 Exercises

FOR EXTRA HELP

Addison-Wesley Math Tutor Center

MathXL

Digital Video Tutor CD 1 Videotape 1

Student's Solutions Manual

MyMathLab

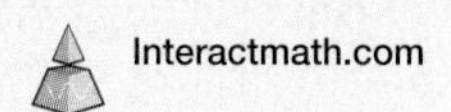

Interactmath.com

Fill in each blank with the correct response.

1. If $x = 3$, then the value of $x + 8$ is ________.

2. If $x = 1$ and $y = 2$, then the value of $5xy$ is ________.

3. "The sum of 13 and x" is represented by the expression ________. If $x = 3$, the value of that expression is ________.

4. Will the equation $x = x + 5$ ever have a solution? ________

5. $2x + 6$ is an ________________ (equation/expression), while $2x + 6 = 8$ is an ________________ (equation/expression).

Exercises 6–10 cover some of the concepts introduced in this section. Give a short explanation for each.

6. Why is $2x^3$ not the same as $2x \cdot 2x \cdot 2x$? Explain, using an exponent to write $2x \cdot 2x \cdot 2x$.

7. If the words *more than* in Example 5(b) were changed to *less than,* how would the equation be changed?

8. Explain in your own words why, when evaluating the expression $4x^2$ for $x = 3$, 3 must be squared *before* multiplying by 4.

9. There are many pairs of values of x and y for which $2x + y$ will equal 6. Name two such pairs and describe how you determined them.

10. Suppose that for the equation $3x - y = 9$, the value of x is given as 4. What would be the corresponding value of y? How do you know this?

Find the numerical value of each expression if **(a)** $x = 4$ *and* **(b)** $x = 6$. *See Example 1.*

11. $4x^2$

(a) **(b)**

12. $5x^2$

(a) **(b)**

13. $\dfrac{3x - 5}{2x}$

(a) **(b)**

14. $\dfrac{4x - 1}{3x}$

(a) **(b)**

15. $\dfrac{6.459x}{2.7}$ (to the nearest thousandth)

(a) **(b)**

16. $\dfrac{.74x^2}{.85}$ (to the nearest thousandth)

(a) **(b)**

17. $3x^2 + x$

(a) **(b)**

18. $2x + x^2$

(a) **(b)**

Find the numerical value of each expression if **(a)** $x = 2$ *and* $y = 1$ *and* **(b)** $x = 1$ *and* $y = 5$. *See Example 2.*

19. $3(x + 2y)$

(a) **(b)**

20. $2(2x + y)$

(a) **(b)**

21. $x + \dfrac{4}{y}$

(a) **(b)**

22. $y + \dfrac{8}{x}$

(a) **(b)**

23. $\dfrac{x}{2} + \dfrac{y}{3}$

(a) **(b)**

24. $\dfrac{x}{5} + \dfrac{y}{4}$

(a) **(b)**

25. $\dfrac{2x + 4y - 6}{5y + 2}$

(a) **(b)**

26. $\dfrac{4x + 3y - 1}{2x + y}$

(a) **(b)**

27. $2y^2 + 5x$

(a) **(b)**

28. $6x^2 + 4y$

(a) **(b)**

29. $\dfrac{3x + y^2}{2x + 3y}$

(a) **(b)**

30. $\dfrac{x^2 + 1}{4x + 5y}$

(a) **(b)**

31. $.841x^2 + .32y^2$

(a) **(b)**

32. $.941x^2 + .2y^2$

(a) **(b)**

Change each word phrase to an algebraic expression. Use x to represent the number. See Example 3.

33. Twelve times a number

34. Thirteen added to a number

35. Two subtracted from a number

36. Eight subtracted from a number

37. One-third of a number, subtracted from seven

38. One-fifth of a number, subtracted from fourteen

39. The difference between twice a number and 6

40. The difference between 6 and half a number

41. 12 divided by the sum of a number and 3

42. The difference between a number and 5, divided by 12

43. The product of 6 and four less than a number

44. The product of 9 and five more than a number

45. In the phrase "four more than the product of a number and 6," does the word *and* signify the operation of addition? Explain.

46. Suppose that the directions on a test read "Solve the following expressions." How would you politely correct the person who wrote these directions?

Decide whether the given number is a solution of the equation. See Example 4.

47. Is 7 a solution of $x - 5 = 12$?

48. Is 10 a solution of $x + 6 = 15$?

49. Is 1 a solution of $5x + 2 = 7$?

50. Is 1 a solution of $3x + 5 = 8$?

51. Is $\frac{1}{5}$ a solution of $6x + 4x + 9 = 11$?

52. Is $\frac{12}{5}$ a solution of $2x + 3x + 8 = 20$?

53. Is 3 a solution of $2y + 3(y - 2) = 14$?

54. Is 2 a solution of $6a + 2(a + 3) = 14$?

55. Is $\frac{1}{3}$ a solution of $\frac{z + 4}{2 - z} = \frac{13}{5}$?

56. Is $\frac{13}{4}$ a solution of $\frac{x + 6}{x - 2} = \frac{37}{5}$?

57. Is 4.3 a solution of $3r^2 - 2 = 53.47$?

58. Is 3.7 a solution of $2x^2 + 1 = 28.38$?

Change each sentence to an equation. Use x to represent the number. See Example 5.

59. The sum of a number and 8 is 18.

60. A number minus three equals 1.

61. Five more than twice a number is 5.

62. The product of 2 and the sum of a number and 5 is 14.

63. Sixteen minus three-fourths of a number is 13.

64. The sum of six-fifths of a number and 2 is 14.

65. Three times a number is equal to 8 more than twice the number.

66. Twelve divided by a number equals $\frac{1}{3}$ times that number.

Identify each as an expression *or an* equation. *See Example 6.*

67. $3x + 2(x - 4)$

68. $5y - (3y + 6)$

69. $7t + 2(t + 1) = 4$

70. $9r + 3(r - 4) = 2$

RELATING CONCEPTS (EXERCISES 71–74) For Individual or Group Work

*A **mathematical model** is an equation that describes the relationship between two quantities. For example, based on data from the U.S. Bureau of Labor Statistics, average hourly earnings of production workers in manufacturing industries in the United States from 1996 through 2002 are approximated by the equation $y = .494x - 974.0$, where x represents the year and y represents the hourly earnings in dollars. Use this model to approximate the hourly earnings during each year. Compare with the actual earnings given in parentheses.*

71. 1997 ($12.49)

72. 2000 ($14.00)

73. 2001 ($14.53)

74. 2002 ($14.95)

1.3 Real Numbers and the Number Line

OBJECTIVES

1. Classify numbers and graph them on number lines.
2. Tell which of two real numbers is less than the other.
3. Find the opposite of a real number.
4. Find the absolute value of a real number.

Study Skills Workbook
Activity 5: Study Cards

OBJECTIVE 1 Classify numbers and graph them on number lines. The numbers used for counting are called the **natural numbers.** We use set braces, { }, to denote the elements of a set.

Natural Numbers

$$\{1, 2, 3, 4, 5, \ldots\}$$

In **Chapter R,** we introduced the set of **whole numbers.**

Whole Numbers

$$\{0, 1, 2, 3, 4, 5, \ldots\}$$

These numbers, along with many others, can be represented on a **number line** like the one in Figure 1. We draw a number line by choosing any point on the line and labeling it 0. Choose any point to the right of 0 and label it 1. The distance between 0 and 1 gives a unit of measure used to locate other points, as shown in Figure 1. The points labeled in Figure 1 correspond to the first few whole numbers.

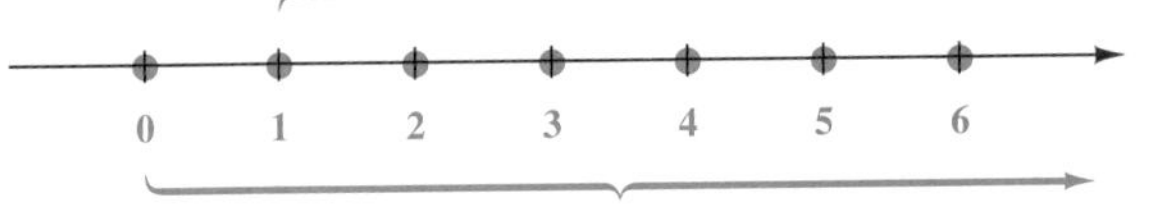

Figure 1

Be sure that number lines are properly labeled.

The natural numbers are located to the right of 0 on the number line. But numbers may also be placed to the left of 0. For each natural number we can place a corresponding number to the left of 0. These numbers, written $-1, -2, -3, -4$, and so on, are shown in Figure 2. Each is the **opposite** or **negative** of a natural number. The natural numbers, their opposites, and 0 form a new set of numbers called the **integers.**

Integers

$$\{\ldots, -3, -2, -1, 0, 1, 2, 3, \ldots\}$$

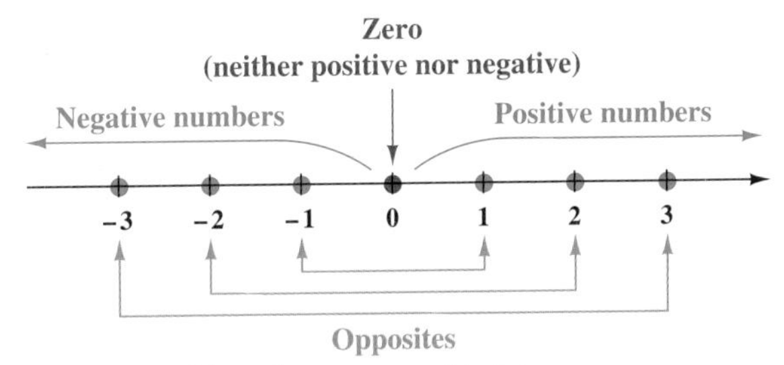

The points correspond to integers.

Figure 2

There are many practical applications of negative numbers. For example, a Fahrenheit temperature on a cold January day might be $-10°$, and a business that spends more than it takes in has a negative "profit."

EXAMPLE 1 Using Negative Numbers in Applications

Use an integer to express the number in each application.

(a) The lowest Fahrenheit temperature ever recorded in meteorological records was 129° below zero at Vostok, Antarctica, on July 21, 1983. (*Source: World Almanac and Book of Facts,* 2004.)
Use $-129°$ because "below zero" indicates a negative number.

(b) The shore surrounding the Dead Sea is 1348 ft below sea level. (*Source: World Almanac and Book of Facts,* 2004.)
Again, "below sea level" indicates a negative number, -1348.

Work Problem 1 at the Side.

1 Use an integer to express the number(s) in each application.

(a) Erin discovers that she has spent \$53 more than she has in her checking account.

(b) The record high Fahrenheit temperature in the United States was 134° in Death Valley, California on July 10, 1913. (*Source: World Almanac and Book of Facts,* 2004.)

(c) A football team gained 5 yd, then lost 10 yd on the next play.

Not all numbers are integers. For example, $\frac{1}{2}$ is not; it is a number halfway between the integers 0 and 1. Also, $3\frac{1}{4}$ is not an integer. These numbers and others that are quotients of integers are **rational numbers.** (The name comes from the word *ratio,* which indicates a quotient.)

Rational Numbers

{numbers that can be written as quotients of integers, with denominators not 0}

Since any integer can be written as the quotient of itself and 1, all integers are also rational numbers. For example, $-5 = \frac{-5}{1}$. A decimal number that comes to an end (terminates), such as .23, is a rational number. For example, $.23 = \frac{23}{100}$. Decimal numbers that repeat in a fixed block of digits, such as $.3333... = .\overline{3}$ and $.454545... = .\overline{45}$, are also rational numbers. For example, $.\overline{3} = \frac{1}{3}$.

As shown in Figures 1 and 2 on the preceding page, to **graph** a number, we place a dot on the number line at the point that corresponds to the number. The number is called the **coordinate** of the point. Think of the graph of a set of numbers as a picture of the set.

2 Graph each number on the number line.

$$-3, \frac{17}{8}, -2.75, 1\frac{1}{2}, -\frac{3}{4}$$

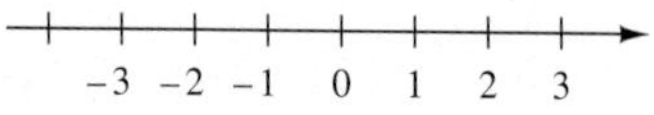

EXAMPLE 2 Graphing Rational Numbers

Graph each number on the number line.

$$-\frac{3}{2}, \quad -\frac{2}{3}, \quad \frac{1}{2}, \quad 1\frac{1}{3}, \quad \frac{23}{8}, \quad 3\frac{1}{4}$$

To locate the improper fractions on the number line, write them as mixed numbers or decimals. The graph is shown in Figure 3.

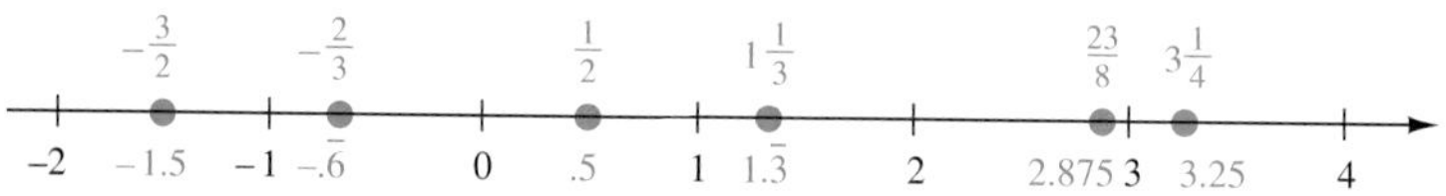

Figure 3

Work Problem 2 at the Side.

ANSWERS

1. (a) -53 (b) 134 (c) 5, -10
2.

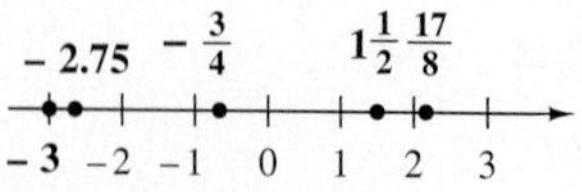

Although many numbers are rational, not all are. For example, a square that measures one unit on a side has a diagonal whose length is the square root of 2, written $\sqrt{2}$. See Figure 4. It can be shown that $\sqrt{2}$ cannot be written as a quotient of integers. Because of this, $\sqrt{2}$ is not rational; it is **irrational.** Other examples of irrational numbers are $\sqrt{3}$, $\sqrt{7}$, $-\sqrt{10}$, and π (the ratio of the circumference of a circle to its diameter).

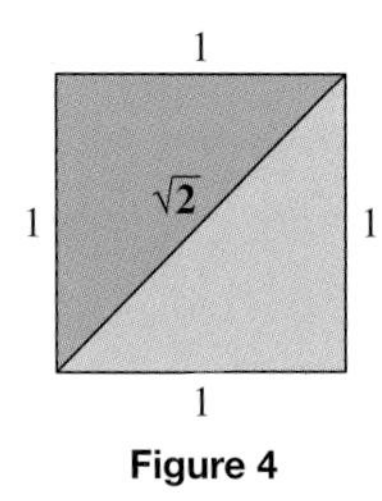

Figure 4

Irrational Numbers
{nonrational numbers represented by points on the number line}

The decimal form of an irrational number neither terminates nor repeats. Irrational numbers are discussed in **Chapter 8.**

Both rational and irrational numbers can be represented by points on the number line and are called **real numbers.**

Real Numbers
{all numbers that are either rational or irrational}

All the numbers mentioned so far are real numbers. The relationships between the various types of numbers are shown in Figure 5. Notice that any real number is either a rational number or an irrational number.

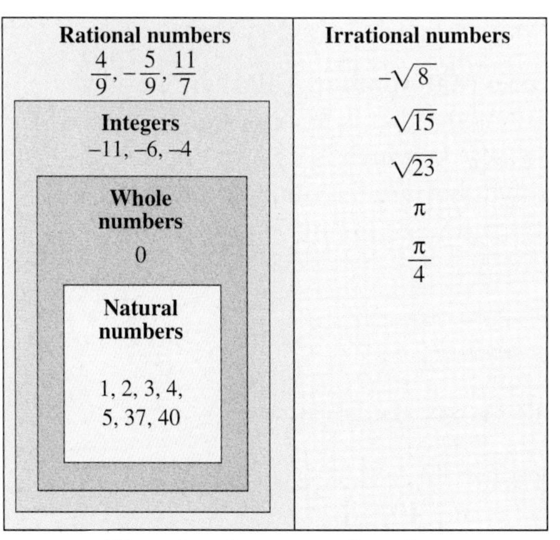

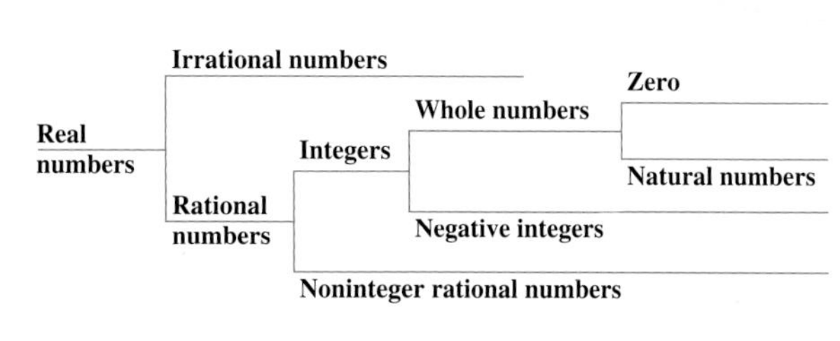

Figure 5

OBJECTIVE 2 Tell which of two real numbers is less than the other. Given any two whole numbers, we can tell which number is less than the other. But what about two negative numbers, as in the set of integers? Moving from 0 to the right along a number line, the positive numbers corresponding to the points on the number line *increase*. For example, $8 < 12$, and 8 is to the left of 12 on a number line. We extend this ordering to all real numbers.

Ordering of the Real Numbers

For any two real numbers a and b, **a is less than b** if a is to the left of b on a number line.

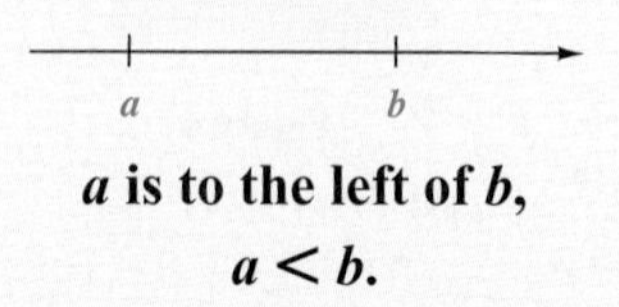

a is to the left of b,
$a < b$.

This means that any negative number is less than 0, and any negative number is less than any positive number. Also, 0 is less than any positive number.

EXAMPLE 3 Determining the Order of Real Numbers

Is it true that $-3 < -1$?

To find out, locate -3 and -1 on a number line, as shown in Figure 6. Because -3 is to the left of -1 on the number line, -3 is less than -1. The statement $-3 < -1$ is true.

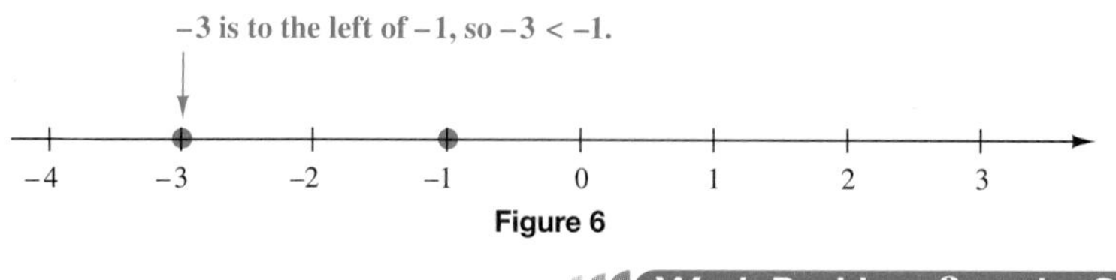

Figure 6

Work Problem 3 at the Side.

3 Tell whether each statement is *true* or *false*.

(a) $-2 < 4$

(b) $6 > -3$

(c) $-9 < -12$

(d) $-4 \geq -1$

(e) $-6 \leq 0$

OBJECTIVE 3 Find the opposite of a real number. Earlier, we saw that every positive integer has a negative integer that is its opposite or negative. This is true for every real number except 0, which is its own opposite.* A characteristic of pairs of opposites is that they are the same distance from 0 on the number line but in opposite directions. See Figure 7.

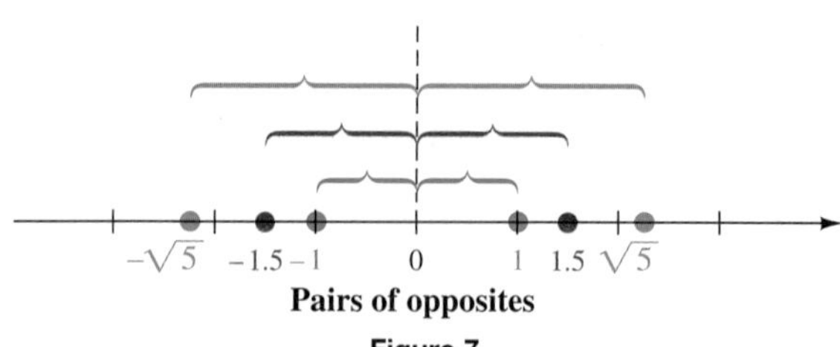

Figure 7

ANSWERS
3. **(a)** true **(b)** true **(c)** false **(d)** false **(e)** true

* The opposite (or negative) of a number is also called the *additive inverse* of the number, as we shall see in **Section 1.7.**

We indicate the opposite of a number by writing the symbol $-$ in front of the number. For example, the opposite of 7 is -7 (read "negative 7"). We could write the opposite of -4 as $-(-4)$, but we know that 4 is the opposite of -4. Since a number can have only one opposite, $-(-4)$ and 4 must represent the same number, so

$$-(-4) = 4.$$

This idea can be generalized.

Double Negative Rule

For any real number a,

$$-(-a) = a.$$

The following chart shows several numbers and their opposites.

Number	Opposite
-4	$-(-4)$, or 4
-3	3
0	0
5	-5
19	-19

The chart suggests the following rule.

Except for 0, the opposite of a number is found by changing the sign of the number.

Work Problem 4 at the Side.

OBJECTIVE 4 Find the absolute value of a real number. As previously mentioned, opposites are numbers the same distance from 0 on the number line but on opposite sides of 0. Another way to say this is to say that opposites have the same *absolute value*. The **absolute value** of a number is the undirected distance between 0 and the number on the number line. The symbol for the absolute value of the number a is $|a|$, read "the absolute value of a." For example, the distance between 2 and 0 on the number line is 2 units, so

$$|2| = 2.$$

Also, the distance between -2 and 0 on the number line is 2, so

$$|-2| = 2.$$

Since distance is a physical measurement, which is never negative, we can make the following statement.

The absolute value of a number can never be negative.

4 Find the opposite of each number.

(a) 6

(b) 15

(c) -9

(d) -12

(e) 0

Answers

4. (a) -6 **(b)** -15 **(c)** 9 **(d)** 12 **(e)** 0

For example,

$$|12| = 12 \quad \text{and} \quad |-12| = 12$$

because both 12 and -12 lie at a distance of 12 units from 0 on the number line. Since the distance of 0 from 0 is 0 units, we have

$$|0| = 0.$$

EXAMPLE 4 Evaluating Absolute Value

Simplify.

(a) $|5| = 5$

(b) $|-5| = 5$

(c) $-|-5| = -(5) = -5$ Replace $|-5|$ with 5.

(d) $-|-13| = -(13) = -13$

(e) $|8 - 5|$

Simplify within the absolute value bars first.

$$|8 - 5| = |3| = 3$$

(f) $-|8 - 5| = -|3| = -3$

(g) $-|12 - 3| = -|9| = -9$

Parts (e)–(g) in Example 4 show that absolute value bars also act as grouping symbols. You must perform any operations within absolute value bars before finding the absolute value.

Work Problem 5 at the Side.

5 Simplify.

(a) $|-6|$

(b) $|9|$

(c) $-|15|$

(d) $-|-9|$

(e) $|9 - 4|$

(f) $-|32 - 2|$

ANSWERS

5. **(a)** 6 **(b)** 9 **(c)** -15 **(d)** -9 **(e)** 5 **(f)** -30

1.3 Exercises

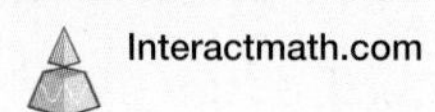

In Exercises 1–6, give an example of a number that satisfies each given condition.

1. An integer between 3.6 and 4.6

2. A rational number between 2.8 and 2.9

3. A whole number that is not positive and is less than 1

4. A whole number greater than 3.5

5. An irrational number that is between $\sqrt{12}$ and $\sqrt{14}$

6. A real number that is neither negative nor positive

*List all numbers from each set that are **(a)** natural numbers, **(b)** whole numbers, **(c)** integers, **(d)** rational numbers, **(e)** irrational numbers, **(f)** real numbers.*

7. $\left\{-9, -\sqrt{7}, -1\frac{1}{4}, -\frac{3}{5}, 0, \sqrt{5}, 3, 5.9, 7\right\}$

8. $\left\{-5.3, -5, -\sqrt{3}, -1, -\frac{1}{9}, 0, 1.2, 4, \sqrt{12}\right\}$

Use an integer to express each number representing a change in the following applications. See Example 1.

9. In all of 2003, there were 1.85 million housing starts. In December 2003, the number of housing starts in the United States increased from the previous month by about 40,000 units. (*Source:* http://money.cnn.com)

10. There were 16,900 fewer Honda Civics produced in the United States in 2002 compared to 2001. (*Source: Ward's AutoInfoBank.*)

11. Between 1995 and 2001, the population of the District of Columbia decreased by about 9000. (*Source:* U.S. Bureau of the Census.)

12. In 2002, Taiwan produced 129,313 more passenger cars than trucks. (*Source:* Automotive News Data Center and Marketing Systems.)

Graph each group of numbers on a number line. See Example 2.

13. $0, 3, -5, -6$

14. $2, 6, -2, -1$

15. $-2, -6, -4, 3, 4$

16. $-5, -3, -2, 0, 4$

17. $\frac{1}{4}, 2\frac{1}{2}, -3\frac{4}{5}, -4, -\frac{13}{8}$

18. $5\frac{1}{4}, \frac{41}{9}, -2\frac{1}{3}, 0, -3\frac{2}{5}$

Select the lesser number in each pair. See Example 3.

19. $-11, -4$

20. $-9, -16$

21. $-21, 1$

22. $-57, 3$

23. $0, -100$

24. $-215, 0$

25. $-\frac{2}{3}, -\frac{1}{4}$

26. $-\frac{3}{8}, -\frac{9}{16}$

Decide whether each statement is true *or* false. *See Example 3.*

27. $8 < -16$

28. $12 < -24$

29. $-3 < -2$

30. $-10 < -9$

For each number, ***(a)*** *find its opposite and* ***(b)*** *find its absolute value.*

31. -2
(a) **(b)**

32. -8
(a) **(b)**

33. 6
(a) **(b)**

34. 11
(a) **(b)**

35. $-\frac{3}{4}$
(a) **(b)**

36. $-\frac{1}{3}$
(a) **(b)**

Simplify. See Example 4.

37. $|-7|$

38. $|-3|$

39. $-|12|$

40. $-|23|$

41. $-\left|-\frac{2}{3}\right|$

42. $-\left|-\frac{4}{5}\right|$

43. $|13 - 4|$

44. $|8 - 7|$

Decide whether each statement is true *or* false.

45. $|-8| < 7$

46. $|-6| \geq -|6|$

47. $4 \leq |4|$

48. $-|-3| > 2$

49. Students often say "The absolute value of a number is always positive." Is this true? If not, explain.

50. If the absolute value of a number is equal to the number itself, what must be true about the number?

To answer the questions in Exercises 51–54, refer to the table, which gives the changes in producer price indexes for two recent periods.

51. What commodity for which period represents the greatest decrease?

52. What commodity for which period represents the least change?

Commodity	Change from 2000 to 2001	Change from 2001 to 2002
Food	.9	−1.4
Shelter	.4	0
Fuel/other utilities	1.8	−13.3
Apparel	−.5	−.8
Private transportation	−5.5	−1.4
Public transportation	−5.5	−2.0
Medical care	.5	.1

Source: U.S. Bureau of Labor Statistics.

53. Which has lesser absolute value, the change for apparel from 2000 to 2001 or from 2001 to 2002?

54. Which has greater absolute value, the change for public transportation from 2000 to 2001 or from 2001 to 2002?

1.4 Adding Real Numbers

OBJECTIVES

1. Add two numbers with the same sign.
2. Add numbers with different signs.
3. Add mentally.
4. Use the order of operations with real numbers.
5. Translate words and phrases that indicate addition.

OBJECTIVE 1 Add two numbers with the same sign. We can use the number line to explain addition of real numbers. Later, we will give the rules for addition. Recall that the answer to an addition problem is called the **sum.**

EXAMPLE 1 Adding with the Number Line

Use the number line to find the sum $2 + 3$.

Add the positive numbers 2 and 3 by starting at 0 and drawing an arrow two units to the *right,* as shown in Figure 8. This arrow represents the number 2 in the sum $2 + 3$. Next, from the right end of this arrow draw another arrow three units to the right. The number below the end of this second arrow is 5, so $2 + 3 = 5$.

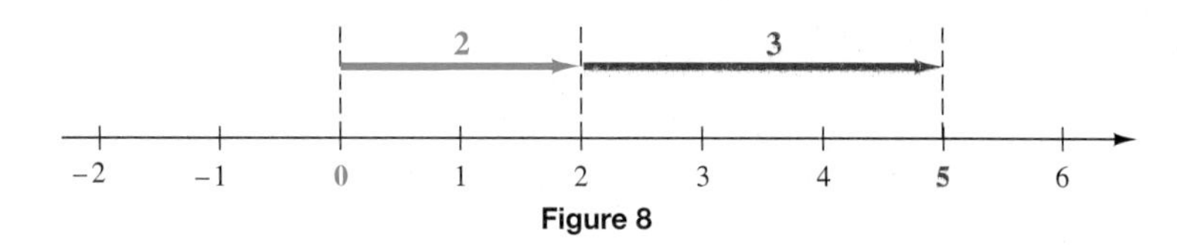

Figure 8

EXAMPLE 2 Adding with the Number Line

Use the number line to find the sum $-2 + (-4)$. (Parentheses are placed around the -4 to avoid the confusing use of $+$ and $-$ next to each other.)

To add the negative numbers -2 and -4 on the number line, we start at 0 and draw an arrow two units to the *left,* as shown in Figure 9. From the left end of this first arrow, we draw a second arrow four units to the left. We draw the arrow to the left to represent the addition of the *negative* number, -4. The number below the end of this second arrow is -6, so $-2 + (-4) = -6$.

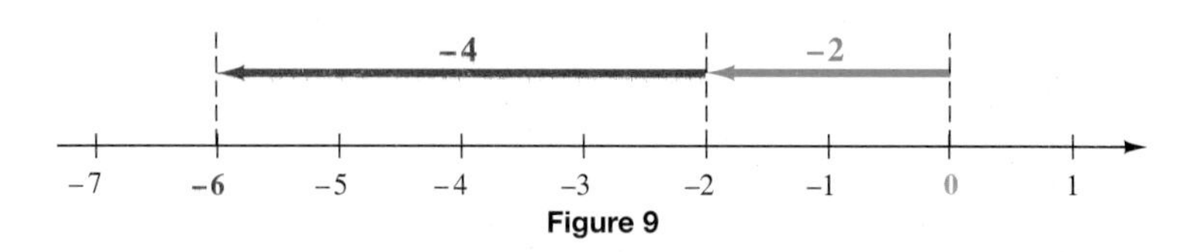

Figure 9

Work Problem 1 at the Side.

1 Use a number line to find each sum.

(a) $1 + 4$

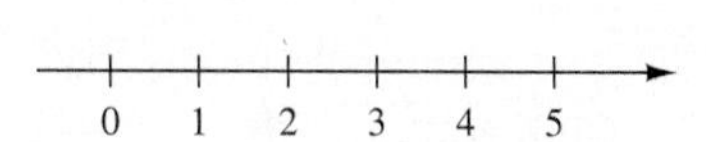

(b) $-2 + (-5)$

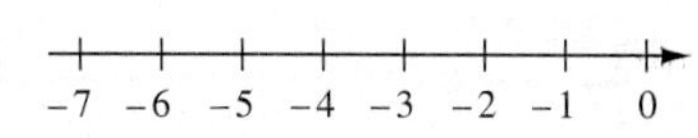

In Example 2, we found that the sum of the two negative numbers -2 and -4 is a negative number whose distance from 0 is the sum of the distance of -2 from 0 and the distance of -4 from 0. That is, ***the sum of two negative numbers is the negative of the sum of their absolute values.***

$$-2 + (-4) = -(|-2| + |-4|) = -(2 + 4) = -6$$

To add two numbers having the same sign, add the absolute values of the numbers. Give the result the same sign as the numbers being added.

Example: $-4 + (-3) = -7$.

ANSWERS

1. (a) $1 + 4 = 5$

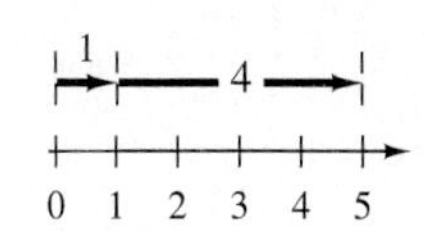

(b) $-2 + (-5) = -7$

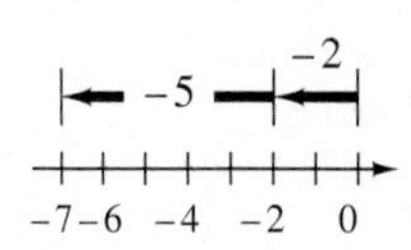

2 Find each sum.

(a) $-7 + (-3)$

(b) $-12 + (-18)$

(c) $-15 + (-4)$

3 Use a number line to find each sum.

(a) $6 + (-3)$

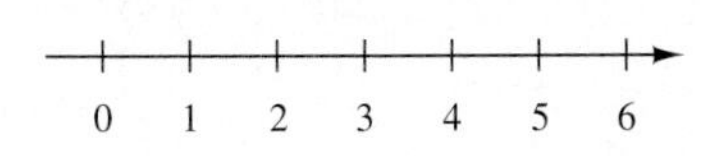

(b) $-5 + 1$

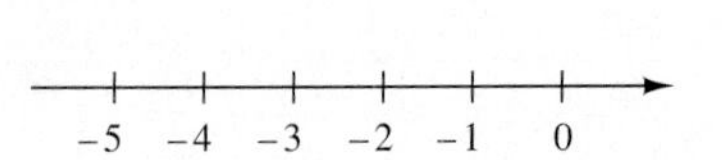

EXAMPLE 3 **Adding Two Negative Numbers**

Find each sum.

(a) $-2 + (-9) = -11$ The sum of two negative numbers is negative.

(b) $-8 + (-12) = -20$

(c) $-15 + (-3) = -18$

Work Problem 2 at the Side.

OBJECTIVE 2 **Add numbers with different signs.** We use the number line again to illustrate the sum of a positive number and a negative number.

EXAMPLE 4 **Adding Numbers with Different Signs**

Use the number line to find the sum $-2 + 5$.

We find the sum $-2 + 5$ on the number line by starting at 0 and drawing an arrow two units to the left. From the left end of this arrow, we draw a second arrow five units to the right, as shown in Figure 10. The number below the end of this second arrow is 3, so $-2 + 5 = 3$.

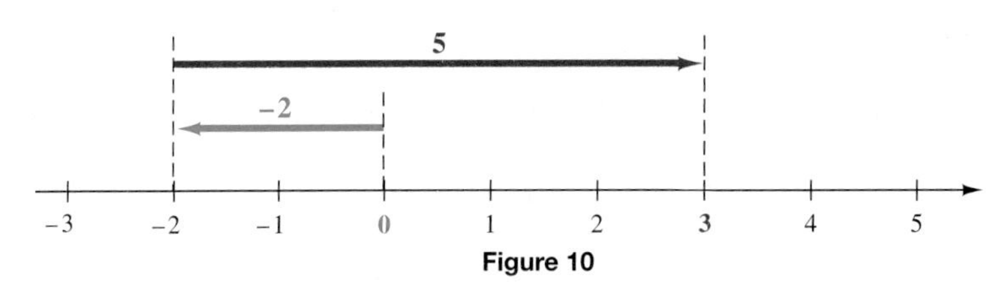

Figure 10

Work Problem 3 at the Side.

Addition of numbers with different signs also can be defined using absolute value.

To add numbers with different signs, find the absolute values of the numbers, and subtract the smaller absolute value from the larger. Give the answer the same sign as the number with the larger absolute value.

Example: $-12 + 6 = -6$.

For example, to add -12 and 5, we find their absolute values:

$$|-12| = 12 \quad \text{and} \quad |5| = 5.$$

Then we find the difference between these absolute values: $12 - 5 = 7$. Since $|-12| > |5|$, the sum will be negative, so

$$-12 + 5 = -7.$$

Calculator Tip The (−) or (+/−) key is used to input a negative number in some scientific calculators. Try using your calculator to add negative numbers.

ANSWERS

2. (a) -10 **(b)** -30 **(c)** -19

3. (a) $6 + (-3) = 3$

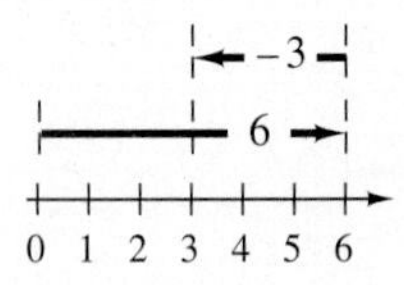

(b) $-5 + 1 = -4$

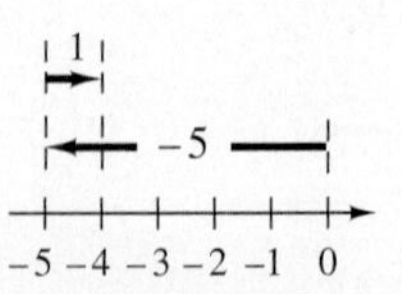

OBJECTIVE 3 Add mentally. While a number line is useful in showing the rules for addition, it is important to be able to find sums mentally.

EXAMPLE 5 Adding a Positive Number and a Negative Number

Check each answer, trying to work the addition mentally. If you have trouble, use a number line.

(a) $7 + (-4) = 3$

(b) $-8 + 12 = 4$

(c) $-\frac{1}{2} + \frac{1}{8} = -\frac{4}{8} + \frac{1}{8} = -\frac{3}{8}$ Remember to find a common denominator first.

(d) $\frac{5}{6} + \left(-1\frac{1}{3}\right) = \frac{5}{6} + \left(-\frac{4}{3}\right) = \frac{5}{6} + \left(-\frac{8}{6}\right) = -\frac{3}{6} = -\frac{1}{2}$

(e) $-4.6 + 8.1 = 3.5$

Work Problem 4 at the Side.

The rules for adding signed numbers are summarized below.

Adding Signed Numbers

Same sign Add the absolute values of the numbers. Give the sum the same sign as the numbers being added.

Different signs Find the absolute values of the numbers, and subtract the smaller absolute value from the larger. Give the answer the sign of the number having the larger absolute value.

OBJECTIVE 4 Use the order of operations with real numbers. Sometimes a problem involves square brackets, [ ]. As we mentioned earlier, brackets are treated just like parentheses. We do the calculations inside the brackets until a single number is obtained. Remember to use the order of operations given in **Section 1.1** for adding more than two numbers.

EXAMPLE 6 Adding with Brackets

Find each sum.

(a) $-3 + [4 + (-8)]$

First work inside the brackets. Follow the order of operations given in **Section 1.1.**

$$-3 + [4 + (-8)] = -3 + (-4) = -7$$

(b) $8 + [(-2 + 6) + (-3)] = 8 + [4 + (-3)] = 8 + 1 = 9$

Work Problem 5 at the Side.

OBJECTIVE 5 Translate words and phrases that indicate addition. We now look at the interpretation of words and phrases that involve addition. Problem solving often requires translating words and phrases into symbols. We began this process with translating simple phrases in **Section 1.1.**

4 Check each answer, trying to work the addition mentally. If you have trouble, use a number line.

(a) $-8 + 2 = -6$

(b) $-15 + 4 = -11$

(c) $17 + (-10) = 7$

(d) $\frac{3}{4} + \left(-1\frac{3}{8}\right) = -\frac{5}{8}$

(e) $-9.5 + 3.8 = -5.7$

5 Find each sum.

(a) $2 + [7 + (-3)]$

(b) $6 + [(-2 + 5) + 7]$

(c) $-9 + [-4 + (-8 + 6)]$

Answers

4. All are correct.

5. (a) 6 (b) 16 (c) -15

6 Write a numerical expression for each phrase, and simplify the expression.

(a) 4 more than -12

(b) The sum of 6 and -7

(c) -12 added to -31

(d) 7 increased by the sum of 8 and -3

7 Solve the problem.

A football team lost 8 yd on first down, lost 5 yd on second down, and then gained 7 yd on third down. How many yards did the team gain or lose altogether?

ANSWERS

6. **(a)** $-12 + 4$; -8 **(b)** $6 + (-7)$; -1
(c) $-31 + (-12)$; -43
(d) $7 + [8 + (-3)]$; 12

7. The team lost 6 yd.

The word *sum* indicates addition. There are other key words and phrases that also indicate addition. Some of these are given in the chart below.

Word or Phrase	Example	Numerical Expression and Simplification
Sum of	The *sum of* -3 and 4	$-3 + 4 = 1$
Added to	5 *added to* -8	$-8 + 5 = -3$
More than	12 *more than* -5	$(-5) + 12 = 7$
Increased by	-6 *increased by* 13	$-6 + 13 = 7$
Plus	3 *plus* 14	$3 + 14 = 17$

EXAMPLE 7 Translating Words and Phrases

Write a numerical expression for each phrase, and simplify the expression.

(a) The **sum of** -8 and 4 and 6

$$-8 + 4 + 6 = (-8 + 4) + 6 = -4 + 6 = 2$$

Notice that parentheses were placed around $-8 + 4$, and this addition was done first, using the order of operations given earlier.

(b) 3 **more than** -5, **increased by** 12

$$-5 + 3 + 12 = (-5 + 3) + 12 = -2 + 12 = 10$$

Work Problem 6 at the Side.

Gains (or increases) and losses (or decreases) sometimes appear in applied problems. When they do, the gains may be interpreted as positive numbers and the losses as negative numbers.

EXAMPLE 8 Interpreting Gains and Losses

The Carolina Panthers football team gained 3 yd on first down, lost 12 yd on second down, and then gained 13 yd on third down. How many yards did the team gain or lose altogether?

The gains are represented by positive numbers and the loss by a negative number.

$$3 + (-12) + 13$$

Add from left to right.

$$\begin{aligned} 3 + (-12) + 13 &= [3 + (-12)] + 13 \\ &= (-9) + 13 \\ &= 4 \end{aligned}$$

The team gained 4 yd altogether.

Work Problem 7 at the Side.

1.4 Exercises

FOR EXTRA HELP

Addison-Wesley Math Tutor Center | MathXL | Digital Video Tutor CD 1 Videotape 1 | Student's Solutions Manual | MyMathLab | Interactmath.com

By the order of operations, what is the first step you would use to simplify each expression?

1. $4[3(-2+5)-1]$

2. $[-4+7(-6+2)]$

3. $9+([-1+(-3)]+5)$

4. $[(-8+4)+(-6)]+5$

Find each sum. See Examples 1–6.

5. $6+(-4)$

6. $8+(-5)$

7. $12+(-15)$

8. $4+(-8)$

9. $-7+(-3)$

10. $-11+(-4)$

11. $-10+(-3)$

12. $-16+(-7)$

13. $-12.4+(-3.5)$

14. $-21.3+(-2.5)$

15. $10+[-3+(-2)]$

16. $13+[-4+(-5)]$

17. $5+[14+(-6)]$

18. $7+[3+(-14)]$

19. $-3+[5+(-2)]$

20. $-7+[10+(-3)]$

21. $-8+[3+(-1)+(-2)]$

22. $-7+[5+(-8)+3]$

23. $\frac{9}{10}+\left(-\frac{3}{5}\right)$

24. $\frac{5}{8}+\left(-\frac{17}{12}\right)$

25. $-\frac{1}{6}+\frac{2}{3}$

26. $-\frac{6}{25}+\frac{19}{20}$

27. $2\frac{1}{2}+\left(-3\frac{1}{4}\right)$

28. $-4\frac{3}{8}+6\frac{1}{2}$

29. $7.8+(-9.4)$

30. $14.7+(-10.1)$

31. $-7.1+[3.3+(-4.9)]$

32. $-9.5+[-6.8+(-1.3)]$

33. $[-8+(-3)]+[-7+(-7)]$

34. $[-5+(-4)]+[9+(-2)]$

35. $\left(-\frac{1}{2}+.25\right)-\left(-\frac{3}{4}+.75\right)$

36. $\left(-\frac{3}{2}-.75\right)-\left(2.25-\frac{1}{2}\right)$

Perform each operation, and then determine whether the statement is true *or* false. *Try to do all work mentally. See Examples 5 and 6.*

37. $-11 + 13 = 13 + (-11)$

38. $16 + (-9) = -9 + 16$

39. $-10 + 6 + 7 = -3$

40. $-12 + 8 + 5 = -1$

41. $\frac{7}{3} + \left(-\frac{1}{3}\right) + \left(-\frac{6}{3}\right) = 0$

42. $-\frac{3}{2} + 1 + \frac{1}{2} = 0$

43. $|-8 + 10| = -8 + (-10)$

44. $|-4 + 6| = -4 + (-6)$

45. $2\frac{1}{5} + \left(-\frac{6}{11}\right) = -\frac{6}{11} + 2\frac{1}{5}$

46. $-1\frac{1}{2} + \frac{5}{8} = \frac{5}{8} + \left(-1\frac{1}{2}\right)$

47. $-7 + [-5 + (-3)] = [(-7) + (-5)] + 3$

48. $6 + [-2 + (-5)] = [(-4) + (-2)] + 5$

RELATING CONCEPTS (EXERCISES 49–52) For Individual or Group Work

Recall the rules for adding signed numbers introduced in this section, and ***work Exercises 49–52 in order.***

49. Suppose that the sum of two numbers is negative, and you know that one of the numbers is positive. What can you conclude about the other number?

50. If you are solving the equation $x + 5 = -7$ from a set of numbers, why could you immediately eliminate any positive numbers as possible solutions? (Remember how you answered Exercise 49.)

51. Suppose that the sum of two numbers is positive, and you know that one of the numbers is negative. What can you conclude about the other number?

52. If you are solving the equation $x + (-8) = 2$ from a set of numbers, why could you immediately eliminate any negative numbers as possible solutions? (Remember how you answered Exercise 51.)

53. In your own words, explain how to add two negative numbers.

54. In your own words, explain how to add a positive number and a negative number. Give two cases.

Write a numerical expression for each phrase, and simplify the expression. See Example 7.

55. The sum of -5 and 12 and 6

56. The sum of -3 and 5 and -12

57. 14 added to the sum of -19 and -4

58. -2 added to the sum of -18 and 11

59. The sum of -4 and -10, increased by 12

60. The sum of -7 and -13, increased by 14

61. $\frac{2}{7}$ more than the sum of $\frac{5}{7}$ and $-\frac{9}{7}$

62. .85 more than the sum of -1.25 and -4.75

Solve each problem. See Example 8.

63. Kramer owed Jerry \$10 for snacks raided from the refrigerator. Kramer later borrowed \$70 from George to finance his latest get-rich scheme. What positive or negative number represents Kramer's financial status?

64. Bonika's checking account balance is \$54.00. She then takes a gamble by writing a check for \$89.00. What is her new balance? (Write the balance as a signed number.)

65. The surface, or rim, of a canyon is at altitude 0. On a hike down into the canyon, a party of hikers stops for a rest at 130 m below the surface. They then descend another 54 m. What is their new altitude? (Write the altitude as a signed number.)

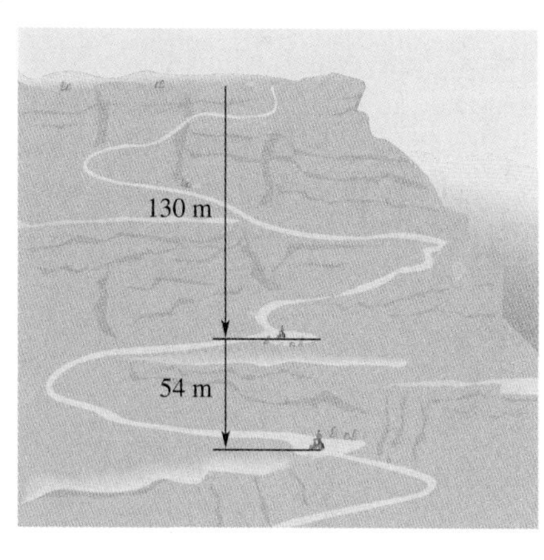

66. A pilot announces to the passengers that the current altitude of their plane is 34,000 ft. Because of some unexpected turbulence, the pilot is forced to descend 2100 ft. What is the new altitude of the plane? (Write the altitude as a signed number.)

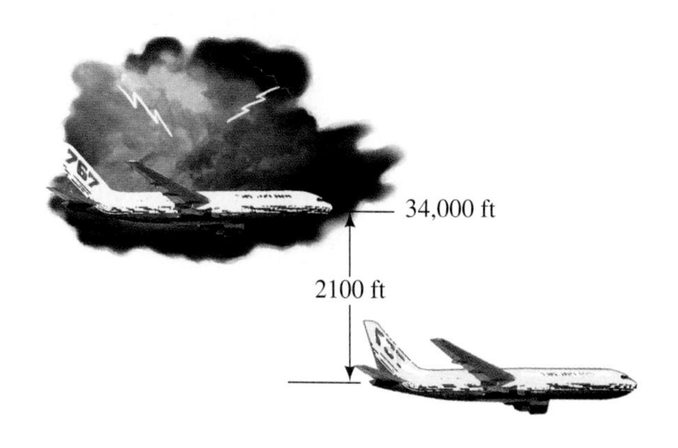

67. On three consecutive passes, Brett Favre of the Green Bay Packers passed for a gain of 6 yd, was sacked for a loss of 12 yd, and passed for a gain of 43 yd. What positive or negative number represents the total net yardage for the plays?

68. On a series of three consecutive running plays, Peyton Manning of the Indianapolis Colts gained 4 yd, lost 3 yd, and lost 2 yd. What positive or negative number represents his total net yardage for the series of plays?

69. The lowest temperature ever recorded in Arkansas was $-29°$F. The highest temperature ever recorded there was 149°F more than the lowest. What was this highest temperature? (*Source: World Almanac and Book of Facts,* 2004.)

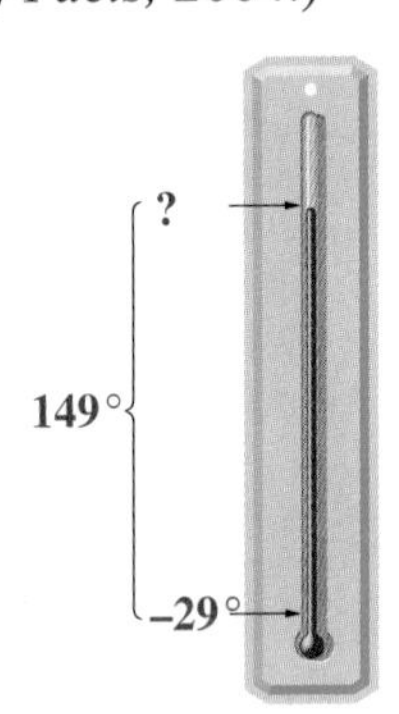

70. On January 23, 1943, the temperature rose 49°F in two minutes in Spearfish, South Dakota. If the starting temperature was $-4°$F, what was the temperature two minutes later?

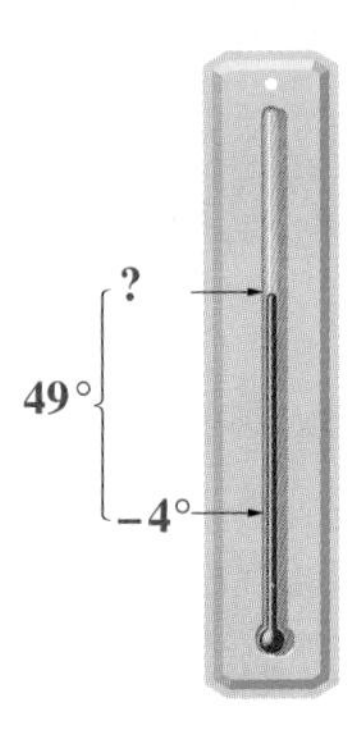

71. Dana Weightman owes \$153 to a credit card company. She makes a \$14 purchase with the card, and then pays \$60 on the account. What is her current balance as a signed number?

72. A female polar bear weighed 660 lb when she entered her winter den. She lost 45 lb during each of the first two months of hibernation, and another 205 lb before leaving the den with her two cubs in March. How much did she weigh when she left the den?

73. Based on census population projections for 2020, New York will lose 5 seats in the U.S. House of Representatives, Pennsylvania will lose 4 seats, and Ohio will lose 3. Write a signed number that represents the total number of seats these three states are projected to lose. (*Source:* Population Reference Bureau.)

74. Michigan is projected to lose 3 seats in the U.S. House of Representatives and Illinois 2 in 2020. The states projected to gain the most seats are California with 9, Texas with 5, Florida with 3, Georgia with 2, and Arizona with 2. Write a signed number that represents the algebraic sum of these changes. (*Source:* Population Reference Bureau.)

1.5 Subtracting Real Numbers

OBJECTIVES

1. Find a difference.
2. Use the definition of subtraction.
3. Work subtraction problems that involve brackets.
4. Translate words and phrases that indicate subtraction.

OBJECTIVE 1 Find a difference. In the operation $a - b$, a is called the **minuend** and b is called the **subtrahend.** As we mentioned earlier, the answer to a subtraction problem is called a **difference.** Differences between signed numbers can be found by using a number line. Addition and subtraction are opposite operations. Thus, because *addition* of a positive number on the number line is shown by drawing an arrow to the *right*, *subtraction* of a positive number is shown by drawing an arrow to the *left*.

EXAMPLE 1 Subtracting with the Number Line

Use the number line to find the difference $7 - 4$.

To find the difference $7 - 4$ on the number line, begin at 0 and draw an arrow 7 units to the *right*. From the right end of this arrow, draw an arrow 4 units to the *left*, as shown in Figure 11. The number at the end of the second arrow shows that $7 - 4 = 3$.

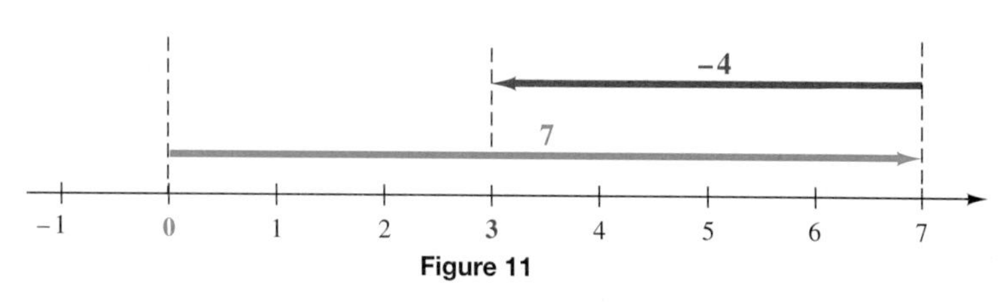

Figure 11

Work Problem 1 at the Side.

1 Use the number line to find each difference.

(a) $5 - 1$

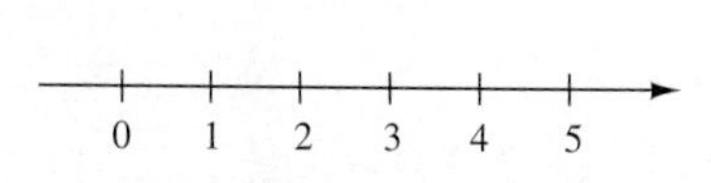

(b) $6 - 2$

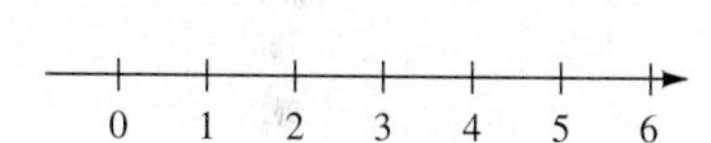

OBJECTIVE 2 Use the definition of subtraction. The procedure used in Example 1 to find $7 - 4$ is exactly the same procedure that would be used to find $7 + (-4)$, so

$$7 - 4 = 7 + (-4).$$

This shows that *subtracting* a positive number from a larger positive number is the same as *adding* the opposite of the smaller number to the larger. We use this idea to define subtraction for all real numbers.

Subtraction

For any real numbers a and b,

$$a - b = a + (-b).$$

Example: $4 - 9 = 4 + (-9) = -5.$

To subtract b from a, add the opposite (or negative) of b to a. In other words, change the subtrahend to its opposite, and add.

Subtracting Signed Numbers

Step 1 Change the subtraction symbol to addition and change the sign of the subtrahend.

Step 2 Add, as in the previous section.

ANSWERS

1. (a) $5 - 1 = 4$

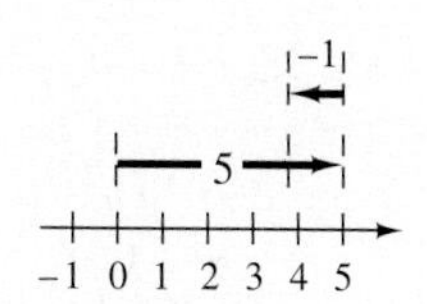

(b) $6 - 2 = 4$

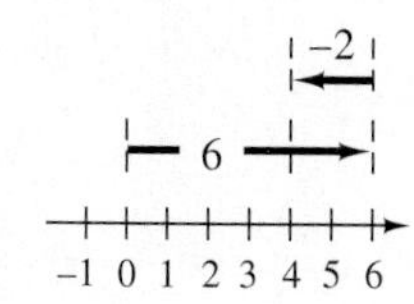

2 Subtract.

(a) $6 - 10$

(b) $-2 - 4$

(c) $3 - (-5)$

(d) $-8 - (-12)$

(e) $\frac{5}{4} - \left(-\frac{3}{7}\right)$

ANSWERS

2. (a) -4 (b) -6 (c) 8 (d) 4 (e) $\frac{47}{28}$

EXAMPLE 2 Using the Definition of Subtraction

Subtract.

No change; Change − to +.; Opposite of 3

(a) $12 - 3 = 12 + (-3) = 9$

In practice, we usually do not change the sign as shown here, since this is the kind of subtraction done in basic arithmetic.

(b) $5 - 7 = 5 + (-7) = -2$

(c) $8 - 15 = 8 + (-15) = -7$

No change; Change − to +.; Opposite of −5

(d) $-3 - (-5) = -3 + (5) = 2$

(e) $\frac{3}{8} - \left(-\frac{4}{5}\right) = \frac{15}{40} - \left(-\frac{32}{40}\right) = \frac{15}{40} + \frac{32}{40} = \frac{47}{40}$

Work Problem 2 at the Side.

Subtraction can be used to reverse the result of an addition problem. For example, if 4 is added to a number and then subtracted from the sum, the original number is the result.

$$12 + 4 = 16 \quad \text{and} \quad 16 - 4 = 12$$

The symbol − has now been used for three purposes:

1. to represent subtraction, as in $9 - 5 = 4$;
2. to represent negative numbers, such as -10, -2, and -3;
3. to represent the opposite (or negative) of a number, as in "the opposite (or negative) of 8 is -8."

We may see more than one use in the same problem, such as $-6 - (-9)$, where -9 is subtracted from -6. The meaning of the symbol depends on its position in the algebraic expression.

OBJECTIVE 3 Work subtraction problems that involve brackets. As before, with problems that have both parentheses and brackets, first do any operations inside the parentheses and brackets. Work from the inside out. Because subtraction is defined in terms of addition, the order of operations from **Section 1.1** can still be used.

EXAMPLE 3 Subtracting with Grouping Symbols

Perform each operation.

(a)
$$\begin{aligned} -6 - [2 - (8 + 3)] &= -6 - [2 - 11] \\ &= -6 - [2 + (-11)] \quad \text{Change } - \text{ to } +. \\ &= -6 - (-9) \\ &= -6 + (9) = 3 \end{aligned}$$

Continued on Next Page

(b) $5 - \left[\left(-\frac{1}{3} - \frac{1}{2}\right) - (4 - 1)\right] = 5 - \left[\left(-\frac{1}{3} + \left(-\frac{1}{2}\right)\right) - 3\right]$

$$= 5 - \left[\left(-\frac{5}{6}\right) - 3\right]$$

$$= 5 - \left[\left(-\frac{5}{6}\right) + (-3)\right]$$

$$= 5 - \left(-\frac{23}{6}\right)$$

$$= 5 + \frac{23}{6} = \frac{53}{6}$$

Work Problem 3 at the Side.

3 Perform each operation.

(a) $2 - [(-3) - (4 + 6)]$

(b) $[(5 - 7) + 3] - 8$

(c) $6 - [(-1 - 4) - 2]$

OBJECTIVE 4 Translate words and phrases that indicate subtraction. Now we translate words and phrases that involve subtraction of real numbers. *Difference* is one of them. Some others are given in the chart below.

Word or Phrase	Example	Numerical Expression and Simplification
Difference between	The *difference between* -3 and -8	$-3 - (-8) = -3 + 8 = 5$
Subtracted from	12 *subtracted from* 18	$18 - 12 = 6$
Less than	6 *less than* 5	$5 - 6 = 5 + (-6) = -1$
Decreased by	9 *decreased by* -4	$9 - (-4) = 9 + 4 = 13$
Minus	-8 *minus* 5	$-8 - 5 = -8 + (-5) = -13$

CAUTION

When you are subtracting two numbers, it is important that you write them in the correct order, because, in general, $a - b \neq b - a$. For example, $5 - 3 \neq 3 - 5$. ***Think carefully before interpreting an expression involving subtraction!*** Subtracting a larger number from a smaller number *always* produces a negative number.

EXAMPLE 4 Translating Words and Phrases

Write a numerical expression for each phrase, and simplify the expression.

(a) The **difference between** -8 and 5

When "difference between" is used, write the numbers in the order they are given.

$$-8 - 5 = -8 + (-5) = -13$$

(b) 4 **subtracted from** the sum of 8 and -3

Here the operation of addition is also used, as indicated by the word *sum*. First, add 8 and -3. Next, subtract 4 from this sum.

$$[8 + (-3)] - 4 = 5 - 4 = 1$$

Continued on Next Page

ANSWERS

3. (a) 15 (b) -7 (c) 13

4 Write a numerical expression for each phrase, and simplify the expression.

(a) The difference between -5 and -12

(b) -2 subtracted from the sum of 4 and -4

(c) 7 less than -2

(d) 9, decreased by 10 less than 7

5 Solve the problem.

The highest elevation in Argentina is Mt. Aconcagua, which is 6960 m above sea level. The lowest point in Argentina is the Valdes Peninsula, 40 m below sea level. Find the difference between the highest and lowest elevations.

(c) 4 **less than** -6

Be careful with order here. 4 must be taken *from* -6, so write -6 first.

$$-6 - 4 = -6 + (-4) = -10$$

Notice that "4 less than -6" differs from "4 *is less than* -6." The statement "4 is less than -6" is symbolized as $4 < -6$ (which is a false statement).

(d) 8, **decreased by** 5 **less than** 12

First, write "5 less than 12" as $12 - 5$. Next, subtract $12 - 5$ from 8.

$$8 - (12 - 5) = 8 - 7 = 1$$

Work Problem 4 at the Side.

We have seen a few applications of signed numbers in earlier sections. The next example involves subtraction of signed numbers.

EXAMPLE 5 Solving a Problem Involving Subtraction

The record high temperature of 134°F in the United States was recorded at Death Valley, California, in 1913. The record low was -80°F, at Prospect Creek, Alaska, in 1971. See Figure 12. What is the difference between these highest and lowest temperatures? (*Source: World Almanac and Book of Facts*, 2004.)

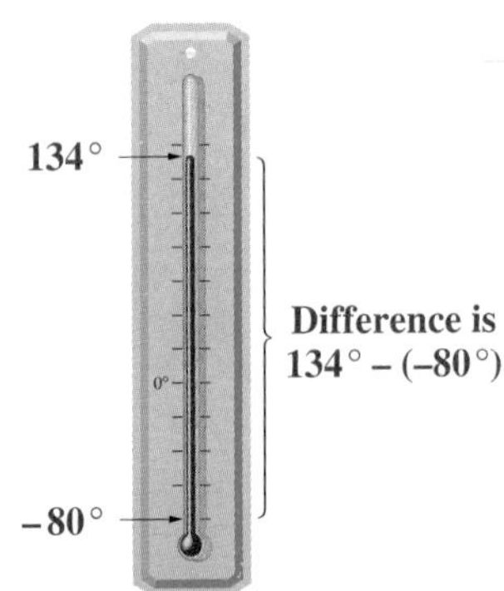

Figure 12

We must subtract the lowest temperature from the highest temperature.

$$134 - (-80) = 134 + 80 \quad \text{Use the definition of subtraction.}$$
$$= 214 \quad \text{Add.}$$

The difference between the two temperatures is 214°F.

Work Problem 5 at the Side.

ANSWERS

4. (a) $-5 - (-12)$; 7
(b) $[4 + (-4)] - (-2)$; 2
(c) $-2 - 7$; -9
(d) $9 - (7 - 10)$; 12

5. 7000 m

1.5 Exercises

FOR EXTRA HELP

Addison-Wesley Math Tutor Center | MathXL | Digital Video Tutor CD 1 Videotape 1 | Student's Solutions Manual | MyMathLab | Interactmath.com

Fill in each blank with the correct response.

1. By the definition of subtraction, in order to perform the subtraction $-6 - (-8)$, we must add the opposite of ________ to ________.

2. By the order of operations, to simplify $8 - [3 - (-4 - 5)]$, the first step is to subtract ________ from ________.

3. "The difference between 7 and 12" translates as ________, while "the difference between 12 and 7" translates as ________.

4. $-9 - (-3) = -9 +$ ______

5. $-8 - 4 = -8 +$ ______

6. $-19 - 22 = -19 +$ ______

Find each difference. See Examples 1–3.

7. $-7 - 3$

8. $-12 - 5$

9. $-10 - 6$

10. $-13 - 16$

11. $7 - (-4)$

12. $9 - (-6)$

13. $6 - (-13)$

14. $13 - (-3)$

15. $-7 - (-3)$

16. $-8 - (-6)$

17. $3 - (4 - 6)$

18. $6 - (7 - 14)$

19. $-3 - (6 - 9)$

20. $-4 - (5 - 12)$

21. $\frac{1}{2} - \left(-\frac{1}{4}\right)$

22. $\frac{1}{3} - \left(-\frac{4}{3}\right)$

23. $-\frac{3}{4} - \frac{5}{8}$

24. $-\frac{5}{6} - \frac{1}{2}$

25. $\frac{5}{8} - \left(-\frac{1}{2} - \frac{3}{4}\right)$

26. $\frac{9}{10} - \left(\frac{1}{8} - \frac{3}{10}\right)$

27. $4.4 - (-9.2)$

28. $6.7 - (-12.6)$

29. $-7.4 - 4.5$

30. $-5.4 - 9.6$

31. $-5.2 - (8.4 - 10.8)$

32. $-9.6 - (3.5 - 12.6)$

33. $[(-3.1) - 4.5] - (.8 - 2.1)$

34. $[(-7.8) - 9.3] - (.6 - 3.5)$

35. $-12 - [(9 - 2) - (-6 - 3)]$

36. $-4 + [(-6 - 9) - (-7 + 4)]$

37. $-8 + [(-3 - 10) - (-4 + 1)]$

38. $\left(-\frac{3}{4} - \frac{5}{2}\right) - \left(-\frac{1}{8} - 1\right)$

39. $\left(-\frac{3}{8} - \frac{2}{3}\right) - \left(-\frac{9}{8} - 3\right)$

40. $[-34.99 + (6.59 - 12.25)] - 8.33$

41. $[-12.25 - (8.34 + 3.57)] - 17.88$

42. Explain in your own words how to subtract signed numbers.

43. We know that, in general, $a - b \neq b - a$. Find two pairs of values for a and b so that $a - b = b - a$.

Simplify each expression. Use the order of operations.

44. $-3 - (-4) - 5$

45. $8 - (-3) - 9 + 6$

46. $-5 - 2 + 4 - 8 - (-6)$

47. Make up a subtraction problem so that the difference between two negative numbers is a negative number.

48. Make up a subtraction problem so that the difference between two negative numbers is a positive number.

Write a numerical expression for each phrase and simplify. See Example 4.

49. The difference between 4 and -8

50. The difference between 7 and -14

51. 8 less than -2

52. 9 less than -13

53. The sum of 9 and -4, decreased by 7

54. The sum of 12 and -7, decreased by 14

55. 12 less than the difference between 8 and -5

56. 19 less than the difference between 9 and -2

Solve each problem. See Example 5.

57. The coldest temperature recorded in Chicago, Illinois, was $-35°$F in 1996. The record low in South Dakota was set in 1936 and was 23°F lower than $-35°$F. What was the record low in South Dakota? (*Source: World Almanac and Book of Facts*, 2004.)

58. No one knows just why humpback whales love to heave their 45-ton bodies out of the water, but leap they do. Mark and Debbie, two researchers based on the island of Maui, noticed that one of their favorite whales, "Pineapple," leaped 15 ft above the surface of the ocean while her mate cruised 12 ft below the surface. What is the difference between these two heights?

59. The top of Mount Whitney, visible from Death Valley, has an altitude of 14,494 ft above sea level. The bottom of Death Valley is 282 ft below sea level. Using 0 as sea level, find the difference between these two elevations. (*Source: World Almanac and Book of Facts*, 2004.)

60. A chemist is running an experiment under precise conditions. At first, she runs it at $-174.6°F$. She then lowers the temperature by 2.3°F. What is the new temperature for the experiment?

61. Ben owed his brother \$10. He later borrowed \$70. What positive or negative number represents his present financial status?

62. Francesca has \$15 in her purse, and Emilio has a debt of \$12. Find the difference between these amounts.

63. For the year 2003, one health club showed a profit of \$86,000, while another showed a loss of \$19,000. Find the difference between these amounts.

64. At 2:00 A.M., a plant worker found that a dial reading was 7.904. At 3:00 A.M., she found the reading to be -3.291. Find the difference between these two readings.

65. J. D. Patin enjoys playing Triominoes every Wednesday night. Last Wednesday, on four successive turns, his scores were -19, 28, -5, and 13. What was his final score for the four turns?

66. Marie Aguillard also enjoys playing Triominoes. On five successive turns, her scores were -13, 15, -12, 24, and 14. What was her total score for the five turns?

67. In August, Alison Romike began with a checking account balance of \$904.89. Her checks and deposits for August are given below:

Checks	Deposits
\$35.84	\$85.00
\$26.14	\$120.76
\$3.12	

Assuming no other transactions, what was her account balance at the end of August?

68. In September, Carter Fenton began with a checking account balance of \$904.89. His checks and deposits for September are given below:

Checks	Deposits
\$41.29	\$80.59
\$13.66	\$276.13
\$84.40	

Assuming no other transactions, what was his account balance at the end of September?

69. A certain Greek mathematician was born in 426 B.C. His father was born 43 years earlier. In what year was his father born?

70. A certain Roman philosopher was born in 325 B.C. Her mother was born 35 years earlier. In what year was her mother born?

71. Kim Falgout owes $870.00 on her MasterCard account. She returns two items costing $35.90 and $150.00 and receives credits for these on the account. Next, she makes a purchase of $82.50, and then two more purchases of $10.00 each. She makes a payment of $500.00. She then incurs a finance charge of $37.23. How much does she still owe?

72. Charles Vosburg owes $679.00 on his Visa account. He returns three items costing $36.89, $29.40, and $113.55 and receives credits for these on the account. Next, he makes purchases of $135.78 and $412.88, and two purchases of $20.00 each. He makes a payment of $400. He then incurs a finance charge of $24.57. How much does he still owe?

73. Dean Baldus enjoys scuba diving. He dives to 34 ft below the surface of a lake. His partner, Jeff Balius, dives to 40 ft below the surface, but then ascends 20 ft. What is the vertical distance between Dean and Jeff?

74. Rhonda Alessi also enjoys scuba diving. She dives to 12 ft below the surface of False River. Her sister, Sandy, dives to 20 ft below the surface, but then ascends 10 ft. What is the vertical distance between Rhonda and Sandy?

75. The height of Mt. Foraker is 17,400 ft, while the depth of the Java Trench is 23,376 ft. What is the vertical distance between the top of Mt. Foraker and the bottom of the Java Trench? (*Source: World Almanac and Book of Facts*, 2004.)

76. The height of Mt. Wilson is 14,246 ft, while the depth of the Cayman Trench is 24,721 ft. What is the vertical distance between the top of Mt. Wilson and the bottom of the Cayman Trench? (*Source: World Almanac and Book of Facts*, 2004.)

The bar graph shown depicts the sales growth changes in various countries for McDonald's revenue during the years 2001 and 2002. The change from one year to the next can be represented by a signed number, found by subtracting the figure for 2001 from the figure for 2002. For example, in Latin America, the sales figure actually increased for the period: $-1.0\% - (-3.9\%) = 2.9\%$.

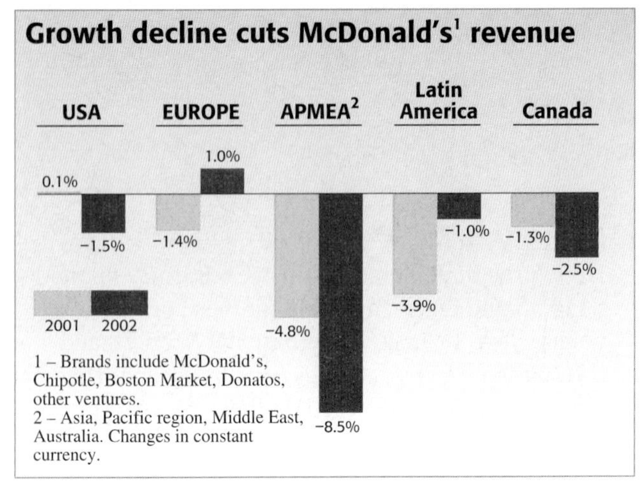

Source: McDonald's Corp.

Use a signed number to answer Exercises 77 and 78.

77. Find the percent change in the United States.

78. Find the percent change in Europe.

Median pricings for existing homes in the United States for the years 1997 through 2002 are shown in the table. Complete the table, determining the change from one year to the next by subtraction.

	Year	Median-Priced Existing Homes	Change from Previous Year
	1997	$121,800	
	1998	$128,400	$6600
79.	1999	$133,300	
80.	2000	$139,000	
81.	2001	$147,800	
82.	2002	$158,100	

Source: National Association of Realtors.

In Exercises 83–86, suppose that x represents a positive number and y represents a negative number. Determine whether the given expression must represent a positive number or a negative number.

83. $x - y$

84. $y - x$

85. $x + |y|$

86. $y - |x|$

1.6 Multiplying and Dividing Real Numbers

In this section we learn how to multiply positive and negative numbers. The result of multiplication is called the **product.** We already know how to multiply positive numbers and that the product of two positive numbers is positive. We also know that the product of 0 and any positive number is 0, and we extend that property to all real numbers.

OBJECTIVES

1. Find the product of numbers with different signs.
2. Find the product of two negative numbers.
3. Use the reciprocal of a number to apply the definition of division.
4. Use the order of operations when multiplying and dividing signed numbers.
5. Evaluate expressions involving variables.
6. Translate words and phrases involving multiplication and division.
7. Translate simple sentences into equations.

Multiplication Property of 0

For any real number a,

$$a \cdot 0 = 0 \cdot a = 0.$$

OBJECTIVE 1 Find the product of numbers with different signs. To define the product of numbers with different signs so that the result is consistent with multiplication of positive numbers, look at the following pattern.

$$3 \cdot 5 = 15$$
$$3 \cdot 4 = 12$$
$$3 \cdot 3 = 9$$
$$3 \cdot 2 = 6$$
$$3 \cdot 1 = 3$$
$$3 \cdot 0 = 0$$
$$3 \cdot (-1) = ?$$

The products decrease by 3.

What should $3(-1)$ equal? Since multiplication can also be considered repeated addition, the product $3(-1)$ represents the sum

$$-1 + (-1) + (-1) = -3,$$

so the product should be -3, which fits the pattern. Also,

$$3(-2) = -2 + (-2) + (-2) = -6.$$

Work Problem 1 at the Side.

The results from Problem 1 maintain the pattern in the list above, which suggests the following rule.

The product of a positive number and a negative number is negative.

Example: $6(-3) = -18.$

EXAMPLE 1 Multiplying a Positive Number and a Negative Number

Find each product using the multiplication rule.

(a) $8(-5) = -(8 \cdot 5) = -40$ **(b)** $-7(2) = -(7 \cdot 2) = -14$

(c) $-9\left(\frac{1}{3}\right) = -3$ **(d)** $-6.2(4.1) = -25.42$

Work Problem 2 at the Side.

1 Find each product by finding the sum of three numbers.

(a) $3(-3)$

(b) $3(-4)$

(c) $3(-5)$

2 Find each product.

(a) $2(-6)$

(b) $7(-8)$

(c) $-9(2)$

(d) $-16\left(\frac{5}{32}\right)$

(e) $4.56(-10)$

ANSWERS

1. (a) -9 (b) -12 (c) -15
2. (a) -12 (b) -56 (c) -18 (d) $-\frac{5}{2}$ (e) -45.6

OBJECTIVE 2 Find the product of two negative numbers. The product of two positive numbers is positive, and the product of a positive number and a negative number is negative. What about the product of two negative numbers? Look at another pattern.

$$\begin{aligned} -5(4) &= -20 \\ -5(3) &= -15 \\ -5(2) &= -10 \\ -5(1) &= -5 \\ -5(0) &= 0 \\ -5(-1) &= ? \end{aligned}$$

The products increase by 5.

The numbers on the left of the equals signs (in color) decrease by 1 for each step down the list. The products on the right increase by 5 for each step down the list. To maintain this pattern, $-5(-1)$ should be 5 more than $-5(0)$, or 5 more than 0, so

$$-5(-1) = 5.$$

The pattern continues with

$$\begin{aligned} -5(-2) &= 10 \\ -5(-3) &= 15 \\ -5(-4) &= 20 \\ -5(-5) &= 25, \end{aligned}$$

and so on. This pattern suggests the next rule.

The product of two negative numbers is positive.

Example: $-5(-4) = 20$.

EXAMPLE 2 Multiplying Two Negative Numbers

Find each product using the multiplication rule.

(a) $-9(-2) = 18$

(b) $-6(-12) = 72$

(c) $-2(4)(-1) = -8(-1) = 8$

(d) $3(-5)(-2) = -15(-2) = 30$

Work Problem 3 at the Side.

Here is a summary of the results for multiplying signed numbers.

Multiplying Signed Numbers

The product of two numbers having the *same* sign is *positive,* and the product of two numbers having *different* signs is *negative.*

OBJECTIVE 3 Use the reciprocal of a number to apply the definition of division. Recall that the result of division is called the **quotient.** In **Section 1.5** we saw that the difference between two numbers is found by adding the opposite of the subtrahend to the minuend. Similarly, the *quotient* of two numbers involves multiplying by the *reciprocal* of the second number.

3 Find each product.

(a) $-5(-6)$

(b) $-7(-3)$

(c) $-8(-5)$

(d) $-11(-2)$

(e) $-17(3)(-7)$

(f) $-41(2)(-13)$

Answers

3. **(a)** 30 **(b)** 21 **(c)** 40 **(d)** 22 **(e)** 357 **(f)** 1066

Reciprocals

Pairs of numbers whose product is 1 are called **reciprocals** of each other.

Since $8 \cdot \frac{1}{8} = \frac{8}{8} = 1$ and $\frac{5}{4} \cdot \frac{4}{5} = \frac{20}{20} = 1$,

the reciprocal of 8 is $\frac{1}{8}$, and that of $\frac{5}{4}$ is $\frac{4}{5}$. The following table shows several numbers and their reciprocals.

Number	Reciprocal
4	$\frac{1}{4}$
-5	$\frac{1}{-5}$ or $-\frac{1}{5}$
$\frac{3}{4}$	$\frac{4}{3}$
$-\frac{5}{8}$	$-\frac{8}{5}$
0	None

By definition, the product of a number and its reciprocal is 1. But the multiplication property of 0 says that the product of 0 and any number is 0. Thus,

0 has no reciprocal.

Work Problem 4 at the Side.

By definition, the quotient of a and b is the product of a and the reciprocal of b.

Division

The quotient $\frac{a}{b}$ of real numbers a and b, with $b \neq 0$, is

$$\frac{a}{b} = a \cdot \frac{1}{b}.$$

Example: $\frac{8}{-4} = 8\left(-\frac{1}{4}\right) = -2.$

This definition indicates that b, the number to divide by, cannot be 0. Since 0 has no reciprocal,

$\frac{a}{0}$ is not a number and *division by 0 is undefined.* If a division problem requires division by 0, write "undefined."

To illustrate, $\frac{6}{2} = 3$ since $2 \cdot 3 = 6$, but there is no number to represent $\frac{6}{0}$, since there is no number that when multiplied by 0 gives 6 as a product.

NOTE

If $a \neq 0$, then $\frac{0}{a} = 0$.

4 Complete the table.

Number	Reciprocal
(a) 6	
(b) -2	
(c) $\frac{2}{3}$	
(d) $-\frac{1}{4}$	
(e) 0	

ANSWERS

4. (a) $\frac{1}{6}$ (b) $\frac{1}{-2} = -\frac{1}{2}$ (c) $\frac{3}{2}$ (d) -4 (e) none

Because division is defined in terms of multiplication, all the rules for multiplying signed numbers also apply to dividing them.

EXAMPLE 3 Using the Definition of Division

Find each quotient.

(a) $\frac{12}{3} = 12 \cdot \frac{1}{3} = 4$

(b) $\frac{5(-2)}{2} = -10 \cdot \frac{1}{2} = -5$

(c) $\frac{-1.47}{-7} = -1.47 \cdot \left(-\frac{1}{7}\right) = .21$

(d) $-\frac{2}{3} \div \left(-\frac{5}{4}\right) = -\frac{2}{3} \cdot \left(-\frac{4}{5}\right) = \frac{8}{15}$

(e) $\frac{-10}{0}$ Undefined

(f) $\frac{0}{13} = 0$ $\quad \frac{0}{a} = 0 \quad (a \neq 0)$

Work Problem 5 at the Side.

When dividing fractions, multiplying by the reciprocal of the divisor works well. However, using the definition of division directly with integers is awkward. It is easier to divide in the usual way, then determine the sign of the answer. The following rule for division can be used instead of multiplying by the reciprocal.

Dividing Signed Numbers

The quotient of two numbers having the *same* sign is *positive;* the quotient of two numbers having *different* signs is *negative.*

Examples: $\frac{-15}{-5} = 3, \quad \frac{15}{-5} = -3, \quad$ and $\quad \frac{-15}{5} = -3.$

EXAMPLE 4 Dividing Signed Numbers

Find each quotient.

(a) $\frac{8}{-2} = -4$

(b) $\frac{-10}{2} = -5$

(c) $\frac{-4.5}{-.09} = 50$

(d) $-\frac{1}{8} \div \left(-\frac{3}{4}\right) = -\frac{1}{8} \cdot \left(-\frac{4}{3}\right) = \frac{1}{6}$

Work Problem 6 at the Side.

From the definitions of multiplication and division of real numbers,

$$\frac{-40}{8} = -40 \cdot \frac{1}{8} = -5 \quad \text{and} \quad \frac{40}{-8} = 40\left(\frac{1}{-8}\right) = -5, \text{ so}$$

$$\frac{-40}{8} = \frac{40}{-8}.$$

Based on this example, the quotient of a positive number and a negative number can be written in any of the following three forms.

For any positive real numbers a and b,

$$\frac{-a}{b} = \frac{a}{-b} = -\frac{a}{b}.$$

5 Find each quotient.

(a) $\frac{42}{7}$

(b) $\frac{-36}{(-2)(-3)}$

(c) $\frac{-12.56}{-.4}$

(d) $\frac{10}{7} \div \left(-\frac{24}{5}\right)$

(e) $\frac{-3}{0}$

(f) $\frac{0}{-53}$

6 Find each quotient.

(a) $\frac{-8}{-2}$

(b) $\frac{-16.4}{2.05}$

(c) $\frac{1}{4} \div \left(-\frac{2}{3}\right)$

(d) $\frac{12}{-4}$

ANSWERS

5. **(a)** 6 **(b)** -6 **(c)** 31.4 **(d)** $-\frac{25}{84}$ **(e)** undefined **(f)** 0

6. **(a)** 4 **(b)** -8 **(c)** $-\frac{3}{8}$ **(d)** -3

Similarly, the quotient of two negative numbers can be expressed as the quotient of two positive numbers.

For any positive real numbers a and b,

$$\frac{-a}{-b} = \frac{a}{b}.$$

OBJECTIVE 4 Use the order of operations when multiplying and dividing signed numbers.

EXAMPLE 5 Using the Order of Operations

Simplify.

(a) $-9(2) - (-3)(2)$

First find all products, working from left to right.

$$\begin{aligned} -9(2) - (-3)(2) &= -18 - (-6) \\ &= -18 + 6 \\ &= -12 \end{aligned}$$

(b) $\begin{aligned} -6(-2) - 3(-4) &= 12 - (-12) \\ &= 12 + 12 \\ &= 24 \end{aligned}$

(c) $\dfrac{5(-2) - 3(4)}{2(1 - 6)}$

Simplify the numerator and denominator separately.

$$\begin{aligned} \frac{5(-2) - 3(4)}{2(1 - 6)} &= \frac{-10 - 12}{2(-5)} && \text{Multiply in numerator. Subtract in denominator.} \\ &= \frac{-22}{-10} && \text{Subtract in numerator. Multiply in denominator.} \\ &= \frac{11}{5} && \text{Write in lowest terms.} \end{aligned}$$

Work Problem 7 at the Side.

The rules for operations with signed numbers are summarized here.

Operations with Signed Numbers

Addition

Same sign Add the absolute values of the numbers. The sum has the same sign as the numbers.

$$-4 + (-6) = -10$$

Different signs Find the absolute values of the numbers, and subtract the smaller absolute value from the larger. Give the sum the sign of the number having the larger absolute value.

$$4 + (-6) = -(6 - 4) = -2$$

(continued)

7 Perform the indicated operations.

(a) $-3(4) - 2(6)$

(b) $-8[-1 - (-4)(-5)]$

(c) $\dfrac{6(-4) - 2(5)}{3(2 - 7)}$

(d) $\dfrac{-6(-8) + 3(9)}{-2[4 - (-3)]}$

ANSWERS

7. **(a)** -24 **(b)** 168 **(c)** $\frac{34}{15}$ **(d)** $-\frac{75}{14}$

Subtraction

Add the opposite of the subtrahend to the minuend.

$$8 - (-3) = 8 + 3 = 11$$

Multiplication and Division

Same sign The product or quotient of two numbers with the same sign is positive.

$$-5(-6) = 30 \quad \text{and} \quad \frac{-36}{-12} = 3$$

Different signs The product or quotient of two numbers with different signs is negative.

$$-5(6) = -30 \quad \text{and} \quad \frac{18}{-6} = -3$$

Division by 0 is undefined.

OBJECTIVE 5 Evaluate expressions involving variables.

EXAMPLE 6 Evaluating Expressions for Numerical Values

Evaluate each expression, given that $x = -1$, $y = -2$, and $m = -3$.

(a) $(3x + 4y)(-2m)$

First substitute the given values for the variables. Then use the order of operations to find the value of the expression.

$(3x + 4y)(-2m) = [3(-1) + 4(-2)][-2(-3)]$ Put parentheses around the number for each variable.

$= [-3 + (-8)][6]$ Find the products.

$= (-11)(6)$ Add inside the brackets.

$= -66$ Multiply.

(b) $2x^2 - 3y^2$

Use parentheses as shown.

$2(-1)^2 - 3(-2)^2 = 2(1) - 3(4)$ Substitute, then apply the exponents.

$= 2 - 12$ Multiply.

$= -10$ Subtract.

(c) $\dfrac{4y^2 + x}{m}$

$\dfrac{4(-2)^2 + (-1)}{-3} = \dfrac{4(4) + (-1)}{-3}$ Substitute, then apply the exponent.

$= \dfrac{16 + (-1)}{-3}$ Multiply.

$= \dfrac{15}{-3}$ Add.

$= -5$ Divide.

Work Problem 8 at the Side.

8 Evaluate each expression.

(a) $2x - 7(y + 1)$ if $x = -4$ and $y = 3$

(b) $2x^2 - 4y^2$ if $x = -2$ and $y = -3$

(c) $\dfrac{4x - 2y}{-3x}$ if $x = 2$ and $y = -1$

Answers

8. **(a)** -36 **(b)** -28 **(c)** $-\frac{5}{3}$

OBJECTIVE 6 Translate words and phrases involving multiplication and division. Just as there are words and phrases that indicate addition or subtraction, certain words and phrases indicate multiplication or division. The chart gives some phrases indicating multiplication.

Word or Phrase	Example	Numerical Expression and Simplification
Product of	The *product of* -5 and -2	$-5(-2) = 10$
Times	13 *times* -4	$13(-4) = -52$
Twice (meaning "2 times")	*Twice* 6	$2(6) = 12$
Of (used with fractions)	$\frac{1}{2}$ *of* 10	$\frac{1}{2}(10) = 5$
Percent of	12% *of* -16	$.12(-16) = -1.92$

EXAMPLE 7 Translating Words and Phrases

Write a numerical expression for each phrase, and simplify the expression.

(a) The product of 12 and the sum of 3 and -6
Here 12 is multiplied by "the sum of 3 and -6."

$$12[3 + (-6)] = 12(-3) = -36$$

(b) Three times the difference between 4 and -11

$$3[4 - (-11)] = 3(4 + 11) = 3(15) = 45$$

(c) Two-thirds of the sum of -5 and -3

$$\frac{2}{3}[-5 + (-3)] = \frac{2}{3}(-8) = -\frac{16}{3}$$

(d) 15% of the difference between 14 and -2
Remember from **Section R.2** that 15% = .15.

$$.15[14 - (-2)] = .15(14 + 2) = .15(16) = 2.4$$

Work Problem 9 at the Side.

9 Write a numerical expression for each phrase and simplify.

(a) The product of 6 and the sum of -5 and -4

(b) Twice the difference between 8 and -4

(c) Three-fifths of the sum of 2 and -7

(d) 20% of the sum of 9 and -4

The word *quotient* refers to the answer in a division problem. In algebra, a quotient is usually represented with a fraction bar; the symbol ÷ is seldom used. When translating an applied problem involving division, use a fraction bar. The chart gives some phrases associated with division.

Word or Phrase	Example	Numerical Expression and Simplification
Quotient of	The *quotient of* -24 and 3	$\frac{-24}{3} = -8$
Divided by	-16 *divided by* -4	$\frac{-16}{-4} = 4$
Ratio of	The *ratio of* 2 to 3	$\frac{2}{3}$

When translating a phrase involving division, we write the first number named as the numerator and the second as the denominator.

ANSWERS

9. **(a)** $6[(-5) + (-4)]; -54$
(b) $2[8 - (-4)]; 24$
(c) $\frac{3}{5}[2 + (-7)]; -3$
(d) $.20[9 + (-4)]; 1$

10 Write a numerical expression for each phrase, and simplify the expression.

(a) The quotient of 20 and the sum of 8 and -3

(b) The product of -9 and 2, divided by the difference between 5 and -1

EXAMPLE 8 Translating Words and Phrases

Write a numerical expression for each phrase, and simplify the expression.

(a) The **quotient** of 14 and the sum of -9 and 2

"Quotient" indicates division. The number 14 is the numerator and "the sum of -9 and 2" is the denominator.

$$\frac{14}{-9+2} = \frac{14}{-7} = -2$$

(b) The product of 5 and -6, **divided by** the difference between -7 and 8

The numerator of the fraction representing the division is obtained by multiplying 5 and -6. The denominator is found by subtracting -7 and 8.

$$\frac{5(-6)}{-7-8} = \frac{-30}{-15} = 2$$

Work Problem 10 at the Side.

11 Write each sentence in symbols, using x to represent the number.

(a) Twice a number is -6.

(b) The difference between -8 and a number is -11.

(c) The sum of 5 and a number is 8.

(d) The quotient of a number and -2 is 6.

OBJECTIVE 7 Translate simple sentences into equations. In this section and the previous two sections, important words and phrases involving the four operations of arithmetic have been introduced. We can use these words and phrases to translate sentences into equations.

EXAMPLE 9 Translating Sentences into Equations

Write each sentence with symbols, using x to represent the number.

(a) Three **times** a number **is** -18.

The word *times* indicates multiplication, and the word *is* translates as the equals sign $(=)$.

$$3x = -18$$

(b) The **sum** of a number and 9 **is** 12.

$$x + 9 = 12$$

(c) The **difference between** a number and 5 **is** 0.

$$x - 5 = 0$$

(d) The **quotient of** 24 and a number **is** -2.

$$\frac{24}{x} = -2$$

Work Problem 11 at the Side.

CAUTION

It is important to recognize the distinction between the types of problems found in Example 8 and Example 9. In Example 8, the phrases translate as *expressions,* while in Example 9, the sentences translate as *equations.* Remember that an equation is a sentence, while an expression is a phrase.

$$\frac{5(-6)}{-7-8} \qquad\qquad 3x = -18$$

↑ Expression ↑ Equation

ANSWERS

10. (a) $\frac{20}{8+(-3)}$; 4
(b) $\frac{-9(2)}{5-(-1)}$; -3

11. (a) $2x = -6$
(b) $-8 - x = -11$
(c) $5 + x = 8$
(d) $\frac{x}{-2} = 6$

1.6 Exercises

FOR EXTRA HELP: Addison-Wesley Math Tutor Center | MathXL | Digital Video Tutor CD 1 Videotape 2 | Student's Solutions Manual | MyMathLab | Interactmath.com

Fill in each blank with one of the following: greater than 0, less than 0, equal to 0.

1. The product or the quotient of two numbers with the same sign is ________________.

2. The product or the quotient of two numbers with different signs is ________________.

3. If three negative numbers are multiplied together, the product is ________________.

4. If two negative numbers are multiplied and then their product is divided by a negative number, the result is ________________.

5. If a negative number is squared and the result is added to a positive number, the final answer is ________________.

6. The reciprocal of a negative number is ________________.

Find each product. See Examples 1 and 2.

7. $-7(4)$ **8.** $-8(5)$ **9.** $-5(-6)$ **10.** $-4(-20)$

11. $-8(0)$ **12.** $0(-12)$ **13.** $-\frac{3}{8}\left(-\frac{20}{9}\right)$ **14.** $-\frac{5}{4}\left(-\frac{6}{25}\right)$

15. $-6.8(.35)$ **16.** $-4.6(.24)$ **17.** $-6\left(-\frac{1}{4}\right)$ **18.** $-8\left(-\frac{1}{2}\right)$

Find each quotient. See Examples 3 and 4.

19. $\frac{-15}{5}$ **20.** $\frac{-18}{6}$ **21.** $\frac{20}{-10}$ **22.** $\frac{28}{-4}$

23. $\frac{-160}{-10}$ **24.** $\frac{-260}{-20}$ **25.** $\frac{0}{-3}$ **26.** $\frac{0}{-5}$

27. $\frac{-10.252}{0}$ **28.** $\frac{-29.584}{0}$ **29.** $\left(-\frac{3}{4}\right) \div \left(-\frac{1}{2}\right)$ **30.** $\left(-\frac{3}{16}\right) \div \left(-\frac{5}{8}\right)$

31. Which expression is undefined?

A. $\frac{5-5}{5+5}$ **B.** $\frac{5+5}{5+5}$ **C.** $\frac{5-5}{5-5}$ **D.** $\frac{5-5}{5}$

32. What is the reciprocal of .4?

Perform each indicated operation. See Example 5.

33. $\frac{-5(-6)}{9-(-1)}$

34. $\frac{-12(-5)}{7-(-5)}$

35. $\frac{-21(3)}{-3-6}$

36. $\frac{-40(3)}{-2-3}$

37. $\frac{-10(2)+6(2)}{-3-(-1)}$

38. $\frac{8(-1)+6(-2)}{-6-(-1)}$

39. $\frac{-27(-2)-(-12)(-2)}{-2(3)-2(2)}$

40. $\frac{-13(-4)-(-8)(-2)}{(-10)(2)-4(-2)}$

41. $\frac{3^2-4^2}{7(-8+9)}$

42. $\frac{5^2-7^2}{2(3+3)}$

43. If x and y are both replaced by negative numbers, is the value of $4x+8y$ positive or negative? What about $4x-8y$?

Evaluate each expression if $x=6$, $y=-4$, and $a=3$. See Example 6.

44. $5x-2y+3a$

45. $6x-5y+4a$

46. $(2x+y)(3a)$

47. $(5x-2y)(-2a)$

48. $\left(\frac{1}{3}x-\frac{4}{5}y\right)\left(-\frac{1}{5}a\right)$

49. $\left(\frac{5}{6}x+\frac{3}{2}y\right)\left(-\frac{1}{3}a\right)$

50. $(-5+x)(-3+y)(3-a)$

51. $(6-x)(5+y)(3+a)$

52. $-2y^2+3a$

53. $5x - 4a^2$

54. $\dfrac{2y^2 - x}{a - 3}$

55. $\dfrac{xy + 9a}{x + y - 2}$

Write a numerical expression for each phrase and simplify. See Examples 7 and 8.

56. The product of -9 and 2, added to 9

57. The product of 4 and -7, added to -12

58. Twice the product of -1 and 6, subtracted from -4

59. Twice the product of -8 and 2, subtracted from -1

60. The product of 12 and the difference between 9 and -8

61. The product of -3 and the difference between 3 and -7

62. Four-fifths of the sum of -8 and -2

63. Three-tenths of the sum of -2 and -28

64. The quotient of -12 and the sum of -5 and -1

65. The quotient of -20 and the sum of -8 and -2

66. The sum of 15 and -3, divided by the product of 4 and -3

67. The sum of -18 and -6, divided by the product of 2 and -4

68. The product of $-\frac{1}{2}$ and $\frac{3}{4}$, divided by $-\frac{2}{3}$

69. The product of $-\frac{2}{3}$ and $-\frac{1}{5}$, divided by $\frac{1}{7}$

Write each sentence with symbols, using x to represent the number. See Example 9.

70. Seven times a number is -42.

71. Nine times a number is -36.

72. The quotient of a number and 3 is -3.

73. The quotient of a number and 4 is -1.

74. $\frac{1}{2}$ less than a number is 2.

75. $\frac{9}{11}$ less than a number is 5.

76. When 15 is divided by a number, the result is -5.

77. When 6 is divided by a number, the result is -3.

RELATING CONCEPTS (EXERCISES 78–83) For Individual or Group Work

To find the average of a group of numbers, we add the numbers and then divide the sum by the number of terms added. ***Work Exercises 78–81 in order,*** *to find the average of* 23, 18, 13, -4, *and* -8. *Then find the averages in Exercises 82 and 83.*

78. Find the sum of the given group of numbers.

79. How many numbers are in the group?

80. Divide your answer for Exercise 78 by your answer for Exercise 79. Give the quotient as a mixed number.

81. What is the average of the given group of numbers?

82. What is the average of all integers between -10 and 14, including both -10 and 14?

83. What is the average of the integers between -15 and -10, including -15 and -10?

Summary Exercises on Operations with Real Numbers

Operations with Signed Numbers

Addition

Same sign Add the absolute values of the numbers. The sum has the same sign as the numbers.

Different signs Find the absolute values of the numbers, and subtract the smaller absolute value from the larger. Give the sum the sign of the number having the larger absolute value.

Subtraction

Add the opposite of the subtrahend to the minuend.

Multiplication and Division

Same sign The product or quotient of two numbers with the same sign is positive.

Different signs The product or quotient of two numbers with different signs is negative.

Division by 0 is undefined.

Perform each indicated operation.

1. $14 - 3 \cdot 10$

2. $-3(8) - 4(-7)$

3. $(3 - 8)(-2) - 10$

4. $-6(7 - 3)$

5. $7 - (-3)(2 - 10)$

6. $-4[(-2)(6) - 7]$

7. $(-4)(7) - (-5)(2)$

8. $-5[-4 - (-2)(-7)]$

9. $40 - (-2)[8 - 9]$

10. $\dfrac{5(-4)}{-7 - (-2)}$

11. $\dfrac{-3 - (-9 + 1)}{-7 - (-6)}$

12. $\dfrac{5(-8 + 3)}{13(-2) + (-7)(-3)}$

13. $\dfrac{6^2 - 8}{-2(2) + 4(-1)}$

14. $\dfrac{16(-8 + 5)}{15(-3) + (-7 - 4)(-3)}$

15. $\dfrac{9(-6) - 3(8)}{4(-7) + (-2)(-11)}$

16. $\dfrac{2^2 + 4^2}{5^2 - 3^2}$

17. $\dfrac{(2 + 4)^2}{(5 - 3)^2}$

18. $\dfrac{4^3 - 3^3}{-5(-4 + 2)}$

19. $\dfrac{-9(-6) + (-2)(27)}{3(8 - 9)}$

20. $|-4(9)| - |-11|$

21. $\dfrac{6(-10 + 3)}{15(-2) - 3(-9)}$

22. $\dfrac{(-9)^2 - 9^2}{3^2 - 5^2}$

23. $\dfrac{(-10)^2 + 10^2}{-10(5)}$

24. $-\dfrac{3}{4} \div \left(-\dfrac{5}{8}\right)$

25. $\dfrac{1}{2} \div \left(-\dfrac{1}{2}\right)$

26. $\dfrac{8^2 - 12}{(-5)^2 + 2(6)}$

27. $\left[\dfrac{5}{8} - \left(-\dfrac{1}{16}\right)\right] + \dfrac{3}{8}$

28. $\left(\dfrac{1}{2} - \dfrac{1}{3}\right) - \dfrac{5}{6}$

29. $-.9(-3.7)$

30. $-5.1(-.2)$

31. $-3^2 - 2^2$

32. $|-2(3) + 4| - |-2|$

33. $40 - (-2)[-5 - 3]$

Evaluate each expression if $x = -2, y = 3,$ *and* $a = 4.$

34. $-x + y - 3a$

35. $(x + 6)^3 - y^3$

36. $(x - y) - (a - 2y)$

37. $\left(\dfrac{1}{2}x + \dfrac{2}{3}y\right)\left(-\dfrac{1}{4}a\right)$

38. $\dfrac{2x + 3y}{a - xy}$

39. $\dfrac{x^2 - y^2}{x^2 + y^2}$

40. $-x^2 + 3y$

41. $\dfrac{-x + 2y}{2x + a}$

42. $\dfrac{2x + a}{-x + 2y}$

1.7 Properties of Real Numbers

If you are asked to find the sum

$$3 + 89 + 97,$$

you might mentally add $3 + 97$ to get 100, and then add $100 + 89$ to get 189. While the order of operations guidelines say to add (or multiply) from left to right, the fact is we may change the order of the terms (or factors) and group them in any way we choose without affecting the sum (or product). This is an example of a shortcut we use in everyday mathematics that is justified by the properties of real numbers introduced in this section. In the following statements, a, b, and c represent real numbers.

OBJECTIVES

1. Use the commutative properties.
2. Use the associative properties.
3. Use the identity properties.
4. Use the inverse properties.
5. Use the distributive property.

OBJECTIVE 1 Use the commutative properties. The word *commute* means to go back and forth. Many people commute to work or to school. If you travel from home to work and follow the same route from work to home, you travel the same distance each time. The **commutative properties** say that if two numbers are added or multiplied in any order, they give the same result.

$$a + b = b + a \quad \text{Addition}$$

$$ab = ba \quad \text{Multiplication}$$

EXAMPLE 1 Using the Commutative Properties

Use a commutative property to complete each statement.

(a) $-8 + 5 = 5 +$ ______

By the commutative property for addition, the missing number is -8 because $-8 + 5 = 5 + (-8)$.

(b) $-2(7) =$ ______ (-2)

By the commutative property for multiplication, the missing number is 7, since $-2(7) = 7(-2)$.

Work Problem 1 at the Side.

1 Complete each statement. Use a commutative property.

(a) $x + 9 = 9 +$ ______

(b) $-12(4) =$ ______ (-12)

(c) $5x = x \cdot$ ______

OBJECTIVE 2 Use the associative properties. When we *associate* one object with another, we tend to think of those objects as being grouped together. The **associative properties** say that when we add or multiply three numbers, we can group them in any manner and get the same answer.

$$(a + b) + c = a + (b + c) \quad \text{Addition}$$

$$(ab)c = a(bc) \quad \text{Multiplication}$$

EXAMPLE 2 Using the Associative Properties

Use an associative property to complete each statement.

(a) $8 + (-1 + 4) = (8 +$ ______ $) + 4$

The missing number is -1.

(b) $[2(-7)]6 = 2$ ______

The missing expression on the right should be $[(-7)6]$.

Work Problem 2 at the Side.

2 Complete each statement. Use an associative property.

(a) $(9 + 10) + (-3)$
$= 9 + [$ ______ $+ (-3)]$

(b) $-5 + (2 + 8)$
$= ($ ______ $) + 8$

(c) $10[-8(-3)] =$ ______

ANSWERS

1. (a) x (b) 4 (c) 5
2. (a) 10 (b) $-5 + 2$ (c) $[10(-8)](-3)$

By the associative property of addition, the sum of three numbers will be the same no matter how the numbers are "associated" in groups. For this reason, parentheses can be left out in many addition problems. For example, both

$$(-1 + 2) + 3 \quad \text{and} \quad -1 + (2 + 3)$$

can be written as

$$-1 + 2 + 3.$$

In the same way, parentheses also can be left out of many multiplication problems.

EXAMPLE 3 Distinguishing between Associative and Commutative Properties

(a) Is $(2 + 4) + 5 = 2 + (4 + 5)$ an example of an associative or a commutative property?

The order of the three numbers is the same on both sides of the equals sign. The only change is in the *grouping*, or association, of the numbers. Therefore, this is an example of an associative property.

(b) Is $6(3 \cdot 10) = 6(10 \cdot 3)$ an example of an associative or a commutative property?

The same numbers, 3 and 10, are grouped on each side. On the left, however, 3 appears first in $(3 \cdot 10)$. On the right, 10 appears first. Since the only change involves the *order* of the numbers, this statement is an example of a commutative property.

(c) Is $(8 + 1) + 7 = 8 + (7 + 1)$ an example of an associative or a commutative property?

In the statement, both the order and the grouping are changed. On the left, the order of the three numbers is 8, 1, and 7. On the right it is 8, 7, and 1. On the left, 8 and 1 are grouped, and on the right, 7 and 1 are grouped. Therefore, both associative and commutative properties are used.

Work Problem 3 at the Side.

3 Decide whether each statement is an example of a commutative property, an associative property, or both.

(a) $2(4 \cdot 6) = (2 \cdot 4)6$

(b) $(2 \cdot 4)6 = (4 \cdot 2)6$

(c) $(2 + 4) + 6 = 4 + (2 + 6)$

We can use commutative and associative properties to simplify expressions.

EXAMPLE 4 Using Commutative and Associative Properties

The commutative and associative properties make it possible to choose pairs of numbers that are easy to add or multiply.

(a)
$$\begin{aligned} 23 + 41 + 2 + 9 + 25 &= (41 + 9) + (23 + 2) + 25 \\ &= 50 + 25 + 25 \\ &= 100 \end{aligned}$$

(b)
$$\begin{aligned} 25(69)(4) &= 25(4)(69) \\ &= 100(69) \\ &= 6900 \end{aligned}$$

Work Problem 4 at the Side.

4 Find the sum.

$5 + 18 + 29 + 31 + 12$

ANSWERS

3. **(a)** associative **(b)** commutative **(c)** both
4. 95

OBJECTIVE 3 Use the identity properties. The identity or value of a real number is left unchanged when identity properties are applied. The **identity properties** say that the sum of 0 and any number equals that number, and the product of 1 and any number equals that number.

$$a + 0 = a \quad \text{and} \quad 0 + a = a \qquad \text{Addition}$$

$$a \cdot 1 = a \quad \text{and} \quad 1 \cdot a = a \qquad \text{Multiplication}$$

The number 0 leaves the identity, or value, of any real number unchanged by addition. For this reason, 0 is called the **identity element for addition** or the **additive identity.** Since multiplication by 1 leaves any real number unchanged, 1 is the **identity element for multiplication** or the **multiplicative identity.**

EXAMPLE 5 Using Identity Properties

These statements are examples of identity properties.

(a) $-3 + 0 = -3$ Addition **(b)** $1 \cdot 25 = 25$ Multiplication

Work Problem 5 at the Side.

We use the identity property for multiplication to write fractions in lowest terms and to find common denominators.

EXAMPLE 6 Using the Identity Element for Multiplication to Simplify Expressions

Simplify each expression.

(a)
$$\frac{49}{35} = \frac{7 \cdot 7}{5 \cdot 7} \qquad \text{Factor.}$$
$$= \frac{7}{5} \cdot \frac{7}{7} \qquad \text{Write as a product.}$$
$$= \frac{7}{5} \cdot 1 \qquad \text{Property of 1}$$
$$= \frac{7}{5} \qquad \text{Identity property}$$

(b)
$$\frac{3}{4} + \frac{5}{24} = \frac{3}{4} \cdot 1 + \frac{5}{24} \qquad \text{Identity property}$$
$$= \frac{3}{4} \cdot \frac{6}{6} + \frac{5}{24} \qquad \text{Use } 1 = \tfrac{6}{6} \text{ to get a common denominator.}$$
$$= \frac{18}{24} + \frac{5}{24} \qquad \text{Multiply.}$$
$$= \frac{23}{24} \qquad \text{Add.}$$

Work Problem 6 at the Side.

5 Use an identity property to complete each statement.

(a) $9 + 0 =$ _____

(b) _____ $+ (-7) = -7$

(c) _____ $\cdot 1 = 5$

6 Use an identity property to simplify each expression.

(a) $\frac{85}{105}$

(b) $\frac{9}{10} - \frac{53}{50}$

ANSWERS

5. (a) 9 (b) 0 (c) 5

6. (a) $\frac{17}{21}$ (b) $-\frac{4}{25}$

OBJECTIVE 4 Use the inverse properties. Each day before you go to work or school, you probably put on your shoes before you leave. Before you go to sleep at night, you probably take them off, and this leads to the same situation that existed before you put them on. These operations from everyday life are examples of inverse operations.

The **inverse properties** of addition and multiplication lead to the additive and multiplicative identities, respectively. The opposite of a, $-a$, is the **additive inverse** of a and the reciprocal of a, $\frac{1}{a}$, is the **multiplicative inverse** of the nonzero number a. The sum of the numbers a and $-a$ is 0, and the product of the nonzero numbers a and $\frac{1}{a}$, is 1.

$$a + (-a) = 0 \quad \text{and} \quad -a + a = 0 \qquad \text{Addition}$$

$$a \cdot \frac{1}{a} = 1 \quad \text{and} \quad \frac{1}{a} \cdot a = 1 \quad (a \neq 0) \qquad \text{Multiplication}$$

EXAMPLE 7 Using Inverse Properties

The following statements are examples of inverse properties.

(a) $\frac{2}{3} \cdot \frac{3}{2} = 1$ Multiplication

(b) $(-5)\left(-\frac{1}{5}\right) = 1$ Multiplication

(c) $-\frac{1}{2} + \frac{1}{2} = 0$ Addition

(d) $4 + (-4) = 0$ Addition

(e) $\frac{1}{2} + (-.5) = 0$ Addition

Work Problem 7 at the Side.

7 Complete each statement so that it is an example of either an identity property or an inverse property. Tell which property is used.

(a) $-6 + \_\_\_\_ = 0$

(b) $\frac{4}{3} \cdot \_\_\_\_ = 1$

(c) $-\frac{1}{9} \cdot \_\_\_\_ = 1$

(d) $275 + \_\_\_\_ = 275$

(e) $-.75 + \frac{3}{4} = \_\_\_\_$

OBJECTIVE 5 Use the distributive property. The everyday meaning of the word *distribute* is "to give out from one to several." An important property of real number operations involves this idea.

Look at the value of the following expressions.

$$2(5 + 8) = 2(13) = 26$$

$$2(5) + 2(8) = 10 + 16 = 26$$

Since both expressions equal 26,

$$2(5 + 8) = 2(5) + 2(8).$$

This result is an example of the *distributive property of multiplication with respect to addition,* the only property involving *both* addition and multiplication. With this property, a product can be changed to a sum or difference. This idea is illustrated by the divided rectangle in Figure 13.

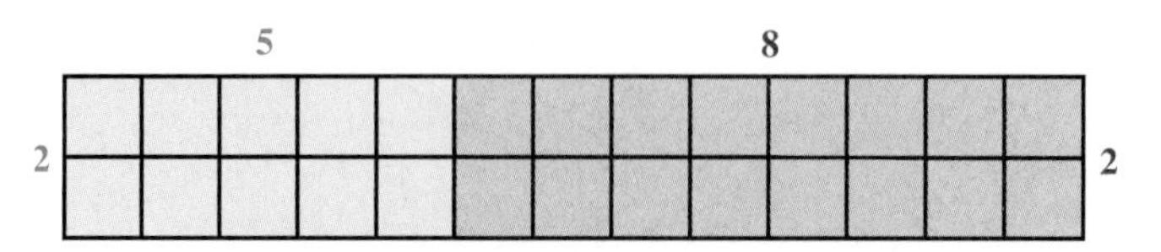

The area of the left part is 2(5) = 10.
The area of the right part is 2(8) = 16.
The total area is 2(5 + 8) = 26 or the total area is 2(5) + 2(8) = 10 + 16 = 26.
Thus, 2(5 + 8) = 2(5) + 2(8).

Figure 13

ANSWERS

7. **(a)** 6; inverse **(b)** $\frac{3}{4}$; inverse **(c)** -9; inverse **(d)** 0; identity **(e)** 0; inverse

The **distributive property** says that multiplying a number a by a sum of numbers $b + c$ gives the same result as multiplying a by b and a by c and then adding the two products.

$$a(b + c) = ab + ac \qquad \text{and} \qquad (b + c)a = ba + ca$$

As the arrows show, the a outside the parentheses is "distributed" over the b and c inside. The distributive property is also valid for subtraction.

$$a(b - c) = ab - ac \qquad \text{and} \qquad (b - c)a = ba - ca$$

The distributive property also can be extended to the sum (or difference) of more than two numbers.

$$a(b + c + d) = ab + ac + ad$$

EXAMPLE 8 Using the Distributive Property

Use the distributive property to rewrite each expression.

(a) $5(9 + 6) = 5 \cdot 9 + 5 \cdot 6$ Distributive property

$= 45 + 30$ Multiply.

$= 75$ Add.

(b) $4(x + 5 + y) = 4x + 4 \cdot 5 + 4y$ Distributive property

$= 4x + 20 + 4y$ Multiply.

(c) $-2(x + 3) = -2x + (-2)(3)$ Distributive property

$= -2x + (-6)$ Multiply.

$= -2x - 6$

(d) $3(k - 9) = 3k - 3 \cdot 9$ Distributive property

$= 3k - 27$ Multiply.

(e) $6 \cdot 8 + 6 \cdot 2$

The distributive property says that $a(b + c) = ab + ac$. This can be reversed to read

$$ab + ac = a(b + c).$$

We use this form of the distributive property to write a sum like $6 \cdot 8 + 6 \cdot 2$ with a common factor (of 6) as a product.

$ab + ac = a(b + c)$ Distributive property

$6 \cdot 8 + 6 \cdot 2 = 6(8 + 2)$ Let $a = 6$, $b = 8$, and $c = 2$.

$= 6(10)$

$= 60$

(f) $8(3r + 11t + 5z) = 8(3r) + 8(11t) + 8(5z)$ Distributive property

$= (8 \cdot 3)r + (8 \cdot 11)t + (8 \cdot 5)z$ Associative property

$= 24r + 88t + 40z$

Work Problem 8 at the Side.

8 Use the distributive property to rewrite each expression.

(a) $2(p + 5)$

(b) $-4(y + 7)$

(c) $5(m - 4)$

(d) $9 \cdot k + 9 \cdot 5$

(e) $3a - 3b$

(f) $7(2y + 7k - 9m)$

ANSWERS

8. **(a)** $2p + 10$ **(b)** $-4y - 28$
(c) $5m - 20$ **(d)** $9(k + 5)$
(e) $3(a - b)$ **(f)** $14y + 49k - 63m$

The symbol $-a$ may be interpreted as $-1 \cdot a$. Similarly, when a negative sign precedes an expression within parentheses, it may also be interpreted as a factor of -1. Thus, we can use the distributive property to remove (or clear) the parentheses from expressions such as $-(2y + 3)$. We do this by first writing $-(2y + 3)$ as $-1 \cdot (2y + 3)$.

$$-(2y + 3) = -1 \cdot (2y + 3) \qquad -a = -1 \cdot a$$
$$= -1 \cdot (2y) + (-1) \cdot (3) \qquad \text{Distributive property}$$
$$= -2y - 3 \qquad \text{Multiply.}$$

EXAMPLE 9 Using the Distributive Property to Remove Parentheses

Write without parentheses.

(a) $-(7r - 8) = -1(7r - 8)$
$$= -1(7r) + (-1)(-8) \qquad \text{Distributive property}$$
$$= -7r + 8 \qquad \text{Multiply.}$$

(b) $-(-9w + 2) = -1(-9w + 2)$
$$= 9w - 2$$

Work Problem 9 at the Side.

9 Write without parentheses.

(a) $-(3k - 5)$

(b) $-(2 - r)$

(c) $-(-5y + 8)$

(d) $-(-z + 4)$

The properties discussed here are the basic properties of real numbers that justify how we add and multiply in algebra. You should know them by name because we will be referring to them frequently. Here is a summary of these properties.

Properties of Addition and Multiplication

For any real numbers a, b, and c, the following properties hold.

Commutative properties $\quad a + b = b + a \qquad ab = ba$

Associative properties $\quad (a + b) + c = a + (b + c)$

$$(ab)c = a(bc)$$

Identity properties There is a real number 0 such that

$$a + 0 = a \quad \text{and} \quad 0 + a = a.$$

There is a real number 1 such that

$$a \cdot 1 = a \quad \text{and} \quad 1 \cdot a = a.$$

Inverse properties For each real number a, there is a single real number $-a$ such that

$$a + (-a) = 0 \quad \text{and} \quad (-a) + a = 0.$$

For each nonzero real number a, there is a single real number $\frac{1}{a}$ such that

$$a \cdot \frac{1}{a} = 1 \quad \text{and} \quad \frac{1}{a} \cdot a = 1.$$

Distributive property $\quad a(b + c) = ab + ac$

$$(b + c)a = ba + ca$$

ANSWERS

9. (a) $-3k + 5$ (b) $-2 + r$ (c) $5y - 8$ (d) $z - 4$

1.7 Exercises

Match each item in Column I with the correct choice from Column II. Choices may be used once, more than once, or not at all.

I	II
1. Identity element for addition	**A.** $(5 \cdot 4) \cdot 3 = 5 \cdot (4 \cdot 3)$
2. Identity element for multiplication	**B.** 0
3. Additive inverse of a	**C.** $-a$
4. Multiplicative inverse, or reciprocal, of the nonzero number a	**D.** -1
5. The only number that has no multiplicative inverse	**E.** $5 \cdot 4 \cdot 3 = 60$
6. An example of an associative property	**F.** 1
7. An example of a commutative property	**G.** $(5 \cdot 4) \cdot 3 = 3 \cdot (5 \cdot 4)$
8. An example of the distributive property	**H.** $5(4 + 3) = 5 \cdot 4 + 5 \cdot 3$
	I. $\frac{1}{a}$

Decide whether each statement is an example of a commutative, associative, identity, or inverse property, or of the distributive property. See Examples 1, 2, 3, and 5–8.

9. $\frac{2}{3}(-4) = -4\left(\frac{2}{3}\right)$

10. $6\left(-\frac{5}{6}\right) = \left(-\frac{5}{6}\right)6$

11. $-6 + (12 + 7) = (-6 + 12) + 7$

12. $(-8 + 13) + 2 = -8 + (13 + 2)$

13. $-6 + 6 = 0$

14. $12 + (-12) = 0$

15. $\left(\frac{2}{3}\right)\left(\frac{3}{2}\right) = 1$

16. $\left(\frac{5}{8}\right)\left(\frac{8}{5}\right) = 1$

17. $2.34 \cdot 1 = 2.34$

18. $-8.456 \cdot 1 = -8.456$

19. $(4 + 17) + 3 = 3 + (4 + 17)$

20. $(-8 + 4) + (-12) = -12 + (-8 + 4)$

21. $6(x + y) = 6x + 6y$

22. $14(t + s) = 14t + 14s$

23. $-\frac{5}{9} = -\frac{5}{9} \cdot \frac{3}{3} = -\frac{15}{27}$

24. $\frac{13}{12} = \frac{13}{12} \cdot \frac{7}{7} = \frac{91}{84}$

25. $5(2x) + 5(3y) = 5(2x + 3y)$

26. $3(5t) - 3(7r) = 3(5t - 7r)$

27. What number(s) satisfy each condition? **(a)** a number that is its own additive inverse **(b)** two numbers that are their own multiplicative inverses

28. The distributive property holds for multiplication with respect to addition. Is there a distributive property for addition with respect to multiplication? That is, does $a + b \cdot c = (a + b)(a + c)$? If not, give an example to show why.

29. Evaluate $25 - (6 - 2)$ and $(25 - 6) - 2$. Use the results to explain why subtraction is or is not associative.

30. Suppose that a classmate shows you the following work.

$$-2(5 - 6) = -2(5) - 2(6) = -10 - 12 = -22$$

The classmate has made a very common error. Explain the error and then work the problem correctly.

Write a new expression that is equal to the given expression, using the given property. Then simplify the new expression if possible. See Examples 1, 2, 5, 7, and 8.

31. $r + 7$; commutative

32. $t + 9$; commutative

33. $s + 0$; identity

34. $w + 0$; identity

35. $-6(x + 7)$; distributive

36. $-5(y + 2)$; distributive

37. $(w + 5) + (-3)$; associative

38. $(b + 8) + (-10)$; associative

39. Explain how the procedure of changing $\frac{3}{4}$ to $\frac{9}{12}$ requires the use of the multiplicative identity element, 1.

Use the properties of this section to simplify each expression. See Example 4.

40. $26 + 8 - 26 + 12$

41. $-\frac{3}{8} + \frac{2}{5} + \frac{8}{5} + \frac{3}{8}$

42. $\frac{9}{7}(-.38)\left(\frac{7}{9}\right)$

Use the distributive property to rewrite each expression. Simplify if possible. See Example 8.

43. $5 \cdot 3 + 5 \cdot 17$

44. $15 \cdot 6 + 5 \cdot 6$

45. $4(t + 3)$

46. $5(w + 4)$

47. $-8(r + 3)$

48. $-11(x + 4)$

49. $-5(y - 4)$

50. $-9(g - 4)$

51. $-\frac{4}{3}(12y + 15z)$

52. $-\frac{2}{5}(10b + 20a)$

53. $8 \cdot z + 8 \cdot w$

54. $4 \cdot s + 4 \cdot r$

55. $7(2v) + 7(5r)$

56. $13(5w) + 13(4p)$

57. $8(3r + 4s - 5y)$

58. $2(5u - 3v + 7w)$

59. $-3(8x + 3y + 4z)$

60. $-5(2x - 5y + 6z)$

Use the distributive property to write each expression without parentheses. See Example 9.

61. $-(4t + 5m)$

62. $-(9x + 12y)$

63. $-(-5c - 4d)$

64. $-(-13x - 15y)$

65. $-(-3q + 5r - 8s)$

66. $-(-4z + 5w - 9y)$

67. "Starting a car" and "driving away in a car" are not commutative. Give an example of another pair of everyday activities that are not commutative.

68. Are "undressing" and "taking a shower" commutative?

69. *True* or *false:* "preparing a meal" and "eating a meal" are commutative.

70. The phrase "dog biting man" has two different meanings, depending on how the words are associated.

(dog biting) man or dog (biting man)

Give another example of a three-word phrase that has different meanings depending on how the words are associated.

71. Use parentheses to show how the associative property can be used to give two different meanings to "foreign sales clerk."

72. Use parentheses to show two different meanings for "hot pink pants."

RELATING CONCEPTS (EXERCISES 73–76) For Individual or Group Work

In ***Section 1.6*** *we used a pattern to see that the product of two negative numbers is a positive number. In the exercises that follow, we show another justification for determining the sign of the product of two negative numbers.* ***Work Exercises 73–76 in order.***

73. Evaluate the expression $-3[5 + (-5)]$ by using the order of operations.

74. Write the expression in Exercise 73 using the distributive property. Do not simplify the products.

75. The product $-3(5)$ should be one of the terms you wrote when answering Exercise 74. Based on the results in **Section 1.6,** what is this product?

76. In Exercise 73, you should have obtained 0 as the answer. Now, consider the following, using the results of Exercises 73 and 75.

$$-3[5 + (-5)] = -3(5) + (-3)(-5)$$
$$0 = -15 + ?$$

The question mark represents the product $-3(-5)$. When added to -15, it must give a sum of 0. Therefore, $-3(-5)$ must equal what?

1.8 Simplifying Expressions

OBJECTIVES

1. Simplify expressions.
2. Identify terms and numerical coefficients.
3. Identify like terms.
4. Combine like terms.
5. Simplify expressions from word phrases.

OBJECTIVE 1 Simplify expressions. We now simplify expressions using the properties of addition and multiplication introduced in **Section 1.7.**

EXAMPLE 1 Simplifying Expressions

Simplify each expression.

(a) $4x + 8 + 9$

Since $8 + 9 = 17$, $4x + 8 + 9 = 4x + 17$.

(b) $4(3m - 2n)$

Use the distributive property.

$$4(3m - 2n) = 4(3m) - 4(2n)$$
$$= 12m - 8n$$

(c) $6 + 3(4k + 5) = 6 + 3(4k) + 3(5)$ Distributive property

$= 6 + 12k + 15$ Multiply.

$= 21 + 12k$ Add.

(d) $5 - (2y - 8) = 5 - 1(2y - 8)$ $-a = -1 \cdot a$

$= 5 - 2y + 8$ Distributive property

$= 13 - 2y$ Add.

NOTE

In Examples 1(c) and 1(d) we mentally used the commutative and associative properties to add in the last step. In practice, these steps are usually left out, but we should realize that they are used whenever the ordering and grouping in a sum are rearranged.

Work Problem 1 at the Side.

1 Simplify each expression.

(a) $9k + 12 - 5$

(b) $7(3p + 2q)$

(c) $2 + 5(3z - 1)$

(d) $-3 - (2 + 5y)$

OBJECTIVE 2 Identify terms and numerical coefficients. A **term** is a number, a variable, or a product or quotient of a number and one or more variables raised to powers. Examples of terms include

$$-9x^2, \quad 15y, \quad -3, \quad 8m^2n, \quad \frac{2}{p}, \quad \text{and} \quad k. \qquad \text{Terms}$$

The **numerical coefficient,** or simply coefficient, of the term $9m$ is 9; the numerical coefficient of $-15x^3y^2$ is -15; the numerical coefficient of x is 1; and the numerical coefficient of 8 is 8. In the expression $\frac{x}{3}$, the numerical coefficient of x is $\frac{1}{3}$ since $\frac{x}{3} = \frac{1x}{3} = \frac{1}{3}x$.

CAUTION

It is important to be able to distinguish between ***terms*** and ***factors.*** For example, in the expression $8x^3 + 12x^2$, there are two *terms*, $8x^3$ and $12x^2$. Terms are separated by a + or − sign. On the other hand, in the one-term expression $(8x^3)(12x^2)$, $8x^3$ and $12x^2$ are *factors*. Factors are multiplied.

ANSWERS

1. (a) $9k + 7$ (b) $21p + 14q$ (c) $15z - 3$ (d) $-5 - 5y$

2 Give the numerical coefficient of each term.

(a) $15q$

(b) $-2m^3$

(c) $-18m^7q^4$

(d) $-r$

(e) $\frac{5x}{4}$

3 Identify each pair of terms as *like* or *unlike*.

(a) $9x, 4x$

(b) $-8y^3, 12y^2$

(c) $5x^2y^4, 5x^4y^2$

(d) $7x^2y^4, -7x^2y^4$

(e) $13kt, 4tk$

4 Combine like terms.

(a) $4k + 7k$

(b) $4r - r$

(c) $5z + 9z - 4z$

(d) $8p + 8p^2$

(e) $5x - 3y + 2x - 5y - 3$

Answers

2. (a) 15 **(b)** -2 **(c)** -18 **(d)** -1 **(e)** $\frac{5}{4}$

3. (a) like **(b)** unlike **(c)** unlike **(d)** like **(e)** like

4. (a) $11k$ **(b)** $3r$ **(c)** $10z$ **(d)** cannot be combined **(e)** $7x - 8y - 3$

Here are some examples of terms and their numerical coefficients.

Term	Numerical Coefficient
$-7y$	-7
$34r^3$	34
$-26x^5yz^4$	-26
$-k$	-1
r	1
$\frac{3x}{8} = \frac{3}{8}x$	$\frac{3}{8}$

Work Problem 2 at the Side.

OBJECTIVE 3 Identify like terms. Terms with exactly the same variables (including the same exponents) are called **like terms.** For example, $9m$ and $4m$ have the same variables and are like terms. Also, $6x^3$ and $-5x^3$ are like terms. The terms $-4y^3$ and $4y^2$ have different exponents and are **unlike terms.** Here are some additional examples.

$5x$ and $-12x$		$3x^2y$ and $5x^2y$	Like terms
$4xy^2$ and $5xy$		$8x^2y^3$ and $7x^3y^2$	Unlike terms

Work Problem 3 at the Side.

OBJECTIVE 4 Combine like terms. Recall the distributive property:

$$x(y + z) = xy + xz.$$

As seen in the previous section, this statement can also be written as

$$xy + xz = x(y + z) \qquad \text{or} \qquad xy + xz = (y + z)x.$$

Thus the distributive property may be used to find the sum or difference of like terms. For example,

$$3x + 5x = (3 + 5)x = 8x.$$

This process is called **combining like terms.**

EXAMPLE 2 Combining Like Terms

Combine like terms in each expression.

(a) $9m + 5m$

Use the distributive property as given above.

$$9m + 5m = (9 + 5)m = 14m$$

(b) $6r + 3r + 2r = (6 + 3 + 2)r = 11r$ Distributive property

(c) $\frac{3}{4}x + x = \frac{3}{4}x + 1x = \left(\frac{3}{4} + 1\right)x = \frac{7}{4}x$ (Note: $x = 1x$.)

(d) $16y^2 - 9y^2 = (16 - 9)y^2 = 7y^2$

(e) $32y + 10y^2$ cannot be combined because $32y$ and $10y^2$ are unlike terms. The distributive property cannot be used here to combine coefficients.

Work Problem 4 at the Side.

CAUTION
Remember that ***only like terms may be combined.***

When an expression involves parentheses, the distributive property is used both "forward" and "backward" to combine like terms, as shown in the following example.

EXAMPLE 3 Simplifying Expressions Involving Like Terms

Simplify each expression.

(a) $14y + 2(6 + 3y) = 14y + 2(6) + 2(3y)$ Distributive property
$= 14y + 12 + 6y$ Multiply.
$= 20y + 12$ Combine like terms.

(b) $9k - 6 - 3(2 - 5k) = 9k - 6 - 3(2) - 3(-5k)$ Distributive property
$= 9k - 6 - 6 + 15k$ Multiply.
$= 24k - 12$ Combine like terms.

(c) $-(2 - r) + 10r = -1(2 - r) + 10r$ $-(2 - r) = -1(2 - r)$
$= -1(2) - 1(-r) + 10r$ Distributive property
$= -2 + r + 10r$ Multiply.
$= -2 + 11r$ Combine like terms.

(d) $5(2a^2 - 6a) - 3(4a^2 - 9) = 10a^2 - 30a - 12a^2 + 27$ Distributive property
$= -2a^2 - 30a + 27$ Combine like terms.

Work Problem 5 at the Side.

5 Simplify.

(a) $10p + 3(5 + 2p)$

(b) $7z - 2 - (1 + z)$

(c) $-(3k^2 + 5k) + 7(k^2 - 4k)$

OBJECTIVE 5 Simplify expressions from word phrases. Earlier we translated words, phrases, and statements into expressions and equations. Now we can simplify translated expressions by combining like terms.

EXAMPLE 4 Translating Words into a Mathematical Expression

Write the following phrase as a mathematical expression and simplify: four times a number, subtracted from the sum of twice the number and 4.

Let x represent the number. The expression is

The sum of twice the number and 4 → $(2x + 4)$; Four times the number → $4x$

$$(2x + 4) - 4x \quad \text{Write with symbols.}$$

which simplifies to

$$-2x + 4. \quad \text{Combine like terms.}$$

Work Problem 6 at the Side.

6 Write each phrase as a mathematical expression, and simplify by combining like terms.

(a) Three times a number, subtracted from the sum of the number and 8

(b) Twice a number added to the sum of 6 and the number

CAUTION
In Example 4, we are dealing with an expression to be simplified, *not* an equation to be solved.

ANSWERS
5. (a) $16p + 15$ (b) $6z - 3$ (c) $4k^2 - 33k$
6. (a) $(x + 8) - 3x$; $-2x + 8$
(b) $2x + (6 + x)$; $3x + 6$

Focus on

Real-Data Applications

Algebraic Expressions and Tuition Costs

Algebraic expressions are useful in real-life scenarios in which the same set of instructions are repeated for different choices of numbers. Below is the description of how tuition and fees are calculated for "Resident of District" students at North Harris Montgomery Community College District (NHMCCD) in Texas for 2003–2004. The information is given in the college's schedule and can be found at the Web site www.nhmccd.edu.*

Fees Required at NHMCCD

[*Residents of the district pay*] tuition at the rate of \$32 per credit hour, a \$6 per credit hour technology fee, a \$2 per credit hour student services fee, and a registration fee of \$12.

For Group Discussion

1. Calculate the tuition and fees for a student who is a resident of the district and who enrolls at NHMCCD for the specified number of credit hours. Let x represent the number of credit hours. Pay attention to the *process* used in your calculations so that you can write the algebraic expression for x credit hours.

(a) 3 credit hours: ________ **(b)** 9 credit hours: ________

(c) 12 credit hours: ________ **(d)** x credit hours: ________ dollars

Write the algebraic expression that represents the tuition and fees for each institution for one semester. Let x represent the number of credit hours. If you have difficulty, first calculate the costs for 3 or 9 credit hours and focus on the process that you used to get the answer.

2. American River College, California (nonresident student) www.losrios.edu*
Enrollment fee \$18 per unit; nonresident tuition \$149 per unit; parking permit \$30 per semester; student representation fee \$1 per semester

3. Austin Community College, Texas (in-state and out-of-district student) www.austin.cc.tx.us*
Tuition \$89 per credit; building fee \$13 per credit; student activity fee \$3 per term; parking fee \$10 per year

4. Anoka Ramsey Community College, Minnesota (in-state resident) www.anokaramsey.edu*
Tuition \$93.12 per credit; technology fee \$5.50 per credit; student activity fee \$4.75 per credit; parking fee \$2.00 per credit; student association fee \$.28 per credit

5. Your own college

---

*Note that URLs sometimes change, although that is unlikely for academic institutions. If the Web address given does not work, use a search engine, such as www.yahoo.com, to find the new URL.

1.8 Exercises

FOR EXTRA HELP

Addison-Wesley Math Tutor Center

MathXL

Digital Video Tutor CD 2 Videotape 2

Student's Solutions Manual

MyMathLab

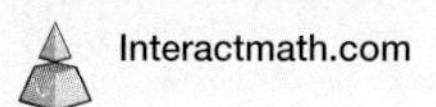

Interactmath.com

In Exercises 1–4, choose the letter of the correct response.

1. Which is true for all real numbers x?

A. $6 + 2x = 8x$ **B.** $6 - 2x = 4x$

C. $6x - 2x = 4x$ **D.** $3 + 8(4x - 6) = 11(4x - 6)$

2. Which is an example of a pair of like terms?

A. $6t, 6w$ **B.** $-8x^2y, 9xy^2$

C. $5ry, 6yr$ **D.** $-5x^2, 2x^3$

3. Which is an example of a term with numerical coefficient 5?

A. $5x^3y^7$ **B.** x^5 **C.** $\frac{x}{5}$ **D.** 5^2xy^3

4. Which is a correct translation for "six times a number, subtracted from the product of eleven and the number" (if x represents the number)?

A. $6x - 11x$ **B.** $11x - 6x$

C. $(11 + x) - 6x$ **D.** $6x - (11 + x)$

Simplify each expression. See Example 1.

5. $3x + 12x$

6. $4y + 9y$

7. $8t - 5t + 2t$

8. $6s - 9s + 4s$

9. $4r + 19 - 8$

10. $7t + 18 - 4$

11. $5 + 2(x - 3y)$

12. $8 + 3(s - 6t)$

13. $-2 - (5 - 3p)$

14. $-10 - (7 - 14r)$

Give the numerical coefficient of each term.

15. $-12k$

16. $-23y$

17. $5m^2$

18. $-3n^6$

19. xw

20. pq

21. $-x$

22. $-t$

23. 74

24. 98

25. Give an example of a pair of like terms with the variable x, such that one of them has a negative numerical coefficient, one has a positive numerical coefficient, and their sum has a positive numerical coefficient.

26. Give an example of a pair of unlike terms such that each term has x as the only variable.

Identify each group of terms as like *or* unlike.

27. $8r, -13r$

28. $-7a, 12a$

29. $5z^4, 9z^3$

30. $8x^5, -10x^3$

31. $4, 9, -24$

32. $7, 17, -83$

33. x, y

34. t, s

35. There is an old saying "You can't add apples and oranges." Explain how this saying can be applied to Objective 3 in this section.

36. Explain how the distributive property is used in combining $6t + 5t$ to get $11t$.

Simplify each expression. See Examples 2 and 3.

37. $-5 - 2(x - 3)$

38. $-8 - 3(2x + 4)$

39. $-\frac{4}{3} + 2t + \frac{1}{3}t - 8 - \frac{8}{3}t$

40. $-\frac{5}{6} + 8x + \frac{1}{6}x - 7 - \frac{7}{6}$

41. $-5.3r + 4.9 - (2r + .7) + 3.2r$

42. $2.7b + 5.8 - (3b + .5) - 4.4b$

43. $2y^2 - 7y^3 - 4y^2 + 10y^3$

44. $9x^4 - 7x^6 + 12x^4 + 14x^6$

45. $13p + 4(4 - 8p)$

46. $5x + 3(7 - 2x)$

47. $-\frac{4}{3}(y - 12) - \frac{1}{6}y$

48. $-\frac{7}{5}(t - 15) - \frac{3}{2}$

49. $-5(5y - 9) + 3(3y + 6)$

50. $-3(2t + 4) + 8(2t - 4)$

Write each phrase as a mathematical expression. Use x to represent the number. Combine like terms when possible. See Example 4.

51. Five times a number, added to the sum of the number and three

52. Six times a number, added to the sum of the number and six

53. A number multiplied by -7, subtracted from the sum of 13 and six times the number

54. A number multiplied by 5, subtracted from the sum of 14 and eight times the number

55. Six times a number added to -4, subtracted from twice the sum of three times the number and 4

56. Nine times a number added to 6, subtracted from triple the sum of 12 and 8 times the number

57. Write the expression $9x - (x + 2)$ using words, as in Exercises 51–56.

58. Write the expression $2(3x + 5) - 2(x + 4)$ using words, as in Exercises 51–56.

RELATING CONCEPTS (EXERCISES 59–62) For Individual or Group Work

A manufacturer has fixed costs of \$1000 *to produce widgets. Each widget costs* \$5 *to make. The fixed cost to produce gadgets is* \$750, *and each gadget costs* \$3 *to make.*
Work Exercises 59–62 in order.

59. Write an expression for the cost to make x widgets. (*Hint:* The cost will be the sum of the fixed cost and the cost per item times the number of items.)

60. Write an expression for the cost to make y gadgets.

61. Write an expression for the total cost to make x widgets and y gadgets.

62. Simplify the expression you wrote in Exercise 61.

Chapter 1

SUMMARY

Study Skills Workbook
Activity 6: Reviewing

KEY TERMS

Section	Term	Definition
1.1	**exponent**	An exponent, or **power**, is a number that indicates how many times a factor is repeated. 3^4 ← Exponent, ↑ Base; Exponential expression
	base	The base is the number that is a repeated factor when written with an exponent.
	exponential expression	A number written with an exponent is an exponential expression.
1.2	**variable**	A variable is a symbol, usually a letter, used to represent an unknown number.
	algebraic expression	An algebraic expression is a collection of numbers, variables, operation symbols, and grouping symbols.
	equation	An equation is a statement that says two expressions are equal.
	solution	A solution of an equation is any value of the variable that makes the equation true.
1.3	**natural numbers**	The set of natural numbers is $\{1, 2, 3, 4, \ldots\}$.
	whole numbers	The set of whole numbers is $\{0, 1, 2, 3, 4, 5, \ldots\}$.
	number line	The number line shows the ordering of the real numbers on an infinite line.
	opposite	The opposite of a number a is the number that is the same distance from 0 on the number line as a, but on the opposite side of 0. This number is also called the **negative** of a or the **additive inverse** of a.
	integers	The set of integers is $\{\ldots, -3, -2, -1, 0, 1, 2, 3, \ldots\}$.
	negative number	A negative number is located to the *left* of 0 on the number line.
	positive number	A positive number is located to the *right* of 0 on the number line.
	rational numbers	A rational number is a number that can be written as the quotient of two integers, with denominator not 0.
	coordinate	The number that corresponds to a point on the number line is the coordinate of that point.
	irrational numbers	An irrational number is a real number that is not a rational number.
	real numbers	Real numbers are numbers that can be represented by points on the number line, or all rational and irrational numbers.
	absolute value	The absolute value of a number is the distance between 0 and the number on the number line.
1.4	**sum**	The answer to an addition problem is called the sum.
1.5	**minuend**	In the operation $a - b$, a is called the minuend.
	subtrahend	In the operation $a - b$, b is called the subtrahend.
	difference	The answer to a subtraction problem is called the difference.
1.6	**product**	The answer to a multiplication problem is called the product.
	quotient	The answer to a division problem is called the quotient.
	reciprocal	Pairs of numbers whose product is 1 are called reciprocals or **multiplicative inverses** of each other.
1.7	**identity element for addition**	When the identity element for addition, which is 0, is added to a number, the number is unchanged.
	identity element for multiplication	When a number is multiplied by the identity element for multiplication, which is 1, the number is unchanged.
1.8	**term**	A term is a number, a variable, or a product or quotient of a number and one or more variables raised to powers.
	numerical coefficient	The numerical factor in a term is its numerical coefficient.
	like terms	Terms with exactly the same variables (including the same exponents) are called like terms.

Number line

Negative numbers | Positive numbers

−3 −2 −1 0 1 2 3

Opposites

New Symbols

Symbol	Meaning	Symbol	Meaning
a^n	n factors of a	$a(b)$, $(a)b$, $(a)(b)$, $a \cdot b$, or ab	a times b
$=$	is equal to	$\frac{a}{b}$, a/b, or $a \div b$	a divided by b
$\neq$	is not equal to	$\{\ \}$	set braces
$<$	is less than	$[\ ]$	square brackets
$\leq$	is less than or equal to	$\lvert x \rvert$	absolute value of x
$>$	is greater than	$\frac{1}{x}$	multiplicative inverse or reciprocal of x $(x \neq 0)$
$\geq$	is greater than or equal to		

Test Your Word Power

See how well you have learned the vocabulary in this chapter. Answers, with examples, follow the Quick Review.

1. The **product** is
- **A.** the answer in an addition problem
- **B.** the answer in a multiplication problem
- **C.** one of two or more numbers that are added to get another number
- **D.** one of two or more numbers that are multiplied to get another number.

2. A number is **prime** if
- **A.** it cannot be factored
- **B.** it has just one factor
- **C.** it has only itself and 1 as factors
- **D.** it has at least two different factors.

3. An **exponent** is
- **A.** a symbol that tells how many numbers are being multiplied
- **B.** a number raised to a power
- **C.** a number that tells how many times a factor is repeated
- **D.** one of two or more numbers that are multiplied.

4. A **variable** is
- **A.** a symbol used to represent an unknown number
- **B.** a value that makes an equation true
- **C.** a solution of an equation
- **D.** the answer in a division problem.

5. An **integer** is
- **A.** a positive or negative number
- **B.** a natural number, its opposite, or zero
- **C.** any number that can be graphed on a number line
- **D.** the quotient of two numbers.

6. A **coordinate** is
- **A.** the number that corresponds to a point on a number line
- **B.** the graph of a number
- **C.** any point on a number line
- **D.** the distance from 0 on a number line.

7. The **absolute value** of a number is
- **A.** the graph of the number
- **B.** the reciprocal of the number
- **C.** the opposite of the number
- **D.** the distance between 0 and the number on a number line.

8. A **term** is
- **A.** a numerical factor
- **B.** a number, a variable, or a product or quotient of numbers and variables raised to powers
- **C.** one of several variables with the same exponents
- **D.** a sum of numbers and variables raised to powers.

9. A **numerical coefficient** is
- **A.** the numerical factor in a term
- **B.** the number of terms in an expression
- **C.** a variable raised to a power
- **D.** the variable factor in a term.

10. The **subtrahend** in $a - b = c$ is
- **A.** a
- **B.** b
- **C.** c
- **D.** $a - b$.

QUICK REVIEW

Concepts	Examples
1.1 *Exponents, Order of Operations, and Inequality* **Order of Operations** Simplify within any parentheses or brackets and above and below fraction bars, using the following steps. *Step 1* Apply all exponents. *Step 2* Multiply or divide from left to right. *Step 3* Add or subtract from left to right.	$\frac{9(2+6)}{2} - 2(2^3+3) = 36 - 2(8+3)$ $= 36 - 2(11)$ $= 36 - 22$ $= 14$
1.2 *Variables, Expressions, and Equations* Evaluate an expression with a variable by substituting a given number for the variable.	Evaluate $2x + y^2$ if $x = 3$ and $y = -4$. $2x + y^2 = 2(3) + (-4)^2$ $= 6 + 16$ $= 22$
Values of a variable that make an equation true are solutions of the equation.	Is 2 a solution of $5x + 3 = 18$? $5(2) + 3 = 18$ $13 = 18$ False 2 is not a solution.
1.3 *Real Numbers and the Number Line* **Ordering Real Numbers**	−3 −2 −1 0 1 2 3 4
a is less than b if a is to the left of b on the number line.	$-2 < 3$ $3 > 0$ $0 < 3$
The opposite or additive inverse of a is $-a$.	$-(5) = -5$ $-(-7) = 7$ $-0 = 0$
The absolute value of a, $\lvert a \rvert$, is the distance between a and 0 on the number line.	$\lvert 13 \rvert = 13$ $\lvert 0 \rvert = 0$ $\lvert -5 \rvert = 5$
1.4 *Adding Real Numbers* To add two numbers with the same sign, add their absolute values. The sum has that same sign.	$9 + 4 = 13$ $-8 + (-5) = -13$
To add two numbers with different signs, subtract their absolute values. The sum has the sign of the number with larger absolute value.	$7 + (-12) = -5$ $-5 + 13 = 8$
1.5 *Subtracting Real Numbers* To subtract signed numbers, change the subtraction symbol to addition, and change the sign of the subtrahend. Add as in the previous section.	$5 - (-2) = 5 + 2 = 7$ $-3 - 4 = -3 + (-4) = -7$ $-2 - (-6) = -2 + 6 = 4$
1.6 *Multiplying and Dividing Real Numbers* The product (or quotient) of two numbers having the *same sign* is *positive*.	$6 \cdot 5 = 30$ $(-7)(-8) = 56$ $\frac{10}{2} = 5$ $\frac{-24}{-6} = 4$
The product (or quotient) of two numbers having *different signs* is *negative*.	$-6(5) = -30$ $6(-5) = -30$ $-18 \div 9 = \frac{-18}{9} = -2$ $49 \div (-7) = \frac{49}{-7} = -7$

(continued)

Concepts	*Examples*
1.6 *Multiplying and Dividing Real Numbers (continued)* To divide a by b, multiply a by the reciprocal of b.	$\dfrac{10}{\frac{2}{3}} = 10 \div \dfrac{2}{3} = 10 \cdot \dfrac{3}{2} = 15$
0 divided by a nonzero number is 0. Division by 0 is undefined.	$\dfrac{0}{5} = 0 \qquad \dfrac{5}{0}$ is undefined.
1.7 *Properties of Real Numbers*	
Commutative Properties $a + b = b + a$ $ab = ba$	$7 + (-1) = -1 + 7$ $5(-3) = (-3)5$
Associative Properties $(a + b) + c = a + (b + c)$ $(ab)c = a(bc)$	$(3 + 4) + 8 = 3 + (4 + 8)$ $[-2(6)]4 = -2[6(4)]$
Identity Properties $a + 0 = a \qquad 0 + a = a$ $a \cdot 1 = a \qquad 1 \cdot a = a$	$-7 + 0 = -7 \qquad 0 + (-7) = -7$ $9 \cdot 1 = 9 \qquad 1 \cdot 9 = 9$
Inverse Properties $a + (-a) = 0 \qquad -a + a = 0$ $a \cdot \dfrac{1}{a} = 1 \qquad \dfrac{1}{a} \cdot a = 1 \ (a \neq 0)$	$7 + (-7) = 0 \qquad -7 + 7 = 0$ $-2\left(-\dfrac{1}{2}\right) = 1 \qquad -\dfrac{1}{2}(-2) = 1$
Distributive Properties $a(b + c) = ab + ac$ $(b + c)a = ba + ca$ $a(b - c) = ab - ac$	$5(4 + 2) = 5(4) + 5(2)$ $(4 + 2)5 = 4(5) + 2(5)$ $9(5 - 4) = 9(5) - 9(4)$
1.8 *Simplifying Expressions* Only like terms may be combined.	$-3y^2 + 6y^2 + 14y^2 = 17y^2$ $-8a^5b^3 + 2a^3b^5 - 6a^5b^3 + 5a^3b^5 = -14a^5b^3 + 7a^3b^5$ $4(3 + 2x) - 6(5 - x) = 12 + 8x - 30 + 6x$ $= 14x - 18$

Answers to Test Your Word Power

1. B; *Example:* The product of 2 and 5, or 2 times 5, is 10.
2. C; *Examples:* 2, 3, 11, 41, 53
3. C; *Example:* In 2^3, the number 3 is the exponent (or power), so 2 is a factor three times; $2^3 = 2 \cdot 2 \cdot 2 = 8$.
4. A; *Examples:* a, b, c
5. B; *Examples:* $-9, 0, 6$
6. A; *Example:* The point graphed three units to the right of 0 on a number line has coordinate 3.
7. D; *Examples:* $|2| = 2$ and $|-2| = 2$
8. B; *Examples:* $6, \frac{x}{2}, -4ab^2$
9. A; *Examples:* The term 3 has numerical coefficient 3, $8z$ has numerical coefficient 8, and $-10x^4y$ has numerical coefficient -10.
10. B; *Example:* In $5 - 3 = 2$, 5 is the minuend, 3 is the subtrahend, and 2 is the difference.

Chapter 1

REVIEW EXERCISES

If you need help with any of these Review Exercises, look in the section indicated in brackets.

[1.1] *Find the value of each exponential expression.*

1. 5^4

2. $(.03)^4$

3. $.21^3$

4. $\left(\frac{5}{2}\right)^3$

Find the value of each expression.

5. $8 \cdot 5 - 13$

6. $5[4^2 + 3(2^3)]$

7. $\frac{7(3^2 - 5)}{16 - 2 \cdot 6}$

8. $\frac{3(9 - 4) + 5(8 - 3)}{2^3 - (5 - 3)}$

Write each word statement in symbols.

9. Thirteen is less than seventeen.

10. Five plus two is not equal to ten.

11. Write $6 < 15$ in words.

12. Construct a false statement that involves addition on the left side, the symbol $\geq$, and division on the right side.

[1.2] *Evaluate each expression if $x = 6$ and $y = 3$.*

13. $2x + 6y$

14. $4(3x - y)$

15. $\frac{x}{3} + 4y$

16. $\frac{x^2 + 3}{3y - x}$

Change each word phrase to an algebraic expression. Use x to represent the number.

17. Six added to a number

18. A number subtracted from eight

19. Nine subtracted from six times a number

20. Three-fifths of a number added to 12

Decide whether the given number is a solution of the equation.

21. $5x + 3(x + 2) = 22; 2$

22. $\frac{x + 5}{3x} = 1; 6$

Change each word sentence to an equation. Use x to represent the number.

23. Six less than twice a number is 10.

24. The product of a number and 4 is 8.

Identify each of the following as either an equation *or an* expression.

25. $5r - 8(r + 7) = 2$

26. $2y + (5y - 9) + 2$

[1.3] *Graph each group of numbers on a number line.*

27. $-4, -\frac{1}{2}, 0, 2.5, 5$

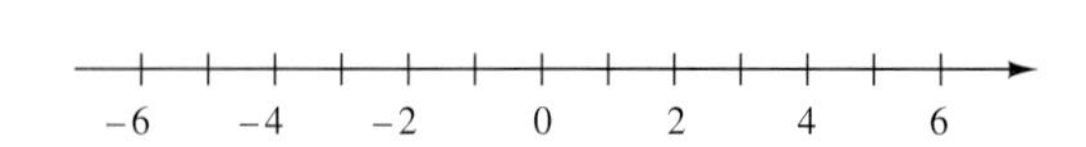

28. $-2, -3, |-3|, |-1|$

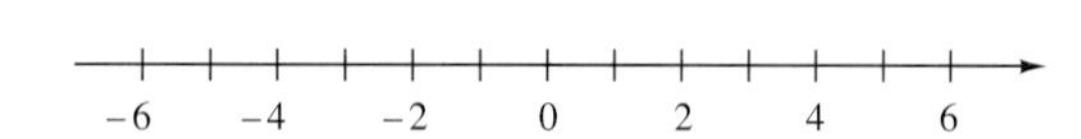

29. $-3\frac{1}{4}, \frac{14}{5}, -1\frac{1}{8}, \frac{5}{6}$

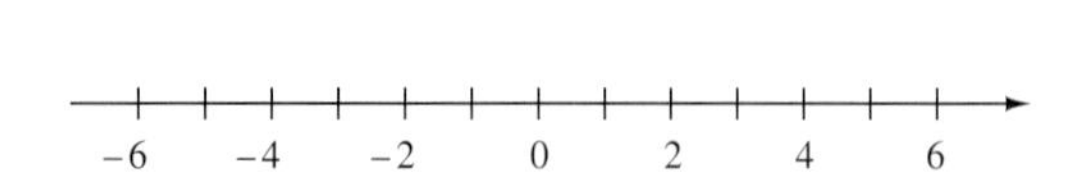

30. $|-4|, -|-3|, -|-5|, -6$

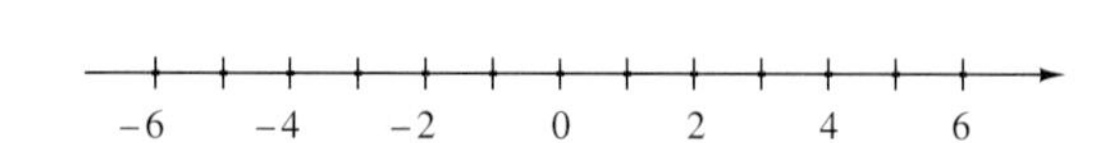

Select the lesser number in each pair.

31. $-10, 5$

32. $-8, -9$

33. $-\frac{2}{3}, -\frac{3}{4}$

34. $0, -|23|$

Decide whether each statement is true *or* false.

35. $12 > -13$

36. $0 > -5$

37. $-9 < -7$

38. $-13 > -13$

Simplify by finding the absolute value.

39. $-|3|$

40. $-|-19|$

41. $-|9 - 2|$

42. $|15 - 6|$

[1.4] *Find each sum.*

43. $-10 + 4$

44. $14 + (-18)$

45. $-8 + (-9)$

46. $\frac{4}{9} + \left(-\frac{5}{4}\right)$

47. $[-6 + (-8) + 8] + [9 + (-13)]$

48. $(-4 + 7) + (-11 + 3) + (-15 + 1)$

Write a numerical expression for each phrase, and simplify the expression.

49. 19 added to the sum of -31 and 12

50. 13 more than the sum of -4 and -8

Solve each problem.

51. Mohammed Hashemi has \$18 in his checking account. He then writes a check for \$26. What negative number represents his balance?

52. The temperature at noon on an August day in Houston was 93°F. After a thunderstorm, it dropped 6°. What was the new temperature?

[1.5] *Find each difference.*

53. $-7 - 4$

54. $-12 - (-11)$

55. $5 - (-2)$

56. $-\frac{3}{7} - \frac{4}{5}$

57. $2.56 - (-7.75)$

58. $(-10 - 4) - (-2)$

59. $(-3 + 4) - (-1)$

60. $|5 - 9| - |-3 + 6|$

Write a numerical expression for each phrase, and simplify the expression.

61. The difference between -4 and -6

62. Five less than the sum of 4 and -8

Solve each problem.

63. In the year 2000, the U.S. budget surplus was \$236.4 billion. By 2004, this had changed to a deficit of \$477.0 billion. By how much had the year 2000 amount decreased? (*Source:* White House; Congressional Budget Office.)

64. The 1994 Women's Olympic Downhill Skiing champion, Katja Seizinger, from West Germany, finished the course in 1 min, 35.93 sec. Seizinger won again in 1998. Her time decreased by 7.04 sec. What was Seizinger's winning time in 1998? (*Source: World Almanac and Book of Facts*, 2004.)

65. Explain in your own words how the subtraction problem $-8 - (-6)$ is performed.

66. Can the difference of two negative numbers be positive? Explain with an example.

The bar graph shows the federal budget outlays for national defense for the years 1993–2001. Use a signed number to represent the change in outlay for each time period. For example, the change from 1995 to 1996 was \$253.3 − \$259.6 = −\$6.3 billion.

67. 2000–2001

68. 1993–1994

69. 1997–1998

70. 1996–1997

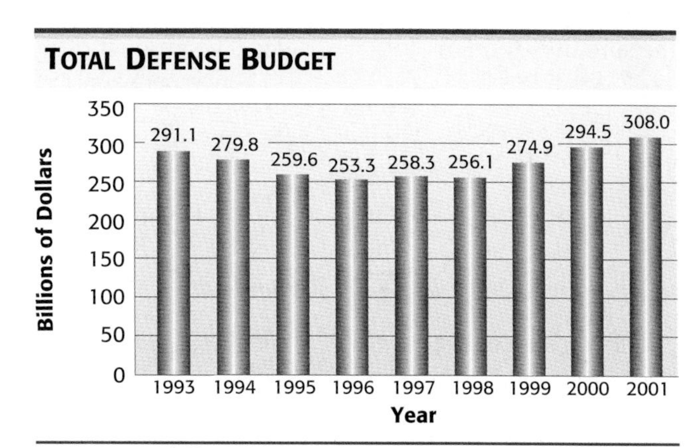

Source: U.S. Office of Management and Budget.

[1.6] *Perform the indicated operations.*

71. $(-12)(-3)$

72. $15(-7)$

73. $\left(-\frac{4}{3}\right)\left(-\frac{3}{8}\right)$

74. $(-4.8)(-2.1)$

75. $5(8 - 12)$

76. $(5 - 7)(8 - 3)$

77. $2(-6) - (-4)(-3)$

78. $3(-10) - 5$

79. $\frac{-36}{-9}$

80. $\frac{220}{-11}$

81. $-\frac{1}{2} \div \frac{2}{3}$

82. $-33.9 \div (-3)$

83. $\frac{-5(3) - 1}{8 - 4(-2)}$

84. $\frac{5(-2) - 3(4)}{-2[3 - (-2)] + 10}$

85. $\frac{10^2 - 5^2}{8^2 + 3^2 - (-2)}$

86. $\frac{4^2 - 8 \cdot 2}{(-1.2)^2 - (-.56)}$

Evaluate each expression if $x = -5$, $y = 4$, *and* $z = -3$.

87. $6x - 4z$

88. $5x + y - z$

89. $5x^2$

90. $z^2(3x - 8y)$

Write a numerical expression for each phrase, and simplify the expression.

91. Nine less than the product of -4 and 5

92. Five-sixths of the sum of 12 and -6

93. The quotient of 12 and the sum of 8 and -4

94. The product of -20 and 12, divided by the difference between 15 and -15

Translate each sentence to an equation, using x to represent the number.

95. The quotient of a number and the sum of the number and 5 is -2.

96. 3 less than 8 times a number is -7.

[1.7] *Decide whether each statement is an example of a commutative, associative, identity, or inverse property, or of the distributive property.*

97. $6 + 0 = 6$

98. $5 \cdot 1 = 5$

99. $-\frac{2}{3}\left(-\frac{3}{2}\right) = 1$

100. $17 + (-17) = 0$

101. $5 + (-9 + 2) = [5 + (-9)] + 2$

102. $w(xy) = (wx)y$

103. $3x + 3y = 3(x + y)$

104. $(1 + 2) + 3 = 3 + (1 + 2)$

Use the distributive property to rewrite each expression. Simplify if possible.

105. $7y + y$

106. $-12(4 - t)$

107. $3(2s) + 3(4y)$

108. $-(-4r + 5s)$

[1.8] *Use the distributive property as necessary and combine like terms.*

109. $16p^2 - 8p^2 + 9p^2$

110. $4r^2 - 3r + 10r + 12r^2$

111. $-8(5k - 6) + 3(7k + 2)$

112. $2s - (-3s + 6)$

113. $-7(2t - 4) - 4(3t + 8) - 19(t + 1)$

114. $3.6t^2 + 9t - 8.1(6t^2 + 4t)$

Translate each phrase into a mathematical expression. Use x to represent the number, and combine like terms when possible.

115. Seven times a number, subtracted from the product of -2 and three times the number

116. The quotient of 9 more than a number and 6 less than the number

117. In Exercise 115, does the word *and* signify addition? Explain.

118. Write the expression $3(4x - 6)$ using words, as in Exercises 115 and 116.

MIXED REVIEW EXERCISES*

Perform the indicated operations.

119. $[(-2) + 7 - (-5)] + [-4 - (-10)]$

120. $\left(-\frac{5}{6}\right)^2$

121. $-|(-7)(-4)| - (-2)$

122. $\frac{6(-4) + 2(-12)}{5(-3) + (-3)}$

123. $\frac{3}{8} - \frac{5}{12}$

124. $\frac{12^2 + 2^2 - 8}{10^2 - (-4)(-15)}$

125. $\frac{8^2 + 6^2}{7^2 + 1^2}$

126. $-16(-3.5) - 7.2(-3)$

127. $2\frac{5}{6} - 4\frac{1}{3}$

128. $-8 + [(-4 + 17) - (-3 - 3)]$

129. $-\frac{12}{5} \div \frac{9}{7}$

130. $(-8 - 3) - 5(2 - 9)$

131. $[-7 + (-2) - (-3)] + [8 + (-13)]$

132. $\frac{15}{2} \cdot \left(-\frac{4}{5}\right)$

Write a numerical expression or an equation for each problem, and simplify any expressions if possible. Use x as the variable, and specify what it represents.

133. In 2003, a company spent \$1400 less on advertising than in the previous year. The total spent for this purpose over these two years was \$25,800.

134. The quotient of a number and 14 less than three times the number

*The order of exercises in this final group does not correspond to the order in which topics occur in the chapter. This random ordering should help you prepare for the chapter test in yet another way.

Chapter 1
TEST

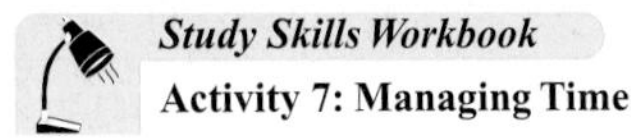

Study Skills Workbook
Activity 7: Managing Time

Decide whether the statement is true *or* false.

1. $4[-20 + 7(-2)] \le -135$ **1.** ______

2. $\left(\frac{1}{2}\right)^2 + \left(\frac{2}{3}\right)^2 = \left(\frac{1}{2} + \frac{2}{3}\right)^2$ **2.** ______

3. Graph the numbers -1, -3, $|-4|$, and $|-1|$ on the number line. **3.**

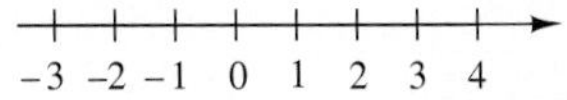

Select the lesser number from each pair.

4. $6, -|-8|$ **4.** ______

5. $-.742, -1.277$ **5.** ______

6. Write in symbols: The quotient of -6 and the sum of 2 and -8. Simplify the expression. **6.** ______

7. If a and b are both negative, is $\frac{a+b}{a \cdot b}$ positive or negative? **7.** ______

Perform the indicated operations whenever possible.

8. $-2 - (5 - 17) + (-6)$ **8.** ______

9. $-5\frac{1}{2} + 2\frac{2}{3}$ **9.** ______

10. $-6.2 - [-7.1 + (2.0 - 3.1)]$ **10.** ______

11. $4^2 + (-8) - (2^3 - 6)$ **11.** ______

12. $(-5)(-12) + 4(-4) + (-8)^2$ **12.** ______

13. $\frac{-7 - |-6 + 2|}{-5 - (-4)}$ **13.** ______

14. ______________

14. $\dfrac{30(-1-2)}{-9[3-(-2)]-12(-2)}$

In Exercises 15 and 16, evaluate each expression if $x = -2$ *and* $y = 4$.

15. ______________

15. $3x - 4y^2$

16. ______________

16. $\dfrac{5x + 7y}{3(x + y)}$

17. ______________

17. The highest Fahrenheit temperature ever recorded in Idaho was 118°, while the lowest was −60°. What is the difference between these highest and lowest temperatures? (*Source: World Almanac and Book of Facts,* 2004.)

Match each example in Column I with a property in Column II.

	I	II
18. ______________	**18.** $3x + 0 = 3x$	**A.** Commutative
19. ______________	**19.** $(5 + 2) + 8 = 8 + (5 + 2)$	**B.** Associative
20. ______________	**20.** $-3(x + y) = -3x + (-3y)$	**C.** Inverse
21. ______________	**21.** $-5 + (3 + 2) = (-5 + 3) + 2$	**D.** Identity
22. ______________	**22.** $-\frac{5}{3}\left(-\frac{3}{5}\right) = 1$	**E.** Distributive

23. ______________

23. Simplify $-2(3x^2 + 4) - 3(x^2 + 2x)$ by using the distributive property and combining like terms.

24. ______________

24. Which properties are used to show that $-(3x + 1) = -3x - 1$?

25. (a) ______________

(b) ______________

(c) ______________

25. Consider the expression $-6[5 + (-2)]$.
(a) Evaluate it by first working within the brackets.
(b) Evaluate it by using the distributive property.
(c) Why must the answers in parts (a) and (b) be the same?

Equations, Inequalities, and Applications

In 1896, 241 competitors (no women among them) from 14 countries gathered in Athens, Greece, for the first modern Olympic Games. In 2004, the XXVIII Olympic Summer Games returned to Athens as a truly international event, attracting 10,500 athletes from 202 countries.

One ceremonial aspect of the games is the flying of the Olympic flag with its five interlocking rings of different colors on a white background. First introduced at the 1920 Games in Antwerp, Belgium, the five rings on the flag symbolize unity among the nations of Africa, the Americas, Asia, Australia, and Europe. (*Source: USA Today,* August 13, 2004; *Microsoft Encarta Encyclopedia 2002.*)

Throughout this chapter we use linear equations to solve applications about the Olympics.

2.1 The Addition Property of Equality

OBJECTIVES

1. **Identify linear equations.**
2. **Use the addition property of equality.**
3. **Simplify equations, and then use the addition property of equality.**

Recall from **Section 1.2** that an *equation* is a statement that two algebraic expressions are equal. The simplest type of equation is a *linear equation*.

OBJECTIVE 1 Identify linear equations.

Linear Equation in One Variable

A **linear equation in one variable** can be written in the form

$$Ax + B = C$$

where A, B, and C are real numbers, with $A \neq 0$.

For example,

$$4x + 9 = 0, \quad 2x - 3 = 5, \quad \text{and} \quad x = 7 \qquad \text{Linear equations}$$

are linear equations in one variable (x). The final two can be written in the specified form using properties developed in this chapter. However,

$$x^2 + 2x = 5, \quad \frac{1}{x} = 6, \quad \text{and} \quad |2x + 6| = 0 \qquad \text{Nonlinear equations}$$

are *not* linear equations.

As we saw in **Section 1.2,** a *solution* of an equation is a number that makes the equation true when it replaces the variable. Equations that have exactly the same solutions are **equivalent equations.** Linear equations are solved by using a series of steps to produce a simpler equivalent equation of the form

$$\mathbf{x = a\ number} \quad \text{or} \quad \mathbf{a\ number = x}.$$

OBJECTIVE 2 Use the addition property of equality. In the equation $x - 5 = 2$, both $x - 5$ and 2 represent the same number because this is the meaning of the equals sign. To solve the equation, we change the left side from $x - 5$ to just x. We do this by adding 5 to $x - 5$. We use 5 because 5 is the opposite (additive inverse) of -5, and $-5 + 5 = 0$. To keep the two sides equal, we must also add 5 to the right side.

$$x - 5 = 2 \qquad \text{Given equation}$$
$$x - 5 + 5 = 2 + 5 \qquad \text{Add 5 to each side.}$$
$$x + 0 = 7 \qquad \text{Additive inverse property}$$
$$x = 7 \qquad \text{Additive identity property}$$

The solution of the given equation is 7. We check by replacing x with 7 in the original equation.

Check:

$$x - 5 = 2 \qquad \text{Original equation}$$
$$7 - 5 \stackrel{?}{=} 2 \qquad \text{Let } x = 7.$$
$$2 = 2 \qquad \text{True}$$

Since the final equation is true, 7 checks as the solution.

To solve the equation, we added the same number to each side. The **addition property of equality** justifies this step.

Addition Property of Equality

If A, B, and C are real numbers, then the equations

$$A = B \quad \text{and} \quad A + C = B + C$$

are equivalent equations.

In words, we can add the same number to each side of an equation without changing the solution.

In the addition property, C represents a real number. This means that any quantity that represents a real number can be added to each side of an equation to change it to an equivalent equation.

NOTE

Equations can be thought of in terms of a balance. Thus, adding the same quantity to each side does not affect the balance. See Figure 1.

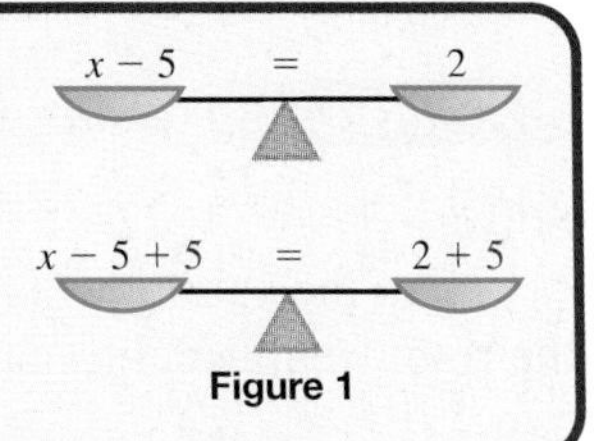

Figure 1

EXAMPLE 1 Using the Addition Property of Equality

Solve $x - 16 = 7$.

Our goal is to get an equivalent equation of the form $x =$ a number. To do this, we use the addition property of equality and add 16 to each side.

$$x - 16 = 7$$
$$x - 16 + \mathbf{16} = 7 + \mathbf{16} \qquad \text{Add 16 to each side.}$$
$$x = 23 \qquad \text{Combine like terms.}$$

Note that we combined the steps that change $x - 16 + 16$ to $x + 0$ and $x + 0$ to x. We check by substituting 23 for x in the *original* equation.

Check:

$$x - 16 = 7 \qquad \text{Original equation}$$
$$\mathbf{23} - 16 = 7 \quad ? \qquad \text{Let } x = 23.$$
$$7 = 7 \qquad \text{True}$$

Since a true statement results, 23 is the solution.

Work Problem 1 at the Side.

EXAMPLE 2 Using the Addition Property of Equality

Solve $x - 2.9 = -6.4$.

We use the addition property of equality to get x alone on the left side.

$$x - 2.9 = -6.4$$
$$x - 2.9 + \mathbf{2.9} = -6.4 + \mathbf{2.9} \qquad \text{Add 2.9 to each side.}$$
$$x = -3.5$$

Check:

$$x - 2.9 = -6.4 \qquad \text{Original equation}$$
$$\mathbf{-3.5} - 2.9 = -6.4 \quad ? \qquad \text{Let } x = -3.5.$$
$$-6.4 = -6.4 \qquad \text{True}$$

Since a true statement results, the solution is -3.5.

Work Problem 2 at the Side.

1 Solve.

(a) $x - 12 = 9$

(b) $x - 25 = -18$

2 Solve.

(a) $x - 3.7 = -8.1$

(b) $a - 4.1 = 6.3$

ANSWERS

1. **(a)** 21 **(b)** 7
2. **(a)** -4.4 **(b)** 10.4

3 Solve.

(a) $-3 = a + 2$

(b) $22 = -16 + r$

The addition property of equality says that the same number may be *added* to each side of an equation. In **Section 1.5,** subtraction was defined as addition of the opposite. Thus, we can also use the following rule when solving an equation.

The same number may be *subtracted* from each side of an equation without changing the solution.

For example, to solve the equation $x + 4 = 10$, we subtract 4 from each side to get $x = 6$.

EXAMPLE 3 Using the Addition Property of Equality

Solve $-7 = x + 22$.

Here the variable x is on the right side of the equation. To get x alone on the right we must eliminate the 22 by subtracting 22 from each side.

$$-7 = x + 22$$

$$-7 - \mathbf{22} = x + 22 - \mathbf{22} \qquad \text{Subtract 22 from each side.}$$

$$-29 = x \quad \text{or} \quad x = -29$$

Check:

$$-7 = x + 22 \qquad \text{Original equation}$$

$$-7 = \mathbf{-29} + 22 \quad ? \qquad \text{Let } x = -29.$$

$$-7 = -7 \qquad \text{True}$$

The check confirms that the solution is -29.

Work Problem 3 at the Side.

CAUTION
The final line of the check does *not* give the solution to the problem, only a confirmation that the solution found is correct.

EXAMPLE 4 Subtracting a Variable Expression

Solve $\frac{3}{5}k + 12 = \frac{8}{5}k$.

To get all terms with variables on the same side of the equation, subtract $\frac{3}{5}k$ from each side.

$$\frac{3}{5}k + 12 = \frac{8}{5}k$$

$$\frac{3}{5}k + 12 - \mathbf{\frac{3}{5}k} = \frac{8}{5}k - \mathbf{\frac{3}{5}k} \qquad \text{Subtract } \tfrac{3}{5}k \text{ from each side.}$$

$$12 = 1k \qquad \tfrac{3}{5}k - \tfrac{3}{5}k = 0;\ \tfrac{8}{5}k - \tfrac{3}{5}k = \tfrac{5}{5}k = 1k$$

$$12 = k \qquad \text{Multiplicative identity property}$$

From now on we will skip the step that changes $1k$ to k. Check the solution by replacing k with 12 in the original equation. The solution is 12.

ANSWERS
3. (a) -5 **(b)** 38

What happens if we solve the equation in Example 4 by first subtracting $\frac{8}{5}k$ from each side?

$$\frac{3}{5}k + 12 = \frac{8}{5}k \qquad \text{Equation from Example 4}$$

$$\frac{3}{5}k + 12 - \frac{8}{5}k = \frac{8}{5}k - \frac{8}{5}k \qquad \text{Subtract } \tfrac{8}{5}k \text{ from each side.}$$

$$12 - k = 0 \qquad \tfrac{3}{5}k - \tfrac{8}{5}k = -\tfrac{5}{5}k = -1k = -k;\ \tfrac{8}{5}k - \tfrac{8}{5}k = 0$$

$$12 - k - 12 = 0 - 12 \qquad \text{Subtract 12 from each side.}$$

$$-k = -12 \qquad \text{Combine like terms; additive inverse}$$

This result gives the value of $-k$, but not of k itself. However, it does say that the additive inverse of k is -12, which means that k must be 12.

$$-k = -12$$

$$k = 12 \qquad \text{Same result as in Example 4}$$

(This result can also be justified using the multiplication property of equality, covered in **Section 2.2.**) We can make the following generalization.

If a is a number and $-x = a$, then $x = -a$.

Work Problem 4 at the Side.

OBJECTIVE 3 Simplify equations, and then use the addition property of equality. Sometimes an equation must be simplified as a first step in its solution.

EXAMPLE 5 Simplifying an Equation before Solving

Solve $3t - 12 + t + 2 = 5 + 3t + 2$.

Begin by combining like terms on each side of the equation to get

$$4t - 10 = 7 + 3t.$$

Next, get all terms that contain variables on the same side of the equation and all terms without variables on the other side. One way to start is to subtract $3t$ from each side.

$$4t - 10 - 3t = 7 + 3t - 3t \qquad \text{Subtract } 3t \text{ from each side.}$$

$$t - 10 = 7 \qquad \text{Combine like terms.}$$

$$t - 10 + 10 = 7 + 10 \qquad \text{Add 10 to each side.}$$

$$t = 17 \qquad \text{Combine like terms.}$$

Check: Substitute 17 for t in the original equation.

$$3t - 12 + t + 2 = 5 + 3t + 2 \qquad \text{Original equation}$$

$$3(17) - 12 + 17 + 2 = 5 + 3(17) + 2 \quad ? \qquad \text{Let } t = 17.$$

$$51 - 12 + 17 + 2 = 5 + 51 + 2 \quad ? \qquad \text{Multiply.}$$

$$58 = 58 \qquad \text{True}$$

The check results in a true statement, so the solution is 17.

Work Problem 5 at the Side.

4 **(a)** Solve $6m = 4 + 5m$.

(b) Solve $\frac{7}{2}m + 1 = \frac{9}{2}m$.

(c) What is the solution of $-x = 6$?

(d) What is the solution of $-x = -12$?

5 Solve.

(a) $4x + 6 + 2x - 3 = 9 + 5x - 4$

(b) $9r + 4r + 6 - 2 = 9r + 4 + 3r$

ANSWERS

4. **(a)** 4 **(b)** 1 **(c)** -6 **(d)** 12

5. **(a)** 2 **(b)** 0

6 Solve.

(a) $4(r + 1) - (3r + 5) = 1$

(b) $-3(m - 4) + 2(5 + 2m) = 29$

EXAMPLE 6 Using the Distributive Property to Simplify an Equation before Solving

Solve $3(2 + 5x) - (1 + 14x) = 6$.

$$3(2 + 5x) - (1 + 14x) = 6$$

$3(2 + 5x) - 1(1 + 14x) = 6$	$-(1 + 14x) = -1(1 + 14x)$
$3(2) + 3(5x) - 1(1) - 1(14x) = 6$	Distributive property
$6 + 15x - 1 - 14x = 6$	Multiply.
$x + 5 = 6$	Combine like terms.
$x + 5 - 5 = 6 - 5$	Subtract 5 from each side.
$x = 1$	Combine like terms.

Check by substituting 1 for x in the original equation. The solution is 1.

CAUTION

Be careful to apply the distributive property correctly in a problem like that in Example 6, or a sign error may result.

Work Problem 6 at the side.

ANSWERS

6. (a) 2 (b) 7

2.1 Exercises

FOR EXTRA HELP Addison-Wesley Math Tutor Center | MathXL | Digital Video Tutor CD 2 Videotape 2 | Student's Solutions Manual | MyMathLab | Interactmath.com

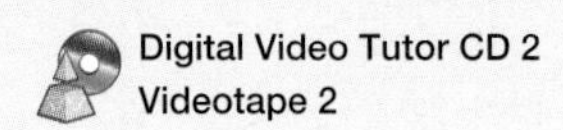

1. Which of the pairs of equations are equivalent equations?

A. $x + 2 = 6$ and $x = 4$ **B.** $10 - x = 5$ and $x = -5$

C. $x + 3 = 9$ and $x = 6$ **D.** $4 + x = 8$ and $x = -4$

2. Decide whether each is an *expression* or an *equation.* If it is an expression, simplify it. If it is an equation, solve it.

(a) $5x + 8 - 4x + 7$ **(b)** $-6y + 12 + 7y - 5$

(c) $5x + 8 - 4x = 7$ **(d)** $-6y + 12 + 7y = -5$

3. Which of the following are not linear equations in one variable?

A. $x^2 - 5x + 6 = 0$ **B.** $x^3 = x$

C. $3x - 4 = 0$ **D.** $7x - 6x = 3 + 9x$

4. Explain how to check a solution of an equation.

Solve each equation, and check your solution. See Examples 1–4.

5. $x - 4 = 8$ **6.** $x - 8 = 9$ **7.** $x - 5 = -8$ **8.** $x - 7 = -9$

9. $r + 9 = 13$ **10.** $t + 6 = 10$ **11.** $x + 26 = 17$ **12.** $x + 45 = 24$

13. $x - 8.4 = -2.1$ **14.** $z - 15.5 = -5.1$ **15.** $t + 12.3 = -4.6$

16. $x + 21.5 = -13.4$ **17.** $7 + r = -3$ **18.** $8 + k = -4$

19. $2 = p + 15$ **20.** $3 = z + 17$ **21.** $-2 = x - 12$

22. $-6 = x - 21$ **23.** $3x = 2x + 7$ **24.** $5x = 4x + 9$

25. $10x + 4 = 9x$ **26.** $8t + 5 = 7t$ **27.** $\frac{9}{7}r - 3 = \frac{2}{7}r$

28. $\frac{8}{5}w - 6 = \frac{3}{5}w$

29. $5.6x + 2 = 4.6x$

30. $9.1x - 5 = 8.1x$

31. $3p + 6 = 10 + 2p$

32. $8x - 4 = -6 + 7x$

Solve each equation, and check your solution. See Examples 5 and 6.

33. $3x + 6 - 10 = 2x - 2$

34. $8k - 4 + 6 = 7k + 1$

35. $6x + 5 + 7x + 3 = 12x + 4$

36. $4x - 3 - 8x + 1 = -5x + 9$

37. $10x + 5x + 7 - 8 = 12x + 3 + 2x$

38. $7p + 4p + 13 - 7 = 7p + 9 + 3p$

39. $5.2q - 4.6 - 7.1q = -.9q - 4.6$

40. $-4.0x + 2.7 - 1.6x = -4.6x + 2.7$

41. $\frac{5}{7}x + \frac{1}{3} = \frac{2}{5} - \frac{2}{7}x + \frac{2}{5}$

42. $\frac{6}{7}s - \frac{3}{4} = \frac{4}{5} - \frac{1}{7}s + \frac{1}{6}$

43. $(5x + 6) - (3 + 4x) = 10$

44. $(8r - 3) - (7r + 1) = -6$

45. $2(p + 5) - (9 + p) = -3$

46. $4(k - 6) - (3k + 2) = -5$

47. $-6(2x + 1) + (13x - 7) = 0$

48. $-5(3w - 3) + (1 + 16w) = 0$

49. $10(-2x + 1) = -19(x + 1)$

50. $2(2 - 3r) = -5(r - 3)$

51. $-2(8p + 2) - 3(2 - 7p) = 2(4 + 2p)$

52. $-5(1 - 2z) + 4(3 - z) = 7(3 + z)$

53. Write an equation where 6 must be added to each side to solve the equation, and the solution is a negative number.

54. Write an equation where $\frac{1}{2}$ must be subtracted from each side, and the solution is a positive number.

2.2 The Multiplication Property of Equality

OBJECTIVES

1 Use the multiplication property of equality.

2 Simplify equations, and then use the multiplication property of equality.

OBJECTIVE 1 **Use the multiplication property of equality.** The addition property of equality alone is not enough to solve some equations, such as $3x + 2 = 17$.

$$3x + 2 = 17$$
$$3x + 2 - 2 = 17 - 2 \quad \text{Subtract 2 from each side.}$$
$$3x = 15 \quad \text{Combine like terms.}$$

Notice that the coefficient of x on the left side is 3, not 1 as desired. We must develop a method that leads to an equation of the form

$$x = \text{a number.}$$

If $3x = 15$, then $3x$ and 15 both represent the same number. Multiplying both $3x$ and 15 by the same number will also result in an equality. The **multiplication property of equality** states that we can multiply each side of an equation by the same nonzero number without changing the solution.

Multiplication Property of Equality

If A, B, and C $(C \neq 0)$ represent real numbers, then the equations

$$A = B \quad \text{and} \quad AC = BC$$

are equivalent equations.

In words, we can multiply each side of an equation by the same nonzero number without changing the solution.

This property can be used to solve $3x = 15$. The $3x$ on the left must be changed to $1x$, or x, instead of $3x$. To isolate x, multiply each side of the equation by $\frac{1}{3}$. We use $\frac{1}{3}$ because $\frac{1}{3}$ is the reciprocal of 3, and $\frac{1}{3} \cdot 3 = \frac{3}{3} = 1$.

$$3x = 15$$
$$\frac{1}{3}(3x) = \frac{1}{3} \cdot 15 \quad \text{Multiply each side by } \tfrac{1}{3}.$$
$$\left(\frac{1}{3} \cdot 3\right)x = \frac{1}{3} \cdot 15 \quad \text{Associative property}$$
$$1x = 5 \quad \text{Multiplicative inverse property}$$
$$x = 5 \quad \text{Multiplicative identity property}$$

The solution of the equation is 5. We can check this result in the original equation. (We usually combine the last two steps shown above.)

Work Problem 1 at the Side.

1 Check that 5 is the solution of $3x = 15$.

Just as the addition property of equality permits *subtracting* the same number from each side of an equation, the multiplication property of equality permits *dividing* each side of an equation by the same nonzero number. For example, the equation $3x = 15$, which we just solved by multiplication, could also be solved by dividing each side by 3, as follows.

$$3x = 15$$
$$\frac{3x}{3} = \frac{15}{3} \quad \text{Divide each side by 3.}$$
$$x = 5$$

ANSWERS

1. Since $3(5) = 15$, the solution of $3x = 15$ is 5.

We can divide each side of an equation by the same nonzero number without changing the solution. Do not, however, divide each side by a variable, since the variable might be equal to 0.

NOTE

In practice, it is usually easier to multiply on each side if the coefficient of the variable is a fraction, and divide on each side if the coefficient is an integer. For example, to solve

$$-\frac{3}{4}x = 12,$$

it is easier to multiply by $-\frac{4}{3}$, the reciprocal of $-\frac{3}{4}$, than to divide by $-\frac{3}{4}$. On the other hand, to solve

$$-5x = -20,$$

it is easier to divide by -5 than to multiply by $-\frac{1}{5}$.

EXAMPLE 1 Dividing Each Side of an Equation by a Nonzero Number

Solve $25p = 30$.

Transform the equation so that p (instead of $25p$) is on the left by using the multiplication property of equality. Divide each side of the equation by 25, the coefficient of p.

$$25p = 30$$

$$\frac{25p}{25} = \frac{30}{25} \qquad \text{Divide by 25.}$$

$$p = \frac{30}{25} = \frac{6}{5} \qquad \text{Write in lowest terms.}$$

To check, substitute $\frac{6}{5}$ for p in the original equation.

Check:

$$25p = 30$$

$$\frac{25}{1}\left(\frac{6}{5}\right) = 30 \quad ? \qquad \text{Let } p = \tfrac{6}{5}.$$

$$30 = 30 \qquad \text{True}$$

The check confirms that the solution is $\frac{6}{5}$.

Work Problem 2 at the Side.

EXAMPLE 2 Solving an Equation with Decimals

Solve $2.1x = 6.09$.

$$2.1x = 6.09$$

$$\frac{2.1x}{\mathbf{2.1}} = \frac{6.09}{\mathbf{2.1}} \qquad \text{Divide each side by 2.1.}$$

$$x = 2.9 \qquad \text{Divide; you may use a calculator.}$$

Check that the solution is 2.9.

Work Problem 3 at the Side.

2 Solve.

(a) $-6p = -14$

(b) $3r = -12$

(c) $-2m = 16$

3 Solve.

(a) $-.7m = -5.04$

(b) $12.5k = -63.75$

ANSWERS

2. (a) $\frac{7}{3}$ (b) -4 (c) -8

3. (a) 7.2 (b) -5.1

In the next two examples, multiplication produces the solution more quickly than division would.

EXAMPLE 3 Using the Multiplication Property of Equality

Solve $\frac{a}{4} = 3$.

Replace $\frac{a}{4}$ by $\frac{1}{4}a$, since dividing by 4 is the same as multiplying by $\frac{1}{4}$. To get a alone on the left, multiply each side by 4, the reciprocal of the coefficient of a.

$$\frac{a}{4} = 3$$

$$\frac{1}{4}a = 3 \qquad \text{Change } \tfrac{a}{4} \text{ to } \tfrac{1}{4}a.$$

$$4 \cdot \frac{1}{4}a = 4 \cdot 3 \qquad \text{Multiply by 4.}$$

$$a = 12 \qquad \text{Multiplicative inverse property; multiplicative identity property}$$

Check:

$$\frac{a}{4} = 3 \qquad \text{Original equation}$$

$$\frac{12}{4} \stackrel{?}{=} 3 \qquad \text{Let } a = 12.$$

$$3 = 3 \qquad \text{True}$$

The solution is 12.

Work Problem 4 at the Side.

EXAMPLE 4 Using the Multiplication Property of Equality

Solve $\frac{3}{4}h = 6$.

Transform the equation so that h is alone on the left by multiplying each side of the equation by $\frac{4}{3}$. Use $\frac{4}{3}$ because $\frac{4}{3} \cdot \frac{3}{4}h = 1 \cdot h = h$.

$$\frac{3}{4}h = 6$$

$$\frac{4}{3}\left(\frac{3}{4}h\right) = \frac{4}{3} \cdot 6 \qquad \text{Multiply by } \tfrac{4}{3}.$$

$$1 \cdot h = \frac{4}{3} \cdot \frac{6}{1} \qquad \text{Multiplicative inverse property}$$

$$h = 8 \qquad \text{Multiplicative identity property; multiply fractions.}$$

Check:

$$\frac{3}{4}h = 6 \qquad \text{Original equation}$$

$$\frac{3}{4}(8) \stackrel{?}{=} 6 \qquad \text{Let } h = 8.$$

$$6 = 6 \qquad \text{True}$$

The solution is 8.

Work Problem 5 at the Side.

4 Solve.

(a) $\frac{y}{5} = 5$

(b) $\frac{p}{4} = -6$

5 Solve.

(a) $-\frac{5}{6}t = -15$

(b) $\frac{3}{4}k = -21$

ANSWERS

4. (a) 25 **(b)** −24

5. (a) 18 **(b)** −28

In **Section 2.1,** we obtained the equation $-k = -12$ in our alternate solution to Example 4. We reasoned that since this equation says that the additive inverse (or opposite) of k is -12, then k must equal 12. We can also use the multiplication property of equality to obtain the same result, as shown in the next example.

EXAMPLE 5 Using the Multiplication Property of Equality When the Coefficient of the Variable Is −1

Solve $-k = -12$.

On the left side, change $-k$ to k by first writing $-k$ as $-1 \cdot k$.

$$-k = -12$$

$$-1 \cdot k = -12 \qquad -k = -1 \cdot k$$

$$-1(-1 \cdot k) = -1(-12) \qquad \text{Multiply by } -1\text{, since } -1(-1) = 1.$$

$$[-1(-1)] \cdot k = 12 \qquad \text{Associative property; multiply.}$$

$$1 \cdot k = 12 \qquad \text{Multiplicative inverse property}$$

$$k = 12 \qquad \text{Multiplicative identity property}$$

Check:

$$-k = -12 \qquad \text{Original equation}$$

$$-(12) \stackrel{?}{=} -12 \qquad \text{Let } k = 12.$$

$$-12 = -12 \qquad \text{True}$$

The solution, 12, checks.

Work Problem 6 at the Side.

OBJECTIVE 2 Simplify equations, and then use the multiplication property of equality. In the next example, it is necessary to simplify the equation before using the multiplication property of equality.

EXAMPLE 6 Simplifying an Equation before Solving

Solve $5m + 6m = 33$.

$$5m + 6m = 33$$

$$11m = 33 \qquad \text{Combine like terms.}$$

$$\frac{11m}{11} = \frac{33}{11} \qquad \text{Divide by 11.}$$

$$1m = 3 \qquad \text{Divide.}$$

$$m = 3 \qquad \text{Multiplicative identity property}$$

The solution, 3, checks.

Work Problem 7 at the Side.

6 Solve.

(a) $-m = 2$

(b) $-p = -7$

7 Solve.

(a) $4r - 9r = 20$

(b) $7m - 5m = -12$

Answers

6. (a) -2 (b) 7

7. (a) -4 (b) -6

2.2 Exercises

FOR EXTRA HELP

Tutor Center Addison-Wesley Math Tutor Center

MathXL MathXL

Digital Video Tutor CD 2 Videotape 2

Student's Solutions Manual

MyMathLab MyMathLab

Interactmath.com

By what number is it necessary to multiply each side of each equation in order to obtain just x on the left side? Do not actually solve these equations.

1. $\frac{2}{3}x = 8$

2. $\frac{4}{5}x = 6$

3. $.1x = 3$

4. $.01x = 8$

5. $-\frac{9}{2}x = -4$

6. $-\frac{8}{3}x = -11$

7. $-x = .36$

8. $-x = .29$

By what number is it necessary to divide each side of each equation in order to obtain just x on the left side? Do not actually solve these equations.

9. $6x = 5$

10. $7x = 10$

11. $-4x = 13$

12. $-13x = 6$

13. $.12x = 48$

14. $.21x = 63$

15. $-x = 23$

16. $-x = 49$

17. A student tried to solve the equation $4x = 8$ by dividing each side by 8. What is a better approach?

18. Which equation does *not* require the use of the multiplication property of equality?

A. $3x - 5x = 6$ **B.** $-\frac{1}{4}x = 12$ **C.** $5x - 4x = 7$ **D.** $\frac{x}{3} = -2$

Solve each equation, and check your solution. See Examples 1–6.

19. $5x = 30$

20. $7x = 56$

21. $2m = 15$

22. $3m = 10$

23. $3a = -15$

24. $5k = -70$

25. $10t = -36$

26. $4s = -34$

27. $-6x = -72$

28. $-8x = -64$

29. $2r = 0$

30. $5x = 0$

31. $-y = 12$

32. $-t = 14$

33. $.2t = 8$

34. $.9x = 18$

35. $-2.1m = 25.62$

36. $-3.9a = 31.2$

37. $\frac{1}{4}y = -12$

38. $\frac{1}{5}p = -3$

39. $\frac{x}{7} = -5$

40. $\frac{k}{8} = -3$

41. $\frac{z}{6} = 12$

42. $\frac{x}{5} = 15$

43. $\frac{2}{7}p = 4$

44. $\frac{3}{8}x = 9$

45. $-\frac{7}{9}c = \frac{3}{5}$

46. $-\frac{5}{6}d = \frac{4}{9}$

47. $4x + 3x = 21$

48. $9x + 2x = 121$

49. $3r - 5r = 10$

50. $9p - 13p = 24$

51. $5m + 6m - 2m = 63$

52. $11r - 5r + 6r = 168$

53. $-6x + 4x - 7x = 0$

54. $-5x + 4x - 8x = 0$

55. $.9w - .5w + .1w = -3$

56. $.5x - .6x + .3x = -1$

57. Write an equation that requires the use of the multiplication property of equality, where each side must be multiplied by $\frac{2}{3}$, and the solution is a negative number.

58. Write an equation that requires the use of the multiplication property of equality, where each side must be divided by 100, and the solution is not an integer.

Write an equation using the information given in the problem. Use x as the variable. Then solve the equation.

59. When a number is multiplied by -4, the result is 10. Find the number.

60. When a number is divided by -5, the result is 2. Find the number.

2.3 More on Solving Linear Equations

OBJECTIVES

1. Learn the four steps for solving a linear equation, and apply them.
2. Solve equations with fractions or decimals as coefficients.
3. Solve equations that have no solution or infinitely many solutions.
4. Write expressions for two related unknown quantities.

OBJECTIVE 1 Learn the four steps for solving a linear equation, and apply them. In this section, we solve more complicated equations using the following four-step method.

Solving Linear Equations

Step 1 **Simplify each side separately.** Clear parentheses using the distributive property, if needed, and combine terms.

Step 2 **Isolate the variable term on one side.** Use the addition property if necessary so that the variable term is on one side of the equation and a number is on the other.

Step 3 **Isolate the variable.** Use the multiplication property if necessary to get the equation in the form $x =$ a number.

Step 4 **Check.** Substitute the proposed solution into the *original* equation to see if a true statement results.

EXAMPLE 1 Using the Four Steps to Solve an Equation

Solve $3r + 4 - 2r - 7 = 4r + 3$.

Our goal is to isolate the variable, r.

Step 1 $3r + 4 - 2r - 7 = 4r + 3$

$r - 3 = 4r + 3$ Combine terms.

Step 2 $r - 3 + 3 = 4r + 3 + 3$ Add 3.

$r = 4r + 6$ Combine terms.

$r - 4r = 4r + 6 - 4r$ Subtract $4r$.

$-3r = 6$ Combine terms.

Step 3 $\frac{-3r}{-3} = \frac{6}{-3}$ Divide by -3.

$r = -2$ $\frac{-3}{-3} = 1$; $1r = r$

Step 4 Substitute -2 for r in the original equation to check.

$3r + 4 - 2r - 7 = 4r + 3$

$3(-2) + 4 - 2(-2) - 7 \stackrel{?}{=} 4(-2) + 3$ Let $r = -2$.

$-6 + 4 + 4 - 7 \stackrel{?}{=} -8 + 3$ Multiply.

$-5 = -5$ True

The solution of the equation is -2.

NOTE

In Step 2 of Example 1, we added and subtracted the terms in such a way that the variable term ended up on the left side of the equation. Choosing differently would have put the variable term on the right side of the equation. Either way, the same solution results.

Work Problem 1 at the Side.

1 Solve.

(a) $5y - 7y + 6y - 9 = 3 + 2y$

(b) $-3k - 5k - 6 + 11 = 2k - 5$

ANSWERS

1. (a) 6 (b) 1

2 Solve.

(a) $7(p - 2) + p = 2p + 4$

(b) $11 + 3(x + 1) = 5x + 16$

EXAMPLE 2 Using the Four Steps to Solve an Equation

Solve $4(k - 3) - k = k - 6$.

Step 1 Clear parentheses using the distributive property.

$$4(k - 3) - k = k - 6$$

$$4(k) + 4(-3) - k = k - 6 \quad \text{Distributive property}$$

$$4k - 12 - k = k - 6 \quad \text{Multiply.}$$

$$3k - 12 = k - 6 \quad \text{Combine terms.}$$

Step 2

$$3k - 12 + 12 = k - 6 + 12 \quad \text{Add 12.}$$

$$3k = k + 6 \quad \text{Combine terms.}$$

$$3k - k = k + 6 - k \quad \text{Subtract } k\text{.}$$

$$2k = 6 \quad \text{Combine terms.}$$

Step 3

$$\frac{2k}{2} = \frac{6}{2} \quad \text{Divide by 2.}$$

$$k = 3$$

Step 4 Check by substituting 3 for k in the original equation.

$$4(k - 3) - k = k - 6$$

$$4(3 - 3) - 3 = 3 - 6 \quad ? \quad \text{Let } k = 3\text{.}$$

$$4(0) - 3 = 3 - 6 \quad ? \quad \text{Work inside the parentheses.}$$

$$-3 = -3 \quad \text{True}$$

The solution of the equation is 3.

Work Problem 2 at the Side.

EXAMPLE 3 Using the Four Steps to Solve an Equation

Solve $8a - (3 + 2a) = 3a + 1$.

Step 1

$$8a - (3 + 2a) = 3a + 1$$

$$8a - 1(3 + 2a) = 3a + 1 \quad \text{Multiplicative identity property}$$

$$8a - 3 - 2a = 3a + 1 \quad \text{Distributive property}$$

$$6a - 3 = 3a + 1 \quad \text{Combine terms.}$$

Step 2

$$6a - 3 + 3 = 3a + 1 + 3 \quad \text{Add 3.}$$

$$6a = 3a + 4 \quad \text{Combine terms.}$$

$$6a - 3a = 3a + 4 - 3a \quad \text{Subtract } 3a\text{.}$$

$$3a = 4 \quad \text{Combine terms.}$$

Step 3

$$\frac{3a}{3} = \frac{4}{3} \quad \text{Divide by 3.}$$

$$a = \frac{4}{3}$$

Step 4 Check that the solution is $\frac{4}{3}$.

ANSWERS

2. (a) 3 (b) -1

CAUTION

Be very careful with signs when solving an equation like the one in Example 3. When clearing parentheses in the expression

$$8a - (3 + 2a),$$

remember that the $-$ sign acts like a factor of -1 and affects the sign of *every* term in the parentheses. Thus,

$$8 - (3 + 2a) = 8 - 3 - 2a.$$

Change to $-$ in *both* terms.

Work Problem 3 at the Side.

3 Solve.

(a) $7m - (2m - 9) = 39$

(b) $4x - (x + 7) = 9$

EXAMPLE 4 Using the Four Steps to Solve an Equation

Solve $4(8 - 3t) = 32 - 8(t + 2)$.

Step 1 $\quad 4(8 - 3t) = 32 - 8(t + 2)$

$32 - 12t = 32 - 8t - 16$ Distributive property

$32 - 12t = 16 - 8t$ Combine terms.

Step 2 $\quad 32 - 12t - \mathbf{32} = 16 - 8t - \mathbf{32}$ Subtract 32.

$-12t = -16 - 8t$ Combine terms.

$-12t + \mathbf{8t} = -16 - 8t + \mathbf{8t}$ Add $8t$.

$-4t = -16$ Combine terms.

Step 3 $\quad \dfrac{-4t}{\mathbf{-4}} = \dfrac{-16}{\mathbf{-4}}$ Divide by -4.

$t = 4$

Step 4 Check this solution in the original equation.

$4(8 - 3t) = 32 - 8(t + 2)$

$4(8 - 3 \cdot \mathbf{4}) = 32 - 8(\mathbf{4} + 2)$? Let $t = 4$.

$4(8 - 12) = 32 - 8(6)$?

$4(-4) = 32 - 48$?

$-16 = -16$ True

The solution, 4, checks.

Work Problem 4 at the Side.

4 Solve.

(a) $2(4 + 3r) = 3(r + 1) + 11$

(b) $2 - 3(2 + 6z) = 4(z + 1) + 18$

OBJECTIVE 2 Solve equations with fractions or decimals as coefficients. We clear an equation of fractions by multiplying each side by the least common denominator (LCD) of all the fractions in the equation. It is a good idea to do this to avoid messy computations.

CAUTION

When clearing an equation of fractions, be sure to multiply every term on each side of the equation by the LCD.

ANSWERS

3. (a) 6 (b) $\frac{16}{3}$

4. (a) 2 (b) $-\frac{13}{11}$

5 Solve $\frac{1}{4}x - 4 = \frac{3}{2}x + \frac{3}{4}x$.

6 Solve $.06(100 - x) + .04x = .05(92)$.

EXAMPLE 5 Solving an Equation with Fractions as Coefficients

Solve $\frac{2}{3}x - \frac{1}{2}x = -\frac{1}{6}x - 2$.

The LCD of all the fractions in the equation is 6, so multiply each side by 6 to clear the fractions.

$$\frac{2}{3}x - \frac{1}{2}x = -\frac{1}{6}x - 2$$

$$6\left(\frac{2}{3}x - \frac{1}{2}x\right) = 6\left(-\frac{1}{6}x - 2\right) \qquad \text{Multiply by 6.}$$

$$6\left(\frac{2}{3}x\right) + 6\left(-\frac{1}{2}x\right) = 6\left(-\frac{1}{6}x\right) + 6(-2) \qquad \text{Distributive property}$$

$$4x - 3x = -x - 12$$

Now use the four steps to solve this equivalent equation.

Step 1 $\quad x = -x - 12 \qquad$ Combine terms.

Step 2 $\quad x + \mathbf{x} = -x - 12 + \mathbf{x} \qquad$ Add x.

$\quad 2x = -12 \qquad$ Combine terms.

Step 3 $\quad \frac{2x}{\mathbf{2}} = \frac{-12}{\mathbf{2}} \qquad$ Divide by 2.

$\quad x = -6$

Step 4 Check by substituting -6 for x in the original equation.

$$\frac{2}{3}(-6) - \frac{1}{2}(-6) = -\frac{1}{6}(-6) - 2 \quad ? \qquad \text{Let } x = -6.$$

$$-4 + 3 = 1 - 2 \quad ?$$

$$-1 = -1 \qquad \text{True}$$

The solution of the equation is -6.

Work Problem 5 at the Side.

EXAMPLE 6 Solving an Equation with Decimals as Coefficients

Solve $.1t + .05(20 - t) = .09(20)$.

The decimals here are expressed as tenths (.1, which equals .10) and hundredths (.05 and .09). Choose the least exponent on 10 needed to eliminate the decimals; in this case, use $10^2 = 100$. A number can be multiplied by 100 by moving the decimal point two places to the right.

$$.10t + .05(20 - t) = .09(20) \qquad .1 = .10$$

$$10t + 5(20 - t) = 9(20) \qquad \text{Multiply by 100.}$$

Step 1 $\quad 10t + 5(20) + 5(-t) = 180 \qquad$ Distributive property

$\quad 10t + 100 - 5t = 180$

$\quad 5t + 100 = 180 \qquad$ Combine terms.

Step 2 $\quad 5t + 100 - \mathbf{100} = 180 - \mathbf{100} \qquad$ Subtract 100.

$\quad 5t = 80 \qquad$ Combine terms.

Step 3 $\quad \frac{5t}{5} = \frac{80}{5} \qquad$ Divide by 5.

$\quad t = 16$

Step 4 Check to confirm that 16 is the solution.

Work Problem 6 at the Side.

Answers
5. -2
6. 70

OBJECTIVE 3 Solve equations that have no solution or infinitely many solutions. Every equation solved so far has had exactly one solution. Sometimes this is not the case, as shown in the next examples.

7 Solve each equation.

(a) $2(x - 6) = 2x - 12$

(b) $3x + 6(x + 1) = 9x - 4$

EXAMPLE 7 Solving an Equation That Has Infinitely Many Solutions

Solve $5x - 15 = 5(x - 3)$.

$5x - 15 = 5(x - 3)$	
$5x - 15 = 5x - 15$	Distributive property
$5x - 15 + \mathbf{15} = 5x - 15 + \mathbf{15}$	Add 15.
$5x = 5x$	Combine terms.
$5x - \mathbf{5x} = 5x - \mathbf{5x}$	Subtract $5x$.
$\mathbf{0 = 0}$	True

The final step leads to an equation that contains no variables ($0 = 0$ in this case). Whenever such a statement is true, as it is in this example, *any* real number is a solution. (Try several replacements for x in the given equation to see that they all satisfy the equation.)

An equation with both sides exactly the same, like $0 = 0$, is called an **identity.** An identity is true for all replacements of the variables. We indicate this by writing ***all real numbers.***

CAUTION

When you are solving an equation like the one in Example 7, do not write "0" as the solution. While 0 is a solution, there are infinitely many other solutions.

EXAMPLE 8 Solving an Equation That Has No Solution

Solve $2x + 3(x + 1) = 5x + 4$.

$2x + 3(x + 1) = 5x + 4$	
$2x + 3x + 3 = 5x + 4$	Distributive property
$5x + 3 = 5x + 4$	Combine terms.
$5x + 3 - \mathbf{5x} = 5x + 4 - \mathbf{5x}$	Subtract $5x$.
$\mathbf{3 = 4}$	False

Again, the variable has disappeared, but this time a false statement ($3 = 4$) results. Whenever this happens in solving an equation, it is a signal that the equation has no solution and we write ***no solution.***

Work Problem 7 at the Side.

ANSWERS

7. **(a)** all real numbers **(b)** no solution

8 One number is 5 more than twice another. If the first number is represented by x, write an expression for the second number.

OBJECTIVE 4 Write expressions for two related unknown quantities. Often we are given a problem in which the sum of two quantities is a particular number, and we are asked to find the values of the two quantities. Example 9 shows how to express the unknown quantities in terms of a single variable.

EXAMPLE 9 Translating a Phrase into an Algebraic Expression

Two numbers have a sum of 23. If one of the numbers is represented by k, find an expression for the other number.

First, suppose that the sum of two numbers is 23, and one of the numbers is 10. How would you find the other number? You would subtract 10 from 23 to get 13.

$$23 - 10 = 13$$

So instead of using 10 as one of the numbers, use k as stated in the problem. The other number would be obtained in the same way. You must subtract k from 23. Therefore, an expression for the other number is

$$23 - k.$$

CAUTION

Since the sum of the two numbers in Example 9 is 23, the expression for the other number must be $23 - k$, *not* $k - 23$. To check, find the sum of the two numbers:

$$k + (23 - k) = 23, \quad \text{as required.}$$

Work Problem 8 at the Side.

ANSWERS

8. $2x + 5$

2.3 Exercises

FOR EXTRA HELP

Tutor Center Addison-Wesley Math Tutor Center

MathXL

Digital Video Tutor CD 2 Videotape 2

Student's Solutions Manual

MyMathLab

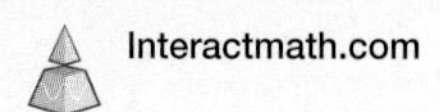
Interactmath.com

Solve each equation, and check your solution. See Examples 1–4, 7, and 8.

1. $5m + 8 = 7 + 4m$

2. $4r + 2 = 3r - 6$

3. $10p + 6 = 12p - 4$

4. $-5x + 8 = -3x + 10$

5. $7r - 5r + 2 = 5r - r$

6. $9p - 4p + 6 = 7p - 3p$

7. $x + 3 = -(2x + 2)$

8. $2x + 1 = -(x + 3)$

9. $4(2x - 1) = -6(x + 3)$

10. $6(3w + 5) = 2(10w + 10)$

11. $6(4x - 1) = 12(2x + 3)$

12. $6(2x + 8) = 4(3x - 6)$

13. $3(2x - 4) = 6(x - 2)$

14. $3(6 - 4x) = 2(-6x + 9)$

15. Which linear equation does *not* have all real numbers as solutions?

A. $5x = 4x + x$ **B.** $2(x + 6) = 2x + 12$ **C.** $\frac{1}{2}x = .5x$ **D.** $3x = 2x$

16. After correctly working through several steps of the solution of a linear equation, a student obtains the equation $7x = 3x$. Then the student divides each side by x to get $7 = 3$ and gives "no solution" as the answer. Is this correct? If not, explain why.

17. Explain in your own words the major steps used in solving a linear equation that does not contain fractions or decimals as coefficients.

18. Explain in your own words the major steps used in solving a linear equation that contains fractions or decimals as coefficients.

Solve each equation, and check your solution. See Examples 5 and 6.

19. $-\frac{2}{7}r + 2r = \frac{1}{2}r + \frac{17}{2}$

20. $\frac{3}{5}t - \frac{1}{10}t = t - \frac{5}{2}$

21. $\frac{1}{9}(x + 18) + \frac{1}{3}(2x + 3) = x + 3$

22. $-\frac{1}{4}(x - 12) + \frac{1}{2}(x + 2) = x + 4$

23. $-\frac{5}{6}q - \left(q - \frac{1}{2}\right) = \frac{1}{4}(q + 1)$

24. $\frac{2}{3}k - \left(k + \frac{1}{4}\right) = \frac{1}{12}(k + 4)$

25. $.30(30) + .15x = .20(30 + x)$

26. $.20(60) + .05x = .10(60 + x)$

27. $.92x + .98(12 - x) = .96(12)$

28. $1.00x + .05(12 - x) = .10(63)$

29. $.02(5000) + .03x = .025(5000 + x)$

30. $.06(10{,}000) + .08x = .072(10{,}000 + x)$

RELATING CONCEPTS (EXERCISES 31–36) For Individual or Group Work

Work Exercises 31–36 in order.

31. Evaluate the term $100ab$ for $a = 2$ and $b = 4$.

32. Will you get the same answer as in Exercise 31 if you evaluate $(100a)b$ for $a = 2$ and $b = 4$? Why or why not?

33. Is the term $(100a)(100b)$ equivalent to $100ab$? Why or why not?

34. If your answer to Exercise 33 is *no,* explain why the distributive property is not involved.

35. The simplest way to solve the equation $.05(x + 2) + .10x = 2.00$ is to begin by multiplying each side by 100. If we do this, the first term on the left becomes $100(.05)(x + 2)$. Is this expression equivalent to $[100(.05)](x + 2)$? Explain. (*Hint:* Compare to Exercises 31 and 32 with $a = .05$ and $b = x + 2$.)

36. Students often want to "distribute" the 100 to both .05 and $(x + 2)$ in the expression $100(.05)(x + 2)$. Is this correct? (*Hint:* See Exercises 34 and 35.)

Solve each equation, and check your solution. See Examples 1–8.

37. $-3(5z + 24) + 2 = 2(3 - 2z) - 4$

38. $-2(2s - 4) - 8 = -3(4s + 4) - 1$

39. $-(6k - 5) - (-5k + 8) = -3$

40. $-(4y + 2) - (-3y - 5) = 3$

41. $\frac{1}{3}(x + 3) + \frac{1}{6}(x - 6) = x + 3$

42. $\frac{1}{2}(x + 2) + \frac{3}{4}(x + 4) = x + 5$

43. $.30(x + 15) + .40(x + 25) = 25$

44. $.10(x + 80) + .20x = 14$

45. $4(x + 3) = 2(2x + 8) - 4$

46. $4(x + 8) = 2(2x + 6) + 20$

47. $8(t - 3) + 4t = 6(2t + 1) - 10$

48. $9(v + 1) - 3v = 2(3v + 1) - 8$

Write the answer to each problem as an algebraic expression. See Example 9.

49. Two numbers have a sum of 12. One number is q. Find the other number.

50. The product of two numbers is 13. One number is k. What is the other number?

51. Mary is a years old. How old will she be in 12 yr? How old was she 5 yr ago?

52. Tom has r quarters. Find the value of the quarters in cents.

53. A bank teller has t dollars in ten-dollar bills. How many ten-dollar bills does the teller have?

54. A plane ticket costs b dollars for an adult and d dollars for a child. Find the total cost of 5 adult and 3 child tickets.

Summary Exercises on Solving Linear Equations

This section of miscellaneous linear equations provides practice in solving all the types introduced in **Sections 2.1–2.3.** Refer to the examples in these sections to review the various solution methods.

Solve each equation, and check your solution.

1. $a + 2 = -3$

2. $2m + 8 = 16$

3. $16.5k = -84.15$

4. $-x = -25$

5. $\frac{4}{5}x = -20$

6. $9x - 7x = -12$

7. $5x - 9 = 4(x - 3)$

8. $\frac{a}{-2} = 8$

9. $-3(t - 5) + 2(7 + 2t) = 36$

10. $\frac{2}{3}x + 8 = \frac{1}{4}x$

11. $.08x + .06(x + 9) = 1.24$

12. $x - 16.2 = 7.5$

13. $4x + 2(3 - 2x) = 6$

14. $-.3x + 2.1(x - 4) = -6.6$

15. $-x = 16$

16. $3(m + 5) - 1 + 2m = 5(m + 2)$

17. $10m - (5m - 9) = 39$

18. $7(p - 2) + p = 2(p + 2)$

19. $-2t + 5t - 9 = 3(t - 4) - 5$

20. $-9z = -21$

21. $.02(50) + .08r = .04(50 + r)$

22. $2.3x + 13.7 = 1.3x + 2.9$

23. $2(3 + 7x) - (1 + 15x) = 2$

24. $6q - 9 = 12 + 3q$

25. $2(5 + 3x) = 3(x + 1) + 13$

26. $r + 9 + 7r = 4(3 + 2r) - 3$

27. $\frac{5}{6}x + \frac{1}{3} = 2x + \frac{2}{3}$

28. $.06x + .09(15 - x) = .07(15)$

29. $\frac{3}{4}(a - 2) - \frac{1}{3}(5 - 2a) = -2$

30. $2 - (m + 4) = 3m + 8$

31. $5.2x - 4.6 - 7.1x = -2.1 - 1.9x - 2.5$

32. $9(2m - 3) - 4(5 + 3m) = 5(4 + m) - 3$

2.4 An Introduction to Applications of Linear Equations

OBJECTIVES

1. Learn the six steps for solving applied problems.
2. Solve problems involving unknown numbers.
3. Solve problems involving sums of quantities.
4. Solve problems involving supplementary and complementary angles.
5. Solve problems involving consecutive integers.

OBJECTIVE 1 Learn the six steps for solving applied problems. We now look at how algebra is used to solve applied problems. Since many meaningful applications of mathematics require concepts that are beyond the level of this book, some of the problems you encounter will seem contrived, and to some extent they are. But the skills you develop in solving simple problems will help you solve more realistic problems in chemistry, physics, biology, business, and other fields.

While there is not one specific method that enables you to solve all kinds of applied problems, the following six-step method is suggested.*

Solving an Applied Problem

Step 1 **Read** the problem, several times if necessary, until you *understand* what is given and what is to be found.

Step 2 **Assign a variable** to represent the unknown value, using diagrams or tables as needed. Write down what the variable represents. Express any other unknown values in terms of the variable.

Step 3 **Write an equation** using the variable expression(s).

Step 4 **Solve** the equation.

Step 5 **State the answer.** Does it seem reasonable?

Step 6 **Check** the answer in the words of the *original* problem.

OBJECTIVE 2 Solve problems involving unknown numbers. Some of the simplest applied problems involve unknown numbers.

> **PROBLEM-SOLVING HINT**
> The third step in solving an applied problem is often the hardest. To translate the problem into an equation, write the given phrases as mathematical expressions. Replace any words that mean *equal* or *same* with an = sign. Other forms of the verb "to be," such as *is, are, was,* and *were,* also translate as an = sign. The = sign leads to an equation to be solved.

EXAMPLE 1 Finding the Value of an Unknown Number

The product of 4, and a number decreased by 7, is 100. What is the number?

Step 1 **Read** the problem carefully. We are asked to find a number.

Step 2 **Assign a variable** to represent the unknown quantity. In this problem, we are asked to find a number, so we write

$$\text{Let } x = \text{the number.}$$

There are no other unknown quantities to find.

Continued on Next Page

* **Appendix A** Strategies for Problem Solving introduces additional methods and tips for solving applied problems.

1 Use the six steps to solve the problem. Give the equation, using x as the variable, and give the answer.

If 5 is added to the product of 9 and a number, the result is 19 less than the number. Find the number.

Step 3 **Write an equation.**

The product of 4,	and	a number	decreased by	7,	is	100.
↓		↓	↓	↓	↓	↓
$4 \cdot$		$(x$	$-$	$7)$	$=$	100

(Because of the commas in the given problem, writing the equation as $4x - 7 = 100$ is incorrect. The equation $4x - 7 = 100$ corresponds to the statement "The product of 4 and a number, decreased by 7, is 100.")

Step 4 **Solve** the equation.

$$4(x - 7) = 100$$

$$4x - 28 = 100 \quad \text{Distributive property}$$

$$4x - 28 + 28 = 100 + 28 \quad \text{Add 28.}$$

$$4x = 128 \quad \text{Combine terms.}$$

$$\frac{4x}{4} = \frac{128}{4} \quad \text{Divide by 4.}$$

$$x = 32$$

Step 5 **State the answer.** The number is 32.

Step 6 **Check.** When 32 is decreased by 7, we get $32 - 7 = 25$. If 4 is multiplied by 25, we get 100, as required. The answer, 32, is correct.

OBJECTIVE 3 Solve problems involving sums of quantities. A common type of problem in elementary algebra involves finding two quantities when the sum of the quantities is known.

PROBLEM-SOLVING HINT
In general, to solve problems involving sums of quantities, choose a variable to represent one of the unknowns and then represent the other quantity in terms of the same variable. (See Example 9 in **Section 2.3.**)

EXAMPLE 2 Finding Numbers of Olympic Medals

In the 2002 Winter Olympics in Salt Lake City, the United States won 10 more medals than Norway. The two countries won a total of 58 medals. How many medals did each country win? (*Source:* U.S. Olympic Committee.)

Step 1 **Read** the problem. We are given information about the total number of medals and asked to find the number each country won.

Step 2 **Assign a variable.**

Let $x =$ the number of medals Norway won.

Then $x + 10 =$ the number of medals the U.S. won.

Continued on Next Page

Answers
1. $9x + 5 = x - 19$; -3

Step 3 **Write an equation.**

The total	is	the number of medals Norway won	plus	the number of medals the U.S. won.
↓	↓	↓	↓	↓
58	$=$	x	$+$	$(x + 10)$

Step 4 **Solve** the equation.

$$58 = 2x + 10 \qquad \text{Combine terms.}$$

$$58 - \mathbf{10} = 2x + 10 - \mathbf{10} \qquad \text{Subtract 10.}$$

$$48 = 2x \qquad \text{Combine terms.}$$

$$\frac{48}{\mathbf{2}} = \frac{2x}{\mathbf{2}} \qquad \text{Divide by 2.}$$

$$24 = x \quad \text{or} \quad x = 24$$

Step 5 **State the answer.** The variable x represents the number of medals Norway won, so Norway won 24 medals. Then the number of medals the United States won is $x + 10 = 24 + 10 = 34$.

Step 6 **Check.** Since the United States won 34 medals and Norway won 24, the total number of medals was $34 + 24 = 58$. Because $34 - 24 = 10$, the United States won 10 more medals than Norway. This information agrees with what is given in the problem, so the answer checks.

PROBLEM-SOLVING HINT

The problem in Example 2 could also be solved by letting x represent the number of medals the United States won. Then $x - 10$ would represent the number of medals Norway won. The equation would be

$$58 = x + (x - 10).$$

The solution of this equation is 34, which is the number of U.S. medals. The number of Norwegian medals would be $34 - 10 = 24$. The answers are the same, whichever approach is used.

Work Problem 2 at the Side.

2 Solve the problem.

On one day of their vacation, Annie drove three times as far as Jim. Altogether they drove 84 mi that day. Find the number of miles driven by each.

EXAMPLE 3 Analyzing a Gasoline/Oil Mixture

A lawn trimmer uses a mixture of gasoline and oil. The mixture contains 16 oz of gasoline for each ounce of oil. If the tank holds 68 oz of the mixture, how many ounces of oil and how many ounces of gasoline does it require when it is full?

Step 1 **Read** the problem. We must find how many ounces of oil and gasoline are needed to fill the tank.

Step 2 **Assign a variable**. Let x = the number of ounces of oil required. Then $16x$ = the number of ounces of gasoline required.

Continued on Next Page

ANSWERS

2. Jim: 21 mi; Annie: 63 mi

A diagram like the following is sometimes helpful.

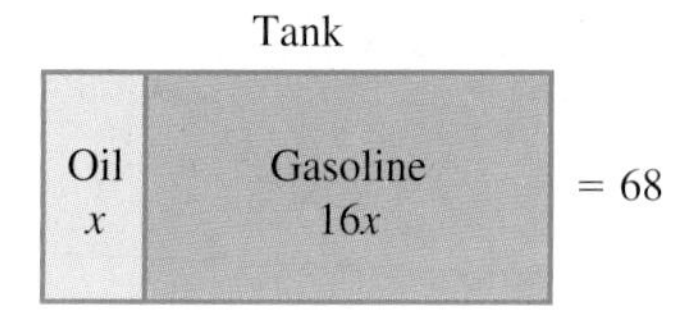

Step 3 **Write an equation.**

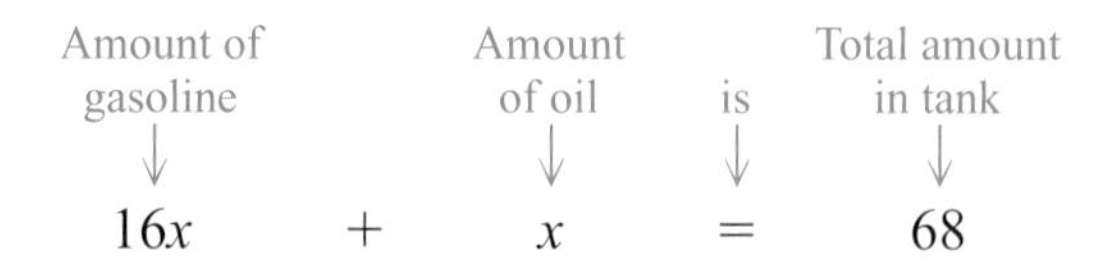

$$16x + x = 68$$

Step 4 **Solve.**

$$17x = 68 \quad \text{Combine terms.}$$

$$\frac{17x}{17} = \frac{68}{17} \quad \text{Divide by 17.}$$

$$x = 4$$

Step 5 **State the answer.** The lawn trimmer requires 4 oz of oil and $16(4) = 64$ oz of gasoline when full.

Step 6 **Check.** Since $4 + 64 = 68$, and 64 is 16 times 4, the answer checks.

Work Problem 3 at the Side.

3 Solve the problem.

At a meeting of the local stamp club, each member brought two nonmembers. If a total of 27 people attended, how many were members and how many were nonmembers?

Meeting

Members x	Nonmembers $2x$	= 27

PROBLEM-SOLVING HINT

Sometimes it is necessary to find three unknown quantities in an applied problem. Frequently the three unknowns are compared in *pairs.* When this happens, it is usually easiest to let the variable represent the unknown found in both pairs. The next example illustrates this.

EXAMPLE 4 Dividing a Board into Pieces

The instructions for a woodworking project call for three pieces of wood. The longest piece must be twice the length of the middle-sized piece, and the shortest piece must be 10 in. shorter than the middle-sized piece. Maria Gonzales has a board 70 in. long that she wishes to use. How long must each piece be?

Step 1 **Read** the problem. Three lengths must be found.

Step 2 **Assign a variable.** Since the middle-sized piece appears in both pairs of comparisons, let x represent the length, in inches, of the middle-sized piece. We have

x = the length of the middle-sized piece,

$2x$ = the length of the longest piece, and

$x - 10$ = the length of the shortest piece.

Continued on Next Page

ANSWERS

3. members: 9; nonmembers: 18

A sketch is helpful here. See Figure 2.

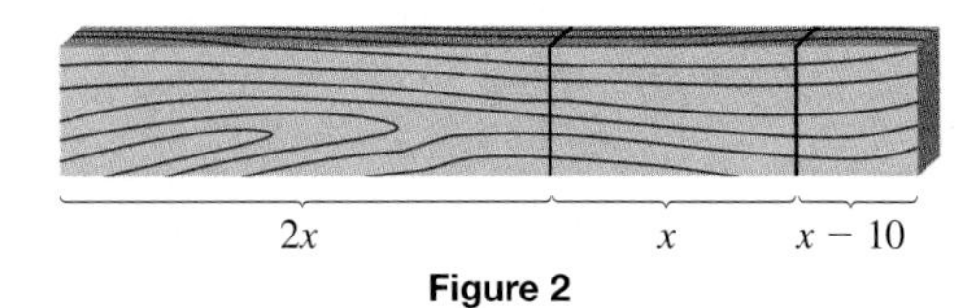

Figure 2

Step 3 **Write an equation.**

Longest		Middle-sized		Shortest	is	Total length
↓		↓		↓	↓	↓
$2x$	$+$	x	$+$	$(x - 10)$	$=$	70

Step 4 **Solve.**

$$4x - 10 = 70 \quad \text{Combine terms.}$$

$$4x - 10 + \mathbf{10} = 70 + \mathbf{10} \quad \text{Add 10.}$$

$$4x = 80 \quad \text{Combine terms.}$$

$$\frac{4x}{\mathbf{4}} = \frac{80}{\mathbf{4}} \quad \text{Divide by 4.}$$

$$x = 20$$

Step 5 **State the answer.** The middle-sized piece is 20 in. long, the longest piece is $2(20) = 40$ in. long, and the shortest piece is $20 - 10 = 10$ in. long.

Step 6 **Check.** The sum of the lengths is 70 in. All conditions of the problem are satisfied.

4 Solve the problem.

A piece of pipe is 50 in. long. It is cut into three pieces. The longest piece is 10 in. longer than the middle-sized piece, and the shortest piece measures 5 in. less than the middle-sized piece. Find the lengths of the three pieces.

OBJECTIVE 4 Solve problems involving supplementary and complementary angles. An angle can be measured by a unit called the degree (°), which is $\frac{1}{360}$ of a complete rotation. Two angles whose sum is 90° are said to be **complementary,** or *complements* of each other. An angle that measures 90° is a **right angle.** Two angles whose sum is 180° are said to be **supplementary,** or supplements of each other. One angle *supplements* the other to form a **straight angle** of 180°. See Figure 3.

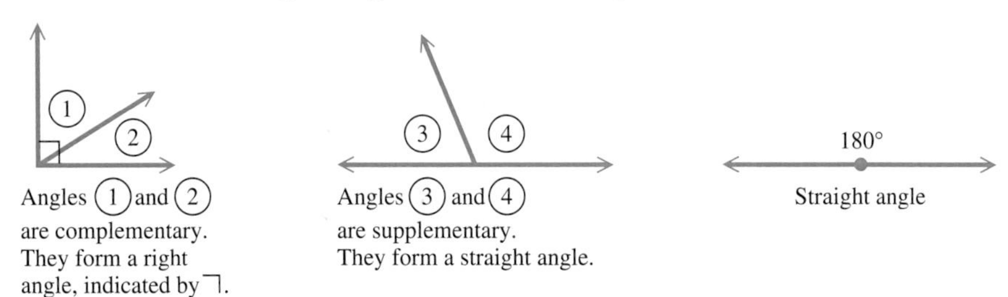

Figure 3

PROBLEM-SOLVING HINT

If x represents the degree measure of an angle, then

$\mathbf{90 - x}$ represents the degree measure of its complement, and

$\mathbf{180 - x}$ represents the degree measure of its supplement.

ANSWERS

4. longest: 25 in.; middle: 15 in.; shortest: 10 in.

5 Find each angle measure.

(a) Fill in the blank below the figure. Then find the supplement of an angle that measures 92°.

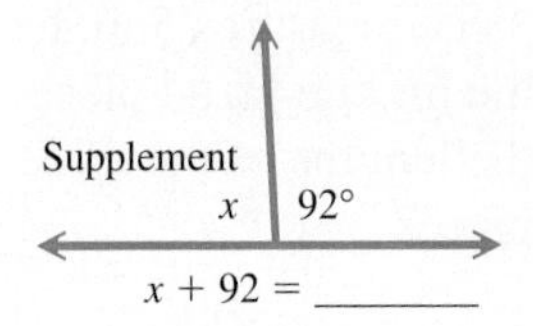

(b) Fill in the blank below the figure. Then find an angle whose complement has twice its measure.

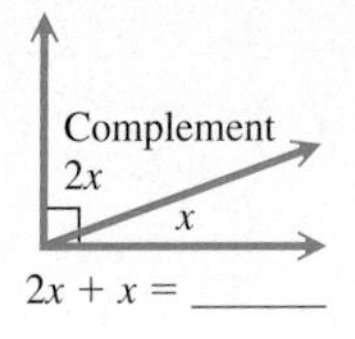

(c) Find the measure of an angle such that twice its complement is 30° less than its supplement.

Answers

5. **(a)** 180; 88° **(b)** 90; 30° **(c)** 30°

EXAMPLE 5 Finding the Measure of an Angle

Find the measure of an angle whose supplement is 10° more than twice its complement.

Step 1 **Read** the problem. We are to find the measure of an angle, given information about its complement and its supplement.

Step 2 **Assign a variable.**

Let $x =$ the degree measure of the angle.

Then $90 - x =$ the degree measure of its complement;

$180 - x =$ the degree measure of its supplement.

We can visualize this information using a sketch. See Figure 4.

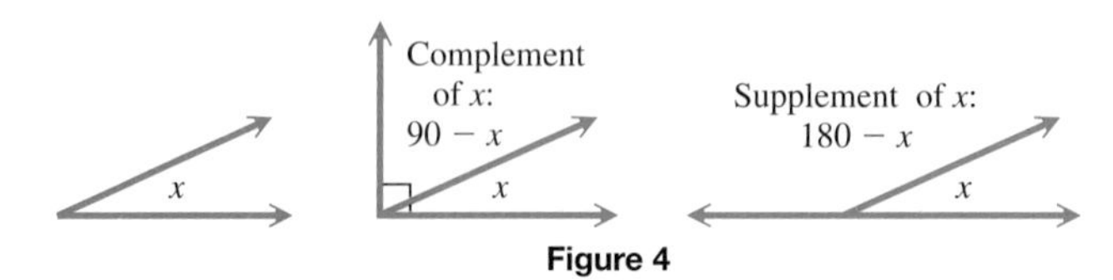

Figure 4

Step 3 **Write an equation.**

Supplement	is	10	more than	twice		its complement.
↓	↓	↓	↓	↓		↓
$180 - x$	$=$	10	$+$	2	$\cdot$	$(90 - x)$

Step 4 **Solve.**

$$180 - x = 10 + 180 - 2x \quad \text{Distributive property}$$
$$180 - x = 190 - 2x \quad \text{Combine terms.}$$
$$180 - x + 2x = 190 - 2x + 2x \quad \text{Add } 2x.$$
$$180 + x = 190 \quad \text{Combine terms.}$$
$$180 + x - 180 = 190 - 180 \quad \text{Subtract 180.}$$
$$x = 10$$

Step 5 **State the answer.** The measure of the angle is 10°.

Step 6 **Check.** The complement of 10° is 80° and the supplement of 10° is 170°. Also, 170° is equal to 10° more than twice 80° (that is, $170 = 10 + 2(80)$ is true). Therefore, the answer is correct.

Work Problem 5 at the Side.

OBJECTIVE 5 Solve problems involving consecutive integers. Two integers that differ by 1 are called **consecutive integers.** For example, 3 and 4, 6 and 7, and -2 and -1 are pairs of consecutive integers. In general, if x represents an integer, $x + 1$ represents the next larger consecutive integer.

Consecutive *even* integers, such as 8 and 10, differ by 2. Similarly, **consecutive *odd* integers,** such as 9 and 11, also differ by two. In general, if x represents an even integer, $x + 2$ represents the next larger consecutive even integer. The same holds true for odd integers; that is, if x is an odd integer, $x + 2$ is the next larger odd integer.

PROBLEM-SOLVING HINT

When solving consecutive integer problems, if $x =$ the first integer, then for any

two consecutive integers, use	$x, \quad x + 1;$
two consecutive *even* integers, use	$x, \quad x + 2;$
two consecutive *odd* integers, use	$x, \quad x + 2.$

6 Solve the problem.

Find two consecutive integers whose sum is -45.

EXAMPLE 6 Finding Consecutive Integers

Two pages that face each other in this book have 277 as the sum of their page numbers. What are the page numbers?

Step 1 **Read** the problem. Because the two pages face each other, they must have page numbers that are consecutive integers.

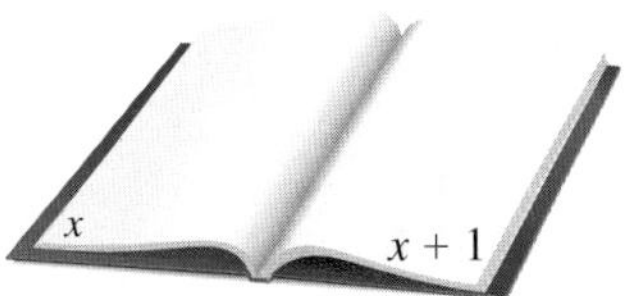

Step 2 **Assign a variable.**

Let $\quad x =$ the first page number.

Then $\quad x + 1 =$ the next page number.

Step 3 **Write an equation.** The sum of the page numbers is 277, so

$$x + (x + 1) = 277.$$

Step 4 **Solve.**

$$2x + 1 = 277 \quad \text{Combine terms.}$$
$$2x = 276 \quad \text{Subtract 1.}$$
$$x = 138 \quad \text{Divide by 2.}$$

Step 5 **State the answer.** The first page number is 138, and the next page number is $138 + 1 = 139$. (Your book is opened to these two pages!)

Step 6 **Check.** The sum of 138 and 139 is 277. The answer is correct.

Work Problem 6 at the Side.

In Example 7, see if you can identify the six steps.

7 Solve the problem.

Find two consecutive even integers such that six times the smaller added to the larger gives a sum of 86.

EXAMPLE 7 Finding Consecutive Odd Integers

If the smaller of two consecutive odd integers is doubled, the result is 7 more than the larger of the two integers. Find the two integers.

Let x be the smaller integer. Since the two numbers are consecutive *odd* integers, then $x + 2$ is the larger. Now write an equation.

If the smaller is doubled,	the result is	7	more than	the larger.
↓	↓	↓	↓	↓
$2x$	$=$	7	$+$	$x + 2$

$$2x = 9 + x \quad \text{Combine terms.}$$
$$x = 9 \quad \text{Subtract } x.$$

The first integer is 9 and the second is $9 + 2 = 11$. To check, we see that when 9 is doubled, we get 18, which is 7 more than the larger odd integer, 11. The answer is correct.

Work Problem 7 at the Side.

ANSWERS

6. $-23, -22$
7. $12, 14$

Focus on

Real-Data Applications

The Magic Number in Sports

The climax of any sports season is the playoffs. Baseball fans eagerly debate predictions of which team will win the pennant for their division. The *magic number* for each first-place team is often used to predict the division winner. The **magic number** is the combined number of wins by the first-place team and losses by the second-place team that would clinch the title for the first-place team.

For Group Discussion

To calculate the magic number, consider the following conditions.

The number of wins for the first-place team (W_1) plus the magic number (M) is one more than the sum of the number of wins to date (W_2) and the number of games remaining in the season (N_2) for the second-place team.

1. First, use the variable definitions to write an equation involving the magic number. Second, solve the equation for the magic number. Write the formula for the magic number.

2. The American League standings with about 10 games left in the 2003 baseball season are shown at the left. There were 162 regulation games in the 2003 season. Find the magic number for each team. The number of games remaining in the season for the second-place team is calculated as

$$N_2 = 162 - (W_2 + L_2).$$

(a) AL East: New York vs Boston

Magic No. _______

(b) AL Central: Minnesota vs Chicago

Magic No. _______

(c) AL West: Oakland vs Seattle

Magic No. _______

American League

East Division	W	L	Pct.	GB
New York	94	58	.618	—
Boston	89	63	.586	5
Toronto	79	73	.520	15
Baltimore	68	83	.450	25½
Tampa Bay	60	92	.395	34
Central Division	**W**	**L**	**Pct.**	**GB**
Minnesota	84	69	.549	—
Chicago	80	72	.526	3½
Kansas City	79	73	.520	4½
Cleveland	66	88	.429	18½
Detroit	38	114	.250	45½
West Division	**W**	**L**	**Pct.**	**GB**
Oakland	92	61	.601	—
Seattle	87	66	.569	5
Anaheim	72	81	.471	20
Texas	68	85	.444	24

Source: USA Today.

[*Note:* For the National League in the 2003 season, Atlanta and San Francisco were runaway winners of the Eastern and Western Divisions, respectively. Chicago, Houston, and St. Louis were locked in a dead heat for the Central Division lead, which was eventually won by Chicago.]

2.4 Exercises

FOR EXTRA HELP

Tutor Center Addison-Wesley Math Tutor Center

MathXL

Digital Video Tutor CD 2 Videotape 2

Student's Solutions Manual

MyMathLab

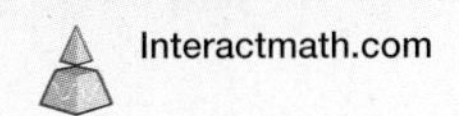

Interactmath.com

1. In your own words, write the general procedure for solving applications as outlined in this section.

2. List some of the words that translate as "=" when writing an equation to solve an applied problem.

3. Suppose that a problem requires you to find the number of cars on a dealer's lot. Which one of the following would not be a reasonable answer? Justify your answer.

   A. 0 **B.** 45 **C.** 1 **D.** $6\frac{1}{2}$

4. Suppose that a problem requires you to find the number of hours a light bulb is on during a day. Which one of the following would not be a reasonable answer? Justify your answer.

   A. 0 **B.** 4.5 **C.** 13 **D.** 25

5. Suppose that a problem requires you to find the distance traveled in miles. Which one of the following would not be a reasonable answer? Justify your answer.

   A. -10 **B.** 1.8 **C.** $10\frac{1}{2}$ **D.** 50

6. Suppose that a problem requires you to find the time in minutes. Which one of the following would not be a reasonable answer? Justify your answer.

   A. 0 **B.** 10.5 **C.** -5 **D.** 90

Solve each problem. See Example 1.

7. The product of 8, and a number increased by 6, is 104. What is the number?

8. The product of 5, and 3 more than twice a number, is 85. What is the number?

9. Two less than three times a number is equal to 14 more than five times the number. What is the number?

10. Nine more than five times a number is equal to 3 less than seven times the number. What is the number?

11. If 2 is subtracted from a number and this difference is tripled, the result is 6 more than the number. Find the number.

12. If 3 is added to a number and this sum is doubled, the result is 2 more than the number. Find the number.

13. The sum of three times a number and 7 more than the number is the same as the difference between -11 and twice the number. What is the number?

14. If 4 is added to twice a number and this sum is multiplied by 2, the result is the same as if the number is multiplied by 3 and 4 is added to the product. What is the number?

Solve each problem. See Example 2.

15. The number of drive-in movie screens has declined steadily in the United States since the 1960s. California and New York were two of the states with the most remaining drive-in movie screens in 2001. California had 11 more screens than New York, and there were 107 screens total in the two states. How many drive-in movie screens remained in each state? (*Source:* National Association of Theatre Owners.)

16. Thursday is the most-watched night for the major broadcast TV networks (ABC, CBS, NBC, and Fox), with 20 million more viewers than Saturday, the least-watched night. The total for the two nights is 102 million viewers. How many viewers of the major networks are there on each of these nights? (*Source:* Nielsen Media Research.)

17. During the 108th session (2002–2003), the U.S. Senate had a total of 99 Democrats and Republicans. There were 3 more Republicans than Democrats. How many Democrats and Republicans were there in the Senate? (*Source: World Almanac and Book of Facts,* 2004.)

18. The total number of Democrats and Republicans in the U.S. House of Representatives during the 108th session was 434. There were 24 more Republicans than Democrats. How many members of each party were there? (*Source: World Almanac and Book of Facts,* 2004.)

19. Bruce Springsteen and the E Street Band generated top revenue on the concert circuit in 2003. Springsteen and second-place Céline Dion together took in \$196.4 million from ticket sales. If Céline Dion took in \$35.4 million less than Bruce Springsteen and the E Street Band, how much revenue did each generate? (*Source: Parade,* February 15, 2004.)

20. The Toyota Camry was the top-selling passenger car in the United States in 2003, followed by the Honda Accord. Honda Accord sales were 35 thousand less than Toyota Camry sales, and 833 thousand of these two cars were sold. How many of each make of car were sold? (*Source:* Ward's Communications.)

21. In the 2002–2003 NBA regular season, the Sacramento Kings won 13 more than twice as many games as they lost. The Kings played 82 games. How many wins and losses did the team have? (*Source:* nba.com)

22. In the 2003 regular baseball season, the Atlanta Braves won 21 less than twice as many games as they lost. They played 162 regular season games. How many wins and losses did the team have? (*Source: World Almanac and Book of Facts,* 2004.)

Solve each problem. See Example 3.

23. The value of a "Mint State-63" (uncirculated) 1950 Jefferson nickel minted at Denver is $\frac{8}{7}$ the value of a similar condition 1945 nickel minted at Philadelphia. Together the total value of the two coins is \$15.00. What is the value of each coin? (*Source:* Yeoman, R., *A Guide Book of United States Coins,* edited by K. Bressett, 56th edition, 2003.)

24. The largest sheep ranch in the world is located in Australia. The number of sheep on the ranch is $\frac{8}{3}$ the number of uninvited kangaroos grazing on the pastureland. Together, herds of these two animals number 88,000. How many sheep and how many kangaroos roam the ranch? (*Source: The Guinness Book of Records.*)

25. In 1988, a dairy in Alberta, Canada, created a sundae with approximately 1 lb of topping for every 83.2 lb of ice cream. The total of the two ingredients weighed approximately 45,225 lb. To the nearest tenth of a pound, how many pounds of ice cream and how many pounds of topping were there? (*Source: The Guinness Book of Records,* 2000.)

26. A husky running the Iditarod (a thousand-mile race between Anchorage and Nome, Alaska) burns $5\frac{3}{8}$ calories in exertion for every 1 calorie burned in thermoregulation in extreme cold. According to one scientific study, a husky in top condition burns an amazing total of 11,200 calories per day. How many calories are burned for exertion, and how many are burned for regulation of body temperature? Round answers to the nearest whole number.

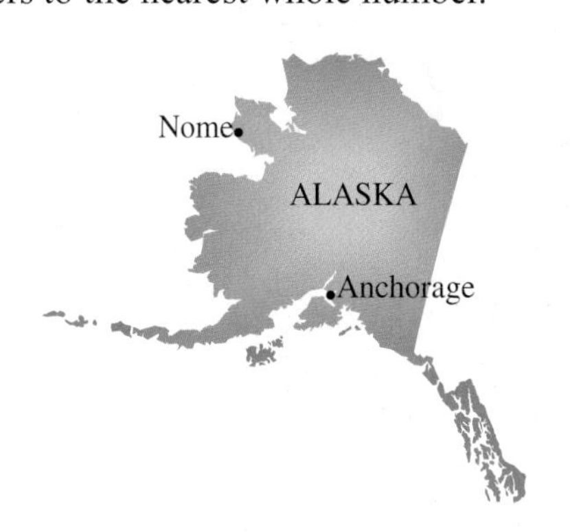

27. In a mixture of concrete, there are 3 lb of cement mix for each pound of gravel. If the mixture contains a total of 140 lb of these two ingredients, how many pounds of gravel are there?

28. A mixture of nuts contains only peanuts and cashews. For every ounce of cashews there are 5 oz of peanuts. If the mixture contains a total of 27 oz, how many ounces of each type of nut does the mixture contain?

Solve each problem. See Example 4.

29. In one day, Akilah Cadet received 13 packages. Federal Express delivered three times as many as Airborne Express, while United Parcel Service delivered 2 fewer than Airborne Express. How many packages did each service deliver to Akilah?

30. In his job at the post office, Eddie Thibodeaux works a 6.5-hr day. He sorts mail, sells stamps, and does supervisory work. One day he sold stamps twice as long as he sorted mail, and he supervised .5 hr longer than he sorted mail. How many hours did he spend at each task?

31. The United States earned 103 medals at the 2004 Summer Olympics in Athens. The number of silver medals earned was 4 more than the number of gold medals. The number of bronze medals earned was 6 less than the number of gold medals. How many of each kind of medal did the United States earn? (*Source: The Gazette,* August 30, 2004.)

32. Nagaraj Nanjappa has a party-length submarine sandwich 59 in. long. He wants to cut it into three pieces so that the middle piece is 5 in. longer than the shortest piece and the shortest piece is 9 in. shorter than the longest piece. How long should the three pieces be?

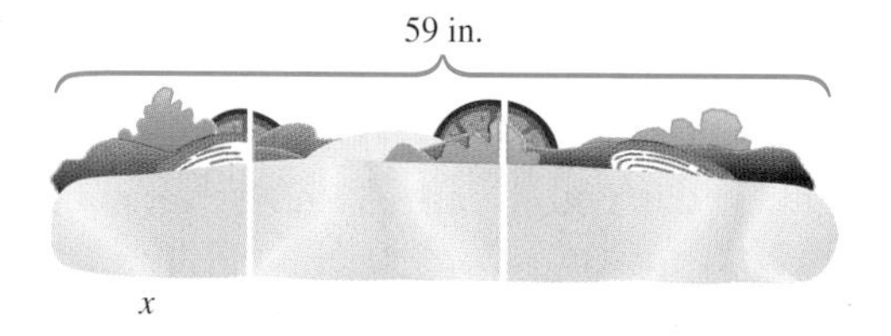

33. Venus is 31.2 million mi farther from the sun than Mercury, while Earth is 57 million mi farther from the sun than Mercury. If the total of the distances from these three planets to the sun is 196.2 million mi, how far away from the sun is Mercury? (All distances given here are mean (*average*) distances.) (*Source: The Universal Almanac.*)

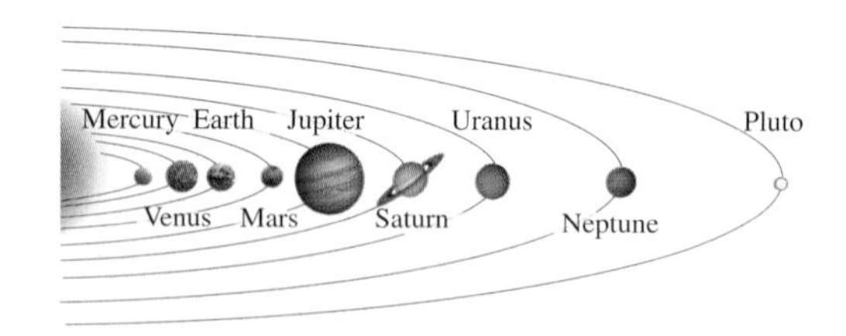

34. Together, Saturn, Jupiter, and Uranus have a total of 113 natural satellites (moons). Jupiter has 30 more satellites than Saturn, and Uranus has 10 fewer satellites than Saturn. How many known satellites does Saturn have? (*Source: World Almanac and Book of Facts,* 2004.)

35. The sum of the measures of the angles of any triangle is 180°. In triangle ABC, angles A and B have the same measure, while the measure of angle C is 60° larger than each of A and B. What are the measures of the three angles?

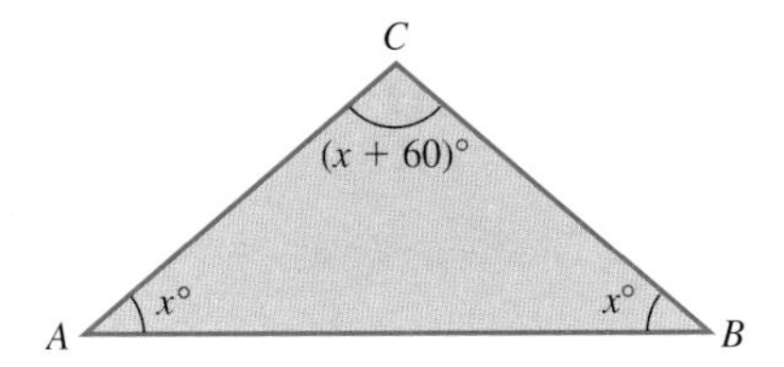

36. In triangle ABC, the measure of angle A is 141° more than the measure of angle B. The measure of angle B is the same as the measure of angle C. Find the measure of each angle. (*Hint:* See Exercise 35.)

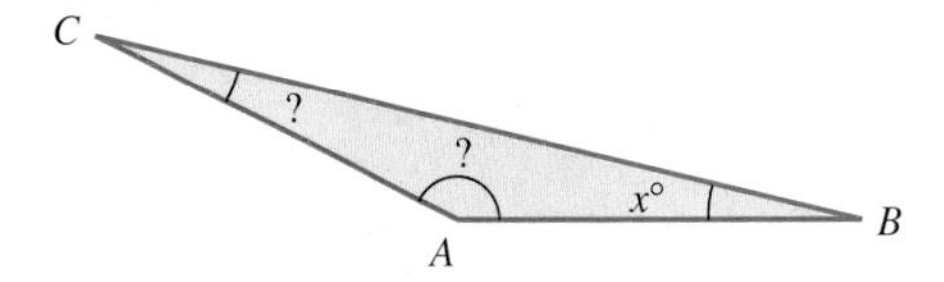

Use the concepts of this section to answer each question.

37. Is there an angle whose supplement is equal to its complement? If so, what is the measure of the angle?

38. Is there an angle that is equal to its supplement? Is there an angle that is equal to its complement? If the answer is yes to either question, give the measure of the angle.

39. If x represents an integer, how can you express the next smaller consecutive integer in terms of x?

40. If x represents an integer, how can you express the next smaller even integer in terms of x?

Solve each problem. See Example 5.

41. Find the measure of an angle whose complement is four times its measure.

42. Find the measure of an angle whose supplement is three times its measure.

43. Find the measure of an angle whose supplement measures 39° more than twice its complement.

44. Find the measure of an angle whose supplement measures 38° less than three times its complement.

45. Find the measure of an angle such that the difference between the measures of its supplement and three times its complement is 10°.

46. Find the measure of an angle such that the sum of the measures of its complement and its supplement is 160°.

Solve each problem. See Examples 6 and 7.

47. The numbers on two consecutively numbered gym lockers have a sum of 137. What are the locker numbers?

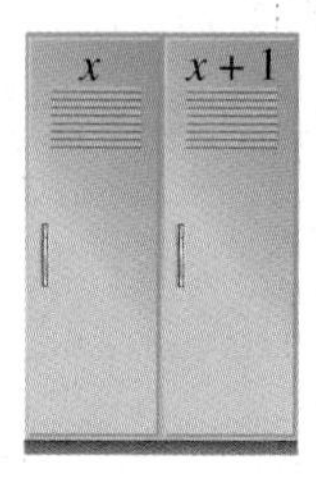

48. The sum of two consecutive checkbook check numbers is 357. Find the numbers.

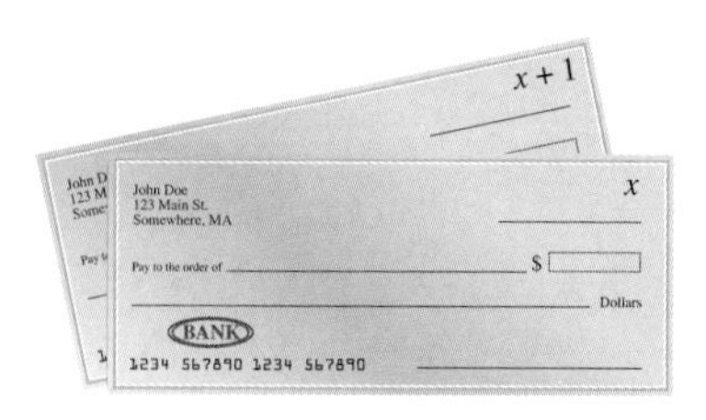

49. Two pages that are back-to-back in this book have 293 as the sum of their page numbers. What are the page numbers?

50. Find two consecutive odd integers such that twice the larger is 17 more than the smaller.

51. Find two consecutive even integers such that the smaller added to three times the larger gives a sum of 46.

52. Two houses on the same side of the street have house numbers that are consecutive even integers. The sum of the integers is 58. What are the two house numbers?

53. When the smaller of two consecutive integers is added to three times the larger, the result is 43. Find the integers.

54. If five times the smaller of two consecutive integers is added to three times the larger, the result is 59. Find the integers.

Apply the ideas of this section to solve Exercises 55 and 56, based on the graphs.

55. In a recent year, the funding for Head Start programs increased by .61 billion dollars from the funding in the previous year. The following year, the increase was .93 billion dollars more. For those three years, the total funding was 16.13 billion dollars. How much was funded in each of these years? (*Source:* U.S. Department of Health and Human Services.)

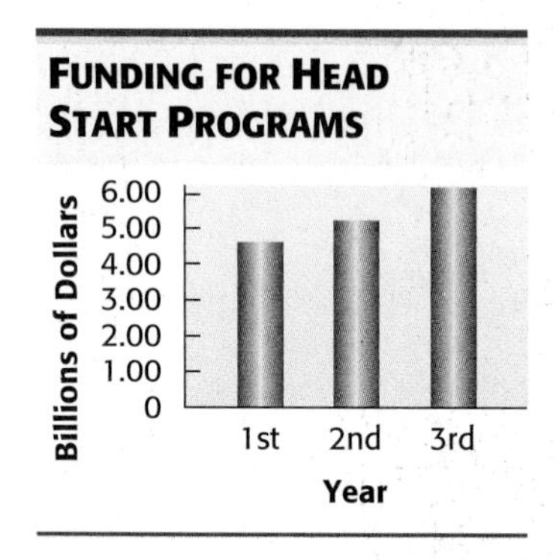

56. According to data provided by the National Safety Council for a recent year, the number of serious injuries per 100,000 participants in football, bicycling, and golf is illustrated in the graph. There were 800 more in bicycling than in golf, and there were 1267 more in football than in bicycling. Altogether there were 3179 serious injuries per 100,000 participants. How many such serious injuries were there in each sport?

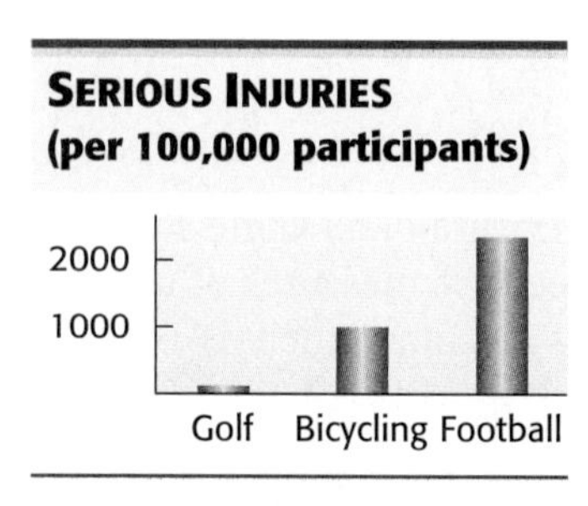

2.5 Formulas and Applications from Geometry

OBJECTIVES

1. Solve a formula for one variable, given the values of the other variables.
2. Use a formula to solve an applied problem.
3. Solve problems involving vertical angles and straight angles.
4. Solve a formula for a specified variable.

Many applied problems can be solved with *formulas*. A **formula** is a mathematical expression in which variables are used to describe a relationship. Formulas exist for geometric figures, for distance, and for money earned on bank savings. The formulas used in this book are given on the inside covers.

OBJECTIVE 1 Solve a formula for one variable, given the values of the other variables. Given the values of all but one of the variables in a formula, we can find the value of the remaining variable. In Example 1, we use the idea of *area*. The **area** of a plane (two-dimensional) geometric figure is a measure of the surface covered by the figure.

EXAMPLE 1 Using Formulas to Evaluate Variables

Find the value of the remaining variable in each formula.

(a) $A = LW$; $A = 64$, $L = 10$

As shown in Figure 5, this formula gives the area of a rectangle with length L and width W.

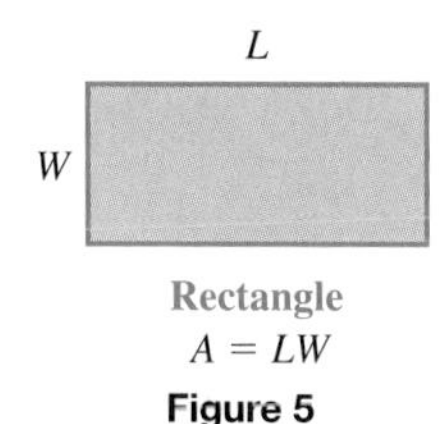

Rectangle
$A = LW$
Figure 5

Substitute the given values into the formula and then solve for W.

$$A = LW$$

$$64 = 10W \quad \text{Let } A = 64 \text{ and } L = 10.$$

$$6.4 = W \quad \text{Divide by 10.}$$

The width is 6.4. Since $10(6.4) = 64$, the given area, the answer checks.

(b) $A = \frac{1}{2}h(b + B)$; $A = 210$, $B = 27$, $h = 10$

This formula gives the area of a trapezoid with parallel sides of lengths b and B and distance h between the parallel sides. See Figure 6.

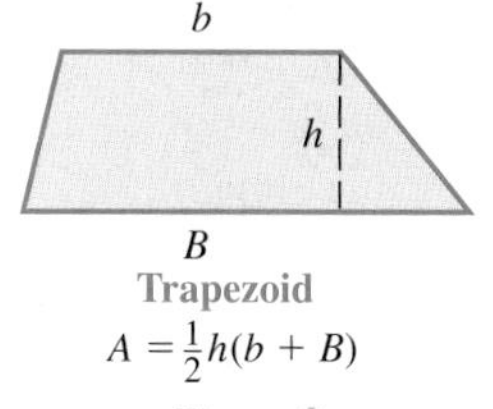

Trapezoid
$A = \frac{1}{2}h(b + B)$
Figure 6

Again, begin by substituting the given values into the formula.

$$A = \frac{1}{2}h(b + B)$$

$$210 = \frac{1}{2}(10)(b + 27) \quad A = 210, h = 10, B = 27$$

Continued on Next Page

Now solve for b.

$$210 = 5(b + 27) \quad \text{Multiply.}$$
$$210 = 5b + 135 \quad \text{Distributive property}$$
$$210 - 135 = 5b + 135 - 135 \quad \text{Subtract 135.}$$
$$75 = 5b \quad \text{Combine terms.}$$
$$\frac{75}{5} = \frac{5b}{5} \quad \text{Divide by 5.}$$
$$15 = b$$

Check that the length of the shorter parallel side, b, is 15.

Work Problem 1 at the Side.

1 Find the value of the remaining variable in each formula.

(a) $I = prt$; $I = \$246$, $r = .06$, $t = 2$

(b) $P = 2L + 2W$; $P = 126$, $W = 25$

OBJECTIVE 2 Use a formula to solve an applied problem. Formulas are often used to solve applied problems. ***It is a good idea to draw a sketch when a geometric figure is involved.*** Example 2 uses the idea of *perimeter*. The **perimeter** of a plane (two-dimensional) geometric figure is the distance around the figure, that is, the sum of the lengths of its sides.

EXAMPLE 2 Finding the Width of a Rectangular Lot

A rectangular lot has perimeter 80 m and length 25 m. Find the width of the lot.

Step 1 **Read.** We are told to find the width of the lot.

Step 2 **Assign a variable.** Let W = the width of the lot in meters. See Figure 7.

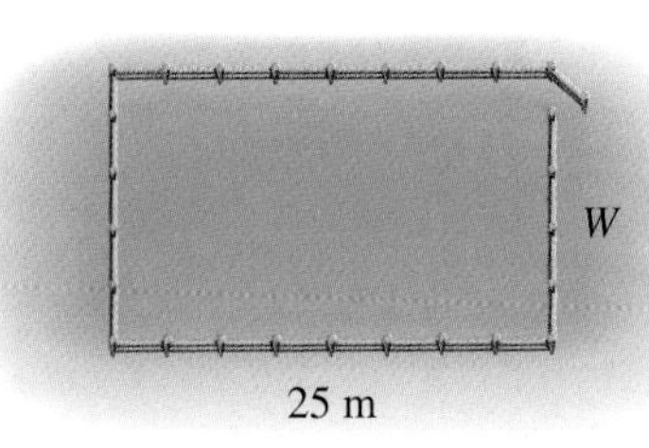

Figure 7

Step 3 **Write an equation.** We use the formula for the perimeter of a rectangle, and find the width by substituting 80 for P and 25 for L.

$$P = 2L + 2W \quad \text{Perimeter formula}$$
$$80 = 2(25) + 2W \quad P = 80, L = 25$$

Step 4 **Solve.**

$$80 = 50 + 2W \quad \text{Multiply.}$$
$$80 - 50 = 50 + 2W - 50 \quad \text{Subtract 50.}$$
$$30 = 2W \quad \text{Combine terms.}$$
$$\frac{30}{2} = \frac{2W}{2} \quad \text{Divide by 2.}$$
$$15 = W$$

Step 5 **State the answer.** The width is 15 m.

Step 6 **Check.** If the width is 15 m and the length is 25 m, the perimeter is $2(25) + 2(15) = 50 + 30 = 80$ m, as required.

Work Problem 2 at the Side.

2 Solve the problem.

A farmer has 800 m of fencing material to enclose a rectangular field. The width of the field is 175 m. Find the length of the field.

Answers

1. (a) $p = \$2050$ (b) $L = 38$
2. 225 m

EXAMPLE 3 **Finding the Height of a Triangular Sail**

The area of a triangular sail of a sailboat is 126 ft^2. (Recall that ft^2 means "square feet.") The base of the sail is 12 ft. Find the height of the sail.

Step 1 **Read.** We must find the height of the triangular sail.

Step 2 **Assign a variable.** Let h = the height of the sail in feet. See Figure 8.

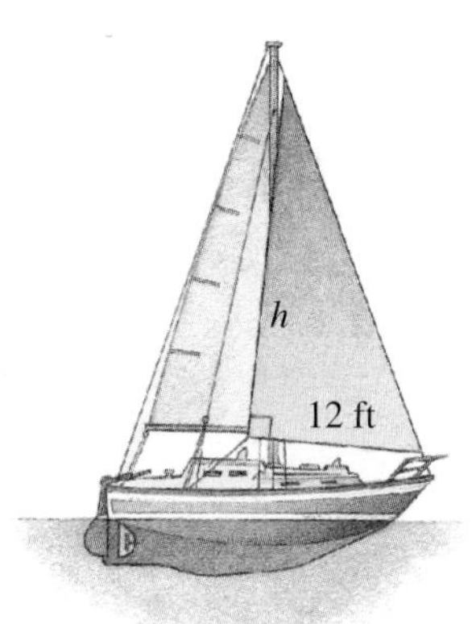

Figure 8

Step 3 **Write an equation.** The formula for the area of a triangle is $A = \frac{1}{2}bh$, where A is the area, b is the base, and h is the height. Using the information given in the problem, substitute 126 for A and 12 for b in the formula.

$$A = \frac{1}{2}bh$$

$$\mathbf{126} = \frac{1}{2}(\mathbf{12})h \qquad A = 126,\ b = 12$$

Step 4 **Solve** the equation.

$$126 = 6h \qquad \text{Multiply.}$$

$$21 = h \qquad \text{Divide by 6.}$$

Step 5 **State the answer.** The height of the sail is 21 ft.

Step 6 **Check** to see that the values $A = 126$, $b = 12$, and $h = 21$ satisfy the formula for the area of a triangle.

Work Problem 3 at the Side.

3 Solve the problem.

The area of a triangle is 120 m^2. The height is 24 m. Find the length of the base of the triangle.

OBJECTIVE 3 **Solve problems involving vertical angles and straight angles.** Figure 9 shows two intersecting lines forming angles that are numbered ①, ②, ③, and ④. Angles ① and ③ lie "opposite" each other. They are called **vertical angles.** Another pair of vertical angles is ② and ④. In geometry, it is shown that ***vertical angles have equal measures.***

Now look at angles ① and ②. When their measures are added, we get the measure of a **straight angle,** which is **180°**. There are three other such pairs of angles: ② and ③, ③ and ④, and ① and ④.

The next example uses these ideas.

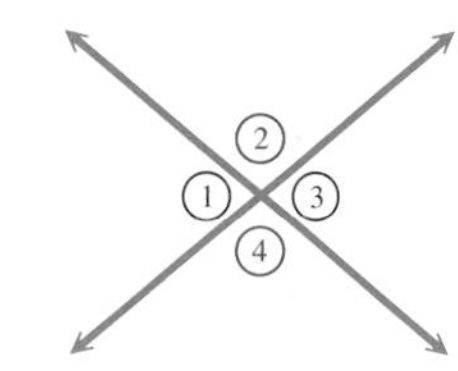

Figure 9

ANSWERS

3. 10 m

4 Find the measure of each marked angle.

(a)

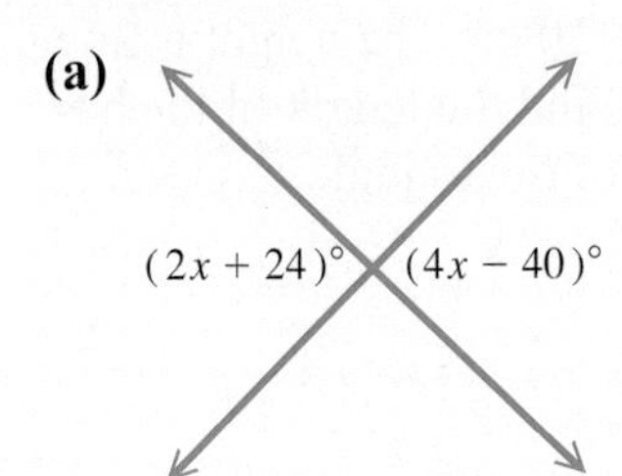

(b)

$(5x + 12)°$ $(3x)°$

EXAMPLE 4 Finding Angle Measures

Refer to the appropriate figure in each part.

(a) Find the measure of each marked angle in Figure 10.

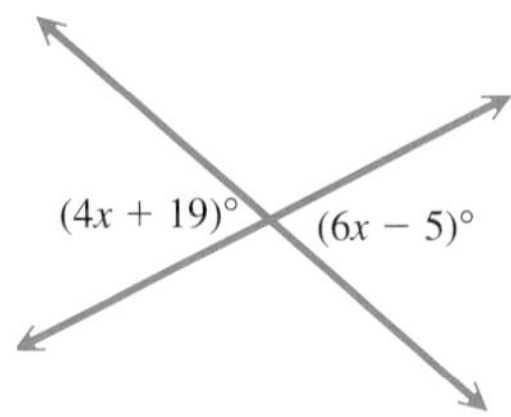

Figure 10

Since the marked angles are vertical angles, they have equal measures.

$$
\begin{aligned}
4x + 19 &= 6x - 5 && \text{Set } 4x + 19 \text{ equal to } 6x - 5. \\
4x + 19 - 4x &= 6x - 5 - 4x && \text{Subtract } 4x. \\
19 &= 2x - 5 \\
19 + 5 &= 2x - 5 + 5 && \text{Add 5.} \\
24 &= 2x \\
\frac{24}{2} &= \frac{2x}{2} && \text{Divide by 2.} \\
12 &= x
\end{aligned}
$$

Since $x = 12$, one angle has measure $4(12) + 19 = \mathbf{67}$ degrees. The other has the same measure, since $6(12) - 5 = \mathbf{67}$ as well. Each angle measures 67°.

(b) Find the measure of each marked angle in Figure 11.

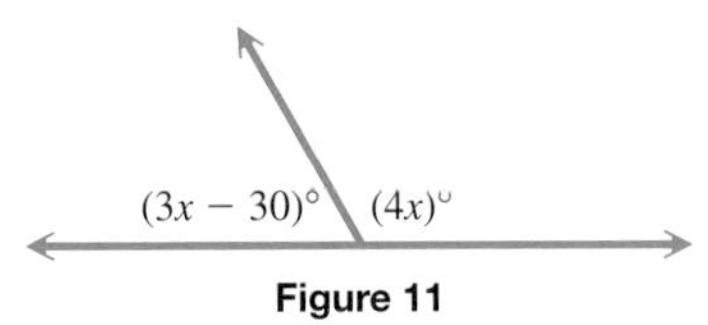

Figure 11

The measures of the marked angles must add to 180° because together they form a straight angle. The equation to solve is

$$
\begin{aligned}
(3x - 30) + 4x &= 180. \\
7x - 30 &= 180 && \text{Combine terms.} \\
7x &= 210 && \text{Add 30.} \\
x &= \mathbf{30} && \text{Divide by 7.}
\end{aligned}
$$

To find the measures of the angles, replace x with 30 in the two expressions.

$$3x - 30 = 3(30) - 30 = 90 - 30 = \mathbf{60}$$

$$4x = 4(30) = \mathbf{120}$$

The two angle measures are 60° and 120°.

CAUTION

In Example 4, the answer is *not* the value of x. Remember to substitute the value of the variable into the expression given for each angle.

Work Problem 4 at the Side.

ANSWERS

4. **(a)** Both measure 88°. **(b)** 117° and 63°

OBJECTIVE 4 Solve a formula for a specified variable. Sometimes it is necessary to solve a large number of problems that use the same formula. For example, a surveying class might need to solve several problems that involve the formula for the area of a rectangle, $A = LW$. Suppose that in each problem the area (A) and the length (L) of a rectangle are given, and the width (W) must be found. Rather than solving for W each time the formula is used, it would be simpler to *rewrite the formula* so that it is solved for W. This process is called **solving for a specified variable** or **solving a literal equation.**

In solving a formula for a specified variable, we treat the specified variable as if it were the *only* variable in the equation, and treat the other variables as if they were numbers. We use the same steps to solve the equation for the specified variable that we have used to solve equations with just one variable.

5 Solve $I = prt$ for t.

EXAMPLE 5 Solving for a Specified Variable

Solve $A = LW$ for W.

Our goal is to get W alone on one side of the equation. To do this, think of undoing what has been done to W. Since W is multiplied by L, undo the multiplication by dividing each side of $A = LW$ by L.

$$A = LW$$

$$\frac{A}{L} = \frac{LW}{L} \qquad \text{Divide by } L.$$

$$\frac{A}{L} = W \quad \text{or} \quad W = \frac{A}{L} \qquad \frac{L}{L} = 1;\ 1W = W$$

The formula is now solved for W.

Work Problem 5 at the Side.

6 Solve $P = a + b + c$ for a.

EXAMPLE 6 Solving for a Specified Variable

Solve $P = 2L + 2W$ for L.

We want to get L alone on one side of the equation. We begin by subtracting $2W$ from each side.

$$P = 2L + 2W$$

$$P - 2W = 2L + 2W - 2W \qquad \text{Subtract } 2W.$$

$$P - 2W = 2L \qquad \text{Combine terms.}$$

$$\frac{P - 2W}{2} = \frac{2L}{2} \qquad \text{Divide by 2.}$$

$$\frac{P - 2W}{2} = L \quad \text{or} \quad L = \frac{P - 2W}{2} \qquad \frac{2}{2} = 1;\ 1L = L$$

The last step gives the formula solved for L, as required.

Work Problem 6 at the Side.

ANSWERS

5. $t = \frac{I}{pr}$
6. $a = P - b - c$

7 **(a)** Solve $A = p + prt$ for t.

(b) Solve $Ax + By = C$ for y.

EXAMPLE 7 Solving for a Specified Variable

Solve $F = \frac{9}{5}C + 32$ for C. (This is the formula for converting temperatures from Celsius to Fahrenheit.)

We need to isolate C on one side of the equation. First undo the addition of 32 to $\frac{9}{5}C$ by subtracting 32 from each side.

$$F = \frac{9}{5}C + 32$$

$$F - 32 = \frac{9}{5}C + 32 - 32 \qquad \text{Subtract 32.}$$

$$F - 32 = \frac{9}{5}C$$

Now multiply each side by $\frac{5}{9}$. Use parentheses on the left.

$$\frac{5}{9}(F - 32) = \frac{5}{9} \cdot \frac{9}{5}C \qquad \text{Multiply by } \tfrac{5}{9}.$$

$$\frac{5}{9}(F - 32) = C \quad \text{or} \quad C = \frac{5}{9}(F - 32)$$

This last result is the formula for converting temperatures from Fahrenheit to Celsius.

Work Problem 7 at the Side.

Answers

7. **(a)** $t = \frac{A - p}{pr}$ **(b)** $y = \frac{C - Ax}{B}$

2.5 Exercises

FOR EXTRA HELP

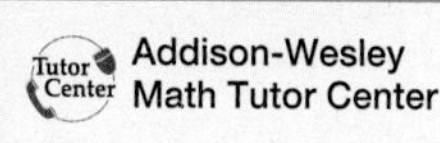

MathXL

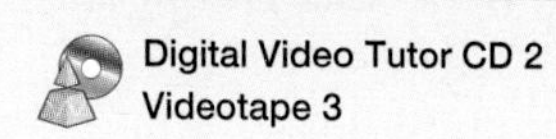

Student's Solutions Manual

1. In your own words, explain what is meant by each term.

(a) Perimeter of a plane geometric figure

(b) Area of a plane geometric figure

2. The distance around a circle is called the ______________ of the circle.

3. If a formula has exactly five variables, how many values would you need to be given in order to find the value of any one variable?

4. The formula for changing Celsius temperature to Fahrenheit is given in Example 7 as $F = \frac{9}{5}C + 32$. Sometimes it is seen as $F = \frac{9C}{5} + 32$. These are both correct. Why is it true that $\frac{9}{5}C$ is equal to $\frac{9C}{5}$?

Decide whether perimeter or area would be used to solve a problem concerning the measure of the quantity.

5. Sod for a lawn

6. Carpeting for a bedroom

7. Baseboards for a living room

8. Fencing for a yard

9. Fertilizer for a garden

10. Tile for a bathroom

11. Determining the cost of planting rye grass in a lawn for the winter

12. Determining the cost of replacing a linoleum floor with a wood floor

In the following exercises a formula is given, along with the values of all but one of the variables in the formula. Find the value of the variable that is not given. (When necessary, use 3.14 as an approximation for π (pi).) See Example 1.

13. $P = 2L + 2W$ (perimeter of a rectangle); $L = 8$, $W = 5$

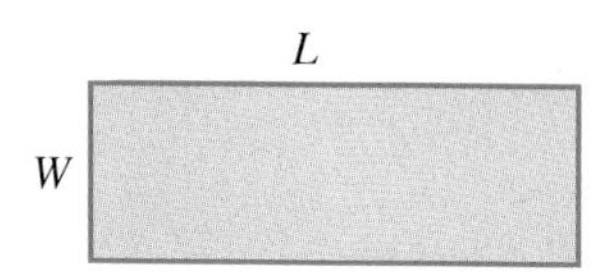

14. $P = 2L + 2W$; $L = 6$, $W = 4$

15. $A = \frac{1}{2}bh$ (area of a triangle); $b = 8$, $h = 16$

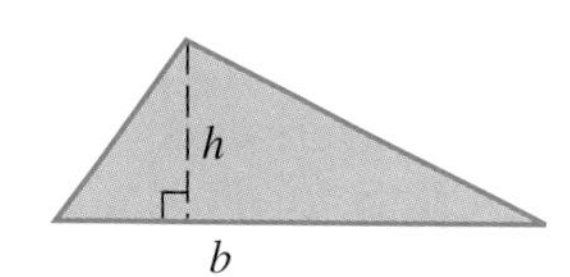

16. $A = \frac{1}{2}bh$; $b = 10$, $h = 14$

17. $P = a + b + c$ (perimeter of a triangle); $P = 12$, $a = 3$, $c = 5$

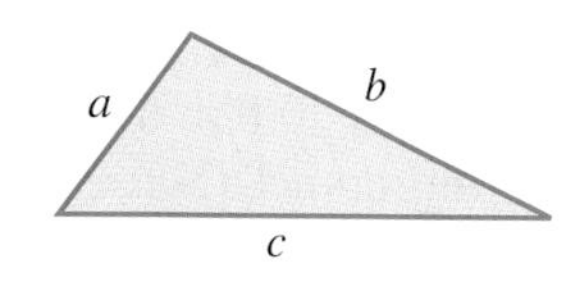

18. $P = a + b + c$; $P = 15$, $a = 3$, $b = 7$

19. $d = rt$ (distance formula); $d = 252, r = 45$

20. $d = rt$; $d = 100, t = 2.5$

21. $I = prt$ (simple interest); $p = 7500, r = .035, t = 6$

22. $I = prt$; $p = 5000, r = .025, t = 7$

23. $C = 2\pi r$ (circumference of a circle); $C = 16.328$

24. $C = 2\pi r$; $C = 8.164$

25. $A = \pi r^2$ (area of a circle); $r = 4$

26. $A = \pi r^2$; $r = 12$

The **volume** *of a three-dimensional object is a measure of the space occupied by the object. For example, we would need to know the volume of a gasoline tank in order to know how many gallons of gasoline it would take to completely fill the tank. In the following exercises, a formula for the volume (V) of a three-dimensional object is given, along with values for the other variables. Evaluate V. (Use 3.14 as an approximation for π.) See Example 1.*

27. $V = LWH$ (volume of a rectangular box); $L = 10, W = 5, H = 3$

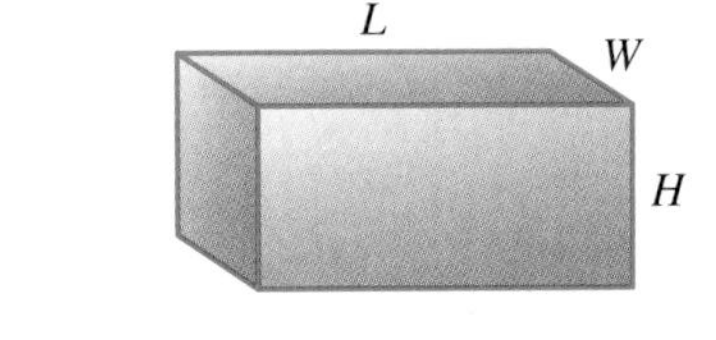

28. $V = LWH$; $L = 12, W = 8, H = 4$

29. $V = \frac{1}{3}Bh$ (volume of a pyramid); $B = 12, h = 13$

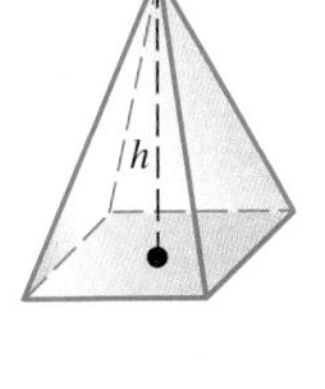

30. $V = \frac{1}{3}Bh$; $B = 36, h = 4$

31. $V = \frac{4}{3}\pi r^3$ (volume of a sphere); $r = 12$

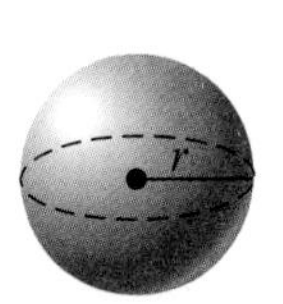

32. $V = \frac{4}{3}\pi r^3$; $r = 6$

Use a formula to write an equation for each application, and then use the problem-solving method of **Section 2.4** *to solve. (Use 3.14 as an approximation for π.)* ***Formulas are found on the inside covers of this book.*** *See Examples 2 and 3.*

33. A prehistoric ceremonial site dating to about 3000 B.C. was discovered at Stanton Drew in southwestern England. The site, which is larger than Stonehenge, is a nearly perfect circle, consisting of nine concentric rings that probably held upright wooden posts. Around this timber temple is a wide, encircling ditch enclosing an area with a diameter of 443 ft. Find this enclosed area to the nearest thousand square feet. (*Source: Archaeology,* vol. 51, no. 1, Jan./Feb. 1998.)

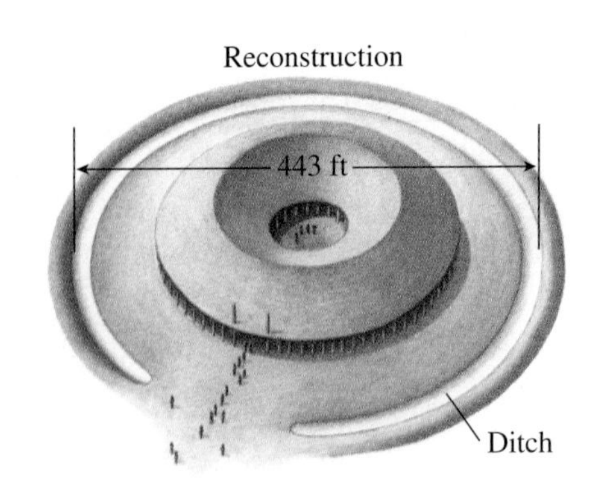

34. The Skydome in Toronto, Canada, is the first stadium with a hard-shell, retractable roof. The steel dome is 630 ft in diameter. To the nearest foot, what is the circumference of this dome? (*Source:* www.4ballparks.com)

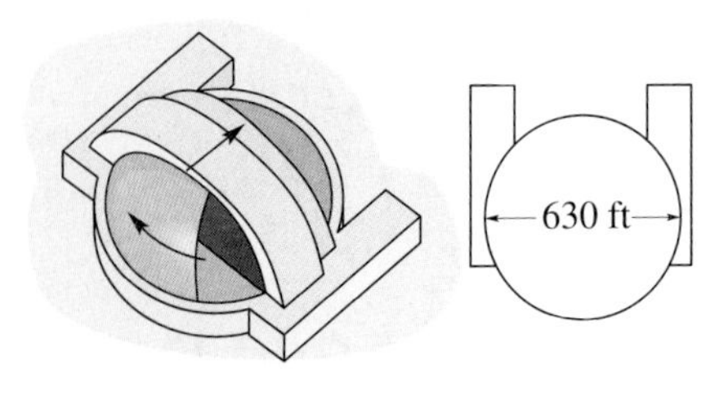

35. The *Daily Banner,* published in Roseberg, Oregon, in the 19th century, had page size 3 in. by 3.5 in. What was the perimeter? What was the area? (*Source: The Guinness Book of Records.*)

36. The newspaper *The Constellation,* printed in 1859 in New York City as part of the Fourth of July celebration, had length 51 in. and width 35 in. What was the perimeter? What was the area? (*Source: The Guinness Book of Records.*)

37. The largest drum ever constructed was made from Japanese cedar and cowhide, with diameter 15.74 ft. It was built in Ishikawa, Japan, and completed on January 1, 2001. What was the area of the circular face of the drum? Round your answer to the nearest hundredth of a square foot. (*Hint:* Use $A = \pi r^2$.) (*Source:* www.guinnessworldrecords.com)

38. What was the circumference of the drum described in Exercise 37? Round your answer to the nearest hundredth of a foot. (*Hint:* Use $C = 2\pi r$.)

39. The survey plat depicted here shows two lots that form a trapezoid. The measures of the parallel sides are 115.80 ft and 171.00 ft. The height of the trapezoid is 165.97 ft. Find the combined area of the two lots. Round your answer to the nearest hundredth of a square foot.

40. Lot A in the figure is in the shape of a trapezoid. The parallel sides measure 26.84 ft and 82.05 ft. The height of the trapezoid is 165.97 ft. Find the area of Lot A. Round your answer to the nearest hundredth of a square foot.

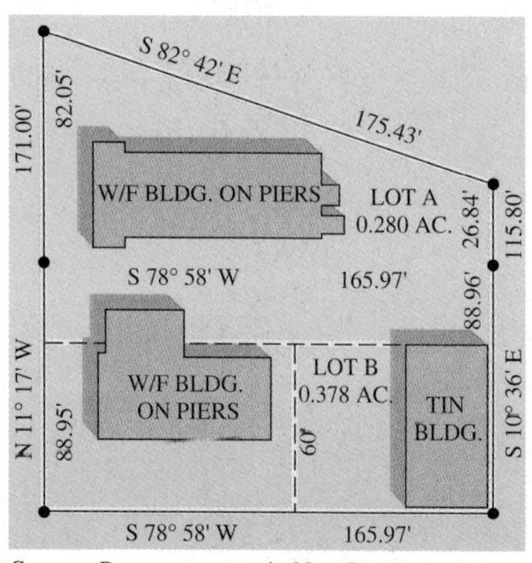

Source: Property survey in New Roads, Louisiana.

41. The U.S. Postal Service requires that any box sent through the mail have length plus girth (distance around) totaling no more than 108 in. The maximum volume that meets this condition is contained by a box with a square end 18 in. on each side. What is the length of the box? What is the maximum volume?

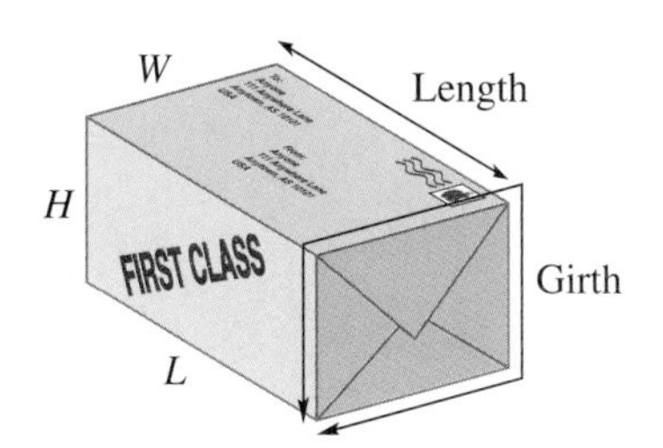

42. The largest box of popcorn was filled by students in Jacksonville, Florida. The box was approximately 40 ft long, $20\frac{2}{3}$ ft wide, and 8 ft high. To the nearest cubic foot, what was the volume of the box? (*Source: The Guinness Book of Records.*)

Find the measure of each marked angle. See Example 4.

43.

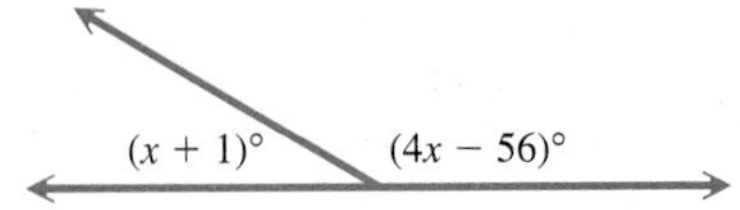

44.

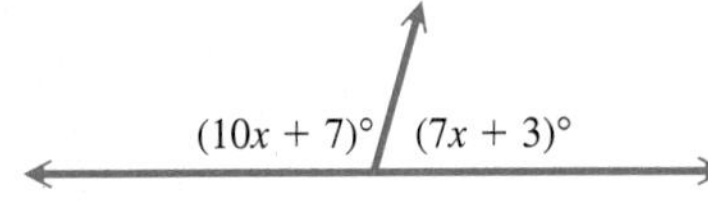

45.

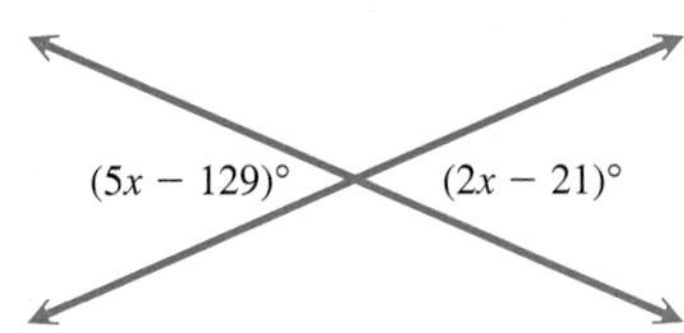

46.

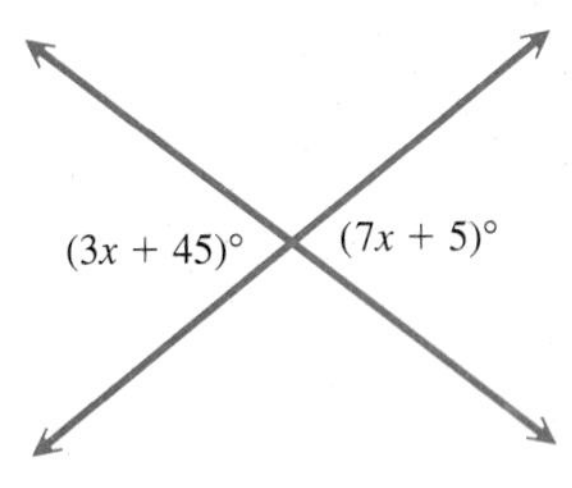

47.

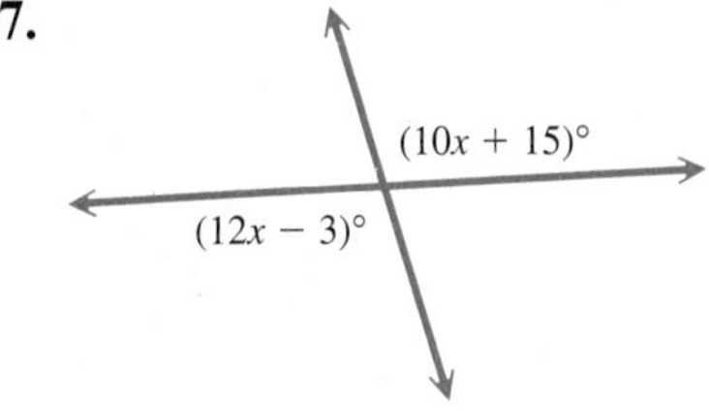

48.

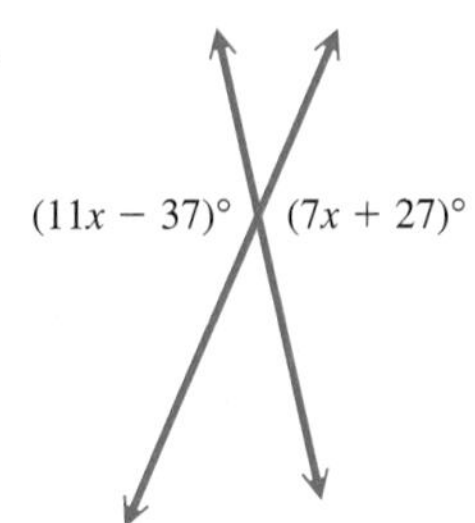

Solve each formula for the specified variable. See Examples 5–7.

49. $d = rt$ for t

50. $d = rt$ for r

51. $V = LWH$ for H

52. $A = LW$ for L

53. $P = a + b + c$ for b

54. $P = a + b + c$ for c

55. $I = prt$ for r

56. $I = prt$ for p

57. $A = \frac{1}{2}bh$ for h

58. $A = \frac{1}{2}bh$ for b

59. $P = 2L + 2W$ for W

60. $A = p + prt$ for r

61. $V = \frac{1}{3}\pi r^2 h$ for h

62. $V = \pi r^2 h$ for h

63. $y = mx + b$ for m

64. $y = mx + b$ for x

65. $M = C(1 + r)$ for r

66. $C = \frac{5}{9}(F - 32)$ for F

2.6 Ratio, Proportion, and Percent

OBJECTIVES

1. Write ratios.
2. Solve proportions.
3. Solve applied problems using proportions.
4. Find percentages and percents.

OBJECTIVE 1 Write ratios. A **ratio** is a comparison of two quantities using a quotient.

Ratio

The ratio of the number a to the number b is written

$$a \text{ to } b, \quad a:b, \quad \text{or} \quad \frac{a}{b}.$$

This last way of writing a ratio is most common in algebra.

EXAMPLE 1 Writing a Word Phrase as a Ratio

Write a ratio for each word phrase.

(a) The ratio of 5 hr to 3 hr is

$$\frac{5 \text{ hr}}{3 \text{ hr}} = \frac{5}{3}.$$

(b) To find the ratio of 6 hr to 3 days, first convert 3 days to hours.

$$3 \text{ days} = 3 \cdot 24$$
$$= 72 \text{ hr}$$

The ratio of 6 hr to 3 days is thus

$$\frac{6 \textbf{ hr}}{3 \textbf{ days}} = \frac{6 \textbf{ hr}}{72 \textbf{ hr}} = \frac{6}{72} = \frac{1}{12}.$$

Work Problem 1 at the Side.

1 Write each ratio.

(a) 9 women to 5 women

(b) 4 in. to 1 ft

An application of ratios is in unit pricing, to see which size of an item offered in different sizes produces the best price per unit. To do this, set up the ratio of the price of the item to the number of units on the label. Then divide to obtain the price per unit.

EXAMPLE 2 Finding the Price per Unit

The Cub Foods supermarket in Coon Rapids, Minnesota, charges the following prices for a jar of extra crunchy peanut butter.

PEANUT BUTTER

Size	Price
18-oz	$1.50
40-oz	$4.14
64-oz	$6.29

Which size is the best buy? That is, which size has the lowest unit price?

Continued on Next Page

ANSWERS

1. (a) $\frac{9}{5}$ (b) $\frac{4}{12} = \frac{1}{3}$

2 Solve the problem.

The local supermarket charges the following prices for a popular brand of pancake syrup.

PANCAKE SYRUP

Size	Price
36-oz	\$3.89
24-oz	\$2.79
12-oz	\$1.89

Which size is the best buy? What is the unit cost for that size?

To find the best buy, write ratios comparing the price for each size jar to the number of units (ounces) per jar. The results in the following table are rounded to the nearest thousandth.

Size	Unit Cost (dollars per ounce)	
18-oz	$\frac{\$1.50}{18} = \$.083$	← The best buy
40-oz	$\frac{\$4.14}{40} = \$.104$	
64-oz	$\frac{\$6.29}{64} = \$.098$	

Because the 18-oz size produces the lowest unit cost, it is the best buy. This example shows that buying the largest size does not always provide the best buy, although this is often true.

Work Problem 2 at the Side.

OBJECTIVE 2 Solve proportions. A ratio is used to compare two numbers or amounts. A **proportion** says that two ratios are equal, so it is a special type of equation. For example,

$$\frac{3}{4} = \frac{15}{20} \qquad \text{Proportion}$$

is a proportion that says that the ratios $\frac{3}{4}$ and $\frac{15}{20}$ are equal. In the proportion

$$\frac{a}{b} = \frac{c}{d} \qquad (b, d \neq 0),$$

a, b, c, and d are the **terms** of the proportion. The a and d terms are called the **extremes,** and the b and c terms are called the **means.** We read the proportion $\frac{a}{b} = \frac{c}{d}$ as "a is to b as c is to d." Multiplying each side of this proportion by the common denominator, bd, gives

$$bd \cdot \frac{a}{b} = bd \cdot \frac{c}{d}$$

$$\frac{b}{b}(d \cdot a) = \frac{d}{d}(b \cdot c) \qquad \text{Associative and commutative properties}$$

$$ad = bc. \qquad \text{Commutative and identity properties}$$

We can also find the products ad and bc by multiplying diagonally.

$$\frac{a}{b} = \frac{c}{d} \qquad bc \qquad ad$$

For this reason, ad and bc are called **cross products.**

If $\frac{a}{b} = \frac{c}{d}$, then the cross products ad and bc are equal.

Also, if $ad = bc$, then $\frac{a}{b} = \frac{c}{d}$ $(b, d \neq 0)$.

From this rule, if $\frac{a}{b} = \frac{c}{d}$ then $ad = bc$; that is, ***the product of the extremes equals the product of the means.***

ANSWERS

2. 36-oz size; \$.108 per oz

NOTE

If $\frac{a}{c} = \frac{b}{d}$, then $ad = cb$, or $ad = bc$. This means that the two proportions are equivalent, and

$$\text{the proportion } \frac{a}{b} = \frac{c}{d} \text{ can always be written as } \frac{a}{c} = \frac{b}{d}.$$

Sometimes one form is more convenient to work with than the other.

Four numbers are used in a proportion. If any three of these numbers are known, the fourth can be found.

EXAMPLE 3 Finding an Unknown in a Proportion

Solve the proportion.

$$\frac{5}{9} = \frac{x}{63}$$

$5 \cdot 63 = 9 \cdot x$ Cross products must be equal.

$315 = 9x$ Multiply.

$35 = x$ Divide by 9.

Check by substituting 35 for x in the proportion. The solution is 35.

Work Problem 3 at the Side.

CAUTION

The cross product method cannot be used directly if there is more than one term on either side.

EXAMPLE 4 Solving an Equation Using Cross Products

Solve the equation.

$$\frac{m-2}{5} = \frac{m+1}{3}$$

$3(m-2) = 5(m+1)$ Find the cross products; use parentheses.

$3m - 6 = 5m + 5$ Distributive property

$3m = 5m + 11$ Add 6.

$-2m = 11$ Subtract $5m$.

$m = -\frac{11}{2}$ Divide by -2.

Check that the solution is $-\frac{11}{2}$.

Work Problem 4 at the Side.

NOTE

When you set cross products equal to each other, you are really multiplying each ratio in the proportion by a common denominator.

3 Solve each proportion.

(a) $\frac{y}{6} = \frac{35}{42}$

(b) $\frac{a}{24} = \frac{15}{16}$

4 Solve each equation.

(a) $\frac{z}{2} = \frac{z+1}{3}$

(b) $\frac{p+3}{3} = \frac{p-5}{4}$

ANSWERS

3. (a) 5 (b) $\frac{45}{2}$

4. (a) 2 (b) -27

5 Solve the problem.

Twelve gal of diesel fuel costs \$20.88. How much would 16.5 gal of the same fuel cost?

OBJECTIVE 3 Solve applied problems using proportions. Proportions are useful in many practical applications. We continue to use the six-step method, although the steps are not numbered here.

EXAMPLE 5 Applying Proportions

After Lee Ann Spahr pumped 5.0 gal of gasoline, the display showing the price read \$9.40. When she finished pumping the gasoline, the price display read \$25.38. How many gallons did she pump?

To solve this problem, set up a proportion, with prices in the numerators and gallons in the denominators. Make sure that the corresponding numbers appear together.

Let $x =$ the number of gallons she pumped.

$$\begin{matrix}\text{Price} \rightarrow \\ \text{Gallons} \rightarrow\end{matrix} \frac{\$9.40}{5.0} = \frac{\$25.38}{x} \begin{matrix}\leftarrow \text{Price} \\ \leftarrow \text{Gallons}\end{matrix}$$

$$9.40x = 5.0(25.38) \quad \text{Cross products}$$

$$9.40x = 126.90 \quad \text{Multiply.}$$

$$x = 13.5 \quad \text{Divide by 9.40.}$$

She pumped a total of 13.5 gal. Check this answer. Notice that the way the proportion was set up uses the fact that the unit price is the same, no matter how many gallons are purchased.

Calculator Tip Using a calculator to perform the arithmetic in Example 5 reduces the possibility of errors.

Work Problem 5 at the Side.

OBJECTIVE 4 Find percentages and percents. Percents are ratios where the second number is always 100. For example, 50% represents the ratio of 50 to 100, 27% represents the ratio of 27 to 100, and so on. We can use the techniques for solving proportions to solve percent problems. Recall from **Section R.2** that the decimal point is moved two places to the left to change a percent to a decimal number. For example, 75% can be written as the decimal .75.

Calculator Tip Many calculators have a percent key that does this automatically.

We can solve a percent problem by writing it as the proportion

$$\frac{amount}{base} = \frac{percent}{100} \quad \text{or} \quad \frac{a}{b} = \frac{p}{100}.$$

The amount, or **percentage,** is compared to the **base** (the whole amount). Since ***percent* means *per 100,*** we compare the numerical value of the percent to 100. Thus, we write 50% as

$$\frac{p}{100} = \frac{50}{100}. \quad p = 50$$

ANSWERS

5. \$28.71

EXAMPLE 6 Finding Percentages

Solve each problem.

(a) Find 15% of 600.
Here, the base is 600, the percent is 15, and we must find the percentage.

$$\frac{a}{b} = \frac{p}{100}$$

$$\frac{a}{600} = \frac{15}{100}$$

$$100a = 600(15) \quad \text{Cross products}$$

$$a = \frac{600(15)}{100} \quad \text{Divide by 100.}$$

$$a = 90$$

Thus, 15% of 600 is 90.

(b) A DVD with a regular price of \$18 is on sale this week at 22% off. Find the amount of the discount and the sale price of the disc.
The discount is 22% of \$18. We want to find a, given b is 18 and p is 22.

$$\frac{a}{b} = \frac{p}{100}$$

$$\frac{a}{18} = \frac{22}{100}$$

$$100a = 18(22) \quad \text{Cross products}$$

$$100a = 396$$

$$a = 3.96 \quad \text{Divide by 100.}$$

The amount of the discount on the DVD is \$3.96, and the sale price is \$18.00 − \$3.96 = \$14.04.

Work Problem 6 at the Side.

6 Solve each problem.

(a) Find 20% of 70.

(b) Find the discount on a television set with a regular price of \$270 if the set is on sale at 25% off. Find the sale price of the set.

EXAMPLE 7 Solving an Applied Percent Problem

A newspaper ad offered a set of tires at a sale price of \$258. The regular price was \$300. What percent of the regular price was the savings?

The savings amounted to \$300 − \$258 = \$42. We can now restate the problem: What percent of 300 is 42? Substitute into the percent proportion. We have $a = 42$, $b = 300$, and p is to be found.

$$\frac{a}{b} = \frac{p}{100}$$

$$\frac{42}{300} = \frac{p}{100}$$

$$300p = 4200 \quad \text{Cross products}$$

$$p = 14 \quad \text{Divide by 300.}$$

The sale price represented a 14% savings.

Work Problem 7 at the Side.

7 Solve each problem.

(a) 90 is what percent of 360?

(b) The interest in 1 yr on deposits of \$11,000 was \$682. What percent interest was paid?

ANSWERS
6. **(a)** 14 **(b)** \$67.50; \$202.50
7. **(a)** 25% **(b)** 6.2%

Focus on

Real-Data Applications

Currency Exchange

When you travel between countries, you need to exchange your U.S. dollars (USD) for the local currency. The exchange rate between currencies changes daily, and you can easily find the updated rates using the Internet. The table shown here was taken from the Bloomberg Currency Calculator Web page.

WESTERN EUROPE CURRENCY RATES

Currency	Currency per 1 unit of USD	
	Symbol	Value
British Pound	GBP	.5456
Euro	EUR	.7968
Danish Krone	DKK	5.9373

Source: Bloomberg L.P.

On February 4, 2004, the currency exchange rate from U.S. dollars to British pounds was given as follows:

\$1.00 U.S. was equivalent to £.5456 (British pounds).

You can set up a proportion to convert dollars to British pounds. For example, suppose you want to determine how many British pounds is equivalent to \$50.

$$\frac{\$1}{£.5456} = \frac{\$50}{£x} \quad \text{or} \quad \frac{1}{.5456} = \frac{50}{x}$$

$$1(x) = .5456(50)$$

$$x = 27.28$$

So at this exchange rate, 27.28 British pounds equals 50 U.S. dollars.

For Group Discussion

1. Based on the currency exchange rates in the table above, find the amount of the local currency equivalent to \$50 U.S. and find the number of U.S. dollars equivalent to 200 units of the local currency. Round to the nearest hundredth as necessary.

   (a) \$50 = ________ Danish Krone and 200 Krone = ________ dollars

   (b) \$50 = ________ Euros and 200 Euros = ________ dollars

2. Set up a proportion to find the number of U.S. dollars equivalent to £1 (British pound).

   £1 (British) was equivalent to \$________ (U.S.).

3. From Problem 2, you should recognize the conversion rate based on £1 as the expression $\frac{1}{.5456}$. What is the mathematical term that describes the relationship between the conversion rates .5456 and $\frac{1}{.5456}$?

2.6 Exercises

FOR EXTRA HELP

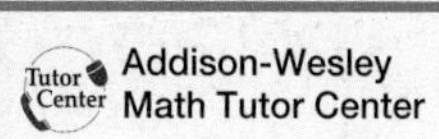
Addison-Wesley Math Tutor Center

MathXL

Digital Video Tutor CD 3 Videotape 3

Student's Solutions Manual

MyMathLab

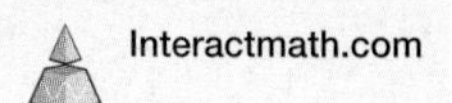
Interactmath.com

1. Match each ratio in Column I with the ratio equivalent to it in Column II.

I	II
(a) 75 to 100	**A.** 80 to 100
(b) 5 to 4	**B.** 50 to 100
(c) $\frac{1}{2}$	**C.** 3 to 4
(d) 4 to 5	**D.** 15 to 12

2. Which one of the following represents a ratio of 3 days to 2 weeks?

A. $\frac{3}{2}$ **B.** $\frac{3}{7}$

C. $\frac{1.5}{1}$ **D.** $\frac{3}{14}$

Write a ratio for each word phrase. In Exercises 7–10, first write the amounts with the same units. Write fractions in lowest terms. See Example 1.

3. 60 ft to 70 ft

4. 40 mi to 30 mi

5. 72 dollars to 220 dollars

6. 120 people to 90 people

7. 30 in. to 8 ft

8. 20 yd to 8 ft

9. 16 min to 1 hr

10. 24 min to 2 hr

Find the best buy (based on price per unit) for each item. Give the unit price to the nearest thousandth for that size. See Example 2. (Source: Cub Foods.)

11. GRANULATED SUGAR

Size	Price
4-lb	$1.78
10-lb	$4.39

12. GROUND COFFEE

Size	Price
13-oz	$2.58
39-oz	$4.44

13. SALAD DRESSING

Size	Price
16-oz	$2.44
32-oz	$2.98
48-oz	$4.95

14. BLACK PEPPER

Size	Price
2-oz	$1.79
4-oz	$2.59
8-oz	$5.59

15. VEGETABLE OIL

Size	Price
16-oz	$1.54
24-oz	$2.08
64-oz	$3.63
128-oz	$5.65

16. MOUTHWASH

Size	Price
8.5-oz	$.99
16.9-oz	$1.87
33.8-oz	$2.49
50.7-oz	$2.99

17. TOMATO KETCHUP

Size	Price
14-oz	$1.39
24-oz	$1.55
36-oz	$1.78
64-oz	$3.99

18. GRAPE JELLY

Size	Price
12-oz	$1.05
18-oz	$1.73
32-oz	$1.84
48-oz	$2.88

19. Explain how percent and ratio are related.

20. Explain the distinction between *ratio* and *proportion.*

Solve each equation. See Examples 3 and 4.

21. $\frac{k}{4} = \frac{175}{20}$

22. $\frac{x}{6} = \frac{18}{4}$

23. $\frac{49}{56} = \frac{z}{8}$

24. $\frac{20}{100} = \frac{z}{80}$

25. $\frac{z}{4} = \frac{z+1}{6}$

26. $\frac{m}{5} = \frac{m-2}{2}$

27. $\frac{3y-2}{5} = \frac{6y-5}{11}$

28. $\frac{2r+8}{4} = \frac{3r-9}{3}$

29. $\frac{5k+1}{6} = \frac{3k-2}{3}$

30. $\frac{x+4}{6} = \frac{x+10}{8}$

31. $\frac{2p+7}{3} = \frac{p-1}{4}$

32. $\frac{3m-2}{5} = \frac{4-m}{3}$

Solve each problem. See Example 5.

33. If 6 gal of premium unleaded gasoline cost \$11.34, how much would it cost to completely fill a 15-gal tank?

34. If sales tax on a \$16.00 compact disc is \$1.32, how much would the sales tax be on a \$120.00 compact disc player?

35. The distance between Kansas City, Missouri, and Denver is 600 mi. On a certain wall map, this is represented by a length of 2.4 ft. On the map, how many feet would there be between Memphis and Philadelphia, two cities that are actually 1000 mi apart?

36. The distance between Singapore and Tokyo is 3300 mi. On a certain wall map, this distance is represented by 11 in. The actual distance between Mexico City and Cairo is 7700 mi. How far apart are they on the same map?

37. As of December 2002, there were 19 vehicles for every 10 U.S. households. How many vehicles were there for the 107 million U.S. households? (*Source:* U.S. Transportation Department.)

38. For the first time, more vehicles are parked in the average American driveway than there are licensed drivers in the house. If there are 18 drivers for every 10 U.S. households, how many drivers are there in the 107 million U.S. households? (*Source:* U.S. Transportation Department.)

39. A chain saw requires a mixture of 2-cycle engine oil and gasoline. According to the directions on a bottle of Oregon 2-cycle Engine Oil, for a 50 to 1 ratio requirement, approximately 2.5 fluid oz of oil are required for 1 gal of gasoline. For 2.75 gal, how many fluid ounces of oil are required?

40. The directions on the bottle mentioned in Exercise 39 indicate that if the ratio requirement is 24 to 1, approximately 5.5 oz of oil are required for 1 gal of gasoline. If gasoline is to be mixed with 22 oz of oil, how much gasoline is to be used?

41. On February 4, 2004, the exchange rate between British pounds and U.S. dollars was 1 pound to \$1.8329. Margaret went to London and exchanged her U.S. currency for British pounds, and received 400 pounds. How much in U.S. dollars did Margaret exchange? (*Source:* Bloomberg L.P.)

42. If 3 U.S. dollars can be exchanged for 3.75 Swiss francs, how many Swiss francs can be obtained for \$49.50? Round to the nearest hundredth. (*Source:* Bloomberg L.P.)

Answer each question about percent. See Example 6.

43. What is 48.6% of 19?

44. What is 26% of 480?

45. What percent of 48 is 96?

46. What percent of 30 is 36?

47. 12% of what number is 3600?

48. 25% of what number is 150?

49. 78.84 is what percent of 292?

50. .392 is what percent of 28?

Use mental techniques to answer the questions in Exercises 51 and 52.

51. The 2000 U.S. Census showed that the population of Alabama was 4,447,000, with 26.0% represented by African-Americans. What is the best estimate of the African-American population in Alabama? (*Source:* U.S. Bureau of the Census.)

A. 500,000 **B.** 750,000
C. 1,000,000 **D.** 1,500,000

52. The 2000 U.S. Census showed that the population of New Mexico was 1,819,000, with 42.1% being Hispanic. What is the best estimate of the Hispanic population of New Mexico? (*Source:* U.S. Bureau of the Census.)

A. 720,000 **B.** 72,000
C. 650,000 **D.** 36,000

Work each problem. Round all money amounts to the nearest dollar and percents to the nearest tenth. See Examples 6 and 7.

53. In 2002, the U.S. civilian labor force consisted of 144,863,000 persons. Of this total, 8,378,000 were unemployed. What was the percent of unemployment? (*Source:* U.S. Bureau of Labor Statistics.)

54. In 2002, the U.S. labor force (excluding agricultural employees, self-employed persons, and the unemployed) consisted of 122,007,000 persons. Of this total, 16,107,000 were union members. What percent of this labor force belonged to unions? (*Source:* U.S. Bureau of Labor Statistics.)

55. As of June 2003, there were 844,000 eateries in the United States. Of these, 177,000 served fast food. What percent of the eateries were *not* fast food establishments? (*Source:* National Restaurant Association.)

56. During 1999 and 2000, the total public and private school enrollment in the United States was 52,885,000. Of this total, 11.4% of the students were enrolled in private schools. How many students were enrolled in public schools? (*Source:* National Center for Education Statistics.)

57. A family of four with a monthly income of \$3800 plans to spend 8% of this amount on entertainment. How much will be spent on entertainment?

58. Quinhon Dac Ho earns \$3200 per month. He wants to save 12% of this amount. How much will he save?

59. The 1916 dime minted in Denver is quite rare. The 1979 edition of *A Guide Book of United States Coins* listed its value in Extremely Fine condition as \$625. The 2003 value had increased to \$2400. What was the percent increase in the value of this coin?

60. Here is a common business problem. If the sales tax rate is 6.5% and I have collected \$3400 in sales tax, how much were my sales?

The Consumer Price Index, issued by the U.S. Bureau of Labor Statistics, provides a means of determining the purchasing power of the U.S. dollar from one year to the next. Using the period from 1982 to 1984 as a measure of 100.0, *the Consumer Price Index for selected years from 1991 to 2003 is shown here. To use the Consumer Price Index to predict a price in a particular year, we can set up a proportion and compare it with a known price in another year, as follows:*

$$\frac{\text{price in year } A}{\text{index in year } A} = \frac{\text{price in year } B}{\text{index in year } B}.$$

Year	Consumer Price Index
1991	136.2
1993	144.5
1995	152.4
1997	160.5
1999	166.6
2001	177.1
2003	183.9

Source: U.S. Bureau of Labor Statistics.

Use the Consumer Price Index figures in the table to find the amount that would be charged for the use of the same amount of electricity that cost \$225 in 1991. Give your answer to the nearest dollar.

61. in 1995 **62.** in 1999 **63.** in 2001 **64.** in 2003

RELATING CONCEPTS (EXERCISES 65–68) For Individual or Group Work

In **Section 2.3** *we solved equations with fractions by first multiplying each side of the equation by the common denominator. A proportion with a variable is this kind of equation.* ***Work Exercises 65–68 in order.*** *The steps justify the method of solving a proportion by cross products.*

65. What is the LCD of the fractions in the equation $\frac{x}{6} = \frac{2}{5}$?

66. Solve the equation in Exercise 65 as follows.

(a) Multiply each side by the LCD. What equation do you get?

(b) Solve the equation from part (a) by dividing each side by the coefficient of x.

67. Solve the equation in Exercise 65 using cross products.

68. Compare your solutions from Exercises 66 and 67. What do you notice?

Summary Exercises on Solving Applied Problems

The following problems are of the various types discussed in this chapter. Solve each problem. The problem-solving steps from ***Section 2.4*** *are repeated here.*

Solving an Applied Problem

Step 1 **Read** the problem, several times if necessary, until you *understand* what is given and what is to be found.

Step 2 **Assign a variable** to represent the unknown value, using diagrams or tables as needed. Write down what the variable represents. Express any other unknown values in terms of the variable.

Step 3 **Write an equation** using the variable expression(s).

Step 4 **Solve** the equation.

Step 5 **State the answer.** Does it seem reasonable?

Step 6 **Check** the answer in the words of the *original* problem.

1. Nevaraz and Smith were opposing candidates in the school board election. Nevaraz received 30 more votes than Smith, with 516 votes cast. How many votes did Smith receive?

2. On an algebra test, the highest grade was 42 points more than the lowest grade. The sum of the two grades was 138. Find the lowest grade.

3. A certain lawn mower uses 3 tanks of gas to cut 10 acres of lawn. How many tanks of gas would be needed for 30 acres?

4. If 2 lb of fertilizer will cover 50 ft^2 of garden, how many pounds would be needed for 225 ft^2?

5. The perimeter of a certain square is seven times the length of a side, decreased by 12. Find the length of a side.

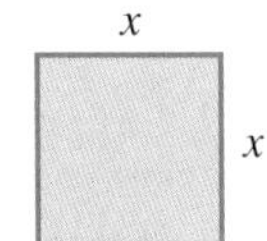

6. The perimeter of a certain rectangle is 16 times the width. The length is 12 cm more than the width. Find the width of the rectangle.

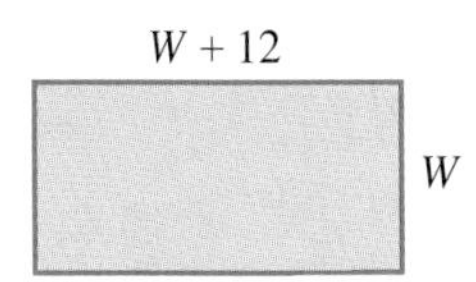

7. Find the measure of an angle whose measure is 70° more than its complement.

8. Find the measure of an angle whose measure is 20° more than its supplement.

9. If 2 is added to five times a number, the result is equal to 5 more than four times the number. Find the number.

10. If four times a number is added to 8, the result is three times the number added to 5. Find the number.

11. The smallest of three consecutive integers is added to twice the largest, producing a result 15 less than four times the middle integer. Find the smallest integer.

12. If the middle of three consecutive even integers is added to 100, the result is 42 more than the sum of the largest integer and twice the smallest. Find the smallest integer.

13. A store has 39 qt of milk, some in pint cartons and some in quart cartons. There are six times as many quart cartons as pint cartons. How many quart cartons are there? (*Hint:* 1 qt $=$ 2 pt.)

14. A rectangular table is three times as long as it is wide. If it were 3 ft shorter and 3 ft wider, it would be square (with all sides equal). How long and how wide is it?

15. Find the measures of the marked angles.

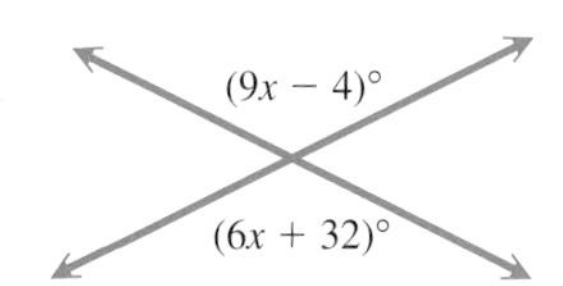

16. Find the measures of the marked angles.

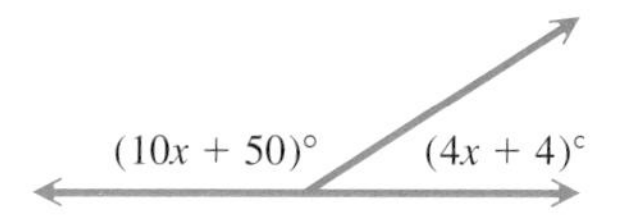

17. Grant Wood painted his most famous work, *American Gothic,* in 1930 on composition board with perimeter 108.44 in. If the width of the painting is 29.88 in., find the length. (*Source: The Gazette*, March 12, 2004.)

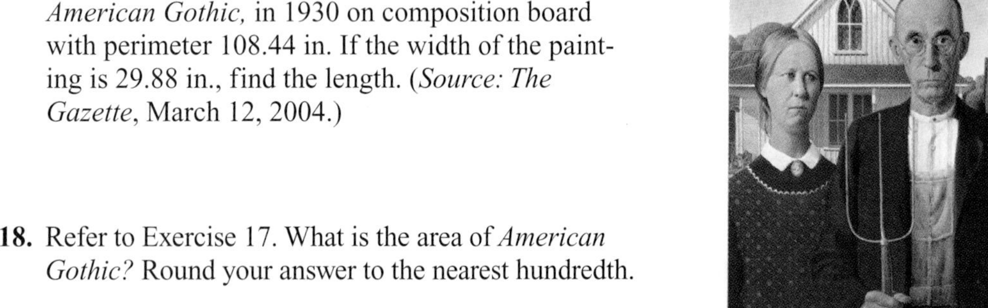

18. Refer to Exercise 17. What is the area of *American Gothic?* Round your answer to the nearest hundredth.

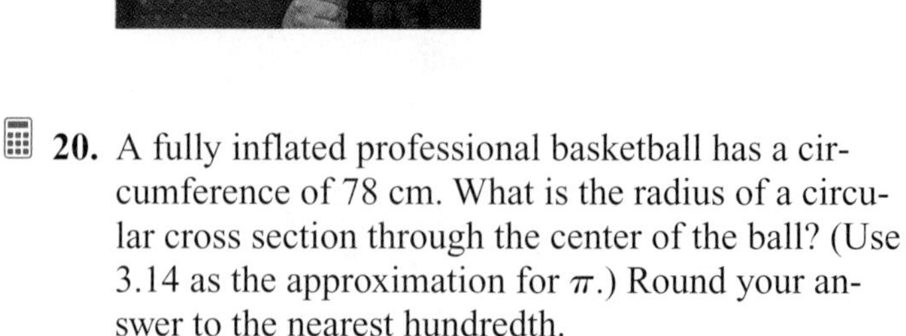

19. If the radius of a certain circle is tripled, and then 8.2 cm are added, the result is the circumference of the circle. Find the radius of the circle. (Use 3.14 as the approximation for π.) Round your answer to the nearest tenth.

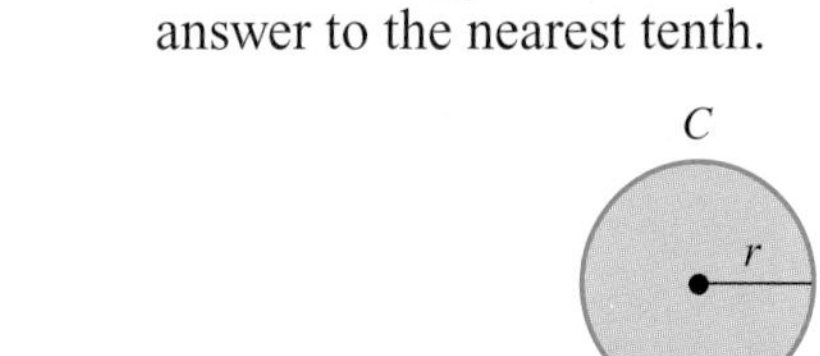

20. A fully inflated professional basketball has a circumference of 78 cm. What is the radius of a circular cross section through the center of the ball? (Use 3.14 as the approximation for π.) Round your answer to the nearest hundredth.

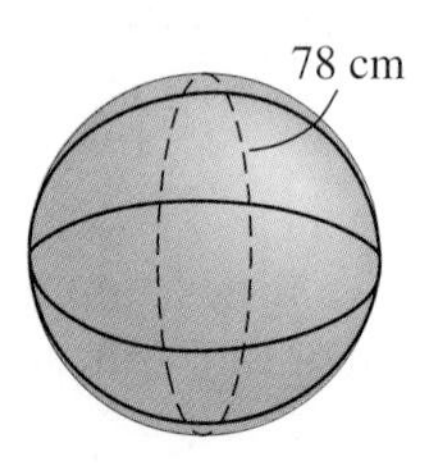

21. Mr. Silvester is 5 yr older than his wife. Five years ago his age was $\frac{4}{3}$ her age. What are their ages now?

22. Chris is 10 yr older than Josh. Next year, Chris will be twice as old as Josh. What are their ages now?

23. An MP-3 player that normally sells for $180 is on sale for $150. What is the percent discount on the player?

24. A grocer marks up cereal boxes by 20%. If a box of Raisin Bran costs the grocer $2.49, what is the grocer's selling price to the nearest cent?

25. Athletes in vigorous training programs can eat 50 calories per day for every 2.2 lb of body weight. To the nearest hundred, how many calories can a 175 lb athlete consume per day? (*Source: The Gazette*, March 23, 2002.)

26. The distance between two cities on a road map is 11 in. The cities are actually 308 mi apart. The distance between two other cities on the map is 15 in. What is the actual distance between those cities?

27. The fountain sculpture *Spoonbridge and Cherry* weighs about 7000 lb. The spoon weighs 200 lb less than five times the weight of the cherry. How much does each part of the sculpture weigh? (*Source:* Minneapolis Sculpture Garden brochure.)

28. In the 2003 Masters Golf Tournament in Augusta, Georgia, Tiger Woods finished with a score of 9 more than the winner, Mike Weir. The sum of their scores was 571. Find their scores. (*Source: World Almanac and Book of Facts*, 2004.)

29. Two slices of bacon contain 85 calories. How many calories are there in twelve slices of bacon?

30. Three ounces of liver contain 22 g of protein. How many ounces of liver provide 121 g of protein?

31. The average cost of a traditional Thanksgiving dinner for 10 people, featuring turkey, stuffing, cranberries, pumpkin pie, and trimmings, was $34.56 in 2002. This price increased 5.0% in 2003. What was the price of this traditional Thanksgiving dinner in 2003? Round your answer to the nearest cent. (*Source:* American Farm Bureau.)

32. Refer to Exercise 31. The cost of a traditional Thanksgiving dinner in 1987, the year this information was first collected, was $26.24. What percent increase is the 2003 cost over the 1987 cost? Round your answer to the nearest tenth. (*Source:* American Farm Bureau.)

33. According to *The Guinness Book of World Records,* the longest recorded voyage in a paddleboat is 2226 mi in 103 days; the boat was propelled down the Mississippi River by the foot power of two boaters. Assuming a constant rate, how far would they have gone in 120 days? Round your answer to the nearest mile.

34. In the 2000 Summer Olympics held in Sydney, Australia, Russian athletes earned 88 medals. Four of every 11 medals were gold. How many gold medals did Russia earn? (*Source: Times Picayune,* October 2, 2000.)

In Exercises 35 and 36, find the best buy based on unit pricing. Give the unit price to the nearest thousandth for that size. (*Source:* Cub Foods.)

35. SPAGHETTI SAUCE

Size	Price
$15\frac{1}{2}$-oz	$1.19
32-oz	$1.69
48-oz	$2.69

36. DISH SOAP

Size	Price
12.6-oz	$1.39
25-oz	$2.44
50-oz	$4.68

2.7 Solving Linear Inequalities

Inequalities are statements with algebraic expressions related by

$<$ "is less than," $\leq$ "is less than or equal to,"

$>$ "is greater than," $\geq$ "is greater than or equal to."

We solve an inequality by finding all real number solutions for it. For example, the solutions of $x \leq 2$ include all *real numbers* that are less than or equal to 2, not just the *integers* less than or equal to 2.

OBJECTIVES

1. Graph the solutions of inequalities on a number line.
2. Use the addition property of inequality.
3. Use the multiplication property of inequality.
4. Solve inequalities using both properties of inequality.
5. Use inequalities to solve applied problems.

OBJECTIVE 1 Graph the solutions of inequalities on a number line. Graphing is a good way to show the solutions of an inequality. To graph all real numbers satisfying $x \leq 2$, we place a closed circle at 2 on a number line and draw an arrow extending from the closed circle to the left (to represent the fact that all numbers less than 2 are also part of the graph). The graph is shown in Figure 12.

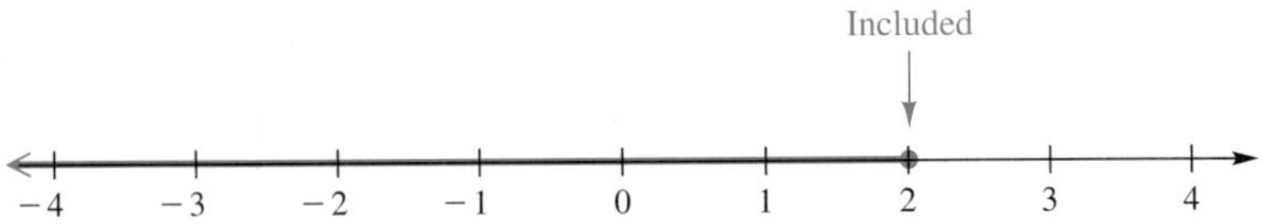

Figure 12

EXAMPLE 1 Graphing the Solutions of an Inequality

Graph $x > -5$.

The statement $x > -5$ says that x can take any value greater than -5, but x cannot equal -5 itself. We show this on a graph by placing an open circle at -5 and drawing an arrow to the right, as in Figure 13. The open circle at -5 shows that -5 is not part of the graph.

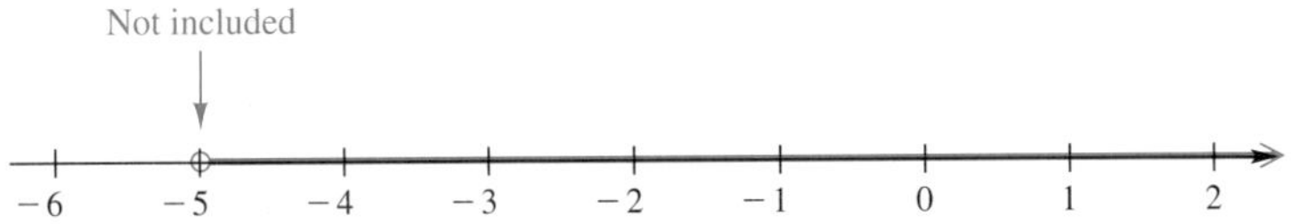

Figure 13

EXAMPLE 2 Graphing the Solutions of an Inequality

Graph $3 > x$.

The statement $3 > x$ means the same as $x < 3$. The graph of $x < 3$ is shown in Figure 14.

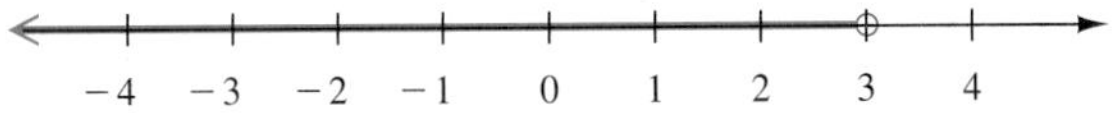

Figure 14

NOTE

To graph an inequality like the one in Example 2, first rewrite it with the variable on the left. Fewer errors occur this way.

Work Problem 1 at the Side.

1 Graph each inequality.

(a) $x \leq 3$

(b) $x > -4$

(c) $-4 \geq x$

(d) $0 < x$

ANSWERS

1. (a)

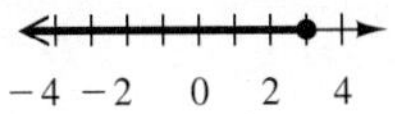

(b)

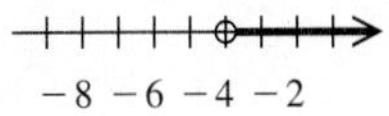

(c)

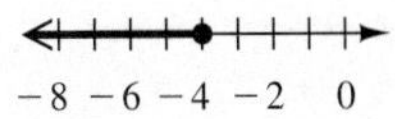

(d)

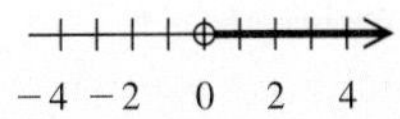

2 Graph each inequality.

(a) $-7 < x < -2$

(b) $-6 < x \leq -4$

EXAMPLE 3 Graphing the Solutions of an Inequality

Graph $-3 \leq x < 2$.

The statement $-3 \leq x < 2$ is read "-3 is less than or equal to x *and* x is less than 2." We graph the solutions of this inequality by placing a closed circle at -3 (because -3 is part of the graph) and an open circle at 2 (because 2 is not part of the graph), then drawing a line segment between the two circles. Notice that the graph includes all points *between* -3 and 2 and includes -3 as well. See Figure 15.

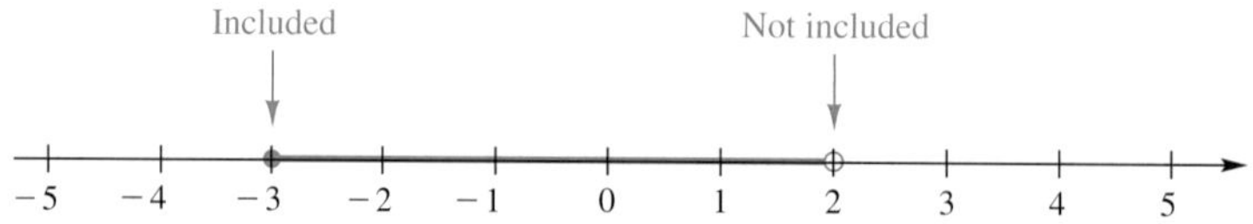

Figure 15

Work Problem 2 at the Side.

OBJECTIVE 2 Use the addition property of inequality. Solving inequalities is similar to solving equations.

Linear Inequality in One Variable

A **linear inequality in one variable** can be written in the form

$$Ax + B < C,$$

where A, B, and C are real numbers, with $A \neq 0$.

Examples of linear inequalities in one variable include

$x + 5 < 2$, $t - 3 \geq 5$, and $2k + 5 \leq 10$. Linear inequalities

(All definitions and rules are also valid for $>$, $\leq$, and $\geq$.)

Consider the inequality $2 < 5$. If 4 is added to each side of this inequality, the result is

$$2 + 4 < 5 + 4$$
$$6 < 9,$$

a true sentence. Now subtract 8 from each side:

$$2 - 8 < 5 - 8$$
$$-6 < -3.$$

The result is again a true sentence. These examples suggest the **addition property of inequality.**

Addition Property of Inequality

For any real numbers A, B, and C, the inequalities

$$A < B \quad \text{and} \quad A + C < B + C$$

have exactly the same solutions.

In words, the same number may be added to each side of an inequality without changing the solutions.

As with the addition property of equality, the same number may be *subtracted* from each side of an inequality.

ANSWERS

2. (a)

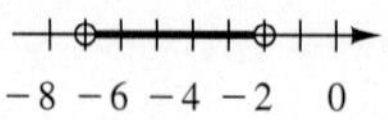

(b)

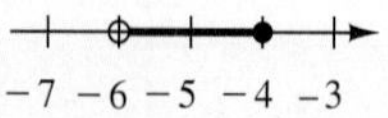

EXAMPLE 4 **Using the Addition Property of Inequality**

Solve $7 + 3k > 2k - 5$.

$$7 + 3k > 2k - 5$$

$$7 + 3k - \mathbf{2k} > 2k - 5 - \mathbf{2k} \quad \text{Subtract } 2k.$$

$$7 + k > -5 \quad \text{Combine terms.}$$

$$7 + k - \mathbf{7} > -5 - \mathbf{7} \quad \text{Subtract 7.}$$

$$k > -12 \quad \text{Combine terms.}$$

A graph of the solutions, $k > -12$, is shown in Figure 16.

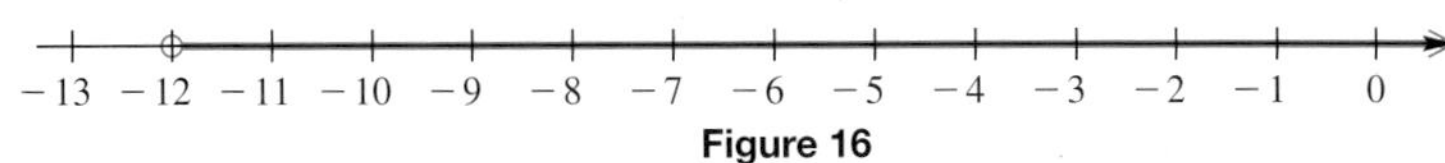

Figure 16

Work Problem 3 at the Side.

3 Solve each inequality, and graph the solutions.

(a) $-1 + 8r < 7r + 2$

(b) $5m - \frac{4}{3} \leq 4m$

OBJECTIVE 3 **Use the multiplication property of inequality.** The addition property of inequality alone cannot be used to solve inequalities such as $4x \geq 28$. These inequalities require the *multiplication property of inequality*. To see how this property works, we look at some examples.

Multiply each side of the inequality $3 < 7$ by the positive number 2.

$$3 < 7$$

$$\mathbf{2}(3) < \mathbf{2}(7) \quad \text{Multiply by 2.}$$

$$6 < 14 \quad \text{True}$$

Now multiply each side of $3 < 7$ by the negative number -5.

$$3 < 7$$

$$\mathbf{-5}(3) < \mathbf{-5}(7) \quad \text{Multiply by } -5.$$

$$-15 < -35 \quad \text{False}$$

To get a true statement when multiplying each side by -5, we must reverse the direction of the inequality symbol.

$$3 < 7$$

$$\mathbf{-5}(3) \mathbf{>} \mathbf{-5}(7) \quad \text{Multiply by } -5\text{; reverse the symbol.}$$

$$-15 > -35 \quad \text{True}$$

Take the inequality $-6 < 2$ as another example. Multiply each side by the positive number 4.

$$-6 < 2$$

$$\mathbf{4}(-6) < \mathbf{4}(2) \quad \text{Multiply by 4.}$$

$$-24 < 8 \quad \text{True}$$

Multiplying each side of $-6 < 2$ by -5 ***and at the same time reversing the direction of the inequality symbol*** gives

$$-6 < 2$$

$$\mathbf{-5}(-6) \mathbf{>} \mathbf{-5}(2) \quad \text{Multiply by } -5\text{; reverse the symbol.}$$

$$30 > -10. \quad \text{True}$$

Work Problem 4 at the Side.

4 **(a)** Multiply each side of $-2 < 8$ by 6 and then by -5. Reverse the direction of the inequality symbol if necessary to make a true statement.

(b) Multiply each side of $-4 > -9$ by 2 and then by -8. Reverse the direction of the inequality symbol if necessary to make a true statement.

ANSWERS

3. (a) $r < 3$

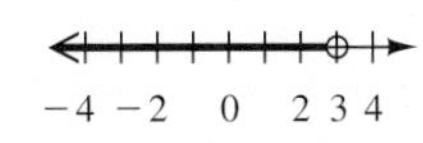

(b) $m \leq \frac{4}{3}$

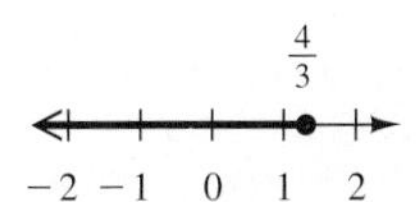

4. (a) $-12 < 48$; $10 > -40$
(b) $-8 > -18$; $32 < 72$

In summary, the **multiplication property of inequality** has two parts.

Multiplication Property of Inequality

For any real numbers A, B, and C $(C \neq 0)$,

1. if C is *positive,* then the inequalities

$$A < B \quad \text{and} \quad AC < BC$$

   have exactly the same solutions;

2. if C is *negative,* then the inequalities

$$A < B \quad \text{and} \quad AC > BC$$

   have exactly the same solutions.

In words, each side of an inequality may be multiplied by the same positive number without changing the solutions. ***If the multiplier is negative, we must reverse the direction of the inequality symbol.***

We can replace $<$ in the multiplication property of inequality with $>$, $\leq$, or $\geq$. As with the multiplication property of equality, the same nonzero number may be divided into each side.

It is important to remember the differences in the multiplication property for positive and negative numbers.

1. When each side of an inequality is multiplied or divided by a positive number, the direction of the inequality symbol *does not change.* (Adding or subtracting terms on each side also does not change the symbol.)
2. When each side of an inequality is multiplied or divided by a negative number, the direction of the symbol *does change.* ***Reverse the direction of the inequality symbol only when multiplying or dividing each side by a negative number.***

EXAMPLE 5 **Using the Multiplication Property of Inequality**

Solve $3r < -18$.

Using the multiplication property of inequality, we divide each side by 3. Since 3 is a positive number, the direction of the inequality symbol *does not* change. ***It does not matter that the number on the right side of the inequality is negative.***

$$3r < -18$$

$$\frac{3r}{3} < \frac{-18}{3} \quad \text{Divide by 3.}$$

$$r < -6$$

The graph of the solutions is shown in Figure 17.

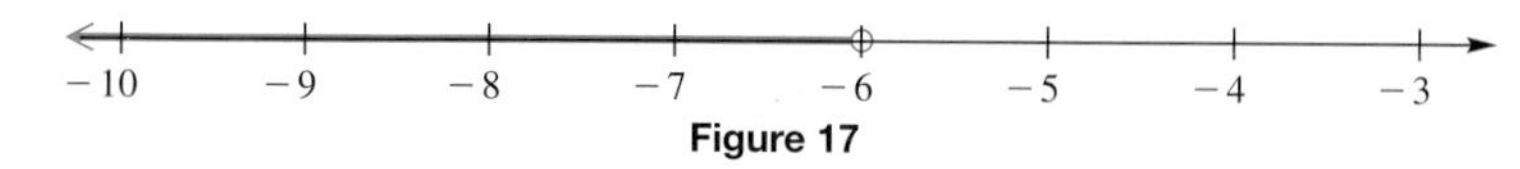

Figure 17

EXAMPLE 6 **Using the Multiplication Property of Inequality**

Solve $-4t \geq 8$.

Here each side of the inequality must be divided by -4, a negative number, which *does* require changing the direction of the inequality symbol.

$$-4t \geq 8$$

$$\frac{-4t}{-4} \leq \frac{8}{-4} \quad \text{Divide by } -4\text{; reverse the symbol.}$$

$$t \leq -2$$

The solutions are graphed in Figure 18.

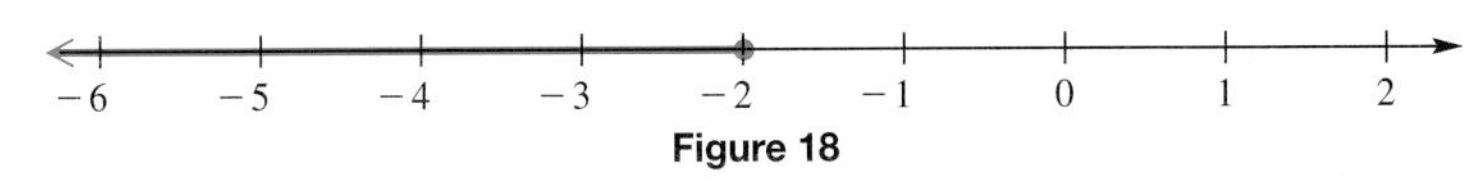

Figure 18

Work Problem 5 at the Side.

5 Solve each inequality. Graph the solutions.

(a) $9x < -18$

(b) $-2r > -12$

(c) $-5p \leq 0$

OBJECTIVE 4 **Solve inequalities using both properties of inequality.** The steps to solve an inequality are summarized below.

Solving Inequalities

Step 1 **Simplify each side separately.** Use the distributive property to clear parentheses and combine terms on each side as needed.

Step 2 **Isolate the variable term on one side.** Use the addition property to get all terms with variables on one side of the inequality and all numbers on the other side.

Step 3 **Isolate the variable.** Use the multiplication property to write the inequality in the form $x < c$ or $x > c$.

EXAMPLE 7 **Solving an Inequality**

Solve $5(k-3) - 7k \geq 4(k-3) + 9$.

Step 1 Clear parentheses; then combine like terms.

$$5(k-3) - 7k \geq 4(k-3) + 9$$

$$5k - 15 - 7k \geq 4k - 12 + 9 \quad \text{Distributive property}$$

$$-2k - 15 \geq 4k - 3 \quad \text{Combine terms.}$$

Step 2 Use the addition property.

$$-2k - 15 - 4k \geq 4k - 3 - 4k \quad \text{Subtract } 4k.$$

$$-6k - 15 \geq -3 \quad \text{Combine terms.}$$

$$-6k - 15 + 15 \geq -3 + 15 \quad \text{Add 15.}$$

$$-6k \geq 12 \quad \text{Combine terms.}$$

Step 3 Divide each side by -6, a negative number. Change the direction of the inequality symbol.

$$\frac{-6k}{-6} \leq \frac{12}{-6} \quad \text{Divide by } -6\text{; reverse the symbol.}$$

$$k \leq -2$$

Continued on Next Page

ANSWERS

5. (a) $x < -2$

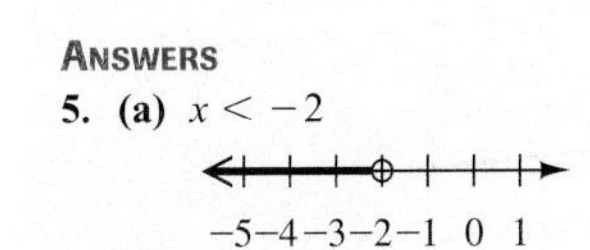

(b) $r < 6$

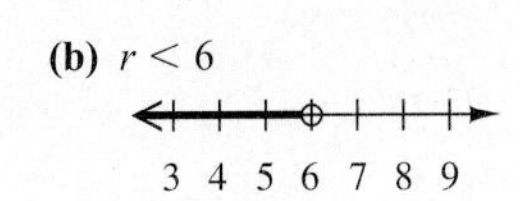

(c) $p \geq 0$

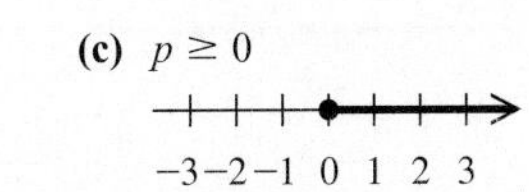

6 Solve each inequality. Graph the solutions.

(a) $5r - r + 2 < 7r - 5$

(b) $4(x - 1) - 3x > -15 - (2x + 1)$

7 Solve the problem.

Maggie has scores of 98, 86, and 88 on her first three tests in algebra. If she wants an average of at least 90 after her fourth test, what score must she make on her fourth test?

ANSWERS

6. (a) $r > \frac{7}{3}$

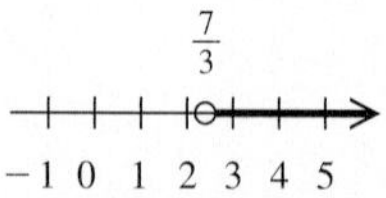

−1 0 1 2 3 4 5

(b) $x > -4$

−6 −5 −4 −3 −2 −1 0 1

7. 88 or more

A graph of the solutions is shown in Figure 19.

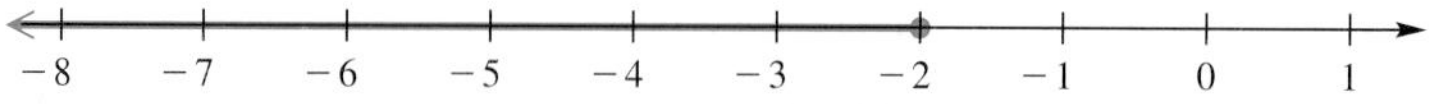

−8 −7 −6 −5 −4 −3 −2 −1 0 1

Figure 19

Work Problem 6 at the Side.

OBJECTIVE 5 Use inequalities to solve applied problems. The chart below gives some of the more common phrases that suggest inequality, along with examples and translations.

Phrase	Example	Inequality
Is greater than	A number *is greater than* 4	$x > 4$
Is less than	A number *is less than* −12	$x < -12$
Is at least	A number *is at least* 6	$x \geq 6$
Is at most	A number *is at most* 8	$x \leq 8$

CAUTION

Do not confuse phrases like "5 less than a number" and statements like "5 *is* less than a number." The first of these is expressed as $x - 5$ while the second is expressed as $5 < x$.

The next application is relevant to anyone who has ever asked, "What score can I make on my next test and have a (particular grade) in this course?" It uses the idea of finding the average of a number of grades. In general, to find the average of n numbers, add the numbers and divide by n.

EXAMPLE 8 Finding an Average Test Score

Brent has test grades of 86, 88, and 78 on his first three tests in geometry. If he wants an average of at least 80 after his fourth test, what score must he make on his fourth test?

Let x represent Brent's score on his fourth test. To find the average of the four scores, add them and divide by 4.

Average ↓ is at least 80. ↓ ↓

$$\frac{86 + 88 + 78 + x}{4} \geq 80$$

$$4\left(\frac{252 + x}{4}\right) \geq 4(80) \quad \text{Add in the numerator; multiply by 4.}$$

$$252 + x \geq 320$$

$$252 - \mathbf{252} + x \geq 320 - \mathbf{252} \quad \text{Subtract 252.}$$

$$x \geq 68 \quad \text{Combine terms.}$$

He must score 68 or more on the fourth test to have an average of *at least* 80.

Work Problem 7 at the Side.

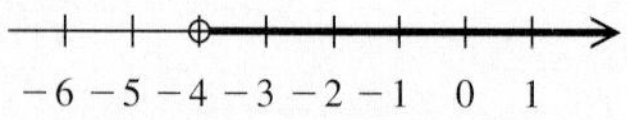

2.7 Exercises

FOR EXTRA HELP Addison-Wesley Math Tutor Center | MathXL | Digital Video Tutor CD 3 Videotape 3 | Student's Solutions Manual | MyMathLab | Interactmath.com

Write an inequality using the variable x that corresponds to each graph of solutions on a number line.

1. −4 −3 −2 −1 0 1 2 3

2. −4 −3 −2 −1 0 1 2 3 4

3. −2 −1 0 1 2 3 4 5

4. −2 −1 0 1 2 3 4 5

5. −1 0 1 2

6.

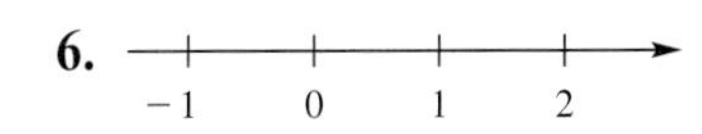

7. −1 0 1 2

8.

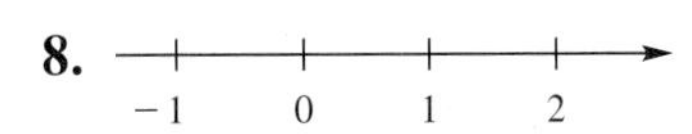

9. How can you determine whether to use an open circle or a closed circle at an endpoint when graphing an inequality on a number line?

10. How does the graph of $t \geq -7$ differ from the graph of $t > -7$?

Graph each inequality on the given number line. See Examples 1–3.

11. $k \leq 4$

12. $r \leq -11$

13. $x > -3$

14. $x > 3$

15. $8 \leq x \leq 10$

16. $3 \leq x \leq 5$

17. $0 < x \leq 10$

18. $-3 \leq x < 5$

19. Why is it *wrong* to write $3 < x < -2$ to indicate that x is between -2 and 3?

20. Your friend tells you that when solving the inequality $6x < -42$, he reversed the direction of the inequality because of the presence of -42. How would you respond?

Solve each inequality, and graph the solutions. See Example 4.

21. $z - 8 \geq -7$

22. $p - 3 \geq -11$

23. $2k + 3 \geq k + 8$

24. $3x + 7 \geq 2x + 11$

25. $3n + 5 < 2n - 6$

26. $5x - 2 < 4x - 5$

27. Under what conditions must the inequality symbol be reversed when using the multiplication property of inequality?

28. Explain the steps you would use to solve the inequality $-5x > 20$.

Solve each inequality, and graph the solutions. See Examples 5 and 6.

29. $3x < 18$

30. $5x < 35$

31. $2x \geq -20$

32. $6m \geq -24$

33. $-8t > 24$

34. $-7x > 49$

35. $-x \geq 0$

36. $-k < 0$

37. $-\frac{3}{4}r < -15$

38. $-\frac{7}{8}t < -14$

39. $-.02x \leq .06$

40. $-.03v \geq -.12$

Solve each inequality, and graph the solutions. See Example 7.

41. $5r + 1 \geq 3r - 9$

42. $6t + 3 < 3t + 12$

43. $6x + 3 + x < 2 + 4x + 4$

44. $-4w + 12 + 9w \geq w + 9 + w$

45. $-x + 4 + 7x \leq -2 + 3x + 6$

46. $14x - 6 + 7x > 4 + 10x - 10$

47. $5(x + 3) - 6x \leq 3(2x + 1) - 4x$

48. $2(x - 5) + 3x < 4(x - 6) + 1$

49. $\frac{2}{3}(p + 3) > \frac{5}{6}(p - 4)$

50. $\frac{7}{9}(n - 4) \leq \frac{4}{3}(n + 5)$

51. $4x - (6x + 1) \leq 8x + 2(x - 3)$

52. $2x - (4x + 3) < 6x + 3(x + 4)$

53. $5(2k + 3) - 2(k - 8) > 3(2k + 4) + k - 2$

54. $2(3z - 5) + 4(z + 6) \geq 2(3z + 2) + 3z - 15$

Solve each application of inequalities. See Example 8.

55. John Douglas has grades of 84 and 98 on his first two history tests. What must he score on his third test so that his average is at least 90?

56. Elizabeth Gainey has scores of 74 and 82 on her first two algebra tests. What must she score on her third test so that her average is at least 80?

57. When 2 is added to the difference between six times a number and 5, the result is greater than 13 added to 5 times the number. Find all such numbers.

58. When 8 is subtracted from the sum of three times a number and 6, the result is less than 4 more than the number. Find all such numbers.

59. The formula for converting Celsius temperature to Fahrenheit is

$$F = \frac{9}{5}C + 32.$$

The Fahrenheit temperature of Providence, Rhode Island, has never exceeded 104°. How would you describe this using Celsius temperature?

60. The formula for converting Fahrenheit temperature to Celsius is

$$C = \frac{5}{9}(F - 32).$$

If the Celsius temperature on a certain day in San Diego, California, is never more than 25°, how would you describe the corresponding Fahrenheit temperature?

61. For what values of x would the rectangle have perimeter of at least 400?

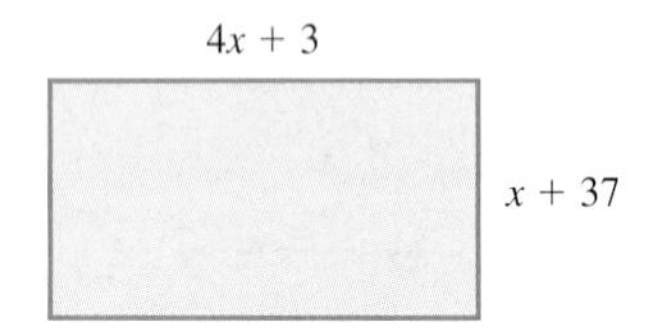

62. For what values of x would the triangle have perimeter of at least 72?

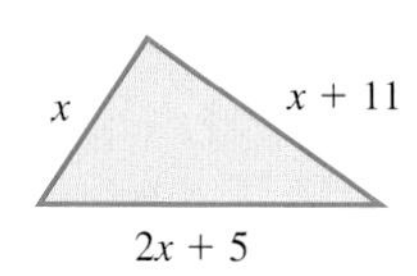

63. A long-distance phone call costs \$2.00 for the first three minutes plus \$.30 per minute for each minute or fractional part of a minute after the first three minutes. If x represents the number of minutes of the length of the call after the first three minutes, then $2 + .30x$ represents the cost of the call. If Jorge has \$5.60 to spend on a call, what is the maximum total time he can use the phone?

64. At the Speedy Gas 'n Go, a car wash costs \$3.00, and gasoline is selling for \$1.50 per gal. Terri Hoelker has \$17.25 to spend, and her car is so dirty that she must have it washed. What is the maximum number of gallons of gasoline that she can purchase?

RELATING CONCEPTS (EXERCISES 65–69) For Individual or Group Work

Work Exercises 65–69 in order, *to see the connection between the solution of an equation and the solutions of the corresponding inequalities. Graph the solutions in Exercises 65–67.*

65. $3x + 2 = 14$

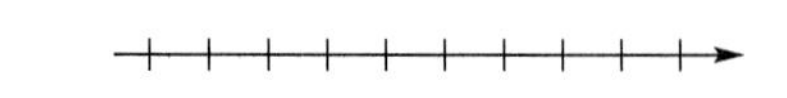

66. $3x + 2 < 14$

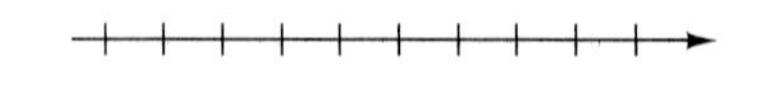

67. $3x + 2 > 14$

68. Now graph all the solutions together on the following number line.

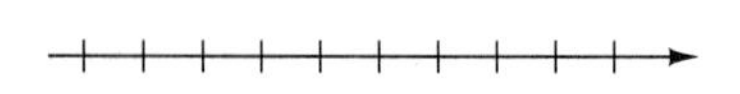

How would you describe the graph?

69. Based on your results from Exercises 65–68, if you were to graph the solutions of

$$-4x + 3 = -1, \quad -4x + 3 > -1,$$
$$\text{and} \quad -4x + 3 < -1$$

on the same number line, what do you think the graph would be?

Chapter 2

SUMMARY

KEY TERMS

Section	Term	Definition
2.1	**linear equation**	A linear equation in one variable is an equation that can be written in the form $Ax + B = C$, where A, B, and C are real numbers, with $A \neq 0$.
	equivalent equations	Equations that have the same solutions are equivalent equations.
2.3	**identity**	An identity is an equation that is true for all replacements of the variable.
2.4	**complementary angles**	Two angles whose measures have a sum of 90° are complementary angles.
	right angle	A right angle measures 90°.
	supplementary angles	Two angles whose measures have a sum of 180° are supplementary angles.
	straight angle	A straight angle measures 180°.
	consecutive integers	Two integers that differ by 1 are consecutive integers.
2.5	**formula**	A formula is a mathematical expression in which variables are used to describe a relationship.
	area	The area of a plane geometric figure is a measure of the surface covered by the figure.
	perimeter	The perimeter of a plane geometric figure is the distance around the figure, that is, the sum of the length of its sides.
	vertical angles	Vertical angles are angles formed by intersecting lines. They have the same measure.
2.6	**ratio**	A ratio is a comparison of two quantities using a quotient.
	proportion	A proportion is a statement that two ratios are equal.
	cross products	The method of cross products provides a way of determining whether a proportion is true. $\frac{a}{b} = \frac{c}{d}$ *ad* and *bc* are cross products.
	terms	In the proportion $\frac{a}{b} = \frac{c}{d}$, a, b, c, and d are the terms. The a and d terms are called the **extremes,** and the b and c terms are called the **means.**
2.7	**inequality**	Inequalities are statements with algebraic expressions related by $<$, $\leq$, $>$, or $\geq$.
	linear inequality	A linear inequality in one variable can be written in the form $Ax + B < C$, $Ax + B \leq C$, $Ax + B > C$, or $Ax + B \geq C$, where A, B, and C are real numbers, with $A \neq 0$.

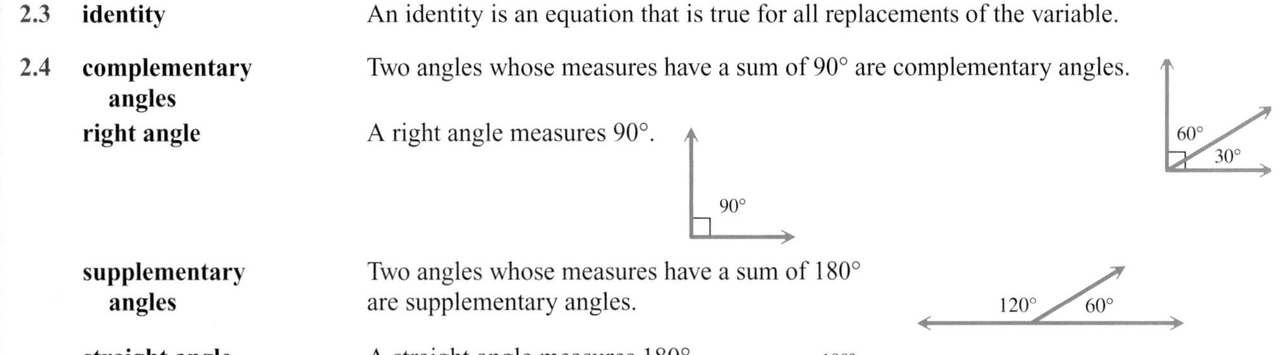

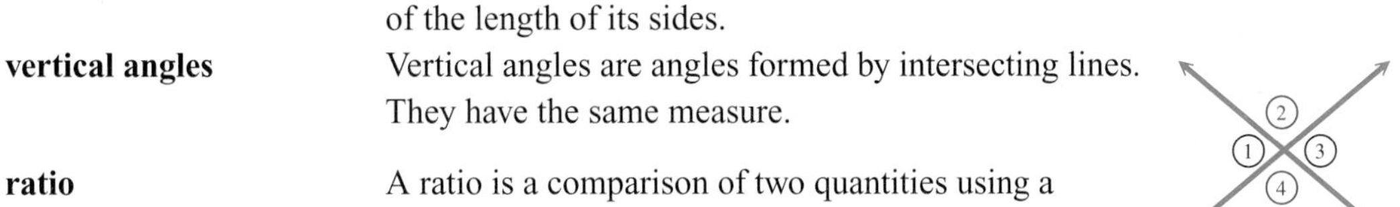

NEW SYMBOLS

$1°$ one degree

a **to** b, $a:b$, **or** $\frac{a}{b}$ the ratio of a to b

TEST YOUR WORD POWER

See how well you have learned the vocabulary in this chapter. Answers, with examples, follow the Quick Review.

1. A **solution** of an equation is a number that
   - **A.** makes an expression undefined
   - **B.** makes the equation false
   - **C.** makes the equation true
   - **D.** makes an expression equal to 0.
2. **Complementary angles** are angles
   - **A.** formed by two parallel lines
   - **B.** whose sum is 90°
   - **C.** whose sum is 180°
   - **D.** formed by perpendicular lines.
3. **Supplementary angles** are angles
   - **A.** formed by two parallel lines
   - **B.** whose sum is 90°
   - **C.** whose sum is 180°
   - **D.** formed by perpendicular lines.
4. A **ratio**
   - **A.** compares two quantities using a quotient
   - **B.** says that two quotients are equal
   - **C.** is a product of two quantities
   - **D.** is a difference between two quantities.
5. A **proportion**
   - **A.** compares two quantities using a quotient
   - **B.** says that two ratios are equal
   - **C.** is a product of two quantities
   - **D.** is a difference between two quantities.
6. An **inequality** is
   - **A.** a statement that two algebraic expressions are equal
   - **B.** a point on a number line
   - **C.** an equation with no solutions
   - **D.** a statement with algebraic expressions related by $<, \leq, >,$ or $\geq$.

QUICK REVIEW

Concepts	*Examples*
2.1 *The Addition Property of Equality* The same number may be added to (or subtracted from) each side of an equation without changing the solution.	Solve. $x - 6 = 12$ $x - 6 + 6 = 12 + 6$ Add 6. $x = 18$ Combine terms.
2.2 *The Multiplication Property of Equality* Each side of an equation may be multiplied (or divided) by the same nonzero number without changing the solution.	Solve. $\frac{3}{4}x = -9$ $\frac{4}{3} \cdot \frac{3}{4}x = \frac{4}{3}(-9)$ Multiply by $\frac{4}{3}$. $x = -12$
2.3 *More on Solving Linear Equations* *Step 1* Simplify each side separately.	Solve. $2x + 2(x + 1) = 14 + x$ $2x + 2x + 2 = 14 + x$ Distributive property $4x + 2 = 14 + x$ Combine terms.
Step 2 Isolate the variable term on one side.	$4x + 2 - x - 2 = 14 + x - x - 2$ Subtract x; subtract 2. $3x = 12$ Combine terms.
Step 3 Isolate the variable.	$\frac{3x}{3} = \frac{12}{3}$ Divide by 3. $x = 4$
Step 4 Check.	*Check:* $2(4) + 2(4 + 1) \stackrel{?}{=} 14 + 4$ Let $x = 4$. $18 = 18$ True The solution is 4.

Concepts	*Examples*
2.4 *An Introduction to Applications of Linear Equations* *Step 1* Read. *Step 2* Assign a variable. *Step 3* Write an equation. *Step 4* Solve the equation. *Step 5* State the answer. *Step 6* Check.	One number is 5 more than another. Their sum is 21. What are the numbers? We are looking for two numbers. Let x represent the smaller number. Then $x + 5$ represents the larger number. $x + (x + 5) = 21$ $2x + 5 = 21$ Combine terms. $2x + 5 - 5 = 21 - 5$ Subtract 5. $2x = 16$ Combine terms. $\frac{2x}{2} = \frac{16}{2}$ Divide by 2. $x = 8$ The numbers are 8 and 13. 13 is 5 more than 8, and $8 + 13 = 21$. It checks.
2.5 *Formulas and Applications from Geometry* To find the value of one of the variables in a formula, given values for the others, substitute the known values into the formula.	Find L if $A = LW$, given that $A = 24$ and $W = 3$. $24 = L \cdot 3$ $A = 24, W = 3$ $\frac{24}{3} = \frac{L \cdot 3}{3}$ Divide by 3. $8 = L$
To solve a formula for one of the variables, isolate that variable by treating the other variables as numbers and using the steps for solving equations.	Solve $P = 2L + 2W$ for W. $P - 2L = 2L + 2W - 2L$ Subtract $2L$. $P - 2L = 2W$ Combine terms. $\frac{P - 2L}{2} = \frac{2W}{2}$ Divide by 2. $\frac{P - 2L}{2} = W$ or $W = \frac{P - 2L}{2}$
2.6 *Ratio, Proportion, and Percent* To write a ratio, express quantities in the same units.	4 ft to 8 in. = 48 in. to 8 in. $= \frac{48}{8} = \frac{6}{1}$
To solve a proportion, use the method of cross products.	Solve $\frac{x}{12} = \frac{35}{60}$. $60x = 12 \cdot 35$ Cross products $60x = 420$ Multiply. $\frac{60x}{60} = \frac{420}{60}$ Divide by 60. $x = 7$
To solve a percent problem, use the proportion $\frac{\text{amount}}{\text{base}} = \frac{\text{percent}}{100}$.	

Concepts	*Examples*
2.7 *Solving Linear Inequalities*	Solve and graph the solutions.
Step 1 Simplify each side separately.	$3(1 - x) + 5 - 2x > 9 - 6$
	$3 - 3x + 5 - 2x > 9 - 6$ Distributive property
	$8 - 5x > 3$ Combine terms.
Step 2 Isolate the variable term on one side.	$8 - 5x - 8 > 3 - 8$ Subtract 8.
	$-5x > -5$ Combine terms.
Step 3 Isolate the variable.	$\frac{-5x}{-5} < \frac{-5}{-5}$ Divide by -5; change $>$ to $<$.
(Be sure to reverse the direction of the inequality symbol when multiplying or dividing by a negative number.)	$x < 1$
	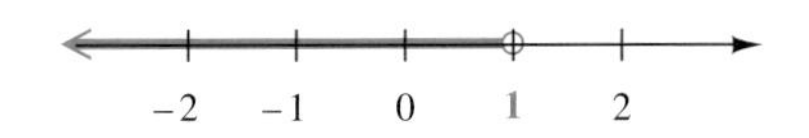

Answers to Test Your Word Power

1. C; *Example:* 8 is the solution of $2x + 5 = 21$.
2. B; *Example:* Angles with measures 35° and 55° are complementary angles.
3. C; *Example:* Angles with measures 112° and 68° are supplementary angles.
4. A; *Example:* $\frac{7 \text{ in.}}{12 \text{ in.}} = \frac{7}{12}$
5. B; *Example:* $\frac{2}{3} = \frac{8}{12}$
6. D; *Examples:* $x < 5, 7 + 2y \geq 11$

Chapter 2

REVIEW EXERCISES

[2.1–2.3] *Solve each equation. Check the solution.*

1. $x - 7 = 2$

2. $4r - 6 = 10$

3. $5x + 8 = 4x + 2$

4. $8t = 7t + \frac{3}{2}$

5. $(4r - 8) - (3r + 12) = 0$

6. $7(2x + 1) = 6(2x - 9)$

7. $-\frac{6}{5}y = -18$

8. $\frac{1}{2}r - \frac{1}{6}r + 3 = 2 + \frac{1}{6}r + 1$

9. $3x - (-2x + 6) = 4(x - 4) + x$

10. $.10(x + 80) + .20x = 8 + .30x$

[2.4] *Solve each problem.*

11. If 7 is added to five times a number, the result is equal to three times the number. Find the number.

12. If 4 is subtracted from twice a number, the result is 36. Find the number.

13. The land area of Hawaii is 5213 mi^2 greater than that of Rhode Island. Together, the areas total 7637 mi^2. What is the area of each state?

14. The height of Seven Falls in Colorado is $\frac{5}{2}$ the height (in feet) of Twin Falls in Idaho. The sum of the heights is 420 ft. Find the height of each.

15. The supplement of an angle measures 10 times the measure of its complement. What is the measure of the angle (in degrees)?

16. Find two consecutive odd integers such that when the smaller is added to twice the larger, the result is 24 more than the larger integer.

[2.5] *A formula is given in each exercise, along with the values for all but one of the variables. Find the value of the variable that is not given. (For Exercises 19 and 20, use* 3.14 *as an approximation for* π.*)*

17. $A = \frac{1}{2}bh$; $A = 44$, $b = 8$

18. $A = \frac{1}{2}h(b + B)$; $b = 3$, $B = 4$, $h = 8$

19. $C = 2\pi r$; $C = 29.83$

20. $V = \frac{4}{3}\pi r^3$; $r = 9$

Solve each formula for the specified variable.

21. $A = LW$ for W

22. $A = \frac{1}{2}h(b + B)$ for h

Find the measure of each marked angle.

23.

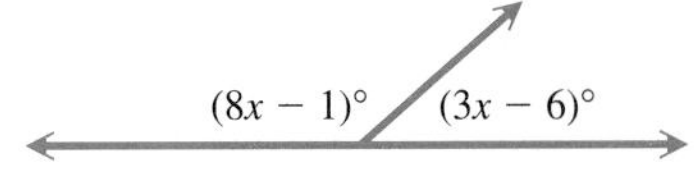

24.

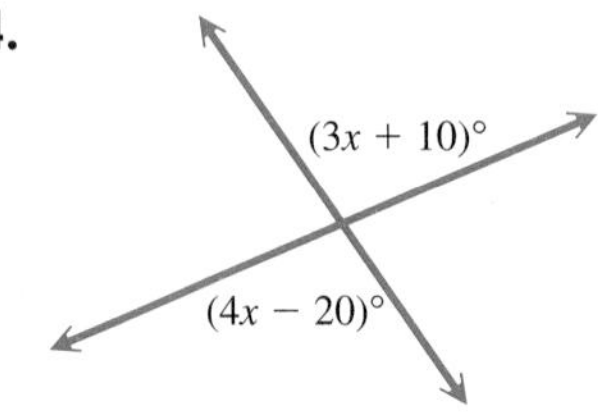

Solve each application of geometry.

25. A cinema screen in Indonesia has length 92.75 ft and width 70.5 ft. What is the perimeter? What is the area? (*Source: The Guinness Book of Records.*)

26. There is a Montezuma cypress in Mexico that is 137 ft tall and has a circumference of about 146.9 ft. What is the diameter of the tree? What is the radius of the tree? Use 3.14 as an approximation for π. Round your answers to the nearest hundredth. (*Source: The Guinness Book of Records*, 2003.)

[2.6] *Write a ratio for each word phrase. Write fractions in lowest terms.*

27. 60 cm to 40 cm

28. 5 days to 2 weeks

29. 90 in. to 10 ft

30. 3 months to 3 yr

Solve each proportion.

31. $\frac{p}{21} = \frac{5}{30}$

32. $\frac{5 + x}{3} = \frac{2 - x}{6}$

33. $\frac{y}{5} = \frac{6y - 5}{11}$

34. Explain how 40% can be expressed as a ratio of two whole numbers.

Solve each problem.

35. If 2 lb of fertilizer will cover 150 ft^2 of lawn, how many pounds would be needed to cover 500 ft^2?

36. If 8 oz of medicine must be mixed with 20 oz of water, how many ounces of medicine must be mixed with 90 oz of water?

37. Biologists tagged 500 fish in Willow Lake on February 18. At a later date they found 7 tagged fish in a sample of 700. Estimate the total number of fish in Willow Lake to the nearest hundred.

38. The distance between two cities on a road map is 32 cm. The two cities are actually 150 km apart. The distance on the map between two other cities is 80 cm. How far apart are these cities?

39. What is 23% of 76?

40. What percent of 12 is 21?

41. 6 is what percent of 18?

42. 36% of what number is 900?

43. Terry paid $18,690, including tax, for her 2003 Chrysler Sebring LX. The sales tax rate was 5%. What was the actual price of the car? (*Source:* Author McGinnis' sales receipt.)

44. Lauren, from the mathematics editorial division of Addison-Wesley, took the mathematics faculty from a community college out to dinner. The bill was $304.75. Lauren added a 15% tip, and paid for the meal with her corporate credit card. What was the total price she paid?

[2.7] *Graph each inequality on the number line provided.*

45. $p \geq -4$

46. $x < 7$

47. $-5 \leq k < 6$

48. $r \geq \frac{1}{2}$

Solve each inequality. Graph the solutions.

49. $x + 6 \geq 3$

50. $5t < 4t + 2$

51. $-6x \leq -18$

52. $8(k - 5) - (2 + 7k) \geq 4$

53. $4x - 3x > 10 - 4x + 7x$

54. $3(2w + 5) + 4(8 + 3w) < 5(3w + 2) + 2w$

55. Justin Sudak has grades of 94 and 88 on his first two calculus tests. What possible scores on a third test will give him an average of at least 90?

56. If nine times a number is added to 6, the result is at most 3. Find all such numbers.

MIXED REVIEW EXERCISES

Solve.

57. $\frac{y}{7} = \frac{y - 5}{2}$

58. $C = \pi d$ for d

59. $-2x > -4$

60. $2k - 5 = 4k + 13$

61. $.05x + .02x = 4.9$

62. $2 - 3(t - 5) = 4 + t$

63. $9x - (7x + 2) = 3x + (2 - x)$

64. $\frac{1}{3}s + \frac{1}{2}s + 7 = \frac{5}{6}s + 5 + 2$

65. On May 13 researchers at Argyle Lake tagged 840 fish. When they returned a few weeks later, their sample of 1000 fish contained 18 that were tagged. Give an approximation of the fish population in Argyle Lake to the nearest hundred.

66. Two-thirds of a number added to the number is 10. What is the number?

67. In the 2002 Winter Olympic Games in Salt Lake City, the United States and Canada earned a total of 51 medals. The United States earned twice as many medals as Canada. How many medals did each country earn? (*Source: The Gazette*, February 2, 2002.)

68. Of the 58 medals earned by the host country, Australia, in the 2000 Summer Olympics, there were 9 more silver than gold medals, and 8 fewer bronze than silver medals. How many of each medal did Australia earn? (*Source: Times Picayune*, October 2, 2000.)

69. The perimeter of a triangle is 96 m. One side is twice as long as another, and the third side is 30 m long. What is the length of the longest side?

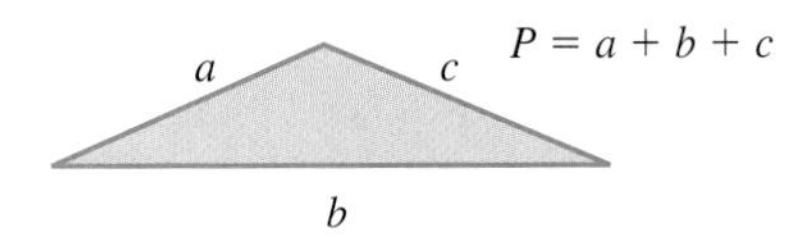

70. The perimeter of a rectangle is 288 ft. The length is 4 ft longer than the width. Find the width.

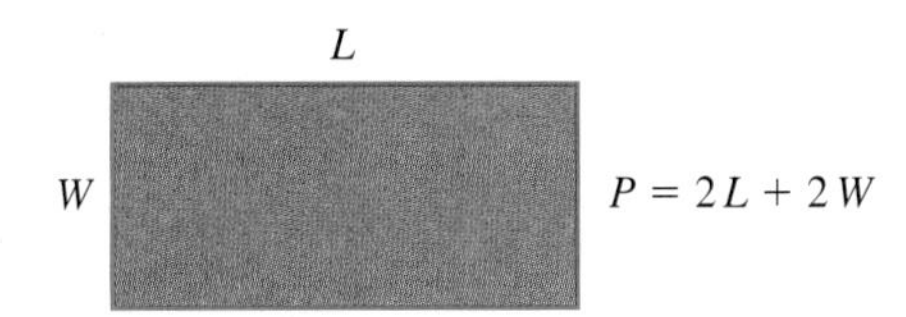

71. Find the best buy. Give the unit price to the nearest thousandth for that size. (*Source:* Cub Foods.)

LAUNDRY DETERGENT

Size	Price
87-oz	\$7.88
131-oz	\$10.98
263-oz	\$19.96

72. If nine pairs of jeans cost \$121.50, find the cost of five pairs. (Assume all are equally priced.)

73. Latarsha has grades of 82 and 96 on her first two English tests. What must she make on her third test so that her average will be at least 90?

74. Find the measure of each marked angle.

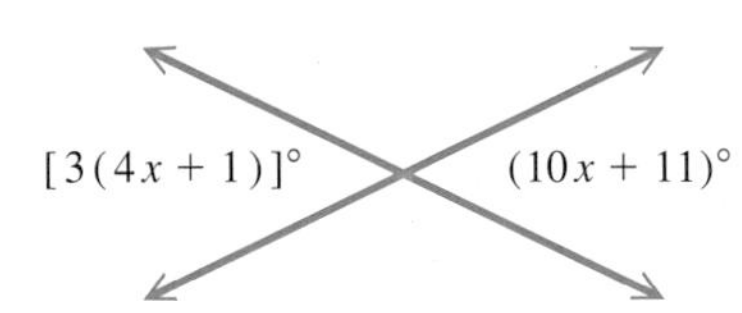

Chapter 2

TEST

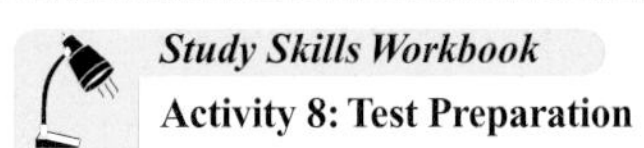

Solve each equation, and check the solution.

1. $3x - 7 = 11$ 1. ________

2. $5x + 9 = 7x + 21$ 2. ________

3. $2 - 3(x - 5) = 3 + (x + 1)$ 3. ________

4. $2.3x + 13.7 = 1.3x + 2.9$ 4. ________

5. $7 - (m - 4) = -3m + 2(m + 1)$ 5. ________

6. $-\frac{4}{7}x = -12$ 6. ________

7. $.06(x + 20) + .08(x - 10) = 4.6$ 7. ________

8. $-8(2x + 4) = -4(4x + 8)$ 8. ________

Solve each problem.

9. **(a)** The San Antonio Spurs beat the New Jersey Nets by 11 points in the final 2003 NBA Championship game held June 15 in San Antonio. The total of the two team scores was 165. What was the final score of the game?

   (b) The high scorer for the Nets during the finals was Jason Kidd. He made 10 fewer field goals than Tim Duncan, the high scorer for the Spurs. The two players made a total of 98 field goals. How many field goals did Kidd make?

   (*Source: World Almanac and Book of Facts*, 2004.)

9. (a) ________
   (b) ________

10. The three largest islands in the Hawaiian island chain are Hawaii (the Big Island), Maui, and Kauai. Together, their areas total 5300 mi^2. The island of Hawaii is 3293 mi^2 larger than the island of Maui, and Maui is 177 mi^2 larger than Kauai. What is the area of each island?

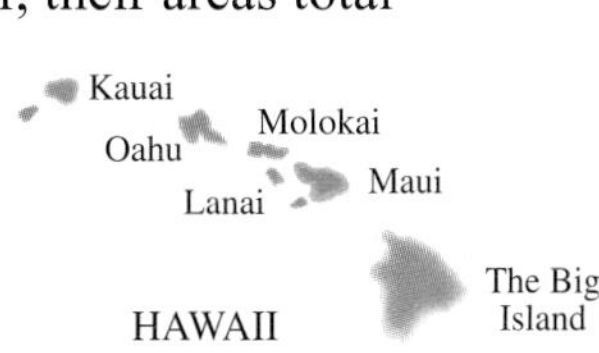

10. ________

11. Find the measure of an angle if its supplement measures 10° more than three times its complement.   11. ________

12. The formula for the perimeter of a rectangle is $P = 2L + 2W$.

(a) Solve for W.

(b) If $P = 116$ and $L = 40$, find the value of W.

Find the measure of each marked angle.

13.

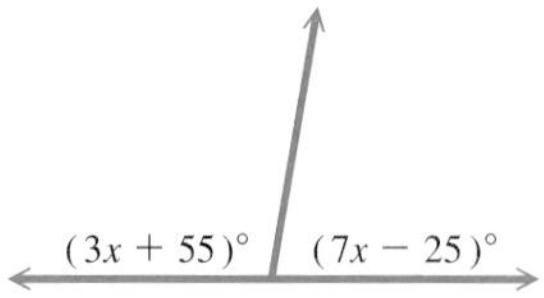

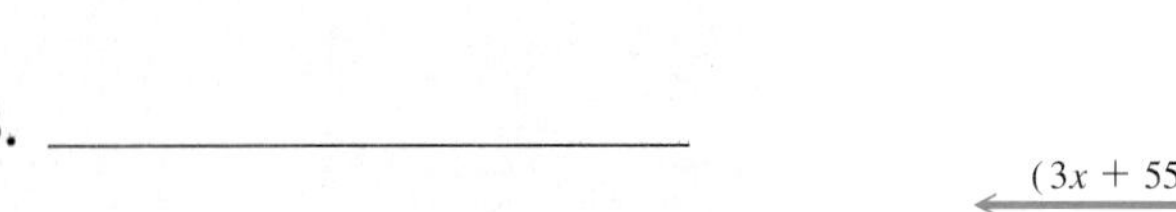

14.

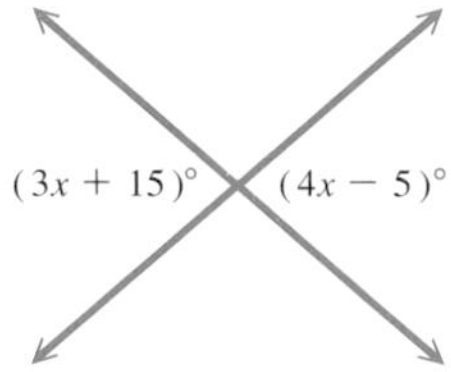

Solve each proportion.

15. $\dfrac{z}{8} = \dfrac{12}{16}$

16. $\dfrac{x + 5}{3} = \dfrac{x - 3}{4}$

17. Find the best buy. Give the unit price to the nearest thousandth for that size.

PROCESSED CHEESE SLICES

Size	Price
12-oz	\$1.99
16-oz	\$2.49
24-oz	\$3.49

18. The distance between Milwaukee and Boston is 1050 mi. On a certain map, this distance is represented by 42 in. On the same map, Seattle and Cincinnati are 92 in. apart. What is the actual distance between Seattle and Cincinnati?

19. The Anaheim Mighty Ducks of the NHL finished the 2002–2003 season with 95 points, up from 69 points in the 2001–2002 season. What was the percent of increase in points? Round your answer to the nearest tenth. (*Source: World Almanac and Book of Facts*, 2004.)

20. Write an inequality involving x that describes the numbers graphed.

(a)

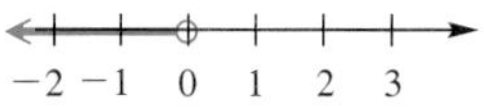

(b)

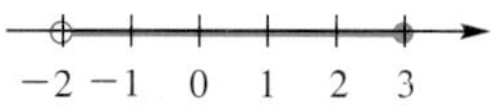

Solve each inequality, and graph the solutions.

21. $-3x > -33$

22. $-.04x \leq .12$

23. $-4x + 2(x - 3) \geq 4x - (3 + 5x) - 7$

24. Shania Johnson has grades of 76 and 81 on her first two algebra tests. If she wants an average of at least 80 after her third test, what score must she make on her third test?

25. Write a short explanation of the additional (extra) rule that must be remembered when solving an inequality (as opposed to solving an equation).

12. (a) ________
(b) ________
13. ________
14. ________
15. ________
16. ________
17. ________
18. ________
19. ________
20. (a) ________
(b) ________
21.
22.
23.
24. ________
25. ________

Cumulative Review Exercises

CHAPTERS R–2

Write each fraction in lowest terms.

1. $\frac{15}{40}$

2. $\frac{108}{144}$

Perform the indicated operations.

3. $\frac{5}{6} + \frac{1}{4} + \frac{7}{15}$

4. $16\frac{7}{8} - 3\frac{1}{10}$

5. $\frac{9}{8} \cdot \frac{16}{3}$

6. $\frac{3}{4} \div \frac{5}{8}$

7. $4.8 + 12.5 + 16.73$

8. $56.3 - 28.99$

9. $67.8(.45)$

10. $236.46 \div 4.2$

11. In making dresses, Earth Works uses $\frac{5}{8}$ yd of trim per dress. How many yards of trim would be used to make 56 dresses?

12. A cook wants to increase a recipe for Quaker Quick Grits that serves 4 to make enough for 10 people. The recipe calls for 3 cups of water. How much water will be needed to serve 10?

13. Buster weighs $27\frac{1}{4}$ lb and Sofi weighs $39\frac{7}{8}$ lb. How much do the two dogs weigh together?

14. A small business owner bought 3 computer workstations of various sizes for $329.99, $379.99, and $439.99 and 3 ergonomic office chairs for $249.99 each. What was the final bill (without tax)? (*Source:* Staples "Furniture Values" catalog, January 2004.)

Tell whether each inequality is true *or* false.

15. $\frac{8(7) - 5(6 + 2)}{3 \cdot 5 + 1} \geq 1$

16. $\frac{4(9 + 3) - 8(4)}{2 + 3 - 3} \geq 2$

Perform the indicated operations.

17. $-11 + 20 + (-2)$

18. $13 + (-19) + 7$

19. $9 - (-4)$

20. $-2(-5)(-4)$

21. $\frac{4 \cdot 9}{-3}$

22. $\frac{8}{7 - 7}$

23. $(-5 + 8) + (-2 - 7)$

24. $(-7 - 1)(-4) + (-4)$

25. $\frac{-3 - (-5)}{1 - (-1)}$

26. $\dfrac{6(-4) - (-2)(12)}{3^2 + 7^2}$

27. $\dfrac{(-3)^2 - (-4)(2^4)}{5 \cdot 2 - (-2)^3}$

28. $\dfrac{-2(5^3) - 6}{4^2 + 2(-5) + (-2)}$

Find the value of each expression when $x = -2$, $y = -4$, and $z = 3$.

29. $xz^3 - 5y^2$

30. $\dfrac{xz - y^3}{-4z}$

Name the property illustrated by each equation.

31. $7(k + m) = 7k + 7m$

32. $3 + (5 + 2) = 3 + (2 + 5)$

33. $7 + (-7) = 0$

34. $3.5(1) = 3.5$

Simplify each expression.

35. $4p - 6 + 3p - 8$

36. $-4(k + 2) + 3(2k - 1)$

Solve each equation, and check the solution.

37. $2r - 6 = 8$

38. $2(p - 1) = 3p + 2$

39. $4 - 5(a + 2) = 3(a + 1) - 1$

40. $2 - 6(z + 1) = 4(z - 2) + 10$

41. $-(m - 1) = 3 - 2m$

42. $\dfrac{x - 2}{3} = \dfrac{2x + 1}{5}$

43. $\dfrac{2x + 3}{5} = \dfrac{x - 4}{2}$

44. $\dfrac{2}{3}x + \dfrac{3}{4}x = -17$

Solve each formula for the indicated variable.

45. $P = a + b + c$ for c

46. $P = 4s$ for s

Solve each inequality. Graph the solutions.

47. $-5z \geq 4z - 18$

48. $6(r - 1) + 2(3r - 5) < -4$

Solve each problem.

49. The small business owner in Exercise 14 paid a sales tax of $6\frac{1}{4}\%$ on his purchase. What was the final bill, including tax, to the nearest cent?

50. A car has a price of \$5000. For trading in her old car, Shannon d'Hemecourt will get 25% off. Find the price of the car with the trade-in.

51. Jennifer Johnston bought textbooks at the college bookstore for \$244.33, including 6% sales tax. What did the books cost?

52. Carter Fenton received a bill from his credit card company for \$104.93. The bill included interest at $1\frac{1}{2}\%$ per month for one month and a \$5.00 late charge. How much did his purchases amount to?

53. The perimeter of a rectangle is 98 cm. The width is 19 cm. Find the length.

?

19 cm

54. The area of a triangle is 104 in.^{2}. The base is 13 in. Find the height.

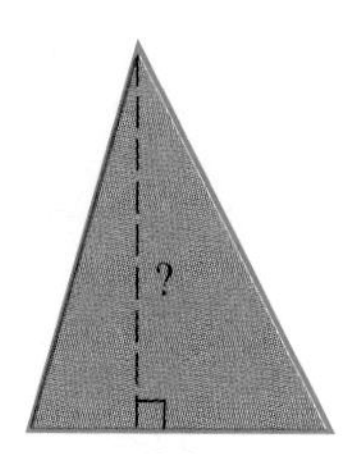

Graphs of Linear Equations and Inequalities in Two Variables

The use and abuse of credit cards is putting college students deeper in debt than ever before. College campuses have become fertile territory as credit card companies pitch their plastic to students at bookstores, student unions, and sporting events. According to federal loan provider Nellie Mae, an estimated 83% of college students have credit cards with an average balance of more than $2300. (*Source:* ISU Extension, *Family Living in Farm Country,* April 2003.)

In Example 6 of Section 3.2 we examine a linear equation that models credit card debt in the United States.

3.1 Reading Graphs; Linear Equations in Two Variables

OBJECTIVES

1. Interpret graphs.
2. Write a solution as an ordered pair.
3. Decide whether a given ordered pair is a solution of a given equation.
4. Complete ordered pairs for a given equation.
5. Complete a table of values.
6. Plot ordered pairs.

We live in an age of information. Graphs provide a quick way to organize and communicate much of this information. They can also be used to analyze data, make predictions, or simply entertain us.

OBJECTIVE 1 Interpret graphs. There are many ways to represent the relationship between two quantities. *Circle graphs, bar graphs*, and *line graphs* are often used for this purpose.

In a **circle graph** or **pie chart,** a circle is used to indicate the total of all the categories represented. The circle is divided into *sectors*, or wedges (like pieces of a pie), whose sizes show the relative magnitudes of the categories. The sum of all the fractional parts of the graph must be 1 (for 1 whole circle).

EXAMPLE 1 Interpreting a Circle Graph

The 2003 market share for satellite-TV home subscribers is shown in the circle graph in Figure 1.

SATELLITE-TV HOME SUBSCRIBERS

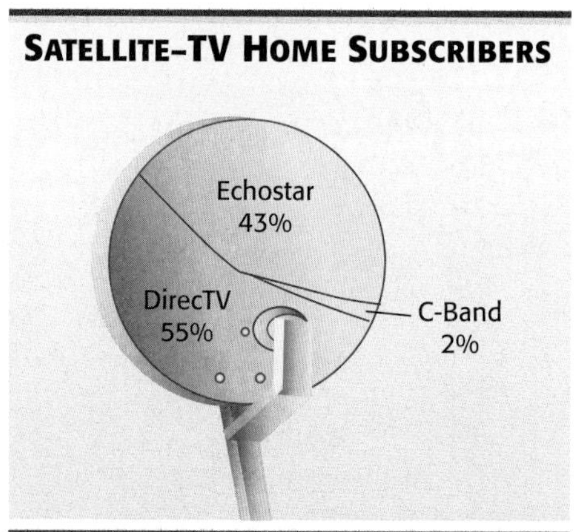

Source: Skyreport.com; *USA Today.*

Figure 1

The number of subscribers reached 22 million in December 2003.

(a) Which provider had the largest share of the home subscriber market in December 2003? What was that share?

In the circle graph, the sector for DirecTV is the largest, so DirecTV had the largest market share, 55%.

(b) Estimate the number of home subscribers to Echostar in December 2003.

A market share of 43% can be rounded to 40%, or .4. We multiply .4 by the total number of subscribers, 22 million. A good estimate for the number of Echostar subscribers would be

$$.4(22) = 8.8 \approx 9 \text{ million.}$$

(c) How many actual home subscribers to Echostar were there?

To find the answer, we multiply the actual percent from the graph for Echostar, 43% or .43, by the number of subscribers, 22 million:

$$.43(22) = 9.46.$$

Thus, 9.46 million homes subscribed to Echostar. This is reasonable given our estimate in part (b).

Work Problem 1 at the Side.

---

1 Refer to the circle graph in Figure 1.

(a) Which provider had the smallest market share in December 2003?

(b) Estimate the number of home subscribers to DirecTV.

(c) How many actual home subscribers to DirecTV were there?

ANSWERS

1. **(a)** C-Band
   (b) 60% of 22 million = 13.2 million
   (c) 12.1 million

A **bar graph** is used to show comparisons. It consists of a series of bars (or simulations of bars) arranged either vertically or horizontally. In a bar graph, values from two categories are paired with each other.

EXAMPLE 2 Interpreting a Bar Graph

The bar graph in Figure 2 illustrates the amount of personal savings, in billions of dollars, accumulated during the years 1997 through 2001.

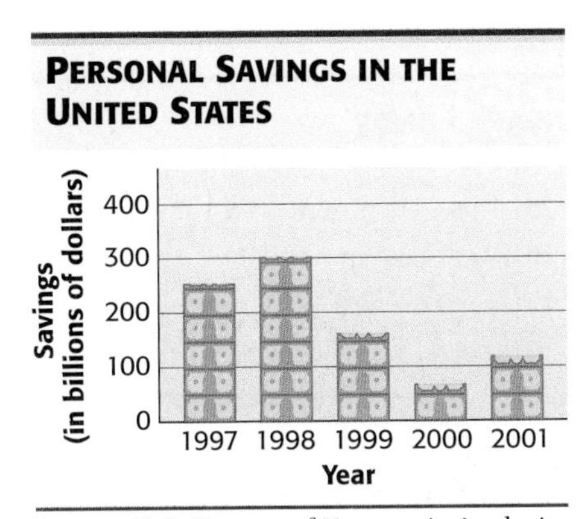

Source: U.S. Bureau of Economic Analysis.

Figure 2

(a) Which year had the greatest amount of savings? Which had the least?

The tallest bar corresponds to 1998, so the largest amount was in 1998. The year 2000, with the shortest bar, had the least.

(b) Which years had amounts greater than \$200 billion?

Locate 200 on the vertical scale and follow the line across to the right. Two years, 1997 and 1998, have bars that extend above the line for 200, so they had amounts greater than \$200 billion.

(c) Estimate the amounts for 1997 and 1998.

Locate the top of the bar for 1997 and move horizontally across to the vertical scale to see that it lies about halfway between 200 and 300. The amount for 1997 was about \$250 billion.

Follow the top of the bar for 1998 across to the vertical scale to see that it is about 300. The amount for 1998 was about \$300 billion.

(d) Estimate the difference between the amounts for the years 1997 and 1998. Interpret this result.

Based on the results of part (c), the difference is approximately

$$\$300 \text{ billion} - \$250 \text{ billion} = \$50 \text{ billion}.$$

The amount of personal savings increased from 1997 to 1998 by about \$50 billion. (However, during the next two years, it decreased each year.)

(e) How did personal savings in 1998 compare to personal savings in 1999?

Comparing the heights of the bars for 1998 and 1999, we see that the bar for 1998 is about twice as tall as the bar for 1999. Therefore, personal savings in 1998 was about twice that for 1999.

Also, if we estimate the amounts of personal savings for 1998 and 1999 as about \$300 billion and about \$150 billion respectively, we see that personal savings in 1998 was about twice that for 1999.

Work Problem 2 at the Side.

2 Refer to the bar graph in Figure 2.

(a) Which years had amounts less than \$200 billion?

(b) Estimate the amounts of personal savings for 2000 and 2001.

ANSWERS

2. (a) 1999, 2000, 2001
(b) \$60 billion; \$120 billion

3 Refer to the line graph in Figure 3.

(a) Which year has the greatest amount of Medicare funds?

(b) Estimate Medicare funds for 2008. Is there a surplus or a deficit in 2008?

(c) About how much will funds decrease from 2006 to 2011?

A **line graph** is used to show changes or trends in data over time. To form a line graph, we connect a series of points representing data with line segments.

EXAMPLE 3 Interpreting a Line Graph

Current projections indicate that funding for Medicare will not cover its costs unless the program changes. The line graph in Figure 3 shows projections for the years 2004 through 2013.

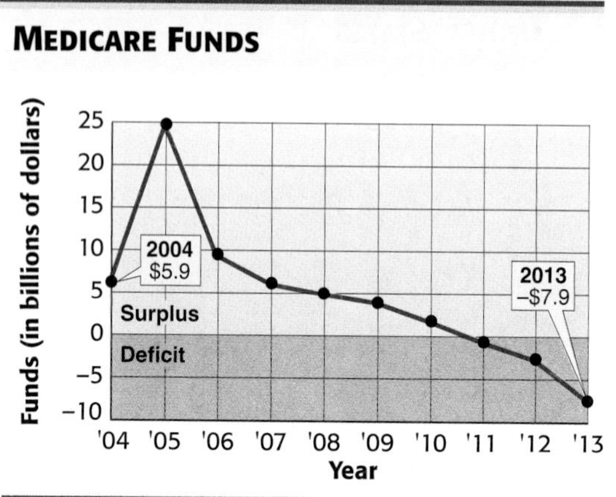

Figure 3

(a) Which is the only period in which Medicare funds are predicted to increase?

Because the graph *rises* from 2004 to 2005 and falls in every other case, funds are predicted to increase between these two years.

(b) What will be the general trend from 2005 to 2013?

Funds will decrease, since the graph *falls* during this period.

(c) In which year will funds first show a deficit?

From 2004 to 2010, the graph is always above 0, but in 2011, it falls below 0 for the first time, indicating a deficit.

(d) Based on the figures shown in the graph, what is the difference in Medicare funds from 2004 to 2013?

$$\underbrace{-\$7.9 \text{ billion}}_{\text{2013 amount}} - \underbrace{\$5.9 \text{ billion}}_{\text{2004 amount}} = \underbrace{-\$13.8 \text{ billion}}_{\text{Difference}}$$

The fund amount will have *decreased* \$13.8 billion (as indicated by the negative sign in −\$13.8).

Work Problem 3 at the Side.

Many everyday situations, such as those illustrated in Examples 2 and 3, involve two quantities that are related. The equations and applications we discussed in **Chapter 2** had only one variable. In this chapter, we extend those ideas to *linear equations in two variables.*

ANSWERS

3. (a) 2005
(b) \$5 billion; surplus
(c) \$10 billion

Linear Equation in Two Variables

A **linear equation in two variables** is an equation that can be written in the form

$$Ax + By = C,$$

where A, B, and C are real numbers and A and B are not both 0.

Some examples of linear equations in two variables in this form, called *standard form*, are

$$3x + 4y = 9, \quad x - y = 0, \quad \text{and} \quad x + 2y = -8.$$ Linear equations in two variables

NOTE

Other linear equations in two variables, such as

$$y = 4x + 5 \quad \text{and} \quad 3x = 7 - 2y,$$

are not written in standard form but could be. We will discuss the forms of linear equations in more detail in **Section 3.4.**

OBJECTIVE 2 Write a solution as an ordered pair. Recall from **Section 1.2** that a *solution* of an equation is a number that makes the equation true when it replaces the variable. For example, the linear equation in one variable $x - 2 = 5$ has solution 7, since replacing x with 7 gives a true statement.

A solution of a linear equation in *two* variables requires *two* numbers, one for each variable. For example, a true statement results when we replace x with 2 and y with 13 in the equation $y = 4x + 5$ since

$$13 = 4(2) + 5. \quad \text{Let } x = 2, y = 13.$$

The pair of numbers $x = 2$ and $y = 13$ gives one solution of the equation $y = 4x + 5$. The phrase "$x = 2$ and $y = 13$" is abbreviated

x-value → $(2, 13)$ ← y-value

Ordered pair

with the x-value, 2, and the y-value, 13, given as a pair of numbers written inside parentheses. ***The x-value is always given first.*** A pair of numbers such as (2, 13) is called an **ordered pair.** As the name indicates, the order in which the numbers are written is important. The ordered pairs (2, 13) and (13, 2) are not the same. The second pair indicates that $x = 13$ and $y = 2$. For two ordered pairs to be equal, their x-coordinates must be equal *and* their y-coordinates must be equal.

Work Problem 4 at the Side.

OBJECTIVE 3 Decide whether a given ordered pair is a solution of a given equation. We substitute the x- and y-values of an ordered pair into a linear equation in two variables to see whether the ordered pair is a solution.

4 Write each solution as an ordered pair.

(a) $x = 5$ and $y = 7$

(b) $y = 6$ and $x = -1$

(c) $y = 4$ and $x = -3$

(d) $x = 3$ and $y = -12$

Answers

4. **(a)** $(5, 7)$ **(b)** $(-1, 6)$ **(c)** $(-3, 4)$ **(d)** $(3, -12)$

5 Decide whether each ordered pair is a solution of the equation $5x + 2y = 20$.

(a) $(0, 10)$

$$5x + 2y = 20$$
$$5(\quad) + 2(\quad) = 20$$
$$\_\_\_\_ + 20 = 20$$
$$\_\_\_\_ = 20$$

Is $(0, 10)$ a solution?

(b) $(2, -5)$

(c) $(3, 2)$

(d) $(-4, 20)$

6 Complete each ordered pair for the equation $y = 2x - 9$.

(a) $(5, \quad)$

$$y = 2(\quad) - 9$$
$$y = \_\_\_\_ - 9$$
$$y = \_\_\_\_$$

The ordered pair is ______.

(b) $(2, \quad)$

(c) $(\quad, 7)$

(d) $(\quad, -13)$

ANSWERS

5. **(a)** 0; 10; 0; 20; yes **(b)** no **(c)** no **(d)** yes

6. **(a)** 5; 10; 1; (5, 1) **(b)** $(2, -5)$ **(c)** $(8, 7)$ **(d)** $(-2, -13)$

EXAMPLE 4 Deciding Whether Ordered Pairs Are Solutions of an Equation

Decide whether each ordered pair is a solution of the equation $2x + 3y = 12$.

(a) $(3, 2)$

To see whether $(3, 2)$ is a solution of the given equation $2x + 3y = 12$, substitute 3 for x and 2 for y in the equation.

$$2x + 3y = 12$$
$$2(3) + 3(2) = 12 \quad ? \qquad \text{Let } x = 3; \text{ let } y = 2.$$
$$6 + 6 = 12 \quad ?$$
$$12 = 12 \qquad \text{True}$$

This result is true, so $(3, 2)$ is a solution of $2x + 3y = 12$.

(b) $(-2, -7)$

$$2x + 3y = 12$$
$$2(-2) + 3(-7) = 12 \quad ? \qquad \text{Let } x = -2; \text{ let } y = -7.$$
$$-4 + (-21) = 12 \quad ?$$
$$-25 = 12 \qquad \text{False}$$

This result is false, so $(-2, -7)$ is *not* a solution of $2x + 3y = 12$.

Work Problem 5 at the Side.

OBJECTIVE 4 Complete ordered pairs for a given equation. Choosing a number for one variable in a linear equation makes it possible to find the value of the other variable.

EXAMPLE 5 Completing Ordered Pairs

Complete each ordered pair for the equation $y = 4x + 5$.

(a) $(7, \quad)$

In this ordered pair, $x = 7$. (Remember that x always comes first.) To find the corresponding value of y, replace x with 7 in the equation.

$$y = 4x + 5$$
$$y = 4(7) + 5 \qquad \text{Let } x = 7.$$
$$y = 28 + 5$$
$$y = 33$$

The ordered pair is $(7, 33)$.

(b) $(\quad, -3)$

In this ordered pair, $y = -3$. Find the value of x by replacing y with -3 in the equation; then solve for x.

$$y = 4x + 5$$
$$-3 = 4x + 5 \qquad \text{Let } y = -3.$$
$$-8 = 4x \qquad \text{Subtract 5 from each side.}$$
$$-2 = x \qquad \text{Divide each side by 4.}$$

The ordered pair is $(-2, -3)$.

Work Problem 6 at the Side.

OBJECTIVE 5 Complete a table of values. Ordered pairs are often displayed in a **table of values.** The table may be written either vertically or horizontally.

EXAMPLE 6 Completing Tables of Values

Complete the table of values for each equation.

(a) $x - 2y = 8$

x	y
2	
10	
	0
	-2

To complete the first two ordered pairs of the table, let $x = 2$ and $x = 10$, respectively.

If	$x = 2,$	If	$x = 10,$
then	$x - 2y = 8$	then	$x - 2y = 8$
becomes	$2 - 2y = 8$	becomes	$10 - 2y = 8$
	$-2y = 6$		$-2y = -2$
	$y = -3.$		$y = 1.$

Now complete the last two ordered pairs by letting $y = 0$ and $y = -2$, respectively.

If	$y = 0,$	If	$y = -2,$
then	$x - 2y = 8$	then	$x - 2y = 8$
becomes	$x - 2(0) = 8$	becomes	$x - 2(-2) = 8$
	$x - 0 = 8$		$x + 4 = 8$
	$x = 8.$		$x = 4.$

The completed table of values follows.

x	y
2	-3
10	1
8	0
4	-2

The corresponding ordered pairs are $(2, -3)$, $(10, 1)$, $(8, 0)$, and $(4, -2)$. Notice that each ordered pair is a solution of the given equation.

(b) $x = 5$

x	y
	-2
	6
	3

The given equation is $x = 5$. No matter which value of y is chosen, the value of x is always 5.

x	y
5	-2
5	6
5	3

The corresponding ordered pairs are $(5, -2)$, $(5, 6)$, and $(5, 3)$.

Work Problem 7 at the Side.

7 Complete the table of values for each equation.

(a) $2x - 3y = 12$

x	y
0	
	0
3	
	-3

(b) $y = 4$

x	y
-3	
2	
5	

ANSWERS

7. (a)

x	y
0	-4
6	0
3	-2
$\frac{3}{2}$	-3

(b)

x	y
-3	4
2	4
5	4

8 Name the quadrant in which each point in the figure is located.

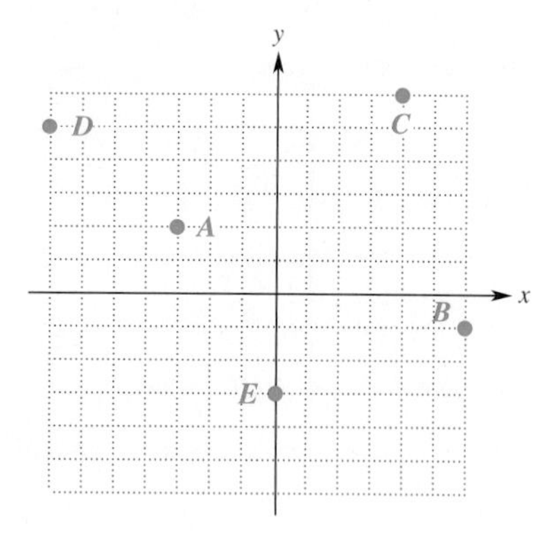

NOTE

We can think of $x = 5$ in Example 6(b) as an equation in two variables by rewriting $x = 5$ as $x + 0y = 5$. This form of the equation shows that for any value of y, the value of x is 5. Similarly, $y = 4$ in Problem 7(b) in the margin on the preceding page is the same as $0x + y = 4$.

OBJECTIVE 6 Plot ordered pairs. In **Section 2.3,** we saw that linear equations in *one* variable had either one, zero, or an infinite number of real number solutions. These solutions could be graphed on *one* number line. Every linear equation in *two* variables has an infinite number of ordered pairs as solutions. Each choice of a number for one variable leads to a particular real number for the other variable.

To graph these solutions, represented as the ordered pairs (x, y), we need *two* number lines, one for each variable. These two number lines are drawn as shown in Figure 4. The horizontal number line is called the ***x*-axis,** and the vertical line is called the ***y*-axis.** Together, the x-axis and y-axis form a **rectangular coordinate system,** also called the **Cartesian coordinate system,** in honor of René Descartes (1596–1650), the French mathematician who is credited with its invention.

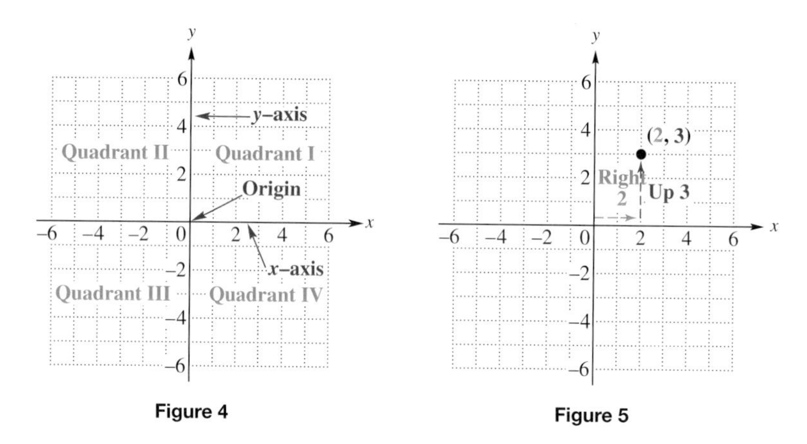

Figure 4

Figure 5

The coordinate system is divided into four regions, called **quadrants.** These quadrants are numbered counterclockwise, as shown in Figure 4. ***Points on the axes themselves are not in any quadrant.*** The point at which the x-axis and y-axis meet is called the **origin.** The origin, which is labeled 0 in Figure 4, is the point corresponding to (0, 0).

Work Problem 8 at the Side.

The x-axis and y-axis determine a **plane,** a flat surface illustrated by a sheet of paper. By referring to the two axes, every point in the plane can be associated with an ordered pair. The numbers in the ordered pair are called the **coordinates** of the point. For example, locate the point associated with the ordered pair (2, 3) by starting at the origin. Since the x-coordinate is 2, go 2 units to the right along the x-axis. Then, since the y-coordinate is 3, turn and go up 3 units on a line parallel to the y-axis. The point (2, 3) is **plotted** in Figure 5. From now on, we will refer to the point with x-coordinate 2 and y-coordinate 3 as the point (2, 3).

ANSWERS

8. *A*, II; *B*, IV; *C*, I; *D*, II; *E*, no quadrant

> **NOTE**
> When we graph on a number line, one number corresponds to each point. On a plane, however, both numbers in the ordered pair are needed to locate a point. The ordered pair is a name for the point.

9 Plot each ordered pair on a coordinate system.

(a) (3, 5) **(b)** (−2, 6)

(c) (−4, 0) **(d)** (−5, −2)

(e) (5, −2) **(f)** (0, −6)

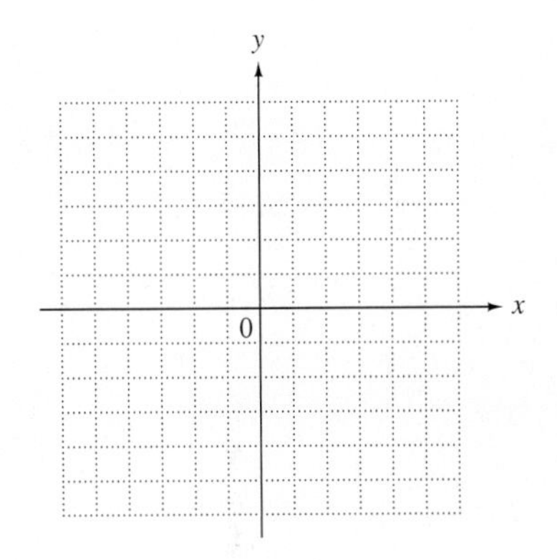

EXAMPLE 7 Plotting Ordered Pairs

Plot each ordered pair on a coordinate system.

(a) (1, 5) **(b)** (−2, 3) **(c)** (−1, −4) **(d)** (7, −2)

(e) $\left(\frac{3}{2}, 2\right)$ **(f)** (5, 0) **(g)** (0, −3)

See Figure 6. In part (c), locate the point (−1, −4) by first going 1 unit to the left along the x-axis. Then turn and go 4 units down, parallel to the y-axis. Plot the point $\left(\frac{3}{2}, 2\right)$ in part (e) by first going $\frac{3}{2}$ (or $1\frac{1}{2}$) units to the right along the x-axis. Then turn and go 2 units up, parallel to the y-axis.

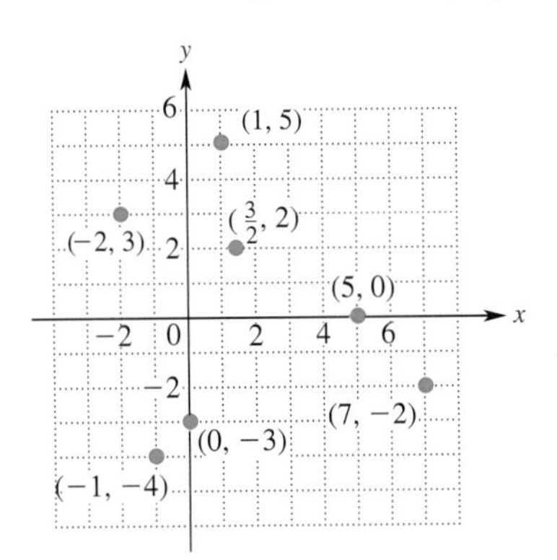

Figure 6

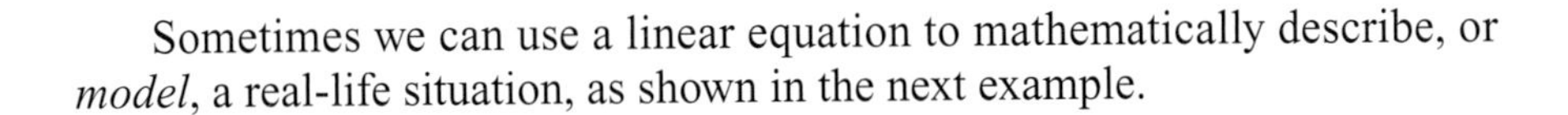

Sometimes we can use a linear equation to mathematically describe, or *model*, a real-life situation, as shown in the next example.

EXAMPLE 8 Completing Ordered Pairs to Estimate Annual Costs of Doctors' Visits

The amount Americans pay annually for doctors' visits increased from 1990 through 2000. This amount can be approximated by the linear equation

$$\underset{\text{Cost}}{y} = 34.3\underset{\text{Year}}{x} - 67{,}693,$$

which relates x, the year, and y, the cost in dollars. (*Source:* U.S. Health Care Financing Administration.)

(a) Complete the table of values for this linear equation.

x (Year)	*y* (Cost)
1990	
1996	
2000	

Continued on Next Page

ANSWERS

9.

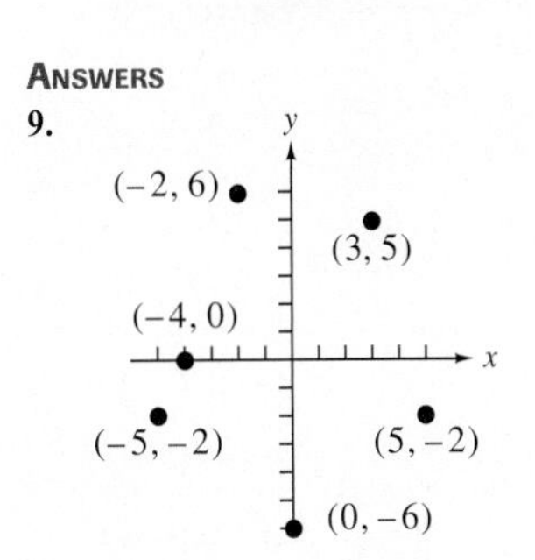

10 Refer to the linear equation in Example 8.

(a) Find the y-value for $x = 1996$. Round to the nearest whole number.

(b) Find the y-value for $x = 2000$. Interpret your result.

To find y when $x = 1990$, substitute into the equation.

$$y = 34.3(\mathbf{1990}) - 67{,}693 \quad \text{Let } x = 1990.$$

$$y = 564 \quad \text{Use a calculator.}$$

This means that in 1990, Americans each spent about \$564 on doctors' visits.

Work Problem 10 at the Side.

Including the results from Problem 10 at the side gives the completed table that follows.

x (Year)	y (Cost)
1990	564
1996	770
2000	907

We can write the results from the table of values as ordered pairs (x, y). Each year x is paired with its cost y:

$$(1990, 564), \quad (1996, 770), \quad \text{and} \quad (2000, 907).$$

(b) Graph the ordered pairs found in part (a).

The ordered pairs are graphed in Figure 7. This graph of ordered pairs of data is called a **scatter diagram.** Notice how the axes are labeled: x represents the year, and y represents the cost in dollars. Different scales are used on the two axes. Here, each square represents two units in the horizontal direction and 100 units in the vertical direction. Because the numbers in the first ordered pair are so large, we show a break in the axes near the origin.

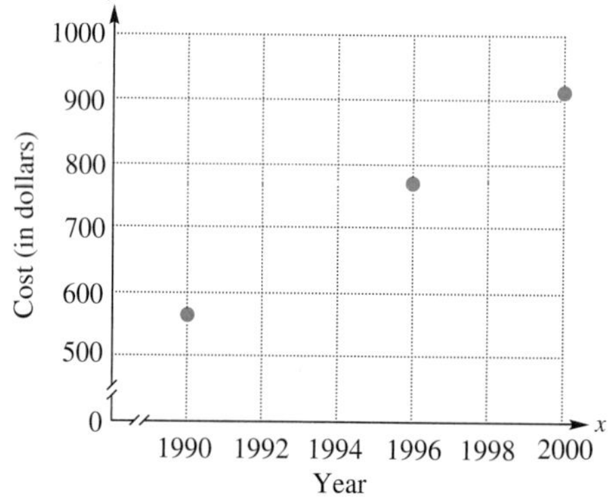

Figure 7

A scatter diagram enables us to tell whether two quantities are related to each other. In Figure 7, the plotted points could be connected to form a straight *line*, so the variables x (year) and y (cost) have a *line*ar relationship. The increase in costs is also reflected.

CAUTION

The equation in Example 8 is valid only for the years 1990 through 2000 because it was based on data for those years. Do not assume that this equation would provide reliable data for other years since the data for those years may not follow the same pattern.

ANSWERS

10. (a) $y = 770$
(b) $y = 907$; In 2000, Americans each spent about \$907 on doctors' visits.

3.1 Exercises

FOR EXTRA HELP

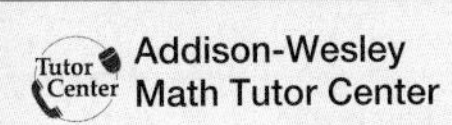
Addison-Wesley Math Tutor Center

MathXL
Digital Video Tutor CD 3 Videotape 3

Student's Solutions Manual

MyMathLab

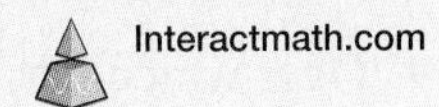
Interactmath.com

On February 13, 2000, Peanuts *fans bid farewell to new segments of this popular comic strip (which is still being printed as* Classic Peanuts*). The circle graph shows the results of a survey of adults to determine their favorite* Peanuts *characters. Use the circle graph to work Exercises 1–4. See Example 1.*

Source: *USA Today*/CNN/Gallup Poll conducted nationwide February 4–6, 2000.

1. Which *Peanuts* character was most popular? What percent of those surveyed named this character as their favorite?

2. A random sample of 5000 adults is surveyed.
   (a) Estimate how many would be expected to name Lucy as their favorite *Peanuts* character. (*Hint:* To estimate, round 8% to 10%.)
   (b) Use the actual figure from the graph to determine how many adults would name Lucy as their favorite character.

3. Regardless of the number of adults surveyed, how many would we expect to name Charlie Brown as their favorite *Peanuts* character compared to Linus?

4. Using the group of 5000 adults, confirm your answer to Exercise 3.

The bar graph compares egg production in millions of eggs for six states in the year 2002. Use the bar graph to work Exercises 5–8. See Example 2.

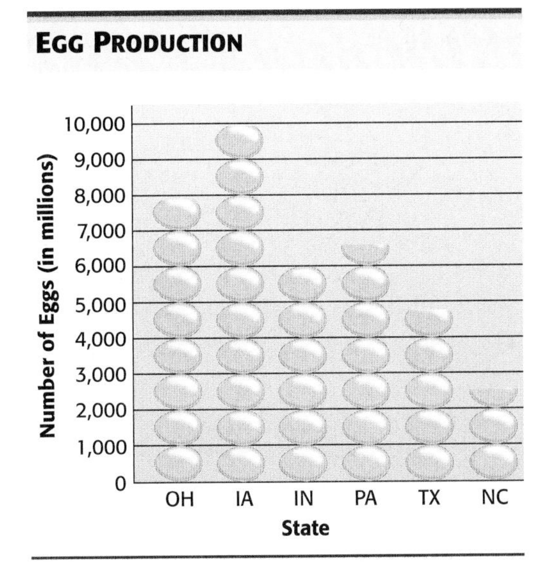

Source: U.S. Department of Agriculture.

5. Name the top two egg-producing states. Estimate their production.

6. Which states had egg production less than 5000 million eggs?

7. Which state had the least production? Estimate this production.

8. How did egg production in Texas (TX) compare to egg production in Iowa (IA)?

The line graph shows the average price, adjusted for inflation, that Americans have paid for a gallon of gasoline for selected years since 1970. Use the line graph to work Exercises 9–12. See Example 3.

9. Over which period of years did the greatest increase in the price of a gallon of gas occur? About how much was this increase?

10. Estimate the price of a gallon of gas during 1985, 1990, 1995, and 2000.

11. Describe the trend in gas prices from 1980 to 1995.

12. During which year(s) did a gallon of gas cost approximately $1.50?

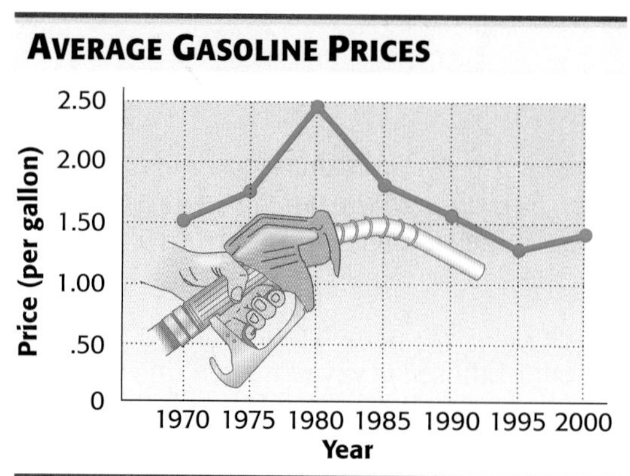

Source: American Petroleum Institute; AP research.

Use the concepts of this section to fill in each blank with the correct response.

13. The symbol (x, y) ______ (does/does not) represent an ordered pair, while the symbols $[x, y]$ and $\{x, y\}$ ______ (do/do not) represent ordered pairs.

14. The point whose graph has coordinates $(-4, 2)$ is in quadrant ______.

15. The point whose graph has coordinates $(0, 5)$ lies on the ______-axis.

16. The ordered pair (4, ______) is a solution of the equation $y = 3$.

17. The ordered pair (______, -2) is a solution of the equation $x = 6$.

18. The ordered pair $(3, 2)$ is a solution of the equation $2x - 5y =$ ______.

Decide whether each ordered pair is a solution of the given equation. See Example 4.

19. $x + y = 9$; $(0, 9)$

20. $x + y = 8$; $(0, 8)$

21. $2x - y = 6$; $(4, 2)$

22. $2x + y = 5$; $(3, -1)$

23. $4x - 3y = 6$; $(2, 1)$

24. $5x - 3y = 15$; $(5, 2)$

25. $y = \frac{2}{3}x$; $(-6, -4)$

26. $y = -\frac{1}{4}x$; $(-8, 2)$

27. $x = -6$; $(5, -6)$

28. $y = 2$; $(2, 4)$

29. Do $(4, -1)$ and $(-1, 4)$ represent the same ordered pair? Explain.

30. Explain why it would be easier to find the corresponding y-value for $x = \frac{1}{3}$ in the equation $y = 6x + 2$ than it would be for $x = \frac{1}{7}$.

Complete each ordered pair for the equation $y = 2x + 7$. See Example 5.

31. (2,) **32.** (0,) **33.** (, 0) **34.** (, −3)

Complete each ordered pair for the equation $y = -4x - 4$. See Example 5.

35. (0,) **36.** (, 0) **37.** (, 16) **38.** (, 24)

Complete each table of values. In Exercises 39–42, write the results as ordered pairs. See Example 6.

39. $2x + 3y = 12$

x	y
0	
	0
	8

( )

40. $4x + 3y = 24$

x	y
0	
	0
	4

41. $3x - 5y = -15$

x	y
0	
	0
	−6

( ) ( )

42. $4x - 9y = -36$

x	y
	0
0	
	8

( )

43. $x = -9$

x	y
	6
	2
	−3

44. $x = 12$

x	y
	3
	8
	0

45. $y = -6$

x	y
8	
4	
−2	

46. $y = -10$

x	y
4	
0	
−4	

47. $x - 8 = 0$

x	y
	8
	3
	0

48. $y + 2 = 0$

x	y
9	
2	
0	

Give the ordered pairs for the points labeled A–F in the figure.

49. *A* **50.** *B* **51.** *C*

52. *D* **53.** *E* **54.** *F*

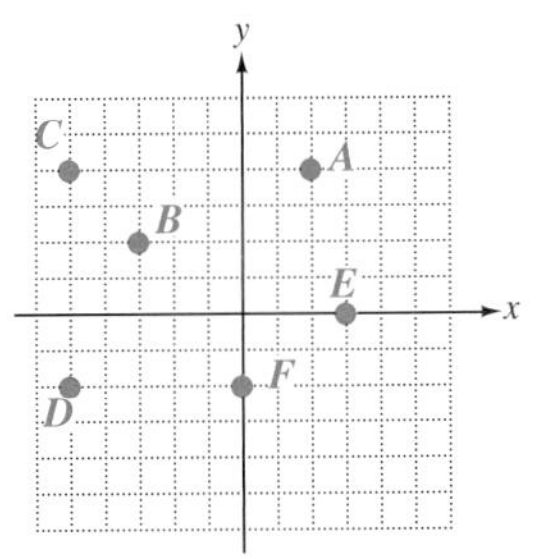

Fill in each blank with the word positive *or the word* negative.

The point with coordinates (x, y) is in

55. quadrant III if x is ________ and y is ________.

56. quadrant II if x is ________ and y is ________.

57. quadrant IV if x is ________ and y is ________.

58. quadrant I if x is ________ and y is ________.

59. A point (x, y) has the property that $xy < 0$. In which quadrant(s) must the point lie? Explain.

60. A point (x, y) has the property that $xy > 0$. In which quadrant(s) must the point lie? Explain.

Plot each ordered pair on the rectangular coordinate system provided. See Example 7.

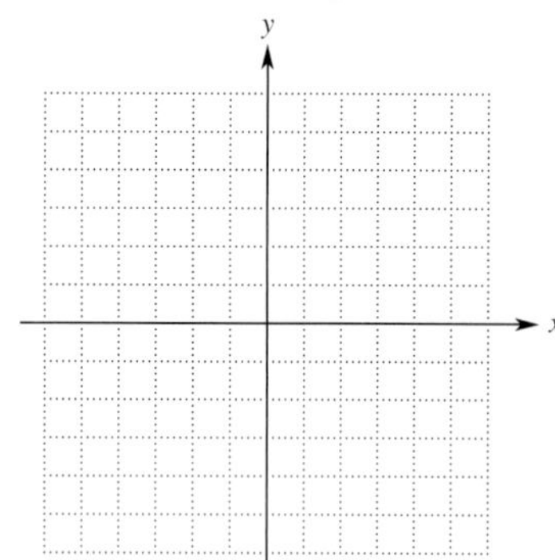

61. $(6, 2)$ **62.** $(5, 3)$ **63.** $(-4, 2)$

64. $(-3, 5)$ **65.** $\left(-\frac{4}{5}, -1\right)$ **66.** $\left(-\frac{3}{2}, -4\right)$

67. $(0, 4)$ **68.** $(0, -3)$ **69.** $(4, 0)$ **70.** $(-3, 0)$

Complete each table of values, and then plot the ordered pairs. See Examples 6 and 7.

71. $x - 2y = 6$

x	y
0	
	0
2	
	−1

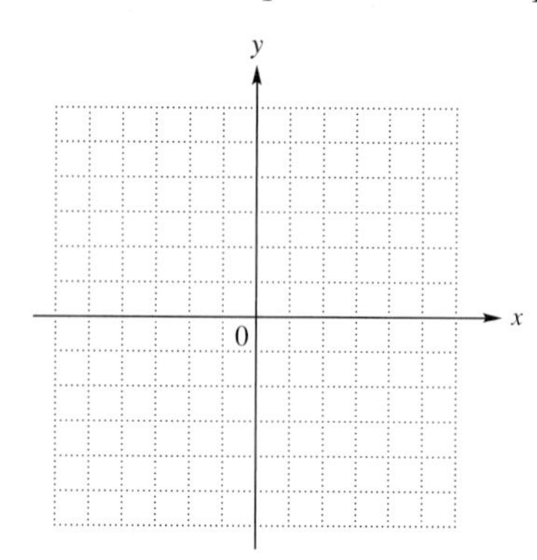

72. $2x - y = 4$

x	y
0	
	0
1	
	−6

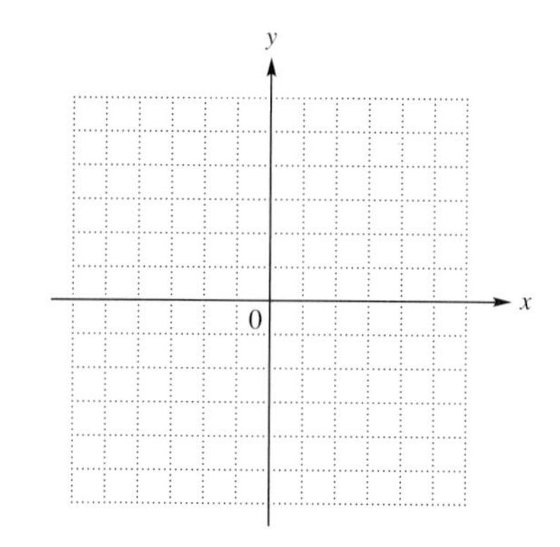

73. $3x - 4y = 12$

x	y
0	
	0
−4	
	−4

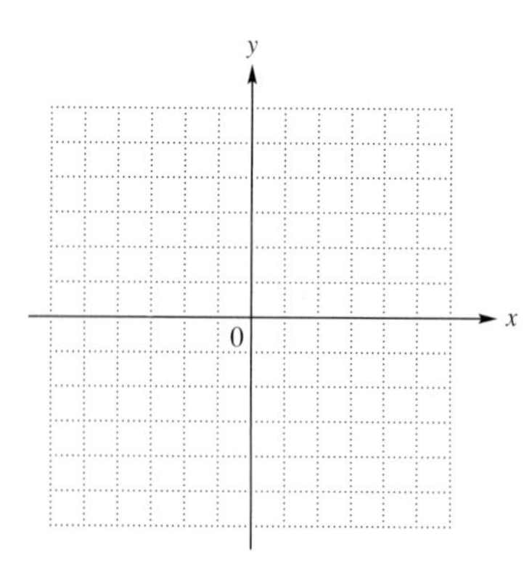

74. $2x - 5y = 10$

x	y
0	
	0
−5	
	−3

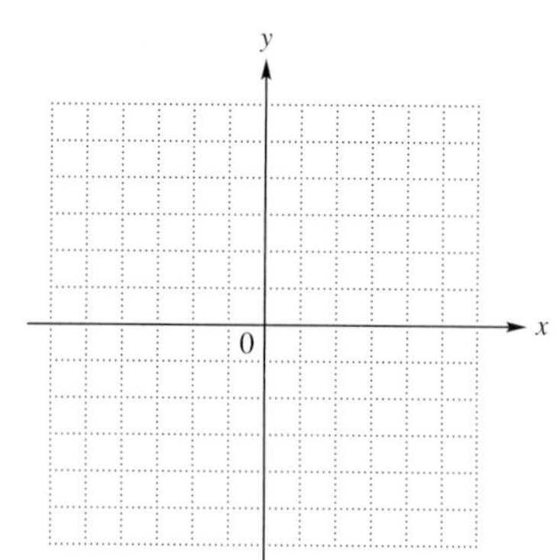

75. $y + 4 = 0$

x	y
0	
5	
−2	
−3	

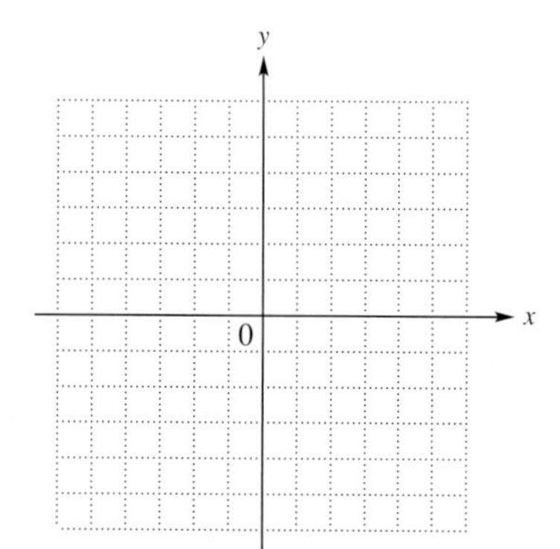

76. $x - 5 = 0$

x	y
	1
	0
	6
	−4

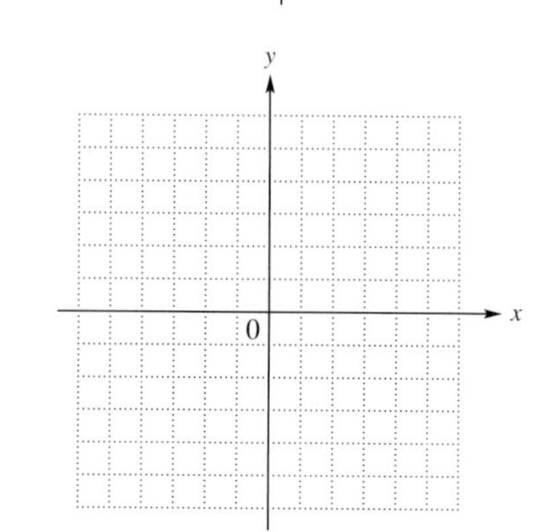

77. Look at the graphs of the ordered pairs in Exercises 71–76. Describe the pattern indicated by the plotted points.

Work each problem. See Example 8.

78. The table shows the amount of e-commerce in billions of dollars.

Year	Amount (in billions)
1996	.7
1997	2.6
1998	7.8
1999	18.9
2000	38.2
2001	45.7

Source: Jupiter Research.

(a) Write the data from the table as ordered pairs (x, y), where x represents the year and y represents the amount of e-commerce in billions of dollars.

(b) What does the ordered pair (2002, 62.0) mean in the context of this discussion?

(c) Make a scatter diagram of the data using the ordered pairs from part (a).

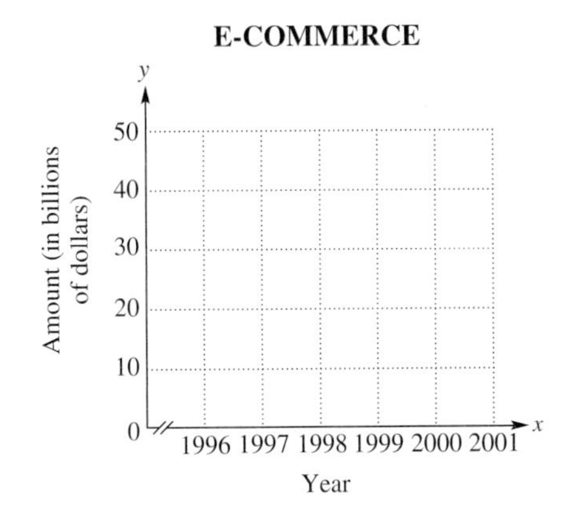

(d) Describe what is happening to the amount of e-commerce during these years.

79. The table shows the rate (in percent) at which 4-year college students (both private and public) graduate within 5 years.

Year	Percent
1996	53.3
1997	52.8
1998	52.1
1999	51.6
2000	51.2
2001	50.9

Source: ACT.

(a) Write the data from the table as ordered pairs (x, y), where x represents the year and y represents graduation percent.

(b) What does the ordered pair (2002, 51.0) mean in the context of this discussion?

(c) Make a scatter diagram of the data using the ordered pairs from part (a).

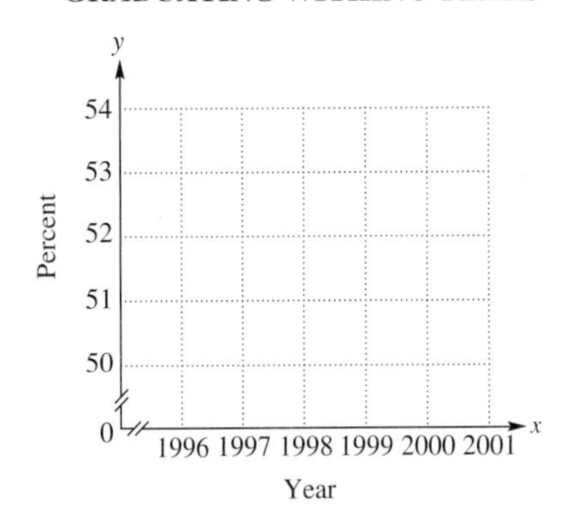

(d) Describe the pattern indicated by the points on the scatter diagram. What is happening to graduation rates for 4-year college students within 5 years?

80. The maximum benefit for the heart from exercising occurs if the heart rate is in the target heart rate zone. The lower limit of this target zone can be approximated by the linear equation

$$y = -.5x + 108,$$

where x represents age and y represents heartbeats per minute. (*Source:* www.fitresource.com)

(a) Complete the table of values for this linear equation.

Age	Heartbeats (per minute)
20	
40	
60	
80	

(b) Write the data from the table of values as ordered pairs.

(c) Make a scatter diagram of the data. Do the points lie in an approximately linear pattern?

TARGET HEART RATE ZONE
(Lower Limit)

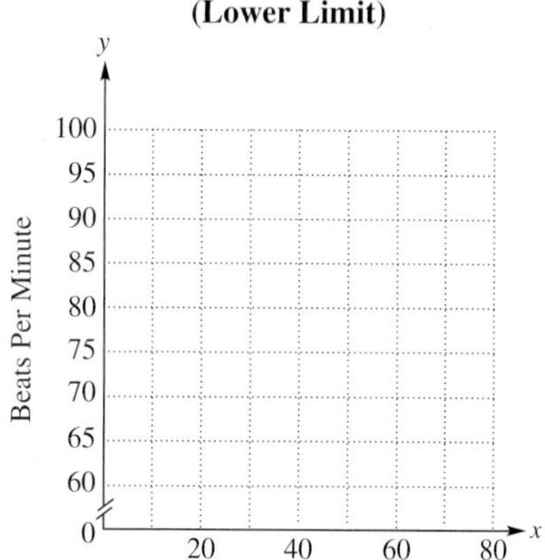

81. (See Exercise 80.) The upper limit of the target heart rate zone can be approximated by the linear equation

$$y = -.8x + 173,$$

where x represents age and y represents heartbeats per minute. (*Source:* www.fitresource.com)

(a) Complete the table of values for this linear equation.

Age	Heartbeats (per minute)
20	
40	
60	
80	

(b) Write the data from the table of values as ordered pairs.

(c) Make a scatter diagram of the data. Describe the pattern indicated by the data.

TARGET HEART RATE ZONE
(Upper Limit)

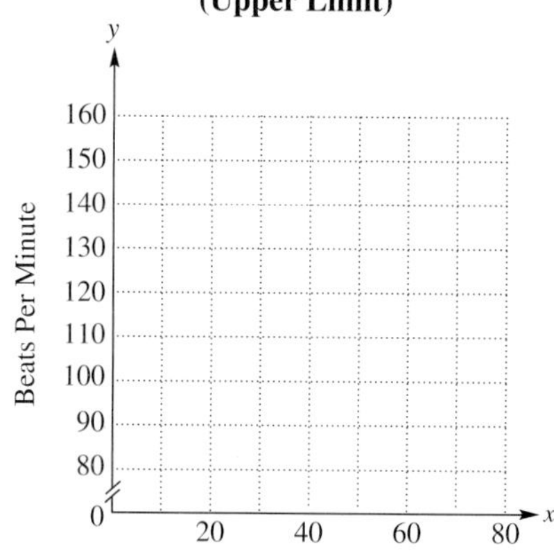

82. Refer to Exercises 80 and 81. What is the target heart rate zone for age 20? age 40?

3.2 Graphing Linear Equations in Two Variables

OBJECTIVES

1. Graph linear equations by plotting ordered pairs.
2. Find intercepts.
3. Graph linear equations where the intercepts coincide.
4. Graph linear equations of the form $y = k$ or $x = k$.
5. Use a linear equation to model data.

OBJECTIVE 1 Graph linear equations by plotting ordered pairs. There are infinitely many ordered pairs that satisfy an equation in two variables. We find ordered pairs that are solutions of the equation $x + 2y = 7$ by choosing as many values of x (or y) as we wish and then completing each ordered pair.

For example, if we choose $x = 1$, then $y = 3$, so the ordered pair (1, 3) is a solution of the equation $x + 2y = 7$.

$$1 + 2(3) = 1 + 6 = 7$$

Work Problem 1 at the Side.

1 Complete each ordered pair for the equation $x + 2y = 7$.

(a) $(-3, \quad)$

(b) $(3, \quad)$

(c) $(-1, \quad)$

(d) $(7, \quad)$

Figure 8 shows a graph of all the ordered pairs found for $x + 2y = 7$ above and in Problem 1 at the side.

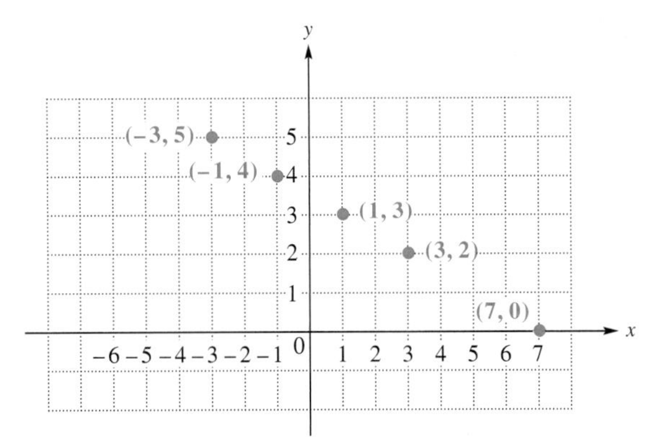

Figure 8

Notice that the points plotted in this figure all appear to lie on a straight line. The line that goes through these points is shown in Figure 9. In fact, all ordered pairs satisfying the equation $x + 2y = 7$ correspond to points that lie on this same straight line. This line gives a "picture" of all the solutions of the equation $x + 2y = 7$. Only a portion of the line is shown here, but it extends indefinitely in both directions, as suggested by the arrowhead on each end of the line.

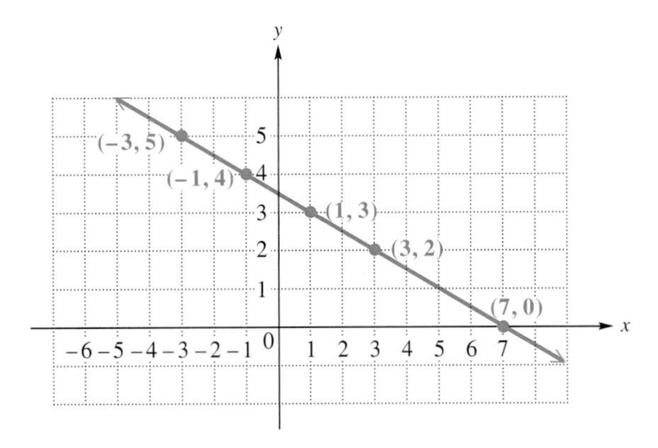

Figure 9

The line in Figure 9 is called the **graph** of the equation $x + 2y = 7$, and the process of plotting the ordered pairs and drawing the line through the corresponding points is called **graphing.** The preceding discussion can be generalized.

ANSWERS
1. (a) $(-3, 5)$ (b) $(3, 2)$ (c) $(-1, 4)$ (d) $(7, 0)$

The graph of any linear equation in two variables is a straight line.

(Notice that the word *line* appears in the term "*line*ar equation.")

Because two distinct points determine a line, a straight line can be graphed by finding any two different points on the line. However, it is a good idea to plot a third point as a check.

EXAMPLE 1 Graphing a Linear Equation

Graph the linear equation $y = -\frac{3}{2}x + 3$.

Although this equation is not in the form $Ax + By = C$, it *could* be written in that form, so it is a linear equation. Two different points on the graph can be found by first letting $x = 0$ and then letting $y = 0$.

If $x = 0$, then

$$y = -\frac{3}{2}x + 3$$

$$y = -\frac{3}{2}(\mathbf{0}) + 3 \qquad \text{Let } x = 0.$$

$$y = 0 + 3$$

$$y = 3.$$

If $y = 0$, then

$$y = -\frac{3}{2}x + 3$$

$$\mathbf{0} = -\frac{3}{2}x + 3 \qquad \text{Let } y = 0.$$

$$\frac{3}{2}x = 3$$

$$x = 2. \qquad \text{Multiply by } \tfrac{2}{3}.$$

This gives the ordered pairs (0, 3) and (2, 0). Find a third point (as a check) by letting x or y equal some other number. For example, let $x = -2$. (Any number could be used, but a multiple of 2 makes multiplying by $-\frac{3}{2}$ easier.)

$$y = -\frac{3}{2}x + 3$$

$$y = -\frac{3}{2}(\mathbf{-2}) + 3 \qquad \text{Let } x = -2.$$

$$y = 3 + 3$$

$$y = 6$$

These three ordered pairs are shown in the table with Figure 10. Plot the corresponding points, then draw a line through them. This line, shown in Figure 10, is the graph of $y = -\frac{3}{2}x + 3$.

x	y
0	3
2	0
−2	6

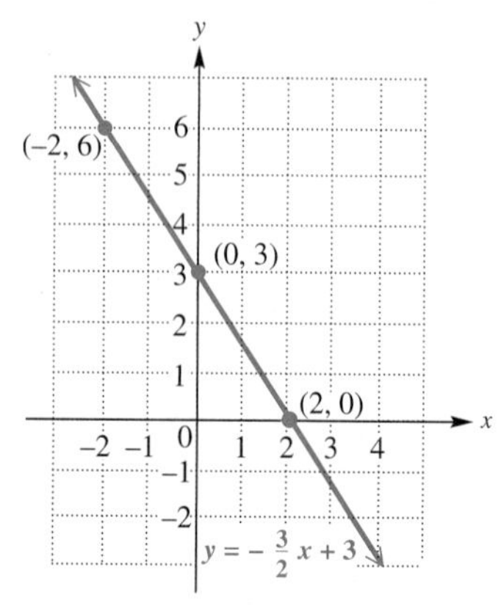

Figure 10

CAUTION

When graphing a linear equation as in Example 1, all three points should lie on the same straight line. If they don't, double-check the ordered pairs you found.

Work Problem 2 at the Side.

EXAMPLE 2 Graphing a Linear Equation

Graph the linear equation $4x = 5y + 20$.

As before, at least two different points are needed to draw the graph. First let $x = 0$ and then let $y = 0$ to complete two ordered pairs.

$4x = 5y + 20$		$4x = 5y + 20$	
$4(\mathbf{0}) = 5y + 20$	Let $x = 0$.	$4x = 5(\mathbf{0}) + 20$	Let $y = 0$.
$0 = 5y + 20$		$4x = 0 + 20$	
$-5y = 20$		$4x = 20$	
$y = -4$		$x = 5$	

The ordered pairs are $(0, -4)$ and $(5, 0)$. Get a third ordered pair (as a check) by choosing some number other than 0 for x or y. We choose $y = 2$. Replacing y with 2 in the equation $4x = 5y + 20$ leads to the ordered pair $\left(\frac{15}{2}, 2\right)$, or $\left(7\frac{1}{2}, 2\right)$.

Plot the three ordered pairs $(0, -4)$, $(5, 0)$, and $\left(7\frac{1}{2}, 2\right)$, and draw a line through them. This line, shown in Figure 11, is the graph of $4x = 5y + 20$.

x	y
0	-4
5	0
$7\frac{1}{2}$	2

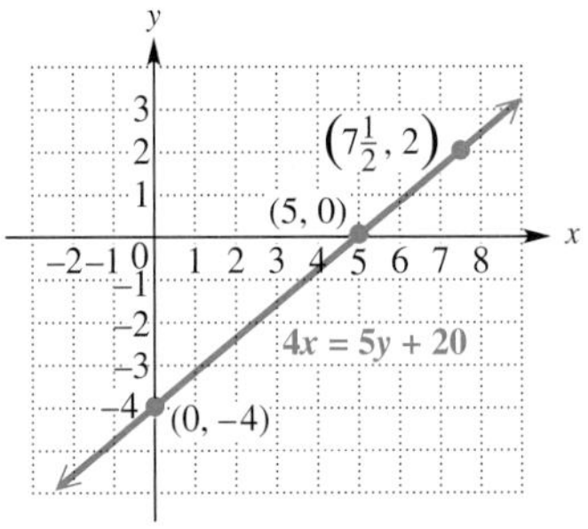

Figure 11

Work Problem 3 at the Side.

OBJECTIVE 2 Find intercepts. In Figure 11, the graph crosses or intersects the y-axis at $(0, -4)$ and the x-axis at $(5, 0)$. For this reason, $(0, -4)$ is called the **y-intercept,** and $(5, 0)$ is called the **x-intercept** of the graph. The intercepts are particularly useful for graphing linear equations. (In general, any point on the y-axis has x-coordinate 0; any point on the x-axis has y-coordinate 0.) The intercepts are found by replacing, in turn, each variable with 0 in the equation and solving for the value of the other variable.

Finding Intercepts

To find the x-intercept, let $y = 0$ in the given equation and solve for x. Then $(x, 0)$ is the x-intercept.

To find the y-intercept, let $x = 0$ in the given equation and solve for y. Then $(0, y)$ is the y-intercept.

2 Complete the table of values, and graph the linear equation.

$x + y = 6$

x	y
0	
	0
2	

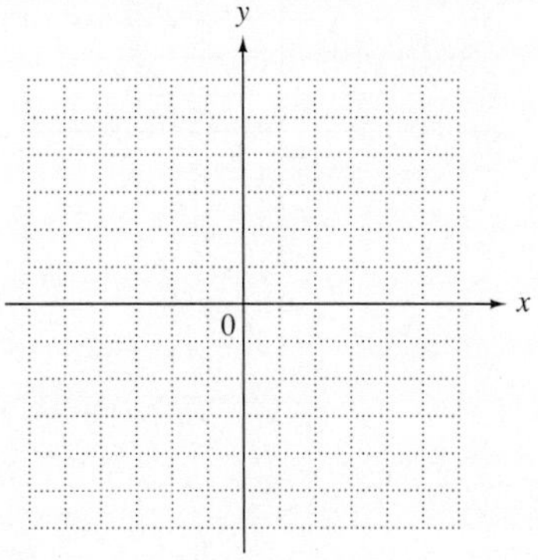

3 Make a table of values, and graph the linear equation.

$2x = 3y + 6$

x	y

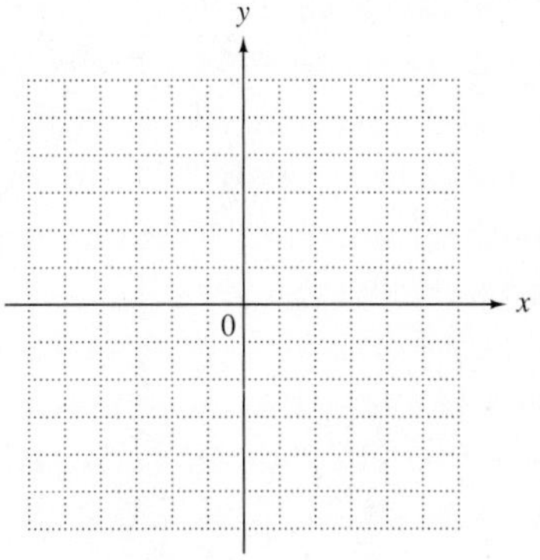

ANSWERS

2.

x	y
0	6
6	0
2	4

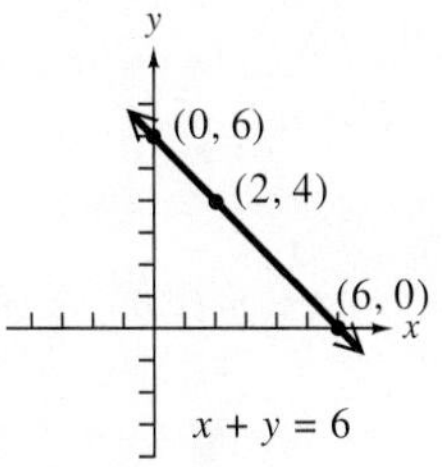

3.

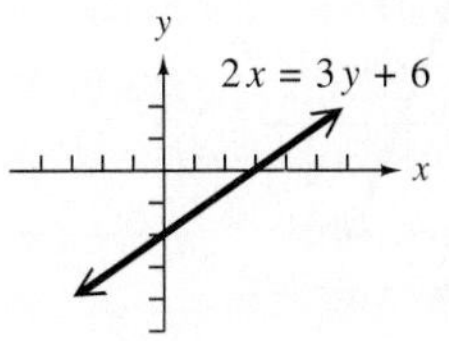

4 Find the intercepts for the graph of $5x + 2y = 10$. Then draw the graph. (Be sure to get a third point as a check.)

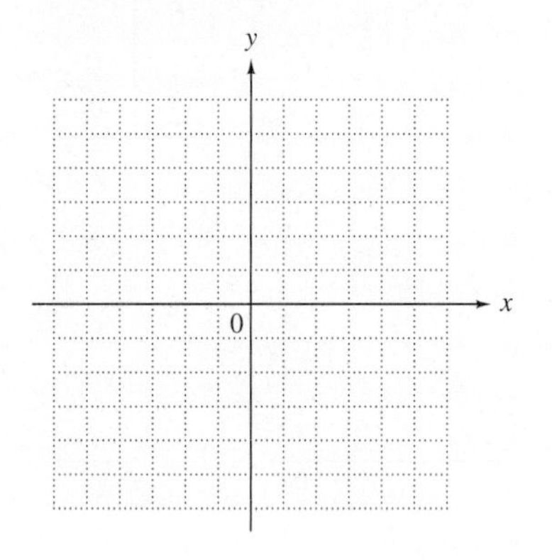

ANSWERS

4. x-intercept $(2, 0)$; y-intercept $(0, 5)$

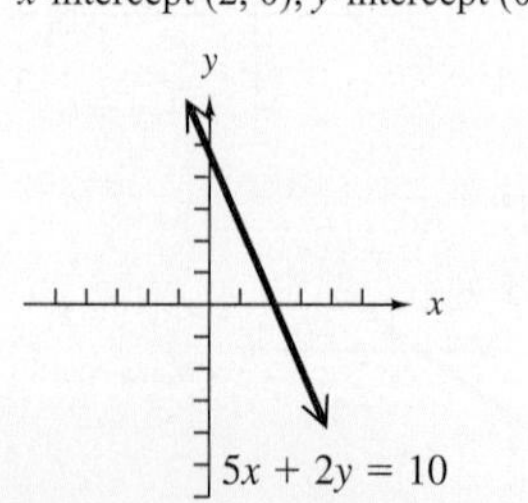

EXAMPLE 3 **Finding Intercep**

Find the intercepts for the graph of 2 $+ y = 4$. Then draw the graph.

To find the y-intercept, let $x =$ (to find the x-intercept, let $y = 0$.

$2x + y = 4$		$2x + y = 4$	
$2(\mathbf{0}) + y = 4$	Let $x = 0$.	$2x + \mathbf{0} = 4$	Let $y = 0$.
$0 + y = 4$		$2x = 4$	
$y = 4$		$x = 2$	

The y-intercept is $(0, 4)$. The x-interc pt is $(2, 0)$. The graph with the two intercepts shown in color is given in F gure 12. Find a third point as a check. For example, choosing $x = 1$ gives $= 2$. Plot $(0, 4)$, $(2, 0)$, and $(1, 2)$ and draw a line through them. This line, s own in Figure 12, is the graph.

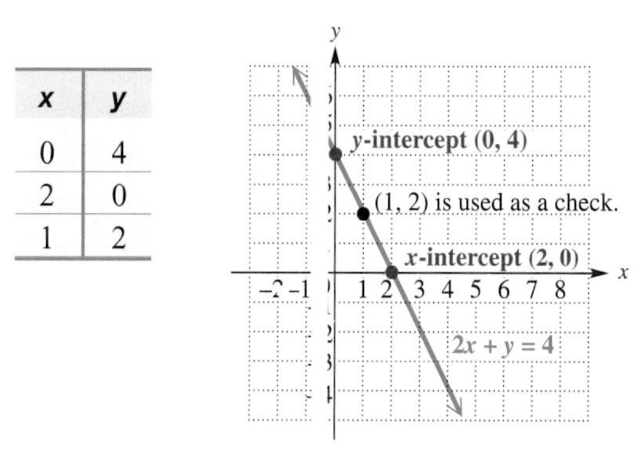

x	y
0	4
2	0
1	2

Figure 12

Work Problem 4 at the Side.

OBJECTIVE 3 **Graph linear equat cns where the intercepts coincide.** In the preceding examples, the x- and y-intercepts were used to help draw the graphs. This is not always possible Example 4 shows what to do when the x- and y-intercepts are the same poi it (that is, coincide).

EXAMPLE 4 **Graphing an Equ tion of the Form $Ax + By = 0$**

Graph the linear equation $x - 3y =$ (.

If we let $x = 0$, then $y = 0$, givin the ordered pair $(0, 0)$. Letting $y = 0$ also gives $(0, 0)$. This is the same or ered pair, so choose two *other* values for x or y. Choosing 2 for y gives $x - 3 \cdot 2 = 0$, or $x = 6$, so another ordered pair is $(6, 2)$. Choosing -6 or x gives -2 for y. The ordered pairs $(-6, -2)$, $(0, 0)$, and $(6, 2)$ are used o sketch the graph in Figure 13.

x	y
0	0
6	2
−6	−2

Figure 13

Example 4 can be generalized as follows.

If A and B are nonzero real numbers, the graph of a linear equation of the form

$$Ax + By = 0$$

passes through the origin (0, 0).

Work Problem 5 at the Side.

OBJECTIVE 4 Graph linear equations of the form $y = k$ or $x = k$. The equation $y = -4$ is a linear equation in which the coefficient of x is 0. (To see this, write $y = -4$ as $0x + y = -4$.) Also, $x = 3$ is a linear equation in which the coefficient of y is 0. These equations lead to horizontal or vertical straight lines, as the next example shows.

EXAMPLE 5 Graphing Equations of the Form $y = k$ and $x = k$

(a) Graph the linear equation $y = -4$.

As the equation states, for any value of x, y is always equal to -4. Three ordered pairs that satisfy the equation are shown in the table of values. Drawing a line through these points gives the horizontal line in Figure 14. The y-intercept is $(0, -4)$; there is no x-intercept.

x	y
-2	-4
0	-4
3	-4

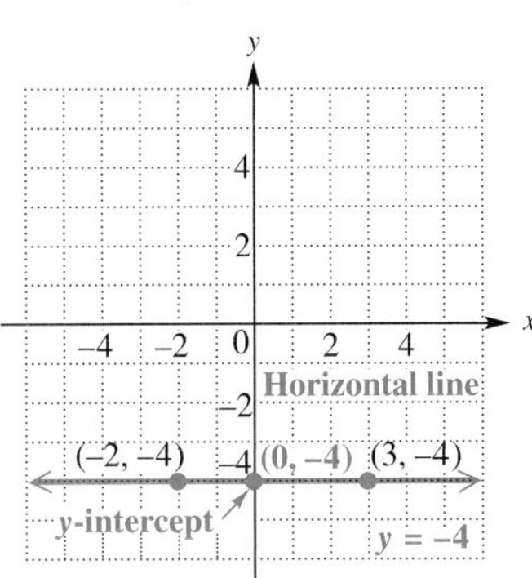

Figure 14

(b) Graph $x - 3 = 0$.

First add 3 to each side of $x - 3 = 0$ to get $x = 3$. All the ordered pairs that satisfy this equation have x-coordinate 3. Any number can be used for y. See Figure 15 for the graph of this vertical line, along with a table of values. The x-intercept is (3, 0); there is no y-intercept.

x	y
3	3
3	0
3	-2

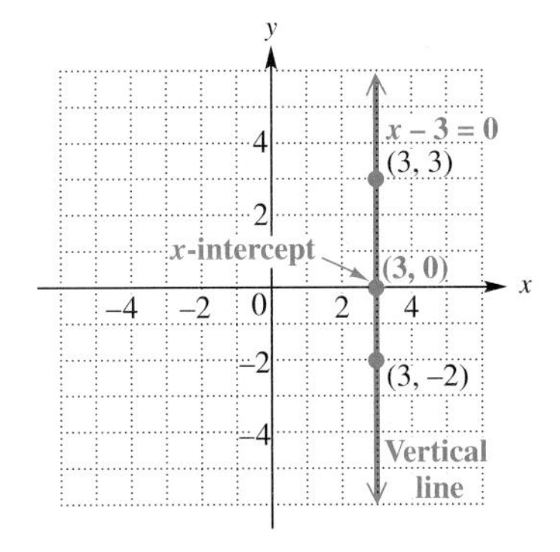

Figure 15

5 Graph each equation.

(a) $2x - y = 0$

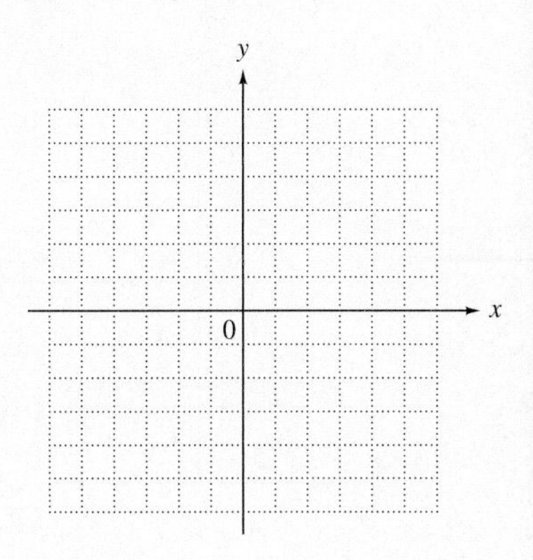

(b) $x = -4y$

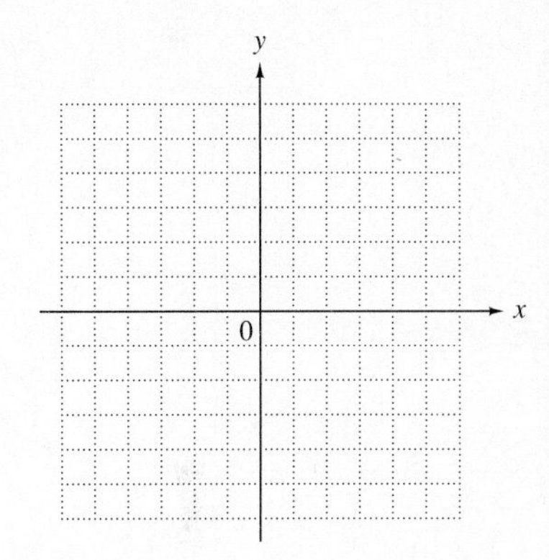

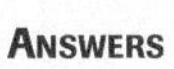

ANSWERS

5. (a)

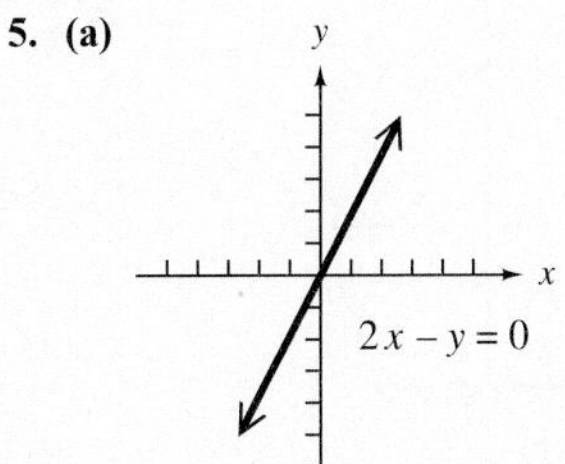

(b)

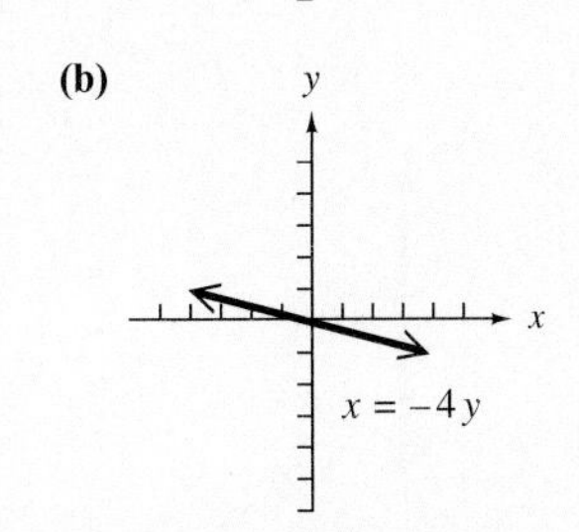

From the results in Example 5, we make the following observations.

Horizontal and Vertical Lines

The graph of the linear equation $y = k$, where k is a real number, is the horizontal line with y-intercept $(0, k)$ and no x-intercept.

The graph of the linear equation $x = k$, where k is a real number, is the vertical line with x-intercept $(k, 0)$ and no y-intercept.

Work Problem 6 at the Side.

The different forms of linear equations from this section and the methods of graphing them are summarized below.

Graphing Linear Equations

Equation	Graphing Method	Example
$y = k$	Draw a horizontal line through $(0, k)$.	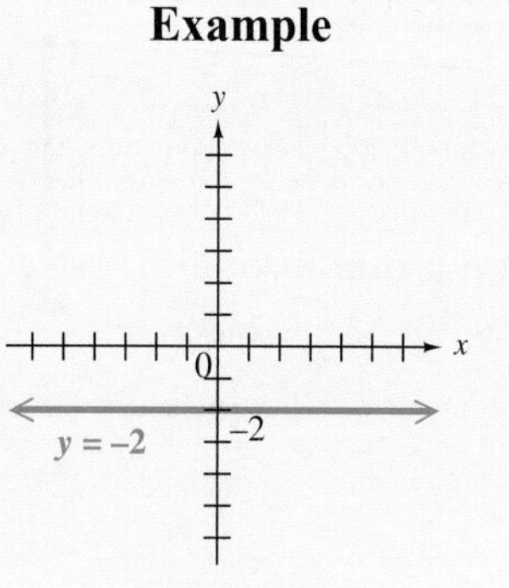
$x = k$	Draw a vertical line through $(k, 0)$.	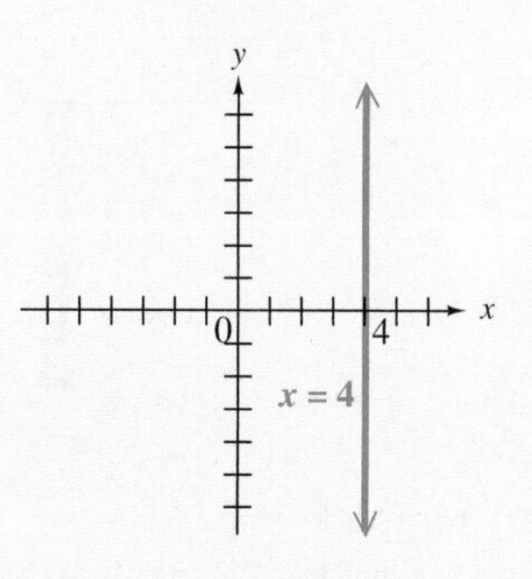
$Ax + By = 0$	Graph passes through $(0, 0)$. To get additional points that lie on the graph, choose any values for x or y, except 0.	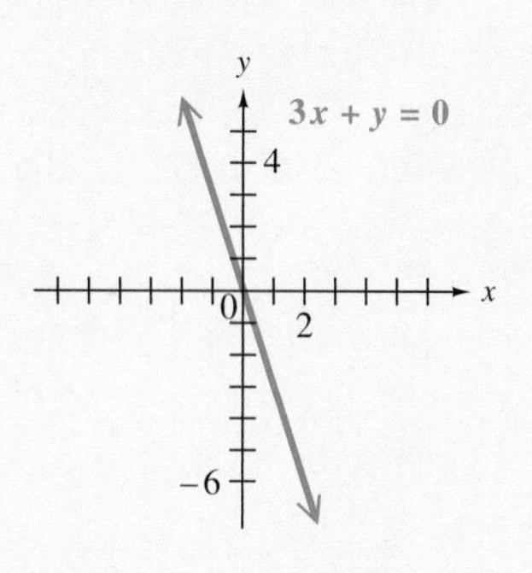

(continued)

6 Graph each equation.

(a) $y = -5$

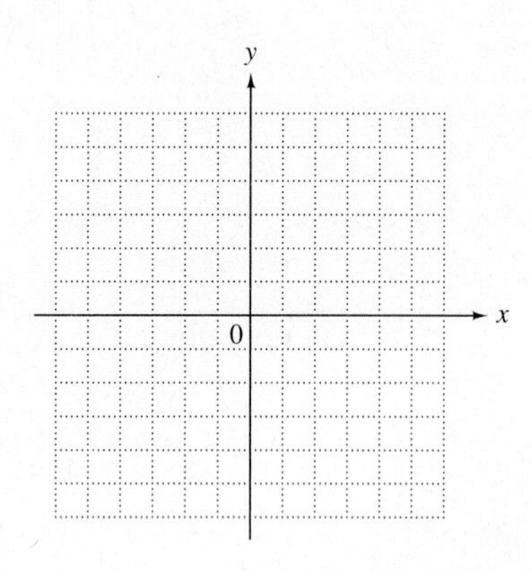

(b) $x + 4 = 6$

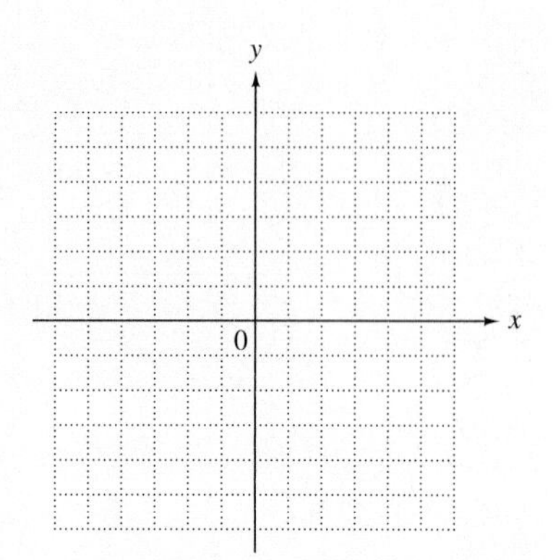

Answers

6. (a)

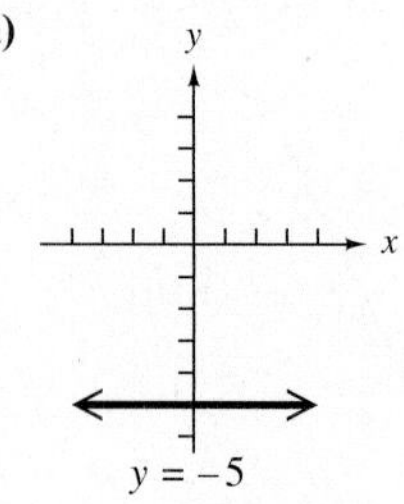

(b)

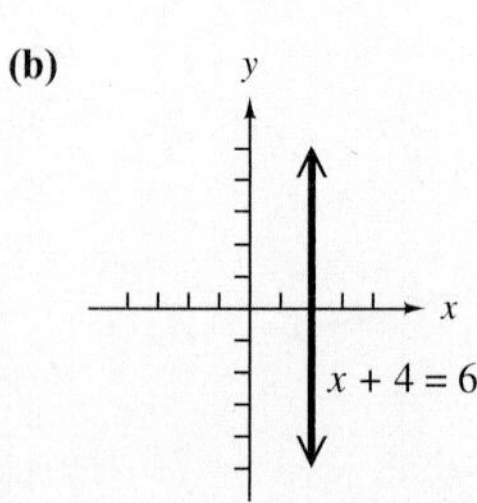

$Ax + By = C$ **$A, B,$ and $C \neq 0$**	Find any two points on the line. A good choice is to find the intercepts. Let $x = 0$, and find the corresponding value of y; then let $y = 0$, and find x. As a check, get a third point by choosing a value of x or y that has not yet been used.	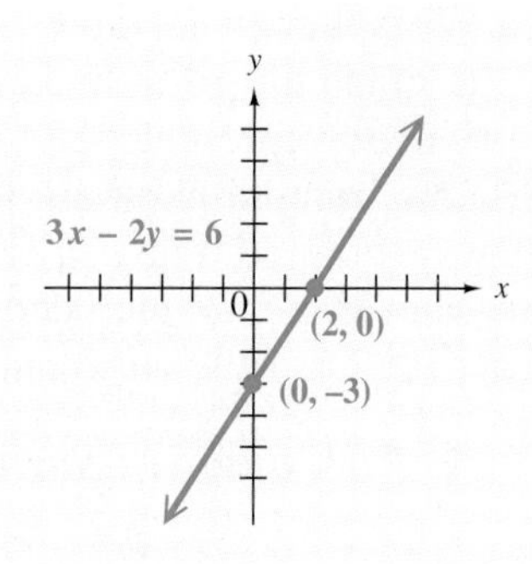

Work Problem 7 at the Side.

> **NOTE**
> Another method of graphing linear equations, using the concepts of slope and y-intercept, will be covered in Objective 2 of **Section 3.4.**

OBJECTIVE 5 Use a linear equation to model data.

EXAMPLE 6 Using a Linear Equation to Model Credit Card Debt

Credit card debt in households in the United States increased steadily during the 1990s. The amount of debt y in billions of dollars can be modeled by the linear equation

$$y = 47.3x + 281,$$

where $x = 0$ represents the year 1992, $x = 1$ represents 1993, and so on. (*Source:* Board of Governors of the Federal Reserve System.)

(a) Use the equation to approximate credit card debt in the years 1992, 1993, and 1999.

Substitute the appropriate value for each year x to find credit card debt in that year.

For 1992: $y = 47.3(0) + 281$ Replace x with 0.

$y = 281$ billion dollars

For 1993: $y = 47.3(1) + 281$ Replace x with 1.

$y = 328.3$ billion dollars

For 1999: $y = 47.3(7) + 281$ Replace x with 7.

$y = 612.1$ billion dollars

Continued on Next Page

7 Match the information about the graphs with the linear equations in A–D.

A. $x = 5$
B. $2x - 5y = 8$
C. $y - 2 = 3$
D. $x + 4y = 0$

(a) The graph of the equation is a horizontal line.

(b) The graph of the equation passes through the origin.

(c) The graph of the equation is a vertical line.

(d) The graph of the equation passes through (9, 2).

ANSWERS
7. (a) C **(b)** D **(c)** A **(d)** B

8 Use the graph and then the equation in Example 6 to approximate credit card debt in 1997.

(b) Write the information from part (a) as three ordered pairs, and use them to graph the given linear equation.

Since x represents the year and y represents the debt in billions of dollars, the ordered pairs are (0, 281), (1, 328.3), and (7, 612.1). Figure 16 shows a graph of these ordered pairs and the line through them. (Note that arrowheads are not included with the graphed line since the data are for the years 1992 to 1999 only, that is, from $x = 0$ to $x = 7$.)

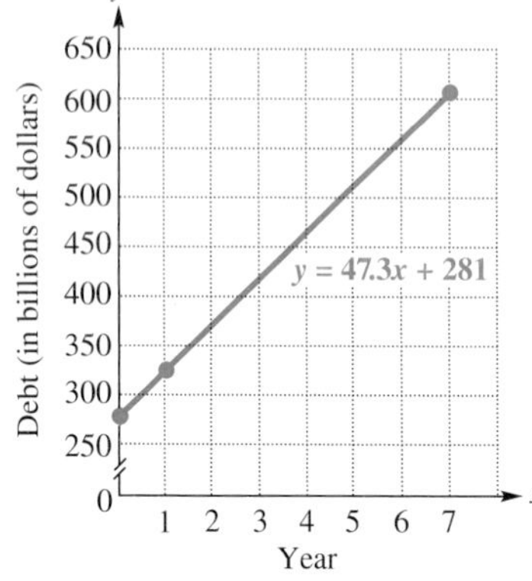

Figure 16

(c) Use the graph and then the equation to approximate credit card debt in 1996.

For 1996, $x = 4$. On the graph, find 4 on the horizontal axis and move up to the graphed line, then across to the vertical axis. It appears that credit card debt in 1996 was about 465 billion dollars.

To use the equation, substitute 4 for x.

$$y = 47.3x + 281$$

$$y = 47.3(4) + 281 \qquad \text{Let } x = 4.$$

$$y = 470.2 \text{ billion dollars}$$

This result for 1996 is close to our estimate using the graph.

Work Problem 8 at the Side.

ANSWERS

8. about 510 billion dollars; 517.5 billion dollars

3.2 Exercises

FOR EXTRA HELP

Addison-Wesley Math Tutor Center | MathXL 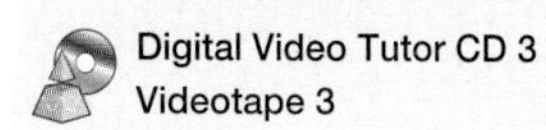 Digital Video Tutor CD 3 Videotape 3  Student's Solutions Manual | MyMathLab | 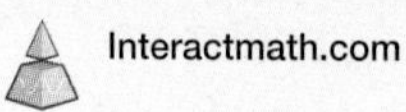 Interactmath.com

Complete the given ordered pairs for each equation. Then graph each equation by plotting the points and drawing a line through them. See Examples 1 and 2.

1. $y = -x + 5$

$(0, \quad), (\quad, 0), (2, \quad)$

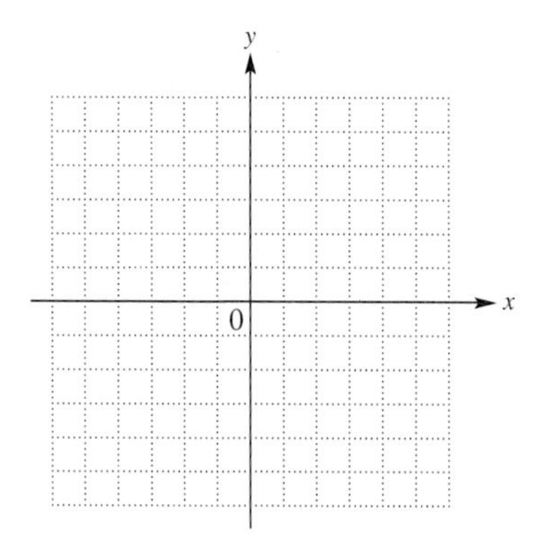

2. $y = x - 2$

$(0, \quad), (\quad, 0), (5, \quad)$

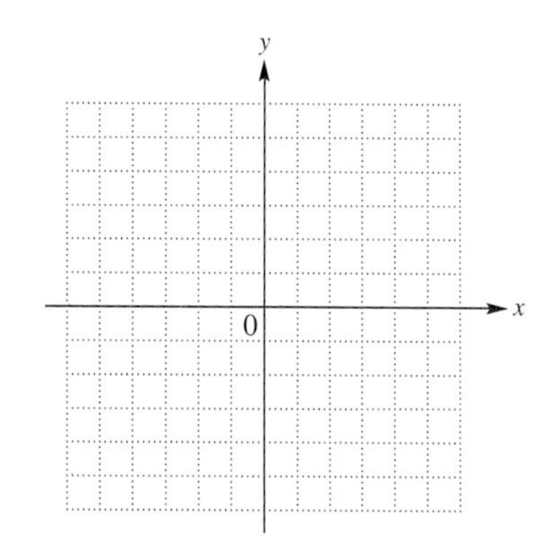

3. $y = \frac{2}{3}x + 1$

$(0, \quad), (3, \quad), (-3, \quad)$

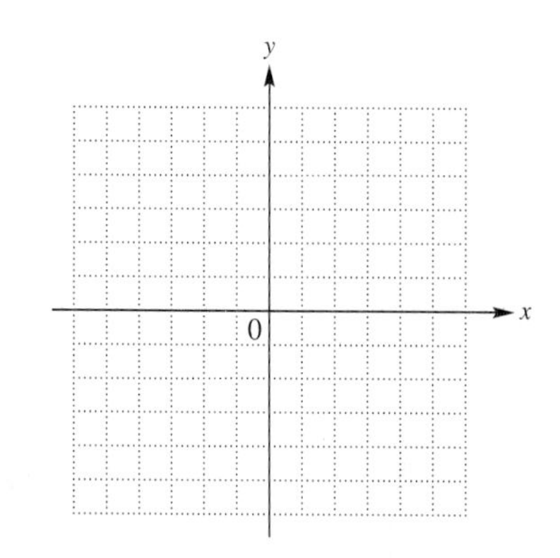

4. $y = -\frac{3}{4}x + 2$

$(0, \quad), (4, \quad), (-4, \quad)$

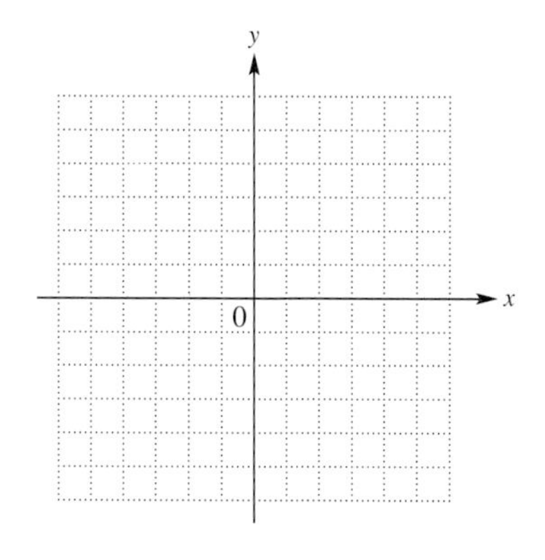

5. $3x = -y - 6$

$(0, \quad), (\quad, 0), \left(-\frac{1}{3}, \quad\right)$

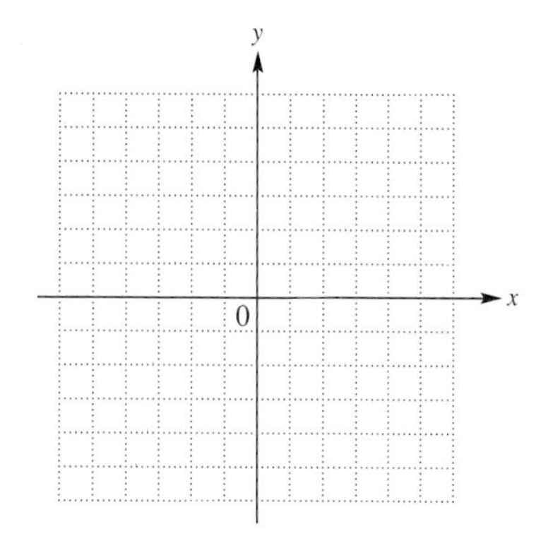

6. $x = 2y + 3$

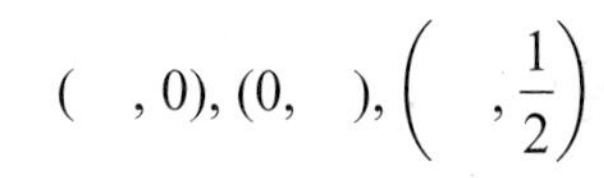

$(\quad, 0), (0, \quad), \left(\quad, \frac{1}{2}\right)$

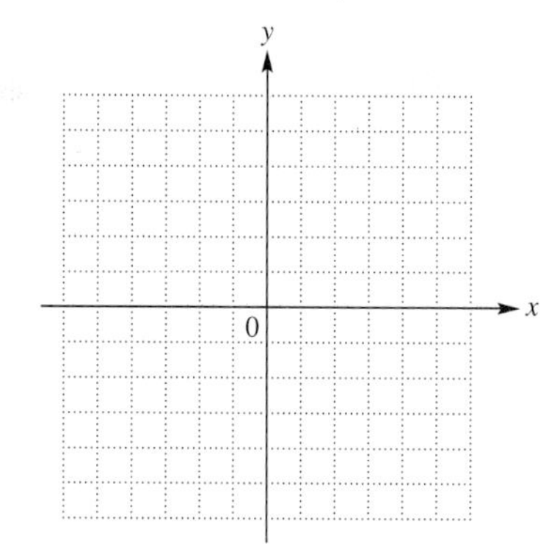

Match the information about each graph in Column I with the correct linear equation in Column II.

I	II
7. The graph of the equation has y-intercept $(0, -4)$.	**A.** $3x + y = -4$
8. The graph of the equation has $(0, 0)$ as x-intercept and y-intercept.	**B.** $x - 4 = 0$
9. The graph of the equation does not have an x-intercept.	**C.** $y = 4x$
10. The graph of the equation has x-intercept $(4, 0)$.	**D.** $y = 4$

Find the intercepts for the graph of each equation. See Example 3.

11. $2x - 3y = 24$

x-intercept:

y-intercept:

12. $-3x + 8y = 48$

x-intercept:

y-intercept:

13. $x + 6y = 0$

x-intercept:

y-intercept:

14. $3x - y = 0$

x-intercept:

y-intercept:

15. A student attempted to graph $4x + 5y = 0$ by finding intercepts. She first let $x = 0$ and found y; then she let $y = 0$ and found x. In both cases, the resulting point was $(0, 0)$. She knew that she needed at least two points to graph the line, but was unsure what to do next because finding intercepts gave her only one point. How would you explain to her what to do next?

16. What is the equation of the x-axis? What is the equation of the y-axis?

Graph each linear equation. See Examples 1–5.

17. $x = y + 2$

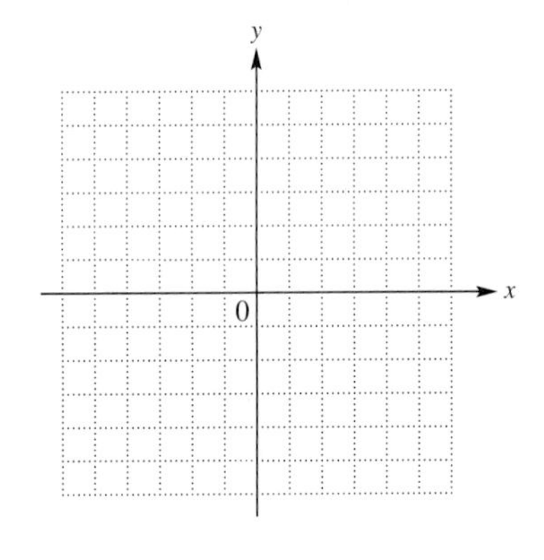

18. $x = -y + 6$

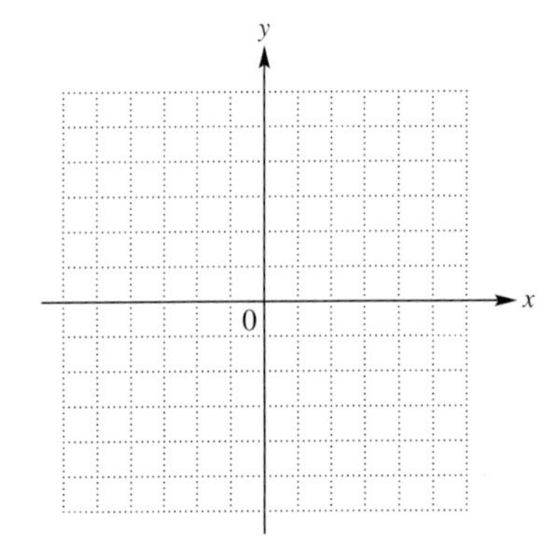

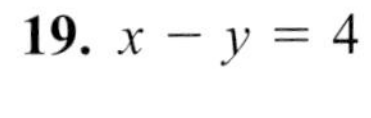

19. $x - y = 4$

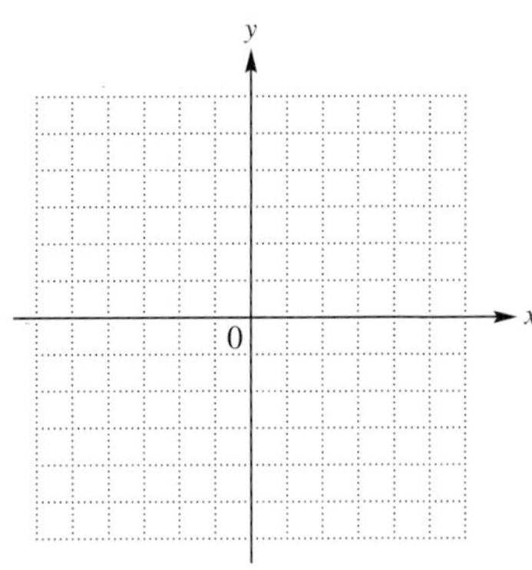

20. $x - y = 5$

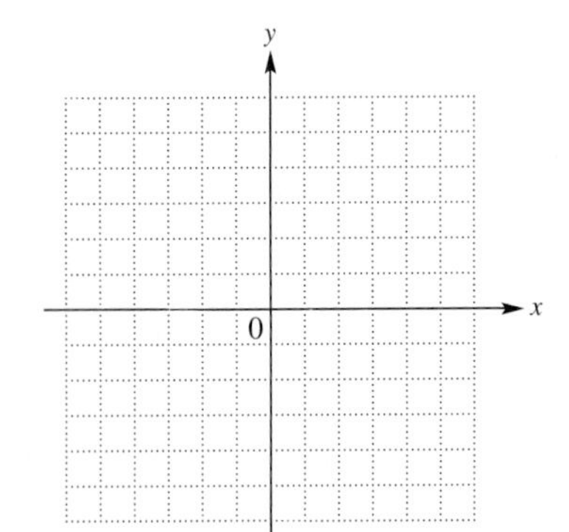

21. $2x + y = 6$

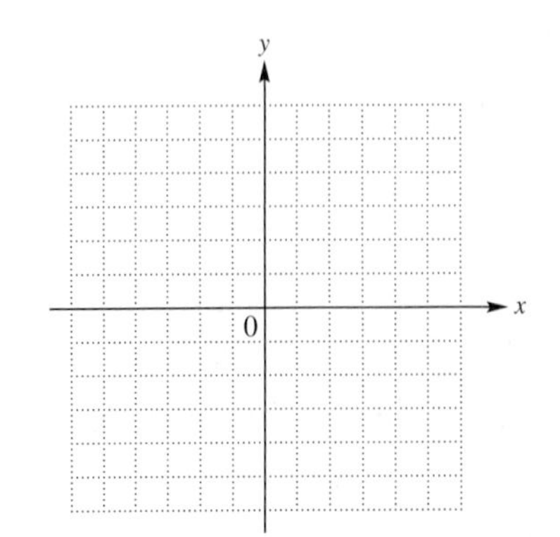

22. $-3x + y = -6$

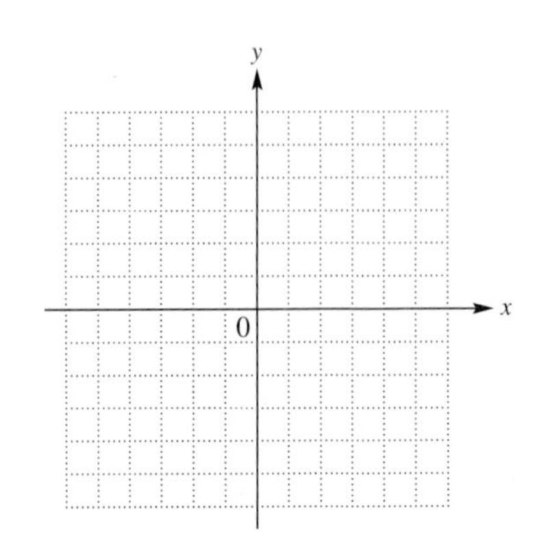

23. $3x + 7y = 14$

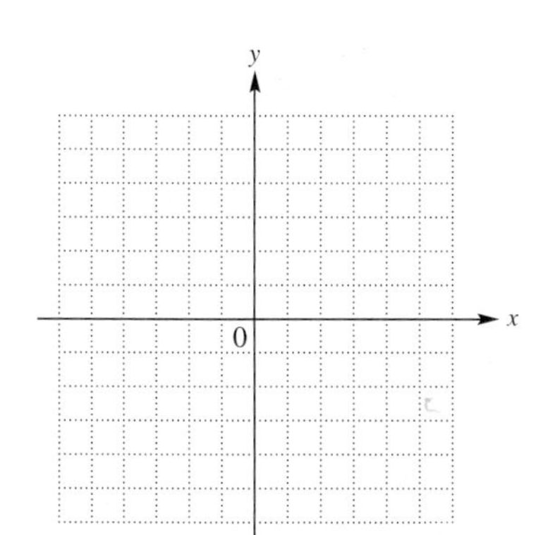

24. $6x - 5y = 18$

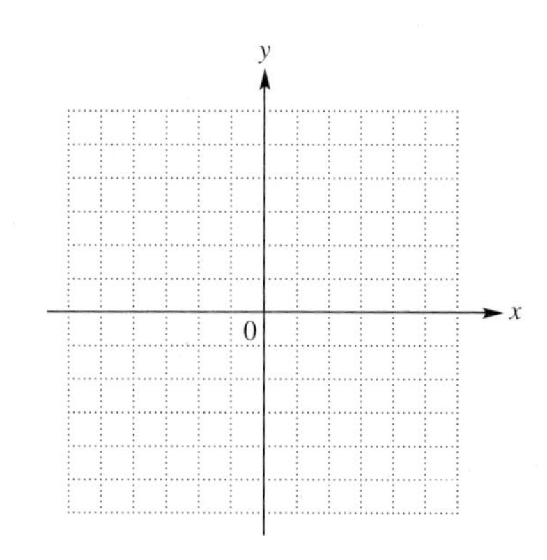

25. $y - 2x = 0$

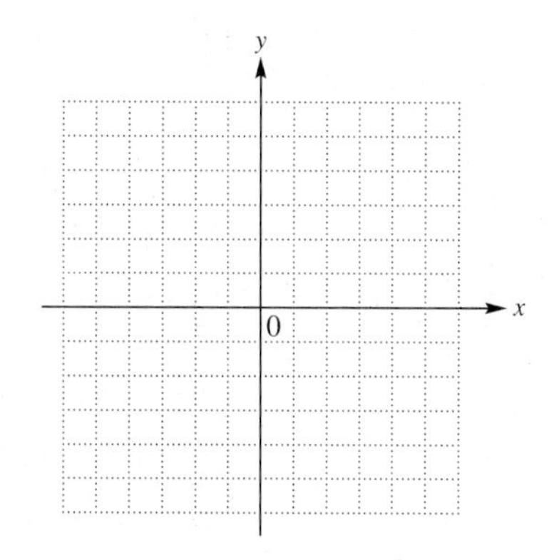

26. $y + 3x = 0$

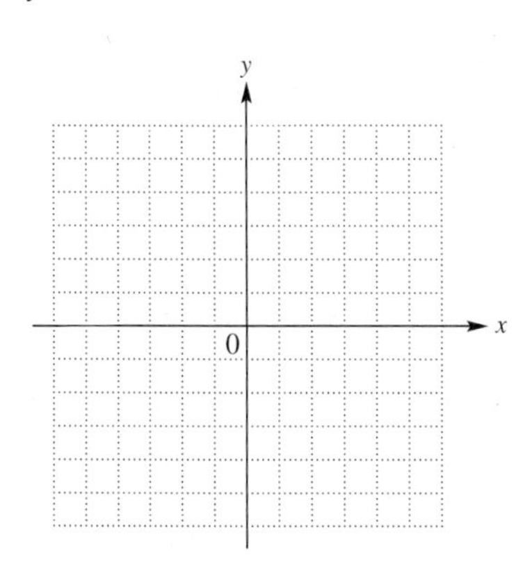

27. $y = -6x$

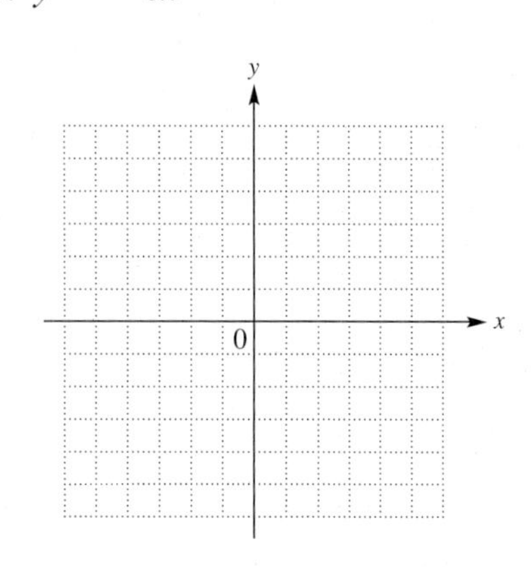

28. $x = 4$

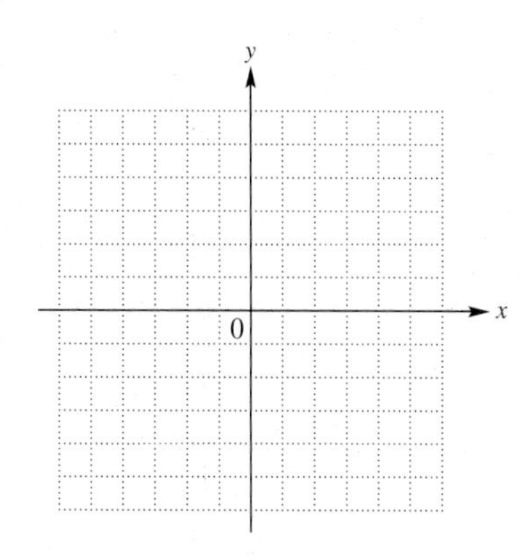

29. $x = -2$

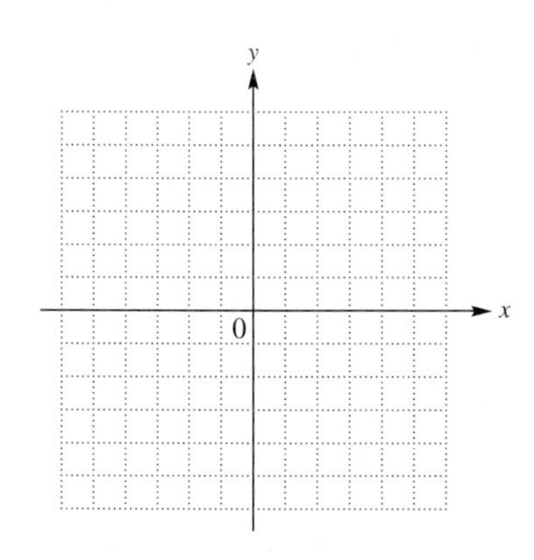

30. $y + 1 = 0$

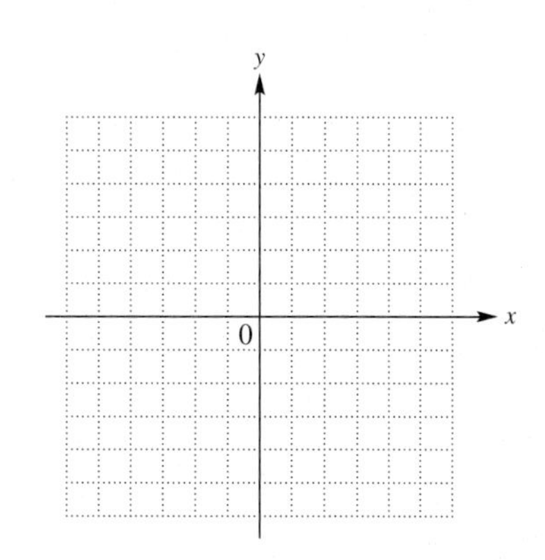

31. $y - 3 = 0$

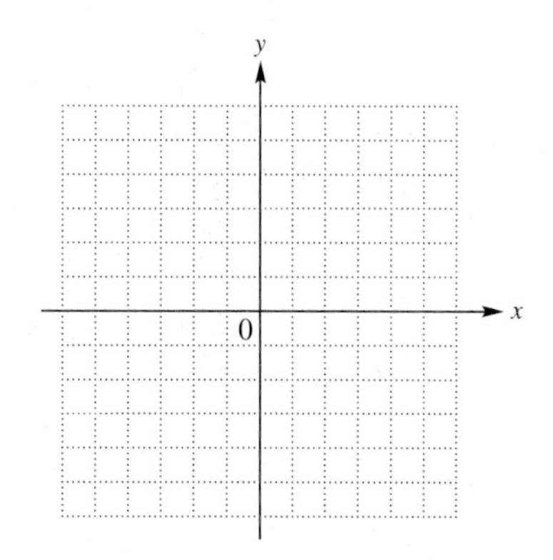

32. Write a few sentences summarizing how to graph a linear equation in two variables.

Solve each problem. See Example 6.

33. The height y (in centimeters) of a woman is related to the length of her radius bone x (from the wrist to the elbow) and is approximated by the linear equation

$$y = 3.9x + 73.5.$$

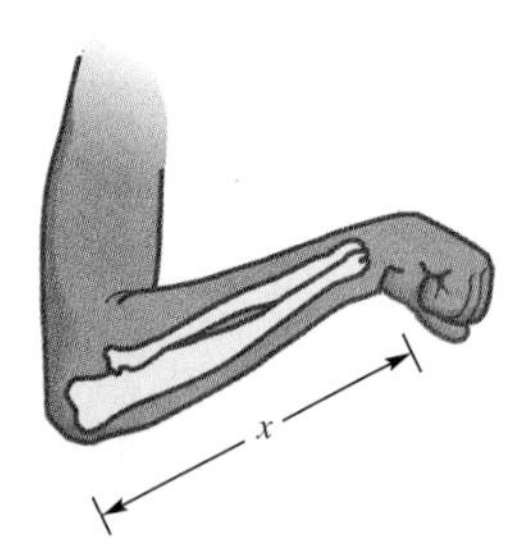

(a) Use the equation to find the approximate heights of women with radius bones of lengths 20 cm, 26 cm, and 22 cm.

(b) Graph the equation using the data from part (a).

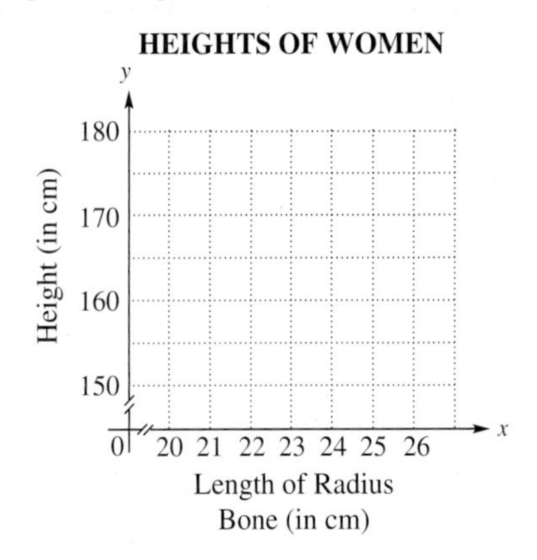

(c) Use the graph to estimate the length of the radius bone in a woman who is 167 cm tall. Then use the equation to find the length of this radius bone to the nearest centimeter. (*Hint:* Substitute for y in the equation.)

34. The weight y (in pounds) of a man taller than 60 in. can be roughly approximated by the linear equation

$$y = 5.5x - 220,$$

where x is the height of the man in inches.

(a) Use the equation to approximate the weights of men whose heights are 62 in., 66 in., and 72 in.

(b) Graph the equation using the data from part (a).

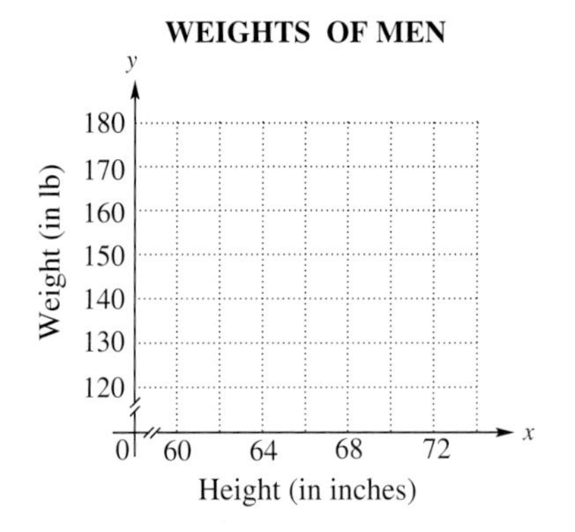

(c) Use the graph to estimate the height of a man who weighs 155 lb. Then use the equation to find the height of this man to the nearest inch. (*Hint:* Substitute for y in the equation.)

35. Refer to **Section 3.1** Exercise 80. Draw a line through the points you plotted in the scatter diagram there.

(a) Use the graph to estimate the lower limit of the target heart rate zone for age 30.

(b) Use the linear equation given there to approximate the lower limit for age 30.

(c) How does the approximation using the equation compare to the estimate from the graph?

36. Refer to **Section 3.1** Exercise 81. Draw a line through the points you plotted in the scatter diagram there.

(a) Use the graph to estimate the upper limit of the target heart rate zone for age 30.

(b) Use the linear equation given there to approximate the upper limit for age 30.

(c) How does the approximation using the equation compare to the estimate from the graph?

37. Use the results of Exercises 35(b) and 36(b) to determine the target heart rate zone for age 30.

38. Should the graphs of the target heart rate zone in the **Section 3.1** exercises be used to estimate the target heart rate zone for ages below 20 or above 80? Why or why not?

39. Per capita consumption of coffee increased for the years 1995 through 2000 as shown in the graph. If $x = 0$ represents 1995, $x = 1$ represents 1996, and so on, per capita consumption y in gallons can be modeled by the linear equation

$$y = 1.13x + 20.67.$$

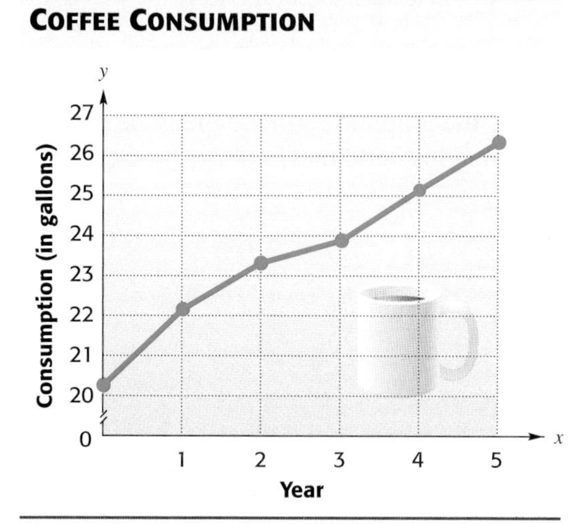

Source: U.S. Department of Agriculture.

(a) Use the equation to approximate consumption in 1995, 1996, and 1998 to the nearest tenth.

(b) Use the graph to estimate consumption for the same years.

(c) How do the approximations using the equation compare to the estimates from the graph?

40. Sporting goods sales y (in billions of dollars) from 1995 through 2000 are modeled by the linear equation

$$y = 3.606x + 41.86,$$

where $x = 5$ corresponds to 1995, $x = 6$ corresponds to 1996, and so on.

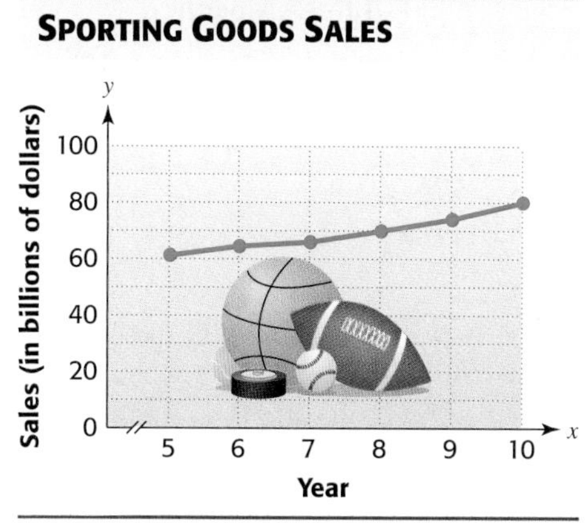

Source: U.S. Bureau of the Census.

(a) Use the equation to approximate sporting goods sales in 1995, 1997, and 2000. Round your answers to the nearest billion dollars.

(b) Use the graph to estimate sales for the same years.

(c) How do the approximations using the equation compare to the estimates using the graph?

41. The graph shows the value of a certain sport-utility vehicle over the first 5 yr of ownership.

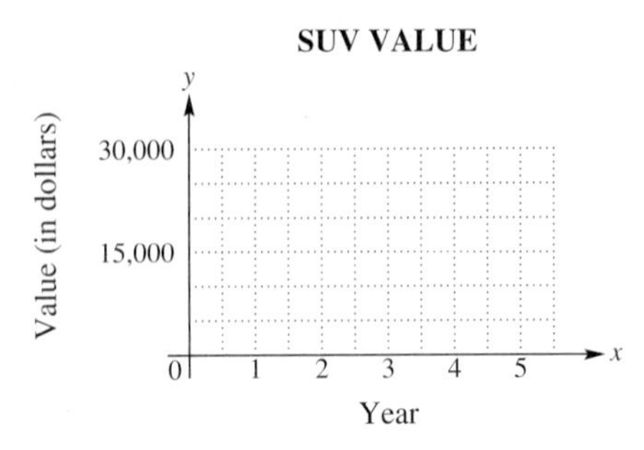

Use the graph to do the following.

(a) Determine the initial value of the SUV.

(b) Find the *depreciation* (loss in value) from the original value after the first 3 yr.

(c) What is the annual or yearly depreciation in each of the first 5 yr?

(d) What does the ordered pair (5, 5000) mean in the context of this problem?

42. Demand for an item is often closely related to its price. As price increases, demand decreases, and as price decreases, demand increases. Suppose demand for a video game is 2000 units when the price is \$40, and demand is 2500 units when the price is \$30.

(a) Let x be the price and y be the demand for the game. Graph the two given pairs of prices and demands.

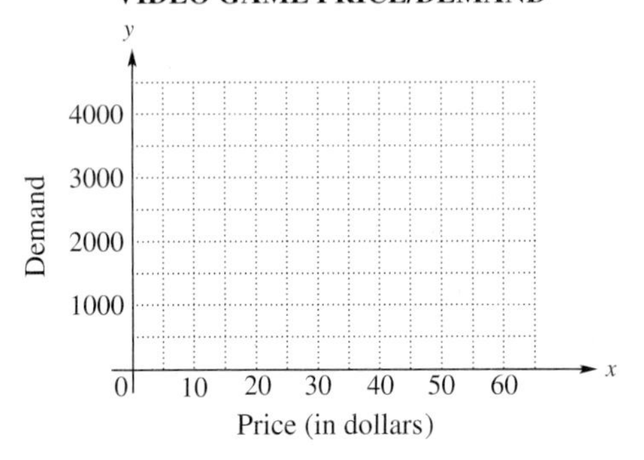

(b) Assume the relationship is linear. Draw a line through the two points from part (a). From your graph, estimate the demand if the price drops to \$20.

(c) Use the graph to estimate the price if the demand is 3500 units.

43. The graph of the linear equation for credit card debt from Example 6,

$$y = 47.3x + 281,$$

where $x = 0$ represents 1992, and so on, and y is in billions of dollars, is shown in the figure. The actual data for 1992 through 1999 is also plotted.

(a) In general, how well does the linear equation model the actual data?

(b) Use the plotted points to estimate the actual credit card debt for 1996. How does it compare to the answer in Example 6(c)?

(c) Should this equation be used to predict credit card debt for the year 2002? Why or why not?

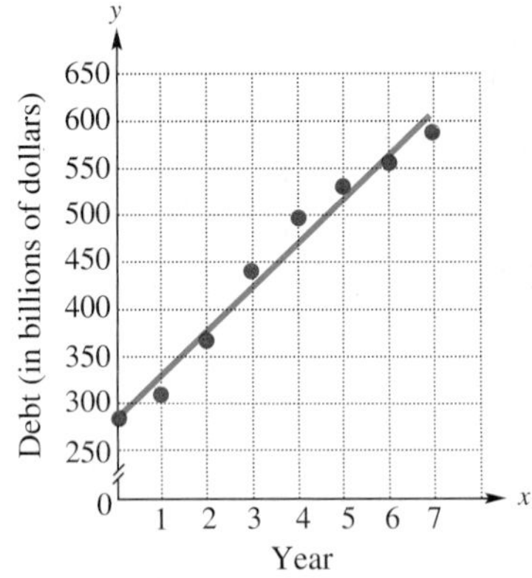

Source: Board of Governors of the Federal Reserve System.

3.3 Slope of a Line

OBJECTIVES

1. Find the slope of a line given two points.
2. Find the slope from the equation of a line.
3. Use slope to determine whether two lines are parallel, perpendicular, or neither.

An important characteristic of the lines we graphed in the previous section is their slant or "steepness", as viewed from left to right. See Figure 17.

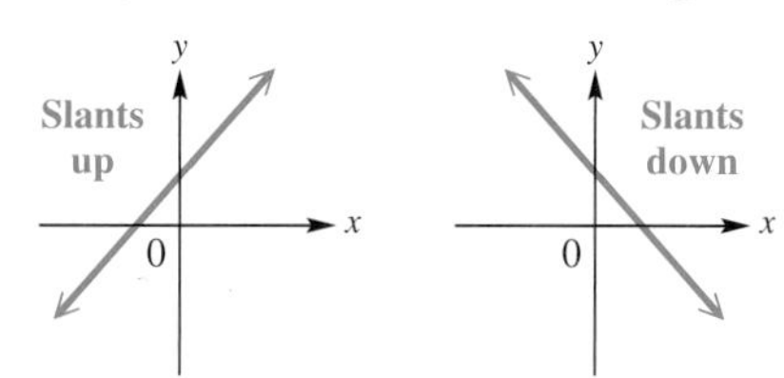

Figure 17

One way to measure the steepness of a line is to compare the vertical change in the line to the horizontal change while moving along the line from one fixed point to another. This measure of steepness is called the *slope* of the line.

OBJECTIVE 1 Find the slope of a line given two points. Figure 18 shows a line through two nonspecific points (x_1, y_1) and (x_2, y_2). (This notation is called **subscript notation.** Read x_1 as "x-sub-one" and x_2 as "x-sub-two.")

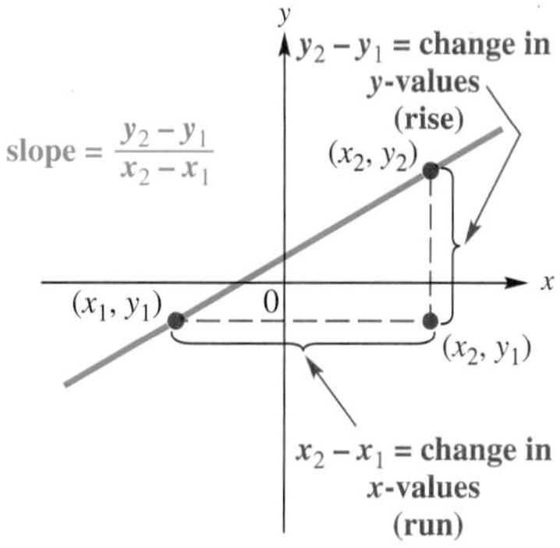

Figure 18

Moving along the line from the point (x_1, y_1) to the point (x_2, y_2) causes y to change by $y_2 - y_1$ units. This is the vertical change or **rise.** Similarly, x changes by $x_2 - x_1$ units, which is the horizontal change or **run.** (In both cases, the change is expressed as a *difference*.) Remember from **Section 2.6** that one way to compare two numbers is by using a ratio. **Slope** is the ratio of the vertical change in y to the horizontal change in x.

EXAMPLE 1 Comparing Rise to Run

Find the slope ratio of the rise to the run for each line shown in Figures 19(a) and 19(b).

We use the two points shown on each line. For the line in Figure 19(a), we find the rise from point Q to point P by determining the vertical change from -4 to 4, the difference $4 - (-4) = \mathbf{8}$. Similarly, we find the run by determining the horizontal change from Q to P, $5 - (-5) = \mathbf{10}$. Slope is the ratio

$$\frac{\text{rise}}{\text{run}} = \frac{\mathbf{8}}{\mathbf{10}} \text{ or } \frac{4}{5}.$$

Continued on Next Page

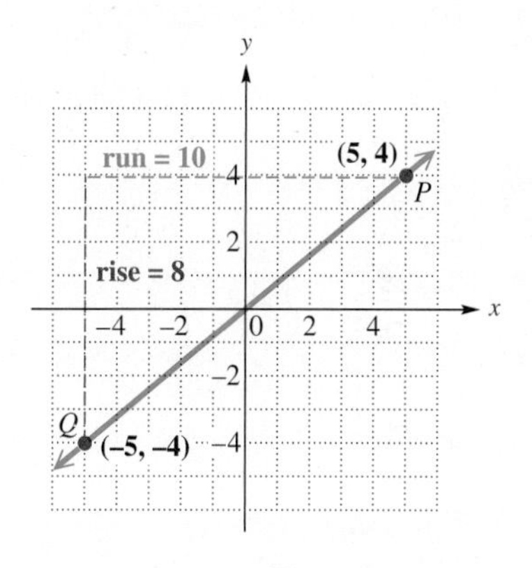

(a)

Figure 19

1 Find the slope ratio of the rise to the run for each line.

(a)

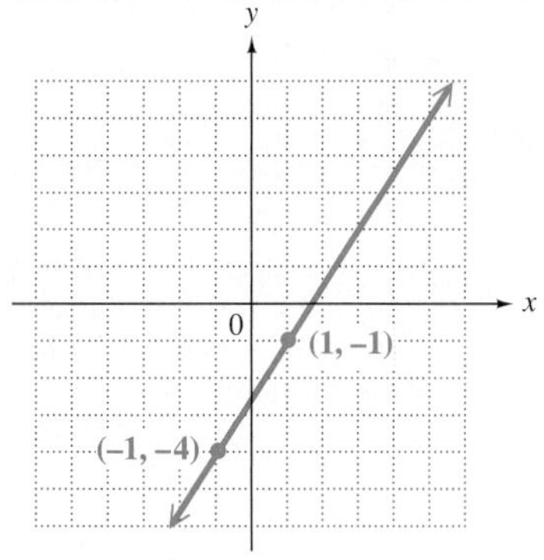

(b)

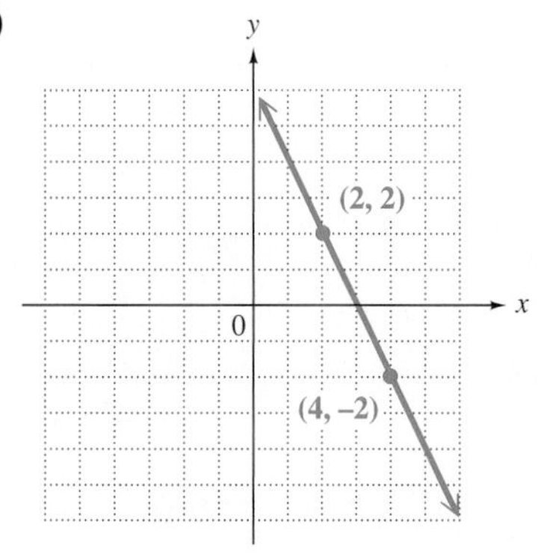

2 Find $\frac{y_2 - y_1}{x_2 - x_1}$ for the following values.

(a) $y_2 = 4, y_1 = -1,$
$x_2 = 3, x_1 = 4$

(b) $x_1 = 3, x_2 = -5,$
$y_1 = 7, y_2 = -9$

(c) $x_1 = 2, x_2 = 7,$
$y_1 = 4, y_2 = 9$

Answers

1. (a) $\frac{3}{2}$ (b) -2

2. (a) -5 (b) 2 (c) 1

Moving from Q to P, the line in Figure 19(b) has rise $3 - (-1) = 4$ and run $2 - 6 = -4$, so the slope is the ratio

$$\frac{\text{rise}}{\text{run}} = \frac{4}{-4} = \frac{1}{-1} \text{ or } -1.$$

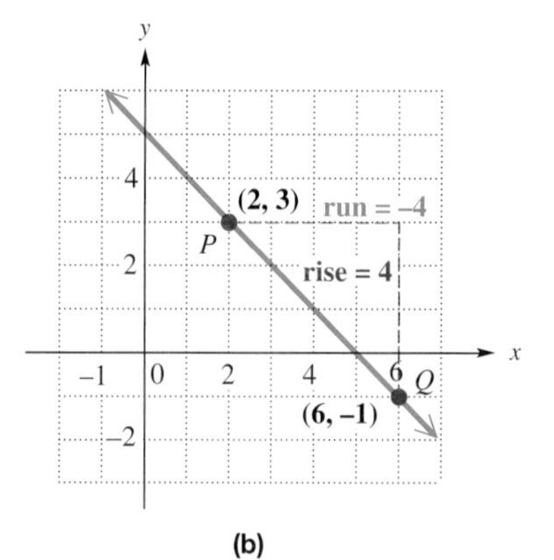

(b)

Figure 19

To confirm our slope ratios, count grid squares from one point on the line to another. For example, starting at the point (5, 0) in Figure 19(b), count up 1 square (the rise in the slope ratio $\frac{1}{-1}$) and then 1 square to the *left* (the run) to arrive at the point (4, 1) on the line.

Work Problem 1 at the Side.

The idea of slope is used in many everyday situations. See Figure 20. For example, a highway with a 10% or $\frac{1}{10}$ grade (or slope) rises 1 m for every 10 m horizontally. Architects specify the pitch of a roof using slope; a $\frac{5}{12}$ roof means that the roof rises 5 ft for every 12 ft in the horizontal direction. The slope of a stairwell also indicates the ratio of the vertical rise to the horizontal run. In the figure, the slope of the stairwell is $\frac{8}{10}$ or $\frac{4}{5}$.

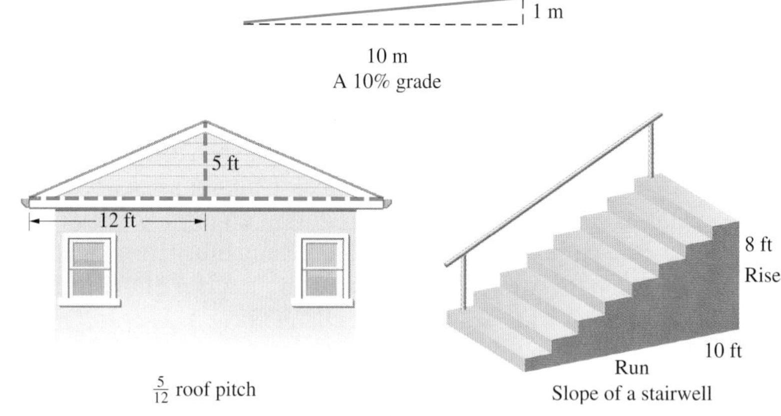

Figure 20

If we know the coordinates of two points on a line, we can find its slope, traditionally designated m, using the slope formula.

Slope Formula

The **slope** of the line through the points (x_1, y_1) and (x_2, y_2) is

$$m = \frac{\textbf{change in } y}{\textbf{change in } x} = \frac{y_2 - y_1}{x_2 - x_1}, \qquad \text{if } x_1 \neq x_2.$$

Work Problem 2 at the Side.

The slope of a line tells how fast y changes for each unit of change in x; that is, the slope gives the rate of change in y for each unit of change in x.

EXAMPLE 2 Using the Slope Formula

Find the slope of each line.

(a) The line through $(1, -2)$ and $(-4, 7)$

Use the slope formula. Let $(-4, 7) = (x_2, y_2)$ and $(1, -2) = (x_1, y_1)$. Then

$$\text{slope } m = \frac{\text{change in } y}{\text{change in } x} = \frac{y_2 - y_1}{x_2 - x_1} = \frac{7 - (-2)}{-4 - 1} = \frac{9}{-5} = -\frac{9}{5}.$$

See Figure 21(a).

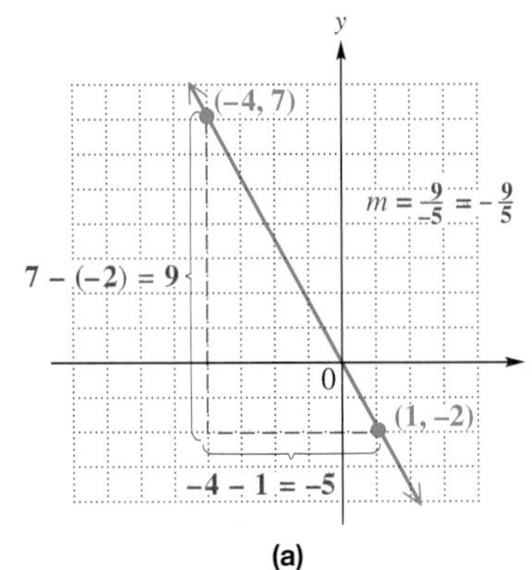

(a)

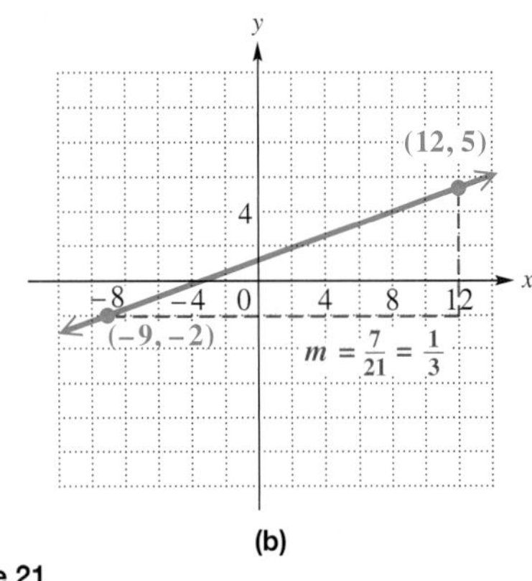

(b)

Figure 21

(b) The line through $(-9, -2)$ and $(12, 5)$

$$m = \frac{y_2 - y_1}{x_2 - x_1} = \frac{5 - (-2)}{12 - (-9)} = \frac{7}{21} = \frac{1}{3}$$

See Figure 21(b). The same slope is obtained by subtracting in reverse order.

$$m = \frac{-2 - 5}{-9 - 12} = \frac{-7}{-21} = \frac{1}{3}$$

CAUTION

It makes no difference which point is (x_1, y_1) or (x_2, y_2); however, it is important to be consistent. Start with the x- and y-values of one point (either one) and subtract the corresponding values of the other point. Also, the slope of a line is the same for *any* two points on the line.

Work Problem 3 at the Side.

In Example 2(a) the slope is negative and the corresponding line in Figure 21(a) falls from left to right. The slope in Example 2(b) is positive and the corresponding line in Figure 21(b) rises from left to right. These facts can be generalized.

3 Find the slope of each line.

(a) Through $(6, -2)$ and $(5, 4)$

(b) Through $(-3, 5)$ and $(-4, -7)$

(c) Through $(6, -8)$ and $(-2, 4)$ (Find this slope in two different ways as in Example 2(b).)

ANSWERS

3. **(a)** -6 **(b)** 12 **(c)** $-\frac{3}{2}; -\frac{3}{2}$

Positive and Negative Slopes

A line with positive slope rises from left to right.

A line with negative slope falls from left to right.

EXAMPLE 3 **Showing that the Slope of a Horizontal Line Is Zero**

Find the slope of the line through $(-8, 4)$ and $(2, 4)$.

$$m = \frac{y_2 - y_1}{x_2 - x_1} = \frac{4 - 4}{-8 - 2} = \frac{0}{-10} = 0 \qquad \text{Zero slope}$$

As shown in Figure 22, the line through the given points is horizontal. ***All horizontal lines have slope 0*** since the difference in y-values is always 0.

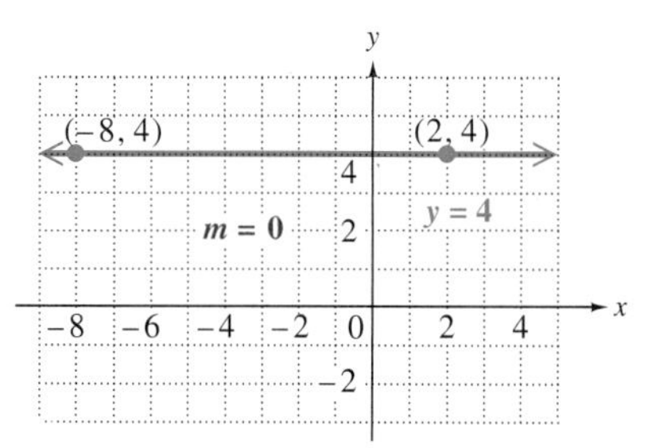

Figure 22

EXAMPLE 4 **Showing that a Vertical Line Has Undefined Slope**

Find the slope of the line through $(6, 2)$ and $(6, -4)$.

$$m = \frac{y_2 - y_1}{x_2 - x_1} = \frac{2 - (-4)}{6 - 6} = \frac{6}{0} \qquad \text{Undefined slope}$$

Because division by 0 is undefined, this line has undefined slope. (This is why the slope formula at the beginning of this section had the restriction $x_1 \neq x_2$.) The graph in Figure 23 shows that this line is vertical. All points on a vertical line have the same x-value, so ***all vertical lines have undefined slope.***

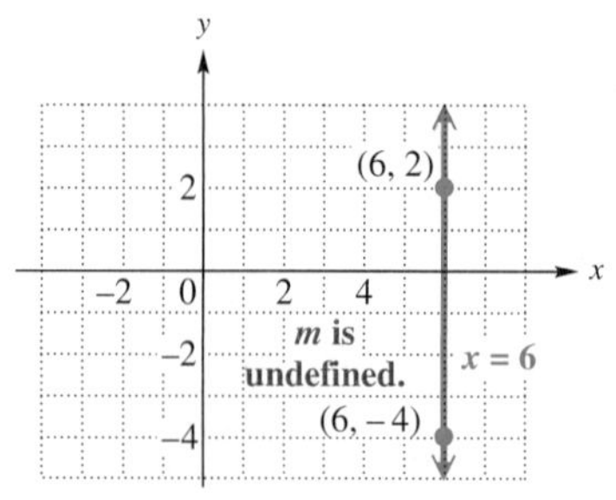

Figure 23

Slopes of Horizontal and Vertical Lines

Horizontal lines, which have equations of the form $y = k$, have **slope 0.**

Vertical lines, which have equations of the form $x = k$, have **undefined slope.**

Work Problem 4 at the Side.

OBJECTIVE 2 Find the slope from the equation of a line. The slope of a line can be found directly from its equation. For example, the slope of the line

$$y = -3x + 5$$

can be found using any two points on the line. We get these two points by first choosing two different values of x and then finding the corresponding values of y. Choose $x = -2$ and $x = 4$.

$y = -3x + 5$	$y = -3x + 5$
$y = -3(-2) + 5$ Let $x = -2$.	$y = -3(4) + 5$ Let $x = 4$.
$y = 6 + 5$	$y = -12 + 5$
$y = 11$	$y = -7$

The ordered pairs are $(-2, 11)$ and $(4, -7)$. Now use the slope formula.

$$m = \frac{11 - (-7)}{-2 - 4} = \frac{18}{-6} = -3$$

The slope, -3, is the same number as the coefficient of x in the equation $y = -3x + 5$. It can be shown that this always happens, *as long as the equation is solved for y.* This fact is used to find the slope of a line from its equation.

Finding the Slope of a Line from Its Equation

Step 1 Solve the equation for y.

Step 2 The slope is given by the coefficient of x.

NOTE

We will see in the next section that the equation $y = -3x + 5$ is written using a special form of the equation of a line,

$$y = mx + b,$$

called *slope-intercept form.*

4 Find the slope of each line.

(a) Through $(2, 5)$ and $(-1, 5)$

(b) Through $(3, 1)$ and $(3, -4)$

(c) With equation $y = -1$

(d) With equation $x - 4 = 0$

ANSWERS

4. **(a)** 0 **(b)** undefined **(c)** 0 **(d)** undefined

5 Find the slope of each line.

(a) $y = -\frac{7}{2}x + 1$

(b) $3x + 2y = 9$

(c) $y + 4 = 0$

(d) $x + 3 = 7$

ANSWERS

5. (a) $-\frac{7}{2}$ (b) $-\frac{3}{2}$ (c) 0 (d) undefined

EXAMPLE 5 Finding Slopes from Equations

Find the slope of each line.

(a) $2x - 5y = 4$

Step 1 Solve the equation for y.

$$2x - 5y = 4$$

$$-5y = -2x + 4 \quad \text{Subtract } 2x.$$

$$y = \frac{2}{5}x - \frac{4}{5} \quad \text{Divide by } -5.$$

Step 2 The slope is given by the coefficient of x, so the slope is $\frac{2}{5}$.

(b) $8x + 4y = 1$

Solve for y.

$$8x + 4y = 1$$

$$4y = -8x + 1 \quad \text{Subtract } 8x.$$

$$y = -2x + \frac{1}{4} \quad \text{Divide by } 4.$$

The slope of this line is given by the coefficient of x, which is -2.

Work Problem 5 at the Side.

OBJECTIVE 3 Use slope to determine whether two lines are parallel, perpendicular, or neither. Two lines in a plane that never intersect are **parallel.** We use slopes to tell whether two lines are parallel. For example, Figure 24 shows the graphs of $x + 2y = 4$ and $x + 2y = -6$. These lines appear to be parallel. Solving for y, we find that both $x + 2y = 4$ and $x + 2y = -6$ have slope $-\frac{1}{2}$. ***Nonvertical parallel lines always have equal slopes.***

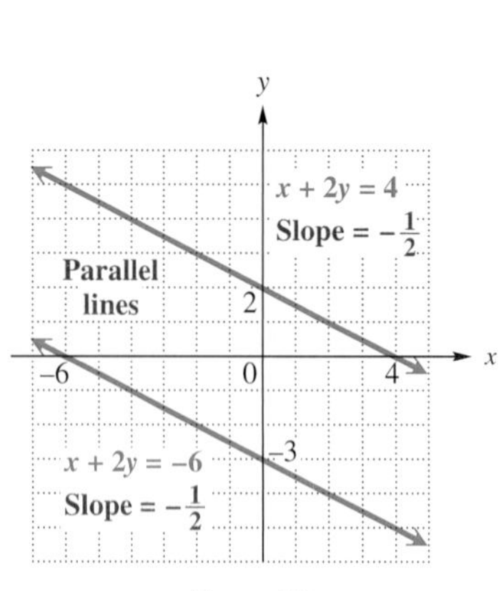

Figure 24

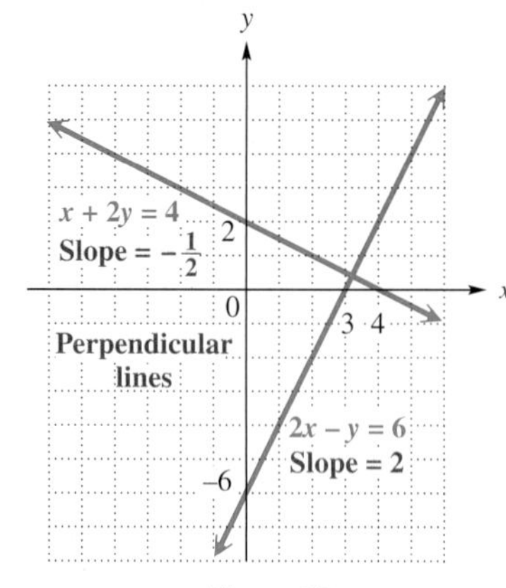

Figure 25

Figure 25 shows the graphs of $x + 2y = 4$ and $2x - y = 6$. These lines appear to be **perpendicular** (that is, they intersect at a 90° angle). Solving for y shows that the slope of $x + 2y = 4$ is $-\frac{1}{2}$, while the slope of $2x - y = 6$ is 2. The product of $-\frac{1}{2}$ and 2 is

$$-\frac{1}{2}(2) = -1.$$

This is true in general; ***the product of the slopes of two perpendicular lines (neither of which is vertical) is always −1.***

Slopes of Parallel and Perpendicular Lines

Two lines with the same slope are parallel.

Two lines whose slopes have a product of -1 are perpendicular.

EXAMPLE 6 Deciding Whether Lines Are Parallel, Perpendicular, or Neither

Decide whether each pair of lines is *parallel, perpendicular,* or *neither.*

(a) $x + 2y = 7$
$-2x + y = 3$

Find the slope of each line by first solving each equation for y.

$$x + 2y = 7$$
$$2y = -x + 7$$
$$y = -\frac{1}{2}x + \frac{7}{2}$$

Slope is $-\frac{1}{2}$.

$$-2x + y = 3$$
$$y = 2x + 3$$

Slope is 2.

Because the slopes are not equal, the lines are not parallel. Check the product of the slopes: $-\frac{1}{2}(2) = -1$. The two lines are perpendicular because the product of their slopes is -1. See Figure 26.

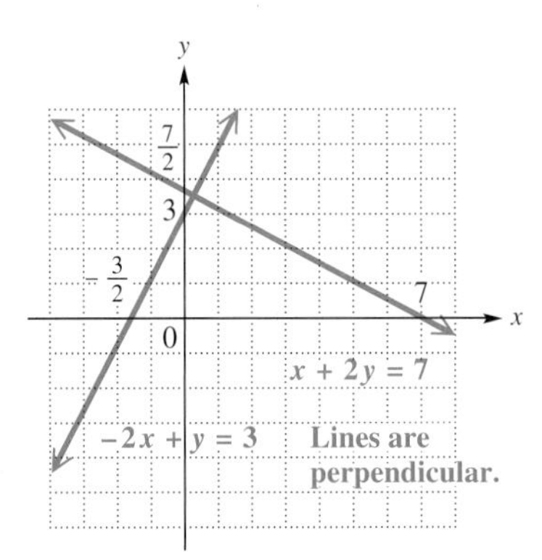

Figure 26

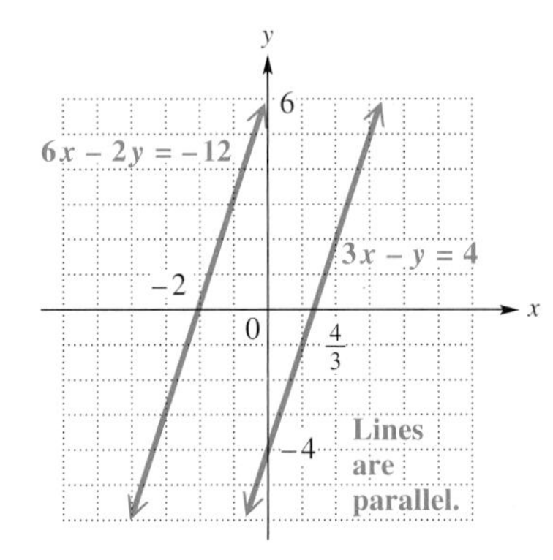

Figure 27

(b) $3x - y = 4$ Solve for y. $\longrightarrow$ $y = 3x - 4$
$6x - 2y = -12$ $\quad$ $y = 3x + 6$

Both lines have slope 3, so the lines are parallel. See Figure 27.

(c) $4x + 3y = 6$ Solve for y. $\longrightarrow$ $y = -\frac{4}{3}x + 2$
$2x - y = 5$ $\quad$ $y = 2x - 5$

Here the slopes are $-\frac{4}{3}$ and 2. Because $-\frac{4}{3} \neq 2$ and $-\frac{4}{3}(2) \neq -1$, these lines are neither parallel nor perpendicular.

(d) $5x - y = 1$ Solve for y. $\longrightarrow$ $y = 5x - 1$
$x - 5y = -10$ $\quad$ $y = \frac{1}{5}x + 2$

The slopes are 5 and $\frac{1}{5}$. The lines are not parallel, nor are they perpendicular. ***(Be careful!*** $5(\frac{1}{5}) = 1$, ***not*** -1.***)***

Work Problem 6 at the Side.

6 Decide whether each pair of lines is *parallel, perpendicular,* or *neither*.

(a) $x + y = 6$
$x + y = 1$

(b) $3x - y = 4$
$x + 3y = 9$

(c) $2x - y = 5$
$2x + y = 3$

(d) $3x - 7y = 35$
$7x - 3y = -6$

ANSWERS

6. **(a)** parallel **(b)** perpendicular **(c)** neither **(d)** neither

Focus on

Real-Data Applications

Linear or Nonlinear? That Is the Question about Windchill

The **windchill factor** measures the cooling effect that the wind has on one's skin. The table gives the windchill factor for various wind speeds and temperatures.

WINDCHILL FACTOR

Wind Speed (mph)	Air Temperature (°Fahrenheit) 35	30	25	20	15	10	5	0	−5	−10	−15	−20	−25	−30	−35
4	35	30	25	20	15	10	5	0	−5	−10	−15	−20	−25	−30	−35
5	32	27	22	16	11	6	0	−5	−10	−15	−21	−26	−31	−36	−42
10	22	16	10	3	−3	−9	−15	−22	−27	−34	−40	−46	−52	−58	−64
15	16	9	2	−5	−11	−18	−25	−31	−38	−45	−51	−58	−65	−72	−78
20	12	4	−3	−10	−17	−24	−31	−39	−46	−53	−60	−67	−74	−81	−88
25	8	1	−7	−15	−22	−29	−36	−44	−51	−59	−66	−74	−81	−88	−96
30	6	−2	−10	−18	−25	−33	−41	−49	−56	−64	−71	−79	−86	−93	−101
35	4	−4	−12	−20	−27	−35	−43	−52	−58	−67	−74	−82	−89	−97	−105
40	3	−5	−13	−21	−29	−37	−45	−53	−60	−69	−76	−84	−92	−100	−107
45	2	−6	−14	−22	−30	−38	−46	−54	−62	−70	−78	−85	−93	−102	−109

Source: USA Today.

The data in the table represents the relationships between two different sets of variable quantities: Windchill versus Air Temperature and Windchill versus Wind Speed. The question is whether either of these relationships is linear, that is, whether the data points, when graphed, could be connected to form a straight line.

Example 1 Windchill versus Air Temperature Choose one measure of wind speed to keep constant, such as 15 mph. Complete the table with Air Temperature (AT) as the *input* and Windchill (WC) as the *output*. Both variables are measured in degrees Fahrenheit.

AT	35	30	25	20	15	10	5	0	−5	−10	−15			
WC	16	9	2	−5	−11	−18	−25	−31	−38					

Example 2 Windchill versus Wind Speed Choose one measure of air temperature to keep constant, such as 10°F. Complete the table with Wind Speed (WS) as the *input* and Windchill (WC) as the *output*. Wind speed is measured in mph.

WS	4	5	10	15	20	25	30		
WC	10	6	−9	−18	−24				

For Group Discussion

1. Refer to the Windchill versus Air Temperature data (Example 1).
   (a) Write the data as ordered pairs.

   (b) On a sheet of graph paper, draw and label a rectangular coordinate system. Use a scale of 5 on the x-axis and the y-axis. Make a scatter diagram of the data. Does the graph represent a linear relationship?
2. Repeat Problem 1 using the Windchill versus Wind Speed data (Example 2).

3.3 Exercises

FOR EXTRA HELP

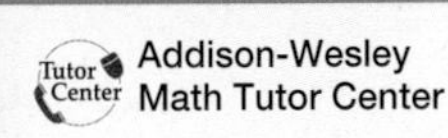
Addison-Wesley Math Tutor Center

MathXL

Digital Video Tutor CD 3 Videotape 3

Student's Solutions Manual

MyMathLab

Interactmath.com

Use the coordinates of the indicated points to find the slope of each line. See Example 1.

1.

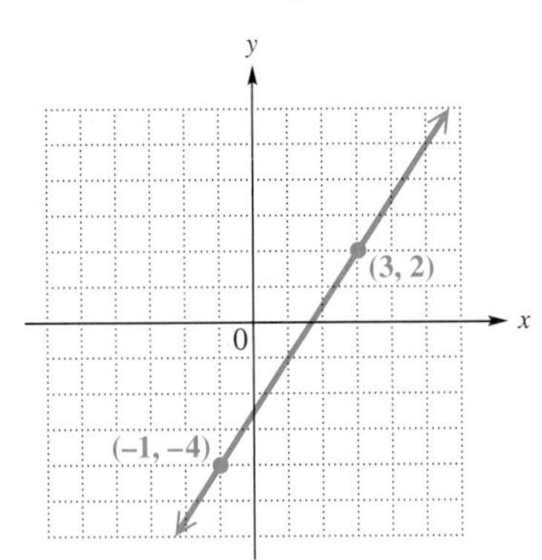

2.

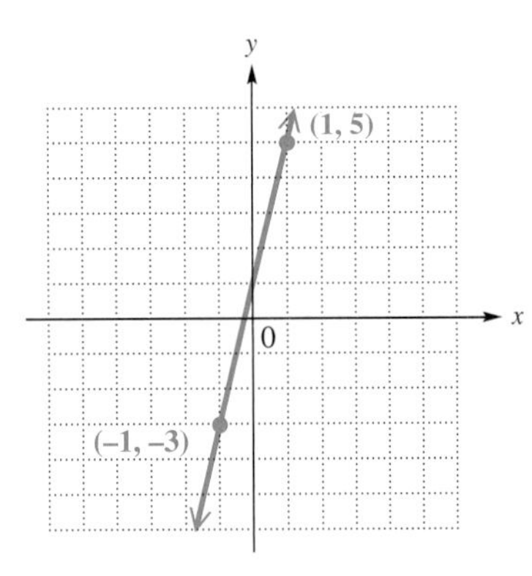

3.

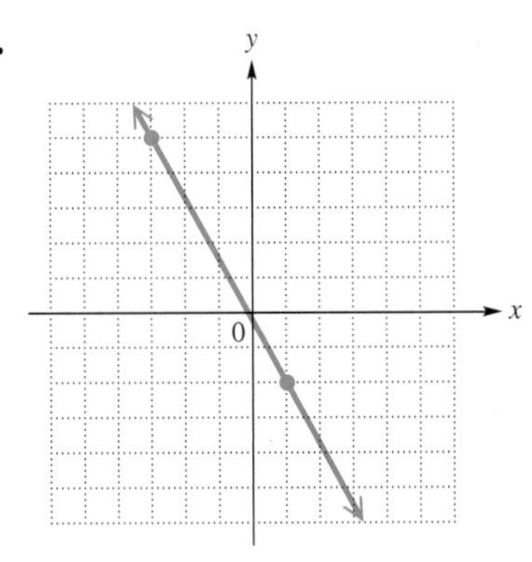

4.

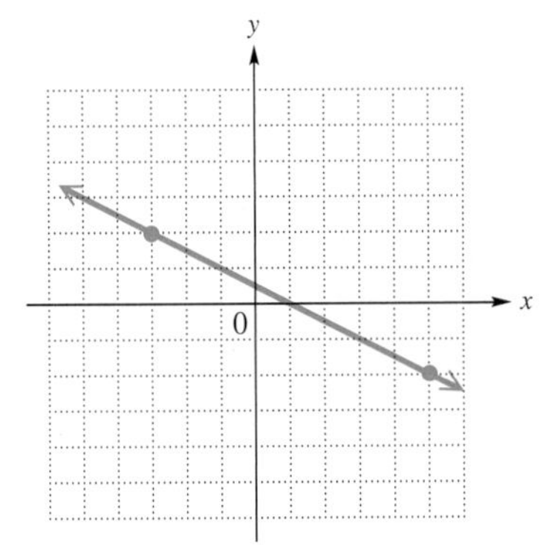

5.

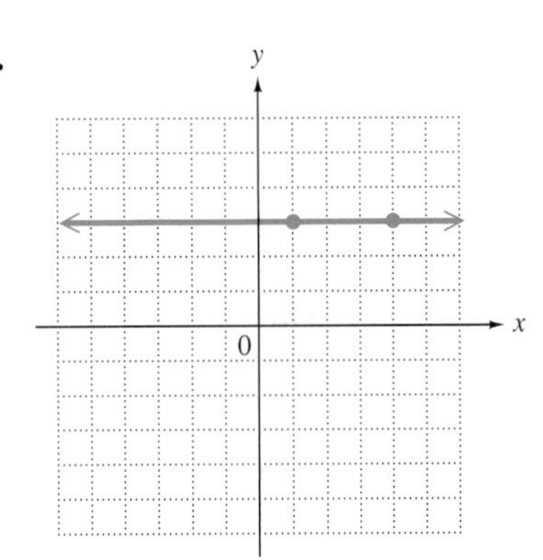

6.

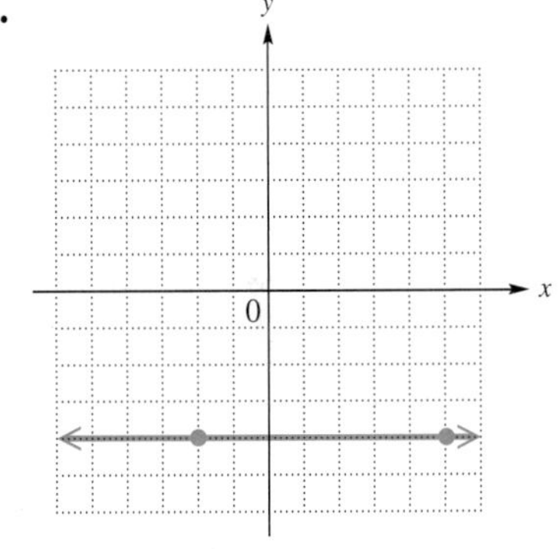

7. In the context of the graph of a straight line, what is meant by "rise"? What is meant by "run"?

8. Look at the graph in Exercise 1, and answer the following.

(a) Start at the point $(-1, -4)$ and count vertically up to the horizontal line that goes through the other plotted point. What is this vertical change? (Remember: "up" means positive, "down" means negative.) ______

(b) From this new position, count horizontally to the other plotted point. What is this horizontal change? (Remember: "right" means positive, "left" means negative.)

______

(c) What is the quotient of the numbers found in parts (a) and (b)? ______
What do we call this number? ____________

9. Refer to Exercise 8. If we were to *start* at the point (3, 2) and *end* at the point $(-1, -4)$, would the answer to part (c) be the same? Explain why or why not.

On the given coordinate system, sketch the graph of a straight line with the indicated slope.

10. Negative

11. Positive

12. Undefined

13. Zero

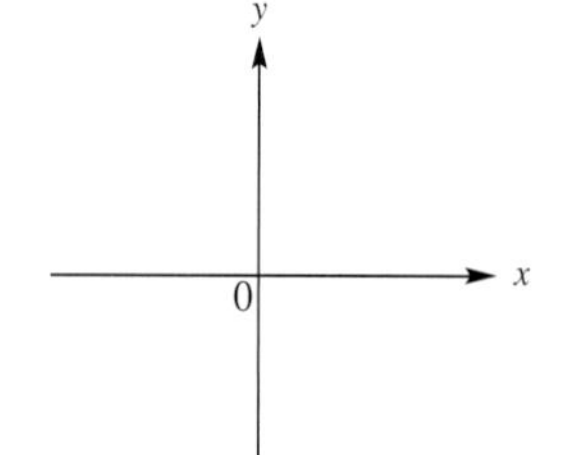

14. Explain in your own words what is meant by the *slope* of a line.

15. A student found the slope of the line through the points (2, 5) and $(-1, 3)$ and got $-\frac{2}{3}$ as his answer. He showed his work as

$$\frac{3-5}{2-(-1)} = \frac{-2}{3} = -\frac{2}{3}.$$

Is he correct? If not, find his error and give the correct slope.

Find the slope of the line through each pair of points. See Examples 1–4.

16. $(4, -1)$ and $(-2, -8)$

17. $(1, -2)$ and $(-3, -7)$

18. $(-8, 0)$ and $(0, -5)$

19. $(0, 3)$ and $(-2, 0)$

20. $(-4, -5)$ and $(-5, -8)$

21. $(-2, 4)$ and $(-3, 7)$

22. $(6, -5)$ and $(-12, -5)$

23. $(4, 3)$ and $(-6, 3)$

24. $(-8, 6)$ and $(-8, -1)$

25. $(-12, 3)$ and $(-12, -7)$

26. $(3.1, 2.6)$ and $(1.6, 2.1)$

27. $\left(-\frac{7}{5}, \frac{3}{10}\right)$ and $\left(\frac{1}{5}, -\frac{1}{2}\right)$

Find the slope of each line. See Example 5.

28. $y = 2x - 3$

29. $y = 5x + 12$

30. $2y = -x + 4$

31. $4y = x + 1$

32. $-6x + 4y = 4$ **33.** $3x - 2y = 3$ **34.** $y = 4$ **35.** $y = 6$

36. $x = 5$ **37.** $x = -2$ **38.** $x + y = 0$ **39.** $x - y = 0$

The figure at the right shows a line that has a positive slope (because it rises from left to right) and a positive y-value for the y-intercept (because it intersects the y-axis above the origin).

For each figure in Exercises 40–45, decide whether ***(a)*** *the slope is* positive, negative, *or* 0 *and whether* ***(b)*** *the y-value of the y-intercept is* positive, negative, *or* 0.

40. (a) ________
(b) ________

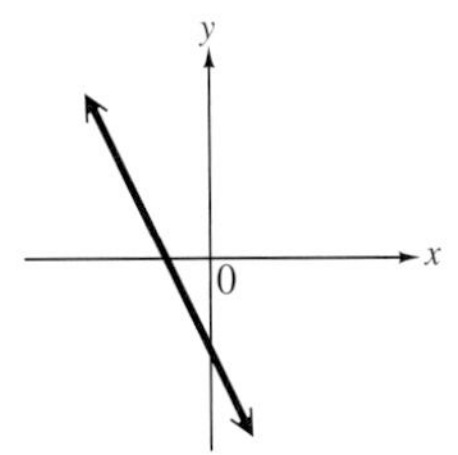

41. (a) ________
(b) ________

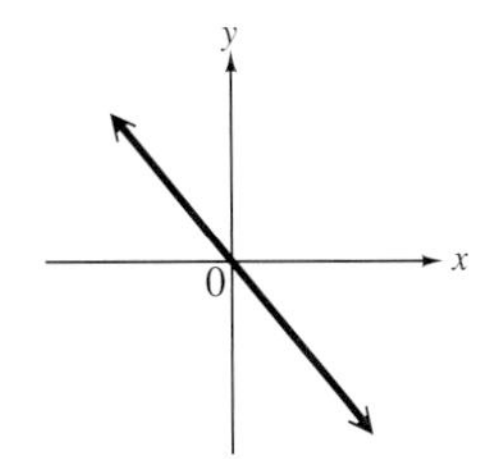

42. (a) ________
(b) ________

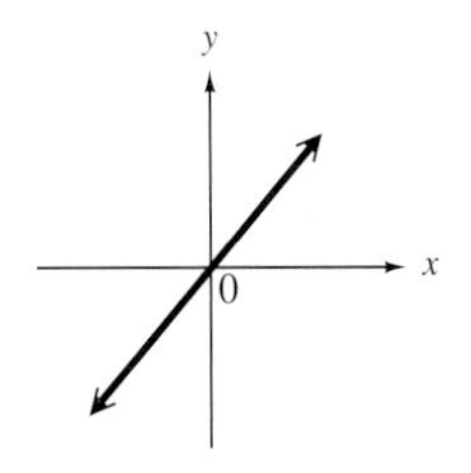

43. (a) ________
(b) ________

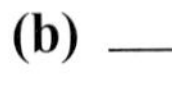

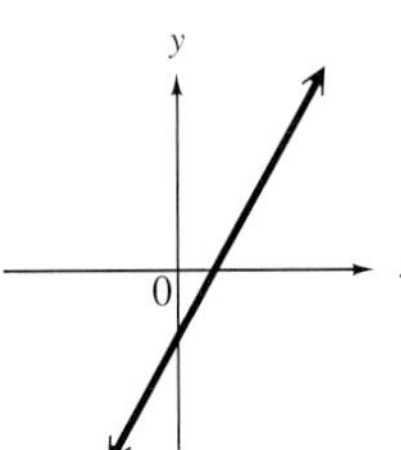

44. (a) ________
(b) ________

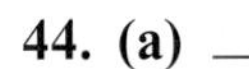

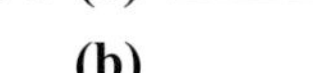

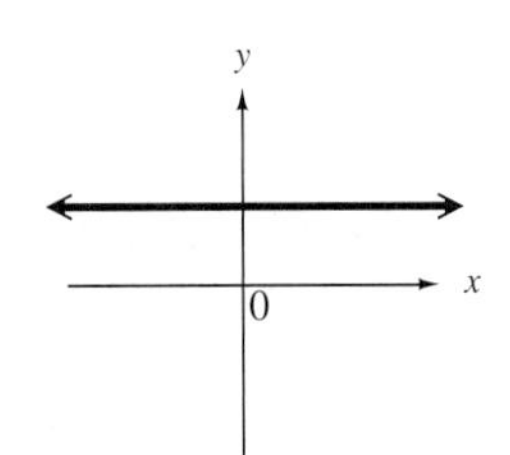

45. (a) ________
(b) ________

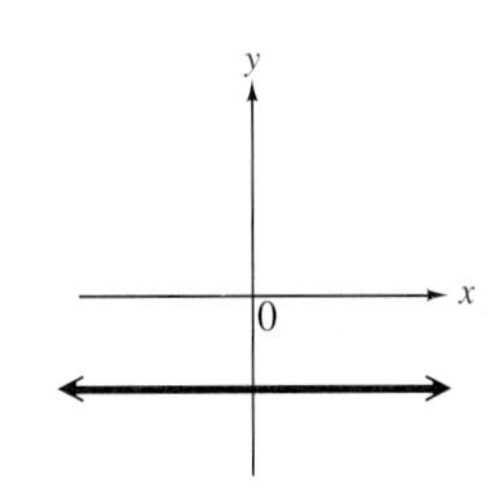

In each pair of equations, give the slope of each line, and then determine whether the two lines are parallel, perpendicular, *or* neither *parallel nor perpendicular. See Example* 6.

46. $2x + 5y = 4$
$4x + 10y = 1$

47. $-4x + 3y = 4$
$-8x + 6y = 0$

48. $8x - 9y = 6$
$8x + 6y = -5$

49. $5x - 3y = -2$
$3x - 5y = -8$

50. $3x - 2y = 6$
$2x + 3y = 3$

51. $3x - 5y = -1$
$5x + 3y = 2$

52. What is the slope (or pitch) of this roof?

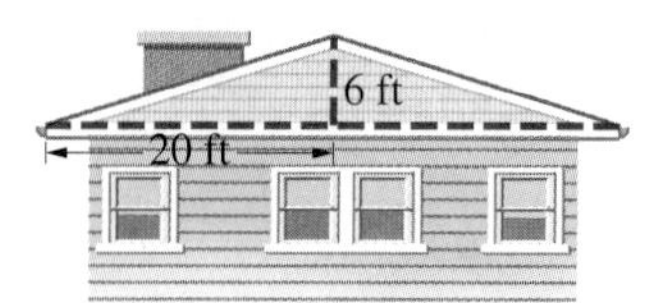

53. What is the slope (or grade) of this hill?

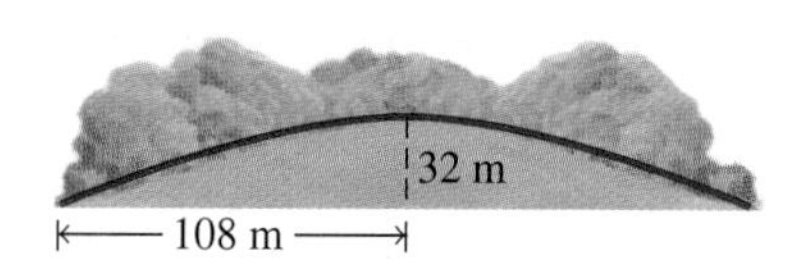

RELATING CONCEPTS (EXERCISES 54–59) For Individual or Group Work

Figure A gives public school enrollment (in thousands) in grades 9–12 in the United States. Figure B gives the (average) number of public school students per computer.

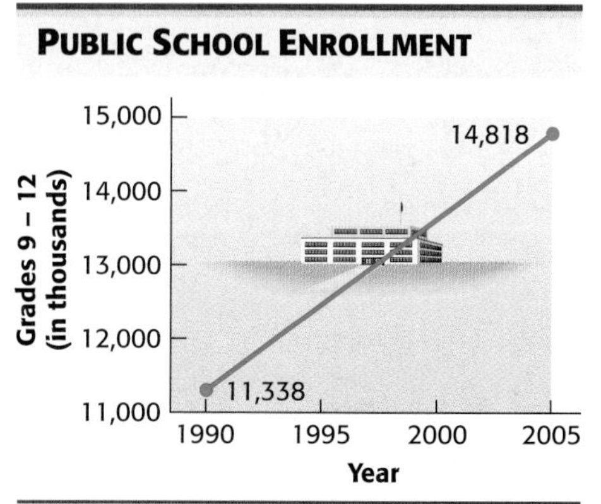

Source: *Digest of Educational Statistics*, annual, and *Projections of Educational Statistics*, annual.

Figure A

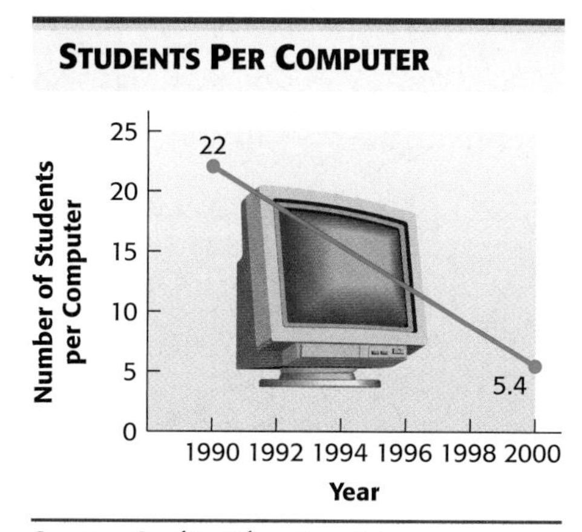

Source: *Quality Education Data.*

Figure B

Work Exercises 54–59 in order.

54. Use the ordered pairs (1990, 11,338) and (2005, 14,818) to find the slope of the line in Figure A.

55. The slope of the line in Figure A is ____________ (positive/negative). This means that during the period represented, enrollment ____________ (increased/decreased).

56. The slope of a line represents its *rate of change*. Based on Figure A, what was the increase in students *per year* during the period shown?

57. Use the given information to find the slope of the line in Figure B.

58. The slope of the line in Figure B is ____________ (positive/negative). This means that during the period represented, the number of students per computer ____________ (increased/decreased).

59. Based on Figure B, what was the decrease in students per computer *per year* during the period shown?

3.4 Equations of Lines

In **Section 3.3**, we found the slope (steepness) of a line from the equation of the line by solving the equation for y. In that form, the slope is the coefficient of x. For example, the slope of the line with equation $y = 2x + 3$ is 2, the coefficient of x. What does the number 3 represent? If $x = 0$, the equation becomes

$$y = 2(0) + 3 = 0 + 3 = 3.$$

Since $y = 3$ corresponds to $x = 0$, $(0, 3)$ is the y-intercept of the graph of $y = 2x + 3$. An equation like $y = 2x + 3$ that is solved for y is said to be in **slope-intercept form** because both the slope and the y-intercept of the line can be read directly from the equation.

OBJECTIVES

1. Write an equation of a line given its slope and y-intercept.
2. Graph a line given its slope and a point on the line.
3. Write an equation of a line given its slope and any point on the line.
4. Write an equation of a line given two points on the line.
5. Find an equation of a line that fits a data set.

Slope-Intercept Form

The slope-intercept form of the equation of a line with slope m and y-intercept $(0, b)$ is

$$y = mx + b.$$

Slope ↑ (points to m); ↑ $(0, b)$ is the y-intercept.

Remember that the intercept in the slope-intercept form is the y-intercept.

NOTE
The slope-intercept form is the most useful form for a linear equation because of the information we can determine from it. It is also the form used by graphing calculators and the one that describes a *linear function*, an important concept in mathematics.

OBJECTIVE 1 Write an equation of a line given its slope and y-intercept. Given the slope and y-intercept of a line, we can use the slope-intercept form to find an equation of the line.

EXAMPLE 1 Finding an Equation of a Line

Find an equation of the line with slope $\frac{2}{3}$ and y-intercept $(0, -1)$.

Here $m = \frac{2}{3}$ and $b = -1$, so an equation is

Slope → m; y-intercept → b

$$y = mx + b$$

$$y = \frac{2}{3}x - 1.$$

Work Problem 1 at the Side.

1 Find an equation of the line with the given slope and y-intercept.

(a) slope $\frac{1}{2}$; y-intercept $(0, -4)$

(b) slope -1; y-intercept $(0, 8)$

(c) slope 3; y-intercept $(0, 0)$

(d) slope 0; y-intercept $(0, 2)$

OBJECTIVE 2 Graph a line given its slope and a point on the line. We can use the slope and y-intercept to graph any line. If a linear equation is given in standard form $Ax + By = C$, it can be graphed using the following procedure.

ANSWERS

1. (a) $y = \frac{1}{2}x - 4$ (b) $y = -x + 8$
(c) $y = 3x$ (d) $y = 2$

2 Graph $3x - 4y = 8$ using the slope and y-intercept.

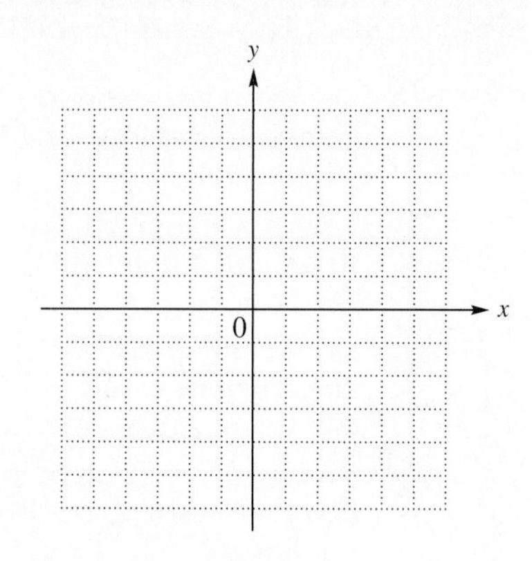

Graphing Using Slope and y-Intercept

Step 1 Solve for y to obtain the $y = mx + b$ form of the equation.

Step 2 Identify the y-intercept from the form just obtained. Graph the point $(0, b)$.

Step 3 The slope of the line is m. Use the geometric interpretation of slope ("rise over run") to find another point on the graph by counting from the y-intercept.

Step 4 Join the two points with a line to obtain the graph of the equation.

EXAMPLE 2 Graphing a Line Using Slope and y-Intercept

Graph $2x - 3y = 3$.

Step 1 Begin by solving for y.

$$2x - 3y = 3 \qquad \text{Given equation}$$

$$-3y = -2x + 3 \qquad \text{Subtract } 2x.$$

$$y = \frac{2}{3}x - 1 \qquad \text{Divide by } -3.$$

Step 2 The y-intercept is $(0, -1)$. Graph this point. See Figure 28.

Step 3 The slope is $\frac{2}{3}$. By the definition of slope,

$$m = \frac{\text{change in } y}{\text{change in } x} = \frac{2}{3}.$$

Counting from the y-intercept 2 units up and 3 units to the right, we obtain another point on the graph, $(3, 1)$.

Step 4 Draw the line through the points $(0, -1)$ and $(3, 1)$ to obtain the graph of $2x - 3y = 3$. See Figure 28.

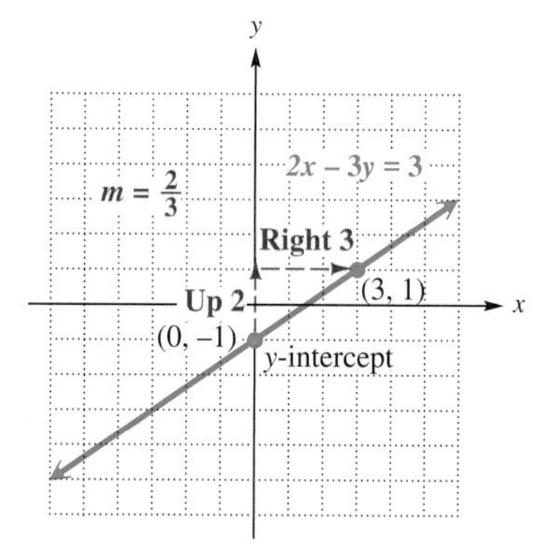

Figure 28

Work Problem 2 at the Side.

The method of Example 2 can be extended to graph a line given its slope and *any* point on the line.

EXAMPLE 3 Graphing a Line Given a Point and the Slope

Graph the line through $(-2, 3)$ with slope -4.

First, locate the point $(-2, 3)$. See Figure 29. Write the slope -4 as

$$m = \frac{\text{change in } y}{\text{change in } x} = \frac{-4}{1}.$$

Continued on Next Page

ANSWERS

2.

y
x
3x – 4y = 8
(4, 1)
(0, –2)

Locate another point on the line by counting 4 units down (because of the negative sign) from $(-2, 3)$ and then 1 unit to the right. Finally, draw the line through this new point P and the given point $(-2, 3)$. See Figure 29.

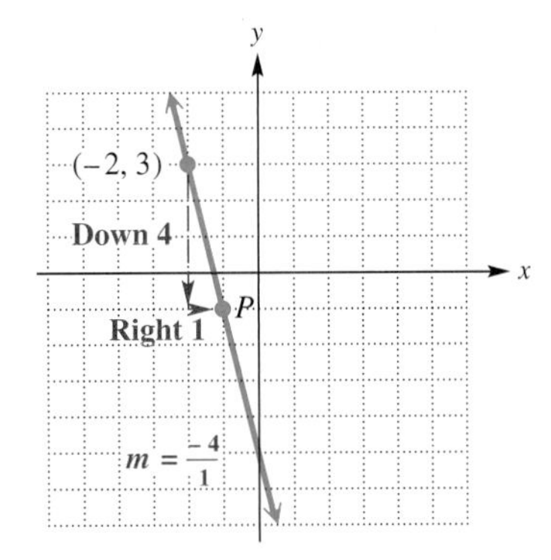

Figure 29

3 Graph the line passing through $(2, -3)$, with slope $-\frac{1}{3}$.

NOTE

In Example 3, we could have written the slope as $\frac{4}{-1}$ instead. In this case, we would move 4 units up from $(-2, 3)$ and then 1 unit to the left (because of the negative sign). Verify that this produces the same line.

Work Problem 3 at the Side.

OBJECTIVE 3 Write an equation of a line given its slope and any point on the line. Let m represent the slope of a line and (x_1, y_1) represent a given point on the line. Let (x, y) represent any other point on the line. Then by the slope formula,

$$\frac{y - y_1}{x - x_1} = m$$

$$y - y_1 = m(x - x_1). \quad \text{Multiply by } x - x_1.$$

This result is the **point-slope form** of the equation of a line.

Point-Slope Form

The point-slope form of the equation of a line with slope m going through (x_1, y_1) is

$$y - y_1 = m(x - x_1).$$

EXAMPLE 4 Using the Point-Slope Form to Write Equations

Find an equation of each line. Write the equation in slope-intercept form.

(a) Through $(-2, 4)$, with slope -3

The given point is $(-2, 4)$ so $x_1 = -2$ and $y_1 = 4$. Also, $m = -3$. Substitute these values into the point-slope form.

$$y - y_1 = m(x - x_1) \quad \text{Point-slope form}$$

$$y - 4 = -3[x - (-2)] \quad \text{Let } x_1 = -2, y_1 = 4, m = -3.$$

$$y - 4 = -3(x + 2)$$

$$y - 4 = -3x - 6 \quad \text{Distributive property}$$

$$y = -3x - 2 \quad \text{Add 4.}$$

Continued on Next Page

ANSWERS

3.

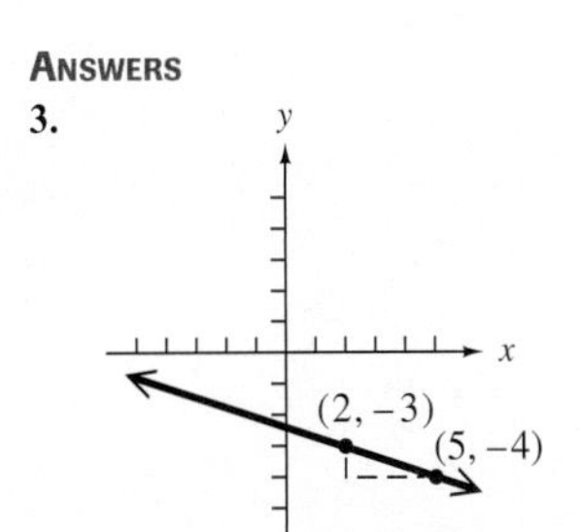

4 Find an equation for each line. Write answers in slope-intercept form.

(a) Through $(-1, 3)$, with slope -2

$$y - y_1 = m(x - x_1)$$
$$y - \_\_\_\_ = \_\_\_\_[x - (\quad)]$$
$$y - 3 = -2(x + \_\_\_\_)$$
$$y - 3 = -2x - \_\_\_\_$$
$$y = \_\_\_\_$$

(b) Through $(5, 2)$, with slope $-\frac{1}{3}$

5 Write an equation in slope-intercept form for the line through each pair of points.

(a) $(-3, 1)$ and $(2, 4)$

(b) $(2, 5)$ and $(-1, 6)$

(b) Through $(4, 2)$, with slope $\frac{3}{5}$

$$y - y_1 = m(x - x_1) \quad \text{Point-slope form}$$
$$y - 2 = \frac{3}{5}(x - 4) \quad \text{Let } x_1 = 4, y_1 = 2, m = \tfrac{3}{5}.$$
$$y - 2 = \frac{3}{5}x - \frac{12}{5} \quad \text{Distributive property}$$
$$y = \frac{3}{5}x - \frac{12}{5} + \frac{10}{5} \quad \text{Add } 2 = \tfrac{10}{5}.$$
$$y = \frac{3}{5}x - \frac{2}{5} \quad \text{Combine terms.}$$

We did not clear fractions after the substitution step because we want the equation in slope-intercept form—that is, solved for y.

Work Problem 4 at the Side.

OBJECTIVE 4 Write an equation of a line given two points on the line. We can also use the point-slope form to find an equation of a line when two points on the line are known.

EXAMPLE 5 Finding the Equation of a Line Given Two Points

Find an equation of the line through the points $(-2, 5)$ and $(3, 4)$. Write the equation in slope-intercept form.

First, find the slope of the line, using the slope formula.

$$\text{slope } m = \frac{y_2 - y_1}{x_2 - x_1} = \frac{5 - 4}{-2 - 3} = \frac{1}{-5} = -\frac{1}{5}$$

Now use either $(-2, 5)$ or $(3, 4)$ and the point-slope form. Using $(3, 4)$ gives

$$y - y_1 = m(x - x_1)$$
$$y - 4 = -\frac{1}{5}(x - 3) \quad \text{Let } x_1 = 3, y_1 = 4, m = -\tfrac{1}{5}.$$
$$y - 4 = -\frac{1}{5}x + \frac{3}{5} \quad \text{Distributive property}$$
$$y = -\frac{1}{5}x + \frac{3}{5} + \frac{20}{5} \quad \text{Add } 4 = \tfrac{20}{5}.$$
$$y = -\frac{1}{5}x + \frac{23}{5}. \quad \text{Combine terms.}$$

The same result would be found using $(-2, 5)$ for (x_1, y_1).

Work Problem 5 at the Side.

Many of the linear equations in **Sections 3.1–3.3** were given in the form $Ax + By = C$, called **standard form,** which we define as follows.

Standard Form

A linear equation is in standard form if it is written as

$$Ax + By = C,$$

where A, B, and C are integers and $A > 0$, $B \neq 0$.

ANSWERS

4. (a) 3; -2; -1; 1; 2; $-2x + 1$
(b) $y = -\frac{1}{3}x + \frac{11}{3}$

5. (a) $y = \frac{3}{5}x + \frac{14}{5}$ (b) $y = -\frac{1}{3}x + \frac{17}{3}$

> **NOTE**
> The preceding definition of standard form is not the same in all texts. A linear equation can be written in this form in many different, equally correct, ways. For example,
>
> $$3x + 4y = 12, \quad 6x + 8y = 24, \quad \text{and} \quad 9x + 12y = 36$$
>
> all represent the same set of ordered pairs. Let us agree that $3x + 4y = 12$ is preferable to the other forms because the greatest common factor of 3, 4, and 12 is 1.

A summary of the types of linear equations follows.

Linear Equations

$x = k$	**Vertical line** Slope is undefined; x-intercept is $(k, 0)$.
$y = k$	**Horizontal line** Slope is 0; y-intercept is $(0, k)$.
$y = mx + b$	**Slope-intercept form** Slope is m; y-intercept is $(0, b)$.
$y - y_1 = m(x - x_1)$	**Point-slope form** Slope is m; line passes through (x_1, y_1).
$Ax + By = C$	**Standard form** Slope is $-\frac{A}{B}$; x-intercept is $\left(\frac{C}{A}, 0\right)$; y-intercept is $\left(0, \frac{C}{B}\right)$.

OBJECTIVE 5 Find an equation of a line that fits a data set. Earlier in this chapter, we gave linear equations that modeled real data, such as annual costs of doctors' visits and amounts of credit card debt, and then used these equations to estimate or predict values. We now develop a procedure to find such an equation if the given set of data fits a linear pattern—that is, its graph consists of points lying close to a straight line.

EXAMPLE 6 Finding an Equation of a Line That Describes Data

The table lists the average annual cost (in dollars) of tuition and fees at public 4-year colleges and universities for selected years. Year 1 represents 1994, year 3 represents 1996, and so on. Plot the data and find an equation that approximates it.

Year	Cost (in dollars)
1	2537
3	2849
5	3110
7	3349
9	3746

Source: U.S. Department of Education.

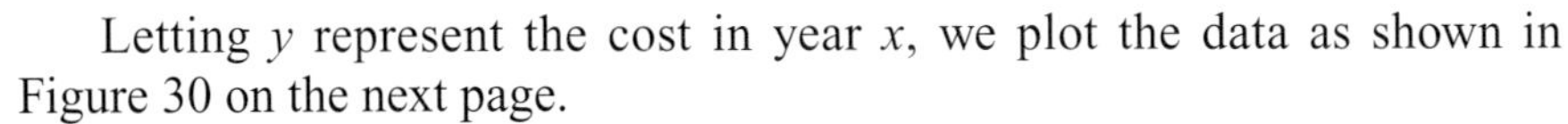

Letting y represent the cost in year x, we plot the data as shown in Figure 30 on the next page.

Continued on Next Page

6 Use the points (1, 2537) and (9, 3746) to find an equation in slope-intercept form that approximates the data of Example 6. How well does this equation approximate the cost in 1998?

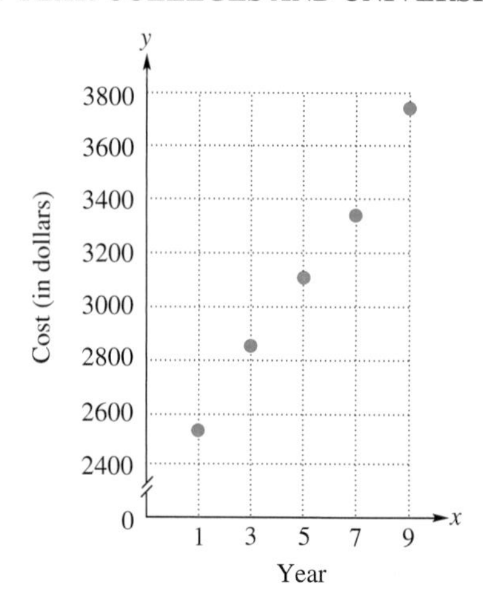

Figure 30

The points appear to lie approximately in a straight line. We can use two of the data pairs and the point-slope form of the equation of a line to get an equation that describes the relationship between the year and the cost. We choose the ordered pairs (3, 2849) and (7, 3349) from the table on the preceding page and find the slope of the line through these points.

$$m = \frac{y_2 - y_1}{x_2 - x_1} = \frac{3349 - 2849}{7 - 3} = \mathbf{125} \qquad \text{Let } (7, 3349) = (x_2, y_2) \text{ and } (3, 2849) = (x_1, y_1).$$

As we might expect, the slope, 125, is positive, indicating that tuition and fees increased $125 each year. Now use this slope and the point (3, 2849) in the point-slope form to find an equation of the line.

$$y - y_1 = m(x - x_1) \qquad \text{Point-slope form}$$
$$y - \mathbf{2849} = \mathbf{125}(x - \mathbf{3}) \qquad \text{Substitute for } x_1, y_1, \text{ and } m.$$
$$y - 2849 = 125x - 375 \qquad \text{Distributive property}$$
$$y = 125x + 2474 \qquad \text{Add 2849.}$$

To see how well this equation approximates the ordered pairs in the data table, let $x = 9$ (for 2002) and find y.

$$y = 125x + 2474 \qquad \text{Equation of the line}$$
$$y = 125(9) + 2474 \qquad \text{Substitute 9 for } x.$$
$$y = 3599$$

The corresponding value in the table for $x = 9$ is 3746, so the equation gives a value that is a bit lower than the actual value. With caution, the equation could be used to predict values for years between 1 and 9 that are not included in the table.

> **NOTE**
> In Example 6, if we had chosen two different data points, we would have gotten a slightly different equation.

ANSWERS

6. $y = 151.125x + 2385.875$; The equation gives $y \approx 3142$ when $x = 5$, which is a very good approximation.

Work Problem 6 at the Side.

3.4 Exercises

FOR EXTRA HELP Addison-Wesley Math Tutor Center | MathXL |  Digital Video Tutor CD 3 Videotape 3 |  Student's Solutions Manual | MyMathLab |  Interactmath.com

Match the correct equation in Column II with the description given in Column I.

I	II
1. Slope $= -2$, through the point $(4, 1)$	**A.** $y = 4x$
2. Slope $= -2$, y-intercept $(0, 1)$	**B.** $y = \frac{1}{4}x$
3. Through the points $(0, 0)$ and $(4, 1)$	**C.** $y = -2x + 1$
4. Through the points $(0, 0)$ and $(1, 4)$	**D.** $y - 1 = -2(x - 4)$

Use the geometric interpretation of slope (rise divided by run, from **Section 3.3***) to find the slope of each line. Then, by identifying the y-intercept from the graph, write the slope-intercept form of the equation of the line.*

5.

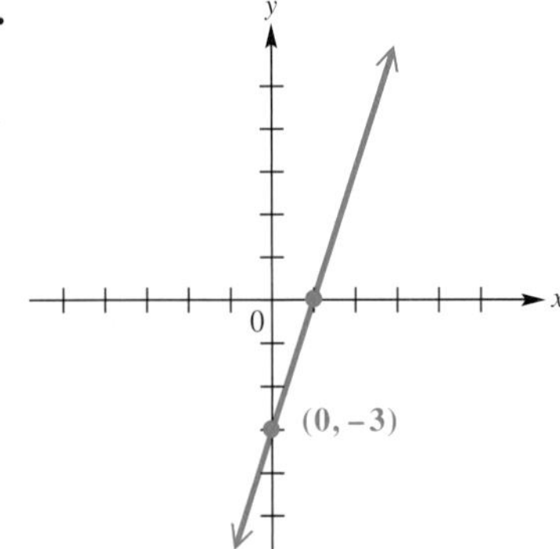

6.

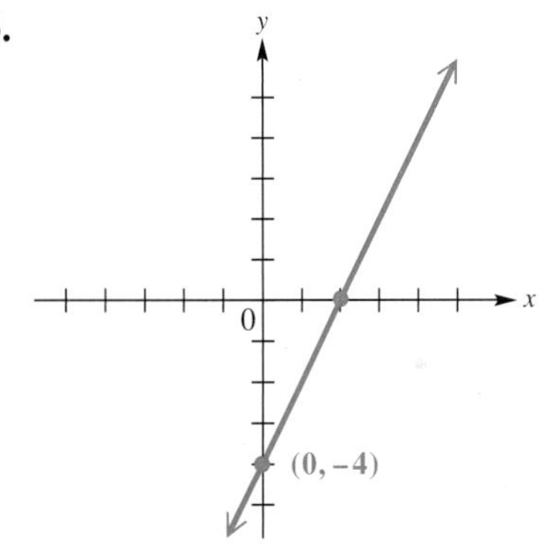

7.

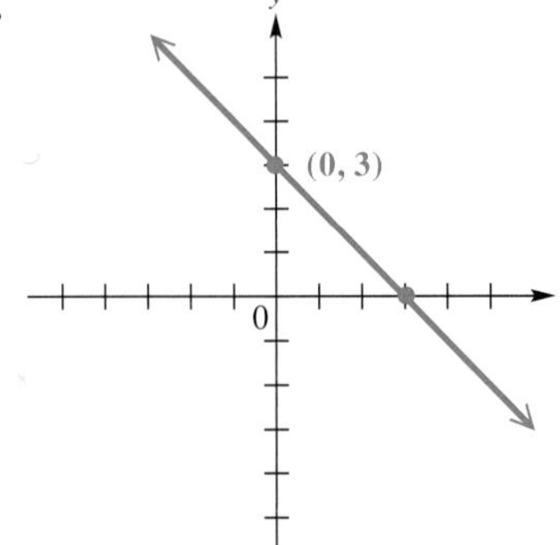

8.

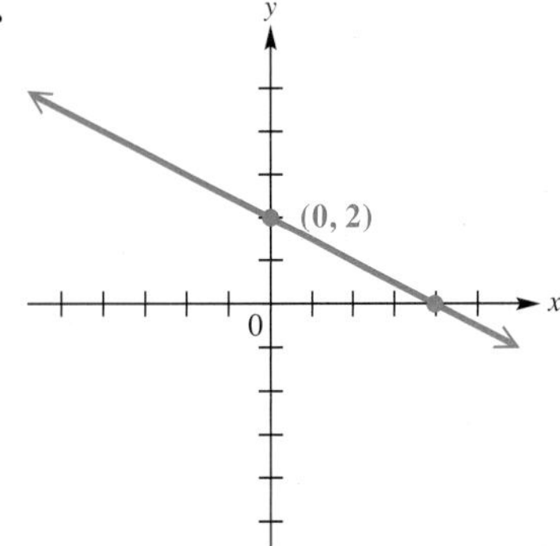

Write the equation of the line with the given slope and y-intercept. See Example 1.

9. slope 4; y-intercept $(0, -3)$

10. slope -5; y-intercept $(0, 6)$

11. slope 0; y-intercept $(0, 3)$

12. slope 3; y-intercept $(0, 0)$

13. Explain why the equation of a vertical line cannot be written in the form $y = mx + b$.

14. Match each equation with the graph that would most closely resemble its graph.

(a) $y = x + 3$

(b) $y = -x + 3$

(c) $y = x - 3$

(d) $y = -x - 3$

A. 

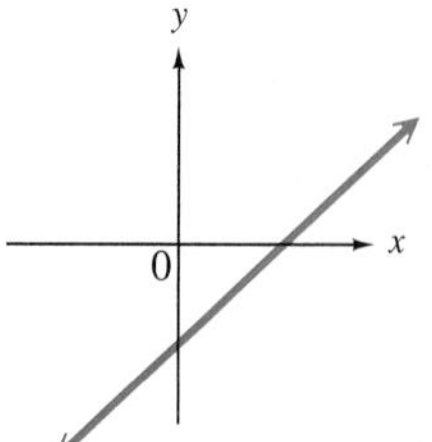

B. 

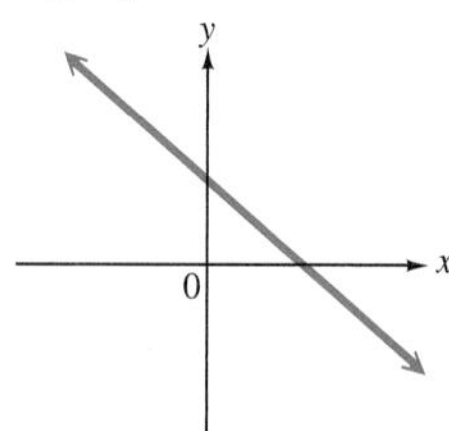

C.

D.

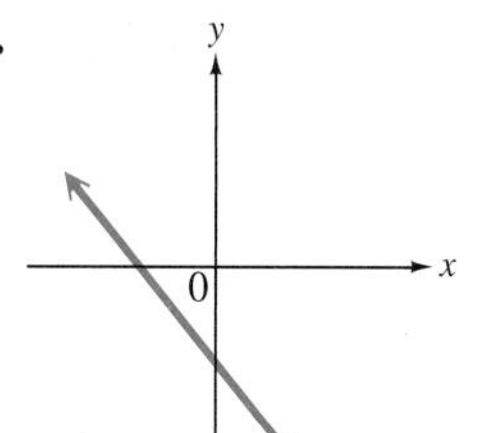

Graph each equation by finding the slope and y-intercept, and using their definitions to find two points on the line. See Example 2.

15. $y = 3x + 2$

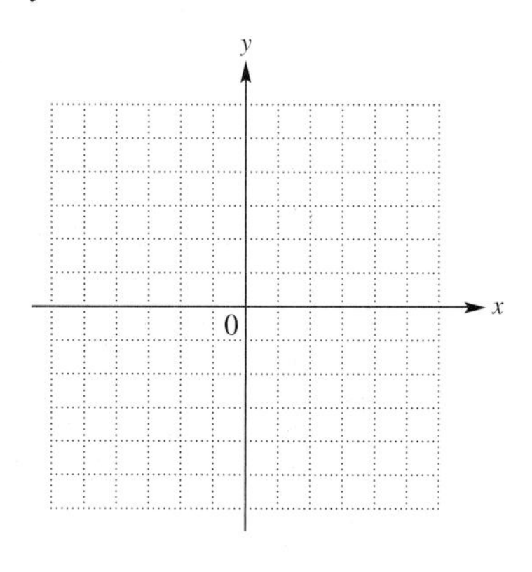

16. $y = 4x - 4$

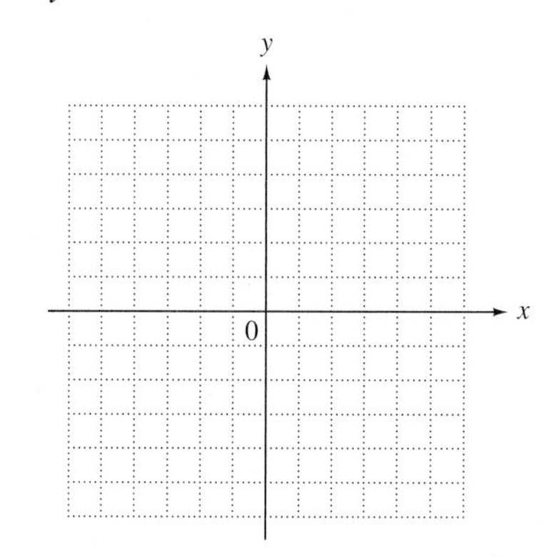

17. $2x + y = -5$

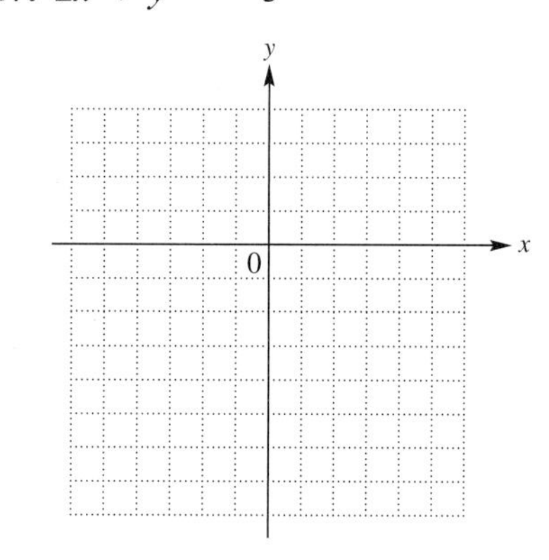

18. $3x + y = -2$

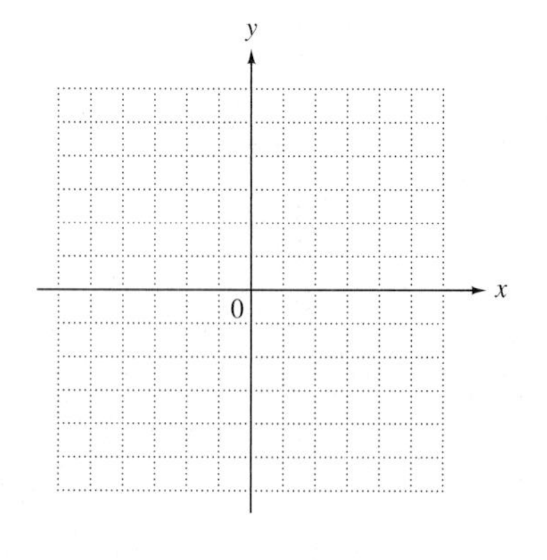

19. $x + 2y = 4$

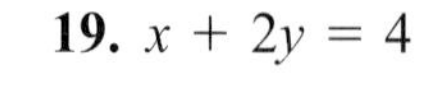

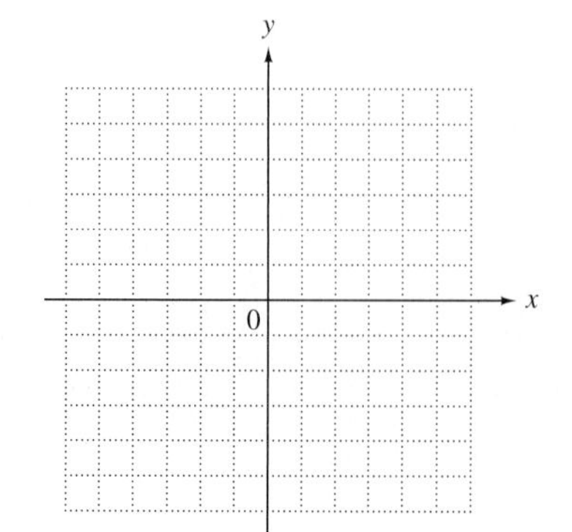

20. $x + 3y = 12$

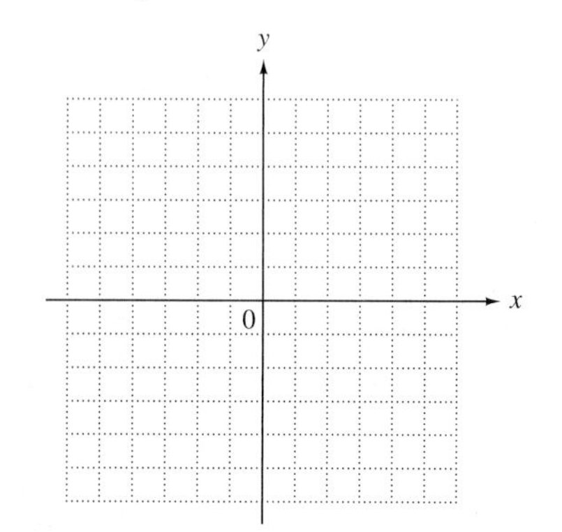

Graph the line through the given point with the given slope. (In Exercises 25–28, recall the types of lines having slope 0 *and undefined slope.) Give the slope-intercept form of the equation of the line if possible. See Example 3.*

21. $(-2, 3), m = \frac{1}{2}$

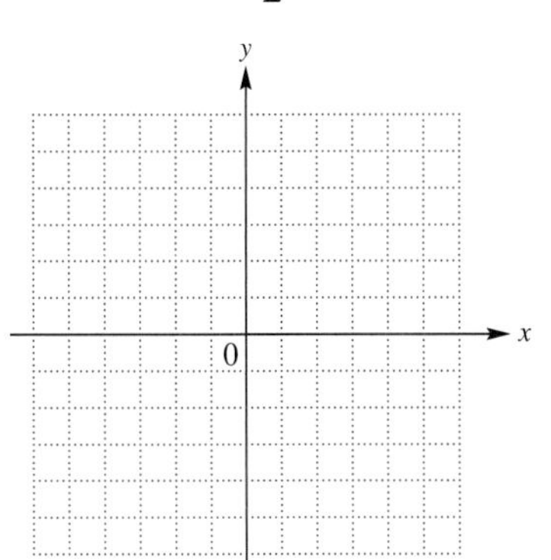

22. $(-4, -1), m = \frac{3}{4}$

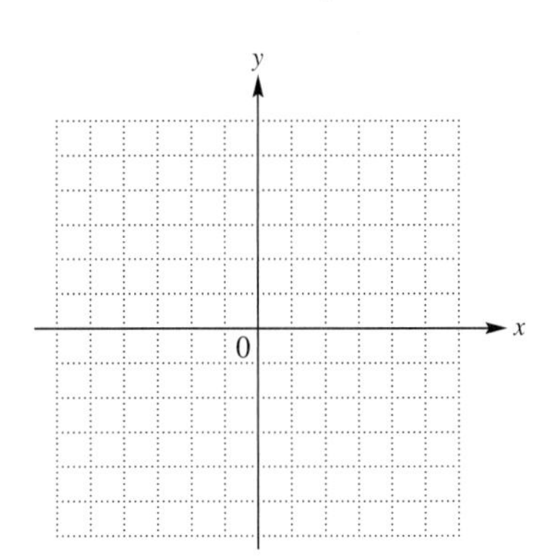

23. $(1, -5), m = -\frac{2}{5}$

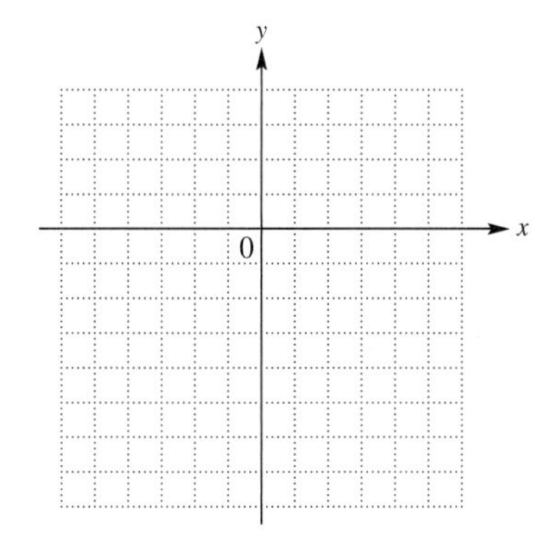

24. $(2, -1), m = -\frac{1}{3}$

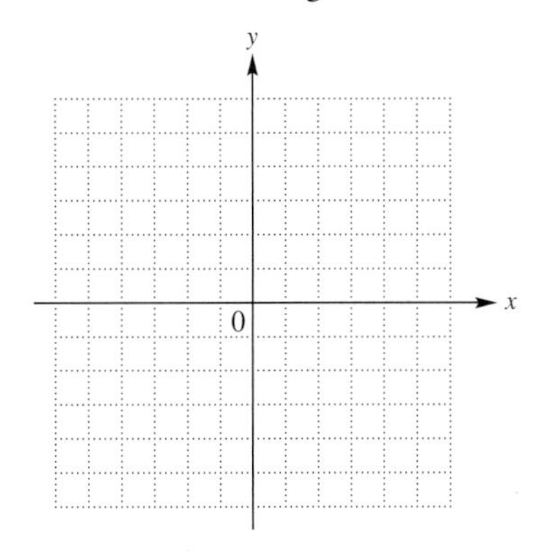

25. $(3, 2), m = 0$

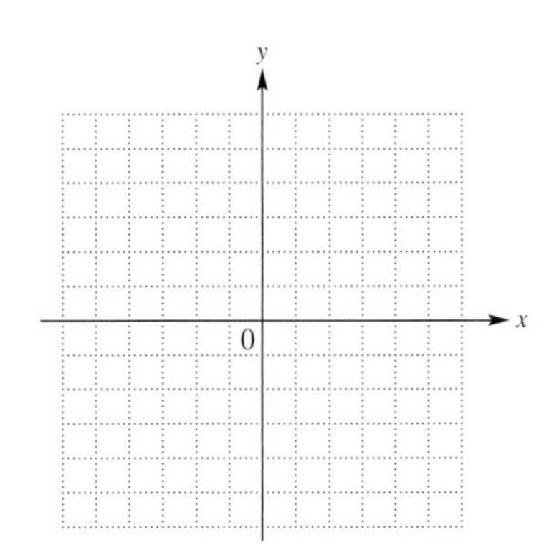

26. $(-2, 3), m = 0$

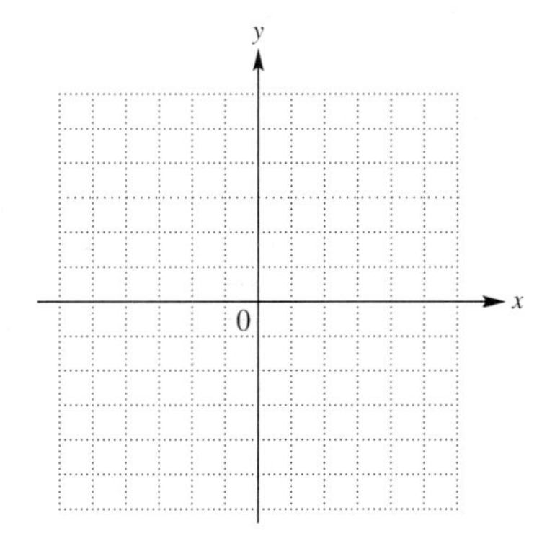

27. $(3, -2)$, undefined slope

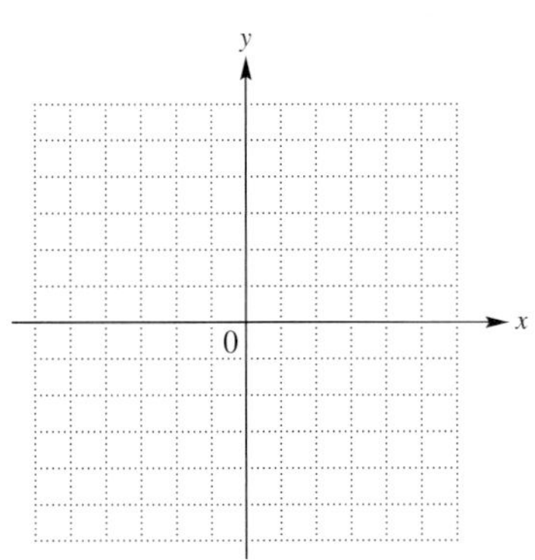

28. $(2, 4)$, undefined slope

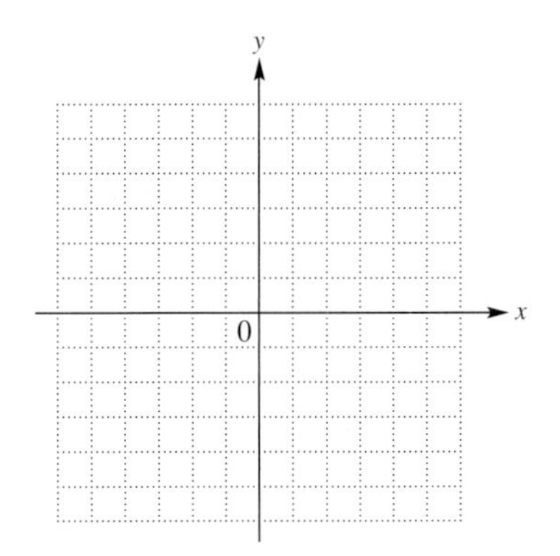

29. $(0, 0), m = \frac{2}{3}$

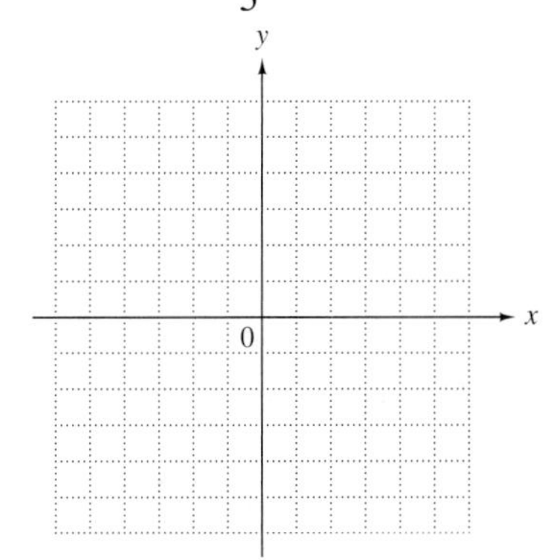

30. (a) What is the common name given to the vertical line whose x-intercept is the origin?

(b) What is the common name given to the line with slope 0 whose y-intercept is the origin?

Write an equation of the line through the given point with the given slope. Write the equation in slope-intercept form. See Example 4.

31. $(4, 1), m = 2$

32. $(2, 7), m = 3$

33. $(3, -10), m = -2$

34. $(2, -5), m = -4$

35. $(-2, 5), m = \frac{2}{3}$

36. $(-4, 1), m = \frac{3}{4}$

Write an equation, in slope-intercept form if possible, of the line through each pair of points. See Example 5.

37. $(8, 5)$ and $(9, 6)$

38. $(4, 10)$ and $(6, 12)$

39. $(-1, -7)$ and $(-8, -2)$

40. $(-2, -1)$ and $(3, -4)$

41. $(0, -2)$ and $(-3, 0)$

42. $(-4, 0)$ and $(0, 2)$

43. $(3, 5)$ and $(3, -2)$

44. $(3, -5)$ and $(-1, -5)$

45. $\left(\frac{1}{2}, \frac{3}{2}\right)$ and $\left(-\frac{1}{4}, \frac{5}{4}\right)$

46. $\left(-\frac{2}{3}, \frac{8}{3}\right)$ and $\left(\frac{1}{3}, \frac{7}{3}\right)$

Write an equation in slope-intercept form of the line satisfying the given conditions.

47. Through $(2, -3)$, parallel to $3x = 4y + 5$

48. Through $(-1, 4)$, perpendicular to $2x + 3y = 8$

49. Perpendicular to $x - 2y = 7$, y-intercept $(0, -3)$

50. Parallel to $5x = 2y + 10$, y-intercept $(0, 4)$

RELATING CONCEPTS (EXERCISES 51–58) For Individual or Group Work

If we think of ordered pairs of the form (C, F), then the two most common methods of measuring temperature, Celsius and Fahrenheit, can be related as follows: When $C = 0, F = 32$, *and when* $C = 100, F = 212$. ***Work Exercises 51–58 in order.***

51. Write two ordered pairs relating these two temperature scales.

52. Find the slope of the line through the two points.

53. Use the point-slope form to find an equation of the line. (Your variables should be C and F rather than x and y.)

54. Write an equation for F in terms of C.

55. Use the equation from Exercise 54 to write an equation for C in terms of F.

56. Use the equation from Exercise 54 to find the Fahrenheit temperature when $C = 30$.

57. Use the equation from Exercise 55 to find the Celsius temperature when $F = 50$.

58. For what temperature is $F = C$?

The cost to produce x items is, in some cases, expressed as $y = mx + b$. *The number b gives the* ***fixed cost*** *(the cost that is the same no matter how many items are produced), and the number m is the* ***variable cost*** *(the cost to produce an additional item). Use this information to work Exercises 59 and 60.*

59. It costs \$400 to start up a business of selling snow cones. Each snow cone costs \$.25 to produce.

(a) What is the fixed cost?

(b) What is the variable cost?

(c) Write the cost equation.

(d) What will be the cost to produce 100 snow cones, based on the cost equation?

(e) How many snow cones will be produced if total cost is \$775?

60. It costs \$2000 to purchase a copier, and each copy costs \$.02 to make.

(a) What is the fixed cost?

(b) What is the variable cost?

(c) Write the cost equation.

(d) What will be the cost to produce 10,000 copies, based on the cost equation?

(e) How many copies will be produced if total cost is \$2600?

Solve each problem. See Example 6.

61. The table lists the average annual cost (in dollars) of tuition and fees at 2-year colleges for selected years, where year 1 represents 1994, year 3 represents 1996, and so on.

Year	Cost (in dollars)
1	1125
3	1239
5	1314
7	1338
9	1379

Source: U.S. Department of Education.

(a) Write five ordered pairs for the data.

(b) Plot the ordered pairs. Do the points lie approximately in a straight line?

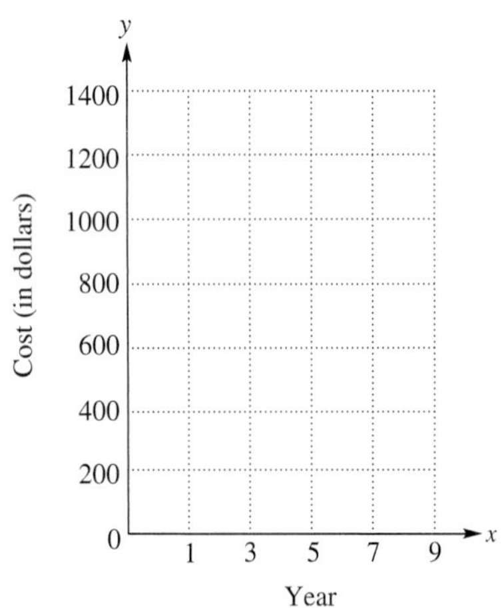

(c) Use the ordered pairs (3, 1239) and (9, 1379) to find the equation of a line that approximates the data. Write the equation in slope-intercept form. (Round the slope to the nearest hundredth and the y-intercept to the nearest whole number.)

(d) Use the equation from part (c) to predict the average annual cost at 2-year colleges in 2004 to the nearest dollar.

62. The table gives heavy-metal nuclear waste (in thousands of metric tons) from spent reactor fuel now stored temporarily at reactor sites, awaiting permanent storage. (*Source:* "Burial of Radioactive Nuclear Waste Under the Seabed," *Scientific American*, January 1998.)

Year *x*	Waste *y*
1995	32
2000*	42
2010*	61
2020*	76

*Estimates by the U.S. Department of Energy.

Let $x = 0$ represent 1995, $x = 5$ represent 2000 (since $2000 - 1995 = 5$), and so on.

(a) For 1995, the ordered pair is (0, 32). Write ordered pairs for the data for the other years given in the table.

(b) Plot the ordered pairs (x, y). Do the points lie approximately in a straight line?

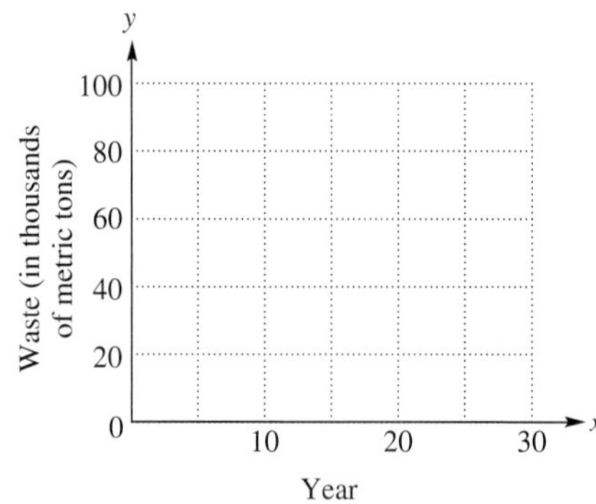

(c) Use the ordered pairs (0, 32) and (25, 76) to find the equation of a line that approximates the other ordered pairs. Write the equation in slope-intercept form.

(d) Use the equation from part (c) to estimate the amount of nuclear waste in 2005. (*Hint:* What is the value of x for 2005?)

Summary Exercises on Graphing Linear Equations

Identify the slope and the y-intercept of the graph of each equation.

1. $3x + y = -6$

2. $2x + y = -4$

3. $-4x - y = 3$

4. $-5x - y = 8$

5. $-3x + 2y = 12$

6. $-5x + 3y = 15$

Graph each equation using its slope and y-intercept.

7. $m = 1, b = -2$

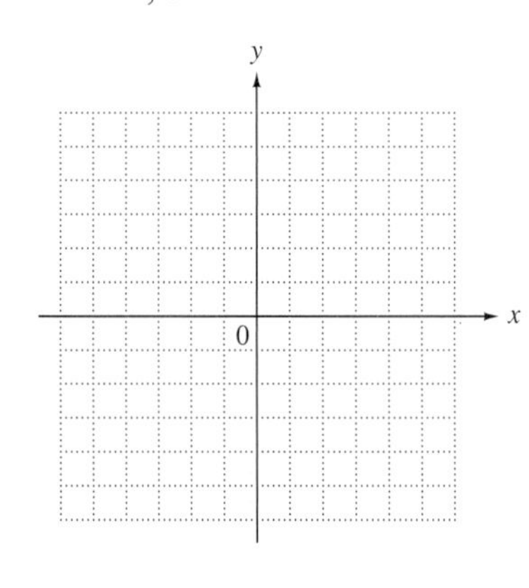

8. $m = 1, b = -4$

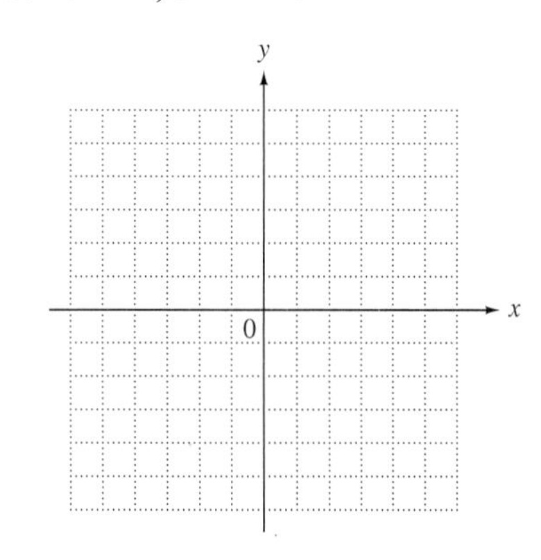

9. $m = -2, b = 6$

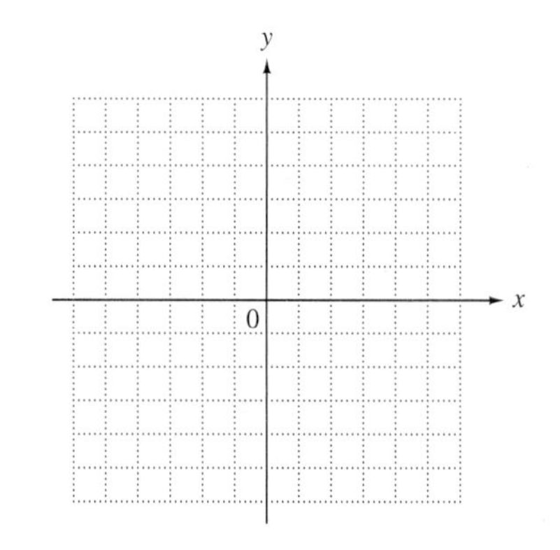

10. $m = -1, b = 6$

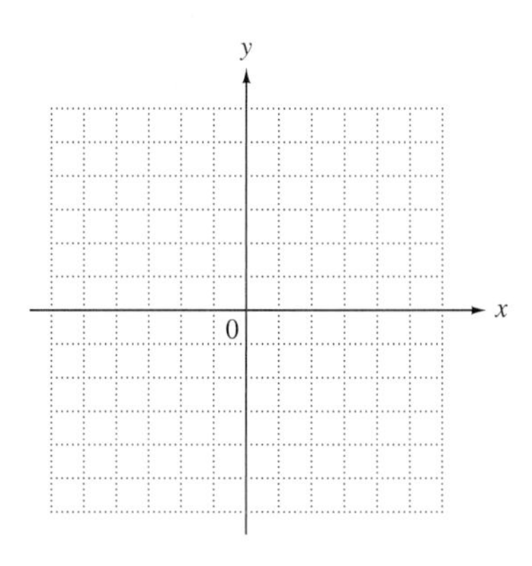

11. $m = -\frac{2}{3}, b = -2$

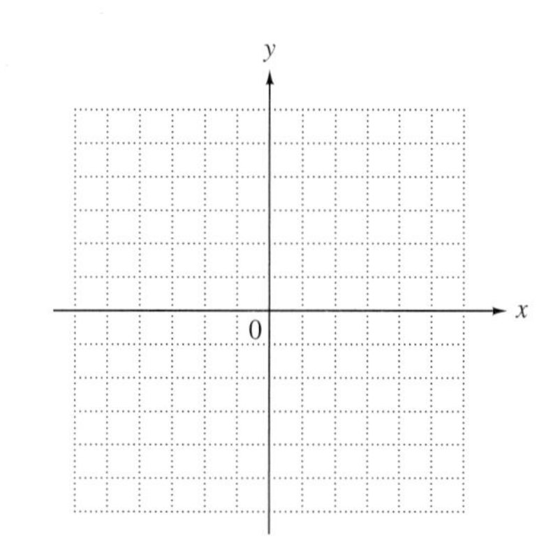

12. $m = -\frac{3}{4}, b = -1$

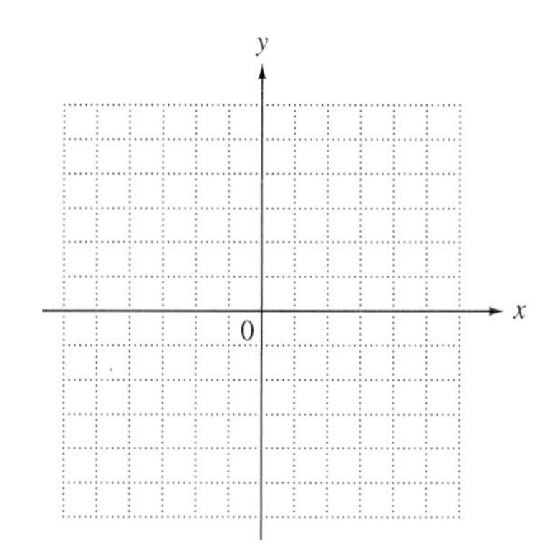

*For each equation **(a)** find the slope, **(b)** find the y-intercept, and **(c)** sketch the graph. Label two points on the graph.*

13. $-x + y = -3$

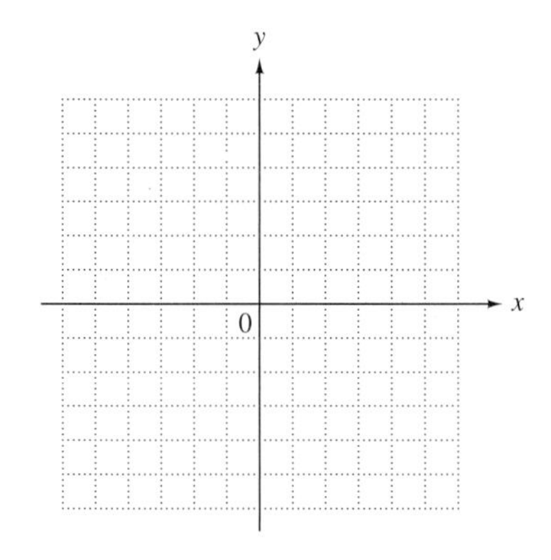

14. $-x + y = -5$

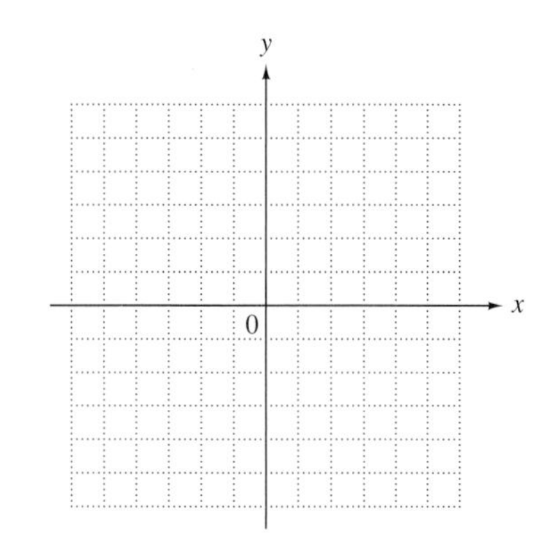

15. $x + 2y = 4$

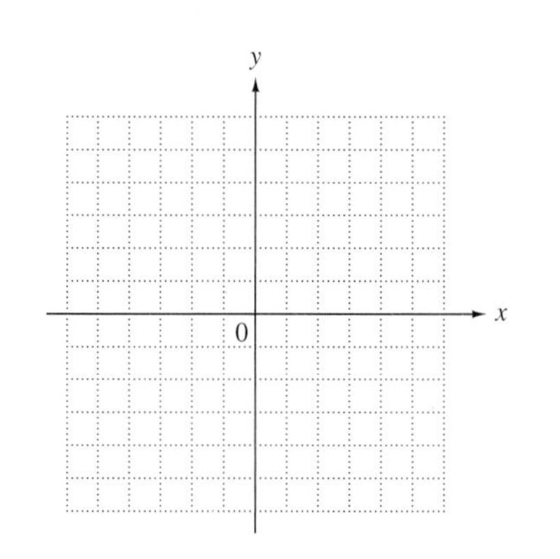

16. $x + 3y = -6$

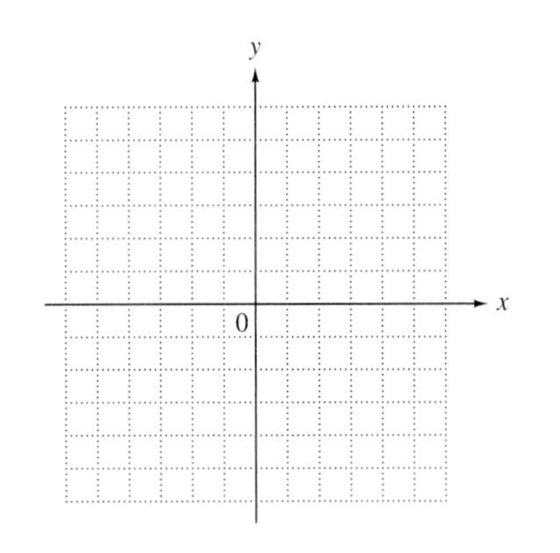

17. $4x - 5y = 20$

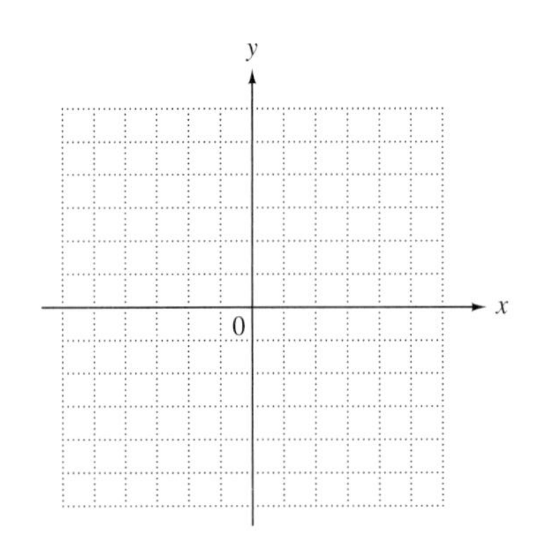

18. $6x - 5y = 30$

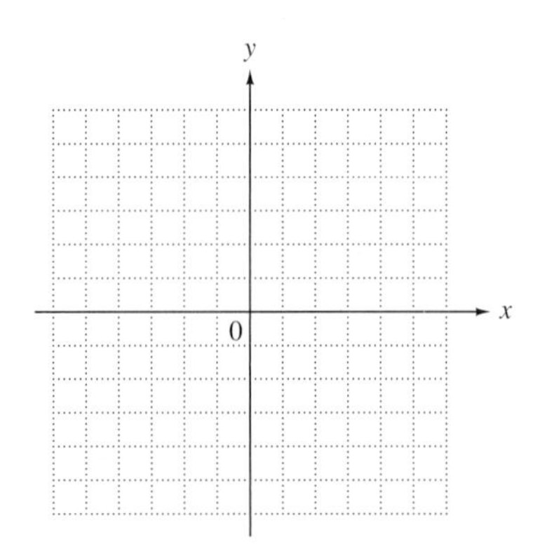

19. $2x + 3y = 12$

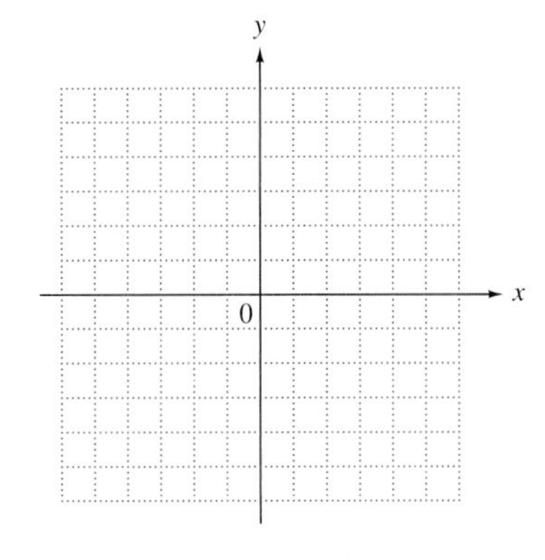

20. $5x + 2y = 10$

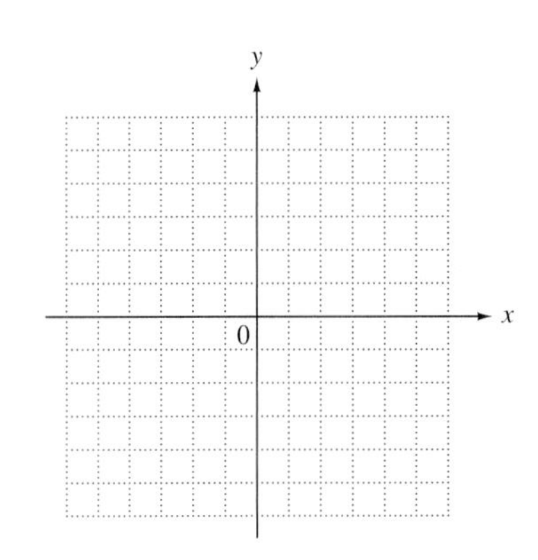

21. $x - 3y = 6$

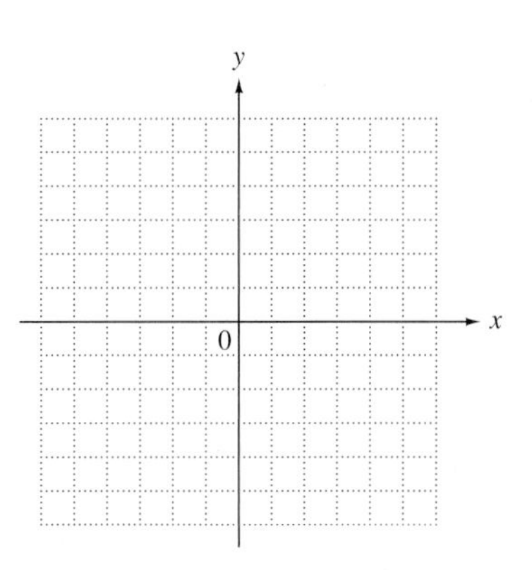

22. $x - 2y = -4$

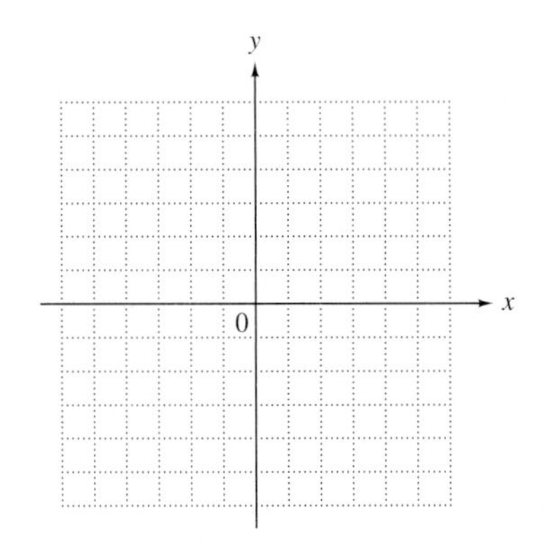

23. $x - 4y = 0$

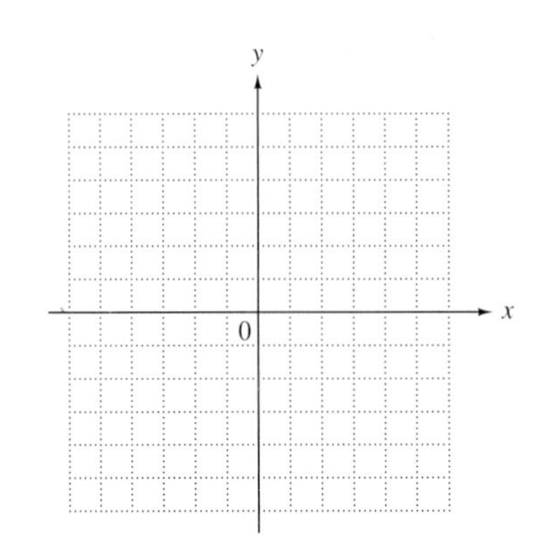

24. $x + 5y = 0$

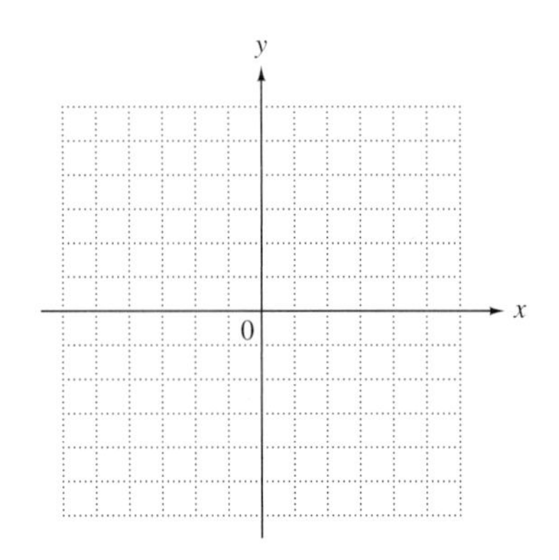

3.5 Graphing Linear Inequalities in Two Variables

OBJECTIVES

1. Graph linear inequalities involving $\leq$ or $\geq$.
2. Graph other linear inequalities.
3. Graph an inequality with boundary through the origin.

In **Section 3.2** we graphed linear equations, such as $2x + 3y = 6$. Now this work is extended to **linear inequalities in two variables,** such as

$$2x + 3y \leq 6.$$

(Recall that $\leq$ is read "is less than or equal to.")

OBJECTIVE 1 Graph linear inequalities involving $\leq$ or $\geq$. The inequality $2x + 3y \leq 6$ means that

$$2x + 3y < 6 \quad \text{or} \quad 2x + 3y = 6.$$

As we found earlier, the graph of $2x + 3y = 6$ is a line. This **boundary line** divides the plane into two regions. The graph of the solutions of the inequality $2x + 3y < 6$ will include only *one* of these regions. We find the required region by solving the given inequality for y.

$$2x + 3y \leq 6$$
$$3y \leq -2x + 6 \qquad \text{Subtract } 2x.$$
$$y \leq -\frac{2}{3}x + 2 \qquad \text{Divide by 3.}$$

By this last statement, ordered pairs in which y is *less than or equal to* $-\frac{2}{3}x + 2$ will be solutions of the inequality. The ordered pairs in which y is equal to $-\frac{2}{3}x + 2$ are on the boundary line, so the ordered pairs in which y *is less than* $-\frac{2}{3}x + 2$ will be *below* that line. (This is because as we move *down* vertically, the y-values *decrease.*) To indicate the solutions, we shade the region below the line, as in Figure 31. The shaded region, along with the line, is the desired graph.

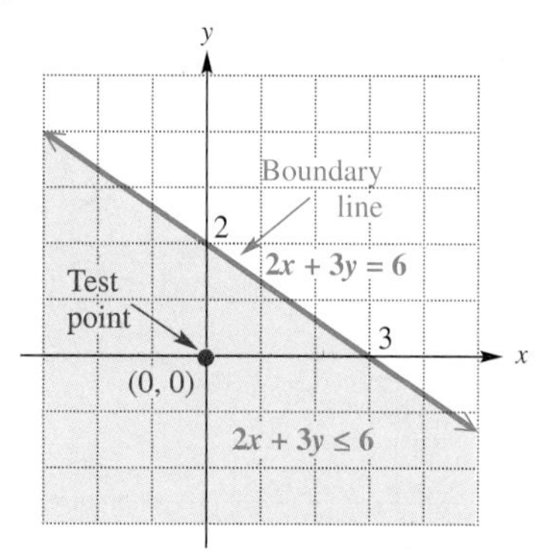

Figure 31

Work Problem 1 at the Side.

1 Shade the appropriate region for each linear inequality.

(a) $x + 2y \geq 6$

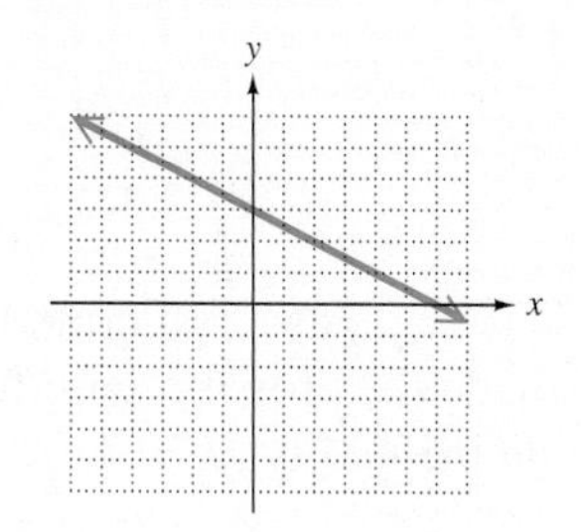

(b) $3x + 4y \leq 12$

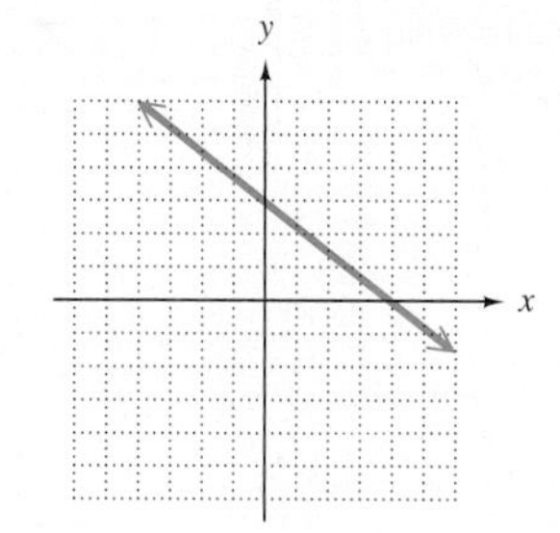

Alternatively, a test point gives a quick way to find the correct region to shade. We choose any point *not* on the boundary line. Because (0, 0) is easy to substitute into an inequality, it is often a good choice, and we will use it here. We substitute 0 for x and 0 for y in the given inequality to see whether the resulting statement is true or false. In the example above,

$$2x + 3y \leq 6$$
$$2(\mathbf{0}) + 3(\mathbf{0}) \overset{?}{\leq} 6 \qquad \text{Let } x = 0 \text{ and } y = 0.$$
$$0 + 0 \overset{?}{\leq} 6$$
$$0 \leq 6. \qquad \text{True}$$

Since the last statement is true, we shade the region that includes the test point (0, 0). This agrees with the result shown in Figure 31.

ANSWERS

1. (a)

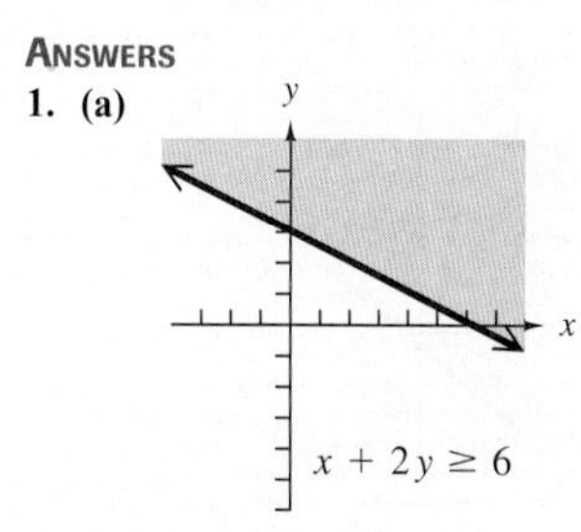

(b)

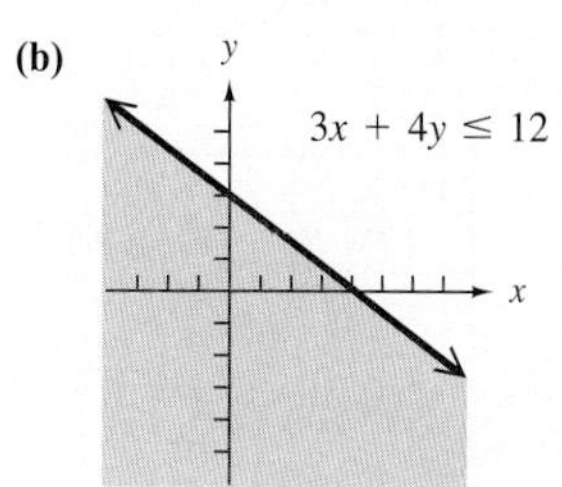

2 Use (0, 0) as a test point to shade the proper region for the inequality $4x - 5y \leq 20$.

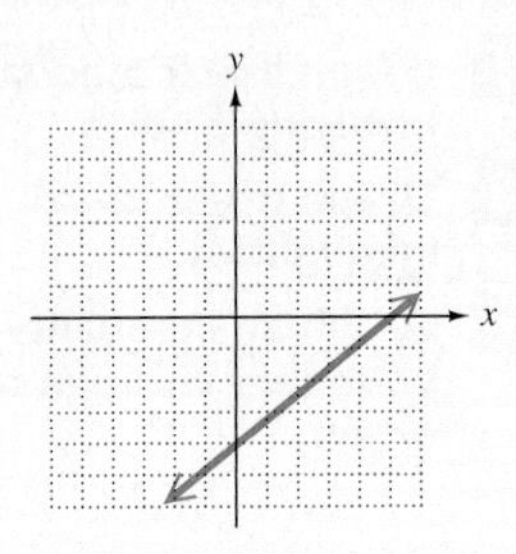

3 Use (1, 1) as a test point to shade the proper region for the inequality $3x + 5y > 15$.

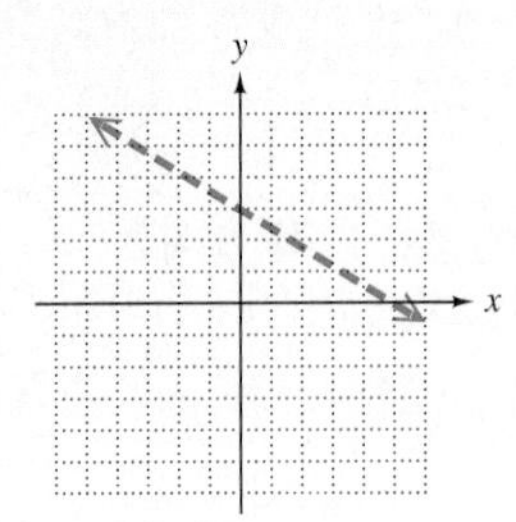

ANSWERS

2.

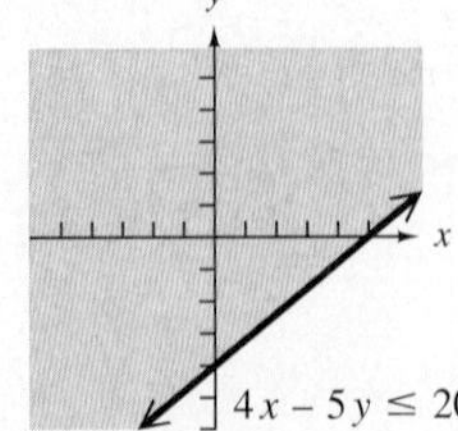

3.

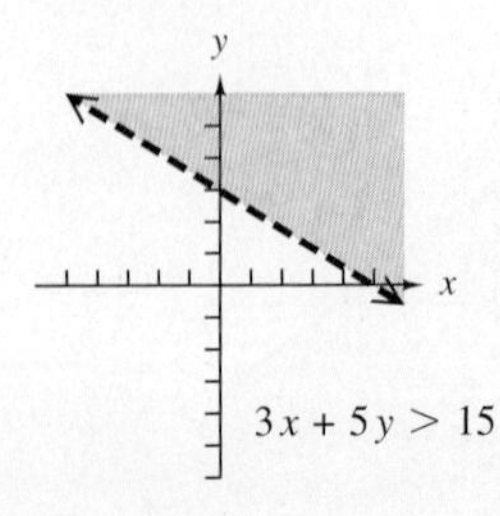

Work Problem 2 at the Side.

OBJECTIVE 2 Graph other linear inequalities. An inequality that does not include the equals sign is graphed in a similar way.

EXAMPLE 1 Graphing a Linear Inequality

Graph the inequality $x - y > 5$.

This inequality does not include the equals sign. Therefore, the points on the line

$$x - y = 5$$

do *not* belong to the graph. However, the line still serves as a boundary for two regions, one of which satisfies the inequality. To graph the inequality, first graph the equation $x - y = 5$. Use a *dashed line* to show that the points on the line are *not* solutions of the inequality $x - y > 5$. Then choose a test point not on the line to see which side of the line satisfies the inequality. We choose $(1, -2)$ this time.

$$x - y > 5$$
$$1 - (-2) > 5 \quad ? \quad \text{Let } x = 1 \text{ and } y = -2.$$
$$3 > 5 \quad \text{False}$$

Because $3 > 5$ is false, the graph of the inequality is the region that does *not* contain $(1, -2)$. We shade the region that does not include the test point $(1, -2)$, as in Figure 32. This shaded region is the desired graph. To check that the proper region is shaded, we select a point in the shaded region and substitute for x and y in the inequality $x - y > 5$. For example, we use $(4, -3)$ from the shaded region as follows.

$$x - y > 5$$
$$4 - (-3) > 5 \quad ? \quad \text{Let } x = 4 \text{ and } y = -3.$$
$$7 > 5 \quad \text{True}$$

This verifies that the correct region is shaded in Figure 32.

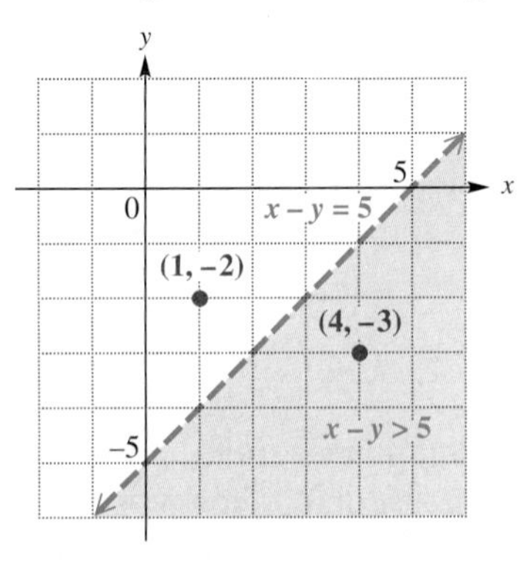

Figure 32

Work Problem 3 at the Side.

A summary of the steps used to graph a linear inequality in two variables follows.

Graphing a Linear Inequality

Step 1 **Graph the boundary.** Graph the line that is the boundary of the region. Use the methods of **Section 3.2.** Draw a solid line if the inequality has $\leq$ or $\geq$; draw a dashed line if the inequality has $<$ or $>$.

Step 2 **Shade the appropriate side.** Use any point not on the line as a test point. Substitute for x and y in the *inequality.* If a true statement results, shade the side containing the test point. If a false statement results, shade the other side.

4 Graph $2x - y \geq -4$.

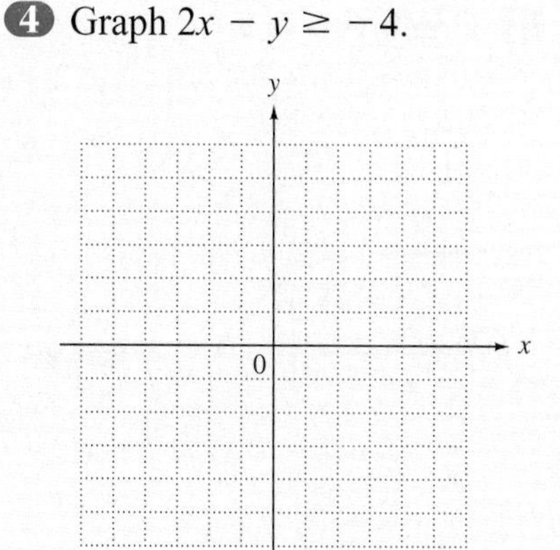

EXAMPLE 2 Graphing a Linear Inequality

Graph the inequality $2x - 5y \geq 10$.

Start by graphing the equation

$$2x - 5y = 10.$$

Use a solid line to show that the points on the line are solutions of the inequality $2x - 5y \geq 10$. Choose any test point not on the line. Again, we choose (0, 0).

$$2x - 5y \geq 10$$

$$2(\mathbf{0}) - 5(\mathbf{0}) \geq 10 \quad ? \quad \text{Let } x = 0 \text{ and } y = 0.$$

$$0 - 0 \geq 10 \quad ?$$

$$0 \geq 10 \quad \text{False}$$

Because $0 \geq 10$ is false, shade the region *not* containing (0, 0). See Figure 33. Verify that a point in the shaded region satisfies the inequality.

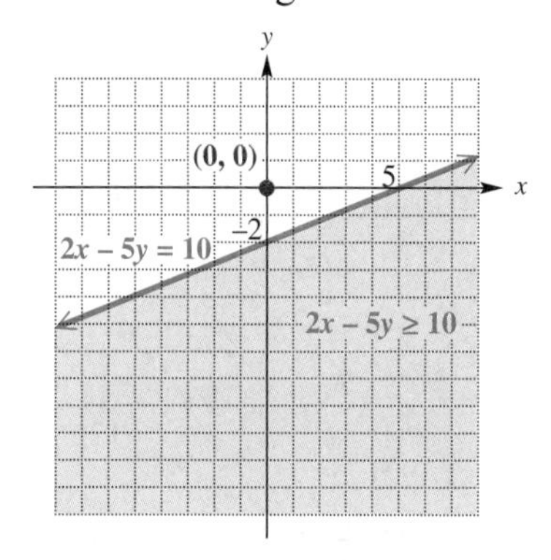

Figure 33

Work Problem 4 at the Side.

EXAMPLE 3 Graphing a Linear Inequality with a Vertical Boundary Line

Graph the inequality $x \leq 3$.

First graph $x = 3$, a vertical line through the point (3, 0). Use a solid line. (Why?) Choose (0, 0) as a test point.

$$x \leq 3$$

$$\mathbf{0} \leq 3 \quad ? \quad \text{Let } x = 0.$$

$$0 \leq 3 \quad \text{True}$$

Continued on Next Page

ANSWERS

4.

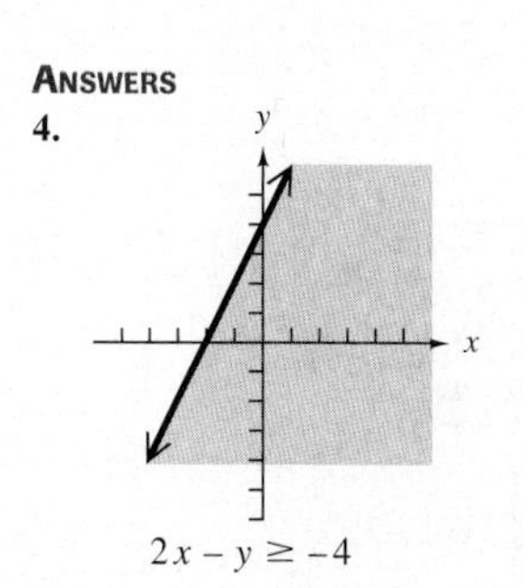

Because $0 \leq 3$ is true, shade the region containing (0, 0), as in Figure 34.

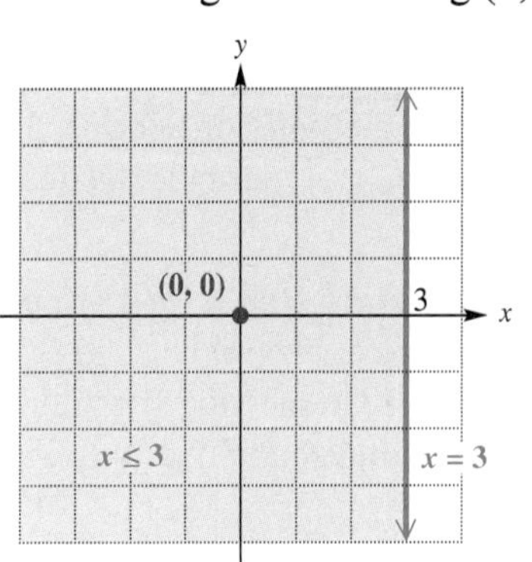

Figure 34

Work Problem 5 at the Side.

5 Graph $y < 4$.

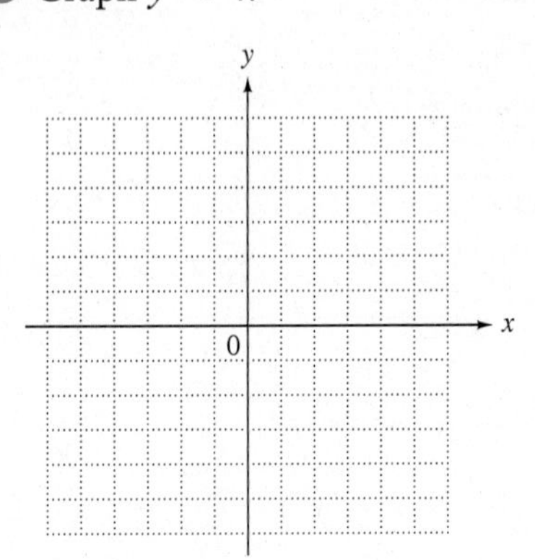

OBJECTIVE 3 Graph an inequality with boundary through the origin. If the graph of an inequality has a boundary line through the origin, (0, 0) cannot be used as a test point.

6 Graph $x \geq -3y$.

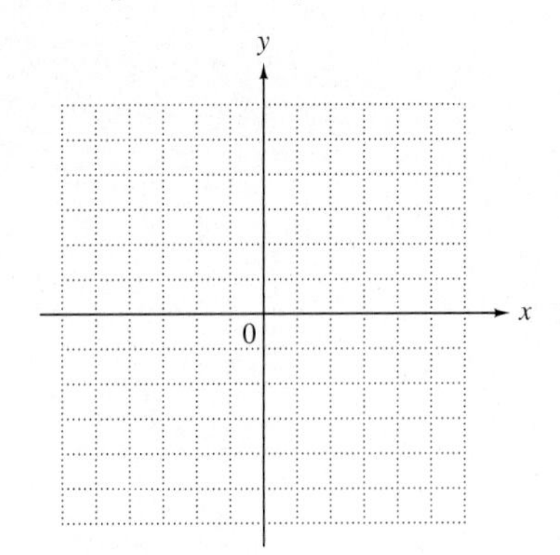

EXAMPLE 4 Graphing a Linear Inequality

Graph the inequality $x \leq 2y$.

We begin by graphing $x = 2y$, using a solid line. Some ordered pairs that can be used to graph this line are (0, 0), (6, 3), and (4, 2). We cannot use (0, 0) as a test point because (0, 0) is on the line $x = 2y$. Instead, we choose a test point off the line, (1, 3).

$$x \leq 2y$$

$$1 \leq 2(3) \quad ? \quad \text{Let } x = 1 \text{ and } y = 3.$$

$$1 \leq 6 \quad \text{True}$$

Because $1 \leq 6$ is true, we shade the side of the graph containing the test point (1, 3). See Figure 35.

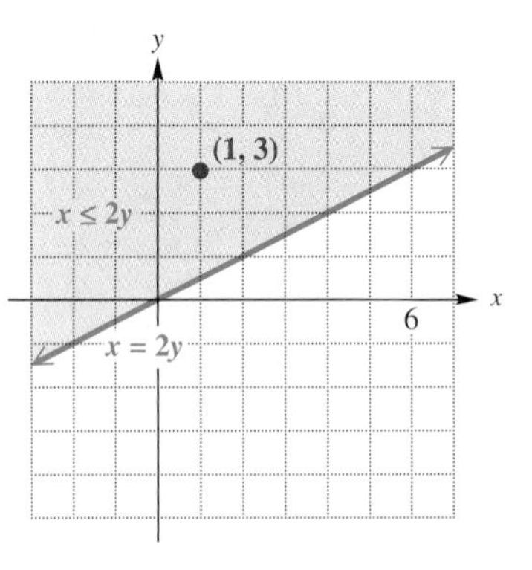

Figure 35

Work Problem 6 at the Side.

ANSWERS

5. $y < 4$

6. $x \geq -3y$

3.5 Exercises

FOR EXTRA HELP

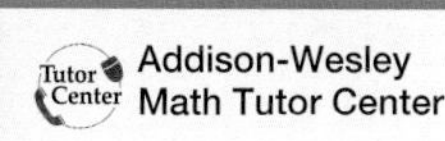
Addison-Wesley Math Tutor Center

MathXL

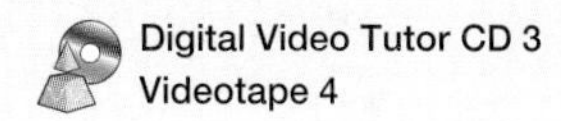
Digital Video Tutor CD 3
Videotape 4

Student's Solutions Manual

MyMathLab

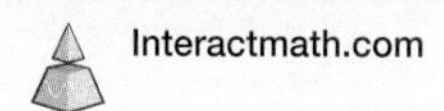
Interactmath.com

Decide whether each statement is true *or* false.

1. The point (4, 0) lies on the graph of $3x - 4y < 12$.

2. The point (4, 0) lies on the graph of $3x - 4y \leq 12$.

3. Both points (4, 1) and (0, 0) lie on the graph of $3x - 2y \geq 0$.

4. The graph of $y > x$ does not contain points in quadrant IV.

The following statements were taken from articles in newspapers or magazines. Each includes a phrase that can be symbolized with one of the inequality symbols $<$, $\leq$, $>$, or $\geq$. In Exercises 5–8, give the inequality symbol for the bold-faced words.

5. According to the Kaiser Family Foundation, in January 2001, (about) one-quarter of Medicare beneficiaries spent **more than** \$2000 a year on drugs.

6. By 1937, a population of as many as a million Attwater's prairie-chickens had been cut to **less than** 9000. (*Source: National Geographic*, March 2002.)

7. In one study, 9 percent of pregnant women had active hepatitis, meaning that **at most** 9 percent of children could get it at birth. (*Source: 2000 Dow Jones & Company, Inc.*)

8. Forty percent of Americans keep **at least** one gun at home. (*Source:* Gallup Organization, quoted in *Reader's Digest*, March 2002.)

In Exercises 9–16, the straight-line boundary has been drawn. Complete each graph by shading the correct region. See Examples 1–4.

9. $x + y \geq 4$

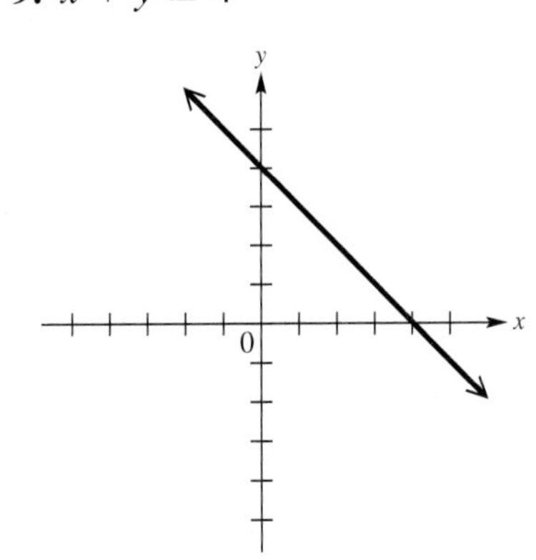

10. $x + y \leq 2$

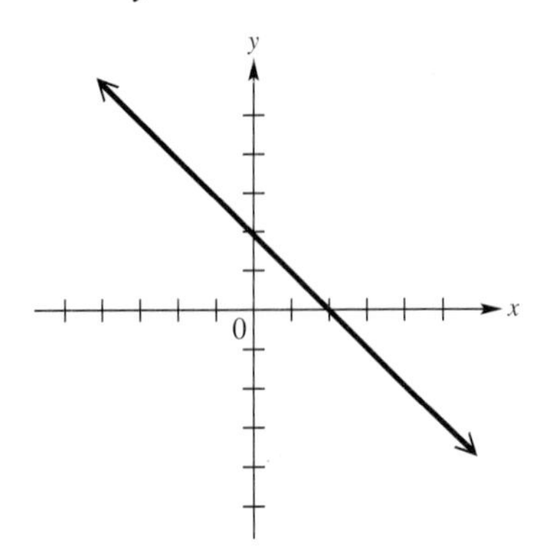

11. $x + 2y \geq 7$

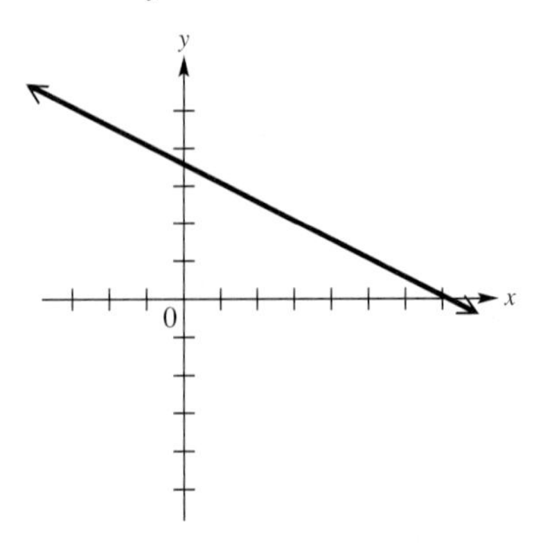

12. $2x + y \geq 5$

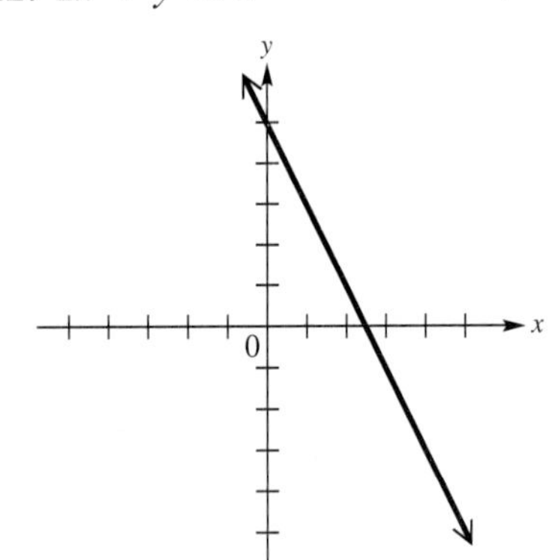

13. $-3x + 4y > 12$

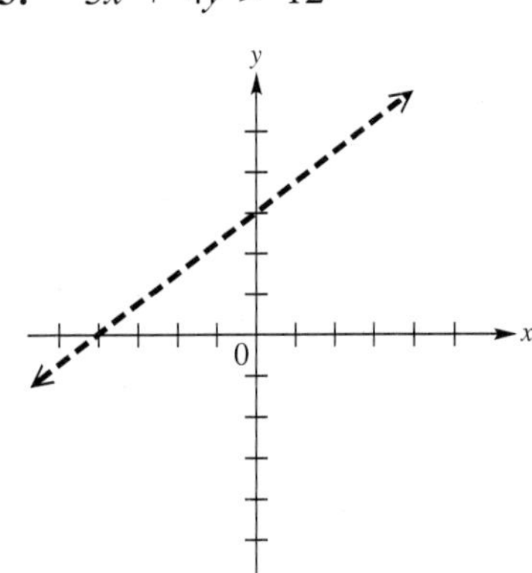

14. $4x - 5y < 20$

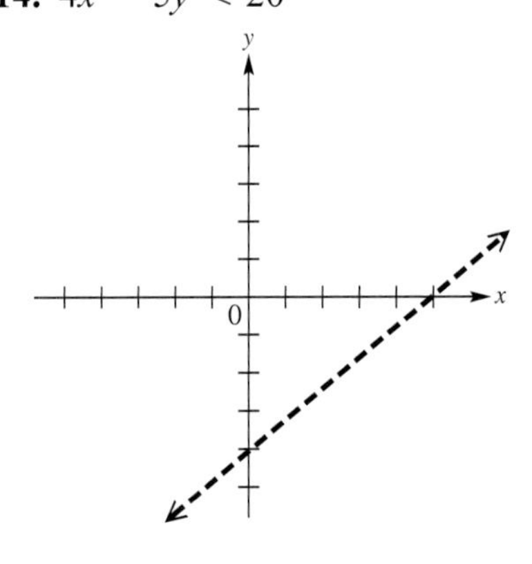

15. $x > 4$

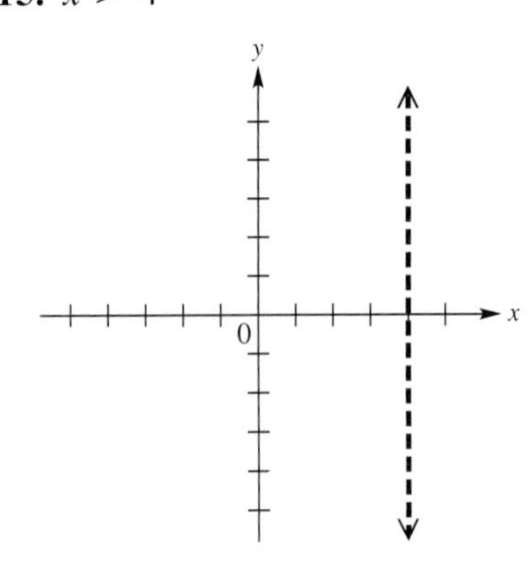

16. $y < -1$

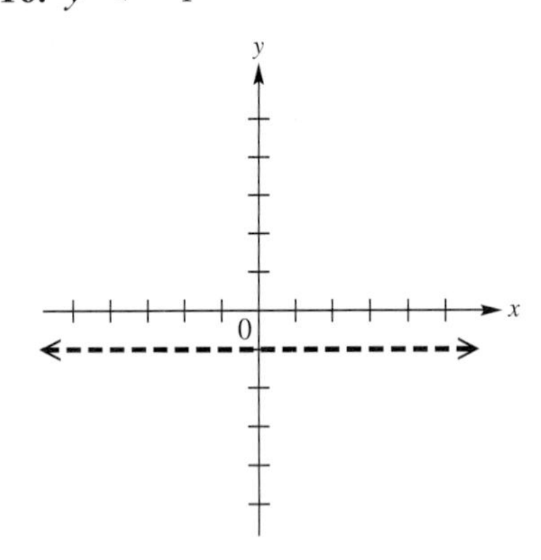

17. Explain how to determine whether to use a dashed line or a solid line when graphing a linear inequality in two variables.

18. Explain why the point (0, 0) is not an appropriate choice for a test point when graphing an inequality whose boundary goes through the origin.

Graph each linear inequality. See Examples 1–4.

19. $x + y \leq 5$

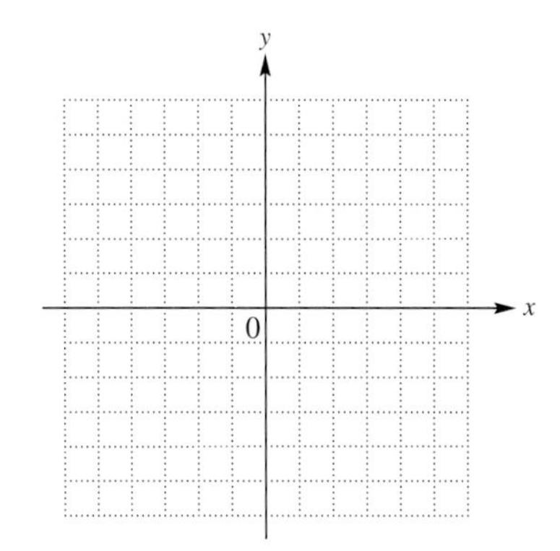

20. $x + y \geq 3$

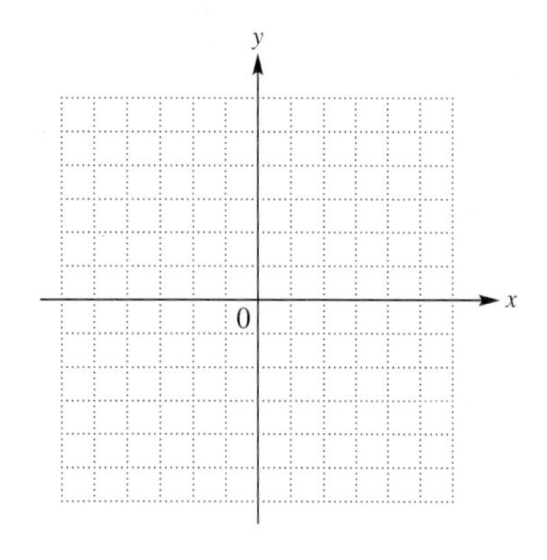

21. $x + 2y < 4$

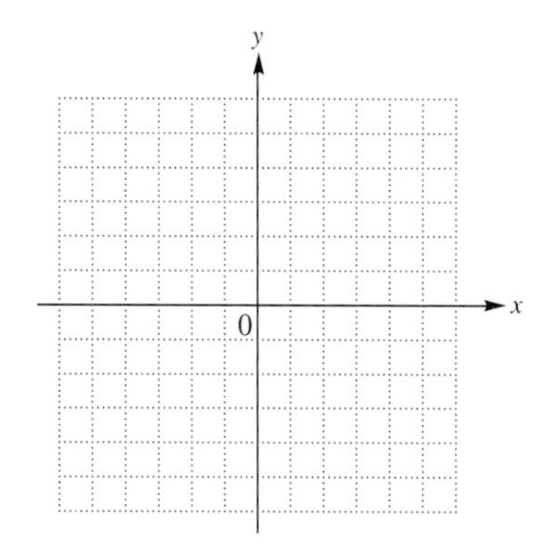

22. $x + 3y > 6$

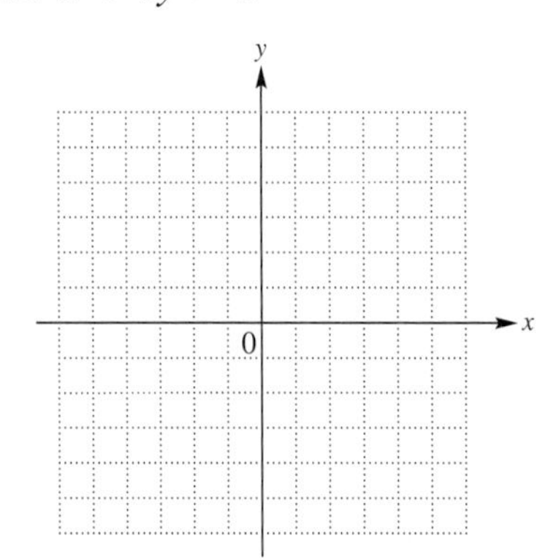

23. $2x + 6 > -3y$

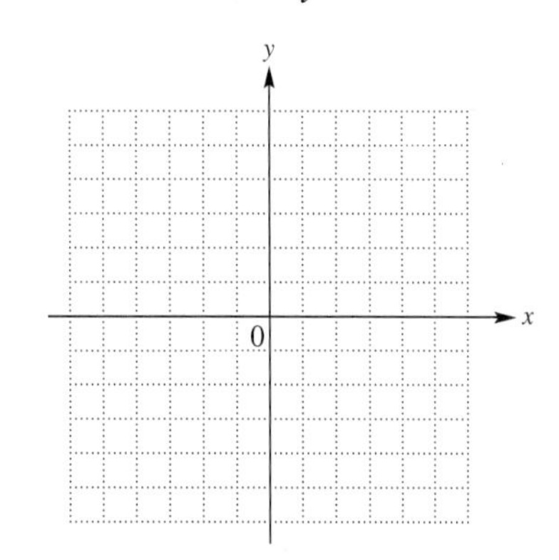

24. $-4y > 3x - 12$

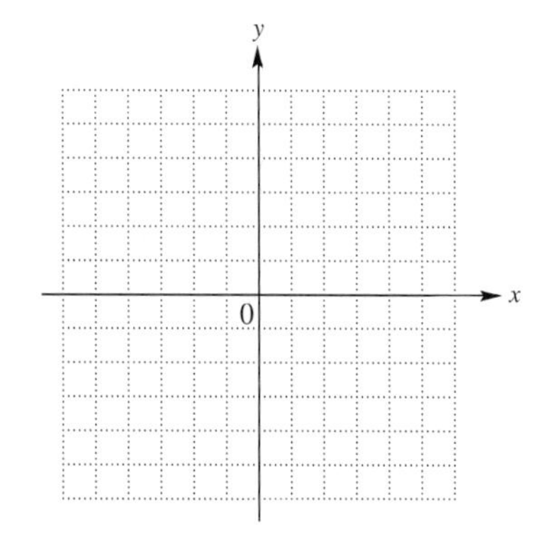

25. $y \geq 2x + 1$

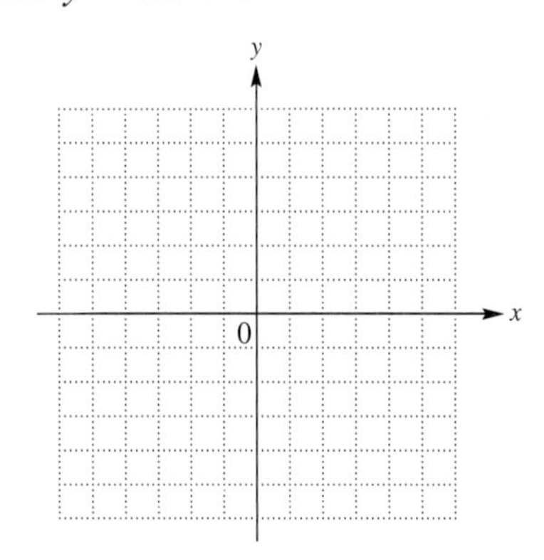

26. $y < -3x + 1$

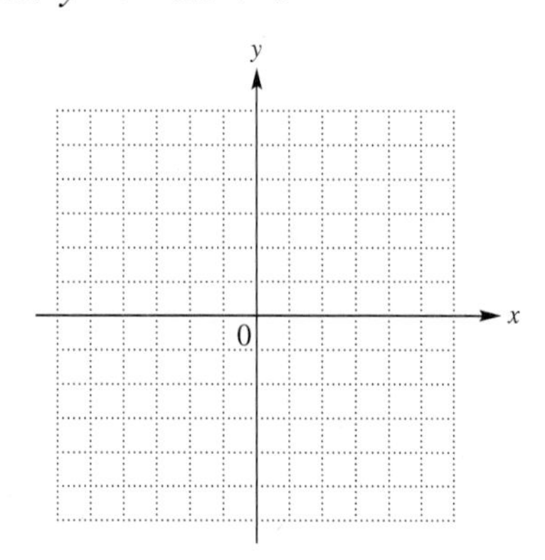

27. $x \leq -2$

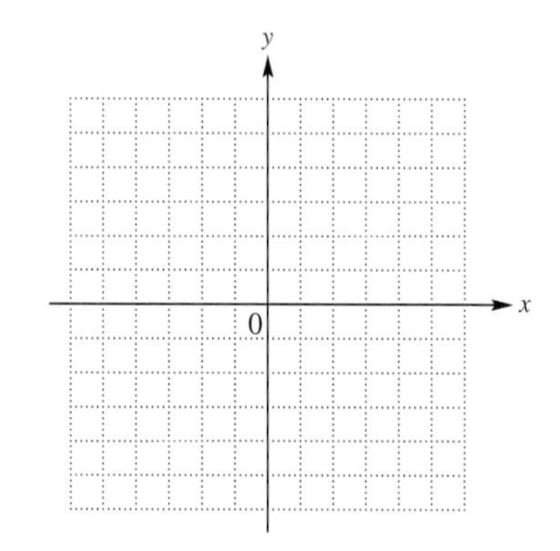

28. $x \geq 1$

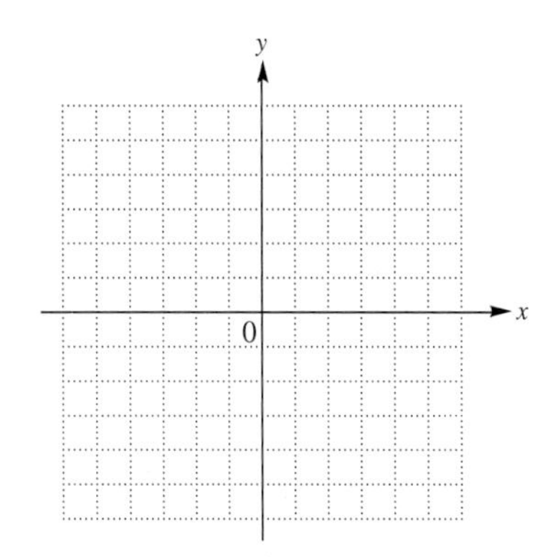

29. $y < 5$

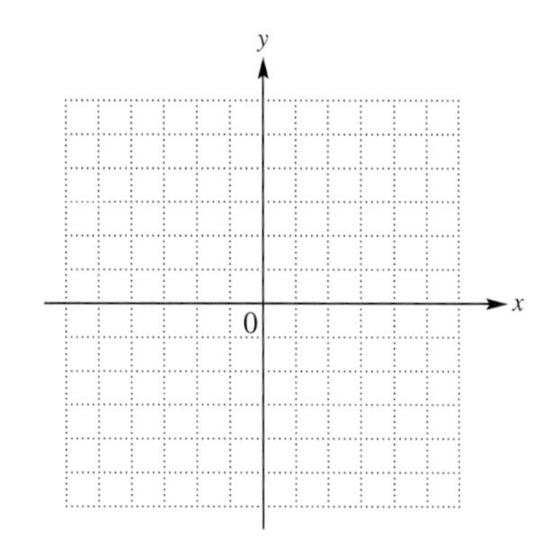

30. $y < -3$

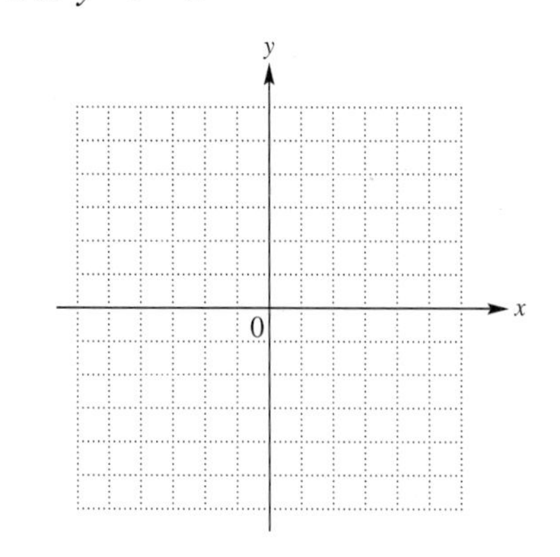

31. $y \geq 4x$

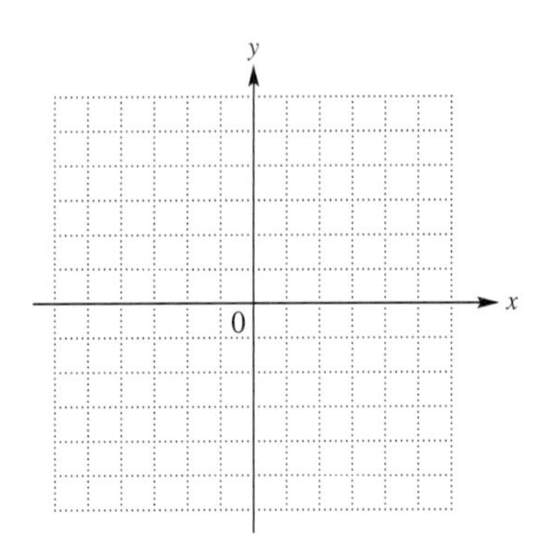

32. $y \leq 2x$

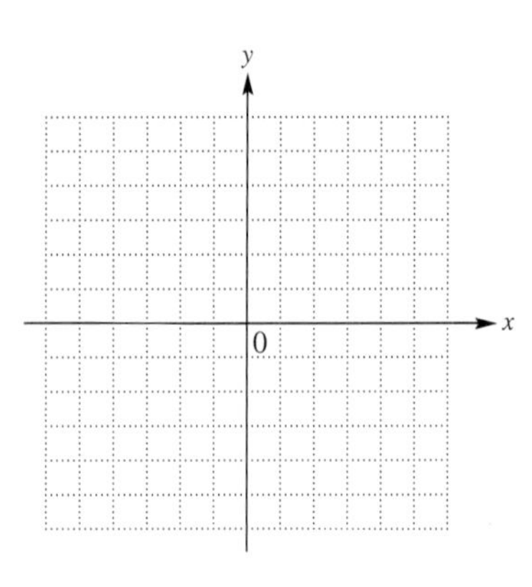

33. Explain why the graph of $y > x$ cannot lie in quadrant IV.

34. Explain why the graph of $y < x$ cannot lie in quadrant II.

Solve each problem. In part (a), $x \geq 0$ and $y \geq 0$, so graph only the part of the inequality in quadrant I.

35. A company will ship x units of merchandise to outlet I and y units of merchandise to outlet II. The company must ship a total of at least 500 units to these two outlets. This can be expressed by writing

$$x + y \geq 500.$$

(a) Graph the inequality.

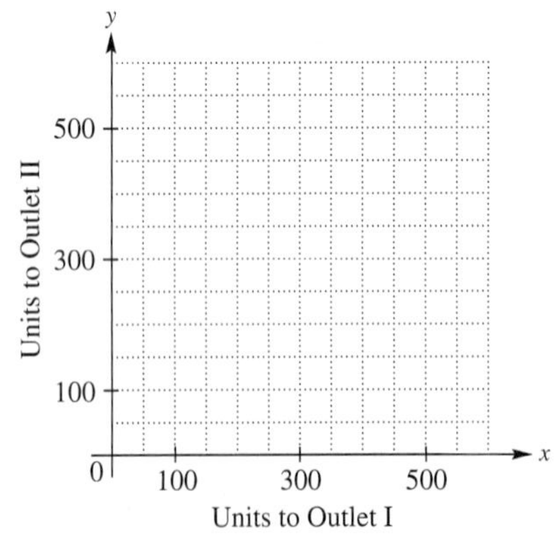

(b) Give two ordered pairs that satisfy the inequality.

36. A toy manufacturer makes stuffed bears and geese. It takes 20 min to sew a bear and 30 min to sew a goose. There is a total of 480 min of sewing time available to make x bears and y geese. These restrictions lead to the inequality

$$20x + 30y \leq 480.$$

(a) Graph the inequality.

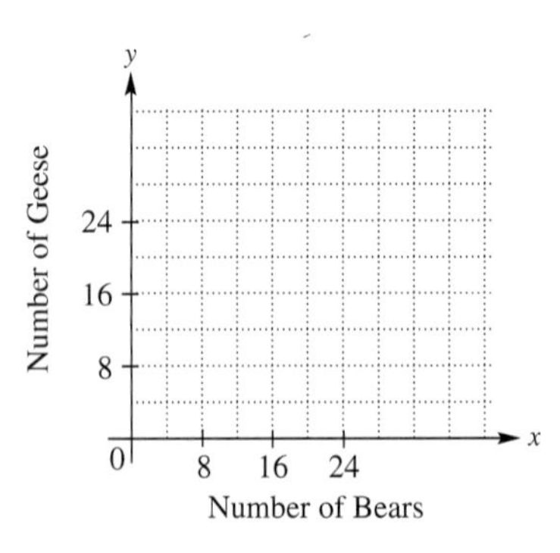

(b) Give two ordered pairs that satisfy the inequality.

Chapter 3

SUMMARY

KEY TERMS

	Term	Definition
3.1	**circle graph**	A circle graph is a circle divided into sectors (or wedges) whose sizes show the relative magnitudes of the categories of data represented.
	bar graph	A bar graph is a series of bars used to show comparisons between two categories of data.
	line graph	A line graph consists of a series of points that are connected with line segments and is used to show changes or trends in data.
	linear equation in two variables	An equation that can be written in the form $Ax + By = C$ is a linear equation in two variables. (A and B are real numbers that cannot both be 0.)
	ordered pair	A pair of numbers written between parentheses in which order is important is called an ordered pair.
	table of values	A table showing selected ordered pairs of numbers that satisfy an equation is called a table of values.
	***x*-axis**	The horizontal axis in a coordinate system is called the x-axis.
	***y*-axis**	The vertical axis in a coordinate system is called the y-axis.
	rectangular (Cartesian) coordinate system	An x-axis and y-axis at right angles form a coordinate system.
	quadrants	A coordinate system divides the plane into four regions called quadrants.
	origin	The point at which the x-axis and y-axis intersect is called the origin.
	plane	A flat surface determined by two intersecting lines is a plane.
	coordinates	The numbers in an ordered pair are called the coordinates of the corresponding point.
	plot	To plot an ordered pair is to find the corresponding point on a coordinate system.
	scatter diagram	A graph of ordered pairs of data is a scatter diagram.
3.2	**graph**	The graph of an equation is the set of all points that correspond to the ordered pairs that satisfy the equation.
	graphing	The process of plotting the ordered pairs that satisfy a linear equation and drawing a line through them is called graphing.
	***y*-intercept**	If a graph intersects the y-axis at k, then the y-intercept is $(0, k)$.
	***x*-intercept**	If a graph intersects the x-axis at k, then the x-intercept is $(k, 0)$.
3.3	**rise**	Rise is the vertical change between two different points on a line.
	run	Run is the horizontal change between two different points on a line.
	slope	The slope of a line is the ratio of the change in y compared to the change in x when moving along the line from one point to another.
	parallel lines	Two lines in a plane that never intersect are parallel.
	perpendicular lines	Perpendicular lines intersect at a 90° angle.
3.5	**linear inequality in two variables**	An inequality that can be written in the form $Ax + By < C$, $Ax + By > C$, $Ax + By \leq C$, or $Ax + By \geq C$ is a linear inequality in two variables.
	boundary line	In the graph of a linear inequality, the boundary line separates the region that satisfies the inequality from the region that does not satisfy the inequality.

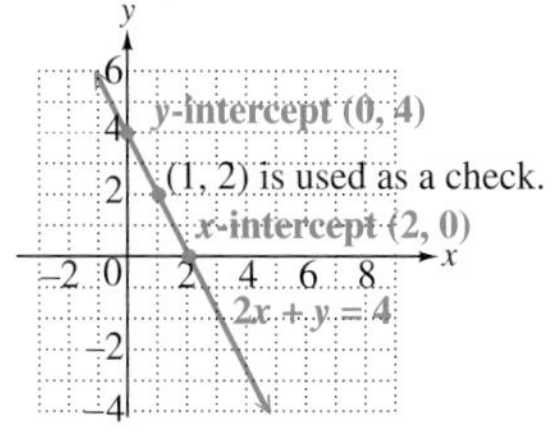

New Symbols

(x, y)	ordered pair	(x_1, y_1)	subscript notation; x-sub-one, y-sub-one	m	slope

Test Your Word Power

See how well you have learned the vocabulary in this chapter. Answers, with examples, follow the Quick Review.

1. An **ordered pair** is a pair of numbers written
   A. in numerical order between brackets
   B. between parentheses or brackets
   C. between parentheses in which order is important
   D. between parentheses in which order does not matter.

2. The **coordinates** of a point are
   A. the numbers in the corresponding ordered pair
   B. the solution of an equation
   C. the values of the x- and y-intercepts
   D. the graph of the point.

3. An **intercept** is
   A. the point where the x-axis and y-axis intersect
   B. a pair of numbers written in parentheses in which order is important
   C. one of the four regions determined by a rectangular coordinate system
   D. a point where a graph intersects the x-axis or the y-axis.

4. The **slope** of a line is
   A. the measure of the run over the rise of the line
   B. the distance between two points on the line
   C. the ratio of the change in y to the change in x along the line
   D. the horizontal change compared to the vertical change of two points on the line.

5. Two lines in a plane are **parallel** if
   A. they represent the same line
   B. they never intersect
   C. they intersect at a 90° angle
   D. one has a positive slope and one has a negative slope.

6. Two lines in a plane are **perpendicular** if
   A. they represent the same line
   B. they never intersect
   C. they intersect at a 90° angle
   D. one has a positive slope and one has a negative slope.

Quick Review

Concepts

3.1 *Reading Graphs; Linear Equations in Two Variables*

Circle graphs, bar graphs, and line graphs are several ways to represent the relationship between two variables.

Examples

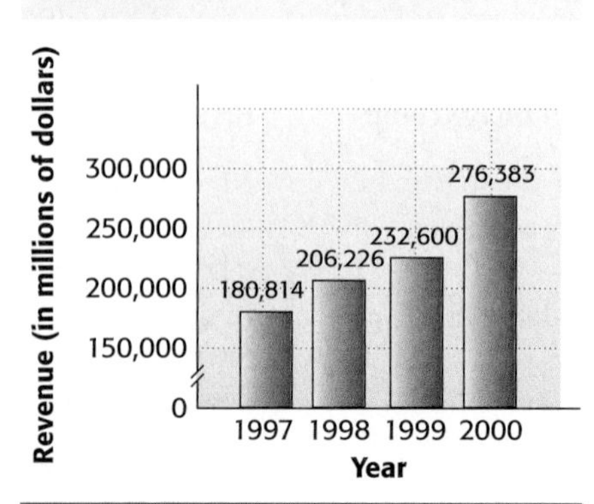

Source: Veronis Suhler Stevenson, *Communications Industry Report,* annual.

The bar graph illustrates communications industry revenue for the years 1997–2000 in millions of dollars.

Concepts	*Examples*

3.1 *Reading Graphs; Linear Equations in Two Variables (continued)*

An ordered pair is a solution of an equation if it makes the equation a true statement.

Is $(2, -5)$ or $(0, -6)$ a solution of $4x - 3y = 18$?

$4(2) - 3(-5) = 23 \neq 18$ — $(2, -5)$ is not a solution.

$4(0) - 3(-6) = 18$ — $(0, -6)$ is a solution.

If a value of either variable in an equation is given, the value of the other variable can be found by substitution.

Complete the ordered pair $(0, \quad)$ for $3x = y + 4$.

$$3(0) = y + 4 \qquad \text{Let } x = 0.$$
$$0 = y + 4$$
$$-4 = y$$

The ordered pair is $(0, -4)$.

To plot the ordered pair $(-3, 4)$, start at the origin, go 3 units to the left, and from there go 4 units up.

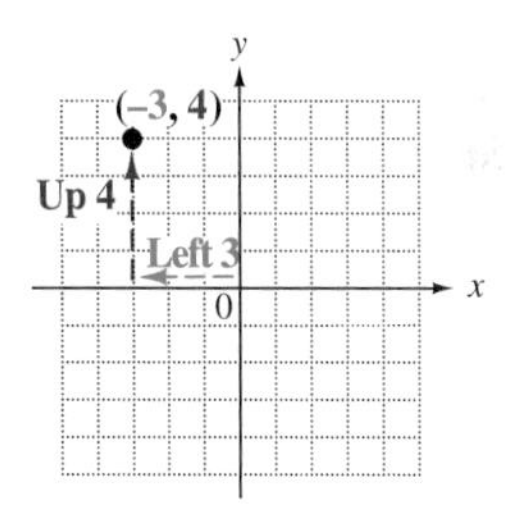

3.2 *Graphing Linear Equations in Two Variables*

To graph a linear equation:

Step 1 Find at least two ordered pairs that are solutions of the equation.

Step 2 Plot the corresponding points.

Step 3 Draw a straight line through the points.

Graph $x - 2y = 4$.

x	y
0	−2
4	0

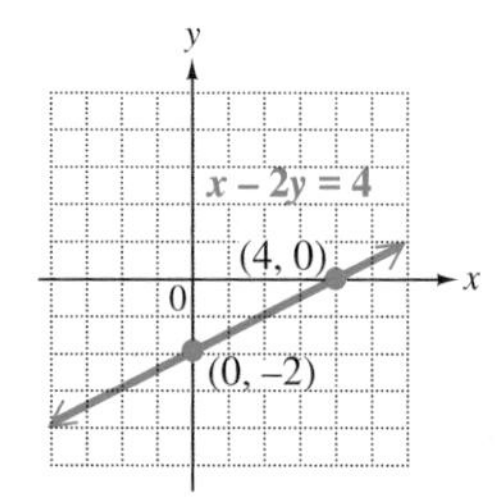

3.3 *Slope of a Line*

The slope of the line through (x_1, y_1) and (x_2, y_2) is

$$m = \frac{\text{change in } y}{\text{change in } x} = \frac{y_2 - y_1}{x_2 - x_1} \quad (x_1 \neq x_2).$$

Horizontal lines have slope 0.

Vertical lines have undefined slope.

To find the slope of a line from its equation, solve for y. The slope is the coefficient of x.

The line through $(-2, 3)$ and $(4, -5)$ has slope

$$m = \frac{-5 - 3}{4 - (-2)} = \frac{-8}{6} = -\frac{4}{3}.$$

The line $y = -2$ has slope 0.

The line $x = 4$ has undefined slope.

Find the slope of the graph of

$$3x - 4y = 12.$$
$$-4y = -3x + 12$$
$$y = \frac{3}{4}x - 3$$

The slope is $\frac{3}{4}$.

Concepts	*Examples*
3.4 *Equations of Lines*	
Slope-Intercept Form $y = mx + b$ m is the slope. $(0, b)$ is the y-intercept.	Find an equation of the line with slope **2** and y-intercept $(0, -5)$. $$y = 2x - 5$$
Point-Slope Form $y - y_1 = m(x - x_1)$ m is the slope. (x_1, y_1) is a point on the line.	Find an equation of the line with slope $-\frac{1}{2}$ through $(-4, 5)$. $$y - 5 = -\frac{1}{2}[x - (-4)]$$ $$y - 5 = -\frac{1}{2}(x + 4)$$ $$y - 5 = -\frac{1}{2}x - 2$$ $$y = -\frac{1}{2}x + 3$$
Standard Form $Ax + By = C$ A, B, and C are integers and $A > 0$, $B \neq 0$.	This equation is written in standard form as $$x + 2y = 6,$$ with $A = 1$, $B = 2$, and $C = 6$.
3.5 *Graphing Linear Inequalities in Two Variables*	
Step 1 Graph the line that is the boundary of the region. Make it solid if the inequality is $\leq$ or $\geq$; make it dashed if the inequality is $<$ or $>$.	Graph $2x + y \leq 5$. Graph the line $2x + y = 5$. Make it solid because the symbol $\leq$ includes equality.
Step 2 Use any point not on the line as a test point. Substitute for x and y in the inequality. If the result is true, shade the side of the line containing the test point; if the result is false, shade the other side.	Use $(1, 0)$ as a test point. $2(1) + 0 \leq 5$? $2 \leq 5$ True Shade the side of the line containing $(1, 0)$. 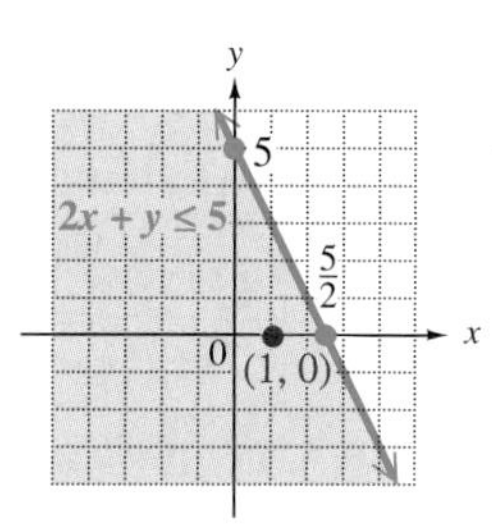

Answers to Test Your Word Power

1. C; *Examples:* (0, 3), (3, 8), (4, 0)
2. A; *Example:* The point associated with the ordered pair (1, 2) has x-coordinate 1 and y-coordinate 2.
3. D; *Example:* The graph of the equation $4x - 3y = 12$ has x-intercept at $(3, 0)$ and y-intercept at $(0, -4)$.
4. C; *Example:* The line through (3, 6) and (5, 4) has slope $\frac{4-6}{5-3} = \frac{-2}{2} = -1$.
5. B; *Example:* See Figure 24 in **Section 3.3.**
6. C; *Example:* See Figure 25 in **Section 3.3.**

Chapter 3

REVIEW EXERCISES

[3.1] *The percents of four-year college students in private institutions who earned a degree within five years of entry between 1996 and 2001 are shown in the graph.*

1. What was the total percent decrease for the period from 1996 to 2001?

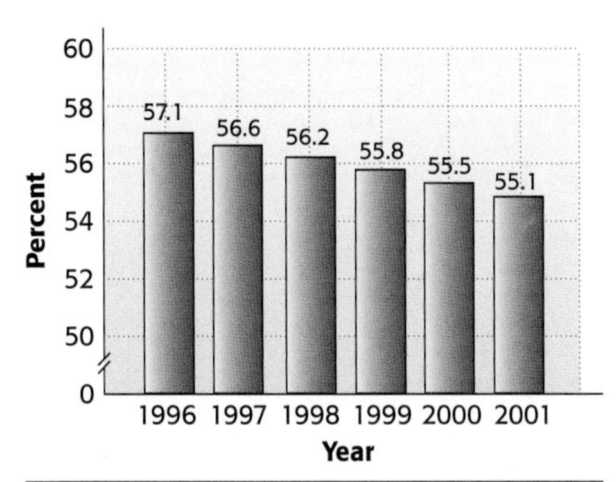

Source: ACT.

2. Write ordered pairs to reflect the data shown in the graph.

3. In what year did the percent show the greatest decrease from the previous year? What was this decrease?

4. Describe the general trend seen in the graph.

Complete the given ordered pairs for each equation.

5. $y = 3x + 2$ $\quad (-1,\ \ ), (0,\ \ ), (\ \ , 5)$

6. $4x + 3y = 6$ $\quad (0,\ \ ), (\ \ , 0), (-2,\ \ )$

7. $x = 3y$ $\quad (0,\ \ ), (8,\ \ ), (\ \ , -3)$

8. $x - 7 = 0$ $\quad (\ \ , -3)\ (\ \ , 0), (\ \ , 5)$

Decide whether each ordered pair is a solution of the given equation.

9. $x + y = 7;\ (2, 5)$

10. $2x + y = 5;\ (-1, 3)$

11. $3x - y = 4;\ \left(\frac{1}{3}, -3\right)$

Plot each ordered pair on the given coordinate system.

12. $(2, 3)$

13. $(-4, 2)$

14. $(3, 0)$

15. $(0, -6)$

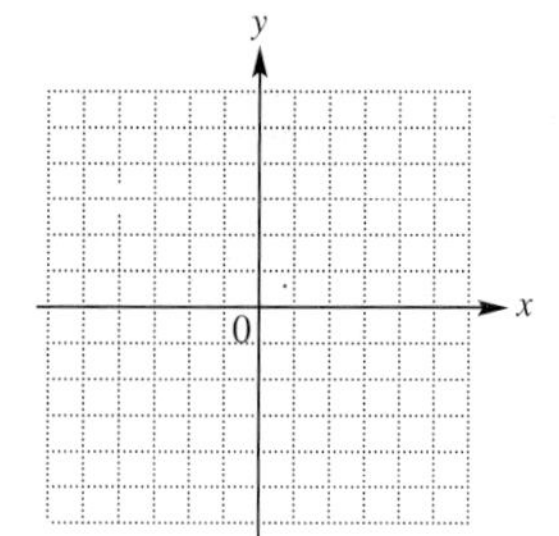

16. If $x > 0$ and $y < 0$, in what quadrant(s) must (x, y) lie? Explain.

17. On what axis does the point $(k, 0)$ lie for any real value of k? the point $(0, k)$? Explain.

Without plotting the given point, name the quadrant in which each point lies.

18. $(-2, 3)$

19. $(-1, -4)$

20. $\left(0, -5\frac{1}{2}\right)$

[3.2] *Find the intercepts for each equation.*

21. $y = 2x + 5$

x-intercept:

y-intercept:

22. $2x + y = -7$

x-intercept:

y-intercept:

23. $3x + 2y = 8$

x-intercept:

y-intercept:

Graph each linear equation.

24. $2x - y = 3$

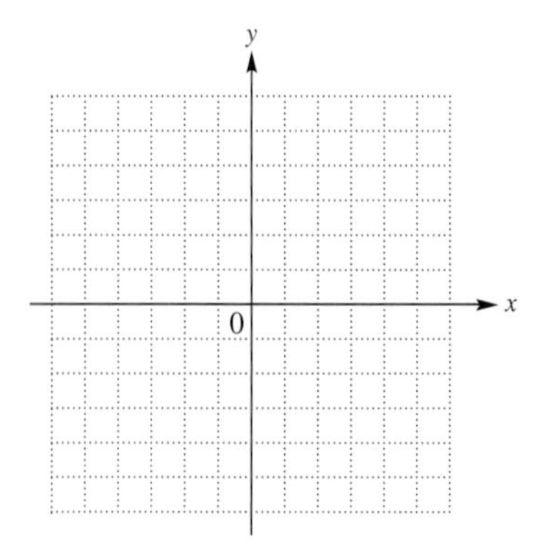

25. $x + 2y = -4$

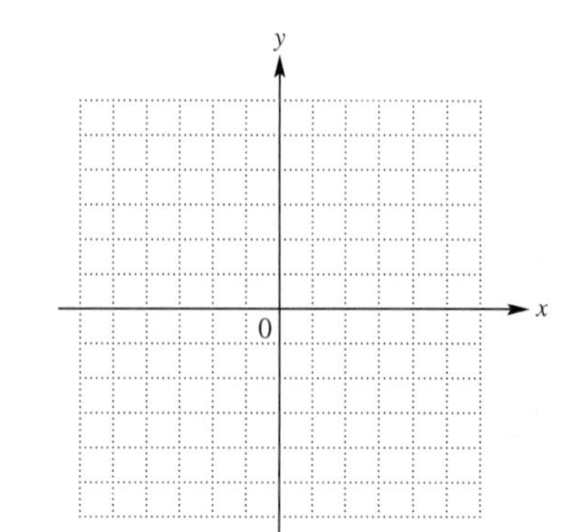

26. $x + y = 0$

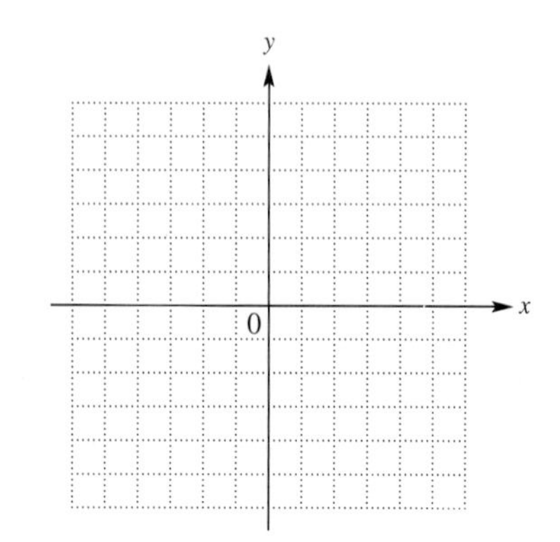

[3.3] *Find the slope of each line.*

27. Through $(2, 3)$ and $(-4, 6)$

28. Through $(0, 0)$ and $(-3, 2)$

29. Through $(0, 6)$ and $(1, 6)$

30. Through $(2, 5)$ and $(2, 8)$

31. $y = 3x - 4$

32. $y = \frac{2}{3}x + 1$

33.

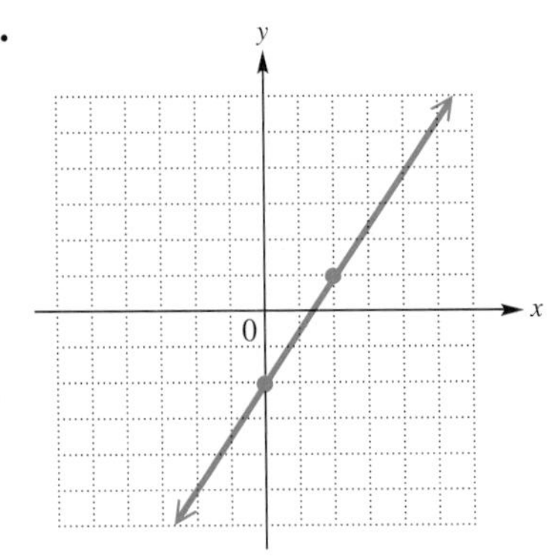

34.

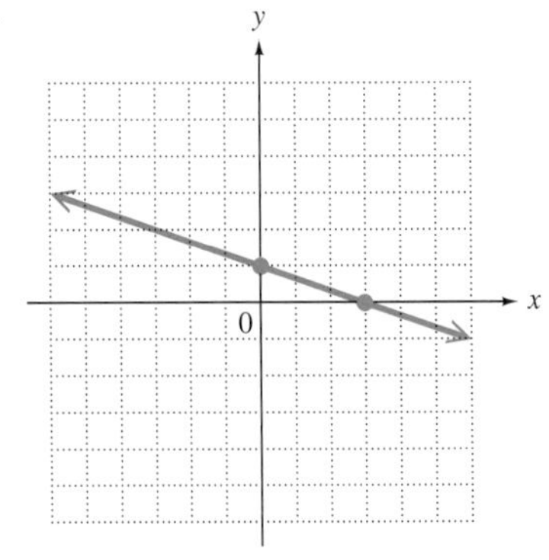

35. $x = 0$

36. $y = 4$

37. The line having these points

x	y
0	1
2	4
6	10

38. (a) A line parallel to the graph of $y = 2x + 3$

(b) A line perpendicular to the graph of $y = -3x + 3$

Decide whether each pair of lines is parallel, perpendicular, *or* neither.

39. $3x + 2y = 6$
$6x + 4y = 8$

40. $x - 3y = 1$
$3x + y = 4$

41. $x - 2y = 8$
$x + 2y = 8$

42. What is the slope of a line perpendicular to a line with undefined slope?

[3.4] *Write an equation in slope-intercept form (if possible) for each line.*

43. $m = -1; b = \frac{2}{3}$

44. The line in Exercise 34

45. Through $(4, -3)$; $m = 1$

46. Through $(-1, 4)$; $m = \frac{2}{3}$

47. Through $(1, -1)$; $m = -\frac{3}{4}$

48. Through $(2, 1)$ and $(-2, 2)$

49. Through $(-4, 1)$ with slope 0

50. Through $\left(\frac{1}{3}, -\frac{3}{4}\right)$ with undefined slope

51. Consider the equation $x + 3y = 15$.

(a) Write it in the form $y = mx + b$.

(b) What is the slope? What is the y-intercept?

(c) Use the slope and the y-intercept to graph the line. Indicate two points on the graph.

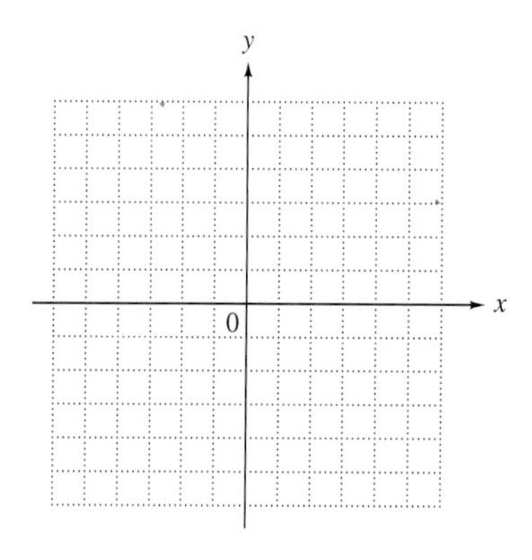

[3.5] *Graph each linear inequality.*

52. $3x + 5y > 9$

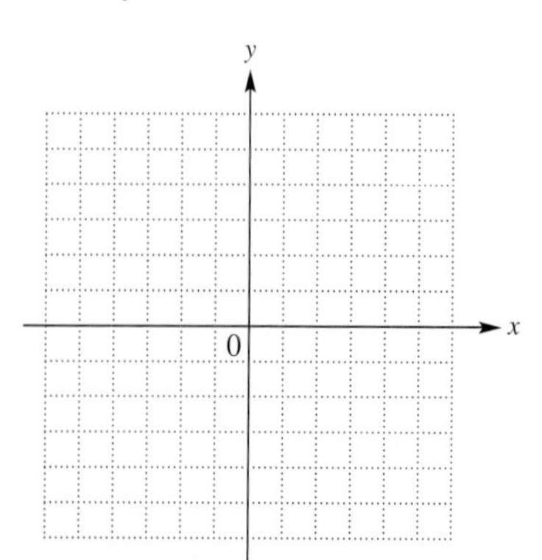

53. $2x - 3y > -6$

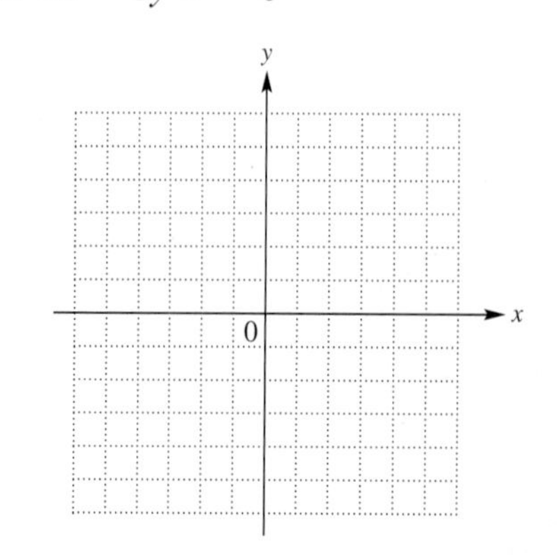

54. $x \geq -4$

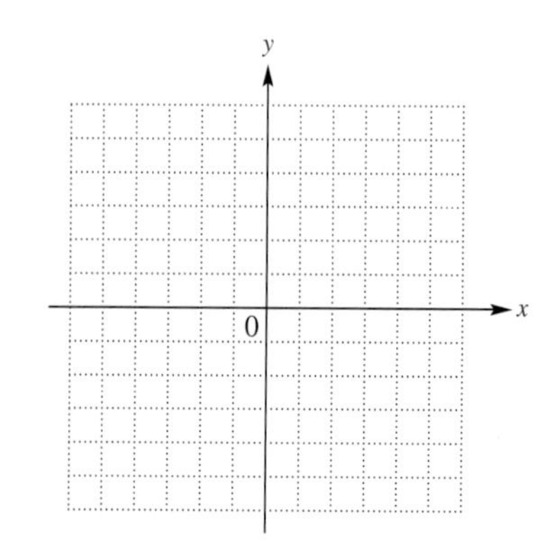

MIXED REVIEW EXERCISES

In Exercises 55–60, match each statement to the appropriate graph or graphs in A–D. Graphs may be used more than once.

A.

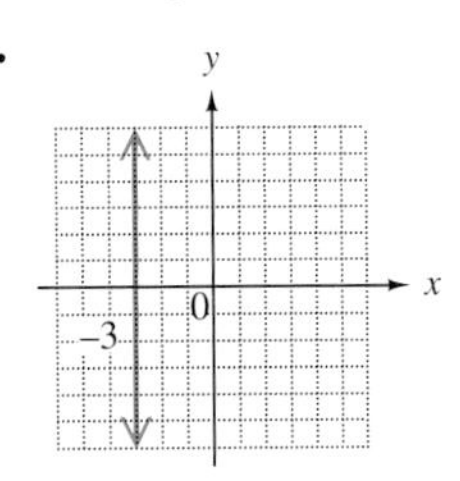

B.

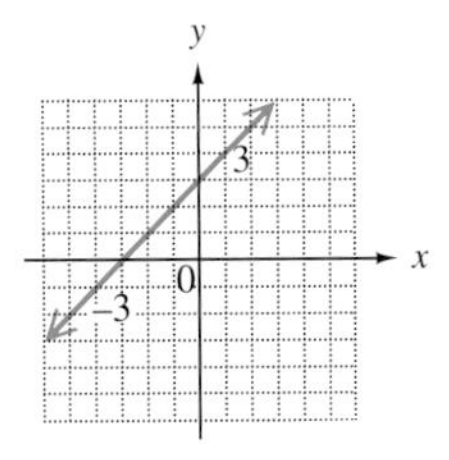

C.

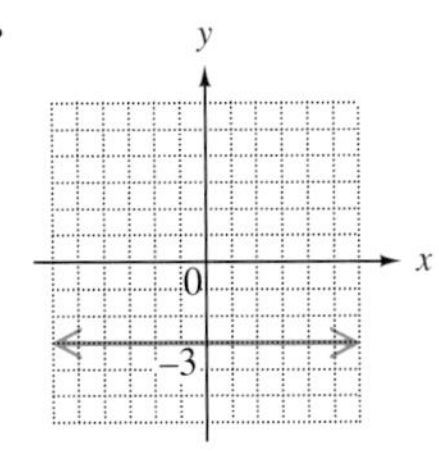

D.

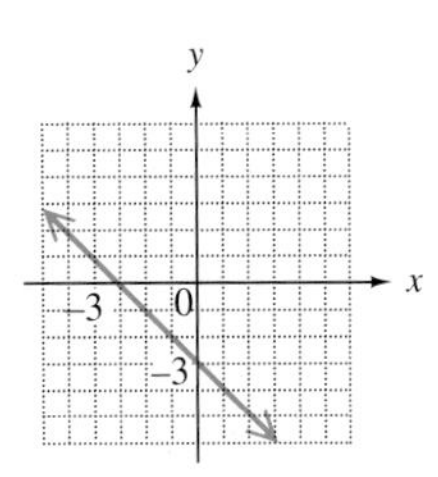

55. The line shown in the graph has undefined slope.

56. The graph of the equation has y-intercept $(0, -3)$.

57. The graph of the equation has x-intercept $(-3, 0)$.

58. The line shown in the graph has negative slope.

59. The graph is that of the equation $y = -3$.

60. The line shown in the graph has slope 1.

Find the intercepts and the slope of each line. Then graph the line.

61. $y = -2x - 5$

x-intercept:

y-intercept:

slope:

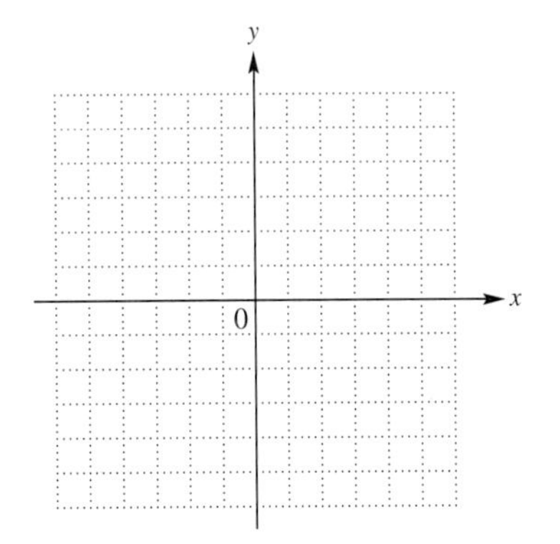

62. $x + 3y = 0$

x-intercept:

y-intercept:

slope:

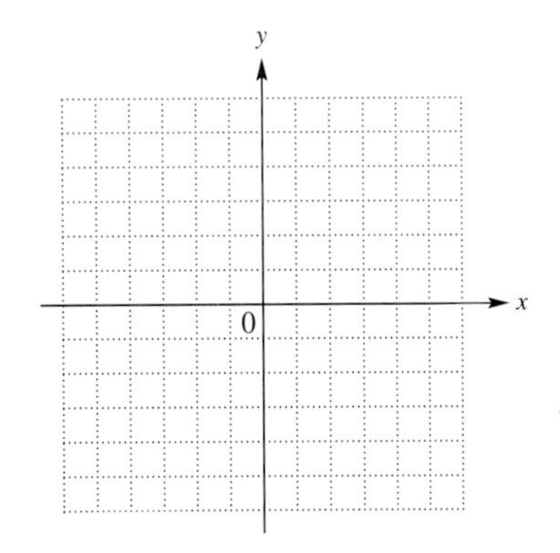

63. $y - 5 = 0$

x-intercept:

y-intercept:

slope:

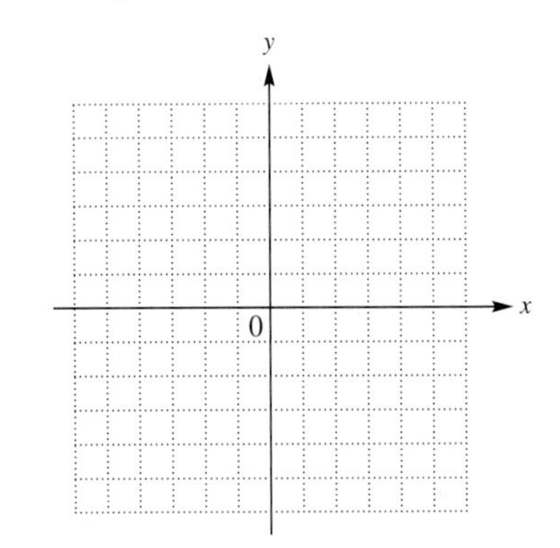

Write an equation in slope-intercept form for each line.

64. $m = -\frac{1}{4}; b = -\frac{5}{4}$

65. Through $(8, 6)$; $m = -3$

66. Through $(3, -5)$ and $(-4, -1)$

Graph each inequality.

67. $y < -4x$

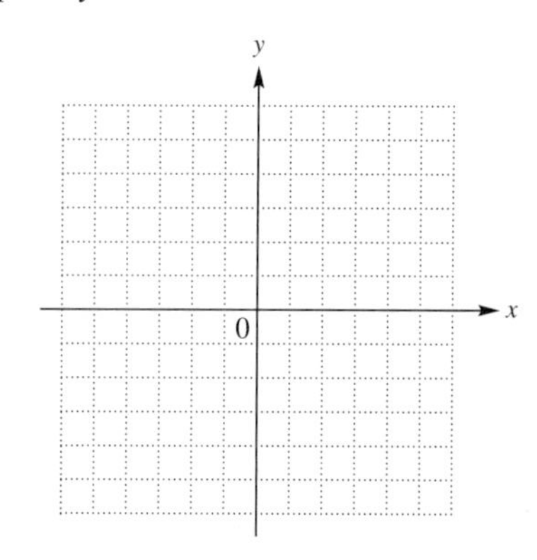

68. $x - 2y \leq 6$

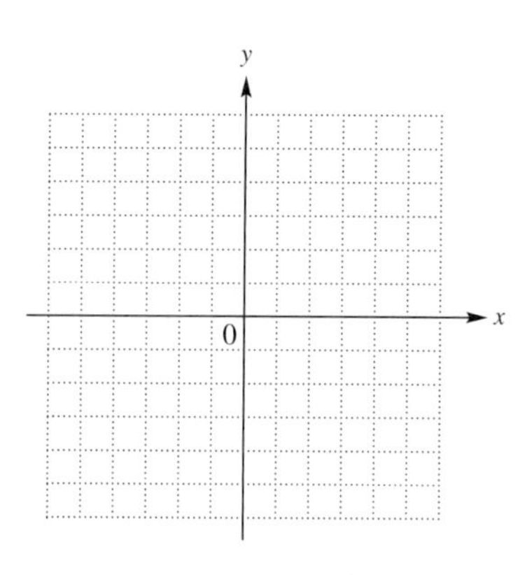

RELATING CONCEPTS (EXERCISES 69–75) For Individual or Group Work

The percents of four-year college students in public schools who earned a degree within five years of entry between 1997 and 2002 are shown in the graph. Use the graph to ***work Exercises 69–75 in order****.*

69. What was the percent decrease from 1997 to 2002?

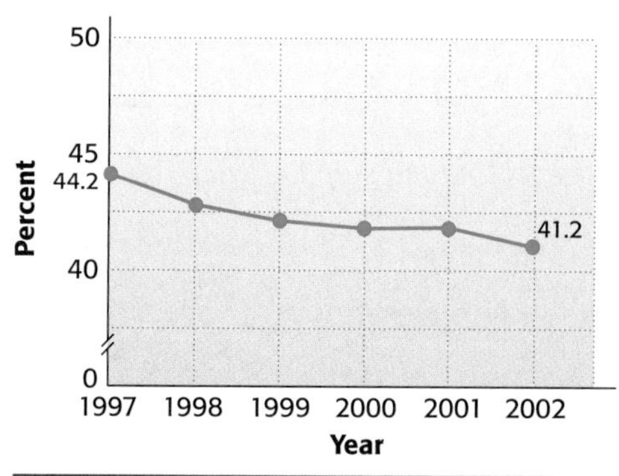

Source: ACT.

70. Since the points of the graph lie approximately in a linear pattern, a straight line can be used to model the data. Will this line have positive or negative slope? Explain.

71. Write two ordered pairs for the data for 1997 and 2002.

72. Use the ordered pairs from Exercise 71 to find the equation of a line that models the data. Write the equation in slope-intercept form.

73. Based on the equation you found in Exercise 72, what is the slope of the line? Does it agree with your answer in Exercise 70?

74. Use the equation from Exercise 72 to approximate the percents for 1998 through 2001, and complete the table. Round your answers to the nearest tenth.

Year	Percent
1998	
1999	
2000	
2001	

75. Use the equation from Exercise 72 to predict the percent for 2003. Can we be sure that this prediction is accurate?

Chapter 3

TEST

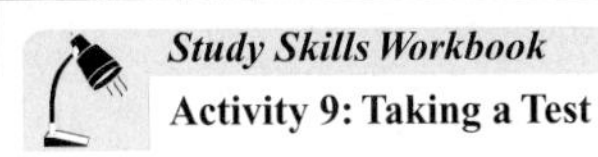

The line graph shows the average prices for a gallon of gasoline in the United States for the years 1998 to 2002. Use the graph to work Exercises 1–3.

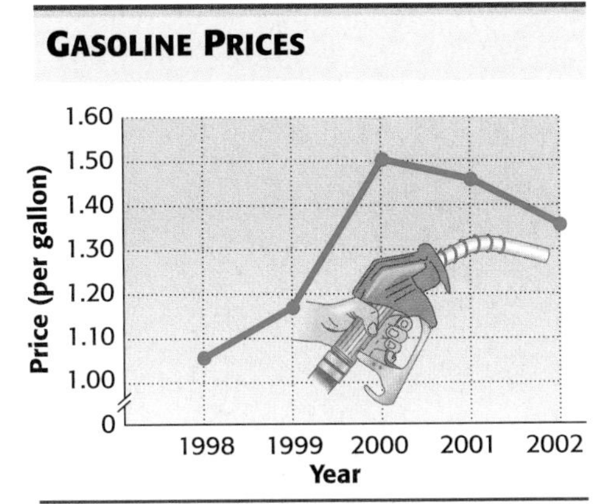

Source: U.S. Department of Energy.

1. About how much did a gallon of gasoline cost in 1998?
2. About how much did the price of a gallon of gasoline increase from 1998 to 2000?
3. During which years did the price decrease?

1. ____________
2. ____________
3. ____________

Graph each linear equation. Give the x- and y-intercepts.

4. $3x + y = 6$

4. x-intercept: ____________
   y-intercept: ____________

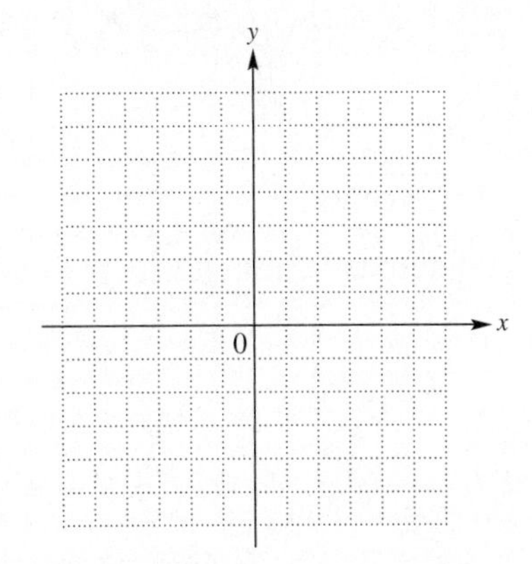

5. $y - 2x = 0$

5. x-intercept: ____________
   y-intercept: ____________

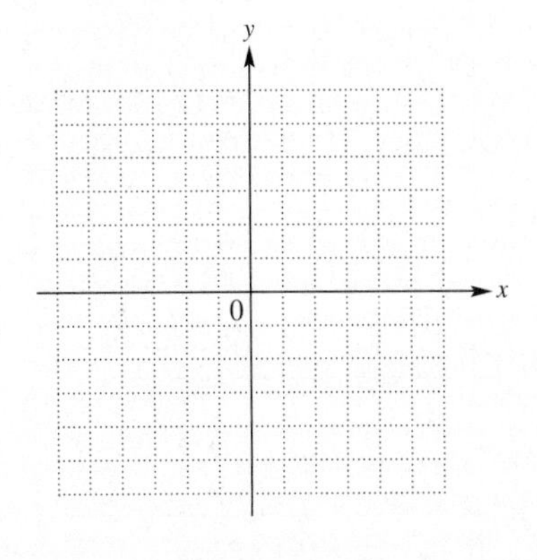

6. x-intercept: ____________

y-intercept: ____________

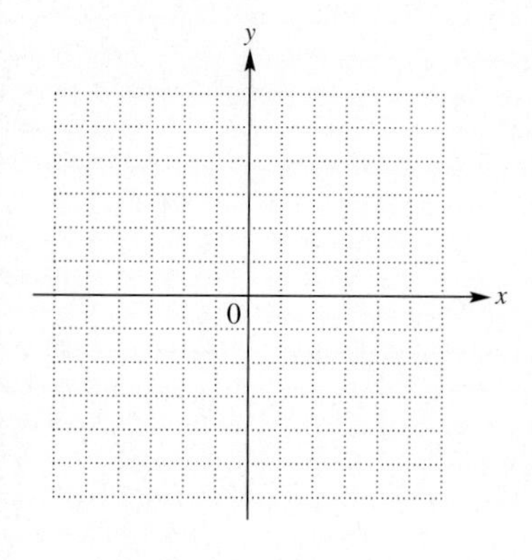

6. $x + 3 = 0$

7. x-intercept: ____________

y-intercept: ____________

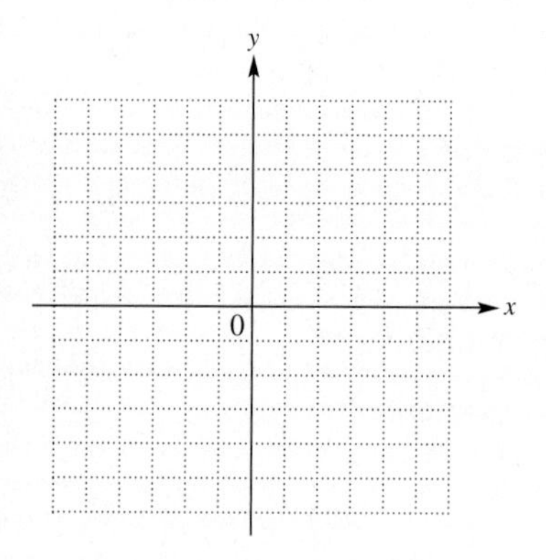

7. $y = 1$

8. x-intercept: ____________

y-intercept: ____________

8. $x - y = 4$

Find the slope of each line.

9. ____________________

9. Through $(-4, 6)$ and $(-1, -2)$

10. ____________________

10. $2x + y = 10$

11. ____________________

11. $x + 12 = 0$

12.

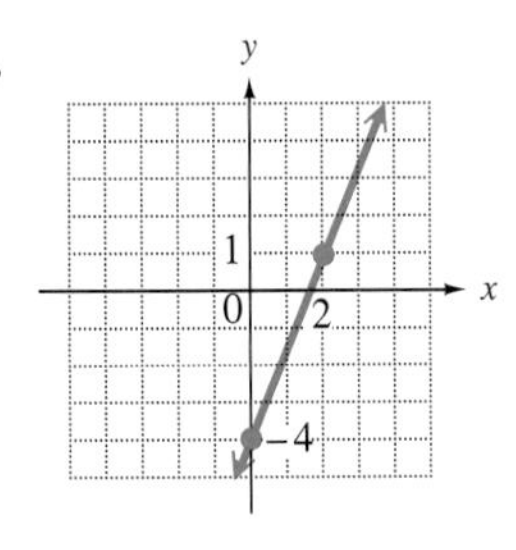

12. ____________________

13. A line parallel to the graph of $y - 4 = 6$

13. ____________________

Write an equation in slope-intercept form for each line.

14. Through $(-1, 4)$; $m = 2$

14. ____________________

15. The line in Exercise 12

15. ____________________

16. Through $(2, -6)$ and $(1, 3)$

16. ____________________

17. x-intercept: $(3, 0)$; y-intercept: $\left(0, \frac{9}{2}\right)$

17. ____________________

Graph each linear inequality.

18. $x + y \leq 3$

18.

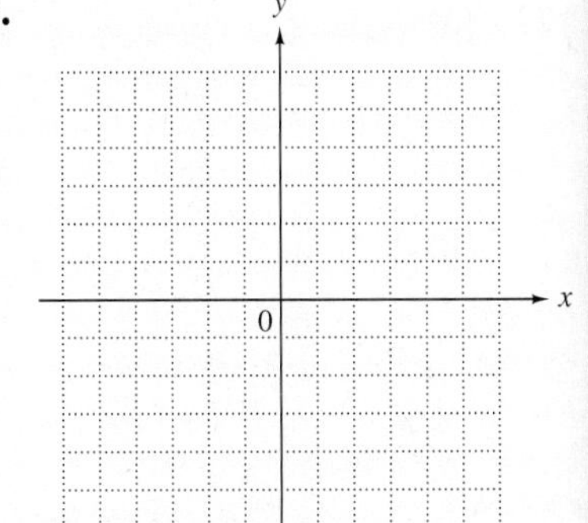

19.

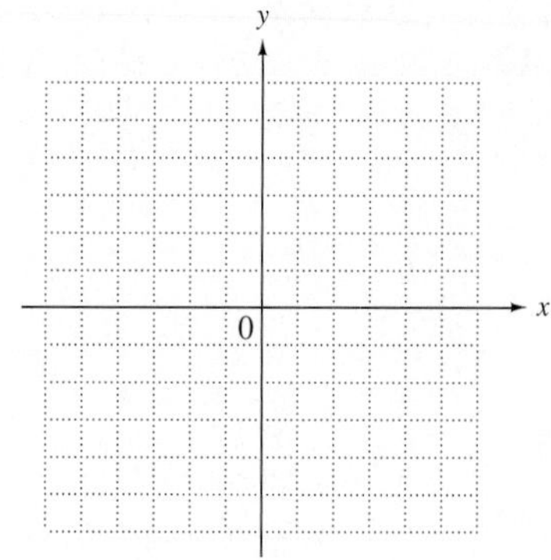

19. $3x - y > 0$

The graph shows total food and drink sales at U.S. restaurants from 1970 through 2000, where 1970 corresponds to $x = 0$. Use the graph to work Exercises 20–22.

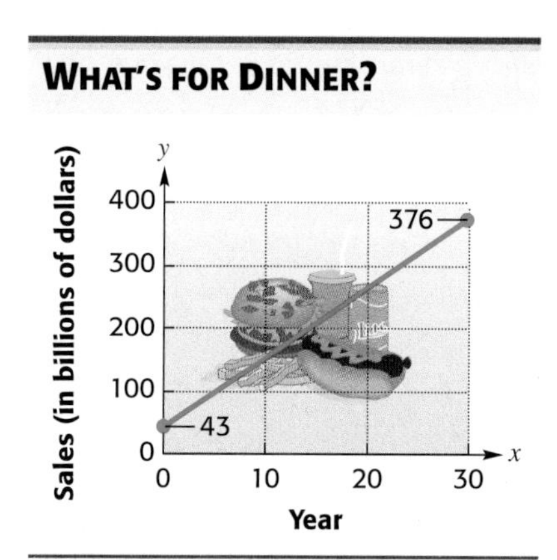

Source: National Restaurant Association.

20. ______________________

20. Is the slope of the line in the graph positive or negative? Explain.

21. ______________________

21. Write two ordered pairs for the data points shown in the graph. Use them to find the slope of the line.

22. The linear equation

$$y = 11.1x + 43$$

approximates food and drink sales y in billions of dollars, where $x = 0$ again represents 1970.

22. (a) ______________________

(a) Use the equation to approximate food and drink sales for 1990 and 1995.

(b) ______________________

(b) What does the ordered pair (30, 376) mean in the context of this problem?

Cumulative Review Exercises

CHAPTERS R–3

Perform the indicated operations.

1. $10\frac{5}{8} - 3\frac{1}{10}$

2. $\frac{3}{4} \div \frac{1}{8}$

3. $5 - (-4) + (-2)$

4. $\frac{(-3)^2 - (-4)(2^4)}{5(2) - (-2)^3}$

5. True or false? $\frac{4(3 - 9)}{2 - 6} \geq 6$

6. Find the value of $xz^3 - 5y^2$ when $x = -2, y = -3,$ and $z = -1$.

7. What property does $3(-2 + x) = -6 + 3x$ illustrate?

8. Simplify $-4p - 6 + 3p + 8$ by combining terms.

Solve.

9. $V = \frac{1}{3}\pi r^2 h$ for h

10. $6 - 3(1 + a) = 2(a + 5) - 2$

11. $-(m - 3) = 5 - 2m$

12. $\frac{y - 2}{3} = \frac{2y + 1}{5}$

Solve each inequality, and graph the solution.

13. $-2.5x < 6.5$

14. $4(x + 3) - 5x < 12$

15. $\frac{2}{3}t - \frac{1}{6}t \leq -2$

Solve each problem.

16. The gap in average annual earnings by level of education continues to increase. Based on the most recent statistics available, a person with a bachelor's degree can expect to earn \$29,200 more each year than someone with a high school diploma. Together the individuals would earn \$102,644. How much can a person at each level of education expect to earn? (*Source*: U.S. Bureau of the Census.)

17. Mount Mayon in the Philippines is the most perfectly shaped conical volcano in the world. Its base is a perfect circle with circumference 80 mi, and it has a height of about 8200 ft. (One mile is 5280 ft.) Find the radius of the circular base to the nearest mile. (*Hint*: This problem has some unneeded information.) (*Source: Microsoft Encarta Encyclopedia 2000.*)

18. The winning times in seconds for the women's 1000 m speed skating event in the Winter Olympics for the years 1960 through 2002 can be closely approximated by the linear equation

$$y = -.5075x + 95.4179,$$

where x is the number of years since 1960. That is, $x = 4$ represents 1964, $x = 8$ represents 1968, and so on. (*Source: World Almanac and Book of Facts,* 2004.)

(a) Use this equation to complete the table of values. Round times to the nearest hundredth of a second.

x	y
12	
28	
34	

(b) Based on this equation, what does the ordered pair (20, 85.27) mean?

19. Baby boomers are expected to inherit \$10.4 trillion from their parents over the next 45 yr, an average of \$50,000 each. The circle graph shows how they plan to spend their inheritances.

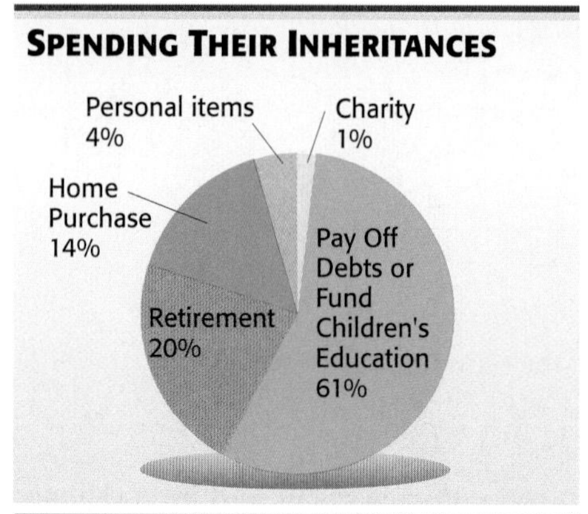

Source: First Interstate Bank Trust and Private Banking Group.

(a) How much of the \$50,000 is expected to go toward home purchase?

(b) How much is expected to go toward retirement?

(c) Use the answer from part (b) to estimate the amount expected to go toward paying off debts or funding children's education.

Consider the linear equation $-3x + 4y = 12$. *Find the following.*

20. The x- and y-intercepts

21. The slope

22. The graph

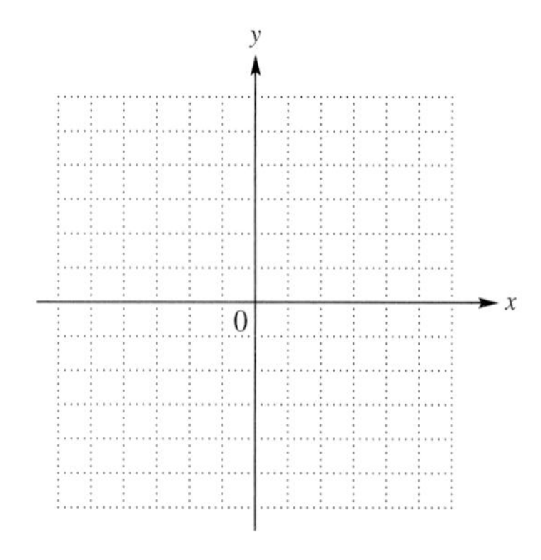

23. The y-value of the point having x-value 4

24. Are the lines with equations $x + 5y = -6$ and $y = 5x - 8$ *parallel, perpendicular,* or *neither?*

Write an equation in slope-intercept form for each line.

25. Through $(2, -5)$ with slope 3

26. Through $(0, 4)$ and $(2, 4)$

Systems of Linear Equations and Inequalities

Although Americans continued their fascination with Hollywood and the movies in 2003, movie attendance and revenues dipped for the first time since 1991. Nonetheless, some 1.52 billion tickets were sold and revenues exceeded $9 billion for the second year in a row. The top box office draws of the year—*The Lord of the Rings: The Return of the King* and *Finding Nemo*—attracted scores of adults and children wishing to get away from it all for a few hours. (*Source:* Exhibitor Relations Co., Nielsen EDI.)

In Exercise 13 of Section 4.4, we use a *system of linear equations* to find out how much money these top films earned.

4.1 Solving Systems of Linear Equations by Graphing

OBJECTIVES

1 Decide whether a given ordered pair is a solution of a system.

2 Solve linear systems by graphing.

3 Solve special systems by graphing.

A **system of linear equations,** often called a **linear system,** consists of two or more linear equations with the same variables. Examples of systems of two linear equations include

$$\begin{aligned} 2x + 3y &= 4 \\ 3x - y &= -5 \end{aligned} \qquad \begin{aligned} x + 3y &= 1 \\ -y &= 4 - 2x \end{aligned} \qquad \begin{aligned} x - y &= 1 \\ y &= 3. \end{aligned}$$ Linear systems

In the system on the right, think of $y = 3$ as an equation in two variables by writing it as $0x + y = 3$.

OBJECTIVE 1 Decide whether a given ordered pair is a solution of a system. A **solution of a system** of linear equations is an ordered pair that makes both equations true at the same time. A solution of an equation is said to *satisfy* the equation.

EXAMPLE 1 Determining Whether an Ordered Pair Is a Solution

Is $(4, -3)$ a solution of each system?

(a) $x + 4y = -8$
$3x + 2y = 6$

To decide whether or not $(4, -3)$ is a solution of the system, substitute 4 for x and -3 for y in each equation.

$x + 4y = -8$		$3x + 2y = 6$	
$4 + 4(-3) = -8$?		$3(4) + 2(-3) = 6$?	
$4 + (-12) = -8$?	Multiply.	$12 + (-6) = 6$?	Multiply.
$-8 = -8$	True	$6 = 6$	True

Because $(4, -3)$ satisfies both equations, it is a solution of the system.

(b) $2x - 5y = -7$
$3x + 4y = 2$

Again, substitute 4 for x and -3 for y in both equations.

$2x + 5y = -7$		$3x + 4y = 2$	
$2(4) + 5(-3) = -7$?		$3(4) + 4(-3) = 2$?	
$8 + (-15) = -7$?	Multiply.	$12 + (-12) = 2$?	Multiply.
$-7 = -7$	True	$0 = 2$	False

The ordered pair $(4, -3)$ is not a solution of this system because it does not satisfy the second equation.

Work Problem 1 at the Side.

OBJECTIVE 2 Solve linear systems by graphing. One way to find the solution of a system of two linear equations is to graph both equations on the same axes. The graph of each line shows points whose coordinates satisfy the equation of that line. Any intersection point would be on both lines and would therefore be a solution of *both* equations. ***Thus, the coordinates of any point where the lines intersect give a solution of the system.***

1 Fill in the blanks, and decide whether the given ordered pair is a solution of the system.

(a) $(2, 5)$

$3x - 2y = -4$
$5x + y = 15$

$3x - 2y = -4$
$3(\_\_\_) - 2(\_\_\_) = -4$

$5x + y = 15$
$5(2) + \_\_\_ = \_\_\_$

$(2, 5)$ ______ a solution. (is/is not)

(b) $(1, -2)$

$x - 3y = 7$
$4x + y = 5$

$(1, -2)$ ______ a solution. (is/is not)

ANSWERS

1. **(a)** 2; 5; 5; 15; is **(b)** is not

The graph in Figure 1 shows that the solution of the system in Example 1(a) is the intersection point $(4, -3)$. Because *two different* straight lines can intersect at no more than one point, there can never be more than one solution for such a system.

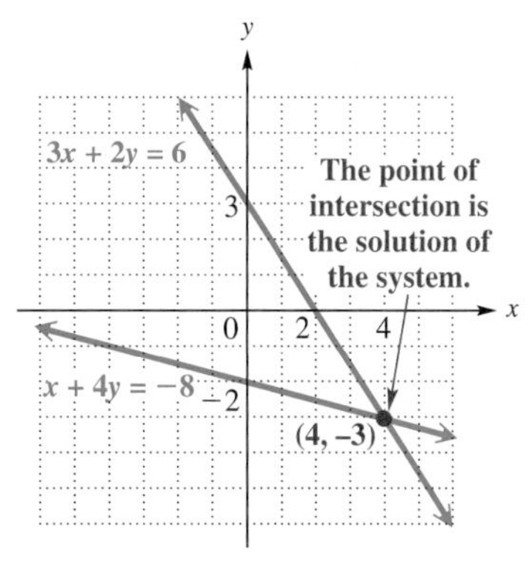

Figure 1

EXAMPLE 2 Solving a System by Graphing

Solve the system of equations by graphing both equations on the same axes.

$$2x + 3y = 4$$
$$3x - y = -5$$

We graph these two equations by plotting several points for each line. Recall from **Section 3.2** that the intercepts are often convenient choices. It is a good idea to find a third ordered pair as a check.

$2x + 3y = 4$

x	y
0	$\frac{4}{3}$
2	0
-2	$\frac{8}{3}$

$3x - y = -5$

x	y
0	5
$-\frac{5}{3}$	0
-2	-1

The lines in Figure 2 suggest that the graphs intersect at the point $(-1, 2)$. We check this by substituting -1 for x and 2 for y in both equations. Because $(-1, 2)$ satisfies both equations, the solution of this system is $(-1, 2)$.

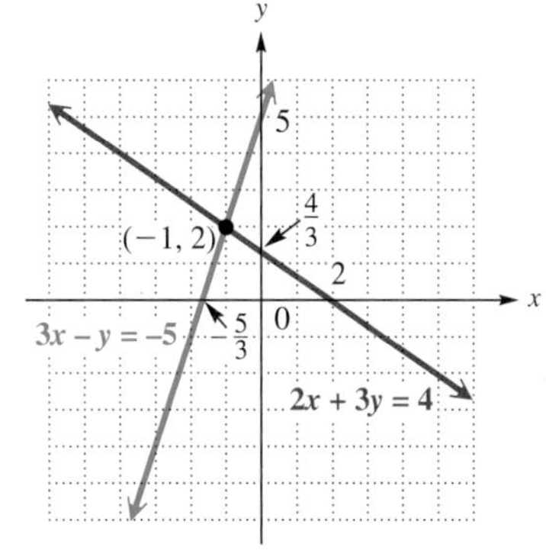

Figure 2

Work Problem 2 at the Side.

NOTE
We can also graph a linear system by writing each equation in the system in slope-intercept form and using the slope and y-intercept to graph each line. For Example 2,

$2x + 3y = 4$ becomes $y = -\frac{2}{3}x + \frac{4}{3}$ y-intercept $(0, \frac{4}{3})$; slope $-\frac{2}{3}$ or $-\frac{2}{3}$

$3x - y = -5$ becomes $y = 3x + 5$. y-intercept $(0, 5)$; slope 3 or $\frac{3}{1}$

Confirm that graphing these equations results in the same lines and the same solution shown in Figure 2.

2 Solve each system of equations by graphing both equations on the same axes. Check your solutions.

(a) $5x - 3y = 9$
$x + 2y = 7$
(One of the lines is already graphed.)

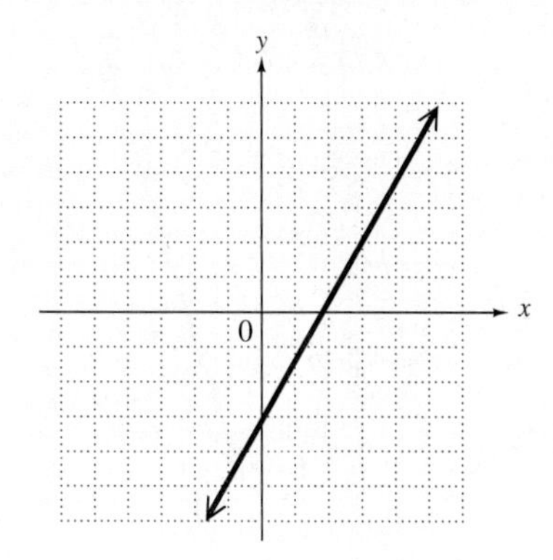

(b) $x + y = 4$
$2x - y = -1$

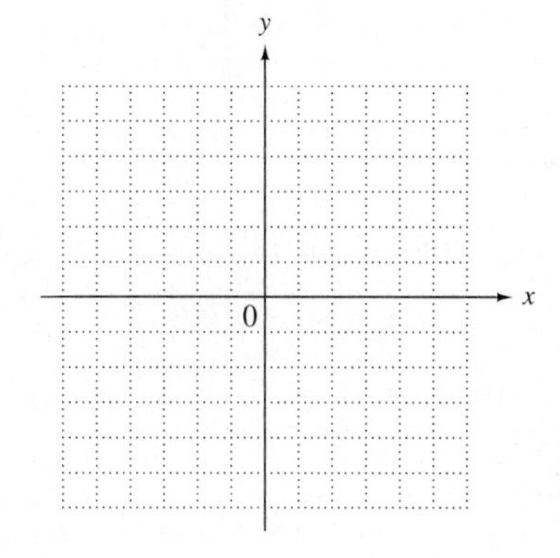

ANSWERS
2. (a) $(3, 2)$ **(b)** $(1, 3)$

3 Solve each system of equations by graphing both equations on the same axes.

(a) $3x - y = 4$
$6x - 2y = 12$
(One of the lines is already graphed.)

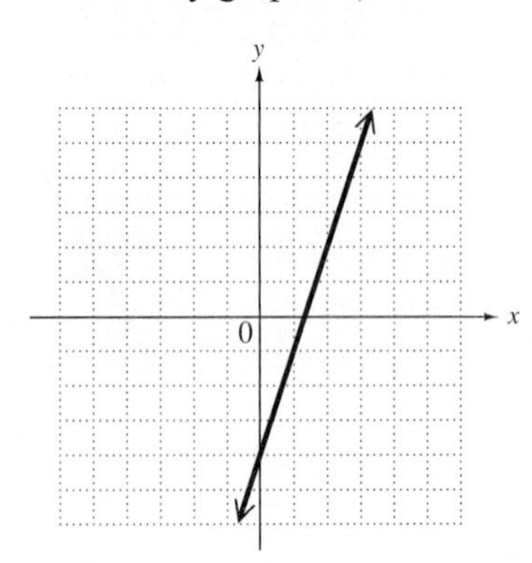

(b) $-x + 3y = 2$
$2x - 6y = -4$

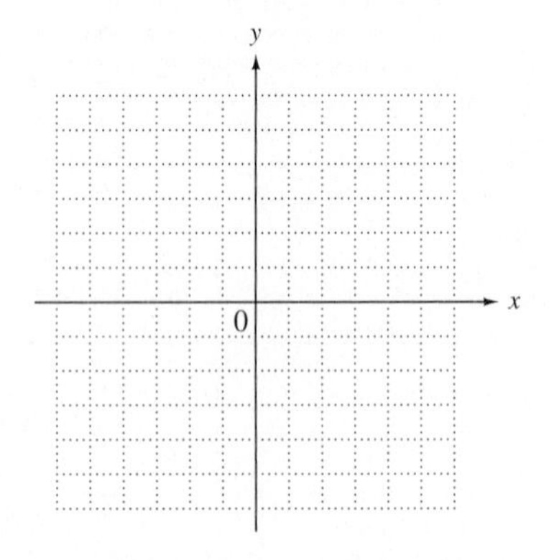

CAUTION
A difficulty with the graphing method of solution is that it may not be possible to determine from the graph the exact coordinates of the point that represents the solution, particularly if these coordinates are not integers. For this reason, algebraic methods of solution are explained later in this chapter. The graphing method does, however, show geometrically how solutions are found and is useful when approximate answers will do.

OBJECTIVE 3 Solve special systems by graphing. Sometimes the graphs of the two equations in a system either do not intersect at all or are the same line, as in the systems in Example 3.

EXAMPLE 3 Solving Special Systems

Solve each system by graphing.

(a) $2x + y = 2$
$2x + y = 8$
The graphs of these lines are shown in Figure 3. The two lines are parallel and have no points in common. For such a system, we will write "***no solution.***"

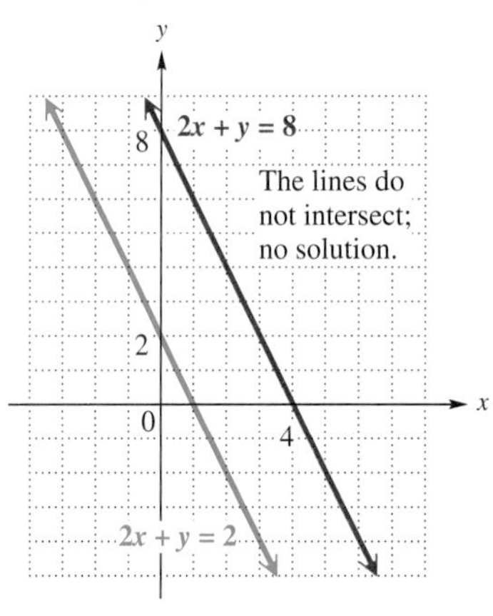

Figure 3

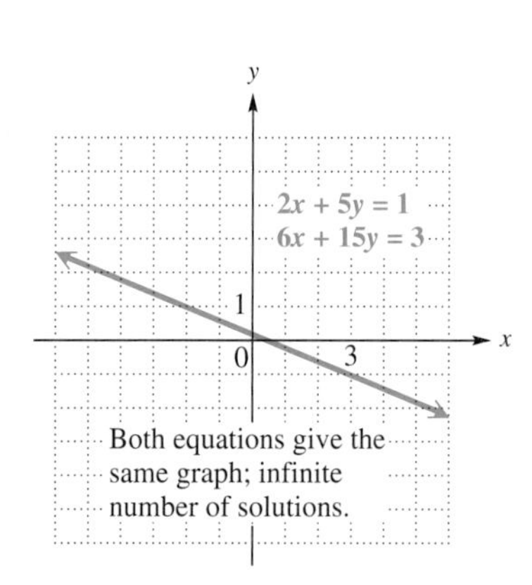

Figure 4

(b) $2x + 5y = 1$
$6x + 15y = 3$
The graphs of these two equations are the same line. See Figure 4. The second equation can be obtained by multiplying each side of the first equation by 3. In this case, every point on the line is a solution of the system, and the solutions are the infinite number of ordered pairs that satisfy the equations. We will write "***infinite number of solutions***" to indicate this case.

Work Problem 3 at the Side.

The system in Example 2 has exactly one solution. A system with at least one solution is called a **consistent system.** A system of equations with no solutions, such as the one in Example 3(a), is called an **inconsistent system.** The equations in Example 2 are **independent equations** with different graphs. The equations of the system in Example 3(b) have the same graph and are equivalent. Because they are different forms of the same equation, these equations are called **dependent equations.**

ANSWERS
3. (a) no solution
(b) infinite number of solutions

Examples 2 and 3 show the three cases that may occur when solving a system of two equations with two variables.

Possible Types of Solutions

1. The graphs intersect at exactly one point, which gives the (single) ordered pair solution of the system. The **system is consistent,** and the **equations are independent.**

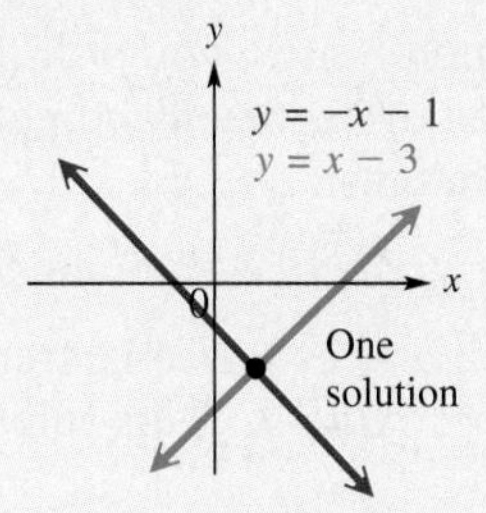

2. The graphs are parallel lines, so there is no solution. The **system is inconsistent.**

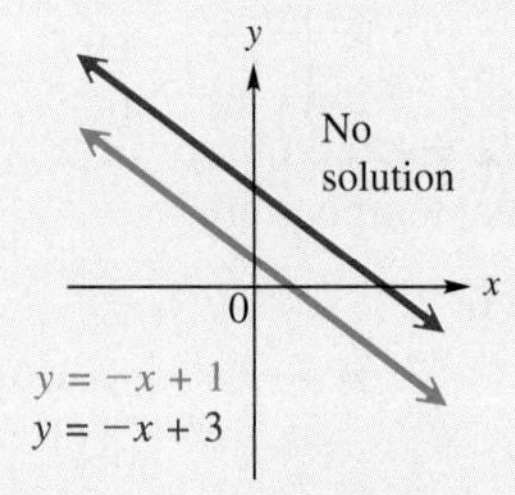

3. The graphs are the same line. The solution is an infinite number of ordered pairs. The **equations are dependent.**

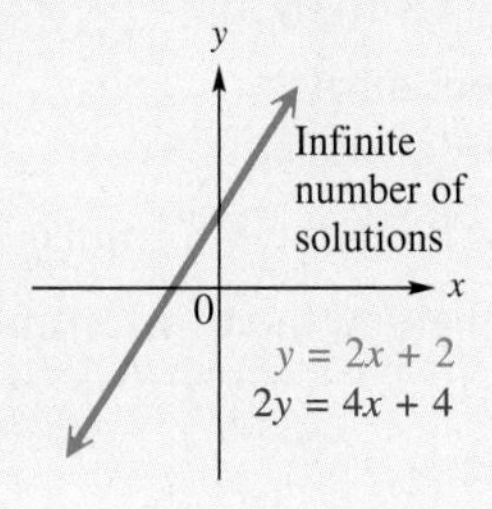

Focus on

Real-Data Applications

Estimating Fahrenheit Temperature

When traveling in countries other than the United States, you will hear the daily high and low temperatures reported in degrees Celsius, instead of degrees Fahrenheit. The following information may help you interpret these Celsius temperatures.

- The linear equation to convert degrees Celsius to degrees Fahrenheit is $F = \frac{9}{5}C + 32$.
- Travel books advise you to use a *rule of thumb* to estimate Fahrenheit temperature that says "*Double the temperature (degrees Celsius) and add 30.*" This rule of thumb is written mathematically as $F = 2C + 30$.

For Group Discussion

Suppose you are interested in knowing for what temperature the rule of thumb and the actual formulas give the same result. You also want to know if the rule of thumb formula is predicting temperatures that are lower or higher than the actual temperature.

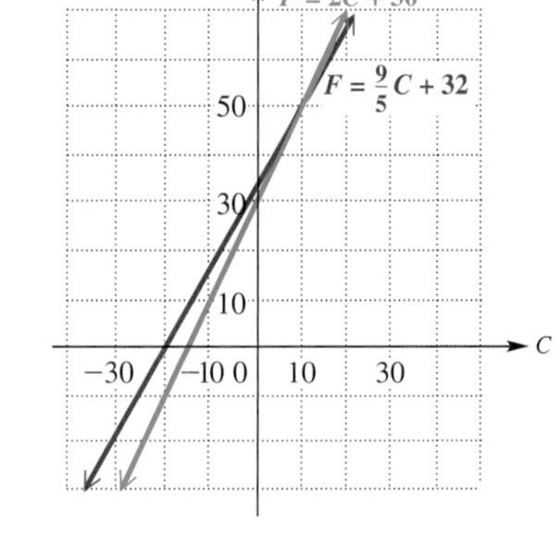

1. The two formulas can be written as the system of equations

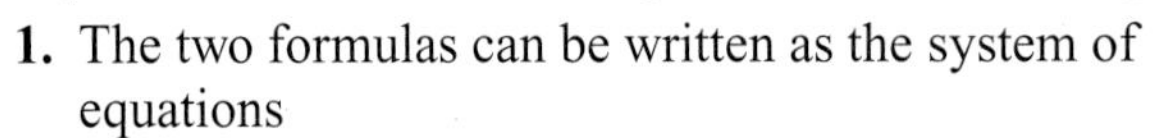

$$F = \frac{9}{5}C + 32$$

$$F = 2C + 30.$$

   (a) Use the graph of the system of equations to find the point of intersection. (*Hint:* To check your answer, use substitution to see if it satisfies both formulas.)

   (b) For what temperature in degrees Celsius do the two formulas agree?

   (c) For what temperature in degrees Fahrenheit do the two formulas agree?

2. **(a)** Complete the table of values to compare the *actual* and the *rule of thumb* formulas for temperature conversion.

°C	°F (*Actual*)	°F (*Rule of Thumb*)
0		
5		
10		
15		
20		
30		

   (b) If the daily low is predicted to be 5°C, then is the rule of thumb estimate too high or too low?

   (c) If the daily high is predicted to be 20°C, then is the rule of thumb estimate too high or too low?

   (d) If the daily high is predicted to be 30°C, then is the rule of thumb estimate too high or too low?

   (e) How many degrees "off" is the rule of thumb estimate for the boiling point of water?

   (f) Comment on the accuracy of using the rule of thumb as an estimate of the actual Fahrenheit temperature.

4.1 Exercises

FOR EXTRA HELP — Tutor Center Addison-Wesley Math Tutor Center | MathXL | Digital Video Tutor CD 4 Videotape 4 | Student's Solutions Manual | MyMathLab | Interactmath.com

1. Which ordered pair could be a solution of the system graphed? Why is it the only valid choice?

A. $(2, 2)$
B. $(-2, 2)$
C. $(-2, -2)$
D. $(2, -2)$

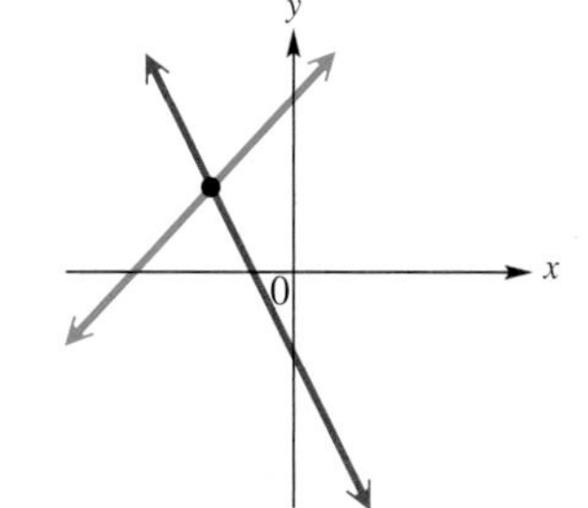

2. Which ordered pair could be a solution of the system graphed? Why is it the only valid choice?

A. $(2, 0)$
B. $(0, 2)$
C. $(-2, 0)$
D. $(0, -2)$

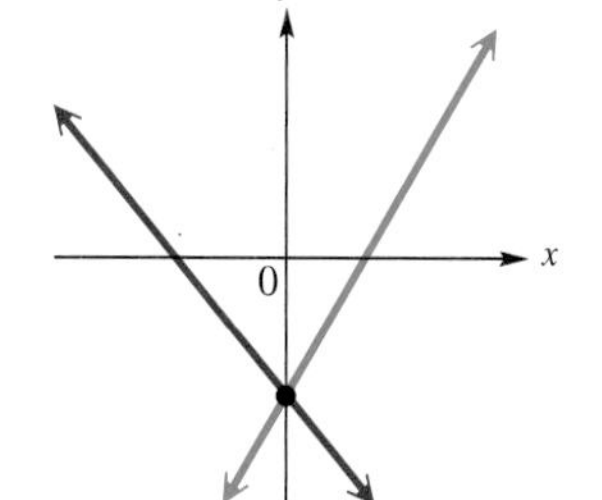

3. How can you tell without graphing that this system has no solution?

$$x + y = 2$$
$$x + y = 4$$

4. Explain why a system of two linear equations cannot have exactly two solutions.

Decide whether the given ordered pair is a solution of the given system. See Example 1.

5. $(2, -3)$
$x + y = -1$
$2x + 5y = 19$

6. $(4, 3)$
$x + 2y = 10$
$3x + 5y = 3$

7. $(-1, -3)$
$3x + 5y = -18$
$4x + 2y = -10$

8. $(-9, -2)$
$2x - 5y = -8$
$3x + 6y = -39$

9. $(7, -2)$
$4x = 26 - y$
$3x = 29 + 4y$

10. $(9, 1)$
$2x = 23 - 5y$
$3x = 24 + 3y$

11. $(6, -8)$
$-2y = x + 10$
$3y = 2x + 30$

12. $(-5, 2)$
$5y = 3x + 20$
$3y = -2x - 4$

Solve each system of equations by graphing. If the two equations produce parallel lines, write no solution. *If the two equations produce the same line, write* infinite number of solutions. *See Examples 2 and 3.*

13. $x - y = 2$
$x + y = 6$

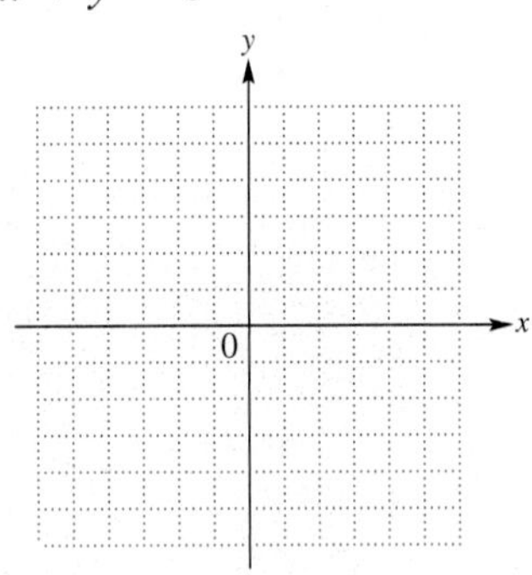

14. $x - y = 3$
$x + y = -1$

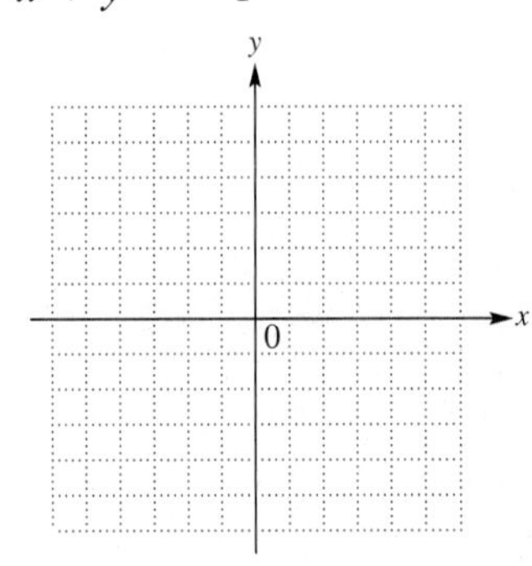

15. $x + y = 4$
$y - x = 4$

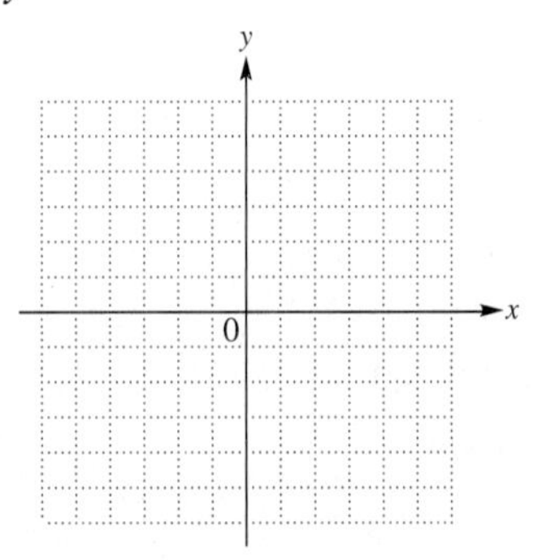

16. $x + y = -5$
$x - y = 5$

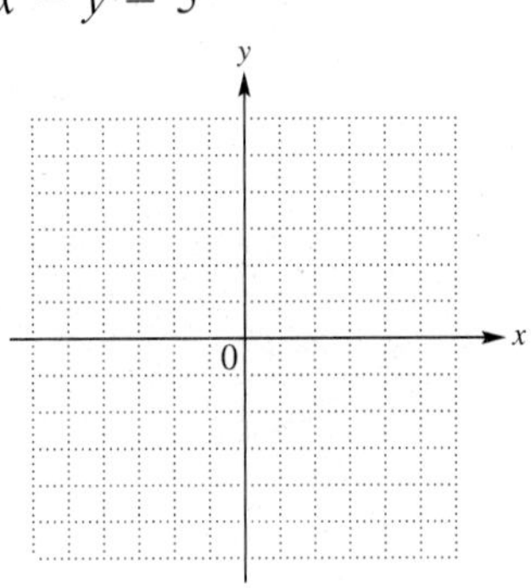

17. $x - 2y = 6$
$x + 2y = 2$

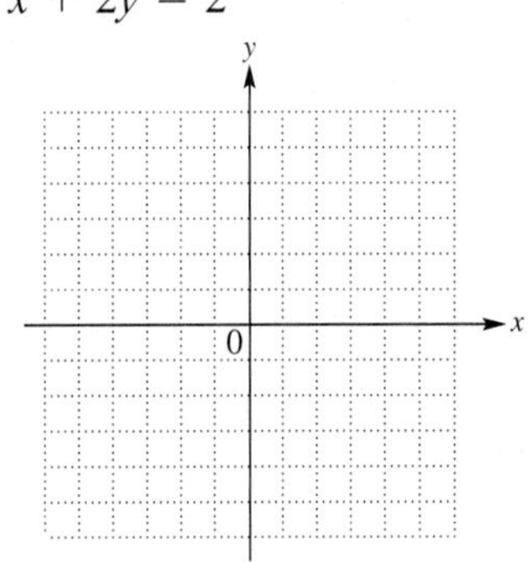

18. $2x - y = 4$
$4x + y = 2$

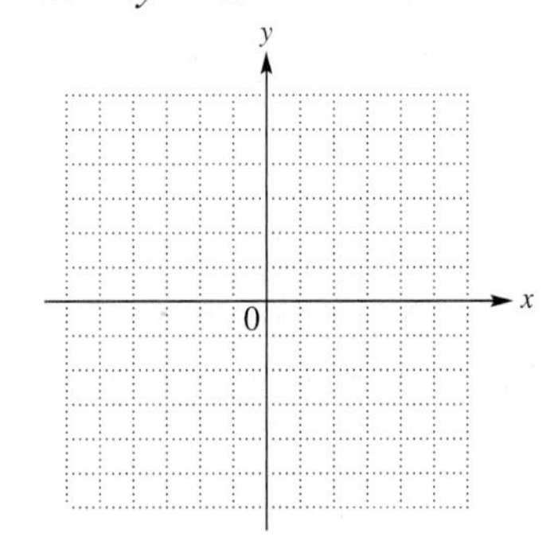

19. $3x - 2y = -3$
$-3x - y = -6$

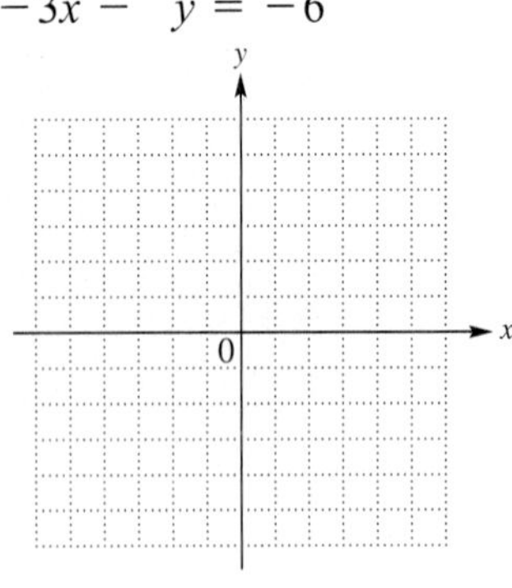

20. $2x - y = 4$
$2x + 3y = 12$

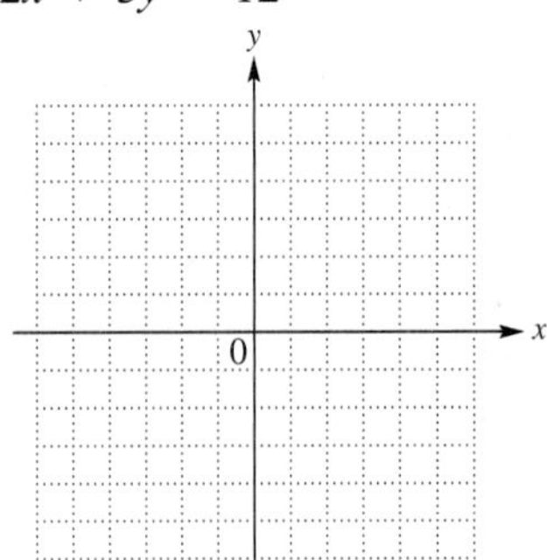

21. $2x - 3y = -6$
$y = -3x + 2$

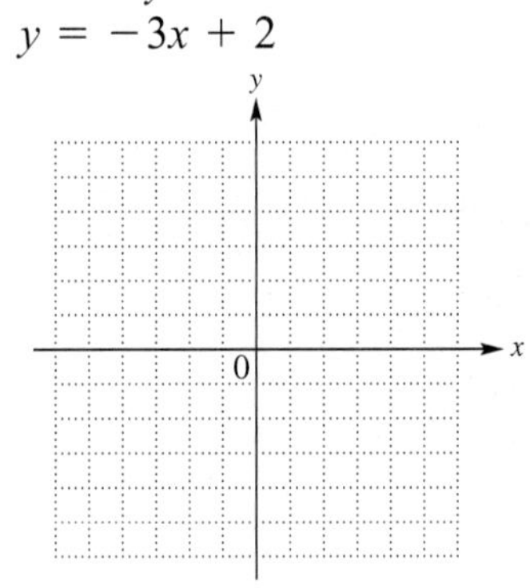

22. $-3x + y = -3$
$y = x - 3$

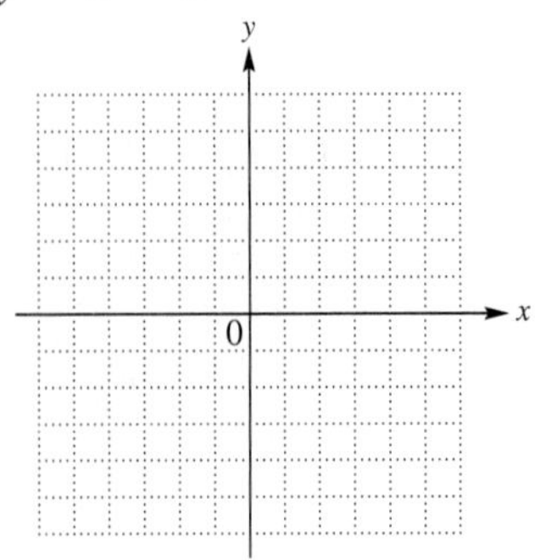

23. $x + 2y = 6$
$2x + 4y = 8$

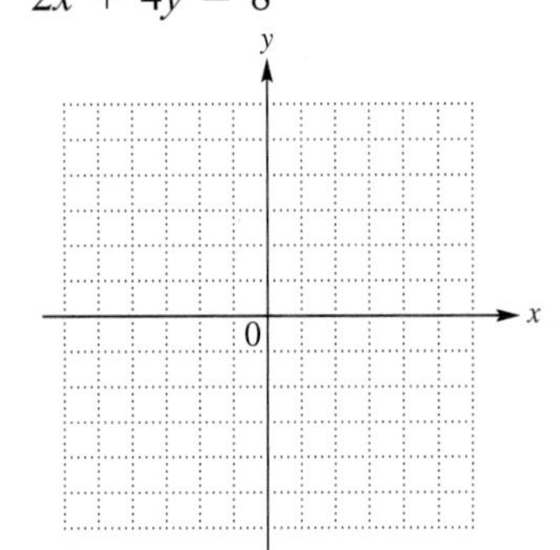

24. $2x - y = 6$
$6x - 3y = 12$

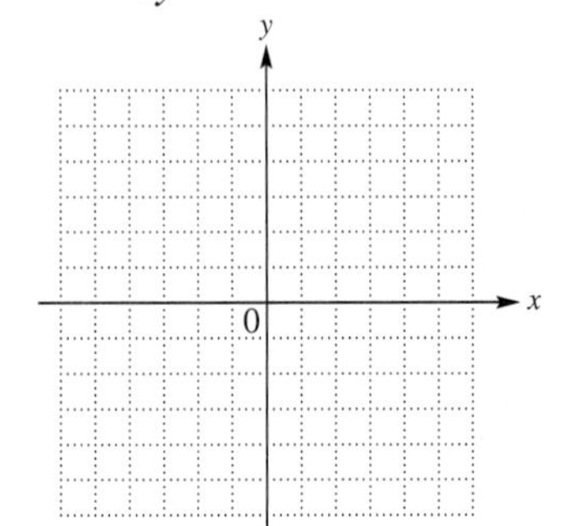

25. $2x - y = 4$
$4x = 2y + 8$

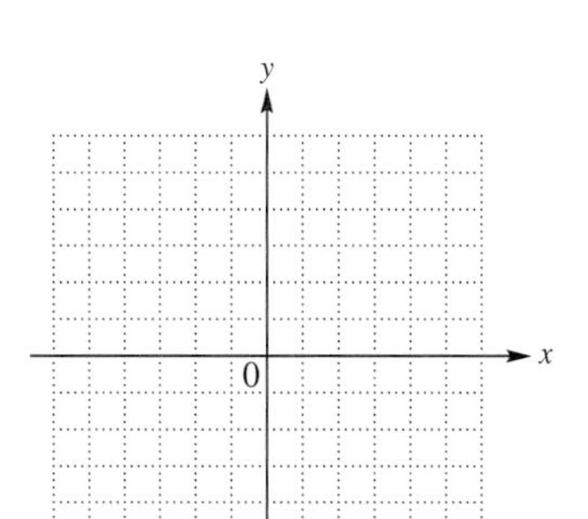

26. $3x = 5 - y$
$6x + 2y = 10$

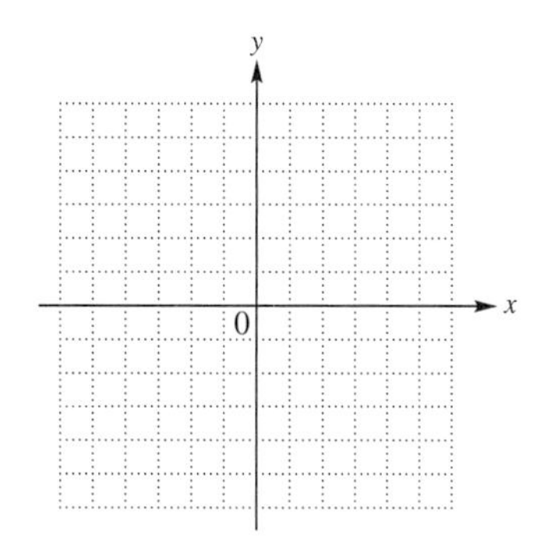

27. $3x - 4y = 24$
$y = -\frac{3}{2}x + 3$

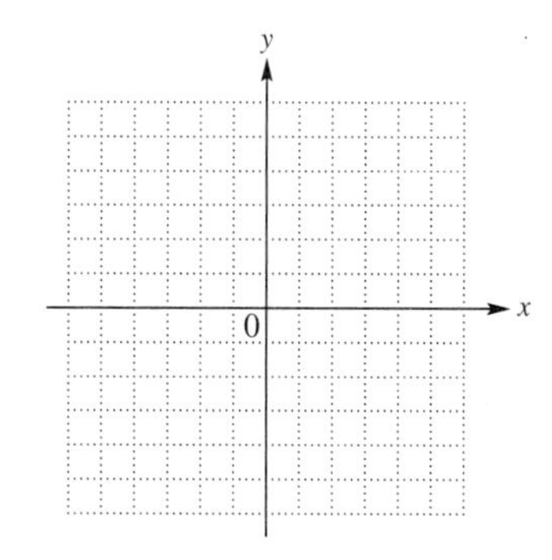

28. $3x - 2y = 12$
$y = -4x + 5$

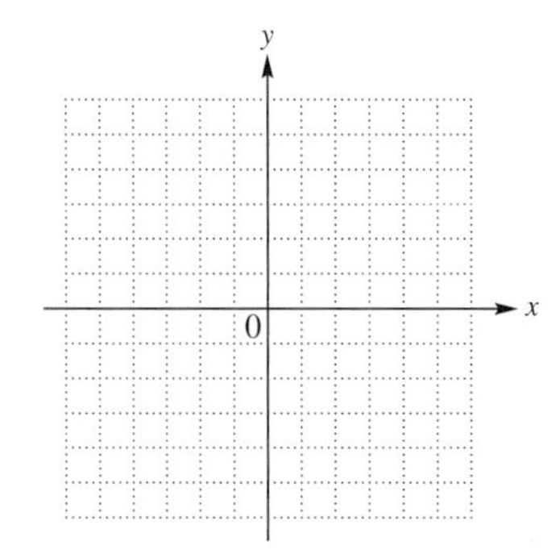

29. $3x = y + 5$
$6x - 5 = 2y$

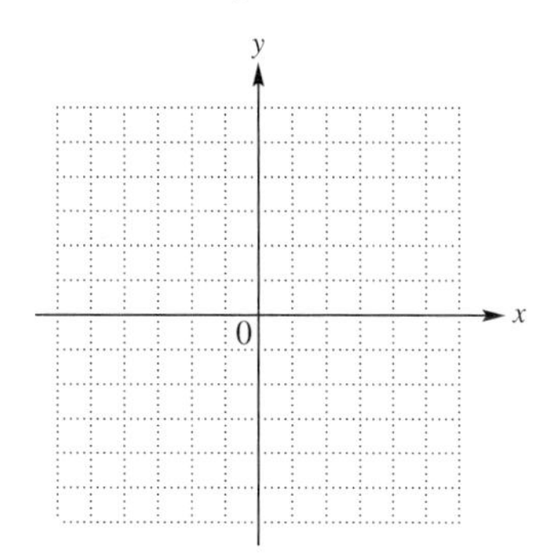

30. $2x = y - 4$
$4x - 2y = -4$

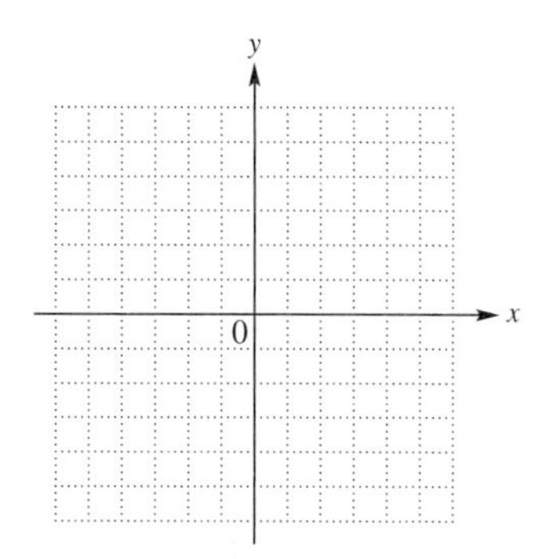

RELATING CONCEPTS (EXERCISES 31–34) For Individual or Group Work

In Exercises 31–33, first write each equation in slope-intercept form. (See the Note following Example 2.) Then use what you learned in ***Chapter 3*** *about slope and the y-intercept to describe the graphs of each system of equations.* ***Work Exercises 31–34 in order.***

31. $3x + 2y = 6$
$-2y = 3x - 5$

32. $2x - y = 4$
$x = .5y + 2$

33. $x - 3y = 5$
$2x + y = 8$

34. Use the results of Exercises 31–33 to determine the number of solutions of each system.

The graph shows network share (the percentage of TV sets in use) for the early evening news programs for the three major broadcast networks from 1986 through 2000.

35. Between what years did the ABC early evening news dominate?

36. During what year did ABC's dominance end? Which network equaled ABC's share that year? What was that share?

37. During what years did ABC and CBS have equal network share? What was the share for each of these years?

WHO'S WATCHING THE EVENING NEWS?

Network Share (percentage of TVs in use)

26, 24, 22, 20, 18, 16, 14, 12, 10

ABC

NBC

CBS

'86 '88 '90 '92 '94 '96 '98 '00

Year

Source: Nielsen Media Research.

38. Which networks most recently had equal share? Write their share as an ordered pair of the form (year, share).

39. Explain one of the drawbacks of solving a system of equations graphically.

40. If the two lines that are the graphs of the equations in a system are parallel, how many solutions does the system have? If the two lines coincide, how many solutions does the system have?

41. Find a system of equations with the solution $(-2, 3)$, and show the graph.

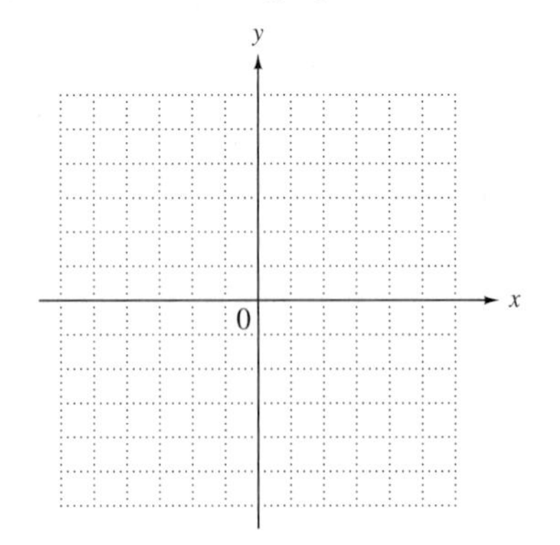

42. Solve the system

$$2x + 3y = 6$$
$$x - 3y = 5$$

by graphing. Can you check your answer? Why or why not?

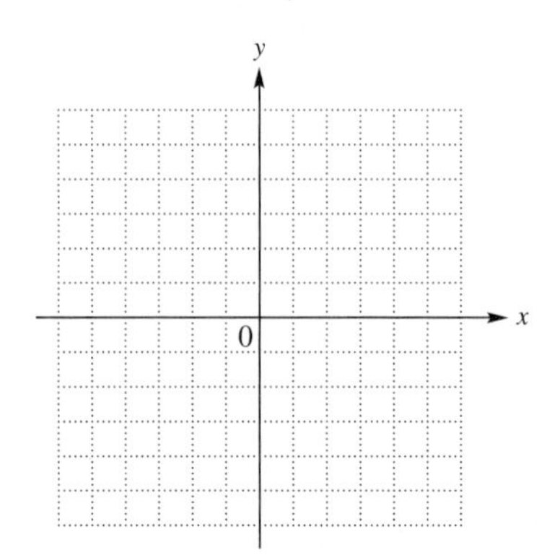

4.2 Solving Systems of Linear Equations by Substitution

OBJECTIVES

1. Solve linear systems by substitution.
2. Solve special systems.
3. Solve linear systems with fractions.

OBJECTIVE 1 Solve linear systems by substitution. Graphing to solve a system of equations has a serious drawback: It is difficult to accurately find a solution such as $\left(\frac{1}{3}, -\frac{5}{6}\right)$ from a graph. One algebraic method for solving a system of equations is the **substitution method.** This method is particularly useful for solving systems where one equation is already solved, or can be solved quickly, for one of the variables.

EXAMPLE 1 Using the Substitution Method

Solve the system

$$3x + 5y = 26$$
$$y = 2x.$$

The second equation is already solved for y. This equation says that $y = 2x$. Substituting $2x$ for y in the first equation gives

$$\begin{aligned} 3x + 5y &= 26 \\ 3x + 5(2x) &= 26 && \text{Let } y = 2x. \\ 3x + 10x &= 26 && \text{Multiply.} \\ 13x &= 26 && \text{Combine terms.} \\ x &= 2. && \text{Divide by 13.} \end{aligned}$$

Because $x = 2$, we find y from the equation $y = 2x$ by substituting 2 for x.

$$y = 2(2) = 4 \qquad \text{Let } x = 2.$$

Check that the solution of the given system is $(2, 4)$ by substituting 2 for x and 4 for y in *both* equations.

Work Problem 1 at the Side.

1. Fill in the blanks to solve by the substitution method. Check your solution.

$$3x + 5y = 69$$
$$y = 4x$$

$$3x + 5(\_\_\_) = 69$$
$$\_\_\_ = 69$$
$$x = \_\_\_$$
$$y = 4(\_\_\_) = \_\_\_$$

The solution is ___.

EXAMPLE 2 Using the Substitution Method

Solve the system

$$2x + 5y = 7$$
$$x = -1 - y.$$

The second equation gives x in terms of y. Substitute $-1 - y$ for x in the first equation.

$$\begin{aligned} 2x + 5y &= 7 \\ 2(-1 - y) + 5y &= 7 && \text{Let } x = -1 - y. \\ -2 - 2y + 5y &= 7 && \text{Distributive property} \\ -2 + 3y &= 7 && \text{Combine terms.} \\ 3y &= 9 && \text{Add 2.} \\ y &= 3 && \text{Divide by 3.} \end{aligned}$$

To find x, substitute 3 for y in the equation $x = -1 - y$ to get

$$x = -1 - 3 = -4.$$

Check that the solution of the given system is $(-4, 3)$.

Work Problem 2 at the Side.

2. Solve by the substitution method. Check your solution.

$$2x + 7y = -12$$
$$x = 3 - 2y$$

ANSWERS

1. $4x$; $23x$; 3; 3; 12; (3, 12)
2. (15, −6)

3 Solve each system by substitution. Check each solution.

(a) Fill in the blanks to solve

$$x + 4y = -1$$
$$2x - 5y = 11.$$

Solve the first equation for x.

$$x = -1 - \_\_\_\_$$

Substitute into the second equation to find y.

$$2(\_\_\_\_) - 5y = 11$$
$$-2 - 8y - 5y = 11$$
$$-2 - \_\_\_\_ y = 11$$
$$\_\_\_\_ y = 13$$
$$y = \_\_\_\_$$

Find x.

$$x = -1 - \_\_\_\_$$
$$x = \_\_\_\_$$

The solution is ____.

(b) $2x + 5y = 4$
$x + y = -1$

ANSWERS

3. **(a)** $4y$; $-1 - 4y$; 13; -13; -1; -4; 3; $(3, -1)$
(b) $(-3, 2)$

CAUTION
Even though we found y first in Example 2, ***the x-coordinate is always written first in the ordered pair solution of a system.***

To solve a system by substitution, follow these steps.

Solving a Linear System by Substitution

Step 1 **Solve one equation for either variable.** If one of the variables has coefficient 1 or -1, choose it.

Step 2 **Substitute** for that variable in the other equation. The result should be an equation with just one variable.

Step 3 **Solve** the equation from Step 2.

Step 4 **Substitute** the result from Step 3 into the equation from Step 1 to find the value of the other variable.

Step 5 **Check** the solution in both of the original equations. Then write the solution as an ordered pair.

EXAMPLE 3 Using the Substitution Method

Use substitution to solve the system

$$2x = 4 - y \quad (1)$$
$$5x + 3y = 10. \quad (2)$$

Step 1 For the substitution method, we must solve one of the equations for either x or y. Because the coefficient of y in equation (1) is -1, we choose equation (1) and solve for y.

$$2x = 4 - y \quad (1)$$
$$2x - 4 = -y \quad \text{Subtract 4.}$$
$$-2x + 4 = y \quad \text{Multiply by } -1.$$

Step 2 Now substitute $-2x + 4$ for y in equation (2).

$$5x + 3y = 10 \quad (2)$$
$$5x + 3(-2x + 4) = 10 \quad \text{Let } y = -2x + 4.$$

Step 3 Now solve the equation from Step 2.

$$5x - 6x + 12 = 10 \quad \text{Distributive property}$$
$$-x + 12 = 10 \quad \text{Combine terms.}$$
$$-x = -2 \quad \text{Subtract 12.}$$
$$x = 2 \quad \text{Multiply by } -1.$$

Step 4 Since $y = -2x + 4$ and $x = 2$, $y = -2(2) + 4 = 0$.

Step 5 Check that $(2, 0)$ is the solution.

$$2x = 4 - y \quad (1)$$
$$2(2) = 4 - 0 \quad ?$$
$$4 = 4 \quad \text{True}$$

$$5x + 3y = 10 \quad (2)$$
$$5(2) + 3(0) = 10 \quad ?$$
$$10 = 10 \quad \text{True}$$

The solution of the system is $(2, 0)$.

Work Problem 3 at the Side.

EXAMPLE 4 Using the Substitution Method

Use substitution to solve the system

$$2x + 3y = 10 \quad (1)$$
$$-3x - 2y = 0. \quad (2)$$

Step 1 To use the substitution method, we must solve one of the equations for one of the variables. We choose equation (1) and solve for x.

$$2x + 3y = 10 \qquad (1)$$
$$2x = 10 - 3y \qquad \text{Subtract } 3y.$$
$$x = 5 - \frac{3}{2}y \qquad \text{Divide by 2.}$$

Step 2 Substitute this expression for x in equation (2).

$$-3x - 2y = 0 \qquad (2)$$
$$-3\left(5 - \frac{3}{2}y\right) - 2y = 0 \qquad \text{Let } x = 5 - \tfrac{3}{2}y.$$

Step 3

$$-15 + \frac{9}{2}y - 2y = 0 \qquad \text{Distributive property}$$
$$-15 + \frac{5}{2}y = 0 \qquad \text{Combine terms.}$$
$$\frac{5}{2}y = 15 \qquad \text{Add 15.}$$
$$y = \frac{30}{5} = 6 \qquad \text{Multiply by } \tfrac{2}{5}.$$

Step 4 Find x by substituting **6** for y in $x = 5 - \frac{3}{2}y$.

$$x = 5 - \frac{3}{2}(6) = -4$$

Step 5 Check that $(-4, 6)$ is the solution.

$2x + 3y = 10$	$-3x - 2y = 0$
$2(-4) + 3(6) = 10$?	$-3(-4) - 2(6) = 0$?
$-8 + 18 = 10$?	$12 - 12 = 0$?
$10 = 10$ True	$0 = 0$ True

The solution of the system is $(-4, 6)$.

4 Solve the system by substitution. Check your solution.

$$3x + 2y = 1$$
$$3x - 4y = -11$$

NOTE

In Example 4, we could have started the solution by solving the second equation for either x or y and then substituting the result into the first equation. The solution would be the same.

Work Problem 4 at the Side.

ANSWERS

4. $(-1, 2)$

OBJECTIVE 2 Solve special systems. We can solve inconsistent systems with graphs that are parallel lines and systems of dependent equations with graphs that are the same line using the substitution method.

EXAMPLE 5 Solving an Inconsistent System by Substitution

Use substitution to solve the system

$$x = 5 - 2y \qquad (1)$$
$$2x + 4y = 6. \qquad (2)$$

Substitute $5 - 2y$ for x in equation (2).

$$2x + 4y = 6 \qquad (2)$$
$$2(5 - 2y) + 4y = 6 \qquad \text{Let } x = 5 - 2y.$$
$$10 - 4y + 4y = 6 \qquad \text{Distributive property}$$
$$\mathbf{10 = 6} \qquad \text{False}$$

This false result means that the equations in the system have graphs that are parallel lines. The system is inconsistent and has no solution. See Figure 5.

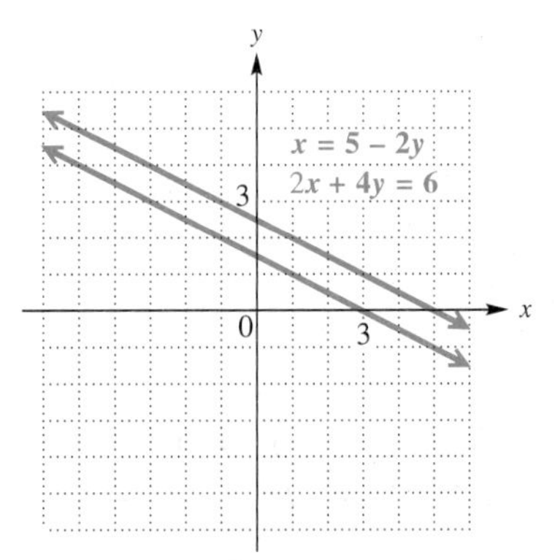

Figure 5

CAUTION
It is a common error to give "false" as the answer to an inconsistent system. The correct response is "no solution."

EXAMPLE 6 Solving a System with Dependent Equations by Substitution

Solve the system by the substitution method.

$$3x - y = 4 \qquad (1)$$
$$-9x + 3y = -12 \qquad (2)$$

Begin by solving equation (1) for y to get $y = 3x - 4$. Substitute $3x - 4$ for y in equation (2) and solve the resulting equation.

$$-9x + 3y = -12 \qquad (2)$$
$$-9x + 3(3x - 4) = -12 \qquad \text{Let } y = 3x - 4.$$
$$-9x + 9x - 12 = -12 \qquad \text{Distributive property}$$
$$\mathbf{0 = 0} \qquad \text{Add 12; combine terms.}$$

Continued on Next Page

This true result means that every solution of one equation is also a solution of the other, so the system has an infinite number of solutions: all the ordered pairs corresponding to points that lie on the common graph. A graph of the equations of this system is shown in Figure 6.

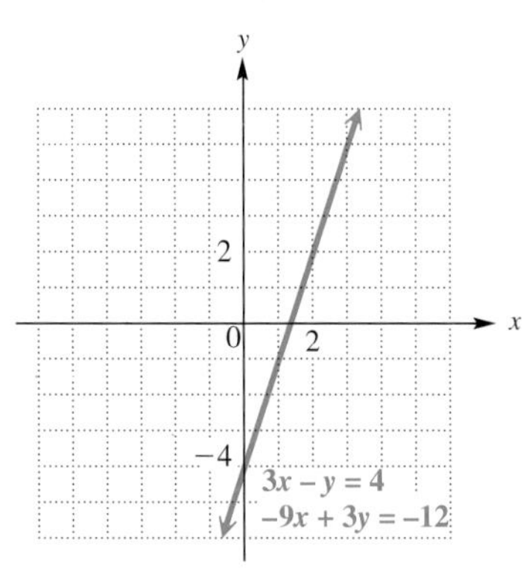

Figure 6

CAUTION
It is a common error to give "true" as the answer to a system of dependent equations. The correct response is "infinite number of solutions."

Work Problem 5 at the Side.

5 Solve each system by substitution.

(a) $8x - y = 4$
$y = 8x + 4$

(b) $7x - 6y = 10$
$-14x + 20 = -12y$

OBJECTIVE 3 Solve linear systems with fractions. When a system includes an equation with fractions as coefficients, eliminate the fractions by multiplying each side of the equation by a common denominator. Then solve the resulting system.

EXAMPLE 7 Using the Substitution Method with Fractions as Coefficients

Solve the system by the substitution method.

$$3x + \frac{1}{4}y = 2 \qquad (1)$$

$$\frac{1}{2}x + \frac{3}{4}y = -\frac{5}{2} \qquad (2)$$

Clear equation (1) of fractions by multiplying each side by 4.

$$4\left(3x + \frac{1}{4}y\right) = 4(2) \qquad \text{Multiply by 4.}$$

$$4(3x) + 4\left(\frac{1}{4}y\right) = 4(2) \qquad \text{Distributive property}$$

$$12x + y = 8 \qquad (3)$$

Now clear equation (2) of fractions by multiplying each side by the common denominator 4.

Continued on Next Page

ANSWERS
5. **(a)** no solution
(b) infinite number of solutions

6 Solve the system by substitution. First clear all fractions. Check your solution.

$$\frac{2}{3}x + \frac{1}{2}y = 6$$
$$\frac{1}{2}x - \frac{3}{4}y = 0$$

$$\frac{1}{2}x + \frac{3}{4}y = -\frac{5}{2} \qquad (2)$$
$$4\left(\frac{1}{2}x + \frac{3}{4}y\right) = 4\left(-\frac{5}{2}\right) \qquad \text{Multiply by 4.}$$
$$4\left(\frac{1}{2}x\right) + 4\left(\frac{3}{4}y\right) = 4\left(-\frac{5}{2}\right) \qquad \text{Distributive property}$$
$$2x + 3y = -10 \qquad (4)$$

The given system of equations has been simplified to the equivalent system

$$12x + y = 8 \qquad (3)$$
$$2x + 3y = -10. \qquad (4)$$

Solve this system by the substitution method. Equation (3) can be solved for y by subtracting $12x$ from each side.

$$12x + y = 8 \qquad (3)$$
$$y = -12x + 8 \qquad \text{Subtract } 12x.$$

Now substitute this result for y in equation (4).

$$2x + 3y = -10 \qquad (4)$$
$$2x + 3(-12x + 8) = -10 \qquad \text{Let } y = -12x + 8.$$
$$2x - 36x + 24 = -10 \qquad \text{Distributive property}$$
$$-34x = -34 \qquad \text{Combine terms; subtract 24.}$$
$$x = 1 \qquad \text{Divide by } -34.$$

Substitute 1 for x in $y = -12x + 8$ to get

$$y = -12(1) + 8 = -4.$$

The solution is $(1, -4)$. Check by substituting 1 for x and -4 for y in both of the original equations.

Work Problem 6 at the Side.

ANSWERS

6. (6, 4)

4.2 Exercises

FOR EXTRA HELP

Addison-Wesley Math Tutor Center

MathXL

Digital Video Tutor CD 4 Videotape 4

Student's Solutions Manual

MyMathLab

Interactmath.com

1. A student solves the system

$$5x - y = 15$$
$$7x + y = 21$$

and finds that $x = 3$, which is the correct value for x. The student gives the solution as "$x = 3$." Is this correct? Explain.

2. When you use the substitution method, how can you tell that a system has

(a) no solution? **(b)** an infinite number of solutions?

Solve each system by the substitution method. Check each solution. See Examples 1–7.

3. $x + y = 12$
$y = 3x$

4. $x + 3y = -28$
$y = -5x$

5. $3x + 2y = 27$
$x = y + 4$

6. $4x + 3y = -5$
$x = y - 3$

7. $3x + 5y = 14$
$x - 2y = -10$

8. $5x + 2y = -1$
$2x - y = -13$

9. $3x + 4 = -y$
$2x + y = 0$

10. $2x - 5 = -y$
$x + 3y = 0$

11. $7x + 4y = 13$
$x + y = 1$

12. $3x - 2y = 19$
$x + y = 8$

13. $3x - y = 5$
$y = 3x - 5$

14. $4x - y = -3$
$y = 4x + 3$

15. $6x - 8y = 6$
$2y = -2 + 3x$

16. $3x + 2y = 6$
$6x = 8 + 4y$

17. $2x + 8y = 3$
$x = 8 - 4y$

18. $2x + 10y = 3$
$x = 1 - 5y$

19. $12x - 16y = 8$
$3x = 4y + 2$

20. $6x + 9y = 6$
$2x = 2 - 3y$

21. $5x + 4y = 40$
$x + y = 1$

22. $x - 5y = 7$
$2x - y = 14$

23. $3x = 6 - 4y$
$9x + 12y = 10$

24. $\frac{1}{3}x - \frac{1}{2}y = \frac{1}{6}$
$3x - 2y = 9$

25. $\frac{1}{5}x + \frac{2}{3}y = -\frac{8}{5}$
$3x - y = 9$

26. $\frac{1}{6}x + \frac{1}{6}y = 2$
$-\frac{1}{2}x - \frac{1}{3}y = -8$

27. $\frac{x}{2} - \frac{y}{3} = 9$
$\frac{x}{5} - \frac{y}{4} = 5$

28. $\frac{x}{3} - \frac{3y}{4} = -\frac{1}{2}$
$\frac{x}{6} + \frac{y}{8} = \frac{3}{4}$

29. $\frac{x}{5} + 2y = \frac{16}{5}$
$\frac{3x}{5} + \frac{y}{2} = -\frac{7}{5}$

RELATING CONCEPTS (EXERCISES 30–33) For Individual or Group Work

A system of linear equations can be used to model the cost and the revenue of a business. ***Work Exercises 30–33 in order.***

30. Suppose that you start a business manufacturing and selling bicycles, and it costs you \$5000 to get started. You determine that each bicycle will cost \$400 to manufacture. Explain why the linear equation $y_1 = 400x + 5000$ gives your *total* cost to manufacture x bicycles (y_1 in dollars).

31. You decide to sell each bike for \$600. What expression in x represents the revenue you will take in if you sell x bikes? Write an equation using y_2 to express your revenue when you sell x bikes (y_2 in dollars).

32. Form a system from the two equations in Exercises 30 and 31, and then solve the system, assuming $y_1 = y_2$, that is, cost = revenue.

33. The value of x from Exercise 32 is the number of bikes it takes to *break even.* Fill in the blanks: When ______ bikes are sold, the break-even point is reached. At that point, you have spent ______ dollars and taken in ______ dollars.

4.3 Solving Systems of Linear Equations by Elimination

OBJECTIVES

1. Solve linear systems by elimination.
2. Multiply when using the elimination method.
3. Use an alternative method to find the second value in a solution.
4. Use the elimination method to solve special systems.

OBJECTIVE 1 Solve linear systems by elimination. An algebraic method that depends on the addition property of equality can be used to solve systems. As mentioned earlier, adding the same quantity to each side of an equation results in equal sums.

$$\text{If} \quad A = B, \quad \text{then} \quad A + C = B + C.$$

This addition can be taken a step further. Adding *equal* quantities, rather than the *same* quantity, to both sides of an equation also results in equal sums.

$$\text{If} \quad A = B \quad \text{and} \quad C = D, \quad \text{then} \quad A + C = B + D.$$

Using the addition property to solve systems is called the **elimination method.** When using this method, the idea is to *eliminate* one of the variables. To do this, one of the variables in the two equations must have coefficients that are opposites.

EXAMPLE 1 Using the Elimination Method

Use the elimination method to solve the system

$$x + y = 5$$
$$x - y = 3.$$

Each equation in this system is a statement of equality, so the sum of the left sides equals the sum of the right sides. Adding in this way gives

$$(x + y) + (x - y) = 5 + 3.$$

Combine terms and simplify to get

$$2x = 8$$
$$x = 4. \quad \text{Divide by 2.}$$

Notice that y has been eliminated. The result, $x = 4$, gives the x-value of the solution of the given system. To find the y-value of the solution, substitute 4 for x in either of the two equations of the system.

Work Problem 1 at the Side.

The solution found at the side, (4, 1), can be checked by substituting 4 for x and 1 for y in both equations of the given system.

Check:

$x + y = 5$		$x - y = 3$	
$4 + 1 = 5$?		$4 - 1 = 3$?	
$5 = 5$	True	$3 = 3$	True

Since both results are true, the solution of the system is (4, 1).

CAUTION

A system is not completely solved until values for both x and y are found. Do not stop after finding the value of only one variable. Remember to write the solution as an ordered pair.

1 **(a)** Substitute 4 for x in the equation $x + y = 5$ to find the value of y.

(b) Give the solution of the system.

ANSWERS

1. (a) $y = 1$ (b) (4, 1)

Work Problem 2 at the Side.

2 Solve each system by the elimination method. Check each solution.

(a) Fill in the blanks to find the solution.

$$x + y = 8$$
$$x - y = 2$$

Add.

$(x + y) + (x - y) = 8 + \_\_\_$

$2\,\_\_\_ = \_\_\_$

$x = \_\_\_$

Find y.

$x - y = 2$

$\_\_\_ - y = 2$

$-y = \_\_\_$

$y = \_\_\_$

The solution is ___.

(b) $3x - y = 7$
$2x + y = 3$

Answers

2. **(a)** 2; x; 10; 5; 5; -3; 3; (5, 3)
(b) (2, -1)

In general, to solve a system by elimination, follow these steps.

Solving a Linear System by Elimination

Step 1 **Write both equations in standard form** $Ax + By = C$.

Step 2 **Transform so that the coefficients of one pair of variable terms are opposites.** Multiply one or both equations by appropriate numbers so that the sum of the coefficients of either the x- or y-terms is 0.

Step 3 **Add** the new equations to eliminate a variable. The sum should be an equation with just one variable.

Step 4 **Solve** the equation from Step 3 for the remaining variable.

Step 5 **Substitute** the result from Step 4 into either of the original equations and solve for the other variable.

Step 6 **Check** the solution in both of the original equations. Then write the solution as an ordered pair.

It does not matter which variable is eliminated first. Usually we choose the one that is more convenient to work with.

EXAMPLE 2 Using the Elimination Method

Solve the system

$$y + 11 = 2x$$
$$5x = y + 26.$$

Step 1 Rewrite both equations in the form $Ax + By = C$ to get the system

$$-2x + y = -11 \quad \text{Subtract } 2x \text{ and } 11.$$
$$5x - y = 26. \quad \text{Subtract } y.$$

Step 2 Because the coefficients of y are 1 and -1, adding will eliminate y. It is not necessary to multiply either equation by a number.

Step 3 Add the two equations. This time we use vertical addition.

$$\begin{array}{rcr} -2x + y & = & -11 \\ 5x - y & = & 26 \\ \hline 3x \quad & = & 15 \end{array} \quad \text{Add in columns.}$$

Step 4 Solve the equation.

$$3x = 15$$
$$x = 5 \quad \text{Divide by 3.}$$

Step 5 Find the value of y by substituting 5 for x in either of the original equations. Choosing the first gives

$$y + 11 = 2x$$
$$y + 11 = 2(5) \quad \text{Let } x = 5.$$
$$y + 11 = 10$$
$$y = -1. \quad \text{Subtract 11.}$$

Continued on Next Page

Step 6 Check the solution by substituting $x = 5$ and $y = -1$ into both of the original equations.

$y + 11 = 2x$		$5x = y + 26$	
$(-1) + 11 = 2(5)$	?	$5(5) = -1 + 26$	?
$10 = 10$	True	$25 = 25$	True

The solution $(5, -1)$ is correct.

Work Problem 3 at the Side.

OBJECTIVE 2 Multiply when using the elimination method. Sometimes we need to multiply each side of one or both equations in a system by some number before adding the equations will eliminate a variable.

EXAMPLE 3 Multiplying Both Equations When Using the Elimination Method

Solve the system

$$2x + 3y = -15 \quad (1)$$
$$5x + 2y = 1. \quad (2)$$

Adding the two equations gives $7x + 5y = -14$, which does not eliminate either variable. However, we can multiply each equation by a suitable number so that the coefficients of one of the two variables are opposites. For example, to eliminate x, multiply each side of equation (1) by 5, and each side of equation (2) by -2.

$$\begin{aligned} 10x + 15y &= -75 && \text{Multiply equation (1) by 5.} \\ -10x - 4y &= -2 && \text{Multiply equation (2) by } -2. \\ \hline 11y &= -77 && \text{Add.} \\ y &= -7 && \text{Divide by 11.} \end{aligned}$$

Substituting -7 for y in either equation (1) or (2) gives $x = 3$. Check that the solution of the system is $(3, -7)$.

Work Problem 4 at the Side.

OBJECTIVE 3 Use an alternative method to find the second value in a solution. Sometimes it is easier to find the value of the second variable in a solution by using the elimination method twice.

EXAMPLE 4 Finding the Second Value Using an Alternative Method

Solve the system

$$4x = 9 - 3y \quad (1)$$
$$5x - 2y = 8. \quad (2)$$

Rearrange the terms in equation (1) so that like terms are aligned in columns. To do this, add $3y$ to each side to get the following system.

$$4x + 3y = 9 \quad (3)$$
$$5x - 2y = 8 \quad (2)$$

One way to proceed is to eliminate y by multiplying each side of equation (3) by 2 and each side of equation (2) by 3, and then adding.

Continued on Next Page

3 Solve each system by the elimination method. Check each solution.

(a) $2x - y = 2$
$4x + y = 10$

(b) $8x - 5y = 32$
$4x + 5y = 4$

4 **(a)** Solve the system in Example 3 by first eliminating the variable y. Check your solution.

(b) Solve

$$6x + 7y = 4$$
$$5x + 8y = -1,$$

and check your solution.

Answers

3. **(a)** $(2, 2)$ **(b)** $\left(3, -\frac{8}{5}\right)$

4. **(a)** $(3, -7)$ **(b)** $(3, -2)$

$$\begin{aligned} 8x + 6y &= 18 && \text{Multiply equation (3) by 2.} \\ 15x - 6y &= 24 && \text{Multiply equation (2) by 3.} \\ \hline 23x &= 42 && \text{Add.} \\ x &= \frac{42}{23} && \text{Divide by 23.} \end{aligned}$$

Substituting $\frac{42}{23}$ for x in one of the given equations would give y, but the arithmetic involved would be messy. Instead, solve for y by starting again with the original equations and eliminating x. Multiply each side of equation (3) by 5 and each side of equation (2) by -4, and then add.

$$\begin{aligned} 20x + 15y &= 45 && \text{Multiply equation (3) by 5.} \\ -20x + 8y &= -32 && \text{Multiply equation (2) by } -4. \\ \hline 23y &= 13 && \text{Add.} \\ y &= \frac{13}{23} && \text{Divide by 23.} \end{aligned}$$

Check that the solution is $\left(\frac{42}{23}, \frac{13}{23}\right)$.

Work Problem 5 at the Side.

5 Solve each system of equations.

(a) $5x = 7 + 2y$
$5y = 5 - 3x$

(b) $3y = 8 + 4x$
$6x = 9 - 2y$

When the value of the first variable is a fraction, the method used in Example 4 helps avoid arithmetic errors. Of course, this method could be used to solve any system of equations.

OBJECTIVE 4 Use the elimination method to solve special systems.

EXAMPLE 5 Using the Elimination Method for an Inconsistent System or Dependent Equations

Solve each system by the elimination method.

(a) $2x + 4y = 5$
$4x + 8y = -9$

Multiply each side of $2x + 4y = 5$ by -2; then add to $4x + 8y = -9$.

$$\begin{aligned} -4x - 8y &= -10 \\ 4x + 8y &= -9 \\ \hline \mathbf{0} &= \mathbf{-19} && \text{False} \end{aligned}$$

The false statement $0 = -19$ shows that the given system has no solution.

(b) $3x - y = 4$
$-9x + 3y = -12$

Multiply each side of the first equation by 3; then add the two equations.

$$\begin{aligned} 9x - 3y &= 12 \\ -9x + 3y &= -12 \\ \hline 0 &= 0 && \text{True} \end{aligned}$$

A true statement occurs when the equations are equivalent. As before, this indicates that every solution of one equation is also a solution of the other; there are an infinite number of solutions. (See **Section 4.2,** Example 6, where the same system was solved using substitution.)

Work Problem 6 at the Side.

6 Solve each system by the elimination method.

(a) $4x + 3y = 10$
$2x + \frac{3}{2}y = 12$

(b) $4x - 6y = 10$
$-10x + 15y = -25$

ANSWERS

5. (a) $\left(\frac{45}{31}, \frac{4}{31}\right)$ (b) $\left(\frac{11}{26}, \frac{42}{13}\right)$
6. (a) no solution
(b) infinite number of solutions

4.3 Exercises

FOR EXTRA HELP

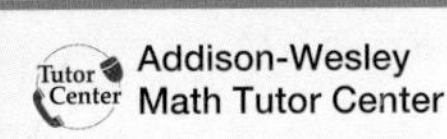
Addison-Wesley Math Tutor Center

MathXL

Digital Video Tutor CD 4 Videotape 4

Student's Solutions Manual

MyMathLab

Interactmath.com

In Exercises 1–3, answer true *or* false *for each statement. If* false, *tell why.*

1. The ordered pair (0, 0) *must* be a solution of a system of the form

$$Ax + By = 0$$
$$Cx + Dy = 0.$$

2. To eliminate the y-terms in the system

$$2x + 12y = 7$$
$$3x + 4y = 1,$$

we should multiply the bottom equation by 3 and then add.

3. The system

$$x + y = 1$$
$$x + y = 2$$

has no solution.

4. Which one of the following systems would be easier to solve using the elimination method? Why?

$$5x - 3y = 7 \qquad 7x + 2y = 4$$
$$2x + 8y = 3 \qquad -7x + 3y = 1$$

Solve each system by the elimination method. Check each solution. See Examples 1 and 2.

5. $x + y = 2$
$2x - y = -5$

6. $3x - y = -12$
$x + y = 4$

7. $2x + y = -5$
$x - y = 2$

8. $2x + y = -15$
$-x - y = 10$

9. $3x + 2y = 0$
$-3x - y = 3$

10. $5x - y = 5$
$-5x + 2y = 0$

11. $6x - y = -1$
$5y = 17 + 6x$

12. $y = 9 - 6x$
$-6x + 3y = 15$

Solve each system by the elimination method. Check each solution. See Examples 3–5.

13. $2x - y = 12$
$3x + 2y = -3$

14. $x + y = 3$
$-3x + 2y = -19$

15. $x + 3y = 19$
$2x - y = 10$

16. $4x - 3y = -19$
$2x + y = 13$

17. $x + 4y = 16$
$3x + 5y = 20$

18. $2x + y = 8$
$5x - 2y = -16$

19. $5x - 3y = -20$
$-3x + 6y = 12$

20. $4x + 3y = -28$
$5x - 6y = -35$

21. $2x - 8y = 0$
$4x + 5y = 0$

22. $3x - 15y = 0$
$6x + 10y = 0$

23. $x + y = 7$
$x + y = -3$

24. $x - y = 4$
$x - y = -3$

25. $-x + 3y = 4$
$-2x + 6y = 8$

26. $6x - 2y = 24$
$-3x + y = -12$

27. $4x - 3y = -19$
$3x + 2y = 24$

28. $\begin{aligned} 5x + 4y &= 12 \\ 3x + 5y &= 15 \end{aligned}$

29. $\begin{aligned} 3x - 7 &= -5y \\ 5x + 4y &= -10 \end{aligned}$

30. $\begin{aligned} 2x + 3y &= 13 \\ 6 + 2y &= -5x \end{aligned}$

31. $\begin{aligned} 2x + 3y &= 0 \\ 4x + 12 &= 9y \end{aligned}$

32. $\begin{aligned} -4x + 3y &= 2 \\ 5x + 3 &= -2y \end{aligned}$

33. $\begin{aligned} 24x + 12y &= -7 \\ 16x - 17 &= 18y \end{aligned}$

34. $\begin{aligned} 9x + 4y &= -3 \\ 6x + 7 &= -6y \end{aligned}$

35. $\begin{aligned} 3x &= 3 + 2y \\ -\frac{4}{3}x + y &= \frac{1}{3} \end{aligned}$

36. $\begin{aligned} 3x &= 27 + 2y \\ x - \frac{7}{2}y &= -25 \end{aligned}$

37. $\begin{aligned} 5x - 2y &= 3 \\ 10x - 4y &= 5 \end{aligned}$

38. $\begin{aligned} 3x - 5y &= 1 \\ 6x - 10y &= 4 \end{aligned}$

39. $\begin{aligned} 6x + 3y &= 0 \\ -18x - 9y &= 0 \end{aligned}$

40. $\begin{aligned} 3x - 5y &= 0 \\ 9x - 15y &= 0 \end{aligned}$

RELATING CONCEPTS (EXERCISES 41–46) For Individual or Group Work

Attending the movies is one of America's favorite forms of entertainment. The graph shows movie attendance from 1991 to 1999. In 1991, attendance was 1141 *million, as represented by the point P*(1991, 1141). *In 1999, attendance was* 1465 *million, as represented by the point Q*(1999, 1465). *We can find an equation of line segment PQ using a system of equations, and then we can use the equation to approximate the attendance in any of the years between 1991 and 1999.* ***Work Exercises 41–46 in order.***

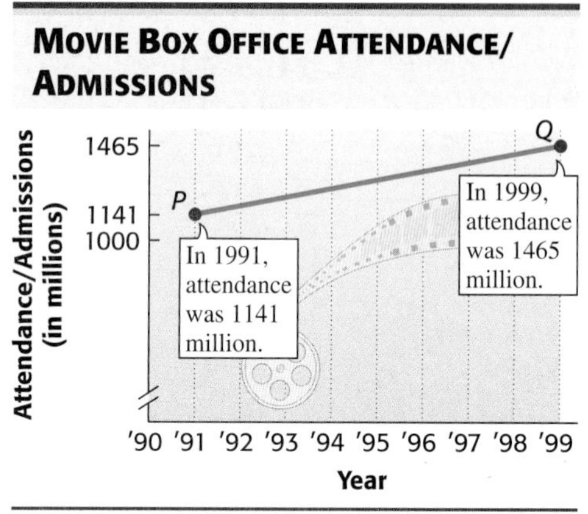

Source: Motion Picture Association of America.

41. The line segment has an equation that can be written in the form $y = ax + b$. Using the coordinates of point P with $x = 1991$ and $y = 1141$, write an equation in the variables a and b.

42. Using the coordinates of point Q with $x = 1999$ and $y = 1465$, write a second equation in the variables a and b.

43. Write the system of equations formed from the two equations in Exercises 41 and 42, and solve the system using the elimination method.

44. What is the equation of the segment PQ?

45. Let $x = 1998$ in the equation of Exercise 44, and solve for y. How does the result compare with the actual figure of 1481 million?

46. The data points for the years 1991 through 1999 do not lie in a perfectly straight line. Explain the pitfalls of relying too heavily on using the equation in Exercise 44 to predict attendance.

Summary Exercises on Solving Systems of Linear Equations

The exercises in this summary include a variety of problems on solving systems of linear equations. Since we do not usually specify the method of solution, use the following guidelines to help you decide whether to use substitution or elimination.

Choosing a Method When Solving a System of Linear Equations

1. If one of the equations of the system is already solved for one of the variables, as in the systems

$$\begin{aligned} 3x + 4y &= 9 \\ y &= 2x - 6 \end{aligned} \quad \text{or} \quad \begin{aligned} -5x + 3y &= 9 \\ x &= 3y - 7, \end{aligned}$$

the substitution method is the better choice.

2. If both equations are in standard $Ax + By = C$ form, as in

$$\begin{aligned} 4x - 11y &= 3 \\ -2x + 3y &= 4, \end{aligned}$$

and none of the variables has coefficient -1 or 1, the elimination method is the better choice.

3. If one or both of the equations are in standard form and the coefficient of one of the variables is -1 or 1, as in the systems

$$\begin{aligned} 3x + y &= -2 \\ -5x + 2y &= 4 \end{aligned} \quad \text{or} \quad \begin{aligned} -x + 3y &= -4 \\ 3x - 2y &= 8, \end{aligned}$$

either method is appropriate.

Use the information in the preceding box to solve each problem.

1. Assuming you want to minimize the amount of work required, tell whether you would use the substitution or elimination method to solve each system. Explain your answers. *Do not actually solve.*

(a) $3x + 2y = 18$
$y = 3x$

(b) $3x + y = -7$
$x - y = -5$

(c) $3x - 2y = 0$
$9x + 8y = 7$

2. Which one of the following systems would be easier to solve using the substitution method? Why?

$$\begin{aligned} 5x - 3y &= 7 \\ 2x + 8y &= 3 \end{aligned} \qquad \begin{aligned} 7x + 2y &= 4 \\ y &= -3x + 1 \end{aligned}$$

In Exercises 3 and 4, ***(a)*** *solve the system by the elimination method,* ***(b)*** *solve the system by the substitution method, and* ***(c)*** *tell which method you prefer for that particular system and why.*

3. $4x - 3y = -8$
$x + 3y = 13$

4. $2x + 5y = 0$
$x = -3y + 1$

Solve each system by the method of your choice. (For Exercises 5–7, see your answers for Exercise 1.)

5. $3x + 2y = 18$
$y = 3x$

6. $3x + y = -7$
$x - y = -5$

7. $3x - 2y = 0$
$9x + 8y = 7$

8. $x + y = 7$
$x = -3 - y$

9. $5x - 4y = 15$
$-3x + 6y = -9$

10. $4x + 2y = 3$
$y = -x$

11. $3x = 7 - y$
$2y = 14 - 6x$

12. $3x - 5y = 7$
$2x + 3y = 30$

13. $3y = 4x + 2$
$5x - 2y = -3$

14. $4x + 3y = 1$
$3x + 2y = 2$

15. $2x - 3y = 7$
$-4x + 6y = 14$

16. $2x + 3y = 10$
$-3x + y = 18$

17. $6x + 5y = 13$
$3x + 3y = 4$

18. $x - 3y = 7$
$4x + y = 5$

Solve each system by any method. First clear all fractions.

19. $\frac{1}{4}x - \frac{1}{5}y = 9$
$y = 5x$

20. $\frac{1}{2}x + \frac{1}{3}y = -\frac{1}{3}$
$\frac{1}{2}x + 2y = -7$

21. $\frac{1}{6}x + \frac{1}{6}y = 1$
$-\frac{1}{2}x - \frac{1}{3}y = -5$

22. $\frac{x}{5} + 2y = \frac{8}{5}$
$\frac{3x}{5} + \frac{y}{2} = -\frac{7}{10}$

23. $\frac{x}{5} + y = \frac{6}{5}$
$\frac{x}{10} + \frac{y}{3} = \frac{5}{6}$

24. $\frac{1}{6}x + \frac{1}{3}y = 8$
$\frac{1}{4}x + \frac{1}{2}y = 12$

4.4 Applications of Linear Systems

OBJECTIVES

1. Solve problems about unknown numbers.
2. Solve problems about quantities and their costs.
3. Solve problems about mixtures.
4. Solve problems about distance, rate (or speed), and time.

Recall from **Section 2.4** the six-step method for solving applied problems. We modify those steps slightly to allow for two variables and two equations.

Solving Applied Problems with Two Variables

Step 1 **Read** the problem, several times if necessary, until you understand what is given and what is to be found.

Step 2 **Assign variables** to represent the unknown values, using diagrams or tables as needed. Write down what each variable represents.

Step 3 **Write two equations** using both variables.

Step 4 **Solve** the system of two equations.

Step 5 **State the answer** to the problem. Is the answer reasonable?

Step 6 **Check** the answer in the words of the original problem.

1 Solve the system

$$x = 1954 + y$$
$$x + y = 24{,}098.$$

OBJECTIVE 1 Solve problems about unknown numbers.

EXAMPLE 1 Solving a Problem about Two Unknown Numbers

In 2000, sales of athletic/sport footwear were \$1954 million more than sales of athletic/sport clothing. Together, total sales for these items were \$24,098 million. (*Source:* National Sporting Goods Association.) What were the sales for each?

Step 1 **Read** the problem carefully. We must find the 2000 sales (in millions of dollars) for athletic/sport clothing and footwear. We know how much more footwear sales were than clothing sales. Also, we know the total sales.

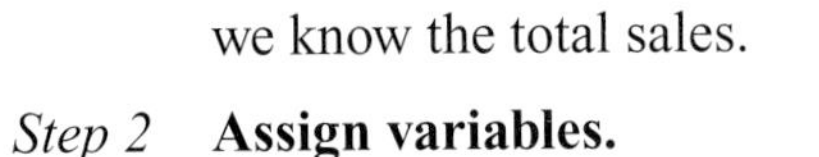

Step 2 **Assign variables.**

Let x = sales of footwear in millions of dollars,

and y = sales of clothing in millions of dollars.

Step 3 **Write two equations.**

$x = 1954 + y$ — Sales of footwear were \$1954 million more than sales of clothing.

$x + y = 24{,}098$ — Total sales were \$24,098 million.

Step 4 **Solve** the system from Step 3. The substitution method works well here since the first equation is already solved for x.

Work Problem 1 at the Side.

Step 5 **State the answer.** Footwear sales were \$13,026 million, and clothing sales were \$11,072 million.

Step 6 **Check** the answer in the original problem. Since

$$13{,}026 - 11{,}072 = 1954 \quad \text{and} \quad 13{,}026 + 11{,}072 = 24{,}098,$$

the answer satisfies the information in the problem.

ANSWERS

1. $x = 13{,}026, y = 11{,}072$

2 Set up a system of equations for the following problem. Do not solve the system.

Two top-grossing Disney movies in 2002 were *Lilo and Stitch* and *The Santa Clause 2*. Together they grossed \$284.2 million. *The Santa Clause 2* grossed \$7.4 million less than *Lilo and Stitch*. How much did each movie gross? (*Source: Variety.*)

Let $x =$ the amount (in millions) that *Lilo and Stitch* grossed,

and $y =$ the amount (in millions) that ____________ grossed.

ANSWERS

2. *The Santa Clause 2*; $x + y = 284.2$
$y = x - 7.4$

CAUTION

If an applied problem asks for *two* values as in Example 1, be sure to give both of them in your answer.

Work Problem 2 at the Side.

OBJECTIVE 2 Solve problems about quantities and their costs. We can also use a linear system to solve an applied problem involving two quantities and their costs.

EXAMPLE 2 Solving a Problem about Quantities and Costs

The all-time top-grossing movie *Titanic** earned more in Europe than in the United States. This may be because average movie prices in Europe exceed those of the United States. (*Source: Parade* magazine, September 13, 1998.)

For example, while the average movie ticket (to the nearest dollar) in 1997–1998 cost \$5 in the United States, it cost an equivalent of \$11 in London. Suppose that a group of 41 Americans and Londoners who paid these average prices spent a total of \$307 for tickets. How many from each country were in the group?

Step 1 **Read** the problem several times.

Step 2 **Assign variables.**

Let $x =$ the number of Americans in the group,
and $y =$ the number of Londoners in the group.

Summarize the information given in the problem in a table. The entries in the first two rows of the Total Value column were found by multiplying the number of tickets sold by the price per ticket.

	Number of Tickets	Price per Ticket in dollars	Total Value
Americans	x	5	$\mathbf{5x}$
Londoners	y	11	$\mathbf{11y}$
Total	41		**307**

Step 3 **Write two equations.** The total number of tickets was 41, so

$$x + y = 41. \qquad \text{Total number of tickets}$$

Since the total value was \$307, the final column leads to

$$5x + 11y = 307. \qquad \text{Total value of tickets}$$

These two equations form the system

$$x + y = 41 \qquad (1)$$
$$5x + 11y = 307. \qquad (2)$$

Step 4 **Solve** the system using the elimination method. To eliminate the x-terms, multiply each side of equation (1) by -5 to get

$$-5x - 5y = -205.$$

Continued on Next Page

* Through September 2003, *Titanic* was still number one. (*Source: Variety.*)

Then add this result to equation (2).

$$\begin{aligned} -5x - 5y &= -205 & \\ 5x + 11y &= 307 & (2) \\ \hline 6y &= 102 & \text{Add.} \\ y &= 17 & \text{Divide by 6.} \end{aligned}$$

Substitute 17 for y in equation (1) to get

$$\begin{aligned} x + y &= 41 & \\ x + 17 &= 41 & \text{Let } y = 17. \\ x &= 24. & \end{aligned}$$

Step 5 **State the answer.** There were 24 Americans and 17 Londoners in the group.

Step 6 **Check.** The sum of 24 and 17 is 41, so the number of moviegoers is correct. Since 24 Americans paid \$5 each and 17 Londoners paid \$11 each, the total of the admission prices is

$$\$5(24) + \$11(17) = \$307,$$

which agrees with the total amount stated in the problem.

Work Problem 3 at the Side.

3 The average movie ticket (to the nearest U.S. dollar) costs \$10 in Geneva and \$8 in Paris. (*Source: Parade* magazine, September 13, 1998.) If a group of 36 people from these two cities paid \$298 for tickets to see *Cheaper by the Dozen,* how many people from each city were there?

(a) Complete the table.

	Number of Tickets Sold	Price (in dollars)	Total Value
Genevans	x		
Parisians	y		
Total			

(b) Write a system of equations.

(c) Solve the system and check your answer in the words of the original problem.

OBJECTIVE 3 Solve problems about mixtures. In **Section 2.6** we solved percent problems using one variable. Many problems about mixtures that involve percent can be solved using a system of two equations in two variables.

EXAMPLE 3 Solving a Mixture Problem Involving Percent

A pharmacist needs 100 L of 50% alcohol solution. She has on hand 30% alcohol solution and 80% alcohol solution, which she can mix. How many liters of each will be required to make the 100 L of 50% alcohol solution?

Step 1 **Read** the problem. Note the percent of each solution and of the mixture.

Step 2 **Assign variables.**

Let x = the number of liters of 30% alcohol needed,

and y = the number of liters of 80% alcohol needed.

Summarize the information in a table. Percents are represented in decimal form.

Percent	Liters of Mixture	Liters of Pure Alcohol
.30	x	$.30x$
.80	y	$.80y$
.50	100	$.50(100)$

Continued on Next Page

ANSWERS

3. (a)

	Number of Tickets Sold	Price (in dollars)	Total Value
Genevans	x	10	$10x$
Parisians	y	8	$8y$
Total	36		298

(b) $x + y = 36$
$10x + 8y = 298$

(c) Genevans: 5; Parisians: 31

4 Solve the system

$$x + \quad y = 100$$
$$.30x + .80y = 50.$$

5 How many liters of 25% alcohol solution must be mixed with 12% solution to get 13 L of 15% solution?

(a) Complete the table.

Percent	Liters	Liters of Pure Alcohol
.25	x	$.25x$
.12	y	
.15	13	

(b) Write a system of equations, and solve it.

6 Solve the problem.

Joe needs 100 cc (cubic centimeters) of 20% acid solution for a chemistry experiment. The lab has on hand only 10% and 25% solutions. How much of each should he mix to get the desired amount of 20% solution?

7 Solve using the distance formula.

A small plane traveled from Warsaw to Rome, averaging 164 mph. The trip took 2.5 hr. What is the distance from Warsaw to Rome?

Answers

4. (60, 40)

5. (a)

Percent	Liters	Liters of Pure Alcohol
.25	x	$.25x$
.12	y	$.12y$
.15	13	$.15(13)$

(b) $x + \quad y = 13$
$.25x + .12y = .15(13)$

3 L of 25%, 10 L of 12%

6. $33\frac{1}{3}$ cc of 10%, $66\frac{2}{3}$ cc of 25%

7. 410 mi

Figure 7 gives an idea of what is actually happening in this problem.

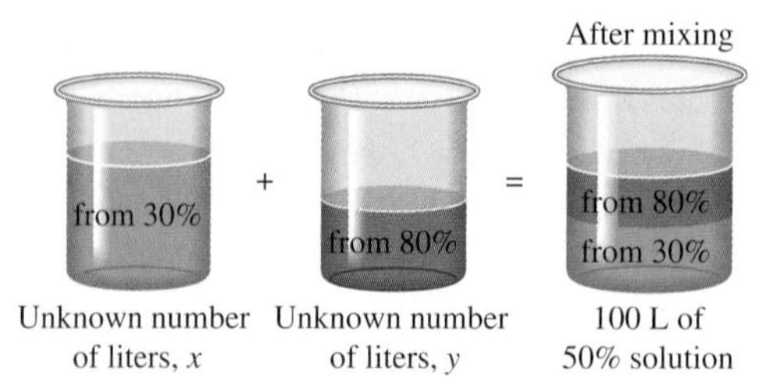

Figure 7

Step 3 **Write two equations.** Since the total number of liters in the final mixture will be 100, the first equation is

$$x + y = 100.$$

To find the amount of pure alcohol in each mixture, multiply the number of liters by the concentration. The amount of pure alcohol in the 30% solution added to the amount of pure alcohol in the 80% solution will equal the amount of pure alcohol in the final 50% solution. This gives the second equation,

$$.30x + .80y = .50(100).$$

These two equations form the system

$$x + \quad y = 100$$
$$.30x + .80y = 50. \qquad .50(100) = 50$$

Step 4 **Solve** this system.

Work Problem 4 at the Side.

Step 5 **State the answer.** From Problem 4 at the side, the pharmacist should use 60 L of the 30% solution and 40 L of the 80% solution.

Step 6 Since $60 + 40 = 100$ and $.30(60) + .80(40) = 50$, this mixture will give the 100 L of 50% solution, as required in the original problem.

Work Problems 5 and 6 at the Side.

OBJECTIVE 4 Solve problems about distance, rate (or speed), and time. If an automobile travels at an average rate of 50 mph for 2 hr, then it travels $50 \times 2 = 100$ mi. This is an example of the basic relationship between distance, rate, and time:

$$\text{distance} = \text{rate} \times \text{time}.$$

This relationship is given by the formula $d = rt$.

Work Problem 7 at the Side.

We can solve problems involving the distance formula with systems of two linear equations. Keep in mind that setting up a table and drawing a sketch will help you solve such problems.

EXAMPLE 4 Solving a Problem about Distance, Rate, and Time

Two executives in cities 400 mi apart drive to a business meeting at a location on the line between their cities. They meet after 4 hr. Find the speed of each car if one car travels 20 mph faster than the other.

Step 1 **Read** the problem carefully.

Step 2 **Assign variables.** Let x = the speed of the faster car, and y = the speed of the slower car.

We use the formula $d = rt$. Since each car travels for 4 hr, the time, t, for each car is 4. See the table. The distance is found by using the formula $d = rt$ and the expressions already entered in the table.

	r	t	d
Faster Car	x	4	$4x$
Slower Car	y	4	$4y$

Find d from $d = rt$.

Figure 8 shows what is happening in the problem.

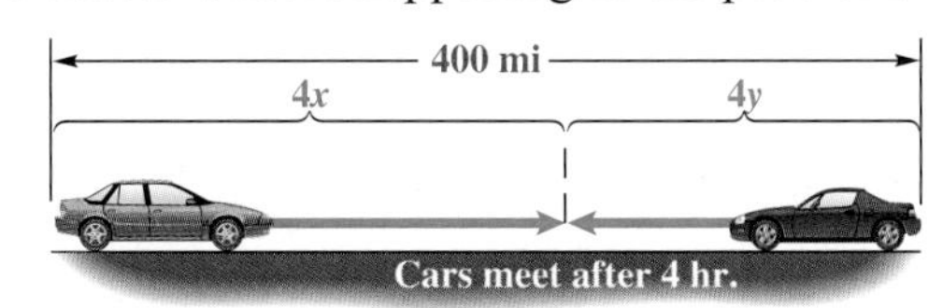

Figure 8

Step 3 **Write two equations.** As shown in the figure, since the total distance traveled by both cars is 400 mi, one equation is

$$4x + 4y = 400.$$

Because the faster car goes 20 mph faster than the slower car, the second equation is

$$x = 20 + y.$$

Step 4 **Solve** the system of equations,

$$4x + 4y = 400 \quad (1)$$
$$x = 20 + y, \quad (2)$$

by substitution. Replace x with $20 + y$ in equation (1) and then solve for y.

$$4(20 + y) + 4y = 400 \quad \text{Let } x = 20 + y.$$
$$80 + 4y + 4y = 400 \quad \text{Distributive property}$$
$$80 + 8y = 400 \quad \text{Combine like terms.}$$
$$8y = 320 \quad \text{Subtract 80.}$$
$$y = 40 \quad \text{Divide by 8.}$$

Since $x = 20 + y$, and $y = 40$,

$$x = 20 + 40 = 60.$$

Step 5 **State the answer.** The speeds of the two cars are 40 mph and 60 mph.

Step 6 **Check** the answer. Since each car travels for 4 hr, total distance is

$$4(60) + 4(40) = 240 + 160 = 400 \text{ mi}, \quad \text{as required.}$$

Work Problem 8 at the Side.

8 Two cars that were 450 mi apart traveled toward each other. They met after 5 hr. If one car traveled twice as fast as the other, what were their speeds?

(a) Complete this table.

	r	t	d
Faster Car	x	5	
Slower Car	y	5	

(b) Write a system, and solve it.

ANSWERS

8. (a)

	r	t	d
Faster Car	x	5	$5x$
Slower Car	y	5	$5y$

(b) $5x + 5y = 450$
$x = 2y$

faster car: $x = 60$ mph
slower car: $y = 30$ mph

9 Solve the system

$$x + y = 320$$
$$x - y = 280.$$

10 Solve the problem.

In 1 hr, Ann can row 2 mi against the current or 10 mi with the current. Find the speed of the current and Ann's speed in still water. (*Hint:* Let $x =$ the speed of the current and $y =$ Ann's speed in still water. Then her rate against the current is $y - x$, and her rate with the current is $y + x$.)

CAUTION
Be careful! ***When you use two variables to solve a problem, you must write two equations.***

EXAMPLE 5 Solving a Problem about Distance, Rate, and Time

A plane flies 560 mi in 1.75 hr traveling with the wind. The return trip against the same wind takes the plane 2 hr. Find the speed of the plane and the speed of the wind.

Step 1 **Read** the problem several times.

Step 2 **Assign variables.**

Let $x =$ the speed of the plane,
and $y =$ the speed of the wind.

The speed (rate) of the plane *with* the wind is $x + y$ mph, and the speed (rate) of the plane *against* the wind is $x - y$ mph. See Figure 9.

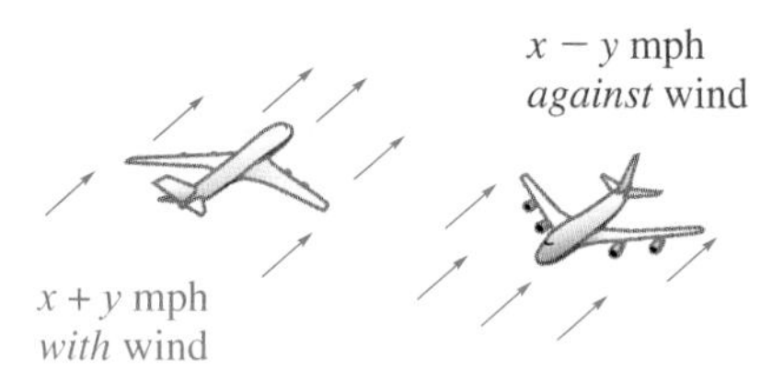

Figure 9

We use this information and the formula $d = rt$ (or $rt = d$) to complete a table.

	r	t	d
With Wind	$x + y$	1.75	560
Against Wind	$x - y$	2	560

Step 3 **Write two equations.** From the table,

$$1.75(x + y) = 560 \xrightarrow{\text{Divide by 1.75.}} x + y = 320 \quad (1)$$

$$2(x - y) = 560 \xrightarrow{\text{Divide by 2.}} x - y = 280. \quad (2)$$

Step 4 **Solve** the system of equations (1) and (2).

Work Problem 9 at the Side.

Step 5 **State the answer.** From Problem 9 at the side, the speed of the plane is 300 mph and the speed of the wind is 20 mph.

Step 6 **Check.** The answer seems reasonable, and true statements result when the values are substituted into the equations of the system.

Work Problem 10 at the Side.

ANSWERS
9. (300, 20)
10. current: 4 mph;
Ann's speed in still water: 6 mph

4.4 Exercises

FOR EXTRA HELP

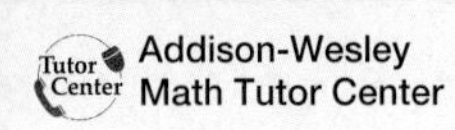

MathXL

MyMathLab

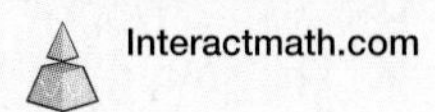

Choose the correct response in Exercises 1–7.

1. Which expression represents the monetary value of x 20-dollar bills?

A. $\frac{x}{20}$ dollars **B.** $\frac{20}{x}$ dollars **C.** $20 + x$ dollars **D.** $20x$ dollars

2. Which expression represents the cost of t pounds of candy that sells for \$1.95 per lb?

A. $\$1.95t$ **B.** $\frac{\$1.95}{t}$ **C.** $\frac{t}{\$1.95}$ **D.** $\$1.95 + t$

3. Which expression represents the amount of interest earned on d dollars at an interest rate of 2%?

A. $2d$ dollars **B.** $.02d$ dollars **C.** $.2d$ dollars **D.** $200d$ dollars

4. Suppose that x liters of a 40% acid solution are mixed with y liters of a 35% solution to obtain 100 L of a 38% solution. One equation in a system for solving this problem is $x + y = 100$. Which one of the following is the other equation?

A. $.35x + .40y = .38(100)$ **B.** $.40x + .35y = .38(100)$

C. $35x + 40y = 38$ **D.** $40x + 35y = .38(100)$

5. According to *Natural History* magazine, the speed of a cheetah is 70 mph. If a cheetah runs for x hours, how many miles does the cheetah cover?

A. $70 + x$ miles **B.** $70 - x$ miles **C.** $\frac{70}{x}$ miles **D.** $70x$ miles

6. What is the speed of a plane that travels at a rate of 560 mph *against* a wind of r mph?

A. $560 + r$ mph **B.** $\frac{560}{r}$ mph **C.** $560 - r$ mph **D.** $r - 560$ mph

7. What is the speed of a plane that travels at a rate of 560 mph *with* a wind of r mph?

A. $\frac{r}{560}$ mph **B.** $560 - r$ mph **C.** $560 + r$ mph **D.** $r - 560$ mph

8. Using the list of steps for solving an applied problem with two variables, write a short paragraph describing the general procedure you will use to solve the problems that follow in this exercise set.

Exercises 9 and 10 are good warm-up problems. In each case, refer to the six-step problem-solving method, fill in the blanks for Steps 2 and 3, and then complete the solution by applying Steps 4–6.

9. The sum of two numbers is 98 and the difference between them is 48. Find the two numbers.

Step 1 **Read** the problem carefully.

Step 2 **Assign variables.**

Let $x =$ the first number and let

$y =$ ________________.

Step 3 **Write two equations.**

First equation: $x + y = 98$

Second equation: __________

10. The sum of two numbers is 201 and the difference between them is 11. Find the two numbers.

Step 1 **Read** the problem carefully.

Step 2 **Assign variables.**

Let $x =$ the first number and let

$y =$ ________________.

Step 3 **Write two equations.**

First equation: $x + y = 201$

Second equation: __________

Write a system of equations for each problem, and then solve the problem.
See Example 1.

11. During 2002, two of the top-grossing concert tours were by the Rolling Stones and Cher. Together the two tours visited 117 cities. Cher visited 51 more cities than the Rolling Stones. How many cities did each tour visit? (*Source:* Pollstar.)

12. In 2003, the two top formats of U.S. commercial radio stations were country and news/talk. There were 288 fewer news/talk stations than country stations, and together they comprised a total of 3912 stations. How many stations of each format were there? (*Source:* M Street Corporation.)

13. The two top-grossing movies of 2003 were *The Lord of the Rings: The Return of the King* and *Finding Nemo. Finding Nemo* grossed \$21.4 million less than *The Lord of the Rings: The Return of the King*, and together the two films took in \$700.8 million. How much did each of these movies earn? (*Source:* Nielsen EDI.)

14. The life span of a \$100 bill is 80 months longer than the life span of a \$1 bill. If the combined life span of a \$1 bill and a \$100 bill is 124 months, what is the life span of each bill? (*Source:* Bureau of Engraving and Printing.)

15. The Terminal Tower in Cleveland, Ohio, is 242 ft shorter than the Key Tower, also in Cleveland. The total of the heights of the two buildings is 1658 ft. Find the heights of the buildings. (*Source: World Almanac and Book of Facts,* 2004.)

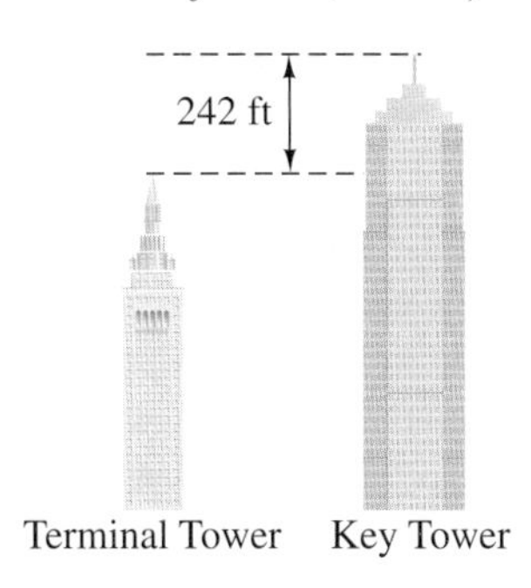

16. In 2003, a total of 943.8 thousand people lived in the cities of Las Vegas, Nevada, and Sacramento, California. Sacramento had 73.4 thousand fewer residents than Las Vegas. What was the population of each city? (*Source:* Bureau of the Census.)

If x units of a product cost C dollars to manufacture and earn revenue of R dollars, the value of x where the expressions for C and R are equal is called the ***break-even quantity,*** *the number of units that produce 0 profit. In Exercises 17 and 18,* ***(a)*** *find the break-even quantity, and* ***(b)*** *decide whether the product should be produced based on whether it will earn a profit. (Profit equals revenue minus cost.)*

17. $C = 85x + 900$; $R = 105x$; no more than 38 units can be sold.

18. $C = 105x + 6000$; $R = 255x$; no more than 400 units can be sold.

Write a system of equations for each problem, and then solve the system. See Example 2.

19. A motel clerk counts his \$1 and \$10 bills at the end of a day. He finds that he has a total of 74 bills having a combined monetary value of \$326. Find the number of bills of each denomination that he has.

Denomination of Bill	Number of Bills	Total Value
\$1	x	
\$10	y	
Totals	74	\$326

20. Letarsha is a bank teller. At the end of a day, she has a total of 69 \$5 and \$10 bills. The total value of the money is \$590. How many of each denomination does she have?

Denomination of Bill	Number of Bills	Total Value
\$5	x	\$5x
\$10	y	
Totals		

21. A newspaper advertised DVDs and CDs. Tracy Sudak went shopping and bought each of her seven nephews a gift, either a DVD of the movie *Miracle* or the latest Linkin Park CD. The DVD cost \$14.95 and the CD cost \$16.88, and she spent a total of \$114.30. How many DVDs and how many CDs did she buy?

22. Terry Wong saw the ad (see Exercise 21) and he, too, went shopping. He bought each of his five nieces a gift, either a DVD of *Home on the Range* or the CD soundtrack to *Scooby Doo 2: Monsters Unleashed.* The DVD cost \$14.99 and the soundtrack cost \$13.88, and he spent a total of \$70.51. How many DVDs and CDs did he buy?

23. Maria Lopez has twice as much money invested at 5% simple annual interest as she does at 4%. If her yearly income from these two investments is \$350, how much does she have invested at each rate?

24. Charles Miller invested his textbook royalty income in two accounts, one paying 3% annual simple interest and the other paying 2% interest. He earned a total of \$11 interest. If he invested three times as much in the 3% account as he did in the 2% account, how much did he invest at each rate?

25. Average movie ticket prices in the United States are, in general, lower than in other countries. It would cost \$77.87 to buy three tickets in Japan plus two tickets in Switzerland. Three tickets in Switzerland plus two tickets in Japan would cost \$73.83. How much does an average movie ticket cost in each of these countries? (*Source:* Business Traveler International.)

26. (See Exercise 25.) Four movie tickets in Germany plus three movie tickets in France would cost \$62.27. Three tickets in Germany plus four tickets in France would cost \$62.19. How much does an average movie ticket cost in each of these countries? (*Source:* Business Traveler International.)

Write a system of equations for each problem, and then solve the system. See Example 3.

27. A 40% dye solution is to be mixed with a 70% dye solution to get 120 L of a 50% solution. How many liters of the 40% and 70% solutions will be needed?

Percent (as a Decimal)	Liters of Solution	Liters of Pure Dye
.40	x	
.70	y	
.50	120	

28. A 90% antifreeze solution is to be mixed with a 75% solution to make 120 L of a 78% solution. How many liters of the 90% and 75% solutions will be used?

Percent (as a Decimal)	Liters of Solution	Liters of Pure Antifreeze
.90	x	
.75	y	
.78	120	

29. A merchant wishes to mix coffee worth \$6 per lb with coffee worth \$3 per lb to get 90 lb of a mixture worth \$4 per lb. How many pounds of the \$6 and the \$3 coffees will be needed?

Dollars per Pound	Pounds	Cost
6	x	
	y	
	90	

30. A grocer wishes to blend candy selling for \$1.20 per lb with candy selling for \$1.80 per lb to get a mixture that will be sold for \$1.40 per lb. How many pounds of the \$1.20 and the \$1.80 candies should be used to get 45 lb of the mixture?

Dollars per Pound	Pounds	Cost
	x	
1.80	y	
	45	

31. How many barrels of pickles worth \$40 per barrel and pickles worth \$60 per barrel must be mixed to obtain 50 barrels of a mixture worth \$48 per barrel?

32. The owner of a nursery wants to mix some fertilizer worth \$70 per bag with some worth \$90 per bag to obtain 40 bags of mixture worth \$77.50 per bag. How many bags of each type should she use?

Write a system of equations for each problem, and then solve the system. See Examples 4 and 5.

33. **RAGBRAI**®, the Des Moines **R**egister's **A**nnual **G**reat **B**icycle **R**ide **A**cross **I**owa, is the longest and oldest touring bicycle ride in the world. Suppose a cyclist began the 490 mi ride on July 25, 2004 in western Iowa at the same time that a car traveling toward it left eastern Iowa. If the bicycle and the car met after 7 hr and the car traveled 40 mph faster than the bicycle, find the average speed of each. (*Source:* www.ragbrai.org)

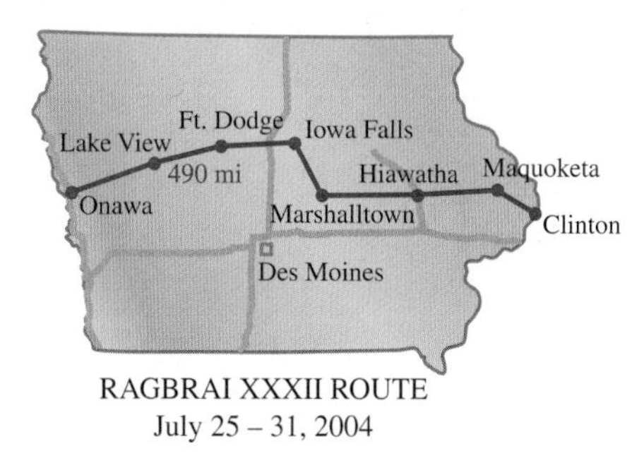

RAGBRAI XXXII ROUTE
July 25 – 31, 2004

34. In 2002, Atlanta's Hartsfield Airport was the nation's busiest. Suppose two planes leave the airport at the same time, one traveling east and the other traveling west. If the planes are 2100 mi apart after 2 hr and one plane travels 50 mph faster than the other, find the speed of each plane. (*Source:* Airports Council International—North America.)

35. Toledo and Cincinnati are 200 mi apart. A car leaves Toledo traveling toward Cincinnati, and another car leaves Cincinnati at the same time, traveling toward Toledo. The car leaving Toledo averages 15 mph faster than the other, and they meet after 1 hr and 36 min. What are the rates of the cars?

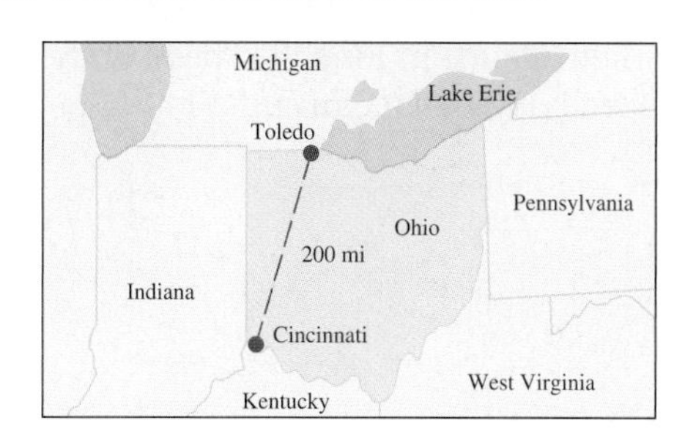

36. Kansas City and Denver are 600 mi apart. Two cars start from these cities, traveling toward each other. They meet after 6 hr. Find the rate of each car if one travels 30 mph slower than the other.

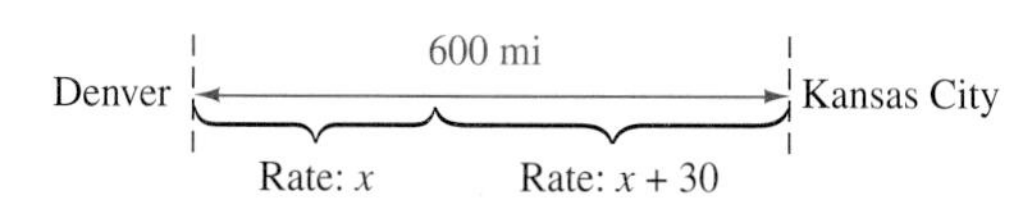

37. At the beginning of a bicycle ride for charity, Roberto and Juana are 30 mi apart. If they leave at the same time and ride in the same direction, Roberto overtakes Juana in 6 hr. If they ride toward each other, they meet in 1 hr. What are their speeds?

38. Mr. Abbot left Farmersville in a plane at noon to travel to Exeter. Mr. Baker left Exeter in his automobile at 2 P.M. to travel to Farmersville. It is 400 mi from Exeter to Farmersville. If the sum of their speeds was 120 mph, and if they crossed paths at 4 P.M., find the speed of each.

39. A boat takes 3 hr to go 24 mi upstream. It can go 36 mi downstream in the same time. Find the speed of the current and the speed of the boat in still water if $x =$ the speed of the boat in still water and $y =$ the speed of the current.

	d	r	t
Downstream	36	$x + y$	
Upstream	24	$x - y$	

40. It takes a boat $1\frac{1}{2}$ hr to go 12 mi downstream, and 6 hr to return. Find the speed of the boat in still water and the speed of the current. Let $x =$ the speed of the boat in still water and $y =$ the speed of the current.

	d	r	t
Downstream	12	$x + y$	$\frac{3}{2}$
Upstream			6

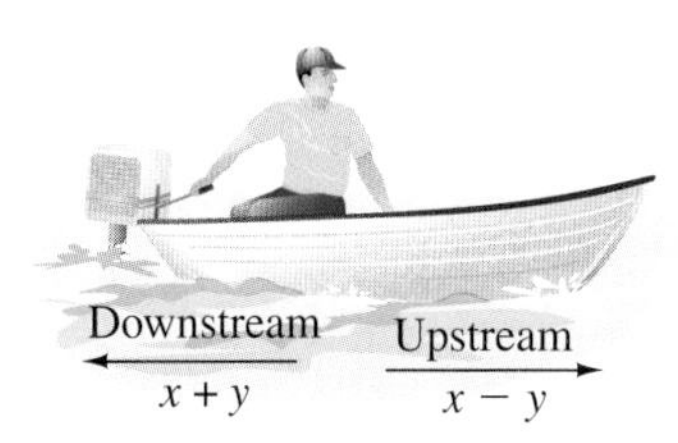

41. If a plane can travel 440 mph against the wind and 500 mph with the wind, find the speed of the wind and the speed of the plane in still air.

440 mph
against wind

500 mph
with wind

42. A small plane travels 200 mph with the wind and 120 mph against it. Find the speed of the wind and the speed of the plane in still air.

4.5 Solving Systems of Linear Inequalities

OBJECTIVE

1 **Solve systems of linear inequalities by graphing.**

We graphed the solutions of a linear inequality in **Section 3.5.** Recall that to graph the solutions of $x + 3y > 12$, for example, we first graph $x + 3y = 12$ by finding and plotting a few ordered pairs that satisfy the equation. Because the points on the line do *not* satisfy the inequality, we use a dashed line. To decide which side of the line includes the points that are solutions, we choose a test point not on the line, such as (0, 0). Substituting these values for x and y in the inequality gives

$$x + 3y > 12$$
$$0 + 3(0) > 12$$
$$0 > 12,$$

a false result. This indicates that the solutions are those points on the side of the line that does not include (0, 0), as shown in Figure 10.

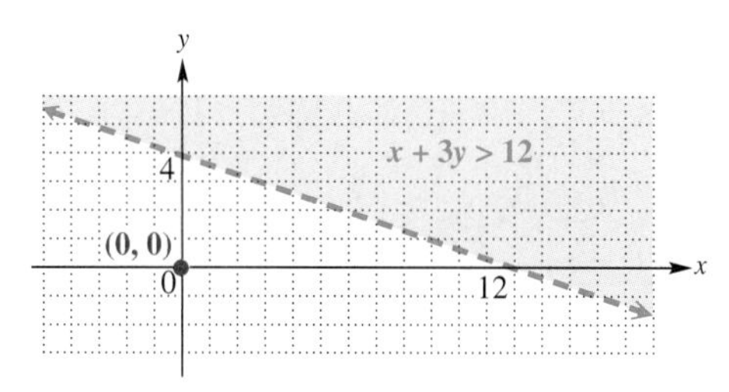

Figure 10

Now we use the same techniques to solve systems of linear inequalities.

OBJECTIVE 1 **Solve systems of linear inequalities by graphing.** A **system of linear inequalities** consists of two or more linear inequalities. The **solution of a system of linear inequalities** includes all points that make all inequalities of the system true at the same time. To solve a system of linear inequalities, use the following steps.

Solving a System of Linear Inequalities

Step 1 **Graph the inequalities.** Graph each inequality using the method of **Section 3.5.**

Step 2 **Choose the intersection.** Indicate the solution of the system by shading the intersection of the graphs (the region where the graphs overlap).

EXAMPLE 1 **Solving a System of Two Linear Inequalities**

Graph the solution of the system

$$3x + 2y \leq 6$$
$$2x - 5y \geq 10.$$

To graph $3x + 2y \leq 6$, graph the solid boundary line $3x + 2y = 6$ and shade the region containing (0, 0), as shown in Figure 11(a) on the next page. Then graph $2x - 5y \geq 10$ with the solid boundary line $2x - 5y = 10$. The test point (0, 0) makes this inequality false, so shade the region on the other side of the boundary line. See Figure 11(b).

Continued on Next Page

1 Graph the solution of the system

$$x - 2y \leq 8$$
$$3x + y \geq 6.$$

To get you started, the graphs of $x - 2y = 8$ and $3x + y = 6$ are shown.

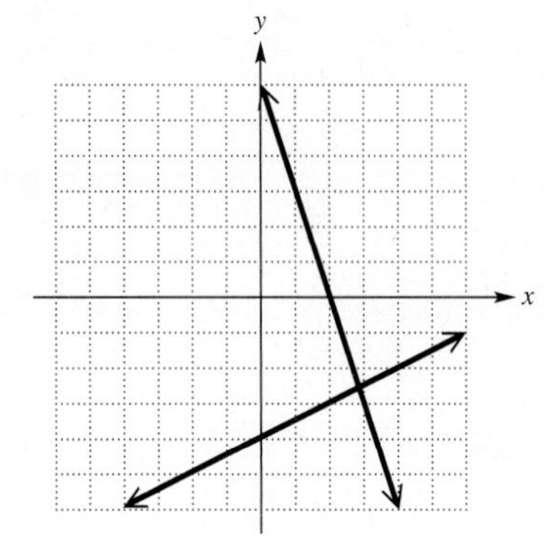

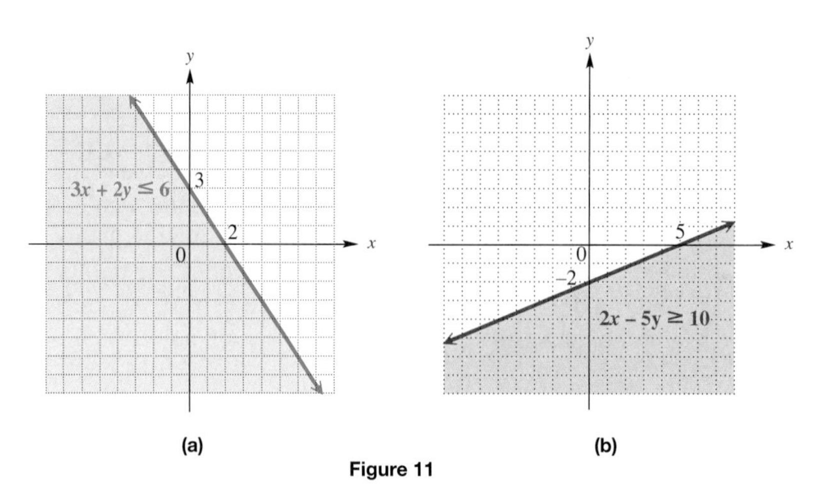

(a) (b)

Figure 11

The solution of this system includes all points in the intersection (overlap) of the graphs of the two inequalities. It includes the shaded region and portions of the two boundary lines shown in Figure 12.

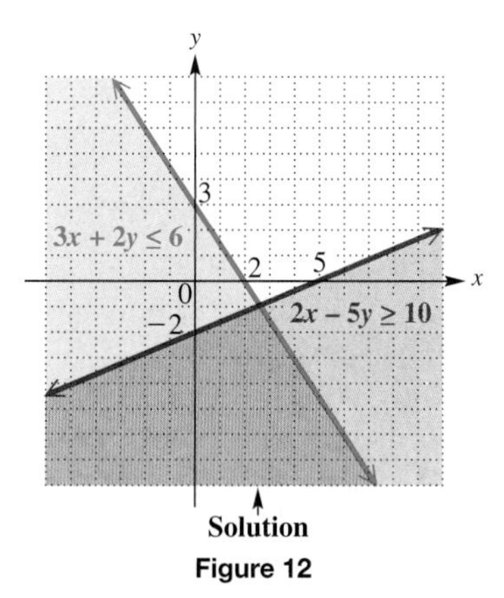

Figure 12

Work Problem 1 at the Side.

NOTE
We usually do all the work on one set of axes. In the following examples, only one graph is shown. Be sure that the region of the final solution is clearly indicated.

EXAMPLE 2 Solving a System of Two Linear Inequalities

Graph the solution of the system

$$x - y > 5$$
$$2x + y < 2.$$

Figure 13 shows the graphs of both $x - y > 5$ and $2x + y < 2$. Dashed lines show that the graphs of the inequalities do not include their boundary lines. The solution of the system is the region with the darkest shading. The solution does not include either boundary line.

Continued on Next Page

ANSWERS

1.

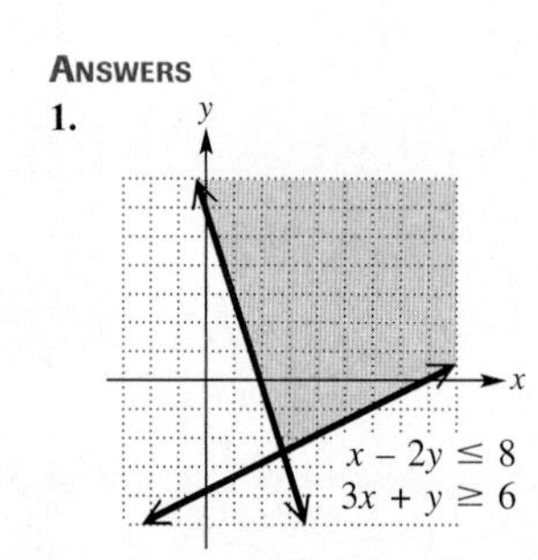

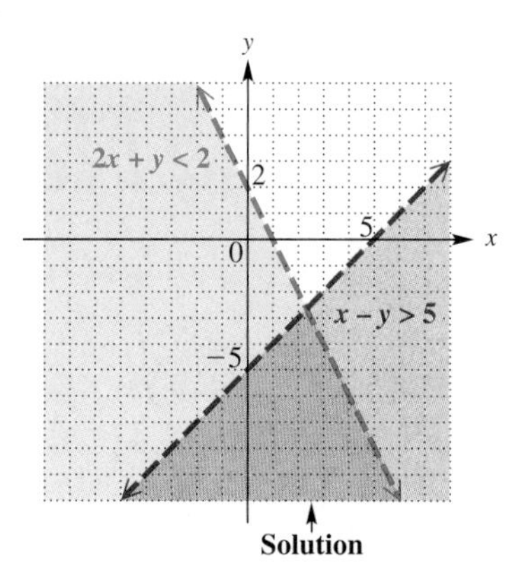

Figure 13

2 Graph the solution of each system.

(a) $x + 2y < 0$
$3x - 4y < 12$

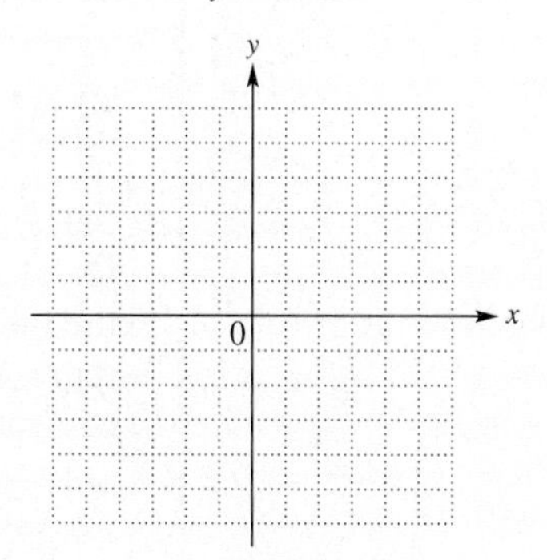

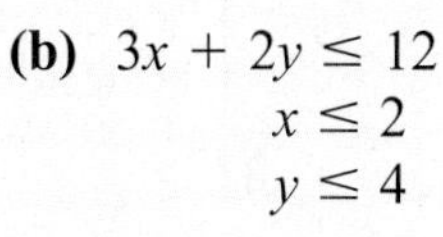

(b) $3x + 2y \leq 12$
$x \leq 2$
$y \leq 4$

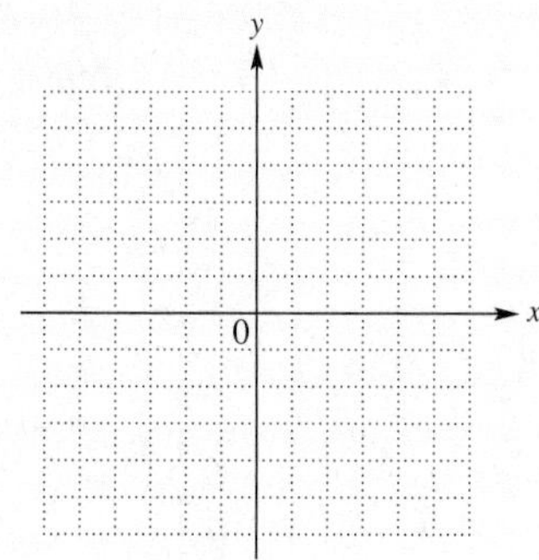

EXAMPLE 3 Solving a System of Three Linear Inequalities

Graph the solution of the system

$$4x - 3y \leq 8$$
$$x \geq 2$$
$$y \leq 4.$$

Recall that $x = 2$ is a vertical line through the point (2, 0), and $y = 4$ is a horizontal line through (0, 4). The graph of the solution is the shaded region in Figure 14, including all boundary lines.

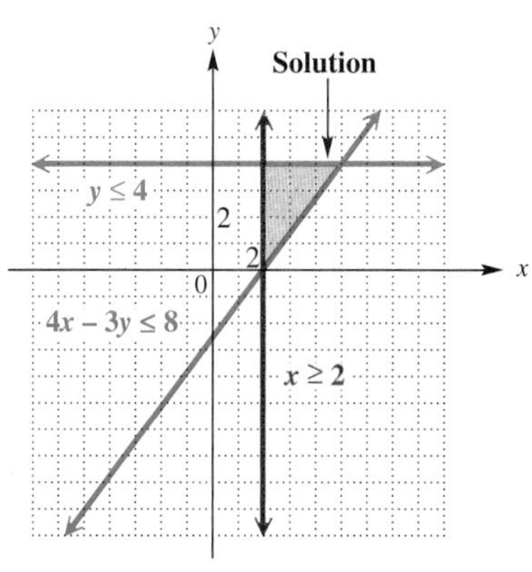

Figure 14

Work Problem 2 at the Side.

Answers

2. (a)

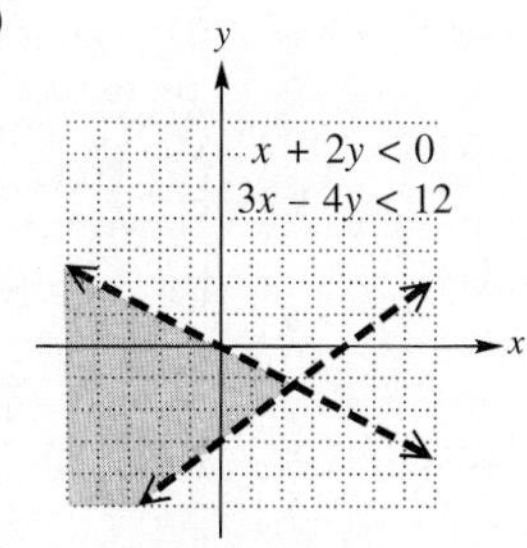

(b)

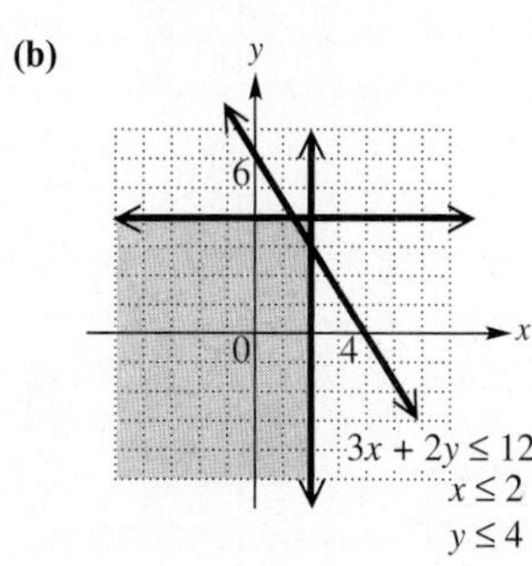

Focus on

Real-Data Applications

Sales of Compact Discs versus Cassettes

Sales of compact discs (CDs) and cassettes for the years 1990 through 2000 are given in the table. The number of years since 1990 is represented by x. Sales of compact discs and cassettes can be modeled by linear equations.

- A linear model for the sales of compact discs is

$$y = 74.523x + 289.3.$$

- A linear model for the sales of cassettes is

$$y = -34.107x + 434.49.$$

y is sales in millions. The actual data and the linear models are graphed below.

Year	x	CD Sales (in millions)	Cassette Sales (in millions)
1990	0	286.5	442.2
1991	1	333.4	360.1
1992	2	407.5	366.4
1993	3	495.4	339.5
1994	4	662.1	345.4
1995	5	722.9	272.6
1996	6	778.9	225.3
1997	7	753.1	172.6
1998	8	847.0	158.5
1999	9	938.9	123.6
2000	10	942.5	76.0

Source: World Almanac and Book of Facts, 2004.

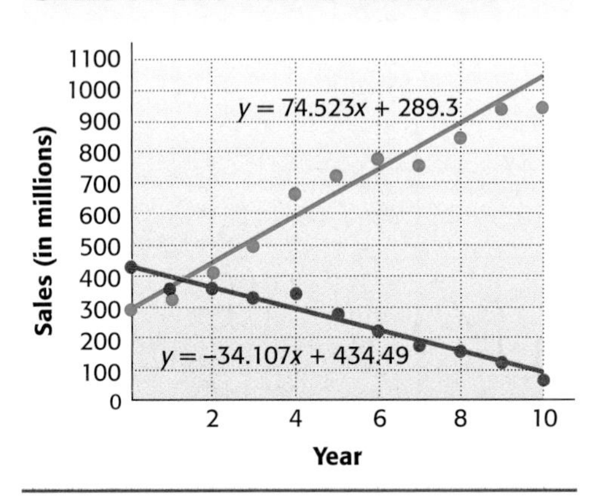

For Group Discussion

1. Shade the region on the graph that corresponds to the solution of the following system.

$$y \le 74.523x + 289.3$$
$$y \ge -34.107x + 434.49$$

Interpret the solution in the context of sales of compact discs and cassettes.

2. Solve the linear inequality $74.523x + 289.3 > -34.107x + 434.49$. Round your answer to the nearest hundredth.

3. What does the solution to the inequality in Problem 2 represent in the context of sales of CDs and cassettes?

4.5 Exercises

FOR EXTRA HELP

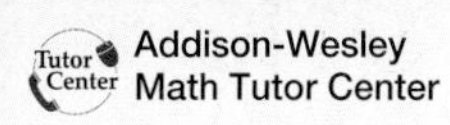

MyMathLab

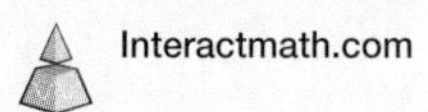

Match each system of inequalities with the correct graph from choices A–D.

1. $x \geq 5$
$y \leq -3$

2. $x \leq 5$
$y \geq -3$

3. $x > 5$
$y < -3$

4. $x < 5$
$y > -3$

A.

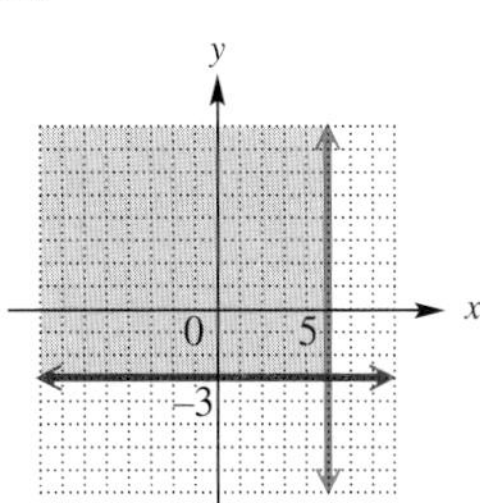

B.

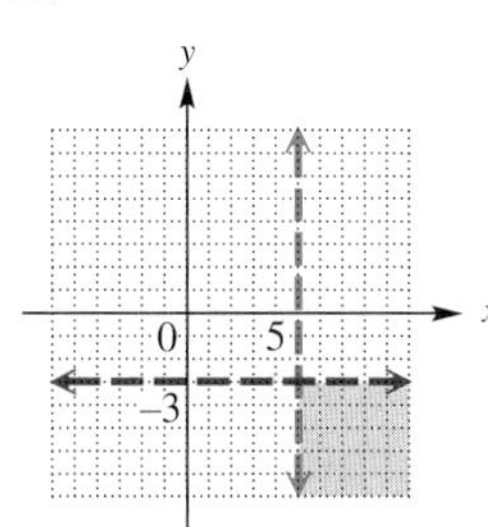

C.

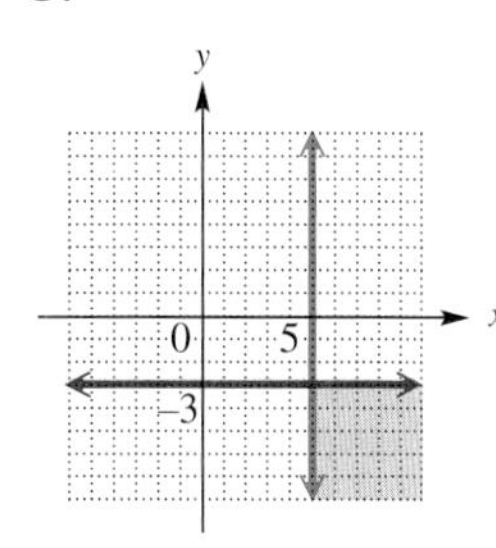

D.

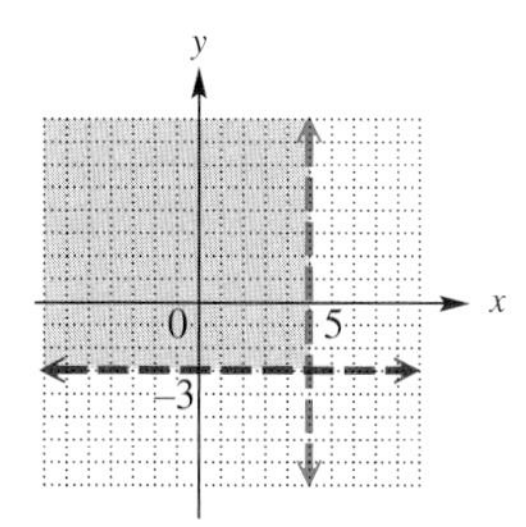

Graph the solution of each system of linear inequalities. See Examples 1–3.

5. $x + y \leq 6$
$x - y \geq 1$

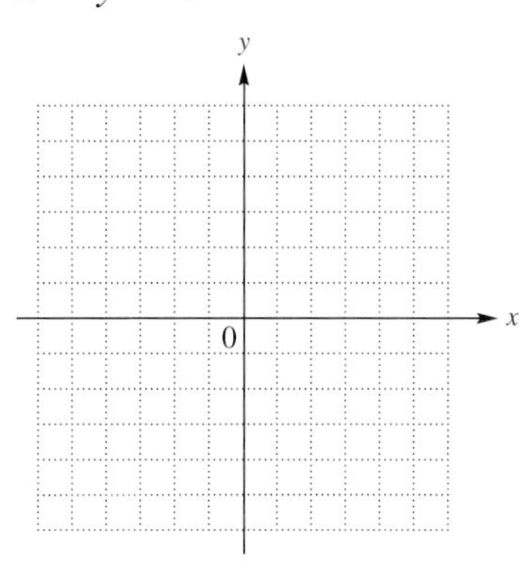

6. $x + y \leq 2$
$x - y \geq 3$

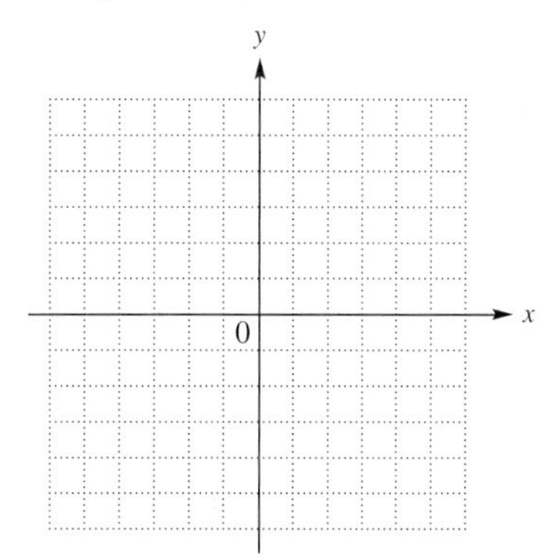

7. $4x + 5y \geq 20$
$x - 2y \leq 5$

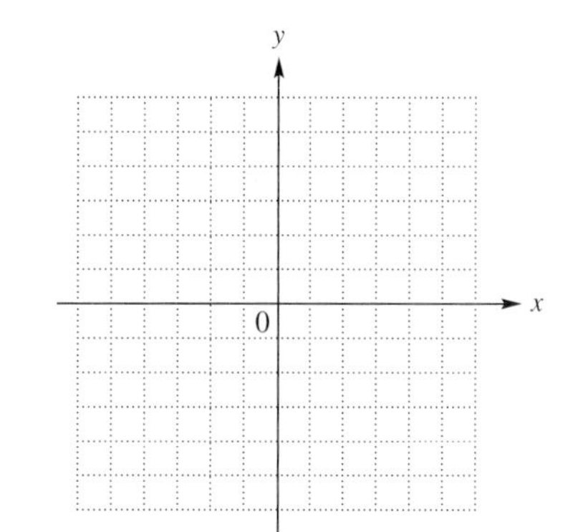

8. $x + 4y \leq 8$
$2x - y \geq 4$

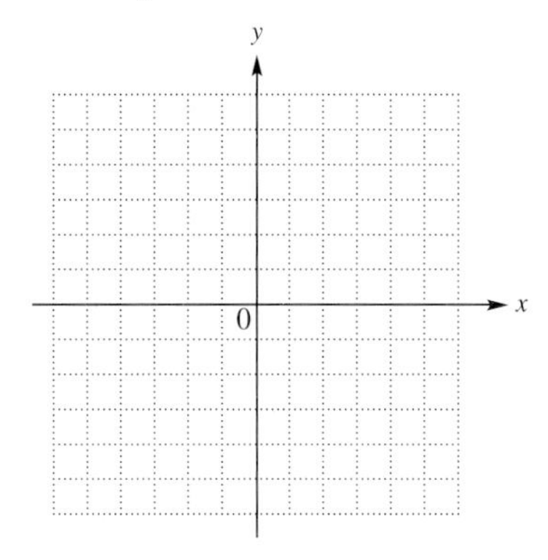

9. $2x + 3y < 6$
$x - y < 5$

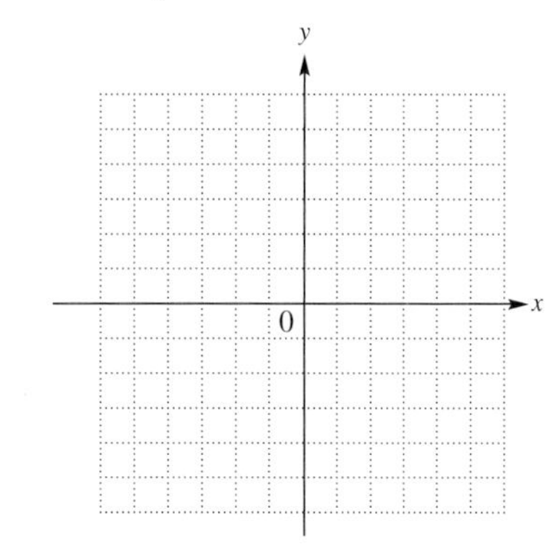

10. $x + 2y < 4$
$x - y < -1$

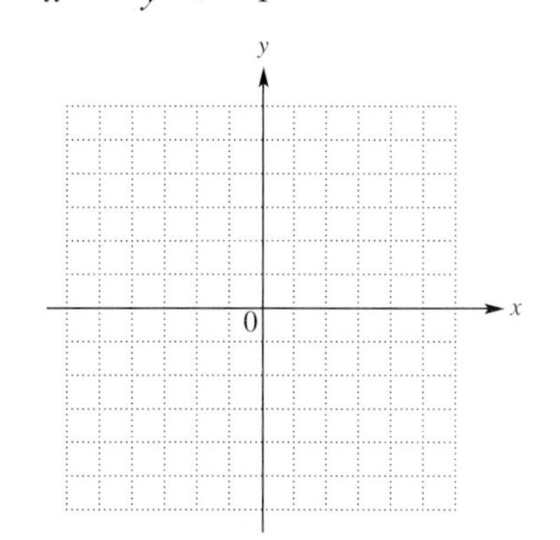

11. $y \leq 2x - 5$
$x < 3y + 2$

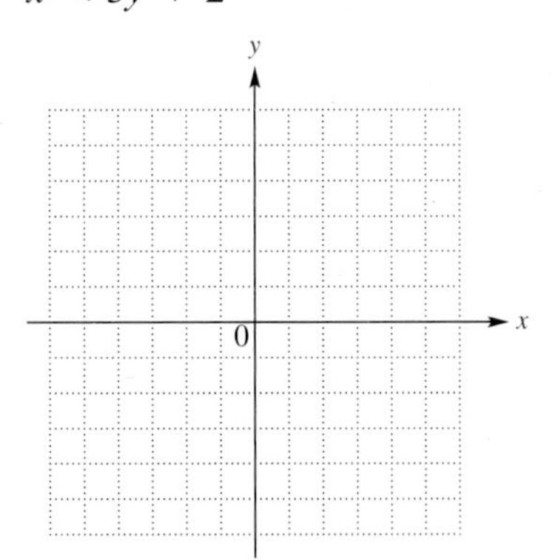

12. $x \geq 2y + 6$
$y > -2x + 4$

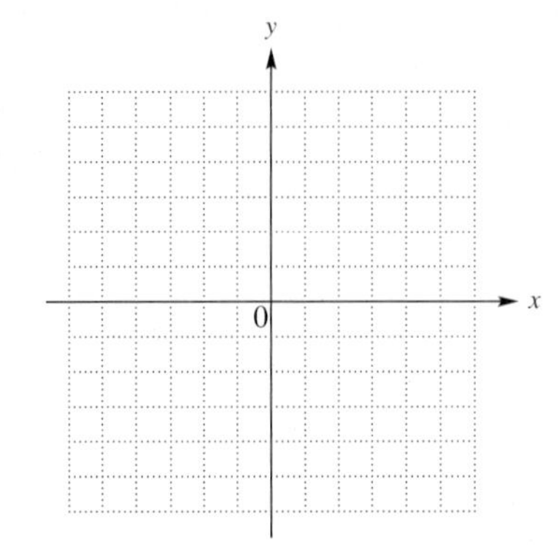

13. $4x + 3y < 6$
$x - 2y > 4$

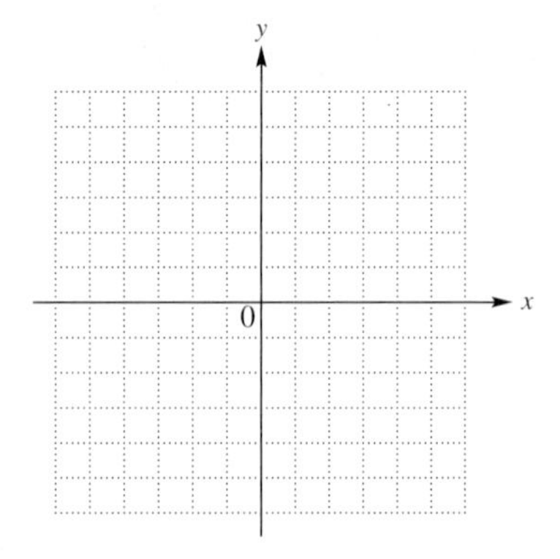

14. $3x + y > 4$
$x + 2y < 2$

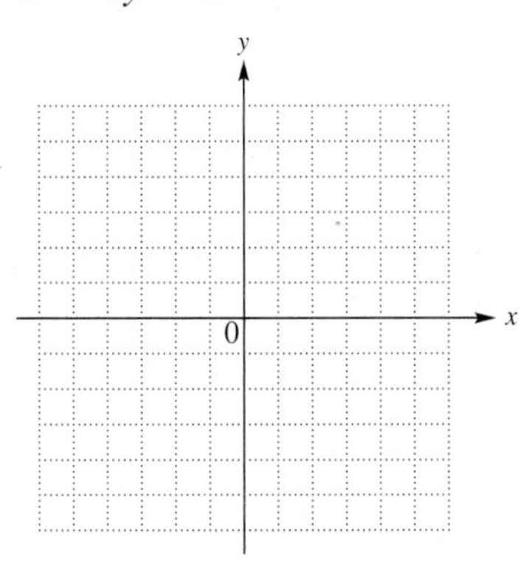

15. $x \leq 2y + 3$
$x + y < 0$

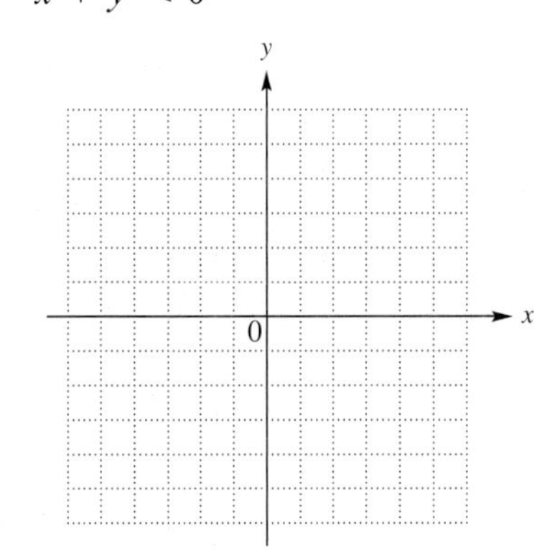

16. $x \leq 4y + 3$
$x + y > 0$

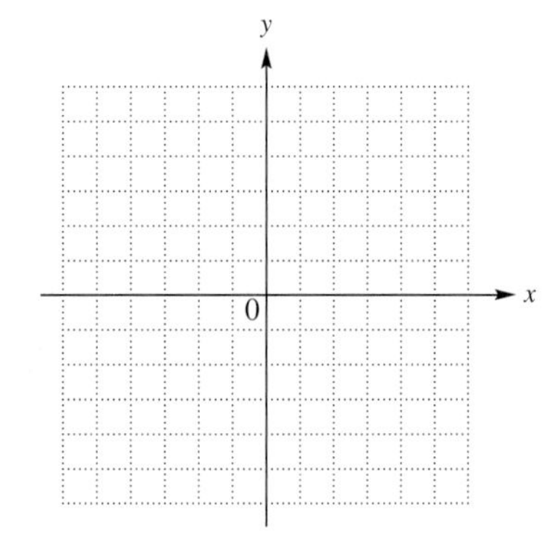

17. $4x + 5y < 8$
$y > -2$
$x > -4$

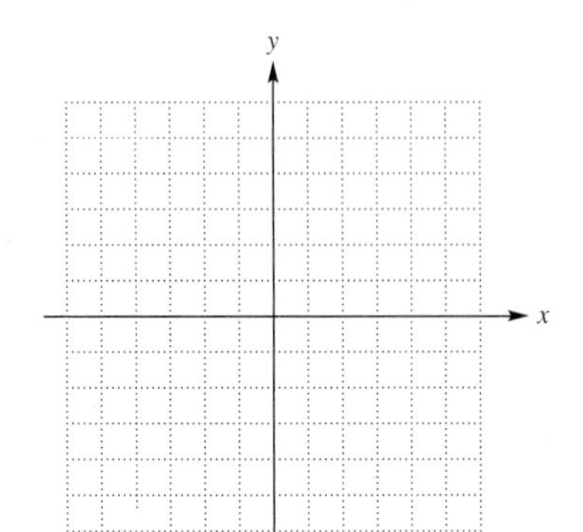

18. $x + y \geq -3$
$x - y \leq 3$
$y \leq 3$

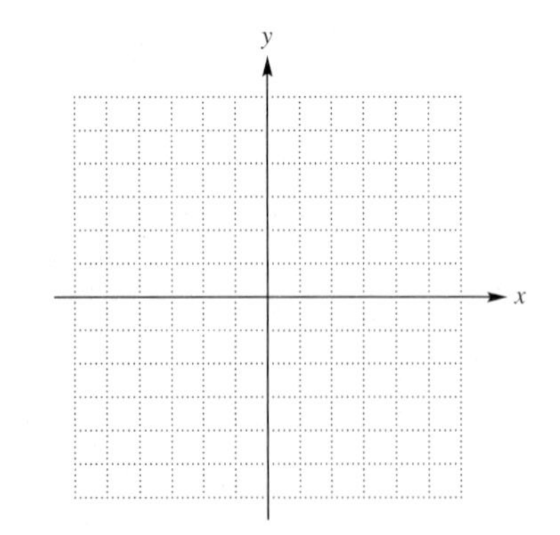

19. $3x - 2y \geq 6$
$x + y \leq 4$
$x \geq 0$
$y \geq -4$

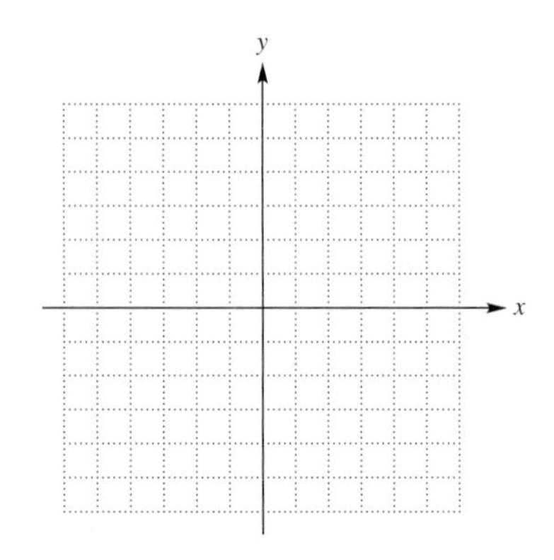

20. Every system of inequalities illustrated in the examples of this section has infinitely many solutions. Explain why this is so. Does this mean that *any* ordered pair is a solution?

Chapter 4

SUMMARY

KEY TERMS

4.1	**system of linear equations**	A system of linear equations (or **linear system**) consists of two or more linear equations with the same variables.
	solution of a system	The solution of a system of linear equations includes all the ordered pairs that make all the equations of the system true at the same time.
	consistent system	A system of equations with at least one solution is a consistent system.
	inconsistent system	An inconsistent system of equations is a system with no solution.
	independent equations	Equations of a system that have different graphs are called independent equations.
	dependent equations	Equations of a system that have the same graph (because they are different forms of the same equation) are called dependent equations.
4.5	**system of linear inequalities**	A system of linear inequalities contains two or more linear inequalities (and no other kinds of inequalities).
	solution of a system of linear inequalities	The solution of a system of linear inequalities includes all points that make all inequalities of the system true at the same time.

TEST YOUR WORD POWER

See how well you have learned the vocabulary in this chapter. Answers, with examples, follow the Quick Review.

1. A **system of linear equations** consists of
- **A.** at least two linear equations with different variables
- **B.** two or more linear equations that have an infinite number of solutions
- **C.** two or more linear equations with the same variables
- **D.** two or more linear inequalities.

2. A **solution of a system** of linear equations is
- **A.** an ordered pair that makes one equation of the system true
- **B.** an ordered pair that makes all the equations of the system true at the same time
- **C.** any ordered pair that makes one or the other or both equations of the system true
- **D.** the set of values that make all the equations of the system false.

3. A **consistent system** is a system of equations
- **A.** with one solution
- **B.** with no solution
- **C.** with two solutions
- **D.** that has parallel lines at its graph.

4. An **inconsistent system** is a system of equations
- **A.** with one solution
- **B.** with no solution
- **C.** with an infinite number of solutions
- **D.** that have the same graph.

5. Dependent equations
- **A.** have different graphs
- **B.** have no solution
- **C.** have one solution
- **D.** are different forms of the same equation.

QUICK REVIEW

Concepts	*Examples*
4.1 *Solving Systems of Linear Equations by Graphing* An ordered pair is a solution of a system if it makes all equations of the system true at the same time.	Is $(4, -1)$ a solution of the system $\begin{aligned} x + y &= 3 \\ 2x - y &= 9 \end{aligned}$? Yes, because $4 + (-1) = 3$ and $2(4) - (-1) = 9$ are both true.
If the graphs of the equations of a system are both sketched on the same axes, then the points of intersection, if any, are solutions of the system. If the graphs of the equations do not intersect (that is, the lines are parallel), then the system has no solution. If the graphs of the equations are the same line, then the system has an infinite number of solutions.	Solve by graphing. $x + y = 5$ $2x - y = 4$ y, x, 5, (3, 2), $x + y = 5$, 0, 2, 5, $2x - y = 4$, −4 The ordered pair (3, 2) satisfies both equations, so (3, 2) is the solution of the system.
4.2 *Solving Systems of Linear Equations by Substitution* *Step 1* Solve one equation for either variable.	Solve by substitution. $x + 2y = -5$ (1) $y = -2x - 1$ (2)
Step 2 Substitute for that variable in the other equation to get an equation in one variable.	Equation (2) is already solved for y. Substitute $-2x - 1$ for y in equation (1).
Step 3 Solve the equation from Step 2.	$x + 2(-2x - 1) = -5$ $x - 4x - 2 = -5$ $-3x - 2 = -5$ $-3x = -3$ $x = 1$
Step 4 Substitute the result into the equation from Step 1 to get the value of the other variable.	To find y, let $x = 1$ in equation (2): $y = -2(1) - 1 = -3.$
Step 5 Check. Write the solution as an ordered pair.	The solution $(1, -3)$ checks.
4.3 *Solving Systems of Linear Equations by Elimination* *Step 1* Write both equations in standard form $Ax + By = C$.	Solve by elimination. $x + 3y = 7$ (1) $3x - y = 1$ (2)
Step 2 If necessary, multiply one or both equations by appropriate numbers so that the sum of the coefficients of either the x- or y-terms is 0.	Multiply equation (1) by -3 to eliminate the x-terms.
Step 3 Add the equations to get an equation with only one variable (or no variable).	$-3x - 9y = -21$ $3x - y = 1$ $-10y = -20$ Add.
Step 4 Solve the equation from Step 3.	$y = 2$ Divide by -10.

Concepts	*Examples*
4.3 *Solving Systems of Linear Equations by Elimination (continued)*	
Step 5 Substitute the solution from Step 4 into either of the original equations to find the value of the remaining variable.	Substitute to get the value of x. $x + 3(2) = 7$ (1) $x + 6 = 7$ $x = 1$
Step 6 Check. Write the solution as an ordered pair.	Since $1 + 3(2) = 7$ and $3(1) - 2 = 1$, the solution $(1, 2)$ checks.
4.4 *Applications of Linear Systems* Use the modified six-step method.	
Step 1 **Read** the problem carefully.	The sum of two numbers is 30. Their difference is 6. Find the numbers.
Step 2 **Assign variables** for each unknown value. Use diagrams or tables as needed.	Let x represent one number. Let y represent the other number.
Step 3 **Write two equations** using both variables.	$x + y = 30$ $x - y = 6$
Step 4 **Solve** the system.	$2x = 36$ Add. $x = 18$ Divide by 2.
Step 5 **State the answer.**	Let $x = 18$ in the first equation: $18 + y = 30$. Solve to get $y = 12$. The numbers are 18 and 12.
Step 6 **Check** the answer in the words of the original problem.	The sum of 18 and 12 is 30, and the difference between 18 and 12 is 6, so the answer checks.
4.5 *Solving Systems of Linear Inequalities* To solve a system of two or more linear inequalities, graph the inequalities on the same axes. (This was explained in **Section 3.5.**) The solution of the system is the intersection (overlap) of the regions of the graphs. The portions of the boundary lines that bound the region of solutions are included for a $\leq$ or $\geq$ inequality and excluded for a $<$ or $>$ inequality.	The shaded region is the solution of the system $2x + 4y \geq 5$ $x \geq 1$. 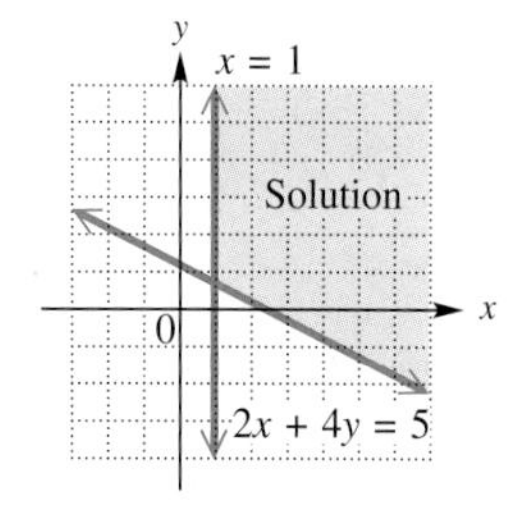

Answers to Test Your Word Power

1. C; *Example:* $2x + y = 7, \quad 3x - y = 3$
2. B; *Example:* The ordered pair (2, 3) satisfies both equations of the system in the Answer 1 example, so it is a solution of the system.
3. A; *Example:* The system in the Answer 1 example is consistent. The graphs of the equations intersect at exactly one point, in this case the solution (2, 3).
4. B; *Example:* The equations of two parallel lines make up an inconsistent system; their graphs never intersect, so there is no solution to the system.
5. D; *Example:* The equations $4x - y = 8$ and $8x - 2y = 16$ are dependent because their graphs are the same line.

Focus on

Real-Data Applications

Systems of Linear Equations and Modeling

A system of linear equations is an efficient tool for finding a linear *model,* or the equation for data that is known to be linear. Recall that the slope-intercept form of the equation of a line is $y = mx + b$. Once we know the values of the slope, m, and the y-intercept, b, we can then write the exact model. If we know two ordered pairs, then we can write a system of linear equations in which the unknown quantities are m and b.

For example, to find the formula to convert from Kelvin (K), the most commonly used thermodynamic temperature scale, as the *input* to degrees Fahrenheit (°F) as the *output*, we only need to know the data presented in the table. The linear equation has the form $F = mK + b$, and ordered pairs have the format (K, F).

	K	°F
Water Freezes	273	32
Water Boils	373	212

The system of linear equations to be solved is

$$32 = m(273) + b$$
$$212 = m(373) + b.$$

For Group Discussion

Use systems of linear equations to find the model for each problem.

1. Solve the system of linear equations given above to find the conversion formula from K to °F.

2. Water freezes at 0°C and boils at 100°C. Write the system of linear equations to find the model for converting from degrees Celsius as the input to degrees Fahrenheit as the output.

3. Suppose you begin a carefully managed weight-loss program in which you expect your weight to decline steadily. (A constant weight loss means that it is reasonable to assume that the relationship between number of weeks on the program and weight is linear.) After two weeks you weigh 179 lb, and after six weeks you weigh 169 lb. Use x to represent the number of weeks on the program and y to represent your weight in pounds at the end of x weeks. Write the linear model for your weight-loss program.

Extension: Suppose a line contains the distinct points (x_1, y_1) and (x_2, y_2), where $x_1 \neq x_2$. Use the method of this activity to derive the slope formula.

Chapter 4

REVIEW EXERCISES

[4.1] *Decide whether the given ordered pair is a solution of the given system.*

1. $(3, 4)$
$4x - 2y = 4$
$5x + y = 19$

2. $(-5, 2)$
$x - 4y = -13$
$2x + 3y = 4$

Solve each system by graphing.

3. $x + y = 4$
$2x - y = 5$

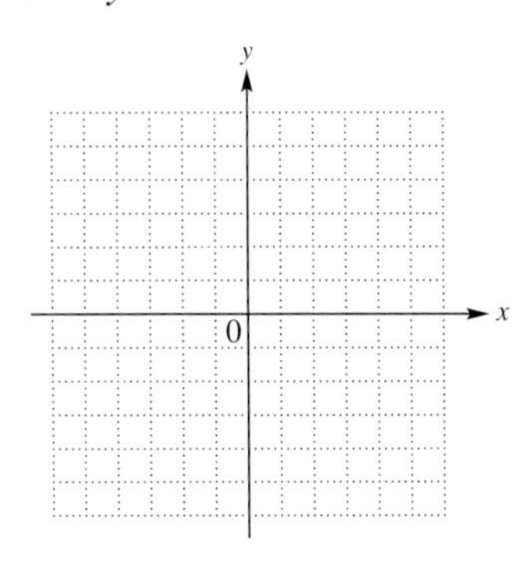

4. $x - 2y = 4$
$2x + y = -2$

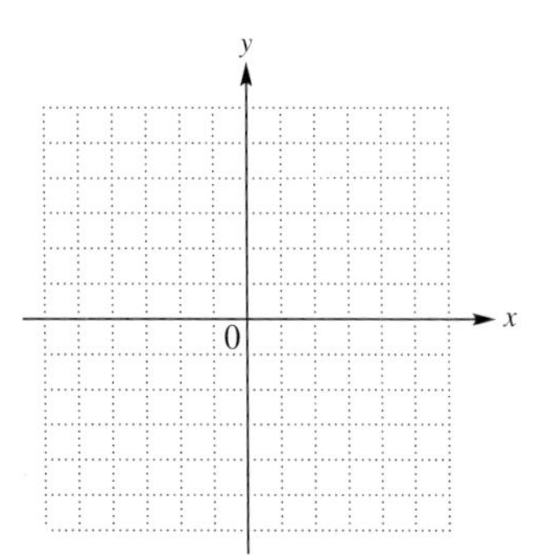

5. $x - 2 = 2y$
$2x - 4y = 4$

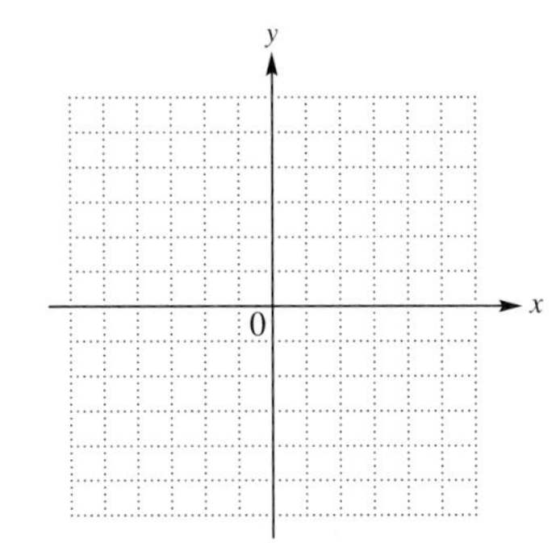

6. $2x + 4 = 2y$
$y - x = -3$

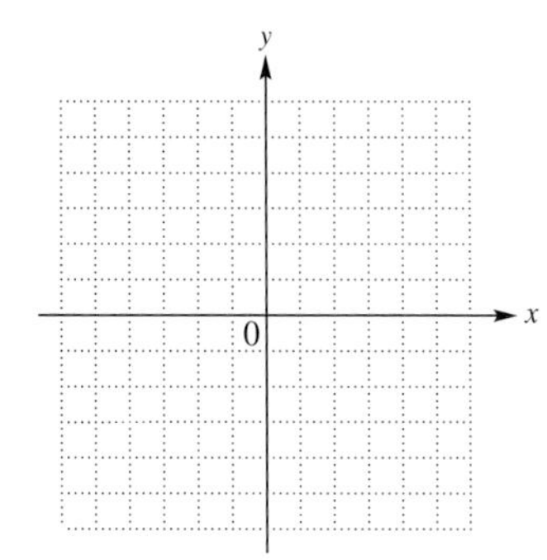

7. When a student was asked to determine whether the ordered pair $(1, -2)$ is a solution of the system

$$x + y = -1$$
$$2x + y = 4,$$

he answered "yes." His reasoning was that the ordered pair satisfies the equation $x + y = -1$; that is, $1 + (-2) = -1$ is true. Why is the student's answer wrong?

[4.2] *Solve each system by the substitution method.*

8. $3x + y = 7$
$x = 2y$

9. $2x - 5y = -19$
$y = x + 2$

10. $4x + 5y = 44$
$x + 2 = 2y$

11. $5x + 15y = 3$
$x + 3y = 2$

[4.3] *Solve each system by the elimination method.*

12. $2x - y = 13$
$x + y = 8$

13. $3x - y = -13$
$x - 2y = -1$

14. $-4x + 3y = 25$
$6x - 5y = -39$

15. $3x - 4y = 9$
$6x - 8y = 18$

16. For the system

$$2x + 12y = 7$$
$$3x + 4y = 1,$$

if we were to multiply the first (top) equation by -3, by what number would we have to multiply the second (bottom) equation in order to

(a) eliminate the x-terms when solving by the elimination method?

(b) eliminate the y-terms when solving by the elimination method?

Solve each system by any method.

17. $x - 2y = 5$
$y = x - 7$

18. $5x - 3y = 11$
$2y = x - 4$

19. $\frac{x}{2} + \frac{y}{3} = 7$
$\frac{x}{4} + \frac{2y}{3} = 8$

20. $\frac{3x}{4} - \frac{y}{3} = \frac{7}{6}$
$\frac{x}{2} + \frac{2y}{3} = \frac{5}{3}$

[4.4] *Solve each problem by using a system of equations.*

21. At the end of 2001, Subway topped McDonald's as the largest restaurant chain in the United States. Subway operated 148 more restaurants than McDonald's, and together the two chains had 26,346 restaurants. How many restaurants did each company operate? (*Source: USA Today,* February 4, 2002.)

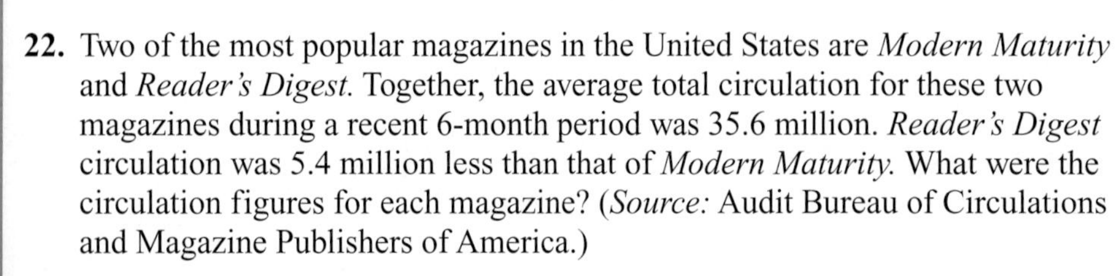

22. Two of the most popular magazines in the United States are *Modern Maturity* and *Reader's Digest.* Together, the average total circulation for these two magazines during a recent 6-month period was 35.6 million. *Reader's Digest* circulation was 5.4 million less than that of *Modern Maturity.* What were the circulation figures for each magazine? (*Source:* Audit Bureau of Circulations and Magazine Publishers of America.)

23. The perimeter of a rectangle is 90 m. Its length is $1\frac{1}{2}$ times its width. Find the length and width of the rectangle.

24. A cashier has 20 bills, all of which are \$10 or \$20 bills. The total value of the money is \$330. How many of each type does the cashier have?

Denomination of Bill	Number of Bills	Total Value
\$10	x	$10x$
\$20		
Totals		\$330

25. Candy that sells for \$1.30 per lb is to be mixed with candy selling for \$.90 per lb to get 100 lb of a mix that will sell for \$1 per lb. How much of each type should be used?

26. A certain plane flying with the wind travels 540 mi in 2 hr. Later, flying against the same wind, the plane travels 690 mi in 3 hr. Find the speed of the plane in still air and the speed of the wind.

27. After taxes, Ms. Cesar's game show winnings were \$18,000. She invested part of it at 3% annual simple interest and the rest at 4%. Her interest income for the first year was \$650. How much did she invest at each rate?

Percent	Amount of Principal	Interest
.03	x	
.04	y	
Totals	\$18,000	

28. A 40% antifreeze solution is to be mixed with a 70% solution to get 90 L of a 50% solution. How many liters of the 40% and 70% solutions will be needed?

Percent	Number of Liters	Amount of Pure Antifreeze
.40	x	
.70	y	
.50	90	

[4.5] *Graph the solution for each system of linear inequalities.*

29. $x + y \geq 2$
$x - y \leq 4$

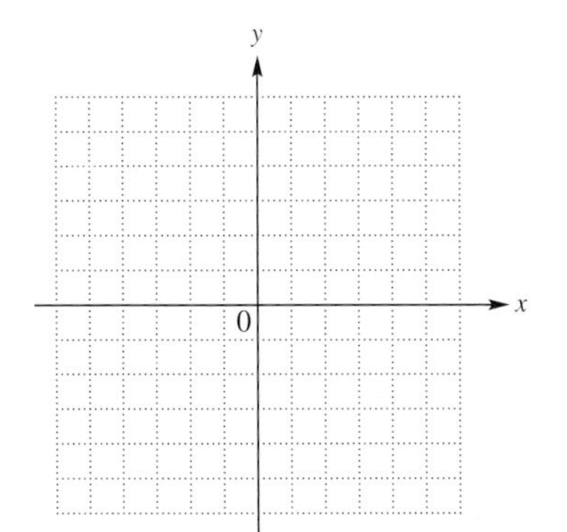

30. $y \geq 2x$
$2x + 3y \leq 6$

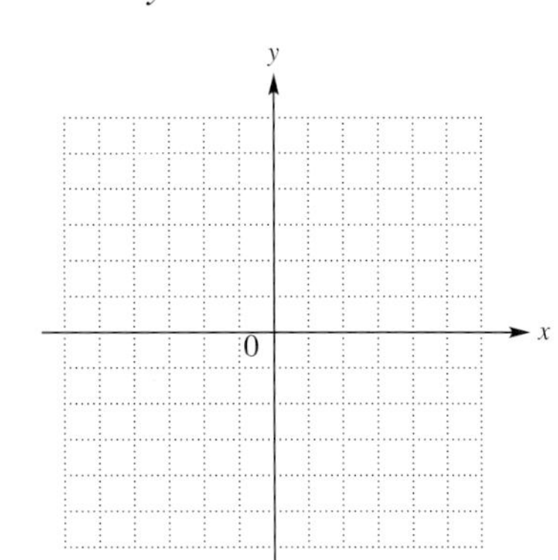

31. $x + y < 3$
$2x > y$

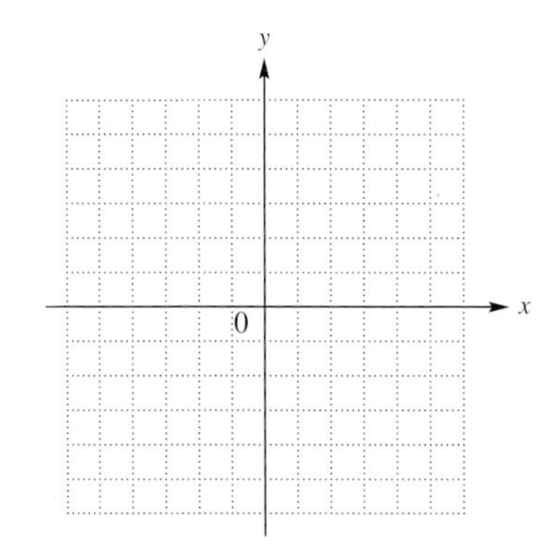

32. Which system of linear inequalities is graphed in the figure?

A. $x \le 3$, $y \le 1$ **B.** $x \le 3$, $y \ge 1$ **C.** $x \ge 3$, $y \le 1$ **D.** $x \ge 3$, $y \ge 1$

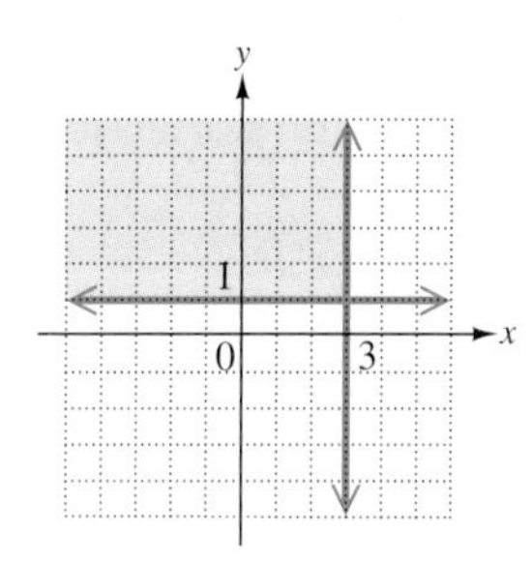

33. Without actually graphing, determine which system of inequalities has no solution.

A. $x \ge 4$, $y \le 3$ **B.** $x + y > 4$, $x + y < 3$ **C.** $x > 2$, $y < 1$ **D.** $x + y > 4$, $x - y < 3$

MIXED REVIEW EXERCISES

Solve each system.

34. $3x + 4y = 6$
$4x - 5y = 8$

35. $\frac{3x}{2} + \frac{y}{5} = -3$
$4x + \frac{y}{3} = -11$

36. $x + 6y = 3$
$2x + 12y = 2$

37. $x + y < 5$
$x - y \ge 2$

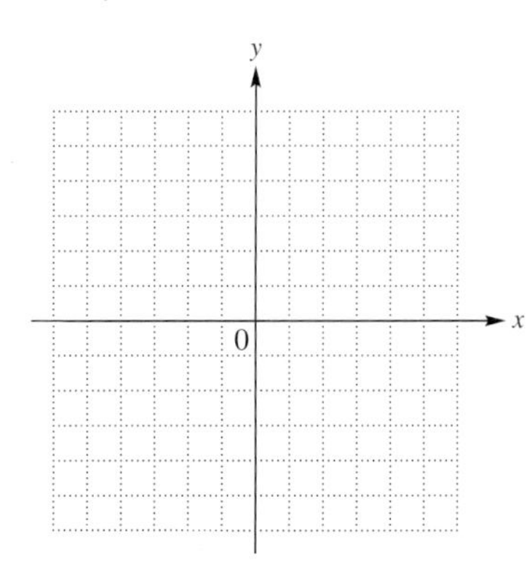

38. $y \le 2x$
$x + 2y > 4$

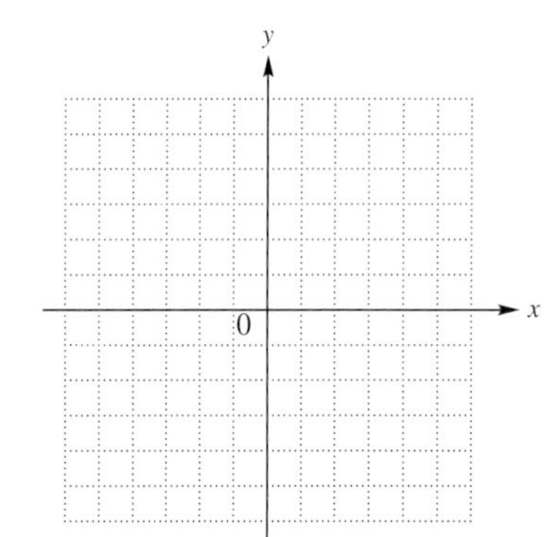

39. $y < -4x$
$y < -2$

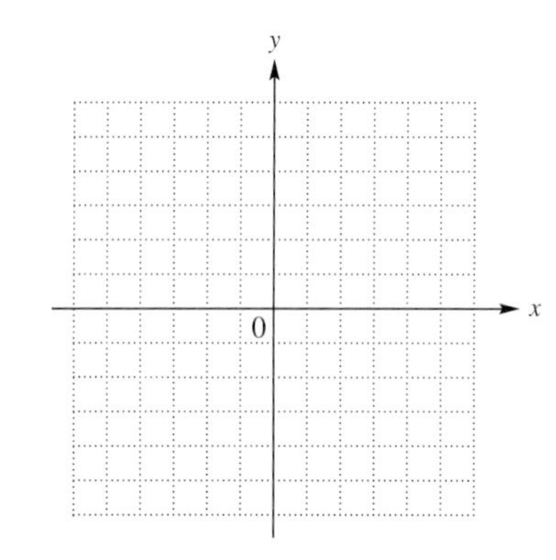

40. The perimeter of an isosceles triangle is 29 in. One side of the triangle is 5 in. longer than each of the two equal sides. Find the lengths of the sides of the triangle.

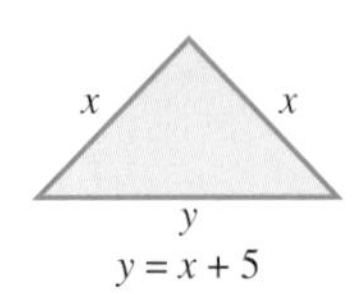

41. In the 2004 National Football League AFC finals, the New England Patriots beat the Indianapolis Colts by 10 points, and the winning score was 4 less than twice the losing score. What was the final score of the game? (*Source:* NFL.)

42. Eboni Perkins compared the monthly payments she would incur for two types of mortgages; fixed-rate and variable-rate. Her observations led to the following graph.

(a) For which years would the monthly payment be more for the fixed-rate mortgage than for the variable-rate mortgage?

(b) In what year would the payments be the same, and what would those payments be?

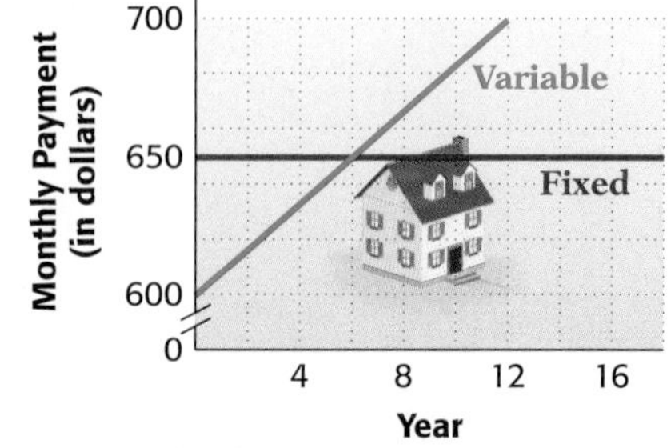

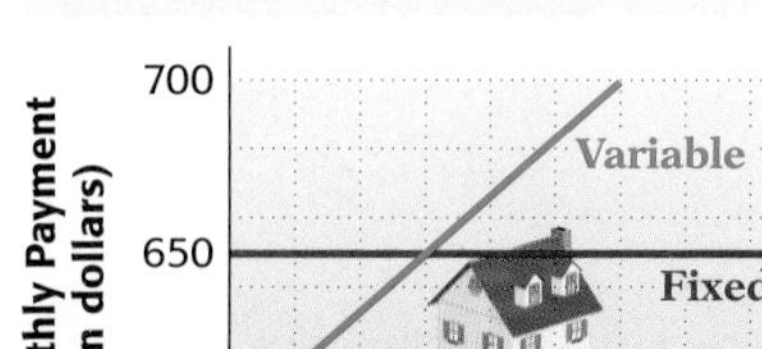

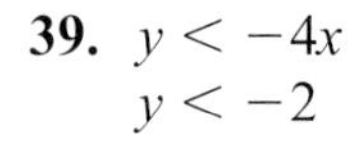

Chapter 4

TEST

Study Skills Workbook
Activity 10: Using Test Results

1. Solve the system by graphing.

$$2x + y = 1$$
$$3x - y = 9$$

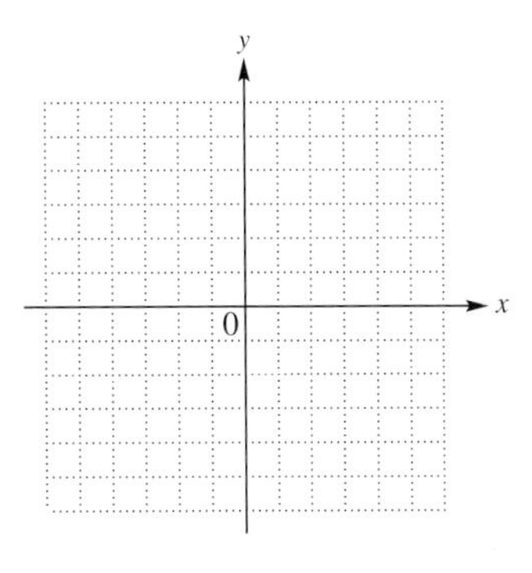

2. Suppose that the graph of a system of two linear equations consists of lines that have the same slope but different y-intercepts. How many solutions does the system have?

Solve each system by the substitution method.

3. $2x + y = -4$
$x = y + 7$

4. $4x + 3y = -35$
$x + y = 0$

Solve each system by the elimination method.

5. $2x - y = 4$
$3x + y = 21$

6. $4x + 2y = 2$
$5x + 4y = 7$

7. $6x - 5y = 0$
$-2x + 3y = 0$

8. $4x + 5y = 2$
$-8x - 10y = 6$

Solve each system by any method.

9. $3x = 6 + y$
$6x - 2y = 12$

10. $\frac{x}{2} - \frac{y}{4} = 7$
$\frac{2x}{3} + \frac{5y}{4} = 3$

1. ______________
2. ______________
3. ______________
4. ______________
5. ______________
6. ______________
7. ______________
8. ______________
9. ______________
10. ______________

Solve each problem.

11. ______________________

11. The distance between Memphis and Atlanta is 782 mi less than the distance between Minneapolis and Houston. Together, the two distances total 1570 mi. How far is it between Memphis and Atlanta? How far is it between Minneapolis and Houston? (*Source: Rand McNally Road Atlas.*)

12. ______________________

12. In 2002, the two most popular amusement parks in the United States were Disneyland and the Magic Kingdom at Walt Disney World. Disneyland had 1.3 million fewer visitors than the Magic Kingdom, and together they had 26.7 million visitors. How many visitors did each park have? (*Source: Amusement Business.*)

13. ______________________

13. A 25% solution of alcohol is to be mixed with a 40% solution to get 50 L of a final mixture that is 30% alcohol. How much of each of the original solutions should be used?

14. ______________________

14. Two cars leave from Perham, Minnesota, and travel in the same direction. One car travels $1\frac{1}{3}$ times as fast as the other. After 3 hr they are 45 mi apart. What are the speeds of the cars?

15.

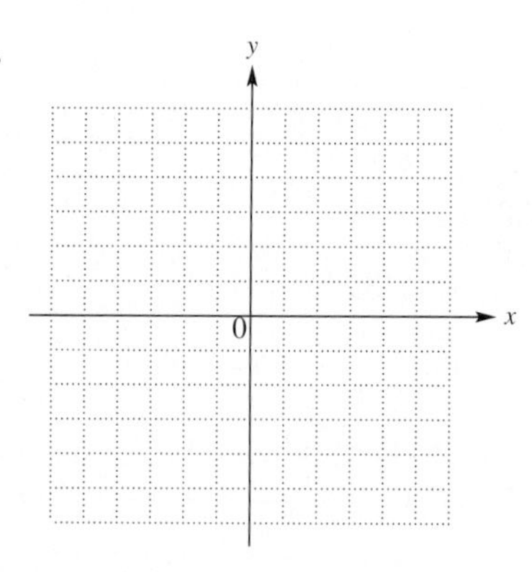

16.

y

x

0

Graph the solution of each system of inequalities.

15. $2x + 7y \leq 14$
$x - y \geq 1$

16. $2x - y > 6$
$4y + 12 \geq -3x$

Cumulative Review Exercises

CHAPTERS R–4

1. List all integer factors of 40.

2. Find the value of the expression if $x = 1$ and $y = 5$.

$$\frac{3x^2 + 2y^2}{10y + 3}$$

Name the property that justifies each statement.

3. $5 + (-4) = (-4) + 5$

4. $r(s - k) = rs - rk$

5. $-\frac{2}{3} + \frac{2}{3} = 0$

6. Evaluate $-2 + 6[3 - (4 - 9)]$.

Solve each linear equation.

7. $2 - 3(6x + 2) = 4(x + 1) + 18$

8. $\frac{3}{2}\left(\frac{1}{3}x + 4\right) = 6\left(\frac{1}{4} + x\right)$

Solve each linear inequality.

9. $-\frac{5}{6}x < 15$

10. $-8 < 2x + 3$

11. No baseball fan should be without a copy of *The Sports Encyclopedia: Baseball 2004* by David S. Neft and Richard M. Cohen. It provides the history of every player, team, and season from 1902–2003, and includes exhaustive statistics. The book has a perimeter of 37.8 in., and its width measures 2.58 in. less than its length. What are its dimensions? (*Source:* www.amazon.com)

Graph each linear equation.

12. $x - y = 4$

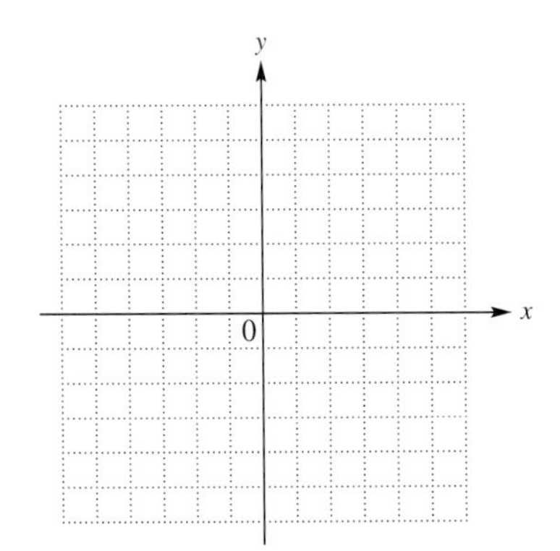

13. $3x + y = 6$

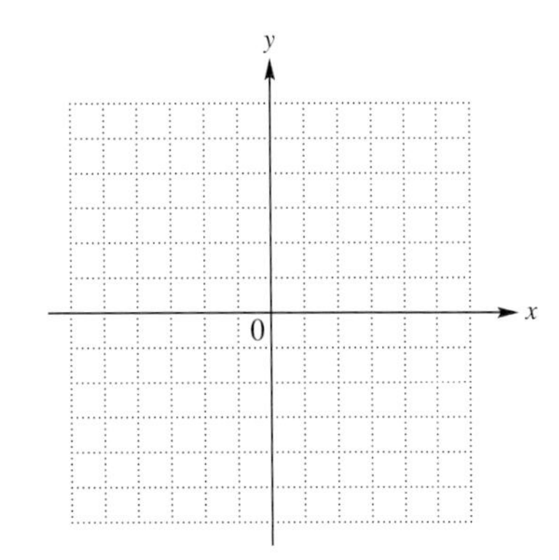

Find the slope of each line.

14. Through $(-5, 6)$ and $(1, -2)$

15. Perpendicular to the line $y = 4x - 3$

Write an equation for each line in slope-intercept form.

16. Through $(-4, 1)$ with slope $\frac{1}{2}$

17. Through the points $(1, 3)$ and $(-2, -3)$

18. (a) Write an equation of the vertical line through $(9, -2)$.

(b) Write an equation of the horizontal line through $(4, -1)$.

Solve each system by any method.

19. $2x - y = -8$
$x + 2y = 11$

20. $4x + 5y = -8$
$3x + 4y = -7$

21. $3x + 4y = 2$
$6x + 8y = 1$

Use a system of equations to solve each problem.

22. Admission prices at a football game were \$6 for adults and \$2 for children. The total value of the tickets sold was \$2528, and 454 tickets were sold. How many adults and how many children attended the game?

Kind of Ticket	Number Sold	Cost of Each (in dollars)	Total Value (in dollars)
Adult	x	6	$6x$
Child	y		
Total	454		

23. The perimeter of a triangle is 53 in. If two sides are of equal length, and the third side measures 4 in. less than each of the equal sides, what are the lengths of the three sides?

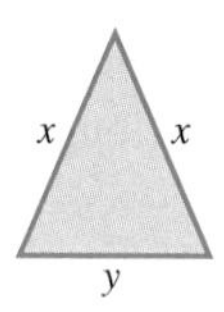

24. Graph the solution of the system

$$x + 2y \le 12$$
$$2x - y \le 8.$$

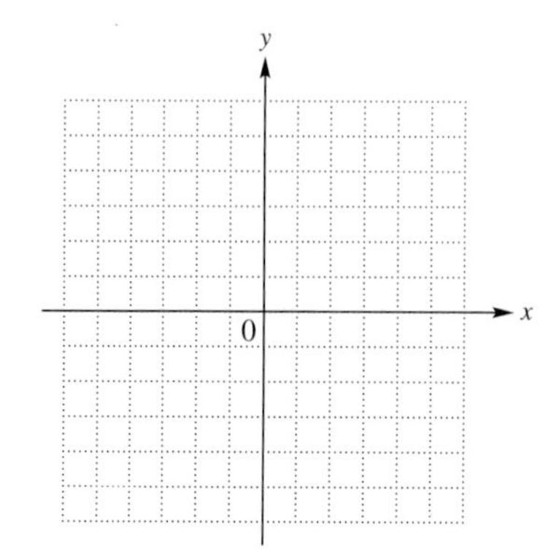

Exponents and Polynomials

Just how much is a trillion? A trillion, which is written 1,000,000,000,000, is an incredibly huge number. A trillion is a million million or a thousand billion. A trillion seconds would last more than 31,000 years, that is, 310 centuries. To be a trillionaire, a person would need a stack of $1000 bills over 67 miles high.

The U.S. budget first exceeded $1 trillion in 1987 and topped $2 trillion for the 2003 fiscal year. (*Source: The Gazette*, February 5, 2002.) In Exercise 1 of Section 5.8, we use exponents to write large numbers like the amount of the NASA budget for the international space station using scientific notation.

5.1 Adding and Subtracting Polynomials

OBJECTIVES

1. **Review combining like terms.**
2. **Know the vocabulary for polynomials.**
3. **Evaluate polynomials.**
4. **Add polynomials.**
5. **Subtract polynomials.**
6. **Add and subtract polynomials with more than one variable.**

Recall from **Section 1.8** that in an expression such as

$$4x^3 + 6x^2 + 5x + 8,$$

the quantities that are added, $4x^3$, $6x^2$, $5x$, and 8 are called *terms.* In the term $4x^3$, the number 4 is called the *numerical coefficient,* or simply the *coefficient,* of x^3. In the same way, 6 is the coefficient of x^2 in the term $6x^2$, 5 is the coefficient of x in the term $5x$, and 8 is the *constant* term.

OBJECTIVE 1 Review combining like terms. In **Section 1.8,** we saw that *like terms* are terms with exactly the same variables, with the same exponents on the variables, such as $2x^2$ and $-5x^2$. Only the coefficients may differ. Like terms are combined, or *added,* by adding their coefficients using the distributive property.

EXAMPLE 1 Adding Like Terms

Simplify each expression by adding like terms.

(a) $-4x^3 + 6x^3 = (-4 + 6)x^3$ Distributive property
$= 2x^3$

(b) $9x^6 - 14x^6 + x^6 = (9 - 14 + 1)x^6$ $x^6 = 1x^6$
$= -4x^6$

(c) $12m^2 + 5m + 4m^2 = (12 + 4)m^2 + 5m$
$= 16m^2 + 5m$

(d) $3x^2y + 4x^2y - x^2y = (3 + 4 - 1)x^2y$
$= 6x^2y$

In Example 1(c), we cannot combine $16m^2$ and $5m$. These two terms are unlike because the exponents on the variables are different. *Unlike terms* have different variables or different exponents on the same variables.

Work Problem 1 at the Side.

1 Add like terms.

(a) $5x^4 + 7x^4$

(b) $9pq + 3pq - 2pq$

(c) $r^2 + 3r + 5r^2$

(d) $8t + 6w$

OBJECTIVE 2 Know the vocabulary for polynomials. A **polynomial in x** is a term or the sum of a finite number of terms of the form ax^n, for any real number a and any whole number n. For example,

$$16x^8 - 7x^6 + 5x^4 - 3x^2 + 4 \quad \text{Polynomial}$$

is a polynomial in x. This polynomial is written in **descending powers,** because the exponents on x decrease from left to right. On the other hand,

$$2x^3 - x^2 + \frac{4}{x} \quad \text{Not a polynomial}$$

is not a polynomial, since a variable appears in a denominator. Of course, we could define a *polynomial* using any variable, not just x, as in Example 1(c). In fact, polynomials may have terms with more than one variable, as in Example 1(d).

Work Problem 2 at the Side.

2 Choose all descriptions that apply for each of the expressions in parts (a)–(d).

A. Polynomial
B. Polynomial written in descending powers
C. Not a polynomial

(a) $3m^5 + 5m^2 - 2m + 1$

(b) $2p^4 + p^6$

(c) $\frac{1}{x} + 2x^2 + 3$

(d) $x - 3$

ANSWERS

1. (a) $12x^4$ (b) $10pq$ (c) $6r^2 + 3r$ (d) cannot be added—unlike terms
2. (a) A and B (b) A (c) C (d) A and B

The **degree of a term** is the sum of the exponents on the variables. A constant term has degree 0. For example, $3x^4$ has degree 4, while $6x^{17}$ has degree 17. The term $5x$ (or $5x^1$) has degree 1, -7 has degree 0, and $2x^2y$ has degree $2 + 1 = 3$ (y has an exponent of 1). The **degree of a polynomial** is the greatest degree of any nonzero term of the polynomial. For example, $3x^4 - 5x^2 + 6$ is of degree 4, the polynomial $5x + 7$ is of degree 1, 3 is of degree 0, and $x^2y + xy - 5xy^2$ is of degree 3.

Three types of polynomials are very common and are given special names. A polynomial with only one term is called a **monomial.** (*Mono*- means "one," as in *mono*rail.) Examples are

$$9m, \quad -6y^5, \quad a^2, \quad \text{and} \quad 6. \qquad \text{Monomials}$$

A polynomial with exactly two terms is called a **binomial.** (*Bi*- means "two," as in *bi*cycle.) Examples are

$$-9x^4 + 9x^3, \quad 8m^2 + 6m, \quad \text{and} \quad 3m^5 - 9m^2. \qquad \text{Binomials}$$

A polynomial with exactly three terms is called a **trinomial.** (*Tri*- means "three," as in *tri*angle.) Examples are

$$9m^3 - 4m^2 + 6, \quad \frac{19}{3}y^2 + \frac{8}{3}y + 5, \quad \text{and} \quad -3m^5 - 9m^2 + 2. \qquad \text{Trinomials}$$

EXAMPLE 2 Classifying Polynomials

For each polynomial, first simplify if possible by combining like terms. Then give the degree and tell whether the polynomial is a *monomial,* a *binomial,* a *trinomial,* or *none of these.*

(a) $2x^3 + 5$
The polynomial cannot be simplified. The degree is 3. The polynomial is a binomial.

(b) $4x - 5x + 2x$
Add like terms to simplify: $4x - 5x + 2x = x$. The degree is 1 (since $x = x^1$). The simplified polynomial is a monomial.

Work Problem 3 at the Side.

OBJECTIVE 3 Evaluate polynomials. A polynomial usually represents different numbers for different values of the variable.

EXAMPLE 3 Evaluating a Polynomial

Find the value of $3x^4 + 5x^3 - 4x - 4$ when $x = -2$ and when $x = 3$.

First, substitute -2 for x.

$$\begin{aligned} 3x^4 + 5x^3 - 4x - 4 &= 3(-2)^4 + 5(-2)^3 - 4(-2) - 4 \\ &= 3 \cdot 16 + 5(-8) - 4(-2) - 4 && \text{Apply exponents.} \\ &= 48 - 40 + 8 - 4 && \text{Multiply.} \\ &= 12 && \text{Add and subtract.} \end{aligned}$$

Next, replace x with 3.

$$\begin{aligned} 3x^4 + 5x^3 - 4x - 4 &= 3(3)^4 + 5(3)^3 - 4(3) - 4 \\ &= 3 \cdot 81 + 5 \cdot 27 - 4(3) - 4 \\ &= 243 + 135 - 12 - 4 \\ &= 362 \end{aligned}$$

3 For each polynomial, first simplify if possible. Then give the degree and tell whether the polynomial is a *monomial, binomial, trinomial,* or *none of these.*

(a) $3x^2 + 2x - 4$

(b) $x^3 + 4x^3$

(c) $x^8 - x^7 + 2x^8$

ANSWERS
3. **(a)** degree 2; trinomial
(b) degree 3; monomial (simplify to $5x^3$)
(c) degree 8; binomial (simplify to $3x^8 - x^7$)

4 Find the value of $2y^3 + 8y - 6$ in each case.

(a) when $y = -1$

(b) when $y = 4$

5 Add each pair of polynomials.

(a)
$$\begin{array}{r} 4x^3 - 3x^2 + 2x \\ \underline{6x^3 + 2x^2 - 3x} \end{array}$$

(b)
$$\begin{array}{r} x^2 - 2x + 5 \\ \underline{4x^2 \qquad\;\; - 2} \end{array}$$

CAUTION

Notice the use of parentheses around the numbers that are substituted for the variable in Example 3. This is particularly important when substituting a negative number for a variable that is raised to a power, so the sign of the product is correct.

Work Problem 4 at the Side.

OBJECTIVE 4 Add polynomials. Polynomials may be added, subtracted, multiplied, and divided.

Adding Polynomials

To add two polynomials, add like terms.

EXAMPLE 4 Adding Polynomials Vertically

(a) Add $6x^3 - 4x^2 + 3$ and $-2x^3 + 7x^2 - 5$.
Write like terms in columns.

$$\begin{array}{r} 6x^3 - 4x^2 + 3 \\ \underline{-2x^3 + 7x^2 - 5} \end{array}$$

Now add, column by column.

$$\begin{array}{rrr} 6x^3 & -4x^2 & 3 \\ \underline{-2x^3} & \underline{7x^2} & \underline{-5} \\ 4x^3 & 3x^2 & -2 \end{array}$$

Add the three sums together.

$$4x^3 + 3x^2 + (-2) = 4x^3 + 3x^2 - 2$$

(b) Add $2x^2 - 4x + 3$ and $x^3 + 5x$.
Write like terms in columns and add column by column.

$$\begin{array}{r} 2x^2 - 4x + 3 \\ \underline{x^3 \qquad\quad + 5x \qquad} \\ x^3 + 2x^2 + \;\; x + 3 \end{array}$$

Leave spaces for missing terms.

Work Problem 5 at the Side.

The polynomials in Example 4 also could be added horizontally.

EXAMPLE 5 Adding Polynomials Horizontally

(a) Add $6x^3 - 4x^2 + 3$ and $-2x^3 + 7x^2 - 5$.
Combine like terms.

$$(6x^3 - 4x^2 + 3) + (-2x^3 + 7x^2 - 5) = 4x^3 + 3x^2 - 2,$$

the same answer found in Example 4(a).

Continued on Next Page

ANSWERS

4. **(a)** -16 **(b)** 154
5. **(a)** $10x^3 - x^2 - x$ **(b)** $5x^2 - 2x + 3$

(b) Add $2x^2 - 4x + 3$ and $x^3 + 5x$.

$$(2x^2 - 4x + 3) + (x^3 + 5x) = 2x^2 - 4x + 3 + x^3 + 5x$$
$$= x^3 + 2x^2 + x + 3 \quad \text{Combine like terms.}$$

Work Problem 6 at the Side.

6 Find each sum.

(a) $(2x^4 - 6x^2 + 7) + (-3x^4 + 5x^2 + 2)$

(b) $(3x^2 + 4x + 2) + (6x^3 - 5x - 7)$

OBJECTIVE 5 Subtract polynomials. In **Section 1.5**, the difference $x - y$ was defined as $x + (-y)$. (We find the difference $x - y$ by adding x and the opposite of y.) For example,

$$7 - 2 = 7 + (-2) = 5 \quad \text{and} \quad -8 - (-2) = -8 + 2 = -6.$$

A similar method is used to subtract polynomials.

Subtracting Polynomials

To subtract two polynomials, change all the signs of the second polynomial and add the result to the first polynomial.

EXAMPLE 6 Subtracting Polynomials

(a) Perform the subtraction $(5x - 2) - (3x - 8)$.
Change the signs in the second polynomial and add like terms.

$$(5x - 2) - (3x - 8) = (5x - 2) + (-3x + 8)$$
$$= 2x + 6$$

(b) Subtract $6x^3 - 4x^2 + 2$ from $11x^3 + 2x^2 - 8$.
Write the problem.

$$(11x^3 + 2x^2 - 8) - (6x^3 - 4x^2 + 2)$$

Change all the signs in the second polynomial and add the two polynomials.

$$(11x^3 + 2x^2 - 8) + (-6x^3 + 4x^2 - 2) = 5x^3 + 6x^2 - 10$$

To check a subtraction problem, use the fact that if

$$a - b = c, \quad \text{then} \quad a = b + c.$$

For example, $6 - 2 = 4$, so we check by writing $6 = 2 + 4$, which is correct. Check the polynomial subtraction above by adding $6x^3 - 4x^2 + 2$ and $5x^3 + 6x^2 - 10$.

$$(6x^3 - 4x^2 + 2) + (5x^3 + 6x^2 - 10) = 11x^3 + 2x^2 - 8$$

Since the sum is $11x^3 + 2x^2 - 8$, the subtraction was performed correctly.

Work Problem 7 at the Side.

7 Subtract, and check your answers by addition.

(a) $(14y^3 - 6y^2 + 2y - 5) - (2y^3 - 7y^2 - 4y + 6)$

(b) Subtract $\left(-\frac{3}{2}y^2 + \frac{4}{3}y + 6\right)$ from $\left(\frac{7}{2}y^2 - \frac{11}{3}y + 8\right)$.

Subtraction also can be done in columns. We will use vertical subtraction in **Section 5.7** when we study polynomial division.

Answers

6. (a) $-x^4 - x^2 + 9$
(b) $6x^3 + 3x^2 - x - 5$

7. (a) $12y^3 + y^2 + 6y - 11$
(b) $5y^2 - 5y + 2$

8 Subtract, using the method of subtracting by columns.

$$(4y^3 - 16y^2 + 2y) - (12y^3 - 9y^2 + 16)$$

9 Perform the indicated operations.

$$(6p^4 - 8p^3 + 2p - 1) - (-7p^4 + 6p^2 - 12) + (p^4 - 3p + 8)$$

10 Add or subtract.

(a) $(3mn + 2m - 4n) + (-mn + 4m + n)$

(b) $(5p^2q^2 - 4p^2 + 2q) - (2p^2q^2 - p^2 - 3q)$

Answers

8. $-8y^3 - 7y^2 + 2y - 16$

9. $14p^4 - 8p^3 - 6p^2 - p + 19$

10. **(a)** $2mn + 6m - 3n$
(b) $3p^2q^2 - 3p^2 + 5q$

EXAMPLE 7 Subtracting Polynomials Vertically

Use the method of subtracting by columns to find

$$(14y^3 - 6y^2 + 2y - 5) - (2y^3 - 7y^2 - 4y + 6).$$

Arrange like terms in columns.

$$\begin{array}{r} 14y^3 - 6y^2 + 2y - 5 \\ 2y^3 - 7y^2 - 4y + 6 \\ \hline \end{array}$$

Change all signs in the second row, and then add.

$$\begin{array}{rl} 14y^3 - 6y^2 + 2y - 5 & \\ -2y^3 + 7y^2 + 4y - 6 & \text{Change signs.} \\ \hline 12y^3 + y^2 + 6y - 11 & \text{Add.} \end{array}$$

Work Problem 8 at the Side.

Either the horizontal or the vertical method may be used for adding or subtracting polynomials.

EXAMPLE 8 Adding and Subtracting More Than Two Polynomials

Perform the indicated operations to simplify the expression

$$(4 - x + 3x^2) - (2 - 3x + 5x^2) + (8 + 2x - 4x^2).$$

Rewrite, changing the subtraction to adding the opposite.

$$\begin{aligned} &(4 - x + 3x^2) - (2 - 3x + 5x^2) + (8 + 2x - 4x^2) \\ &= (4 - x + 3x^2) + (-2 + 3x - 5x^2) + (8 + 2x - 4x^2) \\ &= (2 + 2x - 2x^2) + (8 + 2x - 4x^2) \quad \text{Combine like terms.} \\ &= 10 + 4x - 6x^2 \quad \text{Combine like terms.} \end{aligned}$$

Work Problem 9 at the Side.

OBJECTIVE 6 Add and subtract polynomials with more than one variable. Polynomials in more than one variable are added and subtracted by combining like terms, just as with single-variable polynomials.

EXAMPLE 9 Adding and Subtracting Multivariable Polynomials

Add or subtract as indicated.

(a) $(4a + 2ab - b) + (3a - ab + b)$

$$\begin{aligned} &= 4a + 2ab - b + 3a - ab + b \\ &= 7a + ab \quad \text{Combine like terms.} \end{aligned}$$

(b) $(2x^2y + 3xy + y^2) - (3x^2y - xy - 2y^2)$

$$\begin{aligned} &= 2x^2y + 3xy + y^2 - 3x^2y + xy + 2y^2 \\ &= -x^2y + 4xy + 3y^2 \end{aligned}$$

Work Problem 10 at the Side.

5.1 Exercises

Fill in each blank with the correct response.

1. In the term $7x^5$, the coefficient is ________ and the exponent is ________.

2. The expression $5x^3 - 4x^2$ has ________ term(s).
(how many?)

3. The degree of the term $-4x^8$ is ________.

4. The polynomial $4x^2 - y^2$ ________ an example of a trinomial.
(is/is not)

5. When $x^2 + 10$ is evaluated for $x = 4$, the result is ________.

6. ________ is an example of a monomial with coefficient 5, in the variable x, having degree 9.

For each polynomial, determine the number of terms, and name the coefficient of each term.

7. $6x^4$

8. $-9y^5$

9. t^4

10. s^7

11. $-19r^2 - r$

12. $2y^3 - y$

13. $x + 8x^2$

14. $v - 2v^3$

In each polynomial, combine like terms whenever possible. Write the result with descending powers.

15. $-3m^5 + 5m^5$

16. $-4y^3 + 3y^3$

17. $2r^5 + (-3r^5)$

18. $-19y^2 + 9y^2$

19. $.2m^5 - .5m^2$

20. $-.9y + .9y^2$

21. $-3x^5 + 2x^5 - 4x^5$

22. $6x^3 - 8x^3 + 9x^3$

23. $-4p^7 + 8p^7 + 5p^9$

24. $-3a^8 + 4a^8 - 3a^2$

25. $-4y^2 + 3y^2 - 2y^2 + y^2$

26. $3r^5 - 8r^5 + r^5 + 2r^5$

For each polynomial, first simplify, if possible, and write it with descending powers. Then give the degree of the resulting polynomial, and tell whether it is a monomial, *a* binomial, *a* trinomial, *or* none of these. *See Example 2.*

27. $6x^4 - 9x$

28. $7t^3 - 3t$

29. $5m^4 - 3m^2 + 6m^5 - 7m^3$

30. $6p^5 + 4p^3 - 8p^4 + 10p^2$

31. $\frac{5}{3}x^4 - \frac{2}{3}x^4 + \frac{1}{3}x^2 - 4$

32. $\frac{4}{5}r^6 + \frac{1}{5}r^6 - r^4 + \frac{2}{5}r$

33. $.8x^4 - .3x^4 - .5x^4 + 7$

34. $1.2t^3 - .9t^3 - .3t^3 + 9$

*Find the value of each polynomial **(a)** when $x = 2$ and **(b)** when $x = -1$. See Example 3.*

35. $-2x + 3$

36. $5x - 4$

37. $2x^2 + 5x + 1$

38. $-3x^2 + 14x - 2$

39. $2x^5 - 4x^4 + 5x^3 - x^2$

40. $x^4 - 6x^3 + x^2 + 1$

41. $-4x^5 + x^2$

42. $2x^6 - 4x$

RELATING CONCEPTS (EXERCISES 43–46) For Individual or Group Work

A polynomial can model the distance in feet that a car going approximately 68 mph will skid in t seconds. If we let D represent this distance, then

$$D = 100t - 13t^2.$$

*Each time we evaluate this polynomial for a value of t, we get one and only one output value D. This idea is basic to the concept of a **function**, an important concept in mathematics. Exercises 43–46 illustrate this idea with this polynomial and two others. **Work them in order.***

43. Use the given polynomial to approximate the skidding distance in feet if $t = 5$ sec.

44. Use the polynomial equation $D = 100t - 13t^2$ to find the distance the car will skid in 1 sec. Write an ordered pair of the form (t, D).

45. If gasoline costs \$1.80 per gal, then the monomial $1.80x$ gives the cost, in dollars, of x gallons. How much would 4 gal cost?

46. If it costs \$15 plus \$2 per day to rent a chain saw, the binomial $2x + 15$ gives the cost in dollars to rent the chain saw for x days. How much would it cost to rent the saw for 6 days?

Add or subtract as indicated. See Examples 4 and 7.

47. Add.

$$\begin{array}{r} 3m^2 + 5m \\ \underline{2m^2 - 2m} \end{array}$$

48. Add.

$$\begin{array}{r} 4a^3 - 4a^2 \\ \underline{6a^3 + 5a^2} \end{array}$$

49. Subtract.

$$\begin{array}{r} 12x^4 - x^2 \\ \underline{8x^4 + 3x^2} \end{array}$$

50. Subtract.

$$\begin{array}{r} 13y^5 - y^3 \\ \underline{7y^5 + 5y^3} \end{array}$$

51. Add.

$$\begin{array}{r} \frac{2}{3}x^2 + \frac{1}{5}x + \frac{1}{6} \\ \underline{\frac{1}{2}x^2 - \frac{1}{3}x + \frac{2}{3}} \end{array}$$

52. Add.

$$\begin{array}{r} \frac{4}{7}y^2 - \frac{1}{5}y + \frac{7}{9} \\ \underline{\frac{1}{3}y^2 - \frac{1}{3}y + \frac{2}{5}} \end{array}$$

53. Subtract.

$$\begin{array}{r} 12m^3 - 8m^2 + 6m + 7 \\ \underline{5m^2 \qquad\quad - 4} \end{array}$$

54. Subtract.

$$\begin{array}{r} 5a^4 - 3a^3 + 2a^2 - a + 6 \\ \underline{-6a^4 \qquad\quad - a^2 + a - 1} \end{array}$$

Perform the indicated operations. See Examples 5, 6, and 8.

55. $(2r^2 + 3r - 12) + (6r^2 + 2r)$

56. $(3r^2 + 5r - 6) + (2r - 5r^2)$

57. $(8m^2 - 7m) - (3m^2 + 7m - 6)$

58. $(x^2 + x) - (3x^2 + 2x - 1)$

59. $(16x^3 - x^2 + 3x) + (-12x^3 + 3x^2 + 2x)$

60. $(-2b^6 + 3b^4 - b^2) + (b^6 + 2b^4 + 2b^2)$

61. $(7y^4 + 3y^2 + 2y) - (18y^5 - 5y^3 + y)$

62. $(8t^5 + 3t^3 + 5t) - (19t^4 - 6t^2 + t)$

63. $[(8m^2 + 4m - 7) - (2m^3 - 5m + 2)] - (m^2 + m)$

64. $[(9b^3 - 4b^2 + 3b + 2) - (-2b^3 + b)] - (8b^3 + 6b + 4)$

Find the perimeter of each geometric figure.

65.

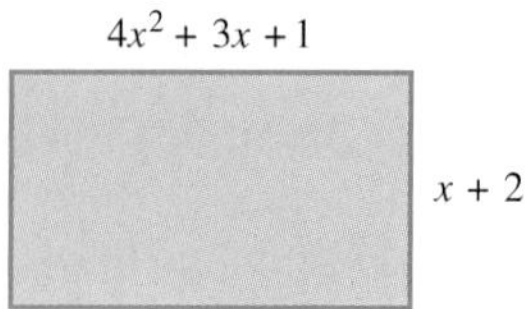

66.

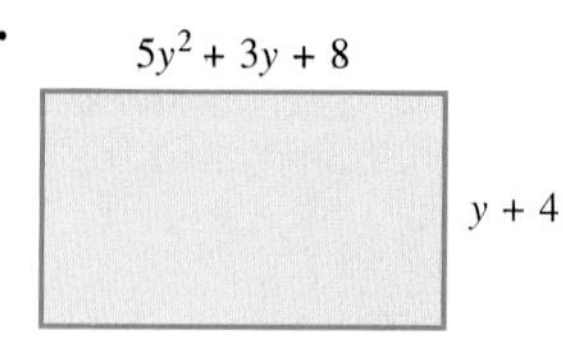

67.

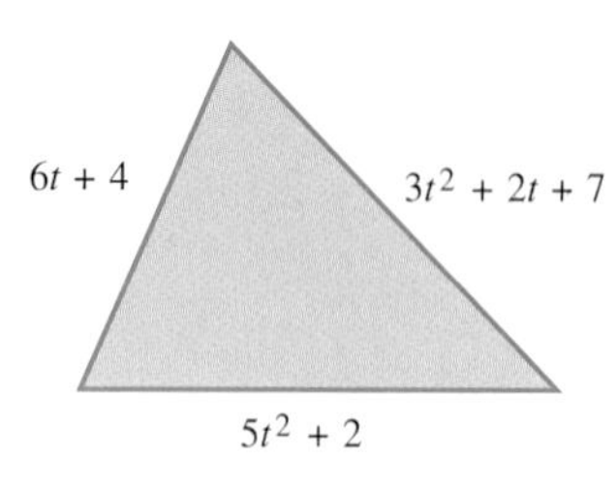

68.

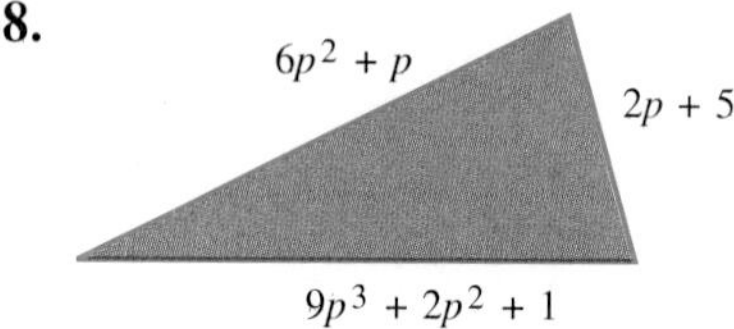

69. Subtract $9x^2 - 3x + 7$ from $-2x^2 - 6x + 4$.

70. Subtract $-5w^3 + 5w^2 - 7$ from $6w^3 + 8w + 5$.

71. Explain why the degree of the term 3^4 is not 4. What is its degree?

72. Can the sum of two polynomials in x, both of degree 3, be of degree 2? If so, give an example.

Add or subtract as indicated. See Example 9.

73. $(9a^2b - 3a^2 + 2b) + (4a^2b - 4a^2 - 3b)$

74. $(4xy^3 - 3x + y) + (5xy^3 + 13x - 4y)$

75. $(2c^4d + 3c^2d^2 - 4d^2) - (c^4d + 8c^2d^2 - 5d^2)$

76. $(3k^2h^3 + 5kh + 6k^3h^2) - (2k^2h^3 - 9kh + k^3h^2)$

77. Subtract.
$$\begin{array}{r} 9m^3n - 5m^2n^2 + 4mn^2 \\ \underline{-3m^3n + 6m^2n^2 + 8mn^2} \end{array}$$

78. Subtract.
$$\begin{array}{r} 12r^5t + 11r^4t^2 - 7r^3t^3 \\ \underline{-8r^5t + 10r^4t^2 + 3r^3t^3} \end{array}$$

5.2 The Product Rule and Power Rules for Exponents

OBJECTIVES

1. Use exponents.
2. Use the product rule for exponents.
3. Use the rule $(a^m)^n = a^{mn}$.
4. Use the rule $(ab)^m = a^m b^m$.
5. Use the rule $\left(\frac{a}{b}\right)^m = \frac{a^m}{b^m}$.
6. Use combinations of the rules for exponents.
7. Use the rules for exponents in an application of geometry.

OBJECTIVE 1 Use exponents. In **Section 1.1** we used exponents to write repeated products. Recall that in the expression 5^2, the number 5 is called the *base* and 2 is called the *exponent* or *power*. The expression 5^2 is called an *exponential expression*. Although we do not usually write a quantity with an exponent of 1, in general, for any quantity a, $a = a^1$.

EXAMPLE 1 Using Exponents

Write $3 \cdot 3 \cdot 3 \cdot 3 \cdot 3$ in exponential form and evaluate.

Since 3 occurs as a factor five times, the base is 3 and the exponent is 5. The exponential expression is 3^5, read "3 to the fifth power" or simply "3 to the fifth." The value is

$$3^5 = 3 \cdot 3 \cdot 3 \cdot 3 \cdot 3 = 243.$$

Work Problem 1 at the Side.

1 Write $2 \cdot 2 \cdot 2 \cdot 2$ in exponential form and evaluate.

EXAMPLE 2 Evaluating Exponential Expressions

Evaluate each exponential expression. Name the base and the exponent.

	Base	Exponent
(a) $5^4 = 5 \cdot 5 \cdot 5 \cdot 5 = 625$	5	4
(b) $-5^4 = -1 \cdot 5^4 = -1 \cdot (5 \cdot 5 \cdot 5 \cdot 5) = -625$	5	4
(c) $(-5)^4 = (-5)(-5)(-5)(-5) = 625$	-5	4

CAUTION

Notice the difference between Examples 2(b) and (c). In -5^4 the lack of parentheses shows that the exponent 4 applies only to the base 5, and not -5. In $(-5)^4$ the parentheses show that the exponent 4 applies to the base -5. In summary, $-a^n$ and $(-a)^n$ are not always the same.

Expression	Base	Exponent	Example
$-a^n$	a	n	$-3^2 = -(3 \cdot 3) = -9$
$(-a)^n$	$-a$	n	$(-3)^2 = (-3)(-3) = 9$

Work Problem 2 at the Side.

2 Evaluate each exponential expression. Name the base and the exponent.

(a) $(-2)^5$ **(b)** -2^5

(c) -4^2 **(d)** $(-4)^2$

OBJECTIVE 2 Use the product rule for exponents. To develop the product rule, we use the definition of an exponential expression.

$$\begin{aligned} 2^4 \cdot 2^3 &= \underbrace{(2 \cdot 2 \cdot 2 \cdot 2)}_{4 \text{ factors}}\underbrace{(2 \cdot 2 \cdot 2)}_{3 \text{ factors}} \\ &= \underbrace{2 \cdot 2 \cdot 2 \cdot 2 \cdot 2 \cdot 2 \cdot 2}_{4 + 3 = 7 \text{ factors}} \\ &= 2^7 \end{aligned}$$

ANSWERS

1. $2^4 = 16$
2. (a) -32; -2; 5 (b) -32; 2; 5 (c) -16; 4; 2 (d) 16; -4; 2

Also,

$$\begin{aligned}6^2 \cdot 6^3 &= (6 \cdot 6)(6 \cdot 6 \cdot 6)\\ &= 6 \cdot 6 \cdot 6 \cdot 6 \cdot 6\\ &= 6^5.\end{aligned}$$

Generalizing from these examples, $2^4 \cdot 2^3 = 2^{4+3} = 2^7$ and $6^2 \cdot 6^3 = 6^{2+3} = 6^5$. In each case, adding the exponents gives the exponent of the product, suggesting the **product rule for exponents.**

Product Rule for Exponents

For any positive integers m and n,

$$a^m \cdot a^n = a^{m+n}.$$

(Keep the same base and add the exponents.)

Example: $6^2 \cdot 6^5 = 6^{2+5} = 6^7$.

CAUTION

Avoid the common error of multiplying the bases when using the product rule.

$$6^2 \cdot 6^5 \neq 36^7$$

Keep the same base and add the exponents.

EXAMPLE 3 Using the Product Rule

Use the product rule for exponents to find each product, if possible.

(a) $6^3 \cdot 6^5 = 6^{3+5} = 6^8$ by the product rule.

(b) $(-4)^7(-4)^2 = (-4)^{7+2} = (-4)^9$ by the product rule.

(c) $x^2 \cdot x = x^2 \cdot x^1 = x^{2+1} = x^3$

(d) $m^4 \cdot m^3 = m^{4+3} = m^7$

(e) $2^3 \cdot 3^2$

The product rule does not apply to the product $2^3 \cdot 3^2$ because the bases are different.

$$2^3 \cdot 3^2 = 8 \cdot 9 = 72$$

(f) $2^3 + 2^4$

The product rule does not apply to $2^3 + 2^4$ because it is a *sum*, not a *product*.

$$2^3 + 2^4 = 8 + 16 = 24$$

CAUTION

The bases of the factors must be the same before we can apply the product rule for exponents.

Work Problem 3 at the Side.

3 Find each product by the product rule, if possible.

(a) $8^2 \cdot 8^5$

(b) $(-7)^5 \cdot (-7)^3$

(c) $y^3 \cdot y$

(d) $4^2 \cdot 3^5$

(e) $6^4 + 6^2$

ANSWERS

3. **(a)** 8^7 **(b)** $(-7)^8$ **(c)** y^4
(d) cannot use the product rule (product: 3888) **(e)** cannot use the product rule

EXAMPLE 4 Using the Product Rule

Multiply $2x^3$ and $3x^7$.

We use the associative and commutative properties and the product rule.

$$2x^3 \cdot 3x^7 = 2 \cdot 3 \cdot x^3 \cdot x^7 = 6x^{10} \qquad 2x^3 = 2 \cdot x^3;\ 3x^7 = 3 \cdot x^7$$

CAUTION

Be sure you understand the difference between *adding* and *multiplying* exponential expressions. For example,

$$8x^3 + 5x^3 = (8 + 5)x^3 = 13x^3, \qquad \text{Add.}$$

but

$$(8x^3)(5x^3) = (8 \cdot 5)x^{3+3} = 40x^6. \qquad \text{Multiply.}$$

Work Problem 4 at the Side.

OBJECTIVE 3 Use the rule $(a^m)^n = a^{mn}$. We can simplify an expression such as $(8^3)^2$ with the product rule for exponents, as follows.

$$(8^3)^2 = (8^3)(8^3) = 8^{3+3} = 8^6$$

The product of the exponents in $(8^3)^2$, $3 \cdot 2$, gives the exponent in 8^6. Also,

$$\begin{aligned}(5^2)^4 &= 5^2 \cdot 5^2 \cdot 5^2 \cdot 5^2 && \text{Definition of exponent}\\ &= 5^{2+2+2+2} && \text{Product rule}\\ &= 5^8,\end{aligned}$$

and $2 \cdot 4 = 8$. These examples suggest **power rule (a) for exponents.**

Power Rule (a) for Exponents

For any positive integers m and n,

$$(a^m)^n = a^{mn}.$$

(Raise a power to a power by multiplying exponents.)

Example: $(3^2)^4 = 3^{2 \cdot 4} = 3^8$.

EXAMPLE 5 Using Power Rule (a)

Use power rule (a) for exponents to simplify each expression.

(a) $(2^5)^3 = 2^{5 \cdot 3} = 2^{15}$ **(b)** $(5^7)^2 = 5^{7 \cdot 2} = 5^{14}$

(c) $(x^2)^5 = x^{2 \cdot 5} = x^{10}$ **(d)** $(n^3)^2 = n^{3 \cdot 2} = n^6$

Work Problem 5 at the Side.

OBJECTIVE 4 Use the rule $(ab)^m = a^m b^m$. We can rewrite the expression $(4x)^3$ as shown below.

$$\begin{aligned}(4x)^3 &= (4x)(4x)(4x) && \text{Definition of exponent}\\ &= 4 \cdot 4 \cdot 4 \cdot x \cdot x \cdot x && \text{Commutative and associative properties}\\ &= 4^3x^3 && \text{Definition of exponent}\end{aligned}$$

This example suggests **power rule (b) for exponents.**

4 Multiply.

(a) $5m^2 \cdot 2m^6$

(b) $3p^5 \cdot 9p^4$

(c) $-7p^5 \cdot (3p^8)$

5 Simplify each expression.

(a) $(5^3)^4$

(b) $(6^2)^5$

(c) $(3^2)^4$

(d) $(a^6)^5$

Answers

4. (a) $10m^8$ **(b)** $27p^9$ **(c)** $-21p^{13}$

5. (a) 5^{12} **(b)** 6^{10} **(c)** 3^8 **(d)** a^{30}

6 Simplify.

(a) $5(mn)^3$

(b) $(3a^2b^4)^5$

(c) $(-5m^2)^3$

Power Rule (b) for Exponents

For any positive integer m,

$$(ab)^m = a^m b^m.$$

(Raise a product to a power by raising each factor to the power.)

Example: $(2p)^5 = 2^5 p^5$.

EXAMPLE 6 Using Power Rule (b)

Use power rule (b) to simplify each expression.

(a) $(3xy)^2 = 3^2x^2y^2$ Power rule (b)

$= 9x^2y^2$

(b) $9(pq)^2 = 9(p^2q^2)$ Power rule (b)

$= 9p^2q^2$

(c) $5(2m^2p^3)^4 = 5[2^4(m^2)^4(p^3)^4]$ Power rule (b)

$= 5(2^4m^8p^{12})$ Power rule (a)

$= 5 \cdot 2^4m^8p^{12}$

$= 80m^8p^{12}$ $5 \cdot 2^4 = 5 \cdot 16 = 80$

(d) $(-5^6)^3 = (-1 \cdot 5^6)^3$ $-a = -1 \cdot a$

$= (-1)^3(5^6)^3$ Power rule (b)

$= -1 \cdot 5^{18}$ Power rule (a)

$= -5^{18}$

CAUTION

Power rule (b) **does not** ***apply to a*** **sum.**

$$(x + 4)^2 \neq x^2 + 4^2$$

Work Problem 6 at the Side.

OBJECTIVE 5 Use the rule $\left(\frac{a}{b}\right)^m = \frac{a^m}{b^m}$. Since the quotient $\frac{a}{b}$ can be written as $a \cdot \frac{1}{b}$, we can use power rule (b), together with some of the properties of real numbers, to get **power rule (c) for exponents.**

Power Rule (c) for Exponents

For any positive integer m,

$$\left(\frac{a}{b}\right)^m = \frac{a^m}{b^m} \quad (b \neq 0).$$

(Raise a quotient to a power by raising both the numerator and the denominator to the power.)

Example: $\left(\frac{5}{3}\right)^2 = \frac{5^2}{3^2}$.

ANSWERS

6. (a) $5m^3n^3$ (b) $3^5a^{10}b^{20}$ or $243a^{10}b^{20}$ (c) -5^3m^6 or $-125m^6$

EXAMPLE 7 Using Power Rule (c)

Simplify each expression.

(a) $\left(\frac{2}{3}\right)^5 = \frac{2^5}{3^5}$ or $\frac{32}{243}$ **(b)** $\left(\frac{m}{n}\right)^4 = \frac{m^4}{n^4}, \quad n \neq 0$

Work Problem 7 at the Side.

7 Simplify. Assume all variables represent nonzero real numbers.

(a) $\left(\frac{5}{2}\right)^4$

(b) $\left(\frac{p}{q}\right)^2$

(c) $\left(\frac{r}{t}\right)^3$

Next we list the rules for exponents discussed in this section. These rules are basic to the study of algebra and should be *memorized.*

Rules for Exponents

For positive integers m and n: *Examples*

		Examples
Product rule	$a^m \cdot a^n = a^{m+n}$	$6^2 \cdot 6^5 = 6^{2+5} = 6^7$
Power rules	**(a)** $(a^m)^n = a^{mn}$	$(3^2)^4 = 3^{2\cdot 4} = 3^8$
	(b) $(ab)^m = a^m b^m$	$(2p)^5 = 2^5 p^5$
	(c) $\left(\frac{a}{b}\right)^m = \frac{a^m}{b^m}$ $(b \neq 0)$.	$\left(\frac{5}{3}\right)^2 = \frac{5^2}{3^2}$

OBJECTIVE 6 Use combinations of the rules for exponents. As shown in the next example, more than one rule may be needed to simplify an expression.

EXAMPLE 8 Using Combinations of Rules

Simplify each expression.

(a)
$$\left(\frac{2}{3}\right)^2 \cdot 2^3 = \frac{2^2}{3^2} \cdot \frac{2^3}{1} \qquad \text{Power rule (c)}$$
$$= \frac{2^2 \cdot 2^3}{3^2 \cdot 1} \qquad \text{Multiply fractions.}$$
$$= \frac{2^5}{3^2} \qquad \text{Product rule}$$

(b)
$$(5x)^3(5x)^4 = (5x)^7 \qquad \text{Product rule}$$
$$= 5^7x^7 \qquad \text{Power rule (b)}$$

(c)
$$(2x^2y^3)^4(3xy^2)^3 = 2^4(x^2)^4(y^3)^4 \cdot 3^3x^3(y^2)^3 \qquad \text{Power rule (b)}$$
$$= 2^4 \cdot x^8 \cdot y^{12} \cdot 3^3 \cdot x^3 \cdot y^6 \qquad \text{Power rule (a)}$$
$$= 2^4 \cdot 3^3x^8x^3y^{12}y^6 \qquad \text{Commutative and associative properties}$$
$$= 16 \cdot 27x^{11}y^{18} \qquad \text{Product rule}$$
$$= 432x^{11}y^{18}$$

Notice that

$$(2x^2y^3)^4 = 2^4x^{2\cdot 4}y^{3\cdot 4}, \quad \textbf{not} \quad (2 \cdot 4)x^{2\cdot 4}y^{3\cdot 4}.$$

Do not multiply the coefficient 2 and the exponent 4.

Continued on Next Page

ANSWERS

7. (a) $\frac{5^4}{2^4}$ or $\frac{625}{16}$ (b) $\frac{p^2}{q^2}$ (c) $\frac{r^3}{t^3}$

8 Simplify.

(a) $(2m)^3(2m)^4$

(b) $\left(\frac{5k^3}{3}\right)^2$

(c) $\left(\frac{1}{5}\right)^4(2x)^2$

(d) $(-3xy^2)^3(x^2y)^4$

9 Find the area.

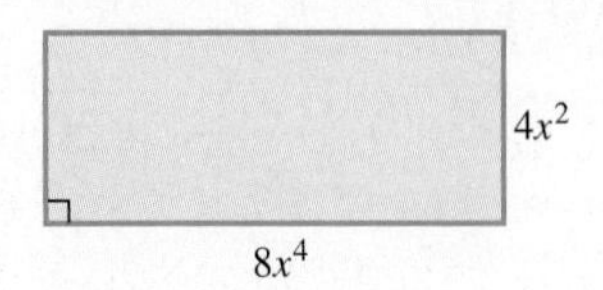

(d) $(-x^3y)^2(-x^5y^4)^3$

Think of the negative sign in each factor as -1.

$$\begin{aligned}(-1x^3y)^2(-1x^5y^4)^3 &= (-1)^2(x^3)^2y^2 \cdot (-1)^3(x^5)^3(y^4)^3 && \text{Power rule (b)}\\ &= (-1)^2(x^6)(y^2)(-1)^3(x^{15})(y^{12}) && \text{Power rule (a)}\\ &= (-1)^5(x^{21})(y^{14}) && \text{Product rule}\\ &= -1x^{21}y^{14}\\ &= -x^{21}y^{14}\end{aligned}$$

Work Problem 8 at the Side.

OBJECTIVE 7 Use the rules for exponents in an application of geometry.

EXAMPLE 9 Using an Area Formula

Find an expression that represents the area of each geometric figure in Figure 1.

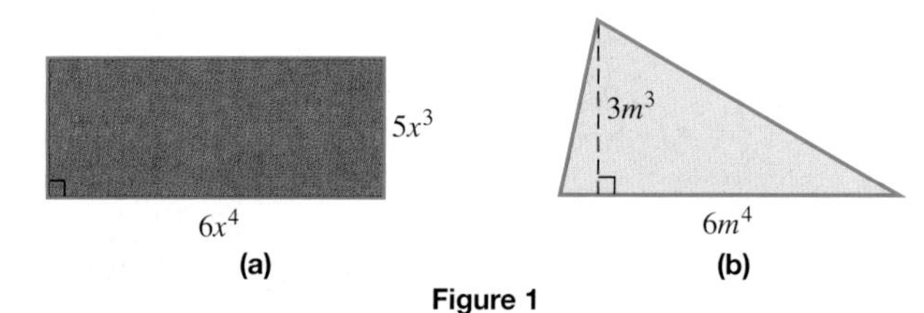

Figure 1

(a) Use the formula for the area of a rectangle, $A = LW$.

$$\begin{aligned}A &= (6x^4)(5x^3) && \text{Area formula}\\ A &= 30x^7 && \text{Product rule}\end{aligned}$$

(b) This is a triangle with base $6m^4$ and height $3m^3$.

$$\begin{aligned}A &= \frac{1}{2}bh && \text{Area formula}\\ &= \frac{1}{2}(6m^4)(3m^3) && \text{Substitute.}\\ &= \frac{1}{2}(18m^7) && \text{Product rule}\\ &= 9m^7\end{aligned}$$

Work Problem 9 at the Side.

ANSWERS

8. (a) 2^7m^7 or $128m^7$ (b) $\frac{5^2k^6}{3^2}$ or $\frac{25k^6}{9}$
(c) $\frac{2^2x^2}{5^4}$ or $\frac{4x^2}{625}$
(d) $-3^3x^{11}y^{10}$ or $-27x^{11}y^{10}$

9. $32x^6$

5.2 Exercises

FOR EXTRA HELP

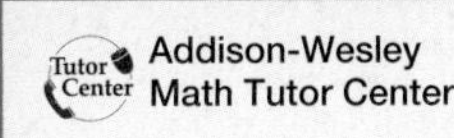

1. What exponent is understood on the base x in the expression xy^2?

2. How are the expressions 3^2, 5^3, and 7^4 read?

Decide whether each statement is true *or* false.

3. $3^3 = 9$

4. $(-2)^4 = 2^4$

5. $(a^2)^3 = a^5$

6. $\left(\frac{1}{4}\right)^2 = \frac{1}{4^2}$

Write each expression using exponents. See Example 1.

7. $(-2)(-2)(-2)(-2)(-2)$

8. $w \cdot w \cdot w \cdot w \cdot w \cdot w$

9. $\left(\frac{1}{2}\right)\left(\frac{1}{2}\right)\left(\frac{1}{2}\right)\left(\frac{1}{2}\right)\left(\frac{1}{2}\right)\left(\frac{1}{2}\right)$

10. $\left(-\frac{1}{4}\right)\left(-\frac{1}{4}\right)\left(-\frac{1}{4}\right)\left(-\frac{1}{4}\right)\left(-\frac{1}{4}\right)$

11. $(-8p)(-8p)$

12. $(-7x)(-7x)(-7x)(-7x)$

13. Explain how the expressions $(-3)^4$ and -3^4 are different.

14. Explain how the expressions $(5x)^3$ and $5x^3$ are different.

Identify the base and the exponent for each exponential expression. In Exercises 15–18, also evaluate the expression. See Example 2.

15. 3^5

16. 2^7

17. $(-3)^5$

18. $(-2)^7$

19. $(-6x)^4$

20. $(-8x)^4$

21. $-6x^4$

22. $-8x^4$

23. Explain why the product rule does not apply to the expression $5^2 + 5^3$. Then evaluate the expression by finding the individual powers and adding the results.

24. Explain why the product rule does not apply to the expression $3^2 \cdot 4^3$. Then evaluate the expression by finding the individual powers and multiplying the results.

Use the product rule to simplify each expression. Write each answer in exponential form. See Examples 3 and 4.

25. $5^2 \cdot 5^6$

26. $3^6 \cdot 3^7$

27. $4^2 \cdot 4^7 \cdot 4^3$

28. $5^3 \cdot 5^8 \cdot 5^2$

29. $(-7)^3(-7)^6$

30. $(-9)^8(-9)^5$

31. $t^3 \cdot t^8 \cdot t^{13}$

32. $n^5 \cdot n^6 \cdot n^9$

33. $(-8r^4)(7r^3)$

34. $(10a^7)(-4a^3)$

35. $(-6p^5)(-7p^5)$

36. $(-5w^8)(-9w^8)$

Use the power rules for exponents to simplify each expression. Write each answer in exponential form. See Examples 5–7.

37. $(4^3)^2$ **38.** $(8^3)^6$ **39.** $(t^4)^5$ **40.** $(y^6)^5$

41. $(7r)^3$ **42.** $(11x)^4$ **43.** $(5xy)^5$ **44.** $(9pq)^6$

45. $8(qr)^3$ **46.** $4(vw)^5$ **47.** $\left(\frac{1}{2}\right)^3$ **48.** $\left(\frac{1}{3}\right)^5$

49. $\left(\frac{a}{b}\right)^3$ $(b \neq 0)$ **50.** $\left(\frac{r}{t}\right)^4$ $(t \neq 0)$ **51.** $\left(\frac{9}{5}\right)^8$ **52.** $\left(\frac{12}{7}\right)^3$

53. $(-2x^2y)^3$ **54.** $(-5m^4p^2)^3$ **55.** $(3a^3b^2)^2$ **56.** $(4x^3y^5)^4$

Use a combination of the rules for exponents introduced in this section to simplify each expression. See Example 8.

57. $\left(\frac{5}{2}\right)^3 \cdot \left(\frac{5}{2}\right)^2$ **58.** $\left(\frac{3}{4}\right)^5 \cdot \left(\frac{3}{4}\right)^6$ **59.** $\left(\frac{9}{8}\right)^3 \cdot 9^2$ **60.** $\left(\frac{8}{5}\right)^4 \cdot 8^3$

61. $(2x)^9(2x)^3$ **62.** $(6y)^5(6y)^8$ **63.** $(-6p)^4(-6p)$ **64.** $(-13q)^3(-13q)$

65. $(6x^2y^3)^5$ **66.** $(5r^5t^6)^7$ **67.** $(x^2)^3(x^3)^5$ **68.** $(y^4)^5(y^3)^5$

69. $(2w^2x^3y)^2(x^4y)^5$ **70.** $(3x^4y^2z)^3(yz^4)^5$ **71.** $(-r^4s)^2(-r^2s^3)^5$

72. $(-ts^6)^4(-t^3s^5)^3$ **73.** $\left(\frac{5a^2b^5}{c^6}\right)^3$ $(c \neq 0)$ **74.** $\left(\frac{6x^3y^9}{z^5}\right)^4$ $(z \neq 0)$

75. $(-5m^3p^4q)^2(p^2q)^3$ **76.** $(-a^4b^5)(-6a^3b^3)^2$ **77.** $(2x^2y^3z)^4(xy^2z^3)^2$

Find the area of each figure. See Example 9.

78.

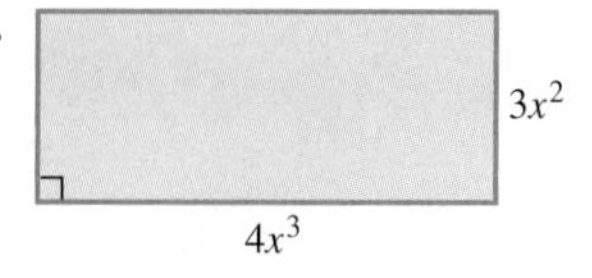

79.

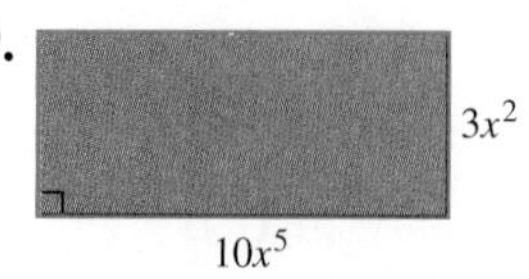

80.

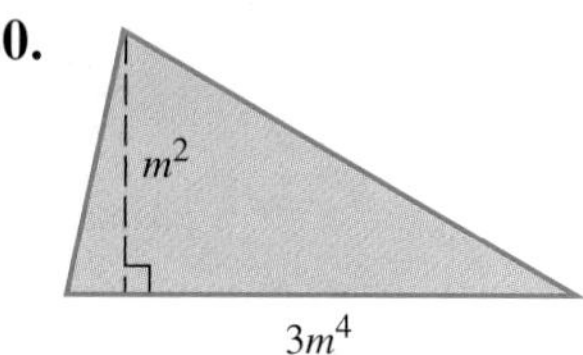

5.3 Multiplying Polynomials

OBJECTIVES

1. Multiply a monomial and a polynomial.
2. Multiply two polynomials.
3. Multiply binomials by the FOIL method.

OBJECTIVE 1 Multiply a monomial and a polynomial. As shown in **Section 5.2,** we find the product of two monomials by using the rules for exponents and the commutative and associative properties. For example,

$$(-8m^6)(-9n^6) = (-8)(-9)(m^6)(n^6) = 72m^6n^6.$$

CAUTION

Do not confuse addition of terms with multiplication of terms.

$$7q^5 + 2q^5 = 9q^5, \quad \text{but} \quad (7q^5)(2q^5) = 7 \cdot 2q^{5+5} = 14q^{10}.$$

To find the product of a monomial and a polynomial with more than one term, we use the distributive property and multiplication of monomials.

EXAMPLE 1 Multiplying a Monomial and a Polynomial

Use the distributive property to find each product.

(a) $4x^2(3x + 5)$

$$4x^2(3x + 5) = 4x^2(3x) + 4x^2(5) \quad \text{Distributive property}$$
$$= 12x^3 + 20x^2 \quad \text{Multiply monomials.}$$

(b) $-8m^3(4m^3 + 3m^2 + 2m - 1)$

$$= -8m^3(4m^3) + (-8m^3)(3m^2)$$
$$+ (-8m^3)(2m) + (-8m^3)(-1) \quad \text{Distributive property}$$
$$= -32m^6 - 24m^5 - 16m^4 + 8m^3 \quad \text{Multiply monomials.}$$

Work Problem 1 at the Side.

1 Find each product.

(a) $5m^3(2m + 7)$

(b) $2x^4(3x^2 + 2x - 5)$

(c) $-4y^2(3y^3 + 2y^2 - 4y + 8)$

OBJECTIVE 2 Multiply two polynomials. We can use the distributive property repeatedly to find the product of any two polynomials. For example, to find the product of the polynomials $x^2 + 3x + 5$ and $x - 4$, think of $x - 4$ as a single quantity and use the distributive property as follows.

$$(x^2 + 3x + 5)(x - 4) = x^2(x - 4) + 3x(x - 4) + 5(x - 4)$$

Now use the distributive property three more times to find the products $x^2(x - 4)$, $3x(x - 4)$, and $5(x - 4)$.

$$x^2(x - 4) + 3x(x - 4) + 5(x - 4)$$
$$= x^2(x) + x^2(-4) + 3x(x) + 3x(-4) + 5(x) + 5(-4)$$
$$= x^3 - 4x^2 + 3x^2 - 12x + 5x - 20 \quad \text{Multiply monomials.}$$
$$= x^3 - x^2 - 7x - 20 \quad \text{Combine terms.}$$

This example suggests the following rule.

Multiplying Polynomials

To multiply two polynomials, multiply each term of the second polynomial by each term of the first polynomial and add the products.

ANSWERS

1. (a) $10m^4 + 35m^3$
   (b) $6x^6 + 4x^5 - 10x^4$
   (c) $-12y^5 - 8y^4 + 16y^3 - 32y^2$

2 Multiply.

(a) $(m^3 - 2m + 1) \cdot (2m^2 + 4m + 3)$

(b) $(6p^2 + 2p - 4)(3p^2 - 5)$

3 Find the product.

$$\begin{array}{r} 3x^2 + 4x - 5 \\ x + 4 \\ \hline \end{array}$$

EXAMPLE 2 Multiplying Two Polynomials

Multiply $(m^2 + 5)(4m^3 - 2m^2 + 4m)$.

Multiply each term of the second polynomial by each term of the first.

$$\begin{aligned}(m^2 + 5)(4m^3 - 2m^2 + 4m) &= m^2(4m^3) + m^2(-2m^2) + m^2(4m) + 5(4m^3) + 5(-2m^2) + 5(4m) \\ &= 4m^5 - 2m^4 + 4m^3 + 20m^3 - 10m^2 + 20m \\ &= 4m^5 - 2m^4 + 24m^3 - 10m^2 + 20m \quad \text{Combine like terms.}\end{aligned}$$

Work Problem 2 at the Side.

When at least one of the factors in a product of polynomials has three or more terms, the multiplication can be performed alternatively by writing one polynomial above the other vertically.

EXAMPLE 3 Multiplying Polynomials Vertically

Multiply $(x^3 + 2x^2 + 4x + 1)(3x + 5)$ using the vertical method.

Write the polynomials as follows.

$$\begin{array}{r} x^3 + 2x^2 + 4x + 1 \\ 3x + 5 \\ \hline \end{array}$$

It is not necessary to line up terms in columns, because any terms may be multiplied (not just like terms). Begin by multiplying each of the terms in the top row by 5.

$$\begin{array}{rl} x^3 + 2x^2 + 4x + 1 & \\ 3x + 5 & \\ \hline 5x^3 + 10x^2 + 20x + 5 & \quad 5(x^3 + 2x^2 + 4x + 1) \end{array}$$

Notice how this process is similar to multiplication of whole numbers. Now multiply each term in the top row by $3x$. Be careful to place like terms in columns, since the final step will involve addition (as in multiplying two whole numbers).

$$\begin{array}{rl} x^3 + 2x^2 + 4x + 1 & \\ 3x + 5 & \\ \hline 5x^3 + 10x^2 + 20x + 5 & \\ 3x^4 + 6x^3 + 12x^2 + 3x & \quad 3x(x^3 + 2x^2 + 4x + 1) \end{array}$$

Add like terms.

$$\begin{array}{r} x^3 + 2x^2 + 4x + 1 \\ 3x + 5 \\ \hline 5x^3 + 10x^2 + 20x + 5 \\ 3x^4 + 6x^3 + 12x^2 + 3x \\ \hline 3x^4 + 11x^3 + 22x^2 + 23x + 5 \end{array}$$

The product is $3x^4 + 11x^3 + 22x^2 + 23x + 5$.

Work Problem 3 at the Side.

ANSWERS

2. (a) $2m^5 + 4m^4 - m^3 - 6m^2 - 2m + 3$
(b) $18p^4 + 6p^3 - 42p^2 - 10p + 20$
3. $3x^3 + 16x^2 + 11x - 20$

EXAMPLE 4 Multiplying Polynomials Vertically

Find the product of $4m^3 - 2m^2 + 4m$ and $\frac{1}{2}m^2 + \frac{5}{2}$.

$$\begin{array}{rl}
4m^3 - 2m^2 + 4m & \\
\frac{1}{2}m^2 + \frac{5}{2} & \\
\hline
10m^3 - 5m^2 + 10m & \text{Terms of top row multiplied by } \tfrac{5}{2} \\
2m^5 - m^4 + 2m^3 \qquad\qquad & \text{Terms of top row multiplied by } \tfrac{1}{2}m^2 \\
\hline
2m^5 - m^4 + 12m^3 - 5m^2 + 10m & \text{Add.}
\end{array}$$

Work Problem 4 at the Side.

4 Find each product.

(a)
$$\begin{array}{r} k^3 - k^2 + k + 1 \\ \frac{2}{3}k - \frac{1}{3} \\ \hline \end{array}$$

(b)
$$\begin{array}{r} a^3 + 3a - 4 \\ 2a^2 + 6a + 5 \\ \hline \end{array}$$

We can use a rectangle to model polynomial multiplication. For example, to find the product

$$(2x + 1)(3x + 2),$$

label a rectangle with each term as shown below on the left. Then put the product of each pair of monomials in the appropriate box as shown on the right.

	$3x$	2
$2x$		
1		

	$3x$	2
$2x$	$6x^2$	$4x$
1	$3x$	2

The product of the binomials is the sum of these four monomial products.

$$\begin{aligned}(2x + 1)(3x + 2) &= 6x^2 + 4x + 3x + 2 \\ &= 6x^2 + 7x + 2\end{aligned}$$

Work Problem 5 at the Side.

5 Use the rectangle method to find each product.

(a) $(4x + 3)(x + 2)$

	x	2
$4x$	$4x^2$	
3		

(b) $(x + 5)(x^2 + 3x + 1)$

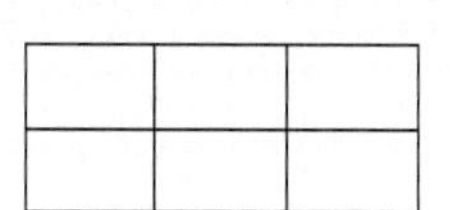

OBJECTIVE 3 Multiply binomials by the FOIL method. In algebra, many of the polynomials to be multiplied are both binomials (with just two terms). For these products, the **FOIL method** reduces the rectangle method to a systematic approach without the rectangle. To develop the FOIL method, we use the distributive property to find $(x + 3)(x + 5)$.

$$\begin{aligned}(x + 3)(x + 5) &= (x + 3)x + (x + 3)5 \\ &= x(x) + 3(x) + x(5) + 3(5) \\ &= x^2 + 3x + 5x + 15 \\ &= x^2 + 8x + 15\end{aligned}$$

Here is where the letters of the word FOIL originate.

$(x + 3)(x + 5)$ Multiply the **First terms**: $x(x)$. F

$(x + 3)(x + 5)$ Multiply the **Outer terms**: $x(5)$. O
This is the **outer product**.

$(x + 3)(x + 5)$ Multiply the **Inner terms**: $3(x)$. I
This is the **inner product**.

$(x + 3)(x + 5)$ Multiply the **Last terms**: $3(5)$. L

The inner product and the outer product should be added mentally so that the three terms of the answer can be written without extra steps as

$$(x + 3)(x + 5) = x^2 + 8x + 15.$$

Answers

4. (a) $\frac{2}{3}k^4 - k^3 + k^2 + \frac{1}{3}k - \frac{1}{3}$
(b) $2a^5 + 6a^4 + 11a^3 + 10a^2 - 9a - 20$

5. (a) $4x^2 + 11x + 6$
(b) $x^3 + 8x^2 + 16x + 5$

6 For the product $(2p - 5)(3p + 7)$, find the following.

(a) Product of first terms

(b) Outer product

(c) Inner product

(d) Product of last terms

(e) Complete product in simplified form

7 Use the FOIL method to find each product.

(a) $(m + 4)(m - 3)$

(b) $(y + 7)(y + 2)$

(c) $(r - 8)(r - 5)$

Answers

6. (a) $2p(3p) = 6p^2$ (b) $2p(7) = 14p$ (c) $-5(3p) = -15p$ (d) $-5(7) = -35$ (e) $6p^2 - p - 35$

7. (a) $m^2 + m - 12$ (b) $y^2 + 9y + 14$ (c) $r^2 - 13r + 40$

A summary of the steps in the FOIL method follows.

Multiplying Binomials by the FOIL Method

Step 1 Multiply the two **F**irst terms of the binomials to get the first term of the answer.

Step 2 Find the **O**uter product and the **I**nner product and combine them (when possible) to get the middle term of the answer.

Step 3 Multiply the two **L**ast terms of the binomials to get the last term of the answer.

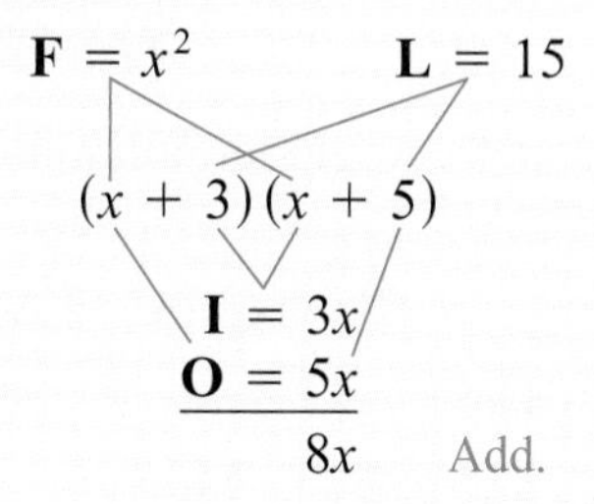

Work Problem 6 at the Side.

EXAMPLE 5 Using the FOIL Method

Use the FOIL method to find the product $(x + 8)(x - 6)$.

Step 1 **F** Multiply the **first** terms.

$$x(x) = x^2$$

Step 2 **O** Find the **outer** product.

$$x(-6) = -6x$$

I Find the **inner** product.

$$8(x) = 8x$$

Add the outer and inner products mentally.

$$-6x + 8x = 2x$$

Step 3 **L** Multiply the **last** terms.

$$8(-6) = -48$$

The product of $x + 8$ and $x - 6$ is the sum of the terms found in the three steps above, so

$$(x + 8)(x - 6) = x^2 + 2x - 48.$$

As a shortcut, this product can be found in the following manner.

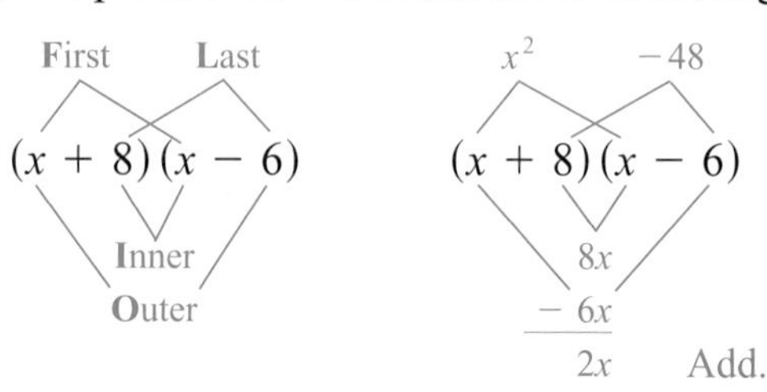

Work Problem 7 at the Side.

It is not possible to add the inner and outer products of the FOIL method if unlike terms result, as shown in the next example.

EXAMPLE 6 Using the FOIL Method

Multiply $(9x - 2)(3y + 1)$.

First	$(9x - 2)(3y + 1)$	$27xy$	
Outer	$(9x - 2)(3y + 1)$	$9x$	Unlike terms
Inner	$(9x - 2)(3y + 1)$	$-6y$	Unlike terms
Last	$(9x - 2)(3y + 1)$	-2	

$$(9x - 2)(3y + 1) = \overset{\text{F}}{27xy} + \overset{\text{O}}{9x} - \overset{\text{I}}{6y} - \overset{\text{L}}{2}$$

Work Problem 8 at the Side.

EXAMPLE 7 Using the FOIL Method

Find each product.

(a) $$(2k + 5y)(k + 3y) = \overset{\text{F}}{2k(k)} + \overset{\text{O}}{2k(3y)} + \overset{\text{I}}{5y(k)} + \overset{\text{L}}{5y(3y)}$$
$$= 2k^2 + 6ky + 5ky + 15y^2$$
$$= 2k^2 + 11ky + 15y^2$$

(b) $(7p + 2q)(3p - q) = 21p^2 - pq - 2q^2$ FOIL

(c) $2x^2(x - 3)(3x + 4) = 2x^2(3x^2 - 5x - 12)$ FOIL

$= 6x^4 - 10x^3 - 24x^2$ Distributive property

Work Problem 9 at the Side.

NOTE

Example 7(c) showed one way to multiply three polynomials. We could have multiplied $2x^2$ and $x - 3$ first, then multiplied that product and $3x + 4$ as follows.

$$2x^2(x - 3)(3x + 4) = (2x^3 - 6x^2)(3x + 4)$$
$$= 6x^4 - 10x^3 - 24x^2$$

8 Find the product.

$(4x - 3)(2y + 5)$

9 Find each product.

(a) $(6m + 5)(m - 4)$

(b) $(3r + 2t)(3r + 4t)$

(c) $y^2(8y + 3)(2y + 1)$

ANSWERS

8. $8xy + 20x - 6y - 15$

9. (a) $6m^2 - 19m - 20$

(b) $9r^2 + 18rt + 8t^2$

(c) $16y^4 + 14y^3 + 3y^2$

Focus on

Real-Data Applications

Algebra as Generalized Arithmetic

The rules of algebra are consistent with those for arithmetic. To learn a method for multiplying binomial factors, such as $(3x + 2)(4x + 1)$, we can observe the method for multiplying two-digit numbers, such as $32 \cdot 41$. The number 32 is shorthand for the expanded number $3 \cdot 10 + 2$, and the number 41 is shorthand for $4 \cdot 10 + 1$. So, multiplying $32 \cdot 41$ is the same as multiplying $(3 \cdot 10 + 2)(4 \cdot 10 + 1)$.

1. An expanded version of the usual algorithm is shown in the first column below. Each partial product, such as 1×2 and 40×30, is shown to clarify how it contributes to the process.
2. In the standard algorithm, it is clear that the 32 represents the sum of 1×2 and 1×30, and 1280 represents the sum of 40×2 and 40×30.
3. When FOIL is used to multiply the numbers, the term $11 \cdot 10$ represents the sum of the partial products 1×30 and 40×2. When we simplify $12 \cdot 10^2 + 11 \cdot 10 + 2$, the result is 1312.
4. When FOIL is used to multiply the binomials, each term exactly matches the corresponding term from the multiplication of the numbers.

Partial Products Multiplication Algorithm	Standard Algorithm	FOIL Method for Numbers	FOIL Method for Variables
32	32	$(3 \cdot 10 + 2)(4 \cdot 10 + 1)$	$(3x + 2)(4x + 1)$
$\times$ 41	$\times$ 41	$12 \cdot 10^2 + 3 \cdot 10 + 8 \cdot 10 + 2$	$12x^2 + 3x + 8x + 2$
$2 = 1 \times 2$	32	$12 \cdot 10^2 + 11 \cdot 10 + 2$	$12x^2 + 11x + 2$
$30 = 1 \times 30$	1280	$1200 + 110 + 2$	
$80 = 40 \times 2$	1312	1312	
$1200 = 40 \times 30$			
1312			

For Group Discussion

Use FOIL to compute each binomial product and corresponding arithmetic product. Verify that the results of the arithmetic product are valid. For the numerical problems, write the correct *signed* product for each term.

1. $(2x + 1)(2x - 3)$ and $(2 \cdot 10 + 1)(2 \cdot 10 - 3)$, which is $21 \cdot 17$
   Does the arithmetic FOIL result simplify to the correct answer?

2. $(4x - 2)(2x + 5)$ and $(4 \cdot 10 - 2)(2 \cdot 10 + 5)$, which is $38 \cdot 25$
   Does the arithmetic FOIL result simplify to the correct answer?

3. $(7x - 3)(5x - 4)$ and $(7 \cdot 10 - 3)(5 \cdot 10 - 4)$, which is $67 \cdot 46$
   Does the arithmetic FOIL result simplify to the correct answer?

4. A mental trick for multiplying two-digit numbers, such as $27 \cdot 18$, follows: "Multiply the ones $(7 \times 8 = 56)$. Write the 6 in the ones place, and carry the 5. Add the inner product (7×1), the outer product (2×8), and the carried 5 $(7 + 16 + 5 = 28)$. Write the 8 in the tens place, and carry the 2. Multiply the tens (2×1), and add to the carried 2. Write 4 in the hundreds place. The answer is 486." Why does the trick work?

5.3 Exercises

FOR EXTRA HELP

Addison-Wesley Math Tutor Center | MathXL | Digital Video Tutor CD 5 Videotape 5 | Student's Solutions Manual | MyMathLab | Interactmath.com

Find each product using the rectangle method shown in the text.

1. $(x + 3)(x + 4)$ **2.** $(x + 5)(x + 2)$ **3.** $(2x + 1)(x^2 + 3x + 2)$ **4.** $(x + 4)(3x^2 + 2x + 1)$

5. In multiplying a monomial by a polynomial, such as in $4x(3x^2 + 7x^3) = 4x(3x^2) + 4x(7x^3)$, the first property that is used is the ____________ property.

6. Match each product in parts (a)–(d) with the correct polynomial in choices A–D.

(a) $(x - 5)(x + 3)$ **(b)** $(x + 5)(x + 3)$ **(c)** $(x - 5)(x - 3)$ **(d)** $(x + 5)(x - 3)$

A. $x^2 + 8x + 15$ **B.** $x^2 - 8x + 15$ **C.** $x^2 - 2x - 15$ **D.** $x^2 + 2x - 15$

Find each product. See Example 1.

7. $-2m(3m + 2)$ **8.** $-5p(6 + 3p)$ **9.** $\frac{3}{4}p(8 - 6p + 12p^3)$

10. $\frac{4}{3}x(3 + 2x + 5x^3)$ **11.** $2y^5(3 + 2y + 5y^4)$ **12.** $2m^4(3m^2 + 5m + 6)$

Find each product. See Examples 2–4.

13. $(6x + 1)(2x^2 + 4x + 1)$ **14.** $(9y - 2)(8y^2 - 6y + 1)$

15. $(4m + 3)(5m^3 - 4m^2 + m - 5)$ **16.** $(y + 4)(3y^3 - 2y^2 + y + 3)$

17. $(2x - 1)(3x^5 - 2x^3 + x^2 - 2x + 3)$ **18.** $(2a + 3)(a^4 - a^3 + a^2 - a + 1)$

19. $(5x^2 + 2x + 1)(x^2 - 3x + 5)$ **20.** $(2m^2 + m - 3)(m^2 - 4m + 5)$

Find each binomial product using the FOIL method. See Examples 5–7.

21. $(n - 2)(n + 3)$ **22.** $(r - 6)(r + 8)$ **23.** $(4r + 1)(2r - 3)$

24. $(5x + 2)(2x - 7)$ **25.** $(3x + 2)(3x - 2)$ **26.** $(7x + 3)(7x - 3)$

27. $(3q + 1)(3q + 1)$

28. $(4w + 7)(4w + 7)$

29. $(3t + 4s)(2t + 5s)$

30. $(8v + 5w)(2v + 3w)$

31. $(-.3t + .4)(t + .6)$

32. $(-.5x + .9)(x - .2)$

33. $\left(x - \frac{2}{3}\right)\left(x + \frac{1}{4}\right)$

34. $\left(-\frac{8}{3} + 3k\right)\left(-\frac{2}{3} - k\right)$

35. $\left(-\frac{5}{4} + 2r\right)\left(-\frac{3}{4} - r\right)$

36. $2m^3(4m - 1)(2m + 3)$

37. $3y^3(2y + 3)(y - 5)$

38. $5t^4(t + 3)(3t - 1)$

RELATING CONCEPTS (EXERCISES 39–44) For Individual or Group Work

Work Exercises 39–44 in order. *(All units are in yards.) Refer to the figure as necessary.*

$3x + 6$

10

39. Find a polynomial that represents the area of the rectangle.

40. Suppose you know that the area of the rectangle is 600 yd^2. Use this information and the polynomial from Exercise 39 to write an equation in x, and solve it.

41. What are the dimensions of the rectangle?

42. Suppose the rectangle represents a lawn and it costs \$3.50 per square yard to lay sod on the lawn. How much will it cost to sod the entire lawn?

43. Use the result of Exercise 41 to find the perimeter of the lawn.

44. Again, suppose the rectangle represents a lawn and it costs \$9.00 per yard to fence the lawn. How much will it cost to fence the lawn?

---

45. Perform the following multiplications: $(x + 4)(x - 4)$; $(y + 2)(y - 2)$; $(r + 7)(r - 7)$. Observe your answers, and explain the pattern that can be found in the answers.

46. Repeat Exercise 45 for the following: $(x + 4)(x + 4)$; $(y - 2)(y - 2)$; $(r + 7)(r + 7)$.

5.4 Special Products

In this section, we develop shortcuts to find certain binomial products that occur frequently.

OBJECTIVES

1. Square binomials.
2. Find the product of the sum and difference of two terms.
3. Find greater powers of binomials.

OBJECTIVE 1 Square binomials. The square of a binomial can be found quickly by using the method shown in Example 1.

EXAMPLE 1 Squaring a Binomial

Find $(m + 3)^2$.

Squaring $m + 3$ by the FOIL method gives

$$(m + 3)(m + 3) = m^2 + 3m + 3m + 9$$
$$= m^2 + 6m + 9.$$

This result has the squares of the first and the last terms of the binomial:

$$m^2 = m^2 \quad \text{and} \quad 3^2 = 9.$$

The middle term, $6m$, is twice the product of the two terms of the binomial, since the outer and inner products are $m(3)$ and $3(m)$, and

$$m(3) + 3(m) = 2(m)(3) = 6m.$$

Work Problem 1 at the Side.

1 Consider the binomial $x + 4$.

(a) What is the first term of the binomial? Square it.

(b) What is the last term of the binomial? Square it.

(c) Find twice the product of the two terms of the binomial.

(d) Find $(x + 4)^2$.

Example 1 suggests the following rules.

Square of a Binomial

The square of a binomial is a trinomial consisting of the square of the first term, plus twice the product of the two terms, plus the square of the last term of the binomial. For a and b,

$$(a + b)^2 = a^2 + 2ab + b^2.$$

Also,

$$(a - b)^2 = a^2 - 2ab + b^2.$$

EXAMPLE 2 Squaring Binomials

Use the rules to square each binomial.

$$(a - b)^2 = a^2 - 2 \cdot a \cdot b + b^2$$

(a) $(5z - 1)^2 = (5z)^2 - 2(5z)(1) + (1)^2$
$= 25z^2 - 10z + 1$ $\qquad (5z)^2 = 5^2z^2 = 25z^2$

(b) $(3b + 5r)^2 = (3b)^2 + 2(3b)(5r) + (5r)^2$
$= 9b^2 + 30br + 25r^2$

(c) $(2a - 9x)^2 = 4a^2 - 36ax + 81x^2$

(d) $\left(4m + \frac{1}{2}\right)^2 = (4m)^2 + 2(4m)\left(\frac{1}{2}\right) + \left(\frac{1}{2}\right)^2$
$= 16m^2 + 4m + \frac{1}{4}$

ANSWERS

1. (a) x; x^2 (b) 4; 16 (c) $8x$ (d) $x^2 + 8x + 16$

2 Find each square by using the rules for the square of a binomial.

(a) $(t + u)^2$

(b) $(2m - p)^2$

(c) $(4p + 3q)^2$

(d) $(5r - 6s)^2$

(e) $\left(3k - \frac{1}{2}\right)^2$

Notice that in the square of a sum, all of the terms are positive, as in Examples 2(b) and (d). In the square of a difference, the middle term is negative, as in Examples 2(a) and (c).

CAUTION

A common error when squaring a binomial is to forget the middle term of the product. In general,

$$(a + b)^2 \neq a^2 + b^2.$$

Work Problem 2 at the Side.

OBJECTIVE 2 Find the product of the sum and difference of two terms. Binomial products of the form $(a + b)(a - b)$ also occur frequently. In these products, one binomial is the sum of two terms, and the other is the difference of the same two terms. For example, the product of $x + 2$ and $x - 2$ is

$$\begin{aligned}(x + 2)(x - 2) &= x^2 - 2x + 2x - 4 \\ &= x^2 - 4.\end{aligned}$$

As this example suggests, the product of $a + b$ and $a - b$ is the difference between two squares.

Product of the Sum and Difference of Two Terms

$$(a + b)(a - b) = a^2 - b^2$$

EXAMPLE 3 Finding the Product of the Sum and Difference of Two Terms

Find each product.

(a) $(x + 4)(x - 4)$

Use the rule for the product of the sum and difference of two terms.

$$\begin{aligned}(x + 4)(x - 4) &= x^2 - 4^2 \\ &= x^2 - 16\end{aligned}$$

(b) $\left(\frac{2}{3} - w\right)\left(\frac{2}{3} + w\right)$

By the commutative property, this product is the same as $\left(\frac{2}{3} + w\right)\left(\frac{2}{3} - w\right)$.

$$\begin{aligned}\left(\frac{2}{3} - w\right)\left(\frac{2}{3} + w\right) &= \left(\frac{2}{3} + w\right)\left(\frac{2}{3} - w\right) \\ &= \left(\frac{2}{3}\right)^2 - w^2 \\ &= \frac{4}{9} - w^2\end{aligned}$$

(c) $x(x + 2)(x - 2) = x(x^2 - 4)$
$= x^3 - 4x$

ANSWERS

2. **(a)** $t^2 + 2tu + u^2$
(b) $4m^2 - 4mp + p^2$
(c) $16p^2 + 24pq + 9q^2$
(d) $25r^2 - 60rs + 36s^2$
(e) $9k^2 - 3k + \frac{1}{4}$

EXAMPLE 4 Finding the Product of the Sum and Difference of Two Terms

Find each product.

$$\begin{matrix} (a & + b) & (a & - b) \\ \downarrow & \downarrow & \downarrow & \downarrow \end{matrix}$$

(a) $(5m + 3)(5m - 3)$

Use the rule for the product of the sum and difference of two terms.

$$(5m + 3)(5m - 3) = (5m)^2 - 3^2$$
$$= 25m^2 - 9$$

(b) $(4x + y)(4x - y) = (4x)^2 - y^2$
$$= 16x^2 - y^2$$

(c) $\left(z - \frac{1}{4}\right)\left(z + \frac{1}{4}\right) = z^2 - \frac{1}{16}$

(d) $2p(p^2 + 3)(p^2 - 3) = 2p(p^4 - 9)$
$$= 2p^5 - 18p$$

Work Problem 3 at the Side.

3 Find each product by using the rule for the sum and difference of two terms.

(a) $(6a + 3)(6a - 3)$

(b) $(10m + 7)(10m - 7)$

(c) $(7p + 2q)(7p - 2q)$

(d) $\left(3r - \frac{1}{2}\right)\left(3r + \frac{1}{2}\right)$

(e) $3x(x^3 - 4)(x^3 + 4)$

The product rules of this section will be important later, particularly in **Chapters 6** and **7.** Therefore, it is important to learn these rules and practice using them.

OBJECTIVE 3 Find greater powers of binomials. The methods used in the previous section and this section can be combined to find greater powers of binomials.

EXAMPLE 5 Finding Greater Powers of Binomials

Find each product.

(a)
$$\begin{aligned} (x + 5)^3 &= (x + 5)^2(x + 5) && a^3 = a^2 \cdot a \\ &= (x^2 + 10x + 25)(x + 5) && \text{Square the binomial.} \\ &= x^3 + 10x^2 + 25x + 5x^2 + 50x + 125 && \text{Multiply polynomials.} \\ &= x^3 + 15x^2 + 75x + 125 && \text{Combine like terms.} \end{aligned}$$

(b)
$$\begin{aligned} (2y - 3)^4 &= (2y - 3)^2(2y - 3)^2 && a^4 = a^2 \cdot a^2 \\ &= (4y^2 - 12y + 9)(4y^2 - 12y + 9) && \text{Square each binomial.} \\ &= 16y^4 - 48y^3 + 36y^2 - 48y^3 + 144y^2 && \text{Multiply polynomials.} \\ &\quad - 108y + 36y^2 - 108y + 81 \\ &= 16y^4 - 96y^3 + 216y^2 - 216y + 81 && \text{Combine like terms.} \end{aligned}$$

(c)
$$\begin{aligned} -2r(r + 2)^3 &= -2r(r + 2)(r + 2)^2 \\ &= -2r(r + 2)(r^2 + 4r + 4) \\ &= -2r(r^3 + 4r^2 + 4r + 2r^2 + 8r + 8) \\ &= -2r(r^3 + 6r^2 + 12r + 8) \\ &= -2r^4 - 12r^3 - 24r^2 - 16r \end{aligned}$$

Work Problem 4 at the Side.

4 Find each product.

(a) $(m + 1)^3$

(b) $(3k - 2)^4$

(c) $-3x(x - 4)^3$

ANSWERS

3. **(a)** $36a^2 - 9$ **(b)** $100m^2 - 49$ **(c)** $49p^2 - 4q^2$ **(d)** $9r^2 - \frac{1}{4}$ **(e)** $3x^7 - 48x$

4. **(a)** $m^3 + 3m^2 + 3m + 1$ **(b)** $81k^4 - 216k^3 + 216k^2 - 96k + 16$ **(c)** $-3x^4 + 36x^3 - 144x^2 + 192x$

Focus on

Real-Data Applications

Using a Rule of Thumb

A Dynamic Homes neighborhood features variations of three models of houses. Each buyer has an option to purchase a concrete patio extension, constructed to his or her choice of size. Each of the three models is designed so that the patio will be in the shape of a large square that is missing a corner square, similar to the diagram shown on the left. The size of the removed corner and the length of the patio vary from model to model, as does the size of the entire patio. To determine the dimensions of the patio, the rectangle of size $b \times (a - b)$ in the middle patio diagram can be rotated and repositioned to form the rectangle shown on the right, with length $a + b$ and width $a - b$.

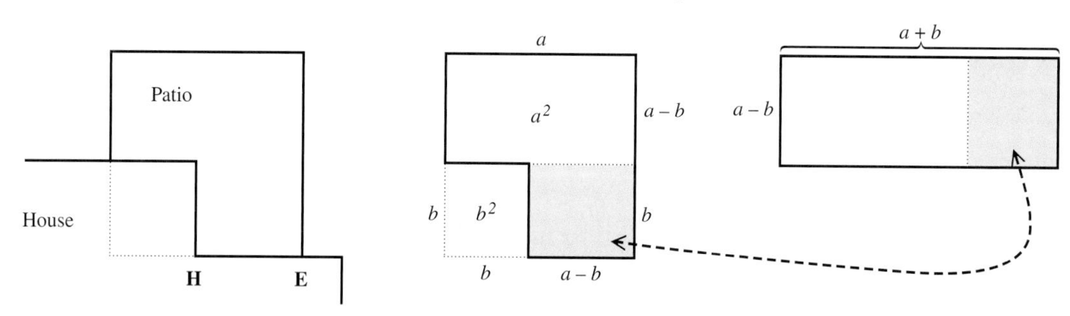

Since every time a house is sold the salesperson must ask the construction supervisor to calculate the cost of the patio extension, the supervisor devises a *rule of thumb* based on the quantity $(a + b)(a - b)$. There are only three possible choices for the value of b, one for each model house. This means that the salesperson needs to know only one additional measurement—the distance from the corner H to the edge of the patio E, which is $a - b$. The thickness of the patio is $\frac{1}{3}$ ft. The cost of concrete is \$58.00 per yd^3. (*Source:* Dunham Price, Inc., Westlake, LA.) There is a 100% markup in price to cover profit as well as costs for ground preparation, building forms, and finishing the concrete, giving a cost factor of \$116.00 per yd^3.

For Group Discussion

1. Why is knowing the values for b and $a - b$ sufficient for finding the quantity $a + b$?

2. Explain why the area of the patio can be written both as $(a + b)(a - b)$ and $a^2 - b^2$.

3. Apply the construction supervisor's rule of thumb to the three examples given in the table. To calculate the total cost for a patio with long side of length a feet, use the table and work from left to right. The second-to-last column is the volume of concrete needed, and the last column is the total cost of the patio. Round to the next tenth.

House Model	Corner Length b ft	Corner Length Doubled $2b$	Distance from Corner H to Patio Edge E $a - b$	Add Previous Two Columns $2b + (a - b)$	Multiply Previous Two Columns $(a + b)(a - b)$	VOLUME Multiply Previous Column by $\frac{1}{3}$ and Divide by 27	COST Multiply Previous Column by Cost Factor of \$116.00
A	5		10				
B	6		12				
C	8		14				

5.4 Exercises

FOR EXTRA HELP

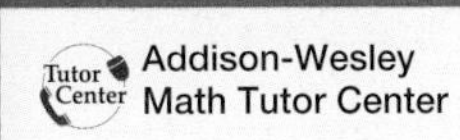

MathXL

MyMathLab

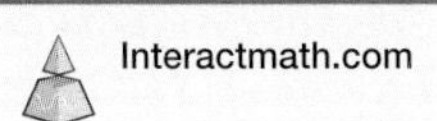

1. Consider the square $(2x + 3)^2$.

(a) What is the square of the first term, $(2x)^2$? ______

(b) What is twice the product of the two terms, $2(2x)(3)$? ______

(c) What is the square of the last term, 3^2? ______

(d) Write the final product, which is a trinomial, using your results from parts (a)–(c). ______

2. Repeat Exercise 1 for the square $(3x - 2)^2$.

Find each square. See Examples 1 and 2.

3. $(a - c)^2$ **4.** $(p - y)^2$ **5.** $(p + 2)^2$ **6.** $(r + 5)^2$

7. $(4x - 3)^2$ **8.** $(5y + 2)^2$ **9.** $(.8t + .7s)^2$ **10.** $(.7z - .3w)^2$

11. $\left(5x + \frac{2}{5}y\right)^2$ **12.** $\left(6m - \frac{4}{5}n\right)^2$ **13.** $\left(4a - \frac{3}{2}b\right)^2$ **14.** $x(2x + 5)^2$

15. $-(4r - 2)^2$ **16.** $-(3y - 8)^2$

17. Consider the product $(7x + 3y)(7x - 3y)$.

(a) What is the product of the first terms, $7x(7x)$? ______

(b) Multiply the outer terms, $7x(-3y)$. Then multiply the inner terms, $3y(7x)$. Add the results. What is this sum? ______

(c) What is the product of the last terms, $3y(-3y)$? ______

(d) Write the complete product using your answers in parts (a) and (c). ______ Why is the sum found in part (b) omitted here?

18. Repeat Exercise 17 for the product $(5x + 7y)(5x - 7y)$.

Find each product. See Examples 3 and 4.

19. $(q + 2)(q - 2)$ **20.** $(x + 8)(x - 8)$ **21.** $(2w + 5)(2w - 5)$ **22.** $(3z + 8)(3z - 8)$

23. $(10x + 3y)(10x - 3y)$ **24.** $(13r + 2z)(13r - 2z)$ **25.** $(2x^2 - 5)(2x^2 + 5)$ **26.** $(9y^2 - 2)(9y^2 + 2)$

27. $\left(7x + \frac{3}{7}\right)\left(7x - \frac{3}{7}\right)$ **28.** $\left(9y + \frac{2}{3}\right)\left(9y - \frac{2}{3}\right)$ **29.** $p(3p + 7)(3p - 7)$ **30.** $q(5q - 1)(5q + 1)$

RELATING CONCEPTS (EXERCISES 31–40) For Individual or Group Work

Special products can be illustrated by using areas of rectangles. Use the figure and **work Exercises 31–36 in order,** *to justify the special product* $(a + b)^2 = a^2 + 2ab + b^2$.

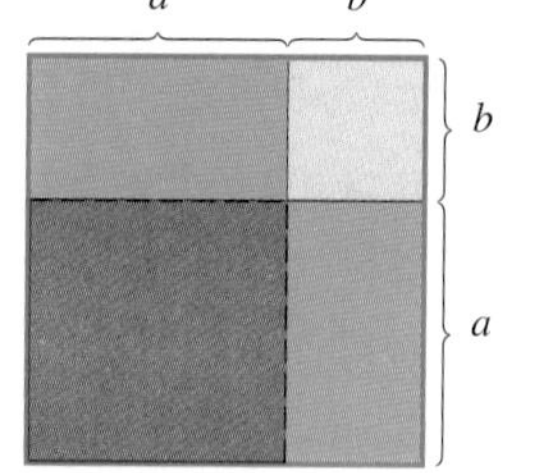

31. Express the area of the large square as the square of a binomial.

32. Give the monomial that represents the area of the red square.

33. Give the monomial that represents the sum of the areas of the blue rectangles.

34. Give the monomial that represents the area of the yellow square.

35. What is the sum of the monomials you obtained in Exercises 32–34?

36. Explain why the binomial square you found in Exercise 31 must equal the polynomial you found in Exercise 35.

To understand how the special product $(a + b)^2 = a^2 + 2ab + b^2$ *can be applied to a purely numerical problem,* **work Exercises 37–40 in order.**

37. Evaluate 35^2 using either traditional paper-and-pencil methods or a calculator.

38. The number 35 can be written as $30 + 5$. Therefore, $35^2 = (30 + 5)^2$. Use the special product for squaring a binomial with $a = 30$ and $b = 5$ to write an expression for $(30 + 5)^2$. Do not simplify at this time.

39. Use the order of operations to simplify the expression you found in Exercise 38.

40. How do the answers in Exercises 37 and 39 compare?

Find each product. See Example 5.

41. $(m - 5)^3$

42. $(p + 3)^3$

43. $(2a + 1)^3$

44. $(3m - 1)^3$

45. $(3r - 2t)^4$

46. $(2z + 5y)^4$

47. $3x^2(x - 3)^3$

48. $4p^3(p + 4)^3$

49. $-8x^2y(x + y)^4$

In Exercises 50 and 51, refer to the figure shown here.

50. Find a polynomial that represents the volume of the cube.

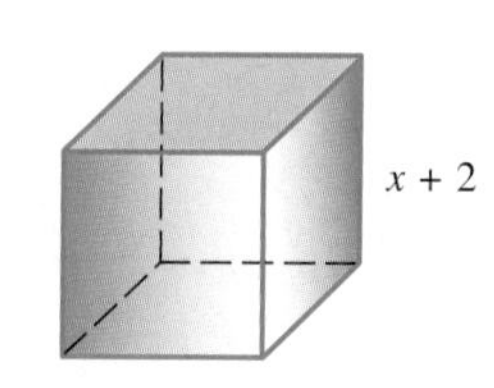

51. If the value of x is 6, what is the volume of the cube?

5.5 Integer Exponents and the Quotient Rule

OBJECTIVES

1. Use 0 as an exponent.
2. Use negative numbers as exponents.
3. Use the quotient rule for exponents.
4. Use combinations of rules.

In all our earlier work, exponents were positive integers. Now we want to develop meaning for exponents that are not positive integers.

Consider the following list of exponential expressions.

$$2^4 = 16$$
$$2^3 = 8$$
$$2^2 = 4$$

Do you see the pattern in the values? Each time we reduce the exponent by 1, the value is divided by 2 (the base). Using this pattern, we can continue the list to smaller and smaller integer exponents.

$$2^1 = 2$$
$$2^0 = 1$$
$$2^{-1} = \frac{1}{2}$$

Work Problem 1 at the Side.

1 Continue the list of exponentials using -2, -3, and -4 as exponents.

$2^{-2} =$ ______

$2^{-3} =$ ______

$2^{-4} =$ ______

From the preceding list and the answers to Problem 1, it appears that we should define 2^0 as 1 and negative exponents as reciprocals.

OBJECTIVE 1 Use 0 as an exponent. We want the definitions of 0 and negative exponents to satisfy the rules for exponents from **Section 5.2.** For example, if $6^0 = 1$,

$$6^0 \cdot 6^2 = 1 \cdot 6^2 = 6^2 \quad \text{and} \quad 6^0 \cdot 6^2 = 6^{0+2} = 6^2,$$

so the product rule is satisfied. Check that the power rules are also valid for a 0 exponent. Thus, we define a 0 exponent as follows.

Zero Exponent

For any nonzero real number a, $\quad a^0 = 1.$

Example: $17^0 = 1$.

2 Evaluate.

(a) 28^0

(b) $(-16)^0$

(c) -7^0

(d) $m^0, \quad m \neq 0$

(e) $-p^0, \quad p \neq 0$

EXAMPLE 1 Using Zero Exponents

Evaluate each exponential expression.

(a) $60^0 = 1$

(b) $(-60)^0 = 1$

(c) $-60^0 = -(1) = -1$

(d) $y^0 = 1, \quad y \neq 0$

(e) $6y^0 = 6(1) = 6, \quad y \neq 0$

(f) $(6y)^0 = 1, \quad y \neq 0$

CAUTION

Look again at Examples 1(b) and (c). In $(-60)^0$, the base is -60 and the exponent is 0. Any nonzero base raised to the exponent 0 is 1. In -60^0, the base is 60. Then $60^0 = 1$, and $-60^0 = -1$.

Work Problem 2 at the Side.

ANSWERS

1. $2^{-2} = \frac{1}{4}$; $2^{-3} = \frac{1}{8}$; $2^{-4} = \frac{1}{16}$

2. (a) 1 (b) 1 (c) -1 (d) 1 (e) -1

OBJECTIVE 2 Use negative numbers as exponents. From the lists at the beginning of this section and margin problem 1, since $2^{-2} = \frac{1}{4}$ and $2^{-3} = \frac{1}{8}$, we can deduce that 2^{-n} should equal $\frac{1}{2^n}$. Is the product rule valid in such cases? For example, if we multiply 6^{-2} by 6^2, we get

$$6^{-2} \cdot 6^2 = 6^{-2+2} = 6^0 = 1.$$

The expression 6^{-2} behaves as if it were the reciprocal of 6^2, because their product is 1. The reciprocal of 6^2 may be written $\frac{1}{6^2}$, leading us to define 6^{-2} as $\frac{1}{6^2}$. This is a particular case of the definition of negative exponents.

Negative Exponents

For any nonzero real number a and any integer n,

$$a^{-n} = \frac{1}{a^n}.$$

Example: $3^{-2} = \frac{1}{3^2}$.

By definition, a^{-n} and a^n are reciprocals, since

$$a^n \cdot a^{-n} = a^n \cdot \frac{1}{a^n} = 1.$$

Since $1^n = 1$, the definition of a^{-n} can also be written

$$a^{-n} = \frac{1}{a^n} = \frac{1^n}{a^n} = \left(\frac{1}{a}\right)^n.$$

For example, $6^{-3} = \left(\frac{1}{6}\right)^3$ and $\left(\frac{1}{3}\right)^{-2} = 3^2$.

EXAMPLE 2 Using Negative Exponents

Simplify by writing each expression with positive exponents.

(a) $3^{-2} = \frac{1}{3^2} = \frac{1}{9}$

(b) $5^{-3} = \frac{1}{5^3} = \frac{1}{125}$

(c) $\left(\frac{1}{2}\right)^{-3} = 2^3 = 8$ $\frac{1}{2}$ and 2 are reciprocals.

Notice that we can change the base to its reciprocal if we also change the sign of the exponent.

(d) $\left(\frac{2}{5}\right)^{-4} = \left(\frac{5}{2}\right)^4$ $\frac{2}{5}$ and $\frac{5}{2}$ are reciprocals.

(e) $\left(\frac{4}{3}\right)^{-5} = \left(\frac{3}{4}\right)^5$

(f) $4^{-1} - 2^{-1} = \frac{1}{4} - \frac{1}{2} = \frac{1}{4} - \frac{2}{4} = -\frac{1}{4}$

Apply the exponents first, then subtract.

Continued on Next Page

(g) $p^{-2} = \frac{1}{p^2}, \quad p \neq 0$

(h) $\frac{1}{x^{-4}}, \quad x \neq 0$

$$\frac{1}{x^{-4}} = \frac{1^{-4}}{x^{-4}} \qquad 1^{-4} = 1$$

$$= \left(\frac{1}{x}\right)^{-4} \qquad \text{Power rule (c)}$$

$$= x^4 \qquad \tfrac{1}{x} \text{ and } x \text{ are reciprocals.}$$

CAUTION

A negative exponent does not indicate a negative number; negative exponents lead to reciprocals.

Expression	**Example**	
a^{-n}	$3^{-2} = \frac{1}{3^2} = \frac{1}{9}$	Not negative
$-a^{-n}$	$-3^{-2} = -\frac{1}{3^2} = -\frac{1}{9}$	Negative

Work Problem 3 at the Side.

The definition of negative exponents allows us to move factors in a fraction if we also change the signs of the exponents. For example,

$$\frac{2^{-3}}{3^{-4}} = \frac{\frac{1}{2^3}}{\frac{1}{3^4}} = \frac{1}{2^3} \div \frac{1}{3^4} = \frac{1}{2^3} \cdot \frac{3^4}{1} = \frac{3^4}{2^3},$$

so that

$$\frac{2^{-3}}{3^{-4}} = \frac{3^4}{2^3}.$$

Changing from Negative to Positive Exponents

For any nonzero numbers a and b, and any integers m and n,

$$\frac{a^{-m}}{b^{-n}} = \frac{b^n}{a^m} \quad \text{and} \quad \left(\frac{a}{b}\right)^{-m} = \left(\frac{b}{a}\right)^m.$$

Examples: $\frac{3^{-5}}{2^{-4}} = \frac{2^4}{3^5}$ and $\left(\frac{4}{5}\right)^{-3} = \left(\frac{5}{4}\right)^3$.

3 Write with positive exponents.

(a) 4^{-3}

(b) 6^{-2}

(c) $\left(\frac{2}{3}\right)^{-2}$

(d) $2^{-1} + 5^{-1}$

(e) $m^{-5}, \quad m \neq 0$

(f) $\frac{1}{z^{-4}}, \quad z \neq 0$

ANSWERS

3. (a) $\frac{1}{4^3}$ **(b)** $\frac{1}{6^2}$ **(c)** $\left(\frac{3}{2}\right)^2$ **(d)** $\frac{1}{2} + \frac{1}{5} = \frac{7}{10}$ **(e)** $\frac{1}{m^5}$ **(f)** z^4

4 Write with only positive exponents. Assume all variables represent nonzero real numbers.

(a) $\dfrac{7^{-1}}{5^{-4}}$

(b) $\dfrac{x^{-3}}{y^{-2}}$

(c) $\dfrac{4h^{-5}}{m^{-2}k}$

(d) p^2q^{-5}

(e) $\left(\dfrac{3m}{p}\right)^{-2}$

ANSWERS

4. (a) $\dfrac{5^4}{7}$ (b) $\dfrac{y^2}{x^3}$ (c) $\dfrac{4m^2}{h^5k}$ (d) $\dfrac{p^2}{q^5}$ (e) $\dfrac{p^2}{3^2m^2}$

EXAMPLE 3 Changing from Negative to Positive Exponents

Write with only positive exponents. Assume all variables represent nonzero real numbers.

(a) $\dfrac{4^{-2}}{5^{-3}} = \dfrac{5^3}{4^2}$

(b) $\dfrac{m^{-5}}{p^{-1}} = \dfrac{p^1}{m^5} = \dfrac{p}{m^5}$

(c) $\dfrac{a^{-2}b}{3d^{-3}} = \dfrac{bd^3}{3a^2}$

Notice that b in the numerator and the coefficient 3 in the denominator were not affected.

(d) $x^3y^{-4} = \dfrac{x^3y^{-4}}{1} = \dfrac{x^3}{y^4}$

(e) $\left(\dfrac{x}{2y}\right)^{-4} = \left(\dfrac{2y}{x}\right)^4 = \dfrac{2^4y^4}{x^4}$

Work Problem 4 at the Side.

CAUTION

Be careful. We cannot change negative exponents to positive exponents using this rule if the exponents occur in a *sum* of terms. For example,

$$\frac{5^{-2} + 3^{-1}}{7 - 2^{-3}}$$

cannot be written with positive exponents using the rule given here. We would have to use the definition of a negative exponent to rewrite this expression with positive exponents, as

$$\frac{\frac{1}{5^2} + \frac{1}{3}}{7 - \frac{1}{2^3}}.$$

OBJECTIVE 3 Use the quotient rule for exponents. We now consider the quotient of exponential expressions with the same base. We know that

$$\frac{6^5}{6^3} = \frac{6 \cdot 6 \cdot 6 \cdot 6 \cdot 6}{6 \cdot 6 \cdot 6} = 6^2.$$

Notice that the difference between the exponents, $5 - 3 = 2$, is the exponent in the quotient. Also,

$$\frac{6^2}{6^4} = \frac{6 \cdot 6}{6 \cdot 6 \cdot 6 \cdot 6} = \frac{1}{6^2} = 6^{-2}.$$

Here, $2 - 4 = -2$. These examples suggest the **quotient rule for exponents.**

Quotient Rule for Exponents

For any nonzero real number a and any integers m and n,

$$\frac{a^m}{a^n} = a^{m-n}.$$

(Keep the same base and subtract the exponents.)

Example: $\frac{5^8}{5^4} = 5^{8-4} = 5^4.$

CAUTION

A common **error** is to write $\frac{5^8}{5^4} = 1^{8-4} = 1^4$. Notice that by the quotient rule, the quotient should have the *same base,* 5. That is,

$$\frac{5^8}{5^4} = 5^{8-4} = 5^4.$$

If you are not sure, use the definition of an exponent to write out the factors:

$$5^8 = 5 \cdot 5 \cdot 5 \cdot 5 \cdot 5 \cdot 5 \cdot 5 \cdot 5 \quad \text{and} \quad 5^4 = 5 \cdot 5 \cdot 5 \cdot 5.$$

Then it is clear that the quotient is 5^4.

EXAMPLE 4 Using the Quotient Rule for Exponents

Simplify, using the quotient rule for exponents. Write answers with positive exponents.

(a) $\frac{5^8}{5^6} = 5^{8-6} = 5^2$

(b) $\frac{4^2}{4^9} = 4^{2-9} = 4^{-7} = \frac{1}{4^7}$

(c) $\frac{5^{-3}}{5^{-7}} = 5^{-3-(-7)} = 5^4$

(d) $\frac{q^5}{q^{-3}} = q^{5-(-3)} = q^8, \quad q \neq 0$

(e) $\frac{3^2x^5}{3^4x^3} = \frac{3^2}{3^4} \cdot \frac{x^5}{x^3} = 3^{2-4} \cdot x^{5-3} = 3^{-2}x^2 = \frac{x^2}{3^2}, \quad x \neq 0$

(f) $\frac{(m+n)^{-2}}{(m+n)^{-4}} = (m+n)^{-2-(-4)} = (m+n)^{-2+4} = (m+n)^2, \; m \neq -n$

(g) $\frac{7x^{-3}y^2}{2^{-1}x^2y^{-5}} = \frac{7 \cdot 2^1y^2y^5}{x^2x^3} = \frac{14y^7}{x^5}, \quad x \neq 0$

Work Problem 5 at the Side.

5 Simplify. Write answers with positive exponents.

(a) $\frac{5^{11}}{5^8}$

(b) $\frac{4^7}{4^{10}}$

(c) $\frac{6^{-5}}{6^{-2}}$

(d) $\frac{8^4m^9}{8^5m^{10}}, \quad m \neq 0$

(e) $\frac{3^{-1}(x+y)^{-3}}{2^{-2}(x+y)^{-4}}, \quad x \neq -y$

ANSWERS

5. (a) 5^3 (b) $\frac{1}{4^3}$ (c) $\frac{1}{6^3}$ (d) $\frac{1}{8m}$ (e) $\frac{4}{3}(x+y)$

The definitions and rules for exponents given in this section and **Section 5.2** are summarized below.

Definitions and Rules for Exponents

For any integers m and n:

		Examples
Product rule	$a^m \cdot a^n = a^{m+n}$	$7^4 \cdot 7^5 = 7^9$
Zero exponent	$a^0 = 1 \quad (a \neq 0)$	$(-3)^0 = 1$
Negative exponent	$a^{-n} = \dfrac{1}{a^n} \quad (a \neq 0)$	$5^{-3} = \dfrac{1}{5^3}$
Quotient rule	$\dfrac{a^m}{a^n} = a^{m-n} \quad (a \neq 0)$	$\dfrac{2^2}{2^5} = 2^{2-5} = 2^{-3} = \dfrac{1}{2^3}$
Power rules	**(a)** $(a^m)^n = a^{mn}$	$(4^2)^3 = 4^6$
	(b) $(ab)^m = a^m b^m$	$(3k)^4 = 3^4 k^4$
	(c) $\left(\dfrac{a}{b}\right)^m = \dfrac{a^m}{b^m} \quad (b \neq 0)$	$\left(\dfrac{2}{3}\right)^2 = \dfrac{2^2}{3^2}$
Negative-to-positive rules	$\dfrac{a^{-m}}{b^{-n}} = \dfrac{b^n}{a^m} \quad (a, b \neq 0)$	$\dfrac{2^{-4}}{5^{-3}} = \dfrac{5^3}{2^4}$
	$\left(\dfrac{a}{b}\right)^{-m} = \left(\dfrac{b}{a}\right)^m$.	$\left(\dfrac{4}{7}\right)^{-2} = \left(\dfrac{7}{4}\right)^2$

OBJECTIVE 4 Use combinations of rules. We sometimes need to use more than one rule to simplify an expression.

EXAMPLE 5 Using a Combination of Rules

Use a combination of the rules for exponents to simplify each expression. Assume all variables represent nonzero real numbers.

(a) $\dfrac{(4^2)^3}{4^5} = \dfrac{4^6}{4^5}$ Power rule (a)

$= 4^{6-5}$ Quotient rule

$= 4^1 = 4$

(b) $(2x)^3(2x)^2 = (2x)^5$ Product rule

$= 2^5x^5$ or $32x^5$ Power rule (b)

(c) $\left(\dfrac{2x^3}{5}\right)^{-4} = \left(\dfrac{5}{2x^3}\right)^4$ Negative-to-positive rule

$= \dfrac{5^4}{2^4x^{12}}$ or $\dfrac{625}{16x^{12}}$ Power rules (a)–(c)

(d) $\left(\dfrac{3x^{-2}}{4^{-1}y^3}\right)^{-3} = \dfrac{3^{-3}x^6}{4^3y^{-9}}$ Power rules (a)–(c)

$= \dfrac{x^6y^9}{4^3 \cdot 3^3}$ or $\dfrac{x^6y^9}{1728}$ Negative-to-positive rule

Continued on Next Page

(e) $\dfrac{(4m)^{-3}}{(3m)^{-4}} = \dfrac{4^{-3}m^{-3}}{3^{-4}m^{-4}}$ Power rule (b)

$= \dfrac{3^4m^4}{4^3m^3}$ Negative-to-positive rule

$= \dfrac{3^4m^{4-3}}{4^3}$ Quotient rule

$= \dfrac{3^4m}{4^3}$ or $\dfrac{81m}{64}$

NOTE

Since the steps can be done in several different orders, there are many equally correct ways to simplify expressions like those in Examples 5(d) and 5(e).

Work Problem 6 at the Side.

6 Simplify. Assume all variables represent nonzero real numbers.

(a) $12^5 \cdot 12^{-7} \cdot 12^6$

(b) $y^{-2} \cdot y^5 \cdot y^{-8}$

(c) $\dfrac{(6x)^{-1}}{(3x^2)^{-2}}$

(d) $\dfrac{3^9 \cdot (x^2y)^{-2}}{3^3 \cdot x^{-4}y}$

Answers

6. **(a)** 12^4 or 20,736 **(b)** $\dfrac{1}{y^5}$ **(c)** $\dfrac{3x^3}{2}$ **(d)** $\dfrac{3^6}{y^3}$ or $\dfrac{729}{y^3}$

Focus on Real-Data Applications

Numbers BIG and SMALL

The ancient Egyptians used the *astonished man* symbol to represent a million. We now use exponents to write very big and very small numbers efficiently. Columbia University professor Edward Kasner asked his nine-year-old nephew to think of a name for the exceedingly large number 10^{100}, which is a 1 followed by 100 zeros. His nephew proclaimed it to be a **googol.** Kasner then called the unimaginably large number 10^{googol} a **googolplex.**

> A googol is 10^{100} or
>
> 10,000,000,000,000,000,000,000,000,000,000,000,
>
> 000,000,000,000,000,000,000,000,000,000,000,
>
> 000,000,000,000,000,000,000,000,000,000,000.

Real-world examples of very big and very small numbers are described in Howard Eves's book *Mathematical Circles Revisited*. A few such examples are listed here.

- The total number of electrons in the universe is, according to an estimate by Sir Arthur Eddington, about 10^{79}.
- The number of grains of sand on the beach at Coney Island, New York, is about 10^{20}.
- The total number of printed words since the Gutenberg Bible appeared is approximately 10^{16}.
- The temperature at the center of an atomic bomb explosion is 2×10^{8} degrees Fahrenheit.
- The diameter of a human hair is about 10^{-4} millimeters.
- The diameter of a nucleus of a cell is about 10^{-6} millimeters (1 micron).
- The probability of winning a lottery by choosing 6 numbers from among 50 is about 6.3×10^{-8}.
- The size of a quark is approximately 10^{-18} millimeters.

Interestingly enough, the Web search engine Google™ is named after a googol. Sergey Brin, president and co-founder of Google, Inc., was a math major. He chose the name Google to describe the vast reach of this search engine. (*Source: The Gazette,* March 2, 2001.)

For Group Discussion

Suppose that you have been offered a new job with "salary negotiable." You present to your new employer the following offer: You will work for 30 days and will be paid 1¢ on day one, 2¢ on day two, 4¢ on day three, 8¢ on day four, and so on, doubling your pay each day for the month. Your employer accepts your proposal, and you begin work on the first day of the next month.

1. On which day will you have received half of your month's wages?
2. When will you have received one-fourth of your total month's wages?
3. What do you think your approximate monthly salary will be?
4. How many days would it take you to become a "googolaire"?

5.5 Exercises

Decide whether each expression is positive, negative, or 0.

1. $(-2)^{-3}$

2. $(-3)^{-2}$

3. -2^4

4. -3^6

5. $\left(\frac{1}{4}\right)^{-2}$

6. $\left(\frac{1}{5}\right)^{-2}$

7. $1 - 5^0$

8. $1 - 7^0$

Each expression is equal to either 0, 1, *or* -1. *Decide which is correct. See Example 1.*

9. $(-4)^0$

10. $(-10)^0$

11. -9^0

12. -5^0

13. $(-2)^0 - 2^0$

14. $(-8)^0 - 8^0$

15. $\frac{0^{10}}{10^0}$

16. $\frac{0^5}{5^0}$

Evaluate each expression. See Examples 1 and 2.

17. $7^0 + 9^0$

18. $8^0 + 6^0$

19. 4^{-3}

20. 5^{-4}

21. $\left(\frac{1}{2}\right)^{-4}$

22. $\left(\frac{1}{3}\right)^{-3}$

23. $\left(\frac{6}{7}\right)^{-2}$

24. $\left(\frac{2}{3}\right)^{-3}$

25. $5^{-1} + 3^{-1}$

26. $6^{-1} + 2^{-1}$

27. $-2^{-1} + 3^{-2}$

28. $(-3)^{-2} + (-4)^{-1}$

RELATING CONCEPTS (EXERCISES 29–32) For Individual or Group Work

In Objective 1, we used the product rule to motivate the definition of a 0 *exponent. We can also use the quotient rule. To see this,* ***work Exercises 29–32 in order.***

29. Consider the expression $\frac{25}{25}$. What is its simplest form?

30. Write the quotient in Exercise 29 using the fact that $25 = 5^2$.

31. Apply the quotient rule for exponents to your answer for Exercise 30. Give the answer as a power of 5.

32. Because your answers for Exercises 29 and 31 both represent $\frac{25}{25}$, they must be equal. Write this equality. What definition does it support?

Use the quotient rule to simplify each expression. Write each expression with positive exponents. Assume that all variables represent nonzero real numbers. See Examples 2–4.

33. $\frac{9^4}{9^5}$

34. $\frac{7^3}{7^4}$

35. $\frac{6^{-3}}{6^2}$

36. $\frac{4^{-2}}{4^3}$

37. $\frac{1}{6^{-3}}$

38. $\frac{1}{5^{-2}}$

39. $\frac{2}{r^{-4}}$

40. $\frac{3}{s^{-8}}$

41. $\frac{4^{-3}}{5^{-2}}$

42. $\frac{6^{-2}}{5^{-4}}$

43. p^5q^{-8}

44. $x^{-8}y^4$

45. $\frac{r^5}{r^{-4}}$

46. $\frac{a^6}{a^{-4}}$

47. $\frac{6^4x^8}{6^5x^3}$

48. $\frac{3^8y^5}{3^{10}y^2}$

49. $\frac{6y^3}{2y}$

50. $\frac{5m^2}{m}$

51. $\frac{3x^5}{3x^2}$

52. $\frac{10p^8}{2p^4}$

Use a combination of the rules for exponents to simplify each expression. Write answers with only positive exponents. Assume that all variables represent nonzero real numbers. See Example 5.

53. $\frac{(7^4)^3}{7^9}$

54. $\frac{(5^3)^2}{5^2}$

55. $x^{-3} \cdot x^5 \cdot x^{-4}$

56. $y^{-8} \cdot y^5 \cdot y^{-2}$

57. $\frac{(3x)^{-2}}{(4x)^{-3}}$

58. $\frac{(2y)^{-3}}{(5y)^{-4}}$

59. $\left(\frac{x^{-1}y}{z^2}\right)^{-2}$

60. $\left(\frac{p^{-4}q}{r^{-3}}\right)^{-3}$

61. $(6x)^4(6x)^{-3}$

62. $(10y)^9(10y)^{-8}$

63. $\frac{(m^7n)^{-2}}{m^{-4}n^3}$

64. $\frac{(m^8n^{-4})^2}{m^{-2}n^5}$

65. $\frac{5x^{-3}}{(4x)^2}$

66. $\frac{-3k^5}{(2k)^2}$

67. $\left(\frac{2p^{-1}q}{3^{-1}m^2}\right)^2$

68. $\left(\frac{4xy^2}{x^{-1}y}\right)^{-2}$

Summary Exercises on the Rules for Exponents

Use the rules for exponents to simplify each expression. Use only positive exponents in your answers. Assume that all variables represent nonzero real numbers.

1. $\left(\frac{6x^2}{5}\right)^{12}$

2. $\left(\frac{rs^2t^3}{3t^4}\right)^6$

3. $(10x^2y^4)^2(10xy^2)^3$

4. $(-2ab^3c)^4(-2a^2b)^3$

5. $\left(\frac{9wx^3}{y^4}\right)^3$

6. $(4x^{-2}y^{-3})^{-2}$

7. $\frac{c^{11}(c^2)^4}{(c^3)^3(c^2)^{-6}}$

8. $\left(\frac{k^4t^2}{k^2t^{-4}}\right)^{-2}$

9. $5^{-1} + 6^{-1}$

10. $\frac{(3y^{-1}z^3)^{-1}(3y^2)}{(y^3z^2)^{-3}}$

11. $\frac{(2xy^{-1})^3}{2^3x^{-3}y^2}$

12. $-8^0 + (-8)^0$

13. $(z^4)^{-3}(z^{-2})^{-5}$

14. $\left(\frac{r^2st^5}{3r}\right)^{-2}$

15. $\frac{(3^{-1}x^{-3}y)^{-1}(2x^2y^{-3})^2}{(5x^{-2}y^2)^{-2}}$

16. $\left(\frac{5x^2}{3x^{-4}}\right)^{-1}$

17. $\left(\frac{-2x^{-2}}{2x^2}\right)^{-2}$

18. $\frac{(x^{-4}y^2)^3(x^2y)^{-1}}{(xy^2)^{-3}}$

19. $\frac{(a^{-2}b^3)^{-4}}{(a^{-3}b^2)^{-2}(ab)^{-4}}$

20. $(2a^{-30}b^{-29})(3a^{31}b^{30})$

21. $5^{-2} + 6^{-2}$

22. $\left(\frac{(x^{47}y^{23})^2}{x^{-26}y^{-42}}\right)^0$

23. $\left(\frac{7a^2b^3}{2}\right)^3$

24. $-(-12^0)$

25. $-(-12)^0$

26. $\frac{0^{12}}{12^0}$

27. $\frac{(2xy^{-3})^{-2}}{(3x^{-2}y^4)^{-3}}$

28. $\left(\frac{a^2b^3c^4}{a^{-2}b^{-3}c^{-4}}\right)^{-2}$

29. $(6x^{-5}z^3)^{-3}$

30. $(2p^{-2}qr^{-3})(2p)^{-4}$

31. $\frac{(xy)^{-3}(xy)^5}{(xy)^{-4}}$

32. $42^0 - (-12)^0$

33. $\frac{(7^{-1}x^{-3})^{-2}(x^4)^{-6}}{7^{-1}x^{-3}}$

34. $\left(\frac{3^{-4}x^{-3}}{3^{-3}x^{-6}}\right)^{-2}$

35. $(5p^{-2}q)^{-3}(5pq^3)^4$

36. $8^{-1} + 6^{-1}$

37. $\left(\frac{4r^{-6}s^{-2}t}{2r^8s^{-4}t^2}\right)^{-1}$

38. $(13x^{-6}y)(13x^{-6}y)^{-1}$

39. $\frac{(8pq^{-2})^4}{(8p^{-2}q^{-3})^3}$

40. $\left(\frac{mn^{-2}p}{m^2np^4}\right)^{-2}\left(\frac{mn^{-2}p}{m^2np^4}\right)^3$

41. $-(-3^0)^0$

42. $5^{-1} - 8^{-1}$

5.6 Dividing a Polynomial by a Monomial

OBJECTIVE 1 Divide a polynomial by a monomial. We add two fractions with a common denominator as follows.

$$\frac{a}{c} + \frac{b}{c} = \frac{a+b}{c}$$

Looking at this statement in reverse gives us a rule for dividing a polynomial by a monomial.

Dividing a Polynomial by a Monomial

To divide a polynomial by a monomial, divide each term of the polynomial by the monomial:

$$\frac{a+b}{c} = \frac{a}{c} + \frac{b}{c} \quad (c \neq 0).$$

Examples: $\frac{2+5}{3} = \frac{2}{3} + \frac{5}{3}$ and $\frac{x+3z}{2y} = \frac{x}{2y} + \frac{3z}{2y}$.

The parts of a division problem are named here.

$$\text{Dividend} \rightarrow \frac{12x^2 + 6x}{6x} = 2x + 1 \leftarrow \text{Quotient}$$
$$\text{Divisor} \rightarrow 6x$$

EXAMPLE 1 Dividing a Polynomial by a Monomial

Divide $5m^5 - 10m^3$ by $5m^2$.

Use the preceding rule, with + replaced by −. Then use the quotient rule.

$$\frac{5m^5 - 10m^3}{5m^2} = \frac{5m^5}{5m^2} - \frac{10m^3}{5m^2} = \mathbf{m^3 - 2m}$$

Check by multiplying: $5m^2(m^3 - 2m) = 5m^5 - 10m^3$.

Because division by 0 is undefined, the quotient

$$\frac{5m^5 - 10m^3}{5m^2}$$

is undefined if $m = 0$. From now on, we assume that no denominators are 0.

Work Problem 1 at the Side.

EXAMPLE 2 Dividing a Polynomial by a Monomial

Divide: $\frac{16a^5 - 12a^4 + 8a^2}{4a^3}$.

Divide each term of $16a^5 - 12a^4 + 8a^2$ by $4a^3$.

$$\frac{16a^5 - 12a^4 + 8a^2}{4a^3} = \frac{16a^5}{4a^3} - \frac{12a^4}{4a^3} + \frac{8a^2}{4a^3}$$

$$= 4a^2 - 3a + \frac{2}{a} \qquad \text{Quotient rule}$$

Continued on Next Page

OBJECTIVE

1 Divide a polynomial by a monomial.

1 Divide.

(a) $\frac{6p^4 + 18p^7}{3p^2}$

(b) $\frac{12m^6 + 18m^5 + 30m^4}{6m^2}$

(c) $(18r^7 - 9r^2) \div (3r)$

ANSWERS

1. (a) $2p^2 + 6p^5$ (b) $2m^4 + 3m^3 + 5m^2$ (c) $6r^6 - 3r$

The quotient $4a^2 - 3a + \frac{2}{a}$ is not a polynomial because of the expression $\frac{2}{a}$, which has a variable in the denominator. While the sum, difference, and product of two polynomials are always polynomials, the quotient of two polynomials may not be.

Again, *check* by multiplying.

$$4a^3\left(4a^2 - 3a + \frac{2}{a}\right) = 4a^3(4a^2) - 4a^3(3a) + 4a^3\left(\frac{2}{a}\right)$$
$$= 16a^5 - 12a^4 + 8a^2$$

Work Problem 2 at the Side.

EXAMPLE 3 Dividing a Polynomial by a Monomial

Divide $12x^4 - 7x^3 + 4x$ by $4x$.

$$\frac{12x^4 - 7x^3 + 4x}{4x} = \frac{12x^4}{4x} - \frac{7x^3}{4x} + \frac{4x}{4x}$$
$$= 3x^3 - \frac{7x^2}{4} + 1 \qquad \text{Quotient rule}$$

Check by multiplying.

CAUTION

In Example 3, notice that the quotient $\frac{4x}{4x} = 1$. It is a common error to leave the 1 out of the answer. Multiplying to check will show that the answer $3x^3 - \frac{7x^2}{4}$ is not correct.

Work Problem 3 at the Side.

EXAMPLE 4 Dividing a Polynomial by a Monomial

Divide the polynomial

$$180x^4y^{10} - 150x^3y^8 + 120x^2y^6 - 90xy^4 + 100y$$

by the monomial $-30xy^2$.

$$\frac{180x^4y^{10} - 150x^3y^8 + 120x^2y^6 - 90xy^4 + 100y}{-30xy^2}$$
$$= \frac{180x^4y^{10}}{-30xy^2} - \frac{150x^3y^8}{-30xy^2} + \frac{120x^2y^6}{-30xy^2} - \frac{90xy^4}{-30xy^2} + \frac{100y}{-30xy^2}$$
$$= -6x^3y^8 + 5x^2y^6 - 4xy^4 + 3y^2 - \frac{10}{3xy}$$

Work Problem 4 at the Side.

2 Divide.

(a) $\dfrac{20x^4 - 25x^3 + 5x}{5x^2}$

(b) $\dfrac{50m^4 - 30m^3 + 20m}{10m^3}$

3 Divide.

(a) $\dfrac{8y^7 - 9y^6 - 11y - 4}{y^2}$

(b) $\dfrac{12p^5 + 8p^4 + 3p^3 - 5p^2}{3p^3}$

4 Divide.

$\dfrac{45x^4y^3 + 30x^3y^2 - 60x^2y}{-15x^2y}$

ANSWERS

2. (a) $4x^2 - 5x + \frac{1}{x}$ (b) $5m - 3 + \frac{2}{m^2}$

3. (a) $8y^5 - 9y^4 - \frac{11}{y} - \frac{4}{y^2}$

(b) $4p^2 + \frac{8p}{3} + 1 - \frac{5}{3p}$

4. $-3x^2y^2 - 2xy + 4$

5.6 Exercises

Fill in each blank with the correct response.

1. In the statement $\dfrac{6x^2 + 8}{2} = 3x^2 + 4$, ________ is the dividend, ________ is the divisor, and ________ is the quotient.

2. The expression $\dfrac{3x + 12}{x}$ is undefined if $x =$ ________.

3. To check the division shown in Exercise 1, multiply ________ by ________ and show that the product is ________.

4. The expression $5x^2 - 3x + 6 + \frac{2}{x}$ ________ (is/is not) a polynomial.

5. Explain why the division problem $\dfrac{16m^3 - 12m^2}{4m}$ can be performed using the method of this section, while the division problem $\dfrac{4m}{16m^3 - 12m^2}$ cannot.

6. Evaluate $\dfrac{5y + 6}{2}$ when $y = 2$. Evaluate $5y + 3$ when $y = 2$. Does $\dfrac{5y + 6}{2}$ equal $5y + 3$?

Perform each division. See Examples 1–4.

7. $\dfrac{60x^4 - 20x^2 + 10x}{2x}$

8. $\dfrac{120x^6 - 60x^3 + 80x^2}{2x}$

9. $\dfrac{20m^5 - 10m^4 + 5m^2}{-5m^2}$

10. $\dfrac{12t^5 - 6t^3 + 6t^2}{-6t^2}$

11. $\dfrac{8t^5 - 4t^3 + 4t^2}{2t}$

12. $\dfrac{8r^4 - 4r^3 + 6r^2}{2r}$

13. $\dfrac{4a^5 - 4a^2 + 8}{4a}$

14. $\dfrac{5t^8 + 5t^7 + 15}{5t}$

15. $\dfrac{12x^5 - 4x^4 + 6x^3}{-6x^2}$

16. $\dfrac{24x^6 - 12x^5 + 30x^4}{-6x^2}$

17. $\dfrac{4x^2 + 20x^3 - 36x^4}{4x^2}$

18. $\dfrac{5x^2 - 30x^4 + 30x^5}{5x^2}$

19. $\dfrac{4x^4 + 3x^3 + 2x}{3x^2}$

20. $\dfrac{5x^4 - 6x^3 + 8x}{3x^2}$

21. $\dfrac{27r^4 - 36r^3 - 6r^2 + 3r - 2}{3r}$

22. $\dfrac{8k^4 - 12k^3 - 2k^2 - 2k - 3}{2k}$

23. $\dfrac{2m^5 - 6m^4 + 8m^2}{-2m^3}$

24. $\dfrac{6r^5 - 8r^4 + 10r^2}{-2r^4}$

25. $(20a^4b^3 - 15a^5b^2 + 25a^3b) \div (-5a^4b)$

26. $(16y^5z - 8y^2z^2 + 12yz^3) \div (-4y^2z^2)$

27. $(120x^{11} - 60x^{10} + 140x^9 - 100x^8) \div (10x^{12})$

28. $(120x^{12} - 84x^9 + 60x^8 - 36x^7) \div (12x^9)$

29. The quotient in Exercise 19 is $\dfrac{4x^2}{3} + x + \dfrac{2}{3x}$. Notice how the third term is written with x in the denominator. Would $\frac{2}{3}x$ be an acceptable form for this term? Explain why or why not. Is $\frac{4}{3}x^2$ an acceptable form for the first term? Why or why not?

30. What expression represents the length of the rectangle?

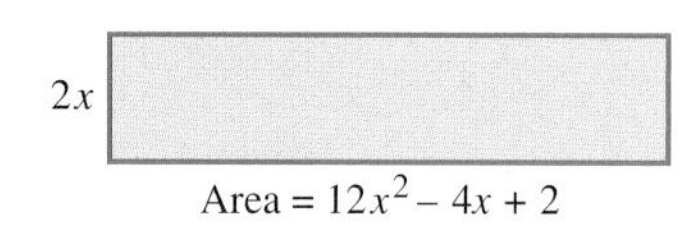

31. What polynomial, when divided by $5x^3$, yields $3x^2 - 7x + 7$ as a quotient?

32. The quotient of a certain polynomial and $-12y^3$ is $6y^3 - 5y^2 + 2y - 3 + \frac{7}{y}$. Find the polynomial.

RELATING CONCEPTS (EXERCISES 33–36) For Individual or Group Work

Our system of numeration is called a decimal system. It is based on powers of ten. In a whole number such as 2846, *each digit is understood to represent the number of powers of ten for its place value. The* 2 *represents two thousands* (2×10^3), *the* 8 *represents eight hundreds* (8×10^2), *the* 4 *represents four tens* (4×10^1), *and the* 6 *represents six ones (or units)* (6×10^0). *In expanded form we write*

$$2846 = (2 \times 10^3) + (8 \times 10^2) + (4 \times 10^1) + (6 \times 10^0).$$

Keeping this information in mind, ***work Exercises 33–36 in order.***

33. Divide 2846 by 2, using paper-and-pencil methods: $2\overline{)2846}$.

34. Write your answer in Exercise 33 in expanded form.

35. Use the methods of this section to divide the polynomial $2x^3 + 8x^2 + 4x + 6$ by 2.

36. Compare your answers in Exercises 34 and 35. How are they similar? How are they different? For what value of x does the answer in Exercise 35 equal the answer in Exercise 34?

5.7 Dividing a Polynomial by a Polynomial

OBJECTIVES

1. **Divide a polynomial by a polynomial.**
2. **Apply division to a geometry problem.**

OBJECTIVE 1 Divide a polynomial by a polynomial. We use a method of "long division" to divide a polynomial by a polynomial (other than a monomial), similar to that used for dividing two whole numbers. Both polynomials must be written in descending powers.

Dividing Whole Numbers	Dividing Polynomials
Step 1	
Divide 6696 by 27.	Divide $8x^3 - 4x^2 - 14x + 15$ by $2x + 3$.
$27\overline{)6696}$	$2x + 3\overline{)8x^3 - 4x^2 - 14x + 15}$
Step 2	
66 divided by 27 = 2; $2 \cdot 27 = \mathbf{54}$.	$8x^3$ divided by $2x = 4x^2$; $4x^2(2x + 3) = \mathbf{8x^3 + 12x^2}$.
$\begin{array}{r} 2 \\ 27\overline{)6696} \\ \underline{\mathbf{54}} \end{array}$	$\begin{array}{r} 4x^2 \qquad\qquad\qquad \\ 2x + 3\overline{)8x^3 - 4x^2 - 14x + 15} \\ \underline{\mathbf{8x^3 + 12x^2}} \qquad\qquad \end{array}$
Step 3	
Subtract; then bring down the next digit.	Subtract; then bring down the next term.
$\begin{array}{r} 2 \\ 27\overline{)6696} \\ \underline{54}\downarrow \\ 129 \end{array}$	$\begin{array}{r} 4x^2 \qquad\qquad\qquad \\ 2x + 3\overline{)8x^3 - 4x^2 - 14x + 15} \\ \underline{8x^3 + 12x^2} \qquad\qquad \\ -16x^2 - 14x \qquad \end{array}$
	(To subtract two polynomials, change the signs of the second and then add.)
Step 4	
129 divided by 27 = 4; $4 \cdot 27 = \mathbf{108}$.	$-16x^2$ divided by $2x = -8x$; $-8x(2x + 3) = \mathbf{-16x^2 - 24x}$.
$\begin{array}{r} 24 \\ 27\overline{)6696} \\ \underline{54} \\ 129 \\ \underline{\mathbf{108}} \end{array}$	$\begin{array}{r} 4x^2 - 8x \qquad\qquad \\ 2x + 3\overline{)8x^3 - 4x^2 - 14x + 15} \\ \underline{8x^3 + 12x^2} \qquad\qquad \\ -16x^2 - 14x \qquad \\ \underline{\mathbf{-16x^2 - 24x}} \qquad \end{array}$
Step 5	
Subtract; then down the next digit.	Subtract; then bring down the next term.
$\begin{array}{r} 24 \\ 27\overline{)6696} \\ \underline{54} \\ 129 \\ \underline{108}\downarrow \\ 216 \end{array}$	$\begin{array}{r} 4x^2 - 8x \qquad\qquad \\ 2x + 3\overline{)8x^3 - 4x^2 - 14x + 15} \\ \underline{8x^3 + 12x^2} \qquad\qquad \\ -16x^2 - 14x \qquad \\ \underline{-16x^2 - 24x} \qquad \\ 10x + 15 \end{array}$

(continued)

Step 6

216 divided by 27 = **8**;
$8 \cdot 27 = \mathbf{216}$.

$$\begin{array}{r} 248 \\ 27\overline{)6696} \\ \underline{54} \\ 129 \\ \underline{108} \\ 216 \\ \underline{\mathbf{216}} \\ 0 \end{array}$$

6696 divided by 27 is 248. The remainder is zero.

$10x$ divided by $2x = \mathbf{5}$;
$5(2x + 3) = \mathbf{10x + 15}$.

$$\begin{array}{r} 4x^2 - 8x + 5 \\ 2x + 3\overline{)8x^3 - 4x^2 - 14x + 15} \\ \underline{8x^3 + 12x^2} \quad\quad\quad\quad \\ -16x^2 - 14x \quad\quad \\ \underline{-16x^2 - 24x} \quad\quad \\ 10x + 15 \\ \underline{\mathbf{10x + 15}} \\ 0 \end{array}$$

$8x^3 - 4x^2 - 14x + 15$ divided by $2x + 3$ is $4x^2 - 8x + 5$. The remainder is zero.

Step 7

Check by multiplying.

$27 \cdot 248 = 6696$

Check by multiplying.

$(2x + 3)(4x^2 - 8x + 5)$
$= 8x^3 - 4x^2 - 14x + 15$

EXAMPLE 1 Dividing a Polynomial by a Polynomial

Divide $5x + 4x^3 - 8 - 4x^2$ by $2x - 1$.

Both polynomials must be written with the exponents in descending order. Rewrite the first polynomial as $4x^3 - 4x^2 + 5x - 8$. Then begin the division process.

Divide $4x^3 - 4x^2 + 5x - 8$ by $2x - 1$.

$$\begin{array}{r} 2x^2 - x + 2 \\ 2x - 1\overline{)4x^3 - 4x^2 + 5x - 8} \\ \underline{4x^3 - 2x^2} \quad\quad\quad\quad \\ -2x^2 + 5x \quad\quad \\ \underline{-2x^2 + x} \quad\quad \\ 4x - 8 \\ \underline{4x - 2} \\ \mathbf{-6} \leftarrow \text{Remainder} \end{array}$$

Step 1 $4x^3$ divided by $2x = \mathbf{2x^2}$; $2x^2(2x - 1) = 4x^3 - 2x^2$.

Step 2 Subtract; bring down the next term.

Step 3 $-2x^2$ divided by $2x = \mathbf{-x}$; $-x(2x - 1) = -2x^2 + x$.

Step 4 Subtract; bring down the next term.

Step 5 $4x$ divided by $2x = \mathbf{2}$; $2(2x - 1) = 4x - 2$.

Step 6 Subtract. The remainder is $\mathbf{-6}$. Write the remainder as the numerator of a fraction that has $2x - 1$ as its denominator. The answer is not a polynomial because of the nonzero remainder.

$$\frac{4x^3 - 4x^2 + 5x - 8}{2x - 1} = 2x^2 - x + 2 + \frac{-6}{2x - 1}$$

Continued on Next Page

Step 7 Check by multiplying.

$$(2x-1)\left(2x^2 - x + 2 + \frac{-6}{2x-1}\right)$$
$$= (2x-1)(2x^2) + (2x-1)(-x) + (2x-1)(2)$$
$$+ (2x-1)\left(\frac{-6}{2x-1}\right)$$
$$= 4x^3 - 2x^2 - 2x^2 + x + 4x - 2 - 6$$
$$= 4x^3 - 4x^2 + 5x - 8$$

Work Problem 1 at the Side.

1 Divide.

(a) $(x^3 + x^2 + 4x - 6) \div (x - 1)$

(b) $\dfrac{p^3 - 2p^2 - 5p + 9}{p + 2}$

EXAMPLE 2 Dividing into a Polynomial with Missing Terms

Divide $x^3 - 1$ by $x - 1$.

Here the polynomial $x^3 - 1$ is missing the x^2-term and the x-term. When terms are missing, use 0 as the coefficient for each missing term. (Zero acts as a placeholder here, just as it does in our number system.)

$$x^3 - 1 = x^3 + 0x^2 + 0x - 1$$

Now divide.

$$\begin{array}{r|l} & x^2 + x + 1 \\ x - 1 & x^3 + 0x^2 + 0x - 1 \\ & \underline{x^3 - x^2} \\ & \quad x^2 + 0x \\ & \quad \underline{x^2 - x} \\ & \qquad x - 1 \\ & \qquad \underline{x - 1} \\ & \qquad\quad 0 \end{array}$$

The remainder is 0. The quotient is $x^2 + x + 1$. *Check* by multiplying.

$$(x - 1)(x^2 + x + 1) = x^3 - 1$$

Work Problem 2 at the Side.

2 Divide.

(a) $\dfrac{r^2 - 5}{r + 4}$

(b) $(x^3 - 8) \div (x - 2)$

EXAMPLE 3 Dividing by a Polynomial with Missing Terms

Divide $x^4 + 2x^3 + 2x^2 - x - 1$ by $x^2 + 1$.

Since $x^2 + 1$ has a missing x term, write it as $x^2 + 0x + 1$. Then go through the division process as follows.

$$\begin{array}{r|l} & x^2 + 2x + 1 \\ x^2 + 0x + 1 & x^4 + 2x^3 + 2x^2 - x - 1 \\ & \underline{x^4 + 0x^3 + x^2} \\ & \quad 2x^3 + x^2 - x \\ & \quad \underline{2x^3 + 0x^2 + 2x} \\ & \qquad x^2 - 3x - 1 \\ & \qquad \underline{x^2 + 0x + 1} \\ & \qquad\quad -3x - 2 \leftarrow \text{Remainder} \end{array}$$

Continued on Next Page

Answers

1. (a) $x^2 + 2x + 6$
   (b) $p^2 - 4p + 3 + \dfrac{3}{p+2}$
2. (a) $r - 4 + \dfrac{11}{r+4}$
   (b) $x^2 + 2x + 4$

3 Divide.

(a)

$(2x^4 + 3x^3 - x^2 + 6x + 5) \div (x^2 - 1)$

(b)

$$\frac{2m^5 + m^4 + 6m^3 - 3m^2 - 18}{m^2 + 3}$$

4 Divide $3x^3 + 7x^2 + 7x + 10$ by $3x + 6$.

5 Divide $x^3 + 4x^2 + 8x + 8$ by $x + 2$.

When the result of subtracting ($-3x - 2$, in this case) is a polynomial of lesser degree than the divisor ($x^2 + 0x + 1$), that polynomial is the remainder. Write the answer as

$$x^2 + 2x + 1 + \frac{-3x - 2}{x^2 + 1}.$$

Multiply to check that this is the correct quotient.

Work Problem 3 at the Side.

EXAMPLE 4 Dividing a Polynomial with a Quotient That Has Fractional Coefficients

Divide $4x^3 + 2x^2 + 3x + 1$ by $4x - 4$.

$$\begin{array}{r} x^2 + \frac{3}{2}x + \frac{9}{4} \\ 4x - 4\overline{)4x^3 + 2x^2 + 3x + 1} \\ \underline{4x^3 - 4x^2 } \\ 6x^2 + 3x \\ \underline{6x^2 - 6x } \\ 9x + 1 \\ \underline{9x - 9} \\ 10 \end{array}$$

The answer is $x^2 + \frac{3}{2}x + \frac{9}{4} + \frac{10}{4x - 4}$.

Work Problem 4 at the Side.

OBJECTIVE 2 Apply division to a geometry problem.

EXAMPLE 5 Using an Area Formula

The area of the rectangle in Figure 2 is given by $x^3 + 4x^2 + 8x + 8$ sq. units and the width by $x + 2$ units. What is its length?

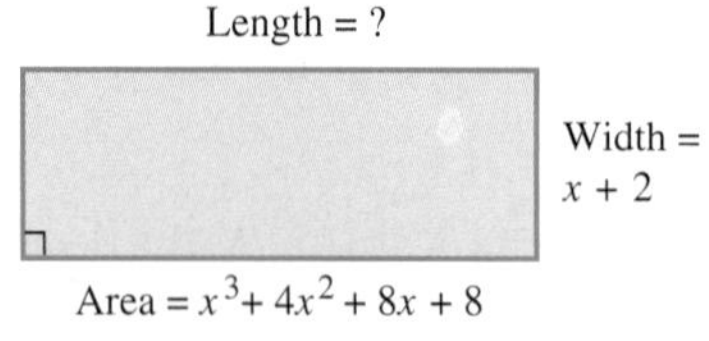

Figure 2

Since $A = LW$, solving for L gives $L = \frac{A}{W}$. Divide $x^3 + 4x^2 + 8x + 8$ by the width, $x + 2$.

Work Problem 5 at the Side.

The quotient from Problem 5, $x^2 + 2x + 4$, represents the length of the rectangle in units.

ANSWERS

3. (a) $2x^2 + 3x + 1 + \frac{9x + 6}{x^2 - 1}$
   (b) $2m^3 + m^2 - 6$
4. $x^2 + \frac{1}{3}x + \frac{5}{3}$
5. $x^2 + 2x + 4$

5.7 Exercises

FOR EXTRA HELP

Tutor Center Addison-Wesley Math Tutor Center

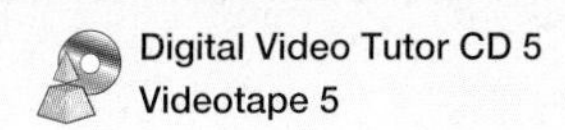

MathXL

Digital Video Tutor CD 5 Videotape 5

Student's Solutions Manual

MyMathLab

Interactmath.com

1. In the division problem $(4x^4 + 2x^3 - 14x^2 + 19x + 10) \div (2x + 5) = 2x^3 - 4x^2 + 3x + 2$, which polynomial is the divisor? Which is the quotient?

2. When dividing one polynomial by another, how do you know when to stop dividing?

3. In dividing $12m^2 - 20m + 3$ by $2m - 3$, what is the first step?

4. In the division in Exercise 3, what is the second step?

Perform each division. See Example 1.

5. $\dfrac{x^2 - x - 6}{x - 3}$

6. $\dfrac{m^2 - 2m - 24}{m - 6}$

7. $\dfrac{2y^2 + 9y - 35}{y + 7}$

8. $\dfrac{2y^2 + 9y + 7}{y + 1}$

9. $\dfrac{p^2 + 2p + 20}{p + 6}$

10. $\dfrac{x^2 + 11x + 16}{x + 8}$

11. $(r^2 - 8r + 15) \div (r - 3)$

12. $(t^2 + 2t - 35) \div (t - 5)$

13. $\dfrac{4a^2 - 22a + 32}{2a + 3}$

14. $\dfrac{9w^2 + 6w + 10}{3w - 2}$

15. $\dfrac{8x^3 - 10x^2 - x + 3}{2x + 1}$

16. $\dfrac{12t^3 - 11t^2 + 9t + 18}{4t + 3}$

Perform each division. See Examples 2–4.

17. $\dfrac{3y^3 + y^2 + 2}{y + 1}$

18. $\dfrac{2r^3 - 6r - 36}{r - 3}$

19. $\dfrac{3k^3 - 4k^2 - 6k + 10}{k^2 - 2}$

20. $\dfrac{5z^3 - z^2 + 10z + 2}{z^2 + 2}$

21. $(x^4 - x^2 - 2) \div (x^2 - 2)$

22. $(r^4 + 2r^2 - 3) \div (r^2 - 1)$

23. $\dfrac{6p^4 - 15p^3 + 14p^2 - 5p + 10}{3p^2 + 1}$

24. $\dfrac{6r^4 - 10r^3 - r^2 + 15r - 8}{2r^2 - 3}$

25. $\dfrac{2x^5 + x^4 + 11x^3 - 8x^2 - 13x + 7}{2x^2 + x - 1}$

26. $\dfrac{4t^5 - 11t^4 - 6t^3 + 5t^2 - t + 3}{4t^2 + t - 3}$

27. $\dfrac{x^4 - 1}{x^2 - 1}$

28. $\dfrac{y^3 + 1}{y + 1}$

29. $(10x^3 + 13x^2 + 4x + 1) \div (5x + 5)$

30. $(6x^3 - 19x^2 - 19x - 4) \div (2x - 8)$

Work each problem. See Example 5.

31. Give the length of the rectangle.

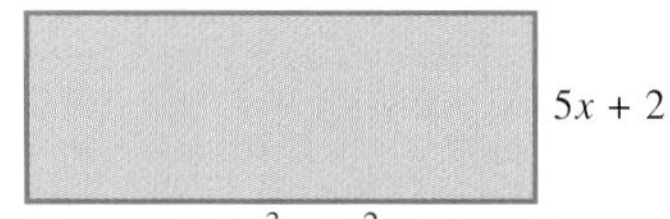

The area is $5x^3 + 7x^2 - 13x - 6$ sq. units.

32. Find the measure of the base of the parallelogram.

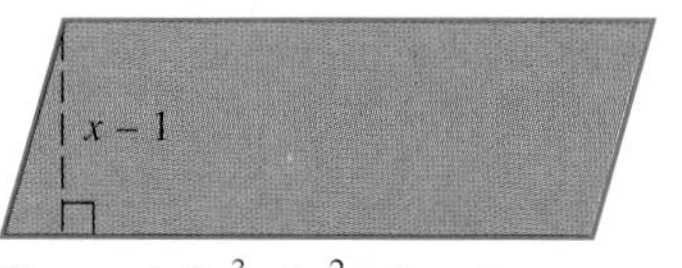

The area is $2x^3 + 2x^2 - 3x - 1$ sq. units.

RELATING CONCEPTS (EXERCISES 33–36) For Individual or Group Work

We can find the value of a polynomial in x for a given value of x by substituting that number for x. Surprisingly, we can accomplish the same thing by division. For example, to find the value of $2x^2 - 4x + 3$ for $x = -3$, we would divide $2x^2 - 4x + 3$ by $x - (-3)$. The remainder will give the value of the polynomial for $x = -3$. ***Work Exercises 33–36 in order.***

33. Find the value of $2x^2 - 4x + 3$ for $x = -3$ by substitution.

34. Divide $2x^2 - 4x + 3$ by $x + 3$. Give the remainder.

35. Compare your answers to Exercises 33 and 34. What do you notice?

36. Choose another polynomial and evaluate it both ways for some value of the variable. Do the answers agree?

5.8 An Application of Exponents: Scientific Notation

OBJECTIVES

1. **Express numbers in scientific notation.**
2. **Convert numbers in scientific notation to numbers without exponents.**
3. **Use scientific notation in calculations.**

OBJECTIVE 1 Express numbers in scientific notation. One example of the use of exponents comes from science. The numbers occurring in science are often extremely large (such as the distance from Earth to the sun, 93,000,000 mi) or extremely small (the wavelength of yellow-green light, approximately .0000006 m). Because of the difficulty of working with many zeros, scientists often express such numbers with exponents. Each number is written as

$$\mathbf{a \times 10^n}, \text{ where } 1 \leq |a| < 10 \text{ and } n \text{ is an integer.}$$

This form is called **scientific notation.** There is always one nonzero digit before the decimal point. This is shown in the following examples.

> **NOTE**
> In work with scientific notation, the times symbol, $\times$, is commonly used.

$3.19 \times 10^1 = 3.19 \times 10 = 31.9$ Decimal point moves 1 place to the right.

$3.19 \times 10^2 = 3.19 \times 100 = 319.$ Decimal point moves 2 places to the right.

$3.19 \times 10^3 = 3.19 \times 1000 = 3190.$ Decimal point moves 3 places to the right.

$3.19 \times 10^{-1} = 3.19 \times .1 = .319$ Decimal point moves 1 place to the left.

$3.19 \times 10^{-2} = 3.19 \times .01 = .0319$ Decimal point moves 2 places to the left.

$3.19 \times 10^{-3} = 3.19 \times .001 = .00319$ Decimal point moves 3 places to the left.

A number in scientific notation is always written with the decimal point after the first nonzero digit and then multiplied by the appropriate power of 10. For example, 35 is written 3.5×10^1, or 3.5×10; 56,200 is written 5.62×10^4, since

$$56{,}200 = 5.62 \times 10{,}000 = 5.62 \times 10^4.$$

To write a number in scientific notation, follow these steps.

Writing a Number in Scientific Notation

Step 1 Move the decimal point to the right of the first nonzero digit.

Step 2 Count the number of places you moved the decimal point.

Step 3 The number of places in Step 2 is the absolute value of the exponent on 10.

Step 4 The exponent on 10 is positive if the original number is greater than the number in Step 1; the exponent is negative if the original number is less than the number in Step 1. If the decimal point is not moved, the exponent is 0.

For negative numbers, follow these steps using the absolute value of the number; then make the result negative.

1 Write each number in scientific notation.

(a) 63,000

(b) 5,870,000

(c) .0571

(d) $-.000062$

EXAMPLE 1 Using Scientific Notation

Write each number in scientific notation.

(a) 93,000,000

Move the decimal point to follow the first nonzero digit. Count the number of places the decimal point was moved.

$$9.3\,000\,000 \qquad \text{7 places}$$

The number will be written in scientific notation as 9.3×10^n. To find the value of n, first compare 9.3 with 93,000,000. Since 93,000,000 is *greater* than 9.3, we must multiply by a *positive* power of 10 so the product 9.3×10^n will equal the larger number.

Since the decimal point was moved 7 places, and since n is positive,

$$93{,}000{,}000 = 9.3 \times 10^7.$$

(b) $463{,}000{,}000{,}000{,}000 = 4.63\,000\,000\,000\,000$ 14 places

$= 4.63 \times 10^{14}$

(c) $3.021 = 3.021 \times 10^0$

(d) .00462

Move the decimal point to the right of the first nonzero digit and count the number of places the decimal point was moved.

$$004.62 \qquad \text{3 places}$$

Because .00462 is *less* than 4.62, the exponent must be *negative*.

$$.00462 = 4.62 \times 10^{-3}$$

(e) $-.0000762 = -7.62 \times 10^{-5}$

Work Problem 1 at the Side.

2 Write without exponents.

(a) 4.2×10^3

(b) 8.7×10^5

(c) 6.42×10^{-3}

OBJECTIVE 2 Convert numbers in scientific notation to numbers without exponents. To convert a number written in scientific notation to a number without exponents, work in reverse. ***Multiplying a number by a positive power of 10 will make the number greater; multiplying by a negative power of 10 will make the number less.***

EXAMPLE 2 Writing Numbers without Exponents

Write each number without exponents.

(a) 6.2×10^3

Since the exponent is positive, make 6.2 greater by moving the decimal point 3 places to the right. It is necessary to attach two 0s.

$$6.2 \times 10^3 = 6.200 = 6200$$

(b) $4.283 \times 10^5 = 4.28300 = 428{,}300$ Move 5 places to the right; attach 0s as necessary.

(c) $-9.73 \times 10^{-2} = -09.73 = -.0973$ Move 2 places to the left.

As these examples show, ***the exponent tells the number of places and the direction that the decimal point is moved.***

Work Problem 2 at the Side.

ANSWERS

1. (a) 6.3×10^4 **(b)** 5.87×10^6 **(c)** 5.71×10^{-2} **(d)** -6.2×10^{-5}

2. (a) 4200 **(b)** 870,000 **(c)** .00642

OBJECTIVE 3 Use scientific notation in calculations. The next example shows how scientific notation can be used with products and quotients.

EXAMPLE 3 Multiplying and Dividing with Scientific Notation

Write each product or quotient without exponents.

(a) $(6 \times 10^3)(5 \times 10^{-4})$

$= (6 \times 5)(10^3 \times 10^{-4})$ Commutative and associative properties

$= 30 \times 10^{-1}$ Product rule for exponents

$= 30. = 3$ Write without exponents.

(b) $\dfrac{6 \times 10^{-5}}{2 \times 10^3} = \dfrac{6}{2} \times \dfrac{10^{-5}}{10^3} = 3 \times 10^{-8} = .00000003$

Work Problem 3 at the Side.

Calculator Tip Calculators usually have a key labeled EE or EXP for scientific notation. See An Introduction to Calculators at the front of this book for more information.

EXAMPLE 4 Applying Scientific Notation

Convert to scientific notation, calculate each computation, then give the result without scientific notation.

(a) In determining helium usage at Kennedy Space Center, the product 70,000(.0283)(1000) must be calculated. (*Source: NASA-AMATYC-NSF Mathematics Explorations II*, Capital Community College, 2000.)

$$70{,}000(.0283)(1000) = (7 \times 10^4)(2.83 \times 10^{-2})(1 \times 10^3)$$
$$= (7 \times 2.83 \times 1)(10^{4-2+3})$$
$$= 19.81 \times 10^5$$
$$= 1{,}981{,}000$$

(b) The ratio of the tidal force exerted by the moon compared to that exerted by the sun is given by

$$\frac{73.5 \times 10^{21} \times (1.5 \times 10^8)^3}{1.99 \times 10^{30} \times (3.84 \times 10^5)^3}.$$

(*Source:* Kastner, Bernice, *Space Mathematics*, NASA.)

$$\frac{7.35 \times 10^1 \times 10^{21} \times 1.5^3 \times 10^{24}}{1.99 \times 10^{30} \times 3.84^3 \times 10^{15}} \approx .22 \times 10^{1+21+24-30-15}$$
$$= .22 \times 10^1$$
$$= 2.2$$

Work Problem 4 at the Side.

3 Simplify, and write without exponents.

(a) $(2.6 \times 10^4)(2 \times 10^{-6})$

(b) $\dfrac{4.8 \times 10^2}{2.4 \times 10^{-3}}$

4 The speed of light is approximately 3.0×10^5 km per sec. (*Source: World Almanac and Book of Facts*.) Write answers without exponents.

(a) How far does light travel in 6.0×10^1 sec?

(b) How many seconds does it take light to travel approximately 1.5×10^8 km from the sun to Earth?

ANSWERS

3. **(a)** .052 **(b)** 200,000
4. **(a)** 18,000,000 km **(b)** 500 sec

Focus on

Real-Data Applications

Earthquake Intensities Measured by the Richter Scale

Charles F. Richter devised a scale in 1935 to compare the intensities, or relative power, of earthquakes. The **intensity** of an earthquake is measured relative to the intensity of a standard **zero-level** earthquake of intensity I_0. The relationship is equivalent to $I = I_0 \times 10^R$, where R is the **Richter scale** measure. For example, if an earthquake has magnitude 5.0 on the Richter scale, then its intensity is calculated as $I = I_0 \times 10^{5.0} = I_0 \times 100{,}000$, which is 100,000 times as intense as a zero-level earthquake. The following diagram illustrates the intensities of earthquakes and their Richter scale magnitudes.

Intensity	I_0	$I_0 \times 10^1$	$I_0 \times 10^2$	$I_0 \times 10^3$	$I_0 \times 10^4$	$I_0 \times 10^5$	$I_0 \times 10^6$	$I_0 \times 10^7$	$I_0 \times 10^8$
Richter Scale	0	1	2	3	4	5	6	7	8

To compare two earthquakes to each other, a ratio of the intensities is calculated. For example, to compare an earthquake that measures 8.0 on the Richter scale to one that measures 5.0, simply find the ratio of the intensities:

$$\frac{\text{intensity } 8.0}{\text{intensity } 5.0} = \frac{I_0 \times 10^{8.0}}{I_0 \times 10^{5.0}} = \frac{10^8}{10^5} = 10^{8-5} = 10^3 = 1000.$$

Therefore an earthquake that measures 8.0 on the Richter scale is 1000 times as intense as one that measures 5.0.

For Group Discussion

The table gives Richter scale measurements for several earthquakes.

Earthquake		Richter Scale Measurement
1960	Concepción, Chile	9.5
1906	San Francisco, California	8.3
1939	Erzincan, Turkey	8.0
1998	Sumatra, Indonesia	7.0
1998	Adana, Turkey	6.3

Source: World Almanac and Book of Facts, 2004.

1. Compare the intensity of the 1939 Erzincan earthquake to the 1998 Sumatra earthquake.

2. Compare the intensity of the 1998 Adana earthquake to the 1906 San Francisco earthquake.

3. Compare the intensity of the 1939 Erzincan earthquake to the 1998 Adana earthquake.

4. Suppose an earthquake measures 7.2 on the Richter scale. How would the intensity of a second earthquake compare if its Richter scale measure differed by $+3.0$? By -1.0?

5.8 Exercises

FOR EXTRA HELP

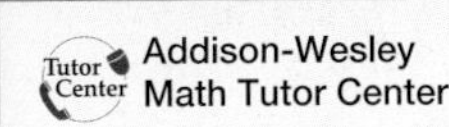
Addison-Wesley Math Tutor Center

MathXL

Digital Video Tutor CD 5 Videotape 5

Student's Solutions Manual

MyMathLab

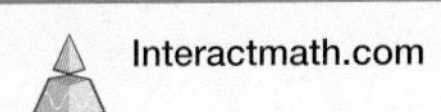
Interactmath.com

Write the numbers (other than dates) mentioned in the following statements in scientific notation.

1. NASA has budgeted \$6,130,900,000 for 2003 and \$5,868,900,000 for 2004 for the international space station. (*Source:* U.S. National Aeronautics and Space Administration.)

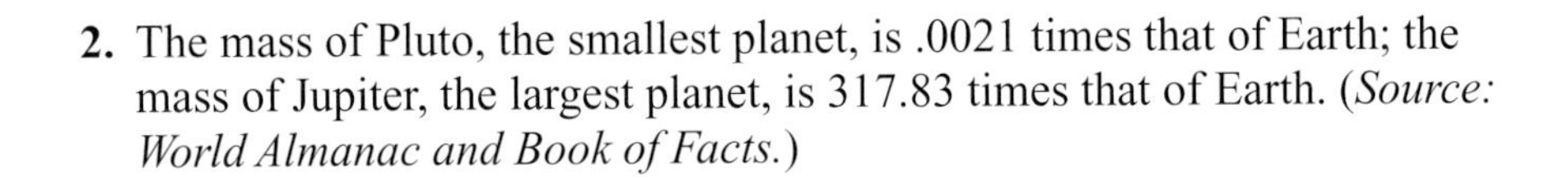

2. The mass of Pluto, the smallest planet, is .0021 times that of Earth; the mass of Jupiter, the largest planet, is 317.83 times that of Earth. (*Source: World Almanac and Book of Facts.*)

3. In 2000, the federal government spent \$69,627,000,000 on research and development. Industry spent \$181,040,000,000 in that same year. (*Source:* U.S. National Science Foundation.)

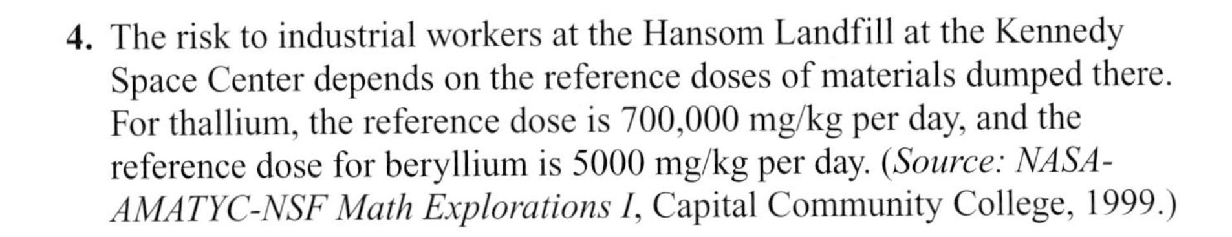

4. The risk to industrial workers at the Hansom Landfill at the Kennedy Space Center depends on the reference doses of materials dumped there. For thallium, the reference dose is 700,000 mg/kg per day, and the reference dose for beryllium is 5000 mg/kg per day. (*Source: NASA-AMATYC-NSF Math Explorations I*, Capital Community College, 1999.)

Determine whether or not the given number is written in scientific notation as defined in Objective 1. If it is not, write it as such.

5. 4.56×10^3

6. 7.34×10^5

7. 5,600,000

8. 34,000

9. .004

10. .0007

11. $.8 \times 10^2$

12. $.9 \times 10^3$

13. Explain in your own words what it means for a number to be written in scientific notation.

14. Explain how to multiply a number by a positive power of ten. Then explain how to multiply a number by a negative power of ten.

Write each number in scientific notation. See Example 1.

15. 5,876,000,000

16. 9,994,000,000

17. 82,350

18. 78,330

19. .000007

20. .0000004

21. $-.00203$

22. $-.0000578$

Write each number without exponents. See Example 2.

23. 7.5×10^5
24. 8.8×10^6
25. 5.677×10^{12}
26. 8.766×10^9

27. -6.21×10^0
28. -8.56×10^0
29. 7.8×10^{-4}

30. 8.9×10^{-5}
31. 5.134×10^{-9}
32. 7.123×10^{-10}

Perform the indicated operations. Write the answers in scientific notation and then without exponents. See Example 3.

33. $(2 \times 10^8) \times (3 \times 10^3)$
34. $(4 \times 10^7) \times (3 \times 10^3)$
35. $(5 \times 10^4) \times (3 \times 10^2)$

36. $(8 \times 10^5) \times (2 \times 10^3)$
37. $(3.15 \times 10^{-4}) \times (2.04 \times 10^8)$
38. $(4.92 \times 10^{-3}) \times (2.25 \times 10^7)$

Perform the indicated operations, and write the answers in scientific notation. See Example 3.

39. $\dfrac{9 \times 10^{-5}}{3 \times 10^{-1}}$
40. $\dfrac{12 \times 10^{-4}}{4 \times 10^{-3}}$
41. $\dfrac{8 \times 10^3}{2 \times 10^2}$

42. $\dfrac{5 \times 10^4}{1 \times 10^3}$
43. $\dfrac{2.6 \times 10^{-3} \times 7.0 \times 10^{-1}}{2 \times 10^2 \times 3.5 \times 10^{-3}}$
44. $\dfrac{9.5 \times 10^{-1} \times 2.4 \times 10^4}{5 \times 10^3 \times 1.2 \times 10^{-2}}$

Work each problem. Give answers without exponents. See Example 4.

45. There are 10^9 social security numbers. The population of the U.S. is about 3×10^8. How many social security numbers are available for each person? (*Source:* U.S. Bureau of the Census.)

46. The number of possible hands in contract bridge is about 6.35×10^{11}. The probability of being dealt one particular hand is $\dfrac{1}{6.35 \times 10^{11}}$. Express this number without scientific notation.

47. The top-grossing movie of all time is *Titanic*. During its first year, box office receipts were about 6×10^8 dollars. That amount represented a fraction of about 9.5×10^{-3} of the total receipts for motion pictures in that year. (*Source:* U.S. Bureau of the Census.) What were the total receipts?

48. In a recent year, the state of Texas had about 1.3×10^6 farms with an average of 7.1×10^2 acres per farm. What was the total number of acres devoted to farmland in Texas that year? (*Source:* National Agricultural Statistics Service, U.S. Department of Agriculture.)

49. The body of a 150-lb person contains about 2.3×10^{-4} lb of copper. How much copper is contained in the bodies of 1200 such people?

50. It takes about 3.6×10^1 sec at a speed of 3.0×10^5 km per sec for light from the sun to reach Venus. (*Source: World Almanac and Book of Facts*, 2004.) How far is Venus from the sun?

Chapter 5

SUMMARY

KEY TERMS

5.1	**polynomial**	A polynomial is a term or the sum of a finite number of terms with whole number exponents.
	descending powers	A polynomial in x is written in descending powers if the exponents on x in its terms are in decreasing order.
	degree of a term	The degree of a term is the sum of the exponents on the variables.
	degree of a polynomial	The degree of a polynomial is the greatest degree of any term of the polynomial.
	monomial	A monomial is a polynomial with one term.
	binomial	A binomial is a polynomial with two terms.
	trinomial	A trinomial is a polynomial with three terms.
5.3	**FOIL**	FOIL is a shortcut method for finding the product of two binomials.
	outer product	The outer product of $(2x + 3)(x - 5)$ is $2x(-5)$.
	inner product	The inner product of $(2x + 3)(x - 5)$ is $3x$.
5.8	**scientific notation**	A number written as $a \times 10^n$, where $1 \leq \lvert a \rvert < 10$ and n is an integer, is in scientific notation.

NEW SYMBOLS

x^{-n} x to the negative n power

TEST YOUR WORD POWER

See how well you have learned the vocabulary in this chapter. Answers, with examples, follow the Quick Review.

1. A **polynomial** is an algebraic expression made up of
   A. a term or a finite product of terms with positive coefficients and exponents
   B. a term or a finite sum of terms with real coefficients and whole number exponents
   C. the product of two or more terms with positive exponents
   D. the sum of two or more terms with whole number coefficients and exponents.

2. The **degree of a term** is
   A. the number of variables in the term
   B. the product of the exponents on the variables
   C. the least exponent on the variables
   D. the sum of the exponents on the variables.

3. A **trinomial** is a polynomial with
   A. only one term
   B. exactly two terms
   C. exactly three terms
   D. more than three terms.

4. A **binomial** is a polynomial with
   A. only one term
   B. exactly two terms
   C. exactly three terms
   D. more than three terms.

5. A **monomial** is a polynomial with
   A. only one term
   B. exactly two terms
   C. exactly three terms
   D. more than three terms.

6. **FOIL** is a method for
   A. adding two binomials
   B. adding two trinomials
   C. multiplying two binomials
   D. multiplying two trinomials.

QUICK REVIEW

Concepts	Examples
5.1 *Adding and Subtracting Polynomials*	
Addition Add like terms.	Add. $\begin{array}{r} 2x^2 + 5x - 3 \\ \underline{5x^2 - 2x + 7} \\ 7x^2 + 3x + 4 \end{array}$
Subtraction Change the signs of the terms in the second polynomial and add to the first polynomial.	Subtract. $(2x^2 + 5x - 3) - (5x^2 - 2x + 7)$ $= (2x^2 + 5x - 3) + (-5x^2 + 2x - 7)$ $= -3x^2 + 7x - 10$
5.2 *The Product Rule and Power Rules for Exponents* For any integers m and n:	
Product rule $a^m \cdot a^n = a^{m+n}$	$2^4 \cdot 2^5 = 2^9$
Power rules (a) $(a^m)^n = a^{mn}$	$(3^4)^2 = 3^8$
(b) $(ab)^m = a^m b^m$	$(6a)^5 = 6^5 a^5$
(c) $\left(\frac{a}{b}\right)^m = \frac{a^m}{b^m} \quad (b \neq 0).$	$\left(\frac{2}{3}\right)^4 = \frac{2^4}{3^4}$
5.3 *Multiplying Polynomials* Multiply each term of the first polynomial by each term of the second polynomial. Then add like terms.	Multiply. $\begin{array}{r} 3x^3 - 4x^2 + 2x - 7 \\ \underline{4x + 3} \\ 9x^3 - 12x^2 + 6x - 21 \\ \underline{12x^4 - 16x^3 + 8x^2 - 28x \quad\;\;} \\ 12x^4 - 7x^3 - 4x^2 - 22x - 21 \end{array}$
FOIL Method	Multiply $(2x + 3)(5x - 4)$.
Step 1 Multiply the two first terms to get the first term of the answer.	$2x(5x) = 10x^2$
Step 2 Find the outer product and the inner product and mentally add them, when possible, to get the middle term of the answer.	$2x(-4) + 3(5x) = 7x$
Step 3 Multiply the two last terms to get the last term of the answer.	$3(-4) = -12$
Add the terms found in Steps 1–3.	$(2x + 3)(5x - 4) = 10x^2 + 7x - 12$
5.4 *Special Products*	
Square of a Binomial $(a + b)^2 = a^2 + 2ab + b^2$ $(a - b)^2 = a^2 - 2ab + b^2$	$(3x + 1)^2 = 9x^2 + 6x + 1$ $(2m - 5n)^2 = 4m^2 - 20mn + 25n^2$
Product of the Sum and Difference of Two Terms $(a + b)(a - b) = a^2 - b^2$	$(4a + 3)(4a - 3) = 16a^2 - 9$

Concepts	*Examples*
5.5 *Integer Exponents and the Quotient Rule* If $a, b \neq 0$, for integers m and n:	
Zero exponent $a^0 = 1$	$15^0 = 1$
Negative exponent $a^{-n} = \frac{1}{a^n}$	$5^{-2} = \frac{1}{5^2} = \frac{1}{25}$
Quotient rule $\frac{a^m}{a^n} = a^{m-n}$	$\frac{4^8}{4^3} = 4^5$
Negative-to-positive rules $\frac{a^{-m}}{b^{-n}} = \frac{b^n}{a^m}$ $\left(\frac{a}{b}\right)^{-m} = \left(\frac{b}{a}\right)^m$.	$\frac{6^{-2}}{7^{-3}} = \frac{7^3}{6^2}$ $\left(\frac{5}{3}\right)^{-4} = \left(\frac{3}{5}\right)^4$
5.6 *Dividing a Polynomial by a Monomial* Divide each term of the polynomial by the monomial: $\frac{a+b}{c} = \frac{a}{c} + \frac{b}{c}$.	Divide. $\frac{4x^3 - 2x^2 + 6x - 8}{2x} = \frac{4x^3}{2x} - \frac{2x^2}{2x} + \frac{6x}{2x} - \frac{8}{2x}$ $= 2x^2 - x + 3 - \frac{4}{x}$
5.7 *Dividing a Polynomial by a Polynomial* Use "long division."	Divide. $3x + 4\overline{)6x^2 - 7x - 21}$ with quotient $2x - 5 + \frac{-1}{3x+4}$ $6x^2 + 8x$ $-15x - 21$ $-15x - 20$ $\mathbf{-1}$ ← Remainder
5.8 *An Application of Exponents: Scientific Notation* To write a number in scientific notation (as $a \times 10^n$), move the decimal point to the right of the first nonzero digit. If the decimal point is moved n places, and this makes the number smaller, n is positive; otherwise, n is negative. If the decimal point is not moved, n is 0.	$247 = 2.47 \times 10^2$ $.0051 = 5.1 \times 10^{-3}$ $4.8 = 4.8 \times 10^0$ $3.25 \times 10^5 = 325{,}000$ $8.44 \times 10^{-6} = .00000844$

Answers to Test Your Word Power

1. B; *Example:* $5x^3 + 2x^2 - 7$
2. D; *Examples:* The term 6 has degree 0, $3x$ has degree 1, $-2x^8$ has degree 8, and $5x^2y^4$ has degree 6.
3. C; *Example:* $2a^2 - 3ab + b^2$
4. B; *Example:* $3t^3 + 5t$
5. A; *Examples:* -5 and $4xy^5$
6. C; *Example:* $(m + 4)(m - 3) = \overset{F}{m(m)} - \overset{O}{3m} + \overset{I}{4m} + \overset{L}{4(-3)} = m^2 + m - 12$

Focus on

Real-Data Applications

Algebra in Euclid's *Elements*

The word *algebra* is derived from *Al-jabr wa'l muqabalah*, a ninth-century treatise written by the Arabic mathematician al-Khwarizmi. The notation that we use today, including the use of letters to represent variables and the symbols $+$ for addition and $-$ for subtraction, was introduced in the sixteenth century by François Viète. In Book II of Euclid's *Elements*, algebraic relationships were written in terms of geometric figures. The following proposition is an example.

Proposition 4: "If a straight line is cut at random, the square on the whole equals the squares on the segments plus twice the rectangle contained by the segments."

This proposition can be viewed geometrically. The straight line segment has length $a + b$. The "square on the whole" is the large outer square, $(a + b)^2$. The "squares on the segments" are a^2 and b^2, and the "rectangle contained by the segments" is ab, as shown in the figure. The algebraic statement equivalent to Proposition 4 is $(a + b)^2 = a^2 + b^2 + 2ab$. You should recognize this as the formula for computing the square of a binomial. We usually write it in the form $(a + b)^2 = a^2 + 2ab + b^2$.

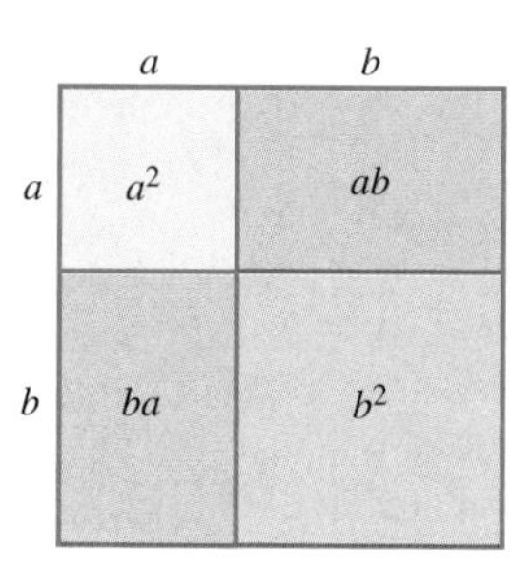

For Group Discussion

Consider the following propositions from Book II of Euclid's *Elements*. An equivalent algebraic statement accompanies each proposition.

Proposition 1: "If there are two straight lines and one of them is cut into any number of segments whatever [2 in the example], then the rectangle contained by the two straight lines equals the sum of the rectangles contained by the uncut straight line and each of the segments."

The equivalent algebraic form is $a(b + c) = ab + ac$.

1. By what name do we know this property?

Proposition 7: "If a straight line is cut at random, then the sum of the square on the whole [a] and that on one of the segments [b] equals twice the rectangle contained by the whole and the said segment plus the square on the remaining segment [$a - b$]."

The equivalent algebraic form is $a^2 + b^2 = 2ab + (a - b)^2$.

2. How is this property written in your textbook?

Proposition 8: An equivalent algebraic form of this proposition is $4ab + (a - b)^2 = (a + b)^2$.

3. Expand both the left and right sides of the formula. Are they the same expression?

Chapter 5

REVIEW EXERCISES

[5.1] *Combine terms where possible in each polynomial. Write the answer in descending powers of the variable. Give the degree of the answer. Identify the polynomial as a* monomial, binomial, trinomial, *or* none of these.

1. $9m^2 + 11m^2 + 2m^2$

2. $-4p + p^3 - p^2 + 8p + 2$

3. $12a^5 - 9a^4 + 8a^3 + 2a^2 - a + 3$

4. $-7y^5 - 8y^4 - y^5 + y^4 + 9y$

Add or subtract as indicated.

5. Add.

$$\begin{array}{r} -2a^3 + 5a^2 \\ -3a^3 - a^2 \\ \hline \end{array}$$

6. Add.

$$\begin{array}{r} 4r^3 - 8r^2 + 6r \\ -2r^3 + 5r^2 + 3r \\ \hline \end{array}$$

7. Subtract.

$$\begin{array}{r} 6y^2 - 8y + 2 \\ -5y^2 + 2y - 7 \\ \hline \end{array}$$

8. Subtract.

$$\begin{array}{r} -12k^4 - 8k^2 + 7k - 5 \\ k^4 + 7k^2 + 11k + 1 \\ \hline \end{array}$$

9. $(2m^3 - 8m^2 + 4) + (8m^3 + 2m^2 - 7)$

10. $(-5y^2 + 3y + 11) + (4y^2 - 7y + 15)$

11. $(6p^2 - p - 8) - (-4p^2 + 2p + 3)$

12. $(12r^4 - 7r^3 + 2r^2) - (5r^4 - 3r^3 + 2r^2 + 1)$

[5.2] *Use the product rule or power rules to simplify each expression. Write the answer in exponential form.*

13. $4^3 \cdot 4^8$

14. $(-5)^6(-5)^5$

15. $(-8x^4)(9x^3)$

16. $(2x^2)(5x^3)(x^9)$

17. $(19x)^5$

18. $(-4y)^7$

19. $5(pt)^4$

20. $\left(\frac{7}{5}\right)^6$

21. $(3x^2y^3)^3$

22. $(t^4)^8(t^2)^5$

23. $(6x^2z^4)^2(x^3yz^2)^4$

24. Explain why the product rule for exponents does not apply to the expression $7^2 + 7^4$.

[5.3] *Find each product.*

25. $5x(2x + 14)$

26. $-3p^3(2p^2 - 5p)$

27. $(3r - 2)(2r^2 + 4r - 3)$

28. $(2y + 3)(4y^2 - 6y + 9)$

29. $(5p^2 + 3p)(p^3 - p^2 + 5)$

30. $(3k - 6)(2k + 1)$

31. $(6p - 3q)(2p - 7q)$

32. $(m^2 + m - 9)(2m^2 + 3m - 1)$

[5.4] *Find each product.*

33. $(a + 4)^2$

34. $(3p - 2)^2$

35. $(2r + 5s)^2$

36. $(r + 2)^3$

37. $(2x - 1)^3$

38. $(6m - 5)(6m + 5)$

39. $(2z + 7)(2z - 7)$

40. $(5a + 6b)(5a - 6b)$

41. $(2x^2 + 5)(2x^2 - 5)$

42. Explain why $(a + b)^2$ is not equal to $a^2 + b^2$.

[5.5] *Evaluate each expression.*

43. $5^0 + 8^0$

44. 2^{-5}

45. $\left(\frac{6}{5}\right)^{-2}$

46. $4^{-2} - 4^{-1}$

Simplify. Write each answer in exponential form, using only positive exponents. Assume all variables represent nonzero numbers.

47. $\frac{6^{-3}}{6^{-5}}$

48. $\frac{x^{-7}}{x^{-9}}$

49. $\frac{p^{-8}}{p^4}$

50. $\frac{r^{-2}}{r^{-6}}$

51. $(2^4)^2$

52. $(9^3)^{-2}$

53. $(5^{-2})^{-4}$

54. $(8^{-3})^4$

55. $\frac{(m^2)^3}{(m^4)^2}$

56. $\frac{y^4 \cdot y^{-2}}{y^{-5}}$

57. $\frac{r^9 \cdot r^{-5}}{r^{-2} \cdot r^{-7}}$

58. $(-5m^3)^2$

59. $(2y^{-4})^{-3}$

60. $\frac{ab^{-3}}{a^4b^2}$

61. $\frac{(6r^{-1})^2 \cdot (2r^{-4})}{r^{-5}(r^2)^{-3}}$

62. $\frac{(2m^{-5}n^2)^3(3m^2)^{-1}}{m^{-2}n^{-4}(m^{-1})^2}$

[5.6] *Perform each division.*

63. $\dfrac{-15y^4}{-9y^2}$

64. $\dfrac{-12x^3y^2}{6xy}$

65. $\dfrac{6y^4 - 12y^2 + 18y}{-6y}$

66. $\dfrac{2p^3 - 6p^2 + 5p}{2p^2}$

67. $(5x^{13} - 10x^{12} + 20x^7 - 35x^5) \div (-5x^4)$

68. $(-10m^4n^2 + 5m^3n^3 + 6m^2n^4) \div (5m^2n)$

[5.7] *Perform each division.*

69. $(2r^2 + 3r - 14) \div (r - 2)$

70. $\dfrac{12m^2 - 11m - 10}{3m - 5}$

71. $\dfrac{10a^3 + 5a^2 - 14a + 9}{5a^2 - 3}$

72. $\dfrac{2k^4 + 4k^3 + 9k^2 - 8}{2k^2 + 1}$

[5.8] *Write each number in scientific notation.*

73. 48,000,000

74. 28,988,000,000

75. .000065

76. .0000000824

Write each number without exponents.

77. 2.4×10^4

78. 7.83×10^7

79. 8.97×10^{-7}

80. 9.95×10^{-12}

Perform the indicated operations and write the answers without exponents.

81. $(2 \times 10^{-3}) \times (4 \times 10^5)$

82. $\dfrac{8 \times 10^4}{2 \times 10^{-2}}$

83. $\dfrac{12 \times 10^{-5} \times 5 \times 10^4}{4 \times 10^3 \times 6 \times 10^{-2}}$

84. $\dfrac{2.5 \times 10^5 \times 4.8 \times 10^{-4}}{7.5 \times 10^8 \times 1.6 \times 10^{-5}}$

85. There are 13 red balls and 39 black balls in a box. Mix them up and draw 13 out one at a time without returning any ball. The probability that the 13 drawings each will produce a red ball is 1.6×10^{-12}. Write the number given in scientific notation without exponents. (*Source:* Warren Weaver, *Lady Luck*, Doubleday & Company, 1963.)

86. According to Campbell, Mitchell, and Reece in *Biology Concepts and Connections* (Benjamin Cummings, 1994, p. 230), "The amount of DNA in a human cell is about 1000 times greater than the DNA in *E. coli*. Does this mean humans have 1000 times as many genes as the 2000 in *E. coli*? The answer is probably no; the human genome is thought to carry between 50,000 and 100,000 genes, which code for various proteins (as well as for tRNA and rRNA)."

Write each number from this quote using scientific notation.

(a) 1000 **(b)** 2000

(c) 50,000 **(d)** 100,000

MIXED REVIEW EXERCISES

Perform the indicated operations. Write with positive exponents. Assume that no denominators are equal to 0.

87. $19^0 - 3^0$

88. $(3p)^4(3p^{-7})$

89. 7^{-2}

90. $(-7 + 2k)^2$

91. $\dfrac{2y^3 + 17y^2 + 37y + 7}{2y + 7}$

92. $\left(\dfrac{6r^{-2}s}{5}\right)^4$

93. $-m^5(8m^2 + 10m + 6)$

94. $\left(\dfrac{1}{2}\right)^{-5}$

95. $(25x^2y^3 - 8xy^2 + 15x^3y) \div (5x)$

96. $(6r^{-2})^{-1}$

97. $(2x + y)^3$

98. $2^{-1} + 4^{-1}$

99. $(a + 2)(a^2 - 4a + 1)$

100. $(5y^3 - 8y^2 + 7) - (-3y^3 + y^2 + 2)$

101. $(2r + 5)(5r - 2)$

102. $(12a + 1)(12a - 1)$

103. Find a polynomial that represents the area of the rectangle shown.

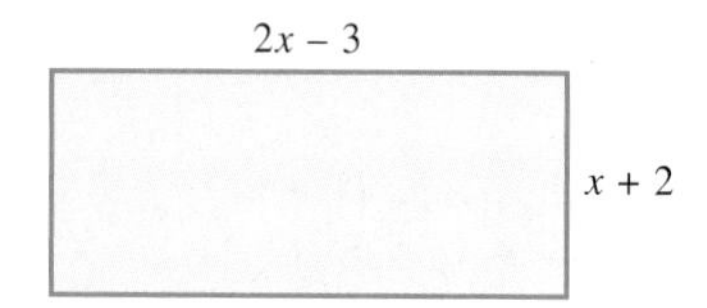

104. If the side of a square has a measure represented by $5x^4 + 2x^2$, what polynomial represents its area?

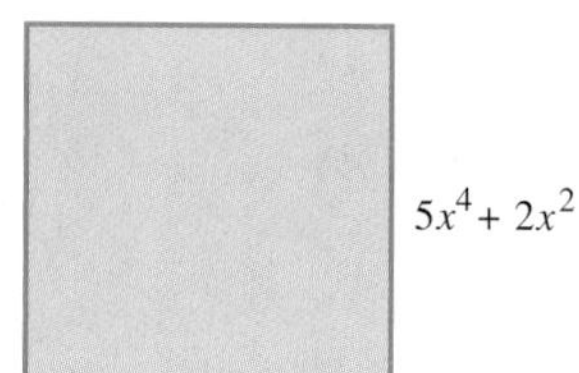

Chapter 5

TEST

Perform the indicated operations.

1. $(5t^4 - 3t^2 + 7t + 3) - (t^4 - t^3 + 3t^2 + 8t + 3)$ 1. ______

2. $(2y^2 - 8y + 8) + (-3y^2 + 2y + 3) - (y^2 + 3y - 6)$ 2. ______

3. Subtract. 3. ______

$$\begin{array}{r} 9t^3 - 4t^2 + 2t + 2 \\ \underline{9t^3 + 8t^2 - 3t - 6} \end{array}$$

Simplify, and write each answer with only positive exponents.

4. $(-2)^3(-2)^2$ 4. ______

5. $\left(\dfrac{6}{m^2}\right)^3, \quad m \neq 0$ 5. ______

6. $3x^2(-9x^3 + 6x^2 - 2x + 1)$ 6. ______

7. $(2r - 3)(r^2 + 2r - 5)$ 7. ______

8. $(t - 8)(t + 3)$ 8. ______

9. $(4x + 3y)(2x - y)$ 9. ______

10. $(5x - 2y)^2$ 10. ______

11. $(10v + 3w)(10v - 3w)$ 11. ______

12. $(x + 1)^3$ 12. ______

Evaluate each expression.

13. 5^{-4} 13. ______

14. $(-3)^0 + 4^0$ 14. ______

15. $4^{-1} + 3^{-1}$ 15. ______

Perform the indicated operations. In Exercises 16 and 17, write each answer using only positive exponents. Assume that variables represent nonzero numbers.

16. ______________________

16. $\dfrac{8^{-1} \cdot 8^4}{8^{-2}}$

17. ______________________

17. $\dfrac{(x^{-3})^{-2}(x^{-1}y)^2}{(xy^{-2})^2}$

18. ______________________

18. $\dfrac{8y^3 - 6y^2 + 4y + 10}{2y}$

19. ______________________

19. $(-9x^2y^3 + 6x^4y^3 + 12xy^3) \div (3xy)$

20. ______________________

20. $\dfrac{2x^2 + x - 36}{x - 4}$

21. ______________________

21. $(3x^3 - x + 4) \div (x - 2)$

Write each number in scientific notation.

22. (a) ______________________

(b) ______________________

22. (a) 344,000,000,000

(b) .00000557

Write each number without exponents.

23. (a) ______________________

(b) ______________________

23. (a) 2.96×10^7

(b) 6.07×10^{-8}

24. ______________________

24. What polynomial expression represents the area of this square?

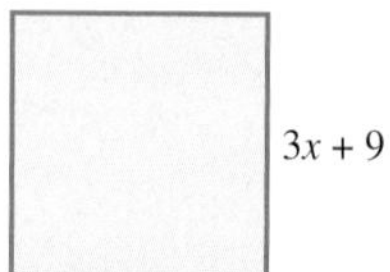

25. ______________________

25. Give an example of this situation: the sum of two fourth-degree polynomials in x is a third-degree polynomial in x.

Cumulative Review Exercises

CHAPTERS R–5

Work each problem.

1. $\frac{2}{3} + \frac{1}{8}$

2. $\frac{7}{4} - \frac{9}{5}$

3. $8.32 - 4.6$

4. 7.21×8.6

5. A retailer has \$34,000 invested in her business. She finds that last year she earned 5.4% on this investment. How much did she earn?

Find the value of each expression if $x = -2$ and $y = 4$.

6. $\frac{4x - 2y}{x + y}$

7. $x^3 - 4xy$

Perform the indicated operations.

8. $\frac{(-13 + 15) - (3 + 2)}{6 - 12}$

9. $-7 - 3[2 + (5 - 8)]$

Decide what property justifies each statement.

10. $(9 + 2) + 3 = 9 + (2 + 3)$

11. $-7 + 7 = 0$

12. $6(4 + 2) = 6(4) + 6(2)$

Solve each equation.

13. $2x - 7x + 8x = 30$

14. $2 - 3(t - 5) = 4 + t$

15. $2(5h + 1) = 10h + 4$

16. $d = rt$ for r

17. $\frac{x}{5} = \frac{x - 2}{7}$

18. $\frac{1}{3}p - \frac{1}{6}p = -2$

19. $.05x + .15(50 - x) = 5.50$

20. $4 - (3x + 12) = (2x - 9) - (5x - 1)$

Solve each problem.

21. A 1-oz mouse takes about 16 times as many breaths as does a 3-ton elephant. (*Source: Dinosaurs, Spitfires, and Sea Dragons*, McGowan, C., Harvard University Press, 1991.) If the two animals take a combined total of 170 breaths per minute, how many breaths does each take during that time period?

22. If a number is subtracted from 8 and this difference is tripled, the result is three times the number. Find this number, and you will learn how many times a dolphin rests during a 24-hr period.

Solve each inequality.

23. $-8x \leq -80$

24. $-2(x + 4) > 3x + 6$

25. $-3 \leq 2x + 5 < 9$

Given $2x - 3y = -6$, find the following.

26. The intercepts of the graph

27. The graph

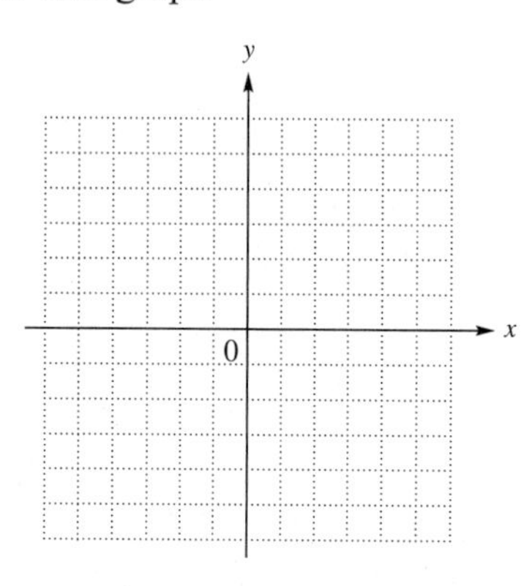

28. The slope of the line

Solve the system using the method indicated.

29. $y = 2x + 5$
$x + y = -4$ (Substitution)

30. $3x + 2y = 2$
$2x + 3y = -7$ (Elimination)

Evaluate each expression.

31. $4^{-1} + 3^0$

32. $2^{-4} \cdot 2^5$

33. $\dfrac{8^{-5} \cdot 8^7}{8^2}$

34. Write with positive exponents only: $\dfrac{(a^{-3}b^2)^2}{(2a^{-4}b^{-3})^{-1}}$.

35. Write in scientific notation: 34,500.

Perform the indicated operations.

36. $(7x^3 - 12x^2 - 3x + 8) + (6x^2 + 4) - (-4x^3 + 8x^2 - 2x - 2)$

37. $6x^5(3x^2 - 9x + 10)$

38. $(7x + 4)(9x + 3)$

39. $(5x + 8)^2$

40. $\dfrac{y^3 - 3y^2 + 8y - 6}{y - 1}$

Factoring and Applications

Wireless communication uses radio waves to carry signals and messages across distances. Cellular phones, one of the most popular forms of wireless communication, have become an invaluable tool for people to stay connected to family, friends, and work while on the go. In 2002 alone, U.S. sales of cellular phones totaled some \$8835 million as 66% of all U.S. households used cell phones. (*Source: Microsoft Encarta Encyclopedia 2002;* Consumer Electronics Association.)

In Exercise 31 of Section 6.7, we use a *quadratic equation* to model the number of cell phones owned by Americans in recent years.

6.1 Factors; The Greatest Common Factor

OBJECTIVES

1. **Find the greatest common factor of a list of numbers.**
2. **Find the greatest common factor of a list of variable terms.**
3. **Factor out the greatest common factor.**
4. **Factor by grouping.**

Recall from **Section R.1** that to **factor** a number means to write it as the product of two or more numbers. The product is called the **factored form** of the number. For example,

$$12 = \underbrace{6 \cdot 2}_{\text{Factored form}}.$$

(Factors: 6 and 2)

Factoring is a process that "undoes" multiplying. We multiply $6 \cdot 2$ to get 12, but we factor 12 by writing it as $6 \cdot 2$.

OBJECTIVE 1 Find the greatest common factor of a list of numbers. An integer that is a factor of two or more integers is a **common factor** of those integers. For example, 6 is a common factor of 18 and 24 because 6 is a factor of both 18 and 24. Other common factors of 18 and 24 are 1, 2, and 3. The **greatest common factor (GCF)** of a list of integers is the largest common factor of those integers. This means 6 is the greatest common factor of 18 and 24, since it is the largest of their common factors.

> **NOTE**
> Factors of a number are also divisors of the number. The greatest common factor is the same as the greatest common divisor.

EXAMPLE 1 Finding the Greatest Common Factor for Numbers

Find the greatest common factor for each list of numbers.

(a) 30, 45

First write each number in prime factored form.

$$30 = 2 \cdot 3 \cdot 5$$
$$45 = 3 \cdot 3 \cdot 5$$

Use each prime the *least* number of times it appears in *all* the factored forms. There is no 2 in the prime factored form of 45, so there will be no 2 in the greatest common factor. The least number of times 3 appears in all the factored forms is 1; the least number of times 5 appears is also 1. From this, the

$$\text{GCF} = 3^1 \cdot 5^1 = 3 \cdot 5 = 15.$$

(b) 72, 120, 432

Find the prime factored form of each number.

$$72 = 2 \cdot 2 \cdot 2 \cdot 3 \cdot 3$$
$$120 = 2 \cdot 2 \cdot 2 \cdot 3 \cdot 5$$
$$432 = 2 \cdot 2 \cdot 2 \cdot 2 \cdot 3 \cdot 3 \cdot 3$$

The least number of times 2 appears in all the factored forms is 3, and the least number of times 3 appears is 1. There is no 5 in the prime factored form of either 72 or 432, so the

$$\text{GCF} = 2^3 \cdot 3^1 = 24.$$

Continued on Next Page

(c) 10, 11, 14
Write the prime factored form of each number.

$$10 = 2 \cdot 5$$
$$11 = 11$$
$$14 = 2 \cdot 7$$

There are no primes common to all three numbers, so the GCF is 1.

Work Problem 1 at the Side.

OBJECTIVE 2 Find the greatest common factor of a list of variable terms. The terms x^4, x^5, x^6, and x^7 have x^4 as the greatest common factor because the least exponent on the variable x is 4.

$$x^4 = 1 \cdot x^4, \quad x^5 = x \cdot x^4, \quad x^6 = x^2 \cdot x^4, \quad x^7 = x^3 \cdot x^4$$

NOTE
The exponent on a variable in the GCF is the *least* exponent that appears on that variable in *all* the terms.

EXAMPLE 2 Finding the Greatest Common Factor for Variable Terms

Find the greatest common factor for each list of terms.

(a) $21m^7, -18m^6, 45m^8$

$$21m^7 = 3 \cdot 7 \cdot m^7$$
$$-18m^6 = -1 \cdot 2 \cdot 3 \cdot 3 \cdot m^6$$
$$45m^8 = 3 \cdot 3 \cdot 5 \cdot m^8$$

First, 3 is the greatest common factor of the coefficients 21, -18, and 45. The least exponent on m is 6, so the

$$\text{GCF} = 3m^6.$$

(b) $x^4y^2, x^7y^5, x^3y^7, y^{15}$

$$x^4y^2, \quad x^7y^5, \quad x^3y^7, \quad y^{15}$$

There is no x in the last term, y^{15}, so x will not appear in the greatest common factor. There is a y in each term, however, and 2 is the least exponent on y. The GCF is y^2.

(c) $-a^2b, -ab^2$

$$-a^2b = -1a^2b = -1 \cdot 1 \cdot a^2b$$
$$-ab^2 = -1ab^2 = -1 \cdot 1 \cdot ab^2$$

The factors of -1 are -1 and 1. Since $1 > -1$, the GCF is $1ab$ or ab.

NOTE
In a list of negative terms, sometimes a negative common factor is preferable (even though it is not the greatest common factor). In Example 2(c), for instance, we might prefer $-ab$ as the common factor. In factoring exercises, either answer will be acceptable.

1 Find the greatest common factor for each list of numbers.

(a) 30, 20, 15

$$30 = 2 \cdot 3 \cdot 5$$
$$20 = 2 \cdot \_\_\_\_ \cdot \_\_\_\_$$
$$15 = 3 \cdot \_\_\_\_$$
$$\text{GCF} = \_\_\_\_$$

(b) 42, 28, 35

(c) 12, 18, 26, 32

(d) 10, 15, 21

ANSWERS
1. (a) 2; 5; 5; 5 (b) 7 (c) 2 (d) 1

2 Find the greatest common factor for each list of terms.

(a) $6m^4, 9m^2, 12m^5$

$6m^4 = 2 \cdot \_\_\_\_ \cdot m^4$

$9m^2 = 3 \cdot \_\_\_\_ \cdot \_\_\_\_$

$12m^5 = 2 \cdot 2 \cdot \_\_\_\_ \cdot \_\_\_\_$

GCF = ____

(b) $-12p^5, -18q^4$

(c) y^4z^2, y^6z^8, z^9

(d) $12p^{11}, 17q^5$

Answers

2. **(a)** $3; 3; m^2; 3; m^5; 3m^2$ **(b)** 6 **(c)** z^2 **(d)** 1

We find the greatest common factor of a list of terms as follows.

Finding the Greatest Common Factor (GCF)

Step 1 **Factor.** Write each number in prime factored form.

Step 2 **List common factors.** List each prime number or each variable that is a factor of every term in the list. (If a prime does not appear in one of the prime factored forms, it cannot appear in the greatest common factor.)

Step 3 **Choose least exponents.** Use as exponents on the common prime factors the *least* exponents from the prime factored forms.

Step 4 **Multiply.** Multiply the primes from Step 3. If there are no primes left after Step 3, the greatest common factor is 1.

Work Problem 2 at the Side.

OBJECTIVE 3 **Factor out the greatest common factor.** The polynomial

$$3m + 12$$

has two terms, $3m$ and 12. The greatest common factor of these two terms is 3. We can write $3m + 12$ so that each term is a product with 3 as one factor.

$$3m + 12 = 3 \cdot m + 3 \cdot 4$$
$$= 3(m + 4) \quad \text{Distributive property}$$

The factored form of $3m + 12$ is $3(m + 4)$. This process is called **factoring out the greatest common factor.**

CAUTION

The polynomial $3m + 12$ is *not* in factored form when written as the *sum*

$$3 \cdot m + 3 \cdot 4. \quad \text{Not in factored form}$$

The *terms* are factored, but the polynomial is not. The factored form of $3m + 12$ is the *product*

$$3(m + 4). \quad \text{In factored form}$$

Writing a polynomial as a product, that is, in factored form, is called **factoring** the polynomial.

EXAMPLE 3 **Factoring Out the Greatest Common Factor**

Factor out the greatest common factor.

(a) $5y^2 + 10y = 5y(y) + 5y(2) \quad \text{GCF} = 5y$

$= 5y(y + 2) \quad$ Distributive property

Check by multiplying: $5y(y + 2) = 5y(y) + 5y(2)$

$= 5y^2 + 10y. \quad$ Original polynomial

Continued on Next Page

(b) $20m^5 + 10m^4 - 15m^3$

$= 5m^3(4m^2) + 5m^3(2m) - 5m^3(3)$ GCF $= 5m^3$

$= 5m^3(4m^2 + 2m - 3)$ Factor out $5m^3$.

Check: $5m^3(4m^2 + 2m - 3) = 20m^5 + 10m^4 - 15m^3$ Original polynomial

(c) $x^5 + x^3 = x^3(x^2) + x^3(1) = x^3(x^2 + 1)$ Don't forget the 1.

(d) $20m^7p^2 - 36m^3p^4 = 4m^3p^2(5m^4) - 4m^3p^2(9p^2)$ GCF $= 4m^3p^2$

$= 4m^3p^2(5m^4 - 9p^2)$

(e) $\frac{1}{6}n^2 + \frac{5}{6}n = \frac{1}{6}n(n) + \frac{1}{6}n(5)$ GCF $= \frac{1}{6}n$

$= \frac{1}{6}n(n + 5)$

CAUTION

Be sure to include the 1 in a problem like Example 3(c). ***Always check that the factored form can be multiplied out to give the original polynomial.***

Work Problem 3 at the Side.

EXAMPLE 4 Factoring Out the Greatest Common Factor

Factor out the greatest common factor.

(a) $a(a + 3) + 4(a + 3)$

The binomial $a + 3$ is the greatest common factor here.

Same

$$a(a + 3) + 4(a + 3) = (a + 3)(a + 4)$$

(b) $x^2(x + 1) - 5(x + 1) = (x + 1)(x^2 - 5)$ Factor out $x + 1$.

Work Problem 4 at the Side.

OBJECTIVE 4 Factor by grouping. When a polynomial has four terms, common factors can sometimes be used to **factor by grouping.**

EXAMPLE 5 Factoring by Grouping

Factor by grouping.

(a) $2x + 6 + ax + 3a$

Group the first two terms and the last two terms, since the first two terms have a common factor of 2 and the last two terms have a common factor of a.

$$2x + 6 + ax + 3a = (2x + 6) + (ax + 3a)$$
$$= 2(x + 3) + a(x + 3)$$

The expression is still not in factored form because it is the *sum* of two terms. Now, however, $x + 3$ is a common factor and can be factored out.

Continued on Next Page

3 Factor out the greatest common factor.

(a) $4x^2 + 6x$

(b) $10y^5 - 8y^4 + 6y^2$

(c) $m^7 + m^9$

(d) $8p^5q^2 + 16p^6q^3 - 12p^4q^7$

(e) $\frac{1}{3}b^2 - \frac{2}{3}b$

(f) $13x^2 - 27$

4 Factor out the greatest common factor.

(a) $r(t - 4) + 5(t - 4)$

(b) $y^2(y + 2) - 3(y + 2)$

(c) $x(x - 1) - 5(x - 1)$

ANSWERS

3. (a) $2x(2x + 3)$
(b) $2y^2(5y^3 - 4y^2 + 3)$
(c) $m^7(1 + m^2)$
(d) $4p^4q^2(2p + 4p^2q - 3q^5)$
(e) $\frac{1}{3}b(b - 2)$
(f) no common factor (except 1)
4. (a) $(t - 4)(r + 5)$
(b) $(y + 2)(y^2 - 3)$
(c) $(x - 1)(x - 5)$

5 Factor by grouping.

(a) $pq + 5q + 2p + 10$

(b) $2xy + 3y + 2x + 3$

(c) $2a^2 - 4a + 3ab - 6b$

(d) $x^3 + 3x^2 - 5x - 15$

$$\begin{aligned} 2x + 6 + ax + 3a &= (2x + 6) + (ax + 3a) && \text{Group terms.} \\ &= 2(x + 3) + a(x + 3) && \text{Factor each group.} \\ &= (x + 3)(2 + a) && \text{Factor out } x + 3. \end{aligned}$$

The final result is in factored form because it is a *product.* Note that the goal in factoring by grouping is to get a common factor, $x + 3$ here, so that the last step is possible. Check by multiplying the binomials using the FOIL method from **Section 5.3.**

Check: $(x + 3)(2 + a) = 2x + ax + 6 + 3a$ FOIL

$= 2x + 6 + ax + 3a,$ Rearrange terms.

which is the original polynomial.

(b)
$$\begin{aligned} 6ax + 24x + a + 4 &= (6ax + 24x) + (a + 4) && \text{Group terms.} \\ &= 6x(a + 4) + \mathbf{1}(a + 4) && \text{Factor each group; remember the 1.} \\ &= (a + 4)(6x + 1) && \text{Factor out } a + 4. \end{aligned}$$

Check: $(a + 4)(6x + 1) = 6ax + a + 24x + 4$ FOIL

$= 6ax + 24x + a + 4,$ Rearrange terms.

which is the original polynomial.

(c)
$$\begin{aligned} 2x^2 - 10x + 3xy - 15y &= (2x^2 - 10x) + (3xy - 15y) && \text{Group terms.} \\ &= 2x(x - 5) + 3y(x - 5) && \text{Factor each group.} \\ &= (x - 5)(2x + 3y) && \text{Factor out the common factor, } x - 5. \end{aligned}$$

Check: $(x - 5)(2x + 3y) = 2x^2 + 3xy - 10x - 15y$ FOIL

$= 2x^2 - 10x + 3xy - 15y$ Original polynomial

(d)
$$\begin{aligned} t^3 + 2t^2 - 3t - 6 &= (t^3 + 2t^2) + (-3t - 6) && \text{Group terms.} \\ &= t^2(t + 2) - 3(t + 2) && \text{Factor out } -3 \text{ so there is a common factor, } t + 2;\ -3(t + 2) = -3t - 6. \\ &= (t + 2)(t^2 - 3) && \text{Factor out } t + 2. \end{aligned}$$

Check: $(t + 2)(t^2 - 3) = t^3 - 3t + 2t^2 - 6$ FOIL

$= t^3 + 2t^2 - 3t - 6$ Original polynomial

CAUTION

Be careful with signs when grouping in a problem like Example 5(d). It is wise to check the factoring in the second step, as shown in the example side comment, before continuing.

Work Problem 5 at the Side.

ANSWERS

5. **(a)** $(p + 5)(q + 2)$
(b) $(2x + 3)(y + 1)$
(c) $(a - 2)(2a + 3b)$
(d) $(x + 3)(x^2 - 5)$

Use these steps to factor a polynomial with four terms by grouping.

Factoring by Grouping

Step 1 **Group terms.** Collect the terms into two groups so that each group has a common factor.

Step 2 **Factor within groups.** Factor out the greatest common factor from each group.

Step 3 **Factor the entire polynomial.** Factor a common binomial factor from the results of Step 2.

Step 4 **If necessary, rearrange terms.** If Step 2 does not result in a common binomial factor, try a different grouping.

EXAMPLE 6 Rearranging Terms Before Factoring by Grouping

Factor by grouping.

(a) $10x^2 - 12y + 15x - 8xy$

Factoring out the common factor of 2 from the first two terms and the common factor of x from the last two terms gives

$$10x^2 - 12y + 15x - 8xy = 2(5x^2 - 6y) + x(15 - 8y).$$

This does not lead to a common factor, so we try rearranging the terms. There is usually more than one way to do this. We try

$$10x^2 - 8xy - 12y + 15x,$$

and group the first two terms and the last two terms as follows.

$$\begin{aligned} 10x^2 - 8xy - 12y + 15x &= 2x(5x - 4y) + 3(-4y + 5x) \\ &= 2x(5x - 4y) + 3(5x - 4y) \\ &= (5x - 4y)(2x + 3) \end{aligned}$$

Check: $(5x - 4y)(2x + 3) = 10x^2 + 15x - 8xy - 12y$ FOIL

$= 10x^2 - 12y + 15x - 8xy$ Original polynomial

(b) $2xy + 12 - 3y - 8x$

We need to rearrange these terms to get two groups that each have a common factor. Trial and error suggests the following grouping.

$2xy + 12 - 3y - 8x = (2xy - 3y) + (-8x + 12)$ Group terms.

$= y(2x - 3) - 4(2x - 3)$ Factor each group; be careful with signs.

$= (2x - 3)(y - 4)$ Factor out $2x - 3$.

Since the quantities in parentheses in the second step must be the same, we factored out -4 rather than 4. *Check* by multiplying.

CAUTION

Use negative signs carefully when grouping, as in Example 6(b), or a sign error will occur. ***Always check by multiplying.***

Work Problem 6 at the Side.

6 Factor by grouping.

(a) $6y^2 - 20w + 15y - 8yw$

(b) $9mn - 4 + 12m - 3n$

Answers

6. (a) $(2y + 5)(3y - 4w)$
(b) $(3m - 1)(3n + 4)$

Focus on Real-Data Applications

Idle Prime Time

A positive integer greater than 1 is a **prime number** if its only factors are 1 and itself. Every positive integer can be written as a product of prime numbers in a unique way, except for the order of the factors. Finding new primes has intrigued people from ancient Greece to modern times. The *Great Internet Mersenne Prime Search* is a consortium headed by George Woltman and Scott Kurowski that has discovered seven world record primes. On May 15, 2004, Josh Findley discovered the prime number, $2^{24,036,583} - 1$. His calculation took over two weeks on his 2.4 GHz Pentium 4 computer. This prime number is nearly a million digits larger than the last prime found and has 7,235,733 decimal digits when written out. (*Source:* http://www.mersenne.org/prime.htm)

Prime numbers are essential in the development of unbreakable codes that, in an era of Internet commerce, ensure security in transmitting and storing computer data.

The oldest known method for finding prime numbers is the Sieve of Eratosthenes, similar to the version shown below. Numbers that are not prime (composite numbers) are eliminated and only the prime numbers are left. Begin with 2. Two is prime but multiples of 2 are not, so delete the remaining numbers in Column 2 and all of Columns 4 and 6. Three is prime, but multiples of 3 are not, so delete the remaining numbers in Column 3. Examine the remaining numbers and eliminate any that are composite (such as 25 or 91). The prime numbers are highlighted.

For Group Discussion

1. **Twin primes** occur in pairs that differ by 2. List all the twin primes from the table.

2. Observe that all prime numbers larger than 3 are in Columns 1 and 5. Each number in Column 5 is 1 less than a multiple of 6, and therefore has the form $6n - 1$. Each number in Column 1 has a similar structure, $6n + 1$. Show that the larger of each of these twin primes, found in the year 2000, has the form $6n + 1$:

$$1,693,965 \times 2^{66,443} \pm 1$$

and

$$4,648,619,711,505 \times 2^{60,000} \pm 1.$$

(*Hint:* Show that the leading term is divisible by both 2 and 3.)

SIEVE OF ERATOSTHENES

Col 1	Col 2	Col 3	Col 4	Col 5	Col 6
1	2	3	4	5	6
7	8	9	10	11	12
13	14	15	16	17	18
19	20	21	22	23	24
25	26	27	28	29	30
31	32	33	34	35	36
37	38	39	40	41	42
43	44	45	46	47	48
49	50	51	52	53	54
55	56	57	58	59	60
61	62	63	64	65	66
67	68	69	70	71	72
73	74	75	76	77	78
79	80	81	82	83	84
85	86	87	88	89	90
91	92	93	94	95	96
97	98	99	100	101	102

3. **Mersenne primes,** named for the 17th-century French monk Marin Mersenne, have the form $2^p - 1$, where p is a prime number. Not all such numbers are prime. Show that $2^{11} - 1$ is composite and $2^5 - 1$ is prime.

4. A **Sophie Germain prime,** named for an 18th-century French mathematician, is an odd prime p for which $2p + 1$ is also prime. For example, 5 is a Sophie Germain prime since 11 $(2 \cdot 5 + 1)$ is prime, but 13 is not since 27 $(2 \cdot 13 + 1)$ is composite. List the Sophie Germain primes from the table.

6.1 Exercises

FOR EXTRA HELP

Tutor Center Addison-Wesley Math Tutor Center

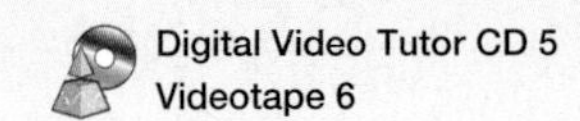

MathXL

Digital Video Tutor CD 5 Videotape 6

Student's Solutions Manual

MyMathLab

Interactmath.com

Find the greatest common factor for each list of numbers. See Example 1.

1. 12, 16 **2.** 18, 24 **3.** 40, 20, 4 **4.** 50, 30, 5

5. 18, 24, 36, 48 **6.** 15, 30, 45, 75 **7.** 4, 9, 12 **8.** 9, 16, 24

Find the greatest common factor for each list of terms. See Example 2.

9. $16y, 24$ **10.** $18w, 27$ **11.** $30x^3, 40x^6, 50x^7$

12. $60z^4, 70z^8, 90z^9$ **13.** $-x^4y^3, -xy^2$ **14.** $-a^4b^5, -a^3b$

15. $42ab^3, -36a, 90b, -48ab$ **16.** $45c^3d, 75c, 90d, -105cd$

Complete each factoring.

17. $9m^4 = 3m^2(\quad)$ **18.** $12p^5 = 6p^3(\quad)$ **19.** $-8z^9 = -4z^5(\quad)$

20. $-15k^{11} = -5k^8(\quad)$ **21.** $6m^4n^5 = 3m^3n(\quad)$ **22.** $27a^3b^2 = 9a^2b(\quad)$

23. $12y + 24 = 12(\quad)$ **24.** $18p + 36 = 18(\quad)$

25. $10a^2 - 20a = 10a(\quad)$ **26.** $15x^2 - 30x = 15x(\quad)$

27. $8x^2y + 12x^3y^2 = 4x^2y(\quad)$ **28.** $18s^3t^2 + 10st = 2st(\quad)$

Factor out the greatest common factor. See Examples 3 and 4.

29. $x^2 - 4x$ **30.** $m^2 - 7m$ **31.** $6t^2 + 15t$ **32.** $8x^2 + 6x$

33. $\frac{1}{4}d^2 - \frac{3}{4}d$ **34.** $\frac{1}{5}z^2 + \frac{3}{5}z$ **35.** $12x^3 + 6x^2$ **36.** $21b^3 - 7b^2$

37. $65y^{10} + 35y^6$ **38.** $100a^5 + 16a^3$ **39.** $11w^3 - 100$

40. $13z^5 - 80$ **41.** $8m^2n^3 + 24m^2n^2$ **42.** $19p^2y - 38p^2y^3$

43. $4x^3 - 10x^2 + 6x$ **44.** $9z^3 - 6z^2 + 12z$ **45.** $13y^8 + 26y^4 - 39y^2$

46. $5x^5 + 25x^4 - 20x^3$

47. $45q^4p^5 + 36qp^6 + 81q^2p^3$

48. $125a^3z^5 + 60a^4z^4 - 85a^5z^2$

49. $c(x + 2) + d(x + 2)$

50. $r(5 - x) + t(5 - x)$

51. $a^2(2a + b) - b(2a + b)$

52. $3x(x^2 + 5) - y(x^2 + 5)$

53. $q(p + 4) - 1(p + 4)$

54. $y^2(x - 4) + 1(x - 4)$

Factor by grouping. See Examples 5 and 6.

55. $5m + mn + 20 + 4n$

56. $ts + 5t + 2s + 10$

57. $6xy - 21x + 8y - 28$

58. $2mn - 8n + 3m - 12$

59. $3xy + 9x + y + 3$

60. $6n + 4mn + 3 + 2m$

61. $7z^2 + 14z - az - 2a$

62. $2b^2 + 3b - 8ab - 12a$

63. $18r^2 + 12ry - 3xr - 2xy$

64. $5m^2 + 15mp - 2mr - 6pr$

65. $w^3 + w^2 + 9w + 9$

66. $y^3 + y^2 + 6y + 6$

67. $3a^3 + 6a^2 - 2a - 4$

68. $10x^3 + 15x^2 - 8x - 12$

69. $16m^3 - 4m^2p^2 - 4mp + p^3$

70. $10t^3 - 2t^2s^2 - 5ts + s^3$

71. $y^2 + 3x + 3y + xy$

72. $m^2 + 14p + 7m + 2mp$

73. $2z^2 + 6w - 4z - 3wz$

74. $2a^2 + 20b - 8a - 5ab$

RELATING CONCEPTS (EXERCISES 75–78) For Individual or Group Work

In many cases, the choice of which pairs of terms to group when factoring by grouping can be made in different ways. To see this for Example 6(b), ***work Exercises 75–78 in order.***

75. Start with the polynomial from Example 6(b), $2xy + 12 - 3y - 8x$, and rearrange the terms as follows: $2xy - 8x - 3y + 12$. What property from **Section 1.7** allows this?

76. Group the first two terms and the last two terms of the rearranged polynomial in Exercise 75. Then factor each group.

77. Is your result from Exercise 76 in factored form? Explain your answer.

78. If your answer to Exercise 77 is *no,* factor the polynomial. Is the result the same as the one shown for Example 6(b)?

6.2 Factoring Trinomials

OBJECTIVES

1 Factor trinomials with a coefficient of 1 for the squared term.

2 Factor trinomials after factoring out the greatest common factor.

Using FOIL, the product of the binomials $k - 3$ and $k + 1$ is

$$(k - 3)(k + 1) = k^2 - 2k - 3. \quad \text{Multiplying}$$

Suppose instead that we are given the polynomial $k^2 - 2k - 3$ and want to rewrite it as the product $(k - 3)(k + 1)$. That is,

$$k^2 - 2k - 3 = (k - 3)(k + 1). \quad \text{Factoring}$$

Recall from **Section 6.1** that this process is called factoring the polynomial. Factoring reverses or "undoes" multiplying.

OBJECTIVE 1 Factor trinomials with a coefficient of 1 for the squared term. When factoring polynomials with integer coefficients, we use only integers in the factors. For example, we can factor $x^2 + 5x + 6$ by finding integers m and n such that

$$x^2 + 5x + 6 = (x + m)(x + n).$$

To find these integers m and n, we first use FOIL to multiply the two binomials on the right side of the equation:

$$(x + m)(x + n) = x^2 + nx + mx + mn.$$

By the distributive property,

$$x^2 + nx + mx + mn = x^2 + (n + m)x + mn.$$

Comparing this result with $x^2 + 5x + 6$ shows that we must find integers m and n having a sum of 5 and a product of 6.

Product of m and n is 6.

$$x^2 + 5x + \mathbf{6} = x^2 + (\mathbf{n + m})x + \mathbf{mn}$$

Sum of m and n is 5.

Because many pairs of integers have a sum of 5, it is best to begin by listing those pairs of integers whose product is 6. Both 5 and 6 are positive, so we consider only pairs in which both integers are positive.

1 **(a)** List all pairs of positive integers whose product is 6.

(b) Find the pair from part (a) whose sum is 5.

Work Problem 1 at the Side.

From Problem 1 at the side, we see that the numbers 1 and 6 and the numbers 2 and 3 both have a product of 6, but only the pair 2 and 3 has a sum of 5. So 2 and 3 are the required integers, and

$$x^2 + 5x + 6 = (x + 2)(x + 3).$$

Check by multiplying the binomials using FOIL. ***Make sure that the sum of the outer and inner products produces the correct middle term.***

Check: $(x + 2)(x + 3) = x^2 + 5x + 6$

$2x$
$3x$
$5x$ Add.

This method of factoring can be used only for trinomials that have 1 as the coefficient of the squared term. Methods for factoring other trinomials will be given in the next two sections.

ANSWERS

1. **(a)** 1, 6; 2, 3 **(b)** 2, 3

2 Factor each trinomial.

(a) $y^2 + 12y + 20$

First complete the given list of numbers.

Factors of 20	Sums of Factors
20, 1	20 + 1 = 21
10, ___	10 + ___ = ___
5, ___	5 + ___ = ___

(b) $x^2 + 9x + 18$

3 Factor each trinomial.

(a) $t^2 - 12t + 32$

First complete the given list of numbers.

Factors of 32	Sums of Factors
−32, −1	−32 + (−1) = −33
−16, ___	−16 + (___) = ___
−8, ___	−8 + (___) = ___

(b) $y^2 - 10y + 24$

Answers

2. (a) 2; 2; 12; 4; 4; 9; $(y + 10)(y + 2)$
   (b) $(x + 3)(x + 6)$
3. (a) −2; −2; −18; −4; −4; −12; $(t - 8)(t - 4)$
   (b) $(y - 6)(y - 4)$

EXAMPLE 1 Factoring a Trinomial with All Positive Terms

Factor $m^2 + 9m + 14$.

Look for two integers whose product is 14 and whose sum is 9. List the pairs of integers whose products are 14. Then examine the sums. Only positive integers are needed since all signs in $m^2 + 9m + 14$ are positive.

Factors of 14	Sums of Factors	
14, 1	14 + 1 = 15	
7, 2	7 + 2 = 9	Sum is 9.

From the list, 7 and 2 are the required integers, since $7 \cdot 2 = 14$ and $7 + 2 = 9$. Thus,

$$m^2 + 9m + 14 = (m + 2)(m + 7).$$

Check: $(m + 2)(m + 7) = m^2 + 7m + 2m + 14$ FOIL

$= m^2 + 9m + 14$ Original polynomial

NOTE

In Example 1, the answer $(m + 2)(m + 7)$ also could have been written

$$(m + 7)(m + 2).$$

Because of the commutative property of multiplication, the order of the factors does not matter. ***Always check by multiplying.***

Work Problem 2 at the Side.

EXAMPLE 2 Factoring a Trinomial with a Negative Middle Term

Factor $x^2 - 9x + 20$.

Find two integers whose product is 20 and whose sum is −9. Since the numbers we are looking for have a *positive product* and a *negative sum,* we consider only pairs of negative integers.

Factors of 20	Sums of Factors	
−20, −1	−20 + (−1) = −21	
−10, −2	−10 + (−2) = −12	
−5, −4	−5 + (−4) = −9	Sum is −9.

The required integers are −5 and −4, so

$$x^2 - 9x + 20 = (x - 5)(x - 4).$$

Check: $(x - 5)(x - 4) = x^2 - 4x - 5x + 20$

$= x^2 - 9x + 20$

Work Problem 3 at the Side.

EXAMPLE 3 Factoring a Trinomial with Two Negative Terms

Factor $p^2 - 2p - 15$.

Find two integers whose product is -15 and whose sum is -2. If these numbers do not come to mind right away, find them (if they exist) by listing all the pairs of integers whose product is -15. Because the last term, -15, is negative, we need pairs of integers with different signs.

Factors of −15	Sums of Factors	
$15, -1$	$15 + (-1) = 14$	
$-15, 1$	$-15 + 1 = -14$	
$5, -3$	$5 + (-3) = 2$	
$\mathbf{-5, 3}$	$\mathbf{-5 + 3 = -2}$	Sum is -2.

The required integers are -5 and 3, so

$$p^2 - 2p - 15 = (p - 5)(p + 3).$$

Check: Multiply $(p - 5)(p + 3)$.

NOTE
In Examples 1–3, notice that we listed factors in descending order (disregarding sign) when we were looking for the required pair of integers. This helps avoid skipping the correct combination.

Work Problem 4 at the Side.

As shown in the next example, some trinomials cannot be factored using only integers. We call such trinomials **prime polynomials.**

EXAMPLE 4 Deciding whether Polynomials Are Prime

Factor each trinomial.

(a) $x^2 - 5x + 12$

As in Example 2, both factors must be negative to give a positive product and a negative sum. First, list all pairs of negative integers whose product is 12. Then examine the sums.

Factors of 12	Sums of Factors
$-12, -1$	$-12 + (-1) = -13$
$-6, -2$	$-6 + (-2) = -8$
$-4, -3$	$-4 + (-3) = -7$

None of the pairs of integers has a sum of -5. Therefore, the trinomial $x^2 - 5x + 12$ *cannot be factored using only integers;* it is a *prime polynomial.*

(b) $k^2 - 8k + 11$

There is no pair of integers whose product is 11 and whose sum is -8, so $k^2 - 8k + 11$ is a prime polynomial.

Work Problem 5 at the Side.

4 Factor each trinomial.

(a) $a^2 - 9a - 22$

(b) $r^2 - 6r - 16$

5 Factor each trinomial, if possible.

(a) $r^2 - 3r - 4$

(b) $m^2 - 2m + 5$

ANSWERS
4. (a) $(a - 11)(a + 2)$ **(b)** $(r - 8)(r + 2)$
5. (a) $(r - 4)(r + 1)$ **(b)** prime

6 Factor each trinomial.

(a) $b^2 - 3ab - 4a^2$

(b) $r^2 - 6rs + 8s^2$

7 Factor each trinomial completely.

(a) $2p^3 + 6p^2 - 8p$

(b) $3x^4 - 15x^3 + 18x^2$

ANSWERS

6. (a) $(b - 4a)(b + a)$
   (b) $(r - 4s)(r - 2s)$
7. (a) $2p(p + 4)(p - 1)$
   (b) $3x^2(x - 3)(x - 2)$

The procedure for factoring a trinomial of the form $x^2 + bx + c$ follows.

Factoring $x^2 + bx + c$

Find two integers whose product is c and whose sum is b.

1. Both integers must be positive if b and c are positive.
2. Both integers must be negative if c is positive and b is negative.
3. One integer must be positive and one must be negative if c is negative.

EXAMPLE 5 Factoring a Trinomial with Two Variables

Factor $z^2 - 2bz - 3b^2$.

Here, the coefficient of z in the middle term is $-2b$, so we need to find two expressions whose product is $-3b^2$ and whose sum is $-2b$. The expressions are $-3b$ and b, so

$$z^2 - 2bz - 3b^2 = (z - 3b)(z + b).$$

Check:
$$\begin{aligned}(z - 3b)(z + b) &= z^2 + zb - 3bz - 3b^2\\ &= z^2 + 1bz - 3bz - 3b^2\\ &= z^2 - 2bz - 3b^2\end{aligned}$$

Work Problem 6 at the Side.

OBJECTIVE 2 Factor trinomials after factoring out the greatest common factor. The trinomial in the next example does not have a coefficient of 1 for the squared term. (In fact, there is no squared term.) However, there may be a common factor.

EXAMPLE 6 Factoring a Trinomial with a Common Factor

Factor $4x^5 - 28x^4 + 40x^3$.

First, factor out the greatest common factor, $4x^3$.

$$4x^5 - 28x^4 + 40x^3 = 4x^3(x^2 - 7x + 10)$$

Now factor $x^2 - 7x + 10$. The integers -5 and -2 have a product of 10 and a sum of -7. The complete factored form is

$$4x^5 - 28x^4 + 40x^3 = 4x^3(x - 5)(x - 2). \quad \text{Include } 4x^3.$$

Check:
$$\begin{aligned}4x^3(x - 5)(x - 2) &= 4x^3(x^2 - 7x + 10)\\ &= 4x^5 - 28x^4 + 40x^3\end{aligned}$$

CAUTION

When factoring, ***always look for a common factor first.*** Remember to include the common factor as part of the answer. As a check, multiplying out the complete factored form should give the original polynomial.

Work Problem 7 at the Side.

6.2 Exercises

FOR EXTRA HELP

Addison-Wesley Math Tutor Center

MathXL

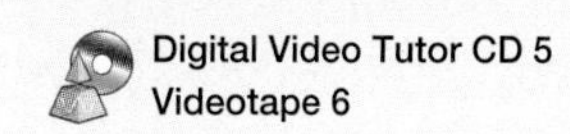

Digital Video Tutor CD 5 Videotape 6

Student's Solutions Manual

MyMathLab

Interactmath.com

1. When factoring a trinomial in x as $(x + a)(x + b)$, what must be true of a and b, if the last term of the trinomial is negative?

2. In Exercise 1, what must be true of a and b if the last term is positive?

3. What is meant by a *prime polynomial?*

4. How can you check your work when factoring a trinomial? Does the check ensure that the trinomial is *completely* factored?

In Exercises 5–8, list all pairs of integers with the given product. Then find the pair whose sum is given. See the tables in Examples 1–4.

5. Product: 12; Sum: 7

6. Product: 18; Sum: 9

7. Product: -24; Sum: -5

8. Product: -36; Sum: -16

9. Which one of the following is the correct factored form of $x^2 - 12x + 32$?

   A. $(x - 8)(x + 4)$ B. $(x + 8)(x - 4)$

   C. $(x - 8)(x - 4)$ D. $(x + 8)(x + 4)$

10. What would be the first step in factoring $2x^3 + 8x^2 - 10x$?

Complete each factoring.

11. $x^2 + 15x + 44 = (x + 4)(\quad)$

12. $r^2 + 15r + 56 = (r + 7)(\quad)$

13. $x^2 - 9x + 8 = (x - 1)(\quad)$

14. $t^2 - 14t + 24 = (t - 2)(\quad)$

15. $y^2 - 2y - 15 = (y + 3)(\quad)$

16. $t^2 - t - 42 = (t + 6)(\quad)$

17. $x^2 + 9x - 22 = (x - 2)(\quad)$

18. $x^2 + 6x - 27 = (x - 3)(\quad)$

19. $y^2 - 7y - 18 = (y + 2)(\quad)$

20. $y^2 - 2y - 24 = (y + 4)(\quad)$

Factor completely. If a polynomial cannot be factored, write prime. *See Examples 1–4.*

21. $y^2 + 9y + 8$

22. $a^2 + 9a + 20$

23. $b^2 + 8b + 15$

24. $x^2 + 6x + 8$

25. $m^2 + m - 20$

26. $p^2 + 4p - 5$

27. $x^2 + 3x - 40$

28. $d^2 + 4d - 45$

29. $y^2 - 8y + 15$

30. $y^2 - 6y + 8$

31. $z^2 - 15z + 56$

32. $x^2 - 13x + 36$

33. $r^2 - r - 30$

34. $q^2 - q - 42$

35. $a^2 - 8a - 48$

36. $m^2 - 10m - 24$

37. $x^2 + 4x + 5$

38. $t^2 + 11t + 12$

Factor completely. See Examples 5 and 6.

39. $r^2 + 3ra + 2a^2$

40. $x^2 + 5xa + 4a^2$

41. $x^2 + 4xy + 3y^2$

42. $p^2 + 9pq + 8q^2$

43. $t^2 - tz - 6z^2$

44. $a^2 - ab - 12b^2$

45. $v^2 - 11vw + 30w^2$

46. $v^2 - 11vx + 24x^2$

47. $4x^2 + 12x - 40$

48. $5y^2 - 5y - 30$

49. $2t^3 + 8t^2 + 6t$

50. $3t^3 + 27t^2 + 24t$

51. $2x^6 + 8x^5 - 42x^4$

52. $4y^5 + 12y^4 - 40y^3$

53. $a^5 + 3a^4b - 4a^3b^2$

54. $z^{10} - 4z^9y - 21z^8y^2$

55. $m^3n - 10m^2n^2 + 24mn^3$

56. $y^3z + 3y^2z^2 - 54yz^3$

57. Use the FOIL method from **Section 5.3** to show that $(2x + 4)(x - 3) = 2x^2 - 2x - 12$. Why, then, is it incorrect to completely factor $2x^2 - 2x - 12$ as $(2x + 4)(x - 3)$?

58. Why is it incorrect to completely factor $3x^2 + 9x - 12$ as the product $(x - 1)(3x + 12)$?

6.3 Factoring Trinomials by Grouping

OBJECTIVE

1 Factor trinomials by grouping when the coefficient of the squared term is not 1.

Trinomials like $2x^2 + 7x + 6$, in which the coefficient of the squared term is *not* 1, are factored with extensions of the methods from the previous sections. One such method uses factoring by grouping from **Section 6.1.**

OBJECTIVE 1 Factor trinomials by grouping when the coefficient of the squared term is not 1. Recall that a trinomial such as $m^2 + 3m + 2$ is factored by finding two integers whose product is 2 and whose sum is 3. To factor $2x^2 + 7x + 6$, we look for two integers whose product is $2 \cdot 6 = 12$ and whose sum is 7.

Sum is 7.

$$\mathbf{2}x^2 + \mathbf{7}x + \mathbf{6}$$

Product is $2 \cdot 6 = 12$.

By considering pairs of positive integers whose product is 12, the necessary integers are found to be 3 and 4. We use these integers to write the middle term, $7x$, as $7x = 3x + 4x$. The trinomial $2x^2 + 7x + 6$ becomes

$$2x^2 + 7x + 6 = 2x^2 + \underbrace{3x + 4x}_{7x} + 6.$$

$$= (2x^2 + 3x) + (4x + 6) \qquad \text{Group terms.}$$

$$= x(2x + 3) + 2(2x + 3) \qquad \text{Factor each group.}$$

Must be the same

$$2x^2 + 7x + 6 = (2x + 3)(x + 2) \qquad \text{Factor out } 2x + 3.$$

Check: $(2x + 3)(x + 2) = 2x^2 + 7x + 6$

In the preceding example, we could have written $7x$ as $4x + 3x$. Factoring by grouping this way would give the same answer.

Work Problem 1 at the Side.

1 **(a)** Factor $2x^2 + 7x + 6$ by writing $7x$ as $4x + 3x$. Complete the following.

$2x^2 + 7x + 6$

$= 2x^2 + 4x + 3x + 6$

$= (2x^2 + \_\_\_) + (3x + \_\_\_)$

$= 2x(x + \_\_\_) + 3(x + \_\_\_)$

$= (\_\_\_)(2x + 3)$

(b) Is the answer in part (a) the same as in the example? (Remember that the order of the factors does not matter.)

EXAMPLE 1 Factoring Trinomials by Grouping

Factor each trinomial.

(a) $6r^2 + r - 1$

We must find two integers with a product of $6(-1) = -6$ and a sum of 1.

Sum is 1.

$$6r^2 + r - 1 = \mathbf{6}r^2 + \mathbf{1}r - \mathbf{1}$$

Product is $6(-1) = -6$.

The integers are -2 and 3. We write the middle term, r, as $-2r + 3r$.

$$6r^2 + r - 1 = 6r^2 - 2r + 3r - 1 \qquad r = -2r + 3r$$

$$= (6r^2 - 2r) + (3r - 1) \qquad \text{Group terms.}$$

$$= 2r(3r - 1) + 1(3r - 1) \qquad \text{The binomials must be the same.}$$

$$= (3r - 1)(2r + 1) \qquad \text{Factor out } 3r - 1.$$

Check: $(3r - 1)(2r + 1) = 6r^2 + r - 1$

Continued on Next Page

ANSWERS

1. (a) $4x$; 6; 2; 2; $x + 2$ (b) yes

2 Factor each trinomial by grouping.

(a) $2m^2 + 7m + 3$

(b) $5p^2 - 2p - 3$

(c) $15k^2 - km - 2m^2$

(b) $12z^2 - 5z - 2$

Look for two integers whose product is $12(-2) = -24$ and whose sum is -5. The required integers are 3 and -8, so

$$\begin{aligned} 12z^2 - 5z - 2 &= 12z^2 + 3z - 8z - 2 && -5z = 3z - 8z \\ &= (12z^2 + 3z) + (-8z - 2) && \text{Group terms.} \\ &= 3z(4z + 1) - 2(4z + 1) && \text{Factor each group; be careful with signs.} \\ &= (4z + 1)(3z - 2). && \text{Factor out } 4z + 1. \end{aligned}$$

Check: $(4z + 1)(3z - 2) = 12z^2 - 5z - 2$

(c) $10m^2 + mn - 3n^2$

Two integers whose product is $10(-3) = -30$ and whose sum is 1 are -5 and 6. Rewrite the trinomial with four terms.

$$\begin{aligned} 10m^2 + mn - 3n^2 &= 10m^2 - 5mn + 6mn - 3n^2 && mn = -5mn + 6mn \\ &= 5m(2m - n) + 3n(2m - n) && \text{Group terms; factor each group.} \\ &= (2m - n)(5m + 3n) && \text{Factor out } 2m - n. \end{aligned}$$

Check by multiplying.

Work Problem 2 at the Side.

3 Factor each trinomial completely.

(a) $4x^2 - 2x - 30$

(b) $18p^4 + 63p^3 + 27p^2$

(c) $6a^2 + 3ab - 18b^2$

EXAMPLE 2 Factoring a Trinomial with a Common Factor by Grouping

Factor $28x^5 - 58x^4 - 30x^3$.

First factor out the greatest common factor, $2x^3$.

$$28x^5 - 58x^4 - 30x^3 = 2x^3(14x^2 - 29x - 15)$$

To factor $14x^2 - 29x - 15$, find two integers whose product is $14(-15) = -210$ and whose sum is -29. Factoring 210 into prime factors gives

$$210 = 2 \cdot 3 \cdot 5 \cdot 7.$$

Combine these prime factors in pairs in different ways, using one positive and one negative to get -210. The factors 6 and -35 have the correct sum, -29. Now rewrite the given trinomial and factor it.

$$\begin{aligned} 28x^5 - 58x^4 - 30x^3 &= 2x^3(14x^2 + 6x - 35x - 15) \\ &= 2x^3[(14x^2 + 6x) + (-35x - 15)] \\ &= 2x^3[2x(7x + 3) - 5(7x + 3)] \\ &= 2x^3[(7x + 3)(2x - 5)] \\ &= 2x^3(7x + 3)(2x - 5) \end{aligned}$$

Check by multiplying.

CAUTION

Remember to include the common factor in the final result.

Work Problem 3 at the Side.

ANSWERS

2. **(a)** $(2m + 1)(m + 3)$
   (b) $(5p + 3)(p - 1)$
   (c) $(5k - 2m)(3k + m)$
3. **(a)** $2(2x + 5)(x - 3)$
   (b) $9p^2(2p + 1)(p + 3)$
   (c) $3(2a - 3b)(a + 2b)$

6.3 Exercises

FOR EXTRA HELP

Addison-Wesley Math Tutor Center | MathXL | Digital Video Tutor CD 6 Videotape 6 | Student's Solutions Manual | MyMathLab | Interactmath.com

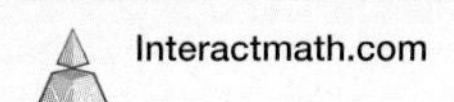

The middle term of each trinomial has been rewritten. Now factor by grouping. See Example 1.

1. $m^2 + 8m + 12$
$= m^2 + 6m + 2m + 12$

2. $x^2 + 9x + 14$
$= x^2 + 7x + 2x + 14$

3. $a^2 + 3a - 10$
$= a^2 + 5a - 2a - 10$

4. $y^2 - 2y - 24$
$= y^2 + 4y - 6y - 24$

5. $10t^2 + 9t + 2$
$= 10t^2 + 5t + 4t + 2$

6. $6x^2 + 13x + 6$
$= 6x^2 + 9x + 4x + 6$

7. $15z^2 - 19z + 6$
$= 15z^2 - 10z - 9z + 6$

8. $12p^2 - 17p + 6$
$= 12p^2 - 9p - 8p + 6$

9. $8s^2 + 2st - 3t^2$
$= 8s^2 - 4st + 6st - 3t^2$

10. $3x^2 - xy - 14y^2$
$= 3x^2 - 7xy + 6xy - 14y^2$

11. $15a^2 + 22ab + 8b^2$
$= 15a^2 + 10ab + 12ab + 8b^2$

12. $25m^2 + 25mn + 6n^2$
$= 25m^2 + 15mn + 10mn + 6n^2$

13. Which pair of integers would be used to rewrite the middle term when factoring $12y^2 + 5y - 2$ by grouping?

A. $-8, 3$ **B.** $8, -3$ **C.** $-6, 4$ **D.** $6, -4$

14. Which pair of integers would be used to rewrite the middle term when factoring $20b^2 - 13b + 2$ by grouping?

A. $10, 3$ **B.** $-10, -3$ **C.** $8, 5$ **D.** $-8, -5$

Complete the steps to factor each trinomial by grouping.

15. $2m^2 + 11m + 12$

(a) Find two integers whose product is _______ · _______ = _______ and whose sum is _______.

(b) The required integers are _______ and _______.

(c) Write the middle term $11m$ as _______ + _______.

(d) Rewrite the given trinomial using four terms.

(e) Factor the polynomial in part (d) by grouping.

(f) Check by multiplying.

16. $6y^2 - 19y + 10$

(a) Find two integers whose product is _______ · _______ = _______ and whose sum is _______.

(b) The required integers are _______ and _______.

(c) Write the middle term $-19y$ as _______ + _______.

(d) Rewrite the given trinomial using four terms.

(e) Factor the polynomial in part (d) by grouping.

(f) Check by multiplying.

Factor each trinomial by grouping. See Examples 1 and 2.

17. $2x^2 + 7x + 3$

18. $3y^2 + 13y + 4$

19. $4r^2 + r - 3$

20. $4r^2 + 3r - 10$

21. $8m^2 - 10m - 3$

22. $20x^2 - 28x - 3$

23. $21m^2 + 13m + 2$

24. $38x^2 + 23x + 2$

25. $6b^2 + 7b + 2$

26. $6w^2 + 19w + 10$

27. $12y^2 - 13y + 3$

28. $15a^2 - 16a + 4$

29. $24x^2 - 42x + 9$

30. $48b^2 - 74b - 10$

31. $2m^3 + 2m^2 - 40m$

32. $3x^3 + 12x^2 - 36x$

33. $32z^5 - 20z^4 - 12z^3$

34. $18x^5 + 15x^4 - 75x^3$

35. $12p^2 + 7pq - 12q^2$

36. $6m^2 - 5mn - 6n^2$

37. $6a^2 - 7ab - 5b^2$

38. $25g^2 - 5gh - 2h^2$

39. $5 - 6x + x^2$

40. $7 + 8x + x^2$

41. On a quiz, a student factored $16x^2 - 24x + 5$ by grouping as follows.

$$\begin{aligned} &16x^2 - 24x + 5 \\ &= 16x^2 - 4x - 20x + 5 \\ &= 4x(4x - 1) - 5(4x - 1) \quad \text{His answer} \end{aligned}$$

He thought his answer was correct since it checked by multiplying. Why was the answer marked wrong? What is the correct factored form?

42. On the same quiz, another student factored $3k^3 - 12k^2 - 15k$ by first factoring out the common factor $3k$ to get $3k(k^2 - 4k - 5)$. Then she wrote

$$\begin{aligned} k^2 - 4k - 5 &= k^2 - 5k + k - 5 \\ &= k(k - 5) + 1(k - 5) \\ &= (k - 5)(k + 1). \quad \text{Her answer} \end{aligned}$$

Why was the answer marked wrong? What is the correct factored form?

6.4 Factoring Trinomials Using FOIL

OBJECTIVE

1 **Factor trinomials using FOIL.**

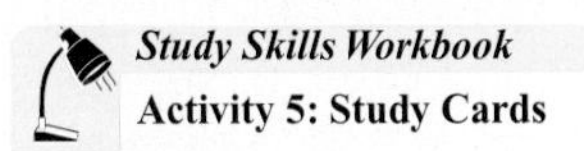

OBJECTIVE 1 Factor trinomials using FOIL. This section shows an alternative method of factoring trinomials in which the coefficient of the squared term is not 1. This method uses trial and error.

To factor $2x^2 + 7x + 6$ (the same trinomial factored at the beginning of **Section 6.3**) by trial and error, we use FOIL backwards. We want to write $2x^2 + 7x + 6$ as the product of two binomials.

$$2x^2 + 7x + 6 = (\qquad)(\qquad)$$

The product of the two first terms of the binomials is $2x^2$. The possible factors of $2x^2$ are $2x$ and x or $-2x$ and $-x$. Since all terms of the trinomial are positive, we consider only positive factors. Thus, we have

$$2x^2 + 7x + 6 = (2x \qquad)(x \qquad).$$

The product of the two last terms, 6, can be factored as $1 \cdot 6$, $6 \cdot 1$, $2 \cdot 3$, or $3 \cdot 2$. Try each pair to find the pair that gives the correct middle term, $7x$.

1 Multiply to decide whether each factored form is correct or incorrect for

$$2x^2 + 7x + 6.$$

(a) $(2x + 1)(x + 6)$

(b) $(2x + 6)(x + 1)$

(c) $(2x + 2)(x + 3)$

Work Problem 1 at the Side.

In part (b) at the side, since $2x + 6 = 2(x + 3)$, the binomial $2x + 6$ has a common factor of 2, while $2x^2 + 7x + 6$ has no common factor other than 1. The product $(2x + 6)(x + 1)$ cannot be correct. (Part (c) also has one binomial factor with a common factor.)

> **NOTE**
> If the original polynomial has no common factor, then none of its binomial factors will either.

Now try the remaining numbers 3 and 2 as factors of 6.

$$(2x + 3)(x + 2) = 2x^2 + 7x + 6 \qquad \text{Correct}$$

$3x$
$4x$
$7x$ Add.

Finally, we see that $2x^2 + 7x + 6$ factors as

$$2x^2 + 7x + 6 = (2x + 3)(x + 2).$$

Check by multiplying: $(2x + 3)(x + 2) = 2x^2 + 7x + 6$.

EXAMPLE 1 Factoring a Trinomial with All Positive Terms Using FOIL

Factor $8p^2 + 14p + 5$.

The number 8 has several possible pairs of factors, but 5 has only 1 and 5 or -1 and -5. For this reason, it is easier to begin by considering the factors of 5. Ignore the negative factors since all coefficients in the trinomial are positive. If $8p^2 + 14p + 5$ can be factored, the factors will have the form

$$(\quad + 5)(\quad + 1).$$

Continued on Next Page

ANSWERS
1. **(a)** incorrect **(b)** incorrect **(c)** incorrect

2 Factor each trinomial.

(a) $2p^2 + 9p + 9$

(b) $6p^2 + 19p + 10$

(c) $8x^2 + 14x + 3$

When factoring $8p^2 + 14p + 5$, the possible pairs of factors of $8p^2$ are $8p$ and p, or $4p$ and $2p$. Try various combinations, checking to see if the middle term is $14p$ in each case.

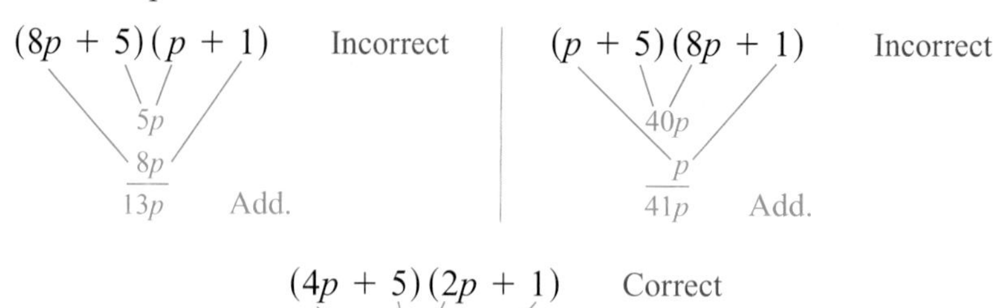

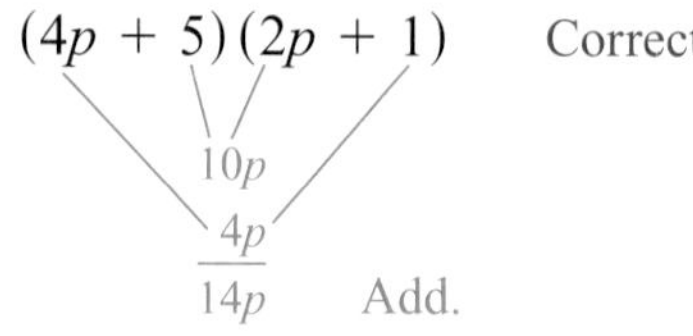

Since $14p$ is the correct middle term,

$$8p^2 + 14p + 5 = (4p + 5)(2p + 1).$$

Check: $(4p + 5)(2p + 1) = 8p^2 + 14p + 5$

Work Problem 2 at the Side.

EXAMPLE 2 **Factoring a Trinomial with a Negative Middle Term Using FOIL**

Factor $6x^2 - 11x + 3$.

Since 3 has only 1 and 3 or -1 and -3 as factors, it is better here to begin by factoring 3. The last term of the trinomial $6x^2 - 11x + 3$ is positive and the middle term has a negative coefficient, so we consider only negative factors. We need two negative factors because the *product* of two negative factors is positive and their *sum* is negative, as required.

Try -3 and -1 as factors of 3:

$$(\quad - 3)(\quad - 1).$$

The factors of $6x^2$ may be either $6x$ and x, or $2x$ and $3x$.

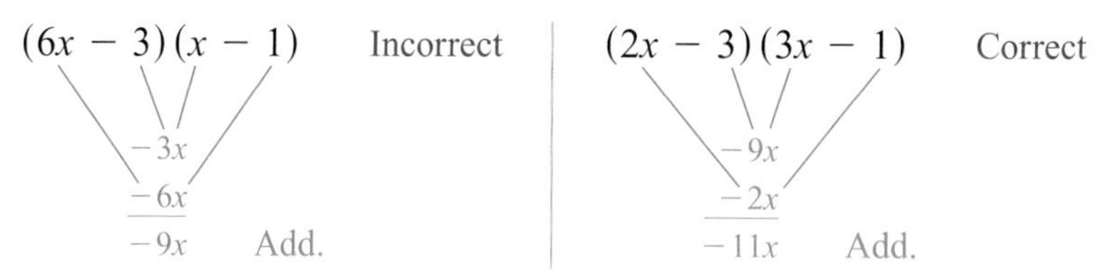

The factors $2x$ and $3x$ produce $-11x$, the correct middle term, so

$$6x^2 - 11x + 3 = (2x - 3)(3x - 1).$$

Check by multiplying.

3 Factor each trinomial.

(a) $4y^2 - 11y + 6$

(b) $9x^2 - 21x + 10$

NOTE

In Example 2, we might also realize that our initial attempt to factor $6x^2 - 11x + 3$ as $(6x - 3)(x - 1)$ *cannot* be correct since $6x - 3$ has a common factor of 3 and the original polynomial does not.

Work Problem 3 at the Side.

ANSWERS

2. (a) $(2p + 3)(p + 3)$
   (b) $(3p + 2)(2p + 5)$
   (c) $(4x + 1)(2x + 3)$
3. (a) $(4y - 3)(y - 2)$
   (b) $(3x - 5)(3x - 2)$

EXAMPLE 3 Factoring a Trinomial with a Negative Last Term Using FOIL

Factor $8x^2 + 6x - 9$.

The integer 8 has several possible pairs of factors, as does -9. Since the last term is negative, one positive factor and one negative factor of -9 are needed. Since the coefficient of the middle term is small, it is wise to avoid large factors such as 8 or 9. We try 4 and 2 as factors of 8, and 3 and -3 as factors of -9, and check the middle term.

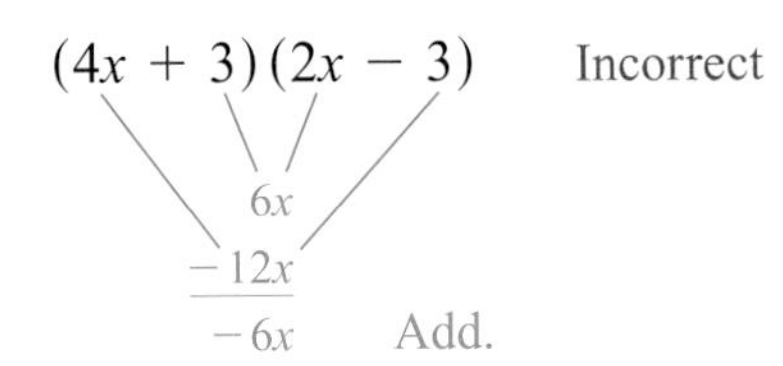

Now we try interchanging 3 and -3, since only the sign of the middle term is incorrect.

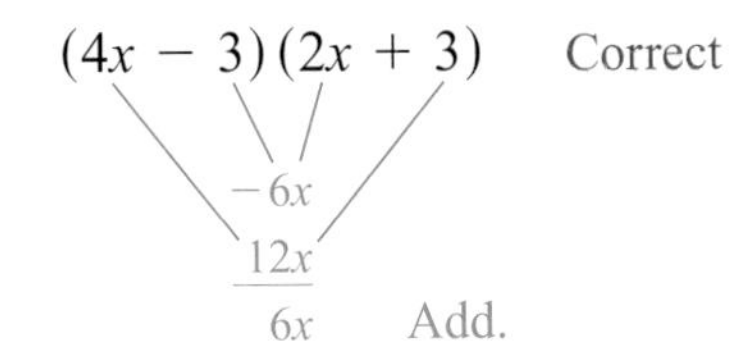

This combination produces $6x$, the correct middle term, so

$$8x^2 + 6x - 9 = (4x - 3)(2x + 3).$$

Work Problem 4 at the Side.

EXAMPLE 4 Factoring a Trinomial with Two Variables

Factor $12a^2 - ab - 20b^2$.

There are several pairs of factors of $12a^2$, including

$$12a \text{ and } a, \quad 6a \text{ and } 2a, \quad \text{and} \quad 4a \text{ and } 3a,$$

just as there are many possible pairs of factors of $-20b^2$, including

$$20b \text{ and } -b, \quad -20b \text{ and } b, \quad 10b \text{ and } -2b,$$
$$-10b \text{ and } 2b, \quad 4b \text{ and } -5b, \quad \text{and} \quad -4b \text{ and } 5b.$$

Once again, since the coefficient of the middle term is small, avoid the larger factors. Try the factors $6a$ and $2a$ and $4b$ and $-5b$.

$$(6a + 4b)(2a - 5b)$$

This cannot be correct, as mentioned before, since $6a + 4b$ has 2 as a common factor, while the given trinomial does not. Try $3a$ and $4a$ with $4b$ and $-5b$.

$$(3a + 4b)(4a - 5b) = 12a^2 + ab - 20b^2 \qquad \text{Incorrect}$$

Here the middle term is ab, rather than $-ab$. Interchange the signs of the last two terms in the factors.

$$(3a - 4b)(4a + 5b) = 12a^2 - ab - 20b^2 \qquad \text{Correct}$$

Work Problem 5 at the Side.

4 Factor each trinomial, if possible.

(a) $6x^2 + 5x - 4$

(b) $6m^2 - 11m - 10$

(c) $4x^2 - 3x - 7$

(d) $3y^2 + 8y - 6$

5 Factor each trinomial.

(a) $2x^2 - 5xy - 3y^2$

(b) $8a^2 + 2ab - 3b^2$

ANSWERS

4. **(a)** $(3x + 4)(2x - 1)$
 (b) $(2m - 5)(3m + 2)$
 (c) $(4x - 7)(x + 1)$
 (d) prime

5. **(a)** $(2x + y)(x - 3y)$
 (b) $(4a + 3b)(2a - b)$

6 Factor each trinomial.

(a) $36z^3 - 6z^2 - 72z$

(b) $-24x^3 + 32x^2y + 6xy^2$

EXAMPLE 5 Factoring Trinomials with Common Factors

Factor each trinomial.

(a) $15y^3 + 55y^2 + 30y$

First factor out the greatest common factor, $5y$.

$$15y^3 + 55y^2 + 30y = 5y(3y^2 + 11y + 6)$$

Now factor $3y^2 + 11y + 6$. Try $3y$ and y as factors of $3y^2$ and 2 and 3 as factors of 6.

$$(3y + 2)(y + 3) = 3y^2 + 11y + 6 \quad \text{Correct}$$

The complete factored form of $15y^3 + 55y^2 + 30y$ is

$$15y^3 + 55y^2 + 30y = 5y(3y + 2)(y + 3).$$

Check by multiplying.

(b) $-24a^3 - 42a^2 + 45a$

The common factor could be $3a$ or $-3a$. If we factor out $-3a$, the first term of the trinomial will be positive, which makes it easier to factor.

$$-24a^3 - 42a^2 + 45a = -3a(8a^2 + 14a - 15) \quad \text{Factor out } -3a.$$

$$= -3a(4a - 3)(2a + 5) \quad \text{Use trial and error.}$$

Check by multiplying.

CAUTION

This caution bears repeating: ***Remember to include the common factor*** in the final factored form.

Work Problem 6 at the Side.

ANSWERS

6. **(a)** $6z(2z - 3)(3z + 4)$
(b) $-2x(6x + y)(2x - 3y)$

6.4 Exercises

FOR EXTRA HELP

Addison-Wesley Math Tutor Center

MathXL

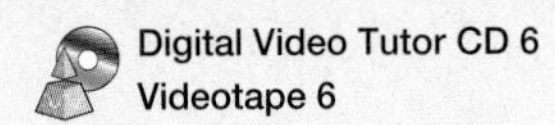

MyMathLab

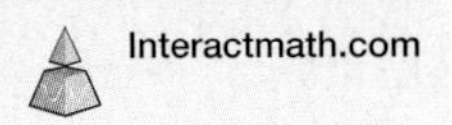

Decide which is the correct factored form of the given polynomial.

1. $2x^2 - x - 1$

A. $(2x - 1)(x + 1)$ **B.** $(2x + 1)(x - 1)$

2. $3a^2 - 5a - 2$

A. $(3a + 1)(a - 2)$ **B.** $(3a - 1)(a + 2)$

3. $4y^2 + 17y - 15$

A. $(y + 5)(4y - 3)$ **B.** $(2y - 5)(2y + 3)$

4. $12c^2 - 7c - 12$

A. $(6c - 2)(2c + 6)$ **B.** $(4c + 3)(3c - 4)$

5. $4k^2 + 13mk + 3m^2$

A. $(4k + m)(k + 3m)$ **B.** $(4k + 3m)(k + m)$

6. $2x^2 + 11x + 12$

A. $(2x + 3)(x + 4)$ **B.** $(2x + 4)(x + 3)$

Complete each factoring.

7. $6a^2 + 7ab - 20b^2 = (3a - 4b)(\qquad)$

8. $9m^2 - 3mn - 2n^2 = (3m + n)(\qquad)$

9. $2x^2 + 6x - 8 = 2(\qquad\qquad)$
$= 2(\qquad)(\qquad)$

10. $3x^2 - 9x - 30 = 3(\qquad\qquad)$
$= 3(\qquad)(\qquad)$

11. $4z^3 - 10z^2 - 6z = 2z(\qquad\qquad)$
$= 2z(\qquad)(\qquad)$

12. $15r^3 - 39r^2 - 18r = 3r(\qquad\qquad)$
$= 3r(\qquad)(\qquad)$

13. For the polynomial $12x^2 + 7x - 12$, 2 is not a common factor. Explain why the binomial $2x - 6$, then, cannot be a factor of the polynomial.

14. Explain how the signs of the last terms of the two binomial factors of a trinomial are determined.

Factor each trinomial completely. See Examples 1–5.

15. $3a^2 + 10a + 7$

16. $7r^2 + 8r + 1$

17. $2y^2 + 7y + 6$

18. $5z^2 + 12z + 4$

19. $15m^2 + m - 2$

20. $6x^2 + x - 1$

21. $12s^2 + 11s - 5$

22. $20x^2 + 11x - 3$

23. $10m^2 - 23m + 12$

24. $6x^2 - 17x + 12$

25. $8w^2 - 14w + 3$

26. $9p^2 - 18p + 8$

27. $20y^2 - 39y - 11$

28. $10x^2 - 11x - 6$

29. $3x^2 - 15x + 16$

30. $2t^2 + 13t - 18$

31. $20x^2 + 22x + 6$

32. $36y^2 + 81y + 45$

33. $40m^2q + mq - 6q$

34. $15a^2b + 22ab + 8b$

35. $15n^4 - 39n^3 + 18n^2$

36. $24a^4 + 10a^3 - 4a^2$

37. $15x^2y^2 - 7xy^2 - 4y^2$

38. $14a^2b^3 + 15ab^3 - 9b^3$

39. $5a^2 - 7ab - 6b^2$

40. $6x^2 - 5xy - y^2$

41. $12s^2 + 11st - 5t^2$

42. $25a^2 + 25ab + 6b^2$

43. $6m^6n + 7m^5n^2 + 2m^4n^3$

44. $12k^3q^4 - 4k^2q^5 - kq^6$

If a trinomial has a negative coefficient for the squared term, such as $-2x^2 + 11x - 12$, *it may be easier to factor by first factoring out the common factor* -1:

$$\begin{aligned} -2x^2 + 11x - 12 &= -1(2x^2 - 11x + 12) \\ &= -1(2x - 3)(x - 4). \end{aligned}$$

Use this method to factor the trinomials in Exercises 45–50.

45. $-x^2 - 4x + 21$

46. $-x^2 + x + 72$

47. $-3x^2 - x + 4$

48. $-5x^2 + 2x + 16$

49. $-2a^2 - 5ab - 2b^2$

50. $-3p^2 + 13pq - 4q^2$

RELATING CONCEPTS (EXERCISES 51–56) For Individual or Group Work

One of the most common problems that beginning algebra students face is this: If an answer obtained doesn't look exactly like the one given in the back of the book, is it necessarily incorrect? Often there are several different equivalent forms of an answer that are all correct. ***Work Exercises 51–56 in order,*** *to see how and why this is possible for factoring problems.*

51. Factor the integer 35 as the product of two prime numbers.

52. Factor the integer 35 as the product of the negatives of two prime numbers.

53. Verify the following factored form: $6x^2 - 11x + 4 = (3x - 4)(2x - 1)$.

54. Verify the following factored form: $6x^2 - 11x + 4 = (4 - 3x)(1 - 2x)$.

55. Compare the two valid factored forms in Exercises 53 and 54. How do the factors in each case compare?

56. Suppose you know that the correct factored form of a particular trinomial is $(7t - 3)(2t - 5)$. Based on your observations in Exercises 51–55, what is another valid factored form?

6.5 Special Factoring Techniques

OBJECTIVES

1. Factor a difference of squares.
2. Factor a perfect square trinomial.

By reversing the rules for multiplication of binomials from **Section 5.4,** we obtain rules for factoring polynomials in certain forms.

OBJECTIVE 1 Factor a difference of squares. The formula for the product of the sum and difference of the same two terms is

$$(a + b)(a - b) = a^2 - b^2.$$

Reversing this rule leads to the following special factoring rule.

Factoring a Difference of Squares

$$a^2 - b^2 = (a + b)(a - b)$$

For example,

$$m^2 - 16 = m^2 - 4^2 = (m + 4)(m - 4).$$

As the next examples show, the following conditions must be true for a binomial to be a difference of squares.

1. Both terms of the binomial must be squares, such as

$$x^2, \quad 9y^2, \quad 25, \quad 1, \quad m^4.$$

2. The terms of the binomial must have different signs (one positive and one negative).

EXAMPLE 1 Factoring Differences of Squares

Factor each binomial, if possible.

$$a^2 - b^2 = (a + b)(a - b)$$

(a) $x^2 - 49 = x^2 - 7^2 = (x + 7)(x - 7)$

(b) $y^2 - m^2 = (y + m)(y - m)$

(c) $z^2 - \frac{9}{16} = z^2 - \left(\frac{3}{4}\right)^2 = \left(z + \frac{3}{4}\right)\left(z - \frac{3}{4}\right)$

(d) $x^2 - 8$

Because 8 is not the square of an integer, this binomial is not a difference of squares. It is a prime polynomial.

(e) $p^2 + 16$

Since $p^2 + 16$ is a *sum* of squares, it is not equal to $(p + 4)(p - 4)$. Also, using FOIL,

$$(p - 4)(p - 4) = p^2 - 8p + 16 \neq p^2 + 16$$

and

$$(p + 4)(p + 4) = p^2 + 8p + 16 \neq p^2 + 16,$$

so $p^2 + 16$ is a prime polynomial.

1 Factor, if possible.

(a) $p^2 - 100$

(b) $x^2 - \frac{25}{36}$

(c) $x^2 + y^2$

(d) $9m^2 - 49$

(e) $64a^2 - 25$

2 Factor completely.

(a) $50r^2 - 32$

(b) $27y^2 - 75$

(c) $25a^2 - 64b^2$

(d) $k^4 - 49$

(e) $81r^4 - 16$

CAUTION
As Example 1(e) suggests, ***after any common factor is removed, a sum of squares cannot be factored.***

EXAMPLE 2 Factoring Differences of Squares

Factor each difference of squares.

$$a^2 - b^2 = (a + b)(a - b)$$

(a) $25m^2 - 16 = (5m)^2 - 4^2 = (5m + 4)(5m - 4)$

(b) $49z^2 - 64 = (7z)^2 - 8^2 = (7z + 8)(7z - 8)$

NOTE
As in previous sections, you should ***always check a factored form by multiplying.***

Work Problem 1 at the Side.

EXAMPLE 3 Factoring More Complex Differences of Squares

Factor completely.

(a) $81y^2 - 36$

First factor out the common factor, 9.

$$\begin{aligned} 81y^2 - 36 &= 9(9y^2 - 4) && \text{Factor out 9.} \\ &= 9[(3y)^2 - 2^2] \\ &= 9(3y + 2)(3y - 2) && \text{Difference of squares} \end{aligned}$$

(b) $9x^2 - 4z^2 = (3x)^2 - (2z)^2 = (3x + 2z)(3x - 2z)$

(c) $p^4 - 36 = (p^2)^2 - 6^2 = (p^2 + 6)(p^2 - 6)$

Neither $p^2 + 6$ nor $p^2 - 6$ can be factored further.

(d) $$\begin{aligned} m^4 - 16 &= (m^2)^2 - 4^2 \\ &= (m^2 + 4)(m^2 - 4) && \text{Difference of squares} \\ &= (m^2 + 4)(m + 2)(m - 2) && \text{Difference of squares again} \end{aligned}$$

CAUTION
Remember to factor again when any of the factors is a difference of squares, as in Example 3(d). Check by multiplying.

Work Problem 2 at the Side.

OBJECTIVE 2 Factor a perfect square trinomial. The expressions 144, $4x^2$, and $81m^6$ are called *perfect squares* because

$$144 = 12^2, \quad 4x^2 = (2x)^2, \quad \text{and} \quad 81m^6 = (9m^3)^2.$$

ANSWERS

1. (a) $(p + 10)(p - 10)$
   (b) $\left(x + \frac{5}{6}\right)\left(x - \frac{5}{6}\right)$
   (c) prime
   (d) $(3m + 7)(3m - 7)$
   (e) $(8a + 5)(8a - 5)$
2. (a) $2(5r + 4)(5r - 4)$
   (b) $3(3y + 5)(3y - 5)$
   (c) $(5a + 8b)(5a - 8b)$
   (d) $(k^2 + 7)(k^2 - 7)$
   (e) $(9r^2 + 4)(3r + 2)(3r - 2)$

A **perfect square trinomial** is a trinomial that is the square of a binomial. For example, $x^2 + 8x + 16$ is a perfect square trinomial because it is the square of the binomial $x + 4$:

$$x^2 + 8x + 16 = (x + 4)(x + 4) = (x + 4)^2.$$

For a trinomial to be a perfect square, *two of its terms must be perfect squares.* For this reason, $16x^2 + 4x + 15$ is not a perfect square trinomial because only the term $16x^2$ is a perfect square.

On the other hand, even if two of the terms are perfect squares, the trinomial may not be a perfect square trinomial. For example, $x^2 + 6x + 36$ has two perfect square terms, but it is not a perfect square trinomial. (Try to find a binomial that can be squared to give $x^2 + 6x + 36$.)

We can multiply to see that the square of a binomial gives one of the following perfect square trinomials.

Factoring Perfect Square Trinomials

$$a^2 + 2ab + b^2 = (a + b)^2$$
$$a^2 - 2ab + b^2 = (a - b)^2$$

The middle term of a perfect square trinomial is always twice the product of the two terms in the squared binomial. (This was shown in **Section 5.4.**) Use this to check any attempt to factor a trinomial that appears to be a perfect square.

EXAMPLE 4 **Factoring a Perfect Square Trinomial**

Factor $x^2 + 10x + 25$.

The term x^2 is a perfect square, and so is 25. Try to factor the trinomial as

$$x^2 + 10x + 25 = (x + 5)^2.$$

To check, take twice the product of the two terms in the squared binomial.

$$2 \cdot x \cdot 5 = 10x$$

Twice; First term of binomial; Last term of binomial

Since $10x$ is the middle term of the trinomial, the trinomial is a perfect square and can be factored as $(x + 5)^2$. Thus,

$$x^2 + 10x + 25 = (x + 5)^2.$$

Work Problem 3 at the Side.

EXAMPLE 5 **Factoring Perfect Square Trinomials**

Factor each trinomial.

(a) $x^2 - 22x + 121$

The first and last terms are perfect squares ($121 = 11^2$ or $(-11)^2$). Check to see whether the middle term of $x^2 - 22x + 121$ is twice the product of the first and last terms of the binomial $x - 11$.

Continued on Next Page

3 Factor each trinomial.

(a) $p^2 + 14p + 49$

(b) $m^2 + 8m + 16$

(c) $x^2 + 2x + 1$

ANSWERS

3. **(a)** $(p + 7)^2$ **(b)** $(m + 4)^2$ **(c)** $(x + 1)^2$

4 Factor each trinomial.

(a) $p^2 - 18p + 81$

(b) $16a^2 + 56a + 49$

(c) $121p^2 + 110p + 100$

(d) $64x^2 - 48x + 9$

(e) $27y^3 + 72y^2 + 48y$

$$2 \cdot x \cdot (-11) = -22x$$

Twice — First term — Last term

Since twice the product of the first and last terms of the binomial is the middle term, $x^2 - 22x + 121$ is a perfect square trinomial and

$$x^2 - 22x + 121 = (x - 11)^2.$$

Same sign

Notice that the sign of the second term in the squared binomial is the same as the sign of the middle term in the trinomial.

(b) $9m^2 - 24m + 16 = (3m)^2 + 2(3m)(-4) + (-4)^2 = (3m - 4)^2$

Twice — First term — Last Term

(c) $25y^2 + 20y + 16$

The first and last terms are perfect squares.

$$25y^2 = (5y)^2 \quad \text{and} \quad 16 = 4^2$$

Twice the product of the first and last terms of the binomial $5y + 4$ is

$$2 \cdot 5y \cdot 4 = 40y,$$

which is not the middle term of $25y^2 + 20y + 16$. This trinomial is not a perfect square. In fact, the trinomial cannot be factored even with the methods of the previous sections; it is a prime polynomial.

(d) $12z^3 + 60z^2 + 75z = 3z(4z^2 + 20z + 25)$ Factor out $3z$.

$$= 3z[(2z)^2 + 2(2z)(5) + 5^2]$$

$$= 3z(2z + 5)^2$$

NOTE

1. The sign of the second term in the squared binomial is always the same as the sign of the middle term in the trinomial.
2. The first and last terms of a perfect square trinomial must be *positive,* because they are squares. For example, the polynomial $x^2 - 2x - 1$ cannot be a perfect square because the last term is negative.
3. Perfect square trinomials can also be factored using grouping or FOIL, although using the method of this section is often easier.

Work Problem 4 at the Side.

The methods of factoring discussed in this section are summarized here.

Special Factoring Rules

Difference of squares	$a^2 - b^2 = (a + b)(a - b)$
Perfect square trinomials	$a^2 + 2ab + b^2 = (a + b)^2$
	$a^2 - 2ab + b^2 = (a - b)^2$

ANSWERS

4. (a) $(p - 9)^2$ (b) $(4a + 7)^2$ (c) prime (d) $(8x - 3)^2$ (e) $3y(3y + 4)^2$

6.5 Exercises

FOR EXTRA HELP

Tutor Center Addison-Wesley Math Tutor Center

MathXL MathXL

Digital Video Tutor CD 6 Videotape 6

Student's Solutions Manual

MyMathLab MyMathLab

Interactmath.com

1. To help you factor a difference of squares, complete the following list of squares.

$1^2 =$ ____ $2^2 =$ ____ $3^2 =$ ____ $4^2 =$ ____ $5^2 =$ ____

$6^2 =$ ____ $7^2 =$ ____ $8^2 =$ ____ $9^2 =$ ____ $10^2 =$ ____

$11^2 =$ ____ $12^2 =$ ____ $13^2 =$ ____ $14^2 =$ ____ $15^2 =$ ____

$16^2 =$ ____ $17^2 =$ ____ $18^2 =$ ____ $19^2 =$ ____ $20^2 =$ ____

2. To use the factoring techniques described in this section, you will sometimes need to recognize fourth powers of integers. Complete the following list of fourth powers.

$1^4 =$ ____ $2^4 =$ ____ $3^4 =$ ____ $4^4 =$ ____ $5^4 =$ ____

3. The following powers of x are all perfect squares: $x^2, x^4, x^6, x^8, x^{10}$. Based on this observation, we may make a conjecture (an educated guess) that if the power of a variable is divisible by ____ (with 0 remainder), then it is a perfect square.

4. Which of the following are differences of squares?

A. $x^2 - 4$ **B.** $y^2 + 9$ **C.** $2a^2 - 25$ **D.** $9m^2 - 1$

Factor each binomial completely. Use your answers in Exercises 1 and 2 as necessary. See Examples 1–3.

5. $y^2 - 25$ **6.** $t^2 - 16$ **7.** $p^2 - \frac{1}{9}$ **8.** $q^2 - \frac{1}{4}$

9. $m^2 - 12$ **10.** $k^2 - 18$ **11.** $9r^2 - 4$ **12.** $4x^2 - 9$

13. $36m^2 - \frac{16}{25}$ **14.** $100b^2 - \frac{4}{49}$ **15.** $36x^2 - 16$ **16.** $32a^2 - 8$

17. $196p^2 - 225$ **18.** $361q^2 - 400$ **19.** $16r^2 - 25a^2$ **20.** $49m^2 - 100p^2$

21. $100x^2 + 49$ **22.** $81w^2 + 16$ **23.** $p^4 - 49$ **24.** $r^4 - 25$

25. $x^4 - 1$ **26.** $y^4 - 16$ **27.** $p^4 - 256$ **28.** $16k^4 - 1$

29. When a student was directed to factor $x^4 - 81$ completely, his teacher did not give him full credit when he answered $(x^2 + 9)(x^2 - 9)$. The student argued that because his answer does indeed give $x^4 - 81$ when multiplied out, he should be given full credit. Was the teacher justified in her grading of this item? Why or why not?

30. The binomial $4x^2 + 16$ is a sum of squares that *can* be factored. How is this binomial factored? When can a sum of squares be factored?

31. In the polynomial $9y^2 + 14y + 25$, the first and last terms are perfect squares. Can the polynomial be factored? If it can, factor it. If it cannot, explain why it is not a perfect square trinomial.

32. Which of the following are perfect square trinomials?

A. $y^2 - 13y + 36$ **B.** $x^2 + 6x + 9$ **C.** $4z^2 - 4z + 1$ **D.** $16m^2 + 10m + 1$

Factor each trinomial completely. It may be necessary to factor out the greatest common factor first. See Examples 4 and 5.

33. $w^2 + 2w + 1$

34. $p^2 + 4p + 4$

35. $x^2 - 8x + 16$

36. $x^2 - 10x + 25$

37. $t^2 + t + \frac{1}{4}$

38. $m^2 + \frac{2}{3}m + \frac{1}{9}$

39. $x^2 - 1.0x + .25$

40. $y^2 - 1.4y + .49$

41. $2x^2 + 24x + 72$

42. $3y^2 - 48y + 192$

43. $16x^2 - 40x + 25$

44. $36y^2 - 60y + 25$

45. $49x^2 - 28xy + 4y^2$

46. $4z^2 - 12zw + 9w^2$

47. $64x^2 + 48xy + 9y^2$

48. $9t^2 + 24tr + 16r^2$

49. $50h^3 - 40h^2y + 8hy^2$

50. $18x^3 + 48x^2y + 32xy^2$

RELATING CONCEPTS (EXERCISES 51–54) For Individual or Group Work

We have seen that multiplication and factoring are reverse processes. We know that multiplication and division are also related. To check a division problem, we multiply the quotient by the divisor to get the dividend. To see how factoring and division are related, ***work Exercises 51–54 in order.***

51. Factor $10x^2 + 11x - 6$.

52. Use long division from **Section 5.7** to divide $10x^2 + 11x - 6$ by $2x + 3$.

53. Could we have predicted the result in Exercise 52 from the result in Exercise 51? Explain.

54. Divide $x^3 - 1$ by $x - 1$. Use your answer to factor $x^3 - 1$.

Summary Exercises on Factoring

As you factor a polynomial, ask yourself these questions to decide on a suitable factoring technique.

Factoring a Polynomial

1. **Is there a common factor?** If so, factor it out.
2. **How many terms are in the polynomial?**
   Two terms: Check to see whether it is a difference of squares.
   Three terms: Is it a perfect square trinomial? If the trinomial is not a perfect square, check to see whether the coefficient of the second-degree term is 1. If so, use the method of **Section 6.2.** If the coefficient of the squared term of the trinomial is not 1, use the general factoring methods of **Sections 6.3** and **6.4.**
   Four terms: Try to factor the polynomial by grouping.
3. **Can any factors be factored further?** If so, factor them.

Factor each polynomial completely. Remember to check by multiplying.

1. $32m^9 + 16m^5 + 24m^3$

2. $2m^2 - 10m - 48$

3. $14k^3 + 7k^2 - 70k$

4. $9z^2 + 64$

5. $6z^2 + 31z + 5$

6. $m^2 - 3mn - 4n^2$

7. $49z^2 - 16y^2$

8. $100n^2r^2 + 30nr^3 - 50n^2r$

9. $16x^2 + 20x$

10. $20 + 5m + 12n + 3mn$

11. $10y^2 - 7yz - 6z^2$

12. $y^4 - 81$

13. $m^2 + 2m - 15$

14. $6y^2 - 5y - 4$

15. $32z^3 + 56z^2 - 16z$

16. $15y^2 + 5y$

17. $z^2 - 12z + 36$

18. $9m^2 - 64$

19. $y^2 - 4yk - 12k^2$

20. $16z^2 - 8z + 1$

21. $6y^2 - 6y - 12$

22. $x^2 + \frac{1}{2}x + \frac{1}{16}$

23. $p^2 - 17p + 66$

24. $a^2 + 17a + 72$

25. $k^2 + 9$

26. $108m^2 - 36m + 3$

27. $z^2 - 3za - 10a^2$

28. $2a^3 + a^2 - 14a - 7$

29. $4k^2 - 12k + 9$

30. $a^2 - 3ab - 28b^2$

31. $16r^2 + 24rm + 9m^2$

32. $3k^2 + 4k - 4$

33. $n^2 - 12n - 35$

34. $a^4 - 625$

35. $16k^2 - 48k + 36$

36. $8k^2 - 10k - 3$

37. $36y^6 - 42y^5 - 120y^4$

38. $5z^3 - 45z^2 + 70z$

39. $8p^2 + 23p - 3$

40. $8k^2 - 2kh - 3h^2$

41. $54m^2 - 24z^2$

42. $4k^2 - 20kz + 25z^2$

43. $6a^2 + 10a - 4$

44. $15h^2 + 11hg - 14g^2$

45. $m^2 - 81$

46. $10z^2 - 7z - 6$

47. $125m^4 - 400m^3n + 195m^2n^2$

48. $9y^2 + 12y - 5$

49. $m^2 - 4m + 4$

50. $36x^2 + 32x + 9$

51. $27p^{10} - 45p^9 - 252p^8$

52. $10m^2 + 25m - 60$

53. $4 - 2q - 6p + 3pq$

54. $k^2 - \frac{64}{121}$

55. $64p^2 - 100m^2$

56. $m^3 + 4m^2 - 6m - 24$

57. $100a^2 - 81y^2$

58. $8a^2 + 23ab - 3b^2$

59. $a^2 + 8a + 16$

60. $4y^2 - 25$

6.6 Solving Quadratic Equations by Factoring

OBJECTIVES

1. Solve quadratic equations by factoring.
2. Solve other equations by factoring.

Galileo Galilei (1564–1642) developed theories to explain physical phenomena and set up experiments to test his ideas. According to legend, Galileo dropped objects of different weights from the Leaning Tower of Pisa to disprove the belief that heavier objects fall faster than lighter objects. He developed a formula for freely falling objects described by $d = 16t^2$, where d is the distance in feet that an object falls (disregarding air resistance) in t seconds, regardless of weight.

The equation $d = 16t^2$ is a *quadratic equation*. A quadratic equation contains a squared term and no terms of higher degree.

Quadratic Equation

A **quadratic equation** is an equation that can be written in the form

$$ax^2 + bx + c = 0,$$

where a, b, and c are real numbers, with $a \neq 0$. The given form is called **standard form.**

$x^2 + 5x + 6 = 0$, $2t^2 - 5t = 3$, $y^2 = 4$ Quadratic equations

In these examples, only $x^2 + 5x + 6 = 0$ is in standard form.

Work Problems 1 and 2 at the Side.

1 Which of the following equations are quadratic equations?

A. $y^2 - 4y - 5 = 0$

B. $x^3 - x^2 + 16 = 0$

C. $2z^2 + 7z = -3$

D. $x + 2y = -4$

2 Write each quadratic equation in standard form.

(a) $x^2 - 3x = 4$

(b) $y^2 = 9y - 8$

Up to now, we have factored *expressions,* including many quadratic expressions of the form $ax^2 + bx + c$. In this section, we use factored quadratic expressions to solve quadratic *equations*.

OBJECTIVE 1 Solve quadratic equations by factoring. We use the **zero-factor property** to solve a quadratic equation by factoring.

Zero-Factor Property

If a and b are real numbers and $ab = 0$, then $a = 0$ or $b = 0$.

In words, if the product of two numbers is 0, then at least one of the numbers must be 0. One number *must* be 0, but both *may* be 0.

EXAMPLE 1 Using the Zero-Factor Property

Solve each equation.

(a) $(x + 3)(2x - 1) = 0$

The product $(x + 3)(2x - 1)$ is equal to 0. By the zero-factor property, the only way that the product of these two factors can be 0 is if at least one of the factors equals 0. Therefore, either $x + 3 = 0$ or $2x - 1 = 0$.

$x + 3 = 0$ or $2x - 1 = 0$ Zero-factor property

$x = -3$ $\quad 2x = 1$ Solve each equation.

$x = \frac{1}{2}$

Continued on Next Page

ANSWERS

1. A, C
2. (a) $x^2 - 3x - 4 = 0$
   (b) $y^2 - 9y + 8 = 0$

3 Solve each equation. Check your solutions.

(a) $(x - 5)(x + 2) = 0$

(b) $(3x - 2)(x + 6) = 0$

(c) $z(2z + 5) = 0$

ANSWERS

3. (a) $-2, 5$ (b) $-6, \frac{2}{3}$ (c) $-\frac{5}{2}, 0$

The given equation, $(x + 3)(2x - 1) = 0$, has two solutions, -3 and $\frac{1}{2}$. *Check* these solutions by substituting -3 for x in the original equation, $(x + 3)(2x - 1) = 0$. Then start over and substitute $\frac{1}{2}$ for x.

If $x = -3$, then

$$(x + 3)(2x - 1) = 0$$
$$(-3 + 3)[2(-3) - 1] = 0 \quad ?$$
$$0(-7) = 0. \quad \text{True}$$

If $x = \frac{1}{2}$, then

$$(x + 3)(2x - 1) = 0$$
$$\left(\frac{1}{2} + 3\right)\left(2 \cdot \frac{1}{2} - 1\right) = 0 \quad ?$$
$$\frac{7}{2}(1 - 1) = 0 \quad ?$$
$$\frac{7}{2} \cdot 0 = 0. \quad \text{True}$$

Both -3 and $\frac{1}{2}$ result in true equations, so they are solutions to the original equation.

(b)
$$y(3y - 4) = 0$$
$$y = 0 \quad \text{or} \quad 3y - 4 = 0 \quad \text{Zero-factor property}$$
$$3y = 4$$
$$y = \frac{4}{3}$$

Check these solutions by substituting each one in the original equation. The solutions are 0 and $\frac{4}{3}$.

Work Problem 3 at the Side.

NOTE

The word *or* as used in Example 1 means "one or the other or both."

In Example 1, each equation to be solved was given with the polynomial in factored form. If the polynomial in an equation is not already factored, first make sure that the equation is in standard form. Then factor.

EXAMPLE 2 Solving Quadratic Equations

Solve each equation.

(a) $x^2 - 5x = -6$

First, write the equation in standard form by adding 6 to each side.

$$x^2 - 5x = -6$$
$$x^2 - 5x + 6 = 0 \quad \text{Add 6.}$$

Now factor $x^2 - 5x + 6$. Find two numbers whose product is 6 and whose sum is -5. These two numbers are -2 and -3, so the equation becomes

$$(x - 2)(x - 3) = 0. \quad \text{Factor.}$$
$$x - 2 = 0 \quad \text{or} \quad x - 3 = 0 \quad \text{Zero-factor property}$$
$$x = 2 \quad \text{or} \quad x = 3 \quad \text{Solve each equation.}$$

Continued on Next Page

Check: If $x = 2$, then

$$2^2 - 5(2) = -6 \quad ?$$
$$4 - 10 = -6 \quad ?$$
$$-6 = -6. \quad \text{True}$$

If $x = 3$, then

$$3^2 - 5(3) = -6 \quad ?$$
$$9 - 15 = -6 \quad ?$$
$$-6 = -6. \quad \text{True}$$

Both solutions check, so the solutions are 2 and 3.

(b) $y^2 = y + 20$

$$y^2 - y - 20 = 0 \quad \text{Write in standard form.}$$
$$(y - 5)(y + 4) = 0 \quad \text{Factor.}$$
$$y - 5 = 0 \quad \text{or} \quad y + 4 = 0 \quad \text{Zero-factor property}$$
$$y = 5 \quad \text{or} \quad y = -4 \quad \text{Solve each equation.}$$

Check the solutions 5 and -4 by substituting in the original equation.

Work Problem 4 at the Side.

4 Solve each equation. Check your solutions.

(a) $m^2 - 3m - 10 = 0$

(b) $r^2 + 2r = 8$

In summary, follow these steps to solve quadratic equations by factoring.

Solving a Quadratic Equation by Factoring

Step 1 **Write the equation in standard form,** that is, with all terms on one side of the equals sign in descending powers of the variable and 0 on the other side.

Step 2 **Factor** completely.

Step 3 **Use the zero-factor property** to set each factor with a variable equal to 0, and solve the resulting equations.

Step 4 **Check** each solution in the original equation.

5 Solve each equation. Check your solutions.

(a) $10a^2 - 5a - 15 = 0$

(b) $4x^2 - 2x = 42$

EXAMPLE 3 Solving a Quadratic Equation with a Common Factor

Solve $4p^2 + 40 = 26p$.

$$4p^2 - 26p + 40 = 0 \quad \text{Standard form}$$
$$2(2p^2 - 13p + 20) = 0 \quad \text{Factor out 2.}$$
$$2p^2 - 13p + 20 = 0 \quad \text{Divide each side by 2.}$$
$$(2p - 5)(p - 4) = 0 \quad \text{Factor.}$$
$$2p - 5 = 0 \quad \text{or} \quad p - 4 = 0 \quad \text{Zero-factor property}$$
$$2p = 5 \qquad p = 4 \quad \text{Solve each equation.}$$
$$p = \frac{5}{2}$$

Check the solutions $\frac{5}{2}$ and 4 by substituting in the original equation.

CAUTION

A common error is to include the common factor 2 as a solution in Example 3. ***Only factors containing* variables *lead to solutions.***

Work Problem 5 at the Side.

ANSWERS

4. (a) $-2, 5$ **(b)** $-4, 2$

5. (a) $-1, \frac{3}{2}$ **(b)** $-3, \frac{7}{2}$

6 Solve each equation. Check your solutions.

(a) $49m^2 - 9 = 0$

(b) $p(4p + 7) = 2$

(c) $m^2 = 3m$

EXAMPLE 4 Solving Quadratic Equations

Solve each equation.

(a) $16m^2 - 25 = 0$

We can factor the left side of the equation as the difference of squares (**Section 6.5**).

$$16m^2 - 25 = 0$$

$$(4m + 5)(4m - 5) = 0 \qquad \text{Factor.}$$

$$4m + 5 = 0 \quad \text{or} \quad 4m - 5 = 0 \qquad \text{Zero-factor property}$$

$$4m = -5 \quad \text{or} \quad 4m = 5 \qquad \text{Solve each equation.}$$

$$m = -\frac{5}{4} \quad \text{or} \quad m = \frac{5}{4}$$

Check the solutions $-\frac{5}{4}$ and $\frac{5}{4}$ in the original equation.

(b) $k(2k + 5) = 3$

We need to write this equation in standard form.

$$k(2k + 5) = 3$$

$$2k^2 + 5k = 3 \qquad \text{Multiply.}$$

$$2k^2 + 5k - 3 = 0 \qquad \text{Subtract 3.}$$

$$(2k - 1)(k + 3) = 0 \qquad \text{Factor.}$$

$$2k - 1 = 0 \quad \text{or} \quad k + 3 = 0 \qquad \text{Zero-factor property}$$

$$2k = 1 \qquad k = -3$$

$$k = \frac{1}{2}$$

Check that the solutions are $\frac{1}{2}$ and -3.

(c) $y^2 = 2y$

$$y^2 - 2y = 0 \qquad \text{Standard form}$$

$$y(y - 2) = 0 \qquad \text{Factor.}$$

$$y = 0 \quad \text{or} \quad y - 2 = 0 \qquad \text{Zero-factor property}$$

$$y = 2$$

Check that the solutions are 0 and 2.

CAUTION

In Example 4(b), the zero-factor property could not be used to solve the equation $k(2k + 5) = 3$ in its given form because of the 3 on the right. ***The zero-factor property applies only to a product that equals 0.***

In Example 4(c), it is tempting to begin by dividing each side of the equation $y^2 = 2y$ by y to get $y = 2$. Note that we do not get the other solution, 0, if we divide by a variable. (We *may* divide each side of an equation by a *nonzero* real number, however. For instance, in Example 3 we divided each side by 2.)

Work Problem 6 at the Side.

ANSWERS

6. (a) $-\frac{3}{7}, \frac{3}{7}$ (b) $-2, \frac{1}{4}$ (c) 0, 3

NOTE
Not all quadratic equations can be solved by factoring. A more general method for solving such equations is given in **Chapter 9.**

OBJECTIVE 2 Solve other equations by factoring. We can also extend the zero-factor property to solve equations that involve more than two factors with variables, as shown in Examples 5 and 6. (These equations are *not* quadratic equations. Why not?)

EXAMPLE 5 Solving an Equation with More Than Two Variable Factors

Solve $6z^3 - 6z = 0$.

$$6z^3 - 6z = 0$$
$$6z(z^2 - 1) = 0 \quad \text{Factor out } 6z.$$
$$6z(z + 1)(z - 1) = 0 \quad \text{Factor } z^2 - 1.$$

By an extension of the zero-factor property, this product can equal 0 only if at least one of the factors equals 0. Write and solve three equations, one for each factor with a variable.

$$6z = 0 \quad \text{or} \quad z + 1 = 0 \quad \text{or} \quad z - 1 = 0$$
$$z = 0 \quad \text{or} \quad z = -1 \quad \text{or} \quad z = 1$$

Check by substituting, in turn, 0, −1, and 1 in the original equation. The solutions are 0, −1, and 1.

Work Problem 7 at the Side.

EXAMPLE 6 Solving an Equation with a Quadratic Factor

Solve $(2x - 1)(x^2 - 9x + 20) = 0$.

$$(2x - 1)(x^2 - 9x + 20) = 0$$
$$(2x - 1)(x - 5)(x - 4) = 0 \quad \text{Factor } x^2 - 9x + 20.$$
$$2x - 1 = 0 \quad \text{or} \quad x - 5 = 0 \quad \text{or} \quad x - 4 = 0 \quad \text{Zero-factor property}$$
$$x = \frac{1}{2} \quad \text{or} \quad x = 5 \quad \text{or} \quad x = 4$$

Check. The solutions are $\frac{1}{2}$, 5, and 4.

Work Problem 8 at the Side.

CAUTION
In Example 6, it would be unproductive to begin by multiplying the two factors together. Keep in mind that the zero-factor property and its extension requires the product of two or more factors to equal 0. Always consider first whether an equation is given in the appropriate form to apply the zero-factor property.

7 Solve each equation. Check your solutions.

(a) $r^3 - 16r = 0$

(b) $x^3 - 3x^2 - 18x = 0$

8 Solve each equation. Check your solutions.

(a) $(m + 3)(m^2 - 11m + 10) = 0$

(b) $(2x + 5)(4x^2 - 9) = 0$

ANSWERS
7. (a) $-4, 0, 4$ (b) $-3, 0, 6$
8. (a) $-3, 1, 10$ (b) $-\frac{5}{2}, -\frac{3}{2}, \frac{3}{2}$

Focus on Real-Data Applications

Factoring Trinomials Made Easy

To factor a trinomial using FOIL, we must find the *Outer* and *Inner* coefficients that sum to give the coefficient of the middle term. Our approach begins with a **key number,** found by multiplying the coefficients of the first and last terms of the trinomial. For the trinomial $6x^2 - x - 2$, for instance, the key number is -12 since $6(-2) = -12$.

Step 1 Display the factors of -12 by entering $Y_1 = -12/X$ in a graphing calculator (Screen 1), and using an automatic table (Screen 2). Factors of -12 are automatically displayed in pairs as $1, -12$; $2, -6$; $3, -4$; $4, -3$; and $6, -2$ (Screen 3). You could scroll up or down to find other factors. Note that $5, -2.4$ and $7, -1.714$ are not acceptable factor pairs since -2.4 and -1.714 are not integers.

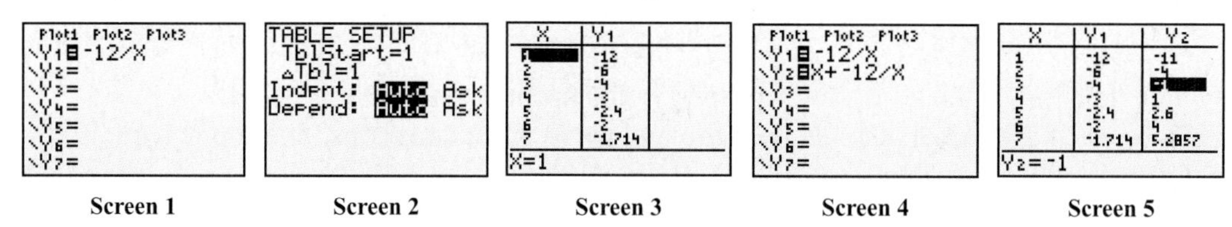

Screen 1 Screen 2 Screen 3 Screen 4 Screen 5

Step 2 Find the pair of factors that sum to the *middle* term coefficient, -1. We can let the calculator do this, too. Enter $Y_2 = X + -12/X$ (Screen 4). In this case, X is one of the factors, and $-12/X$ is the other, so Y_2 will give the sum. Look for -1 in the Y_2 column in Screen 5. (You may have to scroll up or down to find it.)

Step 3 Screen 5 shows that the coefficients of the outer and inner products are 3 and -4. Write $6x^2 - x - 2$ as $6x^2 + 3x - 4x - 2$. Using factoring by grouping,

$$\begin{aligned}(6x^2 + 3x) + (-4x - 2) &= 3x(2x + 1) - 2(2x + 1)\\ &= (2x + 1)(3x - 2).\end{aligned}$$

For Group Discussion

Factor each trinomial given in the column heads of the table. First find the key number, and then use a calculator to find the coefficients of the outer and inner products of FOIL. Finally, factor by grouping.

Trinomial	$3x^2 - 2x - 8$	$2x^2 - 11x + 15$	$10x^2 + 11x - 6$	$4x^2 + 5x + 3$
Key Number	-24 (Why?)			
Outer, Inner Coefficients				
Factor by Grouping				(*Hint:* What does it mean if the middle term coefficient is *not* listed in the Y_2 column?)

6.6 Exercises

FOR EXTRA HELP

Addison-Wesley Math Tutor Center | MathXL | Digital Video Tutor CD 6 Videotape 6 | Student's Solutions Manual | MyMathLab | Interactmath.com

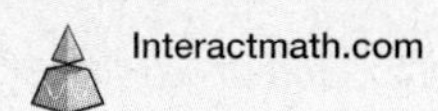

Solve each equation, and check your solutions. See Example 1.

1. $(x + 5)(x - 2) = 0$

2. $(x - 1)(x + 8) = 0$

3. $(2m - 7)(m - 3) = 0$

4. $(6k + 5)(k + 4) = 0$

5. $t(6t + 5) = 0$

6. $w(4w + 1) = 0$

7. $2x(3x - 4) = 0$

8. $6x(4x + 9) = 0$

9. $\left(x + \frac{1}{2}\right)\left(2x - \frac{1}{3}\right) = 0$

10. $\left(a + \frac{2}{3}\right)\left(5a - \frac{1}{2}\right) = 0$

11. $(.5z - 1)(2.5z + 2) = 0$

12. $(.25x + 1)(x - .5) = 0$

13. $(x - 9)(x - 9) = 0$

14. $(2x + 1)(2x + 1) = 0$

15. What is wrong with this "solution"?

$$2x(3x - 4) = 0$$
$$x = 2 \quad \text{or} \quad x = 0 \quad \text{or} \quad 3x - 4 = 0$$
$$x = \frac{4}{3}$$

The solutions are 2, 0, and $\frac{4}{3}$.

16. What is wrong with this "solution"?

$$x(7x - 1) = 0$$
$$7x - 1 = 0 \quad \text{Zero-factor property}$$
$$x = \frac{1}{7}$$

The solution is $\frac{1}{7}$.

Solve each equation, and check your solutions. See Examples 2–6.

17. $y^2 + 3y + 2 = 0$

18. $p^2 + 8p + 7 = 0$

19. $y^2 - 3y + 2 = 0$

20. $r^2 - 4r + 3 = 0$

21. $x^2 = 24 - 5x$

22. $t^2 = 2t + 15$

23. $x^2 = 3 + 2x$

24. $m^2 = 4 + 3m$

25. $z^2 + 3z = -2$

26. $p^2 - 2p = 3$

27. $m^2 + 8m + 16 = 0$

28. $b^2 - 6b + 9 = 0$

29. $3x^2 + 5x - 2 = 0$

30. $6r^2 - r - 2 = 0$

31. $6p^2 = 4 - 5p$

32. $6x^2 = 4 + 5x$

33. $9s^2 + 12s = -4$

34. $36x^2 + 60x = -25$

35. $y^2 - 9 = 0$

36. $m^2 - 100 = 0$

37. $16k^2 - 49 = 0$

38. $4w^2 - 9 = 0$

39. $n^2 = 121$

40. $x^2 = 400$

41. $x^2 = 7x$

42. $t^2 = 9t$

43. $6r^2 = 3r$

44. $10y^2 = -5y$

45. $g(g - 7) = -10$

46. $r(r - 5) = -6$

47. $z(2z + 7) = 4$

48. $b(2b + 3) = 9$

49. $2(y^2 - 66) = -13y$

50. $3(t^2 + 4) = 20t$

51. $5x^3 - 20x = 0$

52. $3x^3 - 48x = 0$

53. $9y^3 - 49y = 0$

54. $16r^3 - 9r = 0$

55. $(2r + 5)(3r^2 - 16r + 5) = 0$

56. $(3m + 4)(6m^2 + m - 2) = 0$

57. $(2x + 7)(x^2 + 2x - 3) = 0$

58. $(x + 1)(6x^2 + x - 12) = 0$

59. Galileo's formula for freely falling objects, $d = 16t^2$, was given at the beginning of this section. The distance d in feet an object falls depends on the time elapsed t in seconds. (This is an example of an important mathematical concept, a *function.*)

(a) Use Galileo's formula and complete the following table. (*Hint:* Substitute each given value into the formula and solve for the unknown value.)

***t* in seconds**	0	1	2	3		
***d* in feet**	0	16			256	576

(b) When $t = 0$, $d = 0$. Explain this in the context of the problem.

(c) When you substituted 256 for d and solved for t, you should have found two solutions, 4 and -4. Why doesn't -4 make sense as an answer?

6.7 Applications of Quadratic Equations

We can use factoring to solve quadratic equations that arise in applications. We follow the same six problem-solving steps given in **Section 2.4.**

OBJECTIVES

1. Solve problems about geometric figures.
2. Solve problems about consecutive integers.
3. Solve problems using the Pythagorean formula.
4. Solve problems using given quadratic models.

Solving an Applied Problem

Step 1 **Read** the problem, several times if necessary, until you *understand* what is given and what is to be found.

Step 2 **Assign a variable** to represent the unknown value, using diagrams or tables as needed. Write down what the variable represents. Express any other unknown values in terms of the variable.

Step 3 **Write an equation** using the variable expression(s).

Step 4 **Solve** the equation.

Step 5 **State the answer.** Does it seem reasonable?

Step 6 **Check** the answer in the words of the original problem.

OBJECTIVE 1 Solve problems about geometric figures. Some of the applied problems in this section require one of the formulas given on the inside covers of the text.

EXAMPLE 1 Solving an Area Problem

The Monroes want to plant a rectangular garden in their yard. The width of the garden will be 4 ft less than its length, and they want it to have an area of 96 ft^2. (ft^2 means square feet.) Find the length and width of the garden.

Step 1 **Read** the problem carefully. We need to find the dimensions of a garden with area 96 ft^2.

Step 2 **Assign a variable.**

Let x = the length of the garden.

Then $x - 4$ = the width. (The width is 4 ft less than the length.)

See Figure 1.

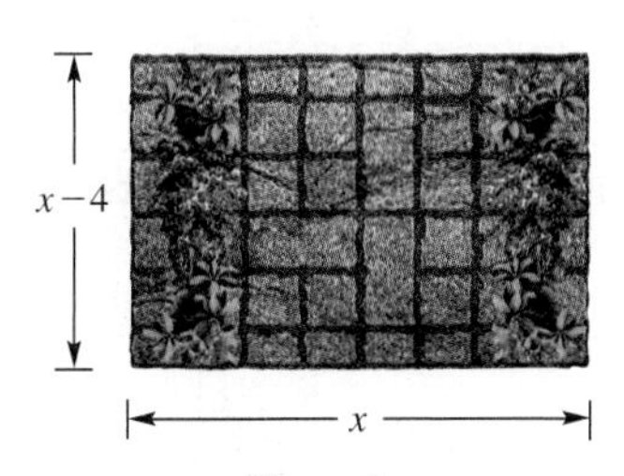

Figure 1

Step 3 **Write an equation.** The area of a rectangle is given by

$$\text{Area} = LW = \text{Length} \times \text{Width}. \qquad \text{Area formula}$$

Substitute 96 for area, x for length, and $x - 4$ for width.

$$A = LW$$

$$96 = x(x - 4) \qquad \text{Let } A = 96, L = x, W = x - 4.$$

Continued on Next Page

1 Solve each problem.

(a) The length of a rectangular room is 2 m more than the width. The area of the floor is 48 m^2. Find the length and width of the room.

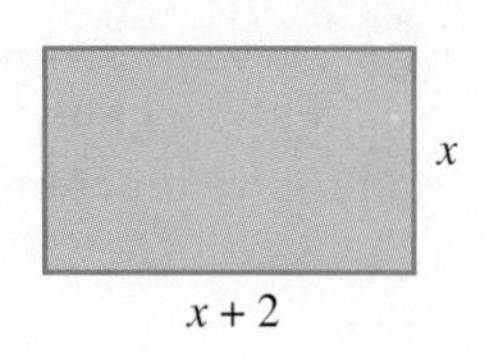

(b) The length of each side of a square is increased by 4 in. The sum of the areas of the original square and the larger square is 106 in^2. What is the length of a side of the original square?

Step 4 **Solve.**

$$96 = x^2 - 4x \quad \text{Distributive property}$$
$$0 = x^2 - 4x - 96 \quad \text{Standard form}$$
$$0 = (x - 12)(x + 8) \quad \text{Factor.}$$
$$x - 12 = 0 \quad \text{or} \quad x + 8 = 0 \quad \text{Zero-factor property}$$
$$x = 12 \quad \text{or} \quad x = -8$$

Step 5 **State the answer.** The solutions are 12 and -8. A rectangle cannot have a side of negative length, so discard -8. The length of the garden will be 12 ft. The width will be $12 - 4 = 8$ ft.

Step 6 **Check.** The width is 4 ft less than the length; the area is $12 \cdot 8 = 96$ ft^2.

CAUTION
When solving applied problems, ***always check solutions*** against physical facts and discard any answers that are not appropriate.

Work Problem 1 at the Side.

OBJECTIVE 2 Solve problems about consecutive integers. Recall from **Section 2.4** that **consecutive integers** are integers that are next to each other on a number line, such as 5 and 6, or -11 and -10. **Consecutive odd integers** are *odd* integers that are next to each other, such as 5 and 7, or -13 and -11. **Consecutive even integers** are defined similarly; for example, 4 and 6 are consecutive even integers, as are -10 and -8.

PROBLEM-SOLVING HINT
In consecutive integer problems, if x represents the first integer, then for

two consecutive integers, use	$x, \quad x + 1;$
three consecutive integers, use	$x, \quad x + 1, \quad x + 2;$
two consecutive even or odd integers, use	$x, \quad x + 2;$
three consecutive even or odd integers, use	$x, \quad x + 2, \quad x + 4.$

EXAMPLE 2 Solving a Consecutive Integer Problem

The product of the numbers on two consecutive post-office boxes is 210. Find the box numbers.

Step 1 **Read** the problem. Note that the boxes are numbered consecutively.

Step 2 **Assign a variable.**

Let x = the first box number.

Then $x + 1$ = the next consecutive box number.

See Figure 2.

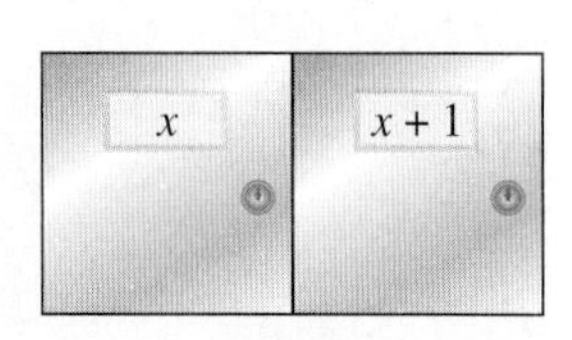

Figure 2

Step 3 **Write an equation.** The product of the box numbers is 210, so

$$x(x + 1) = 210.$$

Continued on Next Page

ANSWERS
1. **(a)** length: 8 m; width: 6 m **(b)** 5 in.

Step 4 **Solve.**

$$x(x + 1) = 210$$
$$x^2 + x = 210 \quad \text{Distributive property}$$
$$x^2 + x - 210 = 0 \quad \text{Standard form}$$
$$(x + 15)(x - 14) = 0 \quad \text{Factor.}$$
$$x + 15 = 0 \quad \text{or} \quad x - 14 = 0 \quad \text{Zero-factor property}$$
$$x = -15 \quad \text{or} \quad x = 14$$

Step 5 **State the answer.** The solutions are -15 and 14. Discard the solution -15 since a box number cannot be negative. When $x = 14$, then $x + 1 = 15$, so the post-office boxes have the numbers 14 and 15.

Step 6 **Check.** The numbers 14 and 15 are consecutive and their product is $14 \cdot 15 = 210$, as required.

Work Problem 2 at the Side.

2 Solve the problem.
The product of the numbers on two consecutive lockers at a health club is 132. Find the locker numbers.

EXAMPLE 3 Solving a Consecutive Integer Problem

The product of two consecutive odd integers is 1 less than five times their sum. Find the integers.

Step 1 **Read** carefully. This problem is a little more complicated.

Step 2 **Assign a variable.** We must find two consecutive *odd* integers.

Let x = the smaller integer.
Then $x + 2$ = the next larger odd integer.

Step 3 **Write an equation.** According to the problem, the product is 1 less than five times the sum.

The product	is	five times the sum	less 1.
↓	↓	↓	↓
$x(x + 2)$	$=$	$5(x + x + 2)$	$- 1$

Step 4 **Solve.**

$$x^2 + 2x = 5x + 5x + 10 - 1 \quad \text{Distributive property}$$
$$x^2 + 2x = 10x + 9 \quad \text{Combine like terms.}$$
$$x^2 - 8x - 9 = 0 \quad \text{Standard form}$$
$$(x - 9)(x + 1) = 0 \quad \text{Factor.}$$
$$x - 9 = 0 \quad \text{or} \quad x + 1 = 0 \quad \text{Zero-factor property}$$
$$x = 9 \quad \text{or} \quad x = -1$$

Step 5 **State the answer.** We need to find two consecutive odd integers.

If $x = 9$ is the smaller, then $x + 2 = 9 + 2 = 11$ is the larger.

If $x = -1$ is the smaller, then $x + 2 = -1 + 2 = 1$ is the larger.

There are two sets of answers here since integers can be positive or negative.

Step 6 **Check.** The product of the first pair of integers is $9 \cdot 11 = 99$. One less than five times their sum is $5(9 + 11) - 1 = 99$. Thus 9 and 11 satisfy the problem. Repeat the check with -1 and 1.

Work Problem 3 at the Side.

3 Solve each problem.

(a) The product of two consecutive even integers is 4 more than two times their sum. Find the integers.

(b) Find three consecutive odd integers such that the product of the smallest and largest is 16 more than the middle integer.

Answers
2. 11 and 12
3. **(a)** 4 and 6 or -2 and 0 **(b)** 3, 5, 7

CAUTION

Do *not* use $x, x + 1, x + 3$, and so on to represent consecutive odd integers. To see why, let $x = 3$. Then $x + 1 = 3 + 1 = 4$ and $x + 3 = 3 + 3 = 6$, and 3, 4, and 6 are not consecutive odd integers.

OBJECTIVE 3 Solve problems using the Pythagorean formula. The next example requires the Pythagorean formula from geometry.

Pythagorean Formula

If a right triangle (a triangle with a 90° angle) has longest side of length c and two other sides of lengths a and b, then

$$a^2 + b^2 = c^2.$$

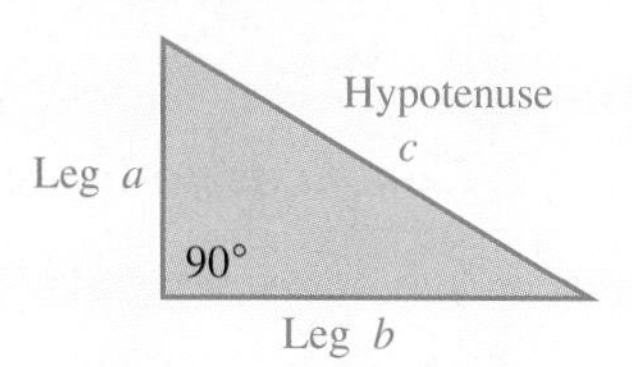

The longest side, the **hypotenuse,** is opposite the right angle. The two shorter sides are the **legs** of the triangle.

EXAMPLE 4 Using the Pythagorean Formula

Ed and Mark leave their office, with Ed traveling north and Mark traveling east. When Mark is 1 mi farther than Ed from the office, the distance between them is 2 mi more than Ed's distance from the office. Find their distances from the office and the distance between them.

Step 1 **Read** the problem again. We must find three distances.

Step 2 **Assign a variable.** Let x represent Ed's distance from the office, $x + 1$ represent Mark's distance from the office, and $x + 2$ represent the distance between them. Place these on a right triangle, as in Figure 3.

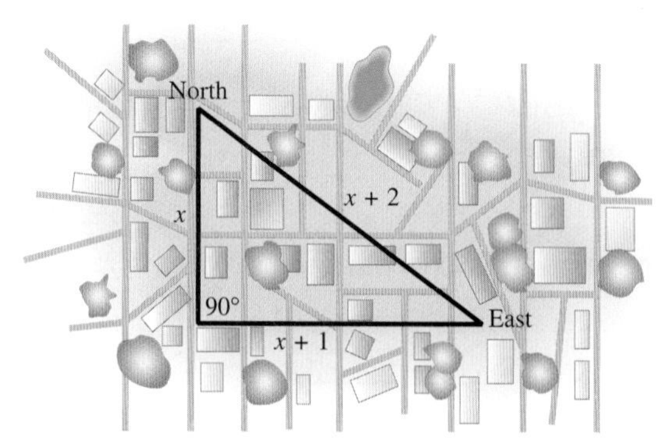

Figure 3

Step 3 **Write an equation.** Substitute into the Pythagorean formula.

$$a^2 + b^2 = c^2$$
$$x^2 + (x + 1)^2 = (x + 2)^2$$

Continued on Next Page

Step 4 **Solve.**

$$x^2 + x^2 + 2x + 1 = x^2 + 4x + 4$$
$$x^2 - 2x - 3 = 0 \quad \text{Standard form}$$
$$(x - 3)(x + 1) = 0 \quad \text{Factor.}$$
$$x - 3 = 0 \quad \text{or} \quad x + 1 = 0 \quad \text{Zero-factor property}$$
$$x = 3 \quad \text{or} \quad x = -1$$

Step 5 **State the answer.** Since -1 cannot represent a distance, 3 is the only possible answer. Ed's distance is 3 mi, Mark's distance is $3 + 1 = 4$ mi, and the distance between them is $3 + 2 = 5$ mi.

Step 6 **Check.** Since $3^2 + 4^2 = 5^2$, the answer is correct.

CAUTION

When solving a problem involving the Pythagorean formula, be sure that the expressions for the sides are properly placed.

$$\textbf{leg}^2 + \textbf{leg}^2 = \textbf{hypotenuse}^2$$

Work Problem 4 at the Side.

OBJECTIVE 4 Solve problems using given quadratic models. In Examples 1–4, we wrote quadratic equations to model, or mathematically describe, various situations and then solved the equations. Now we are given the quadratic models and must use them to determine data.

EXAMPLE 5 Finding the Height of a Ball

A tennis player can hit a ball 180 ft per sec (125 mph). If she hits a ball directly upward, the height h of the ball in feet at time t in seconds is modeled by the quadratic equation

$$h = -16t^2 + 180t + 6.$$

How long will it take for the ball to reach a height of 206 ft?

A height of 206 ft means $h = 206$, so we substitute 206 for h in the equation and then solve for t.

$$\mathbf{206} = -16t^2 + 180t + 6 \quad \text{Let } h = 206.$$
$$-16t^2 + 180t + 6 = 206 \quad \text{Interchange sides.}$$
$$-16t^2 + 180t - 200 = 0 \quad \text{Standard form}$$
$$4t^2 - 45t + 50 = 0 \quad \text{Divide by } -4.$$
$$(4t - 5)(t - 10) = 0 \quad \text{Factor.}$$
$$4t - 5 = 0 \quad \text{or} \quad t - 10 = 0 \quad \text{Zero-factor property}$$
$$t = \frac{5}{4} \quad \text{or} \quad t = 10$$

Since we found two acceptable answers, the ball will be 206 ft above the ground twice (once on its way up and once on its way down)—at $\frac{5}{4}$ sec and at 10 sec. See Figure 4.

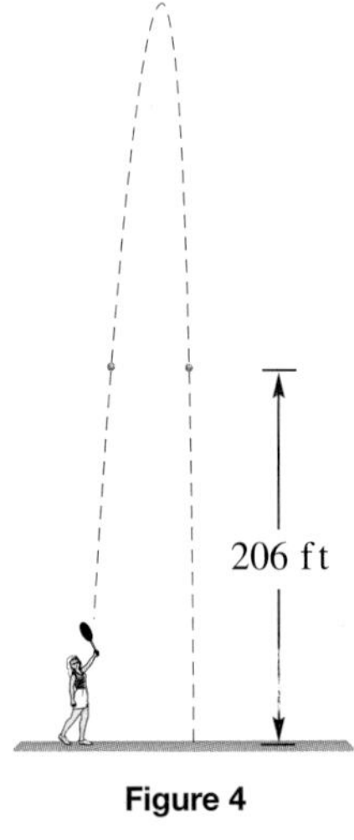

Figure 4

Work Problem 5 at the Side.

4 Solve the problem.

The hypotenuse of a right triangle is 3 in. longer than the longer leg. The shorter leg is 3 in. shorter than the longer leg. Find the lengths of the sides of the triangle.

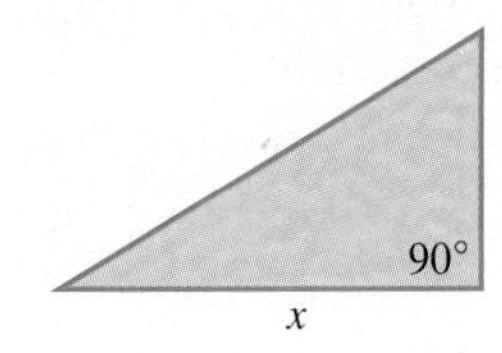

5 Solve the problem.

The number of impulses fired after a nerve has been stimulated is modeled by

$$I = -x^2 + 2x + 60,$$

where x is in milliseconds (ms) after the stimulation. When will 45 impulses occur? Do you get two solutions? Why is only one answer given?

ANSWERS

4. 9 in., 12 in., 15 in.

5. After 5 ms; There are two solutions, -3 and 5; Only one answer makes sense here because a negative answer is not appropriate.

6 Solve the problem.

Use the model in Example 6 to find the annual percent increase in spending on hospital services in 1995. Give your answer to the nearest tenth. How does it compare to the actual data from the table?

EXAMPLE 6 Modeling Increases in Hospital Costs

The annual percent increase y in spending on hospital services in the years 1994–2001 can be modeled by the quadratic equation

$$y = .37x^2 - 4.1x + 12,$$

where $x = 4$ represents 1994, $x = 5$ represents 1995, and so on. (*Source:* Center for Studying Health System Change.)

(a) Use the model to find the annual percent increase to the nearest tenth in 1997.

Since $x = 4$ represents 1994, $x = 7$ represents 1997. Substitute 7 for x in the equation.

$$y = .37(7)^2 - 4.1(7) + 12 \qquad \text{Let } x = 7.$$

$$y = 1.4 \qquad \text{Round to the nearest tenth.}$$

Spending on hospital services increased about 1.4% in 1997.

(b) Repeat part (a) for 2001.

$$y = .37(11)^2 - 4.1(11) + 12 \qquad \text{For 2001, let } x = 11.$$

$$y = 11.7$$

In 2001, spending on hospital services increased about 11.7%.

(c) The model used in parts (a) and (b) was developed using the data in the table below. How do the results in parts (a) and (b) compare to the actual data from the table?

Year	Percent Increase
1994	1.8
1995	.8
1996	.5
1997	1.3
1998	3.4
1999	5.8
2000	7.1
2001	12.0

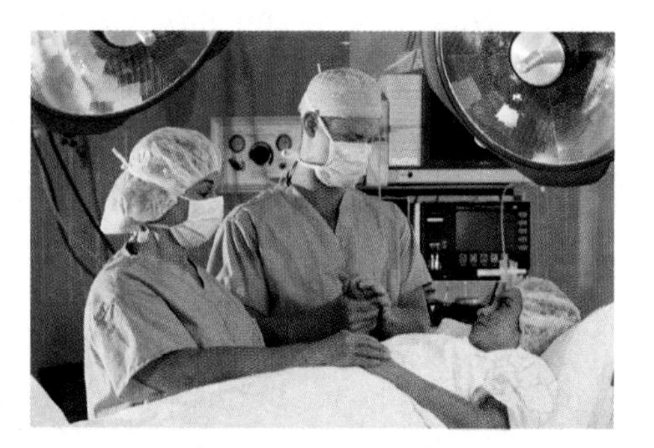

From the table, the actual data for 1997 is 1.3%. Our answer, 1.4%, is slightly high. For 2001, the actual data is 12.0%, so our answer of 11.7% in part (b) is a little low.

Work Problem 6 at the Side.

NOTE

A graph of the quadratic equation from Example 6 is shown in Figure 5. Notice the basic shape of this graph, which follows the general pattern of the data in the table—it decreases from 1994 to 1996 and then increases from 1997 to 2001. We consider such graphs of quadratic equations, called *parabolas,* in **Chapter 9.**

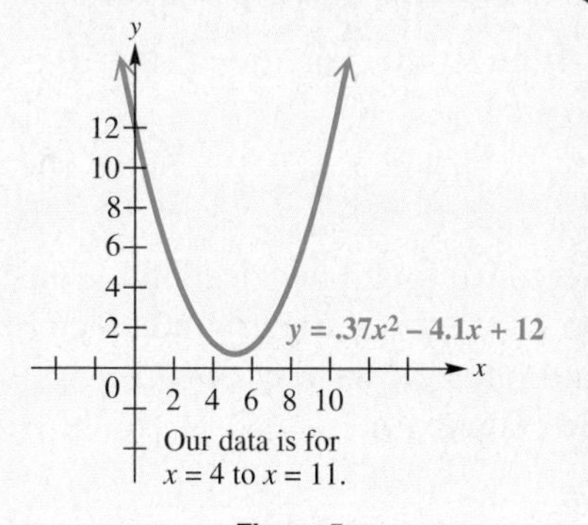

Figure 5

ANSWERS

6. .8%; The actual data for 1995 is .8%, so our answer using the model is the same.

6.7 Exercises

FOR EXTRA HELP

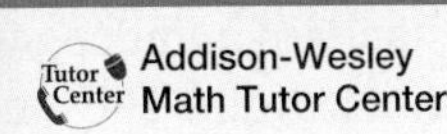

MathXL

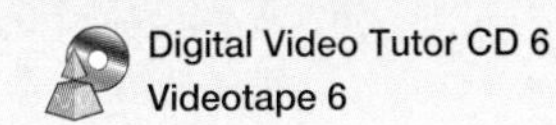

MyMathLab

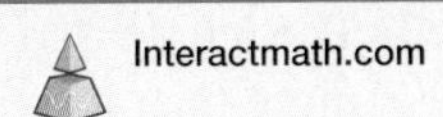

1. To review the six problem-solving steps first introduced in **Section 2.4,** complete each statement.

   Step 1: ________ the problem, several times if necessary, until you understand what is given and what must be found.

   Step 2: Assign a ________ to represent the unknown value.

   Step 3: Write a(n) ________ using the variable expression(s).

   Step 4: ________ the equation.

   Step 5: State the ________.

   Step 6: ________ the answer in the words of the ________ problem.

2. A student solves an applied problem and gets 6 or -3 for the length of the side of a square. Which of these answers is reasonable? Explain.

In Exercises 3–6, a figure and a corresponding geometric formula are given. Using x as the variable, complete Steps 3–6 for each problem. (Refer to the steps in Exercise 1 as needed.)

3.

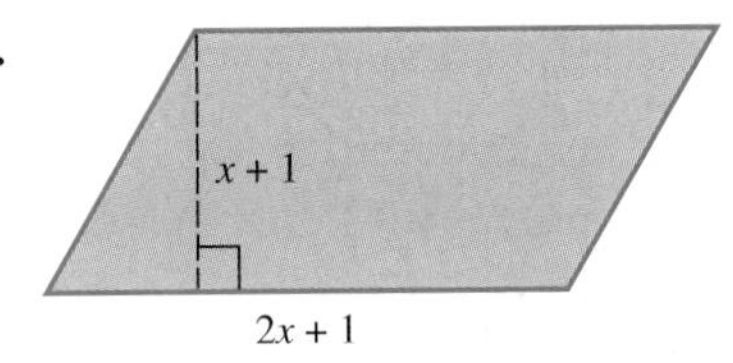

Area of a parallelogram: $A = bh$

The area of this parallelogram is 45 sq. units. Find its base and height.

4.

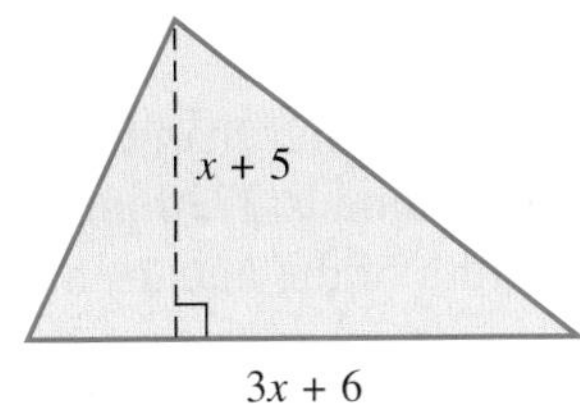

Area of a triangle: $A = \frac{1}{2}bh$

The area of this triangle is 60 sq. units. Find its base and height.

5.

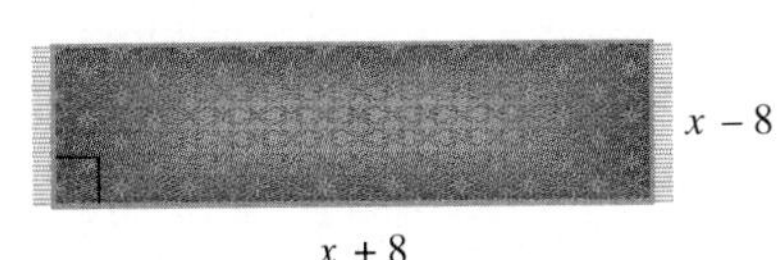

Area of a rectangular rug: $A = LW$

The area of this rug is 80 sq. units. Find its length and width.

6.

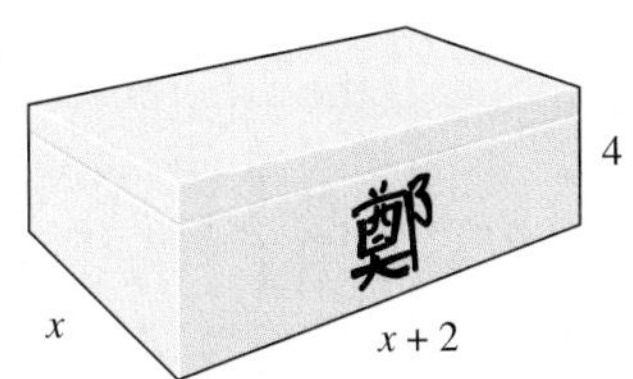

Volume of a rectangular Chinese box: $V = LWH$

The volume of this box is 192 cu. units. Find its length and width.

Solve each problem. Check your answers to be sure they are reasonable. Refer to the formulas on the inside covers. See Example 1.

7. The length of a VHS videocassette shell is 3 in. more than its width. The area of the rectangular top side of the shell is 28 in.^{2}. Find the length and width of the videocassette shell.

8. A plastic box that holds a standard audiocassette has length 4 cm longer than its width. The area of the rectangular top of the box is 77 cm^2. Find the length and width of the box.

9. The dimensions of a Gateway EV700 computer monitor screen are such that its length is 3 in. more than its width. If the length is increased by 1 in. while the width remains the same, the area is increased by 10 in.^{2}. What are the dimensions of the screen? (*Source:* Author's computer.)

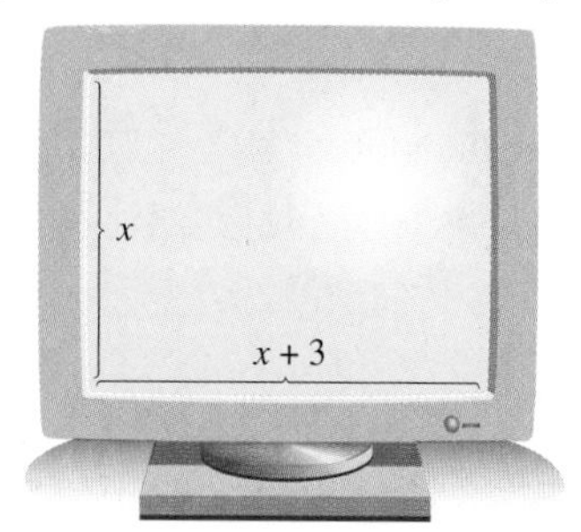

10. The keyboard of the computer in Exercise 9 is 11 in. longer than it is wide. If both its length and width are increased by 2 in., the area of the top of the keyboard is increased by 54 in.^{2}. Find the length and width of the keyboard. (*Source:* Author's computer.)

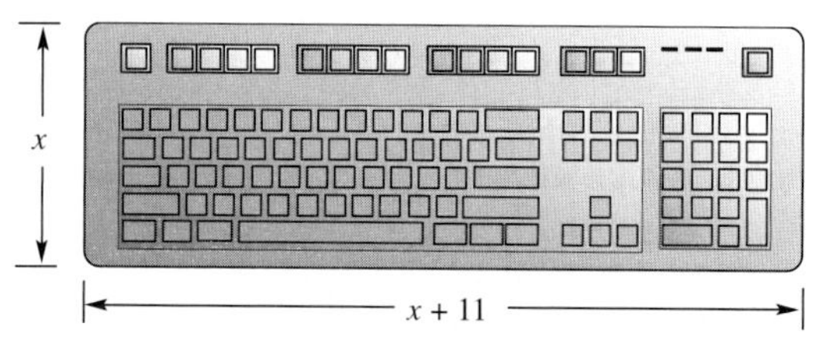

11. A ten-gallon aquarium is 3 in. higher than it is wide. Its length is 21 in., and its volume is 2730 in.^{3}. What are the height and width of the aquarium?

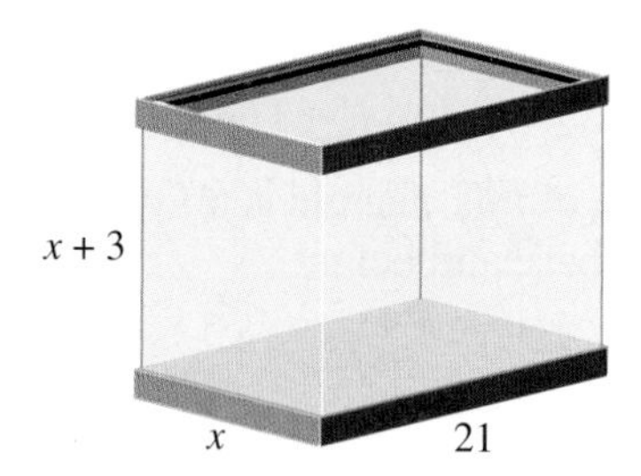

12. A toolbox is 2 ft high, and its width is 3 ft less than its length. If its volume is 80 ft^3, find the length and width of the box.

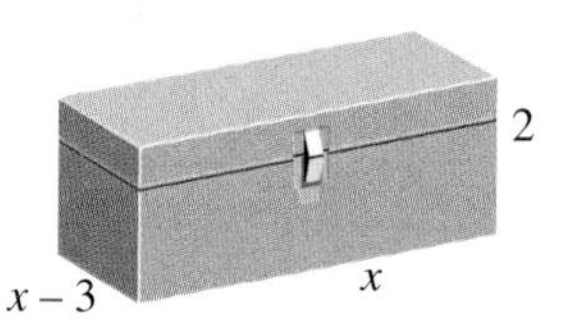

13. A square mirror has sides measuring 2 ft less than the sides of a square painting. If the difference between their areas is 32 ft^2, find the lengths of the sides of the mirror and the painting.

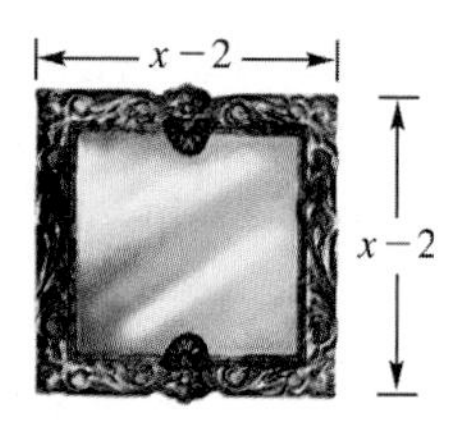

14. The sides of one square have length 3 m more than the sides of a second square. If the area of the larger square is subtracted from 4 times the area of the smaller square, the result is 36 m^2. What are the lengths of the sides of each square?

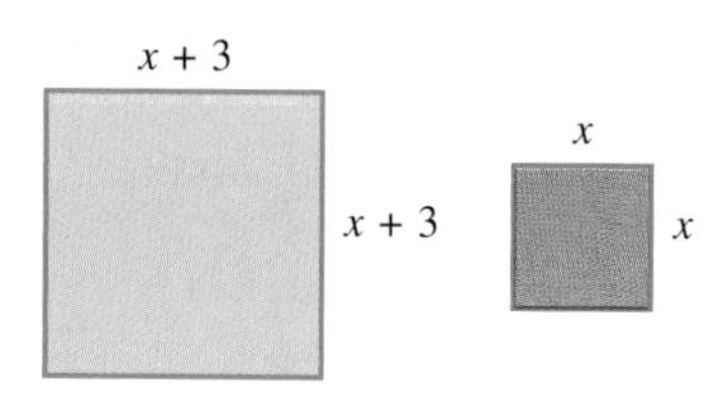

Solve each problem about consecutive integers. See Examples 2 and 3.

15. The product of the numbers on two consecutive volumes of research data is 420. Find the volume numbers.

16. The product of the page numbers on two facing pages of a book is 600. Find the page numbers.

17. The product of two consecutive integers is 11 more than their sum. Find the integers.

18. The product of two consecutive integers is 4 less than four times their sum. Find the integers.

19. Find two consecutive odd integers such that their product is 15 more than three times their sum.

20. Find two consecutive odd integers such that five times their sum is 23 less than their product.

21. Find three consecutive even integers such that the sum of the squares of the smaller two is equal to the square of the largest.

22. Find three consecutive even integers such that the square of the sum of the smaller two is equal to twice the largest.

Use the Pythagorean formula to solve each problem. See Example 4.

23. The hypotenuse of a right triangle is 1 cm longer than the longer leg. The shorter leg is 7 cm shorter than the longer leg. Find the length of the longer leg of the triangle.

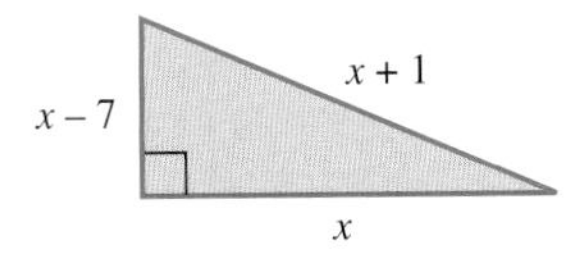

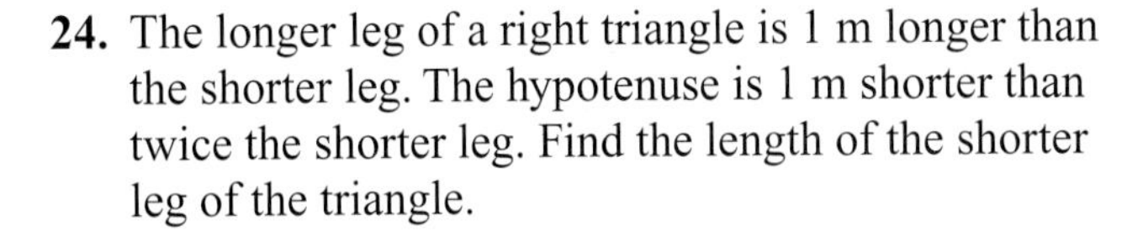

24. The longer leg of a right triangle is 1 m longer than the shorter leg. The hypotenuse is 1 m shorter than twice the shorter leg. Find the length of the shorter leg of the triangle.

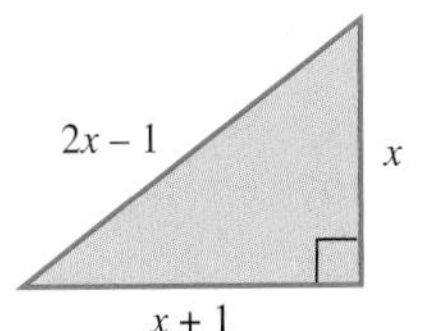

25. Terri works due north of home. Her husband Denny works due east. They leave for work at the same time. By the time Terri is 5 mi from home, the distance between them is 1 mi more than Denny's distance from home. How far from home is Denny?

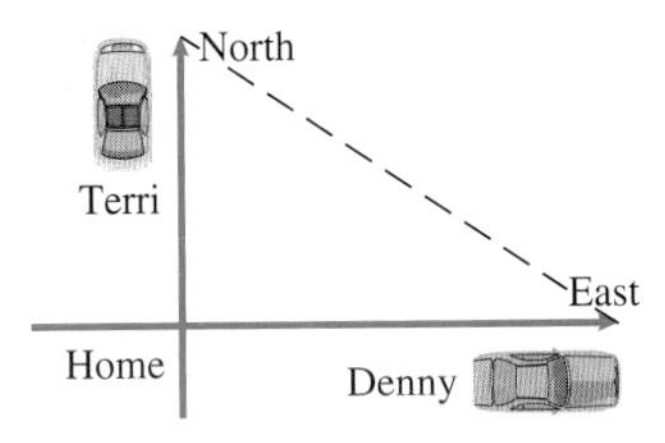

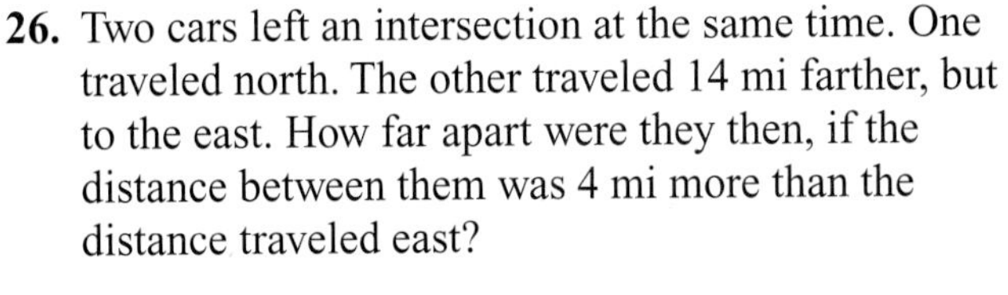

26. Two cars left an intersection at the same time. One traveled north. The other traveled 14 mi farther, but to the east. How far apart were they then, if the distance between them was 4 mi more than the distance traveled east?

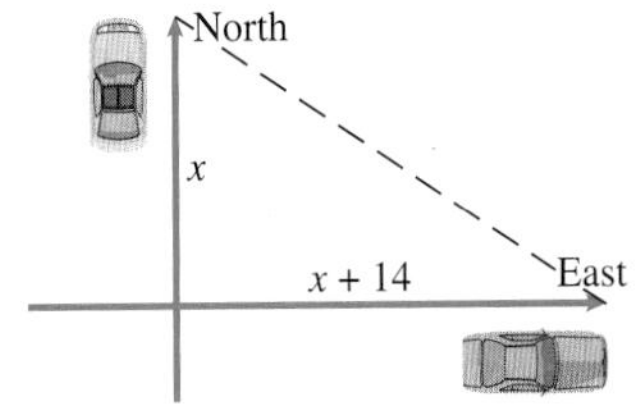

27. A ladder is leaning against a building. The distance from the bottom of the ladder to the building is 4 ft less than the length of the ladder. How high up the side of the building is the top of the ladder if that distance is 2 ft less than the length of the ladder?

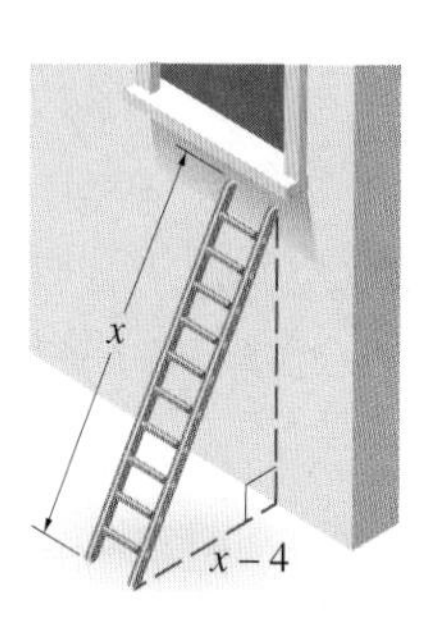

28. A lot has the shape of a right triangle with one leg 2 m longer than the other. The hypotenuse is 2 m less than twice the length of the shorter leg. Find the length of the shorter leg.

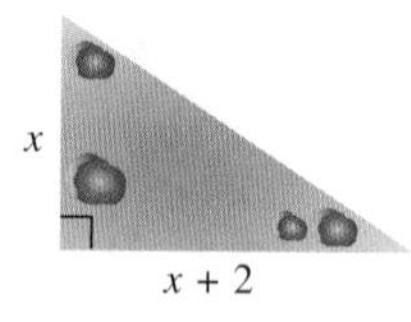

Solve each problem. See Examples 5 and 6.

29. An object propelled from a height of 48 ft with an initial velocity of 32 ft per sec after t seconds has height

$$h = -16t^2 + 32t + 48.$$

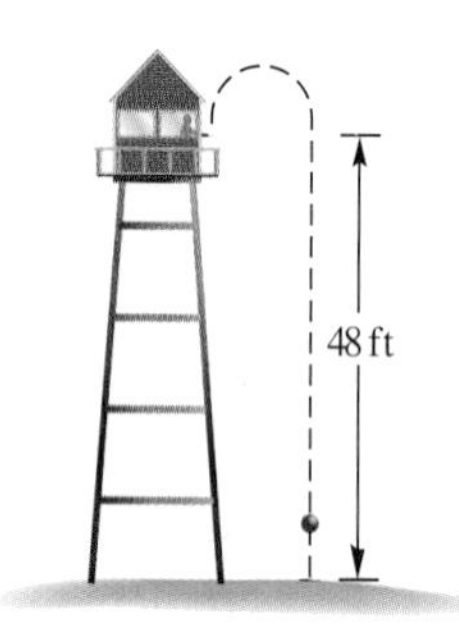

(a) After how many seconds is the height 64 ft? (*Hint:* Let $h = 64$ and solve.)

(b) After how many seconds is the height 60 ft?

(c) After how many seconds does the object hit the ground? (*Hint:* When the object hits the ground, $h = 0$.)

(d) The quadratic equation from part (c) has two solutions, yet only one of them is appropriate for answering the question. Why is this so?

30. If an object is propelled upward from ground level with an initial velocity of 64 ft per sec, its height h in feet t seconds later is

$$h = -16t^2 + 64t.$$

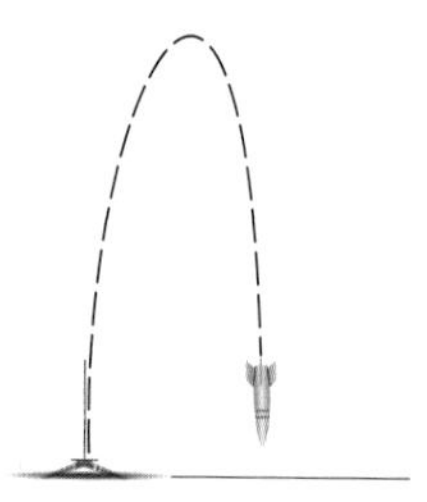

(a) After how many seconds is the height 48 ft?

(b) The object reaches its maximum height 2 sec after it is propelled. What is this maximum height?

(c) After how many seconds does the object hit the ground?

(d) The quadratic equation from part (c) has two solutions, yet only one of them is appropriate for answering the question. Why is this so?

31. The table shows the number of cellular phones (in millions) owned by Americans.

Year	Cellular Phones (in millions)
1990	5
1992	11
1994	24
1996	44
1998	62
2000	109
2002	140

Source: Cellular Telecommunications Industry Association.

We used the data to develop the quadratic equation

$$y = .866x^2 + 1.02x + 5.29,$$

which models the number of cellular phones y (in millions) in the year x, where $x = 0$ represents 1990, $x = 2$ represents 1992, and so on.

(a) Use the model to find the number of cellular phones in 1996 to the nearest tenth. How does the result compare to the actual data in the table?

(b) What value of x corresponds to 2002?

(c) Use the model to find the number of cellular phones in 2002 to the nearest tenth. How does the result compare to the actual data in the table?

(d) Assuming that the trend in the data continues, use the quadratic equation to predict the number of cellular phones in 2005 to the nearest tenth.

RELATING CONCEPTS (EXERCISES 32–40) For Individual or Group Work

The U.S. trade deficit represents the amount by which exports are less than imports. It provides not only a sign of economic prosperity but also a warning of potential decline. The data in the table shows the U.S. trade deficit for 1995 through 2000.

Year	Deficit (in billions of dollars)
1995	97.5
1996	104.3
1997	104.7
1998	164.3
1999	271.3
2000	378.7

Source: U.S. Department of Commerce.

Use the data to ***work Exercises 32–40 in order.***

32. How much did the trade deficit increase from 1999 to 2000? What percent increase is this (to the nearest percent)?

33. The U.S. trade deficit might be approximated by the linear equation

$$y = 40.8x + 66.9,$$

where y is the deficit in billions of dollars. Here $x = 0$ represents 1995, $x = 1$ represents 1996, and so on. Use this equation to approximate the trade deficits in 1997, 1999, and 2000.

34. How do your answers from Exercise 33 compare to the actual data in the table?

35. The trade deficit y (in billions of dollars) might also be approximated by the quadratic equation

$$y = 18.5x^2 - 33.4x + 104,$$

where $x = 0$ again represents 1995, $x = 1$ represents 1996, and so on. Use this equation to approximate the trade deficits in 1997, 1999, and 2000.

36. Compare your answers from Exercise 35 to the actual data in the table. Which equation, the linear one in Exercise 33 or the quadratic one in Exercise 35, models the data better?

37. We can also see graphically why the linear equation is not a very good model for the data. To do so, write the data from the table as a set of ordered pairs (x, y), where x represents the years since 1995 and y represents the trade deficit in billions of dollars.

38. Plot the ordered pairs from Exercise 37 on the graph.

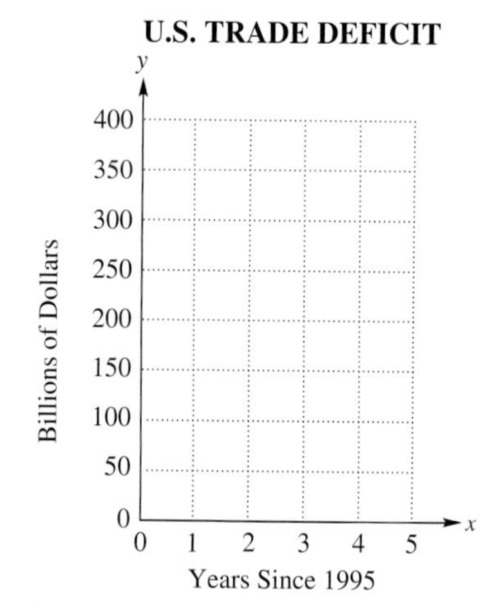

Recall from **Chapter 3** that a linear equation has a straight line for its graph. Do the ordered pairs you plotted lie in a linear pattern?

39. Assuming that the trend in the data continues and since the quadratic equation modeled that data fairly well, use the quadratic equation to predict the trade deficit for the year 2002.

40. The actual trade deficit for 2002 was 417.9 billion dollars.

(a) How does the actual deficit for 2002 compare to your prediction from Exercise 39?

(b) Should the quadratic equation be used to predict the U.S. trade deficit for years after 2000? Explain.

Chapter 6

SUMMARY

KEY TERMS

6.1

factor An expression A is a factor of an expression B if B can be divided by A with 0 remainder.

factored form An expression is in factored form when it is written as a product.

greatest common factor (GCF) The greatest common factor is the largest quantity that is a factor of each of a group of quantities.

factoring The process of writing a polynomial as a product is called factoring.

6.2

prime polynomial A prime polynomial is a polynomial that cannot be factored using only integers.

6.5

perfect square trinomial A perfect square trinomial is a trinomial that can be factored as the square of a binomial.

6.6

quadratic equation A quadratic equation is an equation that can be written in the form $ax^2 + bx + c = 0$, with $a \neq 0$.

standard form The form $ax^2 + bx + c = 0$ is the standard form of a quadratic equation.

6.7

hypotenuse The longest side of a right triangle, opposite the right angle, is the hypotenuse.

legs The two shorter sides of a right triangle are the legs.

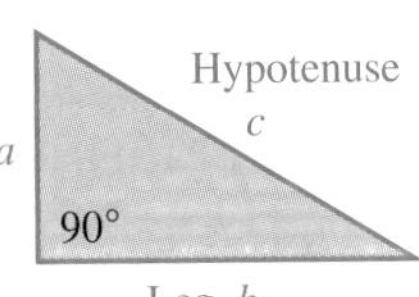

TEST YOUR WORD POWER

See how well you have learned the vocabulary in this chapter. Answers, with examples, follow the Quick Review.

1. **Factoring** is
   A. a method of multiplying polynomials
   B. the process of writing a polynomial as a product
   C. the answer in a multiplication problem
   D. a way to add the terms of a polynomial.

2. A polynomial is in **factored form** when
   A. it is prime
   B. it is written as a sum
   C. the squared term has a coefficient of 1
   D. it is written as a product.

3. The **greatest common factor** of a polynomial is
   A. the least integer that divides evenly into all the terms of the polynomial
   B. the least term that is a factor of all the terms in the polynomial
   C. the greatest term that is a factor of all the terms in the polynomial
   D. the variable that is common to all the terms in the polynomial.

4. A **perfect square trinomial** is a trinomial
   A. that can be factored as the square of a binomial
   B. that cannot be factored
   C. that is multiplied by a binomial
   D. where all terms are perfect squares.

5. A **quadratic equation** is an equation that can be written in the form
   A. $y = mx + b$
   B. $ax^2 + bx + c = 0$ $(a \neq 0)$
   C. $Ax + By = C$
   D. $x = k$.

6. A **hypotenuse** is
   A. either of the two shorter sides of a triangle
   B. the shortest side of a right triangle
   C. the side opposite the right angle in a right triangle
   D. the longest side in any triangle.

Quick Review

Concepts	Examples
6.1 *Factors; The Greatest Common Factor*	
Finding the Greatest Common Factor (GCF)	Find the greatest common factor of $4x^2y$, $-6x^2y^3$, and $2xy^2$.
Step 1 Write each number in prime factored form.	$4x^2y = 2 \cdot 2 \cdot x^2 \cdot y$
Step 2 List each prime number or each variable that is a factor of every term in the list.	$-6x^2y^3 = -1 \cdot 2 \cdot 3 \cdot x^2 \cdot y^3$ $2xy^2 = 2 \cdot x \cdot y^2$
Step 3 Use as exponents on the common prime factors the least exponents from the prime factored forms.	The greatest common factor is $2xy$.
Step 4 Multiply the primes from Step 3.	
Factoring by Grouping	Factor by grouping.
Step 1 Group the terms.	$2a^2 + 2ab + a + b = (2a^2 + 2ab) + (a + b)$
Step 2 Factor out the greatest common factor from each group.	$= 2a(a + b) + 1(a + b)$
Step 3 Factor a common binomial factor from the results of Step 2.	$= (a + b)(2a + 1)$
Step 4 If necessary, rearrange terms and try a different grouping.	
6.2 *Factoring Trinomials*	
To factor $x^2 + bx + c$, find m and n such that $mn = c$ and $m + n = b$.	Factor $x^2 + 6x + 8$.
$mn = c$ ↓ $x^2 + bx + c$ ↑ $m + n = b$	$mn = 8$ ↓ $x^2 + 6x + 8$ ↑ $m + n = 6$
	$m = 2$ and $n = 4$
Then $x^2 + bx + c = (x + m)(x + n)$.	$x^2 + 6x + 8 = (x + 2)(x + 4)$
Check by multiplying.	*Check:* $(x + 2)(x + 4) = x^2 + 4x + 2x + 8$ $= x^2 + 6x + 8$
6.3 *Factoring Trinomials by Grouping*	
To factor $ax^2 + bx + c$ by grouping, first find m and n.	Factor $3x^2 + 14x - 5$. (product of first and last coefficients: -15)
$m + n = b$ ↓ $ax^2 + bx + c$; $mn = ac$ (product of a and c)	Find two integers with a product of $3(-5) = -15$ and a sum of 14. The integers are -1 and 15.
Then factor $ax^2 + mx + nx + b$ by grouping.	$3x^2 + 14x - 5 = 3x^2 - x + 15x - 5$ $= (3x^2 - x) + (15x - 5)$ $= x(3x - 1) + 5(3x - 1)$ $= (3x - 1)(x + 5)$

Concepts	Examples
6.4 *Factoring Trinomials Using FOIL* To factor $ax^2 + bx + c$ by trial and error, use FOIL backwards.	By trial and error, $3x^2 + 14x - 5 = (3x - 1)(x + 5)$.
6.5 *Special Factoring Techniques* **Difference of Squares** $a^2 - b^2 = (a + b)(a - b)$	Factor. $4x^2 - 9 = (2x + 3)(2x - 3)$
Perfect Square Trinomials $a^2 + 2ab + b^2 = (a + b)^2$ $a^2 - 2ab + b^2 = (a - b)^2$	$9x^2 + 6x + 1 = (3x + 1)^2$ $4x^2 - 20x + 25 = (2x - 5)^2$
6.6 *Solving Quadratic Equations by Factoring* **Zero-Factor Property** If a and b are real numbers and $ab = 0$, then $a = 0$ or $b = 0$.	If $(x - 2)(x + 3) = 0$, then $x - 2 = 0$ or $x + 3 = 0$.
Solving a Quadratic Equation by Factoring *Step 1* Write the equation in standard form. *Step 2* Factor. *Step 3* Use the zero-factor property.	Solve $2x^2 = 7x + 15$. $2x^2 - 7x - 15 = 0$ $(2x + 3)(x - 5) = 0$ $2x + 3 = 0$ or $x - 5 = 0$ $2x = -3$ $x = 5$ $x = -\frac{3}{2}$
Step 4 Check.	The solutions $-\frac{3}{2}$ and 5 satisfy the original equation.
6.7 *Applications of Quadratic Equations* **Pythagorean Formula** In a right triangle, the square of the hypotenuse equals the sum of the squares of the legs. $a^2 + b^2 = c^2$ 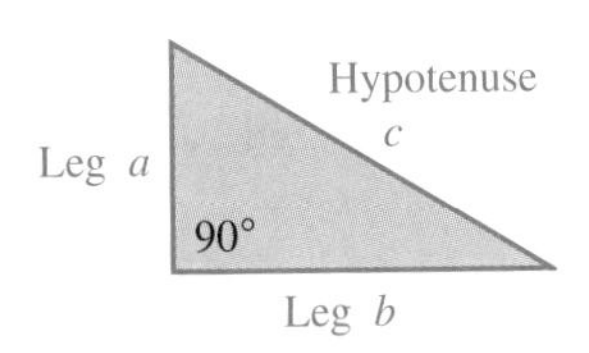 	In a right triangle, one leg measures 2 ft longer than the other. The hypotenuse measures 4 ft longer than the shorter leg. Find the lengths of the three sides of the triangle. Let $x =$ the length of the shorter leg. Then $x^2 + (x + 2)^2 = (x + 4)^2$. Solve this equation to get $x = 6$ or $x = -2$. Discard -2 as a solution. Check that the sides measure 6 ft, $6 + 2 = 8$ ft, and $6 + 4 = 10$ ft.

Answers to Test Your Word Power

1. B; *Example:* $x^2 - 5x - 14 = (x - 7)(x + 2)$
2. D; *Example:* The factored form of $x^2 - 5x - 14$ is $(x - 7)(x + 2)$.
3. C; *Example:* The greatest common factor of $8x^2$, $22xy$, and $16x^3y^2$ is $2x$.
4. A; *Example:* $a^2 + 2a + 1$ is a perfect square trinomial; its factored form is $(a + 1)^2$.
5. B; *Examples:* $y^2 - 3y + 2 = 0$, $x^2 - 9 = 0$, $2m^2 = 6m + 8$
6. C; *Example:* See the triangle included in the Quick Review above for **Section 6.7.**

Focus on

Real-Data Applications

Stopping Distance

The overall *stopping distance* is the sum of the *thinking distance* (how far the car travels once you realize you have to brake) and the *braking distance* (how far the car travels after you apply the brakes).

The data in the table represents three distinct relationships. The *input* is speed in miles per hour for all three relationships. The *output* is thinking distance in feet for the first relationship, braking distance in feet for the second relationship, and overall stopping distance in feet for the third relationship.

Speed	Thinking Distance	Braking Distance	Overall Stopping Distance
20 mph	20 ft	20 ft	40 ft
30 mph	30 ft	45 ft	75 ft
40 mph	40 ft	80 ft	120 ft
50 mph	50 ft	125 ft	175 ft
60 mph	60 ft	180 ft	240 ft
70 mph	70 ft	245 ft	315 ft

Source: Pass Your Driving Theory Test, British School of Motoring (1996).

For Group Discussion

1. In the relationship between thinking distance and speed, the *output* (y) is numerically the same as the *input* (x).

(a) Write the equation that expresses this relationship.

(b) When the speed is doubled from 20 mph to 40 mph, how does the thinking distance change?

(c) Does the same pattern hold true if a 30 mph speed is doubled?

(d) Is the equation linear or quadratic? Explain.

2. The relationship between braking distance and speed is given by the equation $y = \frac{1}{20}x^2$.

(a) Show that this equation corresponds to the table values for speeds of 20 mph and 40 mph.

(b) When the speed is doubled from 20 mph to 40 mph, how does the braking distance change?

(c) Does the same pattern hold true if a 30 mph speed is doubled?

(d) Is the equation linear or quadratic? Explain.

3. The relationship between overall stopping distance and speed is based on the equations in Problems 1 and 2.

(a) Use those results to write the equation that expresses the relationship between overall stopping distance and speed.

(b) Is the equation linear or quadratic? Explain.

(c) A *rule of thumb* for calculating overall stopping distance is to take speed in *tens* of miles per hour, divide by 2, add 1, and multiply the result by the speed in miles per hour. For example, at 40 mph, the rule says: $4 \div 2 + 1 = 3$; $3 \times 40 = 120$ ft. Show why this rule of thumb works. [*Hint:* Factor the right side of the equation in part (a).]

Chapter 6

REVIEW EXERCISES

[6.1] *Factor out the greatest common factor or factor by grouping.*

1. $7t + 14$

2. $60z^3 + 30z$

3. $35x^3 + 70x^2$

4. $100m^2n^3 - 50m^3n^4 + 150m^2n^2$

5. $2xy - 8y + 3x - 12$

6. $6y^2 + 9y + 4xy + 6x$

[6.2] *Factor completely.*

7. $x^2 + 5x + 6$

8. $y^2 - 13y + 40$

9. $q^2 + 6q - 27$

10. $r^2 - r - 56$

11. $r^2 - 4rs - 96s^2$

12. $p^2 + 2pq - 120q^2$

13. $8p^3 - 24p^2 - 80p$

14. $3x^4 + 30x^3 + 48x^2$

15. $m^2 - 3mn - 18n^2$

16. $y^2 - 8yz + 15z^2$

17. $p^7 - p^6q - 2p^5q^2$

18. $3r^5 - 6r^4s - 45r^3s^2$

19. $x^2 + x + 1$

20. $3x^2 + 6x + 6$

[6.3–6.4]

21. To begin factoring $6r^2 - 5r - 6$, what are the possible first terms of the two binomial factors, if we consider only positive integer coefficients?

22. What is the first step you would use to factor $2z^3 + 9z^2 - 5z$?

Factor completely.

23. $2k^2 - 5k + 2$

24. $3r^2 + 11r - 4$

25. $6r^2 - 5r - 6$

26. $10z^2 - 3z - 1$

27. $5t^2 - 11t + 12$

28. $24x^5 - 20x^4 + 4x^3$

29. $-6x^2 + 3x + 30$

30. $10r^3s + 17r^2s^2 + 6rs^3$

[6.5]

31. Which one of the following is a difference of squares?

A. $32x^2 - 1$ **B.** $4x^2y^2 - 25z^2$

C. $x^2 + 36$ **D.** $25y^3 - 1$

32. Which one of the following is a perfect square trinomial?

A. $x^2 + x + 1$ **B.** $y^2 - 4y + 9$

C. $4x^2 + 10x + 25$ **D.** $x^2 - 20x + 100$

Factor completely.

33. $n^2 - 64$

34. $25b^2 - 121$

35. $49y^2 - 25w^2$

36. $144p^2 - 36q^2$

37. $x^2 + 100$

38. $x^2 - \frac{49}{100}$

39. $z^2 + 10z + 25$

40. $r^2 - 12r + 36$

41. $9t^2 - 42t + 49$

42. $16m^2 + 40mn + 25n^2$

43. $54x^3 - 72x^2 + 24x$

44. $x^2 + \frac{2}{3}x + \frac{1}{9}$

[6.6] *Solve each equation, and check the solutions.*

45. $(4t + 3)(t - 1) = 0$

46. $(x + 7)(x - 4)(x + 3) = 0$

47. $x(2x - 5) = 0$

48. $z^2 + 4z + 3 = 0$

49. $m^2 - 5m + 4 = 0$

50. $x^2 = -15 + 8x$

51. $3z^2 - 11z - 20 = 0$

52. $81t^2 - 64 = 0$

53. $y^2 = 8y$

54. $n(n - 5) = 6$

55. $t^2 - 14t + 49 = 0$

56. $t^2 = 12(t - 3)$

57. $(5z + 2)(z^2 + 3z + 2) = 0$

58. $x^2 = 9$

[6.7] *Solve each problem.*

59. The length of a rug is 6 ft more than the width. The area is 40 ft². Find the length and width of the rug.

60. The surface area S of a box is given by

$$S = 2WH + 2WL + 2LH.$$

A treasure chest from a sunken galleon has dimensions as shown in the figure. Its surface area is 650 ft². Find its width.

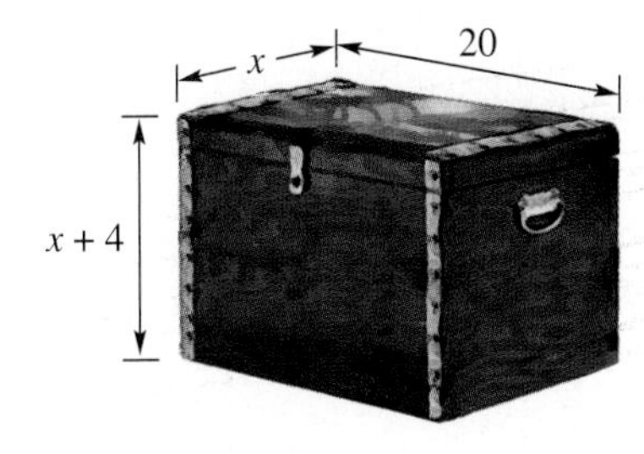

61. The length of a rectangle is three times the width. If the width were increased by 3 m while the length remained the same, the new rectangle would have an area of 30 m². Find the length and width of the original rectangle.

62. The volume of a rectangular box is 120 m³. The width of the box is 4 m, and the height is 1 m less than the length. Find the length and height of the box.

63. The product of two consecutive integers is 29 more than their sum. What are the integers?

64. Two cars left an intersection at the same time. One traveled west, and the other traveled 14 mi less, but to the south. How far apart were they then, if the distance between them was 16 mi more than the distance traveled south?

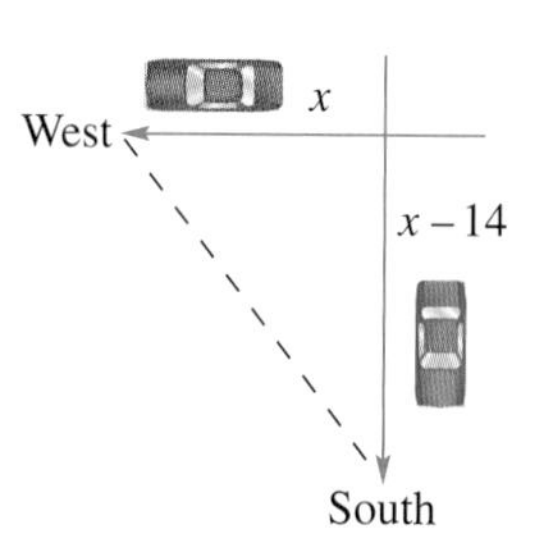

If an object is propelled upward with an initial velocity of 128 *ft per sec, its height h in feet after t seconds is*

$$h = 128t - 16t^2.$$

Find the height of the object after each period of time.

65. 1 sec

66. 2 sec

67. 4 sec

68. For the object described above, when does it return to the ground?

69. Annual revenue in millions of dollars for eBay is shown in the table.

Year	Annual Revenue (in millions of dollars)
1997	5.1
1998	47.4
1999	224.7
2000	431.4
2001	748.8
2002	1214.1

Source: eBay.

Using the data, we developed the quadratic equation

$$y = 47.8x^2 - .135x + 7.65$$

to model eBay revenues y in year x, where $x = 0$ represents 1997, $x = 1$ represents 1998, and so on.

(a) Use the model to find eBay revenue (to the nearest tenth) in 2001. How does your answer compare to the actual data from the table?

(b) Use the model to predict annual revenue (to the nearest tenth) for eBay in 2003.

(c) Revenue for eBay through the third quarter of 2003 was $1516.7 million. Given this information, do you think your prediction in part (b) is reliable? Explain.

MIXED REVIEW EXERCISES

70. Which of the following is *not* factored completely?

A. $3(7t)$ **B.** $3x(7t + 4)$ **C.** $(3 + x)(7t + 4)$ **D.** $3(7t + 4) + x(7t + 4)$

71. Although $(2x + 8)(3x - 4) = 6x^2 + 16x - 32$ is a true statement, the polynomial is not factored completely. Explain why and give the complete factored form.

Factor completely.

72. $z^2 - 11zx + 10x^2$

73. $3k^2 + 11k + 10$

74. $15m^2 + 20mp - 12m - 16p$

75. $y^4 - 625$

76. $6m^3 - 21m^2 - 45m$

77. $24ab^3c^2 - 56a^2bc^3 + 72a^2b^2c$

78. $25a^2 + 15ab + 9b^2$

79. $12x^2yz^3 + 12xy^2z - 30x^3y^2z^4$

80. $2a^5 - 8a^4 - 24a^3$

81. $12r^2 + 8rq - 15q^2$

82. $100a^2 - 9$

83. $49t^2 + 56t + 16$

Solve.

84. $t(t - 7) = 0$

85. $x^2 + 3x = 10$

86. $25x^2 + 20x + 4 = 0$

Solve each problem.

87. A lot is shaped like a right triangle. The hypotenuse is 3 m longer than the longer leg. The longer leg is 6 m longer than twice the length of the shorter leg. Find the lengths of the sides of the lot.

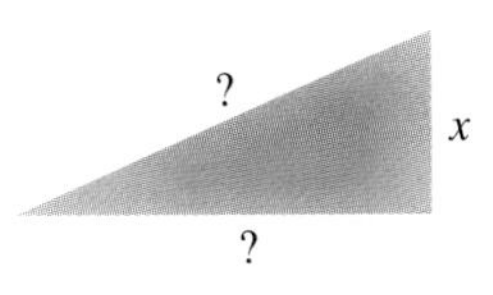

88. A pyramid has a rectangular base with a length that is 2 m more than the width. The height of the pyramid is 6 m, and its volume is 48 m^3. Find the length and width of the base.

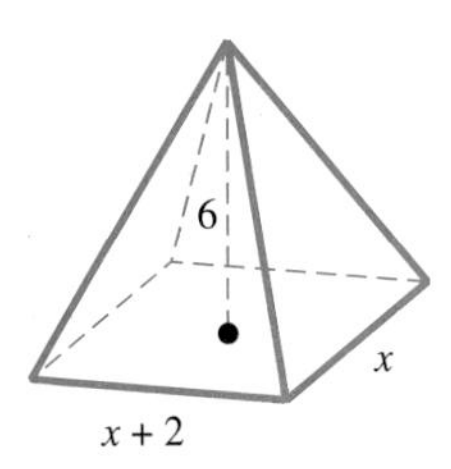

89. The product of the smaller two of three consecutive integers is equal to 23 plus the largest. Find the integers.

90. If an object is dropped, the distance d in feet it falls in t seconds (disregarding air resistance) is given by the quadratic equation

$$d = 16t^2.$$

Find the distance an object would fall in the following times.

(a) 4 sec **(b)** 8 sec

91. The floor plan for a house is a rectangle with length 7 m more than its width. The area is 170 m^2. Find the width and length of the house.

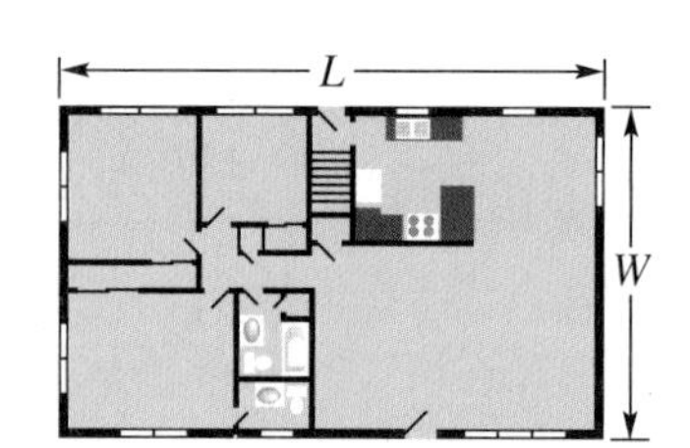

92. The triangular sail of a schooner has an area of 30 m^2. The height of the sail is 4 m more than the base. Find the base of the sail.

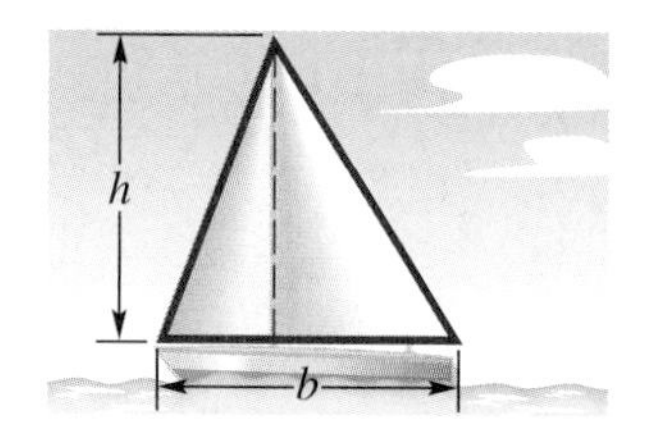

93. The numbers of alternative-fueled vehicles, in thousands, in use for the years 1998–2001 are given in the table.

Year	Number (in thousands)
1998	384
1999	407
2000	432
2001	456

Source: Energy Information Administration, Alternatives to Traditional Fuels, 2001.

Using the data, we developed the quadratic equation

$$y = .25x^2 - 25.65x + 496.6$$

to model the number of vehicles y in year x. Here we used $x = 98$ for 1998, $x = 99$ for 1999, and so on.

(a) What prediction for 2002 is given by the equation?

(b) Why might the prediction for 2002 be unreliable?

Chapter 6
TEST

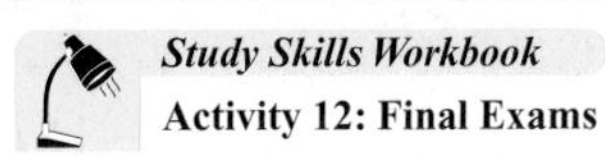

1. Which one of the following is the correct, completely factored form of $2x^2 - 2x - 24$?

A. $(2x + 6)(x - 4)$ **B.** $(x + 3)(2x - 8)$

C. $2(x + 4)(x - 3)$ **D.** $2(x + 3)(x - 4)$

1. ______________

Factor each polynomial completely.

2. $12x^2 - 30x$ **2.** ______________

3. $2m^3n^2 + 3m^3n - 5m^2n^2$ **3.** ______________

4. $2ax - 2bx + ay - by$ **4.** ______________

5. $x^2 - 9x + 14$ **5.** ______________

6. $2x^2 + x - 3$ **6.** ______________

7. $6x^2 - 19x - 7$ **7.** ______________

8. $3x^2 - 12x - 15$ **8.** ______________

9. $10z^2 - 17z + 3$ **9.** ______________

10. $t^2 + 2t + 3$ **10.** ______________

11. $x^2 + \frac{1}{36}$ **11.** ______________

12. $y^2 - 49$ **12.** ______________

13. $9y^2 - 64$ **13.** ______________

14. $x^2 + 16x + 64$ **14.** ______________

15. $4x^2 - 28xy + 49y^2$ **15.** ______________

16. $-2x^2 - 4x - 2$ **16.** ______________

17. $6t^4 + 3t^3 - 108t^2$ **17.** ______________

18. $4r^2 + 10rt + 25t^2$ **18.** ______________

19. ______________________

20. ______________________

21. ______________________

22. ______________________

23. ______________________

24. ______________________

25. ______________________

26. ______________________

27. ______________________

28. ______________________

29. ______________________

30. ______________________

19. $4t^3 + 32t^2 + 64t$

20. $x^4 - 81$

21. Why is $(p + 3)(p + 3)$ *not* the correct factored form of $p^2 + 9$?

Solve each equation.

22. $(x + 3)(x - 9) = 0$

23. $2r^2 - 13r + 6 = 0$

24. $25x^2 - 4 = 0$

25. $x(x - 20) = -100$

26. $t^2 = 3t$

Solve each problem.

27. The length of a rectangular flower bed is 3 ft less than twice its width. The area of the bed is 54 ft². Find the dimensions of the flower bed.

28. Find two consecutive integers such that the square of the sum of the two integers is 11 more than the smaller integer.

29. A carpenter needs to cut a brace to support a wall stud, as shown in the figure. The brace should be 7 ft less than three times the length of the stud. If the brace will be anchored on the floor 15 ft away from the stud, how long should the brace be?

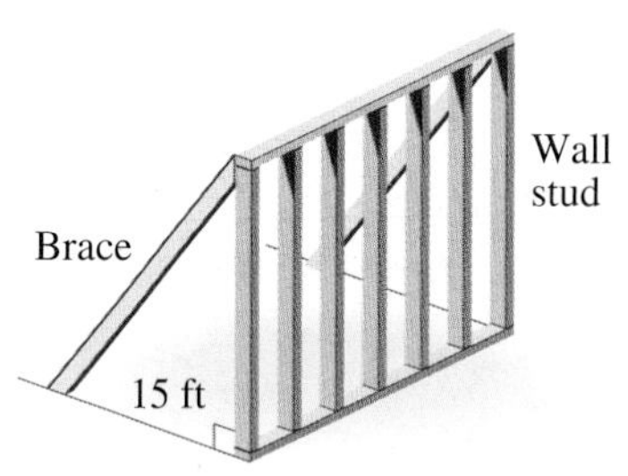

30. The number of U.S. cable TV networks y from 1984 through 2003 can be approximated by the quadratic equation

$$y = 1.06x^2 - 4.77x + 47.9,$$

where $x = 0$ represents 1984, $x = 1$ represents 1985, and so on. (*Source:* National Cable Television Association; FCC annual report.) Use the model to estimate the number of cable TV channels in 2000. Round your answer to the nearest whole number.

Cumulative Review Exercises

CHAPTERS R–6

Solve each equation.

1. $3x + 2(x - 4) = 4(x - 2)$

2. $.3x + .9x = .06$

3. $\frac{2}{3}n - \frac{1}{2}(n - 4) = 3$

4. Solve for P: $A = P + Prt$

5. From a list of "everyday items" often taken for granted, adults were recently surveyed as to those items they wouldn't want to live without. Complete the results shown in the table if 500 adults were surveyed.

Item	Percent That Wouldn't Want to Live Without	Number That Wouldn't Want to Live Without
Toilet paper	69%	
Zipper	42%	
Frozen foods		190
Self-stick note pads		75

(Other items included tape, hairspray, pantyhose, paper clips, and Velcro.)
Source: Market Facts for Kleenex Cottonelle.

Solve each problem.

6. At the 2002 Winter Olympics in Salt Lake City, the top medal winner was Germany with 35. Germany won 9 more silver medals than bronze and 4 fewer gold medals than silver. Find the number of each type of medal won. (*Source: World Almanac and Book of Facts*, 2004.)

7. Although the gender wage gap is narrowing, in 2002 women working full time earned, on average, 77.5 cents for every dollar earned by their male counterparts. How much would a female office worker have earned for a comparable job that paid a male colleague $45,000? (*Source:* Bureau of Labor Statistics.)

8. Find the measures of the marked angles.

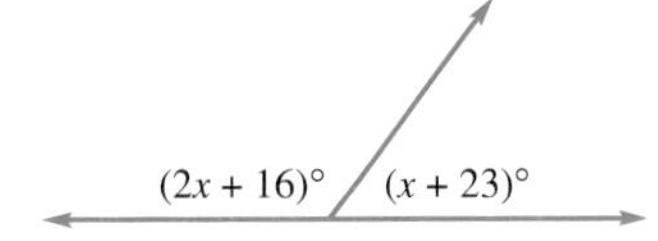

9. Fill in each blank with *positive* or *negative.* The point with coordinates (a, b) is in

(a) quadrant II if a is ________ and b is ________.

(b) quadrant III if a is ________ and b is ________.

Consider the equation $y = 12x + 3$. *Find the following.*

10. The x- and y-intercepts

11. The slope

12. The graph

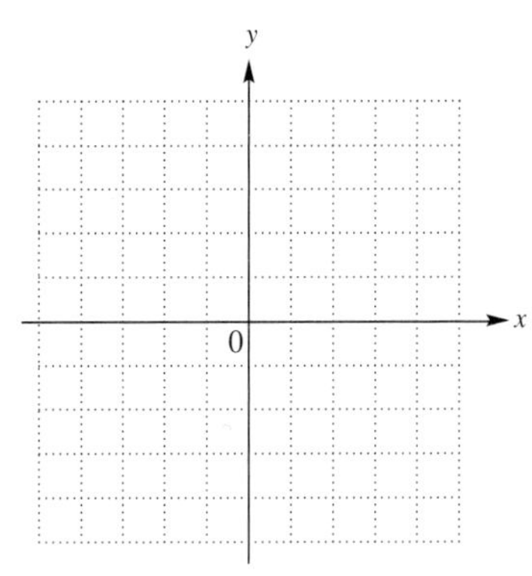

13. The points on the graph show the number of U.S. radio stations in the years 1995–2002, along with the graph of a linear equation that models the data.

(a) Use the ordered pairs shown on the graph to find the slope of the line to the nearest whole number. Interpret the slope.

(b) Use the graph to estimate the number of radio stations in the year 2000. Write your answer as an ordered pair of the form (year, number of radio stations).

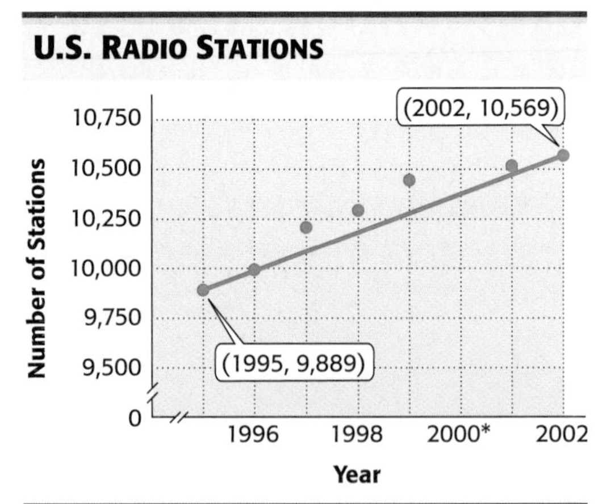

Source: M Street Corporation.
* Data for 2000 unavailable.

Solve each system of equations.

14. $4x - y = -6$
$2x + 3y = 4$

15. $5x + 3y = 10$
$2x + \frac{6}{5}y = 5$

Evaluate each expression.

16. $2^{-3} \cdot 2^5$

17. $\left(\frac{3}{4}\right)^{-2}$

18. $\frac{6^5 \cdot 6^{-2}}{6^3}$

19. $\left(\frac{4^{-3} \cdot 4^4}{4^5}\right)^{-1}$

Simplify each expression and write the answer using only positive exponents. Assume no denominators are 0.

20. $\frac{(p^2)^3p^{-4}}{(p^{-3})^{-1}p}$

21. $\frac{(m^{-2})^3m}{m^5m^{-4}}$

Perform the indicated operations.

22. $(2k^2 + 4k) - (5k^2 - 2) - (k^2 + 8k - 6)$

23. $(9x + 6)(5x - 3)$

24. $(3p + 2)^2$

25. $\frac{8x^4 + 12x^3 - 6x^2 + 20x}{2x}$

Factor completely.

26. $2a^2 + 7a - 4$

27. $10m^2 + 19m + 6$

28. $8t^2 + 10tv + 3v^2$

29. $4p^2 - 12p + 9$

30. $25r^2 - 81t^2$

31. $2pq + 6p^3q + 8p^2q$

Solve each equation.

32. $6m^2 + m - 2 = 0$

33. $8x^2 = 64x$

34. The length of the hypotenuse of a right triangle is 3 m more than twice the length of the shorter leg. The longer leg is 7 m longer than the shorter leg. Find the lengths of the sides.

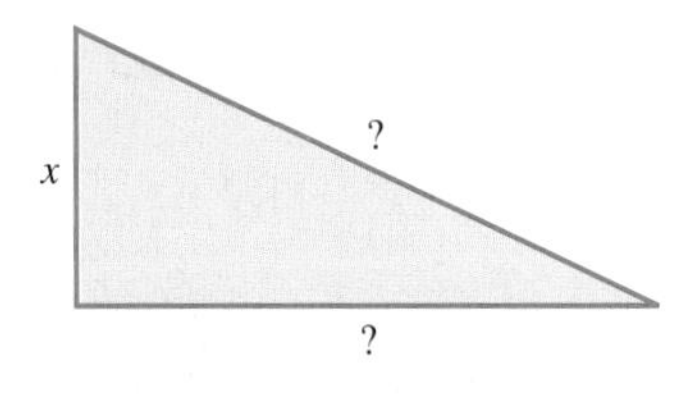

Rational Expressions and Applications

In the 1994 movie *Little Big League,* the young Billy Heywood inherits the Minnesota Twins baseball team and becomes manager. He leads the team to the Division Championship and then to the playoffs. But before the final playoff game, the biggest game of the year, he can't keep his mind on his job because a homework problem is giving him trouble.

If Joe can paint a house in 3 hours, and Sam can paint the same house in 5 hours, how long does it take for them to do it together?

With the help of one of his players, he is able to solve the problem, and the team goes on to victory. Its solution requires the use of algebraic fractions, or *rational expressions*. In Section 7.7 we show how to solve such problems, and in the Chapter Review Exercises, we ask you to solve Billy Heywood's problem.

7.1 The Fundamental Property of Rational Expressions

OBJECTIVES

1. **Find the values of the variable for which a rational expression is undefined.**
2. **Find the numerical value of a rational expression.**
3. **Write rational expressions in lowest terms.**
4. **Recognize equivalent forms of rational expressions.**

The quotient of two integers (with denominator not 0), such as $\frac{2}{3}$ or $-\frac{3}{4}$, is called a rational number. In the same way, the quotient of two polynomials with denominator not equal to 0 is called a *rational expression.*

Rational Expression

A **rational expression** is an expression of the form

$$\frac{P}{Q},$$

where P and Q are polynomials, with $Q \neq 0$.

Examples of rational expressions include

$$\frac{-6x}{x^3 + 8}, \quad \frac{9x}{y + 3}, \quad \text{and} \quad \frac{2m^3}{8}. \quad \text{Rational expressions}$$

Our work with rational expressions will require much of what we learned in **Chapters 5 and 6** on polynomials and factoring, as well as the rules for fractions from **Chapter R.**

OBJECTIVE 1 Find the values of the variable for which a rational expression is undefined. A fraction with denominator 0 is *not* a rational expression because division by 0 is undefined. Be careful when substituting a number in the denominator of a rational expression. For example, in

$$\frac{8x^2}{x - 3}, \leftarrow \text{Denominator cannot equal 0.}$$

the variable x can take on any value except 3. When $x = 3$, the denominator becomes $3 - 3 = \mathbf{0}$, making the expression undefined.

To determine the values for which a rational expression is undefined, use the following procedure.

Determining When a Rational Expression Is Undefined

Step 1 Set the denominator of the rational expression equal to 0.

Step 2 Solve this equation.

Step 3 The solutions of the equation are the values that make the rational expression undefined.

EXAMPLE 1 Finding Values That Make Rational Expressions Undefined

Find any values of the variable for which each rational expression is undefined.

(a) $\frac{p + 5}{3p + 2}$

Remember that the *numerator* may be any number; we must find any value of p that makes the *denominator* equal to 0 since division by 0 is undefined.

Continued on Next Page

Step 1 Set the denominator equal to 0.

$$3p + 2 = 0$$

Step 2 Solve this equation.

$$3p = -2$$

$$p = -\frac{2}{3}$$

Step 3 Since $p = -\frac{2}{3}$ will make the denominator 0, the given expression is undefined for $-\frac{2}{3}$.

(b) $\dfrac{9m^2}{m^2 - 5m + 6}$

Set the denominator equal to 0, and then find the solutions of the equation.

$$m^2 - 5m + 6 = 0$$

$$(m - 2)(m - 3) = 0 \qquad \text{Factor.}$$

$$m - 2 = 0 \quad \text{or} \quad m - 3 = 0 \qquad \text{Zero-factor property}$$

$$m = 2 \quad \text{or} \quad m = 3$$

The original expression is undefined for $m = 2$ and for $m = 3$.

(c) $\dfrac{2r}{r^2 + 1}$

This denominator cannot equal 0 for any value of r because r^2 is always greater than or equal to 0, and adding 1 makes the sum greater than 0. Thus, there are no values for which this rational expression is undefined.

Work Problem 1 at the Side.

1 Find all values for which each rational expression is undefined.

(a) $\dfrac{x + 2}{x - 5}$

(b) $\dfrac{3r}{r^2 + 6r + 8}$

(c) $\dfrac{-5m}{m^2 + 4}$

OBJECTIVE 2 Find the numerical value of a rational expression. We use substitution to evaluate a rational expression for a given value of the variable.

EXAMPLE 2 Evaluating Rational Expressions

Find the numerical value of $\dfrac{3x + 6}{2x - 4}$ for each value of x.

(a) $x = 1$

$$\frac{3x + 6}{2x - 4} = \frac{3(1) + 6}{2(1) - 4} \qquad \text{Let } x = 1.$$

$$= \frac{9}{-2} = -\frac{9}{2}$$

(b) $x = 2$

$$\frac{3x + 6}{2x - 4} = \frac{3(2) + 6}{2(2) - 4} = \frac{12}{0} \qquad \text{Let } x = 2.$$

Substituting 2 for x makes the denominator 0, so the expression is undefined when $x = 2$.

Work Problem 2 at the Side.

2 Find the value of each rational expression when $x = 3$.

(a) $\dfrac{x}{2x + 1}$

(b) $\dfrac{2x + 6}{x - 3}$

Answers

1. (a) 5 (b) $-4, -2$ (c) never undefined
2. (a) $\frac{3}{7}$ (b) undefined

OBJECTIVE 3 Write rational expressions in lowest terms. A fraction such as $\frac{2}{3}$ is said to be in *lowest terms.* How can "lowest terms" be defined? We use the idea of greatest common factor for this definition, which applies to all rational expressions.

Lowest Terms

A rational expression $\frac{P}{Q}$ $(Q \neq 0)$ is in **lowest terms** if the greatest common factor of its numerator and denominator is 1.

The properties of rational numbers also apply to rational expressions. We use the **fundamental property of rational expressions** to write a rational expression in lowest terms.

Fundamental Property of Rational Expressions

If $\frac{P}{Q}$ $(Q \neq 0)$ is a rational expression and if K represents any polynomial, where $K \neq 0$, then

$$\frac{PK}{QK} = \frac{P}{Q}.$$

This property is based on the identity property of multiplication, since

$$\frac{PK}{QK} = \frac{P}{Q} \cdot \frac{K}{K} = \frac{P}{Q} \cdot 1 = \frac{P}{Q}.$$

The next example shows how to write both a rational number and a rational expression in lowest terms. Notice the similarity in the procedures. In both cases, we factor and then divide out the greatest common factor.

EXAMPLE 3 Writing in Lowest Terms

Write each expression in lowest terms.

(a) $\frac{30}{72}$

Begin by factoring.

$$\frac{30}{72} = \frac{2 \cdot 3 \cdot 5}{2 \cdot 2 \cdot 2 \cdot 3 \cdot 3}$$

(b) $\frac{14k^2}{2k^3}$

Write k^2 as $k \cdot k$ and k^3 as $k \cdot k \cdot k$.

$$\frac{14k^2}{2k^3} = \frac{2 \cdot 7 \cdot k \cdot k}{2 \cdot k \cdot k \cdot k}$$

Group any factors common to the numerator and denominator.

$$\frac{30}{72} = \frac{5 \cdot (2 \cdot 3)}{2 \cdot 2 \cdot 3 \cdot (2 \cdot 3)} \qquad \frac{14k^2}{2k^3} = \frac{7(2 \cdot k \cdot k)}{k(2 \cdot k \cdot k)}$$

Use the fundamental property.

$$\frac{30}{72} = \frac{5}{2 \cdot 2 \cdot 3} = \frac{5}{12} \qquad \frac{14k^2}{2k^3} = \frac{7}{k}$$

Work Problem 3 at the Side.

3 Write each rational expression in lowest terms.

(a) $\frac{5x^4}{15x^2}$

(b) $\frac{6p^3}{2p^2}$

ANSWERS

3. (a) $\frac{x^2}{3}$ (b) $3p$

Writing a Rational Expression in Lowest Terms

Step 1 **Factor** the numerator and denominator completely.

Step 2 **Use the fundamental property** to divide out any common factors.

EXAMPLE 4 Writing in Lowest Terms

Write each rational expression in lowest terms.

(a) $\dfrac{3x - 12}{5x - 20}$

Begin by factoring both numerator and denominator. Then use the fundamental property of rational expressions.

$$\frac{3x-12}{5x-20} = \underbrace{\frac{3(x-4)}{5(x-4)}}_{\text{Step 1}} = \underbrace{\frac{3}{5}}_{\text{Step 2}}$$

The given expression is equal to $\frac{3}{5}$ for all values of x, where $x \neq 4$ (since the denominator of the original rational expression is 0 when x is 4).

(b) $\dfrac{m^2 + 2m - 8}{2m^2 - m - 6}$

$$\frac{m^2+2m-8}{2m^2-m-6} = \frac{(m+4)(m-2)}{(2m+3)(m-2)} \quad \text{Factor. (Step 1)}$$

$$= \frac{m+4}{2m+3} \quad \text{Fundamental property (Step 2)}$$

Here $m \neq -\frac{3}{2}$ and $m \neq 2$, since the denominator of the original expression is 0 for these values of m.

From now on, we will write statements of equality of rational expressions with the understanding that they apply only to those real numbers that make neither denominator equal to 0.

CAUTION

Rational expressions cannot be written in lowest terms until after the numerator and denominator have been factored. Only common factors can be divided out, not common terms. For example,

$$\frac{6x+9}{4x+6} = \frac{3(2x+3)}{2(2x+3)} = \frac{3}{2}$$

↑ Divide out the common factor.

$$\frac{6+x}{4x}$$ ← Numerator cannot be factored.

Already in lowest terms

Work Problem 4 at the Side.

4 Write each rational expression in lowest terms.

(a) $\dfrac{4y + 2}{6y + 3}$

(b) $\dfrac{8p + 8q}{5p + 5q}$

(c) $\dfrac{x^2 + 4x + 4}{4x + 8}$

(d) $\dfrac{a^2 - b^2}{a^2 + 2ab + b^2}$

ANSWERS

4. (a) $\frac{2}{3}$ **(b)** $\frac{8}{5}$ **(c)** $\frac{x+2}{4}$ **(d)** $\frac{a-b}{a+b}$

EXAMPLE 5 Writing in Lowest Terms (Factors Are Opposites)

Write $\frac{x - y}{y - x}$ in lowest terms.

At first glance, there does not seem to be any way in which $x - y$ and $y - x$ can be factored to get a common factor. However, $y - x$ can be factored as

$$y - x = -1(-y + x) = -1(x - y).$$

Now, use the fundamental property to simplify.

$$\frac{x - y}{y - x} = \frac{1(x - y)}{-1(x - y)} = \frac{1}{-1} = -1$$

In Example 5, notice that $y - x$ is the opposite of $x - y$. A general rule for this situation follows.

If the numerator and the denominator of a rational expression are opposites, such as in $\frac{x - y}{y - x}$, then the rational expression is equal to -1.

CAUTION
Although x and y appear in both the numerator and denominator in Example 5, we cannot use the fundamental property right away because they are *terms,* not *factors. **Terms are added, while factors are multiplied.***

EXAMPLE 6 Writing in Lowest Terms (Factors Are Opposites)

Write each rational expression in lowest terms.

(a) $\frac{2 - m}{m - 2}$

Since $2 - m$ and $m - 2$ (or $-2 + m$) are opposites,

$$\frac{2 - m}{m - 2} = -1.$$

(b) $\frac{4x^2 - 9}{6 - 4x}$

Factor the numerator and denominator.

$$\frac{4x^2 - 9}{6 - 4x} = \frac{(2x + 3)(2x - 3)}{2(3 - 2x)}$$

$$= \frac{(2x + 3)(2x - 3)}{2(-1)(2x - 3)} \qquad \text{Write } 3 - 2x \text{ as } -1(2x - 3).$$

$$= \frac{2x + 3}{2(-1)} \qquad \text{Fundamental property}$$

$$= \frac{2x + 3}{-2} \quad \text{or} \quad -\frac{2x + 3}{2} \qquad \frac{a}{-b} = -\frac{a}{b}$$

Continued on Next Page

(c) $\dfrac{3 + r}{3 - r}$

The quantity $3 - r$ *is not* the opposite of $3 + r$. This rational expression is already in lowest terms.

Work Problem 5 at the Side.

5 Write each rational expression in lowest terms.

(a) $\dfrac{5 - y}{y - 5}$

(b) $\dfrac{m - n}{n - m}$

(c) $\dfrac{25x^2 - 16}{12 - 15x}$

(d) $\dfrac{9 - k}{9 + k}$

OBJECTIVE 4 Recognize equivalent forms of rational expressions. When working with rational expressions, it is important to be able to recognize equivalent forms of an expression. For example, the common fraction $-\frac{5}{6}$ can also be written as $\frac{-5}{6}$ and as $\frac{5}{-6}$. Consider also the final rational expression from Example 6(b),

$$-\frac{2x + 3}{2}.$$

The $-$ sign representing the -1 factor is in front of the expression, on the same line as the fraction bar. The -1 factor may be placed in front of the expression, in the numerator, or in the denominator. Some other acceptable forms of this rational expression are

$$\frac{-(2x + 3)}{2}, \quad \text{Use parentheses in the numerator.}$$

$$\frac{-2x - 3}{2}, \quad \text{Distribute.}$$

and $$\frac{2x + 3}{-2}. \quad \text{Multiply } -\frac{2x + 3}{2} \text{ by } \frac{-1}{-1}.$$

CAUTION

$\frac{-2x + 3}{2}$ is *not* an equivalent form of $\frac{-(2x + 3)}{2}$. The sign preceding 3 in the numerator of $\frac{-2x + 3}{2}$ should be $-$ rather than $+$. ***Be careful to apply the distributive property correctly.***

EXAMPLE 7 Writing Equivalent Forms of a Rational Expression

Write four equivalent forms of the rational expression

$$-\frac{3x + 2}{x - 6}.$$

If we apply the negative sign to the numerator, we have the equivalent form

$$\frac{-(3x + 2)}{x - 6}.$$

Continued on Next Page

ANSWERS

5. (a) -1 (b) -1 (c) $\dfrac{5x + 4}{-3}$ or $-\dfrac{5x + 4}{3}$
(d) already in lowest terms

By distributing the negative sign in $\frac{-(3x + 2)}{x - 6}$, we have another equivalent form,

$$\frac{-3x - 2}{x - 6}.$$

If we apply the negative sign to the denominator of the given fraction $-\frac{3x + 2}{x - 6}$, we get

$$\frac{3x + 2}{-(x - 6)}$$

or, distributing once again,

$$\frac{3x + 2}{-x + 6}.$$

CAUTION

Recall that $-\frac{5}{6} \neq \frac{-5}{-6}$. Thus, in Example 7, it would be incorrect to distribute the negative sign in $-\frac{3x + 2}{x - 6}$ to *both* the numerator *and* the denominator. This would lead to the *opposite* of the original expression.

Work Problem 6 at the Side.

6 Decide whether each rational expression is equivalent to

$$-\frac{2x - 6}{x + 3}.$$

(a) $\frac{-(2x - 6)}{x + 3}$

(b) $\frac{-2x + 6}{x + 3}$

(c) $\frac{-2x - 6}{x + 3}$

(d) $\frac{2x - 6}{-(x + 3)}$

(e) $\frac{2x - 6}{-x - 3}$

(f) $\frac{2x - 6}{x - 3}$

ANSWERS

6. **(a)** equivalent **(b)** equivalent **(c)** not equivalent **(d)** equivalent **(e)** equivalent **(f)** not equivalent

7.1 Exercises

FOR EXTRA HELP

Addison-Wesley Math Tutor Center | MathXL | Digital Video Tutor CD 6 Videotape 7 | Student's Solutions Manual | MyMathLab | Interactmath.com

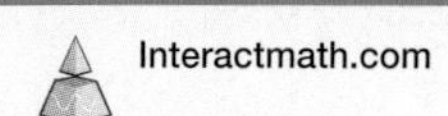

1. Fill in each blank with the correct response.

(a) The rational expression $\frac{x+5}{x-3}$ is undefined when $x =$ ____, and is equal to 0 when $x =$ ____.

(b) The rational expression $\frac{p-q}{q-p}$ is undefined when $p =$ ____, and in all other cases when written in lowest terms is equal to ____.

2. Make the correct choice for each blank.

(a) $\frac{4-r^2}{4+r^2}$ ______ (is/is not) equal to -1.

(b) $\frac{5+2x}{3-x}$ and $\frac{-5-2x}{x-3}$ ______ (are/are not) equivalent rational expressions.

3. Define *rational expression* in your own words, and give an example.

4. Give an example of a rational expression that is not in lowest terms, and then show the steps required to write it in lowest terms.

Find any value(s) for which each rational expression is undefined. See Example 1.

5. $\frac{2}{5y}$ **6.** $\frac{7}{3z}$ **7.** $\frac{4x^2}{3x-5}$ **8.** $\frac{2x^3}{3x-4}$

9. $\frac{m+2}{m^2+m-6}$ **10.** $\frac{r-5}{r^2-5r+4}$ **11.** $\frac{3x}{x^2+2}$ **12.** $\frac{4q}{q^2+9}$

Find the numerical value of each rational expression when **(a)** $x = 2$ *and* **(b)** $x = -3$. *See Example 2.*

13. $\frac{5x-2}{4x}$ **14.** $\frac{3x+1}{5x}$ **15.** $\frac{2x^2-4x}{3x}$ **16.** $\frac{4x^2-1}{5x}$

17. $\frac{(-3x)^2}{4x+12}$ **18.** $\frac{(-2x)^3}{3x+9}$ **19.** $\frac{5x+2}{2x^2+11x+12}$ **20.** $\frac{7-3x}{3x^2-7x+2}$

21. If 2 is substituted for x in the rational expression $\frac{x-2}{x^2-4}$, the result is $\frac{0}{0}$. We often hear the statement "Any number divided by itself is 1." Does this mean that this expression is equal to 1 for $x = 2$? If not, explain.

22. For $x \neq 2$, the rational expression $\frac{2(x-2)}{x-2}$ is equal to 2. Can $\frac{2x-2}{x-2}$ also be simplified to 2? Explain.

Write each rational expression in lowest terms. See Examples 3 and 4.

23. $\frac{18r^3}{6r}$

24. $\frac{27p^2}{3p}$

25. $\frac{4(y-2)}{10(y-2)}$

26. $\frac{15(m-1)}{9(m-1)}$

27. $\frac{(x+1)(x-1)}{(x+1)^2}$

28. $\frac{(t+5)(t-3)}{(t-1)(t+5)}$

29. $\frac{7m+14}{5m+10}$

30. $\frac{8z-24}{4z-12}$

31. $\frac{m^2-n^2}{m+n}$

32. $\frac{a^2-b^2}{a-b}$

33. $\frac{12m^2-3}{8m-4}$

34. $\frac{20p^2-45}{6p-9}$

35. $\frac{3m^2-3m}{5m-5}$

36. $\frac{6t^2-6t}{2t-2}$

37. $\frac{9r^2-4s^2}{9r+6s}$

38. $\frac{16x^2-9y^2}{12x-9y}$

39. $\frac{2x^2-3x-5}{2x^2-7x+5}$

40. $\frac{3x^2+8x+4}{3x^2-4x-4}$

41. $\frac{zw+4z-3w-12}{zw+4z+5w+20}$

42. $\frac{km+4k+4m+16}{km+4k+5m+20}$

Write each rational expression in lowest terms. See Examples 5 and 6.

43. $\frac{6-t}{t-6}$

44. $\frac{2-k}{k-2}$

45. $\frac{m^2-1}{1-m}$

46. $\frac{a^2-b^2}{b-a}$

47. $\frac{q^2-4q}{4q-q^2}$

48. $\frac{z^2-5z}{5z-z^2}$

Write four equivalent expressions for each expression. See Example 7.

49. $-\frac{x+4}{x-3}$

50. $-\frac{x+6}{x-1}$

51. $-\frac{2x-3}{x+3}$

52. $-\frac{5x-6}{x+4}$

53. $\frac{-3x+1}{5x-6}$

54. $\frac{-2x-9}{3x+1}$

55. The area of the rectangle is represented by x^4+10x^2+21. What is the width? (*Hint*: Use $W=\frac{A}{L}$.)

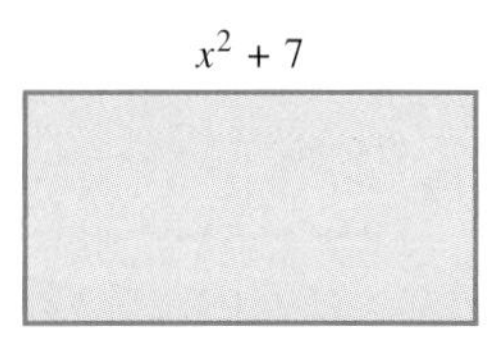

56. The volume of the box is represented by

$$(x^2+8x+15)(x+4).$$

Find the polynomial that represents the area of the bottom of the box.

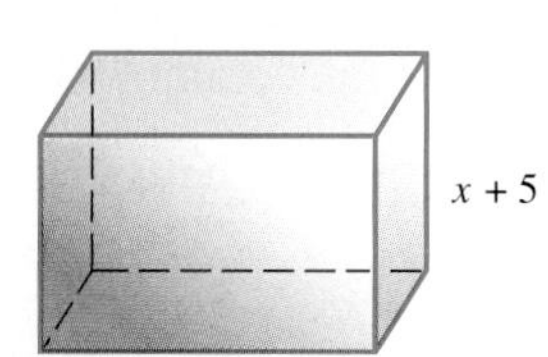

7.2 Multiplying and Dividing Rational Expressions

OBJECTIVES

1. Multiply rational expressions.
2. Find reciprocals.
3. Divide rational expressions.

OBJECTIVE 1 Multiply rational expressions. The product of two fractions is found by multiplying the numerators and multiplying the denominators. Rational expressions are multiplied in the same way.

Multiplying Rational Expressions

The product of the rational expressions $\frac{P}{Q}$ and $\frac{R}{S}$ is

$$\frac{P}{Q} \cdot \frac{R}{S} = \frac{PR}{QS}.$$

In words: To multiply rational expressions, multiply the numerators and multiply the denominators.

In the following example, the parallel discussion with rational numbers and rational expressions lets you compare the steps.

EXAMPLE 1 Multiplying Rational Expressions

Multiply. Write each answer in lowest terms.

(a) $\frac{3}{10} \cdot \frac{5}{9}$ **(b)** $\frac{6}{x} \cdot \frac{x^2}{12}$

Indicate the product of the numerators and the product of the denominators.

$$\frac{3}{10} \cdot \frac{5}{9} = \frac{3 \cdot 5}{10 \cdot 9} \qquad \frac{6}{x} \cdot \frac{x^2}{12} = \frac{6 \cdot x^2}{x \cdot 12}$$

Leave the products in factored form because common factors are needed to write the product in lowest terms. Factor the numerator and denominator to further identify any common factors. Then use the fundamental property to write each product in lowest terms.

$$\frac{3}{10} \cdot \frac{5}{9} = \frac{3 \cdot 5}{2 \cdot 5 \cdot 3 \cdot 3} = \frac{1}{6} \qquad \frac{6}{x} \cdot \frac{x^2}{12} = \frac{6 \cdot x \cdot x}{2 \cdot 6 \cdot x} = \frac{x}{2}$$

Work Problem 1 at the Side.

EXAMPLE 2 Multiplying Rational Expressions

Multiply $\frac{x+y}{2x} \cdot \frac{x^2}{(x+y)^2}$. Write the answer in lowest terms.

Use the definition of multiplication. Indicate the products in the first step so that common factors are easily identified.

$$\frac{x+y}{2x} \cdot \frac{x^2}{(x+y)^2} = \frac{(x+y)x^2}{2x(x+y)^2}$$ Multiply numerators. Multiply denominators.

$$= \frac{(x+y)x \cdot x}{2x(x+y)(x+y)}$$ Factor; identify common factors.

$$= \frac{x}{2(x+y)}$$ $\frac{(x+y)x}{x(x+y)} = 1$; lowest terms

Work Problem 2 at the Side.

1 Multiply. Write each answer in lowest terms.

(a) $\frac{2}{7} \cdot \frac{5}{10}$

(b) $\frac{3m^2}{2} \cdot \frac{10}{m}$

(c) $\frac{8p^2q}{3} \cdot \frac{9}{q^2p}$

2 Multiply. Write each answer in lowest terms.

(a) $\frac{a+b}{5} \cdot \frac{30}{2(a+b)}$

(b) $\frac{3(p-q)}{q^2} \cdot \frac{q}{2(p-q)^2}$

ANSWERS

1. (a) $\frac{1}{7}$ (b) $15m$ (c) $\frac{24p}{q}$

2. (a) 3 (b) $\frac{3}{2q(p-q)}$

3 Multiply. Write each answer in lowest terms.

(a) $\dfrac{x^2+7x+10}{3x+6} \cdot \dfrac{6x-6}{x^2+2x-15}$

(b) $\dfrac{m^2+4m-5}{m+5} \cdot \dfrac{m^2+8m+15}{m-1}$

4 Find each reciprocal.

(a) $\dfrac{6b^5}{3r^2b}$

(b) $\dfrac{t^2-4t}{t^2+2t-3}$

Answers

3. (a) $\dfrac{2(x-1)}{x-3}$ (b) $(m+5)(m+3)$

4. (a) $\dfrac{3r^2b}{6b^5}$ (b) $\dfrac{t^2+2t-3}{t^2-4t}$

EXAMPLE 3 Multiplying Rational Expressions

Multiply. Write the answer in lowest terms.

$$\frac{x^2+3x}{x^2-3x-4} \cdot \frac{x^2-5x+4}{x^2+2x-3} = \frac{(x^2+3x)(x^2-5x+4)}{(x^2-3x-4)(x^2+2x-3)} \quad \text{Definition of multiplication}$$

$$= \frac{x(x+3)(x-4)(x-1)}{(x-4)(x+1)(x+3)(x-1)} \quad \text{Factor.}$$

$$= \frac{x}{x+1} \quad \text{Lowest terms}$$

The quotients

$$\frac{x+3}{x+3}, \quad \frac{x-4}{x-4}, \quad \text{and} \quad \frac{x-1}{x-1}$$

are all equal to 1, justifying the final product $\dfrac{x}{x+1}$.

Work Problem 3 at the Side.

OBJECTIVE 2 Find reciprocals. If the product of two rational expressions is 1, the rational expressions are called **reciprocals** (or **multiplicative inverses**) of each other. The reciprocal of a rational expression is found by interchanging the numerator and the denominator. For example,

$$\frac{2x-1}{x-5} \quad \text{has reciprocal} \quad \frac{x-5}{2x-1}.$$

EXAMPLE 4 Finding Reciprocals of Rational Expressions

Find the reciprocal of each rational expression.

(a) $\dfrac{4p^3}{9q}$

Interchange the numerator and denominator. The reciprocal is $\dfrac{9q}{4p^3}$.

(b) $\dfrac{k^2-9}{k^2-k-20}$ has reciprocal $\dfrac{k^2-k-20}{k^2-9}$.

Work Problem 4 at the Side.

OBJECTIVE 3 Divide rational expressions. To develop a method for dividing rational numbers and rational expressions, consider the following problem. Suppose that you have $\frac{7}{8}$ gal of milk and you wish to find how many quarts you have. Since 1 qt is $\frac{1}{4}$ gal, you must ask yourself, "How many $\frac{1}{4}$s are there in $\frac{7}{8}$?" This would be interpreted as

$$\frac{7}{8} \div \frac{1}{4} \quad \text{or} \quad \frac{\frac{7}{8}}{\frac{1}{4}}$$

since the fraction bar means division.

The fundamental property of rational expressions discussed earlier can be applied to rational number values of P, Q, and K. With $P = \frac{7}{8}$, $Q = \frac{1}{4}$, and $K = 4$ (K is the reciprocal of $Q = \frac{1}{4}$),

$$\frac{P}{Q} = \frac{P \cdot K}{Q \cdot K} = \frac{\frac{7}{8} \cdot 4}{\frac{1}{4} \cdot 4} = \frac{\frac{7}{8} \cdot 4}{1} = \frac{7}{8} \cdot \frac{4}{1}.$$

So, to divide $\frac{7}{8}$ by $\frac{1}{4}$, we must multiply $\frac{7}{8}$ by the reciprocal of $\frac{1}{4}$, namely $\frac{4}{1}$ or 4. Since $\frac{7}{8}(4) = \frac{7}{2}$, there are $\frac{7}{2}$ or $3\frac{1}{2}$ qt in $\frac{7}{8}$ gal.

The preceding discussion illustrates the rule for dividing common fractions. To divide $\frac{a}{b}$ by $\frac{c}{d}$, multiply $\frac{a}{b}$ by the reciprocal of $\frac{c}{d}$. Division of rational expressions is defined in the same way.

Dividing Rational Expressions

If $\frac{P}{Q}$ and $\frac{R}{S}$ are any two rational expressions, with $\frac{R}{S} \neq 0$, then

$$\frac{P}{Q} \div \frac{R}{S} = \frac{P}{Q} \cdot \frac{S}{R} = \frac{PS}{QR}.$$

In words: To divide one rational expression by another rational expression, multiply the first rational expression (dividend) by the reciprocal of the second rational expression (divisor).

The next example shows the division of two rational numbers and the division of two rational expressions.

EXAMPLE 5 Dividing Rational Expressions

Divide. Write each answer in lowest terms.

(a) $\frac{5}{8} \div \frac{7}{16}$ (b) $\frac{y}{y+3} \div \frac{4y}{y+5}$

Multiply the first expression by the reciprocal of the second.

(a)

$$\frac{5}{8} \div \frac{7}{16} = \frac{5}{8} \cdot \frac{16}{7} \quad \text{Reciprocal of } \tfrac{7}{16}$$
$$= \frac{5 \cdot 16}{8 \cdot 7}$$
$$= \frac{5 \cdot 8 \cdot 2}{8 \cdot 7}$$
$$= \frac{10}{7}$$

(b)

$$\frac{y}{y+3} \div \frac{4y}{y+5}$$
$$= \frac{y}{y+3} \cdot \frac{y+5}{4y} \quad \text{Reciprocal of } \tfrac{4y}{y+5}$$
$$= \frac{y(y+5)}{(y+3)(4y)}$$
$$= \frac{y+5}{4(y+3)}$$

Work Problem 5 at the Side.

5 Divide. Write each answer in lowest terms.

(a) $\frac{3}{4} \div \frac{5}{16}$

(b) $\frac{r}{r-1} \div \frac{3r}{r+4}$

(c) $\frac{6x-4}{3} \div \frac{15x-10}{9}$

ANSWERS

5. (a) $\frac{12}{5}$ (b) $\frac{r+4}{3(r-1)}$ (c) $\frac{6}{5}$

6 Divide. Write each answer in lowest terms.

(a) $\frac{5a^2b}{2} \div \frac{10ab^2}{8}$

(b) $\frac{(3t)^2}{w} \div \frac{3t^2}{5w^4}$

EXAMPLE 6 Dividing Rational Expressions

Divide. Write the answer in lowest terms.

$$\frac{(3m)^2}{(2p)^3} \div \frac{6m^3}{16p^2} = \frac{(3m)^2}{(2p)^3} \cdot \frac{16p^2}{6m^3}$$ Multiply by the reciprocal.

$$= \frac{9m^2}{8p^3} \cdot \frac{16p^2}{6m^3}$$ Power rule for exponents

$$= \frac{9 \cdot 16m^2p^2}{8 \cdot 6p^3m^3}$$ Multiply numerators. Multiply denominators.

$$= \frac{3}{mp}$$ Lowest terms

Work Problem 6 at the Side.

7 Divide. Write each answer in lowest terms.

(a) $\frac{y^2 + 4y + 3}{y + 3} \div \frac{y^2 - 4y - 5}{y - 3}$

(b) $\frac{4x(x + 3)}{2x + 1} \div \frac{-x^2(x + 3)}{4x^2 - 1}$

EXAMPLE 7 Dividing Rational Expressions

Divide. Write the answer in lowest terms.

$$\frac{x^2 - 4}{(x + 3)(x - 2)} \div \frac{(x + 2)(x + 3)}{-2x}$$

$$= \frac{x^2 - 4}{(x + 3)(x - 2)} \cdot \frac{-2x}{(x + 2)(x + 3)}$$ Multiply by the reciprocal.

$$= \frac{(x + 2)(x - 2)}{(x + 3)(x - 2)} \cdot \frac{-2x}{(x + 2)(x + 3)}$$ Factor.

$$= \frac{-2x(x + 2)(x - 2)}{(x + 3)(x - 2)(x + 2)(x + 3)}$$ Multiply numerators. Multiply denominators.

$$= \frac{-2x}{(x + 3)^2} \quad \text{or} \quad -\frac{2x}{(x + 3)^2}$$ Lowest terms

Work Problem 7 at the Side.

8 Divide. Write each answer in lowest terms.

(a) $\frac{ab - a^2}{a^2 - 1} \div \frac{a - b}{a - 1}$

(b) $\frac{x^2 - 9}{2x + 6} \div \frac{9 - x^2}{4x - 12}$

EXAMPLE 8 Dividing Rational Expressions (Factors Are Opposites)

Divide. Write the answer in lowest terms.

$$\frac{m^2 - 4}{m^2 - 1} \div \frac{2m^2 + 4m}{1 - m}$$

$$= \frac{m^2 - 4}{m^2 - 1} \cdot \frac{1 - m}{2m^2 + 4m}$$ Multiply by the reciprocal.

$$= \frac{(m + 2)(m - 2)}{(m + 1)(m - 1)} \cdot \frac{1 - m}{2m(m + 2)}$$ Factor; $1 - m$ and $m - 1$ differ only in sign.

$$= \frac{-1(m - 2)}{2m(m + 1)}$$ From **Section 7.1**, $\frac{1 - m}{m - 1} = -1$.

$$= \frac{2 - m}{2m(m + 1)}$$ Distribute -1 in the numerator.

Work Problem 8 at the Side.

Answers

6. (a) $\frac{2a}{b}$ (b) $15w^3$
7. (a) $\frac{y - 3}{y - 5}$ (b) $-\frac{4(2x - 1)}{x}$
8. (a) $\frac{-a}{a + 1}$ (b) $\frac{-2x + 6}{3 + x}$

In summary, follow these steps to multiply or divide rational expressions.

Multiplying or Dividing Rational Expressions

Step 1 **Note the operation.** If the operation is division, use the definition of division to rewrite as multiplication.

Note: Steps 2 and 3 may be interchanged. It is a matter of personal preference.

Step 2 **Factor** all numerators and denominators completely.

Step 3 **Multiply** numerators and multiply denominators.

Step 4 **Write in lowest terms** using the fundamental property.

Focus on

Real-Data Applications

Is 5 or 10 Minutes Really Worth the Risk?

A poignant e-mail entitled *The Drive Home* recounted the story of the habitual speeder named Jack, who was stopped by a policeman. The policeman happened to be his friend. Rather than writing a speeding ticket, the policeman handed Jack a note that described the loss of his daughter to a speeding driver. Jack was deeply affected and changed his attitude toward life on the road. The e-mail then gave some facts about speed.

- For a trip of 20 miles in a 40 mile-per-hour (mph) zone, speeding by 10 mph will save 6 minutes.
- For a trip of 50 miles in a 55 mph zone, speeding by 10 mph will save 8.3 minutes and speeding by 15 mph will save 11.6 minutes.

The question is, "Is 5 or 10 minutes really worth the risk?"

The relationship between time, distance, and rate is given by $t = \frac{d}{r}$. In the example, the time it takes to travel 20 miles in a 40 mph zone is $\frac{20 \text{ mi}}{40 \text{ mph}} = \frac{1}{2}$ hr, or 30 minutes. At a speed of 50 mph, the time is $\frac{20 \text{ mi}}{50 \text{ mph}} = \frac{2}{5}$ hr, or 24 minutes, since $\frac{2}{5}(60 \text{ min}) = 24 \text{ min}$. The author's claim that only 6 minutes were saved is valid.

For Group Discussion

1. Verify the claims that traveling 50 miles at 65 mph in a 55 mph zone saves only 8.3 minutes and traveling at 70 mph saves only 11.6 minutes.

2. To save 10 minutes in a trip of 20 miles in a 40 mph zone, by how many miles per hour would Jack have to exceed the speed limit? (*Hint:* To save 10 minutes, the time for the trip would have to be only 20 minutes, which is $\frac{20}{60} = \frac{1}{3}$ hr. Determine the fraction of the form $\frac{20}{40 + x}$ that is equivalent to $\frac{1}{3}$ and then determine the value for x.)

3. Suppose you have to drive 10 miles to school in a 30 mph zone and you are running late. To save 5 minutes on your trip, at what speed would you have to travel? If stopped by a policeman, would you likely be given a ticket?

7.2 Exercises

FOR EXTRA HELP

Addison-Wesley Math Tutor Center | MathXL | Digital Video Tutor CD 6 Videotape 7 |  Student's Solutions Manual | MyMathLab |  Interactmath.com

1. Match each multiplication problem in Column I with the correct product in Column II.

	I		II
(a)	$\dfrac{5x^3}{10x^4} \cdot \dfrac{10x^7}{2x}$	**A.**	$\dfrac{2}{5x^5}$
(b)	$\dfrac{10x^4}{5x^3} \cdot \dfrac{10x^7}{2x}$	**B.**	$\dfrac{5x^5}{2}$
(c)	$\dfrac{5x^3}{10x^4} \cdot \dfrac{2x}{10x^7}$	**C.**	$\dfrac{1}{10x^7}$
(d)	$\dfrac{10x^4}{5x^3} \cdot \dfrac{2x}{10x^7}$	**D.**	$10x^7$

2. Match each division problem in Column I with the correct quotient in Column II.

	I		II
(a)	$\dfrac{5x^3}{10x^4} \div \dfrac{10x^7}{2x}$	**A.**	$\dfrac{5x^5}{2}$
(b)	$\dfrac{10x^4}{5x^3} \div \dfrac{10x^7}{2x}$	**B.**	$10x^7$
(c)	$\dfrac{5x^3}{10x^4} \div \dfrac{2x}{10x^7}$	**C.**	$\dfrac{2}{5x^5}$
(d)	$\dfrac{10x^4}{5x^3} \div \dfrac{2x}{10x^7}$	**D.**	$\dfrac{1}{10x^7}$

Multiply. Write each answer in lowest terms. See Examples 1 and 2.

3. $\dfrac{10m^2}{7} \cdot \dfrac{14}{15m}$

4. $\dfrac{36z^3}{6z} \cdot \dfrac{28}{z^2}$

5. $\dfrac{16y^4}{18y^5} \cdot \dfrac{15y^5}{y^2}$

6. $\dfrac{20x^5}{-2x^2} \cdot \dfrac{8x^4}{35x^3}$

7. $\dfrac{2(c+d)}{3} \cdot \dfrac{18}{6(c+d)^2}$

8. $\dfrac{4(y-2)}{x} \cdot \dfrac{3x}{6(y-2)^2}$

Find the reciprocal of each rational expression. See Example 4.

9. $\dfrac{3p^3}{16q}$

10. $\dfrac{6x^4}{9y^2}$

11. $\dfrac{r^2+rp}{7}$

12. $\dfrac{16}{9a^2+36a}$

13. $\dfrac{z^2+7z+12}{z^2-9}$

14. $\dfrac{p^2-4p+3}{p^2-3p}$

Divide. Write each answer in lowest terms. See Examples 5, 6, and 7.

15. $\dfrac{9z^4}{3z^5} \div \dfrac{3z^2}{5z^3}$

16. $\dfrac{35q^8}{9q^5} \div \dfrac{25q^6}{10q^5}$

17. $\dfrac{4t^4}{2t^5} \div \dfrac{(2t)^3}{-6}$

18. $\dfrac{-12a^6}{3a^2} \div \dfrac{(2a)^3}{27a}$

19. $\dfrac{3}{2y-6} \div \dfrac{6}{y-3}$

20. $\dfrac{4m+16}{10} \div \dfrac{3m+12}{18}$

21. Explain in your own words how to multiply rational expressions.

22. Explain in your own words how to divide rational expressions.

Multiply or divide. Write each answer in lowest terms. See Examples 3, 7, and 8.

23. $\dfrac{5x-15}{3x+9} \cdot \dfrac{4x+12}{6x-18}$

24. $\dfrac{8r+16}{24r-24} \cdot \dfrac{6r-6}{3r+6}$

25. $\dfrac{2-t}{8} \div \dfrac{t-2}{6}$

26. $\dfrac{4}{m-2} \div \dfrac{16}{2-m}$

27. $\dfrac{5-4x}{5+4x} \cdot \dfrac{4x+5}{4x-5}$

28. $\dfrac{5-x}{5+x} \cdot \dfrac{x+5}{x-5}$

29. $\dfrac{6(m-2)^2}{5(m+4)^2} \cdot \dfrac{15(m+4)}{2(2-m)}$

30. $\dfrac{7(q-1)}{3(q+1)^2} \cdot \dfrac{6(q+1)}{3(1-q)^2}$

31. $\dfrac{p^2+4p-5}{p^2+7p+10} \div \dfrac{p-1}{p+4}$

32. $\dfrac{z^2-3z+2}{z^2+4z+3} \div \dfrac{z-1}{z+1}$

33. $\dfrac{2k^2-k-1}{2k^2+5k+3} \div \dfrac{4k^2-1}{2k^2+k-3}$

34. $\dfrac{2m^2-5m-12}{m^2+m-20} \div \dfrac{4m^2-9}{m^2+4m-5}$

35. $\dfrac{2k^2+3k-2}{6k^2-7k+2} \cdot \dfrac{4k^2-5k+1}{k^2+k-2}$

36. $\dfrac{2m^2-5m-12}{m^2-10m+24} \div \dfrac{4m^2-9}{m^2-9m+18}$

37. $\dfrac{m^2+2mp-3p^2}{m^2-3mp+2p^2} \div \dfrac{m^2+4mp+3p^2}{m^2+2mp-8p^2}$

38. $\dfrac{r^2+rs-12s^2}{r^2-rs-20s^2} \div \dfrac{r^2-2rs-3s^2}{r^2+rs-30s^2}$

39. $\left(\dfrac{x^2+10x+25}{x^2+10x} \cdot \dfrac{10x}{x^2+15x+50}\right) \div \dfrac{x+5}{x+10}$

40. $\left(\dfrac{m^2-12m+32}{8m} \cdot \dfrac{m^2-8m}{m^2-8m+16}\right) \div \dfrac{m-8}{m-4}$

41. Consider the division problem $\dfrac{x-6}{x+4} \div \dfrac{x+7}{x+5}$. We know that division by 0 is undefined, so the restrictions on x are $x \neq -4$, $x \neq -5$, and $x \neq -7$. Why is the last restriction needed?

42. If the rational expression $\dfrac{5x^2y^3}{2pq}$ represents the area of a rectangle and $\dfrac{2xy}{p}$ represents the length, what rational expression represents the width?

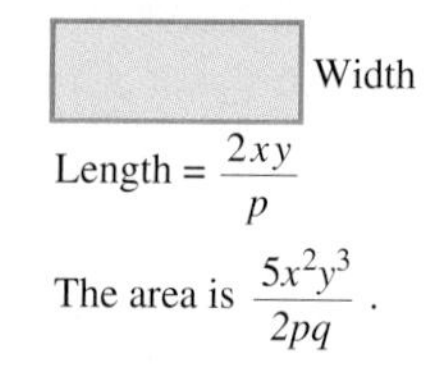

7.3 Least Common Denominators

OBJECTIVES

1. **Find the least common denominator for a group of fractions.**
2. **Rewrite rational expressions with given denominators.**

OBJECTIVE 1 Find the least common denominator for a group of fractions. Just as with common fractions, adding or subtracting rational expressions (to be discussed in the next section) often requires a **least common denominator (LCD),** the simplest expression that is divisible by all denominators. For example, the least common denominator for $\frac{2}{9}$ and $\frac{5}{12}$ is 36 because 36 is the smallest positive number divisible by both 9 and 12.

Least common denominators can often be found by inspection. For example, the LCD for $\frac{1}{6}$ and $\frac{2}{3m}$ is $6m$. In other cases, the LCD can be found by a procedure similar to that used in **Section 6.1** for finding the greatest common factor.

Finding the Least Common Denominator (LCD)

Step 1 **Factor** each denominator into prime factors.

Step 2 **List each different denominator factor** the *greatest* number of times it appears in any of the denominators.

Step 3 **Multiply** the denominator factors from Step 2 to get the LCD.

When each denominator is factored into prime factors, every prime factor must be a factor of the least common denominator.

In Example 1, we find the LCD for both numerical and algebraic denominators.

EXAMPLE 1 Finding the LCD

Find the LCD for each pair of fractions.

(a) $\frac{1}{24}, \frac{7}{15}$ **(b)** $\frac{1}{8x}, \frac{3}{10x}$

Step 1 Write each denominator in factored form with numerical coefficients in prime factored form.

(a)
$$24 = 2 \cdot 2 \cdot 2 \cdot 3 = 2^3 \cdot 3$$
$$15 = 3 \cdot 5$$

(b)
$$8x = 2 \cdot 2 \cdot 2 \cdot x = 2^3 \cdot x$$
$$10x = 2 \cdot 5 \cdot x$$

Step 2 We find the LCD by taking each different factor the *greatest* number of times it appears as a factor in any of the denominators.

(a) The factor 2 appears three times in one product and not at all in the other, so the greatest number of times 2 appears is three. The greatest number of times both 3 and 5 appear is one.

(b) Here 2 appears three times in one product and once in the other, so the greatest number of times 2 appears is three. The greatest number of times 5 appears is one, and the greatest number of times x appears in either product is one.

Step 3 (a)
$$\text{LCD} = 2 \cdot 2 \cdot 2 \cdot 3 \cdot 5$$
$$= 2^3 \cdot 3 \cdot 5$$
$$= 120$$

(b)
$$\text{LCD} = 2 \cdot 2 \cdot 2 \cdot 5 \cdot x$$
$$= 2^3 \cdot 5 \cdot x$$
$$= 40x$$

Work Problem 1 at the Side.

1 Find the LCD for each pair of fractions.

(a) $\frac{7}{10}, \frac{1}{25}$

(b) $\frac{7}{20p}, \frac{11}{30p}$

(c) $\frac{4}{5x}, \frac{12}{10x}$

ANSWERS

1. (a) 50 **(b)** $60p$ **(c)** $10x$

2 Find the LCD.

(a) $\frac{4}{16m^3n}, \frac{5}{9m^5}$

(b) $\frac{3}{25a^2}, \frac{2}{10a^3b}$

EXAMPLE 2 Finding the LCD

Find the LCD for $\frac{5}{6r^2}$ and $\frac{3}{4r^3}$.

Step 1 Factor each denominator.

$$6r^2 = 2 \cdot 3 \cdot r^2$$
$$4r^3 = 2^2 \cdot r^3$$

Step 2 The greatest number of times 2 appears is two, the greatest number of times 3 appears is one, and the greatest number of times r appears is three; therefore,

Step 3 $\text{LCD} = 2^2 \cdot 3 \cdot r^3 = 12r^3.$

Work Problem 2 at the Side.

3 Find the LCD.

(a) $\frac{7}{3a}, \frac{11}{a^2 - 4a}$

(b) $\frac{2m}{m^2 - 3m + 2}, \frac{5m - 3}{m^2 + 3m - 10}, \frac{4m + 7}{m^2 + 4m - 5}$

(c) $\frac{6}{x - 4}, \frac{3x - 1}{4 - x}$

EXAMPLE 3 Finding the LCD

Find the LCD.

(a) $\frac{6}{5m}, \frac{4}{m^2 - 3m}$

Factor each denominator.

$$5m = 5 \cdot m$$
$$m^2 - 3m = m(m - 3)$$

Use each different factor the greatest number of times it appears.

$$\text{LCD} = 5 \cdot m \cdot (m - 3) = 5m(m - 3)$$

Because m is not a *factor* of $m - 3$, both factors, m and $m - 3$, must appear in the LCD.

(b) $\frac{1}{r^2 - 4r - 5}, \frac{3}{r^2 - r - 20}, \frac{1}{r^2 - 10r + 25}$

$$r^2 - 4r - 5 = (r - 5)(r + 1)$$
$$r^2 - r - 20 = (r - 5)(r + 4)$$
$$r^2 - 10r + 25 = (r - 5)^2$$

Factor each denominator.

Use each different factor the greatest number of times it appears as a factor. The LCD is

$$(r - 5)^2(r + 1)(r + 4).$$

(c) $\frac{1}{q - 5}, \frac{3}{5 - q}$

The expressions $q - 5$ and $5 - q$ are opposites of each other because

$$-(q - 5) = -q + 5 = 5 - q.$$

Therefore, either $q - 5$ or $5 - q$ can be used as the LCD.

Work Problem 3 at the Side.

OBJECTIVE 2 Rewrite rational expressions with given denominators. Once the LCD has been found, the next step in preparing to add or subtract two rational expressions is to use the fundamental property to write equivalent rational expressions. The next example shows how to do this with both numerical and algebraic fractions.

ANSWERS

2. **(a)** $144m^5n$ **(b)** $50a^3b$
3. **(a)** $3a(a - 4)$
 (b) $(m - 1)(m - 2)(m + 5)$
 (c) either $x - 4$ or $4 - x$

EXAMPLE 4 Writing Rational Expressions with Given Denominators

Rewrite each rational expression with the indicated denominator.

(a) $\dfrac{3}{8} = \dfrac{}{40}$ **(b)** $\dfrac{9k}{25} = \dfrac{}{50k}$

For each example, first factor the denominator on the right. Then compare the denominator on the left with the one on the right to decide what factors are missing. (It may be necessary to factor both denominators.)

$$\frac{3}{8} = \frac{}{5 \cdot 8}$$

A factor of 5 is missing. Using the fundamental property, multiply $\frac{3}{8}$ by $\frac{5}{5}$.

$$\frac{3}{8} = \frac{3}{8} \cdot \frac{5}{5} = \frac{15}{40}$$

$\frac{5}{5} = 1$

$$\frac{9k}{25} = \frac{}{25 \cdot 2k}$$

Factors of 2 and k are missing. Multiply by $\frac{2k}{2k}$.

$$\frac{9k}{25} = \frac{9k}{25} \cdot \frac{2k}{2k} = \frac{18k^2}{50k}$$

$\frac{2k}{2k} = 1$

Work Problem 4 at the Side.

4 Rewrite each rational expression with the indicated denominator.

(a) $\dfrac{3}{4} = \dfrac{}{36}$

(b) $\dfrac{7k}{5} = \dfrac{}{30p}$

EXAMPLE 5 Writing Rational Expressions with Given Denominators

Rewrite each rational expression with the indicated denominator.

(a)
$$\frac{8}{3x+1} = \frac{}{12x+4}$$

Factor the denominator on the right.

$$\frac{8}{3x+1} = \frac{}{4(3x+1)} \qquad \text{Factor.}$$

The missing factor is 4, so multiply the fraction on the left by $\frac{4}{4}$.

$$\frac{8}{3x+1} \cdot \frac{4}{4} = \frac{32}{12x+4} \qquad \text{Fundamental property}$$

(b)
$$\frac{12p}{p^2+8p} = \frac{}{p^3+4p^2-32p}$$

Factor $p^2 + 8p$ as $p(p+8)$. Compare with the denominator on the right, which factors as $p(p+8)(p-4)$. The factor $p-4$ is missing, so multiply $\frac{12p}{p(p+8)}$ by $\frac{p-4}{p-4}$.

$$\frac{12p}{p^2+8p} = \frac{12p}{p(p+8)} \cdot \frac{p-4}{p-4} \qquad \text{Fundamental property}$$

$$= \frac{12p(p-4)}{p(p+8)(p-4)} \qquad \text{Multiplication of rational expressions}$$

$$= \frac{12p^2-48p}{p^3+4p^2-32p} \qquad \text{Multiply the factors.}$$

ANSWERS

4. (a) $\dfrac{27}{36}$ (b) $\dfrac{42kp}{30p}$

5 Rewrite each rational expression with the indicated denominator.

(a) $\dfrac{9}{2a + 5} = \dfrac{}{6a + 15}$

(b) $\dfrac{5k + 1}{k^2 + 2k} = \dfrac{}{k^3 + k^2 - 2k}$

NOTE

In the next section we add and subtract rational expressions, which sometimes requires the steps illustrated in Examples 4 and 5. While it is beneficial to leave the denominator in factored form, we multiplied the factors in the denominator in Example 5(b),

$$\frac{12p(p - 4)}{p(p + 8)(p - 4)},$$

to give the answer,

$$\frac{12p^2 - 48p}{p^3 + 4p^2 - 32p},$$

in the same form as the original problem.

Work Problem 5 at the Side.

Answers

5. (a) $\dfrac{27}{6a + 15}$ (b) $\dfrac{(5k + 1)(k - 1)}{k^3 + k^2 - 2k}$

7.3 Exercises

FOR EXTRA HELP

Addison-Wesley Math Tutor Center

MathXL

Digital Video Tutor CD 6 Videotape 7

Student's Solutions Manual

MyMathLab

Interactmath.com

Choose the correct response in Exercises 1–4.

1. Suppose that the greatest common factor of a and b is 1. Then the least common denominator for $\frac{1}{a}$ and $\frac{1}{b}$ is

A. a **B.** b **C.** ab **D.** 1.

2. If a is a factor of b, then the least common denominator for $\frac{1}{a}$ and $\frac{1}{b}$ is

A. a **B.** b **C.** ab **D.** 1.

3. The least common denominator for $\frac{11}{20}$ and $\frac{1}{2}$ is

A. 40 **B.** 2 **C.** 20 **D.** none of these.

4. Suppose that we wish to write the fraction $\dfrac{1}{(x-4)^2(y-3)}$ with denominator $(x-4)^3(y-3)^2$. We must multiply both the numerator and the denominator by

A. $(x-4)(y-3)$ **B.** $(x-4)^2$

C. $x-4$ **D.** $(x-4)^2(y-3)$.

Find the least common denominator for the fractions in each list. See Examples 1–3.

5. $\dfrac{2}{15}, \dfrac{3}{10}, \dfrac{7}{30}$

6. $\dfrac{5}{24}, \dfrac{7}{12}, \dfrac{9}{28}$

7. $\dfrac{3}{x^4}, \dfrac{5}{x^7}$

8. $\dfrac{2}{y^5}, \dfrac{3}{y^6}$

9. $\dfrac{5}{36q}, \dfrac{17}{24q}$

10. $\dfrac{4}{30p}, \dfrac{9}{50p}$

11. $\dfrac{6}{21r^3}, \dfrac{8}{12r^5}$

12. $\dfrac{9}{35t^2}, \dfrac{5}{49t^6}$

13. If the denominators of two fractions in prime factored form are $2^3 \cdot 3$ and $2^2 \cdot 5$, what is the factored form of their LCD?

14. Suppose two rational expressions have denominators $(t+4)^3(t-3)$ and $(t+4)^2(t+8)$. Find the factored form of their LCD. What is the similarity between the answers for this problem and for Exercise 13?

15. If two denominators have greatest common factor equal to 1, how can you easily find their least common denominator?

16. Suppose two fractions have denominators a^k and a^r, where k and r are natural numbers, with $k > r$. What is their least common denominator?

Find the least common denominator for each group of fractions. See Examples 1–3.

17. $\dfrac{9}{28m^2}, \dfrac{3}{12m-20}$

18. $\dfrac{15}{27a^3}, \dfrac{8}{9a-45}$

19. $\dfrac{7}{5b-10}, \dfrac{11}{6b-12}$

20. $\dfrac{3}{7x^2+21x}, \dfrac{1}{5x^2+15x}$

21. $\dfrac{5}{c-d}, \dfrac{8}{d-c}$

22. $\dfrac{4}{y-x}, \dfrac{7}{x-y}$

23. $\dfrac{3}{k^2 + 5k}, \dfrac{2}{k^2 + 3k - 10}$

24. $\dfrac{1}{z^2 - 4z}, \dfrac{4}{z^2 - 3z - 4}$

25. $\dfrac{5}{p^2 + 8p + 15}, \dfrac{3}{p^2 - 3p - 18}, \dfrac{2}{p^2 - p - 30}$

26. $\dfrac{10}{y^2 - 10y + 21}, \dfrac{2}{y^2 - 2y - 3}, \dfrac{5}{y^2 - 6y - 7}$

Rewrite each rational expression with the given denominator. See Examples 4 and 5.

27. $\dfrac{4}{11} = \dfrac{}{55}$

28. $\dfrac{6}{7} = \dfrac{}{42}$

29. $\dfrac{-5}{k} = \dfrac{}{9k}$

30. $\dfrac{-3}{q} = \dfrac{}{6q}$

31. $\dfrac{13}{40y} = \dfrac{}{80y^3}$

32. $\dfrac{5}{27p} = \dfrac{}{108p^4}$

33. $\dfrac{5t^2}{6r} = \dfrac{}{42r^4}$

34. $\dfrac{8y^2}{3x} = \dfrac{}{30x^3}$

35. $\dfrac{5}{2(m + 3)} = \dfrac{}{8(m + 3)}$

36. $\dfrac{7}{4(y - 1)} = \dfrac{}{16(y - 1)}$

37. $\dfrac{-4t}{3t - 6} = \dfrac{}{12 - 6t}$

38. $\dfrac{-7k}{5k - 20} = \dfrac{}{40 - 10k}$

39. $\dfrac{14}{z^2 - 3z} = \dfrac{}{z(z - 3)(z - 2)}$

40. $\dfrac{12}{x(x + 4)} = \dfrac{}{x(x + 4)(x - 9)}$

41. $\dfrac{2(b - 1)}{b^2 + b} = \dfrac{}{b^3 + 3b^2 + 2b}$

42. $\dfrac{3(c + 2)}{c(c - 1)} = \dfrac{}{c^3 - 5c^2 + 4c}$

7.4 Adding and Subtracting Rational Expressions

OBJECTIVES

1. Add rational expressions having the same denominator.
2. Add rational expressions having different denominators.
3. Subtract rational expressions.

To add and subtract rational expressions, we find least common denominators and write equivalent fractions with the LCD.

OBJECTIVE 1 Add rational expressions having the same denominator. We find the sum of two rational expressions with the same procedure that we used for adding two fractions in **Section R.1.**

Adding Rational Expressions

If $\frac{P}{Q}$ and $\frac{R}{Q}$ $(Q \neq 0)$ are rational expressions, then

$$\frac{P}{Q} + \frac{R}{Q} = \frac{P+R}{Q}.$$

In words: To add rational expressions with the same denominator, add the numerators and keep the same denominator.

The first example shows how the addition of rational expressions compares with that of rational numbers.

EXAMPLE 1 Adding Rational Expressions with the Same Denominator

Add. Write each answer in lowest terms.

(a) $\frac{4}{9} + \frac{2}{9}$ (b) $\frac{3x}{x+1} + \frac{3}{x+1}$

The denominators are the same, so the sum is found by adding the two numerators and keeping the same (common) denominator.

$$\frac{4}{9} + \frac{2}{9} = \frac{4+2}{9} = \frac{6}{9} = \frac{2}{3}$$

$$\frac{3x}{x+1} + \frac{3}{x+1} = \frac{3x+3}{x+1} = \frac{3(x+1)}{x+1} = 3$$

Work Problem 1 at the Side.

1 Add. Write each answer in lowest terms.

(a) $\frac{7}{15} + \frac{3}{15}$

(b) $\frac{3}{y+4} + \frac{2}{y+4}$

(c) $\frac{x}{x+y} + \frac{1}{x+y}$

(d) $\frac{a}{a+b} + \frac{b}{a+b}$

(e) $\frac{x^2}{x+1} + \frac{x}{x+1}$

OBJECTIVE 2 Add rational expressions having different denominators. We use the following steps to add two rational expressions with different denominators.

Adding with Different Denominators

Step 1 **Find the least common denominator (LCD).**

Step 2 **Rewrite each rational expression** as an equivalent rational expression with the LCD as the denominator.

Step 3 **Add** the numerators to get the numerator of the sum. The LCD is the denominator of the sum.

Step 4 **Write in lowest terms** using the fundamental property.

ANSWERS

1. (a) $\frac{2}{3}$ (b) $\frac{5}{y+4}$ (c) $\frac{x+1}{x+y}$ (d) 1 (e) x

2 Add. Write each answer in lowest terms.

(a) $\frac{1}{10}+\frac{1}{15}$

(b) $\frac{6}{5x}+\frac{9}{2x}$

(c) $\frac{m}{3n}+\frac{2}{7n}$

EXAMPLE 2 Adding Rational Expressions with Different Denominators

Add. Write each answer in lowest terms.

(a) $\frac{1}{12}+\frac{7}{15}$ **(b)** $\frac{2}{3y}+\frac{1}{4y}$

Step 1 First find the LCD using the methods of the previous section.

$$\text{LCD} = 2^2 \cdot 3 \cdot 5 = 60 \qquad \text{LCD} = 2^2 \cdot 3 \cdot y = 12y$$

Step 2 Now rewrite each rational expression as a fraction with the LCD (either 60 or $12y$) as the denominator.

$$\frac{1}{12}+\frac{7}{15}=\frac{1(5)}{12(5)}+\frac{7(4)}{15(4)} = \frac{5}{60}+\frac{28}{60}$$

$$\frac{2}{3y}+\frac{1}{4y}=\frac{2(4)}{3y(4)}+\frac{1(3)}{4y(3)} = \frac{8}{12y}+\frac{3}{12y}$$

Step 3 Since the fractions now have common denominators, add the numerators.

Step 4 Write in lowest terms if necessary.

$$\frac{5}{60}+\frac{28}{60}=\frac{5+28}{60} = \frac{33}{60}=\frac{11}{20}$$

$$\frac{8}{12y}+\frac{3}{12y}=\frac{8+3}{12y} = \frac{11}{12y}$$

Work Problem 2 at the Side.

EXAMPLE 3 Adding Rational Expressions

Add. Write the answer in lowest terms.

$$\frac{2x}{x^2-1}+\frac{-1}{x+1}$$

Step 1 Since the denominators are different, find the LCD.

$$x^2-1=(x+1)(x-1)$$

$$x+1 \text{ is prime.}$$

The LCD is $(x+1)(x-1)$.

Step 2 Rewrite each rational expression as a fraction with common denominator $(x+1)(x-1)$.

$$\frac{2x}{x^2-1}+\frac{-1}{x+1}=\frac{2x}{(x+1)(x-1)}+\frac{-1(x-1)}{(x+1)(x-1)}$$ Multiply the second fraction by $\frac{x-1}{x-1}$.

$$=\frac{2x}{(x+1)(x-1)}+\frac{-x+1}{(x+1)(x-1)}$$ Distributive property

Step 3 $$=\frac{2x-x+1}{(x+1)(x-1)}$$ Add numerators; keep the same denominator.

$$=\frac{x+1}{(x+1)(x-1)}$$ Combine like terms in the numerator.

Continued on Next Page

ANSWERS

2. (a) $\frac{1}{6}$ (b) $\frac{57}{10x}$ (c) $\frac{7m+6}{21n}$

Step 4 $= \dfrac{1(x+1)}{(x+1)(x-1)}$ Identity property for multiplication

$= \dfrac{1}{x-1}$ Fundamental property

Work Problem 3 at the Side.

EXAMPLE 4 Adding Rational Expressions

Add. Write the answer in lowest terms.

$$\frac{2x}{x^2+5x+6} + \frac{x+1}{x^2+2x-3}$$

$$= \frac{2x}{(x+2)(x+3)} + \frac{x+1}{(x+3)(x-1)} \quad \text{Factor the denominators.}$$

The LCD is $(x+2)(x+3)(x-1)$. Use the fundamental property.

$$= \frac{2x(x-1)}{(x+2)(x+3)(x-1)} + \frac{(x+1)(x+2)}{(x+2)(x+3)(x-1)}$$

$$= \frac{2x(x-1)+(x+1)(x+2)}{(x+2)(x+3)(x-1)} \quad \text{Add numerators; keep the same denominator.}$$

$$= \frac{2x^2-2x+x^2+3x+2}{(x+2)(x+3)(x-1)} \quad \text{Multiply.}$$

$$= \frac{3x^2+x+2}{(x+2)(x+3)(x-1)} \quad \text{Combine like terms.}$$

It is usually more convenient to leave the denominator in factored form. The numerator cannot be factored here, so the expression is in lowest terms.

Work Problem 4 at the Side.

EXAMPLE 5 Adding Rational Expressions with Denominators That Are Opposites

Add. Write the answer in lowest terms.

$$\frac{y}{y-2} + \frac{8}{2-y}$$

One way to get a common denominator is to multiply the second expression by -1 in both the numerator and the denominator, giving $y-2$ as a common denominator.

$$\frac{y}{y-2} + \frac{8}{2-y} = \frac{y}{y-2} + \frac{8(-1)}{(2-y)(-1)} \quad \text{Fundamental property}$$

$$= \frac{y}{y-2} + \frac{-8}{y-2} \quad \text{Distributive property}$$

$$= \frac{y-8}{y-2} \quad \text{Add numerators; keep the same denominator.}$$

If we had chosen to use $2-y$ as the common denominator, the final answer would be in the form $\frac{8-y}{2-y}$, which is equivalent to $\frac{y-8}{y-2}$.

Work Problem 5 at the Side.

3 Add. Write each answer in lowest terms.

(a) $\dfrac{2p}{3p+3} + \dfrac{5p}{2p+2}$

(b) $\dfrac{4}{y^2-1} + \dfrac{6}{y+1}$

(c) $\dfrac{-2}{p+1} + \dfrac{4p}{p^2-1}$

4 Add. Write each answer in lowest terms.

(a) $\dfrac{2k}{k^2-5k+4} + \dfrac{3}{k^2-1}$

(b) $\dfrac{4m}{m^2+3m+2} + \dfrac{2m-1}{m^2+6m+5}$

5 Add. Write the answer in lowest terms.

$\dfrac{m}{2m-3n} + \dfrac{n}{3n-2m}$

ANSWERS

3. (a) $\dfrac{19p}{6(p+1)}$ **(b)** $\dfrac{2(3y-1)}{(y+1)(y-1)}$ **(c)** $\dfrac{2}{p-1}$

4. (a) $\dfrac{(2k-3)(k+4)}{(k-4)(k-1)(k+1)}$ **(b)** $\dfrac{6m^2+23m-2}{(m+2)(m+1)(m+5)}$

5. $\dfrac{m-n}{2m-3n}$ or $\dfrac{n-m}{3n-2m}$

6 Subtract. Write each answer in lowest terms.

(a) $\dfrac{3}{m^2} - \dfrac{2}{m^2}$

(b) $\dfrac{x}{2x+3} - \dfrac{3x+4}{2x+3}$

OBJECTIVE 3 Subtract rational expressions. To subtract rational expressions, use the following rule.

Subtracting Rational Expressions

If $\frac{P}{Q}$ and $\frac{R}{Q}$ $(Q \neq 0)$ are rational expressions, then

$$\frac{P}{Q} - \frac{R}{Q} = \frac{P-R}{Q}.$$

In words: To subtract rational expressions with the same denominator, subtract the numerators and keep the same denominator.

EXAMPLE 6 Subtracting Rational Expressions with the Same Denominator

Subtract. Write the answer in lowest terms.

Use parentheses around the quantity being subtracted.

$$\frac{2m}{m-1} - \frac{m+3}{m-1} = \frac{2m-(m+3)}{m-1}$$ Subtract numerators; keep the same denominator.

$$= \frac{2m-m-3}{m-1}$$ Distributive property

$$= \frac{m-3}{m-1}$$ Combine like terms.

CAUTION

Sign errors often occur in subtraction problems like the one in Example 6. Remember that the numerator of the fraction being subtracted must be treated as a single quantity. ***Be sure to use parentheses after the subtraction sign*** to avoid this common error.

Work Problem 6 at the Side.

EXAMPLE 7 Subtracting Rational Expressions with Different Denominators

Subtract. Write the answer in lowest terms.

$$\frac{9}{x-2} - \frac{3}{x}$$

The LCD is $x(x-2)$.

$$\frac{9}{x-2} - \frac{3}{x} = \frac{9x}{x(x-2)} - \frac{3(x-2)}{x(x-2)}$$ Rewrite each expression with the LCD.

$$= \frac{9x-3(x-2)}{x(x-2)}$$ Subtract numerators; keep the same denominator.

Continued on Next Page

ANSWERS

6. (a) $\dfrac{1}{m^2}$ (b) $\dfrac{-2(x+2)}{2x+3}$

$$= \frac{9x - 3x + 6}{x(x - 2)}$$ Distributive property; Be careful with signs.

$$= \frac{6x + 6}{x(x - 2)}$$ Combine like terms.

$$= \frac{6(x + 1)}{x(x - 2)}$$ Factor.

NOTE

We factor the final numerator in Example 7 to get an answer in the form $\frac{6(x+1)}{x(x-2)}$. The fundamental property does not apply, since there are no common factors. The answer is in lowest terms.

Work Problem 7 at the Side.

EXAMPLE 8 Subtracting Rational Expressions with Denominators That Are Opposites

Subtract. Write the answer in lowest terms.

$$\frac{3x}{x - 5} - \frac{2x - 25}{5 - x}$$

The denominators are opposites, so either may be used as the common denominator. We will choose $x - 5$.

$$\frac{3x}{x - 5} - \frac{2x - 25}{5 - x} = \frac{3x}{x - 5} - \frac{2x - 25}{5 - x} \cdot \frac{-1}{-1}$$ Fundamental property

$$= \frac{3x}{x - 5} - \frac{-2x + 25}{x - 5}$$ Multiply.

$$= \frac{3x - (-2x + 25)}{x - 5}$$ Subtract numerators; use parentheses.

$$= \frac{3x + 2x - 25}{x - 5}$$ Distributive property

$$= \frac{5x - 25}{x - 5}$$ Combine like terms.

$$= \frac{5(x - 5)}{x - 5}$$ Factor.

$$= 5$$ Lowest terms

Work Problem 8 at the Side.

EXAMPLE 9 Subtracting Rational Expressions

Subtract. Write the answer in lowest terms.

$$\frac{6x}{x^2 - 2x + 1} - \frac{1}{x^2 - 1}$$

Begin by factoring the denominators.

$$x^2 - 2x + 1 = (x - 1)^2 \quad \text{and} \quad x^2 - 1 = (x - 1)(x + 1)$$

Continued on Next Page

7 Subtract. Write each answer in lowest terms.

(a) $\frac{1}{k + 4} - \frac{2}{k}$

(b) $\frac{6}{a + 2} - \frac{1}{a - 3}$

8 Subtract. Write each answer in lowest terms.

(a) $\frac{5}{x - 1} - \frac{3x}{1 - x}$

(b) $\frac{2y}{y - 2} - \frac{1 + y}{2 - y}$

ANSWERS

7. (a) $\frac{-k - 8}{k(k + 4)}$ **(b)** $\frac{5(a - 4)}{(a + 2)(a - 3)}$

8. (a) $\frac{5 + 3x}{x - 1}$ **(b)** $\frac{3y + 1}{y - 2}$

From the factored denominators, identify the LCD,

$$(x-1)^2(x+1).$$

Use the factor $x-1$ twice because it appears twice in the first denominator.

$$\frac{6x}{(x-1)^2} - \frac{1}{(x-1)(x+1)}$$

$$= \frac{6x(x+1)}{(x-1)^2(x+1)} - \frac{1(\mathbf{x-1})}{(\mathbf{x-1})(x-1)(x+1)} \quad \text{Fundamental property}$$

$$= \frac{6x(x+1) - 1(x-1)}{(x-1)^2(x+1)} \quad \text{Subtract numerators.}$$

$$= \frac{6x^2 + 6x - x + 1}{(x-1)^2(x+1)} \quad \text{Distributive property}$$

$$= \frac{6x^2 + 5x + 1}{(x-1)^2(x+1)} \text{ or } \frac{(2x+1)(3x+1)}{(x-1)^2(x+1)} \quad \text{Combine like terms.}$$

Verify that the final expression is in lowest terms.

Work Problem 9 at the Side.

9 Subtract. Write each answer in lowest terms.

(a) $\frac{4y}{y^2-1} - \frac{5}{y^2+2y+1}$

(b) $\frac{3r}{r^2-5r} - \frac{4}{r^2-10r+25}$

EXAMPLE 10 Subtracting Rational Expressions

Subtract. Write the answer in lowest terms.

$$\frac{q}{q^2-4q-5} - \frac{3}{2q^2-13q+15}$$

To find the LCD, factor each denominator.

$$q^2-4q-5 = (q+1)(q-5)$$
$$2q^2-13q+15 = (q-5)(2q-3)$$

The LCD is $(q+1)(q-5)(2q-3)$. Rewrite each rational expression with the LCD, using the fundamental property.

$$\frac{q}{(q+1)(q-5)} - \frac{3}{(q-5)(2q-3)}$$

$$= \frac{q(\mathbf{2q-3})}{(q+1)(q-5)(\mathbf{2q-3})} - \frac{3(\mathbf{q+1})}{(\mathbf{q+1})(q-5)(2q-3)}$$

$$= \frac{q(2q-3) - 3(q+1)}{(q+1)(q-5)(2q-3)} \quad \text{Subtract numerators.}$$

$$= \frac{2q^2-3q-3q-3}{(q+1)(q-5)(2q-3)} \quad \text{Distributive property}$$

$$= \frac{2q^2-6q-3}{(q+1)(q-5)(2q-3)} \quad \text{Combine like terms.}$$

Verify that the final expression is in lowest terms.

Work Problem 10 at the Side.

10 Subtract. Write each answer in lowest terms.

(a) $\frac{2}{p^2-5p+4} - \frac{3}{p^2-1}$

(b) $\frac{q}{2q^2+5q-3} - \frac{3q+4}{3q^2+10q+3}$

Answers

9. (a) $\frac{4y^2-y+5}{(y+1)^2(y-1)}$

(b) $\frac{3r-19}{(r-5)^2}$

10. (a) $\frac{14-p}{(p-4)(p-1)(p+1)}$

(b) $\frac{-3q^2-4q+4}{(2q-1)(q+3)(3q+1)}$

7.4 Exercises

FOR EXTRA HELP

Addison-Wesley Math Tutor Center | MathXL | Digital Video Tutor CD 7 Videotape 7 | Student's Solutions Manual | MyMathLab | Interactmath.com

Match the expression in Column I with the correct sum or difference in Column II.

Study Skills Workbook **Activity 11: Mind Maps**

I	II
1. $\frac{x}{x+6}+\frac{6}{x+6}$	**A.** 2
2. $\frac{2x}{x-6}-\frac{12}{x-6}$	**B.** $\frac{x-6}{x+6}$
3. $\frac{6}{x-6}-\frac{x}{x-6}$	**C.** -1
4. $\frac{6}{x+6}-\frac{x}{x+6}$	**D.** $\frac{6+x}{6x}$
5. $\frac{x}{x+6}-\frac{6}{x+6}$	**E.** 1
6. $\frac{1}{x}+\frac{1}{6}$	**F.** 0
7. $\frac{1}{6}-\frac{1}{x}$	**G.** $\frac{x-6}{6x}$
8. $\frac{1}{6x}-\frac{1}{6x}$	**H.** $\frac{6-x}{x+6}$

Note: When adding and subtracting rational expressions, several different equivalent forms of the answer often exist. If your answer does not look exactly like the one given in the back of the book, check to see whether you have written an equivalent form.

Add or subtract. Write each answer in lowest terms. See Examples 1 and 6.

9. $\frac{4}{m}+\frac{7}{m}$

10. $\frac{5}{p}+\frac{11}{p}$

11. $\frac{a+b}{2}-\frac{a-b}{2}$

12. $\frac{x-y}{2}-\frac{x+y}{2}$

13. $\frac{x^2}{x+5}+\frac{5x}{x+5}$

14. $\frac{t^2}{t-3}+\frac{-3t}{t-3}$

15. $\frac{y^2-3y}{y+3}+\frac{-18}{y+3}$

16. $\frac{r^2-8r}{r-5}+\frac{15}{r-5}$

17. Explain with an example how to add or subtract rational expressions with the same denominator.

18. Explain with an example how to add or subtract rational expressions with different denominators.

Add or subtract. Write each answer in lowest terms. See Examples 2, 3, 4, and 7.

19. $\frac{z}{5} + \frac{1}{3}$

20. $\frac{p}{8} + \frac{3}{5}$

21. $\frac{5}{7} - \frac{r}{2}$

22. $\frac{10}{9} - \frac{z}{3}$

23. $-\frac{3}{4} - \frac{1}{2x}$

24. $-\frac{5}{8} - \frac{3}{2a}$

25. $\frac{x+1}{6} + \frac{3x+3}{9}$

26. $\frac{2x-6}{4} + \frac{x+5}{6}$

27. $\frac{x+3}{3x} + \frac{2x+2}{4x}$

28. $\frac{x+2}{5x} + \frac{6x+3}{3x}$

29. $\frac{2}{x+3} + \frac{1}{x}$

30. $\frac{3}{x-4} + \frac{2}{x}$

31. $\frac{x}{x-2} + \frac{4}{x+2}$

32. $\frac{2x}{x-1} + \frac{3}{x+1}$

33. $\frac{t}{t+2} + \frac{5-t}{t} - \frac{4}{t^2+2t}$

34. $\frac{2p}{p-3} + \frac{2+p}{p} - \frac{-6}{p^2-3p}$

35. What are the two possible LCDs that could be used for the sum

$$\frac{10}{m-2}+\frac{5}{2-m}?$$

36. If one form of the correct answer to a sum or difference of rational expressions is $\frac{4}{k-3}$, what would be an alternative form of the answer if the denominator is $3-k$?

Add or subtract. Write each answer in lowest terms. See Examples 5 and 8.

37. $\frac{4}{x-5}+\frac{6}{5-x}$

38. $\frac{10}{m-2}+\frac{5}{2-m}$

39. $\frac{-1}{1-y}+\frac{3-4y}{y-1}$

40. $\frac{-4}{p-3}-\frac{p+1}{3-p}$

41. $\frac{2}{x-y^2}+\frac{7}{y^2-x}$

42. $\frac{-8}{p-q^2}+\frac{3}{q^2-p}$

43. $\frac{x}{5x-3y}-\frac{y}{3y-5x}$

44. $\frac{t}{8t-9s}-\frac{s}{9s-8t}$

45. $\frac{3}{4p-5}+\frac{9}{5-4p}$

46. $\frac{8}{3-7y}-\frac{2}{7y-3}$

In each subtraction problem, the rational expression that follows the subtraction sign has a numerator with more than one term. Be very careful with signs and find each difference. See Examples 6–10.

47. $\frac{2m}{m-n}-\frac{5m+n}{2m-2n}$

48. $\frac{5p}{p-q}-\frac{3p+1}{4p-4q}$

49. $\dfrac{5}{x^2 - 9} - \dfrac{x + 2}{x^2 + 4x + 3}$

50. $\dfrac{1}{a^2 - 1} - \dfrac{a - 1}{a^2 + 3a - 4}$

51. $\dfrac{2q + 1}{3q^2 + 10q - 8} - \dfrac{3q + 5}{2q^2 + 5q - 12}$

52. $\dfrac{4y - 1}{2y^2 + 5y - 3} - \dfrac{y + 3}{6y^2 + y - 2}$

Perform the indicated operations. See Examples 1–10.

53. $\dfrac{4}{r^2 - r} + \dfrac{6}{r^2 + 2r} - \dfrac{1}{r^2 + r - 2}$

54. $\dfrac{6}{k^2 + 3k} - \dfrac{1}{k^2 - k} + \dfrac{2}{k^2 + 2k - 3}$

55. $\dfrac{x + 3y}{x^2 + 2xy + y^2} + \dfrac{x - y}{x^2 + 4xy + 3y^2}$

56. $\dfrac{m}{m^2 - 1} + \dfrac{m - 1}{m^2 + 2m + 1}$

57. $\dfrac{r + y}{18r^2 + 9ry - 2y^2} + \dfrac{3r - y}{36r^2 - y^2}$

58. $\dfrac{2x - z}{2x^2 + xz - 10z^2} - \dfrac{x + z}{x^2 - 4z^2}$

59. Refer to the rectangle in the figure.

(a) Find an expression that represents its perimeter. Give the simplified form.

(b) Find an expression that represents its area. Give the simplified form.

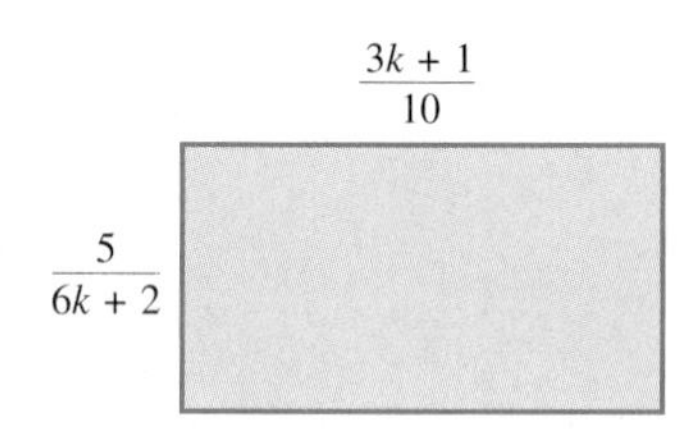

60. Refer to the triangle in the figure. Find an expression that represents its perimeter.

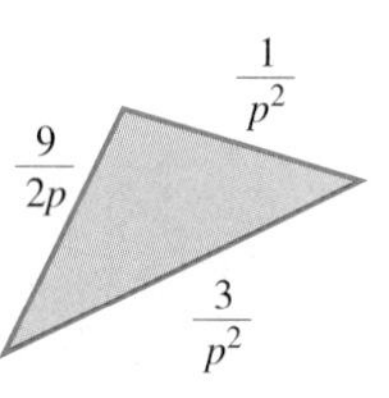

7.5 Complex Fractions

OBJECTIVES

1. Simplify a complex fraction by writing it as a division problem (Method 1).
2. Simplify a complex fraction by multiplying numerator and denominator by the least common denominator (Method 2).

The quotient of two mixed numbers in arithmetic, such as $2\frac{1}{2} \div 3\frac{1}{4}$, can be written as a fraction:

$$2\frac{1}{2} \div 3\frac{1}{4} = \frac{2\frac{1}{2}}{3\frac{1}{4}} = \frac{2 + \frac{1}{2}}{3 + \frac{1}{4}}.$$

The last expression is the quotient of expressions that involve fractions. In algebra, some rational expressions also have fractions in the numerator, or denominator, or both.

Complex Fraction

A rational expression with one or more fractions in the numerator, denominator, or both, is called a **complex fraction.**

Examples of complex fractions include

$$\frac{2 + \frac{1}{2}}{3 + \frac{1}{4}}, \quad \frac{\frac{3x^2 - 5x}{6x^2}}{2x - \frac{1}{x}}, \quad \text{and} \quad \frac{3 + x}{5 - \frac{2}{x}}. \qquad \text{Complex fractions}$$

The parts of a complex fraction are named as follows.

$$\frac{\frac{2}{p} - \frac{1}{q}}{\frac{3}{p} + \frac{5}{q}}$$

- $\frac{2}{p} - \frac{1}{q}$ ← Numerator of complex fraction
- ← Main fraction bar
- $\frac{3}{p} + \frac{5}{q}$ ← Denominator of complex fraction

OBJECTIVE 1 Simplify a complex fraction by writing it as a division problem (Method 1). Since the main fraction bar represents division in a complex fraction, one method of simplifying a complex fraction involves division.

Method 1

To simplify a complex fraction:

Step 1 Write both the numerator and denominator as single fractions.

Step 2 Change the complex fraction to a division problem.

Step 3 Perform the indicated division.

Once again, the first example shows complex fractions from both arithmetic and algebra.

1 Simplify each complex fraction using Method 1.

(a) $\dfrac{\frac{2}{5}+\frac{1}{4}}{\frac{1}{2}+\frac{1}{3}}$

(b) $\dfrac{6+\frac{1}{x}}{5-\frac{2}{x}}$

(c) $\dfrac{9-\frac{4}{p}}{\frac{2}{p}+1}$

2 Simplify each complex fraction using Method 1.

(a) $\dfrac{\frac{rs^2}{t}}{\frac{r^2s}{t^2}}$

(b) $\dfrac{\frac{m^2n^3}{p}}{\frac{m^4n}{p^2}}$

EXAMPLE 1 Simplifying Complex Fractions (Method 1)

Simplify each complex fraction.

(a) $\dfrac{\frac{2}{3}+\frac{5}{9}}{\frac{1}{4}+\frac{1}{12}}$ (b) $\dfrac{6+\frac{3}{x}}{\frac{x}{4}+\frac{1}{8}}$

Step 1 First, write each numerator as a single fraction.

$$\frac{2}{3}+\frac{5}{9}=\frac{2(3)}{3(3)}+\frac{5}{9}=\frac{6}{9}+\frac{5}{9}=\frac{11}{9}$$

$$6+\frac{3}{x}=\frac{6}{1}+\frac{3}{x}=\frac{6x}{x}+\frac{3}{x}=\frac{6x+3}{x}$$

Do the same thing with each denominator.

$$\frac{1}{4}+\frac{1}{12}=\frac{1(3)}{4(3)}+\frac{1}{12}=\frac{3}{12}+\frac{1}{12}=\frac{4}{12}$$

$$\frac{x}{4}+\frac{1}{8}=\frac{x(2)}{4(2)}+\frac{1}{8}=\frac{2x}{8}+\frac{1}{8}=\frac{2x+1}{8}$$

Step 2 The original complex fraction can now be written as follows.

$$\frac{\frac{11}{9}}{\frac{4}{12}} \qquad \frac{\frac{6x+3}{x}}{\frac{2x+1}{8}}$$

Step 3 Now use the definition of division and multiply by the reciprocal. Then write in lowest terms using the fundamental property.

$$\frac{11}{9}\div\frac{4}{12}=\frac{11}{9}\cdot\frac{12}{4}=\frac{11\cdot 3\cdot 4}{3\cdot 3\cdot 4}=\frac{11}{3}$$

$$\frac{6x+3}{x}\div\frac{2x+1}{8}=\frac{6x+3}{x}\cdot\frac{8}{2x+1}=\frac{3(2x+1)}{x}\cdot\frac{8}{2x+1}=\frac{24}{x}$$

Work Problem 1 at the Side.

EXAMPLE 2 Simplifying a Complex Fraction (Method 1)

Simplify the complex fraction.

$$\frac{\frac{xp}{q^3}}{\frac{p^2}{qx^2}}$$

Here the numerator and denominator are already single fractions, so use the definition of division and then the fundamental property.

$$\frac{xp}{q^3}\div\frac{p^2}{qx^2}=\frac{xp}{q^3}\cdot\frac{qx^2}{p^2}=\frac{x^3}{q^2p}$$

Work Problem 2 at the Side.

ANSWERS

1. (a) $\frac{39}{50}$ (b) $\frac{6x+1}{5x-2}$ (c) $\frac{9p-4}{2+p}$

2. (a) $\frac{st}{r}$ (b) $\frac{n^2p}{m^2}$

EXAMPLE 3 Simplifying a Complex Fraction (Method 1)

Simplify the complex fraction.

$$\frac{\frac{3}{x+2}-4}{\frac{2}{x+2}+1} = \frac{\frac{3}{x+2}-\frac{4(x+2)}{x+2}}{\frac{2}{x+2}+\frac{1(x+2)}{x+2}}$$ Write both second terms with a denominator of $x + 2$.

$$= \frac{\frac{3-4(x+2)}{x+2}}{\frac{2+1(x+2)}{x+2}}$$ Subtract in the numerator. Add in the denominator.

$$= \frac{\frac{3-4x-8}{x+2}}{\frac{2+x+2}{x+2}}$$ Distributive property

$$= \frac{\frac{-5-4x}{x+2}}{\frac{4+x}{x+2}}$$ Combine like terms.

$$= \frac{-5-4x}{x+2} \cdot \frac{x+2}{4+x}$$ Multiply by the reciprocal.

$$= \frac{-5-4x}{4+x}$$ Lowest terms

CAUTION
Be aware that

$$\frac{\frac{a}{b}+\frac{c}{d}}{\frac{e}{f}+\frac{g}{h}} \neq \left(\frac{a}{b}+\frac{c}{d}\right)\cdot\left(\frac{f}{e}+\frac{h}{g}\right).$$

Work Problem 3 at the Side.

OBJECTIVE 2 Simplify a complex fraction by multiplying numerator and denominator by the least common denominator (Method 2). Since any expression can be multiplied by a form of 1 to get an equivalent expression, we may multiply both the numerator and the denominator of a complex fraction by the same nonzero expression to get an equivalent complex fraction. If we choose the expression to be the LCD of all the fractions within the complex fraction, the complex fraction will be simplified. This is Method 2.

3 Simplify by Method 1.

$$\frac{\frac{2}{x-1}+\frac{1}{x+1}}{\frac{3}{x-1}-\frac{4}{x+1}}$$

ANSWERS

3. $\frac{3x+1}{-x+7}$

4 Simplify by Method 2.

(a) $\dfrac{\frac{2}{3}-\frac{1}{4}}{\frac{4}{9}+\frac{1}{2}}$

(b) $\dfrac{2-\frac{6}{a}}{3+\frac{4}{a}}$

(c) $\dfrac{\frac{p}{5-p}}{\frac{4p}{2p+1}}$

Method 2

To simplify a complex fraction:

Step 1 Find the LCD of all fractions within the complex fraction.

Step 2 Multiply both the numerator and the denominator of the complex fraction by this LCD using the distributive property as necessary. Write in lowest terms.

In the next example, Method 2 is used to simplify the complex fractions from Example 1.

EXAMPLE 4 Simplifying Complex Fractions (Method 2)

Simplify each complex fraction.

(a) $\dfrac{\frac{2}{3}+\frac{5}{9}}{\frac{1}{4}+\frac{1}{12}}$ **(b)** $\dfrac{6+\frac{3}{x}}{\frac{x}{4}+\frac{1}{8}}$

Step 1 Find the LCD for all denominators in the complex fraction.

The LCD for 3, 9, 4, and 12 is 36. | The LCD for x, 4, and 8 is $8x$.

Step 2 Multiply the numerator and denominator of the complex fraction by the LCD.

$$\frac{\frac{2}{3}+\frac{5}{9}}{\frac{1}{4}+\frac{1}{12}}=\frac{36\left(\frac{2}{3}+\frac{5}{9}\right)}{36\left(\frac{1}{4}+\frac{1}{12}\right)}=\frac{36\left(\frac{2}{3}\right)+36\left(\frac{5}{9}\right)}{36\left(\frac{1}{4}\right)+36\left(\frac{1}{12}\right)}=\frac{24+20}{9+3}=\frac{44}{12}=\frac{4\cdot 11}{4\cdot 3}=\frac{11}{3}$$

$$\frac{6+\frac{3}{x}}{\frac{x}{4}+\frac{1}{8}}=\frac{8x\left(6+\frac{3}{x}\right)}{8x\left(\frac{x}{4}+\frac{1}{8}\right)}=\frac{8x(6)+8x\left(\frac{3}{x}\right)}{8x\left(\frac{x}{4}\right)+8x\left(\frac{1}{8}\right)}=\frac{48x+24}{2x^2+x}=\frac{24(2x+1)}{x(2x+1)}=\frac{24}{x}$$

Work Problem 4 at the Side.

EXAMPLE 5 Simplifying a Complex Fraction (Method 2)

Simplify the complex fraction.

$$\frac{\frac{3}{5m}-\frac{2}{m^2}}{\frac{9}{2m}+\frac{3}{4m^2}}$$

Continued on Next Page

ANSWERS

4. (a) $\frac{15}{34}$ (b) $\frac{2a-6}{3a+4}$ (c) $\frac{2p+1}{4(5-p)}$

The LCD for $5m$, m^2, $2m$, and $4m^2$ is $20m^2$. Multiply the numerator and denominator by $20m^2$.

$$\frac{\frac{3}{5m}-\frac{2}{m^2}}{\frac{9}{2m}+\frac{3}{4m^2}}=\frac{20m^2\left(\frac{3}{5m}-\frac{2}{m^2}\right)}{20m^2\left(\frac{9}{2m}+\frac{3}{4m^2}\right)}$$

$$=\frac{20m^2\left(\frac{3}{5m}\right)-20m^2\left(\frac{2}{m^2}\right)}{20m^2\left(\frac{9}{2m}\right)+20m^2\left(\frac{3}{4m^2}\right)} \qquad \text{Distributive property}$$

$$=\frac{12m-40}{90m+15}$$

Work Problem 5 at the Side.

5 Simplify by Method 2.

$$\frac{\frac{2}{5x}-\frac{3}{x^2}}{\frac{7}{4x}+\frac{1}{2x^2}}$$

Either of the two methods shown in this section can be used to simplify a complex fraction. You may want to choose one method and stick with it to eliminate confusion. However, some students prefer to use Method 1 for problems like Example 2, which is the quotient of two fractions. They prefer Method 2 for problems like Examples 1, 3, 4, and 5, which have sums or differences in the numerators or denominators or both.

EXAMPLE 6 **Deciding on a Method and Simplifying Complex Fractions**

Simplify.

(a) $$\frac{\frac{1}{y}+\frac{2}{y+2}}{\frac{4}{y}-\frac{3}{y+2}}$$

Although either method will work, we will use Method 2 since there are sums and differences in the numerator and denominator. The LCD is $y(y+2)$. Multiply the numerator and denominator by the LCD.

$$\frac{\frac{1}{y}+\frac{2}{y+2}}{\frac{4}{y}-\frac{3}{y+2}}=\frac{\left(\frac{1}{y}+\frac{2}{y+2}\right)y(y+2)}{\left(\frac{4}{y}-\frac{3}{y+2}\right)y(y+2)}$$

$$=\frac{1(y+2)+2y}{4(y+2)-3y} \qquad \text{Distributive property; fundamental property}$$

$$=\frac{y+2+2y}{4y+8-3y} \qquad \text{Distributive property}$$

$$=\frac{3y+2}{y+8} \qquad \text{Combine like terms.}$$

Continued on Next Page

ANSWERS

5. $\frac{8x-60}{35x+10}$

6 Simplify. Use either method.

(a) $$\frac{\frac{1}{x}+\frac{2}{x-1}}{\frac{2}{x}-\frac{4}{x-1}}$$

(b) $$\frac{1-\frac{2}{x}-\frac{15}{x^2}}{1+\frac{5}{x}+\frac{6}{x^2}}$$

(c) $$\frac{\frac{2x+3}{x-4}}{\frac{4x^2-9}{x^2-16}}$$

(b) $$\frac{1-\frac{2}{x}-\frac{3}{x^2}}{1-\frac{5}{x}+\frac{6}{x^2}}$$

Use Method 2.

$$\frac{1-\frac{2}{x}-\frac{3}{x^2}}{1-\frac{5}{x}+\frac{6}{x^2}}=\frac{\left(1-\frac{2}{x}-\frac{3}{x^2}\right)x^2}{\left(1-\frac{5}{x}+\frac{6}{x^2}\right)x^2}$$ Multiply numerator and denominator by the LCD, x^2.

$$=\frac{x^2-2x-3}{x^2-5x+6}$$ Distributive property

$$=\frac{(x-3)(x+1)}{(x-3)(x-2)}$$ Factor.

$$=\frac{x+1}{x-2}$$ Lowest terms

(c) $$\frac{\frac{x+2}{x-3}}{\frac{x^2-4}{x^2-9}}$$

Since this is simply a quotient of two rational expressions, we will use Method 1.

$$\frac{\frac{x+2}{x-3}}{\frac{x^2-4}{x^2-9}}=\frac{x+2}{x-3}\div\frac{x^2-4}{x^2-9}$$

$$=\frac{x+2}{x-3}\cdot\frac{x^2-9}{x^2-4}$$ Definition of division

$$=\frac{x+2}{x-3}\cdot\frac{(x+3)(x-3)}{(x+2)(x-2)}$$ Factor.

$$=\frac{x+3}{x-2}$$ Multiply.

Work Problem 6 at the Side.

Answers

6. **(a)** $\frac{3x+1}{-2x-2}$ **(b)** $\frac{x-5}{x+2}$ **(c)** $\frac{x+4}{2x-3}$

7.5 Exercises

Note: In many problems involving complex fractions, several different equivalent forms of the answer often exist. If your answer does not look exactly like the one given in the back of the book, check to see whether you have written an equivalent form.

1. Consider the complex fraction $\dfrac{\frac{1}{2}-\frac{1}{3}}{\frac{5}{6}-\frac{1}{12}}$. Answer each part, outlining Method 1 for simplifying this complex fraction.

(a) To combine the terms in the numerator, we must find the LCD of $\frac{1}{2}$ and $\frac{1}{3}$. What is this LCD? Determine the simplified form of the numerator of the complex fraction.

(b) To combine the terms in the denominator, we must find the LCD of $\frac{5}{6}$ and $\frac{1}{12}$. What is this LCD? Determine the simplified form of the denominator of the complex fraction.

(c) Now use the results from parts (a) and (b) to write the complex fraction as a division problem using the symbol $\div$.

(d) Perform the operation from part (c) to obtain the final simplification.

2. Consider the same complex fraction given in Exercise 1, $\dfrac{\frac{1}{2}-\frac{1}{3}}{\frac{5}{6}-\frac{1}{12}}$. Answer each part, outlining Method 2 for simplifying this complex fraction.

(a) We must determine the LCD of all the fractions within the complex fraction. What is this LCD?

(b) Multiply every term in the complex fraction by the LCD found in part (a), but at this time do not combine the terms in the numerator and the denominator.

(c) Now combine the terms from part (b) to obtain the simplified form of the complex fraction.

Simplify each complex fraction. Use either method. See Examples 1–6.

3. $\dfrac{-\frac{4}{3}}{\frac{2}{9}}$

4. $\dfrac{-\frac{5}{6}}{\frac{5}{4}}$

5. $\dfrac{\frac{p}{q^2}}{\frac{p^2}{q}}$

6. $\dfrac{\frac{a}{x}}{\frac{a^2}{2x}}$

7. $\dfrac{\frac{x}{y^2}}{\frac{x^2}{y}}$

8. $\dfrac{\frac{p^4}{r}}{\frac{p^2}{r^2}}$

9. $\dfrac{\frac{4a^4b^3}{3a}}{\frac{2ab^4}{b^2}}$

10. $\dfrac{\frac{2r^4t^2}{3t}}{\frac{5r^2t^5}{3r}}$

11. $\dfrac{\dfrac{m+2}{3}}{\dfrac{m-4}{m}}$

12. $\dfrac{\dfrac{q-5}{q}}{\dfrac{q+5}{3}}$

13. $\dfrac{\dfrac{2}{x}-3}{\dfrac{2-3x}{2}}$

14. $\dfrac{6+\dfrac{2}{r}}{\dfrac{3r+1}{4}}$

15. $\dfrac{\dfrac{1}{x}+x}{\dfrac{x^2+1}{8}}$

16. $\dfrac{\dfrac{3}{m}-m}{\dfrac{3-m^2}{4}}$

17. $\dfrac{a-\dfrac{5}{a}}{a+\dfrac{1}{a}}$

18. $\dfrac{q+\dfrac{1}{q}}{q+\dfrac{4}{q}}$

19. $\dfrac{\dfrac{1}{2}+\dfrac{1}{p}}{\dfrac{2}{3}+\dfrac{1}{p}}$

20. $\dfrac{\dfrac{3}{4}-\dfrac{1}{r}}{\dfrac{1}{5}+\dfrac{1}{r}}$

21. $\dfrac{\dfrac{t}{t+2}}{\dfrac{4}{t^2-4}}$

22. $\dfrac{\dfrac{m}{m+1}}{\dfrac{3}{m^2-1}}$

23. $\dfrac{\dfrac{1}{k+1}-1}{\dfrac{1}{k+1}+1}$

24. $\dfrac{\dfrac{2}{p-1}+2}{\dfrac{3}{p-1}-2}$

25. $\dfrac{2+\dfrac{1}{x}-\dfrac{28}{x^2}}{3+\dfrac{13}{x}+\dfrac{4}{x^2}}$

26. $\dfrac{4-\dfrac{11}{x}-\dfrac{3}{x^2}}{2-\dfrac{1}{x}-\dfrac{15}{x^2}}$

27. $\dfrac{\dfrac{1}{m-1}+\dfrac{2}{m+2}}{\dfrac{2}{m+2}-\dfrac{1}{m-3}}$

28. $\dfrac{\dfrac{5}{r+3}-\dfrac{1}{r-1}}{\dfrac{2}{r+2}+\dfrac{3}{r+3}}$

29. $2-\dfrac{2}{2+\dfrac{2}{2+2}}$

30. $3-\dfrac{2}{4+\dfrac{2}{4-2}}$

7.6 Solving Equations with Rational Expressions

In **Section 2.3** we solved equations with fractions as coefficients. By using the multiplication property of equality, we cleared the fractions by multiplying by the LCD. We continue this work here.

OBJECTIVES

1. Distinguish between operations with rational expressions and equations with terms that are rational expressions.
2. Solve equations with rational expressions.
3. Solve a formula for a specified variable.

OBJECTIVE 1 Distinguish between operations with rational expressions and equations with terms that are rational expressions. Before solving equations with rational expressions, you must understand the difference between sums and differences of terms with rational coefficients, or rational *expressions,* and *equations* with terms that are rational expressions. Sums and differences lead to *expressions,* while *equations* are solved.

EXAMPLE 1 Distinguishing between Expressions and Equations

Identify each of the following as an *expression* or an *equation.* Then simplify the expression or solve the equation.

(a) $\frac{3}{4}x - \frac{2}{3}x$

This is a difference of two terms. It represents an *expression* since there is no equals sign. Find the LCD, write each coefficient with this LCD, and combine like terms.

$$\frac{3}{4}x - \frac{2}{3}x = \frac{9}{12}x - \frac{8}{12}x \quad \text{Get a common denominator.}$$

$$= \frac{1}{12}x \quad \text{Combine like terms.}$$

(b) $\frac{3}{4}x - \frac{2}{3}x = \frac{1}{2}$

Because of the equals sign, this is an *equation* to be solved. We proceed as in **Section 2.3,** using the multiplication property of equality to clear fractions. The LCD is 12.

$$\frac{3}{4}x - \frac{2}{3}x = \frac{1}{2}$$

$$12\left(\frac{3}{4}x - \frac{2}{3}x\right) = 12\left(\frac{1}{2}\right) \quad \text{Multiply by 12.}$$

$$12\left(\frac{3}{4}x\right) - 12\left(\frac{2}{3}x\right) = 12\left(\frac{1}{2}\right) \quad \text{Distributive property}$$

$$9x - 8x = 6 \quad \text{Multiply.}$$

$$x = 6 \quad \text{Combine like terms.}$$

Continued on Next Page

Check:

$$\frac{3}{4}x - \frac{2}{3}x = \frac{1}{2} \quad \text{Original equation}$$

$$\frac{3}{4}(6) - \frac{2}{3}(6) \stackrel{?}{=} \frac{1}{2} \quad \text{Let } x = 6.$$

$$\frac{9}{2} - 4 \stackrel{?}{=} \frac{1}{2} \quad \text{Multiply.}$$

$$\frac{1}{2} = \frac{1}{2} \quad \text{True}$$

The check shows that 6 is the solution of the equation.

The ideas of Example 1 can be summarized as follows.

When adding or subtracting rational expressions, the LCD must be kept throughout the simplification. When solving an equation, the LCD is used to multiply each side so that denominators are eliminated.

Work Problem 1 at the Side.

OBJECTIVE 2 Solve equations with rational expressions. When an equation involves fractions as in Example 1(b), we use the multiplication property of equality to clear it of fractions. Choose as multiplier the LCD of all denominators in the fractions of the equation.

EXAMPLE 2 Solving an Equation with Rational Expressions

Solve $\frac{x}{3} + \frac{x}{4} = 10 + x$. Check the solution.

We begin by multiplying each side of the equation by 12, the LCD.

$$12\left(\frac{x}{3} + \frac{x}{4}\right) = 12(10 + x)$$

$$12\left(\frac{x}{3}\right) + 12\left(\frac{x}{4}\right) = 12(10) + 12x \quad \text{Distributive property}$$

$$4x + 3x = 120 + 12x$$

$$7x = 120 + 12x \quad \text{Combine like terms.}$$

$$-5x = 120 \quad \text{Subtract } 12x.$$

$$x = -24 \quad \text{Divide by } -5.$$

Check:

$$\frac{x}{3} + \frac{x}{4} = 10 + x \quad \text{Original equation}$$

$$\frac{-24}{3} + \frac{-24}{4} \stackrel{?}{=} 10 - 24 \quad \text{Let } x = -24.$$

$$-8 - 6 \stackrel{?}{=} -14$$

$$-14 = -14 \quad \text{True}$$

The solution is -24.

Work Problem 2 at the Side.

1 Identify each as an *expression* or an *equation*. Then perform the operation to simplify the expression or solve the equation.

(a) $\frac{x}{3} + \frac{x}{5} = 7 + x$

(b) $\frac{2x}{3} - \frac{4x}{9}$

2 Solve each equation, and check your solutions.

(a) $\frac{x}{5} + 3 = \frac{3}{5}$

(b) $\frac{x}{2} - \frac{x}{3} = \frac{5}{6}$

Answers

1. **(a)** equation; -15 **(b)** expression; $\frac{2x}{9}$

2. **(a)** -12 **(b)** 5

CAUTION
Note that the use of the LCD here is different from its use in the previous section. Here, we use the multiplication property of equality to multiply each side of an *equation* by the LCD. Earlier, we used the fundamental property to multiply a *fraction* by another fraction that had the LCD as both its numerator and denominator. Be careful not to confuse these two methods.

EXAMPLE 3 Solving an Equation with Rational Expressions

Solve $\frac{p}{2} - \frac{p-1}{3} = 1$.

$$6\left(\frac{p}{2} - \frac{p-1}{3}\right) = 6 \cdot 1 \quad \text{Multiply by the LCD, 6.}$$

$$6\left(\frac{p}{2}\right) - 6\left(\frac{p-1}{3}\right) = 6 \quad \text{Distributive property}$$

$$3p - 2(p-1) = 6$$

Be very careful to put parentheses around $p - 1$; otherwise, you may find an incorrect solution. Continue simplifying and solve.

$$3p - 2p + 2 = 6 \quad \text{Distributive property}$$

$$p + 2 = 6 \quad \text{Combine like terms.}$$

$$p = 4 \quad \text{Subtract 2.}$$

Check to see that 4 is correct by replacing p with 4 in the original equation.

Work Problem 3 at the Side.

When solving an equation that has a variable in the denominator, remember that the number 0 cannot be used as a denominator. Therefore, ***the solution cannot be a number that will make the denominator equal* 0.**

EXAMPLE 4 Solving an Equation with Rational Expressions

Solve $\frac{x}{x-2} = \frac{2}{x-2} + 2$. Check the proposed solution.

The common denominator is $x - 2$. (*Note:* Because $x = 2$ makes a denominator in the given equation equal 0, x cannot equal 2.) Solve the equation by multiplying each side of the equation by $x - 2$.

$$(x-2)\left(\frac{x}{x-2}\right) = (x-2)\left(\frac{2}{x-2} + 2\right)$$

$$(x-2)\left(\frac{x}{x-2}\right) = (x-2)\left(\frac{2}{x-2}\right) + (x-2)(2)$$

$$x = 2 + 2x - 4$$

$$x = -2 + 2x \quad \text{Combine like terms.}$$

$$-x = -2 \quad \text{Subtract } 2x.$$

$$x = 2 \quad \text{Divide by } -1.$$

Continued on Next Page

3 Solve each equation, and check your solutions.

(a) $\frac{k}{6} - \frac{k+1}{4} = -\frac{1}{2}$

(b) $\frac{2m-3}{5} - \frac{m}{3} = -\frac{6}{5}$

ANSWERS
3. (a) 3 **(b)** -9

4 Solve the equation, and check your solution.

$$1 - \frac{2}{x+1} = \frac{2x}{x+1}$$

Check: The proposed solution is 2. If we substitute 2 in the original equation, we get

$$\frac{x}{x-2} = \frac{2}{x-2} + 2 \qquad \text{Original equation}$$

$$\frac{2}{2-2} = \frac{2}{2-2} + 2 \qquad ?$$

$$\frac{2}{0} = \frac{2}{0} + 2. \qquad ?$$

Notice that 2 makes both denominators equal 0. Because 0 cannot be the denominator, there is no solution.

While it is always a good idea to check solutions to guard against arithmetic and algebraic errors, ***it is essential to check proposed solutions when variables appear in denominators in the original equation.*** Some students like to determine which numbers cannot be solutions *before* solving the equation.

Work Problem 4 at the Side.

The steps used to solve an equation with rational expressions follow.

Solving An Equation with Rational Expressions

Step 1 **Multiply each side of the equation by the LCD.** (This clears the equation of fractions.)

Step 2 **Solve** the resulting equation.

Step 3 **Check** each proposed solution by substituting it in the original equation. Reject any that cause a denominator to equal 0.

EXAMPLE 5 Solving an Equation with Rational Expressions

Solve $\frac{2}{x^2 - x} = \frac{1}{x^2 - 1}$. Check the proposed solution.

Step 1 Begin by finding the LCD.

$$\frac{2}{x(x-1)} = \frac{1}{(x+1)(x-1)} \qquad \text{Factor the denominators to find the LCD.}$$

Since $x^2 - x$ can be factored as $x(x-1)$, and $x^2 - 1$ can be factored as $(x+1)(x-1)$, the LCD is $x(x+1)(x-1)$.

Step 2 Notice that 0, -1, and 1 cannot be solutions of this equation. Multiply each side of the equation by $x(x+1)(x-1)$.

$$x(x+1)(x-1)\frac{2}{x(x-1)} = x(x+1)(x-1)\frac{1}{(x+1)(x-1)}$$

$$2(x+1) = x$$

$$2x + 2 = x \qquad \text{Distributive property}$$

$$2 = -x \qquad \text{Subtract } 2x.$$

$$x = -2 \qquad \text{Multiply by } -1\text{; rewrite.}$$

Continued on Next Page

ANSWERS

4. no solution (When the equation is solved, -1 is found. However, because $x = -1$ leads to a 0 denominator in the original equation, there is no solution.)

Step 3 The proposed solution is -2, which does not make any denominator equal 0.

Check:

$$\frac{2}{x^2 - x} = \frac{1}{x^2 - 1} \qquad \text{Original equation}$$

$$\frac{2}{(-2)^2 - (-2)} \stackrel{?}{=} \frac{1}{(-2)^2 - 1} \qquad \text{Let } x = -2.$$

$$\frac{2}{4 + 2} \stackrel{?}{=} \frac{1}{4 - 1}$$

$$\frac{1}{3} = \frac{1}{3} \qquad \text{True}$$

The solution is indeed -2.

Work Problem 5 at the Side.

5 Solve each equation, and check your solutions.

(a) $\dfrac{4}{x^2 - 3x} = \dfrac{1}{x^2 - 9}$

(b) $\dfrac{2}{p^2 - 2p} = \dfrac{3}{p^2 - p}$

EXAMPLE 6 Solving an Equation with Rational Expressions

Solve $\dfrac{2m}{m^2 - 4} + \dfrac{1}{m - 2} = \dfrac{2}{m + 2}$.

Factor the first denominator on the left.

$$\frac{2m}{(m + 2)(m - 2)} + \frac{1}{m - 2} = \frac{2}{m + 2}$$

Multiply by the LCD, $(m + 2)(m - 2)$. (Notice that -2 and 2 cannot be solutions.)

$$(m + 2)(m - 2)\left(\frac{2m}{(m + 2)(m - 2)} + \frac{1}{m - 2}\right) = (m + 2)(m - 2)\frac{2}{m + 2}$$

$$(m + 2)(m - 2)\frac{2m}{(m + 2)(m - 2)} + (m + 2)(m - 2)\frac{1}{m - 2} = (m + 2)(m - 2)\frac{2}{m + 2}$$

$$2m + m + 2 = 2(m - 2)$$

$$3m + 2 = 2m - 4 \qquad \text{Combine like terms; distributive property}$$

$$m + 2 = -4 \qquad \text{Subtract } 2m.$$

$$m = -6 \qquad \text{Subtract 2.}$$

Check to see that -6 is a solution of the given equation.

Work Problem 6 at the Side.

6 Solve each equation, and check your solutions.

(a) $\dfrac{2p}{p^2 - 1} = \dfrac{2}{p + 1} - \dfrac{1}{p - 1}$

(b) $\dfrac{8r}{4r^2 - 1} = \dfrac{3}{2r + 1} + \dfrac{3}{2r - 1}$

EXAMPLE 7 Solving an Equation with Rational Expressions

Solve $\dfrac{1}{x - 1} + \dfrac{1}{2} = \dfrac{2}{x^2 - 1}$.

The denominator $x^2 - 1$ factors as $(x + 1)(x - 1)$. Multiply each side of the equation by the LCD, $2(x + 1)(x - 1)$. (Notice that -1 and 1 cannot be solutions.)

Continued on Next Page

Answers
5. (a) -4 (b) 4
6. (a) -3 (b) 0

7 Solve the equation, and check your solution.

$$\frac{2}{3x+1} - \frac{1}{x} = \frac{-6x}{3x+1}$$

$$2(x+1)(x-1)\left(\frac{1}{x-1} + \frac{1}{2}\right) = 2(x+1)(x-1)\frac{2}{(x+1)(x-1)}$$

$$2(x+1)(x-1)\frac{1}{x-1} + \mathbf{2(x+1)(x-1)}\frac{1}{2}$$

$$= 2(x+1)(x-1)\frac{2}{(x+1)(x-1)}$$

$$2(x+1) + (x+1)(x-1) = 4$$

$$2x + 2 + x^2 - 1 = 4 \quad \text{Distributive property}$$

$$x^2 + 2x + 1 = 4 \quad \text{Combine like terms.}$$

$$x^2 + 2x - 3 = 0 \quad \text{Subtract 4.}$$

$$(x+3)(x-1) = 0 \quad \text{Factor.}$$

Solving this equation suggests that $x = -3$ or $x = 1$. But 1 makes a denominator of the original equation equal 0, so 1 is not a solution. However, -3 is a solution, as shown by substituting -3 for x in the original equation.

Check:

$$\frac{1}{x-1} + \frac{1}{2} = \frac{2}{x^2-1} \quad \text{Original equation}$$

$$\frac{1}{-3-1} + \frac{1}{2} \stackrel{?}{=} \frac{2}{(-3)^2-1} \quad \text{Let } x = -3.$$

$$\frac{1}{-4} + \frac{1}{2} \stackrel{?}{=} \frac{2}{9-1} \quad \text{Simplify.}$$

$$\frac{1}{4} = \frac{1}{4} \quad \text{True}$$

The check shows that -3 is a solution.

Work Problem 7 at the Side.

8 Solve each equation, and check your solutions.

(a) $\frac{1}{x-2} + \frac{1}{5} = \frac{2}{5(x^2-4)}$

(b) $\frac{6}{5a+10} - \frac{1}{a-5} = \frac{4}{a^2-3a-10}$

EXAMPLE 8 Solving an Equation with Rational Expressions

Solve $\frac{1}{k^2+4k+3} + \frac{1}{2k+2} = \frac{3}{4k+12}$.

Factor the three denominators to get the common denominator, $4(k+1)(k+3)$. (Notice that -1 and -3 cannot be solutions.)

$$4(k+1)(k+3)\left(\frac{1}{(k+1)(k+3)} + \frac{1}{2(k+1)}\right)$$

$$= 4(k+1)(k+3)\frac{3}{4(k+3)} \quad \text{Multiply by the LCD.}$$

$$4(k+1)(k+3)\frac{1}{\mathbf{(k+1)(k+3)}} + 2 \cdot \mathbf{2(k+1)}(k+3)\frac{1}{\mathbf{2(k+1)}}$$

$$= 4(k+1)(k+3)\frac{3}{\mathbf{4(k+3)}}$$

$$4 + 2(k+3) = 3(k+1) \quad \text{Simplify.}$$

$$4 + 2k + 6 = 3k + 3 \quad \text{Distributive property}$$

$$2k + 10 = 3k + 3 \quad \text{Combine like terms.}$$

$$7 = k \quad \text{Subtract } 2k \text{ and 3.}$$

Check to see that 7 is a solution of the given equation.

Work Problem 8 at the Side.

ANSWERS

7. $\frac{1}{2}$

8. **(a)** $-4, -1$ **(b)** 60

OBJECTIVE 3 Solve a formula for a specified variable. Solving a formula for a specified variable was first discussed in **Section 2.5.** ***Remember to treat the variable for which you are solving as if it were the only variable, and all others as if they were constants.***

9 Solve $z = \dfrac{x}{x + y}$ for y.

EXAMPLE 9 Solving for a Specified Variable

Solve $a = \dfrac{v - w}{t}$ for v.

Our goal is to get v alone on one side of the equation.

$$a = \frac{v - w}{t} \qquad \text{Given equation}$$

$$at = v - w \qquad \text{Multiply by } t.$$

$$at + w = v \qquad \text{Add } w.$$

or

$$v = at + w$$

To check this, substitute $at + w$ for v in the original equation. The final result will be the identity $a = a$, indicating that the result obtained is correct.

Work Problem 9 at the Side.

EXAMPLE 10 Solving for a Specified Variable

Solve the formula $\dfrac{1}{a} = \dfrac{1}{b} + \dfrac{1}{c}$ for c.

The LCD of all the fractions in the equation is abc, so multiply each side by abc.

$$\frac{1}{a} = \frac{1}{b} + \frac{1}{c}$$

$$abc\left(\frac{1}{a}\right) = abc\left(\frac{1}{b} + \frac{1}{c}\right) \qquad \text{Multiply by the LCD, } abc.$$

$$abc\left(\frac{1}{a}\right) = abc\left(\frac{1}{b}\right) + abc\left(\frac{1}{c}\right) \qquad \text{Distributive property}$$

$$bc = ac + ab$$

Since we are solving for c, transform so that all terms with c are on one side of the equation. Do this by subtracting ac from each side.

$$bc - ac = ab \qquad \text{Subtract } ac.$$

Factor out the common factor c on the left.

$$c(b - a) = ab \qquad \text{Factor out } c.$$

Finally, divide each side by the coefficient of c, which is $b - a$.

$$c = \frac{ab}{b - a}$$

ANSWERS

9. $y = \dfrac{x - zx}{z}$

10 Solve $\frac{2}{x} = \frac{1}{y} + \frac{1}{z}$ for z.

CAUTION

Students often have trouble in the step that involves factoring out the variable for which they are solving. In Example 10, we had to factor out c on the left side so that we could divide both sides by $b - a$.

$$bc - ac = ab$$
$$c(b - a) = ab$$
$$c = \frac{ab}{b - a}$$

When solving an equation for a specified variable, ***be sure that the specified variable appears alone on only one side of the equals sign in the final equation.***

Work Problem 10 at the Side.

Answers

10. $z = \frac{xy}{2y - x}$ or $z = \frac{-xy}{x - 2y}$

7.6 Exercises

FOR EXTRA HELP

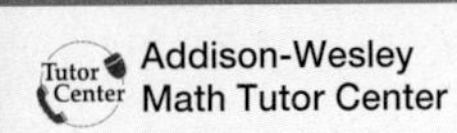

MathXL

MyMathLab

Identify each as an expression *or an* equation. *Then simplify the expression or solve the equation. See Example 1.*

1. $\frac{7}{8}x + \frac{1}{5}x$

2. $\frac{4}{7}x + \frac{3}{5}x$

3. $\frac{7}{8}x + \frac{1}{5}x = 1$

4. $\frac{4}{7}x + \frac{3}{5}x = 1$

5. $\frac{3}{5}y - \frac{7}{10}y$

6. $\frac{3}{5}y - \frac{7}{10}y = 1$

7. Explain how the LCD is used in a different way when adding and subtracting rational expressions compared to solving equations with rational expressions.

8. If we multiply each side of the equation $\frac{6}{x+5} = \frac{6}{x+5}$ by $x + 5$, we get $6 = 6$. Are all real numbers solutions of this equation? Explain.

Solve each equation, and check your solutions. See Examples 2 and 3.

9. $\frac{2}{3}x + \frac{1}{2}x = -7$

10. $\frac{1}{4}x - \frac{1}{3}x = 1$

11. $\frac{p}{3} - \frac{p}{6} = 4$

12. $\frac{x}{15} + \frac{x}{5} = 4$

13. $\frac{3x}{5} - 6 = x$

14. $\frac{5t}{4} + t = 9$

15. $\frac{4m}{7} + m = 11$

16. $a - \frac{3a}{2} = 1$

17. $\frac{z-1}{4} = \frac{z+3}{3}$

18. $\frac{r-5}{2} = \frac{r+2}{3}$

19. $\frac{3p+6}{8} = \frac{3p-3}{16}$

20. $\frac{2z+1}{5} = \frac{7z+5}{15}$

21. $\dfrac{2x+3}{-6}=\dfrac{3}{2}$

22. $\dfrac{4y+3}{6}=\dfrac{5}{2}$

23. $\dfrac{q+2}{3}+\dfrac{q-5}{5}=\dfrac{7}{3}$

24. $\dfrac{b+7}{8}-\dfrac{b-2}{3}=\dfrac{4}{3}$

25. $\dfrac{t}{6}+\dfrac{4}{3}=\dfrac{t-2}{3}$

26. $\dfrac{x}{2}=\dfrac{5}{4}+\dfrac{x-1}{4}$

27. $\dfrac{3m}{5}-\dfrac{3m-2}{4}=\dfrac{1}{5}$

28. $\dfrac{8p}{5}=\dfrac{3p-4}{2}+\dfrac{5}{2}$

29. What values of x would have to be rejected as possible solutions of the equation $\dfrac{1}{x-4}=\dfrac{3}{2x}$?

30. What is wrong with the following problem? "Solve $\dfrac{2}{3x}+\dfrac{1}{5x}$."

Solve each equation, and check your solutions. See Examples 4–8.

31. $\dfrac{2x+3}{x}=\dfrac{3}{2}$

32. $\dfrac{5-2y}{y}=\dfrac{1}{4}$

33. $\dfrac{k}{k-4}-5=\dfrac{4}{k-4}$

34. $\dfrac{-5}{a+5}=\dfrac{a}{a+5}+2$

35. $\dfrac{3}{x-1}+\dfrac{2}{4x-4}=\dfrac{7}{4}$

36. $\dfrac{2}{p+3}+\dfrac{3}{8}=\dfrac{5}{4p+12}$

37. $\dfrac{y}{3y+3}=\dfrac{2y-3}{y+1}-\dfrac{2y}{3y+3}$

38. $\dfrac{2k+3}{k+1}-\dfrac{3k}{2k+2}=\dfrac{-2k}{2k+2}$

39. $\dfrac{2}{m}=\dfrac{m}{5m+12}$

40. $\frac{x}{4-x} = \frac{2}{x}$

41. $\frac{-2}{z+5} + \frac{3}{z-5} = \frac{20}{z^2-25}$

42. $\frac{3}{r+3} - \frac{2}{r-3} = \frac{-12}{r^2-9}$

43. $\frac{3y}{y^2+5y+6} = \frac{5y}{y^2+2y-3} - \frac{2}{y^2+y-2}$

44. $\frac{x+4}{x^2-3x+2} - \frac{5}{x^2-4x+3} = \frac{x-4}{x^2-5x+6}$

45. $\frac{5x}{14x+3} = \frac{1}{x}$

46. $\frac{m}{8m+3} = \frac{1}{3m}$

47. $\frac{2}{z-1} - \frac{5}{4} = \frac{-1}{z+1}$

48. $\frac{5}{p-2} = 7 - \frac{10}{p+2}$

49. If you are solving a formula for the letter k, and your steps lead to the equation $kr - mr = km$, what would be your next step?

50. If you are solving a formula for the letter k, and your steps lead to the equation $kr - km = mr$, what would be your next step?

Solve each formula for the specified variable. See Example 9.

51. $m = \frac{kF}{a}$ for F

52. $I = \frac{kE}{R}$ for E

53. $m = \frac{kF}{a}$ for a

54. $I = \frac{kE}{R}$ for R

55. $I = \frac{E}{R + r}$ for R

56. $I = \frac{E}{R + r}$ for r

57. $h = \frac{2A}{B + b}$ for A

58. $d = \frac{2S}{n(a + L)}$ for S

59. $d = \frac{2S}{n(a + L)}$ for a

60. $h = \frac{2A}{B + b}$ for B

Solve each equation for the specified variable. See Example 10.

61. $\frac{2}{r} + \frac{3}{s} + \frac{1}{t} = 1$ for t

62. $\frac{5}{p} + \frac{2}{q} + \frac{3}{r} = 1$ for r

63. $\frac{1}{a} - \frac{1}{b} - \frac{1}{c} = 2$ for c

64. $\frac{-1}{x} + \frac{1}{y} + \frac{1}{z} = 4$ for y

65. $9x + \frac{3}{z} = \frac{5}{y}$ for z

66. $-3t - \frac{4}{p} = \frac{6}{s}$ for p

Summary Exercises on Rational Expressions and Equations

We have performed the four operations of arithmetic with rational expressions and solved equations with rational expressions. The exercises in this summary include a mixed variety of problems of these types.

Students often confuse *simplifying* rational expressions with the *solution of equations* with rational expressions. For example, the four possible operations to simplify the rational expressions $\frac{1}{x}$ and $\frac{1}{x-2}$ are performed as follows.

Add:

$$\frac{1}{x}+\frac{1}{x-2}=\frac{1(x-2)}{x(x-2)}+\frac{x(1)}{x(x-2)}$$ Write with a common denominator.

$$=\frac{x-2+x}{x(x-2)}$$ Add numerators; keep the same denominator.

$$=\frac{2x-2}{x(x-2)}$$ Combine like terms.

Subtract:

$$\frac{1}{x}-\frac{1}{x-2}=\frac{1(x-2)}{x(x-2)}-\frac{x(1)}{x(x-2)}$$ Write with a common denominator.

$$=\frac{x-2-x}{x(x-2)}$$ Subtract numerators; keep the same denominator.

$$=\frac{-2}{x(x-2)}$$ Combine like terms.

Multiply:

$$\frac{1}{x}\cdot\frac{1}{x-2}=\frac{1}{x(x-2)}$$ Multiply numerators and multiply denominators.

Divide:

$$\frac{1}{x}\div\frac{1}{x-2}=\frac{1}{x}\cdot\frac{x-2}{1}=\frac{x-2}{x}$$ Multiply by the reciprocal of the divisor.

On the other hand, consider the *equation*

$$\frac{1}{x}+\frac{1}{x-2}=\frac{3}{4}.$$

Neither 0 nor 2 can be a solution of this equation, since each will cause a denominator to equal 0. We use the multiplication property of equality to multiply each side by the LCD, $4x(x-2)$, leading to an equation with no denominators.

$$4x(x-2)\frac{1}{x}+4x(x-2)\frac{1}{x-2}=4x(x-2)\frac{3}{4}$$

$$4(x-2)+4x=3x(x-2)$$

$$4x-8+4x=3x^2-6x \quad \text{Distributive property}$$

$$0=3x^2-14x+8 \quad \text{Standard form}$$

$$0=(3x-2)(x-4) \quad \text{Factor.}$$

$$x-4=0 \quad \text{or} \quad 3x-2=0 \quad \text{Zero-factor property}$$

$$x=4 \quad \text{or} \quad x=\frac{2}{3}$$

Both $\frac{2}{3}$ and 4 are solutions since neither makes a denominator equal 0.

In conclusion, remember the following points when working exercises involving rational expressions.

Points to Remember When Working with Rational Expressions

1. The fundamental property is applied only after numerators and denominators have been *factored.*
2. When adding and subtracting rational expressions, the common denominator must be kept throughout the problem and in the final result.
3. Always look to see if the answer is in lowest terms; if it is not, use the fundamental property.
4. When solving equations, the LCD is used to clear the equation of fractions. Multiply each side by the LCD. (Notice how this differs from the use of the LCD in Point 2.)
5. When solving equations with rational expressions, reject any proposed solution that causes an original denominator to equal 0.

For each exercise, indicate "expression" if an expression is to be simplified or "equation" if an equation is to be solved. Then simplify the expression or solve the equation.

1. $\frac{4}{p} + \frac{6}{p}$

2. $\frac{x^3y^2}{x^2y^4} \cdot \frac{y^5}{x^4}$

3. $\frac{1}{x^2 + x - 2} \div \frac{4x^2}{2x - 2}$

4. $\frac{8}{m - 5} = 2$

5. $\frac{2y^2 + y - 6}{2y^2 - 9y + 9} \cdot \frac{y^2 - 2y - 3}{y^2 - 1}$

6. $\frac{2}{k^2 - 4k} + \frac{3}{k^2 - 16}$

7. $\frac{x - 4}{5} = \frac{x + 3}{6}$

8. $\frac{3t^2 - t}{6t^2 + 15t} \div \frac{6t^2 + t - 1}{2t^2 - 5t - 25}$

9. $\frac{4}{p + 2} + \frac{1}{3p + 6}$

10. $\frac{1}{y} + \frac{1}{y - 3} = -\frac{5}{4}$

11. $\frac{3}{t - 1} + \frac{1}{t} = \frac{7}{2}$

12. $\frac{6}{y} - \frac{2}{3y}$

13. $\frac{5}{4z} - \frac{2}{3z}$

14. $\frac{k + 2}{3} = \frac{2k - 1}{5}$

15. $\frac{1}{m^2 + 5m + 6} + \frac{2}{m^2 + 4m + 3}$

16. $\frac{2k^2 - 3k}{20k^2 - 5k} \div \frac{2k^2 - 5k + 3}{4k^2 + 11k - 3}$

17. $\frac{2}{x + 1} + \frac{5}{x - 1} = \frac{10}{x^2 - 1}$

18. $\frac{x}{x - 2} + \frac{3}{x + 2} = \frac{8}{x^2 - 4}$

7.7 Applications of Rational Expressions

OBJECTIVES

1. Solve problems about numbers.
2. Solve problems about distance, rate, and time.
3. Solve problems about work.

In **Section 7.6** we solved equations with rational expressions; now we can solve applications that involve this type of equation. The six-step problem-solving method of **Section 2.4** still applies.

OBJECTIVE 1 Solve problems about numbers. We begin with an example about an unknown number.

EXAMPLE 1 Solving a Problem about an Unknown Number

If the same number is added to both the numerator and the denominator of the fraction $\frac{2}{5}$, the result is equivalent to $\frac{2}{3}$. Find the number.

Step 1 **Read** the problem carefully. We are trying to find a number.

Step 2 **Assign a variable.** Here, let x = the number added to the numerator and the denominator.

Step 3 **Write an equation.** The fraction

$$\frac{2+x}{5+x}$$

represents the result of adding the same number to both the numerator and the denominator. Since this result is equivalent to $\frac{2}{3}$, the equation is

$$\frac{2+x}{5+x}=\frac{2}{3}.$$

Step 4 **Solve** this equation by multiplying each side by the LCD, $3(5+x)$.

$$3(5+x)\frac{2+x}{5+x}=3(5+x)\frac{2}{3}$$

$$3(2+x)=2(5+x)$$

$$6+3x=10+2x \qquad \text{Distributive property}$$

$$x=4 \qquad \text{Subtract } 2x\text{; subtract 6.}$$

Step 5 **State the answer.** The number is 4.

Step 6 **Check** the solution in the words of the original problem. If 4 is added to both the numerator and the denominator of $\frac{2}{5}$, the result is $\frac{6}{9}=\frac{2}{3}$, as required.

Work Problem 1 at the Side.

1 Solve each problem.

(a) A certain number is added to the numerator and subtracted from the denominator of $\frac{5}{8}$. The new fraction equals the reciprocal of $\frac{5}{8}$. Find the number.

(b) The denominator of a fraction is 1 more than the numerator. If 6 is added to the numerator and subtracted from the denominator, the result is $\frac{15}{4}$. Find the original fraction.

OBJECTIVE 2 Solve problems about distance, rate, and time. If an automobile travels at an average rate of 65 mph for 2 hr, then it travels $65 \times 2 = 130$ mi. Recall from **Section 4.4** that this is an example of the basic relationship between distance, rate, and time given by the formula $d = rt$. By solving, in turn, for r and t in the formula, we obtain two other equivalent forms of the formula. The three forms are given below.

Distance, Rate, and Time Relationship

$$d=rt \qquad r=\frac{d}{t} \qquad t=\frac{d}{r}$$

ANSWERS

1. (a) 3 (b) $\frac{9}{10}$

2 Solve each problem.

(a) The world record in the men's 100-m dash was set in 1999 by Maurice Green, who ran it in 9.79 sec. What was his speed in meters per second? (*Source:* http://english.sydneylink.com)

(b) The world record for the women's 3000-m run was set by Junxia Wang in 1993. Her speed was 6.173 m per sec. What was her time in seconds?

(c) A small plane flew from Chicago to St. Louis averaging 145 mph. The trip took 2 hr. What is the distance between Chicago and St. Louis?

ANSWERS

2. (a) 10.21 m per sec **(b)** 486 sec **(c)** 290 mi

The next example illustrates the uses of these formulas.

EXAMPLE 2 Finding Distance, Rate, or Time

(a) The speed of sound is 1088 ft per sec at sea level at 32°F. In 5 sec under these conditions, sound travels

$$\underset{\text{Rate}}{1088} \times \underset{\text{Time}}{5} = \underset{\text{Distance}}{5440 \text{ ft.}}$$

Here, we found distance given rate and time, using $d = rt$.

(b) The winner of the first Indianapolis 500 race (in 1911) was Ray Harroun, driving a Marmon Wasp at an average speed of 74.602 mph. (*Source: World Almanac and Book of Facts,* 2004.)

To complete the 500 mi, it took him

$$\frac{500 \leftarrow \text{Distance}}{74.602 \leftarrow \text{Rate}} = 6.70 \text{ hr (rounded)}. \leftarrow \text{Time}$$

Here, we found time given rate and distance using $t = \frac{d}{r}$. To convert .70 hr to minutes, multiply by 60 to get $.70(60) = 42$. It took Harroun about 6 hr, 42 min to complete the race.

(c) At the 2004 Olympic Games in Athens, Greece, Dutch swimmer Inge de Bruijn won the women's 50-m freestyle swimming event in 24.58 sec. (*Source:* www.olympics.com)

Her rate was

$$\text{Rate} = \frac{\text{Distance} \rightarrow 50}{\text{Time} \rightarrow 24.58} = 2.03 \text{ m per sec (rounded)}.$$

Work Problem 2 at the Side.

PROBLEM-SOLVING HINT

Many applied problems use the formulas just discussed. The next two examples show how to solve typical applications of the formula $d = rt$. A helpful strategy for solving such problems is to ***first make a sketch*** showing what is happening in the problem. ***Then make a table*** using the information given, along with the unknown quantities. The table will help you organize the information, and the sketch will help you set up the equation.

EXAMPLE 3 Solving a Motion Problem about Distance, Rate, and Time

Two cars leave Baton Rouge, Louisiana, at the same time and travel east on Interstate 10. One travels at a constant speed of 55 mph and the other travels at a constant speed of 63 mph. In how many hours will the distance between them be 24 mi?

Continued on Next Page

Step 1 **Read** the problem. We are trying to find the time when the distance between the cars will be 24 mi.

Step 2 **Assign a variable.** Since we are looking for time, let t = the number of hours until the distance between them is 24 mi. The sketch in Figure 1 shows what is happening in the problem.

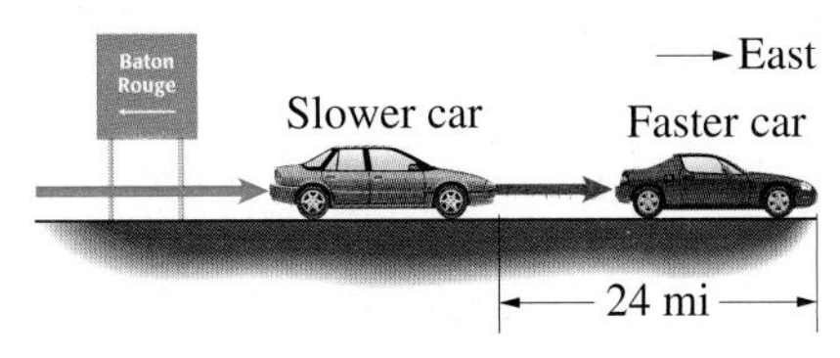

Figure 1

Now, construct a table like the one that follows. Fill in the information given in the problem, and use t for the time traveled by each car. Multiply rate by time to get the expressions for distances traveled.

	Rate	× Time	= Distance
Faster Car	63	t	$63t$
Slower Car	55	t	$55t$

Difference is 24 mi.

The quantities $63t$ and $55t$ represent the two distances. Refer to Figure 1, and notice that the *difference* between the larger distance and the smaller distance is 24 mi.

Step 3 **Write an equation.**

$$63t - 55t = 24$$

Step 4 **Solve.**

$$\begin{aligned} 63t - 55t &= 24 \\ 8t &= 24 \quad \text{Combine like terms.} \\ t &= 3 \quad \text{Divide by 8.} \end{aligned}$$

Step 5 **State the answer.** It will take the cars 3 hr to be 24 mi apart.

Step 6 **Check.** After 3 hr the faster car will have traveled $63 \times 3 = 189$ mi, and the slower car will have traveled $55 \times 3 = 165$ mi. Since $189 - 165 = 24$, the conditions of the problem are satisfied.

PROBLEM-SOLVING HINT

In motion problems like the one in Example 3, once you have filled in two pieces of information in each row of the table, you should automatically fill in the third piece of information, using the appropriate form of the formula relating distance, rate, and time. Set up the equation based on your sketch and the information in the table.

Work Problem 3 at the Side.

3 Solve each problem.

(a) From a point on a straight road, Lupe and Maria ride bicycles in opposite directions. Lupe rides 10 mph and Maria rides 12 mph. In how many hours will they be 55 mi apart?

(b) At a given hour, two steamboats leave a city in the same direction on a straight canal. One travels at 18 mph, and the other travels at 25 mph. In how many hours will the boats be 35 mi apart?

ANSWERS

3. (a) $2\frac{1}{2}$ hr **(b)** 5 hr

EXAMPLE 4 **Solving a Problem about Distance, Rate, and Time**

The Tickfaw River has a current of 3 mph. A motorboat takes as long to go 12 mi downstream as to go 8 mi upstream. What is the speed of the boat in still water?

Step 1 **Read** the problem again. We are looking for the speed of the boat in still water.

Step 2 **Assign a variable.** Let $x =$ the speed of the boat in still water. Because the current pushes the boat when the boat is going downstream, the speed of the boat downstream will be the sum of the speed of the boat and the speed of the current, $x + 3$ mph. Because the current slows down the boat when the boat is going upstream, the boat's speed upstream is given by the difference between the speed of the boat in still water and the speed of the current, $x - 3$ mph. See Figure 2.

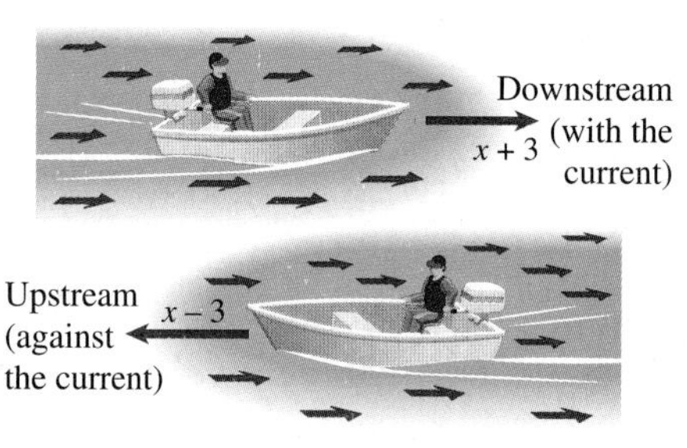

Figure 2

This information is summarized in the following table.

	d	r	t
Downstream	12	$x + 3$	
Upstream	8	$x - 3$	

Fill in the column representing time by using the formula $t = \frac{d}{r}$. Then the time upstream is the distance divided by the rate, or

$$t = \frac{d}{r} = \frac{8}{x - 3},$$

and the time downstream is also the distance divided by the rate, or

$$t = \frac{d}{r} = \frac{12}{x + 3}.$$

Now complete the table.

	d	r	t
Downstream	12	$x + 3$	$\frac{12}{x + 3}$
Upstream	8	$x - 3$	$\frac{8}{x - 3}$

Times are equal.

Step 3 **Write an equation.** According to the original problem, the time upstream equals the time downstream. The two times from the table must therefore be equal, giving the equation

$$\frac{12}{x + 3} = \frac{8}{x - 3}.$$

Continued on Next Page

Step 4 **Solve.** Begin by multiplying each side by $(x + 3)(x - 3)$.

$$(x+3)(x-3)\frac{12}{x+3} = (x+3)(x-3)\frac{8}{x-3}$$

$$12(x-3) = 8(x+3)$$

$$12x - 36 = 8x + 24 \quad \text{Distributive property}$$

$$4x = 60 \quad \text{Subtract } 8x\text{; add 36.}$$

$$x = 15 \quad \text{Divide by 4.}$$

Step 5 **State the answer.** The speed of the boat in still water is 15 mph.

Step 6 **Check.** First find the speed of the boat downstream, which is $15 + 3 = 18$ mph. Traveling 12 mi would take

$$t = \frac{d}{r} = \frac{12}{18} = \frac{2}{3} \text{ hr.}$$

The speed of the boat upstream is $15 - 3 = 12$ mph, and traveling 8 mi would take

$$t = \frac{d}{r} = \frac{8}{12} = \frac{2}{3} \text{ hr.}$$

The time upstream equals the time downstream, as required.

Work Problem 4 at the Side.

4 Solve each problem.

(a) A boat can go 20 mi against the current in the same time it can go 60 mi with the current. The current is flowing at 4 mph. Find the speed of the boat with no current.

(b) An airplane, maintaining a constant airspeed, takes as long to go 450 mi with the wind as it does to go 375 mi against the wind. If the wind is blowing at 15 mph, what is the speed of the plane?

OBJECTIVE 3 Solve problems about work. Suppose that you can mow your lawn in 4 hr. Then after 1 hr, you will have mowed $\frac{1}{4}$ of the lawn. After 2 hr, you will have mowed $\frac{2}{4}$ or $\frac{1}{2}$ of the lawn, and so on. This idea is generalized as follows.

Rate of Work

If a job can be completed in t units of time, then the rate of work is

$$\frac{1}{t} \textbf{ job per unit of time.}$$

PROBLEM-SOLVING HINT

The relationship between problems involving work and problems involving distance is a very close one. Recall that the formula $d = rt$ says that distance traveled is equal to rate of travel multiplied by time traveled. Similarly, the fractional part of a job accomplished is equal to the rate of work multiplied by the time worked. In the lawn mowing example, after 3 hr, the fractional part of the job done is

$$\underbrace{\frac{1}{4}}_{\text{Rate of work}} \cdot \underbrace{3}_{\text{Time worked}} = \underbrace{\frac{3}{4}}_{\text{Fractional part of job done}}.$$

After 4 hr, $\frac{1}{4}(4) = 1$ whole job has been done.

ANSWERS

4. **(a)** 8 mph **(b)** 165 mph

EXAMPLE 5 Solving a Problem about Work Rates

With spraying equipment, Mateo can paint the woodwork in a small house in 8 hr. His assistant, Chet, needs 14 hr to complete the same job painting by hand. If both Mateo and Chet work together, how long will it take them to paint the woodwork?

Step 1 **Read** the problem again. We are looking for time working together.

Step 2 **Assign a variable.** Let $x =$ the number of hours it will take for Mateo and Chet to paint the woodwork, working together.

Certainly, x will be less than 8, since Mateo alone can complete the job in 8 hr. Begin by making a table as shown. Remember that based on the previous discussion, Mateo's rate alone is $\frac{1}{8}$ job per hour, and Chet's rate is $\frac{1}{14}$ job per hour.

	Rate	Time Working Together	Fractional Part of the Job Done When Working Together
Mateo	$\frac{1}{8}$	x	$\frac{1}{8}x$
Chet	$\frac{1}{14}$	x	$\frac{1}{14}x$

Sum is 1 whole job.

Step 3 **Write an equation.** Since together Mateo and Chet complete 1 whole job, we must add their individual fractional parts and set the sum equal to 1.

$$\underbrace{\text{Fractional part done by Mateo}}_{\frac{1}{8}x} + \underbrace{\text{Fractional part done by Chet}}_{\frac{1}{14}x} = \underbrace{\text{1 whole job}}_{1}$$

Step 4 **Solve.**

$$56\left(\frac{1}{8}x + \frac{1}{14}x\right) = 56(1) \qquad \text{Multiply by the LCD, 56.}$$

$$56\left(\frac{1}{8}x\right) + 56\left(\frac{1}{14}x\right) = 56(1) \qquad \text{Distributive property}$$

$$7x + 4x = 56$$

$$11x = 56 \qquad \text{Combine like terms.}$$

$$x = \frac{56}{11} \qquad \text{Divide by 11.}$$

Step 5 **State the answer.** Working together, Mateo and Chet can paint the woodwork in $\frac{56}{11}$ hr, or $5\frac{1}{11}$ hr.

Step 6 **Check** to be sure the answer is correct.

NOTE

An alternative approach in work problems is to consider the part of the job that can be done in 1 hr. For instance, in Example 5 Mateo can do the entire job in 8 hr, and Chet can do it in 14 hr. Thus, their work rates, as we saw in Example 5, are $\frac{1}{8}$ and $\frac{1}{14}$, respectively. Since it takes them x hr to complete the job when working together, in 1 hr they can paint $\frac{1}{x}$ of the woodwork. The amount painted by Mateo in 1 hr plus the amount painted by Chet in 1 hr must equal the amount they can do together. This leads to the equation

Amount by Chet ↓

$$\text{Amount by Mateo} \rightarrow \frac{1}{8} + \frac{1}{14} = \frac{1}{x}. \leftarrow \text{Amount together}$$

Compare this with the equation in Example 5. Multiplying each side by $56x$ leads to

$$7x + 4x = 56,$$

the same equation found in the third line of Step 4 in the example. The same solution results.

Work Problem 5 at the Side.

5 Solve each problem.

(a) Michael can paint a room, working alone, in 8 hr. Lindsay can paint the same room, working alone, in 6 hr. How long will it take them if they work together?

(b) Roberto can detail his Camaro in 2 hr working alone. His brother Marco can do the job in 3 hr working alone. How long would it take them if they worked together?

Answers

5. **(a)** $3\frac{3}{7}$ hr **(b)** $1\frac{1}{5}$ hr

Focus on

Real-Data Applications

Upward Mobility*

As a struggling college student you have been driving the "Wimp," a 1989 Honda CRX, which has one saving grace—it gets 30 miles per gallon in the city. Now that you are graduating and are being recruited for your dream job, your first major purchase will be a new truck or sport-utility vehicle. In addition to car payments, you also must consider increased gasoline costs.

Based on past experience, you anticipate driving 15,000 miles a year. The new vehicle requires premium unleaded gasoline at $1.40 per gallon, instead of the $1.35 cost for regular unleaded that you use now. Currently, you use 500 gallons of gasoline per year, and you spend $675 per year on gasoline.

Gasoline usage $\quad \dfrac{15{,}000 \text{ mi}}{1 \text{ yr}} \div \dfrac{30 \text{ mi}}{1 \text{ gal}} = \dfrac{15{,}000 \text{ mi}}{1 \text{ yr}} \cdot \dfrac{1 \text{ gal}}{30 \text{ mi}} = 500 \text{ gal per yr}$

Current gasoline costs $\quad \dfrac{500 \text{ gal}}{1 \text{ yr}} \cdot \dfrac{\$1.35 \text{ gal}}{1 \text{ gal}} = \$675 \text{ per yr}$

You test drove a 4.8 Liter, V8, Chevy Tahoe that is rated to get 15 miles to the gallon (mpg). To calculate the additional costs for gasoline, you must first compute the cost for gasoline usage in the Tahoe and then subtract your current gasoline costs. Observe the *process* so that you can copy it to devise a **cost equation** that you can use to evaluate costs for all the other vehicles that you want to test drive. The *variable* quantity (the number of miles per gallon) is shown in blue.

Chevy Tahoe gasoline usage $\quad \dfrac{15{,}000 \text{ mi}}{1 \text{ yr}} \div \dfrac{15 \text{ mi}}{1 \text{ gal}} = \dfrac{15{,}000 \text{ mi}}{1 \text{ yr}} \cdot \dfrac{1 \text{ gal}}{15 \text{ mi}}$

Projected gasoline cost increase $\quad \dfrac{15{,}000 \text{ gal}}{15 \text{ yr}} \cdot \dfrac{\$1.40}{1 \text{ gal}} - \$675 = \dfrac{(15{,}000)(1.40)}{15} - 675$

For Group Discussion

1. Write a cost equation that computes the additional costs, y, for gasoline for a truck or SUV that gets x mpg. Assume that the new vehicle requires premium gasoline. (*Hint:* Replace the variable quantity in the process above with x.)

2. Evaluate the cost equation for the following vehicles to predict increased gasoline costs.
   (a) Honda CR-V, 4 L, 23 mpg
   (b) Ford Expedition, 5.4 L, V8, 14 mpg
   (c) Chevy Yukon, 5.3 L, V8, 12 mpg

3. A calculator graph of the cost equation is shown here.

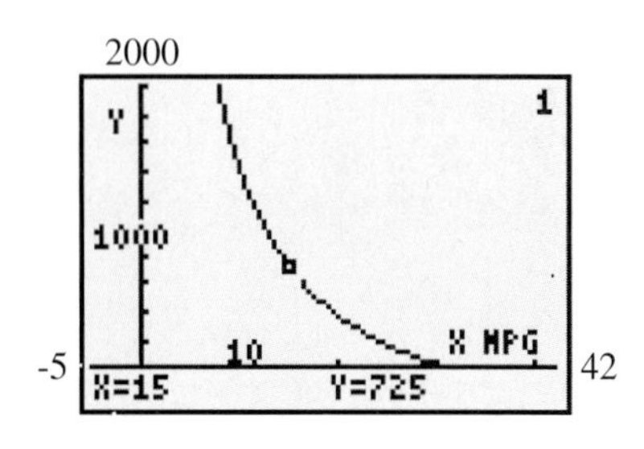

   (a) What can you conclude about the effect of *decreasing* mileage rating on the additional gasoline costs?

   (b) The graph crosses the x-axis at approximately $x = 31$. Explain why.

* Based on *Driving Rationally* by Patricia Stone, Tomball College. Original vehicle models and gasoline prices are used.

7.7 Exercises

FOR EXTRA HELP

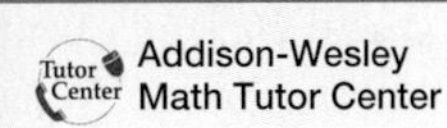
Addison-Wesley Math Tutor Center

MathXL

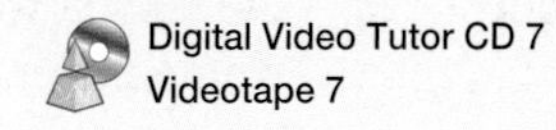
Digital Video Tutor CD 7 Videotape 7

Student's Solutions Manual

MyMathLab

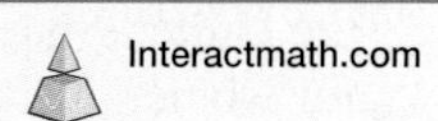
Interactmath.com

Use Steps 2 and 3 of the six-step problem solving method to set up the equation you would use to solve each problem. (Remember that Step 1 is to read the problem carefully.) Do not actually solve the equation. See Example 1.

1. The numerator of the fraction $\frac{5}{6}$ is increased by an amount so that the value of the resulting fraction is equivalent to $\frac{13}{3}$. By what amount was the numerator increased?

(a) Let $x =$ ______________. *(Step 2)*

(b) Write an expression for "the numerator of the fraction $\frac{5}{6}$ is increased by an amount."

(c) Set up an equation to solve the problem.

(Step 3)

2. If the same number is added to the numerator and subtracted from the denominator of $\frac{23}{12}$, the resulting fraction is equivalent to $\frac{3}{2}$. What is the number?

(a) Let $x =$ ______________. *(Step 2)*

(b) Write an expression for "a number is added to the numerator of $\frac{23}{12}$." Then write an expression for "the same number is subtracted from the denominator of $\frac{23}{12}$."

(c) Set up an equation to solve the problem.

(Step 3)

Use the six-step method to solve each problem. See Example 1.

3. In a certain fraction, the denominator is 4 less than the numerator. If 3 is added to both the numerator and the denominator, the resulting fraction is equivalent to $\frac{3}{2}$. What was the original fraction?

4. In a certain fraction, the denominator is 6 more than the numerator. If 3 is added to both the numerator and the denominator, the resulting fraction is equivalent to $\frac{5}{7}$. What was the original fraction?

5. The denominator of a certain fraction is three times the numerator. If 2 is added to the numerator and subtracted from the denominator, the resulting fraction is equivalent to 1. What was the original fraction?

6. The numerator of a certain fraction is four times the denominator. If 6 is added to both the numerator and the denominator, the resulting fraction is equivalent to 2. What was the original fraction?

7. One-sixth of a number is 5 more than the same number. What is the number?

8. One-third of a number is 2 more than one-sixth of the same number. What is the number?

9. A quantity, its $\frac{3}{4}$, its $\frac{1}{2}$, and its $\frac{1}{3}$, added together, become 93. What is the quantity? (*Source: Rhind Mathematical Papyrus.*)

10. A quantity, its $\frac{2}{3}$, its $\frac{1}{2}$, and its $\frac{1}{7}$, added together, become 33. What is the quantity? (*Source: Rhind Mathematical Papyrus.*)

Solve each problem. See Example 2.

11. At the Tyson Foods Invitational, Gail Devers of the United States won the indoor 60-m hurdle event for women in 7.10 sec. What was her rate? (*Source:* www.usatoday/sports.com)

12. In the 2002 Winter Games, Catriona LeMay Doan of Canada won the 500-m speed skating event for women. Her rate was 13.4048 m per sec. What was her time (to the nearest hundredth of a second)? (*Source: Sports Illustrated Sports Almanac.*)

13. The winner of the 2003 Daytona 500 (mile) race was Michael Waltrip, who drove his Chevrolet to victory with a rate of 133.870 mph. What was his time? (*Source: World Almanac and Book of Facts,* 2004.)

14. In 2003, Gil de Ferran drove his Dallara-Toyota to victory in the Indianapolis 500 (mile) race. His rate was 156.291 mph. What was his time? (*Source: World Almanac and Book of Facts,* 2004.)

15. Meseret Defar of Ethiopia won the women's 5000-m race in the 2004 Olympics with a time of 14.761 min. What was her rate? (*Source:* www.olympics.com)

16. The winner of the women's 1500-m race in the 2004 Olympics was Kelly Holmes of Great Britain with a time of 3.965 min. What was her rate? (*Source:* www.olympics.com)

Set up the equation you would use to solve each problem. Do not actually solve the equation. See Examples 3 and 4.

17. Luvenia can row 4 mph in still water. It takes as long to row 8 mi upstream as 24 mi downstream. How fast is the current? (Let x = speed of the current.)

	d	*r*	*t*
Upstream	8	$4 - x$	
Downstream	24	$4 + x$	

18. Julio flew his airplane 500 mi against the wind in the same time it took him to fly it 600 mi with the wind. If the speed of the wind was 10 mph, what was the average speed of his plane? (Let x = speed of the plane in still air.)

	d	*r*	*t*
Against the Wind	500	$x - 10$	
With the Wind	600	$x + 10$	

Solve each problem. See Examples 3 and 4.

19. If a migrating hawk travels m mph in still air, what is its rate when it flies into a steady headwind of 5 mph? What is its rate with a tailwind of 5 mph?

20. Suppose Stephanie walks D mi at R mph in the same time that Wally walks d mi at r mph. Give an equation relating D, R, d, and r.

21. A plane flies 350 mi with the wind in the same time that it can fly 310 mi against the wind. The plane has a still-air speed of 165 mph. Find the speed of the wind.

22. A boat can go 20 mi against a current in the same time that it can go 60 mi with the current. The current is 4 mph. Find the speed of the boat in still water.

23. The distance from Seattle, Washington, to Victoria, British Columbia, is about 148 mi by ferry. It takes about 4 hr less time to travel by the same ferry to Vancouver, British Columbia, a distance of about 74 mi. What is the average speed of the ferry?

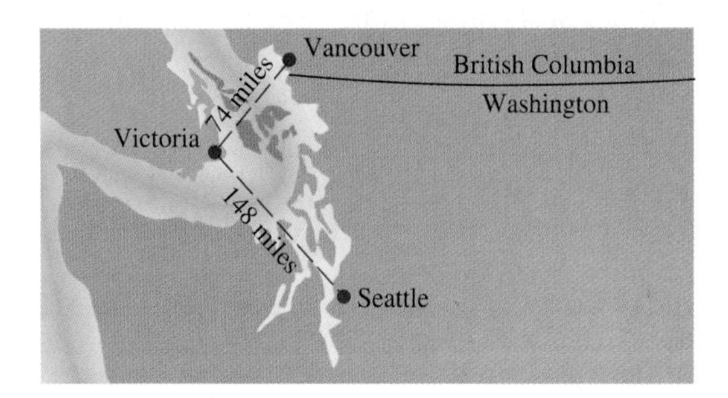

24. Alexis Glaser flew from Dallas to Indianapolis at 180 mph and then flew back at 150 mph. The trip at the slower speed took 1 hr longer than the trip at the higher speed. Find the distance between the two cities.

In Exercises 25 and 26, set up the equation you would use to solve each problem. Do not actually solve the equation. See Example 5.

25. Edwin Bedford can tune up his Chevy in 2 hr working alone. His son, Beau, can do the job in 3 hr working alone. How long would it take them if they worked together? (Let t represent the time working together.)

	r	t	w
Edwin		t	
Beau		t	

26. Working alone, Jorge can paint a room in 8 hr. Caterina can paint the same room working alone in 6 hr. How long will it take them if they work together? (Let x represent the time working together.)

	r	t	w
Jorge		x	
Caterina		x	

Solve each problem. See Example 5.

27. Lea can groom the horses in her boarding stable in 5 hr, while Tran needs 4 hr. How long will it take them to groom the horses if they work together?

28. Geraldo and Luisa Hernandez operate a small laundry. Luisa, working alone, can clean a day's laundry in 9 hr. Geraldo can clean a day's laundry in 8 hr. How long would it take them if they work together?

29. Todd's copier can do a printing job in 7 hr. Scott's copier can do the same job in 12 hr. How long would it take to do the job using both copiers?

30. A pump can pump the water out of a flooded basement in 10 hr. A smaller pump takes 12 hr. How long would it take to pump the water from the basement using both pumps?

31. Hilda can paint a room in 6 hr. Working together with Brenda, they can paint the room in $3\frac{3}{4}$ hr. How long would it take Brenda to paint the room by herself?

32. Grant can completely mess up his room in 15 min. If his cousin Wade helps him, they can completely mess up the room in $8\frac{4}{7}$ min. How long would it take Wade to mess up the room by himself?

33. An inlet pipe can fill a swimming pool in 9 hr, and an outlet pipe can empty the pool in 12 hr. Through an error, both pipes are left open. How long will it take to fill the pool?

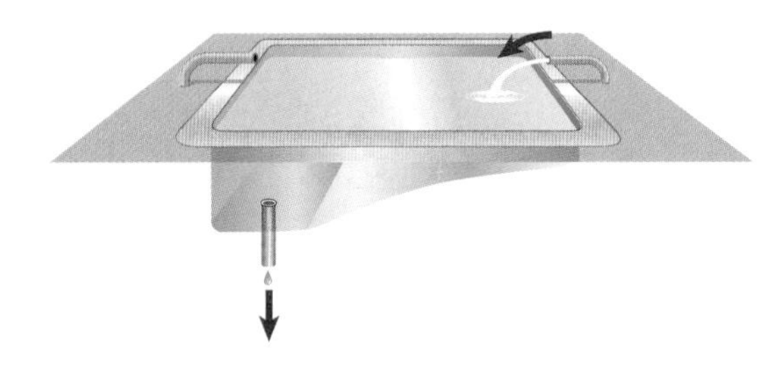

34. One pipe can fill a swimming pool in 6 hr, and another pipe can do it in 9 hr. How long will it take the two pipes working together to fill the pool $\frac{3}{4}$ full?

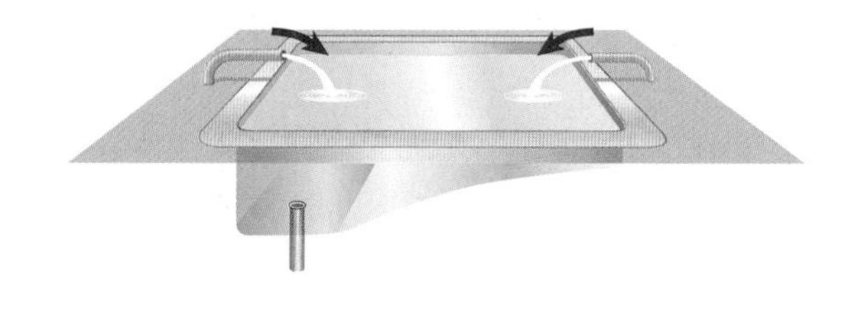

35. Refer to Exercise 33. Assume the error was discovered after both pipes had been running for 3 hr, and the outlet pipe was then closed. How much more time would then be required to fill the pool? (*Hint:* Consider how much of the job had been done when the error was discovered.)

36. A cold water faucet can fill a sink in 12 min, and a hot water faucet can fill it in 15 min. The drain can empty the sink in 25 min. If both faucets are on and the drain is open, how long will it take to fill the sink?

37. Refer to the table in Exercise 17. Suppose that a student made the error of interchanging the positions of the expressions $4 - x$ and $4 + x$, but used the correct method of setting up the equation, applying the formula $t = \frac{d}{r}$. Solve the equation the student used, and explain how the student should immediately know that there is something wrong in the setup.

7.8 Variation

OBJECTIVE 1 Solve problems about direct variation. Suppose that gasoline costs \$1.80 per gal. Then 1 gal costs \$1.80, 2 gal cost 2(\$1.80) = \$3.60, 3 gal cost 3(\$1.80) = \$5.40, and so on. Each time, the total cost is obtained by multiplying the number of gallons by the price per gallon. In general, if k equals the price per gallon and x equals the number of gallons, then the total cost y is equal to kx. Notice that as the number of gallons increases, the total cost increases.

The preceding discussion is an example of variation. Equations with fractions often result when discussing variation. As in the gasoline example, two variables *vary directly* if one is a constant multiple of the other.

OBJECTIVES

1. Solve problems about direct variation.
2. Solve problems about inverse variation.

Direct Variation

y varies directly as x if there exists a constant k such that

$$y = kx.$$

The constant k in the equation for direct variation is a numerical value, such as 1.80 in the gasoline price discussion.

EXAMPLE 1 Using Direct Variation

Suppose y varies directly as x, and $y = 20$ when $x = 4$. Find y when $x = 9$.

Since y varies directly as x, there is a constant k such that $y = kx$. We know that $y = 20$ when $x = 4$. Substituting these values into $y = kx$ and solving for k gives

$$y = kx$$
$$20 = k \cdot 4$$
$$k = 5.$$

Since $y = kx$ and $k = 5$,

$$y = 5x. \qquad \text{Let } k = 5.$$

When $x = 9$,

$$y = 5x = 5 \cdot 9 = 45. \qquad \text{Let } x = 9.$$

Thus, $y = 45$ when $x = 9$.

Work Problem 1 at the Side.

1. Solve each problem.

(a) If z varies directly as t, and $z = 11$ when $t = 4$, find z when $t = 32$.

(b) The circumference of a circle varies directly as the radius. A circle with a radius of 7 cm has a circumference of 43.96 cm. Find the circumference if the radius is 11 cm.

OBJECTIVE 2 Solve problems about inverse variation. Another common type of variation, in which the value of one variable increases while the value of another decreases, is called *inverse variation.* For example, an increase in the supply of an item causes a decrease in the price of the item.

ANSWERS

1. (a) 88 (b) 69.08 cm

2 Solve the problem.

Suppose z varies inversely as t, and $z = 8$ when $t = 2$. Find z when $t = 32$.

Inverse Variation

y* varies inversely as *x if there exists a constant k such that

$$y = \frac{k}{x}.$$

EXAMPLE 2 **Using Inverse Variation**

Suppose y varies inversely as x, and $y = 3$ when $x = 8$. Find y when $x = 6$.

Since y varies inversely as x, there is a constant k such that $y = \frac{k}{x}$. We know that $y = 3$ when $x = 8$, so we can find k.

$$y = \frac{k}{x}$$

$$3 = \frac{k}{8}$$

$$k = 24$$

Since $y = \frac{24}{x}$, we let $x = 6$ and solve for y.

$$y = \frac{24}{x} = \frac{24}{6} = 4$$

Therefore, when $x = 6$, $y = 4$.

Work Problem 2 at the Side.

3 Solve the problem.

The current in a simple electrical circuit varies inversely as the resistance. If the current is 80 amps when the resistance is 10 ohms, find the current if the resistance is 16 ohms.

EXAMPLE 3 **Using Inverse Variation**

In the manufacturing of a certain medical syringe, the cost of producing the syringe varies inversely as the number produced. If 10,000 syringes are produced, the cost is \$2 per unit. Find the cost per unit to produce 25,000 syringes.

Let x = the number of syringes produced

and c = the cost per unit.

Since c varies inversely as x, there is a constant k such that

$$c = \frac{k}{x}.$$

Find k by replacing c with 2 and x with 10,000.

$$2 = \frac{k}{10{,}000}$$

$$20{,}000 = k \qquad \text{Multiply by 10,000.}$$

Since $c = \frac{k}{x}$,

$$c = \frac{20{,}000}{25{,}000} = .80. \qquad \text{Let } k = 20{,}000 \text{ and } x = 25{,}000.$$

The cost per unit to make 25,000 syringes is \$.80.

Work Problem 3 at the Side.

Answers

2. $\frac{1}{2}$
3. 50 amps

7.8 Exercises

FOR EXTRA HELP Addison-Wesley Math Tutor Center · MathXL · Digital Video Tutor CD 7 Videotape 8 · Student's Solutions Manual · MyMathLab · Interactmath.com

1. **(a)** If the constant of variation is positive and y varies directly as x, then as x increases, y ________________ (increases/decreases).

(b) If the constant of variation is positive and y varies inversely as x, then as x increases, y ________________ (increases/decreases).

2. Bill Veeck was the owner of several major league baseball teams in the 1950s and 1960s. He was known to often sit in the stands and enjoy games with his paying customers. Here is a quote attributed to him:

"I have discovered in 20 years of moving around a ballpark, that the knowledge of the game is usually in inverse proportion to the price of the seats."

Explain in your own words the meaning of this statement. (To prove his point, Veeck once allowed the fans to vote on managerial decisions.)

Solve each problem involving direct variation. See Example 1.

3. If z varies directly as x, and $z = 30$ when $x = 8$, find z when $x = 4$.

4. If y varies directly as x, and $x = 27$ when $y = 6$, find x when $y = 2$.

5. If d varies directly as r, and $d = 200$ when $r = 40$, find d when $r = 60$.

6. If d varies directly as t, and $d = 150$ when $t = 3$, find d when $t = 5$.

Solve each problem involving inverse variation. See Example 2.

7. If z varies inversely as x, and $z = 50$ when $x = 2$, find z when $x = 25$.

8. If x varies inversely as y, and $x = 3$ when $y = 8$, find y when $x = 4$.

9. If m varies inversely as r, and $m = 12$ when $r = 8$, find m when $r = 16$.

10. If p varies inversely as q, and $p = 7$ when $q = 6$, find p when $q = 2$.

Solve each variation problem. See Examples 1–3.

11. For a given base, the area of a triangle varies directly as its height. Find the area of a triangle with a height of 6 in., if the area is 10 in.^{2} when the height is 4 in.

12. The interest on an investment varies directly as the rate of interest. If the interest is \$48 when the interest rate is 5%, find the interest when the rate is 4.2%.

13. Hooke's law for an elastic spring states that the distance a spring stretches varies directly with the force applied. If a force of 75 lb stretches a certain spring 16 in., how much will a force of 200 lb stretch the spring?

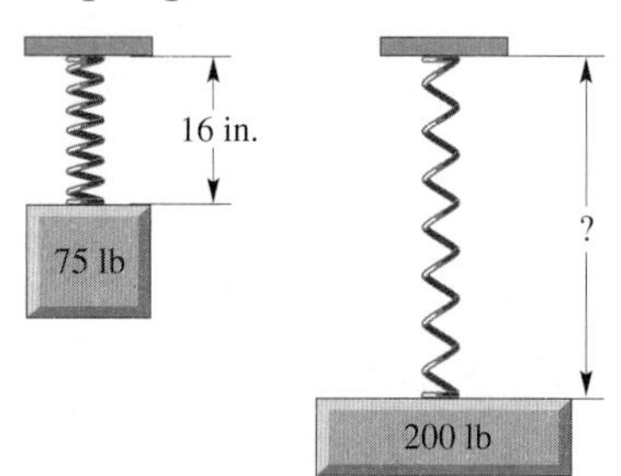

14. The pressure exerted by water at a given point varies directly with the depth of the point beneath the surface of the water. Water exerts 4.34 lb per in.^{2} for every 10 ft traveled below the water's surface. What is the pressure exerted on a scuba diver at 20 ft?

15. For a constant area, the length of a rectangle varies inversely as the width. The length of a rectangle is 27 ft when the width is 10 ft. Find the width of a rectangle with the same area if the length is 18 ft.

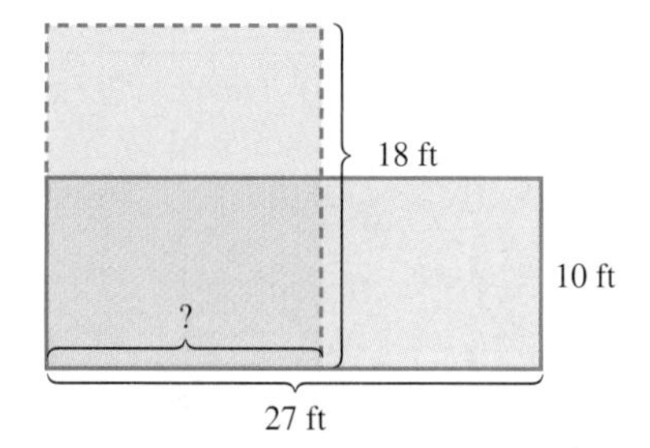

16. Over a specified distance, speed varies inversely with time. If a Dodge Viper on a test track goes a certain distance in one-half minute at 160 mph, what speed is needed to go the same distance in three-fourths minute?

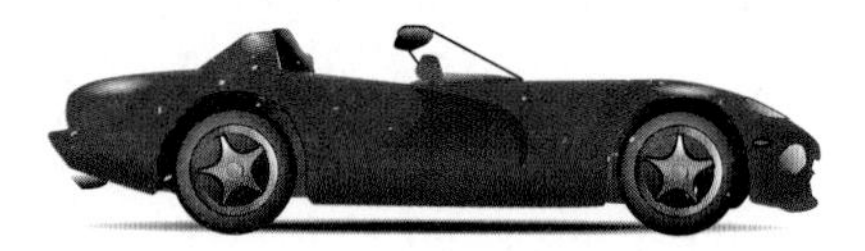

17. If the temperature is constant, the pressure of a gas in a container varies inversely as the volume of the container. If the pressure is 10 lb per ft^2 in a container with volume 3 ft^3, what is the pressure in a container with volume 1.5 ft^3?

18. The current in a simple electrical circuit varies inversely as the resistance. If the current is 20 amps when the resistance is 5 ohms, find the current when the resistance is 8 ohms.

19. In the inversion of raw sugar, the rate of change of the amount of raw sugar varies directly as the amount of raw sugar remaining. The rate is 200 kg per hr when there are 800 kg left. What is the rate of change per hour when only 100 kg are left?

20. The force required to compress a spring varies directly as the change in the length of the spring. If a force of 12 lb is required to compress a certain spring 3 in., how much force is required to compress the spring 5 in.?

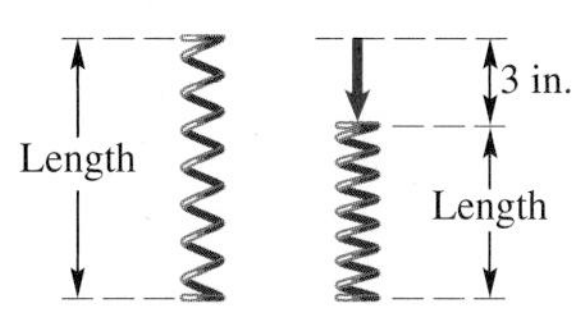

*Use personal experience or intuition to determine whether the situation suggests direct or inverse variation.**

21. The rate and the distance traveled by a pickup truck in 3 hr

22. The number of different lottery tickets you buy and your probability of winning that lottery

23. The number of days from now until December 25 and the magnitude of the frenzy of Christmas shopping

24. The amount of pressure put on the accelerator of a car and the speed of the car

25. The amount of gasoline that you pump and the amount of empty space left in your tank

26. The surface area of a balloon and its diameter

27. The amount of gasoline you pump and the amount you will pay

28. The number of days until the end of the baseball season and the number of home runs that Alex Rodriguez has

* The authors thank Linda Kodama of Kapi'olani Community College for suggesting the inclusion of exercises of this type.

*Two triangles are **similar** if they have the same shape (but not necessarily the same size). Similar triangles have side lengths that vary directly. The figure shows two similar triangles. Notice that the ratios of the corresponding sides are all equal to $\frac{3}{2}$:*

$$\frac{3}{2} = \frac{3}{2} \qquad \frac{4.5}{3} = \frac{3}{2} \qquad \frac{6}{4} = \frac{3}{2}.$$

2 3 4 3 4.5 6

If we know that two triangles are similar, we can set up a direct variation equation to solve for the length of an unknown side.

Find the length x, given that the pair of triangles are similar.

29.

3 3 2

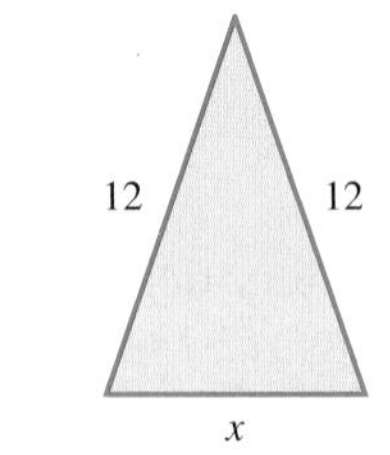

30.

3 5 x

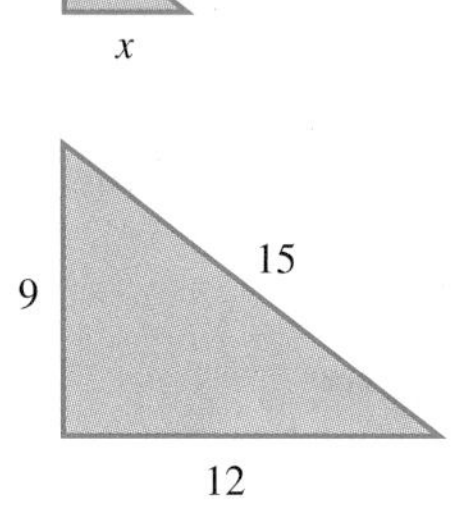

31.

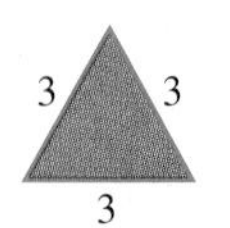

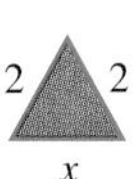

32.

$\frac{4}{3}$ x 2

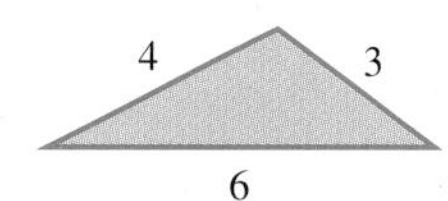

Use similar triangles and direct variation to solve each problem. (Source: The Guinness Book of World Records.)

33. One of the tallest candles ever constructed was exhibited at the 1897 Stockholm Exhibition. If it cast a shadow 5 ft long at the same time a vertical pole 32 ft high cast a shadow 2 ft long, how tall was the candle?

34. An enlarged version of the chair used by George Washington at the Constitutional Convention casts a shadow 18 ft long at the same time a vertical pole 12 ft high casts a shadow 4 ft long. How tall is the chair?

Chapter 7

SUMMARY

KEY TERMS

7.1	**rational expression**	The quotient of two polynomials with denominator not 0 is called a rational expression.
	lowest terms	A rational expression is written in lowest terms if the greatest common factor of its numerator and denominator is 1.
7.3	**least common denominator (LCD)**	The simplest expression that is divisible by all denominators is called the least common denominator.
7.5	**complex fraction**	A rational expression with one or more fractions in the numerator, denominator, or both, is called a complex fraction.
7.8	**direct variation**	y varies directly as x if there is a constant k such that $y = kx$.
	inverse variation	y varies inversely as x if there is a constant k such that $y = \frac{k}{x}$.

TEST YOUR WORD POWER

See how well you have learned the vocabulary in this chapter. Answers, with examples, follow the Quick Review.

1. A **rational expression** is
- **A.** an algebraic expression made up of a term or the sum of a finite number of terms with real coefficients and whole number exponents
- **B.** a polynomial equation of degree 2
- **C.** an expression with one or more fractions in the numerator, denominator, or both
- **D.** the quotient of two polynomials with denominator not 0.

2. A **complex fraction** is
- **A.** an algebraic expression made up of a term or the sum of a finite number of terms with real coefficients and whole number exponents
- **B.** a polynomial equation of degree 2
- **C.** a rational expression with one or more fractions in the numerator, denominator, or both
- **D.** the quotient of two polynomials with denominator not 0.

3. In a given set of fractions, the **least common denominator** is
- **A.** the smallest denominator of all the denominators
- **B.** the smallest expression that is divisible by all the denominators
- **C.** the largest integer that evenly divides the numerator and denominator of all the fractions
- **D.** the largest denominator of all the denominators.

4. If two positive quantities x and y are in **direct variation,** and the constant of variation is positive, then
- **A.** as x increases, y decreases
- **B.** as x increases, y increases
- **C.** as x increases, y remains constant
- **D.** as x decreases, y remains constant.

Quick Review

Concepts	Examples
7.1 *The Fundamental Property of Rational Expressions* To find the value(s) for which a rational expression is undefined, set the denominator equal to 0 and solve the equation.	Find the values for which the expression is undefined. $$\frac{x-4}{x^2-16}$$ $$x^2-16=0$$ $$(x-4)(x+4)=0$$ $$x-4=0 \quad \text{or} \quad x+4=0$$ $$x=4 \quad \text{or} \quad x=-4$$ The rational expression is undefined for 4 or -4.
Writing a Rational Expression in Lowest Terms	Write $\frac{x^2-1}{(x-1)^2}$ in lowest terms.
Step 1 Factor the numerator and denominator.	$$\frac{x^2-1}{(x-1)^2}=\frac{(x-1)(x+1)}{(x-1)(x-1)}=\frac{x+1}{x-1}$$
Step 2 Use the fundamental property to divide out common factors from the numerator and denominator.	
There are often several different equivalent forms of a rational expression.	Give two equivalent forms of $-\frac{x-1}{x+2}$. Distribute the $-$ sign in the numerator to get $\frac{-(x-1)}{x+2}$ or $\frac{-x+1}{x+2}$; do so in the denominator to get $\frac{x-1}{-x-2}$. (There are other forms as well.)
7.2 *Multiplying and Dividing Rational Expressions* **Multiplying Rational Expressions**	Multiply. $\frac{3x+9}{x-5}\cdot\frac{x^2-3x-10}{x^2-9}$
Step 1 Factor.	$$=\frac{3(x+3)}{x-5}\cdot\frac{(x-5)(x+2)}{(x+3)(x-3)}$$
Step 2 Multiply numerators and multiply denominators.	$$=\frac{3(x+3)(x-5)(x+2)}{(x-5)(x+3)(x-3)}$$
Step 3 Write in lowest terms.	$$=\frac{3(x+2)}{x-3}$$
Dividing Rational Expressions	Divide. $\frac{2x+1}{x+5}\div\frac{6x^2-x-2}{x^2-25}$
Step 1 Multiply the first rational expression by the reciprocal of the second.	$$=\frac{2x+1}{x+5}\cdot\frac{x^2-25}{6x^2-x-2}$$
Step 2 Factor.	$$=\frac{2x+1}{x+5}\cdot\frac{(x+5)(x-5)}{(2x+1)(3x-2)}$$
Steps 3 and 4 Multiply the numerators and the denominators, and write in lowest terms.	$$=\frac{x-5}{3x-2}$$

Concepts	*Examples*
7.3 *Least Common Denominators* **Finding the LCD**	Find the LCD for $\frac{3}{k^2 - 8k + 16}$ and $\frac{1}{4k^2 - 16k}$.
Step 1 Factor each denominator into prime factors.	$k^2 - 8k + 16 = (k - 4)^2$ $4k^2 - 16k = 4k(k - 4)$
Step 2 List each different factor the greatest number of times it appears.	$\text{LCD} = (k - 4)^2 \cdot 4 \cdot k$
Step 3 Multiply the factors from Step 2 to get the LCD.	$= 4k(k - 4)^2$
Writing a Rational Expression with a Specified Denominator	Find the numerator: $\frac{5}{2z^2 - 6z} = \frac{}{4z^3 - 12z^2}$.
Step 1 Factor both denominators.	$\frac{5}{2z(z - 3)} = \frac{}{4z^2(z - 3)}$
Step 2 Decide what factors the denominator must be multiplied by to equal the specified denominator.	$2z(z - 3)$ must be multiplied by $2z$.
Step 3 Multiply the rational expression by that factor divided by itself (multiply by 1).	$\frac{5}{2z(z - 3)} \cdot \frac{2z}{2z} = \frac{10z}{4z^2(z - 3)} = \frac{10z}{4z^3 - 12z^2}$
7.4 *Adding and Subtracting Rational Expressions* **Adding Rational Expressions**	Add. $\frac{2}{3m + 6} + \frac{m}{m^2 - 4}$
Step 1 Find the LCD.	$3m + 6 = 3(m + 2)$ $m^2 - 4 = (m + 2)(m - 2)$ The LCD is $3(m + 2)(m - 2)$.
Step 2 Rewrite each rational expression with the LCD as denominator.	$= \frac{2(m - 2)}{3(m + 2)(m - 2)} + \frac{3m}{3(m + 2)(m - 2)}$
Step 3 Add the numerators to get the numerator of the sum. The LCD is the denominator of the sum.	$= \frac{2m - 4 + 3m}{3(m + 2)(m - 2)}$
Step 4 Write in lowest terms.	$= \frac{5m - 4}{3(m + 2)(m - 2)}$
Subtracting Rational Expressions Follow the same steps as for addition, but subtract in Step 3.	Subtract. $\frac{6}{k + 4} - \frac{2}{k}$ The LCD is $k(k + 4)$. $\frac{6k}{(k + 4)k} - \frac{2(k + 4)}{k(k + 4)} = \frac{6k - 2(k + 4)}{k(k + 4)}$ $= \frac{6k - 2k - 8}{k(k + 4)}$ Be careful with signs when subtracting the numerators. $= \frac{4k - 8}{k(k + 4)}$ or $\frac{4(k - 2)}{k(k + 4)}$

Concepts	Examples

7.5 *Complex Fractions*

Simplifying Complex Fractions

Simplify.

Method 1 Simplify the numerator and denominator separately. Then divide the simplified numerator by the simplified denominator.

Method 1
$$\frac{\frac{1}{a} - a}{1 - a} = \frac{\frac{1}{a} - \frac{a^2}{a}}{1 - a} = \frac{\frac{1 - a^2}{a}}{1 - a}$$
$$= \frac{1 - a^2}{a} \cdot \frac{1}{1 - a}$$
$$= \frac{(1 - a)(1 + a)}{a(1 - a)} = \frac{1 + a}{a}$$

Method 2 Multiply the numerator and denominator of the complex fraction by the LCD of all the denominators in the complex fraction. Write in lowest terms.

Method 2
$$\frac{\frac{1}{a} - a}{1 - a} = \frac{\left(\frac{1}{a} - a\right)a}{(1 - a)a} = \frac{\frac{a}{a} - a^2}{(1 - a)a}$$
$$= \frac{1 - a^2}{(1 - a)a} = \frac{(1 + a)(1 - a)}{(1 - a)a}$$
$$= \frac{1 + a}{a}$$

7.6 *Solving Equations with Rational Expressions*

Solving Equations with Rational Expressions

Solve $\frac{x}{x - 3} + \frac{4}{x + 3} = \frac{18}{x^2 - 9}$.

Step 1 Find the LCD of all denominators in the equation.

The LCD is $(x - 3)(x + 3)$. Note that 3 and -3 cannot be solutions, as they cause a denominator to equal 0.

$$\frac{x}{x - 3} + \frac{4}{x + 3} = \frac{18}{(x - 3)(x + 3)}$$ Factor.

Step 2 Multiply each side of the equation by this LCD.

$$x(x + 3) + 4(x - 3) = 18$$ Multiply by $(x - 3)(x + 3)$.

Step 3 Solve the resulting equation, which should have no fractions.

$$x^2 + 3x + 4x - 12 = 18$$ Distributive property

$$x^2 + 7x - 30 = 0$$ Subtract 18; standard form

$$(x - 3)(x + 10) = 0$$ Factor.

$x - 3 = 0$ or $x + 10 = 0$ Zero-factor property

Step 4 Check each proposed solution, and reject any value that causes an original denominator to equal 0.

Reject $\rightarrow x = 3$ or $x = -10$

The only solution is -10.

Concepts	*Examples*

7.7 *Applications of Rational Expressions*

Solving Problems about Distance

Use the six-step method.

Step 1 **Read** the problem carefully.

On a trip from Sacramento to Monterey, Marge traveled at an average speed of 60 mph. The return trip, at an average speed of 64 mph, took $\frac{1}{4}$ hr less. How far did she travel between the two cities?

Step 2 **Assign a variable.** Use a table to identify distance, rate, and time. Solve $d = rt$ for the unknown quantity in the table.

Let $x =$ the unknown distance.

	d	r	$t = \frac{d}{r}$
Going	x	60	$\frac{x}{60}$
Returning	x	64	$\frac{x}{64}$

Step 3 **Write an equation.** From the wording in the problem, decide the relationship between the quantities. Use those expressions to write an equation.

Since the time for the return trip was $\frac{1}{4}$ hr less, the time going equals the time returning plus $\frac{1}{4}$.

$$\frac{x}{60} = \frac{x}{64} + \frac{1}{4}$$

Step 4 **Solve** the equation.

$$16x = 15x + 240 \quad \text{Multiply by 960.}$$

$$x = 240 \quad \text{Subtract } 15x.$$

Step 5 **State the answer.**

She traveled 240 mi.

Step 6 **Check** the solution.

The trip there took $\frac{240}{60} = 4$ hr, while the return trip took $\frac{240}{64} = 3\frac{3}{4}$ hr, which is $\frac{1}{4}$ hr less time. The solution checks.

Solving Problems about Work

Step 1 **Read** the problem carefully.

It takes the regular mail carrier 6 hr to cover her route. A substitute takes 8 hr to cover the same route. How long would it take them to cover the route together?

Step 2 **Assign a variable.** State what the variable represents. Put the information from the problem in a table. If a job is done in t units of time, then the rate is $\frac{1}{t}$.

Let $x =$ the number of hours to cover the route together.

The rate of the regular carrier is $\frac{1}{6}$ job per hour; the rate of the substitute is $\frac{1}{8}$ job per hour. Multiply rate by time to get the fractional part of the job done.

	Rate	Time	Part of the Job Done
Regular	$\frac{1}{6}$	x	$\frac{1}{6}x$
Substitute	$\frac{1}{8}$	x	$\frac{1}{8}x$

Step 3 **Write an equation.** The sum of the fractional parts should equal 1 (whole job).

The equation is $\frac{1}{6}x + \frac{1}{8}x = 1$.

Step 4 **Solve** the equation.

The solution of the equation is $\frac{24}{7}$. The solution checks, because $\frac{1}{6}\left(\frac{24}{7}\right) + \frac{1}{8}\left(\frac{24}{7}\right) = 1$ is true.

Steps 5 and 6 **State the answer** and **check** the solution.

It would take them $\frac{24}{7}$ or $3\frac{3}{7}$ hr to cover the route together.

Concepts	*Examples*
7.8 *Variation*	
Solving Variation Problems	If y varies inversely as x, and $y = 4$ when $x = 9$, find y when $x = 6$.
Step 1 Write the variation equation. Use $y = kx$ for direct variation, $y = \frac{k}{x}$ for inverse variation.	The equation for inverse variation is $y = \frac{k}{x}$.
Step 2 Find k by substituting the given values of x and y into the equation.	$4 = \frac{k}{9}$ $k = 36$
Step 3 Write the equation with the value of k from Step 2 and the given value of x or y. Solve for the remaining variable.	$y = \frac{36}{x}$ $k = 36$ $y = \frac{36}{6}$ Let $x = 6$. $y = 6$

Answers to Test Your Word Power

1. D; *Examples:* $-\frac{3}{4y}$, $\frac{5x^3}{x+2}$, $\frac{a+3}{a^2-4a-5}$
2. C; *Examples:* $\frac{\frac{2}{3}}{\frac{4}{7}}$, $\frac{x-\frac{1}{y}}{x+\frac{1}{y}}$, $\frac{\frac{2}{a+1}}{a^2-1}$
3. B; *Examples:* The least common denominator of $\frac{1}{2}$, $\frac{1}{3}$, and $\frac{1}{4}$ is 12. The least common denominator of $\frac{1}{x}$ and $\frac{1}{x+1}$ is $x(x+1)$.
4. B; *Example:* The equation $y = 3x$ represents direct variation. When $x = 2$, $y = 6$. If x increases to 3, then y increases to $3(3) = 9$.

Chapter 7

REVIEW EXERCISES

[7.1] *Find the value(s) of the variable for which each rational expression is undefined.*

1. $\dfrac{4}{x - 3}$

2. $\dfrac{y + 3}{2y}$

3. $\dfrac{m - 2}{m^2 - 2m - 3}$

4. $\dfrac{2k + 1}{3k^2 + 17k + 10}$

*Find the numerical value of each rational expression when **(a)** $x = -2$ and **(b)** $x = 4$.*

5. $\dfrac{x^2}{x - 5}$

6. $\dfrac{4x - 3}{5x + 2}$

7. $\dfrac{3x}{x^2 - 4}$

8. $\dfrac{x - 1}{x + 2}$

Write each rational expression in lowest terms.

9. $\dfrac{5a^3b^3}{15a^4b^2}$

10. $\dfrac{m - 4}{4 - m}$

11. $\dfrac{4x^2 - 9}{6 - 4x}$

12. $\dfrac{4p^2 + 8pq - 5q^2}{10p^2 - 3pq - q^2}$

Write four equivalent expressions for each fraction.

13. $-\dfrac{4x - 9}{2x + 3}$

14. $\dfrac{8 - 3x}{3 + 6x}$

[7.2] *Find each product or quotient. Write each answer in lowest terms.*

15. $\dfrac{8x^2}{12x^5} \cdot \dfrac{6x^4}{2x}$

16. $\dfrac{9m^2}{(3m)^4} \div \dfrac{6m^5}{36m}$

17. $\dfrac{x - 3}{4} \cdot \dfrac{5}{2x - 6}$

18. $\dfrac{2r + 3}{r - 4} \cdot \dfrac{r^2 - 16}{6r + 9}$

19. $\dfrac{3q + 3}{5 - 6q} \div \dfrac{4q + 4}{2(5 - 6q)}$

20. $\dfrac{y^2 - 6y + 8}{y^2 + 3y - 18} \div \dfrac{y - 4}{y + 6}$

21. $\dfrac{2p^2 + 13p + 20}{p^2 + p - 12} \cdot \dfrac{p^2 + 2p - 15}{2p^2 + 7p + 5}$

22. $\dfrac{3z^2 + 5z - 2}{9z^2 - 1} \cdot \dfrac{9z^2 + 6z + 1}{z^2 + 5z + 6}$

[7.3] *Find the least common denominator for each list of fractions.*

23. $\frac{1}{8}, \frac{5}{12}, \frac{7}{32}$

24. $\frac{4}{9y}, \frac{7}{12y^2}, \frac{5}{27y^4}$

25. $\frac{1}{m^2 + 2m}, \frac{4}{m^2 + 7m + 10}$

26. $\frac{3}{x^2 + 4x + 3}, \frac{5}{x^2 + 5x + 4}, \frac{2}{x^2 + 7x + 12}$

Rewrite each rational expression with the given denominator.

27. $\frac{5}{8} = \frac{}{56}$

28. $\frac{10}{k} = \frac{}{4k}$

29. $\frac{3}{2a^3} = \frac{}{10a^4}$

30. $\frac{9}{x - 3} = \frac{}{18 - 6x}$

31. $\frac{-3y}{2y - 10} = \frac{}{50 - 10y}$

32. $\frac{4b}{b^2 + 2b - 3} = \frac{}{(b + 3)(b - 1)(b + 2)}$

[7.4] *Add or subtract as indicated. Write each answer in lowest terms.*

33. $\frac{10}{x} + \frac{5}{x}$

34. $\frac{6}{3p} - \frac{12}{3p}$

35. $\frac{9}{k} - \frac{5}{k - 5}$

36. $\frac{4}{y} + \frac{7}{7 + y}$

37. $\frac{m}{3} - \frac{2 + 5m}{6}$

38. $\frac{12}{x^2} - \frac{3}{4x}$

39. $\frac{5}{a - 2b} + \frac{2}{a + 2b}$

40. $\frac{4}{k^2 - 9} - \frac{k + 3}{3k - 9}$

41. $\frac{8}{z^2 + 6z} - \frac{3}{z^2 + 4z - 12}$

42. $\frac{11}{2p - p^2} - \frac{2}{p^2 - 5p + 6}$

[7.5] *Simplify each complex fraction.*

43. $\dfrac{\frac{a^4}{b^2}}{\frac{a^3}{b}}$

44. $\dfrac{\frac{y - 3}{y}}{\frac{y + 3}{4y}}$

45. $\dfrac{\frac{3m + 2}{m}}{\frac{2m - 5}{6m}}$

46. $\dfrac{\frac{1}{p} - \frac{1}{q}}{\frac{1}{q - p}}$

47. $\dfrac{x + \frac{1}{w}}{x - \frac{1}{w}}$

48. $\dfrac{\frac{1}{r + t} - 1}{\frac{1}{r + t} + 1}$

[7.6] *Solve each equation. Check your solutions.*

49. $\frac{k}{5} - \frac{2}{3} = \frac{1}{2}$

50. $\frac{4 - z}{z} + \frac{3}{2} = \frac{-4}{z}$

51. $\frac{x}{2} - \frac{x - 3}{7} = -1$

52. $\frac{3y - 1}{y - 2} = \frac{5}{y - 2} + 1$

53. $\frac{3}{m - 2} + \frac{1}{m - 1} = \frac{7}{m^2 - 3m + 2}$

Solve for the specified variable.

54. $m = \frac{Ry}{t}$ for t

55. $x = \frac{3y - 5}{4}$ for y

56. $\frac{1}{r} - \frac{1}{s} = \frac{1}{t}$ for t

[7.7] *Solve each problem. Use the six-step method.*

57. One-fourth of a number is 9 less than the same number. What is the number?

58. In a certain fraction, the denominator is 5 less than the numerator. If 5 is added to both the numerator and the denominator, the resulting fraction is equivalent to $\frac{5}{4}$. Find the original fraction.

59. The denominator of a certain fraction is six times the numerator. If 3 is added to the numerator and subtracted from the denominator, the resulting fraction is equivalent to $\frac{2}{5}$. Find the original fraction.

60. On August 18, 1996, Scott Sharp won the True Value 200-mi Indy race driving a Ford with an average speed of 130.934 mph. What was his time? (*Source: Sports Illustrated Sports Almanac.*)

61. A man can plant his garden in 5 hr, working alone. His daughter can do the same job in 8 hr. How long would it take them if they worked together?

62. At a given hour, two steamboats leave a city in the same direction on a straight canal. One travels at 18 mph, and the other travels at 25 mph. In how many hours will the boats be 70 mi apart?

[7.8] *Solve each problem.*

63. If a parallelogram has a fixed area, the height varies inversely as the base. A parallelogram has a height of 8 cm and a base of 12 cm. Find the height if the base is changed to 24 cm.

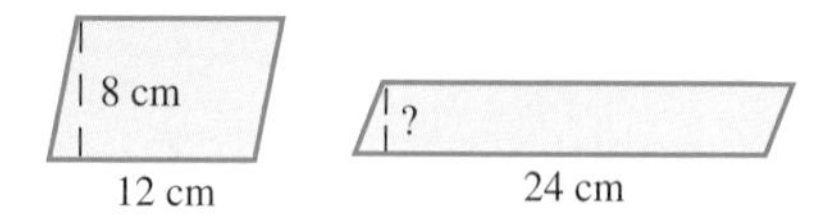

64. If y varies directly as x, and $x = 12$ when $y = 5$, find x when $y = 3$.

MIXED REVIEW EXERCISES

Perform the indicated operations.

65. $\dfrac{4}{m-1} - \dfrac{3}{m+1}$

66. $\dfrac{8p^5}{5} \div \dfrac{2p^3}{10}$

67. $\dfrac{r-3}{8} \div \dfrac{3r-9}{4}$

68. $\dfrac{\dfrac{5}{x} - 1}{\dfrac{5-x}{3x}}$

69. $\dfrac{4}{z^2 - 2z + 1} - \dfrac{3}{z^2 - 1}$

Solve.

70. $F = \dfrac{k}{d - D}$ for d

71. $\dfrac{2}{z} - \dfrac{z}{z+3} = \dfrac{1}{z+3}$

72. Anne Kelly flew her plane 400 km with the wind in the same time it took her to go 200 km against the wind. The speed of the wind is 50 km per hr. Find the speed of the plane in still air.

73. "If Joe can paint a house in 3 hours, and Sam can paint the same house in 5 hours, how long does it take for them to do it together?" (From the movie *Little Big League*.)

74. In rectangles of constant area, length and width vary inversely. When the length is 24, the width is 2. What is the width when the length is 12?

75. If w varies inversely as z, and $w = 16$ when $z = 3$, find w when $z = 2$.

Chapter 7

TEST

Study Skills Workbook
Activity 12: Final Exams

1. Find any values for which $\dfrac{3x - 1}{x^2 - 2x - 8}$ is undefined.

2. Find the numerical value of $\dfrac{6r + 1}{2r^2 - 3r - 20}$ when

(a) $r = -2$ and **(b)** $r = 4$.

3. Write four rational expressions equivalent to $-\dfrac{6x - 5}{2x + 3}$.

Write each rational expression in lowest terms.

4. $\dfrac{-15x^6y^4}{5x^4y}$

5. $\dfrac{6a^2 + a - 2}{2a^2 - 3a + 1}$

Multiply or divide. Write each answer in lowest terms.

6. $\dfrac{5(d - 2)}{9} \div \dfrac{3(d - 2)}{5}$

7. $\dfrac{6k^2 - k - 2}{8k^2 + 10k + 3} \cdot \dfrac{4k^2 + 7k + 3}{3k^2 + 5k + 2}$

8. $\dfrac{4a^2 + 9a + 2}{3a^2 + 11a + 10} \div \dfrac{4a^2 + 17a + 4}{3a^2 + 2a - 5}$

Find the least common denominator for each list of fractions.

9. $\dfrac{-3}{10p^2}, \dfrac{21}{25p^3}, \dfrac{-7}{30p^5}$

10. $\dfrac{r + 1}{2r^2 + 7r + 6}, \dfrac{-2r + 1}{2r^2 - 7r - 15}$

Rewrite each rational expression with the given denominator.

11. $\dfrac{15}{4p} = \dfrac{}{64p^3}$

12. $\dfrac{3}{6m - 12} = \dfrac{}{42m - 84}$

Add or subtract. Write each answer in lowest terms.

13. $\dfrac{4x + 2}{x + 5} + \dfrac{-2x + 8}{x + 5}$

14. $\dfrac{-4}{y + 2} + \dfrac{6}{5y + 10}$

15. $\dfrac{x + 1}{3 - x} - \dfrac{x^2}{x - 3}$

16. $\dfrac{3}{2m^2 - 9m - 5} - \dfrac{m + 1}{2m^2 - m - 1}$

1. ______
2. (a) ______
   (b) ______
3. ______
4. ______
5. ______
6. ______
7. ______
8. ______
9. ______
10. ______
11. ______
12. ______
13. ______
14. ______
15. ______
16. ______

Simplify each complex fraction.

17. ____________________

17. $\dfrac{\dfrac{2p}{k^2}}{\dfrac{3p^2}{k^3}}$

18. ____________________

18. $\dfrac{\dfrac{1}{x+3}-1}{1+\dfrac{1}{x+3}}$

19. ____________________

19. Solve the equation $\dfrac{2x}{x-3}+\dfrac{1}{x+3}=\dfrac{-6}{x^2-9}$. Be sure to check your answer(s).

20. ____________________

20. Solve the formula $F=\dfrac{k}{d-D}$ for D.

Solve each problem.

21. ____________________

21. If the same number is added to the numerator and subtracted from the denominator of $\frac{5}{6}$, the resulting fraction is equivalent to $\frac{1}{10}$. What is the number?

22. ____________________

22. A boat goes 7 mph in still water. It takes as long to go 20 mi upstream as 50 mi downstream. Find the speed of the current.

23. ____________________

23. A man can paint a room in his house, working alone, in 5 hr. His wife can do the job in 4 hr. How long will it take them to paint the room if they work together?

24. ____________________

24. If x varies directly as y, and $x = 12$ when $y = 4$, find x when $y = 9$.

25. ____________________

25. Under certain conditions, the length of time that it takes for fruit to ripen during the growing season varies inversely as the average maximum temperature during the season. If it takes 25 days for fruit to ripen with an average maximum temperature of 80°F, find the number of days it would take at 75°F. Round your answer to the nearest whole number.

Cumulative Review Exercises

CHAPTERS R–7

1. Evaluate $3 + 4\left(\frac{1}{2} - \frac{3}{4}\right)$.

Solve. Graph the solutions in Exercises 5 and 6.

2. $3(2y - 5) = 2 + 5y$

3. $A = \frac{1}{2}bh$ for b

4. $\frac{2 + m}{2 - m} = \frac{3}{4}$

5. $5y \leq 6y + 8$

6. $5m - 9 > 2m + 3$

Sketch each graph.

7. $y = -3x + 2$

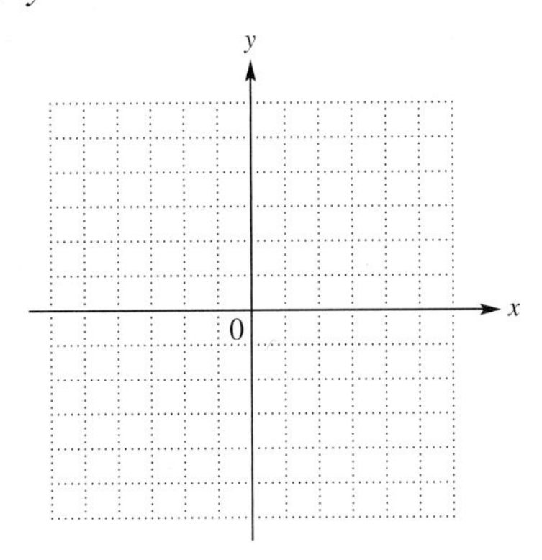

8. $y \geq 2x + 3$

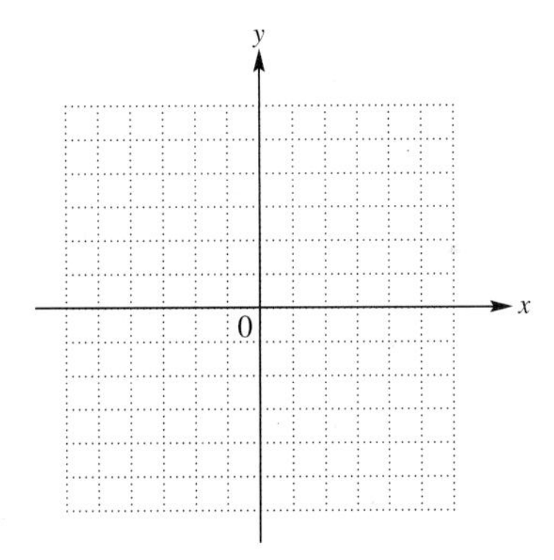

9. Solve by elimination.

$$3x + 4y = 5$$
$$6x + 7y = 8$$

10. Solve by substitution.

$$y = -3x + 1$$
$$x + 2y = -3$$

Simplify each expression. Write with only positive exponents.

11. $\frac{(2x^3)^{-1} \cdot x}{2^3x^5}$

12. $\frac{(m^{-2})^3m}{m^5m^{-4}}$

13. $\frac{2p^3q^4}{8p^5q^3}$

Perform the indicated operations.

14. $(2k^2 + 3k) - (k^2 + k - 1)$

15. $8x^2y^2(9x^4y^5)$

16. $(2a - b)^2$

17. $(y^2 + 3y + 5)(3y - 1)$

18. $\frac{12p^3 + 2p^2 - 12p + 4}{2p - 2}$

Factor completely.

19. $8t^2 + 10tv + 3v^2$

20. $8r^2 - 9rs + 12s^2$

21. $16x^4 - 1$

Solve each equation.

22. $r^2 = 2r + 15$

23. $(r - 5)(2r + 1)(3r - 2) = 0$

Solve each problem.

24. One number is 4 more than another. The product of the numbers is 2 less than the smaller number. Find the smaller number.

25. The length of a rectangle is 2 m less than twice the width. The area is 60 m^2. Find the width of the rectangle.

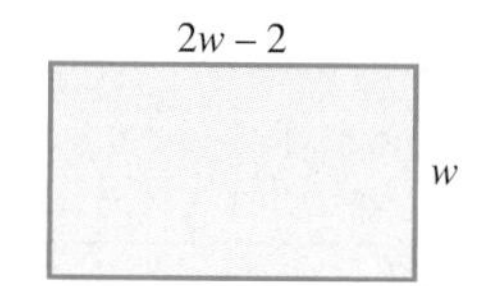

26. For what value(s) of t is $\frac{2 + t}{t^2 - 4}$ undefined?

27. One of the following is equal to 1 for *all* real numbers. Which one is it?

A. $\frac{k^2 + 2}{k^2 + 2}$ **B.** $\frac{4 - m}{4 - m}$ **C.** $\frac{2x + 9}{2x + 9}$ **D.** $\frac{x^2 - 1}{x^2 - 1}$

28. Which one of the following rational expressions is *not* equivalent to $\frac{4 - 3x}{7}$?

A. $-\frac{-4 + 3x}{7}$ **B.** $-\frac{4 - 3x}{-7}$ **C.** $\frac{-4 + 3x}{-7}$ **D.** $\frac{-(3x + 4)}{7}$

Perform each operation, and write the answer in lowest terms.

29. $\frac{5}{q} - \frac{1}{q}$

30. $\frac{3}{7} + \frac{4}{r}$

31. $\frac{4}{5q - 20} - \frac{1}{3q - 12}$

32. $\frac{2}{k^2 + k} - \frac{3}{k^2 - k}$

33. $\frac{7z^2 + 49z + 70}{16z^2 + 72z - 40} \div \frac{3z + 6}{4z^2 - 1}$

34. Simplify the complex fraction $\frac{\frac{4}{a} + \frac{5}{2a}}{\frac{7}{6a} - \frac{1}{5a}}$.

Solve each equation. Check your solutions.

35. $\frac{r + 2}{5} = \frac{r - 3}{3}$

36. $\frac{1}{x} = \frac{1}{x + 1} + \frac{1}{2}$

Solve each problem.

37. On a business trip, Monica traveled to her destination at an average speed of 60 mph. Coming home, her average speed was 50 mph, and the trip took $\frac{1}{2}$ hr longer. How far did she travel each way?

38. Chandler can weed the yard in 3 hr. Ross can weed the yard in 2 hr. How long would it take them if they worked together?

Roots and Radicals

The London Eye opened on New Year's Eve in 1999. This unique Ferris wheel features 32 observation capsules and has a diameter of 135 meters. Located on the bank of the Thames River, it faces the Houses of Parliament and is the sixth tallest structure in London. (*Source:* www.londoneye.com)

In the Focus on Real-Data Applications at the end of Section 8.6, we use a formula involving *radicals,* the subject of this chapter, to determine the truth of the claim that passengers on the London Eye can see Windsor Castle, 25 miles away.

8.1 Evaluating Roots

OBJECTIVES

1. Find square roots.
2. Decide whether a given root is rational, irrational, or not a real number.
3. Find decimal approximations for irrational square roots.
4. Use the Pythagorean formula.
5. Find higher roots.

In **Section 1.1,** we discussed the idea of the *square* of a number. Recall that squaring a number means multiplying the number by itself.

$$\text{If } a = 8, \quad \text{then} \quad a^2 = 8 \cdot 8 = 64.$$

$$\text{If } a = -4, \quad \text{then} \quad a^2 = (-4)(-4) = 16.$$

$$\text{If } a = -\frac{1}{2}, \quad \text{then} \quad a^2 = \left(-\frac{1}{2}\right)\left(-\frac{1}{2}\right) = \frac{1}{4}.$$

In this chapter, the opposite process is considered.

$$\text{If } a^2 = 64, \quad \text{then} \quad a = ?.$$

$$\text{If } a^2 = 16, \quad \text{then} \quad a = ?.$$

$$\text{If } a^2 = \frac{1}{4}, \quad \text{then} \quad a = ?.$$

OBJECTIVE 1 Find square roots. To find a in the three preceding statements, we must find a number that when multiplied by itself results in the given number. The number a is called a **square root** of the number a^2.

EXAMPLE 1 Finding All Square Roots of a Number

Find all square roots of 49.

To find a square root of 49, think of a number that when multiplied by itself gives 49. One square root is 7 because $7 \cdot 7 = 49$. Another square root of 49 is -7 because $(-7)(-7) = 49$. The number 49 has two square roots, 7 and -7; one is positive, and one is negative.

Work Problem 1 at the Side.

1 Find all square roots.

(a) 100

(b) 25

(c) 36

(d) $\frac{25}{36}$

The **positive** or **principal square root** of a number is written with the symbol $\sqrt{\ }$. For example, the positive square root of 121 is 11, written

$$\sqrt{121} = 11.$$

The symbol $-\sqrt{\ }$ is used for the **negative square root** of a number. For example, the negative square root of 121 is -11, written

$$-\sqrt{121} = -11.$$

The symbol $\sqrt{\ }$, called a **radical sign,** always represents the positive square root (except that $\sqrt{0} = 0$). The number inside the radical sign is called the **radicand,** and the entire expression, radical sign and radicand, is called a **radical.**

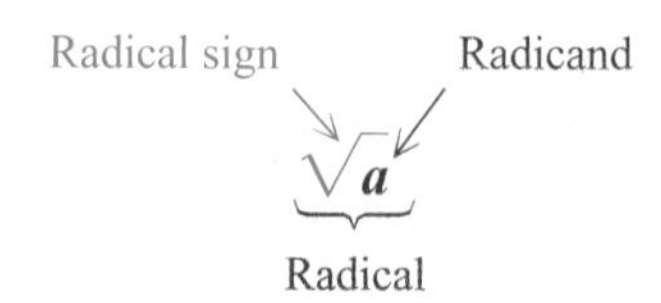

An algebraic expression containing a radical is called a **radical expression.**

Radicals have a long mathematical history. The radical sign $\sqrt{\ }$ has been used since sixteenth-century Germany and was probably derived from the letter *R*. The radical symbol in the margin comes from the Latin word for root, *radix*. It was first used by Leonardo da Pisa (Fibonnaci) in 1220.

Early radical symbol

ANSWERS

1. (a) $10, -10$ (b) $5, -5$ (c) $6, -6$ (d) $\frac{5}{6}, -\frac{5}{6}$

Our discussion of square roots is summarized as follows.

Square Roots of *a*

If a is a positive real number, then

$\sqrt{a}$ is the positive or principal square root of a,

and $-\sqrt{a}$ is the negative square root of a.

For nonnegative a,

$$\sqrt{a} \cdot \sqrt{a} = (\sqrt{a})^2 = a \quad \text{and} \quad -\sqrt{a} \cdot (-\sqrt{a}) = (-\sqrt{a})^2 = a.$$

Also, $\sqrt{0} = 0$.

Calculator Tip Most calculators have a square root key, usually labeled $\sqrt{x}$, that allows us to find the square root of a number. On some models, the square root key must be used in conjunction with the key marked INV or 2nd.

EXAMPLE 2 Finding Square Roots

Find each square root.

(a) $\sqrt{144}$

The radical $\sqrt{144}$ represents the positive or principal square root of 144. Think of a positive number whose square is 144.

$$12^2 = 144, \quad \text{so} \quad \sqrt{144} = 12.$$

(b) $-\sqrt{1024}$

This symbol represents the negative square root of 1024. A calculator with a square root key can be used to find $\sqrt{1024} = 32$. Then, $-\sqrt{1024} = -32$.

(c) $\sqrt{\frac{4}{9}} = \frac{2}{3}$ **(d)** $-\sqrt{\frac{16}{49}} = -\frac{4}{7}$ **(e)** $\sqrt{.81} = .9$

Work Problem 2 at the Side.

As noted above, when the square root of a positive real number is squared, the result is that positive real number. (Also, $(\sqrt{0})^2 = 0$.)

EXAMPLE 3 Squaring Radical Expressions

Find the *square* of each radical expression.

(a) $\sqrt{13}$

$(\sqrt{13})^2 = 13$ Definition of square root

(b) $-\sqrt{29}$

$(-\sqrt{29})^2 = 29$ The square of a *negative* number is positive.

(c) $\sqrt{p^2 + 1}$

$(\sqrt{p^2 + 1})^2 = p^2 + 1$

Work Problem 3 at the Side.

2 Find each square root.

(a) $\sqrt{16}$

(b) $-\sqrt{169}$

(c) $-\sqrt{225}$

(d) $\sqrt{729}$

(e) $\sqrt{\frac{36}{25}}$

(f) $\sqrt{.49}$

3 Find the *square* of each radical expression.

(a) $\sqrt{41}$

(b) $-\sqrt{39}$

(c) $\sqrt{2x^2 + 3}$

ANSWERS

2. **(a)** 4 **(b)** -13 **(c)** -15 **(d)** 27 **(e)** $\frac{6}{5}$ **(f)** .7

3. **(a)** 41 **(b)** 39 **(c)** $2x^2 + 3$

OBJECTIVE 2 Decide whether a given root is rational, irrational, or not a real number. All numbers with square roots that are rational are called **perfect squares.**

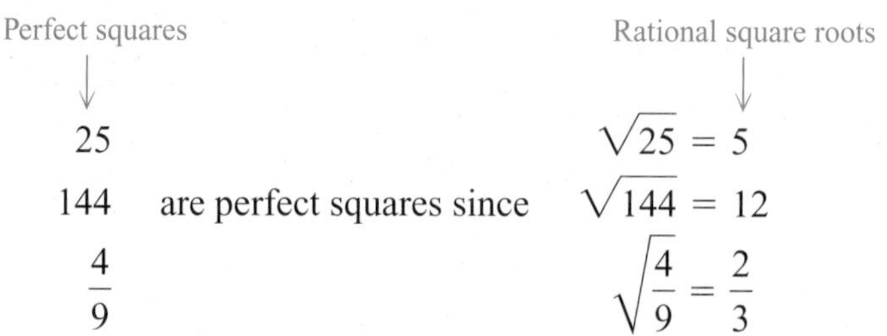

Perfect squares		Rational square roots
25		$\sqrt{25} = 5$
144	are perfect squares since	$\sqrt{144} = 12$
$\frac{4}{9}$		$\sqrt{\frac{4}{9}} = \frac{2}{3}$

A number that is not a perfect square has a square root that is not a rational number. For example, $\sqrt{5}$ is not a rational number because it cannot be written as the ratio of two integers. Its decimal equivalent (or approximation) neither terminates nor repeats. However, $\sqrt{5}$ is a real number and corresponds to a point on the number line. As mentioned in **Section 1.3,** a real number that is not rational is called an **irrational number.** The number $\sqrt{5}$ is irrational. Many square roots of integers are irrational.

If a is a *positive* real number that is *not* a perfect square, then $\sqrt{a}$ is irrational.

Not every number has a real number square root. For example, there is no real number that can be squared to obtain -36. (The square of a real number can never be negative.) Because of this, $\sqrt{-36}$ ***is not a real number.***

If a is a *negative* real number, then $\sqrt{a}$ is *not* a real number.

CAUTION
Be careful not to confuse $\sqrt{-36}$ and $-\sqrt{36}$. $\sqrt{-36}$ is not a real number since there is no real number that can be squared to obtain -36. However, $-\sqrt{36}$ is the negative square root of 36, which is -6.

EXAMPLE 4 Identifying Types of Square Roots

Tell whether each square root is *rational, irrational,* or *not a real number.*

(a) $\sqrt{17}$
Because 17 is not a perfect square, $\sqrt{17}$ is irrational.

(b) $\sqrt{64}$
The number 64 is a perfect square, 8^2, so $\sqrt{64} = 8$ is a rational number.

(c) $\sqrt{-25}$
There is no real number whose square is -25. Therefore, $\sqrt{-25}$ is not a real number.

Work Problem 4 at the Side.

4 Tell whether each square root is *rational, irrational,* or *not a real number.*

(a) $\sqrt{9}$

(b) $\sqrt{7}$

(c) $\sqrt{\frac{4}{9}}$

(d) $\sqrt{72}$

(e) $\sqrt{-43}$

ANSWERS
4. **(a)** rational **(b)** irrational **(c)** rational **(d)** irrational **(e)** not a real number

NOTE

Not all irrational numbers are square roots of integers. For example, π (approximately 3.14159) is an irrational number that is not a square root of any integer.

OBJECTIVE 3 Find decimal approximations for irrational square roots. Even if a number is irrational, a decimal that approximates the number can be found using a calculator. For example, if we use a calculator to find $\sqrt{10}$, the display might show 3.16227766, which is only an *approximation* of $\sqrt{10}$, not an exact rational value.

EXAMPLE 5 Approximating Irrational Square Roots

Find a decimal approximation for each square root. Round answers to the nearest thousandth.

(a) $\sqrt{11}$

Using the square root key of a calculator gives $3.31662479 \approx 3.317$, where $\approx$ means "is approximately equal to."

(b) $\sqrt{39} \approx 6.245$ Use a calculator. **(c)** $-\sqrt{740} \approx -27.203$

Work Problem 5 at the Side.

5 Find a decimal approximation for each square root. Round answers to the nearest thousandth.

(a) $\sqrt{28}$

(b) $\sqrt{63}$

(c) $-\sqrt{190}$

(d) $\sqrt{1000}$

OBJECTIVE 4 Use the Pythagorean formula. Many applications of square roots use the Pythagorean formula. Recall from **Section 6.7** that by this formula if c is the length of the hypotenuse of a right triangle, and a and b are the lengths of the two legs, as shown in Figure 1, then

$$a^2 + b^2 = c^2.$$

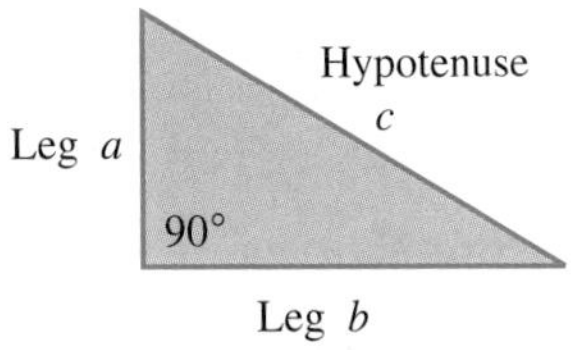

Figure 1

EXAMPLE 6 Using the Pythagorean Formula

Find the length of the unknown side of each right triangle with sides a, b, and c, where c is the hypotenuse.

(a) $a = 3, b = 4$

Use the Pythagorean formula to find c^2 first.

$$c^2 = a^2 + b^2$$
$$c^2 = 3^2 + 4^2 \quad \text{Let } a = 3 \text{ and } b = 4.$$
$$c^2 = 9 + 16 \quad \text{Square.}$$
$$c^2 = 25 \quad \text{Add.}$$

Since the length of a side of a triangle must be a positive number, find the positive square root of 25 to get c.

$$c = \sqrt{25} = 5$$

Continued on Next Page

Answers

5. (a) 5.292 (b) 7.937 (c) −13.784 (d) 31.623

6 Find the length of the unknown side in each right triangle. Give any decimal approximations to the nearest thousandth.

(a) $a = 7, b = 24$

(b) $c = 15, b = 13$

(c) 8, 11, ?

(b) $c = 9, b = 5$

Substitute the given values in the Pythagorean formula. Then solve for a^2.

$$c^2 = a^2 + b^2$$

$$9^2 = a^2 + 5^2 \quad \text{Let } c = 9 \text{ and } b = 5.$$

$$81 = a^2 + 25 \quad \text{Square.}$$

$$56 = a^2 \quad \text{Subtract 25.}$$

Use a calculator to find $a = \sqrt{56} \approx 7.483$.

CAUTION

Be careful not to make the common mistake of thinking that $\sqrt{a^2 + b^2}$ equals $a + b$. As Example 6(a) shows, $\sqrt{9 + 16} = \sqrt{25} = 5$. However, $\sqrt{9} + \sqrt{16} = 3 + 4 = 7$. Since $5 \neq 7$, in general,

$$\sqrt{a^2 + b^2} \neq a + b.$$

Work Problem 6 at the Side.

The Pythagorean formula can be used to solve applied problems that involve right triangles. Use the same six problem-solving steps that we have been using throughout the text.

EXAMPLE 7 Using the Pythagorean Formula to Solve an Application

A ladder 10 ft long leans against a wall. The foot of the ladder is 6 ft from the base of the wall. How high up the wall does the top of the ladder rest?

Step 1 **Read** the problem again.

Step 2 **Assign a variable.** As shown in Figure 2, a right triangle is formed with the ladder as the hypotenuse. Let a represent the height of the top of the ladder when measured straight down to the ground.

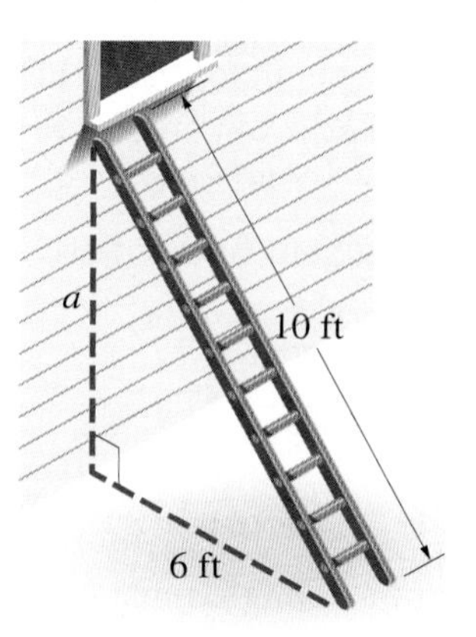

Figure 2

Continued on Next Page

Answers

6. **(a)** 25 **(b)** $\sqrt{56} \approx 7.483$ **(c)** $\sqrt{57} \approx 7.550$

Step 3 **Write an equation** using the Pythagorean formula.

$$c^2 = a^2 + b^2$$

$$10^2 = a^2 + 6^2 \quad \text{Let } c = 10 \text{ and } b = 6.$$

Step 4 **Solve.**

$$100 = a^2 + 36 \quad \text{Square.}$$

$$64 = a^2 \quad \text{Subtract 36.}$$

$$\sqrt{64} = a$$

$$a = 8 \quad \sqrt{64} = 8$$

Choose the positive square root of 64 since a represents a length.

Step 5 **State the answer.** The top of the ladder rests 8 ft up the wall.

Step 6 **Check.** From Figure 2, we see that we must have

$$8^2 + 6^2 = 10^2 \quad ?$$

$$64 + 36 = 100. \quad \text{True}$$

The check confirms that the top of the ladder rests 8 ft up the wall.

Work Problem 7 at the Side.

7 A rectangle has dimensions 5 ft by 12 ft. Find the length of its diagonal.

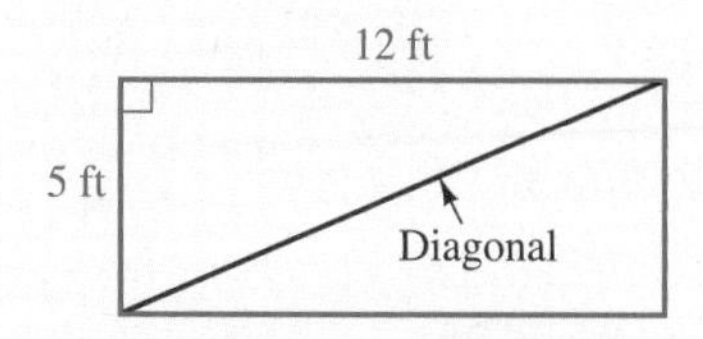

(Note that the diagonal divides the rectangle into two right triangles with itself as the hypotenuse.)

OBJECTIVE 5 Find higher roots. Finding the square root of a number is the inverse (reverse) of squaring a number. In a similar way, there are inverses to finding the cube of a number, or finding the fourth or higher power of a number. These inverses are the **cube root,** written $\sqrt[3]{a}$, and the **fourth root,** written $\sqrt[4]{a}$. Similar symbols are used for higher roots. In general, we have the following.

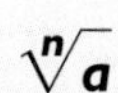

The nth root of a is written $\sqrt[n]{a}$.

In $\sqrt[n]{a}$, the number n is the **index** or **order** of the radical.

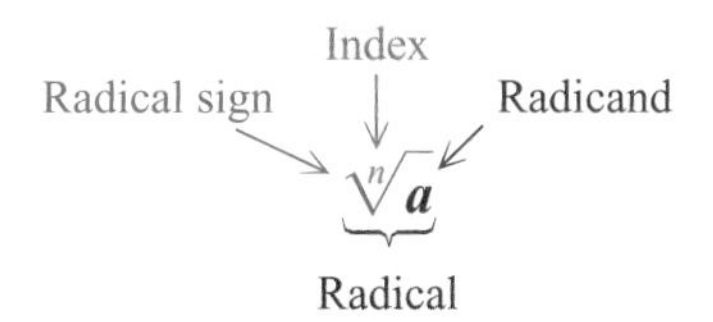

We could write $\sqrt[2]{a}$ instead of $\sqrt{a}$, but the simpler symbol $\sqrt{a}$ is customary since the square root is the most commonly used root.

Calculator Tip A calculator that has a key marked $\sqrt[x]{y}$, x^y, or y^x (again perhaps in conjunction with the INV or 2nd key) can be used to find higher roots.

When working with cube roots or fourth roots, it is helpful to memorize the first few *perfect cubes* ($1^3 = 1, 2^3 = 8, 3^3 = 27$, and so on) and the first few *perfect fourth powers* ($1^4 = 1, 2^4 = 16, 3^4 = 81$, and so on).

Work Problem 8 at the Side.

8 Complete the following list of perfect cubes and perfect fourth powers.

Perfect Cubes	Perfect Fourth Powers
$1^3 = 1$	$1^4 = 1$
$2^3 = 8$	$2^4 = 16$
$3^3 = 27$	$3^4 = 81$
$4^3 =$ ____	$4^4 =$ ____
$5^3 =$ ____	$5^4 =$ ____
$6^3 =$ ____	$6^4 =$ ____
$7^3 =$ ____	$7^4 =$ ____
$8^3 =$ ____	$8^4 =$ ____
$9^3 =$ ____	$9^4 =$ ____
$10^3 =$ ____	$10^4 =$ ____

ANSWERS

7. 13 ft

8. Perfect cubes: 64; 125; 216; 343; 512; 729; 1000
Perfect fourth powers: 256; 625; 1296; 2401; 4096; 6561; 10,000

9 Find each cube root.

(a) $\sqrt[3]{27}$

(b) $\sqrt[3]{64}$

(c) $\sqrt[3]{-125}$

10 Find each root.

(a) $\sqrt[4]{81}$

(b) $\sqrt[4]{-81}$

(c) $-\sqrt[4]{625}$

(d) $\sqrt[5]{243}$

(e) $\sqrt[5]{-243}$

ANSWERS

9. **(a)** 3 **(b)** 4 **(c)** -5

10. **(a)** 3 **(b)** not a real number **(c)** -5 **(d)** 3 **(e)** -3

EXAMPLE 8 Finding Cube Roots

Find each cube root.

(a) $\sqrt[3]{8}$

Look for a number that can be cubed to give 8. Because $2^3 = 8$, $\sqrt[3]{8} = 2$.

(b) $\sqrt[3]{-8} = -2$ because $(-2)^3 = -8$.

(c) $\sqrt[3]{216} = 6$ because $6^3 = 216$.

Notice in Example 8(b) that we can find the cube root of a negative number. (Contrast this with the square root of a negative number, which is not real.) In fact, the cube root of a positive number is positive, and the cube root of a negative number is negative. ***There is only one real number cube root for each real number.***

Work Problem 9 at the Side.

When a radical has an ***even index*** (square root, fourth root, and so on), ***the radicand must be nonnegative*** to yield a real number root. Also,

$$\sqrt{a}, \sqrt[4]{a}, \sqrt[6]{a}, \text{ and so on are positive (principal) roots;}$$

$$-\sqrt{a}, -\sqrt[4]{a}, -\sqrt[6]{a}, \text{ and so on are negative roots.}$$

EXAMPLE 9 Finding Higher Roots

Find each root.

(a) $\sqrt[4]{16}$

$\sqrt[4]{16} = 2$ because 2 is positive and $2^4 = 16$.

(b) $-\sqrt[4]{16}$

From part (a), $\sqrt[4]{16} = 2$, so the negative root $-\sqrt[4]{16} = -2$.

(c) $\sqrt[4]{-16}$

For a real number fourth root, the radicand must be nonnegative. There is no real number that equals $\sqrt[4]{-16}$.

(d) $-\sqrt[5]{32}$

First find $\sqrt[5]{32}$. Because 2 is the number whose fifth power is 32, $\sqrt[5]{32} = 2$. Since $\sqrt[5]{32} = 2$, it follows that

$$-\sqrt[5]{32} = -2.$$

(e) $\sqrt[5]{-32}$

Because $(-2)^5 = -32$, $\sqrt[5]{-32} = -2$.

Work Problem 10 at the Side.

8.1 Exercises

FOR EXTRA HELP

Addison-Wesley Math Tutor Center

MathXL

Digital Video Tutor CD 8 Videotape 8

Student's Solutions Manual

MyMathLab

Interactmath.com

Decide whether each statement is true *or* false. *If* false, *tell why.*

1. Every positive number has two real square roots.

2. A negative number has negative square roots.

3. Every nonnegative number has two real square roots.

4. The positive square root of a positive number is its principal square root.

5. The cube root of every real number has the same sign as the number itself.

6. Every positive number has three real cube roots.

Find all square roots of each number. See Example 1.

7. 9 **8.** 16 **9.** 64 **10.** 100 **11.** 169

12. 225 **13.** $\frac{25}{196}$ **14.** $\frac{81}{400}$ **15.** 900 **16.** 1600

Find each square root. See Examples 2 and 4(c).

17. $\sqrt{1}$ **18.** $\sqrt{4}$ **19.** $\sqrt{49}$ **20.** $\sqrt{81}$ **21.** $-\sqrt{256}$

22. $-\sqrt{196}$ **23.** $-\sqrt{\frac{144}{121}}$ **24.** $-\sqrt{\frac{49}{36}}$ **25.** $\sqrt{.64}$ **26.** $\sqrt{.16}$

27. $\sqrt{-121}$ **28.** $\sqrt{-64}$ **29.** $-\sqrt{-49}$ **30.** $-\sqrt{-100}$

Find the square of each radical expression. See Example 3.

31. $\sqrt{100}$ **32.** $\sqrt{36}$ **33.** $-\sqrt{19}$ **34.** $-\sqrt{99}$

35. $\sqrt{\frac{2}{3}}$ **36.** $\sqrt{\frac{5}{7}}$ **37.** $\sqrt{3x^2 + 4}$ **38.** $\sqrt{9y^2 + 3}$

What must be true about the value of a for each statement in Exercises 39–42 to be true?

39. $\sqrt{a}$ represents a positive number.

40. $-\sqrt{a}$ represents a negative number.

41. $\sqrt{a}$ is not a real number.

42. $-\sqrt{a}$ is not a real number.

Write rational, irrational, *or* not a real number *for each number. If a number is rational, give its exact value. If a number is irrational, give a decimal approximation to the nearest thousandth. Use a calculator as necessary. See Examples 4 and 5.*

43. $\sqrt{25}$

44. $\sqrt{169}$

45. $\sqrt{29}$

46. $\sqrt{33}$

47. $-\sqrt{64}$

48. $-\sqrt{81}$

49. $-\sqrt{300}$

50. $-\sqrt{500}$

51. $\sqrt{-29}$

52. $\sqrt{-47}$

53. $\sqrt{1200}$

54. $\sqrt{1500}$

Work Exercises 55 and 56 without using a calculator.

55. Choose the best estimate for the length and width (in meters) of this rectangle.

A. 11 by 6 **B.** 11 by 7 **C.** 10 by 7 **D.** 10 by 6

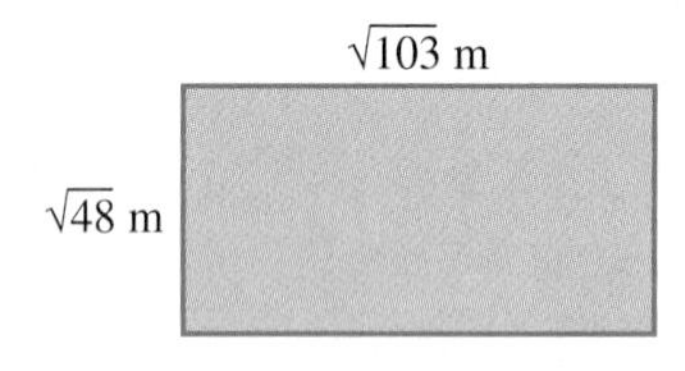

56. Choose the best estimate for the base and height (in feet) of this triangle.

A. $b = 8, h = 5$ **B.** $b = 8, h = 4$

C. $b = 9, h = 5$ **D.** $b = 9, h = 4$

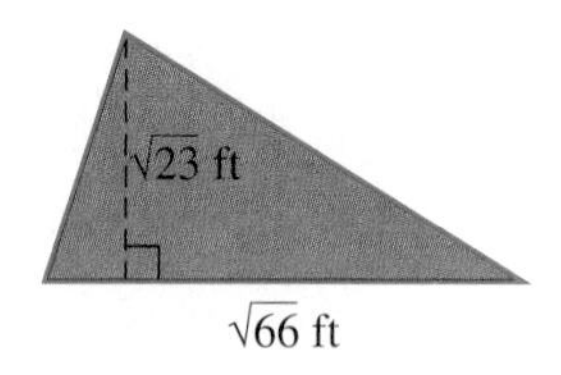

Find the length of the unknown side of each right triangle with sides a, b, and c, where c is the hypotenuse. See Figure 1 and Example 6. Give any decimal approximations to the nearest thousandth.

57. $a = 8, b = 15$

58. $a = 24, b = 10$

59. $a = 6, c = 10$

60. $b = 12, c = 13$

61. $a = 11, b = 4$

62. $a = 13, b = 9$

Solve each problem. See Example 7.

63. The diagonal of a rectangle measures 25 cm. The width of the rectangle is 7 cm. Find the length of the rectangle.

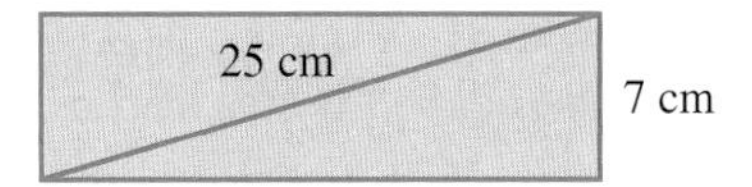

64. The length of a rectangle is 40 m, and the width is 9 m. Find the measure of the diagonal of the rectangle.

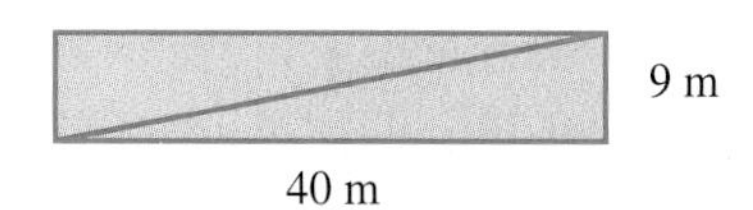

65. Tyler is flying a kite on 100 ft of string. How high is it above his hand (vertically) if the horizontal distance between Tyler and the kite is 60 ft?

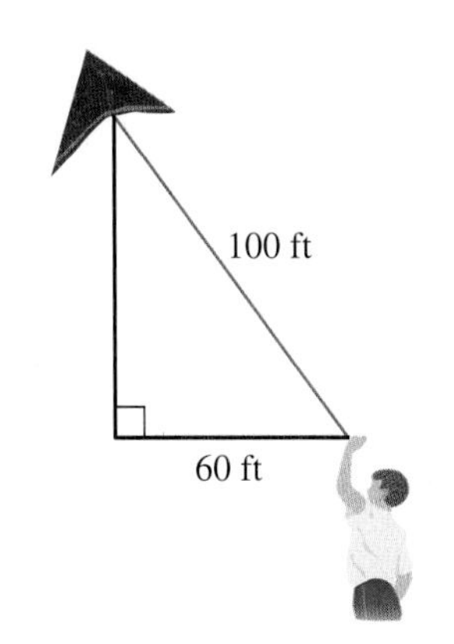

66. A guy wire is attached to the mast of a short-wave transmitting antenna. It is attached 96 ft above ground level. If the wire is staked to the ground 72 ft from the base of the mast, how long is the wire?

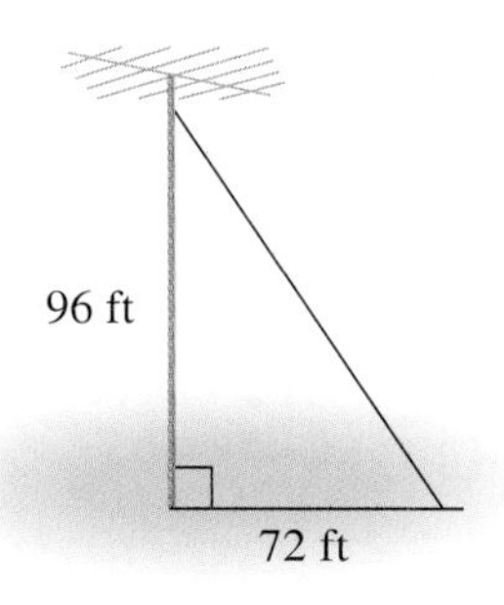

67. A surveyor measured the distances shown in the figure. Find the distance across the lake between points R and S.

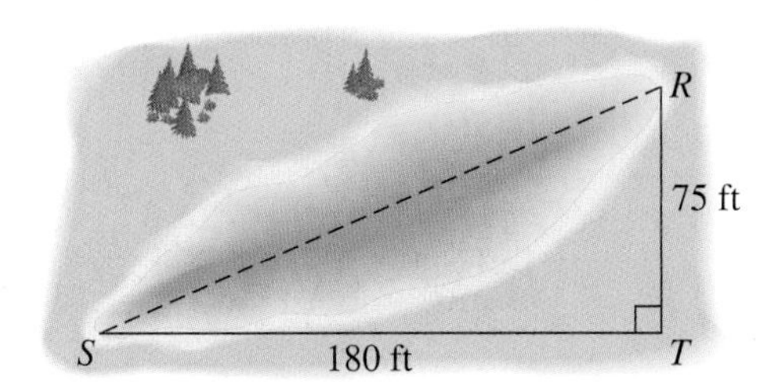

68. A boat is being pulled toward a dock with a rope attached at water level. When the boat is 24 ft from the dock, 30 ft of rope is extended. What is the height of the dock above the water?

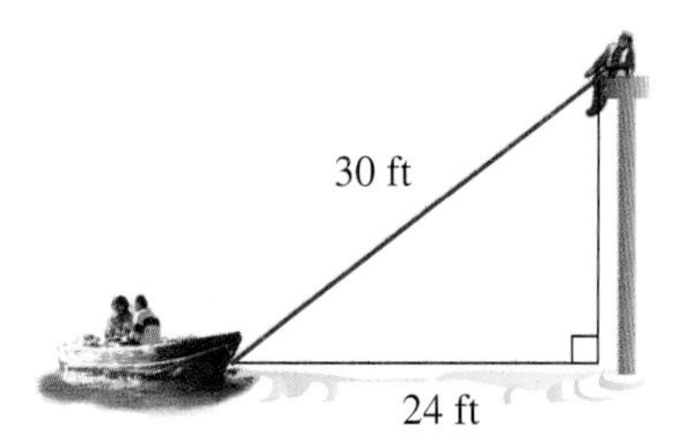

69. What is the value of x (to the nearest thousandth) in the figure?

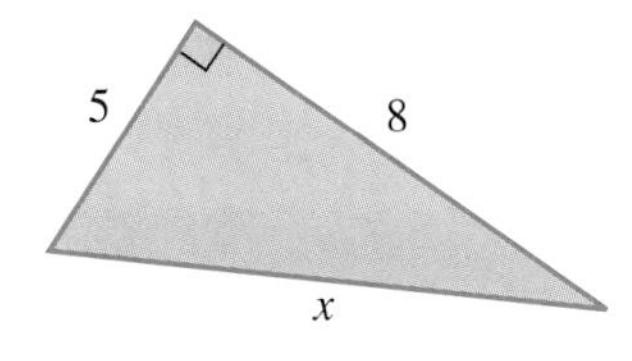

70. What is the value of y (to the nearest thousandth) in the figure?

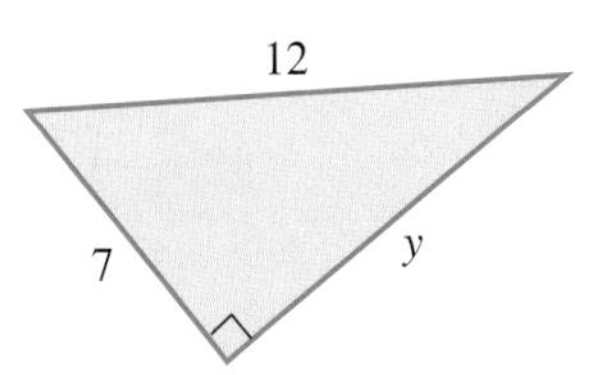

71. One of the authors of this text took this photo of a broken tree in a field near his home. The vertical distance from the base of the broken tree to the point of the break is 6.5 ft. The length of the broken part is 15 ft. How far along the ground (to the nearest tenth) is it from the base of the tree to the point where the broken part touches the ground?

72. A television set is "sized" according to the diagonal measurement of the viewing screen. One of the authors of this text purchased a 19-in. TV, so the TV measures 19 in. from one corner of the viewing screen diagonally to the other corner. The viewing screen is 15.5 in. wide. Find the height of the viewing screen (to the nearest tenth). (*Source:* Phillips Magnavox color television 19PR21C1.)

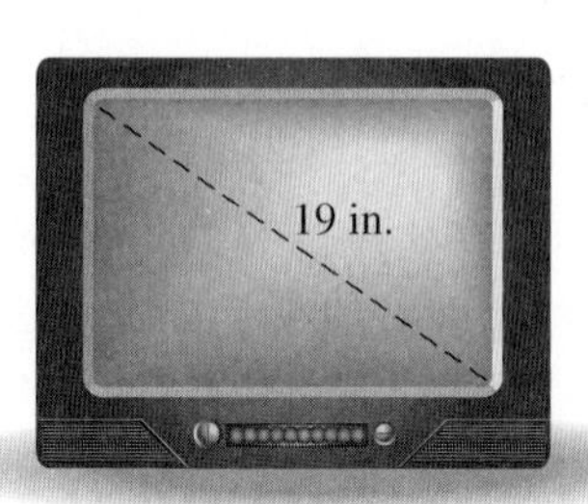

73. Use specific values for a and b different from those given in the "Caution" following Example 6 to show that $\sqrt{a^2 + b^2} \neq a + b$.

74. Why would the values $a = 0$ and $b = 1$ *not* be satisfactory in Exercise 73?

Find each root. See Examples 8 and 9.

75. $\sqrt[3]{1}$

76. $\sqrt[3]{8}$

77. $\sqrt[3]{125}$

78. $\sqrt[3]{1000}$

79. $\sqrt[3]{-27}$

80. $\sqrt[3]{-64}$

81. $\sqrt[3]{-216}$

82. $\sqrt[3]{-343}$

83. $-\sqrt[3]{8}$

84. $-\sqrt[3]{216}$

85. $\sqrt[4]{256}$

86. $\sqrt[4]{625}$

87. $\sqrt[4]{1296}$

88. $\sqrt[4]{10{,}000}$

89. $\sqrt[4]{-1}$

90. $\sqrt[4]{-625}$

91. $-\sqrt[4]{81}$

92. $-\sqrt[4]{256}$

93. $\sqrt[5]{-1024}$

94. $\sqrt[5]{-100{,}000}$

8.2 Multiplying, Dividing, and Simplifying Radicals

OBJECTIVES

1. Multiply radicals.
2. Simplify radicals using the product rule.
3. Simplify radicals using the quotient rule.
4. Simplify radicals involving variables.
5. Simplify higher roots.

OBJECTIVE 1 Multiply radicals. We develop several rules for finding products and quotients of radicals in this section. To illustrate the rule for products, notice that

$$\sqrt{4} \cdot \sqrt{9} = 2 \cdot 3 = \mathbf{6} \quad \text{and} \quad \sqrt{4 \cdot 9} = \sqrt{36} = \mathbf{6},$$

showing that

$$\sqrt{4} \cdot \sqrt{9} = \sqrt{4 \cdot 9}.$$

This result is a particular case of the **product rule for radicals.**

Product Rule for Radicals

For nonnegative real numbers a and b,

$$\sqrt{a} \cdot \sqrt{b} = \sqrt{a \cdot b} \quad \text{and} \quad \sqrt{a \cdot b} = \sqrt{a} \cdot \sqrt{b}.$$

In words, the product of two radicals is the radical of the product, and the radical of a product is the product of the radicals.

EXAMPLE 1 Using the Product Rule to Multiply Radicals

Use the product rule for radicals to find each product.

(a) $\sqrt{2} \cdot \sqrt{3} = \sqrt{2 \cdot 3} = \sqrt{6}$ Product rule

(b) $\sqrt{7} \cdot \sqrt{5} = \sqrt{35}$ Product rule

(c) $\sqrt{11} \cdot \sqrt{a} = \sqrt{11a}$ Assume $a \geq 0$.

Work Problem 1 at the Side.

1 Use the product rule for radicals to find each product.

(a) $\sqrt{6} \cdot \sqrt{11}$

(b) $\sqrt{2} \cdot \sqrt{5}$

(c) $\sqrt{10} \cdot \sqrt{r}, \quad r \geq 0$

OBJECTIVE 2 Simplify radicals using the product rule. A square root radical is *simplified* when no perfect square factor remains under the radical sign. This is accomplished by using the product rule.

EXAMPLE 2 Using the Product Rule to Simplify Radicals

Simplify each radical.

(a) $\sqrt{20}$

Because 20 has a perfect square factor of 4, we can write

$$\begin{aligned} \sqrt{20} &= \sqrt{4 \cdot 5} && \text{4 is a perfect square.} \\ &= \sqrt{4} \cdot \sqrt{5} && \text{Product rule} \\ &= 2\sqrt{5}. && \sqrt{4} = 2 \end{aligned}$$

Thus, $\sqrt{20} = 2\sqrt{5}$. Because 5 has no perfect square factor (other than 1), $2\sqrt{5}$ is called the *simplified form* of $\sqrt{20}$. Note that $2\sqrt{5}$ represents a product, where the factors are 2 and $\sqrt{5}$.

We could also factor 20 into prime factors and look for pairs of like factors. Each pair of like factors produces one factor outside the radical. Thus,

$$\sqrt{20} = \sqrt{2 \cdot 2 \cdot 5} = 2\sqrt{5}.$$

Continued on Next Page

ANSWERS

1. (a) $\sqrt{66}$ (b) $\sqrt{10}$ (c) $\sqrt{10r}$

2 Simplify each radical.

(a) $\sqrt{8}$

(b) $\sqrt{27}$

(c) $\sqrt{50}$

(d) $\sqrt{60}$

(e) $\sqrt{30}$

(b) $\sqrt{72}$

Begin by looking for the *largest* perfect square factor of 72. This number is 36, so

$$\sqrt{72} = \sqrt{36 \cdot 2} \quad \text{36 is a perfect square.}$$
$$= \sqrt{36} \cdot \sqrt{2} \quad \text{Product rule}$$
$$= 6\sqrt{2}. \quad \sqrt{36} = 6$$

We could also factor 72 into its prime factors and look for pairs of like factors.

$$\sqrt{72} = \sqrt{2 \cdot 2 \cdot 2 \cdot 3 \cdot 3} = 2 \cdot 3 \cdot \sqrt{2} = 6\sqrt{2}$$

In either case, we obtain $6\sqrt{2}$ as the simplified form of $\sqrt{72}$. However, our work is simpler if we begin with the largest perfect square factor.

(c)
$$\sqrt{300} = \sqrt{100 \cdot 3} \quad \text{100 is a perfect square.}$$
$$= \sqrt{100} \cdot \sqrt{3} \quad \text{Product rule}$$
$$= 10\sqrt{3} \quad \sqrt{100} = 10$$

(d) $\sqrt{15}$

The number 15 has no perfect square factors (except 1), so $\sqrt{15}$ cannot be simplified further.

Work Problem 2 at the Side.

Sometimes the product rule can be used to simplify a product, as Example 3 shows.

EXAMPLE 3 Multiplying and Simplifying Radicals

Find each product and simplify.

(a)
$$\sqrt{9} \cdot \sqrt{75} = 3\sqrt{75} \quad \sqrt{9} = 3$$
$$= 3\sqrt{25 \cdot 3} \quad \text{25 is a perfect square.}$$
$$= 3\sqrt{25} \cdot \sqrt{3} \quad \text{Product rule}$$
$$= 3 \cdot 5 \cdot \sqrt{3} \quad \sqrt{25} = 5$$
$$= 15\sqrt{3} \quad \text{Multiply.}$$

Notice that we could have used the product rule to get $\sqrt{9} \cdot \sqrt{75} = \sqrt{675}$, and then simplified. However, the product rule as used here allows us to obtain the final answer without using a large number like 675.

(b)
$$\sqrt{8} \cdot \sqrt{12} = \sqrt{8 \cdot 12} \quad \text{Product rule}$$
$$= \sqrt{4 \cdot 2 \cdot 4 \cdot 3} \quad \text{Factor; 4 is a perfect square.}$$
$$= \sqrt{4} \cdot \sqrt{4} \cdot \sqrt{2 \cdot 3} \quad \text{Product rule}$$
$$= 2 \cdot 2 \cdot \sqrt{6} \quad \sqrt{4} = 2$$
$$= 4\sqrt{6} \quad \text{Multiply.}$$

Continued on Next Page

Answers

2. (a) $2\sqrt{2}$ (b) $3\sqrt{3}$ (c) $5\sqrt{2}$ (d) $2\sqrt{15}$ (e) cannot be simplified further

(c) $2\sqrt{3} \cdot 3\sqrt{6} = 2 \cdot 3 \cdot \sqrt{3 \cdot 6}$ Product rule

$= 6\sqrt{18}$ Multiply.

$= 6\sqrt{9 \cdot 2}$ Factor; 9 is a perfect square.

$= 6\sqrt{9} \cdot \sqrt{2}$ Product rule

$= 6 \cdot 3 \cdot \sqrt{2}$ $\sqrt{9} = 3$

$= 18\sqrt{2}$ Multiply.

NOTE

We could also simplify Example 3(b) as follows.

$$\sqrt{8} \cdot \sqrt{12} = \sqrt{4 \cdot 2} \cdot \sqrt{4 \cdot 3}$$
$$= 2\sqrt{2} \cdot 2\sqrt{3}$$
$$= 2 \cdot 2 \cdot \sqrt{2} \cdot \sqrt{3}$$
$$= 4\sqrt{6} \quad \text{Same result}$$

There is often more than one way to find such a product.

Work Problem 3 at the Side.

OBJECTIVE 3 Simplify radicals using the quotient rule. The **quotient rule for radicals** is very similar to the product rule. It, too, can be used either way.

Quotient Rule for Radicals

If a and b are nonnegative real numbers and $b \neq 0$, then

$$\sqrt{\frac{a}{b}} = \frac{\sqrt{a}}{\sqrt{b}} \quad \text{and} \quad \frac{\sqrt{a}}{\sqrt{b}} = \sqrt{\frac{a}{b}}.$$

In words, the radical of a quotient is the quotient of the radicals, and the quotient of two radicals is the radical of the quotient.

EXAMPLE 4 Using the Quotient Rule to Simplify Radicals

Use the quotient rule to simplify each radical.

(a) $\sqrt{\frac{25}{9}} = \frac{\sqrt{25}}{\sqrt{9}} = \frac{5}{3}$ Quotient rule

(b) $\frac{\sqrt{288}}{\sqrt{2}} = \sqrt{\frac{288}{2}} = \sqrt{144} = 12$ Quotient rule

(c) $\sqrt{\frac{3}{4}} = \frac{\sqrt{3}}{\sqrt{4}} = \frac{\sqrt{3}}{2}$ Quotient rule

Work Problem 4 at the Side.

3 Find each product and simplify.

(a) $\sqrt{3} \cdot \sqrt{15}$

(b) $\sqrt{10} \cdot \sqrt{50}$

(c) $\sqrt{12} \cdot \sqrt{2}$

(d) $\sqrt{7} \cdot \sqrt{14}$

(e) $3\sqrt{5} \cdot 4\sqrt{10}$

4 Use the quotient rule to simplify each radical.

(a) $\sqrt{\frac{81}{16}}$

(b) $\frac{\sqrt{192}}{\sqrt{3}}$

(c) $\sqrt{\frac{10}{49}}$

ANSWERS

3. (a) $3\sqrt{5}$ (b) $10\sqrt{5}$ (c) $2\sqrt{6}$ (d) $7\sqrt{2}$ (e) $60\sqrt{2}$

4. (a) $\frac{9}{4}$ (b) 8 (c) $\frac{\sqrt{10}}{7}$

5 Simplify $\dfrac{8\sqrt{50}}{4\sqrt{5}}$.

6 Simplify.

(a) $\sqrt{\dfrac{5}{6}} \cdot \sqrt{120}$

(b) $\sqrt{\dfrac{3}{8}} \cdot \sqrt{\dfrac{7}{2}}$

Answers

5. $2\sqrt{10}$

6. (a) 10 **(b)** $\dfrac{\sqrt{21}}{4}$

EXAMPLE 5 Using the Quotient Rule to Divide Radicals

Simplify $\dfrac{27\sqrt{15}}{9\sqrt{3}}$.

Use multiplication of fractions and the quotient rule as follows.

$$\frac{27\sqrt{15}}{9\sqrt{3}} = \frac{27}{9} \cdot \frac{\sqrt{15}}{\sqrt{3}} = \frac{27}{9} \cdot \sqrt{\frac{15}{3}} = 3\sqrt{5}$$

Work Problem 5 at the Side.

Some problems require both the product and quotient rules.

EXAMPLE 6 Using Both the Product and Quotient Rules

Simplify $\sqrt{\dfrac{3}{5}} \cdot \sqrt{\dfrac{1}{5}}$.

$$\sqrt{\frac{3}{5}} \cdot \sqrt{\frac{1}{5}} = \sqrt{\frac{3}{5} \cdot \frac{1}{5}} \quad \text{Product rule}$$

$$= \sqrt{\frac{3}{25}} \quad \text{Multiply fractions.}$$

$$= \frac{\sqrt{3}}{\sqrt{25}} \quad \text{Quotient rule}$$

$$= \frac{\sqrt{3}}{5} \quad \sqrt{25} = 5$$

Work Problem 6 at the Side.

OBJECTIVE 4 Simplify radicals involving variables. Radicals can also involve variables, such as $\sqrt{x^2}$. Simplifying such radicals can get a little tricky. If x represents a nonnegative number, then $\sqrt{x^2} = x$. If x represents a negative number, then $\sqrt{x^2} = -x$, the opposite of x (which is positive). For example,

$$\sqrt{5^2} = 5, \qquad \text{but} \qquad \sqrt{(-5)^2} = \sqrt{25} = 5, \quad \text{the } \textit{opposite} \text{ of } -5.$$

This means that the square root of a squared number is always nonnegative. We can use absolute value to express this.

$\sqrt{a^2}$

For any real number a,

$$\sqrt{a^2} = |a|.$$

The product and quotient rules apply when variables appear under the radical sign, as long as the variables represent only *nonnegative* real numbers. ***To avoid negative radicands, variables under radical signs are assumed to be nonnegative in this text.*** Therefore, absolute value bars are not necessary, since for $x \geq 0$, $|x| = x$.

EXAMPLE 7 Simplifying Radicals Involving Variables

Simplify each radical. Assume that all variables represent nonnegative real numbers.

(a) $\sqrt{x^4} = x^2$ since $(x^2)^2 = x^4$

(b) $\sqrt{25m^6} = \sqrt{25} \cdot \sqrt{m^6}$ Product rule

$= 5m^3$ $(m^3)^2 = m^6$

(c) $\sqrt{8p^{10}} = \sqrt{4 \cdot 2 \cdot p^{10}}$ 4 is a perfect square

$= \sqrt{4} \cdot \sqrt{2} \cdot \sqrt{p^{10}}$ Product rule

$= 2 \cdot \sqrt{2} \cdot p^5$ $(p^5)^2 = p^{10}$

$= 2p^5\sqrt{2}$

(d) $\sqrt{r^9} = \sqrt{r^8 \cdot r}$

$= \sqrt{r^8} \cdot \sqrt{r}$ Product rule

$= r^4\sqrt{r}$ $(r^4)^2 = r^8$

(e) $\sqrt{\frac{5}{x^2}} = \frac{\sqrt{5}}{\sqrt{x^2}}$ Quotient rule

$= \frac{\sqrt{5}}{x}$ $x \neq 0$

NOTE

A quick way to find the square root of a variable raised to an even power is to divide the exponent by the index, 2. For example,

$$\sqrt{x^6} = x^3 \quad \text{and} \quad \sqrt{x^{10}} = x^5.$$

$6 \div 2 = 3$ $10 \div 2 = 5$

Work Problem 7 at the Side.

OBJECTIVE 5 Simplify higher roots. The product and quotient rules for radicals also work for other roots. To simplify cube roots, look for factors that are *perfect cubes*. A **perfect cube** is a number with a rational cube root. For example, $\sqrt[3]{64} = 4$, and because 4 is a rational number, 64 is a perfect cube. Higher roots are handled in a similar manner.

Properties of Radicals

For all real numbers where the indicated roots exist,

$$\sqrt[n]{a} \cdot \sqrt[n]{b} = \sqrt[n]{ab} \quad \text{and} \quad \frac{\sqrt[n]{a}}{\sqrt[n]{b}} = \sqrt[n]{\frac{a}{b}} \quad (b \neq 0).$$

7 Simplify each radical. Assume that all variables represent nonnegative real numbers.

(a) $\sqrt{x^8}$

(b) $\sqrt{36y^6}$

(c) $\sqrt{100p^{12}}$

(d) $\sqrt{12z^2}$

(e) $\sqrt{a^5}$

(f) $\sqrt{\frac{10}{n^4}}$, $n \neq 0$

ANSWERS

7. (a) x^4 (b) $6y^3$ (c) $10p^6$ (d) $2z\sqrt{3}$ (e) $a^2\sqrt{a}$ (f) $\frac{\sqrt{10}}{n^2}$

8 Simplify each radical.

(a) $\sqrt[3]{108}$

(b) $\sqrt[4]{160}$

(c) $\sqrt[4]{\dfrac{16}{625}}$

9 Simplify each radical.

(a) $\sqrt[3]{z^9}$

(b) $\sqrt[3]{8x^6}$

(c) $\sqrt[3]{54t^5}$

(d) $\sqrt[3]{\dfrac{a^{15}}{64}}$

ANSWERS

8. (a) $3\sqrt[3]{4}$ (b) $2\sqrt[4]{10}$ (c) $\dfrac{2}{5}$

9. (a) z^3 (b) $2x^2$ (c) $3t\sqrt[3]{2t^2}$ (d) $\dfrac{a^5}{4}$

EXAMPLE 8 Simplifying Higher Roots

Simplify each radical.

(a) $\sqrt[3]{32} = \sqrt[3]{\mathbf{8} \cdot 4}$ 8 is a perfect cube.

$= \sqrt[3]{\mathbf{8}} \cdot \sqrt[3]{4}$ Product rule

$= 2\sqrt[3]{4}$

(b) $\sqrt[4]{32} = \sqrt[4]{\mathbf{16} \cdot 2}$ 16 is a perfect fourth power.

$= \sqrt[4]{\mathbf{16}} \cdot \sqrt[4]{2}$ Product rule

$= 2\sqrt[4]{2}$

(c) $\sqrt[3]{\dfrac{27}{125}} = \dfrac{\sqrt[3]{27}}{\sqrt[3]{125}} = \dfrac{3}{5}$ Quotient rule

Work Problem 8 at the Side.

Higher roots of radicals involving variables can also be simplified. To simplify cube roots with variables, use the fact that for any real number a,

$$\sqrt[3]{a^3} = a.$$

This is true whether a is positive or negative. (Why?)

EXAMPLE 9 Simplifying Cube Roots Involving Variables

Simplify each radical.

(a) $\sqrt[3]{m^6} = m^2$ $(m^2)^3 = m^6$

(b) $\sqrt[3]{27x^{12}} = \sqrt[3]{27} \cdot \sqrt[3]{x^{12}}$ Product rule

$= 3x^4$ $3^3 = 27$; $(x^4)^3 = x^{12}$

(c) $\sqrt[3]{32a^4} = \sqrt[3]{8a^3 \cdot 4a}$ 8 is a perfect cube.

$= \sqrt[3]{8a^3} \cdot \sqrt[3]{4a}$ Product rule

$= 2a\sqrt[3]{4a}$ $(2a)^3 = 8a^3$

(d) $\sqrt[3]{\dfrac{y^3}{125}} = \dfrac{\sqrt[3]{y^3}}{\sqrt[3]{125}}$ Quotient rule

$= \dfrac{y}{5}$

Work Problem 9 at the Side.

8.2 Exercises

FOR EXTRA HELP

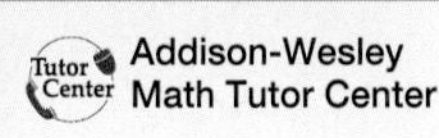
Addison-Wesley Math Tutor Center

MathXL

Digital Video Tutor CD 8 Videotape 8

Student's Solutions Manual

MyMathLab

Interactmath.com

Decide whether each statement is true *or* false. *If* false, *show why.*

1. $\sqrt{(-6)^2} = -6$

2. $\sqrt[3]{(-6)^3} = -6$

Use the product rule for radicals to find each product. See Example 1.

3. $\sqrt{3} \cdot \sqrt{5}$

4. $\sqrt{3} \cdot \sqrt{7}$

5. $\sqrt{2} \cdot \sqrt{11}$

6. $\sqrt{2} \cdot \sqrt{15}$

7. $\sqrt{6} \cdot \sqrt{7}$

8. $\sqrt{5} \cdot \sqrt{6}$

9. $\sqrt{13} \cdot \sqrt{r}, r \geq 0$

10. $\sqrt{19} \cdot \sqrt{k}, k \geq 0$

11. Which one of the following radicals is simplified? See Example 2.

A. $\sqrt{47}$ **B.** $\sqrt{45}$ **C.** $\sqrt{48}$ **D.** $\sqrt{44}$

12. If p is a prime number, is $\sqrt{p}$ in simplified form? Explain your answer.

Simplify each radical. See Example 2.

13. $\sqrt{45}$

14. $\sqrt{27}$

15. $\sqrt{24}$

16. $\sqrt{44}$

17. $\sqrt{90}$

18. $\sqrt{56}$

19. $\sqrt{75}$

20. $\sqrt{18}$

21. $\sqrt{125}$

22. $\sqrt{80}$

23. $\sqrt{145}$

24. $\sqrt{110}$

25. $\sqrt{160}$

26. $\sqrt{128}$

27. $-\sqrt{700}$

28. $-\sqrt{600}$

Find each product and simplify. See Example 3.

29. $\sqrt{3} \cdot \sqrt{18}$

30. $\sqrt{3} \cdot \sqrt{21}$

31. $\sqrt{12} \cdot \sqrt{48}$

32. $\sqrt{50} \cdot \sqrt{72}$

33. $\sqrt{12} \cdot \sqrt{30}$

34. $\sqrt{30} \cdot \sqrt{24}$

35. $2\sqrt{10} \cdot 3\sqrt{2}$

36. $5\sqrt{6} \cdot 2\sqrt{10}$

37. $5\sqrt{3} \cdot 2\sqrt{15}$

38. $4\sqrt{6} \cdot 3\sqrt{2}$

39. Simplify the product $\sqrt{8} \cdot \sqrt{32}$ in two ways. First, multiply 8 by 32 and simplify the square root of this product. Second, simplify $\sqrt{8}$, simplify $\sqrt{32}$, and then multiply. How do the answers compare? Make a conjecture (an educated guess) about whether the correct answer can always be obtained using either method when simplifying a product such as this.

40. Simplify the radical $\sqrt{288}$ in two ways. First, factor 288 as $144 \cdot 2$ and then simplify. Second, factor 288 as $48 \cdot 6$ and then simplify. How do the answers compare? Make a conjecture concerning the quickest way to simplify such a radical.

Simplify each radical expression. See Examples 4–6.

41. $\sqrt{\dfrac{16}{225}}$

42. $\sqrt{\dfrac{9}{100}}$

43. $\sqrt{\dfrac{7}{16}}$

44. $\sqrt{\dfrac{13}{25}}$

45. $\dfrac{\sqrt{75}}{\sqrt{3}}$

46. $\dfrac{\sqrt{200}}{\sqrt{2}}$

47. $\sqrt{\dfrac{5}{2}} \cdot \sqrt{\dfrac{125}{8}}$

48. $\sqrt{\dfrac{8}{3}} \cdot \sqrt{\dfrac{512}{27}}$

49. $\dfrac{30\sqrt{10}}{5\sqrt{2}}$

50. $\dfrac{50\sqrt{20}}{2\sqrt{10}}$

Simplify each radical. Assume that all variables represent nonnegative real numbers. See Example 7.

51. $\sqrt{m^2}$ **52.** $\sqrt{k^2}$ **53.** $\sqrt{y^4}$ **54.** $\sqrt{s^4}$

55. $\sqrt{36z^2}$ **56.** $\sqrt{49n^2}$ **57.** $\sqrt{400x^6}$ **58.** $\sqrt{900y^8}$

59. $\sqrt{18x^8}$ **60.** $\sqrt{20r^{10}}$ **61.** $\sqrt{45c^{14}}$ **62.** $\sqrt{50d^{20}}$

63. $\sqrt{z^5}$ **64.** $\sqrt{y^3}$ **65.** $\sqrt{a^{13}}$ **66.** $\sqrt{p^{17}}$

67. $\sqrt{64x^7}$ **68.** $\sqrt{25t^{11}}$ **69.** $\sqrt{x^6y^{12}}$ **70.** $\sqrt{a^8b^{10}}$

71. $\sqrt{81m^4n^2}$ **72.** $\sqrt{100c^4d^6}$ **73.** $\sqrt{\frac{7}{x^{10}}},\quad x \neq 0$ **74.** $\sqrt{\frac{14}{z^{12}}},\quad z \neq 0$

75. $\sqrt{\frac{y^4}{100}}$ **76.** $\sqrt{\frac{w^8}{144}}$ **77.** $\sqrt{\frac{x^6}{y^8}},\quad y \neq 0$ **78.** $\sqrt{\frac{a^4}{b^6}},\quad b \neq 0$

Simplify each radical. See Example 8.

79. $\sqrt[3]{40}$ **80.** $\sqrt[3]{48}$ **81.** $\sqrt[3]{54}$ **82.** $\sqrt[3]{135}$

83. $\sqrt[3]{128}$ **84.** $\sqrt[3]{192}$ **85.** $\sqrt[4]{80}$ **86.** $\sqrt[4]{243}$

87. $\sqrt[3]{\frac{8}{27}}$ **88.** $\sqrt[3]{\frac{64}{125}}$ **89.** $\sqrt[3]{-\frac{216}{125}}$ **90.** $\sqrt[3]{-\frac{1}{64}}$

Simplify each radical. See Example 9.

91. $\sqrt[3]{p^3}$

92. $\sqrt[3]{w^3}$

93. $\sqrt[3]{x^9}$

94. $\sqrt[3]{y^{18}}$

95. $\sqrt[3]{64z^6}$

96. $\sqrt[3]{125a^{15}}$

97. $\sqrt[3]{343a^9b^3}$

98. $\sqrt[3]{216m^3n^6}$

99. $\sqrt[3]{16t^5}$

100. $\sqrt[3]{24x^4}$

101. $\sqrt[3]{\frac{m^{12}}{8}}$

102. $\sqrt[3]{\frac{n^9}{27}}$

The volume of a cube is found with the formula $V = s^3$, where s is the length of an edge of the cube. Use this information in Exercises 103 and 104.

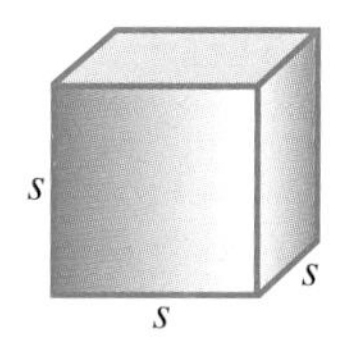

103. A container in the shape of a cube has a volume of 216 cm^3. What is the depth of the container?

104. A cube-shaped box must be constructed to contain 128 ft^3. What should the dimensions (height, width, and length) of the box be?

The volume of a sphere is found with the formula $V = \frac{4}{3}\pi r^3$, where r is the length of the radius of the sphere. Use this information in Exercises 105 and 106.

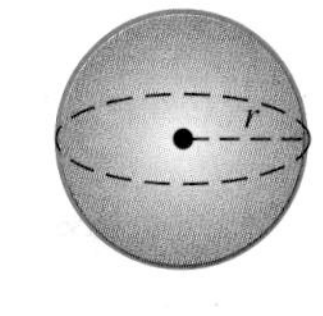

105. A ball in the shape of a sphere has a volume of 288π in.^{3}. What is the radius of the ball?

106. Suppose that the volume of the ball described in Exercise 105 is multiplied by 8. How is the radius affected?

Work Exercises 107 and 108 without using a calculator.

107. Choose the best estimate for the area (in square inches) of this rectangle.

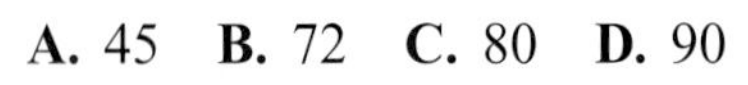

A. 45 **B.** 72 **C.** 80 **D.** 90

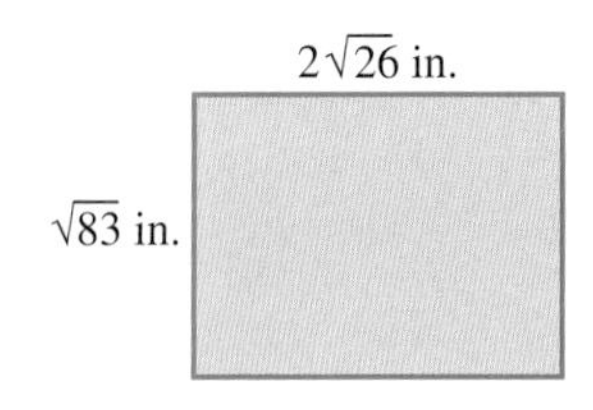

108. Choose the best estimate for the area (in square feet) of the triangle.

A. 20 **B.** 40 **C.** 60 **D.** 80

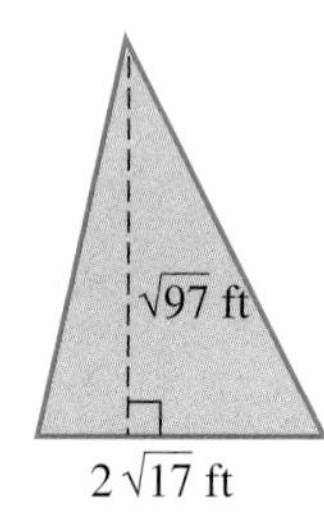

8.3 Adding and Subtracting Radicals

OBJECTIVES

1. Add and subtract radicals.
2. Simplify radical sums and differences.
3. Simplify more complicated radical expressions.

OBJECTIVE 1 Add and subtract radicals. We add or subtract radicals by using the distributive property. For example,

$$8\sqrt{3} + 6\sqrt{3} = (8 + 6)\sqrt{3} = 14\sqrt{3}.$$

Also, $$2\sqrt{11} - 7\sqrt{11} = -5\sqrt{11}.$$

Only **like radicals,** those that are ***multiples of the same root of the same number,*** can be combined in this way. In the examples above, $8\sqrt{3}$ and $6\sqrt{3}$ are like radicals, as are $2\sqrt{11}$ and $-7\sqrt{11}$. On the other hand, examples of **unlike radicals** are

$2\sqrt{5}$ and $2\sqrt{3}$, Radicands are different.

as well as $2\sqrt{3}$ and $2\sqrt[3]{3}$. Indexes are different.

Work Problem 1 at the Side.

1 Indicate whether the radicals in each pair are *like* or *unlike*.

(a) $5\sqrt{6}$ and $4\sqrt{6}$

(b) $2\sqrt{3}$ and $3\sqrt{2}$

(c) $\sqrt{10}$ and $\sqrt[3]{10}$

(d) $7\sqrt{2x}$ and $8\sqrt{2x}$

(e) $\sqrt{3y}$ and $\sqrt{6y}$

EXAMPLE 1 Adding and Subtracting Like Radicals

Add or subtract, as indicated.

(a) $3\sqrt{6} + 5\sqrt{6} = (3 + 5)\sqrt{6} = 8\sqrt{6}$ Distributive property

(b) $5\sqrt{10} - 7\sqrt{10} = (5 - 7)\sqrt{10} = -2\sqrt{10}$

(c) $\sqrt{7} + 2\sqrt{7} = 1\sqrt{7} + 2\sqrt{7} = (1 + 2)\sqrt{7} = 3\sqrt{7}$

(d) $\sqrt{5} + \sqrt{5} = 1\sqrt{5} + 1\sqrt{5} = 2\sqrt{5}$

(e) $\sqrt{3} + \sqrt{7}$ cannot be added using the distributive property.

Work Problem 2 at the Side.

2 Add or subtract, as indicated.

(a) $8\sqrt{5} + 2\sqrt{5}$

(b) $-4\sqrt{3} + 9\sqrt{3}$

(c) $12\sqrt{11} - 3\sqrt{11}$

(d) $\sqrt{15} + \sqrt{15}$

(e) $2\sqrt{7} + 2\sqrt{10}$

OBJECTIVE 2 Simplify radical sums and differences. Sometimes one or more radical expressions in a sum or difference must first be simplified. Any like radicals that result can then be added or subtracted.

EXAMPLE 2 Adding and Subtracting Radicals That Must Be Simplified

Add or subtract, as indicated.

(a)
$$\begin{aligned} 3\sqrt{2} + \sqrt{8} &= 3\sqrt{2} + \sqrt{4 \cdot 2} && \text{Factor.} \\ &= 3\sqrt{2} + \sqrt{4} \cdot \sqrt{2} && \text{Product rule} \\ &= 3\sqrt{2} + 2\sqrt{2} && \sqrt{4} = 2 \\ &= 5\sqrt{2} && \text{Add like radicals.} \end{aligned}$$

(b)
$$\begin{aligned} \sqrt{18} - \sqrt{27} &= \sqrt{9 \cdot 2} - \sqrt{9 \cdot 3} && \text{Factor.} \\ &= \sqrt{9} \cdot \sqrt{2} - \sqrt{9} \cdot \sqrt{3} && \text{Product rule} \\ &= 3\sqrt{2} - 3\sqrt{3} && \sqrt{9} = 3 \end{aligned}$$

Since $\sqrt{2}$ and $\sqrt{3}$ are unlike radicals, the answer cannot be simplified.

Continued on Next Page

ANSWERS

1. (a) like (b) unlike (c) unlike (d) like (e) unlike
2. (a) $10\sqrt{5}$ (b) $5\sqrt{3}$ (c) $9\sqrt{11}$ (d) $2\sqrt{15}$ (e) cannot be added

3 Add or subtract, as indicated.

(a) $\sqrt{8} + 4\sqrt{2}$

(b) $\sqrt{27} + \sqrt{12}$

(c) $5\sqrt{200} - 6\sqrt{18}$

4 Simplify each radical expression. Assume that all variables represent non-negative real numbers.

(a) $\sqrt{7} \cdot \sqrt{21} + 2\sqrt{27}$

(b) $\sqrt{3r} \cdot \sqrt{6} + \sqrt{8r}$

(c) $y\sqrt{72} - \sqrt{18y^2}$

(d) $\sqrt[3]{81x^4} + 5\sqrt[3]{24x^4}$

(c)
$$\begin{aligned} 2\sqrt{12} + 3\sqrt{75} &= 2(\sqrt{4} \cdot \sqrt{3}) + 3(\sqrt{25} \cdot \sqrt{3}) && \text{Product rule} \\ &= 2(2\sqrt{3}) + 3(5\sqrt{3}) && \sqrt{4} = 2;\ \sqrt{25} = 5 \\ &= 4\sqrt{3} + 15\sqrt{3} && \text{Multiply.} \\ &= 19\sqrt{3} && \text{Add like radicals.} \end{aligned}$$

Work Problem 3 at the Side

OBJECTIVE 3 Simplify more complicated radical expressions. When simplifying, the order of operations from **Section 1.1** still applies.

EXAMPLE 3 Simplifying Radical Expressions

Simplify each radical expression. Assume that all variables represent non-negative real numbers.

(a)
$$\begin{aligned} \sqrt{5} \cdot \sqrt{15} + 4\sqrt{3} &= \sqrt{5 \cdot 15} + 4\sqrt{3} && \text{Product rule} \\ &= \sqrt{75} + 4\sqrt{3} && \text{Multiply.} \\ &= \sqrt{25 \cdot 3} + 4\sqrt{3} && \text{25 is a perfect square.} \\ &= \sqrt{25} \cdot \sqrt{3} + 4\sqrt{3} && \text{Product rule} \\ &= 5\sqrt{3} + 4\sqrt{3} && \sqrt{25} = 5 \\ &= 9\sqrt{3} && \text{Add like radicals.} \end{aligned}$$

(b)
$$\begin{aligned} \sqrt{2} \cdot \sqrt{6k} + \sqrt{27k} &= \sqrt{12k} + \sqrt{27k} && \text{Product rule} \\ &= \sqrt{4 \cdot 3k} + \sqrt{9 \cdot 3k} && \text{Factor.} \\ &= \sqrt{4} \cdot \sqrt{3k} + \sqrt{9} \cdot \sqrt{3k} && \text{Product rule} \\ &= 2\sqrt{3k} + 3\sqrt{3k} && \sqrt{4} = 2;\ \sqrt{9} = 3 \\ &= 5\sqrt{3k} && \text{Add like radicals.} \end{aligned}$$

(c)
$$\begin{aligned} 3x\sqrt{50} + \sqrt{2x^2} &= 3x\sqrt{25 \cdot 2} + \sqrt{x^2 \cdot 2} && \text{Factor.} \\ &= 3x\sqrt{25} \cdot \sqrt{2} + \sqrt{x^2} \cdot \sqrt{2} && \text{Product rule} \\ &= 3x \cdot 5\sqrt{2} + x\sqrt{2} && \sqrt{25} = 5;\ \sqrt{x^2} = x \\ &= 15x\sqrt{2} + x\sqrt{2} && \text{Multiply.} \\ &= 16x\sqrt{2} && \text{Add like radicals.} \end{aligned}$$

(d)
$$\begin{aligned} 2\sqrt[3]{32m^3} - \sqrt[3]{108m^3} &= 2\sqrt[3]{(8m^3)4} - \sqrt[3]{(27m^3)4} && \text{Factor.} \\ &= 2(2m)\sqrt[3]{4} - 3m\sqrt[3]{4} && \sqrt[3]{8m^3} = 2m;\ \sqrt[3]{27m^3} = 3m \\ &= 4m\sqrt[3]{4} - 3m\sqrt[3]{4} && \text{Multiply.} \\ &= m\sqrt[3]{4} && \text{Subtract like radicals.} \end{aligned}$$

CAUTION

A sum or difference of radicals can be simplified only if the radicals are like radicals. Thus, $\sqrt{5} + 3\sqrt{5} = 4\sqrt{5}$, but $\sqrt{5} + 5\sqrt{3}$ cannot be simplified further. Also, $2\sqrt{3} + 5\sqrt[3]{3}$ cannot be simplified further.

Work Problem 4 at the Side.

ANSWERS

3. (a) $6\sqrt{2}$ **(b)** $5\sqrt{3}$ **(c)** $32\sqrt{2}$

4. (a) $13\sqrt{3}$ **(b)** $5\sqrt{2r}$ **(c)** $3y\sqrt{2}$ **(d)** $13x\sqrt[3]{3x}$

8.3 Exercises

FOR EXTRA HELP

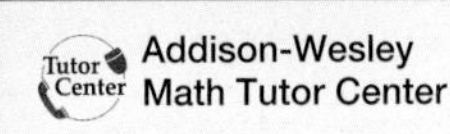

Math XL
MathXL

MyMathLab
MyMathLab

Fill in each blank with the correct response.

1. $5\sqrt{2} + 6\sqrt{2} = (5 + 6)\sqrt{2} = 11\sqrt{2}$ is an example of the ____________ property.

2. Like radicals have the same ____________ of the same ____________.

3. $\sqrt{5} + 5\sqrt{3}$ cannot be simplified because the ____________ are different.

4. $4\sqrt[3]{2} + 3\sqrt{2}$ cannot be simplified because the ____________ are different.

Simplify and add or subtract wherever possible. See Examples 1 and 2.

5. $14\sqrt{7} - 19\sqrt{7}$

6. $16\sqrt{2} - 18\sqrt{2}$

7. $\sqrt{17} + 4\sqrt{17}$

8. $5\sqrt{19} + \sqrt{19}$

9. $6\sqrt{7} - \sqrt{7}$

10. $11\sqrt{14} - \sqrt{14}$

11. $\sqrt{45} + 4\sqrt{20}$

12. $\sqrt{24} + 6\sqrt{54}$

13. $5\sqrt{72} - 3\sqrt{50}$

14. $6\sqrt{18} - 5\sqrt{32}$

15. $-5\sqrt{32} + 2\sqrt{98}$

16. $-4\sqrt{75} + 3\sqrt{12}$

17. $5\sqrt{7} - 3\sqrt{28} + 6\sqrt{63}$

18. $3\sqrt{11} + 5\sqrt{44} - 8\sqrt{99}$

19. $2\sqrt{8} - 5\sqrt{32} - 2\sqrt{48}$

20. $5\sqrt{72} - 3\sqrt{48} + 4\sqrt{128}$

21. $4\sqrt{50} + 3\sqrt{12} - 5\sqrt{45}$

22. $6\sqrt{18} + 2\sqrt{48} + 6\sqrt{28}$

23. $\frac{1}{4}\sqrt{288} + \frac{1}{6}\sqrt{72}$

24. $\frac{2}{3}\sqrt{27} + \frac{3}{4}\sqrt{48}$

Find the perimeter of each figure.

25.

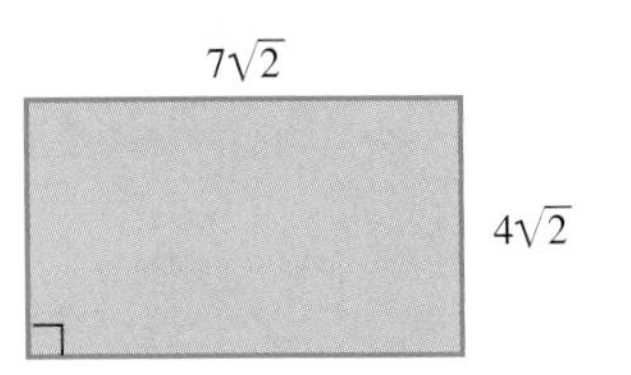

26.

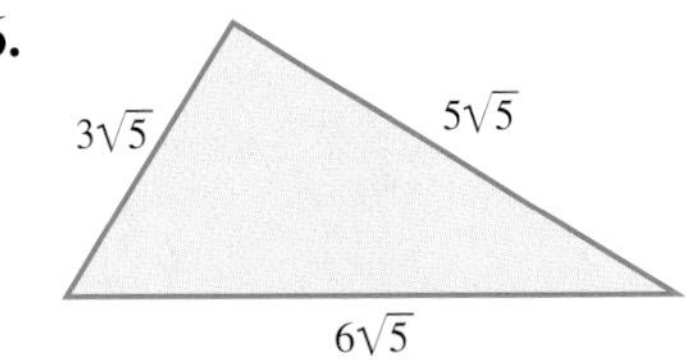

Perform the indicated operations. Assume that all variables represent nonnegative real numbers. See Example 3.

27. $\sqrt{6} \cdot \sqrt{2} + 9\sqrt{3}$

28. $4\sqrt{15} \cdot \sqrt{3} + 4\sqrt{5}$

29. $\sqrt{9x} + \sqrt{49x} - \sqrt{25x}$

30. $\sqrt{4a} - \sqrt{16a} + \sqrt{100a}$

31. $\sqrt{6x^2} + x\sqrt{24}$

32. $\sqrt{75x^2} + x\sqrt{108}$

33. $3\sqrt{8x^2} - 4x\sqrt{2} - x\sqrt{8}$

34. $\sqrt{2b^2} + 3b\sqrt{18} - b\sqrt{200}$

35. $-8\sqrt{32k} + 6\sqrt{8k}$

36. $4\sqrt{12x} + 2\sqrt{27x}$

37. $2\sqrt{125x^2z} + 8x\sqrt{80z}$

38. $\sqrt{48x^2y} + 5x\sqrt{27y}$

39. $4\sqrt[3]{16} - 3\sqrt[3]{54}$

40. $5\sqrt[3]{128} + 3\sqrt[3]{250}$

41. $6\sqrt[3]{8p^2} - 2\sqrt[3]{27p^2}$

42. $8k\sqrt[3]{54k} + 6\sqrt[3]{16k^4}$

43. $5\sqrt[4]{m^3} + 8\sqrt[4]{16m^3}$

44. $5\sqrt[4]{m^5} + 3\sqrt[4]{81m^5}$

RELATING CONCEPTS (EXERCISES 45–48) For Individual or Group Work

Adding and subtracting like radicals is no different than adding and subtracting like terms. ***Work Exercises 45–48 in order.***

45. Combine like terms: $5x^2y + 3x^2y - 14x^2y$.

46. Combine like terms: $5(p - 2q)^2(a + b) + 3(p - 2q)^2(a + b) - 14(p - 2q)^2(a + b)$.

47. Combine like radicals: $5a^2\sqrt{xy} + 3a^2\sqrt{xy} - 14a^2\sqrt{xy}$.

48. Compare your answers in Exercises 45–47. How are they alike? How are they different?

8.4 Rationalizing the Denominator

OBJECTIVES

1. Rationalize denominators with square roots.
2. Write radicals in simplified form.
3. Rationalize denominators with cube roots.

OBJECTIVE 1 Rationalize denominators with square roots. Although calculators now make it fairly easy to divide by a radical in an expression such as $\frac{1}{\sqrt{2}}$, it is sometimes easier to work with radical expressions if the denominators do not contain any radicals. For example, the radical in the denominator of $\frac{1}{\sqrt{2}}$ can be eliminated by multiplying the numerator and denominator by $\sqrt{2}$, since $\sqrt{2} \cdot \sqrt{2} = \sqrt{4} = 2$.

$$\frac{1}{\sqrt{2}} = \frac{1 \cdot \sqrt{2}}{\sqrt{2} \cdot \sqrt{2}} = \frac{\sqrt{2}}{2} \qquad \text{Multiply by } \tfrac{\sqrt{2}}{\sqrt{2}} = 1.$$

This process of changing the denominator from a radical (irrational number) to a rational number is called **rationalizing the denominator.** The value of the radical expression is not changed; only the form is changed, because the expression has been multiplied by 1 in the form of $\frac{\sqrt{2}}{\sqrt{2}}$.

EXAMPLE 1 Rationalizing Denominators

Rationalize each denominator.

(a) $\frac{9}{\sqrt{6}}$

We must eliminate the radical in the denominator.

$$\frac{9}{\sqrt{6}} = \frac{9 \cdot \sqrt{6}}{\sqrt{6} \cdot \sqrt{6}} \qquad \text{Multiply by } \tfrac{\sqrt{6}}{\sqrt{6}} = 1.$$

$$= \frac{9\sqrt{6}}{6} \qquad \sqrt{6} \cdot \sqrt{6} = \sqrt{36} = 6$$

$$= \frac{3\sqrt{6}}{2} \qquad \text{Lowest terms}$$

(b) $\frac{12}{\sqrt{8}}$

The denominator could be rationalized by multiplying by $\sqrt{8}$. However, first simplifying the denominator is more direct.

$$\sqrt{8} = \sqrt{4} \cdot \sqrt{2} = 2\sqrt{2}$$

Then multiply the numerator and denominator by $\sqrt{2}$.

$$\frac{12}{\sqrt{8}} = \frac{12}{2\sqrt{2}} \qquad \sqrt{8} = 2\sqrt{2}$$

$$= \frac{12 \cdot \sqrt{2}}{2\sqrt{2} \cdot \sqrt{2}} \qquad \text{Multiply by } \tfrac{\sqrt{2}}{\sqrt{2}} = 1.$$

$$= \frac{12 \cdot \sqrt{2}}{2 \cdot 2} \qquad \sqrt{2} \cdot \sqrt{2} = \sqrt{4} = 2$$

$$= \frac{12\sqrt{2}}{4} \qquad \text{Multiply.}$$

$$= 3\sqrt{2} \qquad \text{Lowest terms}$$

1 Rationalize each denominator.

(a) $\frac{3}{\sqrt{5}}$

(b) $\frac{-6}{\sqrt{11}}$

(c) $-\frac{\sqrt{7}}{\sqrt{2}}$

(d) $\frac{20}{\sqrt{18}}$

2 Simplify.

(a) $\sqrt{\frac{16}{11}}$

(b) $\sqrt{\frac{5}{18}}$

(c) $\sqrt{\frac{8}{32}}$

Answers

1. (a) $\frac{3\sqrt{5}}{5}$ **(b)** $\frac{-6\sqrt{11}}{11}$ **(c)** $-\frac{\sqrt{14}}{2}$ **(d)** $\frac{10\sqrt{2}}{3}$

2. (a) $\frac{4\sqrt{11}}{11}$ **(b)** $\frac{\sqrt{10}}{6}$ **(c)** $\frac{1}{2}$

NOTE

In Example 1(b), we could also have rationalized the original denominator $\sqrt{8}$ by multiplying by $\sqrt{2}$, since $\sqrt{8} \cdot \sqrt{2} = \sqrt{16} = 4$.

$$\frac{12}{\sqrt{8}} = \frac{12 \cdot \sqrt{2}}{\sqrt{8} \cdot \sqrt{2}} = \frac{12\sqrt{2}}{\sqrt{16}} = \frac{12\sqrt{2}}{4} = 3\sqrt{2}$$

Both approaches are correct.

Work Problem 1 at the Side.

OBJECTIVE 2 Write radicals in simplified form. A radical is considered to be in simplified form if the following three conditions are met.

Simplified Form of a Radical

1. The radicand contains no factor (except 1) that is a perfect square (when dealing with square roots), a perfect cube (when dealing with cube roots), and so on.
2. The radicand has no fractions.
3. No denominator contains a radical.

EXAMPLE 2 Simplifying a Radical

Simplify $\sqrt{\frac{27}{5}}$.

This violates condition 2. To begin, use the quotient rule for radicals.

$$\sqrt{\frac{27}{5}} = \frac{\sqrt{27}}{\sqrt{5}} \quad \text{Quotient rule}$$

$$= \frac{\sqrt{27} \cdot \sqrt{5}}{\sqrt{5} \cdot \sqrt{5}} \quad \text{Rationalize the denominator.}$$

$$= \frac{\sqrt{27} \cdot \sqrt{5}}{5} \quad \sqrt{5} \cdot \sqrt{5} = \sqrt{25} = 5$$

$$= \frac{\sqrt{9 \cdot 3} \cdot \sqrt{5}}{5} \quad \text{Factor.}$$

$$= \frac{\sqrt{9} \cdot \sqrt{3} \cdot \sqrt{5}}{5} \quad \text{Product rule}$$

$$= \frac{3 \cdot \sqrt{3} \cdot \sqrt{5}}{5} \quad \sqrt{9} = 3$$

$$= \frac{3\sqrt{15}}{5} \quad \text{Product rule}$$

Work Problem 2 at the Side.

EXAMPLE 3 Simplifying a Product of Radicals

Simplify $\sqrt{\frac{5}{8}} \cdot \sqrt{\frac{1}{6}}$

Use both the product and quotient rules.

$$\sqrt{\frac{5}{8}} \cdot \sqrt{\frac{1}{6}} = \sqrt{\frac{5}{8} \cdot \frac{1}{6}} \quad \text{Product rule}$$

$$= \sqrt{\frac{5}{48}} \quad \text{Multiply fractions.}$$

$$= \frac{\sqrt{5}}{\sqrt{48}} \quad \text{Quotient rule}$$

First simplify the denominator and then rationalize it.

$$= \frac{\sqrt{5}}{\sqrt{16} \cdot \sqrt{3}} \quad \text{Product rule}$$

$$= \frac{\sqrt{5}}{4\sqrt{3}} \quad \sqrt{16} = 4$$

$$= \frac{\sqrt{5} \cdot \sqrt{3}}{4\sqrt{3} \cdot \sqrt{3}} \quad \text{Rationalize the denominator.}$$

$$= \frac{\sqrt{15}}{4 \cdot 3} \quad \text{Product rule; } \sqrt{3} \cdot \sqrt{3} = 3$$

$$= \frac{\sqrt{15}}{12} \quad \text{Multiply.}$$

Work Problem 3 at the Side.

3 Simplify.

(a) $\sqrt{\frac{1}{2}} \cdot \sqrt{\frac{5}{6}}$

(b) $\sqrt{\frac{1}{10}} \cdot \sqrt{20}$

(c) $\sqrt{\frac{5}{8}} \cdot \sqrt{\frac{24}{10}}$

EXAMPLE 4 Simplifying a Quotient of Radicals

Simplify $\frac{\sqrt{4x}}{\sqrt{y}}$. Assume that x and y represent positive real numbers.

Multiply the numerator and denominator by $\sqrt{y}$.

$$\frac{\sqrt{4x}}{\sqrt{y}} = \frac{\sqrt{4x} \cdot \sqrt{y}}{\sqrt{y} \cdot \sqrt{y}} \quad \text{Rationalize the denominator.}$$

$$= \frac{\sqrt{4xy}}{y} \quad \text{Product rule; } \sqrt{y} \cdot \sqrt{y} = y$$

$$= \frac{2\sqrt{xy}}{y} \quad \sqrt{4} = 2$$

Work Problem 4 at the Side.

4 Simplify $\frac{\sqrt{5p}}{\sqrt{q}}$. Assume that p and q represent positive real numbers.

Answers

3. **(a)** $\frac{\sqrt{15}}{6}$ **(b)** $\sqrt{2}$ **(c)** $\frac{\sqrt{6}}{2}$

4. $\frac{\sqrt{5pq}}{q}$

5 Simplify $\sqrt{\frac{5r^2t^2}{7}}$. Assume that r and t represent nonnegative real numbers.

EXAMPLE 5 Simplifying a Radical Quotient

Simplify $\sqrt{\frac{2x^2y}{3}}$. Assume that x and y represent nonnegative real numbers.

$$\sqrt{\frac{2x^2y}{3}} = \frac{\sqrt{2x^2y}}{\sqrt{3}} \quad \text{Quotient rule}$$

$$= \frac{\sqrt{2x^2y} \cdot \sqrt{3}}{\sqrt{3} \cdot \sqrt{3}} \quad \text{Rationalize the denominator.}$$

$$= \frac{\sqrt{6x^2y}}{3} \quad \text{Product rule; } \sqrt{3} \cdot \sqrt{3} = 3$$

$$= \frac{\sqrt{x^2}\sqrt{6y}}{3} \quad \text{Product rule}$$

$$= \frac{x\sqrt{6y}}{3} \quad \sqrt{x^2} = x, \text{ since } x \geq 0.$$

Work Problem 5 at the Side.

OBJECTIVE 3 Rationalize denominators with cube roots. A denominator with a cube root is rationalized by changing the radicand in the denominator to a perfect cube, as shown in the next example.

EXAMPLE 6 Rationalizing Denominators with Cube Roots

Rationalize each denominator.

(a) $\sqrt[3]{\frac{3}{2}}$

First write the expression as a quotient of radicals. Then multiply the numerator and denominator by the appropriate number of factors of 2 to make the denominator a perfect cube. This will eliminate the radical in the denominator. Here, multiply by $\sqrt[3]{2^2}$.

$$\sqrt[3]{\frac{3}{2}} = \frac{\sqrt[3]{3}}{\sqrt[3]{2}} = \frac{\sqrt[3]{3} \cdot \sqrt[3]{2^2}}{\sqrt[3]{2} \cdot \sqrt[3]{2^2}} = \frac{\sqrt[3]{3 \cdot 2^2}}{\sqrt[3]{2^3}} = \frac{\sqrt[3]{12}}{2} \qquad \sqrt[3]{2^3} = \sqrt[3]{8} = 2$$

Denominator is a perfect cube.

(b) $\frac{\sqrt[3]{3}}{\sqrt[3]{4}}$

Since $\sqrt[3]{4} \cdot \sqrt[3]{2} = \sqrt[3]{2^2} \cdot \sqrt[3]{2} = \sqrt[3]{2^3} = 2$, multiply the numerator and denominator by $\sqrt[3]{2}$.

$$\frac{\sqrt[3]{3}}{\sqrt[3]{4}} = \frac{\sqrt[3]{3} \cdot \sqrt[3]{2}}{\sqrt[3]{2^2} \cdot \sqrt[3]{2}} = \frac{\sqrt[3]{6}}{\sqrt[3]{2^3}} = \frac{\sqrt[3]{6}}{2}$$

Continued on Next Page

ANSWERS

5. $\frac{rt\sqrt{35}}{7}$

(c) $\dfrac{\sqrt[3]{2}}{\sqrt[3]{3x^2}}, \quad x \neq 0$

Multiply the numerator and denominator by the appropriate number of factors of 3 and of x to get a perfect cube in the denominator. Here, multiply by $\sqrt[3]{3^2x}$ (that is, $\sqrt[3]{9x}$) since $\sqrt[3]{3x^2} \cdot \sqrt[3]{3^2x} = \sqrt[3]{(3x)^3} = 3x$.

$$\frac{\sqrt[3]{2}}{\sqrt[3]{3x^2}} = \frac{\sqrt[3]{2} \cdot \sqrt[3]{3^2x}}{\sqrt[3]{3x^2} \cdot \sqrt[3]{3^2x}} = \frac{\sqrt[3]{18x}}{\sqrt[3]{(3x)^3}} = \frac{\sqrt[3]{18x}}{3x}$$

Denominator is a perfect cube.

CAUTION

A common error in a problem like the one in Example 6(a) is to multiply by $\sqrt[3]{2}$ instead of $\sqrt[3]{2^2}$. Doing this would give a denominator of $\sqrt[3]{2} \cdot \sqrt[3]{2} = \sqrt[3]{4}$. Because 4 is not a perfect cube, the denominator is still not rationalized.

Work Problem 6 at the Side.

6 Rationalize each denominator.

(a) $\sqrt[3]{\dfrac{5}{7}}$

(b) $\dfrac{\sqrt[3]{5}}{\sqrt[3]{9}}$

(c) $\dfrac{\sqrt[3]{4}}{\sqrt[3]{25y}}, \quad y \neq 0$

ANSWERS

6. (a) $\dfrac{\sqrt[3]{245}}{7}$ (b) $\dfrac{\sqrt[3]{15}}{3}$ (c) $\dfrac{\sqrt[3]{20y^2}}{5y}$

Focus on

Real-Data Applications

The Golden Ratio—A Star Number

The **Golden Ratio,** the number $\frac{1 + \sqrt{5}}{2}$, is called phi, ϕ. The number has been known since ancient times and is called the *sacred ratio* in the Rhind Papyrus from 1600 B.C. The Egyptians used ϕ to build the Great Pyramids, and the ancient Greeks used ϕ in art and architecture, striving for the proportion that was most pleasing to the eye.

The Golden Ratio is widespread in the star formed by connecting the vertices (corners) of a regular pentagon inscribed in a circle. In the figure shown, ABCDE forms a regular pentagon. Each of the following ratios forms the Golden Ratio, ϕ.

$$\frac{AC}{AY} \qquad \frac{AY}{AX}$$

Using symmetry, you should be able to identify other similar relationships.

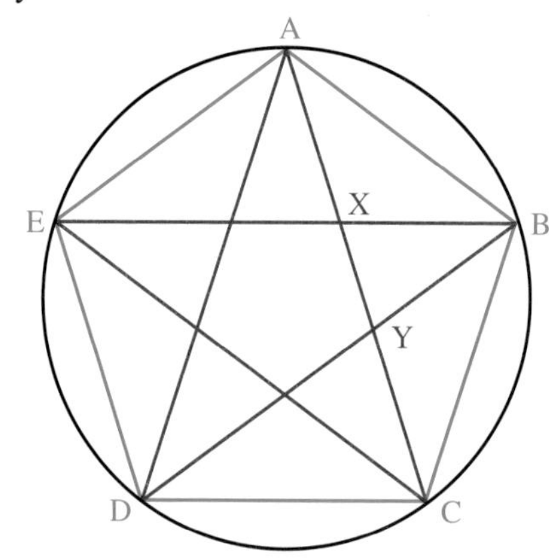

For Group Discussion

1. List two more ratios in the star that form the Golden Ratio. You may label additional points.

2. The Golden Ratio has some curious properties. Use $\phi = \frac{1 + \sqrt{5}}{2}$ to find the following quantities. Write your answers in exact form (using radicals). Rationalize denominators when appropriate.

   (a) $\frac{1}{\phi}$

   (b) $\phi - 1$

   (c) ϕ^2

   (d) $\phi + 1$

3. Based on your results from Problem 2, which forms in (a)–(d) are equivalent?

8.4 Exercises

FOR EXTRA HELP

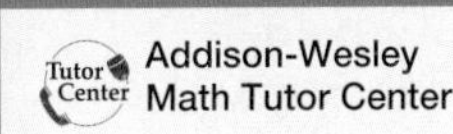

MathXL | Digital Video Tutor CD 9 Videotape 8 | Student's Solutions Manual | MyMathLab | Interactmath.com

Rationalize each denominator. See Examples 1 and 2.

1. $\dfrac{8}{\sqrt{2}}$

2. $\dfrac{12}{\sqrt{3}}$

3. $\dfrac{-\sqrt{11}}{\sqrt{3}}$

4. $\dfrac{-\sqrt{13}}{\sqrt{5}}$

5. $\dfrac{7\sqrt{3}}{\sqrt{5}}$

6. $\dfrac{4\sqrt{6}}{\sqrt{5}}$

7. $\dfrac{24\sqrt{10}}{16\sqrt{3}}$

8. $\dfrac{18\sqrt{15}}{12\sqrt{2}}$

9. $\dfrac{16}{\sqrt{27}}$

10. $\dfrac{24}{\sqrt{18}}$

11. $\dfrac{-3}{\sqrt{50}}$

12. $\dfrac{-5}{\sqrt{75}}$

13. $\dfrac{63}{\sqrt{45}}$

14. $\dfrac{27}{\sqrt{32}}$

15. $\dfrac{\sqrt{24}}{\sqrt{8}}$

16. $\dfrac{\sqrt{36}}{\sqrt{18}}$

17. $\sqrt{\dfrac{1}{2}}$

18. $\sqrt{\dfrac{1}{3}}$

19. $\sqrt{\dfrac{13}{5}}$

20. $\sqrt{\dfrac{17}{11}}$

21. When we rationalize the denominator in an expression such as $\dfrac{4}{\sqrt{3}}$, we multiply both the numerator and denominator by $\sqrt{3}$. By what number are we actually multiplying the given expression, and what property of real numbers justifies the fact that our result is equal to the given expression?

22. In Example 1(a), we show algebraically that $\dfrac{9}{\sqrt{6}}$ is equal to $\dfrac{3\sqrt{6}}{2}$. Support this result numerically by finding the decimal approximation of $\dfrac{9}{\sqrt{6}}$ on your calculator, and then finding the decimal approximation of $\dfrac{3\sqrt{6}}{2}$. What do you notice?

Simplify each product of radicals. See Example 3.

23. $\sqrt{\dfrac{7}{13}} \cdot \sqrt{\dfrac{13}{3}}$

24. $\sqrt{\dfrac{19}{20}} \cdot \sqrt{\dfrac{20}{3}}$

25. $\sqrt{\dfrac{21}{7}} \cdot \sqrt{\dfrac{21}{8}}$

26. $\sqrt{\dfrac{5}{8}} \cdot \sqrt{\dfrac{5}{6}}$

27. $\sqrt{\dfrac{1}{12}} \cdot \sqrt{\dfrac{1}{3}}$

28. $\sqrt{\dfrac{1}{8}} \cdot \sqrt{\dfrac{1}{2}}$

29. $\sqrt{\dfrac{2}{9}} \cdot \sqrt{\dfrac{9}{2}}$

30. $\sqrt{\dfrac{4}{3}} \cdot \sqrt{\dfrac{3}{4}}$

Simplify each radical. Assume that all variables represent positive real numbers. See Examples 4 and 5.

31. $\dfrac{\sqrt{7}}{\sqrt{x}}$

32. $\dfrac{\sqrt{19}}{\sqrt{y}}$

33. $\dfrac{\sqrt{4x^3}}{\sqrt{y}}$

34. $\dfrac{\sqrt{9t^3}}{\sqrt{s}}$

35. $\sqrt{\dfrac{5x^3z}{6}}$

36. $\sqrt{\dfrac{3st^3}{5}}$

37. $\sqrt{\dfrac{9a^2r^5}{7t}}$

38. $\sqrt{\dfrac{16x^3y^2}{13z}}$

39. Which one of the following would be an appropriate choice for multiplying the numerator and denominator of $\dfrac{\sqrt[3]{2}}{\sqrt[3]{5}}$ by in order to rationalize the denominator?

A. $\sqrt[3]{5}$ **B.** $\sqrt[3]{25}$ **C.** $\sqrt[3]{2}$ **D.** $\sqrt[3]{4}$

40. In Example 6(b), we multiplied the numerator and denominator of $\dfrac{\sqrt[3]{3}}{\sqrt[3]{4}}$ by $\sqrt[3]{2}$ to rationalize the denominator. Suppose we had chosen to multiply by $\sqrt[3]{16}$ instead. Would we have obtained the correct answer after all simplifications were done?

Rationalize each denominator. Assume that variables in the denominator represent nonzero real numbers. See Example 6.

41. $\sqrt[3]{\dfrac{3}{2}}$

42. $\sqrt[3]{\dfrac{2}{5}}$

43. $\dfrac{\sqrt[3]{4}}{\sqrt[3]{7}}$

44. $\dfrac{\sqrt[3]{5}}{\sqrt[3]{10}}$

45. $\sqrt[3]{\dfrac{3}{4y^2}}$

46. $\sqrt[3]{\dfrac{3}{25x^2}}$

47. $\dfrac{\sqrt[3]{7m}}{\sqrt[3]{36n}}$

48. $\dfrac{\sqrt[3]{11p}}{\sqrt[3]{49q}}$

*In Exercises 49 and 50, **(a)** give the answer as a simplified radical and **(b)** use a calculator to give the answer correct to the nearest thousandth.*

49. The period p of a pendulum is the time it takes for it to swing from one extreme to the other and back again. The value of p in seconds is given by

$$p = k \cdot \sqrt{\frac{L}{g}},$$

where L is the length of the pendulum, g is the acceleration due to gravity, and k is a constant. Find the period when $k = 6$, $L = 9$ ft, and $g = 32$ ft per sec per sec.

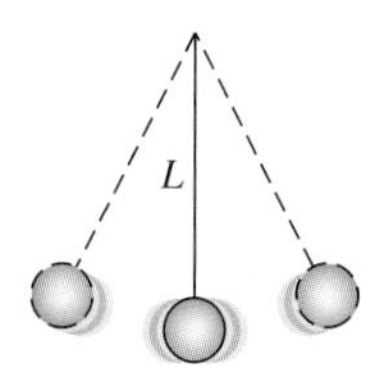

50. The velocity v of a meteorite approaching Earth is given by

$$v = \frac{k}{\sqrt{d}}$$

km per sec, where d is its distance from the center of Earth and k is a constant. What is the velocity of a meteorite that is 6000 km away from the center of Earth, if $k = 450$?

8.5 More Simplifying and Operations with Radicals

OBJECTIVES

1. Simplify products of radical expressions.
2. Use conjugates to rationalize denominators of radical expressions.
3. Write radical expressions with quotients in lowest terms.

The conditions for which a radical is in simplest form were listed in **Section 8.4.** Below is a set of guidelines to follow when you are simplifying radical expressions.

Simplifying Radical Expressions

1. If a radical represents a rational number, use that rational number in place of the radical.

Examples: $\sqrt{49} = 7; \sqrt{\frac{169}{9}} = \frac{13}{3}.$

2. If a radical expression contains products of radicals, use the product rule for radicals, $\sqrt[n]{x} \cdot \sqrt[n]{y} = \sqrt[n]{xy}$, to get a single radical.

Examples: $\sqrt{3} \cdot \sqrt{2} = \sqrt{6}; \sqrt[3]{5} \cdot \sqrt[3]{x} = \sqrt[3]{5x}.$

3. If a radicand has a factor that is a perfect square, express the radical as the product of the positive square root of the perfect square and the remaining radical factor. A similar statement applies to higher roots.

Examples: $\sqrt{20} = \sqrt{4 \cdot 5} = \sqrt{4} \cdot \sqrt{5} = 2\sqrt{5};$

$\sqrt[3]{16} = \sqrt[3]{8 \cdot 2} = \sqrt[3]{8} \cdot \sqrt[3]{2} = 2\sqrt[3]{2}.$

4. If a radical expression contains sums or differences of radicals, use the distributive property to combine like radicals.

Examples: $3\sqrt{2} + 4\sqrt{2} = 7\sqrt{2};$

$3\sqrt{2} + 4\sqrt{3}$ cannot be simplified further.

5. Rationalize any denominator containing a radical.

Examples: $\frac{5}{\sqrt{3}} = \frac{5 \cdot \sqrt{3}}{\sqrt{3} \cdot \sqrt{3}} = \frac{5\sqrt{3}}{3};$

$\sqrt{\frac{3}{2}} = \frac{\sqrt{3}}{\sqrt{2}} = \frac{\sqrt{3} \cdot \sqrt{2}}{\sqrt{2} \cdot \sqrt{2}} = \frac{\sqrt{6}}{2};$

$\sqrt[3]{\frac{1}{4}} = \frac{\sqrt[3]{1}}{\sqrt[3]{4}} = \frac{\sqrt[3]{1} \cdot \sqrt[3]{2}}{\sqrt[3]{4} \cdot \sqrt[3]{2}} = \frac{\sqrt[3]{2}}{\sqrt[3]{8}} = \frac{\sqrt[3]{2}}{2}.$

OBJECTIVE 1 Simplify products of radical expressions. Use the above guidelines.

EXAMPLE 1 Multiplying Radical Expressions

Find each product, and simplify.

(a) $\sqrt{5}(\sqrt{8} - \sqrt{32})$

Start by simplifying $\sqrt{8}$ and $\sqrt{32}$.

$$\sqrt{8} = 2\sqrt{2} \quad \text{and} \quad \sqrt{32} = 4\sqrt{2}.$$

Continued on Next Page

1 Find each product, and simplify.

(a) $\sqrt{7}(\sqrt{2} + \sqrt{5})$

(b) $\sqrt{2}(\sqrt{8} + \sqrt{20})$

(c) $(\sqrt{2} + 5\sqrt{3})(\sqrt{3} - 2\sqrt{2})$

(d) $(\sqrt{2} - \sqrt{5})(\sqrt{10} + \sqrt{2})$

$$\begin{aligned} \sqrt{5}(\sqrt{8} - \sqrt{32}) &= \sqrt{5}(2\sqrt{2} - 4\sqrt{2}) && \sqrt{8} = 2\sqrt{2};\ \sqrt{32} = 4\sqrt{2} \\ &= \sqrt{5}(-2\sqrt{2}) && \text{Subtract like radicals.} \\ &= -2\sqrt{5 \cdot 2} && \text{Product rule} \\ &= -2\sqrt{10} && \text{Multiply.} \end{aligned}$$

(b) $(\sqrt{3} + 2\sqrt{5})(\sqrt{3} - 4\sqrt{5})$

We can find the products of sums of radicals in the same way that we found the product of binomials in **Section 5.3,** using the FOIL method.

$$\begin{aligned} &(\sqrt{3} + 2\sqrt{5})(\sqrt{3} - 4\sqrt{5}) \\ &= \underbrace{\sqrt{3}(\sqrt{3})}_{\text{First}} + \underbrace{\sqrt{3}(-4\sqrt{5})}_{\text{Outer}} + \underbrace{2\sqrt{5}(\sqrt{3})}_{\text{Inner}} + \underbrace{2\sqrt{5}(-4\sqrt{5})}_{\text{Last}} \\ &= 3 - 4\sqrt{15} + 2\sqrt{15} - 8 \cdot 5 && \text{Product rule} \\ &= 3 - 2\sqrt{15} - 40 && \text{Add like radicals; multiply.} \\ &= -37 - 2\sqrt{15} && \text{Combine terms.} \end{aligned}$$

(c) $(\sqrt{3} + \sqrt{21})(\sqrt{3} - \sqrt{7})$

$$\begin{aligned} &= \sqrt{3}(\sqrt{3}) + \sqrt{3}(-\sqrt{7}) + \sqrt{21}(\sqrt{3}) \\ &\quad + \sqrt{21}(-\sqrt{7}) && \text{FOIL} \\ &= 3 - \sqrt{21} + \sqrt{63} - \sqrt{147} && \text{Product rule} \\ &= 3 - \sqrt{21} + \sqrt{9} \cdot \sqrt{7} - \sqrt{49} \cdot \sqrt{3} && \text{9 and 49 are perfect squares.} \\ &= 3 - \sqrt{21} + 3\sqrt{7} - 7\sqrt{3} && \sqrt{9} = 3;\ \sqrt{49} = 7 \end{aligned}$$

Since there are no like radicals, no terms can be combined.

Work Problem 1 at the Side.

The special products of binomials discussed in **Section 5.4** can be applied to radicals. Example 2 uses the rules for the square of a binomial,

$$(a + b)^2 = a^2 + 2ab + b^2 \quad \text{and} \quad (a - b)^2 = a^2 - 2ab + b^2.$$

EXAMPLE 2 Using Special Products with Radicals

Find each product.

(a) $(\sqrt{10} - 7)^2$

Follow the second pattern given above. Let $a = \sqrt{10}$ and $b = 7$.

$$\begin{aligned} (\sqrt{10} - 7)^2 &= (\sqrt{10})^2 - 2(\sqrt{10})(7) + 7^2 \\ &= 10 - 14\sqrt{10} + 49 && (\sqrt{10})^2 = 10;\ 7^2 = 49 \\ &= 59 - 14\sqrt{10} && \text{Combine terms.} \end{aligned}$$

(b)
$$\begin{aligned} (2\sqrt{3} + 4)^2 &= (2\sqrt{3})^2 + 2(2\sqrt{3})(4) + 4^2 && a = 2\sqrt{3};\ b = 4 \\ &= 12 + 16\sqrt{3} + 16 && (2\sqrt{3})^2 = 4 \cdot 3 = 12 \\ &= 28 + 16\sqrt{3} \end{aligned}$$

Continued on Next Page

Answers

1. (a) $\sqrt{14} + \sqrt{35}$
(b) $4 + 2\sqrt{10}$
(c) $11 - 9\sqrt{6}$
(d) $2\sqrt{5} + 2 - 5\sqrt{2} - \sqrt{10}$

(c) $(5 - \sqrt{x})^2 = 5^2 - 2(5)(\sqrt{x}) + (\sqrt{x})^2$

$= 25 - 10\sqrt{x} + x, \quad x \geq 0$

CAUTION

Be careful! In Examples 2(a) and (b),

$$59 - 14\sqrt{10} \neq 45\sqrt{10} \quad \text{and} \quad 28 + 16\sqrt{3} \neq 44\sqrt{3}.$$

Only like radicals can be combined.

Work Problem 2 at the Side.

Example 3 uses the rule for the product of the sum and difference of two terms,

$$(a + b)(a - b) = a^2 - b^2.$$

EXAMPLE 3 Using a Special Product with Radicals

Find each product.

(a) $(4 + \sqrt{3})(4 - \sqrt{3})$

Follow the pattern given above. Let $a = 4$ and $b = \sqrt{3}$.

$$(4 + \sqrt{3})(4 - \sqrt{3}) = 4^2 - (\sqrt{3})^2$$

$= 16 - 3 \qquad 4^2 = 16; (\sqrt{3})^2 = 3$

$= 13$

(b) $(\sqrt{x} - \sqrt{6})(\sqrt{x} + \sqrt{6}) = (\sqrt{x})^2 - (\sqrt{6})^2$

$= x - 6, \quad x \geq 0 \qquad (\sqrt{x})^2 = x; (\sqrt{6})^2 = 6$

Work Problem 3 at the Side.

Notice that the results in Example 3 do not contain radicals. The pairs of expressions being multiplied, $4 + \sqrt{3}$ and $4 - \sqrt{3}$, and $\sqrt{x} - \sqrt{6}$ and $\sqrt{x} + \sqrt{6}$, are called **conjugates** of each other.

OBJECTIVE 2 Use conjugates to rationalize denominators of radical expressions. Conjugates similar to those in Example 3 can be used to rationalize the denominators in more complicated quotients, such as

$$\frac{2}{4 - \sqrt{3}}.$$

By Example 3(a), if this denominator, $4 - \sqrt{3}$, is multiplied by $4 + \sqrt{3}$, then the product $(4 - \sqrt{3})(4 + \sqrt{3})$ is the rational number 13. Multiplying the numerator and denominator of the quotient by $4 + \sqrt{3}$ gives

$$\frac{2}{4 - \sqrt{3}} = \frac{2(4 + \sqrt{3})}{(4 - \sqrt{3})(4 + \sqrt{3})} = \frac{2(4 + \sqrt{3})}{13}.$$

The denominator has now been rationalized and contains no radicals.

2 Find each product. Simplify the answers.

(a) $(\sqrt{5} - 3)^2$

(b) $(4\sqrt{2} + 5)^2$

(c) $(6 + \sqrt{m})^2, \quad m \geq 0$

3 Find each product. Simplify the answers.

(a) $(3 + \sqrt{5})(3 - \sqrt{5})$

(b) $(\sqrt{3} - 2)(\sqrt{3} + 2)$

(c) $(\sqrt{5} + \sqrt{3})(\sqrt{5} - \sqrt{3})$

(d) $(\sqrt{10} - \sqrt{y})(\sqrt{10} + \sqrt{y}),$ $y \geq 0$

ANSWERS

2. **(a)** $14 - 6\sqrt{5}$ **(b)** $57 + 40\sqrt{2}$ **(c)** $36 + 12\sqrt{m} + m$

3. **(a)** 4 **(b)** -1 **(c)** 2 **(d)** $10 - y$

4 Rationalize each denominator.

(a) $\dfrac{5}{4+\sqrt{2}}$

(b) $\dfrac{\sqrt{5}+3}{2-\sqrt{5}}$

(c) $\dfrac{1}{\sqrt{6}+\sqrt{3}}$

(d) $\dfrac{7}{5-\sqrt{x}}$

Using Conjugates to Simplify Radical Expressions

To simplify a radical expression with two terms in the denominator, where at least one of those terms is a radical, multiply both the numerator and the denominator by the conjugate of the denominator.

EXAMPLE 4 Using Conjugates to Rationalize Denominators

Simplify by rationalizing each denominator.

(a) $\dfrac{5}{3+\sqrt{5}}$

We can eliminate the radical in the denominator by multiplying both the numerator and denominator by $3-\sqrt{5}$, the conjugate of the denominator.

$$\frac{5}{3+\sqrt{5}} = \frac{5(3-\sqrt{5})}{(3+\sqrt{5})(3-\sqrt{5})} \qquad \text{Multiply by } \frac{3-\sqrt{5}}{3-\sqrt{5}} = 1.$$

$$= \frac{5(3-\sqrt{5})}{3^2-(\sqrt{5})^2} \qquad (a+b)(a-b) = a^2-b^2$$

$$= \frac{5(3-\sqrt{5})}{9-5} \qquad 3^2 = 9;\ (\sqrt{5})^2 = 5$$

$$= \frac{5(3-\sqrt{5})}{4} \qquad \text{Subtract.}$$

(b) $\dfrac{6+\sqrt{2}}{\sqrt{2}-5}$

Multiply the numerator and denominator by $\sqrt{2}+5$, the conjugate of the denominator.

$$\frac{6+\sqrt{2}}{\sqrt{2}-5} = \frac{(6+\sqrt{2})(\sqrt{2}+5)}{(\sqrt{2}-5)(\sqrt{2}+5)} \qquad \text{Multiply by } \frac{\sqrt{2}+5}{\sqrt{2}+5} = 1.$$

$$= \frac{6\sqrt{2}+30+2+5\sqrt{2}}{2-25} \qquad \text{FOIL; } (a+b)(a-b) = a^2-b^2$$

$$= \frac{11\sqrt{2}+32}{-23} \qquad \text{Combine terms.}$$

$$= \frac{-11\sqrt{2}-32}{23} \qquad \frac{a}{-b} = \frac{-a}{b}$$

(c) $$\frac{4}{3-\sqrt{x}} = \frac{4(3+\sqrt{x})}{(3-\sqrt{x})(3+\sqrt{x})} \qquad \text{Multiply by } \frac{3+\sqrt{x}}{3+\sqrt{x}} = 1.$$

$$= \frac{4(3+\sqrt{x})}{9-x} \qquad 3^2 = 9;\ (\sqrt{x})^2 = x$$

(We assume here that $x > 0$ and $x \neq 9$.)

Work Problem 4 at the Side.

ANSWERS

4. (a) $\dfrac{5(4-\sqrt{2})}{14}$ (b) $-11-5\sqrt{5}$ (c) $\dfrac{\sqrt{6}-\sqrt{3}}{3}$ (d) $\dfrac{7(5+\sqrt{x})}{25-x}$

OBJECTIVE 3 Write radical expressions with quotients in lowest terms.

EXAMPLE 5 Writing a Radical Quotient in Lowest Terms

Write $\frac{3\sqrt{3} + 9}{12}$ in lowest terms.

Factor the numerator and denominator, and then use the fundamental property from **Section 7.1** to divide out common factors.

$$\frac{3\sqrt{3} + 9}{12} = \frac{3(\sqrt{3} + 3)}{3(4)} = 1 \cdot \frac{\sqrt{3} + 3}{4} = \frac{\sqrt{3} + 3}{4}$$

CAUTION

An expression like the one in Example 5 can only be simplified by factoring a common factor from the denominator and *each* term of the numerator. For example,

$$\frac{4 + 8\sqrt{5}}{4} \neq 1 + 8\sqrt{5}.$$

First factor to get $\frac{4 + 8\sqrt{5}}{4} = \frac{4(1 + 2\sqrt{5})}{4} = 1 + 2\sqrt{5}.$

Work Problem 5 at the Side.

5 Write each quotient in lowest terms.

(a) $\frac{5\sqrt{3} - 15}{10}$

(b) $\frac{12 + 8\sqrt{5}}{16}$

ANSWERS

5. (a) $\frac{\sqrt{3} - 3}{2}$ (b) $\frac{3 + 2\sqrt{5}}{4}$

Focus on

Real-Data Applications

Spaceship Earth—A Geodesic Sphere

Geodesic domes became a popular design base for houses in the 1960s. The original patent was awarded to R. Buckminster Fuller in 1951. For the same square footage of interior space, a geodesic dome has less surface area and, thus, both reduces energy costs and is less prone to storm damage. One of the most famous geodesic domes is the full geodesic sphere Spaceship Earth, which is a major attraction at Epcot Center at Walt Disney World, Florida.

The sphere is made up of pyramids, shaped from three equilateral **facets,** or faces. A set of four pyramids forms a **panel** that is also an equilateral triangle. There are 954 panels, each supporting 12 facets. Some of the facets are removed for support beams, so there are actually only 11,324 facets on the sphere. The structure of the panels and the facets are easily seen in a close-up view of Spaceship Earth.

The outside **surface area** of Spaceship Earth is approximately 150,000 ft^2. To envision the idea of surface area, think about the task of painting the geodesic sphere. The painted surface is the surface area.

Each of the facets is an equilateral triangle. The area A of an equilateral triangle that has a side of length s is given by the formula

$$A = \frac{\sqrt{3}}{4}s^2.$$

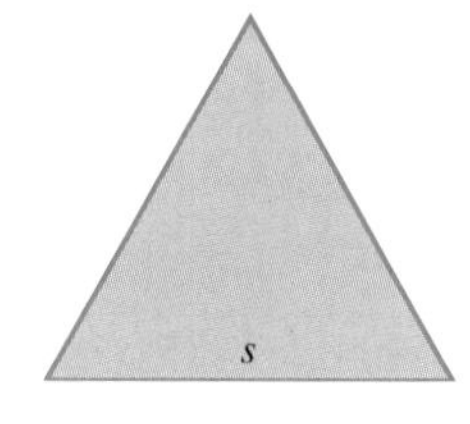

For Group Discussion

1. Use the given data for the outside surface area and the number of facets to find the surface area of one of Spaceship Earth's triangular facets. Round your answer to the nearest hundredth of a square foot. (Record the actual result for use in calculations for subsequent problems.)

2. Use the formula for the area of an equilateral triangle to find the length of the side of one of the triangular facets, to the nearest tenth of a foot.

3. How long is the side of one of the equilateral triangular panels?

4. Derive the formula for the area of an equilateral triangle. (*Hint:* Envision the right triangle that makes up half of the equilateral triangle, and use the Pythagorean formula to write an equation that relates the height h and the side s. Then use the formula for the area of a triangle.)

8.5 Exercises

FOR EXTRA HELP

Addison-Wesley Math Tutor Center

MathXL

Digital Video Tutor CD 9
Videotape 8

Student's Solutions Manual

MyMathLab

Interactmath.com

In Exercises 1–4, perform the operations mentally, and write the answers without doing intermediate steps.

1. $\sqrt{49} + \sqrt{36}$

2. $\sqrt{100} - \sqrt{81}$

3. $\sqrt{2} \cdot \sqrt{8}$

4. $\sqrt{8} \cdot \sqrt{8}$

Simplify each expression. Use the five guidelines given in this section. Assume that all variables represent nonnegative real numbers. See Examples 1–3.

5. $\sqrt{5}\left(\sqrt{3} - \sqrt{7}\right)$

6. $\sqrt{7}\left(\sqrt{10} + \sqrt{3}\right)$

7. $2\sqrt{5}\left(\sqrt{2} + 3\sqrt{5}\right)$

8. $3\sqrt{7}\left(2\sqrt{7} + 4\sqrt{5}\right)$

9. $3\sqrt{14} \cdot \sqrt{2} - \sqrt{28}$

10. $7\sqrt{6} \cdot \sqrt{3} - 2\sqrt{18}$

11. $\left(2\sqrt{6} + 3\right)\left(3\sqrt{6} + 7\right)$

12. $\left(4\sqrt{5} - 2\right)\left(2\sqrt{5} - 4\right)$

13. $\left(5\sqrt{7} - 2\sqrt{3}\right)\left(3\sqrt{7} + 4\sqrt{3}\right)$

14. $\left(2\sqrt{10} + 5\sqrt{2}\right)\left(3\sqrt{10} - 3\sqrt{2}\right)$

15. $\left(8 - \sqrt{7}\right)^2$

16. $\left(6 - \sqrt{11}\right)^2$

17. $\left(2\sqrt{7} + 3\right)^2$

18. $\left(4\sqrt{5} + 5\right)^2$

19. $\left(\sqrt{a} + 1\right)^2$

20. $\left(\sqrt{y} + 4\right)^2$

21. $\left(5 - \sqrt{2}\right)\left(5 + \sqrt{2}\right)$

22. $\left(3 - \sqrt{5}\right)\left(3 + \sqrt{5}\right)$

23. $(\sqrt{8} - \sqrt{7})(\sqrt{8} + \sqrt{7})$

24. $(\sqrt{12} - \sqrt{11})(\sqrt{12} + \sqrt{11})$

25. $(\sqrt{y} - \sqrt{10})(\sqrt{y} + \sqrt{10})$

26. $(\sqrt{t} - \sqrt{13})(\sqrt{t} + \sqrt{13})$

27. $(\sqrt{2} + \sqrt{3})(\sqrt{6} - \sqrt{2})$

28. $(\sqrt{3} + \sqrt{5})(\sqrt{15} - \sqrt{5})$

29. $(\sqrt{10} - \sqrt{5})(\sqrt{5} + \sqrt{20})$

30. $(\sqrt{6} - \sqrt{3})(\sqrt{3} + \sqrt{18})$

31. $(\sqrt{5} + \sqrt{30})(\sqrt{6} + \sqrt{3})$

32. $(\sqrt{10} - \sqrt{20})(\sqrt{2} - \sqrt{5})$

33. $(\sqrt{5} - \sqrt{10})(\sqrt{x} - \sqrt{2})$

34. $(\sqrt{x} + \sqrt{6})(\sqrt{10} + \sqrt{3})$

35. In Example 1(b), the original expression simplifies to $-37 - 2\sqrt{15}$. Students often try to simplify such expressions by combining -37 and -2 to get $-39\sqrt{15}$, which is incorrect. Explain why.

36. If you try to rationalize the denominator of $\frac{2}{4 + \sqrt{3}}$ by multiplying the numerator and denominator by $4 + \sqrt{3}$, what problem arises? What should you multiply by?

Rationalize each denominator. Write quotients in lowest terms. Assume that all variables represent nonnegative real numbers. See Examples 4 and 5.

37. $\frac{1}{3 + \sqrt{2}}$

38. $\frac{1}{4 - \sqrt{3}}$

39. $\frac{14}{2 - \sqrt{11}}$

40. $\frac{19}{5 - \sqrt{6}}$

41. $\frac{\sqrt{2}}{2-\sqrt{2}}$

42. $\frac{\sqrt{7}}{7-\sqrt{7}}$

43. $\frac{\sqrt{5}}{\sqrt{2}+\sqrt{3}}$

44. $\frac{\sqrt{3}}{\sqrt{2}+\sqrt{3}}$

45. $\frac{\sqrt{5}+2}{2-\sqrt{3}}$

46. $\frac{\sqrt{7}+3}{4-\sqrt{5}}$

47. $\frac{12}{\sqrt{x}+1}$

48. $\frac{10}{\sqrt{x}-4}$

49. $\frac{3}{7-\sqrt{x}}$

50. $\frac{1}{6+\sqrt{z}}$

Write each quotient in lowest terms. See Example 5.

51. $\frac{6\sqrt{11}-12}{6}$

52. $\frac{12\sqrt{5}-24}{12}$

53. $\frac{2\sqrt{3}+10}{16}$

54. $\frac{4\sqrt{6}+24}{20}$

55. $\frac{12-\sqrt{40}}{4}$

56. $\frac{9-\sqrt{72}}{12}$

RELATING CONCEPTS (EXERCISES 57–62) For Individual or Group Work

Work Exercises 57–62 in order, *to see why a common student error is indeed an error.*

57. Use the distributive property to write $6(5 + 3x)$ as a sum.

58. Your answer in Exercise 57 should be $30 + 18x$. Why can't we combine these two terms to get $48x$?

59. Repeat Exercise 14 from earlier in this exercise set.

60. Your answer in Exercise 59 should be $30 + 18\sqrt{5}$. Many students will, in error, try to combine these terms to get $48\sqrt{5}$. Why is this wrong?

61. Write the expression similar to $30 + 18x$ that simplifies to $48x$. Then write the expression similar to $30 + 18\sqrt{5}$ that simplifies to $48\sqrt{5}$.

62. Write a short paragraph explaining the similarities between combining like terms and combining like radicals.

Solve each problem.

63. The radius of the circular top or bottom of a tin can with a surface area S and a height h is given by

$$r = \frac{-h + \sqrt{h^2 + .64S}}{2}.$$

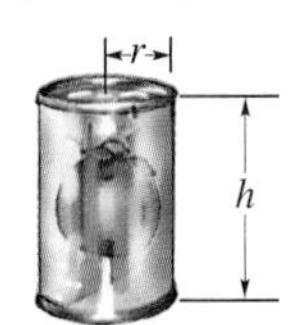

What radius should be used to make a can with a height of 12 in. and a surface area of 400 in.^{2}?

64. If an investment of P dollars grows to A dollars in 2 yr, the annual rate of return on the investment is given by

$$r = \frac{\sqrt{A} - \sqrt{P}}{\sqrt{P}}.$$

Rationalize the denominator. Then find the annual rate of return r (as a percent) if \$50,000 increases to \$58,320.

Summary Exercises on Operations with Radicals

Perform all indicated operations and express each answer in simplest form. Assume that all variables represent positive real numbers.

1. $5\sqrt{10} - 8\sqrt{10}$

2. $\sqrt{5}(\sqrt{5} - \sqrt{3})$

3. $(1 + \sqrt{3})(2 - \sqrt{6})$

4. $\sqrt{98} - \sqrt{72} + \sqrt{50}$

5. $(3\sqrt{5} - 2\sqrt{7})^2$

6. $\dfrac{3}{\sqrt{6}}$

7. $\sqrt[3]{16t^2} - \sqrt[3]{54t^2} + \sqrt[3]{128t^2}$

8. $\dfrac{8}{\sqrt{7} - \sqrt{5}}$

9. $\dfrac{1 + \sqrt{2}}{1 - \sqrt{2}}$

10. $(1 + \sqrt[3]{3})(1 - \sqrt[3]{3} + \sqrt[3]{9})$

11. $(\sqrt{3} + 6)(\sqrt{3} - 6)$

12. $\dfrac{1}{\sqrt{t} + \sqrt{3}}$

13. $\sqrt[3]{8x^3y^5z^6}$

14. $\dfrac{12}{\sqrt[3]{9}}$

15. $\dfrac{5}{\sqrt{6} - 1}$

16. $\sqrt{\dfrac{2}{3x}}$

17. $\dfrac{6\sqrt{3}}{5\sqrt{12}}$

18. $\dfrac{8\sqrt{50}}{2\sqrt{25}}$

19. $\dfrac{-4}{\sqrt[3]{4}}$

20. $\dfrac{\sqrt{6} - \sqrt{5}}{\sqrt{6} + \sqrt{5}}$

21. $\sqrt{75x} - \sqrt{12x}$

22. $(5 + 3\sqrt{3})^2$

23. $(\sqrt{7} - \sqrt{6})(\sqrt{7} + \sqrt{6})$

24. $\sqrt[3]{\dfrac{16}{81}}$

25. $x\sqrt[4]{x^5} - 3\sqrt[4]{x^9} + x^2\sqrt[4]{x}$

26. $\sqrt{6} + \sqrt{6}$

27. $\sqrt{14} + \sqrt{17}$

28. $9\sqrt{24} - 2\sqrt{54} + 3\sqrt{20}$

29. $\sqrt{\dfrac{3}{4}} \cdot \sqrt{\dfrac{1}{5}}$

30. $\dfrac{5}{\sqrt{5}}$

31. $\sqrt[3]{24} + 6\sqrt[3]{81}$

32. $\dfrac{8}{4 - \sqrt{x}}$

33. $\sqrt[3]{4}\,(\sqrt[3]{2} - 3)$

34. $\sqrt{32x} - \sqrt{18x}$

35. $\sqrt{\dfrac{5}{8}}$

36. $(7 + \sqrt{x})^2$

37. A biologist has shown that the number of different plant species S on a Galápagos Island is related to the area of the island, A (in square miles), by

$$S = 28.6\sqrt[3]{A}.$$

How many plant species (to the nearest whole number) would exist on such an island with the following areas?

(a) 8 mi^2 **(b)** 27,000 mi^2

8.6 Solving Equations with Radicals

OBJECTIVES

1. Solve radical equations.
2. Identify equations with no solutions.
3. Solve equations by squaring a binomial.

A **radical equation** is an equation with a variable in the radicand, such as

$$\sqrt{x+1} = 3 \quad \text{or} \quad 3\sqrt{x} = \sqrt{8x+9}. \qquad \text{Radical equations}$$

OBJECTIVE 1 Solve radical equations. The addition and multiplication properties of equality are not enough to solve radical equations. We need a new property, called the *squaring property*.

Squaring Property of Equality

If each side of a given equation is squared, all solutions of the original equation are *among* the solutions of the squared equation.

CAUTION

Be very careful with the squaring property: Using this property can give a new equation with *more* solutions than the original equation. For example, starting with the equation $x = 4$ and squaring each side gives

$$x^2 = 4^2 \quad \text{or} \quad x^2 = 16.$$

This last equation, $x^2 = 16$, has *two* solutions, 4 or -4, while the original equation, $x = 4$, has only *one* solution, 4. Because of this possibility, checking is more than just a guard against algebraic errors when solving an equation with radicals. It is an essential part of the solution process. ***All potential solutions from the squared equation must be checked in the original equation.***

EXAMPLE 1 Using the Squaring Property of Equality

Solve $\sqrt{p+1} = 3$.

Use the squaring property of equality to square each side of the equation.

$$(\sqrt{p+1})^2 = 3^2$$

$$p + 1 = 9 \qquad (\sqrt{p+1})^2 = p + 1$$

$$p = 8 \qquad \text{Subtract 1.}$$

Now check this potential solution in the original equation.

Check:

$$\sqrt{p+1} = 3$$

$$\sqrt{8+1} = 3 \quad ? \quad \text{Let } p = 8.$$

$$\sqrt{9} = 3 \quad ?$$

$$3 = 3 \qquad \text{True}$$

Because this statement is true, 8 is the solution of $\sqrt{p+1} = 3$. In this case the equation obtained by squaring had just one solution, which also satisfied the original equation.

Work Problem 1 at the Side.

1 Solve each equation. Be sure to check your solutions.

(a) $\sqrt{k} = 3$

(b) $\sqrt{x-2} = 4$

(c) $\sqrt{9-t} = 4$

ANSWERS

1. (a) 9 (b) 18 (c) -7

2 Solve each equation.

(a) $\sqrt{3x+9} = 2\sqrt{x}$

(b) $5\sqrt{x} = \sqrt{20x+5}$

EXAMPLE 2 Using the Squaring Property with a Radical on Each Side

Solve $3\sqrt{x} = \sqrt{x+8}$.

Squaring each side gives

$$(3\sqrt{x})^2 = (\sqrt{x+8})^2$$
$$3^2(\sqrt{x})^2 = (\sqrt{x+8})^2 \qquad (ab)^2 = a^2b^2$$
$$9x = x + 8 \qquad (\sqrt{x})^2 = x;\ (\sqrt{x+8})^2 = x+8$$
$$8x = 8 \qquad \text{Subtract } x.$$
$$x = 1. \qquad \text{Divide by 8.}$$

Check:

$$3\sqrt{x} = \sqrt{x+8} \qquad \text{Original equation}$$
$$3\sqrt{1} \stackrel{?}{=} \sqrt{1+8} \qquad \text{Let } x = 1.$$
$$3(1) \stackrel{?}{=} \sqrt{9}$$
$$3 = 3 \qquad \text{True}$$

The solution of $3\sqrt{x} = \sqrt{x+8}$ is 1.

CAUTION
Do not write the final result obtained in the check as the solution. In Example 2, the solution is 1, *not* 3.

Work Problem 2 at the Side.

OBJECTIVE 2 Identify equations with no solutions. Not all radical equations have solutions, as shown in Examples 3 and 4.

EXAMPLE 3 Using the Squaring Property When One Side Is Negative

Solve $\sqrt{x} = -3$.

Square each side of the equation.

$$(\sqrt{x})^2 = (-3)^2$$
$$x = 9 \longleftarrow \text{Potential solution}$$

Check:

$$\sqrt{x} = -3$$
$$\sqrt{9} \stackrel{?}{=} -3 \qquad \text{Let } x = 9.$$
$$3 = -3 \qquad \text{False}$$

Because the statement $3 = -3$ is false, the number 9 is *not* a solution of the given equation and is said to be an **extraneous solution;** it must be discarded. In fact, $\sqrt{x} = -3$ has no solution.

NOTE
Because $\sqrt{x}$ represents the *principal* or *nonnegative* square root of x in Example 3, we might have seen immediately that there is no solution.

ANSWERS
2. (a) 9 (b) 1

EXAMPLE 4 **Using the Squaring Property with a Quadratic Equation**

Solve $p = \sqrt{p^2 + 5p + 10}$.

Square each side.

$$p^2 = (\sqrt{p^2 + 5p + 10})^2$$

$p^2 = p^2 + 5p + 10$	$(\sqrt{p^2 + 5p + 10})^2 = p^2 + 5p + 10$
$0 = 5p + 10$	Subtract p^2.
$-10 = 5p$	Subtract 10.
$p = -2$	Divide by 5.

Check this potential solution in the original equation.

Check:

$p = \sqrt{p^2 + 5p + 10}$		
$-2 = \sqrt{(-2)^2 + 5(-2) + 10}$	?	Let $p = -2$.
$-2 = \sqrt{4 - 10 + 10}$	?	
$-2 = 2$		False

Because $p = -2$ leads to a false result, the equation has no solution.

Work Problem 3 at the Side.

3 Solve each equation. (*Hint:* In part (a), subtract 4 from each side.)

(a) $\sqrt{x} + 4 = 0$

(b) $x = \sqrt{x^2 - 4x - 16}$

OBJECTIVE 3 **Solve equations by squaring a binomial.** The next examples use the following rules from **Section 5.4.**

$$(a + b)^2 = a^2 + 2ab + b^2$$

and

$$(a - b)^2 = a^2 - 2ab + b^2.$$

By these patterns, for example,

$$(x - 3)^2 = x^2 - 2x(3) + 3^2$$
$$= x^2 - 6x + 9.$$

Work Problem 4 at the Side.

4 Square each expression.

(a) $w - 5$

(b) $2k - 5$

(c) $3m - 2p$

EXAMPLE 5 **Using the Squaring Property When One Side Has Two Terms**

Solve $\sqrt{2x - 3} = x - 3$.

Square each side, using the preceding result to square the binomial on the right side of the equation.

$$(\sqrt{2x - 3})^2 = (x - 3)^2$$
$$2x - 3 = x^2 - 6x + 9$$

This equation is quadratic because of the x^2-term. As shown in **Section 6.6,** to solve this equation requires that one side be equal to 0. Subtract $2x$ and add 3 on each side, giving

$0 = x^2 - 8x + 12.$	Standard form
$0 = (x - 6)(x - 2)$	Factor.
$x - 6 = 0$ or $x - 2 = 0$	Zero-factor property
$x = 6$ or $x = 2$	Solve.

Continued on Next Page

ANSWERS

3. **(a)** no solution **(b)** no solution

4. **(a)** $w^2 - 10w + 25$ **(b)** $4k^2 - 20k + 25$ **(c)** $9m^2 - 12mp + 4p^2$

5 Solve each equation.

(a) $\sqrt{6w+6} = w + 1$

(b) $2u - 1 = \sqrt{10u + 9}$

Check *both* of these potential solutions in the original equation.

Check:

If $x = 6$, then

$$\sqrt{2x-3} = x - 3$$
$$\sqrt{2(6)-3} = 6 - 3 \quad ?$$
$$\sqrt{12-3} = 3 \quad ?$$
$$\sqrt{9} = 3 \quad ?$$
$$3 = 3. \quad \text{True}$$

If $x = 2$, then

$$\sqrt{2x-3} = x - 3$$
$$\sqrt{2(2)-3} = 2 - 3 \quad ?$$
$$\sqrt{4-3} = -1 \quad ?$$
$$\sqrt{1} = -1 \quad ?$$
$$1 = -1. \quad \text{False}$$

Only 6 is a valid solution of the equation; 2 is extraneous.

Work Problem 5 at the Side.

Sometimes we must write an equation in a different form before squaring each side. For example, suppose we want to solve $3\sqrt{x} - 1 = 2x$. Squaring each side gives

$$\left(3\sqrt{x} - 1\right)^2 = (2x)^2$$
$$9x - 6\sqrt{x} + 1 = 4x^2,$$

a more complicated equation that still contains a radical. In a case like this it would be better to rewrite the original equation so that the radical is alone on one side of the equals sign, as shown in Example 6.

EXAMPLE 6 Rewriting an Equation before Using the Squaring Property

Solve $3\sqrt{x} - 1 = 2x$.

Isolate the radical by adding 1 to each side.

$$3\sqrt{x} = 2x + 1$$
$$\left(3\sqrt{x}\right)^2 = (2x+1)^2 \qquad \text{Square each side.}$$
$$9x = 4x^2 + 4x + 1$$
$$0 = 4x^2 - 5x + 1 \qquad \text{Subtract } 9x.$$
$$0 = (4x-1)(x-1) \qquad \text{Factor.}$$
$$4x - 1 = 0 \quad \text{or} \quad x - 1 = 0 \qquad \text{Zero-factor property}$$
$$x = \frac{1}{4} \quad \text{or} \quad x = 1 \qquad \text{Solve.}$$

Check:

If $x = \frac{1}{4}$, then

$$3\sqrt{x} - 1 = 2x$$
$$3\sqrt{\frac{1}{4}} - 1 = 2\left(\frac{1}{4}\right) \quad ?$$
$$\frac{1}{2} = \frac{1}{2}. \quad \text{True}$$

If $x = 1$, then

$$3\sqrt{x} - 1 = 2x$$
$$3\sqrt{1} - 1 = 2(1) \quad ?$$
$$2 = 2. \quad \text{True}$$

Both solutions check, so the solutions to the original equation are $\frac{1}{4}$ and 1.

ANSWERS
5. (a) 5, −1 (b) 4

CAUTION
Errors often occur when each side of an equation is squared. For instance, in Example 6 when each side of

$$3\sqrt{x} = 2x + 1$$

is squared, ***the entire binomial on the right must be squared.*** Here, $(2x + 1)^2 = 4x^2 + 4x + 1$. It would be incorrect to square the $2x$ and the 1 separately to get $4x^2 + 1$.

Work Problem 6 at the Side.

Some radical equations require squaring twice, as in the next example.

EXAMPLE 7 Using the Squaring Property Twice

Solve $\sqrt{21 + x} = 3 + \sqrt{x}$.

$$(\sqrt{21 + x})^2 = (3 + \sqrt{x})^2 \quad \text{Square each side.}$$
$$21 + x = 9 + 6\sqrt{x} + x$$
$$12 = 6\sqrt{x} \quad \text{Subtract 9; subtract } x.$$
$$2 = \sqrt{x} \quad \text{Divide by 6.}$$
$$2^2 = (\sqrt{x})^2 \quad \text{Square each side again.}$$
$$4 = x$$

Check: If $x = 4$, then

$$\sqrt{21 + x} = 3 + \sqrt{x} \quad \text{Original equation}$$
$$\sqrt{21 + 4} = 3 + \sqrt{4} \quad ?$$
$$5 = 5. \quad \text{True}$$

The solution is 4.

Work Problem 7 at the Side.

In summary, use the following steps to solve a radical equation.

Solving a Radical Equation

Step 1 **Isolate a radical.** Arrange the terms so that a radical is alone on one side of the equation.

Step 2 **Square each side.**

Step 3 **Combine like terms.**

Step 4 **Repeat Steps 1–3, if necessary.** If there is still a term with a radical, repeat Steps 1–3.

Step 5 **Solve the equation.** Find all potential solutions.

Step 6 **Check.** All potential solutions *must* be checked in the original equation.

6 Solve each equation.

(a) $\sqrt{x} - 3 = x - 15$

(b) $\sqrt{z + 5} + 2 = z + 5$

7 Solve each equation.

(a) $\sqrt{p + 1} - \sqrt{p - 4} = 1$

(b) $\sqrt{2x + 1} + \sqrt{x + 4} = 3$

ANSWERS
6. (a) 16 **(b)** -1
7. (a) 8 **(b)** 0

Focus on

Real-Data Applications

On a Clear Day

The Empire State Building, the Eiffel Tower, and the world's tallest buildings evoke images of sitting on top of the world. The prospect of viewing sights from such lofty heights stirs our imaginations. But how far can we really see? On a clear day, the maximum distance in kilometers that you can see from a tall building is given by the formula

$$\textbf{sight distance} = \textbf{111.7}\sqrt{\textbf{height of building in kilometers}}.$$

(*Source: A Sourcebook of Applications of School Mathematics*, NCTM, 1980.)

For Group Discussion

On a clear day, how far could you see from the top of each of these famous buildings and structures? Round answers to the nearest mile. Recall that **1 ft ≈ .3048 m** and **1 km ≈ .621371 mi.**

1. The London Eye, which opened on New Year's Eve 1999, is a unique form of a Ferris wheel that features 32 observation capsules. It is located on the bank of the Thames River facing the Houses of Parliament and is the sixth tallest structure in London, with a diameter of 135 m. (*Source:* www.londoneye.com) Does the formula justify the claim that on a clear day, passengers on the London Eye can see Windsor Castle, 25 mi away?

2. The Empire State Building opened in 1931 on 5th Avenue in New York City. The building is 1250 ft high. (The antenna reaches to 1454 ft.) The observation deck, located on the 102nd floor, is at a height of 1050 ft. (*Source:* www.esbnyc.com) How far could you see on a clear day from the observation deck?

3. The twin Petronas Towers in Kuala Lumpur, Malaysia, are 1483 ft high (including the spires). (*Source: World Almanac and Book of Facts*, 2004.) How far would one of the builders have been able to see on a clear day from the top of a spire?

4. The Khufu Pyramid in Giza (also known as Cheops Pyramid) was built in about 2566 B.C. to a height, at that time, of 482 ft. It is now only about 450 ft high. (*Source:* www.touregypt.net/cheop.htm) How far would one of the original builders of the pyramid have been able to see from the top of the pyramid?

8.6 Exercises

FOR EXTRA HELP

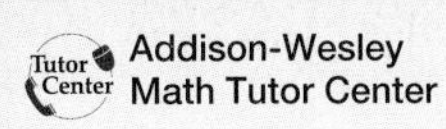

MathXL

MyMathLab

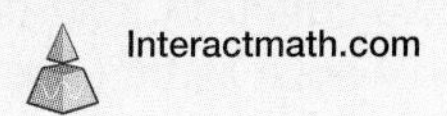

Solve each equation. See Examples 1–4.

1. $\sqrt{x} = 7$

2. $\sqrt{k} = 10$

3. $\sqrt{t + 2} = 3$

4. $\sqrt{x + 7} = 5$

5. $\sqrt{r - 4} = 9$

6. $\sqrt{k - 12} = 3$

7. $\sqrt{4 - t} = 7$

8. $\sqrt{9 - s} = 5$

9. $\sqrt{2t + 3} = 0$

10. $\sqrt{5x - 4} = 0$

11. $\sqrt{3x - 8} = -2$

12. $\sqrt{6x + 4} = -3$

13. $\sqrt{w} - 4 = 7$

14. $\sqrt{t} + 3 = 10$

15. $\sqrt{10x - 8} = 3\sqrt{x}$

16. $\sqrt{17t - 4} = 4\sqrt{t}$

17. $5\sqrt{x} = \sqrt{10x + 15}$

18. $4\sqrt{z} = \sqrt{20z - 16}$

19. $\sqrt{3x - 5} = \sqrt{2x + 1}$

20. $\sqrt{5x + 2} = \sqrt{3x + 8}$

21. $k = \sqrt{k^2 - 5k - 15}$

22. $s = \sqrt{s^2 - 2s - 6}$

23. $7x = \sqrt{49x^2 + 2x - 10}$

24. $6x = \sqrt{36x^2 + 5x - 5}$

25. The first step in solving the equation $\sqrt{2x + 1} = x - 7$ is to square each side of the equation. Errors often occur in solving equations such as this one when the right side of the equation is squared incorrectly. What is the square of the right side?

26. What is wrong with the following "solution"?

$$\begin{aligned} -\sqrt{x - 1} &= -4 & \\ -(x - 1) &= 16 & \text{Square each side.} \\ -x + 1 &= 16 & \text{Distributive property} \\ -x &= 15 & \text{Subtract 1.} \\ x &= -15 & \text{Multiply by } -1. \end{aligned}$$

Solve each equation. See Examples 5 and 6.

27. $\sqrt{2x + 1} = x - 7$

28. $\sqrt{3x + 3} = x - 5$

29. $\sqrt{3k + 10} + 5 = 2k$

30. $\sqrt{4t + 13} + 1 = 2t$

31. $\sqrt{5x + 1} - 1 = x$

32. $\sqrt{x + 1} - x = 1$

33. $\sqrt{6t + 7} + 3 = t + 5$

34. $\sqrt{10x + 24} = x + 4$

35. $x - 4 - \sqrt{2x} = 0$

36. $x - 3 - \sqrt{4x} = 0$

37. $\sqrt{x} + 6 = 2x$

38. $\sqrt{k} + 12 = k$

Solve each equation. See Example 7.

39. $\sqrt{x + 1} - \sqrt{x - 4} = 1$

40. $\sqrt{2x + 3} + \sqrt{x + 1} = 1$

41. $\sqrt{x} = \sqrt{x - 5} + 1$

42. $\sqrt{2x} = \sqrt{x + 7} - 1$

43. $\sqrt{3x + 4} - \sqrt{2x - 4} = 2$

44. $\sqrt{1 - x} + \sqrt{x + 9} = 4$

45. $\sqrt{2x + 11} + \sqrt{x + 6} = 2$

46. $\sqrt{x + 9} + \sqrt{x + 16} = 7$

Solve each problem.

47. The square root of the sum of a number and 4 is 5. Find the number.

48. A certain number is the same as the square root of the product of 8 and the number. Find the number.

49. Three times the square root of 2 equals the square root of the sum of some number and 10. Find the number.

50. The negative square root of a number equals that number decreased by 2. Find the number.

Solve each problem. Give answers to the nearest tenth.

51. To estimate the speed at which a car was traveling at the time of an accident, a police officer drives the car involved in the accident under conditions similar to those during which the accident took place and then skids to a stop. If the car is driven at 30 mph, then the speed at the time of the accident is given by

$$s = 30\sqrt{\frac{a}{p}},$$

where a is the length of the skid marks left at the time of the accident and p is the length of the skid marks in the police test. Find s for the following values of a and p.

(a) $a = 862$ ft; $p = 156$ ft

(b) $a = 382$ ft; $p = 96$ ft

(c) $a = 84$ ft; $p = 26$ ft

52. A formula for calculating the distance, d, one can see from an airplane to the horizon on a clear day is

$$d = 1.22\sqrt{x},$$

where x is the altitude of the plane in feet and d is given in miles.

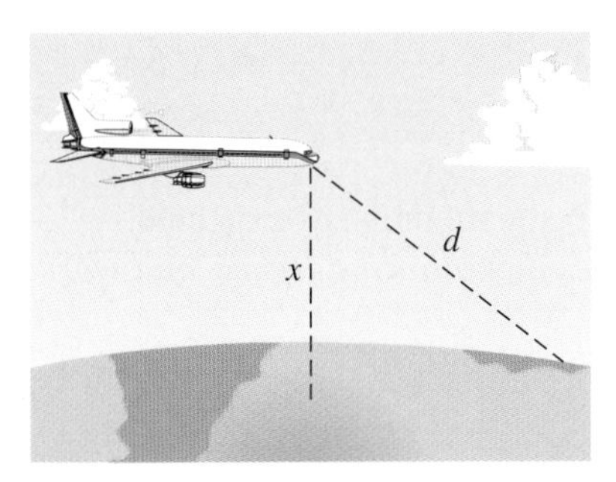

How far can one see to the horizon in a plane flying at the following altitudes?

(a) 15,000 ft

(b) 18,000 ft

(c) 24,000 ft

53. A surveyor wants to find the height of a building. At a point 110.0 ft from the base of the building he sights to the top of the building and finds the distance to be 193.0 ft. How high is the building?

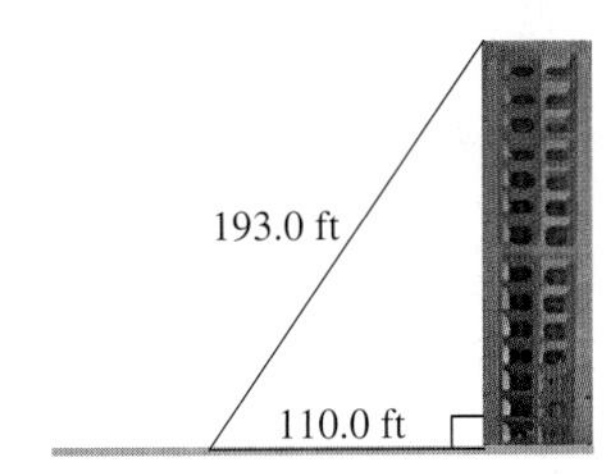

54. Two towns are separated by dense woods. To go from Town B to Town A, it is necessary to travel due west for 19.0 mi, then turn due north and travel for 14.0 mi. How far apart are the towns?

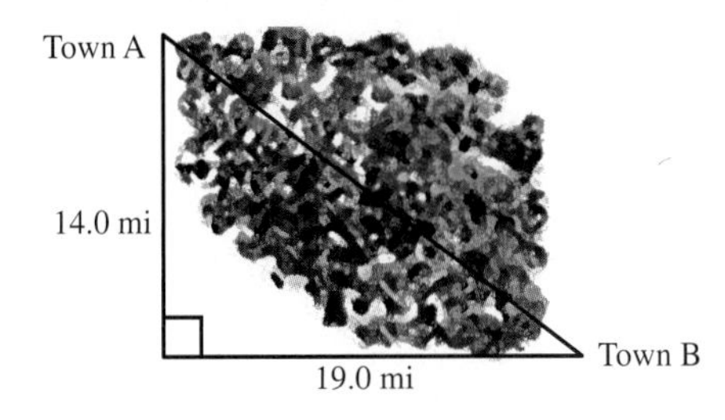

RELATING CONCEPTS (EXERCISES 55–60) For Individual or Group Work

The most common formula for the area of a triangle is $A = \frac{1}{2}bh$, where b is the length of the base and h is the height. What if the height is not known? What if we know only the lengths of the sides? Another formula, known as **Heron's formula,** allows us to calculate the area of a triangle if we know the lengths of the sides a, b, and c. First let s equal the **semiperimeter,** which is one-half the perimeter.

$$s = \frac{1}{2}(a + b + c)$$

The area A is given by the formula

$$A = \sqrt{s(s - a)(s - b)(s - c)}.$$

For example, the familiar 3–4–5 right triangle has area

$$A = \frac{1}{2}(3)(4) = 6 \text{ square units,}$$

using the familiar formula. Using Heron's formula, $s = \frac{1}{2}(3 + 4 + 5) = 6$, and

$$\begin{aligned} A &= \sqrt{6(6 - 3)(6 - 4)(6 - 5)} \\ &= \sqrt{6 \cdot 3 \cdot 2 \cdot 1} \\ &= \sqrt{36} = 6 \end{aligned}$$

The area is 6 square units, as expected.

Consider the following figure, and ***work Exercises 55–60 in order****.*

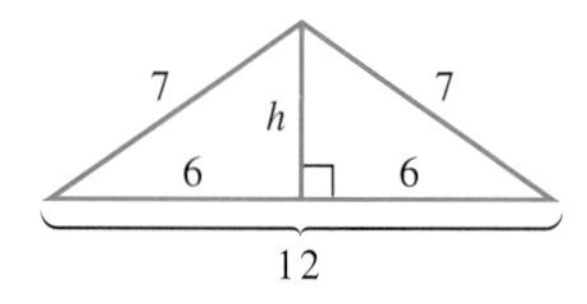

55. The lengths of the sides of the entire triangle are 7, 7, and 12. Find the semiperimeter s.

56. Now use Heron's formula to find the area of the entire triangle. Write it as a simplified radical.

57. Find the value of h by using the Pythagorean formula.

58. Find the area of each of the congruent right triangles forming the entire triangle by using the formula $A = \frac{1}{2}bh$.

59. Double your result from Exercise 58 to determine the area of the entire triangle.

60. How do your answers in Exercises 56 and 59 compare? (*Note:* They should be equal, since the area of the entire triangle is unique.)

Chapter 8

SUMMARY

KEY TERMS

8.1	**square root**	The number b is a square root of a if $b^2 = a$.
	principal square root	The positive square root of a number is its principal square root.
	radicand	The number or expression inside a radical sign is called the radicand.
	radical	A radical sign with a radicand is called a radical.
	radical expression	An algebraic expression containing a radical is called a radical expression.
	perfect square	A number with a rational square root is called a perfect square.
	irrational number	A real number that is not rational is called an irrational number.
	cube root	The number b is a cube root of a if $b^3 = a$.
	index (order)	In a radical of the form $\sqrt[n]{a}$, the number n is the index or order.
8.2	**perfect cube**	A number with a rational cube root is called a perfect cube.
8.3	**like radicals**	Like radicals are multiples of the same root of the same number.
8.4	**rationalizing the denominator**	The process of changing the denominator of a fraction from a radical (irrational number) to an expression not involving a radical is called rationalizing the denominator.
8.5	**conjugate**	The conjugate of $a + b$ is $a - b$.
8.6	**radical equation**	An equation with a variable in the radicand is a radical equation.
	extraneous solution	A potential solution that does not satisfy the given equation is an extraneous solution.

Radical sign → $\sqrt[n]{a}$ ← Radicand; Index points to n; Radical is the whole $\sqrt[n]{a}$.

NEW SYMBOLS

$\sqrt{\ }$ radical sign $\approx$ is approximately equal to $\sqrt[3]{a}$ cube root of a $\sqrt[n]{a}$ nth root of a

TEST YOUR WORD POWER

See how well you have learned the vocabulary in this chapter. Answers, with examples, follow the Quick Review.

1. A **square root** of a number is
   - **A.** the number raised to the second power
   - **B.** the number under a radical sign
   - **C.** a number that when multiplied by itself gives the original number
   - **D.** the inverse of the number.

2. A **radicand** is
   - **A.** the index of a radical
   - **B.** the number or expression inside the radical sign
   - **C.** the positive root of a number
   - **D.** the radical sign.

3. A **radical** is
   - **A.** a symbol that indicates the nth root
   - **B.** an algebraic expression containing a square root
   - **C.** the positive nth root of a number
   - **D.** a radical sign and the number or expression inside it.

4. The **principal root** of a positive number with even index n is
   - **A.** the positive nth root of the number
   - **B.** the negative nth root of the number
   - **C.** the square root of the number
   - **D.** the cube root of the number.

5. An **irrational number** is
   - **A.** the quotient of two integers, with denominator not 0
   - **B.** a decimal number that neither terminates nor repeats
   - **C.** the principal square root of a number
   - **D.** a nonreal number.

(continued)

TEST YOUR WORD POWER (CONTINUED)

6. The **Pythagorean formula** states that, in a right triangle,
   A. the sum of the measures of the angles is 180°
   B. the sum of the lengths of the two shorter sides equals the length of the longest side
   C. the longest side is opposite the right angle
   D. the square of the length of the longest side equals the sum of the squares of the lengths of the two shorter sides.

7. **Like radicals** are
   A. radicals in simplest form
   B. algebraic expressions containing radicals
   C. multiples of the same root of the same number
   D. radicals with the same index.

8. **Rationalizing the denominator** is the process of
   A. eliminating fractions from a radical expression
   B. changing the denominator of a fraction from a radical to an expression not involving a radical.
   C. clearing a radical expression of radicals
   D. multiplying radical expressions.

9. The **conjugate** of $a + b$ is
   A. $a - b$
   B. $a \cdot b$
   C. $a \div b$
   D. $(a + b)^2$.

10. An **extraneous solution** is a value
    A. that makes an equation false and must be discarded
    B. that makes an equation true
    C. that makes an expression equal 0
    D. that checks in the original equation.

QUICK REVIEW

Concepts	*Examples*
8.1 *Evaluating Roots* If a is a positive real number, then $\sqrt{a}$ is the positive or principal square root of a; $-\sqrt{a}$ is the negative square root of a; $\sqrt{0} = 0$. If a is a negative real number, then $\sqrt{a}$ is not a real number.	$\sqrt{49} = 7$ $-\sqrt{81} = -9$ $\sqrt{-25}$ is not a real number.
If a is a positive rational number, then $\sqrt{a}$ is rational if a is a perfect square. $\sqrt{a}$ is irrational if a is not a perfect square.	$\sqrt{\frac{4}{9}}$ and $\sqrt{16}$ are rational. $\sqrt{\frac{2}{3}}$ and $\sqrt{21}$ are irrational.
Every real number has exactly one real cube root.	$\sqrt[3]{27} = 3$; $\sqrt[3]{-8} = -2$
Pythagorean Formula If c is the length of the longest side (hypotenuse) of a right triangle and a and b are the lengths of the shorter sides (legs), then $a^2 + b^2 = c^2$.	Find b for the triangle in the figure. $10^2 + b^2 = (2\sqrt{61})^2$ $100 + b^2 = 4(61)$ $100 + b^2 = 244$ $b^2 = 144$ $b = 12$

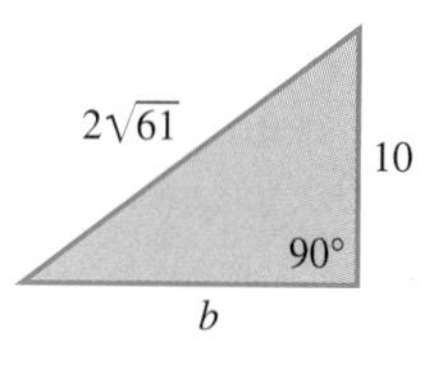

Concepts	*Examples*
8.2 *Multiplying, Dividing, and Simplifying Radicals*	
Product Rule for Radicals For nonnegative real numbers a and b, $\sqrt{a} \cdot \sqrt{b} = \sqrt{ab}$ and $\sqrt{ab} = \sqrt{a} \cdot \sqrt{b}$.	$\sqrt{5} \cdot \sqrt{7} = \sqrt{5 \cdot 7} = \sqrt{35}$ $\sqrt{8} \cdot \sqrt{2} = \sqrt{16} = 4$ $\sqrt{48} = \sqrt{16 \cdot 3} = \sqrt{16} \cdot \sqrt{3} = 4\sqrt{3}$
Quotient Rule for Radicals If a and b are nonnegative real numbers and $b \neq 0$, then $\sqrt{\frac{a}{b}} = \frac{\sqrt{a}}{\sqrt{b}}$ and $\frac{\sqrt{a}}{\sqrt{b}} = \sqrt{\frac{a}{b}}$.	$\sqrt{\frac{25}{64}} = \frac{\sqrt{25}}{\sqrt{64}} = \frac{5}{8}$; $\frac{\sqrt{8}}{\sqrt{2}} = \sqrt{\frac{8}{2}} = \sqrt{4} = 2$
If all indicated roots are real, then $\sqrt[n]{a} \cdot \sqrt[n]{b} = \sqrt[n]{ab}$ and $\frac{\sqrt[n]{a}}{\sqrt[n]{b}} = \sqrt[n]{\frac{a}{b}}$ $(b \neq 0)$.	$\sqrt[3]{5} \cdot \sqrt[3]{3} = \sqrt[3]{15}$; $\frac{\sqrt[4]{12}}{\sqrt[4]{4}} = \sqrt[4]{\frac{12}{4}} = \sqrt[4]{3}$
8.3 *Adding and Subtracting Radicals* Add and subtract like radicals by using the distributive property. ***Only like radicals can be combined in this way.***	$2\sqrt{5} + 4\sqrt{5} = (2 + 4)\sqrt{5}$ $= 6\sqrt{5}$ $\sqrt{8} + \sqrt{32} = 2\sqrt{2} + 4\sqrt{2}$ $= 6\sqrt{2}$
8.4 *Rationalizing the Denominator* The denominator of a radical can be rationalized by multiplying both the numerator and denominator by a number that will eliminate the radical from the denominator.	$\frac{2}{\sqrt{3}} = \frac{2 \cdot \sqrt{3}}{\sqrt{3} \cdot \sqrt{3}} = \frac{2\sqrt{3}}{3}$ $\sqrt[3]{\frac{5}{121}} = \frac{\sqrt[3]{5} \cdot \sqrt[3]{11}}{\sqrt[3]{11^2} \cdot \sqrt[3]{11}} = \frac{\sqrt[3]{55}}{11}$
8.5 *More Simplifying and Operations with Radicals* When appropriate, use the rules for adding and multiplying polynomials to simplify radical expressions.	$\sqrt{6}(\sqrt{5} - \sqrt{7}) = \sqrt{30} - \sqrt{42}$ $(\sqrt{3} + 1)(\sqrt{3} - 2) = 3 - 2\sqrt{3} + \sqrt{3} - 2$ FOIL $= 1 - \sqrt{3}$ Combine terms.
The formulas $(a + b)^2 = a^2 + 2ab + b^2$, $(a - b)^2 = a^2 - 2ab + b^2$, and $(a + b)(a - b) = a^2 - b^2$ are useful when simplifying radical expressions.	$(\sqrt{13} - \sqrt{2})^2 = (\sqrt{13})^2 - 2(\sqrt{13})(\sqrt{2}) + (\sqrt{2})^2$ $= 13 - 2\sqrt{26} + 2$ $= 15 - 2\sqrt{26}$ $(\sqrt{5} + \sqrt{3})(\sqrt{5} - \sqrt{3}) = 5 - 3 = 2$

(continued)

Concepts	Examples
8.5 *More Simplifying and Operations with Radicals (continued)* Any denominators with radicals should be rationalized.	$\frac{3}{\sqrt{6}} = \frac{3 \cdot \sqrt{6}}{\sqrt{6} \cdot \sqrt{6}} = \frac{3\sqrt{6}}{6} = \frac{\sqrt{6}}{2}$
If a radical expression contains two terms in the denominator and at least one of those terms is a square root radical, multiply both the numerator and denominator by the conjugate of the denominator.	$\frac{6}{\sqrt{7}-\sqrt{2}} = \frac{6(\sqrt{7}+\sqrt{2})}{(\sqrt{7}-\sqrt{2})(\sqrt{7}+\sqrt{2})}$ $= \frac{6(\sqrt{7}+\sqrt{2})}{7-2}$ Multiply. $= \frac{6(\sqrt{7}+\sqrt{2})}{5}$ Subtract.
8.6 *Solving Equations with Radicals* **Solving a Radical Equation** *Step 1* Isolate a radical. *Step 2* Square each side. (By the squaring property of equality, all solutions of the original equation are *among* the solutions of the squared equation.)	Solve $\sqrt{2x-3} + x = 3$ $\sqrt{2x-3} = 3 - x$ Isolate the radical. $(\sqrt{2x-3})^2 = (3-x)^2$ Square each side. $2x - 3 = 9 - 6x + x^2$
Step 3 Combine like terms. *Step 4* If there is still a term with a radical, repeat Steps 1–3. *Step 5* Solve the equation for potential solutions.	$0 = x^2 - 8x + 12$ Standard form $0 = (x-2)(x-6)$ Factor. $x - 2 = 0$ or $x - 6 = 0$ Zero-factor property. $x = 2$ or $x = 6$ Solve.
Step 6 Check all potential solutions from Step 5 in the original equation.	A check is essential here. Verify that 2 is the only solution. (6 is extraneous.)

ANSWERS TO TEST YOUR WORD POWER

1. C; *Examples:* 6 is a square root of 36 since $6^2 = 6 \cdot 6 = 36$; -6 is also a square root of 36.
2. B; *Example:* In $\sqrt{3xy}$, $3xy$ is the radicand.
3. D; *Examples:* $\sqrt{144}$, $\sqrt{4xy^2}$, and $\sqrt{4 + t^2}$
4. A; *Examples:* $\sqrt{36} = 6$, $\sqrt[4]{81} = 3$, and $\sqrt[6]{64} = 2$
5. B; *Examples:* π, $\sqrt{2}$, $-\sqrt{5}$
6. D; *Example:* In a right triangle where $a = 6$, $b = 8$, and $c = 10$, $6^2 + 8^2 = 10^2$.
7. C; *Examples:* $\sqrt{7}$ and $3\sqrt{7}$ are like radicals; so are $2\sqrt[3]{6k}$ and $5\sqrt[3]{6k}$.
8. B; *Example:* To rationalize the denominator of $\frac{5}{\sqrt{3}+1}$, multiply the numerator and denominator by $\sqrt{3} - 1$ to get $\frac{5(\sqrt{3}-1)}{2}$.
9. A; *Example:* The conjugate of $\sqrt{3} + 1$ is $\sqrt{3} - 1$.
10. A; *Example:* The potential solution 2 is extraneous when $\sqrt{5q-1} + 3 = 0$ is solved using the method of **Section 8.6.**

Chapter 8

REVIEW EXERCISES

[8.1] *Find all square roots of each number.*

1. 49 **2.** 81 **3.** 196 **4.** 121 **5.** 225 **6.** 729

Find each root.

7. $\sqrt{16}$ **8.** $-\sqrt{.36}$ **9.** $\sqrt[3]{1000}$ **10.** $\sqrt[4]{81}$

11. $\sqrt{-8100}$ **12.** $-\sqrt{4225}$ **13.** $\sqrt{\dfrac{49}{36}}$ **14.** $\sqrt{\dfrac{100}{81}}$

Match each radical in Column I with the equivalent choice in Column II. Choices may be used once, more than once, or not at all.

I	II
15. $\sqrt{64}$	**A.** 4
16. $-\sqrt{64}$	**B.** 8
17. $\sqrt{-64}$	**C.** -4
18. $\sqrt[3]{64}$	**D.** Not a real number
19. $\sqrt[3]{-64}$	**E.** 16
20. $-\sqrt[3]{-64}$	**F.** -8

21. Find the value of x.

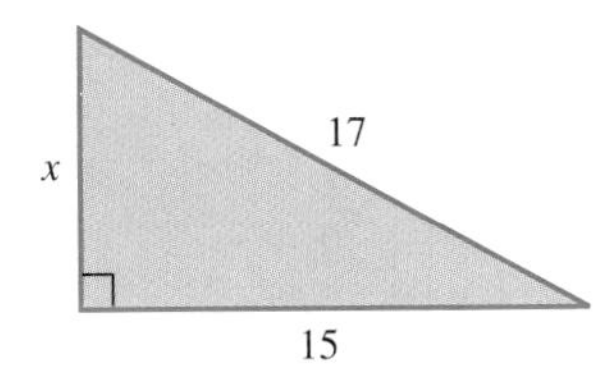

22. A Gateway EV700 computer monitor has viewing screen dimensions as shown in the figure. Find the diagonal measure of the viewing screen to the nearest tenth. (*Source:* Author's computer.)

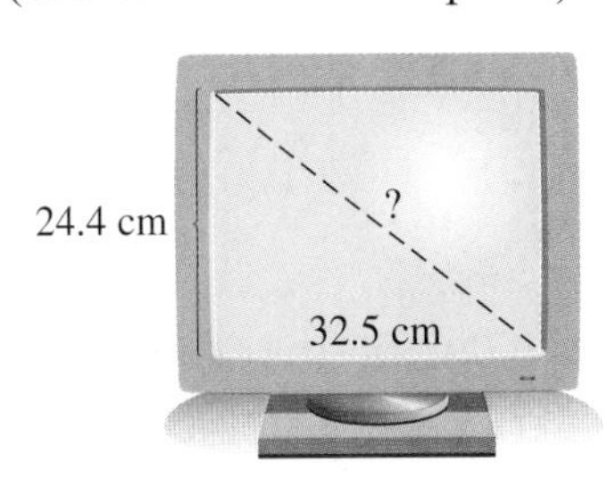

Write rational, irrational, *or* not a real number *for each number. If a number is rational, give its exact value. If a number is irrational, give a decimal approximation for the number. Round approximations to the nearest thousandth.*

23. $\sqrt{23}$ **24.** $\sqrt{169}$ **25.** $-\sqrt{25}$ **26.** $\sqrt{-4}$

[8.2] *Simplify each expression.*

27. $\sqrt{2} \cdot \sqrt{7}$ **28.** $\sqrt{5} \cdot \sqrt{15}$ **29.** $-\sqrt{27}$ **30.** $\sqrt{48}$

31. $\sqrt{160}$ **32.** $\sqrt{12} \cdot \sqrt{27}$ **33.** $\sqrt{32} \cdot \sqrt{48}$ **34.** $\sqrt{50} \cdot \sqrt{125}$

35. $\sqrt{\frac{9}{4}}$ **36.** $-\sqrt{\frac{121}{400}}$ **37.** $\sqrt{\frac{7}{169}}$ **38.** $\sqrt{\frac{1}{6}} \cdot \sqrt{\frac{5}{6}}$

39. $\sqrt{\frac{2}{5}} \cdot \sqrt{\frac{2}{45}}$ **40.** $\frac{3\sqrt{10}}{\sqrt{5}}$ **41.** $\frac{24\sqrt{12}}{6\sqrt{3}}$ **42.** $\frac{8\sqrt{150}}{4\sqrt{75}}$

Simplify each expression. Assume that all variables represent nonnegative real numbers.

43. $\sqrt{p} \cdot \sqrt{p}$ **44.** $\sqrt{k} \cdot \sqrt{m}$ **45.** $\sqrt{r^{18}}$ **46.** $\sqrt{x^{10}y^{16}}$

47. $\sqrt{x^9}$ **48.** $\sqrt{\frac{36}{p^2}}$, $p \neq 0$ **49.** $\sqrt{a^{15}b^{21}}$

50. $\sqrt{121x^6y^{10}}$ **51.** $\sqrt[3]{y^6}$ **52.** $\sqrt[3]{216x^{15}}$

53. Use a calculator to find approximations for $\sqrt{.5}$ and $\frac{\sqrt{2}}{2}$. Based on your results, do you think that these two expressions represent the same number? If so, verify it *algebraically*.

[8.3] *Simplify and combine terms where possible.*

54. $\sqrt{11} + \sqrt{11}$

55. $3\sqrt{2} + 6\sqrt{2}$

56. $3\sqrt{75} + 2\sqrt{27}$

57. $4\sqrt{12} + \sqrt{48}$

58. $4\sqrt{24} - 3\sqrt{54} + \sqrt{6}$

59. $2\sqrt{7} - 4\sqrt{28} + 3\sqrt{63}$

60. $\frac{2}{5}\sqrt{75} + \frac{3}{4}\sqrt{160}$

61. $\frac{1}{3}\sqrt{18} + \frac{1}{4}\sqrt{32}$

62. $\sqrt{15} \cdot \sqrt{2} + 5\sqrt{30}$

Simplify each expression. Assume that all variables represent nonnegative real numbers.

63. $\sqrt{4x} + \sqrt{36x} - \sqrt{9x}$

64. $\sqrt{16p} + 3\sqrt{p} - \sqrt{49p}$

65. $\sqrt{20m^2} - m\sqrt{45}$

66. $3k\sqrt{8k^2n} + 5k^2\sqrt{2n}$

[8.4] *Perform the indicated operations, and write all answers in simplest form. Rationalize all denominators. Assume that all variables represent nonnegative real numbers.*

67. $\frac{10}{\sqrt{3}}$

68. $\frac{8\sqrt{2}}{\sqrt{5}}$

69. $\frac{12}{\sqrt{24}}$

70. $\sqrt{\frac{2}{5}}$

71. $\sqrt{\frac{5}{14}} \cdot \sqrt{28}$

72. $\sqrt{\frac{2}{7}} \cdot \sqrt{\frac{1}{3}}$

73. $\sqrt{\frac{r^2}{16x}}, \quad x \neq 0$

74. $\sqrt[3]{\frac{1}{3}}$

Solve each problem.

75. The radius r of a cone in terms of its volume V is given by the formula

$$r = \sqrt{\frac{3V}{\pi h}}.$$

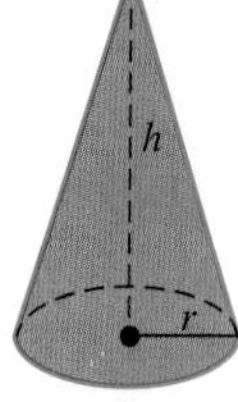

Rationalize the denominator of the radical expression.

76. The radius r of a sphere in terms of its surface area S is given by the formula

$$r = \sqrt{\frac{S}{4\pi}}.$$

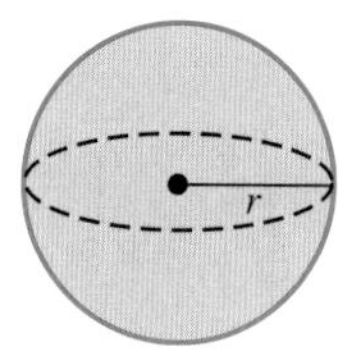

Rationalize the denominator of the radical expression.

[8.5] *Simplify each expression.*

77. $-\sqrt{3}(\sqrt{5} + \sqrt{27})$

78. $3\sqrt{2}(\sqrt{3} + 2\sqrt{2})$

79. $(2\sqrt{3} - 4)(5\sqrt{3} + 2)$

80. $(\sqrt{7} + 2\sqrt{6})(\sqrt{12} - \sqrt{2})$

81. $(2\sqrt{3} + 5)(2\sqrt{3} - 5)$

82. $(\sqrt{x} + 2)^2, \quad x \geq 0$

Rationalize each denominator.

83. $\dfrac{1}{2 + \sqrt{5}}$

84. $\dfrac{2}{\sqrt{2} - 3}$

85. $\dfrac{3}{1 + \sqrt{x}}, \quad x \geq 0$

86. $\dfrac{\sqrt{8}}{\sqrt{2} + 6}$

87. $\dfrac{\sqrt{5} - 1}{\sqrt{2} + 3}$

88. $\dfrac{2 + \sqrt{6}}{\sqrt{3} - 1}$

Write each quotient in lowest terms.

89. $\dfrac{15 + 10\sqrt{6}}{15}$

90. $\dfrac{3 + 9\sqrt{7}}{12}$

91. $\dfrac{6 + \sqrt{192}}{2}$

[8.6] *Solve each equation.*

92. $\sqrt{x} + 5 = 0$

93. $\sqrt{k + 1} = 7$

94. $\sqrt{5t + 4} = 3\sqrt{t}$

95. $\sqrt{2p + 3} = \sqrt{5p - 3}$

96. $\sqrt{4x + 1} = x - 1$

97. $\sqrt{13 + 4t} = t + 4$

98. $\sqrt{2 - x} + 3 = x + 7$

99. $\sqrt{x} - x + 2 = 0$

100. $\sqrt{x + 2} - \sqrt{x - 3} = 1$

MIXED REVIEW EXERCISES

Simplify each expression if possible. Assume that all variables represent nonnegative real numbers.

101. $\sqrt{3} \cdot \sqrt{27}$

102. $2\sqrt{27} + 3\sqrt{75} - \sqrt{300}$

103. $\sqrt{\frac{121}{t^2}}, \quad t \neq 0$

104. $\frac{1}{5 + \sqrt{2}}$

105. $\sqrt{\frac{1}{3}} \cdot \sqrt{\frac{24}{5}}$

106. $\sqrt{50y^2}$

107. $\sqrt[3]{-125}$

108. $-\sqrt{5}(\sqrt{2} + \sqrt{75})$

109. $\sqrt{\frac{16r^3}{3s}}, \quad s \neq 0$

110. $\frac{12 + 6\sqrt{13}}{12}$

111. $-\sqrt{162} + \sqrt{8}$

112. $(\sqrt{5} - \sqrt{2})^2$

113. $(6\sqrt{7} + 2)(4\sqrt{7} - 1)$

114. $-\sqrt{121}$

115. $\sqrt{98}$

Solve.

116. $\sqrt{x + 2} = x - 4$

117. $\sqrt{k} + 3 = 0$

118. $\sqrt{1 + 3t} - t = -3$

119. The *fall speed,* in miles per hour, of a vehicle running off the road into a ditch is given by the formula

$$S = \frac{2.74D}{\sqrt{h}},$$

where D is the horizontal distance traveled from the level surface to the bottom of the ditch and h is the height (or depth) of the ditch. What is the fall speed (to the nearest tenth) of a vehicle that traveled 32 ft horizontally into a ditch 5 ft deep?

RELATING CONCEPTS (EXERCISES 120–124) For Individual or Group Work

In **Chapter 3** *we plotted points in the rectangular coordinate plane. In all cases our points had coordinates that were rational numbers. However, ordered pairs may have irrational coordinates as well. Consider the points* $A(2\sqrt{14}, 5\sqrt{7})$ *and* $B(-3\sqrt{14}, 10\sqrt{7})$. ***Work Exercises 120–124 in order.***

120. Write an expression that represents the slope of the line containing points A and B. Do not simplify yet.

121. Simplify the numerator and the denominator in the expression from Exercise 120 by combining like radicals.

122. Write the fraction from Exercise 121 as the square root of a fraction in lowest terms.

123. Rationalize the denominator of the expression found in Exercise 122.

124. Based on your answer in Exercise 123, does line AB rise or fall from left to right?

Chapter 8
TEST

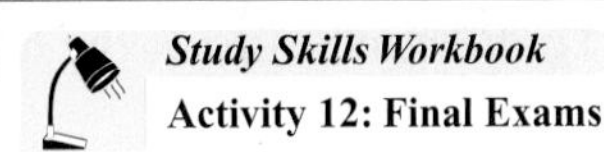

On this test, assume that all variables represent nonnegative real numbers.

1. Find all square roots of 196. **1.** ______

2. Consider $\sqrt{142}$. **2. (a)** ______ **(b)** ______

(a) Determine whether it is rational or irrational.

(b) Find a decimal approximation to the nearest thousandth.

3. If $\sqrt{a}$ is not a real number, then what kind of number must a be? **3.** ______

Simplify where possible.

4. $\sqrt[3]{216}$ **4.** ______

5. $-\sqrt{27}$ **5.** ______

6. $\sqrt{\dfrac{128}{25}}$ **6.** ______

7. $\sqrt[3]{32}$ **7.** ______

8. $\dfrac{20\sqrt{18}}{5\sqrt{3}}$ **8.** ______

9. $3\sqrt{28} + \sqrt{63}$ **9.** ______

10. $3\sqrt{27x} - 4\sqrt{48x} + 2\sqrt{3x}$ **10.** ______

11. $\sqrt{32x^2y^3}$ **11.** ______

12. $(6 - \sqrt{5})(6 + \sqrt{5})$ **12.** ______

13. $(2 - \sqrt{7})(3\sqrt{2} + 1)$ **13.** ______

14. $(\sqrt{5} + \sqrt{6})^2$ **14.** ______

15. (a) ____________

(b) ____________

16. ____________

17. ____________

18. ____________

19. ____________

20. ____________

21. ____________

22. ____________

23. ____________

24. ____________

25. ____________

Solve each problem.

15. Find the measure of the unknown leg of this right triangle.

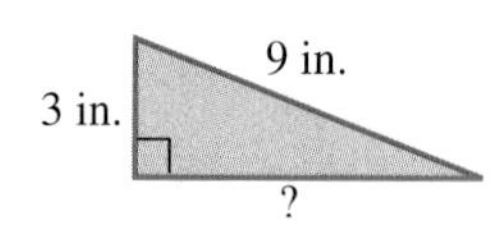

(a) Give its length in simplified radical form.

(b) Round the answer to the nearest thousandth.

16. In electronics, the impedance Z of an alternating current circuit is given by the formula

$$Z = \sqrt{R^2 + X^2},$$

where R is the resistance and X is the reactance, both in ohms. Find the value of the impedance Z if $R = 40$ ohms and $X = 30$ ohms. (*Source:* Cooke, N., and J. Orleans, *Mathematics Essential to Electricity and Radio*, McGraw-Hill, 1943.)

Rationalize each denominator.

17. $\dfrac{5\sqrt{2}}{\sqrt{7}}$

18. $\sqrt{\dfrac{2}{3x}}$, $x \neq 0$

19. $\dfrac{-2}{\sqrt[3]{4}}$

20. $\dfrac{-3}{4 - \sqrt{3}}$

21. Write in lowest terms: $\dfrac{\sqrt{12} + 3\sqrt{128}}{6}$.

Solve each equation.

22. $\sqrt{p} + 4 = 0$

23. $\sqrt{x + 1} = 5 - x$

24. $3\sqrt{x} - 2 = x$

25. What is wrong with the following "solution"?

$$\begin{aligned} \sqrt{2x + 1} + 5 &= 0 & \\ \sqrt{2x + 1} &= -5 & \text{Subtract 5.} \\ 2x + 1 &= 25 & \text{Square both sides.} \\ 2x &= 24 & \text{Subtract 1.} \\ x &= 12 & \text{Divide by 2.} \end{aligned}$$

The solution is 12.

Cumulative Review Exercises

CHAPTERS R–8

Simplify each expression.

1. $3(6+7)+6\cdot 4-3^2$

2. $\dfrac{3(6+7)+3}{2(4)-1}$

3. $|-6|-|-3|$

Solve each equation or inequality.

4. $5(k-4)-k=k-11$

5. $-\dfrac{3}{4}x \le 12$

6. $5z+3-4>2z+9+z$

7. The Pro Rodeo Cowboy All-Around Champion in 2001, Cody Ohl, earned \$22,422 more than Trevor Brazile, the 2002 champion. The two champions earned a total of \$570,416. How much did each champion earn? (*Source: World Almanac and Book of Facts*, 2004.)

Graph.

8. $-4x+5y=-20$

9. $x=2$

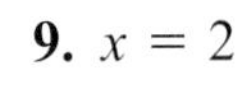

10. $2x-5y>10$

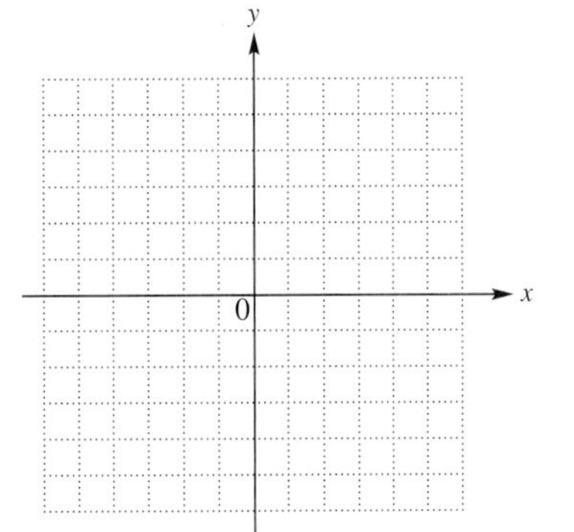

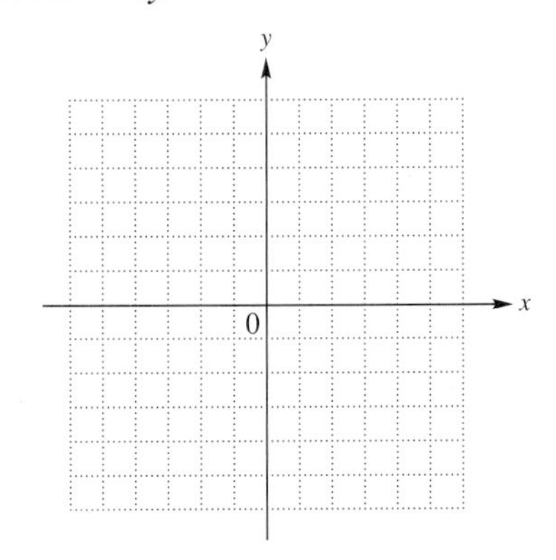

The graph shows a linear equation that models federal spending on each of the two major-party political conventions, in millions of dollars.

11. Use the ordered pairs shown on the graph to find the slope of the line to the nearest hundredth. Interpret the slope.

12. Use the slope from Exercise 11 and the ordered pair (2000, 13.5) to find the equation of the line that models the data.

13. Use the equation from Exercise 12 to project convention spending for 2004. Round your answer to the nearest tenth.

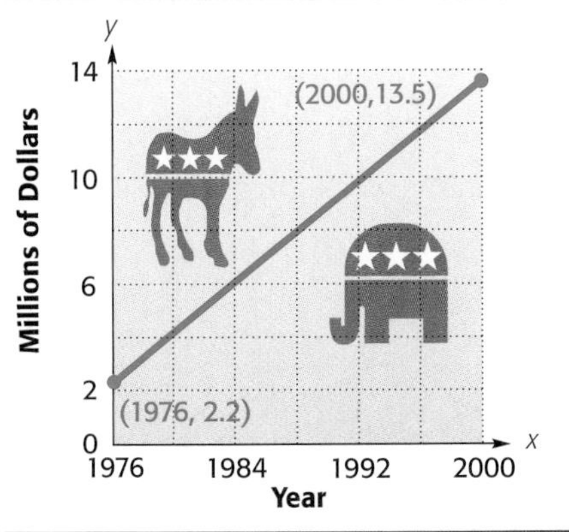

Source: Federal Election Commission.

Solve each system of equations.

14. $4x - y = 19$
$3x + 2y = -5$

15. $2x - y = 6$
$3y = 6x - 18$

16. Des Moines and Chicago are 345 mi apart. Two cars start from these cities traveling toward each other. They meet after 3 hr. The car from Chicago has an average speed 7 mph faster than the other car. Find the average speed of each car. (*Source: State Farm Road Atlas.*)

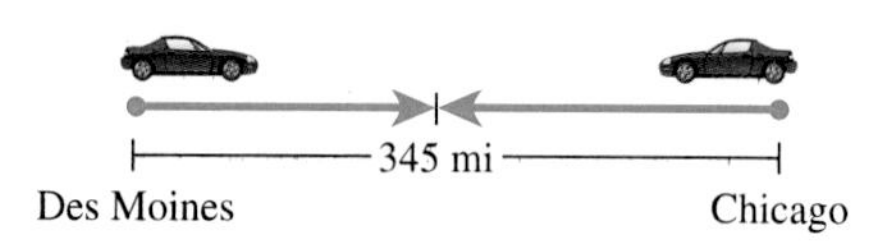

Simplify and write each expression without negative exponents. Assume that variables represent positive numbers.

17. $(3x^6)(2x^2y)^2$

18. $\left(\dfrac{3^2y^{-2}}{2^{-1}y^3}\right)^{-3}$

19. Subtract $7x^3 - 8x^2 + 4$ from $10x^3 + 3x^2 - 9$.

20. Divide $\dfrac{8t^3 - 4t^2 - 14t + 15}{2t + 3}$.

21. The percentage of U.S. households investing in mutual funds has soared since 1980. As of September 30, 2003, mutual-fund assets totaled $\$6.9 \times 10^{12}$. (*Source:* Investment Company Institute (U.S.).) Write the total mutual-fund assets without scientific notation.

Factor each polynomial completely.

22. $m^2 + 12m + 32$

23. $25t^4 - 36$

24. $12a^2 + 4ab - 5b^2$

25. $81z^2 + 72z + 16$

Solve each quadratic equation.

26. $x^2 - 7x = -12$

27. $(x + 4)(x - 1) = -6$

Perform the indicated operations. Express answers in lowest terms.

28. $\dfrac{x^2 - 3x - 4}{x^2 + 3x} \cdot \dfrac{x^2 + 2x - 3}{x^2 - 5x + 4}$

29. $\dfrac{t^2 + 4t - 5}{t + 5} \div \dfrac{t - 1}{t^2 + 8t + 15}$

30. $\dfrac{2}{x + 3} - \dfrac{4}{x - 1}$

Simplify each expression if possible. Assume all variables represent nonnegative real numbers.

31. $\sqrt{27} - 2\sqrt{12} + 6\sqrt{75}$

32. $\dfrac{2}{\sqrt{3} + \sqrt{5}}$

33. $\sqrt{200x^2y^5}$

34. $(3\sqrt{2} + 1)(4\sqrt{2} - 3)$

35. Solve $\sqrt{x} + 2 = x - 10$.

Quadratic Equations

In this chapter we develop methods of solving quadratic equations. The graphs of such equations in two variables, called *parabolas,* have many applications. For example, the Parkes radio telescope, pictured here, has a *parabolic* dish shape with a diameter of 210 ft and a depth of 32 ft. (*Source:* Mar, J. and Liebowitz, H., *Structure Technology for Large Radio and Radar Telescope Systems,* The MIT Press, 1969.)

In Section 9.4 we use a graph to model this parabolic shape and find its quadratic equation.

9.1 Solving Quadratic Equations by the Square Root Property

OBJECTIVES

1. **Solve equations of the form $x^2 = k$, where $k > 0$.**
2. **Solve equations of the form $(ax + b)^2 = k$, where $k > 0$.**
3. **Use formulas involving squared variables.**

In **Section 6.6** we solved quadratic equations by factoring. However, since not all quadratic equations can be solved by factoring, we need to develop other methods. In this chapter we do just that. Recall that a *quadratic equation* is an equation that can be written in the form

$$ax^2 + bx + c = 0 \qquad \text{Standard form}$$

for real numbers a, b, and c, with $a \neq 0$. As we saw in **Section 6.6,** to solve $x^2 + 4x + 3 = 0$ by the zero-factor property, we begin by factoring the left side and then setting each factor equal to 0.

$$x^2 + 4x + 3 = 0$$
$$(x + 3)(x + 1) = 0 \qquad \text{Factor.}$$
$$x + 3 = 0 \quad \text{or} \quad x + 1 = 0 \qquad \text{Zero-factor property}$$
$$x = -3 \quad \text{or} \quad x = -1 \qquad \text{Solve each equation.}$$

The solutions are -3 and -1.

OBJECTIVE 1 Solve equations of the form $x^2 = k$, where $k > 0$. We can solve equations such as $x^2 = 9$ by factoring as follows.

$$x^2 = 9$$
$$x^2 - 9 = 0 \qquad \text{Subtract 9.}$$
$$(x + 3)(x - 3) = 0 \qquad \text{Factor.}$$
$$x + 3 = 0 \quad \text{or} \quad x - 3 = 0 \qquad \text{Zero-factor property}$$
$$x = -3 \quad \text{or} \quad x = 3 \qquad \text{Solve each equation.}$$

We might also solve $x^2 = 9$ by noticing that x must be a number whose square is 9. Thus, $x = \sqrt{9} = 3$ or $x = -\sqrt{9} = -3$. This approach is generalized as the **square root property of equations.**

Square Root Property of Equations

If k is a positive number and if $a^2 = k$, then

$$a = \sqrt{k} \quad \text{or} \quad a = -\sqrt{k}.$$

EXAMPLE 1 Solving Quadratic Equations of the Form $x^2 = k$

Solve each equation. Write radicals in simplified form.

(a) $x^2 = 16$

By the square root property, if $x^2 = 16$, then

$$x = \sqrt{16} = 4 \quad \text{or} \quad x = -\sqrt{16} = -4.$$

An abbreviation for $x = 4$ or $x = -4$ is $x = \pm 4$ (read "positive or negative 4"). Check each solution by substituting it for x in the original equation.

(b) $z^2 = 5$

The solutions $\sqrt{5}$ and $-\sqrt{5}$ may be written as $\pm\sqrt{5}$.

Continued on Next Page

(c)

$$5m^2 - 32 = 8$$

$$5m^2 = 40 \qquad \text{Add 32.}$$

$$m^2 = 8 \qquad \text{Divide by 5.}$$

$$m = \sqrt{8} \quad \text{or} \quad m = -\sqrt{8} \qquad \text{Square root property}$$

$$m = 2\sqrt{2} \quad \text{or} \quad m = -2\sqrt{2} \qquad \sqrt{8} = \sqrt{4} \cdot \sqrt{2} = 2\sqrt{2}$$

The solutions are $\pm 2\sqrt{2}$.

(d) $x^2 = -4$

Because -4 is a negative number and because the square of a real number cannot be negative, there is no real number solution for this equation. (The square root property cannot be used because of the requirement that k must be positive.)

Work Problem 1 at the Side.

OBJECTIVE 2 Solve equations of the form $(ax + b)^2 = k$, where $k > 0$. In each equation in Example 1, the exponent 2 had a single variable as its base. The square root property can be extended to solve equations where the base is a binomial, as shown in the next example.

EXAMPLE 2 Solving Quadratic Equations of the Form $(x + b)^2 = k$

Solve each equation.

(a) $(x - 3)^2 = 16$

Apply the square root property, using $x - 3$ as the base.

$$(x - 3)^2 = 16$$

$$x - 3 = \sqrt{16} \quad \text{or} \quad x - 3 = -\sqrt{16}$$

$$x - 3 = 4 \quad \text{or} \quad x - 3 = -4 \qquad \sqrt{16} = 4$$

$$x = 7 \quad \text{or} \quad x = -1 \qquad \text{Add 3.}$$

Check both answers in the original equation.

$$(x - 3)^2 = 16 \qquad\qquad (x - 3)^2 = 16$$

$$(7 - 3)^2 = 16 \ ? \quad \text{Let } x = 7. \qquad\qquad (-1 - 3)^2 = 16 \ ? \quad \text{Let } x = -1.$$

$$4^2 = 16 \ ? \qquad\qquad (-4)^2 = 16 \ ?$$

$$16 = 16 \quad \text{True} \qquad\qquad 16 = 16 \quad \text{True}$$

The solutions are 7 and -1.

(b)

$$(x + 1)^2 = 6$$

$$x + 1 = \sqrt{6} \quad \text{or} \quad x + 1 = -\sqrt{6} \qquad \text{Square root property}$$

$$x = -1 + \sqrt{6} \quad \text{or} \quad x = -1 - \sqrt{6} \qquad \text{Add } -1.$$

Check:

$$\left(-1 + \sqrt{6} + 1\right)^2 = \left(\sqrt{6}\right)^2 = 6;$$

$$\left(-1 - \sqrt{6} + 1\right)^2 = \left(-\sqrt{6}\right)^2 = 6.$$

The solutions are $-1 + \sqrt{6}$ and $-1 - \sqrt{6}$.

Work Problem 2 at the Side.

1 Solve each equation. Write radicals in simplified form.

(a) $x^2 = 49$

(b) $x^2 = 11$

(c) $2x^2 + 8 = 32$

(d) $x^2 = -9$

2 Solve each equation.

(a) $(x + 2)^2 = 36$

(b) $(x - 4)^2 = 3$

ANSWERS

1. **(a)** $-7, 7$ **(b)** $-\sqrt{11}, \sqrt{11}$ **(c)** $-2\sqrt{3}, 2\sqrt{3}$ **(d)** no real number solution
2. **(a)** $-8, 4$ **(b)** $4 + \sqrt{3}, 4 - \sqrt{3}$

3 Solve $(2x - 5)^2 = 18$.

4 Solve each equation.

(a) $(5x + 1)^2 = 7$

(b) $(7x - 1)^2 = -1$

5 Use the formula in Example 5 to approximate the length of a bass weighing 2.80 lb and having girth 11 in.

ANSWERS

3. $\frac{5 + 3\sqrt{2}}{2}, \frac{5 - 3\sqrt{2}}{2}$

4. (a) $\frac{-1 + \sqrt{7}}{5}, \frac{-1 - \sqrt{7}}{5}$

(b) no real number solution

5. approximately 17.48 in.

EXAMPLE 3 Solving a Quadratic Equation of the Form $(ax + b)^2 = k$

Solve $(3r - 2)^2 = 27$.

$3r - 2 = \sqrt{27}$ or $3r - 2 = -\sqrt{27}$ Square root property

$3r - 2 = 3\sqrt{3}$ or $3r - 2 = -3\sqrt{3}$ $\sqrt{27} = \sqrt{9} \cdot \sqrt{3} = 3\sqrt{3}$

$3r = 2 + 3\sqrt{3}$ or $3r = 2 - 3\sqrt{3}$ Add 2.

$r = \frac{2 + 3\sqrt{3}}{3}$ or $r = \frac{2 - 3\sqrt{3}}{3}$ Divide by 3.

The solutions are $\frac{2 + 3\sqrt{3}}{3}$ and $\frac{2 - 3\sqrt{3}}{3}$.

Work Problem 3 at the Side.

CAUTION
The solutions in Example 3 are fractions that cannot be simplified, since 3 is *not* a common factor in the numerator.

EXAMPLE 4 Recognizing a Quadratic Equation with No Real Solution

Solve $(x + 3)^2 = -9$.

Because the square root of -9 is not a real number, there is no real number solution for this equation.

Work Problem 4 at the Side.

OBJECTIVE 3 Use formulas involving squared variables.

EXAMPLE 5 Finding the Length of a Bass

The formula $w = \frac{L^2 g}{1200}$ is used to approximate the weight of a bass, in pounds, given its length L and its girth g, where both are measured in inches. Approximate the length of a bass weighing 2.20 lb and having girth 10 in. (*Source: Sacramento Bee*, November 29, 2000.)

$w = \frac{L^2 g}{1200}$ Given formula

$2.20 = \frac{L^2 \cdot 10}{1200}$ $w = 2.20, g = 10$

$2640 = 10L^2$ Multiply by 1200.

$L^2 = 264$ Divide by 10; interchange the sides.

$L \approx 16.25$ Use a calculator; $L > 0$.

The length of the bass is a little more than 16 in. (We discard the negative solution -16.25 since L represents length.)

Work Problem 5 at the Side.

9.1 Exercises

FOR EXTRA HELP

Tutor Center Addison-Wesley Math Tutor Center

MathXL MathXL

Digital Video Tutor CD 9 Videotape 9

Student's Solutions Manual

MyMathLab MyMathLab

Interactmath.com

Decide whether each statement is true *or* false. *If* false, *tell why.*

1. If k is a prime number, then $x^2 = k$ has two irrational solutions.

2. If k is a positive perfect square, then $x^2 = k$ has two rational solutions.

3. If k is a positive integer, then $x^2 = k$ must have two rational solutions.

4. If $-10 < k < 0$, then $x^2 = k$ has no real solution.

5. If $-10 < k < 10$, then $x^2 = k$ has no real solution.

6. If k is an integer greater than 24 and less than 26, then $x^2 = k$ has two solutions, -5 and 5.

Solve each equation by using the square root property. Write all radicals in simplest form. See Example 1.

7. $x^2 = 81$

8. $x^2 = 121$

9. $k^2 = 14$

10. $m^2 = 22$

11. $t^2 = 48$

12. $x^2 = 54$

13. $x^2 = \frac{25}{4}$

14. $m^2 = \frac{36}{121}$

15. $x^2 = -100$

16. $x^2 = -64$

17. $z^2 = 2.25$

18. $w^2 = 56.25$

19. $r^2 - 3 = 0$

20. $x^2 - 13 = 0$

21. $7x^2 = 4$

22. $3x^2 = 10$

23. $5x^2 + 4 = 8$

24. $4x^2 - 3 = 7$

25. $3x^2 - 8 = 64$

26. $2x^2 + 7 = 61$

Solve each equation by using the square root property. Express all radicals in simplest form. See Examples 2–4.

27. $(x - 3)^2 = 25$

28. $(x - 7)^2 = 16$

29. $(x + 5)^2 = -13$

30. $(x + 2)^2 = -17$

31. $(x - 8)^2 = 27$

32. $(x - 5)^2 = 40$

33. $(3x + 2)^2 = 49$

34. $(5x + 3)^2 = 36$

35. $(4x - 3)^2 = 9$

36. $(7x - 5)^2 = 25$

37. $(5 - 2x)^2 = 30$

38. $(3 - 2x)^2 = 70$

39. $(3x + 1)^2 = 18$

40. $(5x + 6)^2 = 75$

41. $\left(\frac{1}{2}x + 5\right)^2 = 12$

42. $\left(\frac{1}{3}x + 4\right)^2 = 27$

43. $(4x - 1)^2 - 48 = 0$

44. $(2x - 5)^2 - 180 = 0$

45. Michael solved the equation in Exercise 37 and wrote his solutions as $\frac{5 + \sqrt{30}}{2}, \frac{5 - \sqrt{30}}{2}$.

Lindsay solved the same equation and wrote her solutions as $\frac{-5 + \sqrt{30}}{-2}, \frac{-5 - \sqrt{30}}{-2}$.

The teacher gave them both full credit. Explain why both students were correct, although their answers seem to differ.

46. In the solutions found in Example 3 of this section, why is it not valid to simplify by dividing out the 3s in the numerators and denominators?

Solve each problem. See Example 5.

47. One expert at marksmanship can hold a silver dollar at forehead level, drop it, draw his gun, and shoot the coin as it passes waist level. The distance traveled by a falling object is given by

$$d = 16t^2,$$

where d is the distance (in feet) the object falls in t seconds. If the coin falls about 4 ft, use the formula to estimate the time that elapses between the dropping of the coin and the shot.

48. The illumination produced by a light source depends on the distance from the source. For a particular light source, this relationship can be expressed as

$$d^2 = \frac{4050}{I},$$

where d is the distance from the source (in feet) and I is the amount of illumination in foot-candles. How far from the source is the illumination equal to 50 foot-candles?

49. The area A of a circle with radius r is given by the formula

$$A = \pi r^2.$$

If a circle has area 81π in.^{2}, what is its radius?

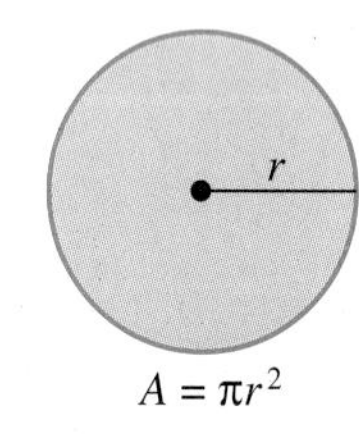

50. The surface area S of a sphere with radius r is given by the formula

$$S = 4\pi r^2.$$

If a sphere has surface area 36π ft^2, what is its radius?

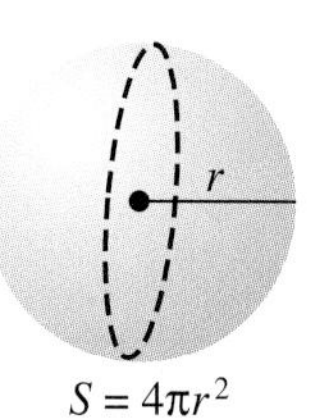

The amount A that P dollars invested at an annual rate of interest r will grow to in 2 yr is $A = P(1 + r)^2$.

51. At what interest rate will \$100 grow to \$110.25 in 2 yr?

52. At what interest rate will \$500 grow to \$572.45 in 2 yr?

9.2 Solving Quadratic Equations by Completing the Square

OBJECTIVES

1. Solve quadratic equations by completing the square when the coefficient of the squared term is 1.
2. Solve quadratic equations by completing the square when the coefficient of the squared term is not 1.
3. Simplify an equation before solving.
4. Solve applied problems that require quadratic equations.

OBJECTIVE 1 Solve quadratic equations by completing the square when the coefficient of the squared term is 1. The methods we have studied so far are not enough to solve the equation

$$x^2 + 6x + 7 = 0.$$

If we could write the equation in the form $(x + 3)^2$ equals a constant, we could solve it with the square root property discussed in **Section 9.1.** To do that, we need to have a perfect square trinomial on one side of the equation.

Recall from **Section 6.5** that a perfect square trinomial has the form

$$x^2 + 2kx + k^2 \quad \text{or} \quad x^2 - 2kx + k^2,$$

where k represents a positive number.

EXAMPLE 1 Creating Perfect Square Trinomials

Complete each trinomial so it is a perfect square.

(a) $x^2 + \_\_\_\_ + 16$

Here $k^2 = 16$, so $k = 4$ and $2kx = 2(4)(x) = 8x$. The perfect square trinomial is $x^2 + 8x + 16$.

(b) $4m^2 - \_\_\_\_ + 9$

Replace x^2 by $4m^2 = (2m)^2$ and k^2 by $9 = 3^2$, to get the middle term

$$2(3)(2m) = 12m.$$

The perfect square trinomial is $4m^2 - 12m + 9$.

(c) $x^2 + 18x + \_\_\_\_$

Here the middle term $18x$ must equal $2kx$.

$$2kx = 18x$$
$$k = 9 \qquad \text{Divide by } 2x.$$

Thus, $k = 9$ and $k^2 = 81$. The required trinomial is $x^2 + 18x + 81$.

Work Problem 1 at the Side.

1 Complete each trinomial so it is a perfect square.

(a) $x^2 + \_\_\_\_ + 36$

(b) $25x^2 - \_\_\_\_ + 4$

(c) $x^2 + 14x + \_\_\_\_$

EXAMPLE 2 Rewriting an Equation to Use the Square Root Property

Solve $x^2 + 6x + 7 = 0$.

Start by subtracting 7 from each side of the equation.

$$x^2 + 6x = -7$$

The quantity on the left side of $x^2 + 6x = -7$ must be made into a perfect square trinomial. Here $2kx = 6x$, so $k = 3$ and $k^2 = 9$. The expression $x^2 + 6x + 9$ is a perfect square, since

$$x^2 + 6x + 9 = (x + 3)^2.$$

Therefore, if 9 is added to each side, the equation will have a perfect square trinomial on the left side, as needed.

$$x^2 + 6x + 9 = -7 + 9 \qquad \text{Add 9.}$$
$$(x + 3)^2 = 2 \qquad \text{Factor.}$$

Continued on Next Page

ANSWERS

1. (a) $12x$ (b) $20x$ (c) 49

2 Solve $x^2 - 4x - 1 = 0$.

Now use the square root property to complete the solution.

$$(x + 3)^2 = 2$$

$$x + 3 = \sqrt{2} \quad \text{or} \quad x + 3 = -\sqrt{2}$$

$$x = -3 + \sqrt{2} \quad \text{or} \quad x = -3 - \sqrt{2}$$

The solutions of the original equation are $-3 + \sqrt{2}$ and $-3 - \sqrt{2}$. Check by substituting $-3 + \sqrt{2}$ and $-3 - \sqrt{2}$ for x in the original equation.

Work Problem 2 at the Side.

The process of changing the form of the equation in Example 2 from

$$x^2 + 6x + 7 = 0 \quad \text{to} \quad (x + 3)^2 = 2$$

is called **completing the square.** Completing the square changes only the form of the equation. To see this, multiply out the left side of $(x + 3)^2 = 2$ and combine terms. Then subtract 2 from each side to see that the result is $x^2 + 6x + 7 = 0$.

Look again at the original equation,

$$x^2 + 6x + 7 = 0.$$

Note that 9 is the square of half the coefficient of x, 6.

$$\frac{1}{2} \cdot 6 = 3 \quad \text{and} \quad 3^2 = 9$$

Coefficient of x (6)

To complete the square in Example 2, we added 9 to each side.

3 Solve by completing the square.

(a) $x^2 + 4x = 1$

(b) $z^2 + 6z - 3 = 0$

EXAMPLE 3 Completing the Square to Solve a Quadratic Equation

Complete the square to solve $x^2 - 8x = 5$.

To complete the square on $x^2 - 8x$, take half the coefficient of x and square it.

$$\frac{1}{2}(-8) = -4 \quad \text{and} \quad (-4)^2 = 16$$

Coefficient of x (-8)

Add the result, 16, to each side of the equation.

$x^2 - 8x = 5$	Given equation
$x^2 - 8x + 16 = 5 + 16$	Add 16.
$(x - 4)^2 = 21$	Factor the left side as the square of a binomial.

Now apply the square root property.

$x - 4 = \sqrt{21}$	or	$x - 4 = -\sqrt{21}$	Square root property
$x = 4 + \sqrt{21}$	or	$x = 4 - \sqrt{21}$	Add 4.

A check indicates that the solutions are $4 + \sqrt{21}$ and $4 - \sqrt{21}$.

Work Problem 3 at the Side.

ANSWERS

2. $2 + \sqrt{5}, 2 - \sqrt{5}$
3. (a) $-2 + \sqrt{5}, -2 - \sqrt{5}$
(b) $-3 + 2\sqrt{3}, -3 - 2\sqrt{3}$

OBJECTIVE 2 Solve quadratic equations by completing the square when the coefficient of the squared term is not 1. If an equation has the form

$$ax^2 + bx + c = 0, \quad \text{where } a \neq 1,$$

then to obtain 1 as the coefficient of x^2, we first divide each side of the equation by a.

EXAMPLE 4 Solving a Quadratic Equation by Completing the Square

Solve $4x^2 + 16x = 9$.

Before completing the square, the coefficient of x^2 must be 1, not 4. We make the coefficient 1 by dividing each side of the equation by 4.

$$4x^2 + 16x = 9$$

$$x^2 + 4x = \frac{9}{4} \quad \text{Divide by 4.}$$

Next, we complete the square by taking half the coefficient of x, or $\frac{1}{2}(4) = 2$, and squaring the result: $2^2 = 4$. We add 4 to each side of the equation, combine terms on the right side, and factor on the left.

$$x^2 + 4x + 4 = \frac{9}{4} + 4 \quad \text{Add 4.}$$

$$x^2 + 4x + 4 = \frac{25}{4} \quad \text{Combine terms.}$$

$$(x + 2)^2 = \frac{25}{4} \quad \text{Factor.}$$

Use the square root property and solve for x.

$$x + 2 = \sqrt{\frac{25}{4}} \quad \text{or} \quad x + 2 = -\sqrt{\frac{25}{4}} \quad \text{Square root property}$$

$$x + 2 = \frac{5}{2} \quad \text{or} \quad x + 2 = -\frac{5}{2} \quad \text{Take square roots.}$$

$$x = -2 + \frac{5}{2} \quad \text{or} \quad x = -2 - \frac{5}{2} \quad \text{Add } -2.$$

$$x = \frac{1}{2} \quad \text{or} \quad x = -\frac{9}{2} \quad \text{Combine terms.}$$

Check:

$$4x^2 + 16x = 9$$

$$4\left(\frac{1}{2}\right)^2 + 16\left(\frac{1}{2}\right) = 9 \quad ?$$

$$4\left(\frac{1}{4}\right) + 8 = 9 \quad ?$$

$$1 + 8 = 9 \quad ?$$

$$9 = 9 \quad \text{True}$$

$$4x^2 + 16x = 9$$

$$4\left(-\frac{9}{2}\right)^2 + 16\left(-\frac{9}{2}\right) = 9 \quad ?$$

$$4\left(\frac{81}{4}\right) - 72 = 9 \quad ?$$

$$81 - 72 = 9 \quad ?$$

$$9 = 9 \quad \text{True}$$

The two solutions are $\frac{1}{2}$ and $-\frac{9}{2}$.

4 Solve by completing the square.

(a) $9x^2 + 18x = -5$

(b) $4t^2 - 24t + 11 = 0$

The steps in solving a quadratic equation $ax^2 + bx + c = 0$ by completing the square are summarized here.

Solving a Quadratic Equation by Completing the Square

Step 1 **Be sure the squared term has coefficient 1.** If the coefficient of the squared term is 1, proceed to Step 2. If the coefficient of the squared term is not 1 but some other nonzero number a, divide each side of the equation by a.

Step 2 **Write in correct form.** Make sure that all terms with variables are on one side of the equals sign and that all constants are on the other side.

Step 3 **Complete the square.** Take half the coefficient of the first-degree term, and square the result. Add the square to each side of the equation. Factor the variable side, and combine terms on the other side.

Step 4 **Solve.** Apply the square root property to solve the equation.

Work Problem 4 at the Side.

EXAMPLE 5 Solving a Quadratic Equation by Completing the Square

Solve $2x^2 - 7x = 9$.

Step 1 Divide each side of the equation by 2 to get a coefficient of 1 for the x^2-term.

$$x^2 - \frac{7}{2}x = \frac{9}{2} \qquad \text{Divide by 2.}$$

(Step 2 is not needed for this equation.)

Step 3 Now take half the coefficient of x and square it. Half of $-\frac{7}{2}$ is $-\frac{7}{4}$, and $\left(-\frac{7}{4}\right)^2 = \frac{49}{16}$. Add $\frac{49}{16}$ to each side of the equation, and write the left side as a perfect square.

$$x^2 - \frac{7}{2}x + \frac{49}{16} = \frac{9}{2} + \frac{49}{16} \qquad \text{Add } \frac{49}{16}.$$

$$\left(x - \frac{7}{4}\right)^2 = \frac{121}{16} \qquad \text{Factor on the left; add on the right.}$$

Step 4 Use the square root property.

$$x - \frac{7}{4} = \sqrt{\frac{121}{16}} \quad \text{or} \quad x - \frac{7}{4} = -\sqrt{\frac{121}{16}}$$

$$x = \frac{7}{4} + \frac{11}{4} \quad \text{or} \quad x = \frac{7}{4} - \frac{11}{4} \qquad \text{Add } \frac{7}{4};\ \sqrt{\frac{121}{16}} = \frac{11}{4}.$$

$$x = \frac{18}{4} = \frac{9}{2} \quad \text{or} \quad x = -\frac{4}{4} = -1$$

Check that the solutions are $\frac{9}{2}$ and -1.

Work Problem 5 at the Side.

5 Solve by completing the square.

(a) $3x^2 + 5x = 2$

(b) $2x^2 - 4x - 1 = 0$

Answers

4. (a) $-\frac{1}{3}, -\frac{5}{3}$ (b) $\frac{11}{2}, \frac{1}{2}$

5. (a) $-2, \frac{1}{3}$ (b) $\frac{2 + \sqrt{6}}{2}, \frac{2 - \sqrt{6}}{2}$

EXAMPLE 6 Solving a Quadratic Equation by Completing the Square

Solve $4p^2 + 8p + 5 = 0$.

$$4p^2 + 8p + 5 = 0$$

$$p^2 + 2p + \frac{5}{4} = 0 \qquad \text{Divide by 4.}$$

$$p^2 + 2p = -\frac{5}{4} \qquad \text{Subtract } \tfrac{5}{4}.$$

The coefficient of p is 2. Take half of 2, square the result, and add this square to each side: $\left[\frac{1}{2}(2)\right]^2 = 1$. The left side can then be written as a perfect square.

$$p^2 + 2p + 1 = -\frac{5}{4} + 1 \qquad \text{Add 1.}$$

$$(p + 1)^2 = -\frac{1}{4} \qquad \text{Factor; add.}$$

The square root of $-\frac{1}{4}$ is not a real number, so the square root property does not apply. This equation has no real number solution.*

Work Problem 6 at the Side.

6 Solve $5x^2 + 3x + 1 = 0$ by completing the square.

OBJECTIVE 3 Simplify an equation before solving. The next example shows how to simplify a quadratic equation before solving it.

EXAMPLE 7 Simplifying an Equation before Completing the Square

Solve $(x + 3)(x - 1) = 2$.

$$(x + 3)(x - 1) = 2 \qquad \text{Given equation}$$

$$x^2 + 2x - 3 = 2 \qquad \text{Use FOIL.}$$

$$x^2 + 2x = 5 \qquad \text{Add 3.}$$

$$x^2 + 2x + 1 = 5 + 1 \qquad \text{Add } \left[\tfrac{1}{2}(2)\right]^2 = 1.$$

$$(x + 1)^2 = 6 \qquad \text{Factor on the left; add on the right.}$$

$$x + 1 = \sqrt{6} \quad \text{or} \quad x + 1 = -\sqrt{6} \qquad \text{Square root property}$$

$$x = -1 + \sqrt{6} \quad \text{or} \quad x = -1 - \sqrt{6} \qquad \text{Add } -1.$$

The solutions are $-1 + \sqrt{6}$ and $-1 - \sqrt{6}$.

Work Problem 7 at the Side.

7 Solve each equation.

(a) $r(r - 3) = -1$

(b) $(x + 2)(x + 1) = 5$

NOTE

The solutions given in Example 7 are *exact*. In applications, decimal solutions are more appropriate. Using the square root key of a calculator, $\sqrt{6} \approx 2.449$. Evaluating the two solutions gives

$$x \approx 1.449 \quad \text{and} \quad x \approx -3.449.$$

ANSWERS

6. no real number solution

7. (a) $\frac{3 + \sqrt{5}}{2}, \frac{3 - \sqrt{5}}{2}$

(b) $\frac{-3 + \sqrt{21}}{2}, \frac{-3 - \sqrt{21}}{2}$

* The equation in Example 6 has no *real number* solution. In the context of another number system, called the *complex numbers,* however, this equation does have solutions. The complex numbers include numbers whose squares are negative. These numbers are discussed in intermediate and college algebra courses.

8 Suppose a ball is propelled upward with an initial velocity of 128 ft per sec. Its height at time t (in seconds) is given by

$$s = -16t^2 + 128t,$$

where s is in feet. At what times will it be 48 ft above the ground? Give answers to the nearest tenth.

OBJECTIVE 4 Solve applied problems that require quadratic equations. There are many practical applications of quadratic equations. The next example illustrates an application from physics.

EXAMPLE 8 Solving a Velocity Problem

If a ball is propelled into the air from ground level with an initial velocity of 64 ft per sec, its height s (in feet) in t seconds is given by the formula

$$s = -16t^2 + 64t.$$

How long will it take the ball to reach a height of 48 ft?

Since s represents the height, substitute **48** for s in the formula to get

$$48 = -16t^2 + 64t.$$

We solve this equation for time, t, by completing the square. First we divide each side by -16. We also interchange the sides of the equation.

$-3 = t^2 - 4t$	Divide by -16.
$t^2 - 4t = -3$	Interchange the sides.
$t^2 - 4t + 4 = -3 + 4$	Add $\left[\frac{1}{2}(-4)\right]^2 = 4$.
$(t - 2)^2 = 1$	Factor.
$t - 2 = 1$ or $t - 2 = -1$	Square root property
$t = 3$ or $t = 1$	Add 2.

You may wonder how we can get two correct answers for the time required for the ball to reach a height of 48 ft. The ball reaches that height twice, once on the way up and again on the way down. So it takes 1 sec to reach 48 ft on the way up, and then after 3 sec, the ball reaches 48 ft again on the way down.

Work Problem 8 at the Side.

ANSWERS
8. .4 and 7.6 sec

9.2 Exercises

FOR EXTRA HELP

Tutor Center Addison-Wesley Math Tutor Center

MathXL

Digital Video Tutor CD 9 Videotape 9

Student's Solutions Manual

MyMathLab

Interactmath.com

Complete each trinomial so that it is a perfect square. See Example 1.

1. $x^2 + \_\_\_\_ + 25$

2. $9x^2 + \_\_\_\_ + 4$

3. $z^2 - 14z + \_\_\_\_$

4. $4a^2 - 32a + \_\_\_\_$

5. Which step is an appropriate way to begin solving the quadratic equation

$$2x^2 - 4x = 9$$

by completing the square?

A. Add 4 to each side of the equation.

B. Factor the left side as $2x(x - 2)$.

C. Factor the left side as $x(2x - 4)$.

D. Divide each side by 2.

6. In Example 3 of **Section 6.6,** we solved the quadratic equation

$$4p^2 - 26p + 40 = 0$$

by factoring. If we were to solve by completing the square, would we get the same solutions, $\frac{5}{2}$ and 4?

Find the number that should be added to each expression to make it a perfect square. See Examples 1–3.

7. $x^2 + 20x$

8. $z^2 + 18z$

9. $x^2 - 5x$

10. $m^2 - 9m$

11. $r^2 + \frac{1}{2}r$

12. $s^2 - \frac{1}{3}s$

Solve each equation by completing the square. See Examples 2 and 3.

13. $x^2 - 4x = -3$

14. $x^2 - 2x = 8$

15. $x^2 + 5x + 6 = 0$

16. $x^2 + 6x + 5 = 0$

17. $x^2 + 2x - 5 = 0$

18. $x^2 + 4x + 1 = 0$

19. $t^2 + 6t + 9 = 0$

20. $k^2 - 8k + 16 = 0$

21. $x^2 + x - 1 = 0$

22. $x^2 + x - 3 = 0$

Solve each equation by completing the square. See Examples 4–7.

23. $4x^2 + 4x - 3 = 0$

24. $9x^2 + 3x - 2 = 0$

25. $2x^2 - 4x = 5$

26. $2x^2 - 6x = 3$

27. $2p^2 - 2p + 3 = 0$

28. $3q^2 - 3q + 4 = 0$

29. $3k^2 + 7k = 4$

30. $2k^2 + 5k = 1$

31. $(x + 3)(x - 1) = 5$

32. $(y - 8)(y + 2) = 24$

33. $-x^2 + 2x = -5$

34. $-r^2 + 3r = -2$

Solve each problem. See Example 8.

35. If an object is propelled upward from ground level with an initial velocity of 96 ft per sec, its height s (in feet) in t seconds is given by the formula $s = -16t^2 + 96t$. In how many seconds will the object be at a height of 80 ft?

36. How much time will it take the object described in Exercise 35 to be at a height of 100 ft? Round your answers to the nearest tenth.

37. If an object is propelled upward on the surface of Mars from ground level with an initial velocity of 104 ft per sec, its height s (in feet) in t seconds is given by the formula $s = -13t^2 + 104t$. How long will it take for the object to be at a height of 195 ft?

38. How long will it take the object in Exercise 37 to return to the surface? (*Hint:* When it returns to the surface, $s = 0$.)

39. A farmer has a rectangular cattle pen with perimeter 350 ft and area 7500 ft^2. What are the dimensions of the pen? (*Hint:* Use the figure to set up the equation.)

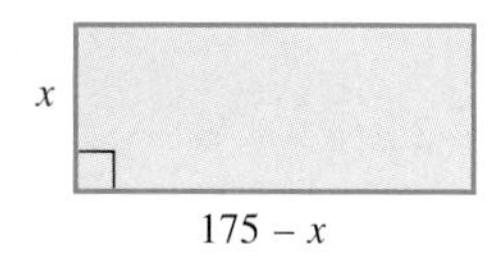

40. The base of a triangle measures 1 m more than three times the height of the triangle. Its area is 15 m^2. Find the lengths of the base and the height.

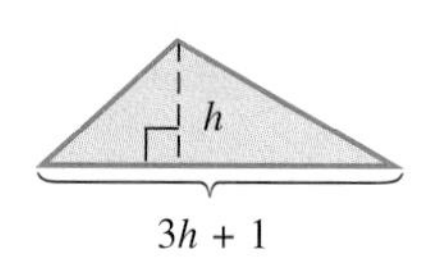

9.3 Solving Quadratic Equations by the Quadratic Formula

We can solve any quadratic equation by completing the square, but the method is tedious. In this section we complete the square on the general quadratic equation

$$ax^2 + bx + c = 0, \quad a \neq 0,$$

to obtain the *quadratic formula,* a formula that gives the solution(s) for any quadratic equation.

OBJECTIVES

1. Identify the values of a, b, and c in a quadratic equation.
2. Use the quadratic formula to solve quadratic equations.
3. Solve quadratic equations with only one solution.
4. Solve quadratic equations with fractions as coefficients.

NOTE
In $ax^2 + bx + c = 0$, there is a restriction that a is not zero. If it were, the equation would be linear, not quadratic.

OBJECTIVE 1 Identify the values of *a*, *b*, and *c* in a quadratic equation. The first step in solving a quadratic equation by this new method is to identify the values of a, b, and c in the standard form of the quadratic equation.

EXAMPLE 1 Identifying Values of *a*, *b*, and *c* in Quadratic Equations

Match the coefficients of each quadratic equation with the letters a, b, and c of the standard quadratic equation $ax^2 + bx + c = 0$.

(a) $2x^2 + 3x - 5 = 0$ (with a, b, c marked above 2, 3, -5)
In this example $a = 2$, $b = 3$, and $c = -5$.

(b) $-x^2 + 2 = 6x$
First rewrite the equation with 0 on one side to match the standard form $ax^2 + bx + c = 0$.

$$-x^2 + 2 = 6x$$
$$-x^2 - 6x + 2 = 0 \qquad \text{Subtract } 6x.$$

Here, $a = -1$, $b = -6$, and $c = 2$. (Notice that the coefficient of x^2 is understood to be -1.)

(c) $5x^2 - 12 = 0$
The x-term is missing, so write the equation as

$$5x^2 + 0x - 12 = 0.$$

Then $a = 5$, $b = 0$, and $c = -12$.

(d) $(2x - 7)(x + 4) = -23$
Write the equation in standard form.

$$(2x - 7)(x + 4) = -23$$
$$2x^2 + x - 28 = -23 \qquad \text{Use FOIL on the left.}$$
$$2x^2 + x - 5 = 0 \qquad \text{Add 23.}$$

Now, identify the values: $a = 2$, $b = 1$, and $c = -5$.

Work Problem 1 at the Side.

1 Match the coefficients of each quadratic equation with the letters a, b, and c of the standard quadratic equation $ax^2 + bx + c = 0$.

(a) $5x^2 + 2x - 1 = 0$

(b) $3x^2 = x - 2$

(c) $9x^2 - 13 = 0$

(d) $(3x + 2)(x - 1) = 8$

ANSWERS
1. (a) $a = 5, b = 2, c = -1$
(b) $a = 3, b = -1, c = 2$
(c) $a = 9, b = 0, c = -13$
(d) $a = 3, b = -1, c = -10$

OBJECTIVE 2 Use the quadratic formula to solve quadratic equations. To develop the quadratic formula, we follow the steps for completing the square on $ax^2 + bx + c = 0$ $(a > 0)$ given in **Section 9.2.** For comparison, we also show the corresponding steps for solving $2x^2 + x - 5 = 0$ (from Example 1(d)).

Step 1 Transform so that the coefficient of the squared term equals 1.

$2x^2 + x - 5 = 0$		$ax^2 + bx + c = 0$	Standard form
$x^2 + \frac{1}{2}x - \frac{5}{2} = 0$	Divide by 2.	$x^2 + \frac{b}{a}x + \frac{c}{a} = 0$	Divide by a.

Step 2 Write so that the variable terms are alone on the left side.

$x^2 + \frac{1}{2}x = \frac{5}{2}$	Add $\frac{5}{2}$.	$x^2 + \frac{b}{a}x = -\frac{c}{a}$	Subtract $\frac{c}{a}$.

Step 3 Add the square of half the coefficient of x to each side, factor the left side, and combine terms on the right.

$x^2 + \frac{1}{2}x + \frac{1}{16} = \frac{5}{2} + \frac{1}{16}$	Add $\frac{1}{16}$.	$x^2 + \frac{b}{a}x + \frac{b^2}{4a^2} = -\frac{c}{a} + \frac{b^2}{4a^2}$	Add $\frac{b^2}{4a^2}$.
$\left(x + \frac{1}{4}\right)^2 = \frac{41}{16}$	Factor; add on right.	$\left(x + \frac{b}{2a}\right)^2 = \frac{b^2 - 4ac}{4a^2}$	Factor; add on right.

Step 4 Use the square root property to complete the solution.

$x + \frac{1}{4} = \pm\sqrt{\frac{41}{16}}$	$x + \frac{b}{2a} = \pm\sqrt{\frac{b^2 - 4ac}{4a^2}}$
$x + \frac{1}{4} = \pm\frac{\sqrt{41}}{4}$	$x + \frac{b}{2a} = \pm\frac{\sqrt{b^2 - 4ac}}{2a}$
$x = -\frac{1}{4} \pm \frac{\sqrt{41}}{4}$	$x = -\frac{b}{2a} \pm \frac{\sqrt{b^2 - 4ac}}{2a}$
$x = \frac{-1 \pm \sqrt{41}}{4}$	$x = \frac{-b \pm \sqrt{b^2 - 4ac}}{2a}$

The final result in the column on the right is called the **quadratic formula.** (It is also valid for $a < 0$.) ***It is a key result that should be memorized.*** Notice that there are two values, one for the + sign and one for the − sign.

Quadratic Formula

The solutions of the quadratic equation $ax^2 + bx + c = 0$, $a \neq 0$, are

$$\mathbf{x = \frac{-b + \sqrt{b^2 - 4ac}}{2a}} \quad \text{and} \quad \mathbf{x = \frac{-b - \sqrt{b^2 - 4ac}}{2a}},$$

or in compact form,

$$\mathbf{x = \frac{-b \pm \sqrt{b^2 - 4ac}}{2a}}.$$

CAUTION

Notice in the quadratic formula that the fraction bar is under $-b$ as well as the radical. When using this formula, ***be sure to find the values of*** $-b \pm \sqrt{b^2 - 4ac}$ ***first, then divide those results by the value of*** $2a$.

2 Solve by using the quadratic formula.

(a) $2x^2 + 3x - 5 = 0$

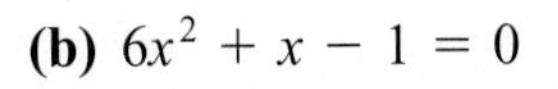

(b) $6x^2 + x - 1 = 0$

EXAMPLE 2 Solving a Quadratic Equation by the Quadratic Formula

Use the quadratic formula to solve $2x^2 - 7x - 9 = 0$. (This equation was solved by completing the square in **Section 9.2.** See Example 5.)

Match the coefficients of the variables with those of the standard quadratic equation

$$ax^2 + bx + c = 0.$$

Here, $a = 2$, $b = -7$, and $c = -9$. Substitute these numbers into the quadratic formula, and simplify the result.

$$x = \frac{-b \pm \sqrt{b^2 - 4ac}}{2a}$$

$$x = \frac{-(-7) \pm \sqrt{(-7)^2 - 4(2)(-9)}}{2(2)} \qquad \text{Let } a = 2,\ b = -7,\ c = -9.$$

$$x = \frac{7 \pm \sqrt{49 + 72}}{4}$$

$$x = \frac{7 \pm \sqrt{121}}{4}$$

$$x = \frac{7 \pm 11}{4} \qquad \sqrt{121} = 11$$

Find the two individual solutions by first using the plus sign, and then using the minus sign:

$$x = \frac{7 + 11}{4} = \frac{18}{4} = \frac{9}{2} \quad \text{or} \quad x = \frac{7 - 11}{4} = \frac{-4}{4} = -1.$$

Check:

$$2x^2 - 7x - 9 = 0$$
$$2\left(\frac{9}{2}\right)^2 - 7\left(\frac{9}{2}\right) - 9 = 0 \quad ?$$
$$\frac{81}{2} - \frac{63}{2} - 9 = 0 \quad ?$$
$$0 = 0 \quad \text{True}$$

$$2x^2 - 7x - 9 = 0$$
$$2(-1)^2 - 7(-1) - 9 = 0 \quad ?$$
$$2 + 7 - 9 = 0 \quad ?$$
$$0 = 0 \quad \text{True}$$

The solutions are $\frac{9}{2}$ and -1, the same solutions obtained by completing the square in **Section 9.2.**

Work Problem 2 at the Side.

ANSWERS

2. **(a)** $1, -\frac{5}{2}$ **(b)** $-\frac{1}{2}, \frac{1}{3}$

3 Solve $x^2 + 1 = -8x$ by the quadratic formula.

EXAMPLE 3 **Rewriting a Quadratic Equation before Using the Quadratic Formula**

Solve $x^2 = 2x + 1$.

Find a, b, and c by rewriting the equation in standard form (with 0 on one side). Add $-2x - 1$ to each side of the equation to get

$$x^2 - 2x - 1 = 0.$$

Then $a = 1$, $b = -2$, and $c = -1$. Substitute these values into the quadratic formula.

$$x = \frac{-b \pm \sqrt{b^2 - 4ac}}{2a}$$

$$x = \frac{-(-2) \pm \sqrt{(-2)^2 - 4(1)(-1)}}{2(1)} \qquad \text{Let } a = 1, b = -2, c = -1.$$

$$x = \frac{2 \pm \sqrt{4 + 4}}{2}$$

$$x = \frac{2 \pm \sqrt{8}}{2}$$

$$x = \frac{2 \pm 2\sqrt{2}}{2} \qquad \sqrt{8} = \sqrt{4} \cdot \sqrt{2} = 2\sqrt{2}$$

Write these solutions in lowest terms by factoring $2 \pm 2\sqrt{2}$ as $2(1 \pm \sqrt{2})$.

$$x = \frac{2 \pm 2\sqrt{2}}{2} = \frac{2(1 \pm \sqrt{2})}{2} = 1 \pm \sqrt{2}$$

The two solutions of the original equation are $1 + \sqrt{2}$ and $1 - \sqrt{2}$.

Work Problem 3 at the Side.

4 Solve $9x^2 - 12x + 4 = 0$.

OBJECTIVE 3 **Solve quadratic equations with only one solution.** In the quadratic formula, when the quantity under the radical, $b^2 - 4ac$, equals 0, the equation has just one rational number solution. In this case, the trinomial $ax^2 + bx + c$ is a perfect square.

EXAMPLE 4 **Solving a Quadratic Equation with Only One Solution**

Solve $4x^2 + 25 = 20x$.

Write the equation as

$$4x^2 - 20x + 25 = 0. \qquad \text{Subtract } 20x.$$

Here, $a = 4$, $b = -20$, and $c = 25$. By the quadratic formula,

$$x = \frac{-(-20) \pm \sqrt{(-20)^2 - 400}}{2(4)} = \frac{20 \pm 0}{8} = \frac{5}{2}.$$

In this case, $b^2 - 4ac = 0$, and the trinomial $4x^2 - 20x + 25$ is a perfect square. There is just one solution, $\frac{5}{2}$.

Work Problem 4 at the Side.

ANSWERS

3. $-4 + \sqrt{15}, -4 - \sqrt{15}$
4. $\frac{2}{3}$

NOTE
The single solution of the equation in Example 4 is a rational number. If all solutions of a quadratic equation are rational, the equation can be solved by factoring as well.

OBJECTIVE 4 Solve quadratic equations with fractions as coefficients. It is usually easier to clear quadratic equations of fractions before solving them. This is accomplished by multiplying all terms in the equation by the LCD of the fractions.

EXAMPLE 5 Solving a Quadratic Equation with Fractions as Coefficients

Solve $\frac{1}{10}t^2 = \frac{2}{5}t - \frac{1}{2}$.

Eliminate the denominators by multiplying each side of the equation by the least common denominator, 10.

$$10\left(\frac{1}{10}t^2\right) = 10\left(\frac{2}{5}t - \frac{1}{2}\right)$$

$$10\left(\frac{1}{10}t^2\right) = 10\left(\frac{2}{5}t\right) - 10\left(\frac{1}{2}\right) \quad \text{Distributive property}$$

$$t^2 = 4t - 5 \quad \text{Multiply.}$$

$$t^2 - 4t + 5 = 0 \quad \text{Standard form.}$$

From this form, we see that $a = 1$, $b = -4$, and $c = 5$. Use the quadratic formula to complete the solution.

$$t = \frac{-(-4) \pm \sqrt{(-4)^2 - 4(1)(5)}}{2(1)}$$

$$t = \frac{4 \pm \sqrt{16 - 20}}{2}$$

$$t = \frac{4 \pm \sqrt{-4}}{2}$$

The radical $\sqrt{-4}$ is not a real number, so the equation has no real number solution.

Work Problem 5 at the Side.

5 Solve.

(a) $x^2 - \frac{4}{3}x + \frac{2}{3} = 0$

(b) $x^2 - \frac{9}{5}x = \frac{2}{5}$

Answers

5. **(a)** no real number solution
(b) $-\frac{1}{5}, 2$

Focus on

Real-Data Applications

Estimating Interest Rates

Banks offer savings accounts in the form of money market funds that pay variable interest rates, depending on the status of the prime interest rate in effect at the beginning of each month. The prime interest rate is the rate that the Federal Reserve pays its largest customers.

The compound interest formula, $A = P(1 + r)^n$, computes the amount A that P dollars invested at a periodic rate of interest r will grow to in n periods. The period for compound interest can be a year (annual), six months (semi-annual), three months (quarterly), one month (monthly), or even daily. The periodic interest rate is the annual rate divided by the number of periods per year. For example, to find the monthly rate, you divide the annual rate by 12. Similarly, if you know the monthly rate, you can multiply it by 12 to find the annual rate. If you borrow money to purchase a car or a house, the quantities P, r, and n are typically determined for the entire purchase.

Suppose you are saving money on a month-to-month basis at your bank. On March 1 you open a money market fund account with a deposit of \$250; on April 1 you deposit \$175; and on May 1 you deposit \$300. Your deposit slip on May 1 shows a balance of \$726.20. You are curious about the average interest rate earned from the date you opened the account. The investment can be thought of as three separate transactions, added together.

March 1:	The investment earns interest for 2 months (March, April).	$A = 250(1 + r)^2$
April 1:	The investment earns interest for 1 month (April).	$A = 175(1 + r)$
May 1:	The investment has not yet earned interest.	$A = 300$

The total investment value can be written as a quadratic equation:

$$250(1 + r)^2 + 175(1 + r) + 300 = 726.20.$$

To solve this equation, you could expand and collect like terms, but the algebra is tedious. A better option is to replace $1 + r$ with x and solve the quadratic equation

$$250x^2 + 175x + 300 = 726.20.$$

Once you have found x, you simply subtract 1 to get r, the monthly interest rate.

For Group Discussion

1. Consider the quadratic equation $250x^2 + 175x + 300 = 726.20$.

(a) Use the quadratic formula to solve the equation for x. Remember to write it in standard form. (*Hint:* x must have a value between 1 and 2 since x is $1 + r$ and r is an interest rate.)

(b) Calculate the periodic (monthly) rate, r, and the annual rate, rounded to ten-thousandths. Write the annual interest rate as a percent.

2. After talking to an investment advisor, you decide to invest in another fund. On the first of June, July, and August you again invest \$250, \$175, and \$300, respectively. After making the August 1 deposit, your account value is \$732.96. To find the average interest rate, you must now solve

$$250x^2 + 175x + 300 = 732.96.$$

(a) Use the quadratic formula to solve the equation for x.

(b) Calculate the periodic (monthly) rate, r, and the annual rate, rounded to ten-thousandths. Write the annual interest rate as a percent.

9.3 Exercises

FOR EXTRA HELP

Tutor Center Addison-Wesley Math Tutor Center | MathXL | Digital Video Tutor CD 9 Videotape 9 | Student's Solutions Manual | MyMathLab | Interactmath.com

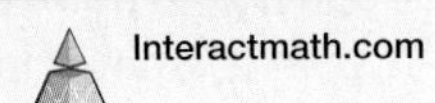

Write each equation in the form $ax^2 + bx + c = 0$, if necessary. Then identify the values of a, b, and c. Do not actually solve the equation. See Example 1.

1. $4x^2 + 5x - 9 = 0$

$a =$ ______ $b =$ ______ $c =$ ______

2. $8x^2 + 3x - 4 = 0$

$a =$ ______ $b =$ ______ $c =$ ______

3. $3x^2 = 4x + 2$

$a =$ ______ $b =$ ______ $c =$ ______

4. $5x^2 = 3x - 6$

$a =$ ______ $b =$ ______ $c =$ ______

5. $3x^2 = -7x$

$a =$ ______ $b =$ ______ $c =$ ______

6. $9x^2 = 8x$

$a =$ ______ $b =$ ______ $c =$ ______

Use the quadratic formula to solve each equation. Write all radicals in simplified form, and write all answers in lowest terms. See Examples 2–4.

7. $k^2 = -12k + 13$

8. $r^2 = 8r + 9$

9. $p^2 - 4p + 4 = 0$

10. $9x^2 + 6x + 1 = 0$

11. $2x^2 + 12x = -5$

12. $5m^2 + m = 1$

13. $2x^2 = 5 + 3x$

14. $2z^2 = 30 + 7z$

15. $6x^2 + 6x = 0$

16. $4n^2 - 12n = 0$

17. $-2x^2 = -3x + 2$

18. $-x^2 = -5x + 20$

19. $3x^2 + 5x + 1 = 0$

20. $6x^2 - 6x + 1 = 0$

21. $7x^2 = 12x$

22. $9r^2 = 11r$

23. $x^2 - 24 = 0$

24. $z^2 - 96 = 0$

25. $25x^2 - 4 = 0$

26. $16x^2 - 9 = 0$

27. $3x^2 - 2x + 5 = 10x + 1$

28. $4x^2 - x + 4 = x + 7$

29. $2x^2 + x + 5 = 0$

30. $3x^2 + 2x + 8 = 0$

31. If we apply the quadratic formula and find that the value of $b^2 - 4ac$ is negative, what can we conclude?

32. If we were to solve the quadratic equation $-2x^2 - 4x + 3 = 0$, we might choose to use $a = -2$, $b = -4$, and $c = 3$. On the other hand, we might decide to multiply each side by -1 to begin, obtaining the equation $2x^2 + 4x - 3 = 0$, and then use $a = 2$, $b = 4$, and $c = -3$. Show that in either case, we obtain the same solutions.

Use the quadratic formula to solve each equation. See Example 5.

33. $\frac{3}{2}k^2 - k - \frac{4}{3} = 0$

34. $\frac{2}{5}x^2 - \frac{3}{5}x - 1 = 0$

35. $\frac{1}{2}x^2 + \frac{1}{6}x = 1$

36. $\frac{2}{3}t^2 - \frac{4}{9}t = \frac{1}{3}$

37. $.5x^2 = x + .5$

38. $.25x^2 = -1.5x - 1$

39. $\frac{3}{8}x^2 - x + \frac{17}{24} = 0$

40. $\frac{1}{3}x^2 + \frac{8}{9}x + \frac{7}{9} = 0$

Solve each problem.

41. A frog is sitting on a stump 3 ft above the ground. He hops off the stump and lands on the ground 4 ft away. During his leap, his height h is given by the equation

$$h = -.5x^2 + 1.25x + 3,$$

where x is the distance in feet from the base of the stump, and h is in feet. How far was the frog from the base of the stump when he was 1.25 ft above the ground?

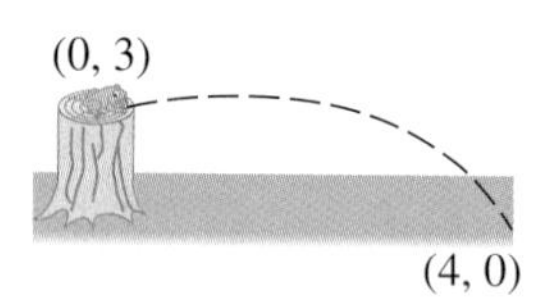

42. An astronaut on the moon throws a baseball upward. The height h of the ball, in feet, x seconds after he throws it, is given by the equation

$$h = -2.7x^2 + 30x + 6.5.$$

After how many seconds is the ball 12 ft above the moon's surface? Give answer(s) to the nearest tenth.

43. A rule for estimating the number of board feet of lumber that can be cut from a log depends on the diameter of the log. To find the diameter d required to get 9 board feet of lumber, we use the equation

$$\left(\frac{d-4}{4}\right)^2 = 9.$$

Solve this equation for d. Are both answers reasonable?

44. An old Babylonian problem asks for the length of the side of a square, given that the area of the square minus the length of a side is 870. Find the length of the side. (*Source:* Eves, H., *An Introduction to the History of Mathematics,* Sixth Edition, Saunders College Publishing, 1990.)

Summary Exercises on Quadratic Equations

Four algebraic methods have now been introduced for solving quadratic equations written in the form $ax^2 + bx + c = 0$. The following chart shows some advantages and some disadvantages of each method.

Method	Advantages	Disadvantages
1. Factoring	It is usually the fastest method.	Not all equations can be solved by factoring. Some factorable polynomials are difficult to factor.
2. Square root property	It is the simplest method for solving equations of the form $(ax + b)^2$ = a number.	Few equations are given in this form.
3. Completing the square	It can always be used. (Also, the procedure is useful in other areas of mathematics.)	It requires more steps than other methods.
4. Quadratic formula	It can always be used.	It is more difficult than factoring because of the $\sqrt{b^2 - 4ac}$ expression.

Solve each quadratic equation by the method of your choice.

1. $x^2 = 36$

2. $x^2 + 3x = -1$

3. $x^2 - \frac{100}{81} = 0$

4. $81t^2 = 49$

5. $z^2 - 4z + 3 = 0$

6. $w^2 + 3w + 2 = 0$

7. $z(z - 9) = -20$

8. $x^2 + 3x - 2 = 0$

9. $(3k - 2)^2 = 9$

10. $(2s - 1)^2 = 10$

11. $(x + 6)^2 = 121$

12. $(5k + 1)^2 = 36$

13. $(3r - 7)^2 = 24$

14. $(7p - 1)^2 = 32$

15. $(5x - 8)^2 = -6$

16. $2t^2 + 1 = t$

17. $-2x^2 = -3x - 2$

18. $-2x^2 + x = -1$

19. $8z^2 = 15 + 2z$

20. $3k^2 = 3 - 8k$

21. $0 = -x^2 + 2x + 1$

22. $3x^2 + 5x = -1$

23. $5x^2 - 22x = -8$

24. $x(x + 6) + 4 = 0$

25. $(x + 2)(x + 1) = 10$

26. $16x^2 + 40x + 25 = 0$

27. $4x^2 = -1 + 5x$

28. $2p^2 = 2p + 1$

29. $3x(3x + 4) = 7$

30. $5x - 1 + 4x^2 = 0$

31. $\frac{x^2}{2} + \frac{7x}{4} + \frac{11}{8} = 0$

32. $t(15t + 58) = -48$

33. $9k^2 = 16(3k + 4)$

34. $\frac{1}{5}x^2 + x + 1 = 0$

35. $x^2 - x + 3 = 0$

36. $4x^2 - 11x + 8 = -2$

37. $-3x^2 + 4x = -4$

38. $z^2 - \frac{5}{12}z = \frac{1}{6}$

39. $5k^2 + 19k = 2k + 12$

40. $\frac{1}{2}x^2 - x = \frac{15}{2}$

41. $x^2 - \frac{4}{15} = -\frac{4}{15}x$

42. $(x + 2)(x - 4) = 16$

9.4 Graphing Quadratic Equations

OBJECTIVES

1. Graph quadratic equations.
2. Find the vertex of a parabola.
3. Solve an application involving a parabola.

OBJECTIVE 1 Graph quadratic equations. In **Chapter 3** we saw that the graph of a linear equation in two variables is a straight line that represents all the solutions of the equation. Quadratic equations in two variables, of the form $y = ax^2 + bx + c$, are graphed in this section.

The simplest quadratic equation is $y = x^2$ (or $y = 1x^2 + 0x + 0$). The graph of this equation is not a straight line, as seen in Example 1 that follows. The equation can be graphed in much the same way that straight lines were graphed, by finding ordered pairs that satisfy the equation $y = x^2$.

EXAMPLE 1 Graphing a Quadratic Equation

Graph $y = x^2$.

Select several values for x; then find the corresponding y-values. For example, selecting $x = 2$ gives

$$y = 2^2 = 4,$$

and so the point (2, 4) is on the graph of $y = x^2$. (Recall that in an ordered pair such as (2, 4), the x-value comes first and the y-value second.)

Work Problem 1 at the Side.

If the points from Problem 1 at the side are plotted on a coordinate system and a smooth curve is drawn through them, the graph is as shown in Figure 1. The table of values completed in Problem 1 is shown with the graph.

x	y
3	9
2	4
1	1
0	0
−1	1
−2	4
−3	9

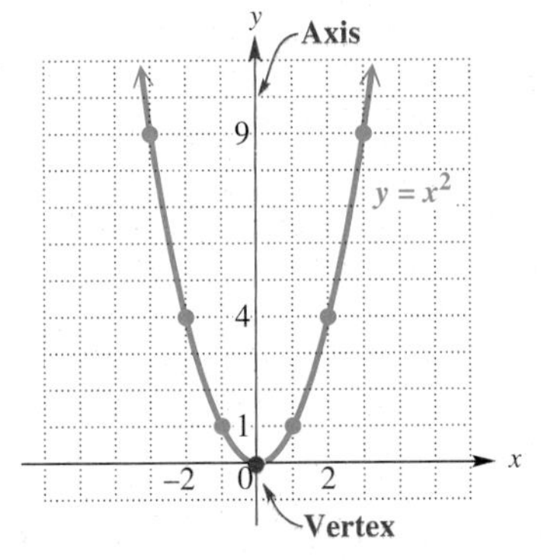

Figure 1

Work Problem 2 at the Side.

1. Complete the table of values for $y = x^2$.

x	y
3	
2	4
1	
0	
−1	
−2	
−3	

2. Graph $y = \frac{1}{2}x^2$ by first completing the table of values.

x	y
−2	
−1	
0	
1	
2	

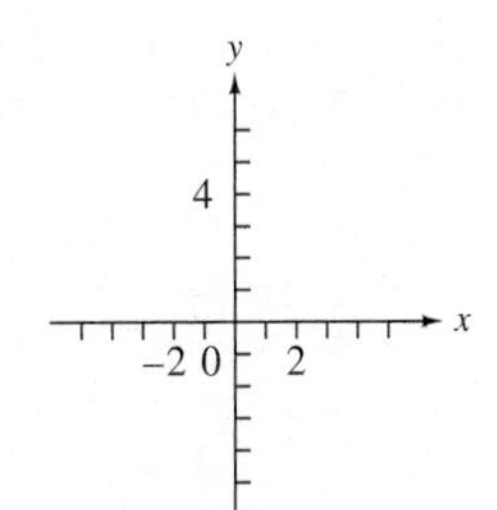

The curve in Figure 1 is called a **parabola.** The point (0, 0), the *lowest* point on this graph, is called the **vertex** of the parabola. The vertical line through the vertex (the y-axis here) is called the **axis** of the parabola. The axis of a parabola is a **line of symmetry** for the graph, because if the graph is folded on this line, the two halves will coincide.

Every equation of the form

$$y = ax^2 + bx + c,$$

with $a \neq 0$, has a graph that is a parabola. Because of its many useful properties, the parabola occurs frequently in real-life applications. For example, if an object is thrown into the air, the path that the object follows is a parabola (ignoring wind resistance). The cross sections of radar, spotlight, and telescope reflectors also form parabolas.

ANSWERS

1.

x	y
3	9
2	4
1	1
0	0
−1	1
−2	4
−3	9

2.

x	y
−2	2
−1	$\frac{1}{2}$
0	0
1	$\frac{1}{2}$
2	2

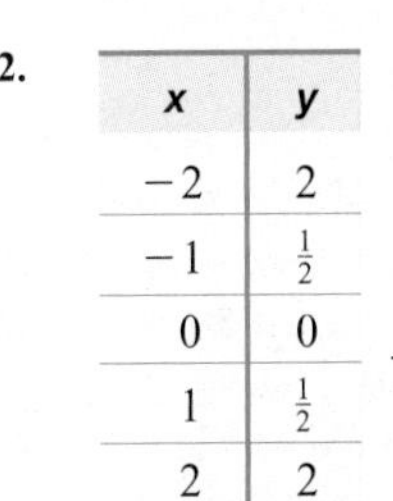

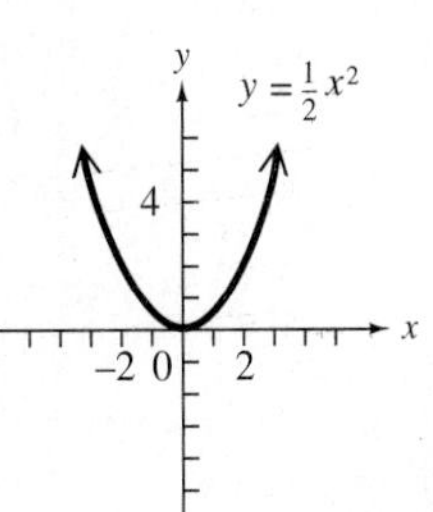

3 Complete each ordered pair for $y = -x^2 - 1$.

$(-2, \quad), \quad (-1, \quad),$
$(1, \quad), \quad (2, \quad)$

4 Graph each equation, and identify each vertex.

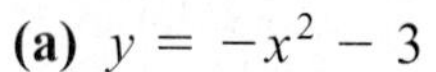

(a) $y = -x^2 - 3$

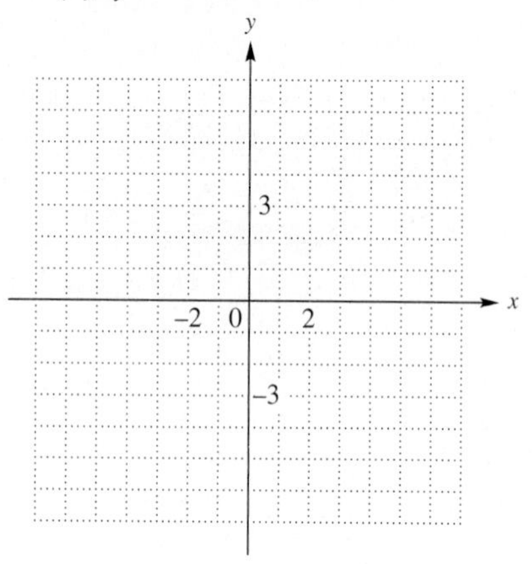

(b) $y = x^2 + 3$

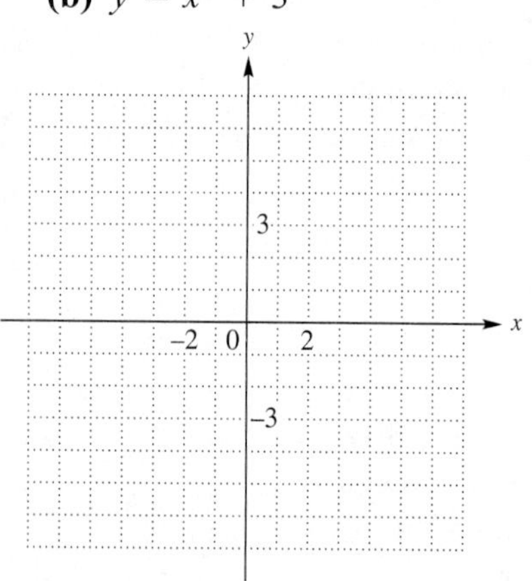

EXAMPLE 2 Graphing a Parabola

Graph $y = -x^2 - 1$.

Find several ordered pairs. To begin, check for intercepts. Let $x = 0$ to find the y-intercept.

$$y = -x^2 - 1 = -0^2 - 1 = -1,$$

giving the ordered pair $(0, -1)$. Let $y = 0$ to find the x-intercepts.

$$y = -x^2 - 1$$
$$0 = -x^2 - 1$$
$$x^2 = -1$$

This equation has no real number solution, so there are no x-intercepts. Choose additional x-values near $x = 0$ to find other ordered pairs.

Work Problem 3 at the Side.

The ordered pair $(0, -1)$ and the ordered pairs from Problem 3 at the side are listed in the table shown with Figure 2. Plot these points and connect them with a smooth curve as shown in Figure 2. The vertex of this parabola is $(0, -1)$. The graph opens downward because x^2 has a negative coefficient, so the vertex is the *highest* point of the graph.

x	y
-2	-5
-1	-2
0	-1
1	-2
2	-5

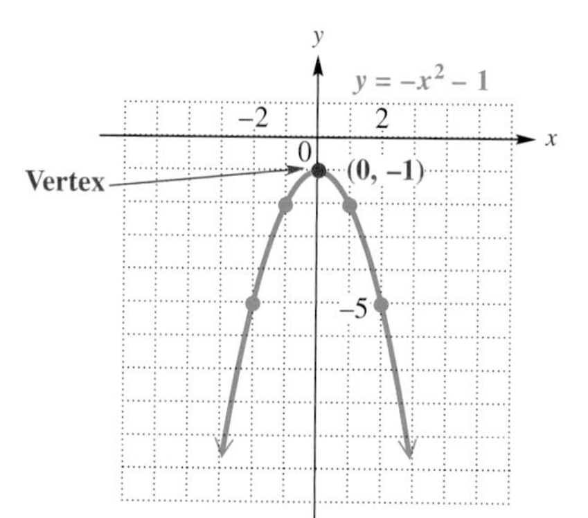

Figure 2

Work Problem 4 at the Side.

OBJECTIVE 2 Find the vertex of a parabola. The vertex is the most important point to locate when you are graphing a quadratic equation. The next example shows how to find the vertex in a more general case.

EXAMPLE 3 Finding the Vertex to Graph a Parabola

Graph $y = x^2 - 2x - 3$.

We want to find the vertex of the graph. ***If a parabola has two x-intercepts, the vertex is exactly halfway between them.*** Therefore, we begin by finding the x-intercepts. We let $y = 0$ in the equation, and solve for x.

$$0 = x^2 - 2x - 3$$
$$0 = (x + 1)(x - 3) \quad \text{Factor.}$$
$$x + 1 = 0 \quad \text{or} \quad x - 3 = 0 \quad \text{Zero-factor property}$$
$$x = -1 \quad \text{or} \quad x = 3$$

There are two x-intercepts, $(-1, 0)$ and $(3, 0)$.

Continued on Next Page

ANSWERS

3. $(-2, -5), (-1, -2), (1, -2), (2, -5)$

4. (a)

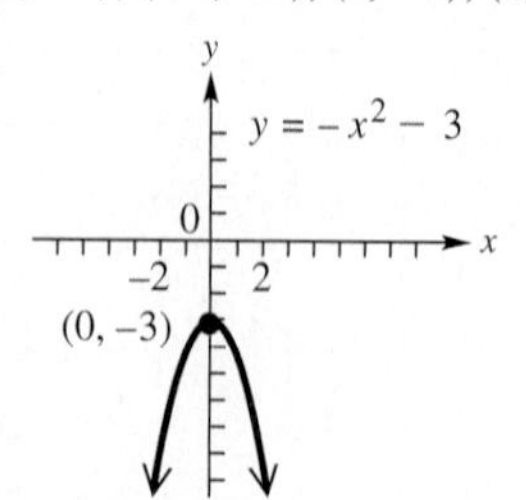

(b)

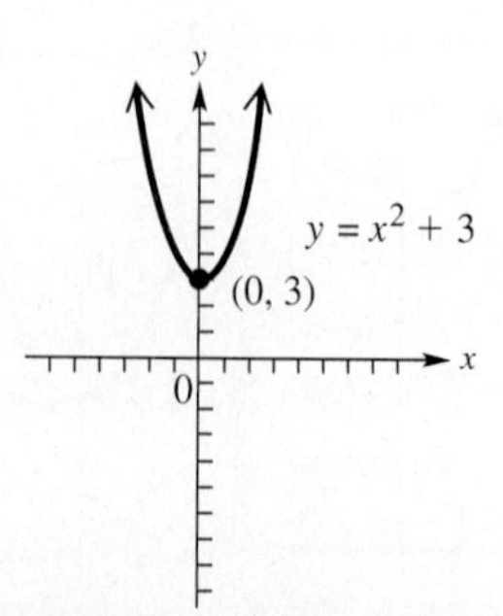

As previously mentioned, the x-value of the vertex is halfway between the x-values of the two x-intercepts. Thus, it is $\frac{1}{2}$ their sum.

$$x = \frac{1}{2}(-1 + 3) = \frac{1}{2}(2) = 1$$

We find the corresponding y-value by substituting 1 for x in the equation.

$$y = 1^2 - 2(1) - 3 = -4$$

The vertex is $(1, -4)$. Here, the axis is the line $x = 1$.

Now we find the y-intercept.

$$y = 0^2 - 2(0) - 3 \qquad \text{Let } x = 0.$$
$$y = -3$$

The y-intercept is $(0, -3)$.

We plot the three intercepts and the vertex, and find additional ordered pairs as needed. For example, if $x = 2$, then

$$y = 2^2 - 2(2) - 3 = -3,$$

leading to the ordered pair $(2, -3)$. A table with the ordered pairs we have found is shown with the graph in Figure 3.

x	y	
-2	5	
-1	0	← x-intercept
0	-3	← y-intercept
1	-4	← Vertex
2	-3	
3	0	← x-intercept
4	5	

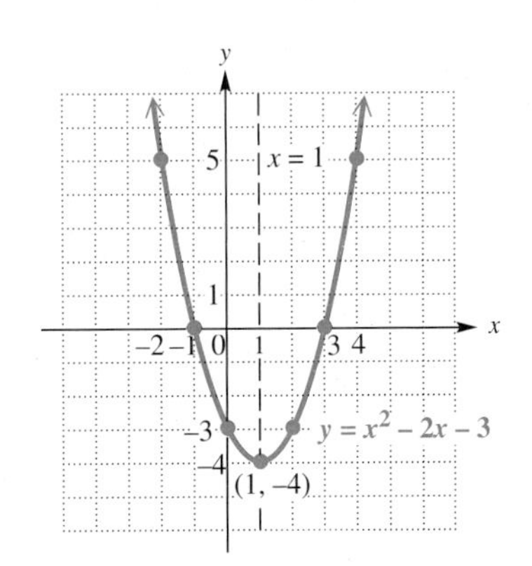

Figure 3

Work Problem 5 at the Side.

We can generalize from Example 3. The x-coordinates of the x-intercepts for $y = ax^2 + bx + c$, by the quadratic formula, are

$$x = \frac{-b + \sqrt{b^2 - 4ac}}{2a} \quad \text{and} \quad x = \frac{-b - \sqrt{b^2 - 4ac}}{2a}.$$

Thus, the x-value of the vertex is

$$x = \frac{1}{2}\left(\frac{-b + \sqrt{b^2 - 4ac}}{2a} + \frac{-b - \sqrt{b^2 - 4ac}}{2a}\right)$$

$$x = \frac{1}{2}\left(\frac{-b + \sqrt{b^2 - 4ac} - b - \sqrt{b^2 - 4ac}}{2a}\right)$$

$$x = \frac{1}{2}\left(\frac{-2b}{2a}\right)$$

$$x = -\frac{b}{2a}.$$

5 Graph $y = x^2 + 2x - 8$

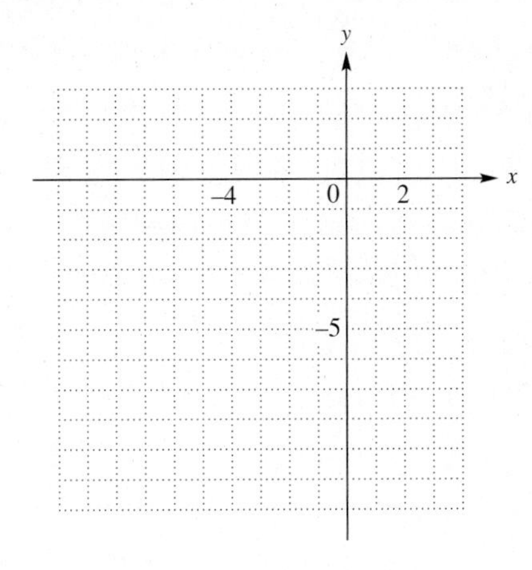

ANSWERS

5.

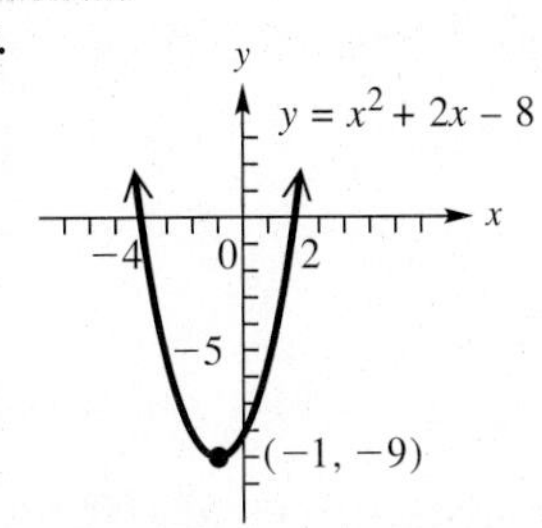

6 Complete the following ordered pairs for $y = x^2 - 4x + 1$.

(5,), (1,), (4,), (3,), (−1,)

For the equation in Example 3, $y = x^2 - 2x - 3$, $a = 1$, and $b = -2$. Thus, the x-value of the vertex is

$$x = -\frac{b}{2a} = -\frac{-2}{2(1)} = 1,$$

which is the same x-value for the vertex we found in Example 3. (The x-value of the vertex is $x = -\frac{b}{2a}$, even if the graph has no x-intercepts.) A procedure for graphing quadratic equations follows.

Graphing the Parabola $y = ax^2 + bx + c$

Step 1 **Find the vertex.** Let $x = -\frac{b}{2a}$, and find the corresponding y-value by substituting for x in the equation.

Step 2 **Find the y-intercept.**

Step 3 **Find the x-intercepts** (if they exist).

Step 4 **Plot** the intercepts and the vertex.

Step 5 **Find and plot additional ordered pairs** near the vertex and intercepts as needed, using the symmetry about the axis of the parabola.

EXAMPLE 4 Graphing a Parabola

Graph $y = x^2 - 4x + 1$.

The x-value of the vertex is

$$x = -\frac{b}{2a} = -\frac{-4}{2(1)} = 2.$$

The y-value of the vertex is

$$y = 2^2 - 4(2) + 1 = \mathbf{-3},$$

so the vertex is $(2, -3)$. The axis is the line $x = 2$. Now find the intercepts. Let $x = 0$ in $y = x^2 - 4x + 1$ to get the y-intercept $(0, 1)$. Let $y = 0$ to get the x-intercepts. If $y = 0$, the equation is $0 = x^2 - 4x + 1$, which cannot be solved by factoring. Use the quadratic formula to solve for x.

$$x = \frac{4 \pm \sqrt{16 - 4}}{2} \qquad \text{Let } a = 1, b = -4, c = 1.$$

$$x = \frac{4 \pm \sqrt{12}}{2}$$

$$x = \frac{4 \pm 2\sqrt{3}}{2} \qquad \sqrt{12} = \sqrt{4} \cdot \sqrt{3} = 2\sqrt{3}$$

$$x = \frac{2(2 \pm \sqrt{3})}{2} = 2 \pm \sqrt{3} \qquad \text{Factor; lowest terms}$$

Use a calculator to find that the x-intercepts are $(3.7, 0)$ and $(.3, 0)$, with x-values approximated to the nearest tenth.

Work Problem 6 at the Side.

Continued on Next Page

ANSWERS

6. (5, 6), (1, −2), (4, 1), (3, −2), (−1, 6)

Plot the intercepts, vertex, and the points found in Problem 6. Connect these points with a smooth curve. The graph is shown in Figure 4.

x	y
-1	6
0	1
$2 - \sqrt{3} \approx .3$	0
1	-2
2	-3
3	-2
$2 + \sqrt{3} \approx 3.7$	0
4	1
5	6

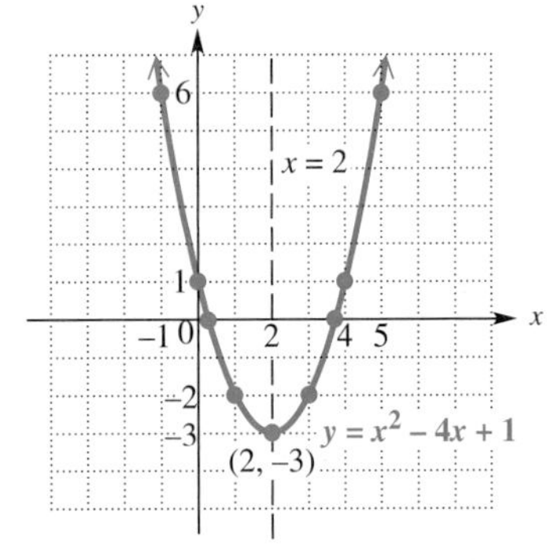

Figure 4

Work Problem 7 at the Side.

OBJECTIVE 3 Solve an application involving a parabola. Parabolic shapes are found all around us. Satellite dishes that receive television signals are becoming more popular each year. Radio telescopes use parabolic reflectors to track incoming signals. The final example discusses how to describe a cross section of a parabolic dish using an equation.

EXAMPLE 5 Finding the Equation of a Parabolic Satellite Dish

The Parkes radio telescope has a parabolic dish shape with a diameter of 210 ft and a depth of 32 ft. (*Source:* Mar, J. and Liebowitz, H., *Structure Technology for Large Radio and Radar Telescope Systems,* The MIT Press, 1969.) Figure 5(a) shows a diagram of such a dish. Figure 5(b) shows how a cross section of the dish can be modeled by a graph, with the vertex of the parabola at the origin of a coordinate system. Find the equation of this graph.

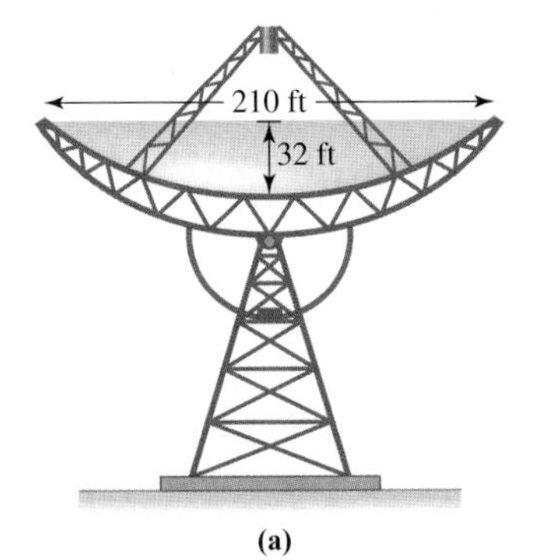

(a)

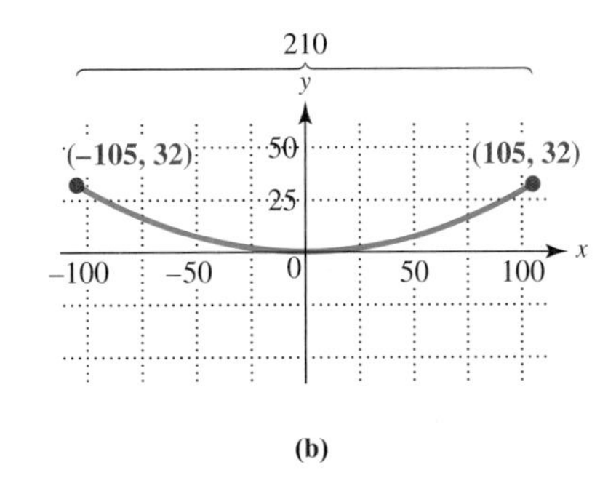

(b)

Figure 5

Continued on Next Page

7 Graph each parabola. Identify the vertex and y-intercept, and give the coordinates of the x-intercepts to the nearest tenth.

(a) $y = x^2 - 3x - 3$

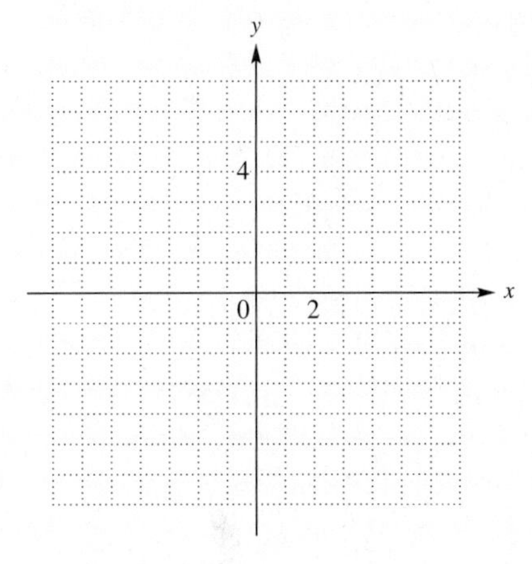

(b) $y = -x^2 + 2x + 4$

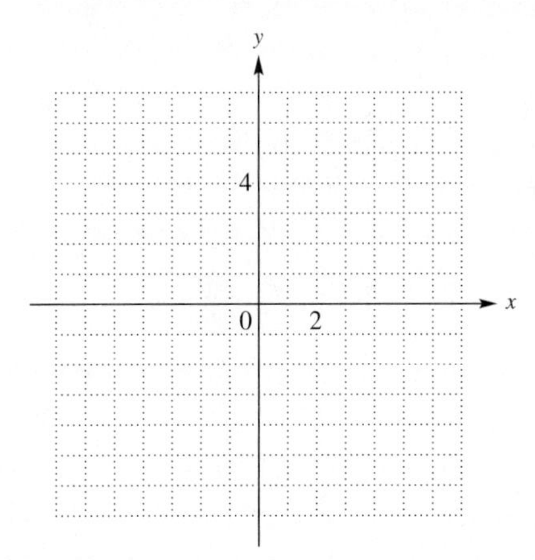

ANSWERS

7. (a)

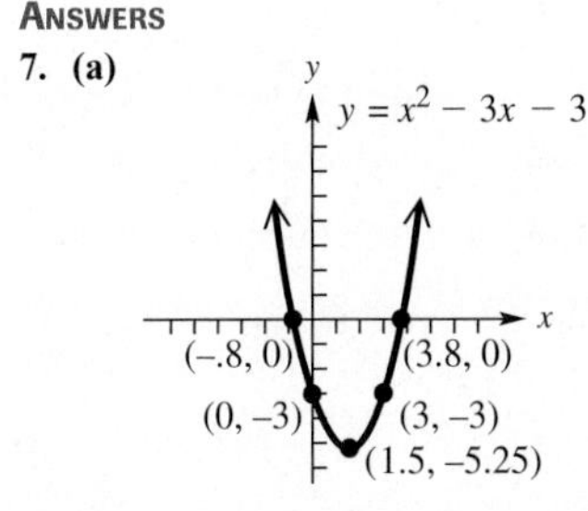

(b)

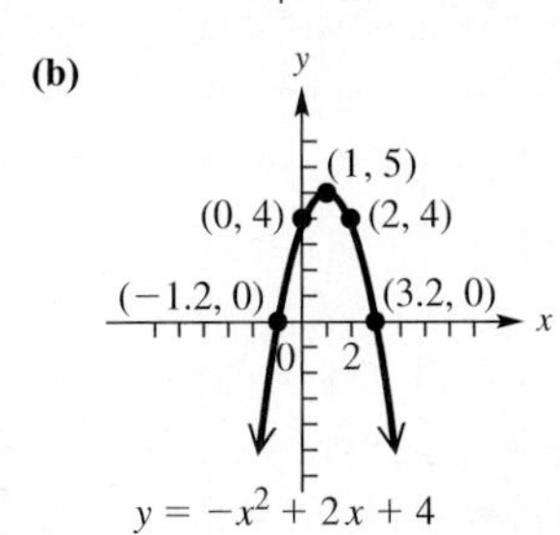

8 Suppose that a radio telescope has a parabolic dish shape with a diameter of 350 ft and a depth of 48 ft. Find the equation of the graph of a cross section.

Because the vertex is at the origin, the equation will be of the form $y = ax^2$. (See Example 1 and Problem 1, for example.) As shown in Figure 5(b), one point on the graph has coordinates (105, 32). Letting $x = 105$ and $y = 32$, we can solve for a.

$$y = ax^2 \quad \text{General equation}$$
$$32 = a(105)^2 \quad \text{Substitute for } x \text{ and } y.$$
$$32 = 11{,}025a \quad 105^2 = 11{,}025$$
$$a = \frac{32}{11{,}025} \quad \text{Divide by } 11{,}025.$$

Thus the equation is $y = \frac{32}{11{,}025}x^2$.

Work Problem 8 at the Side.

ANSWERS

8. $y = \frac{48}{30{,}625}x^2$

9.4 Exercises

FOR EXTRA HELP

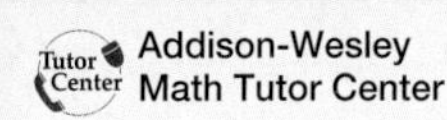

MathXL

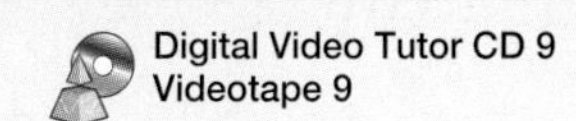

MyMathLab

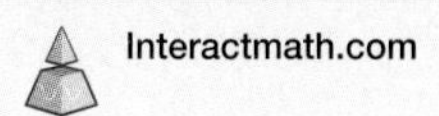

1. In your own words, explain what is meant by the vertex of a parabola.

2. In your own words, explain what is meant by the line of symmetry of a parabola that opens upward or downward.

Graph each equation. Give the coordinates of the vertex in each case. See Examples 1–4.

3. $y = 2x^2$

4. $y = 3x^2$

5. $y = x^2 - 4$

6. $y = x^2 - 6$

7. $y = -x^2 + 2$

8. $y = -x^2 + 4$

9. $y = (x + 3)^2$

10. $y = (x - 4)^2$

11. $y = x^2 + 2x + 3$

12. $y = x^2 - 4x + 3$

13. $y = -x^2 + 6x - 5$

14. $y = -x^2 - 4x - 3$

15. Based on your work in Exercises 3–14, what seems to be the direction in which the parabola $y = ax^2 + bx + c$ opens if $a > 0$? if $a < 0$?

16. See Examples 1–3. How many real solutions does a quadratic equation have if its corresponding graph has

(a) no x-intercepts

(b) one x-intercept

(c) two x-intercepts?

Solve each problem. See Example 5.

17. The U.S. Naval Research Laboratory designed a giant radio telescope that had a diameter of 300 ft and maximum depth of 44 ft. The graph on the right below describes a cross section of this telescope. Find the equation of this parabola. (*Source:* Mar, J. and Liebowitz, H., *Structure Technology for Large Radio and Radar Telescope Systems*, The MIT Press, 1969.)

18. Suppose the telescope in Exercise 17 had a diameter of 400 ft and maximum depth of 50 ft. Find the equation of this parabola.

y
(–150, 44)
(150, 44)
50
–150 –100 –50 0 50 100 150
x

9.5 Introduction to Functions

If gasoline costs \$1.85 per gal and you buy 1 gal, then you must pay $\$1.85(1) = \$1.85$. If you buy 2 gal, your cost is $\$1.85(2) = \$3.70$; for 3 gal, your cost is $\$1.85(3) = \$5.55$, and so on. Generalizing, if x represents the number of gallons, then the cost is $\$1.85x$. If we let y represent the cost, then the equation $y = 1.85x$ *relates* the number of gallons, x, to the cost in dollars, y. The ordered pairs (x, y) that satisfy this equation form a *relation.*

OBJECTIVES

1. Understand the definition of a relation.
2. Understand the definition of a function.
3. Decide whether an equation defines a function.
4. Use function notation.
5. Apply the function concept in an application.

OBJECTIVE 1 Understand the definition of a relation. In an ordered pair (x, y), x and y are called the **components** of the ordered pair. Any set of ordered pairs is called a **relation.*** The set of all first components in the ordered pairs of a relation is the **domain** of the relation, and the set of all second components in the ordered pairs is the **range** of the relation.

EXAMPLE 1 Using Ordered Pairs to Define Relations

(a) The relation $\{(0, 1), (2, 5), (3, 8), (4, 2)\}$ has domain $\{0, 2, 3, 4\}$ and range $\{1, 2, 5, 8\}$. The correspondence between the elements of the domain and the elements of the range is shown in Figure 6.

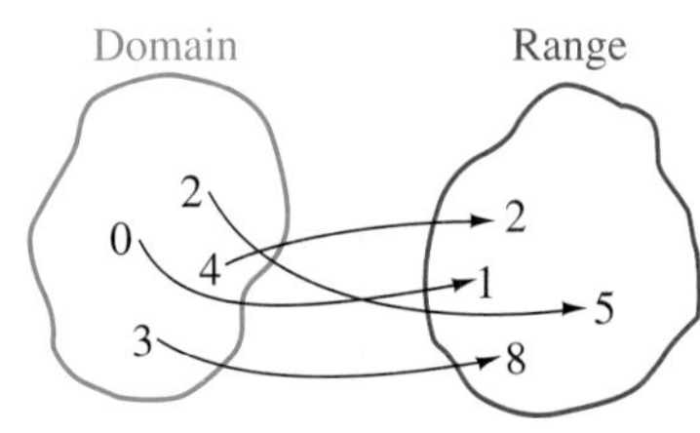

Figure 6

(b) The relation $\{(3, 5), (3, 6), (3, 7), (3, 8)\}$ has domain $\{3\}$ and range $\{5, 6, 7, 8\}$.

Work Problem 1 at the Side.

1 Give the domain and the range of each relation.

(a) $\{(5, 10), (15, 20), (25, 30), (35, 40)\}$

(b) $\{(1, 4), (2, 4), (3, 4)\}$

OBJECTIVE 2 Understand the definition of a function. We now investigate an important type of relation, called a *function.*

Function

A **function** is a set of ordered pairs in which each first component corresponds to exactly one second component.

By this definition, the relation in Example 1(a) is a function, but the relation in Example 1(b) is not a function. Notice, however, that if the components of the ordered pairs in Example 1(b) were interchanged, giving the relation

$$\{(5, 3), (6, 3), (7, 3), (8, 3)\},$$

then the relation *would* be a function; in this case, each domain element (first component) corresponds to *exactly one* range element (second component).

* It is standard notation to use set braces around the ordered pairs that form a relation. You may want to refer to **Appendix B,** which covers set notation.

ANSWERS

1. **(a)** domain: $\{5, 15, 25, 35\}$; range: $\{10, 20, 30, 40\}$
   (b) domain: $\{1, 2, 3\}$; range: $\{4\}$

2 Decide whether each relation is a function.

(a) $\{(-2, 8), (-1, 1), (0, 0), (1, 1), (2, 8)\}$

(b) $\{(5, 2), (5, 1), (5, 0)\}$

ANSWERS
2. **(a)** function **(b)** not a function

EXAMPLE 2 Determining Whether Relations Are Functions

Determine whether each relation is a function.

(a) $\{(-2, 4), (-1, 1), (0, 0), (1, 1), (2, 4)\}$

Notice that each first component appears once and only once. Because of this, the relation is a function.

(b) $\{(9, 3), (9, -3), (4, 2)\}$

The first component 9 appears in two ordered pairs, and corresponds to two different second components. Therefore, this relation is not a function.

Work Problem 2 at the Side.

The simple relations given in Examples 1 and 2 were defined by listing the ordered pairs or by showing the correspondence with a figure. Most useful functions have an infinite number of ordered pairs and are usually defined with equations that tell how to get the second components, given the first. We have been using equations with x and y as the variables, where x represents the first component (input) and y the second component (output) in the ordered pairs.

Here are some everyday examples of functions.

1. The **cost** y in dollars charged by an express mail company is a function of the **weight in pounds** x determined by the equation $y = 1.5(x - 1) + 9$.
2. In one state, the sales tax is 6% of the price of an item. The **tax** y on a particular item is a function of the **price** x, because $y = .06x$.
3. The **distance** d traveled by a car moving at a constant speed of 45 mph is a function of the **time** t. Thus, $d = 45t$.

The function concept can be illustrated by an input-output "machine," as seen in Figure 7. It shows how the express mail company equation $y = 1.5(x - 1) + 9$ provides an output (the cost, represented by y) for a given input (the weight in pounds, given by x).

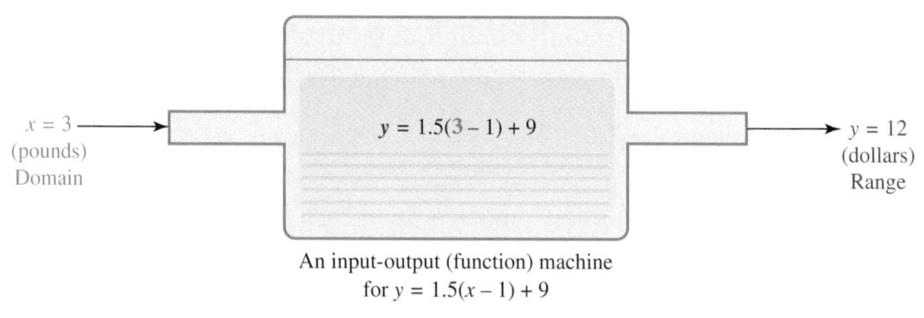

An input-output (function) machine for $y = 1.5(x - 1) + 9$

Figure 7

OBJECTIVE 3 Decide whether an equation defines a function. Given the graph of an equation, the definition of a function can be used to decide whether or not the graph represents a function. By the definition of a function, each x-value must lead to exactly one y-value. In Figure 8(a) on the next page, the indicated x-value leads to two y-values, so this graph is not the graph of a function. A vertical line can be drawn that intersects this graph in more than one point.

(a)

(b)

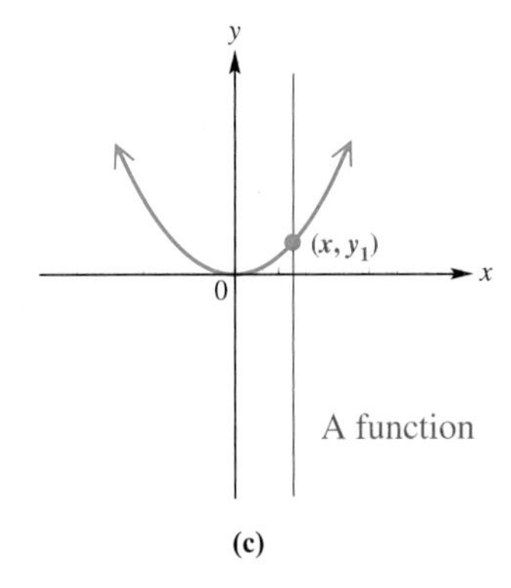

(c)

Figure 8

On the other hand, in Figures 8(b) and 8(c) any vertical line will intersect each graph in no more than one point. Because of this, the graphs in Figures 8(b) and 8(c) are graphs of functions. This idea leads to the **vertical line test** for a function.

Vertical Line Test

If a vertical line intersects a graph in more than one point, the graph is not the graph of a function.

As Figure 8(b) suggests, any nonvertical line is the graph of a function. For this reason, any linear equation of the form $y = mx + b$ defines a function. (Recall that a vertical line has undefined slope.) Also, any vertical parabola, as in Figure 8(c), is the graph of a function, so any quadratic equation of the form $y = ax^2 + bx + c$ $(a \neq 0)$ defines a function.

EXAMPLE 3 Deciding Whether Relations Define Functions

Decide whether each relation graphed or defined is a function.

(a)

Because there are two ordered pairs with first component -4, this is not the graph of a function.

Continued on Next Page

3 Decide whether each relation graphed or defined is a function.

(a)

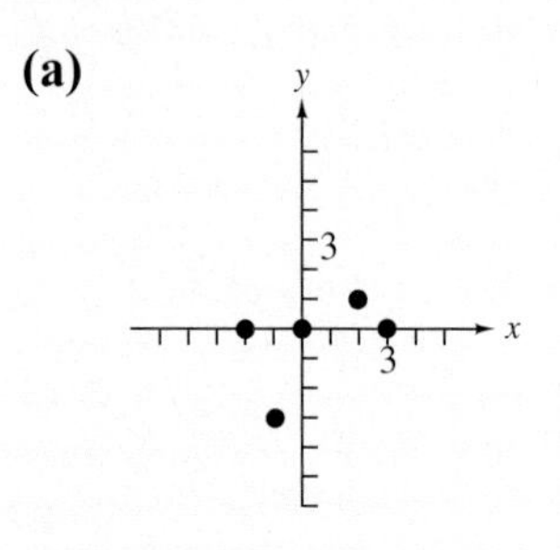

(b)

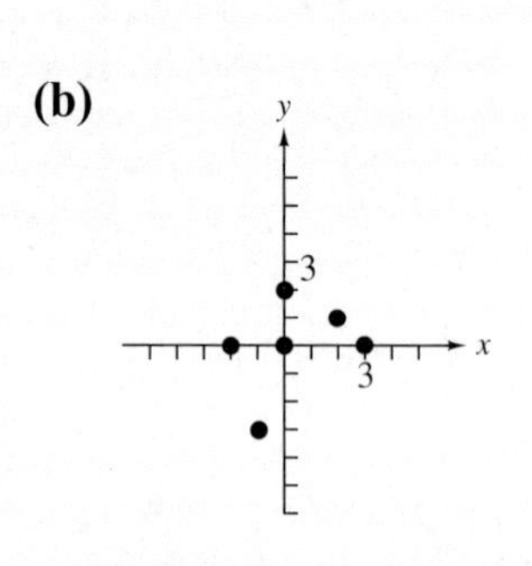

(c)

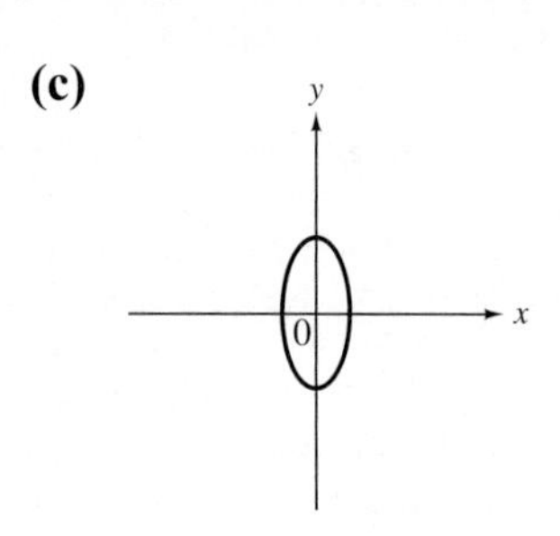

(d)

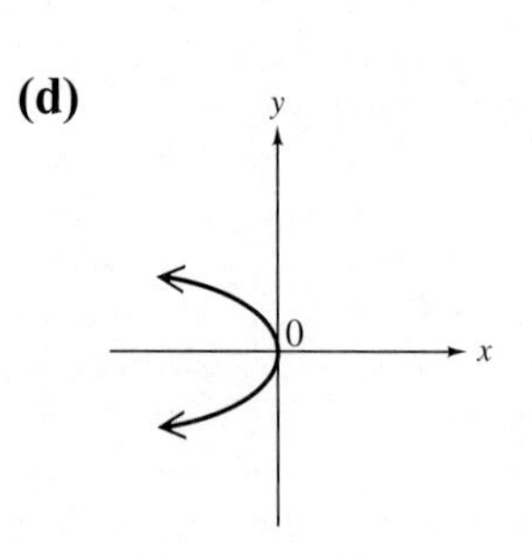

(e) $y = 3$

ANSWERS

3. **(a)** function **(b)** not a function **(c)** not a function **(d)** not a function **(e)** function

---

(b)

Every first component is paired with one and only one second component, and so no vertical line intersects the graph in more than one point. Therefore, this is the graph of a function.

(c) $y = 2x - 9$

This linear equation is in the form $y = mx + b$. Since the graph of this equation is a line that is not vertical, the equation defines a function.

(d)

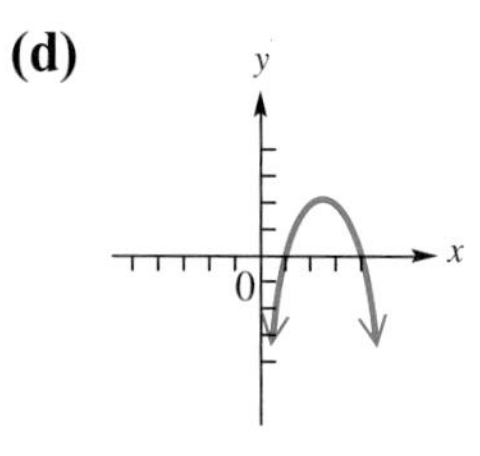

Use the vertical line test. Any vertical line will cross the graph of a vertical parabola just once, so this is the graph of a function.

(e)

The vertical line test shows that this graph is not the graph of a function; a vertical line could cross the graph twice.

(f) $x = 4$

The graph of $x = 4$ is a vertical line, so the equation does not define a function.

Work Problem 3 at the Side.

OBJECTIVE 4 Use function notation. The letters f, g, and h are commonly used to name functions. For example, the function $y = 3x + 5$ may be written

$$f(x) = 3x + 5,$$

where $f(x)$ is read "f of x." The notation $f(x)$ is another way of writing y in a function. For the function defined by $f(x) = 3x + 5$, if $x = 7$ then

$$f(7) = 3 \cdot 7 + 5 \qquad \text{Let } x = 7.$$
$$= 21 + 5 = 26.$$

Read this result, $f(7) = 26$, as "f of 7 equals 26." The notation $f(7)$ means the value of y when x is 7. The statement $f(7) = 26$ says that the value of y is 26 when x is 7. It also indicates that the point (7, 26) lies on the graph of f.

To find $f(-3)$, substitute -3 for x.

$$f(-3) = 3(-3) + 5 \qquad \text{Let } x = -3.$$
$$= -9 + 5 = -4$$

CAUTION

The notation $f(x)$ does *not* mean f times x. ***The symbol f(x) represents the function value of x. It also represents the y-value that corresponds to x.***

Function Notation

In the notation $f(x)$,

	f	is the name of the function,
	x	is the domain value,
and	$f(x)$	is the range value y for the domain value x.

EXAMPLE 4 Using Function Notation

For the function defined by $f(x) = x^2 - 3$, find the following.

(a) $f(2)$

Substitute 2 for x.

$$f(x) = x^2 - 3$$
$$f(2) = 2^2 - 3 \quad \text{Let } x = 2.$$
$$= 4 - 3 = 1$$

(b) $f(0) = 0^2 - 3 = 0 - 3 = -3$

(c) $f(-3) = (-3)^2 - 3 = 9 - 3 = 6$

Work Problem 4 at the Side.

OBJECTIVE 5 Apply the function concept in an application. Because a function assigns to each element in its domain exactly one element in its range, the function concept is used in real-data applications where two quantities are related. Our final example discusses such an application, using a table of values.

EXAMPLE 5 Applying the Function Concept to Population

Asian-American populations (in millions) are shown in the table.

ASIAN-AMERICAN POPULATION

Year	Population (in millions)
1996	9.7
1998	10.5
2000	11.2
2002	12

Source: U.S. Bureau of the Census.

(a) Use the table to write a set of ordered pairs that defines a function f.

If we choose the years as the domain elements and the populations as the range elements, the information in the table can be written as a set of four ordered pairs. In set notation, the function f is

$$f = \{(1996, 9.7), (1998, 10.5), (2000, 11.2), (2002, 12)\}.$$

(b) What is the domain of f? What is the range?

The domain is the set of years, or x-values:

$$\{1996, 1998, 2000, 2002\}.$$

The range is the set of populations, in millions, or y-values:

$$\{9.7, 10.5, 11.2, 12\}.$$

Continued on Next Page

4 For $f(x) = 6x - 2$, find each function value.

(a) $f(-1)$

(b) $f(0)$

(c) $f(1)$

ANSWERS

4. **(a)** -8 **(b)** -2 **(c)** 4

5 The numbers of U.S. children (in millions) educated at home for selected years are given in the table.

School Year	Number of Children
1997	1.1
1998	1.2
1999	1.3
2000	1.5
2001	1.7

Source: National Home Education Research Institute, Salem, OR.

(a) Write a set of ordered pairs that defines a function f for the data.

(b) Give the domain and range of f.

(c) Find $f(1999)$.

(d) In what year did the number of children equal 1.5 million?

ANSWERS

5. **(a)** $f = \{(1997, 1.1), (1998, 1.2), (1999, 1.3), (2000, 1.5), (2001, 1.7)\}$
(b) domain: $\{1997, 1998, 1999, 2000, 2001\}$; range: $\{1.1, 1.2, 1.3, 1.5, 1.7\}$
(c) 1.3 million
(d) 2000

(c) Find $f(1996)$ and $f(2000)$.

We repeat the table and the set of ordered pairs from part (a),

ASIAN-AMERICAN POPULATION

Year	Population (in millions)
1996	9.7
1998	10.5
2000	11.2
2002	12

Source: U.S. Bureau of the Census.

$$f = \{(1996, 9.7), (1998, 10.5), (2000, 11.2), (2002, 12)\}$$

Thus, $f(1996) = 9.7$ million and $f(2000) = 11.2$ million.

(d) For what x-value does $f(x)$ equal 12 million? 10.5 million?

We use the table or the ordered pairs found in part (a) to determine $f(2002) = 12$ million and $f(1998) = 10.5$ million.

Work Problem 5 at the Side.

9.5 Exercises

FOR EXTRA HELP

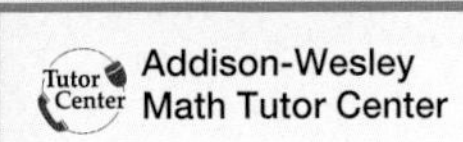

Addison-Wesley Math Tutor Center

MathXL

Digital Video Tutor CD 9
Videotape 9

Student's Solutions Manual

MyMathLab

Interactmath.com

Complete the following table for the function defined by $f(x) = x + 2$.

	x	$x + 2$	$f(x)$	(x, y)
	0	2	2	(0, 2)
1.	1			
2.	2			
3.	3			
4.	4			

5. Describe the graph of function f in Exercises 1–4 if the domain is $\{0, 1, 2, 3, 4\}$.

6. Describe the graph of function f in Exercises 1–4 if the domain is the set of all real numbers.

Determine whether each relation is a function. Give the domain and the range in Exercises 7–12. See Examples 1–3.

7. $\{(-4, 3), (-2, 1), (0, 5), (-2, -8)\}$

8. $\{(3, 7), (1, 4), (0, -2), (-1, -1), (-2, 5)\}$

9.

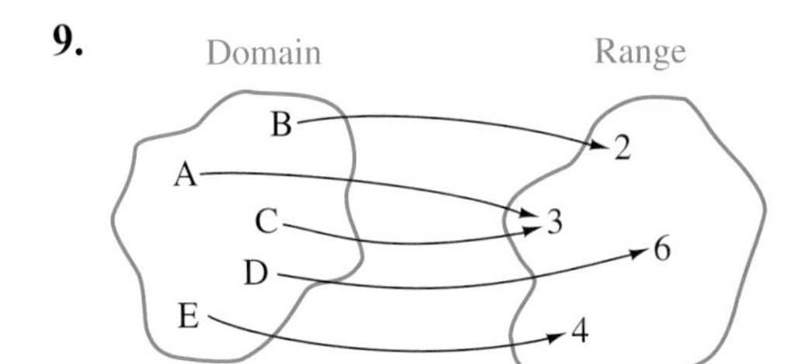

10.

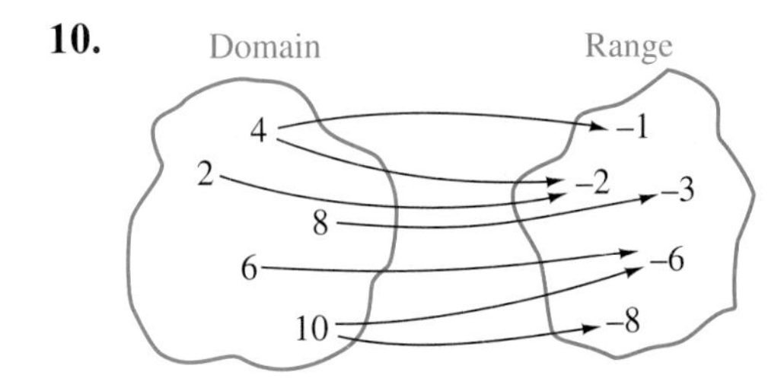

11.

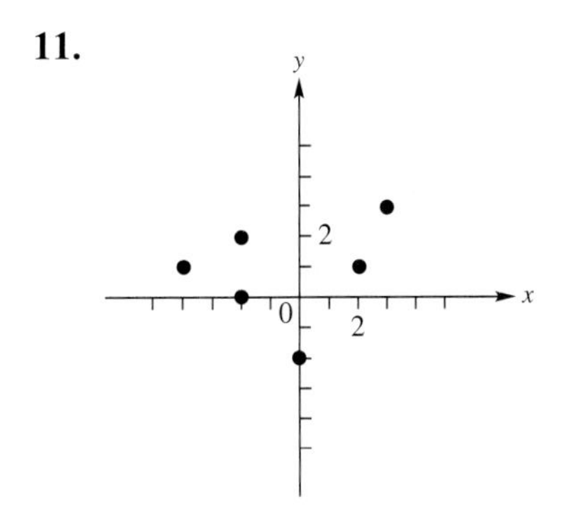

12.

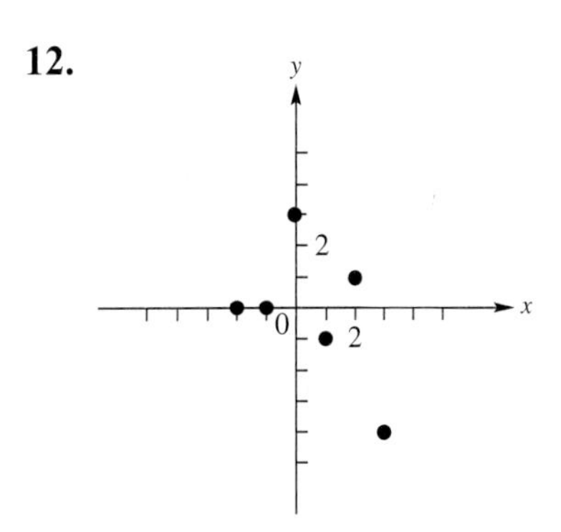

13.

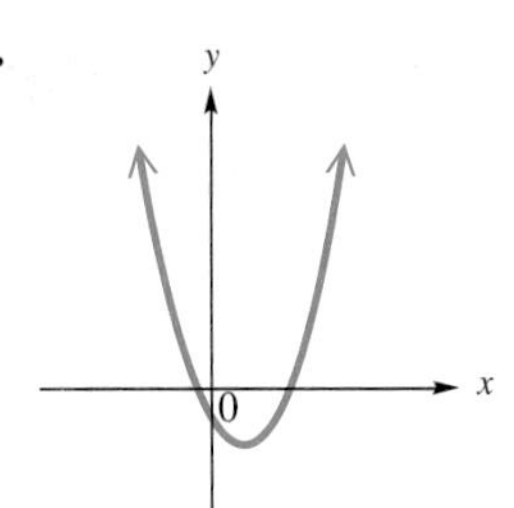

14.

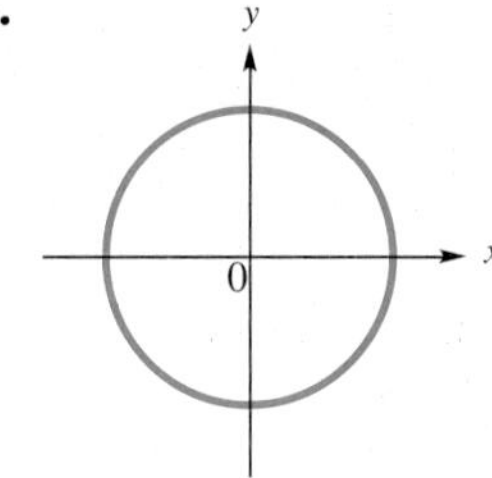

15.

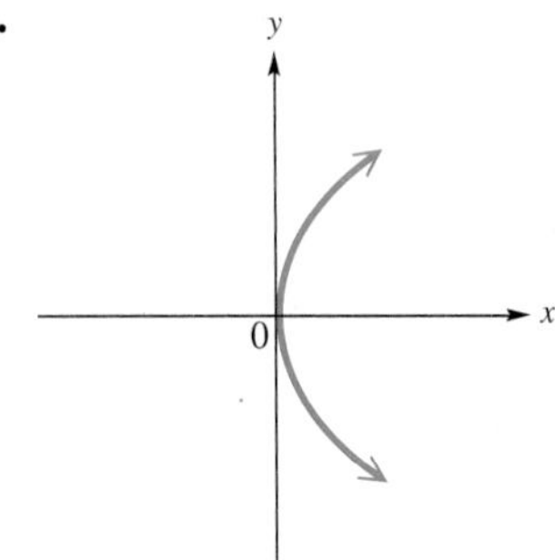

16.

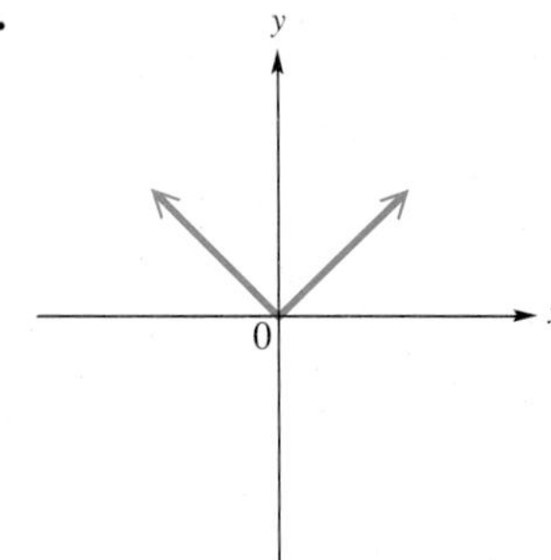

Decide whether each equation defines y as a function of x. (Remember that to be a function, every value of x must give one and only one value of y.) See Example 3.

17. $y = 5x + 3$ **18.** $y = -7x + 12$ **19.** $x = |y|$ **20.** $x = y^2$

RELATING CONCEPTS (EXERCISES 21–24) For Individual or Group Work

*A function defined by $f(x) = 3x - 4$, called a **linear function** because its graph is a straight line, can be graphed by replacing $f(x)$ with y and then using the methods described earlier. Let us assume that some function is written in the form $f(x) = mx + b$, for particular values of m and b. **Work Exercises 21–24 in order.***

21. If $f(2) = 4$, name the coordinates of one point on the line.

22. If $f(-1) = -4$, name the coordinates of another point on the line.

23. Use the results of Exercises 21 and 22 to find the slope of the line.

24. Use the slope-intercept form of the equation of a line to write the function in the form $f(x) = mx + b$.

*For each function f, find **(a)** $f(2)$, **(b)** $f(0)$, and **(c)** $f(-3)$. See Example 4.*

25. $f(x) = 4x + 3$

26. $f(x) = -3x + 5$

27. $f(x) = x^2 - x + 2$

28. $f(x) = x^3 + x$

29. $f(x) = |x|$

30. $f(x) = |x + 7|$

The number of U.S. foreign-born residents has grown by more than 43% since 1990. The graph shows the number of such residents (in millions) for selected years. Use the information in the graph for Exercises 31–35. See Example 5.

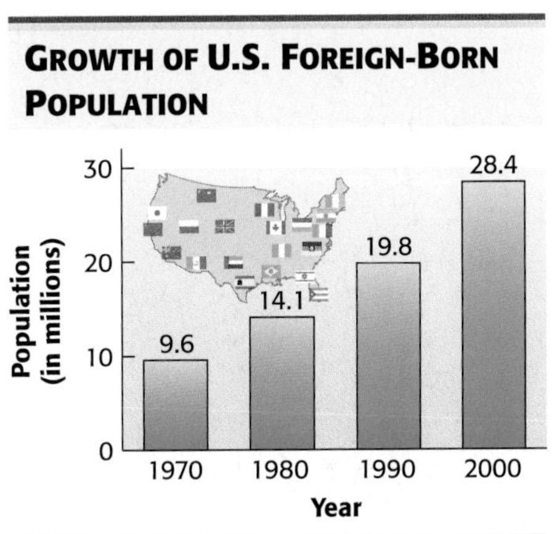

Source: U.S. Bureau of the Census.

31. Write the information in the graph as a set of ordered pairs. Does this set define a function?

32. Suppose that g is the name given to this relation. Give the domain and range of g.

33. Find $g(1980)$ and $g(1990)$.

34. For what value of x does $g(x) = 28.4$ (million)?

35. Suppose $g(2002) = 30.3$ (million). What does this tell you in the context of the application?

RELATING CONCEPTS (EXERCISES 36–40) For Individual or Group Work

The data give the percent of U.S. active-duty female military personnel during selected years from 1987 to 2003. Use the information in the table to ***work Exercises 36–40 in order.***

ACTIVE-DUTY FEMALE MILITARY PERSONNEL

Year	Percent
1987	10.2
1993	11.6
2000	14.4
2003	15.1

Source: U.S. Department of Defense.

36. Plot the ordered pairs (year, percent) from the table. Do the points suggest that a linear function would give a reasonable approximation of the data?

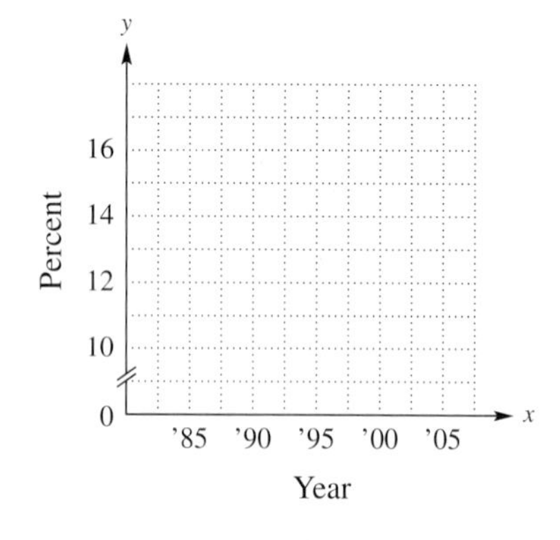

37. Use the first and last data pairs in the table to write an equation relating the data.

38. Use your equation from Exercise 37 to approximate the percent of active-duty female military personnel in 1993 and 2000. Round to the nearest tenth of a percent.

39. Use the second and third data pairs to write an equation relating the data.

40. Use your equation from Exercise 39 to approximate the number of active-duty female military personnel in 1987 and 2003. Which of the results in Exercise 38 and this exercise give better approximations?

Chapter 9
SUMMARY

KEY TERMS

9.4

parabola The graph of the quadratic equation $y = ax^2 + bx + c$ is called a parabola.

vertex The vertex of a parabola that opens upward or downward is the lowest or highest point on the graph.

axis The axis of a parabola that opens upward or downward is a vertical line through the vertex.

line of symmetry If a graph is folded on its line of symmetry, the two sides coincide.

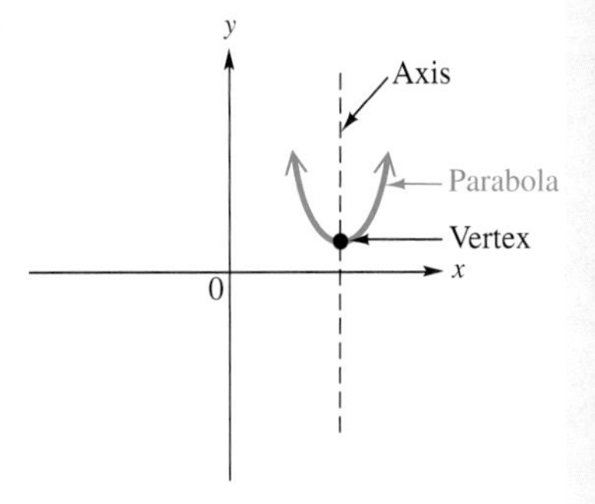

9.5

components In an ordered pair (x, y), x and y are the components.

relation Any set of ordered pairs is called a relation.

domain The set of all first components in the ordered pairs of a relation is the domain of the relation.

range The set of all second components in the ordered pairs of a relation is the range of the relation.

function A function is a set of ordered pairs in which each first component corresponds to exactly one second component.

NEW SYMBOLS

$\pm$ positive or negative

$f(x)$ function f of x

TEST YOUR WORD POWER

See how well you have learned the vocabulary in this chapter. Answers, with examples, follow the Quick Review.

1. A **relation** is
   - **A.** any set of ordered pairs
   - **B.** a set of ordered pairs in which each first component corresponds to exactly one second component
   - **C.** two sets of ordered pairs that are related
   - **D.** a graph of ordered pairs.

2. The **domain** of a relation is
   - **A.** the set of all x- and y-values in the ordered pairs of the relation
   - **B.** the difference between the components in an ordered pair of the relation
   - **C.** the set of all first components in the ordered pairs of the relation
   - **D.** the set of all second components in the ordered pairs of the relation.

3. The **range** of a relation is
   - **A.** the set of all x- and y-values in the ordered pairs of the relation
   - **B.** the difference between the components in an ordered pair of the relation
   - **C.** the set of all first components in the ordered pairs of the relation
   - **D.** the set of all second components in the ordered pairs of the relation.

4. A **function** is
   - **A.** any set of ordered pairs
   - **B.** a set of ordered pairs in which each first component corresponds to exactly one second component
   - **C.** two sets of ordered pairs that are related
   - **D.** a graph of ordered pairs.

QUICK REVIEW

Concepts	*Examples*
9.1 *Solving Quadratic Equations by the Square Root Property* **Square Root Property of Equations** If k is positive, and if $a^2 = k$, then $a = \sqrt{k}$ or $a = -\sqrt{k}$.	Solve $(2x + 1)^2 = 5$. $2x + 1 = \sqrt{5}$ or $2x + 1 = -\sqrt{5}$ $2x = -1 + \sqrt{5}$ or $2x = -1 - \sqrt{5}$ $x = \dfrac{-1 + \sqrt{5}}{2}$ or $x = \dfrac{-1 - \sqrt{5}}{2}$ The solutions are $\dfrac{-1 + \sqrt{5}}{2}$ and $\dfrac{-1 - \sqrt{5}}{2}$.
9.2 *Solving Quadratic Equations by Completing the Square* **Solving a Quadratic Equation by Completing the Square** *Step 1* If the coefficient of the squared term is 1, go to Step 2. If it is not 1, divide each side of the equation by this coefficient. *Step 2* Make sure that all variable terms are on one side of the equation, and all constant terms are on the other. *Step 3* Take half the coefficient of x, square it, and add the square to each side of the equation. Factor the variable side, and combine terms on the other side. *Step 4* Use the square root property to solve the equation.	Solve $2x^2 + 4x - 1 = 0$. $x^2 + 2x - \dfrac{1}{2} = 0$ Divide by 2. $x^2 + 2x = \dfrac{1}{2}$ Add $\frac{1}{2}$. $x^2 + 2x + 1 = \dfrac{1}{2} + 1$ $\left[\frac{1}{2}(2)\right]^2 = 1$ $(x + 1)^2 = \dfrac{3}{2}$ Factor; combine like terms. $x + 1 = \sqrt{\dfrac{3}{2}}$ or $x + 1 = -\sqrt{\dfrac{3}{2}}$ $x + 1 = \dfrac{\sqrt{3} \cdot \sqrt{2}}{\sqrt{2} \cdot \sqrt{2}}$ or $x + 1 = -\dfrac{\sqrt{3} \cdot \sqrt{2}}{\sqrt{2} \cdot \sqrt{2}}$ $x + 1 = \dfrac{\sqrt{6}}{2}$ or $x + 1 = -\dfrac{\sqrt{6}}{2}$ $x = -1 + \dfrac{\sqrt{6}}{2}$ or $x = -1 - \dfrac{\sqrt{6}}{2}$ $x = \dfrac{-2 + \sqrt{6}}{2}$ or $x = \dfrac{-2 - \sqrt{6}}{2}$ The solutions are $\dfrac{-2 + \sqrt{6}}{2}$ and $\dfrac{-2 - \sqrt{6}}{2}$.

Concepts	*Examples*
9.3 *Solving Quadratic Equations by the Quadratic Formula*	
Quadratic Formula The solutions of $ax^2 + bx + c = 0$ $(a \neq 0)$ are $$x = \frac{-b \pm \sqrt{b^2 - 4ac}}{2a}.$$	Solve $3x^2 - 4x - 2 = 0$. $$x = \frac{-(-4) \pm \sqrt{(-4)^2 - 4(3)(-2)}}{2(3)}$$ $$x = \frac{4 \pm \sqrt{16 + 24}}{6}$$ $$x = \frac{4 \pm \sqrt{40}}{6} = \frac{4 \pm 2\sqrt{10}}{6}$$ $$x = \frac{2(2 \pm \sqrt{10})}{2(3)} = \frac{2 \pm \sqrt{10}}{3}$$ The solutions are $\frac{2 + \sqrt{10}}{3}$ and $\frac{2 - \sqrt{10}}{3}$.
9.4 *Graphing Quadratic Equations*	
To graph $y = ax^2 + bx + c$: *Step 1* Find the vertex: $x = -\frac{b}{2a}$; find y by substituting this value for x in the equation.	Graph $y = 2x^2 - 5x - 3$. $$x = -\frac{b}{2a} = -\frac{-5}{2(2)} = \frac{5}{4}$$ $$y = 2\left(\frac{5}{4}\right)^2 - 5\left(\frac{5}{4}\right) - 3 = 2\left(\frac{25}{16}\right) - \frac{25}{4} - 3$$ $$= \frac{25}{8} - \frac{50}{8} - \frac{24}{8} = -\frac{49}{8}$$ The vertex is $\left(\frac{5}{4}, -\frac{49}{8}\right)$.
Step 2 Find the y-intercept.	$$y = 2(0)^2 - 5(0) - 3 = -3$$ The y-intercept is $(0, -3)$.
Step 3 Find the x-intercepts (if they exist).	$$0 = 2x^2 - 5x - 3$$ $$0 = (2x + 1)(x - 3)$$ $2x + 1 = 0$ or $x - 3 = 0$ $2x = -1$ or $x = 3$ $x = -\frac{1}{2}$ The x-intercepts are $\left(-\frac{1}{2}, 0\right)$ and $(3, 0)$.
Step 4 Plot the intercepts and the vertex. *Step 5* Find and plot additional ordered pairs near the vertex and intercepts as needed.	See table and graph below.

x	y
$-\frac{1}{2}$	0
0	-3
1	-6
$\frac{5}{4}$	$-\frac{49}{8}$
2	-5
3	0

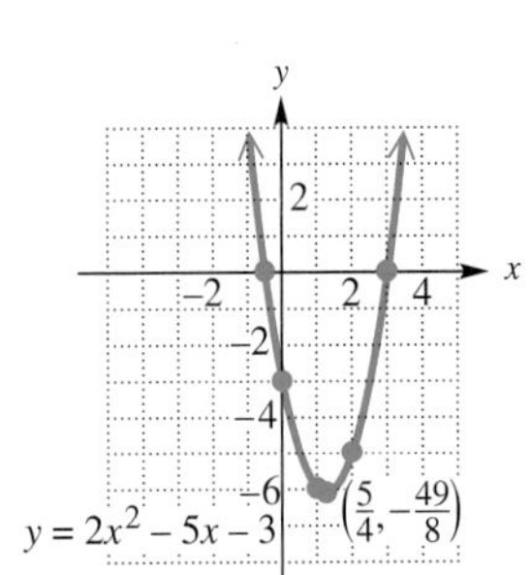

Concepts	Examples
9.5 *Introduction to Functions* **Vertical Line Test** If a vertical line intersects a graph in more than one point, the graph is not the graph of a function.	By the vertical line test, the graph shown is not the graph of a function. 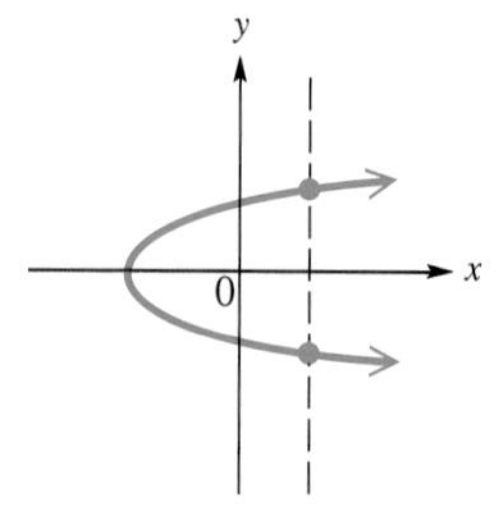
	The function $\{(10, 5), (20, 15), (30, 25)\}$ has domain $\{10, 20, 30\}$ and range $\{5, 15, 25\}$.
	If $f(x) = 2x + 7$, then $f(3) = 2(3) + 7 = 13$.

Answers to Test Your Word Power

1. A; *Example:* $\{(0, 2), (2, 4), (3, 6), (-1, 3)\}$
2. C; *Example:* The domain in the relation given in Problem 1 is the set of x-values, that is, $\{0, 2, 3, -1\}$.
3. D; *Example:* The range of the relation given in Problem 1 is the set of y-values, that is, $\{2, 4, 6, 3\}$.
4. B; *Example:* The relation given in Problem 1 is a function since each x-value corresponds to exactly one y-value.

Chapter 9
REVIEW EXERCISES

[9.1] *In Exercises 1–8, solve each equation by using the square root property. Express all radicals in simplest form.*

1. $y^2 = 144$

2. $x^2 = 37$

3. $m^2 = 128$

4. $(k + 2)^2 = 25$

5. $(r - 3)^2 = 10$

6. $(2p + 1)^2 = 14$

7. $(3k + 2)^2 = -3$

8. $(3x + 5)^2 = 0$

[9.2] *Solve each equation by completing the square.*

9. $m^2 + 6m + 5 = 0$

10. $p^2 + 4p = 7$

11. $-x^2 + 5 = 2x$

12. $2x^2 - 3 = -8x$

13. $4(x^2 + 7x) + 29 = -20$

14. $(4x + 1)(x - 1) = -7$

Solve each problem.

15. If an object is propelled upward on Earth from a height of 50 ft, with an initial velocity of 32 ft per sec, then its height after t sec is given by $h = -16t^2 + 32t + 50$, where h is in feet. After how many seconds will it reach a height of 30 ft?

16. Find the lengths of the three sides of the right triangle shown.

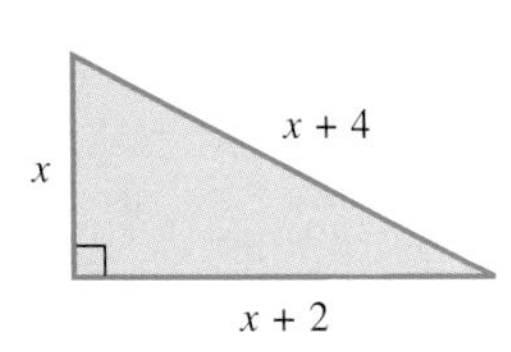

17. What must be added to $x^2 + kx$ to make it a perfect square?

[9.3]

18. Consider the equation $x^2 - 9 = 0$.

(a) Solve the equation by factoring.

(b) Solve the equation by the square root property.

(c) Solve the equation by the quadratic formula.

(d) Compare your answers. If a quadratic equation can be solved by both factoring and the quadratic formula, should we always get the same results? Explain.

Solve each equation by using the quadratic formula.

19. $-4x^2 - 2x + 7 = 0$

20. $2x^2 + 8 = 4x + 11$

21. $x(5x - 1) = 1$

22. $\frac{1}{4}x^2 = 2 - \frac{3}{4}x$

23. $\frac{1}{2}x^2 + 3x = 5$

24. Why is this not the statement of the quadratic formula for $ax^2 + bx + c = 0$?

$$x = -b \pm \frac{\sqrt{b^2 - 4ac}}{2a}$$

[9.4] *Sketch the graph of each equation. Identify each vertex.*

25. $y = -3x^2$

26. $y = -x^2 + 5$

27. $y = x^2 - 2x + 1$

28. $y = -x^2 + 2x + 3$

29. $y = x^2 + 4x + 2$

30. $y = (x + 4)^2$

31. Refer to Example 5 and Exercise 17 in **Section 9.4.** Suppose that a telescope has a diameter of 200 ft and a maximum depth of 30 ft. Find the equation for a cross section of the parabolic dish.

[9.5] *Decide whether each relation is or is not a function. In Exercises 32 and 33, give the domain and the range.*

32. $\{(-2, 4), (0, 8), (2, 5), (2, 3)\}$

33. $\{(8, 3), (7, 4), (6, 5), (5, 6), (4, 7)\}$

34.

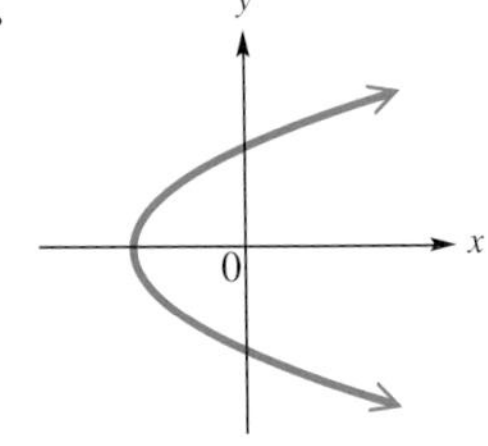

35.

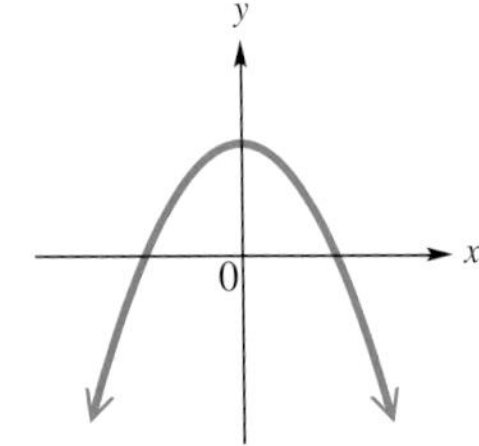

36. $2x + 3y = 12$

37. $y = x^2$

38. $x = 2|y|$

*Find **(a)** $f(2)$ and **(b)** $f(-1)$.*

39. $f(x) = 3x + 2$

40. $f(x) = 2x^2 - 1$

41. $f(x) = |x + 3|$

Becky and Brad are the owners of Cole's Baseball Cards. They have found that the price y, in dollars, of a particular Brad Radke baseball card depends on the demand x, in hundreds, for the card, according to the function defined by

$$y = -x^2 + 12x - 26.$$

42. What demand produces a price of \$6 for the card?

43. Find the vertex of the parabola $y = -x^2 + 12x - 26$.

44. Give the demand and price that correspond to the vertex.

MIXED REVIEW EXERCISES

Solve by any method.

45. $(2t - 1)(t + 1) = 54$

46. $(2p + 1)^2 = 100$

47. $(k + 2)(k - 1) = 3$

48. $6t^2 + 7t - 3 = 0$

49. $2x^2 + 3x + 2 = x^2 - 2x$

50. $x^2 + 2x + 5 = 7$

51. $m^2 - 4m + 10 = 0$

52. $k^2 - 9k + 10 = 0$

53. $(5x + 6)^2 = 0$

54. $\frac{1}{2}r^2 = \frac{7}{2} - r$

55. $x^2 + 4x = 1$

56. $7x^2 - 8 = 5x^2 + 8$

Chapter 9
TEST

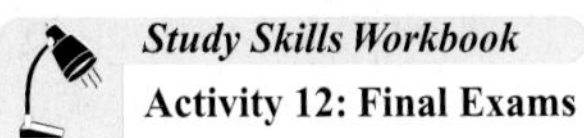
Study Skills Workbook
Activity 12: Final Exams

Solve by using the square root property.

1. $x^2 = 39$ **1.** ____________

2. $(x + 3)^2 = 64$ **2.** ____________

3. $(4x + 3)^2 = 24$ **3.** ____________

Solve by completing the square.

4. $x^2 - 4x = 6$ **4.** ____________

5. $2x^2 + 12x - 3 = 0$ **5.** ____________

Solve by the quadratic formula.

6. $2x^2 + 5x - 3 = 0$ **6.** ____________

7. $3w^2 + 2 = 6w$ **7.** ____________

8. $4x^2 + 8x + 11 = 0$ **8.** ____________

9. $t^2 - \frac{5}{3}t + \frac{1}{3} = 0$ **9.** ____________

Solve by the method of your choice.

10. $p^2 - 2p - 1 = 0$ **10.** ____________

11. $(2x + 1)^2 = 18$ **11.** ____________

12. $(x - 5)(2x - 1) = 1$ **12.** ____________

13. $t^2 + 25 = 10t$ **13.** ____________

14. ____________________

15. ____________________

16. ____________________

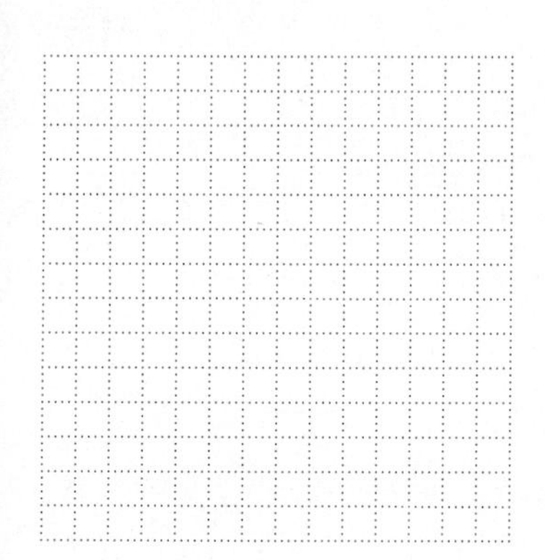

17. ____________________

18. (a) ____________________

(b) ____________________

19. ____________________

20. ____________________

Solve each problem.

14. If an object is propelled into the air from ground level with an initial velocity of 64 ft per sec, its height s (in feet) after t seconds is given by the formula $s = -16t^2 + 64t$. After how many seconds will the object reach a height of 64 ft?

15. Find the lengths of the three sides of the right triangle.

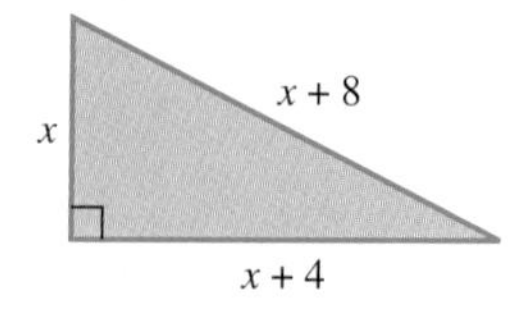

Sketch the graph of each equation. Identify each vertex.

16. $y = (x - 3)^2$

17. $y = -x^2 - 2x - 4$

18. Decide whether each relation represents a function. If it does, give the domain and the range.

(a) $\{(2, 3), (2, 4), (2, 5)\}$

(b) $\{(0, 2), (1, 2), (2, 2)\}$

19. Use the vertical line test to determine whether the graph is that of a function.

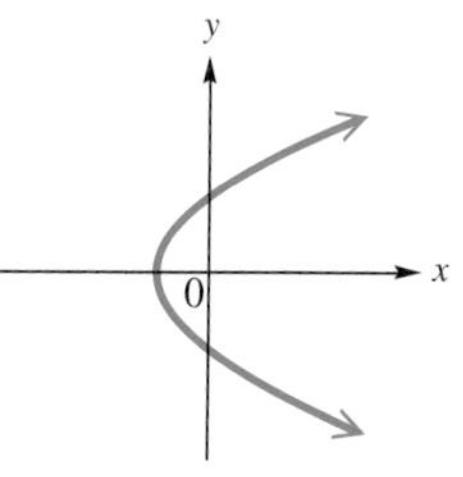

20. If $f(x) = 3x + 7$, find $f(-2)$.

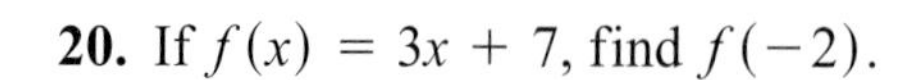

Cumulative Review Exercises

CHAPTERS R–9

Note: **This cumulative review exercise set may be considered as a final examination for the course.**

Perform the indicated operations, wherever possible.

1. $\dfrac{-4 \cdot 3^2 + 2 \cdot 3}{2 - 4 \cdot 1}$

2. $-9 - (-8)(2) + 6 - (6 + 2)$

3. $|-3| - |1 - 6|$

4. $-4r + 14 + 3r - 7$

5. $13k - 4k + k - 14k + 2k$

6. $5(4m - 2) - (m + 7)$

Solve each equation.

7. $6x - 5 = 13$

8. $3k - 9k - 8k + 6 = -64$

9. $2(m - 1) - 6(3 - m) = -4$

10. The perimeter of a basketball court is 288 ft. The width of the court is 44 ft less than the length. What are the dimensions of the court?

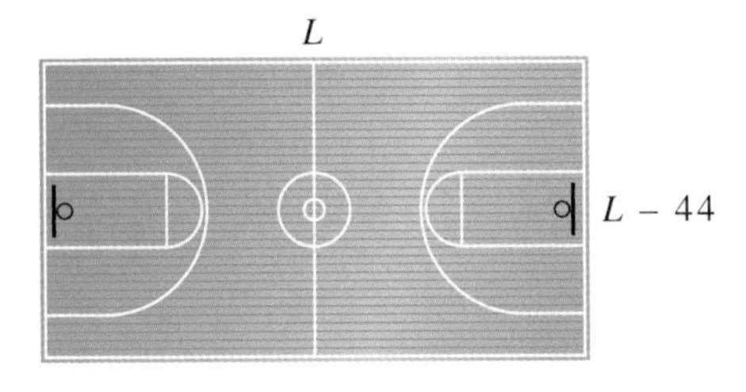

11. Find the measures of the marked angles.

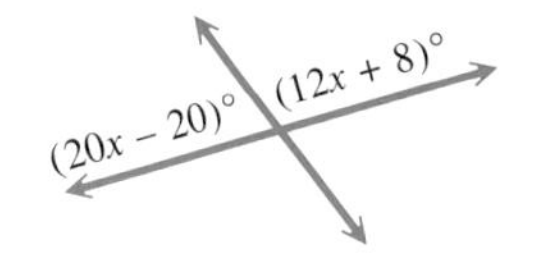

12. Solve the formula $P = 2L + 2W$ for L.

Solve each inequality. Graph the solutions.

13. $-8m < 16$

14. $-9p + 2(8 - p) - 6 \geq 4p - 50$

Graph the following.

15. $2x + 3y = 6$

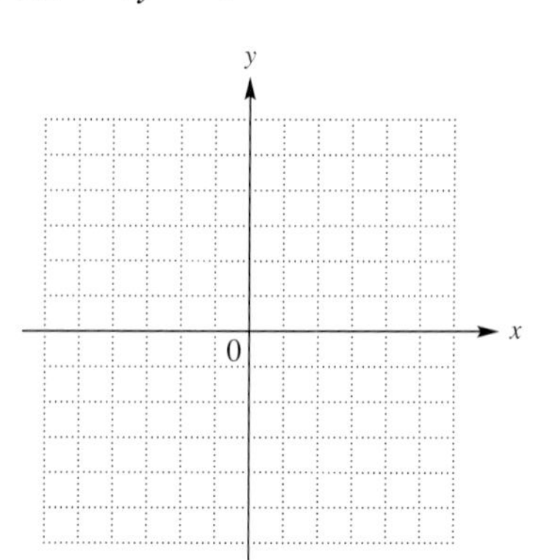

16. $y = 3$

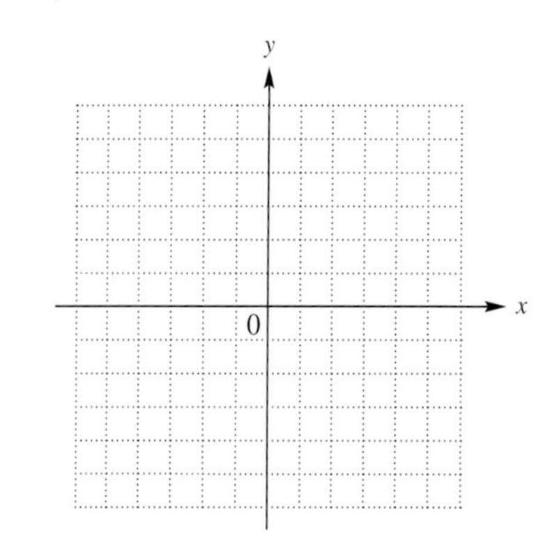

17. $2x - 5y < 10$

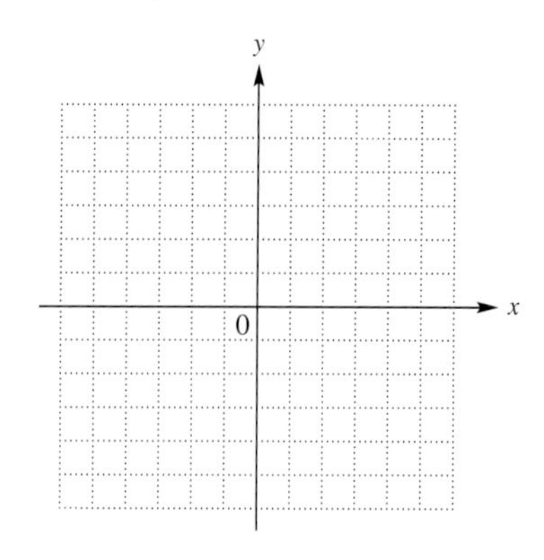

18. Find the slope of the line through $(-1, 4)$ and $(5, 2)$.

19. Write an equation of a line with slope 2 and y-intercept $(0, 3)$. Give it in the form $Ax + By = C$.

Solve each system of equations.

20. $2x + y = -4$
$-3x + 2y = 13$

21. $3x - 5y = 8$
$-6x + 10y = 16$

22. Based on prices quoted in a Pay-as-You-Go Internet site in February 2004, you could purchase 3 Motorola cell phones and 2 Kyocera cell phones for \$379.95. You could also purchase 2 Motorola cell phones and 3 Kyocera cell phones for \$369.95. Find the price for a single phone of each model.

Graph the solutions of the system of inequalities.

23. $2x + y \leq 4$
$x - y > 2$

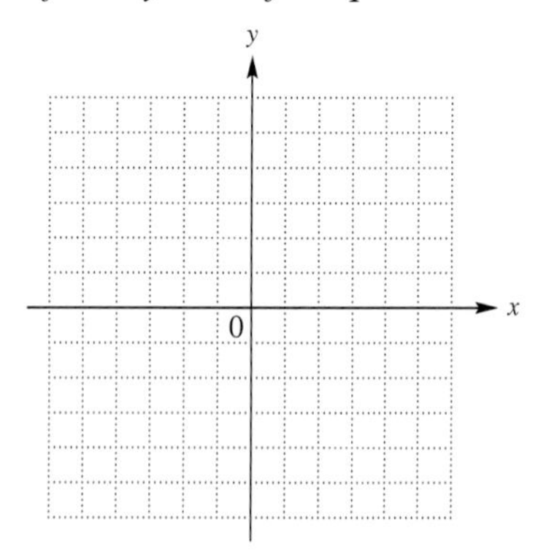

Simplify. Write answers with positive exponents.

24. $(3^2 \cdot x^{-4})^{-1}$

25. $\left(\frac{b^{-3}c^4}{b^5c^3}\right)^{-2}$

26. $\left(\frac{5}{3}\right)^{-3}$

Perform the indicated operations.

27. $(5x^5 - 9x^4 + 8x^2) - (9x^2 + 8x^4 - 3x^5)$

28. $(2x - 5)(x^3 + 3x^2 - 2x - 4)$

29. $(5t + 9)^2$

30. $\dfrac{3x^3 + 10x^2 - 7x + 4}{x + 4}$

Factor as completely as possible.

31. $16x^3 - 48x^2y$

32. $16x^4 - 1$

33. $2a^2 - 5a - 3$

34. $25m^2 - 20m + 4$

Solve each equation.

35. $x^2 + 3x - 54 = 0$

36. $3x^2 = x + 4$

37. The length of a rectangle is 2.5 times its width. The area is 1000 m^2. Find the length.

Perform the indicated operations. Write all answers in lowest terms.

38. $\dfrac{2}{a - 3} \div \dfrac{5}{2a - 6}$

39. $\dfrac{1}{k} - \dfrac{2}{k - 1}$

40. $\dfrac{2}{a^2 - 4} + \dfrac{3}{a^2 - 4a + 4}$

41. $\dfrac{6 + \dfrac{1}{x}}{3 - \dfrac{1}{x}}$

42. Solve $\dfrac{1}{x+3} + \dfrac{1}{x} = \dfrac{7}{10}$.

Simplify each expression as completely as possible.

43. $\sqrt{100}$

44. $\dfrac{6\sqrt{6}}{\sqrt{5}}$

45. $3\sqrt{5} - 2\sqrt{20} + \sqrt{125}$

46. $\sqrt[3]{16a^3b^4} - \sqrt[3]{54a^3b^4}$

Solve.

47. $\sqrt{x+2} = x - 4$

48. $2a^2 - 2a = 1$

49. Graph the parabola $y = x^2 - 4$. Identify the vertex. What are the x-intercepts?

50. **(a)** Is $\{(0, 4), (1, 2), (3, 5)\}$ a function?

(b) Give the domain and the range of the relation in part (a).

(c) If $f(x) = -2x + 7$, find $f(-2)$.

(d) Is this the graph of a function?

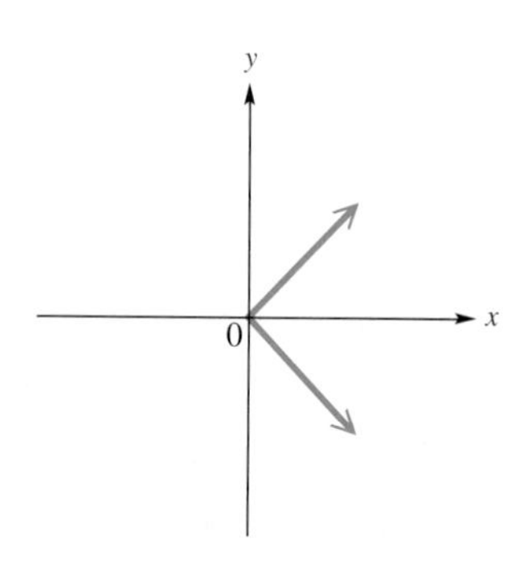

Appendix A Strategies for Problem Solving

Appendix A Strategies for Problem Solving

OBJECTIVE

1 Learn additional problem-solving strategies.

OBJECTIVE 1 **Learn additional problem-solving strategies.** In **Section 2.4,** we introduce a six-step method for problem solving that we use throughout this text. This method is based on a problem-solving process developed by Hungarian native George Polya, among whose many publications is the modern classic *How to Solve It.*

George Polya (1887–1985)

Polya's Four-Step Process for Problem Solving

Step 1 **Understand the problem.** You cannot solve a problem if you do not understand what you are asked to find. The problem must be read and analyzed carefully. You may need to read it several times. After you have done so, ask yourself, "What must I find?"

Step 2 **Devise a plan.** There are many ways to attack a problem. Decide what plan is appropriate for the particular problem you are solving.

Step 3 **Carry out the plan.** Once you know how to approach the problem, carry out your plan. You may run into "dead ends" and unforeseen roadblocks, but be persistent. If you are able to solve a problem without a struggle, it isn't much of a problem, is it?

Step 4 **Look back and check.** Check your answer to see that it is reasonable. Does it satisfy the conditions of the problem? Have you answered all the questions the problem asks? Can you solve the problem a different way and come up with the same answer?

Work Problem 1 at the Side.

1 Compare the six-step problem-solving method given in **Section 2.4** with Polya's four steps.

In Step 2 of Polya's problem-solving process, we are told to devise a plan. The box on the next page lists some strategies that may prove useful.

Answers

1. Step 1 compares to Polya's first step, Steps 2 and 3 compare to his second step, Step 4 compares to his third step, and Step 6 compares to his fourth step.

Problem-Solving Strategies

Make a table or a chart.
Look for a pattern.
Solve a similar simpler problem.
Draw a sketch.
Write an equation and solve it.
If a formula applies, use it.
Work backward.
Guess and check.
Use trial and error.
Use common sense.
Look for a "catch" if an answer seems too obvious or impossible.

A particular problem solution may involve one or more of the strategies listed here, and you should try to be creative in your problem-solving techniques. The examples that follow illustrate some of these strategies.

The problem in Example 1 is a famous one in the history of mathematics and first appeared in *Liber Abaci*, a book written by the Italian mathematician Leonardo Pisano (also known as Fibonacci) in the year 1202. We apply Polya's four-step process to solve it.

EXAMPLE 1 Using a Table or a Chart

A man put a pair of rabbits in a cage. During the first month the rabbits produced no offspring, but each month thereafter produced one new pair of rabbits. If each new pair thus produced reproduces in the same manner, how many pairs of rabbits will there be at the end of one year?

Step 1 **Understand the problem.** After several readings, we can reword the problem as follows:

How many pairs of rabbits will the man have at the end of one year if he starts with one pair, and they reproduce this way: During the first month of life, each pair produces no new rabbits, but each month thereafter each pair produces one new pair?

Step 2 **Devise a plan.** Since there is a definite pattern to how the rabbits will reproduce, we can construct a table as shown below. Once the table is completed, the final entry in the final column is our answer.

Month	Number of Pairs at Start	Number of New Pairs Produced	Number of Pairs at End of Month
1st			
2nd			
3rd			
4th			
5th			
6th			
7th			
8th			
9th			
10th			
11th			
12th			

Continued on Next Page

Step 3 **Carry out the plan.** At the start of the first month there is only one pair of rabbits. No new pairs are produced during the first month, so there is $1 + 0 = 1$ pair present at the end of the first month. This pattern continues throughout the table. We add the number in the first column of numbers to the number in the second column to get the number in the third.

Month	Number of Pairs at Start	+ Number of New Pairs Produced	= Number of Pairs at End of Month
1st	1	0	1
2nd	1	1	2
3rd	2	1	3
4th	3	2	5
5th	5	3	8
6th	8	5	13
7th	13	8	21
8th	21	13	34
9th	34	21	55
10th	55	34	89
11th	89	55	144
12th	144	89	**233**

$1 + 0 = 1$
$1 + 1 = 2$
$2 + 1 = 3$
.
.
.
$144 + 89 = \mathbf{233}$

There will be 233 pairs of rabbits at the end of one year.

Step 4 **Look back and check.** This problem can be checked by going back and making sure that we have interpreted it correctly, which we have. Double-check the arithmetic. We have answered the question posed by the problem, so the problem is solved.

NOTE
The sequence shown in color in the table in Example 1 is called the *Fibonacci sequence,* and many of its interesting properties are investigated in other mathematics courses.

Work Problem 2 at the Side.

In the remaining examples of this section, we use Polya's process but we do not list the steps specifically as we did in Example 1.

EXAMPLE 2 Working Backward

Rob Zwettler goes to the racetrack with his buddies on a weekly basis. One week he tripled his money, but then lost \$12. He took his money back the next week, doubled it, but then lost \$40. The following week he tried again, taking his money back with him. He quadrupled it, and then played well enough to take that much home with him, a total of \$224. How much did he start with the first week?

This problem asks us to find Rob's starting amount, given information about his winnings and losses. We also know his final amount. The method of working backward can be applied quite easily.

Continued on Next Page

2 Refer to Example 1, and observe the sequence of numbers in color, the Fibonacci sequence. Choose any four consecutive terms. Multiply the first one chosen by the fourth, and then write the product. Now multiply the two middle terms and write the product. Repeat this process a few more times. What do you notice when the two products are compared?

Answers
2. The products will always differ by 1.

Since his final amount was \$224 and this represents four times the amount he started with on the third week, we *divide* \$224 by 4 to find that he started the third week with \$56. Before he lost \$40 the second week, he had this \$56 plus the \$40 he lost, giving him \$96. This represented double what he started with, so he started with \$96 *divided by* 2, or \$48, the second week. Repeating this process once more for the first week, before his \$12 loss he had

$$\$48 + \$12 = \$60,$$

which represents triple what he started with. Therefore, he started with

$$\$60 \div 3, \quad \text{or} \quad \$20.$$

To check our answer, \$20, observe the following equations that depict the winnings and losses:

First week: $(3 \times \$20) - \$12 = \$60 - \$12 = \mathbf{\$48}$

Second week: $(2 \times \mathbf{\$48}) - \$40 = \$96 - \$40 = \$56$

Third week: $(4 \times \$56) = \mathbf{\$224}.$ His final amount

Work Problem 3 at the Side.

Recall that $5^2 = 5 \cdot 5 = 25$, that is, 5 squared is 25. Thus, 25 is called a **perfect square.** (See **Section 6.5.**) Other perfect squares include

$$4, \quad 9, \quad 16, \quad 36, \quad \text{and so on.} \qquad \text{Perfect squares}$$

EXAMPLE 3 Using Trial and Error

The mathematician Augustus De Morgan lived in the nineteenth century. He once made the following statement: "I was x years old in the year x^2." In what year was De Morgan born?

We must find the year of De Morgan's birth. The problem tells us that he lived in the nineteenth century, which is another way of saying that he lived during the 1800s. One year of his life was a perfect square, so we must find a number between 1800 and 1900 that is a perfect square. Use trial and error.

$$42^2 = 42 \cdot 42 = 1764$$
$$43^2 = 43 \cdot 43 = \mathbf{1849} \qquad \text{1849 is between 1800 and 1900.}$$
$$44^2 = 44 \cdot 44 = 1936$$

The only natural number whose square is between 1800 and 1900 is 43, since $43^2 = 1849$. Therefore, De Morgan was 43 yr old in 1849. The final step in solving the problem is to subtract 43 from 1849 to find the year of his birth:

$$1849 - 43 = \mathbf{1806}. \qquad \text{He was born in 1806.}$$

While the following check may seem unorthodox, it works: Look up De Morgan's birth date in a book on mathematics history, such as *An Introduction to the History of Mathematics*, Sixth Edition, by Howard W. Eves.

Work Problem 4 at the Side.

As mentioned above, $5^2 = 25$. The inverse (opposite) of squaring a number is called taking the **square root.** (See **Section 8.1.**) We indicate the positive square root using a **radical sign** $\sqrt{}$. Thus, $\sqrt{25} = 5$. Also,

$$\sqrt{4} = 2, \quad \sqrt{9} = 3, \quad \sqrt{16} = 4, \quad \text{and so on.} \qquad \text{Square roots}$$

3 Solve each problem.

(a) Phyllis Crittenden bought a book for \$10 and then spent half her remaining money on a train ticket. She then bought lunch for \$4 and spent half her remaining money at a bazaar. She left the bazaar with \$20. How much money did she start with?

(b) If a, b, and c are digits for which

$$\begin{array}{r} 7\ a\ 2 \\ -\ 4\ 8\ b \\ \hline c\ 7\ 3, \end{array}$$

then $a + b + c =$ ____.

A. 14 **B.** 15 **C.** 16 **D.** 17 **E.** 18.
(*Source: Mathematics Teacher* calendar, September 22, 1999.)

4 Solve each problem.

(a) Assuming that he lives that long, one of the authors of this book will be 76 yr old in the year x^2, where x is a counting number. In what year was he born?

(b) Place each of the digits 1, 2, 3, 4, 5, 6, 7, and 8 in separate boxes so that boxes that share common corners do not contain successive digits.
(*Source: Mathematics Teacher* calendar, November 29, 1997.)

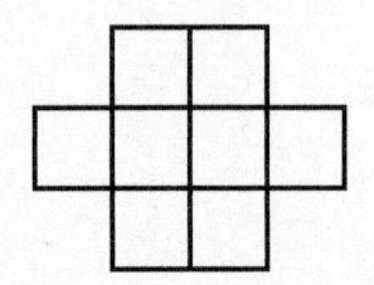

ANSWERS
3. **(a)** \$98 **(b)** D
4. **(a)** 1949 **(b)** Here is one possible solution.

	3	5	
7	1	8	2
	4	6	

The next problem dates back to Hindu mathematics, circa 850.

EXAMPLE 4 Guessing and Checking

One-fourth of a herd of camels was seen in the forest. Twice the square root of that herd had gone to the mountain slopes, and 3 times 5 camels remained on the riverbank. What is the numerical measure of that herd of camels?

The numerical measure of the herd of camels must be a natural number. Since the problem mentions "one-fourth of a herd" and "the square root of that herd," the number of camels must be both a multiple of 4 and a perfect square, so no fractions will be encountered. The smallest natural number that satisfies both conditions is 4. We write an equation where x represents the numerical measure of the herd, and then substitute 4 for x to see if it is a solution.

"one-fourth of the herd" + "twice the square root of that herd" + "3 times 5 camels" = "the numerical measure of the herd"

$$\frac{1}{4}x + 2\sqrt{x} + 3 \cdot 5 = x$$

$$\frac{1}{4}(4) + 2\sqrt{4} + 3 \cdot 5 = 4 \quad \text{Let } x = 4.$$

$$1 + 4 + 15 = 4 \quad ? \quad \sqrt{4} = 2$$

$$20 \neq 4$$

Since 4 is not the solution, try 16, the next perfect square that is a multiple of 4.

$$\frac{1}{4}(16) + 2\sqrt{16} + 3 \cdot 5 = 16 \quad \text{Let } x = 16.$$

$$4 + 8 + 15 = 16 \quad ? \quad \sqrt{16} = 4$$

$$27 \neq 16$$

Since 16 is not a solution, try 36.

$$\frac{1}{4}(36) + 2\sqrt{36} + 3 \cdot 5 = 36 \quad \text{Let } x = 36.$$

$$9 + 12 + 15 = 36 \quad ? \quad \sqrt{36} = 6$$

$$36 = 36$$

We see that 36 is the numerical measure of the herd. Check in the words of the problem: "One-fourth of 36, plus twice the square root of 36, plus 3 times 5" gives 9 plus 12 plus 15, which equals 36.

Work Problem 5 at the Side.

5 Solve each problem.

(a) I am thinking of a positive number. If I square it, double the result, take half of that result, and then add 12, I get 21. What is my number?

(b) The same author mentioned in Margin Problem 4(a) graduated from high school in the year that satisfies these conditions:

(1) The sum of the digits is 23;

(2) The hundreds digit is 3 more than the tens digit;

(3) No digit is an 8.

In what year did he graduate?

EXAMPLE 5 Considering a Similar Simpler Problem and Looking for a Pattern

The digit farthest to the right in a natural number is called the *ones* or *units* digit, since it tells how many ones are contained in the number when grouping by tens is considered. What is the ones (or units) digit in 2^{4000}?

Recall that 2^{4000} means that 2 is used as a factor 4000 times:

$$2^{4000} = \underbrace{2 \times 2 \times 2 \times \cdots \times 2.}_{4000 \text{ factors}}$$

Certainly, we are not expected to evaluate this number. To answer the question, we examine some smaller powers of 2 and then look for a pattern. We start with the exponent 1 and look at the first twelve powers of 2.

Continued on Next Page

ANSWERS
5. (a) 3 (b) 1967

6 Solve each problem.

(a) What is the units digit in 7^{491}?

(b) What is the 103rd digit in the decimal representation for $\frac{1}{11}$?

7 Solve each problem.

(a) What is the maximum number of small squares in which we may place a cross ($\times$) and not have any row, column, or diagonal completely filled with crosses? Illustrate your answer.

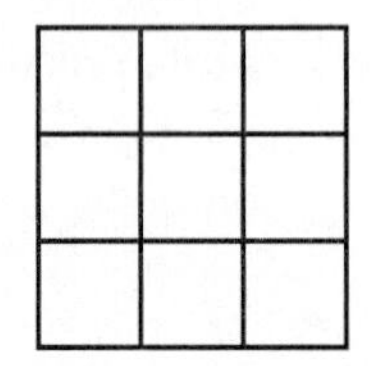

(b) By drawing two straight lines, divide the face of a clock into three regions such that the numbers in the regions have the same total. (*Source: Mathematics Teacher* calendar, October 28, 1998.)

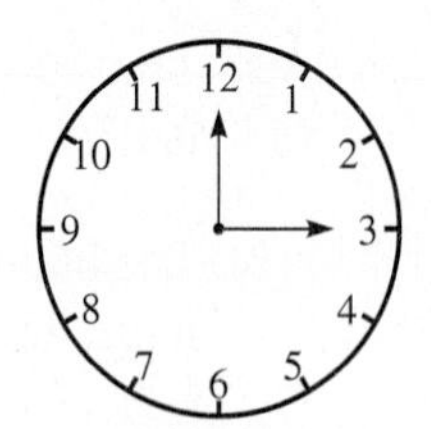

Answers

6. (a) 3 **(b)** 0

7. (a) 6

	X	X
X		X
X	X	

One of several possibilities

(b) Each region has a sum of 26.

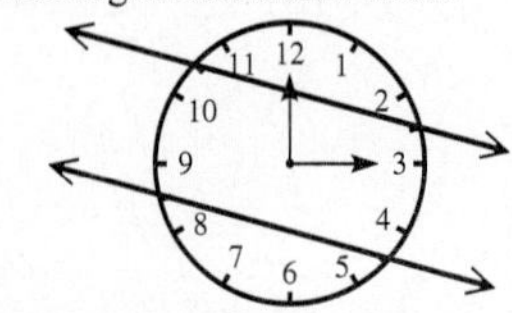

$2^1 = 2$	$2^5 = 32$	$2^9 = 512$
$2^2 = 4$	$2^6 = 64$	$2^{10} = 1024$
$2^3 = 8$	$2^7 = 128$	$2^{11} = 2048$
$2^4 = 16$	$2^8 = 256$	$2^{12} = 4096$

Notice that in each of the four rows above, the ones digit is the same. The final row, which contains the exponents 4, 8, and 12, has the ones digit 6. Each of these exponents is divisible by 4, and since 4000 is divisible by 4, we observe the pattern to predict that the units digit in 2^{4000} is **6**.

The units digit for any other power of 2 can be found if we divide the exponent by 4 and compare the remainder to the preceding list of powers. For example, to find the units digit of 2^{543}, we divide 543 by 4 to get a quotient of 135 and a remainder of 3. The units digit is the same as that of 2^3, which is 8.

Work Problem 6 at the Side.

EXAMPLE 6 Drawing a Sketch

An array of nine dots is arranged in a 3 $\times$ 3 square, as shown in Figure 1. Is it possible to join the dots with exactly four straight lines if you are not allowed to pick up your pencil from the paper and may not trace over a line that has already been drawn? If so, show how.

Figure 1

Figure 2 shows three attempts. In each case, something is wrong. In the first sketch, one dot is not joined. In the second, the figure cannot be drawn without picking up your pencil from the paper or tracing over a line that has already been drawn. In the third figure, all dots have been joined, but you have used five lines as well as retraced over the figure.

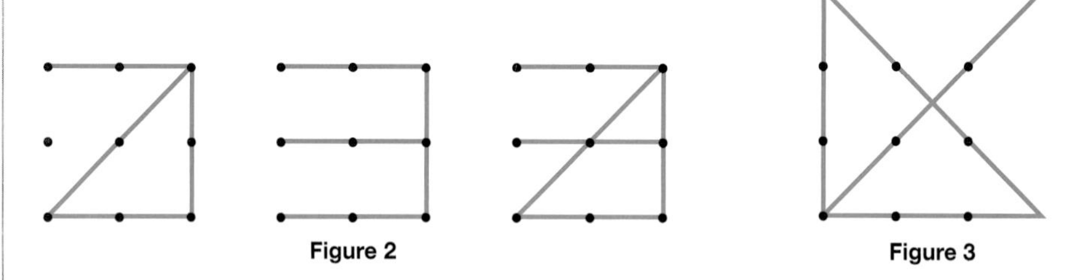

Figure 2

Figure 3

However, the conditions of the problem can be satisfied, as shown in Figure 3. We "went outside of the box," which was not prohibited by the conditions of the problem. This is an example of creative thinking—we used a strategy that is usually not considered at first, since our initial attempts involved "staying within the confines" of the figure.

Work Problem 7 at the Side.

The final example falls into a category of problems that involve a "catch." Some of these problems seem too easy or perhaps impossible at first, because we tend to overlook an obvious situation. We must look carefully at the use of language in such problems. And, of course, we should never forget to use common sense.

EXAMPLE 7 Using Common Sense

Two currently minted U.S. coins together have a total value of \$1.05. One is not a dollar. What are the two coins?

Our initial reaction might be, "The only way to have two such coins with a total of \$1.05 is to have a nickel and a dollar, but the problem says that one of them is not a dollar." This statement is indeed true. What we must realize here is that the one that is not a dollar is the nickel, and the *other* coin is a dollar! So the two coins are a dollar and a nickel.

Work Problem 8 at the Side.

8 Solve each problem.

(a) Which is correct? Three cubed *is* nine or three cubed *are* nine?

(b) If you take 7 bowling pins from 10 bowling pins, what do you have?

(c) If it takes $7\frac{1}{2}$ min to boil an egg, how long does it take to boil 5 eggs?

ANSWERS

8. **(a)** Neither is correct, since $3^3 = 27$.
(b) 7 bowling pins
(c) $7\frac{1}{2}$ min (Boil them all at the same time.)

Appendix A Exercises

FOR EXTRA HELP

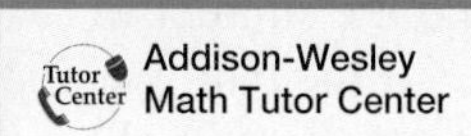
Addison-Wesley Math Tutor Center

MathXL

Student's Solutions Manual

MyMathLab

Interactmath.com

Exercises 1–13 are from the popular monthly calendar feature in the journal Mathematics Teacher. *The authors wish to thank the many journal contributors for permission to use these problems. Original calendar dates are included.*

Use the various problem-solving strategies to solve each problem. In many cases, there is more than one possible approach, so be creative.

1. You are working in a store that has been very careless with the stock. Three boxes of socks are each incorrectly labeled. The labels say *red socks, green socks,* and *red and green socks*. How can you relabel the boxes correctly by taking only one sock out of one box, without looking inside the boxes? (October 22, 2001)

2. Three dice with faces numbered 1 through 6 are stacked as shown. Seven of the eighteen faces are visible, leaving eleven faces hidden on the back, on the bottom, and between faces. The total number of dots not visible in this view is ______.

A. 21

B. 22

C. 31

D. 41

E. 53

(September 17, 2001)

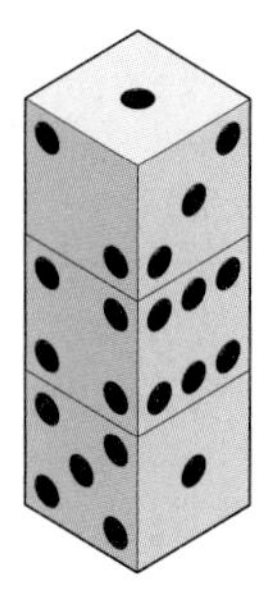

3. At his birthday party, Mr. Green would not directly tell how old he was. He said, "If you add the year of my birth to this year, subtract the year of my tenth birthday and the year of my fiftieth birthday, and then add my present age, the result is eighty." How old was Mr. Green? (December 14, 1997)

4. Today is your first day driving a city bus. When you leave downtown, you have 23 passengers. At the first stop, 3 people exit and 5 people get on the bus. At the second stop, 11 people exit and 8 people get on the bus. At the third stop, 5 people exit and 10 people get on. How old is the bus driver? (April 1, 2002)

5. You and a friend are playing tick-tack-toe, where three in a row *loses*. You are O. If you want to win, what must your next move be? (October 21, 2001)

X	X	O
O		
	X	

6. How can you connect each square with the triangle that has the same number? Lines cannot cross, enter a square or triangle, or go outside the diagram. (October 15, 1999)

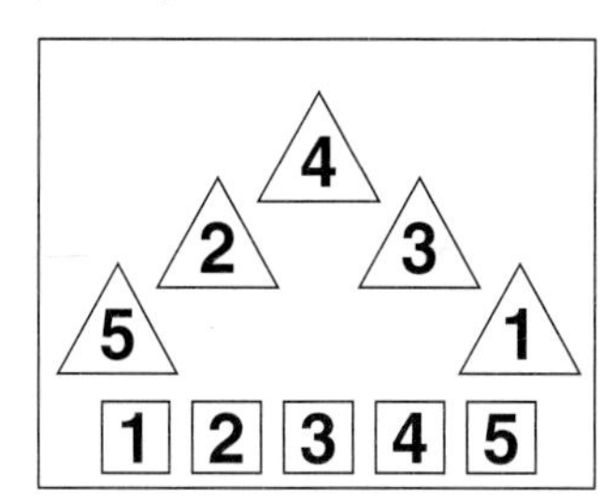

7. Pat and Chris have the same birthday. Pat is twice as old as Chris was when Pat was as old as Chris is now. If Pat is now 24 yr old, how old is Chris? (December 3, 2001)

8. Balls numbered 1 through 6 are arranged in a *difference triangle*. Note that in any row, the difference between the larger and the smaller of two successive balls is the number of the ball that appears below them. Arrange balls numbered 1 through 10 in a difference triangle. (May 6, 1998)

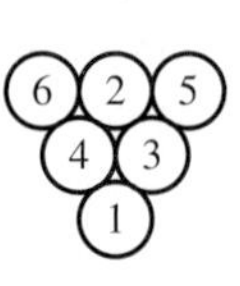

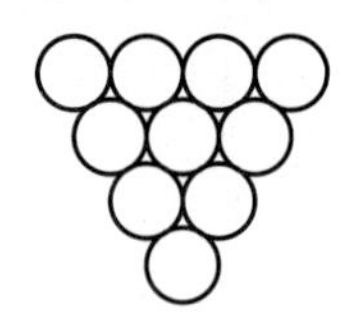

9. Only one of these numbers is a perfect square. Which one is it? (October 8, 1997)

329,476 389,372 964,328
326,047 724,203

10. While traveling to his grandmother's for Christmas, George fell asleep halfway through the journey. When he awoke, he still had to travel half the distance that he had traveled while sleeping. For what part of the entire journey had he been asleep? (December 25, 1998)

11. You have brought two unmarked buckets to a stream. The buckets hold 7 gal and 3 gal of water, respectively. How can you obtain exactly 5 gal of water to take home? (October 19, 1997)

12. Chip and Dale collected 32 acorns on Monday and stored them with their acorn supply. After Chip fell asleep, Dale ate half the acorns. This pattern continued through Friday night, with 32 acorns being added and half the supply being eaten. On Saturday morning, Chip counted the acorns and found that they had only 35. How many acorns had they started with on Monday morning? (March 12, 1997)

13. What are the final two digits of 7^{1997}? (November 29, 1997)

14. If you raise 3 to the 324th power, what is the units digit of the result?

15. A frog is at the bottom of a 20-ft well. Each day it crawls up 4 ft but each night it slips back 3 ft. After how many days will the frog reach the top of the well?

16. A lily pad grows so that each day it doubles its size. On the twentieth day of its life, it completely covers a pond. On what day was the pond half covered?

17. Some children are standing in a circular arrangement. They are evenly spaced and arranged in numerical order. The fourth child is standing directly opposite the twelfth child. How many children are there in the circle?

18. A *perfect number* is a natural number that is equal to the sum of all its counting number divisors except itself. For example, 28 is a perfect number, since its divisors other than itself are 1, 2, 4, 7, and 14, and $1 + 2 + 4 + 7 + 14 = 28$. What is the least perfect number?

19. Draw a diagram that satisfies the following description, using the minimum number of birds: "Two birds above a bird, two birds below a bird, and a bird between two birds."

20. Donna is taller than David but shorter than Bill. Dan is shorter than Bob. What is the first letter in the name of the tallest person?

21. A *magic square* is a square array of numbers that has the property that the sum of the numbers in any row, column, or diagonal is the same. Fill in the square below so that it becomes a magic square, and all digits 1, 2, 3, . . . , 9 are used exactly once.

6		8
	5	
		4

22. Refer to Exercise 21. Complete the magic square below so that all counting numbers 1, 2, 3, . . . , 16 are used exactly once, and the sum in each row, column, or diagonal is 34.

6			9
	15		14
11		10	
16		13	

23. What is the minimum number of pitches that a baseball player who pitches a complete game can make in a regulation 9-inning baseball game?

24. What is the least natural number whose written name in the English language has its letters in alphabetical order?

25. You have eight coins. Seven are genuine and one is a fake, which weighs a little less than the other seven. You have a balance scale, which you may use only three times. Tell how to locate the bad coin in three weighings. Then show how to detect the bad coin in only *two* weighings.

26. A person must take a wolf, a goat, and some cabbage across a river. The rowboat to be used has room for the person plus either the wolf, the goat, or the cabbage. If the person takes the cabbage in the boat, the wolf will eat the goat. While the wolf crosses in the boat, the cabbage will be eaten by the goat. The goat and cabbage are safe only when the person is present. Even so, the person gets everything across the river. Explain how. (This problem dates back to around the year 750.)

27. When the diagram shown is folded to form a cube, what letter is opposite the face marked Z?

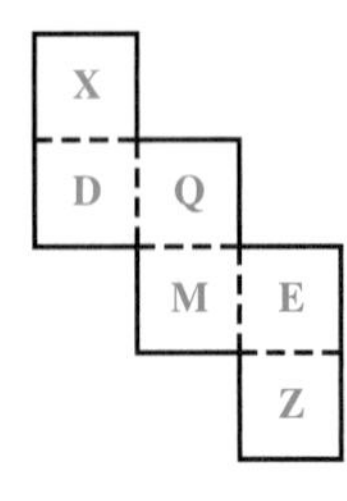

28. Draw a square in the following figure so that no two cats share the same region.

29. (This is an ancient Hindu problem.) Beautiful maiden with beaming eyes, tell me . . . which is the number that when multiplied by 3, then increased by $\frac{3}{4}$ the product, then divided by 7, diminished by $\frac{1}{3}$ of the quotient, multiplied by itself, diminished by 52, by the extraction of the square root, addition of 8, and division by 10 gives the number 2?

30. A teenager's age increased by 2 gives a perfect square. Her age decreased by 10 gives the square root of that perfect square. She is 5 yr older than her brother. How old is her brother?

31. Draw the following figure without picking up your pencil from the paper and without tracing over a line you have already drawn.

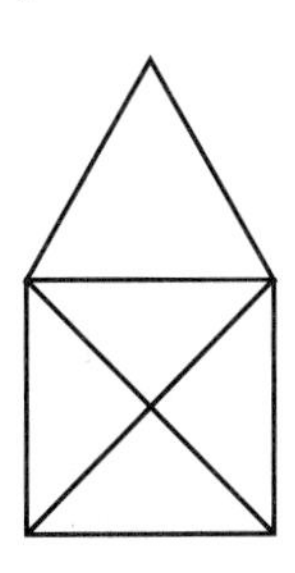

32. Repeat Exercise 31 for the figure shown here.

33. James, Dan, Jessica, and Cathy form a pair of married couples. Their ages are 36, 31, 30, and 29. Jessica is married to the oldest person in the group. James is older than Jessica but younger than Cathy. Who is married to whom, and what are their ages?

34. If a year has two consecutive months with Friday the thirteenth, what months must they be?

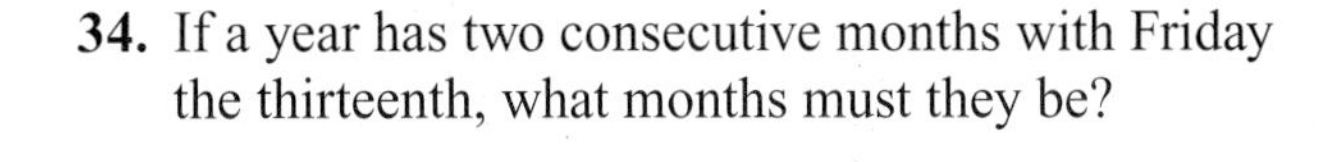

35. The brother of the chief executive officer (CEO) of a major industrial firm died. The man who died had no brother. How is this possible?

36. Some months have 30 days and some have 31 days. How many months have 28 days?

37. How much dirt is there in a cubical hole, 6 ft on each side?

38. Becky's mother has three daughters. She named her first daughter Penny and her second daughter Nichole. What did she name her third daughter?

39. Place one of the arithmetic operations $+$, $-$, $\times$, or $\div$ between each pair of successive numbers on the left side of this equation to make it true. Any operation may be used more than once or not at all.

$$1 \quad 2 \quad 3 \quad 4 \quad 5 \quad 6 \quad 7 \quad 8 \quad 9 = 100$$

40. In the addition problem below, some digits are missing as indicated by the blanks. If the problem is done correctly, what is the sum of the missing digits?

$$\begin{array}{rrrr} & \_ & 3 & 5 \\ & 8 & \_ & 6 \\ + & 1 & 4 & \_ \\ \hline \_ & 4 & 0 & 8 \end{array}$$

41. Fill in the blanks so that the multiplication problem below uses all digits 0, 1, 2, 3, . . . , 9 exactly once, and is correctly worked.

$$\begin{array}{r} \_\;0\;2 \\ \times \qquad 3\;\_ \\ \hline \_\;5,\;\_\;\_\;\_ \end{array}$$

42. Based on your knowledge of elementary arithmetic, describe the pattern that can be observed when the following operations are performed:

$$9 \times 1, \quad 9 \times 2, \quad 9 \times 3, \quad \ldots, \quad 9 \times 9.$$

(*Hint:* Add the digits in the answers. What do you notice?)

43. How many triangles are in the following figure?

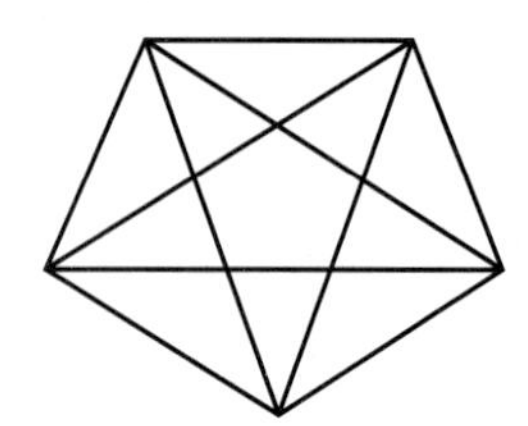

44. How many squares are in the following figure?

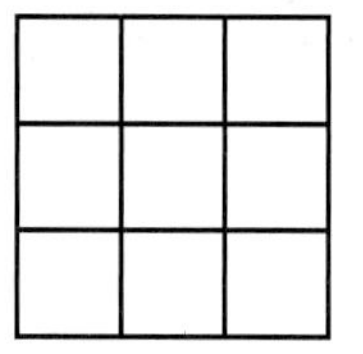

45. Volumes 1 and 2 of *The Complete Works of Wally Smart* are standing in numerical order from left to right on your bookshelf. Volume 1 has 450 pages and Volume 2 has 475 pages. Excluding the covers, how many pages are between page 1 of Volume 1 and page 475 of Volume 2?

46. At a hardware store, I can buy 1 for \$.75 and I can buy 68356 for \$3.75. What am I buying?

47. Eve said to Adam, "If you give me one dollar, then we will have the same amount of money." Adam then replied, "Eve, if you give me one dollar, I will have double the amount of money you are left with." How much does each have?

48. A drawer contains 20 black socks and 20 white socks. If the light is off and you reach into the drawer to get your socks, what is the minimum number of socks you must pull out in order to be sure that you have a matching pair?

Appendix B
Sets

Appendix B Sets

OBJECTIVE 1 List the elements of a set. A **set** is a collection of things. The objects in a set are called the **elements** (or **members**) of the set. A set is represented by listing its elements between **set braces,** { }. The order in which the elements of a set are listed is unimportant.

EXAMPLE 1 Listing the Elements of Sets

Represent each set by listing the elements.

(a) The set of states in the United States that border on the Pacific Ocean = {California, Oregon, Washington, Hawaii, Alaska}.

(b) The set of all counting numbers less than 6 = {1, 2, 3, 4, 5}.

Work Problem 1 at the Side.

OBJECTIVE 2 Learn the vocabulary and symbols used to discuss sets. Capital letters are used to name sets. To state that 5 is an element of

$$S = \{1, 2, 3, 4, 5\},$$

write $5 \in S$. The statement $6 \notin S$ means that 6 is not an element of S.

A set with no elements is called the **empty set,** or the **null set**. The symbols $\emptyset$ or { } are used for the empty set. If we let A be the set of all cats that fly, then A is the empty set.

$$A = \emptyset \quad \text{or} \quad A = \{\ \}$$

CAUTION
Do not make the common error of writing the empty set as $\{\emptyset\}$.

In any discussion of sets, there is some set that includes all the elements under consideration. This set is called the **universal set** for that situation. For example, if the discussion is about presidents of the United States, then the set of all presidents of the United States is the universal set. The universal set is denoted U.

OBJECTIVES

1. List the elements of a set.
2. Learn the vocabulary and symbols used to discuss sets.
3. Decide whether a set is finite or infinite.
4. Decide whether a given set is a subset of another set.
5. Find the complement of a set.
6. Find the union and the intersection of two sets.

1 Represent each set by listing the elements.

(a) The set of states whose names begin with the letter O

(b) The set of letters of the alphabet that follow T

(c) The set of even natural numbers less than 10

(d) The set of odd counting numbers between 15 and 20

ANSWERS
(a) {Oregon, Ohio, Oklahoma}
(b) {U, V, W, X, Y, Z}
(c) {2, 4, 6, 8}
(d) {17, 19}

OBJECTIVE 3 Decide whether a set is finite or infinite. In Example 1, there are five elements in the set in part (a) and five in part (b). If the number of elements in a set is either 0 or a counting number, then the set is a **finite set**. On the other hand, the set of natural numbers, for example, is an **infinite set**, because there is no final natural number. We can list the elements of the set of natural numbers as

$$N = \{1, 2, 3, 4, \ldots\},$$

where the three dots indicate that the set continues indefinitely. Not all infinite sets can be listed in this way. For example, there is no way to list the elements in the set of all real numbers between 1 and 2.

EXAMPLE 2 Distinguishing between Finite and Infinite Sets

List the elements of each set, if possible. Decide whether each set is *finite* or *infinite*.

(a) The set of all integers
One way to list the elements is $\{\ldots, -2, -1, 0, 1, 2, \ldots\}$. The set is infinite.

(b) The set of all natural numbers between 0 and 5
List the elements of this set as $\{1, 2, 3, 4\}$. The set is finite.

(c) The set of all irrational numbers
This is an infinite set whose elements cannot be listed.

Work Problem 2 at the Side.

Two sets are equal if they have exactly the same elements. Thus, the set of natural numbers and the set of positive integers are equal sets. Also, the sets $\{1, 2, 4, 7\}$ and $\{4, 2, 7, 1\}$ are equal. The order of the elements does not make a difference.

OBJECTIVE 4 Decide whether a given set is a subset of another set. If all elements of a set A are also elements of a new set B, then we say A is a **subset** of B, written $A \subseteq B$. We use the symbol $A \nsubseteq B$ to mean that A is not a subset of B.

EXAMPLE 3 Using Subset Notation

Let $A = \{1, 2, 3, 4\}$, $B = \{1, 4\}$, and $C = \{1\}$. Then

$$B \subseteq A, \quad C \subseteq A, \quad \text{and} \quad C \subseteq B,$$

but

$$A \nsubseteq B, \quad A \nsubseteq C, \quad \text{and} \quad B \nsubseteq C.$$

Work Problem 3 at the Side.

The set $M = \{a, b\}$ has four subsets: $\{a, b\}$, $\{a\}$, $\{b\}$, and $\emptyset$. The empty set is defined to be a subset of any set. How many subsets does $N = \{a, b, c\}$ have? There is one subset with 3 elements: $\{a, b, c\}$. There are three subsets with 2 elements:

$$\{a, b\}, \quad \{a, c\}, \quad \text{and} \quad \{b, c\}.$$

There are three subsets with 1 element:

$$\{a\}, \quad \{b\}, \quad \text{and} \quad \{c\}.$$

There is one subset with 0 elements: $\emptyset$. Thus, set N has eight subsets.

2 List the elements of each set, if possible. Decide whether each set is *finite* or *infinite*.

(a) The set of whole numbers

(b) The set of odd natural numbers between 10 and 20

(c) The set of integers greater than 3

(d) The set of rational numbers

3 Let

$$A = \{2, 4, 6, 8, 10, 12\},$$
$$B = \{2, 4, 8, 10\}, \text{ and}$$
$$C = \{4, 10, 12\}.$$

Tell whether each statement is *true* or *false*.

(a) $B \subseteq A$

(b) $C \subseteq B$

(c) $A \nsubseteq C$

(d) $B \nsubseteq C$

ANSWERS

2. **(a)** $\{0, 1, 2, 3, \ldots\}$; infinite
(b) $\{11, 13, 15, 17, 19\}$; finite
(c) $\{4, 5, 6, \ldots\}$; infinite
(d) This is an infinite set whose elements cannot be listed.

3. **(a)** true **(b)** false **(c)** true **(d)** true

The following generalization can be made.

Number of Subsets of a Set
A set with n elements has 2^n subsets.

To illustrate the relationships between sets, **Venn diagrams** are often used. A rectangle represents the universal set, U. The sets under discussion are represented by regions within the rectangle. The Venn diagram in Figure 1 shows that $B \subseteq A$.

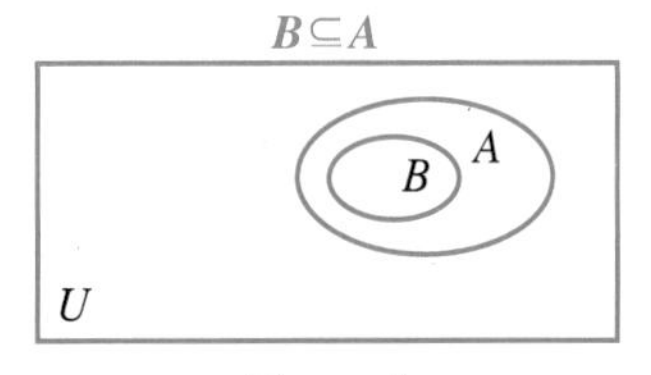

Figure 1

OBJECTIVE 5 Find the complement of a set. For every set A, there is a set A', the **complement** of A, that contains all the elements of U that are not in A. The shaded region in the Venn diagram in Figure 2 represents A'.

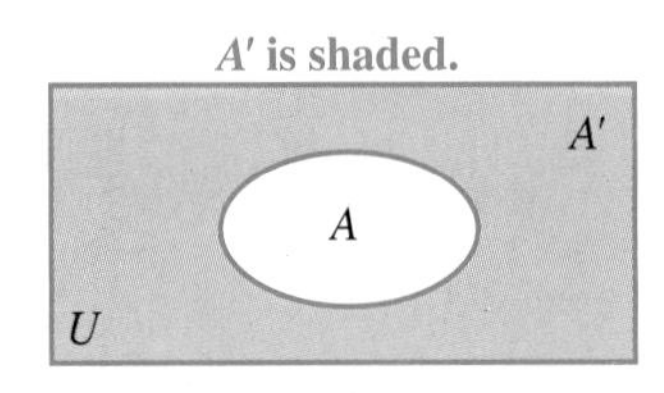

Figure 2

EXAMPLE 4 Determining Complements of Sets

Given $U = \{a, b, c, d, e, f, g\}$, $A = \{a, b, c\}$, $B = \{a, d, f, g\}$, and $C = \{d, e\}$, then

$$A' = \{d, e, f, g\}, \quad B' = \{b, c, e\}, \quad \text{and} \quad C' = \{a, b, c, f, g\}.$$

Work Problem 4 at the Side.

4 Let
$U = \{0, 1, 2, 3, 4, 5, 6, 7, 8\}$,
$M = \{0, 2, 4, 6, 8\}$,
$N = \{1, 3, 5, 7\}$, and
$Q = \{0, 1, 2, 3, 4\}$.

List the elements in each set.

(a) M'

(b) N'

(c) Q'

OBJECTIVE 6 Find the union and the intersection of two sets. The **union** of two sets A and B, written $A \cup B$, is the set of all elements of A together with all elements of B. Thus, for the sets in Example 4,

$$A \cup B = \{a, b, c, d, f, g\} \qquad A = \{a, b, c\}, B = \{a, d, f, g\}$$

and

$$A \cup C = \{a, b, c, d, e\}. \qquad A = \{a, b, c\}, C = \{d, e\}$$

In Figure 3 the shaded region is the union of sets A and B.

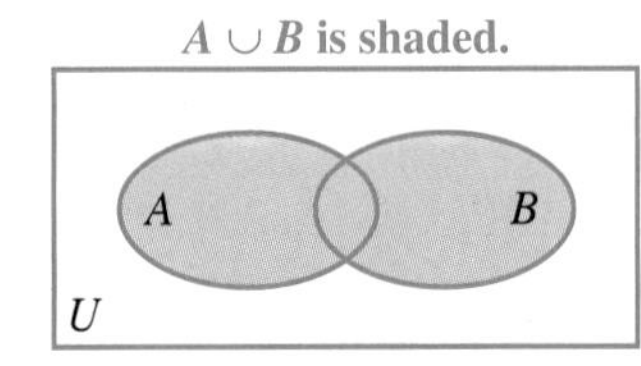

Figure 3

EXAMPLE 5 Finding the Union of Two Sets

If $M = \{2, 5, 7\}$ and $N = \{1, 2, 3, 4, 5\}$, then

$$M \cup N = \{1, 2, 3, 4, 5, 7\}.$$

Work Problem 5 at the Side.

5 Using the sets given in Margin Problem 4, find the following.

(a) $M \cup N$

(b) $N \cup Q$

ANSWERS

4. **(a)** $\{1, 3, 5, 7\}$ **(b)** $\{0, 2, 4, 6, 8\}$ **(c)** $\{5, 6, 7, 8\}$

5. **(a)** $\{0, 1, 2, 3, 4, 5, 6, 7, 8\} = U$ **(b)** $\{0, 1, 2, 3, 4, 5, 7\}$

6 Using the sets given in Margin Problem 4, find the following.

(a) $M \cap Q$

(b) $N \cap Q$

(c) $M \cap N$

7 Let

$U = \{1, 2, 3, 4, 6, 8, 10\}$,
$A = \{1, 3, 4, 6\}$,
$B = \{2, 4, 6, 8, 10\}$, and
$C = \{2, 8\}$.

Find the following.

(a) $B \cup C$

(b) $A \cap B$

(c) $A \cap C$

(d) A'

(e) $A \cup B$

Answers

6. **(a)** $\{0, 2, 4\}$ **(b)** $\{1, 3\}$ **(c)** $\emptyset$

7. **(a)** $\{2, 4, 6, 8, 10\} = B$ **(b)** $\{4, 6\}$ **(c)** $\emptyset$ **(d)** $\{2, 8, 10\}$ **(e)** $\{1, 2, 3, 4, 6, 8, 10\} = U$

The **intersection** of two sets A and B, written $A \cap B$, is the set of all elements that belong to both A and B. For example, if

$$A = \{\text{Jose, Ellen, Marge, Kevin}\}$$

and

$$B = \{\text{Jose, Patrick, Ellen, Sue}\},$$

then

$$A \cap B = \{\text{Jose, Ellen}\}.$$

The shaded region in Figure 4 represents the intersection of sets A and B.

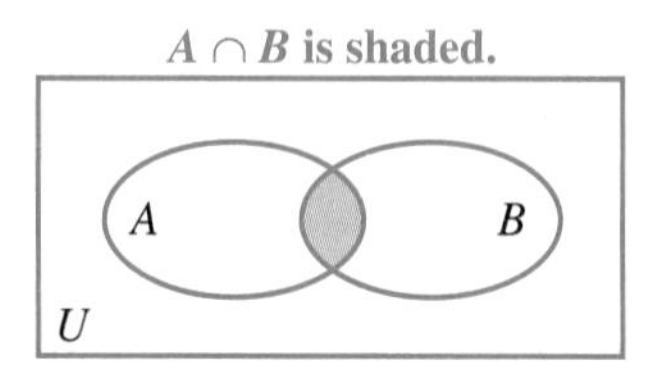

Figure 4

EXAMPLE 6 Finding the Intersection of Two Sets

Suppose that

$P = \{3, 9, 27\}$, $Q = \{2, 3, 10, 18, 27, 28\}$, and $R = \{2, 10, 28\}$.

List the elements in each set.

(a) $P \cap Q = \{3, 27\}$ **(b)** $Q \cap R = \{2, 10, 28\} = R$ **(c)** $P \cap R = \emptyset$

Work Problem 6 at the Side.

Sets like P and R in Example 6 that have no elements in common are called **disjoint sets.** The Venn diagram in Figure 5 shows a pair of disjoint sets.

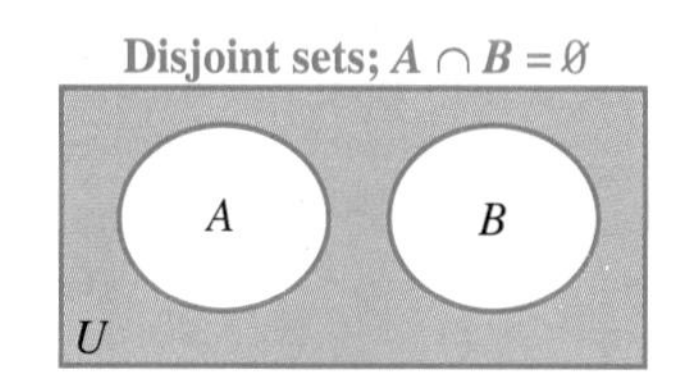

Figure 5

EXAMPLE 7 Using Set Operations

Let

$$U = \{2, 5, 7, 10, 14, 20\},$$
$$A = \{2, 10, 14, 20\},$$
$$B = \{5, 7\}, \text{ and}$$
$$C = \{2, 5, 7\}.$$

Find the following.

(a) $A \cup B = \{2, 5, 7, 10, 14, 20\} = U$

(b) $A \cap B = \emptyset$ **(c)** $B \cup C = \{2, 5, 7\} = C$

(d) $B \cap C = \{5, 7\} = B$ **(e)** $A' = \{5, 7\} = B$

Work Problem 7 at the Side.

Appendix B Exercises

FOR EXTRA HELP

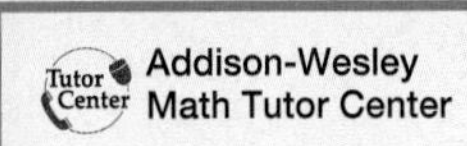
Addison-Wesley Math Tutor Center

MathXL

Student's Solutions Manual

MyMathLab

Interactmath.com

List the elements of each set. See Examples 1 and 2.

1. The set of all natural numbers less than 8

2. The set of all integers between 4 and 10

3. The set of seasons

4. The set of months of the year

5. The set of women presidents of the United States

6. The set of all living humans who are more than 200 yr old

7. The set of letters of the alphabet between K and M

8. The set of letters of the alphabet between D and H

9. The set of positive even integers

10. The set of all multiples of 5

11. Which of the sets described in Exercises 1–10 are infinite sets?

12. Which of the sets described in Exercises 1–10 are finite sets?

Tell whether each statement is true *or* false.

13. $5 \in \{1, 2, 5, 8\}$

14. $6 \in \{1, 2, 3, 4, 5\}$

15. $2 \in \{1, 3, 5, 7, 9\}$

16. $1 \in \{6, 2, 5, 1\}$

17. $7 \notin \{2, 4, 6, 8\}$

18. $7 \notin \{1, 3, 5, 7\}$

19. $\{2, 4, 9, 12, 13\} = \{13, 12, 9, 4, 2\}$

20. $\{7, 11, 4\} = \{7, 11, 4, 0\}$

Let

$$A = \{1, 3, 4, 5, 7, 8\}, \quad B = \{2, 4, 6, 8\}, \quad C = \{1, 3, 5, 7\},$$
$$D = \{1, 2, 3\}, \quad E = \{3, 7\}, \quad \text{and} \quad U = \{1, 2, 3, 4, 5, 6, 7, 8, 9, 10\}.$$

Tell whether each statement is true *or* false. *See Examples 3, 5, 6, and 7.*

21. $A \subseteq U$ **22.** $D \subseteq A$ **23.** $\emptyset \subseteq A$ **24.** $\{1, 2\} \subseteq D$ **25.** $C \subseteq A$

26. $A \subseteq C$ **27.** $D \subseteq B$ **28.** $E \subseteq C$ **29.** $D \nsubseteq E$ **30.** $E \nsubseteq A$

31. There are exactly 4 subsets of E.

32. There are exactly 8 subsets of D.

33. There are exactly 12 subsets of C.

34. There are exactly 16 subsets of B.

35. $\{4, 6, 8, 12\} \cap \{6, 8, 14, 17\} = \{6, 8\}$

36. $\{2, 5, 9\} \cap \{1, 2, 3, 4, 5\} = \{2, 5\}$

37. $\{3, 1, 0\} \cap \{0, 2, 4\} = \{0\}$

38. $\{4, 2, 1\} \cap \{1, 2, 3, 4\} = \{1, 2, 3\}$

39. $\{3, 9, 12\} \cap \emptyset = \{3, 9, 12\}$

40. $\{3, 9, 12\} \cup \emptyset = \emptyset$

41. $\{4, 9, 11, 7, 3\} \cup \{1, 2, 3, 4, 5\} = \{1, 2, 3, 4, 5, 7, 9, 11\}$

42. $\{1, 2, 3\} \cup \{1, 2, 3\} = \{1, 2, 3\}$

43. $\{3, 5, 7, 9\} \cup \{4, 6, 8\} = \emptyset$

44. $\{5, 10, 15, 20\} \cup \{5, 15, 30\} = \{5, 15\}$

Let

$$U = \{\text{a, b, c, d, e, f, g, h}\}, \quad A = \{\text{a, b, c, d, e, f}\}, \quad B = \{\text{a, c, e}\}, \quad C = \{\text{a, f}\}, \quad \text{and} \quad D = \{\text{d}\}.$$

List the elements in each set. See Examples 4–7.

45. A' **46.** B' **47.** C' **48.** D'

49. $A \cap B$ **50.** $B \cap A$ **51.** $A \cap D$ **52.** $B \cap D$

53. $B \cap C$ **54.** $A \cup B$ **55.** $B \cup D$ **56.** $B \cup C$

57. $C \cup B$ **58.** $C \cup D$ **59.** $A \cap \emptyset$ **60.** $B \cup \emptyset$

61. Name every pair of disjoint sets among A–D above.

Appendix C Mean, Median, and Mode

Appendix C Mean, Median, and Mode

OBJECTIVES

1. Find the mean of a list of numbers.
2. Find a weighted mean.
3. Find the median.
4. Find the mode.

OBJECTIVE 1 **Find the mean of a list of numbers.** Making sense of a long list of numbers can be difficult. So when we analyze data, one of the first things to look for is a *measure of central tendency*—a single number that we can use to represent the entire list of numbers. One such measure is the *average* or **mean.** The mean can be found with the following formula.

Finding the Mean (Average)

$$\text{mean} = \frac{\text{sum of all values}}{\text{number of values}}$$

EXAMPLE 1 **Finding the Mean (Average)**

David had test scores of 84, 90, 95, 98, and 88. Find his mean (average) score.

Use the formula for finding the mean. Add up all the test scores and then divide the sum by the number of tests.

$$\text{mean} = \frac{84 + 90 + 95 + 98 + 88}{5} \quad \begin{matrix}\leftarrow \text{Sum of test scores} \\ \leftarrow \text{Number of tests}\end{matrix}$$

$$\text{mean} = \frac{455}{5}$$

$$\text{mean} = 91 \quad \text{Divide.}$$

David has a mean (average) score of 91.

Work Problem 1 at the Side.

1 Tanya had test scores of 96, 98, 84, 88, 82, and 92. Find her mean (average) score.

EXAMPLE 2 **Applying the Mean (Average)**

The sales of photo albums at Sarah's Card Shop for each day last week were \$86, \$103, \$118, \$117, \$126, \$158, and \$149. Find the mean daily sales of photo albums.

To find the mean, add all the daily sales amounts and then divide the sum by the number of days (7).

Continued on Next Page

ANSWERS
1. 90

2 Find the mean monthly bill for monthly long-distance phone bills of \$25.12, \$42.58, \$76.19, \$32.00, \$81.11, \$26.41, \$19.76, \$59.32, \$71.18, and \$21.03.

$$\text{mean} = \frac{\$86 + \$103 + \$118 + \$117 + \$126 + \$158 + \$149}{7} \quad \begin{array}{l}\leftarrow \text{Sum of sales} \\ \leftarrow \text{Number of days}\end{array}$$

$$\text{mean} = \frac{\$857}{7}$$

$$\text{mean} \approx \$122.43 \quad \text{Nearest cent}$$

Work Problem 2 at the Side.

OBJECTIVE 2 Find a weighted mean. Some items in a list of data might appear more than once. In this case, we find a **weighted mean,** in which each value is "weighted" by multiplying it by the number of times it occurs.

EXAMPLE 3 Finding a Weighted Mean

The table shows the amount of contribution and the number of times the amount was given (frequency) to a food pantry. Find the weighted mean.

Contribution Value	Frequency
\$ 3	4 ← 4 people each contributed \$3.
\$ 5	2
\$ 7	1
\$ 8	5
\$ 9	3
\$10	2
\$12	1
\$13	2

In most cases, the same amount was given by more than one person. For example, \$3 was given by four people, and \$8 was given by five people. Other amounts, such as \$12, were given by only one person.

To find the mean, multiply each contribution value by its frequency. Then add the products. Next, add the numbers in the *frequency* column to find the total number of values, that is, the total number of people who contributed money.

Value	Frequency	Product
\$ 3	4	(\$3 · 4) = \$12
\$ 5	2	(\$5 · 2) = \$10
\$ 7	1	(\$7 · 1) = \$ 7
\$ 8	5	(\$8 · 5) = \$40
\$ 9	3	(\$9 · 3) = \$27
\$10	2	(\$10 · 2) = \$20
\$12	1	(\$12 · 1) = \$12
\$13	2	(\$13 · 2) = \$26
Totals	20	\$154

Finally, divide the totals.

$$\text{mean} = \frac{\$154}{20} = \$7.70$$

The mean contribution to the food pantry was \$7.70.

Work Problem 3 at the Side.

3 Alice Westmoreland works downtown. Some days she can park in cheap lots that charge \$6 or \$7. Other days she has to park in lots that charge \$9 or \$10. Last month she kept track of the amount she spent each day for parking and the number of days she spent that amount. Find her average daily parking cost.

Parking Fee	Frequency
\$ 6	2
\$ 7	6
\$ 8	3
\$ 9	4
\$10	6

ANSWERS
2. \$45.47
3. \$8.29 (to the nearest cent)

A common use of the weighted mean is to find a student's *grade point average (GPA),* as shown in the next example.

EXAMPLE 4 Applying the Weighted Mean

Find the GPA (grade point average) for a student who earned the following grades last semester. Assume A = 4, B = 3, C = 2, D = 1, and F = 0. The number of credits determines how many times the grade is counted (the frequency).

Course	Credits	Grade	Credits · Grade
Mathematics	4	A (= 4)	$4 \cdot 4 = 16$
Speech	3	C (= 2)	$3 \cdot 2 = 6$
English	3	B (= 3)	$3 \cdot 3 = 9$
Computer Science	2	A (= 4)	$2 \cdot 4 = 8$
Theater	2	D (= 1)	$2 \cdot 1 = 2$
Totals	14		41

It is common to round grade point averages to the nearest hundredth, so

$$\text{GPA} = \frac{41}{14} \approx 2.93.$$

Work Problem 4 at the Side.

OBJECTIVE 3 Find the median. Because it can be affected by extremely high or low numbers, the mean is often a poor indicator of central tendency for a list of numbers. In cases like this, another measure of central tendency, called the *median,* can be used. The **median** divides a group of numbers in half; half the numbers lie above the median, and half lie below the median.

Find the median by listing the numbers *in order* from *smallest* to *largest.* If the list contains an *odd* number of items, the median is the *middle number.*

EXAMPLE 5 Finding the Median

Find the median for this list of prices.

\$7, \$23, \$15, \$6, \$18, \$12, \$24

First arrange the numbers in numerical order from least to greatest.

Least → 6, 7, 12, 15, 18, 23, 24 ← Greatest

Next, find the middle number in the list.

6, 7, 12, **15**, 18, 23, 24

Three are below. (6, 7, 12) — Middle number (15) — Three are above. (18, 23, 24)

The median price is \$15.

Work Problem 5 at the Side.

If a list contains an *even* number of items, there is no single middle number. In this case, the median is defined as the mean (average) of the *middle two* numbers.

4 Find the GPA (grade point average) for a student who earned the following grades. Round to the nearest hundredth.

Course	Credits	Grade
Mathematics	5	A (= 4)
English	3	C (= 2)
Biology	4	B (= 3)
History	3	B (= 3)

5 Find the median for the following number of customers helped each hour at an order desk.

35, 33, 27, 31, 39, 50, 59, 25, 30

Answers
4. 3.13
5. 33

6 Find the median for this list of measurements.

178 ft, 261 ft, 126 ft, 189 ft, 121 ft, 195 ft

7 Find the mode for each list of numbers.

(a) Ages of part-time employees (in years): 28, 16, 22, 28, 34, 22, 28

(b) Total points on a screening exam: 312, 219, 782, 312, 219, 426, 507, 600

(c) Monthly commissions of salespeople: \$1706, \$1289, \$1653, \$1892, \$1301, \$1782

EXAMPLE 6 Finding the Median

Find the median for this list of ages, in years.

74, 7, 15, 13, 25, 28, 47, 59, 32, 68

First arrange the numbers in numerical order from least to greatest. Then, because the list has an even number of ages, find the middle *two* numbers.

Least → 7, 13, 15, 25, 28, 32, 47, 59, 68, 74 ← Greatest

(28, 32: Middle two numbers)

The median age is the mean of the two middle numbers.

$$\text{median} = \frac{28 + 32}{2} = \frac{60}{2} = 30 \text{ yr}$$

Work Problem 6 at the Side.

OBJECTIVE 4 Find the mode. Another statistical measure is the **mode,** which is the number that occurs *most often* in a list of numbers. For example, if the test scores for 10 students were

74, 81, 39, 74, 82, 80, 100, 92, 74, and 85,

then the mode is 74. Three students earned a score of 74, so 74 appears more times on the list than any other score. It is *not* necessary to place the numbers in numerical order when looking for the mode although that may help you find it more easily.

A list can have two modes; such a list is sometimes called **bimodal.** If no number occurs more frequently than any other number in a list, then the list has *no mode.*

EXAMPLE 7 Finding the Mode

Find the mode for each list of numbers.

(a) 51, 32, 49, 51, 49, 90, 49, 60, 17, 60

The number 49 occurs three times, which is more often than any other number. Therefore, 49 is the mode.

(b) 482, 485, 483, 485, 487, 487, 489, 486

Because both 485 and 487 occur twice, each is a mode. This list is bimodal.

(c) 10,708; 11,519; 10,972; 12,546; 13,905; 12,182

No number occurs more than once. This list has no mode.

Work Problem 7 at the Side.

Measures of Central Tendency

The **mean** is the sum of all the values divided by the number of values. It is the mathematical *average.*

The **median** is the middle number in a group of values that are listed from least to greatest. It divides a group of numbers in half.

The **mode** is the value that occurs most often in a group of values.

ANSWERS

6. 183.5 ft

7. **(a)** 28 **(b)** bimodal; 219 and 312 **(c)** no mode

Appendix C Exercises

FOR EXTRA HELP

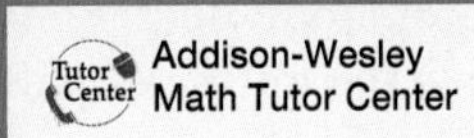

MathXL

MyMathLab

Find the mean for each list of numbers. Round answers to the nearest tenth when necessary. See Example 1.

1. Ages of infants at the child care center (in months): 4, 9, 6, 4, 7, 10, 9

2. Monthly electric bills: $53, $77, $38, $29, $49, $48

3. Final exam scores: 92, 51, 59, 86, 68, 73, 49, 80

4. Quiz scores: 18, 25, 21, 8, 16, 13, 23, 19

5. Annual salaries: $31,900; $32,850; $34,930; $39,712; $38,340; $60,000

6. Numbers of people attending baseball games: 27,500; 18,250; 17,357; 14,298; 33,110

Solve each problem. See Examples 2 and 3.

7. The Athletic Shoe Store sold shoes at the following prices: $75.52, $36.15, $58.24, $21.86, $47.68, $106.57, $82.72, $52.14, $28.60, $72.92. Find the mean shoe sales amount.

8. In one evening, a waitress collected the following checks from her dinner customers: $30.10, $42.80, $91.60, $51.20, $88.30, $21.90, $43.70, $51.20. Find the mean dinner check amount.

9. The table shows the face value (policy amount) of life insurance policies sold and the number of policies sold for each amount by the New World Life Company during one week. Find the weighted mean amount for the policies sold.

Policy Amount	Number of Policies Sold
$ 10,000	6
$ 20,000	24
$ 25,000	12
$ 30,000	8
$ 50,000	5
$100,000	3
$250,000	2

10. Detroit Metro-Sales Company prepared the following table showing the gasoline mileage obtained by each of the cars in their automobile fleet. Find the weighted mean to determine the miles per gallon for the fleet of cars.

Miles per Gallon	Number of Autos
15	5
20	6
24	10
30	14
32	5
35	6
40	4

Find each weighted mean. Round answers to the nearest tenth when necessary. See Example 3.

11.

Quiz Scores	Frequency
3	4
5	2
6	5
8	5
9	2

12.

Credits per Student	Frequency
9	3
12	5
13	2
15	6
18	1

13.

Hours Worked	Frequency
12	4
13	2
15	5
19	3
22	1
23	5

14.

Students per Class	Frequency
25	1
26	2
29	5
30	4
32	3
33	5

Find the GPA (grade point average) for students earning the following grades. Assume A = 4, B = 3, C = 2, D = 1, *and* F = 0. *Round answers to the nearest hundredth. See Example 4.*

15.

Course	Credits	Grade
Biology	4	B
Biology Lab	2	A
Mathematics	5	C
Health	1	F
Psychology	3	B

16.

Course	Credits	Grade
Chemistry	3	A
English	3	B
Mathematics	4	B
Theater	2	C
Astronomy	3	C

17. Look again at the grades in Exercise 15. Find the student's GPA in each of these situations.

(a) The student earned a B instead of an F in the 1-credit class.

(b) The student earned a B instead of a C in the 5-credit class.

(c) Both (a) and (b) happened.

18. List the credits for the courses you're taking at this time. List the lowest grades you think you will earn in each class and find your GPA. Then list the highest grades you think you will earn and find your GPA.

Find the median for each list of numbers. See Examples 5 and 6.

19. Number of e-mail messages received: 9, 12, 14, 15, 23, 24, 28

20. Deliveries by a newspaper distributor: 99, 108, 109, 123, 126, 129, 146, 168, 170

21. Students enrolled in algebra each semester: 328, 549, 420, 592, 715, 483

22. Number of cars in the parking lot each day: 520, 523, 513, 1283, 338, 509, 290, 420

23. Number of computer service calls taken each day: 51, 48, 96, 40, 47, 23, 95, 56, 34, 48

24. Number of gallons of paint sold per week: 1072, 1068, 1093, 1042, 1056, 205, 1009, 1081

The table lists the cruising speed and distance flown without refueling for several types of larger airplanes used to carry passengers. Use the table to answer Exercises 25–28.

Type of Airplane	Cruising Speed (miles per hour)	Distance without Refueling (miles)
747-400	565	7650
747-200	558	6450
DC-9	505	1100
DC-10	550	5225
727	530	1550
757	530	2875

Source: Northwest Airlines *WorldTraveler*.

25. What is the mean distance flown without refueling, to the nearest mile?

26. Find the mean cruising speed.

27. **(a)** Find the median distance.

(b) Is the median distance similar to the mean distance from Exercise 25? Explain why or why not.

28. **(a)** Find the median cruising speed.

(b) Is the median speed similar to the mean speed from Exercise 26? Explain why or why not.

Find the mode(s) for each list of numbers. Indicate whether a list is bimodal or has no mode. See Example 7.

29. Number of samples taken each hour: 3, 8, 5, 1, 7, 6, 8, 4, 5, 8

30. Monthly water bills: $21, $32, $46, $32, $49, $32, $49, $25, $32

31. Ages of retirees (in years): 74, 68, 68, 68, 75, 75, 74, 74, 70, 77

32. Patients admitted to the hospital each week: 30, 19, 25, 78, 36, 20, 45, 85, 38

33. The number of boxes of candy sold by each child: 5, 9, 17, 3, 2, 8, 19, 1, 4, 20, 10, 6

34. The weights of soccer players (in pounds): 158, 161, 165, 162, 165, 157, 163, 162

Appendix D Factoring Sums and Differences of Cubes

Appendix D Factoring Sums and Differences of Cubes

OBJECTIVES

1. Factor the difference of cubes.
2. Factor the sum of cubes.

OBJECTIVE 1 Factor the difference of cubes. The difference of squares was factored in **Section 6.5;** we can also factor the **difference of cubes.** Use the following pattern.

Difference of Cubes

$$a^3 - b^3 = (a - b)(a^2 + ab + b^2)$$

This pattern should be memorized. Multiply on the right to see that the pattern gives the correct factors, as shown in the margin.

$$\begin{array}{r} a^2 + ab + b^2 \\ a - b \\ \hline - a^2b - ab^2 - b^3 \\ a^3 + a^2b + ab^2 \quad\quad \\ \hline a^3 \quad\quad\quad\quad\quad - b^3 \end{array}$$

Notice the pattern of the terms in the factored form of $a^3 - b^3$.

- $a^3 - b^3 =$ (a binomial factor)(a trinomial factor)
- The binomial factor has the difference of the cube roots of the given terms.
- The terms in the trinomial factor are all positive.
- What you write in the binomial factor determines the trinomial factor.

$$a^3 - b^3 = (a - b)(\underbrace{a^2}_{\text{First term squared}} + \underbrace{ab}_{\text{positive product of the terms}} + \underbrace{b^2}_{\text{second term squared}})$$

EXAMPLE 1 Factoring Differences of Cubes

Factor each difference of cubes.

(a) $m^3 - 125$

Use the pattern for the difference of cubes.

$$a^3 - b^3 = (a - b)(a^2 + ab + b^2)$$

$$m^3 - 125 = m^3 - 5^3 = (m - 5)(m^2 + 5m + 5^2)$$
$$= (m - 5)(m^2 + 5m + 25)$$

Continued on Next Page

1 Factor each difference of cubes.

(a) $t^3 - 64$

(b) $2x^3 - 54$

(c) $8k^3 - y^3$

2 Factor each sum of cubes.

(a) $x^3 + 8$

(b) $64y^3 + 1$

(c) $27m^3 + 343n^3$

ANSWERS

1. **(a)** $(t - 4)(t^2 + 4t + 16)$
   (b) $2(x - 3)(x^2 + 3x + 9)$
   (c) $(2k - y)(4k^2 + 2ky + y^2)$
2. **(a)** $(x + 2)(x^2 - 2x + 4)$
   (b) $(4y + 1)(16y^2 - 4y + 1)$
   (c) $(3m + 7n)(9m^2 - 21mn + 49n^2)$

(b) $8p^3 - 27$
Since $8p^3 = (2p)^3$ and $27 = 3^3$,

$$\begin{aligned} 8p^3 - 27 &= (2p)^3 - 3^3 \\ &= (2p - 3)[(2p)^2 + (2p)3 + 3^2] \\ &= (2p - 3)(4p^2 + 6p + 9). \end{aligned}$$

(c) $4m^3 - 32n^3 = 4(m^3 - 8n^3)$ Factor out the common factor.

$$\begin{aligned} &= 4[m^3 - (2n)^3] \quad 8n^3 = (2n)^3 \\ &= 4(m - 2n)[m^2 + m(2n) + (2n)^2] \\ &= 4(m - 2n)(m^2 + 2mn + 4n^2) \end{aligned}$$

Work Problem 1 at the Side.

CAUTION

A common error in factoring a difference of cubes, such as $a^3 - b^3 = (a - b)(a^2 + ab + b^2)$, is to try to factor $a^2 + ab + b^2$. It is easy to confuse this factor with a perfect square trinomial, $a^2 + 2ab + b^2$. It is unusual to be able to further factor an expression of the form $a^2 + ab + b^2$.

OBJECTIVE 2 Factor the sum of cubes. A sum of squares, such as $m^2 + 25$, cannot be factored using real numbers, but a **sum of cubes** can be factored by the following pattern, ***which should be memorized.***

Sum of Cubes

$$a^3 + b^3 = (a + b)(a^2 - ab + b^2)$$

The only difference between the pattern for the *sum* of cubes and the pattern for the *difference* of cubes is the positive and negative signs.

$$a^3 - b^3 = (a - b)(a^2 + ab + b^2) \quad \text{Difference of cubes}$$

Positive; Same sign; Opposite sign

$$a^3 + b^3 = (a + b)(a^2 - ab + b^2) \quad \text{Sum of cubes}$$

Positive; Same sign; Opposite sign

EXAMPLE 2 Factoring Sums of Cubes

Factor each sum of cubes.

(a)
$$\begin{aligned} k^3 + 27 &= k^3 + 3^3 \\ &= (k + 3)(k^2 - 3k + 3^2) \\ &= (k + 3)(k^2 - 3k + 9) \end{aligned}$$

(b)
$$\begin{aligned} 8m^3 + 125p^3 &= (2m)^3 + (5p)^3 \\ &= (2m + 5p)[(2m)^2 - (2m)(5p) + (5p)^2] \\ &= (2m + 5p)(4m^2 - 10mp + 25p^2) \end{aligned}$$

Work Problem 2 at the Side.

Appendix D Exercises

FOR EXTRA HELP

Addison-Wesley Math Tutor Center | MathXL | Student's Solutions Manual | MyMathLab | Interactmath.com

1. To help you factor the sum or difference of cubes, complete the following list of cubes.

$1^3 =$ ____ $2^3 =$ ____ $3^3 =$ ____ $4^3 =$ ____ $5^3 =$ ____

$6^3 =$ ____ $7^3 =$ ____ $8^3 =$ ____ $9^3 =$ ____ $10^3 =$ ____

2. The following powers of x are all perfect cubes: $x^3, x^6, x^9, x^{12}, x^{15}$. Based on this observation, we may make a conjecture that if the power of a variable is divisible by ____ (with 0 remainder), then we have a perfect cube.

3. Which of the following are differences of cubes?

A. $9x^3 - 125$ **B.** $x^3 - 16$ **C.** $x^3 - 1$ **D.** $8x^3 - 27y^3$

4. Which of the following are sums of cubes?

A. $x^3 + 1$ **B.** $x^3 + 36$ **C.** $12x^3 + 27$ **D.** $64x^3 + 216y^3$

Factor. Use your answers in Exercises 1 and 2 as necessary. See Examples 1 and 2.

5. $a^3 + 1$

6. $m^3 + 8$

7. $a^3 - 1$

8. $m^3 - 8$

9. $p^3 + q^3$

10. $w^3 + z^3$

11. $y^3 - 216$

12. $x^3 - 343$

13. $k^3 + 1000$

14. $p^3 + 512$

15. $27x^3 - 1$

16. $64y^3 - 27$

17. $125a^3 + 8$

18. $216b^3 + 125$

19. $y^3 - 8x^3$

20. $w^3 - 216z^3$

21. $27a^3 - 64b^3$

22. $125m^3 - 8n^3$

23. $8p^3 + 729q^3$

24. $27x^3 + 1000y^3$

25. $16t^3 - 2$

26. $3p^3 - 81$

27. $40w^3 + 135$

28. $32z^3 + 500$

29. $x^3 + y^6$

30. $p^9 + q^3$

31. $125k^3 - 8m^9$

32. $125c^6 - 216d^3$

RELATING CONCEPTS (EXERCISES 33–40) For Individual or Group Work

A binomial may be both a difference of squares and a difference of cubes. One example of such a binomial is $x^6 - 1$. Using the techniques of ***Section 6.5,*** *one factoring method will give the complete factored form, while the other will not.* ***Work Exercises 33–40 in order,*** *to determine the method to use.*

33. Factor $x^6 - 1$ as the difference of two squares.

34. The factored form obtained in Exercise 33 consists of a difference of cubes multiplied by a sum of cubes. Factor each binomial further.

35. Now start over and factor $x^6 - 1$ as a difference of cubes.

36. The factored form obtained in Exercise 35 consists of a binomial that is a difference of squares and a trinomial. Factor the binomial further.

37. Compare your results in Exercises 34 and 36. Which one of these is the completely factored form?

38. Verify that the trinomial in the factored form in Exercise 36 is the product of the two trinomials in the factored form in Exercise 34.

39. Use the results of Exercises 33–38 to complete the following statement:

In general, if I must choose between factoring first using the method for a difference of squares or the method for a difference of cubes, I should choose the ________________ method to eventually obtain the complete factored form.

40. Find the *complete* factored form of $x^6 - 729$ using the knowledge you have gained in Exercises 33–39.

Answers to Selected Exercises

In this section we provide the answers that we think most students will obtain when they work the exercises using the methods explained in the text. If your answer does not look exactly like the one given here, it is not necessarily wrong. In many cases there are equivalent forms of the answer that are correct. For example, if the answer section shows $\frac{3}{4}$ and your answer is .75, you have obtained the right answer but written it in a different (yet equivalent) form. Unless the directions specify otherwise, .75 is just as valid an answer as $\frac{3}{4}$.

In general, if your answer does not agree with the one given in the text, see whether it can be transformed into the other form. If it can, then it is the correct answer. If you still have doubts, talk with your instructor.

DIAGNOSTIC PRETEST

(page xxix)

1. 18 **2.** $\frac{253}{24}$ or $10\frac{13}{24}$ **3.** 28.322 **4. (a)** .0099 **(b)** 472% **5.** $-|35|$ **6.** 56 **7.** 0 **8.** 4 **9.** -2 **10.** New York: 8,084,316; Los Angeles: 3,798,981 **11.** 58°, 122° **12.** $x \leq 5$

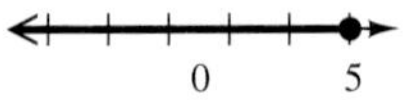

13. x-intercept: $(5, 0)$; y-intercept: $(0, -2)$

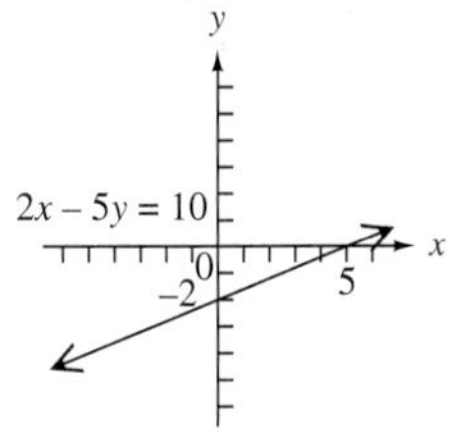

14. -4 **15.** $y = -x + 1$
16.

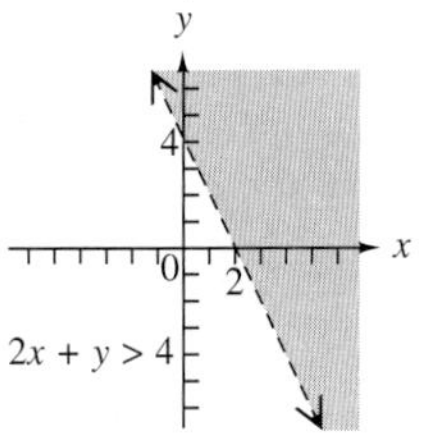

17. (2, 3) **18.** no solution **19.** Marla: 61 mph; Rick: 53 mph
20.

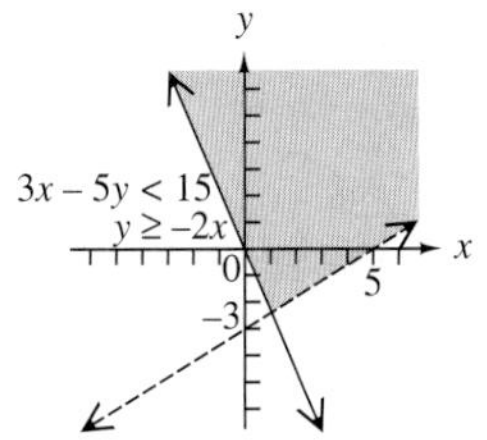

21. $-2m^3 - 9m - 5$ **22.** $49z^2 + 42zw + 9w^2$ **23.** $-\frac{3}{10}$ **24. (a)** 4.45×10^8 **(b)** .000234 **25.** $(3x - 4)(x + 2)$ **26.** $(4n + 7)(4n - 7)$ **27.** $-3, 5$ **28.** 15 in. by 11 in.

29. $\frac{x + 5}{x + 4}$ **30.** $-\frac{1}{3r}$ **31.** $\frac{z^3 + 2z^2 + 3z}{(z - 3)(z + 3)}$ **32.** $y + 1$ **33.** 0 **34.** $12 - 2\sqrt{35}$ **35.** $2\sqrt{10}$ **36.** 27 **37.** $3 + 2\sqrt{5}, 3 - 2\sqrt{5}$ **38.** $\frac{-3 + \sqrt{17}}{4}, \frac{-3 - \sqrt{17}}{4}$

39. vertex: $(0, 5)$

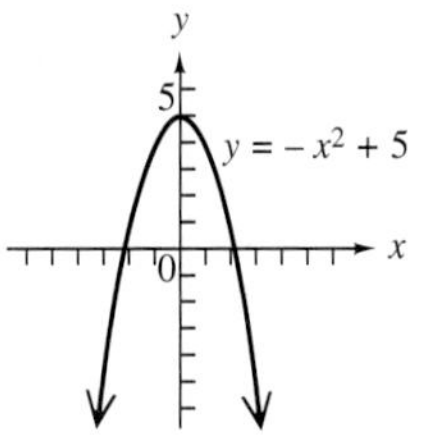

40. function; domain: $\{-2, -1, 0, 1, 2\}$; range: $\{0, 1, 4\}$

CHAPTER R PREALGEBRA REVIEW

Section R.1 (page 9)

1. true **3.** false; The fraction $\frac{17}{51}$ can be simplified to $\frac{1}{3}$. **5.** false; *Product* indicates multiplication, so the product of 8 and 2 is 16. **7.** prime **9.** composite **11.** composite **13.** neither **15.** $2 \cdot 3 \cdot 5$ **17.** $2 \cdot 2 \cdot 3 \cdot 3 \cdot 7$ **19.** $2 \cdot 2 \cdot 31$ **21.** 29 **23.** $\frac{1}{2}$ **25.** $\frac{5}{6}$ **27.** $\frac{1}{5}$ **29.** $\frac{6}{5}$ **31.** A **33.** $\frac{24}{35}$ **35.** $\frac{6}{25}$ **37.** $\frac{6}{5}$ or $1\frac{1}{5}$ **39.** $\frac{232}{15}$ or $15\frac{7}{15}$ **41.** $\frac{10}{3}$ or $3\frac{1}{3}$ **43.** 12 **45.** $\frac{1}{16}$ **47.** $\frac{84}{47}$ or $1\frac{37}{47}$ **49.** Multiply the first fraction (the dividend) by the reciprocal of the second fraction (the divisor) to divide two fractions. **51.** $\frac{2}{3}$ **53.** $\frac{8}{9}$ **55.** $\frac{27}{8}$ or $3\frac{3}{8}$ **57.** $\frac{17}{36}$ **59.** $\frac{11}{12}$ **61.** $\frac{4}{3}$ or $1\frac{1}{3}$ **63.** 6 cups **65.** $618\frac{3}{4}$ ft **67.** $\frac{9}{16}$ in. **69.** $\frac{5}{16}$ in. **71.** $16\frac{5}{8}$ yd **73.** $\frac{1}{20}$ **75.** more than $1\frac{1}{25}$ million

Section R.2 (page 21)

1. (a) 6 **(b)** 9 **(c)** 1 **(d)** 7 **(e)** 4 **3. (a)** 46.25 **(b)** 46.2 **(c)** 46 **(d)** 50 **5.** C **7.** B **9.** $\frac{4}{10}$ **11.** $\frac{64}{100}$ **13.** $\frac{138}{1000}$ **15.** $\frac{3805}{1000}$ **17.** 139; 143.094 **19.** 27; 25.61 **21.** 10; 15.33 **23.** 82; 81.716 **25.** 17; 15.211 **27.** .006; .006 **29.** 90; 116.48 **31.** 6; 7.15 **33.** 2; 2.05 **35.** 6000; 5711.6 **37.** .2; .162 **39.** To add or subtract decimals, line up the decimal points in a column, add or subtract as usual, and move the decimal point straight down in the sum or difference. **41.** .125 **43.** .25 **45.** $.\overline{5}$; .556 **47.** $.1\overline{6}$; .167 **49.** To convert a decimal to a percent, move the decimal point two places to the right and attach a percent symbol (%). **51.** .54 **53.** 1.17 **55.** .024 **57.** .0625 **59.** .008 **61.** 75% **63.** .4% **65.** 128% **67.** 30% **69.** 75% **71.** $83.\overline{3}\%$ **73. (a)** $\frac{2}{32}$ **(b)** $\frac{2}{32}$

CHAPTER 1 THE REAL NUMBER SYSTEM

Section 1.1 (page 29)

1. true **3.** false; Using the guidelines for order of operations gives $4 + 3(8 - 2) = 4 + 3(6) = 4 + 18 = 22$. **5.** false; The correct translation is $4 = 16 - 12$. **7.** 49 **9.** 144 **11.** 64 **13.** 1000 **15.** 81 **17.** 1024 **19.** $\frac{16}{81}$ **21.** .000064 **23.** The 4 would be applied last because we work first inside the parentheses. **25.** 58 **27.** 13 **29.** 32 **31.** 19 **33.** $\frac{49}{30}$ **35.** 12 **37.** 36.14 **39.** 26 **41.** 4 **43.** 95 **45.** 12 **47.** 14 **49.** $\frac{19}{2}$ **51.** false **53.** true **55.** true **57.** false **59.** false **61.** true **63.** $15 = 5 + 10$ **65.** $9 > 5 - 4$ **67.** $16 \neq 19$ **69.** $2 \leq 3$ **71.** Seven is less than nineteen. True **73.** One-third is not equal to three-tenths. True **75.** Eight is greater than or equal to eleven. False **77.** $30 > 5$ **79.** $3 \leq 12$ **81.** Alaska, Texas, California, Idaho **83.** Alaska, Texas, California, Idaho, Missouri

Section 1.2 (page 35)

1. 11 **3.** $13 + x$; 16 **5.** expression; equation **7.** The equation would be $5x - 9 = 49$. **9.** Answers will vary. Two such pairs are $x = 0, y = 6$ and $x = 1, y = 4$. To find a pair, choose one number, substitute it for a variable, then calculate the value for the other variable. **11. (a)** 64 **(b)** 144 **13. (a)** $\frac{7}{8}$ **(b)** $\frac{13}{12}$ **15. (a)** 9.569 **(b)** 14.353 **17. (a)** 52 **(b)** 114 **19. (a)** 12 **(b)** 33 **21. (a)** 6 **(b)** $\frac{9}{5}$ **23. (a)** $\frac{4}{3}$ **(b)** $\frac{13}{6}$ **25. (a)** $\frac{2}{7}$ **(b)** $\frac{16}{27}$ **27. (a)** 12 **(b)** 55 **29. (a)** 1 **(b)** $\frac{28}{17}$ **31. (a)** 3.684 **(b)** 8.841 **33.** $12x$ **35.** $x - 2$ **37.** $7 - \frac{1}{3}x$ **39.** $2x - 6$ **41.** $\frac{12}{x+3}$ **43.** $6(x - 4)$ **45.** The word *and* does not signify addition here. In the phrase "the product of a number and 6," *and* connects two quantities to be multiplied. **47.** no **49.** yes **51.** yes **53.** no **55.** yes **57.** yes **59.** $x + 8 = 18$ **61.** $2x + 5 = 5$ **63.** $16 - \frac{3}{4}x = 13$ **65.** $3x = 2x + 8$ **67.** expression **69.** equation **71.** \$12.52; more by \$.03 **72.** \$14.00; same **73.** \$14.49; less by \$.04 **74.** \$14.99; more by \$.04

Section 1.3 (page 45)

1. 4 **3.** 0 **5.** One example is $\sqrt{13}$. There are others. **7. (a)** 3, 7 **(b)** 0, 3, 7 **(c)** $-9, 0, 3, 7$ **(d)** $-9, -1\frac{1}{4}, -\frac{3}{5}, 0, 3, 5.9, 7$ **(e)** $-\sqrt{7}, \sqrt{5}$ **(f)** All are real numbers. **9.** 40,000 **11.** -9000 **13.**

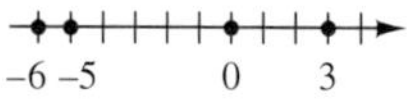

15.

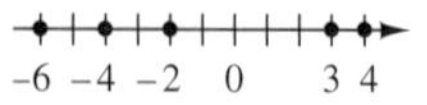

17.

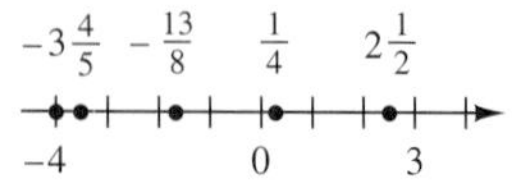

19. -11 **21.** -21 **23.** -100 **25.** $-\frac{2}{3}$ **27.** false **29.** true **31. (a)** 2 **(b)** 2 **33. (a)** -6 **(b)** 6 **35. (a)** $\frac{3}{4}$ **(b)** $\frac{3}{4}$ **37.** 7 **39.** -12 **41.** $-\frac{2}{3}$ **43.** 9 **45.** false **47.** true **49.** No; the statement is false for one number, 0. **51.** fuel/other utilities from 2001 to 2002 **53.** 2000 to 2001

Section 1.4 (page 51)

1. Add -2 and 5. **3.** Add -1 and -3. **5.** 2 **7.** -3 **9.** -10 **11.** -13 **13.** -15.9 **15.** 5 **17.** 13 **19.** 0 **21.** -8 **23.** $\frac{3}{10}$ **25.** $\frac{1}{2}$ **27.** $-\frac{3}{4}$ **29.** -1.6 **31.** -8.7 **33.** -25 **35.** $-\frac{1}{4}$ **37.** true **39.** false **41.** true **43.** false **45.** true **47.** false **49.** It must be negative and have the larger absolute value. **50.** The sum of a positive number and 5 cannot be -7. **51.** It must be positive and have the larger absolute value. **52.** The sum of a negative number and -8 cannot be 2. **53.** Add the absolute values of the numbers. The sum will be negative. **55.** $-5 + 12 + 6$; 13 **57.** $[-19 + (-4)] + 14$; -9 **59.** $[-4 + (-10)] + 12$; -2 **61.** $\left[\frac{5}{7} + \left(-\frac{9}{7}\right)\right] + \frac{2}{7}$; $-\frac{2}{7}$ **63.** $-\$80$ **65.** -184 m **67.** 37 yd **69.** 120°F **71.** $-\$107$ **73.** -12

Section 1.5 (page 59)

1. -8; -6 **3.** $7 - 12$; $12 - 7$ **5.** -4 **7.** -10 **9.** -16 **11.** 11 **13.** 19 **15.** -4 **17.** 5 **19.** 0 **21.** $\frac{3}{4}$ **23.** $-\frac{11}{8}$ **25.** $\frac{15}{8}$ **27.** 13.6 **29.** -11.9 **31.** -2.8 **33.** -6.3 **35.** -28 **37.** -18 **39.** $\frac{37}{12}$ **41.** -42.04 **43.** For example, let $a = 1, b = 1$ or let $a = 2, b = 2$. In general, choose $a = b$. **45.** 8 **47.** For example, $-8 - (-2) = -6$. **49.** $4 - (-8)$; 12 **51.** $-2 - 8$; -10 **53.** $[9 + (-4)] - 7$; -2 **55.** $[8 - (-5)] - 12$; 1 **57.** -58°F **59.** 14,776 ft **61.** $-\$80$ **63.** \$105,000 **65.** 17 **67.** \$1045.55 **69.** 469 B.C. **71.** \$323.83 **73.** 14 ft **75.** 40,776 ft **77.** -1.6% **79.** \$4900 **81.** \$8800 **83.** positive **85.** positive

Section 1.6 (page 73)

1. greater than 0 **3.** less than 0 **5.** greater than 0 **7.** -28 **9.** 30 **11.** 0 **13.** $\frac{5}{6}$ **15.** -2.38 **17.** $\frac{3}{2}$ **19.** -3 **21.** -2 **23.** 16 **25.** 0 **27.** undefined **29.** $\frac{3}{2}$ **31.** C **33.** 3 **35.** 7 **37.** 4 **39.** -3 **41.** -1 **43.** negative; impossible to tell **45.** 68 **47.** -228 **49.** 1 **51.** 0 **53.** -6 **55.** undefined **57.** $-12 + 4(-7)$; -40 **59.** $-1 - 2(-8)(2)$; 31 **61.** $-3[3 - (-7)]$; -30 **63.** $\frac{3}{10}[-2 + (-28)]$; -9 **65.** $\frac{-20}{-8 + (-2)}$; 2 **67.** $\frac{-18 + (-6)}{2(-4)}$; 3 **69.** $\frac{-\frac{2}{3}\left(-\frac{1}{5}\right)}{\frac{1}{7}}$; $\frac{14}{15}$ **71.** $9x = -36$ **73.** $\frac{x}{4} = -1$ **75.** $x - \frac{9}{11} = 5$ **77.** $\frac{6}{x} = -3$ **78.** 42 **79.** 5 **80.** $8\frac{2}{5}$ **81.** $8\frac{2}{5}$ **82.** 2 **83.** $-12\frac{1}{2}$

Summary Exercises on Operations with Real Numbers (page 77)

1. -16 **2.** 4 **3.** 0 **4.** -24 **5.** -17 **6.** 76 **7.** -18 **8.** 90 **9.** 38 **10.** 4 **11.** -5 **12.** 5 **13.** $-\frac{7}{2}$ or $-3\frac{1}{2}$ **14.** 4

15. 13 **16.** $\frac{5}{4}$ or $1\frac{1}{4}$ **17.** 9 **18.** $\frac{37}{10}$ or $3\frac{7}{10}$ **19.** 0 **20.** 25 **21.** 14 **22.** 0 **23.** -4 **24.** $\frac{6}{5}$ or $1\frac{1}{5}$ **25.** -1 **26.** $\frac{52}{37}$ or $1\frac{15}{37}$ **27.** $\frac{17}{16}$ or $1\frac{1}{16}$ **28.** $-\frac{2}{3}$ **29.** 3.33 **30.** 1.02 **31.** -13 **32.** 0 **33.** 24 **34.** -7 **35.** 37 **36.** -3 **37.** -1 **38.** $\frac{1}{2}$ **39.** $-\frac{5}{13}$ **40.** 5 **41.** undefined **42.** 0

Section 1.7 (page 85)

1. B **3.** C **5.** B **7.** G **9.** commutative property **11.** associative property **13.** inverse property **15.** inverse property **17.** identity property **19.** commutative property **21.** distributive property **23.** identity property **25.** distributive property **27. (a)** 0 **(b)** 1, -1 **29.** $25 - (6 - 2) = 25 - 4 = 21$ and $(25 - 6) - 2 = 19 - 2 = 17$. Since these results are different, subtraction is not associative. **31.** $7 + r$ **33.** s **35.** $-6x + (-6)7$; $-6x - 42$ **37.** $w + [5 + (-3)]$; $w + 2$ **39.** We must multiply $\frac{3}{4}$ by 1 in the form $\frac{3}{3}$: $\frac{3}{4} \cdot \frac{3}{3} = \frac{9}{12}$. **41.** 2 **43.** $5(3 + 17)$; 100 **45.** $4t + 12$ **47.** $-8r - 24$ **49.** $-5y + 20$ **51.** $-16y - 20z$ **53.** $8(z + w)$ **55.** $7(2v + 5r)$ **57.** $24r + 32s - 40y$ **59.** $-24x - 9y - 12z$ **61.** $-4t - 5m$ **63.** $5c + 4d$ **65.** $3q - 5r + 8s$ **67.** Answers will vary. For example, "putting on your socks" and "putting on your shoes" **69.** false **71.** (foreign sales) clerk; foreign (sales clerk) **73.** 0 **74.** $-3(5) + (-3)(-5)$ **75.** -15 **76.** The product $-3(-5)$ must equal 15, since it is the additive inverse of -15.

Section 1.8 (page 93)

1. C **3.** A **5.** $15x$ **7.** $5t$ **9.** $4r + 11$ **11.** $5 + 2x - 6y$ **13.** $-7 + 3p$ **15.** -12 **17.** 5 **19.** 1 **21.** -1 **23.** 74 **25.** Answer will vary. For example, $-3x$ and $4x$ **27.** like **29.** unlike **31.** like **33.** unlike **35.** We cannot "add" two unlike terms to obtain a single term, so we must be able to identify like terms in order to combine them. **37.** $1 - 2x$ **39.** $-\frac{1}{3}t - \frac{28}{3}$ **41.** $-4.1r + 4.2$ **43.** $-2y^2 + 3y^3$ **45.** $-19p + 16$ **47.** $-\frac{3}{2}y + 16$ **49.** $-16y + 63$ **51.** $(x + 3) + 5x$; $6x + 3$ **53.** $(13 + 6x) - (-7x)$; $13 + 13x$ **55.** $2(3x + 4) - (-4 + 6x)$; 12 **57.** Wording may vary. One example is "the difference between 9 times a number and the sum of the number and 2." **59.** $1000 + 5x$ (dollars) **60.** $750 + 3y$ (dollars) **61.** $1000 + 5x + 750 + 3y$ (dollars) **62.** $1750 + 5x + 3y$ (dollars)

Chapter 1 Review Exercises (page 99)

1. 625 **2.** .00000081 **3.** .009261 **4.** $\frac{125}{8}$ **5.** 27 **6.** 200 **7.** 7 **8.** $\frac{20}{3}$ **9.** $13 < 17$ **10.** $5 + 2 \neq 10$ **11.** Six is less than fifteen. **12.** Answers will vary. One example is $2 + 5 \geq \frac{16}{2}$. **13.** 30 **14.** 60 **15.** 14 **16.** 13 **17.** $x + 6$ **18.** $8 - x$ **19.** $6x - 9$ **20.** $12 + \frac{3}{5}x$ **21.** yes **22.** no **23.** $2x - 6 = 10$ **24.** $4x = 8$ **25.** equation **26.** expression

27. 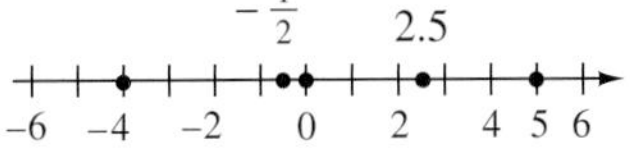

28. 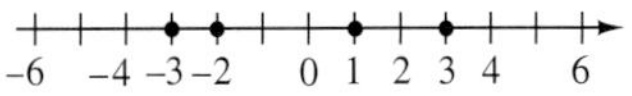

29. 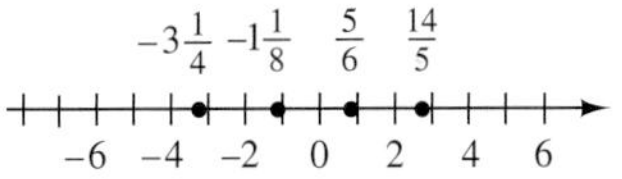

30. 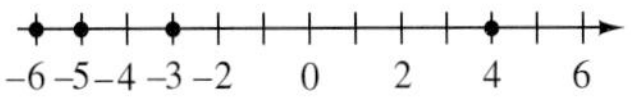

31. -10 **32.** -9 **33.** $-\frac{3}{4}$ **34.** $-|23|$ **35.** true **36.** true **37.** true **38.** false **39.** -3 **40.** -19 **41.** -7 **42.** 9 **43.** -6 **44.** -4 **45.** -17 **46.** $-\frac{29}{36}$ **47.** -10 **48.** -19 **49.** $(-31 + 12) + 19$; 0 **50.** $[-4 + (-8)] + 13$; 1 **51.** $-\$8$ **52.** 87°F **53.** -11 **54.** -1 **55.** 7 **56.** $-\frac{43}{35}$ **57.** 10.31 **58.** -12 **59.** 2 **60.** 1 **61.** $-4 - (-6)$; 2 **62.** $[4 + (-8)] - 5$; -9 **63.** \$713.4 billion **64.** 1 min, 28.89 sec **65.** The first step is to change subtracting -6 to adding its opposite, 6. So the problem becomes $-8 + 6$. This sum is -2. **66.** Yes; for example, $-2 - (-6) = -2 + 6 = 4$, a positive number. **67.** \$13.5 billion **68.** $-\$11.3$ billion **69.** $-\$2.2$ billion **70.** \$5.0 billion **71.** 36 **72.** -105 **73.** $\frac{1}{2}$ **74.** 10.08 **75.** -20 **76.** -10 **77.** -24 **78.** -35 **79.** 4 **80.** -20 **81.** $-\frac{3}{4}$ **82.** 11.3 **83.** -1 **84.** undefined **85.** 1 **86.** 0 **87.** -18 **88.** -18 **89.** 125 **90.** -423 **91.** $-4(5) - 9$; -29 **92.** $\frac{5}{6}[12 + (-6)]$; 5 **93.** $\frac{12}{8 + (-4)}$; 3 **94.** $\frac{-20(12)}{15 - (-15)}$; -8 **95.** $\frac{x}{x + 5} = -2$ **96.** $8x - 3 = -7$ **97.** identity property **98.** identity property **99.** inverse property **100.** inverse property **101.** associative property **102.** associative property **103.** distributive property **104.** commutative property **105.** $(7 + 1)y$; $8y$ **106.** $-12 \cdot 4 - 12(-t)$; $-48 + 12t$ **107.** $3(2s + 4y)$; $6s + 12y$ **108.** $-1(-4r) + (-1)(5s)$; $4r - 5s$ **109.** $17p^2$ **110.** $16r^2 + 7r$ **111.** $-19k + 54$ **112.** $5s - 6$ **113.** $-45t - 23$ **114.** $-45t^2 - 23.4t$ **115.** $-2(3x) - 7x$; $-13x$ **116.** $\frac{x + 9}{x - 6}$ **117.** No. The use of *and* there indicates the two quantities that are to be multiplied. **118.** Answers may vary. For example, "3 times the difference between 4 times a number and 6" **119.** 16 **120.** $\frac{25}{36}$ **121.** -26 **122.** $\frac{8}{3}$ **123.** $-\frac{1}{24}$ **124.** $\frac{7}{2}$ **125.** 2 **126.** 77.6 **127.** $-1\frac{1}{2}$ **128.** 11 **129.** $-\frac{28}{15}$ **130.** 24 **131.** -11 **132.** -6 **133.** $2x - 1400 = 25{,}800$; x represents the amount spent in 2002. **134.** $\frac{x}{3x - 14}$; x represents the number.

Chapter 1 Test (page 105)

1. true **2.** false **3.** 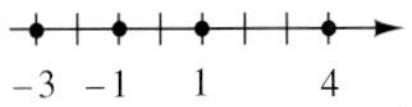

4. $-|-8|$ (or -8) **5.** -1.277 **6.** $\frac{-6}{2 + (-8)}$; 1 **7.** negative **8.** 4 **9.** $-2\frac{5}{6}$ **10.** 2 **11.** 6 **12.** 108 **13.** 11 **14.** $\frac{30}{7}$

15. -70 **16.** 3 **17.** 178°F **18.** D **19.** A **20.** E **21.** B **22.** C **23.** $-9x^2 - 6x - 8$ **24.** identity and distributive properties **25. (a)** -18 **(b)** -18 **(c)** The distributive property tells us that the two methods produce equal results.

Chapter 2 Equations, Inequalities, and Applications

Section 2.1 (page 113)

1. A and C **3.** A and B **5.** 12 **7.** -3 **9.** 4 **11.** -9 **13.** 6.3 **15.** -16.9 **17.** -10 **19.** -13 **21.** 10 **23.** 7 **25.** -4 **27.** 3 **29.** -2 **31.** 4 **33.** 2 **35.** -4 **37.** 4 **39.** 0 **41.** $\frac{7}{15}$ **43.** 7 **45.** -4 **47.** 13 **49.** 29 **51.** 18 **53.** Answers will vary. One example is $x - 6 = -8$.

Section 2.2 (page 119)

1. $\frac{3}{2}$ **3.** 10 **5.** $-\frac{2}{9}$ **7.** -1 **9.** 6 **11.** -4 **13.** .12 **15.** -1 **17.** To get x alone on the left side, divide each side by 4, the coefficient of x. **19.** 6 **21.** $\frac{15}{2}$ **23.** -5 **25.** $-\frac{18}{5}$ **27.** 12 **29.** 0 **31.** -12 **33.** 40 **35.** -12.2 **37.** -48 **39.** -35 **41.** 72 **43.** 14 **45.** $-\frac{27}{35}$ **47.** 3 **49.** -5 **51.** 7 **53.** 0 **55.** -6 **57.** Answers will vary. One example is $\frac{3}{2}x = -6$. **59.** $-4x = 10$; $-\frac{5}{2}$

Section 2.3 (page 127)

1. -1 **3.** 5 **5.** 1 **7.** $-\frac{5}{3}$ **9.** -1 **11.** no solution **13.** all real numbers **15.** D **17.** Simplify each side separately. Use the addition property to get all variable terms on one side of the equation and all numbers on the other, and then combine terms. Use the multiplication property to get the equation in the form $x =$ a number. Check the solution. **19.** 7 **21.** 0 **23.** $\frac{3}{25}$ **25.** 60 **27.** 4 **29.** 5000 **31.** 800 **32.** Yes, you will get $(100 \cdot 2) \cdot 4 = 800$. This is a result of the associative property of multiplication. **33.** No, because $(100a)(100b) = 10{,}000ab \neq 100ab$. **34.** The distributive property involves the operation of addition as well. **35.** Yes; the associative property of multiplication is used. **36.** no **37.** $-\frac{72}{11}$ **39.** 0 **41.** -6 **43.** 15 **45.** all real numbers **47.** no solution **49.** $12 - q$ **51.** $a + 12$; $a - 5$ **53.** $\frac{t}{10}$

Summary Exercises on Solving Linear Equations (page 131)

1. -5 **2.** 4 **3.** -5.1 **4.** 25 **5.** -25 **6.** -6 **7.** -3 **8.** -16 **9.** 7 **10.** $-\frac{96}{5}$ **11.** 5 **12.** 23.7 **13.** all real numbers **14.** 1 **15.** -16 **16.** no solution **17.** 6 **18.** 3 **19.** no solution **20.** $\frac{7}{3}$ **21.** 25 **22.** -10.8 **23.** 3 **24.** 7 **25.** 2 **26.** all real numbers **27.** $-\frac{2}{7}$ **28.** 10 **29.** $\frac{14}{17}$ **30.** $-\frac{5}{2}$ **31.** all real numbers **32.** 64

Section 2.4 (page 141)

1. The procedure should include the following steps: read the problem carefully; assign a variable to represent the unknown to be found; write down variable expressions for any other unknown quantities; translate into an equation; solve the equation; state the answer; check your solution. **3.** D; there cannot be a fractional number of cars. **5.** A; distance cannot be negative. **7.** 7 **9.** -8 **11.** 6 **13.** -3 **15.** California: 59 screens; New York: 48 screens **17.** Democrats: 48; Republicans: 51 **19.** Bruce Springsteen and the E Street Band: \$115.9 million; Céline Dion: \$80.5 million **21.** wins: 59; losses: 23 **23.** 1950 Denver nickel: \$8.00; 1945 Philadelphia nickel: \$7.00 **25.** ice cream: 44,687.9 lb; topping: 537.1 lb **27.** 35 lb **29.** Airborne Express: 3; Federal Express: 9; United Parcel Service: 1 **31.** gold: 35; silver: 39; bronze: 29 **33.** 36 million mi **35.** A and B: 40°; C: 100° **37.** no **39.** $x - 1$ **41.** 18° **43.** 39° **45.** 50° **47.** 68, 69 **49.** 146, 147 **51.** 10, 12 **53.** 10, 11 **55.** \$4.66 billion, \$5.27 billion, \$6.20 billion

Section 2.5 (page 153)

1. (a) The perimeter of a plane geometric figure is the distance around the figure. **(b)** The area of a plane geometric figure is the measure of the surface covered or enclosed by the figure. **3.** four **5.** area **7.** perimeter **9.** area **11.** area **13.** $P = 26$ **15.** $A = 64$ **17.** $b = 4$ **19.** $t = 5.6$ **21.** $I = 1575$ **23.** $r = 2.6$ **25.** $A = 50.24$ **27.** $V = 150$ **29.** $V = 52$ **31.** $V = 7234.56$ **33.** about 154,000 ft^2 **35.** perimeter: 13 in.; area: 10.5 in.^{2} **37.** 194.48 ft^2 **39.** 23,800.10 ft^2 **41.** length: 36 in.; volume: 11,664 in.^{3} **43.** 48°, 132° **45.** 51°, 51° **47.** 105°, 105° **49.** $t = \frac{d}{r}$ **51.** $H = \frac{V}{LW}$ **53.** $b = P - a - c$ **55.** $r = \frac{I}{pt}$ **57.** $h = \frac{2A}{b}$ **59.** $W = \frac{P - 2L}{2}$ or $W = \frac{P}{2} - L$ **61.** $h = \frac{3V}{\pi r^2}$ **63.** $m = \frac{y - b}{x}$ **65.** $r = \frac{M - C}{C}$ or $r = \frac{M}{C} - 1$

Section 2.6 (page 163)

1. (a) C **(b)** D **(c)** B **(d)** A **3.** $\frac{6}{7}$ **5.** $\frac{18}{55}$ **7.** $\frac{5}{16}$ **9.** $\frac{4}{15}$ **11.** 10-lb size; \$.439 **13.** 32-oz size; \$.093 **15.** 128-oz size; \$.044 **17.** 36-oz size; \$.049 **19.** A percent is a ratio where the basis of comparison is 100. For example, 27% represents the ratio of 27 to 100. **21.** 35 **23.** 7 **25.** 2 **27.** -1 **29.** 5 **31.** $-\frac{31}{5}$ **33.** \$28.35 **35.** 4 ft **37.** 203.3 million vehicles **39.** 6.875 fluid oz **41.** \$733.16 **43.** 9.234 **45.** 200% **47.** 30,000 **49.** 27% **51.** C **53.** 5.8% **55.** 79.0% **57.** \$304 **59.** 284% **61.** \$252 **63.** \$293 **65.** 30 **66. (a)** $5x = 12$ **(b)** $\frac{12}{5}$ **67.** $\frac{12}{5}$ **68.** Both methods give the same solution.

Summary Exercises on Solving Applied Problems (page 167)

1. 243 votes **2.** 48 **3.** 9 tanks **4.** 9 lb **5.** 4 **6.** 2 cm **7.** 80° **8.** 100° **9.** 3 **10.** -3 **11.** 15 **12.** 28 **13.** 36 quart cartons **14.** length: 9 ft; width: 3 ft **15.** 104°, 104° **16.** 140°, 40° **17.** 24.34 in. **18.** 727.28 in.^{2} **19.** 2.5 cm **20.** 12.42 cm **21.** Mr. Silvester: 25 yr old; Mrs. Silvester: 20 yr old **22.** Chris: 19 yr old; Josh: 9 yr old **23.** $16\frac{2}{3}\%$ **24.** \$2.99

25. 4000 calories **26.** 420 mi **27.** cherry: 1200 lb; spoon: 5800 lb **28.** Mike Weir: 281; Tiger Woods: 290 **29.** 510 calories **30.** $16\frac{1}{2}$ oz **31.** \$36.29 **32.** 38.3% **33.** approximately 2593 mi **34.** 32 gold medals **35.** 32-oz size; \$.053 **36.** 50-oz size; \$.094

Section 2.7 (page 177)

1. $x > -4$ **3.** $x \leq 4$ **5.** $-1 < x < 2$ **7.** $-1 < x \leq 2$

9. Use an open circle if the symbol is > or <. Use a closed circle if the symbol is ≥ or ≤.

11. 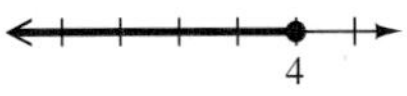 4 **13.** 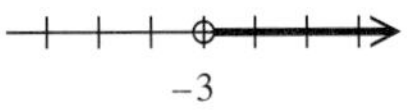 -3

15.  8 10 **17.** 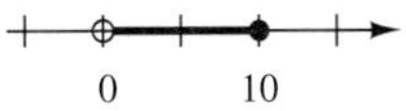 0 10

19. It would imply that $3 < -2$, which is false.

21. $z \geq 1$ 1

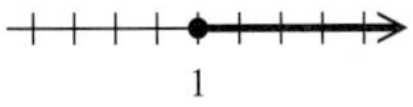

23. $k \geq 5$ 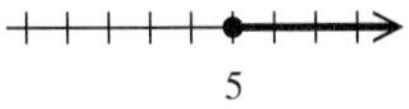 5

25. $n < -11$ -11

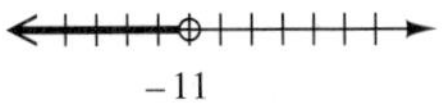

27. It must be reversed when multiplying or dividing by a negative number.

29. $x < 6$ 6

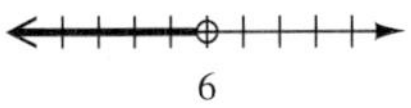

31. $x \geq -10$ -10

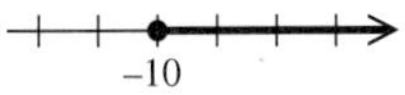

33. $t < -3$ -3

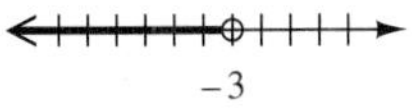

35. $x \leq 0$ 0

37. $r > 20$ 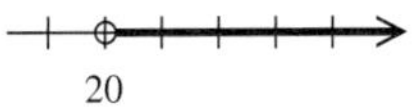 20

39. $x \geq -3$ -3

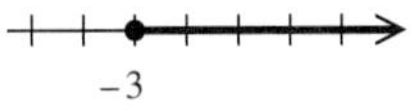

41. $r \geq -5$ -5

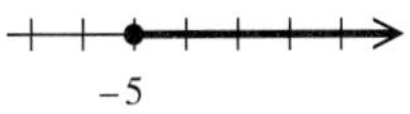

43. $x < 1$ 1

45. $x \leq 0$ 0

47. $x \geq 4$  4

49. $p < 32$ 32

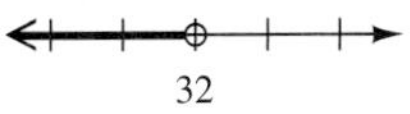

51. $x \geq \frac{5}{12}$ 0 $\frac{5}{12}$

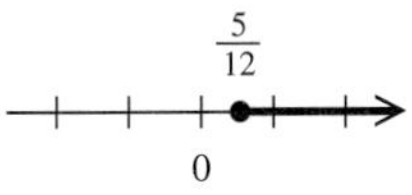

53. $k > -21$ -21

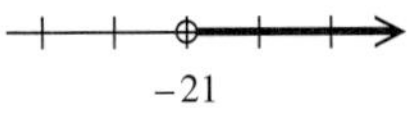

55. 88 or more **57.** all numbers greater than 16 **59.** It has never exceeded 40°C. **61.** 32 or greater **63.** 15 min

65. 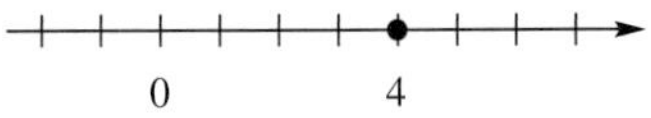 0 4

66. 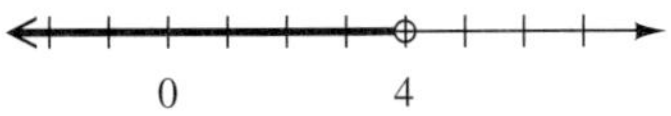 0 4

67. 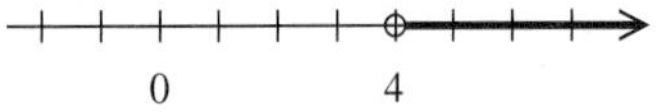 0 4

68. It is the set of all real numbers.
0 4

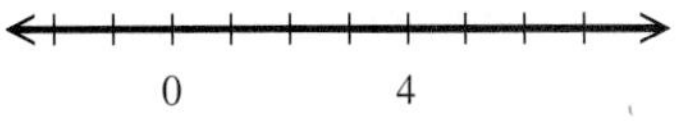

69. The graph would be the set of all real numbers.

Chapter 2 Review Exercises (page 185)

1. 9 **2.** 4 **3.** -6 **4.** $\frac{3}{2}$ **5.** 20 **6.** $-\frac{61}{2}$ **7.** 15 **8.** 0 **9.** no solution **10.** all real numbers **11.** $-\frac{7}{2}$ **12.** 20

13. Hawaii: 6425 mi²; Rhode Island: 1212 mi² **14.** Seven Falls: 300 ft; Twin Falls: 120 ft **15.** 80° **16.** 11, 13 **17.** $h = 11$ **18.** $A = 28$ **19.** $r = 4.75$ **20.** $V = 3052.08$ **21.** $W = \frac{A}{L}$ **22.** $h = \frac{2A}{b + B}$ **23.** 135°, 45° **24.** 100°, 100°

25. perimeter: 326.5 ft; area: 6538.875 ft² **26.** diameter: 46.78 ft; radius: 23.39 ft **27.** $\frac{3}{2}$ **28.** $\frac{5}{14}$ **29.** $\frac{3}{4}$ **30.** $\frac{1}{12}$ **31.** $\frac{7}{2}$ **32.** $-\frac{8}{3}$ **33.** $\frac{25}{19}$ **34.** 40% means $\frac{40}{100}$ or $\frac{2}{5}$. It is the same as the ratio of 2 to 5. **35.** $6\frac{2}{3}$ lb **36.** 36 oz **37.** 50,000 fish **38.** 375 km **39.** 17.48 **40.** 175% **41.** $33\frac{1}{3}$% **42.** 2500 **43.** \$17,800 **44.** \$350.46

45.  -4 **46.** 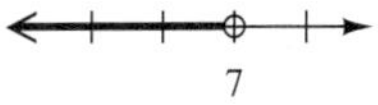 7

47.  -5 6 **48.** 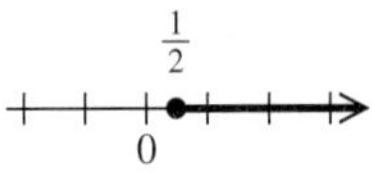 0 $\frac{1}{2}$

49. $x \geq -3$ -3

50. $t < 2$ 2

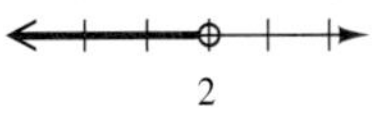

51. $x \geq 3$ 3

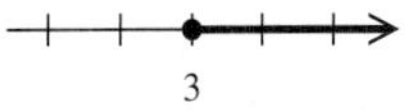

52. $k \geq 46$ 46

53. $x < -5$ -5

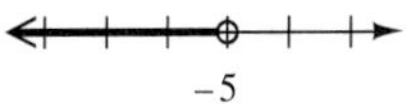

54. $w < -37$ -37

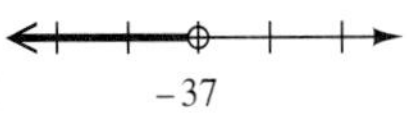

55. 88 or more **56.** all numbers less than or equal to $-\frac{1}{3}$

57. 7 **58.** $d = \frac{C}{\pi}$ **59.** $x < 2$ **60.** -9 **61.** 70 **62.** $\frac{13}{4}$ **63.** no solution **64.** all real numbers **65.** 46,700 fish **66.** 6 **67.** United States: 34; Canada: 17 **68.** gold: 16; silver: 25; bronze: 17 **69.** 44 m **70.** 70 ft **71.** 263-oz size; \$.076 **72.** \$67.50 **73.** 92 or more **74.** 51°, 51°

Chapter 2 Test (page 189)

1. 6 **2.** -6 **3.** $\frac{13}{4}$ **4.** -10.8 **5.** no solution **6.** 21 **7.** 30 **8.** all real numbers **9. (a)** Spurs: 88; Nets: 77 **(b)** 44 field goals **10.** Hawaii: 4021 mi²; Maui: 728 mi²; Kauai: 551 mi²

11. 50° **12. (a)** $W = \frac{P - 2L}{2}$ or $W = \frac{P}{2} - L$ **(b)** 18 **13.** 100°, 80° **14.** 75°, 75° **15.** 6 **16.** −29 **17.** 24-oz size; \$.145 **18.** 2300 mi **19.** 37.7% **20. (a)** $x < 0$ **(b)** $-2 < x \leq 3$

21. $x < 11$

11

22. $x \geq -3$

−3

23. $x \leq 4$

4

24. 83 or more **25.** When an inequality is multiplied or divided by a negative number, the direction of the symbol must be reversed.

Cumulative Review Exercises: Chapters R–2 (page 191)

1. $\frac{3}{8}$ **2.** $\frac{3}{4}$ **3.** $\frac{31}{20}$ **4.** $\frac{551}{40}$ or $13\frac{31}{40}$ **5.** 6 **6.** $\frac{6}{5}$ **7.** 34.03 **8.** 27.31 **9.** 30.51 **10.** 56.3 **11.** 35 yd **12.** $7\frac{1}{2}$ cups **13.** $67\frac{1}{8}$ lb **14.** \$1899.94 **15.** true **16.** true **17.** 7 **18.** 1 **19.** 13 **20.** −40 **21.** −12 **22.** undefined **23.** −6 **24.** 28 **25.** 1 **26.** 0 **27.** $\frac{73}{18}$ **28.** −64 **29.** −134 **30.** $-\frac{29}{6}$ **31.** distributive property **32.** commutative property **33.** inverse property **34.** identity property **35.** $7p - 14$ **36.** $2k - 11$ **37.** 7 **38.** −4 **39.** −1 **40.** $-\frac{3}{5}$ **41.** 2 **42.** −13 **43.** 26 **44.** −12 **45.** $c = P - a - b$ **46.** $s = \frac{P}{4}$

47. $z \leq 2$

2

48. $r < 1$

1

49. \$2018.69 **50.** \$3750 **51.** \$230.50 **52.** \$98.45 **53.** 30 cm **54.** 16 in.

Chapter 3 Graphs of Linear Equations and Inequalities in Two Variables

Section 3.1 (page 203)

1. Snoopy; 31% **3.** Since 26% is twice as much as 13%, we can expect twice as many adults to favor Charlie Brown. **5.** Ohio (OH): about 8000 million eggs; Iowa (IA): about 10,000 million eggs **7.** North Carolina (NC); about 2500 million eggs **9.** from 1975 to 1980; about \$.75 **11.** The price of a gallon of gas was decreasing. **13.** does; do not **15.** y **17.** 6 **19.** yes **21.** yes **23.** no **25.** yes **27.** no **29.** No. For two ordered pairs (x, y) to be equal, the x-values must be equal and the y-values must be equal. Here we have $4 \neq -1$ and $-1 \neq 4$. **31.** 11 **33.** $-\frac{7}{2}$ **35.** −4 **37.** −5 **39.** 4; 6; −6; (0, 4); (6, 0); (−6, 8) **41.** 3; −5; −15; (0, 3); (−5, 0); (−15, −6) **43.** −9; −9; −9 **45.** −6; −6; −6 **47.** 8; 8; 8 **49.** (2, 4) **51.** (−5, 4) **53.** (3, 0) **55.** negative; negative **57.** positive; negative **59.** If $xy < 0$, then either $x < 0$ and $y > 0$ or $x > 0$ and $y < 0$. If $x < 0$ and $y > 0$, then the point lies in quadrant II. If $x > 0$ and $y < 0$, then the point lies in quadrant IV.

61.–70.

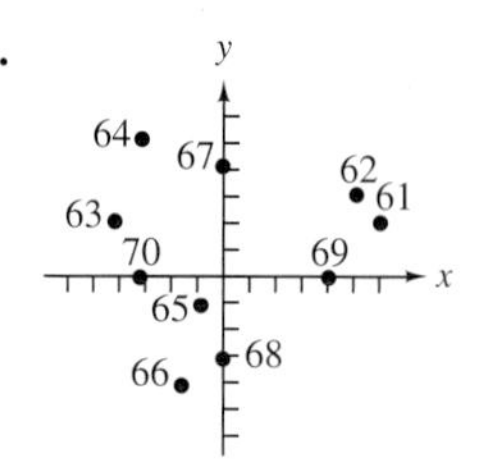

71. −3; 6; −2; 4

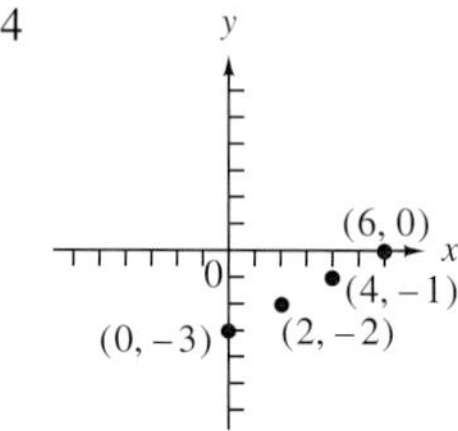

73. $-3; 4; -6; -\frac{4}{3}$

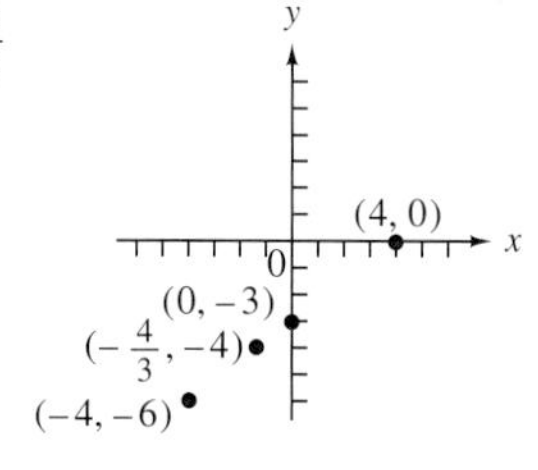

75. −4; −4; −4; −4

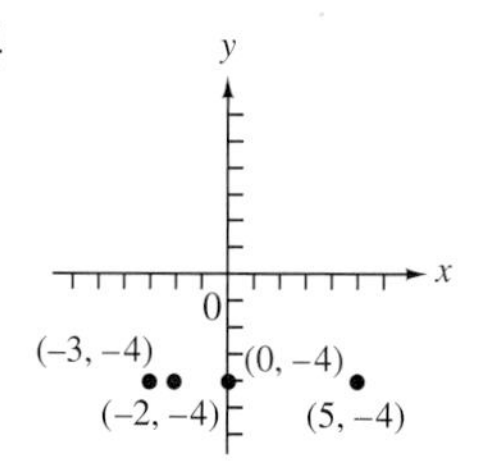

77. The points in each graph appear to lie on a straight line. **79. (a)** (1996, 53.3), (1997, 52.8), (1998, 52.1), (1999, 51.6), (2000, 51.2), (2001, 50.9) **(b)** (2002, 51.0) indicates that 51.0 percent of college students in 2002 graduated within 5 years.

(c)

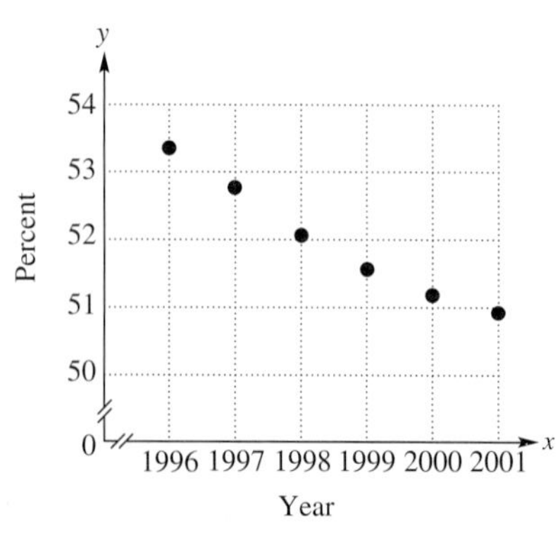

(d) The points appear to be approximated by a straight line. Graduation rates for 4-year college students within 5 years are decreasing.

81. (a) 157, 141, 125, 109 **(b)** (20, 157), (40, 141), (60, 125), (80, 109)

81. (c)

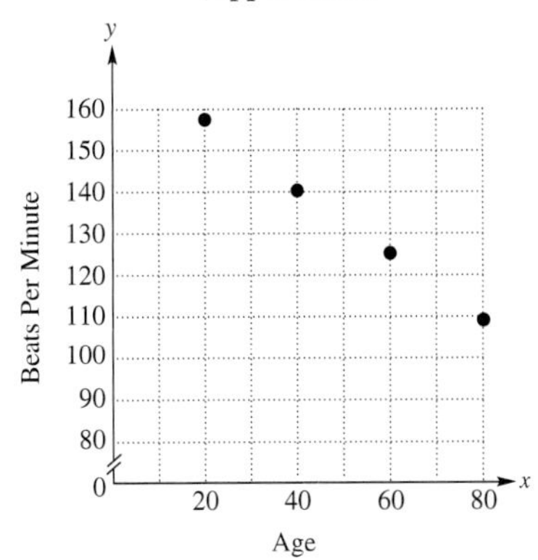

(d) The points lie in a linear pattern.

Section 3.2 (page 217)

1. 5; 5; 3

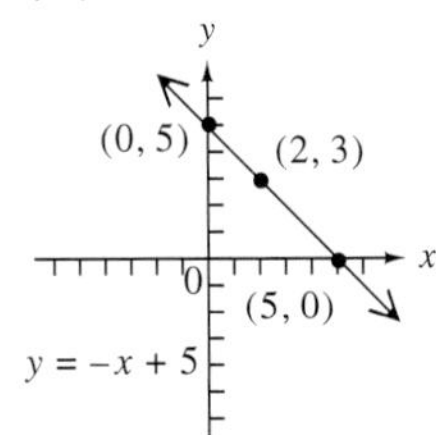

3. 1; 3; −1

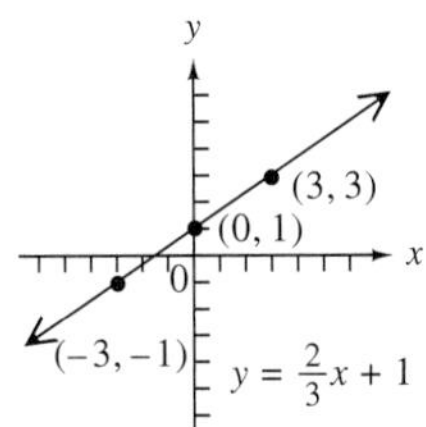

5. −6; −2; −5

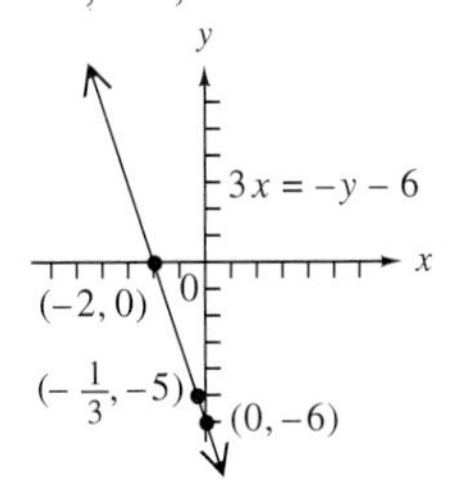

7. A **9.** D **11.** (12, 0); (0, −8) **13.** (0, 0); (0, 0)
15. Choose a value *other than* 0 for either x or y. For example, if $x = -5, y = 4$.

17.

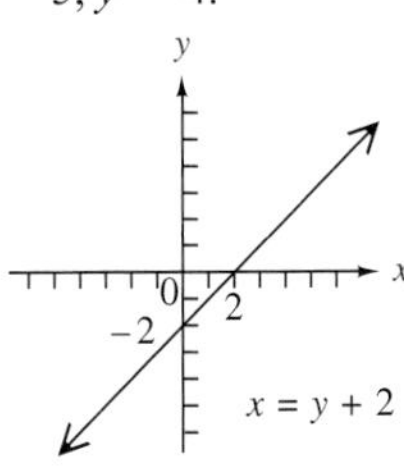

19.

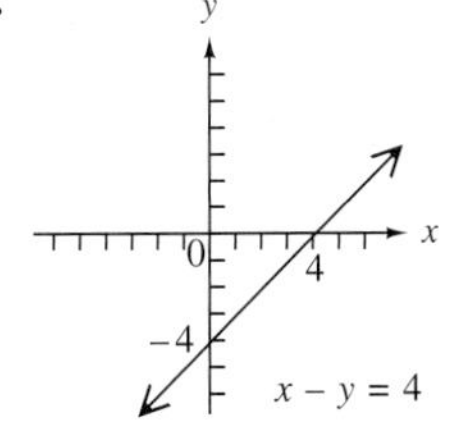

21.

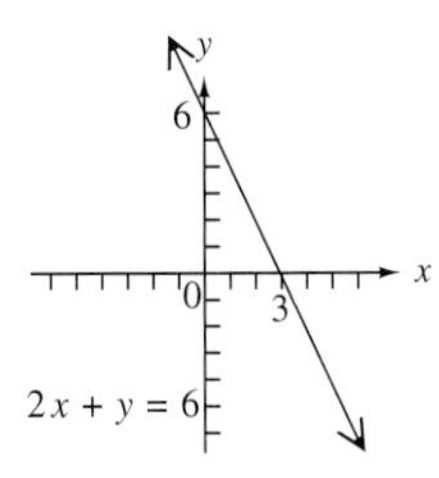

23.

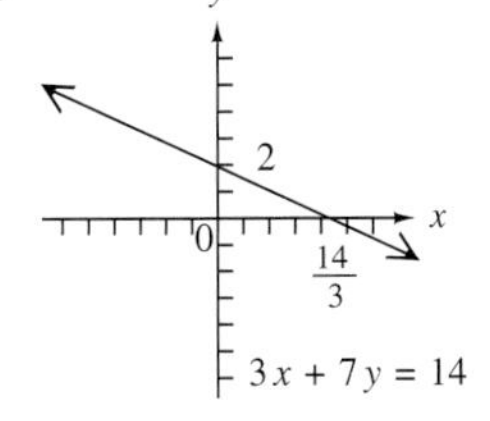

25.

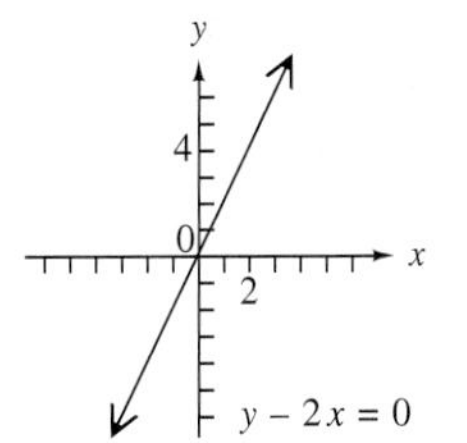

27.

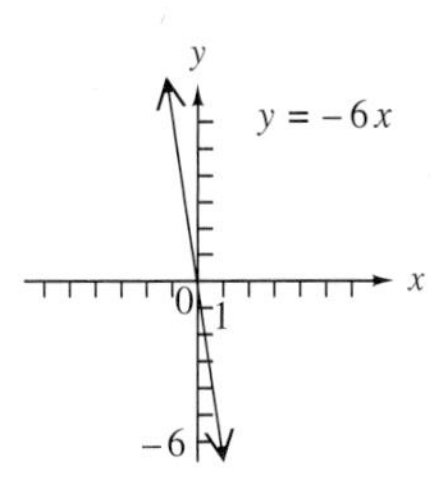

29.

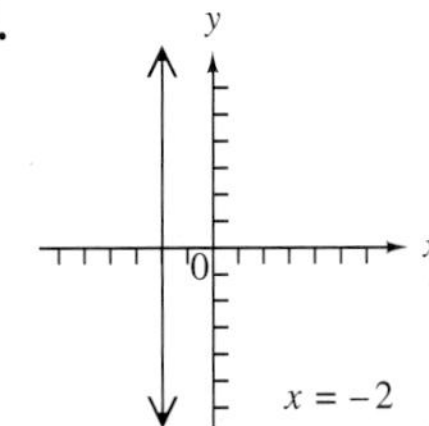

31.

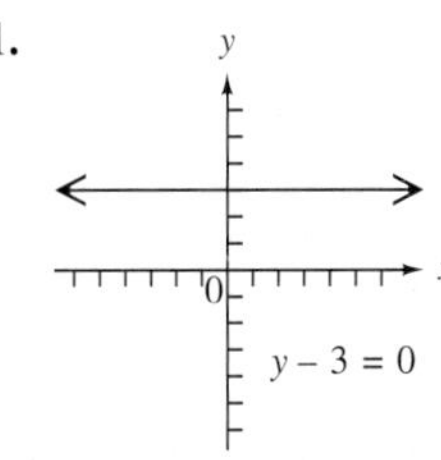

33. (a) 151.5 cm, 174.9 cm, 159.3 cm
(b)

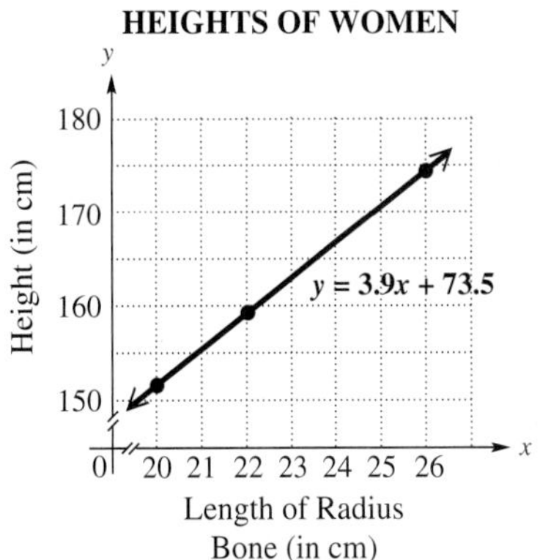

(c) 24 cm; 24 cm **35. (a)** 93 **(b)** 93 **(c)** They are the same. **37.** between 93 and 149 **39. (a)** 1995: 20.7; 1996: 21.8; 1998: 24.1 (all in gallons) **(b)** 1995: 20.3; 1996: 22.1; 1998: 23.9 (all in gallons) **(c)** They differ by .4, .3, and .2 gallon, respectively. **41. (a)** \$30,000 **(b)** \$15,000 **(c)** \$5000 **(d)** After 5 yr, the SUV has a value of \$5000. **43. (a)** The equation is a fairly good model. **(b)** The actual debt for 1996 is about \$500 billion. It is about \$30 billion less than the amount given by the equation. **(c)** No. Data for future years might not follow the same pattern, so the linear equation would not be a reliable model.

Section 3.3 (page 231)

1. $\frac{3}{2}$ **3.** $-\frac{7}{4}$ **5.** 0 **7.** Rise is the vertical change between two different points on a line. Run is the horizontal change between two different points on a line. **9.** Yes, the answer would be the same. It doesn't matter which point you start with. The slope would be expressed as the quotient of −6 and −4, which simplifies to $\frac{3}{2}$.
10.–13. Answers will vary.

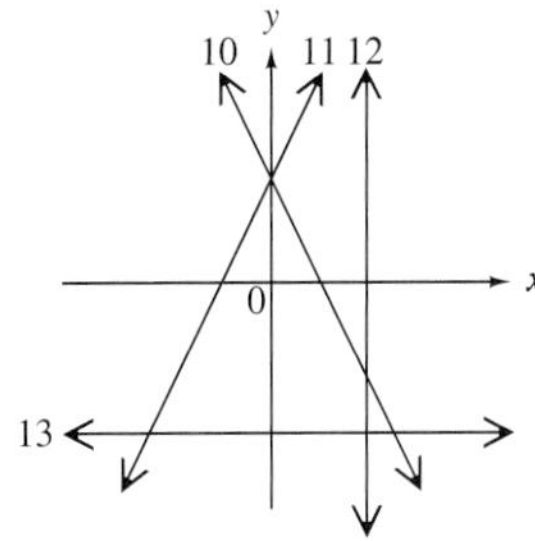

15. His answer is incorrect. Because he found the difference $3 - 5 = -2$ in the numerator, he should have subtracted in the same order in the denominator to get $-1 - 2 = -3$. The correct slope is $\frac{-2}{-3} = \frac{2}{3}$.

17. $\frac{5}{4}$ **19.** $\frac{3}{2}$ **21.** -3 **23.** 0 **25.** undefined **27.** $-\frac{1}{2}$
29. 5 **31.** $\frac{1}{4}$ **33.** $\frac{3}{2}$ **35.** 0 **37.** undefined **39.** 1
41. (a) negative **(b)** 0 **43. (a)** positive **(b)** negative
45. (a) 0 **(b)** negative **47.** $\frac{4}{3}; \frac{4}{3}$; parallel **49.** $\frac{5}{3}; \frac{3}{5}$; neither
51. $\frac{3}{5}; -\frac{5}{3}$; perpendicular **53.** $\frac{8}{27}$ **54.** 232 thousand or 232,000
55. positive; increased **56.** 232,000 students **57.** -1.66
58. negative; decreased **59.** 1.66 students per computer

Section 3.4 (page 241)

1. D **3.** B **5.** $y = 3x - 3$ **7.** $y = -x + 3$ **9.** $y = 4x - 3$
11. $y = 3$ **13.** A vertical line has undefined slope, so there is no value for m. Also, there is no y-intercept, so there can be no value for b.

15.

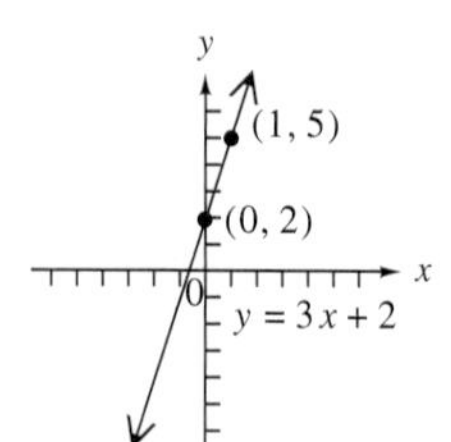

17.

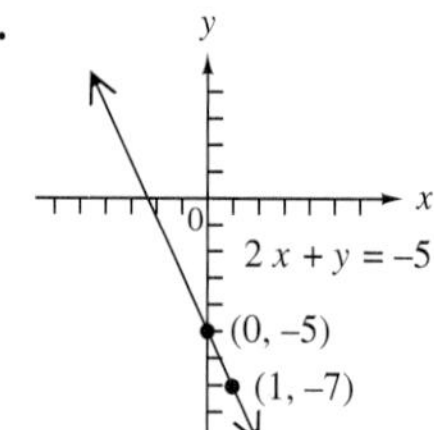

19.

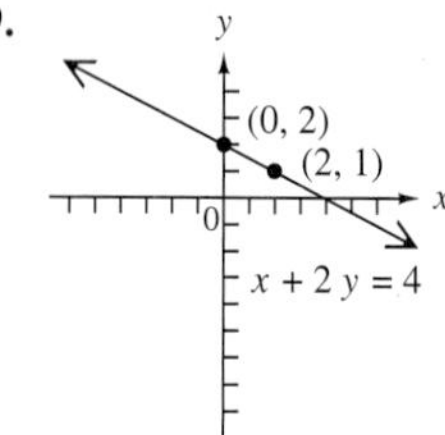

21. $y = \frac{1}{2}x + 4$

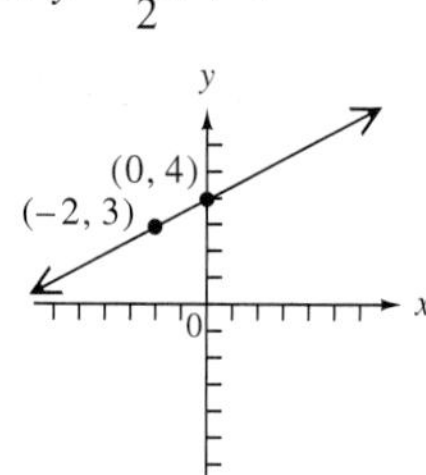

23. $y = -\frac{2}{5}x - \frac{23}{5}$

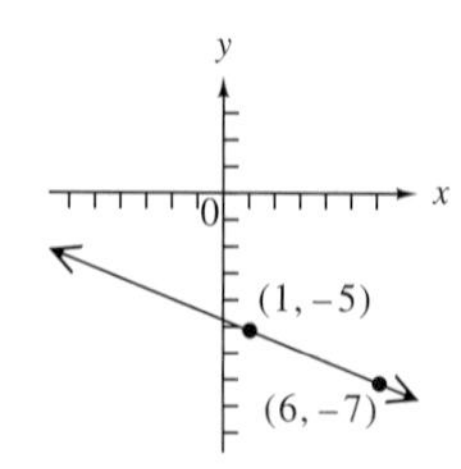

25. $y = 2$

27. $x = 3$ (no slope-intercept form)

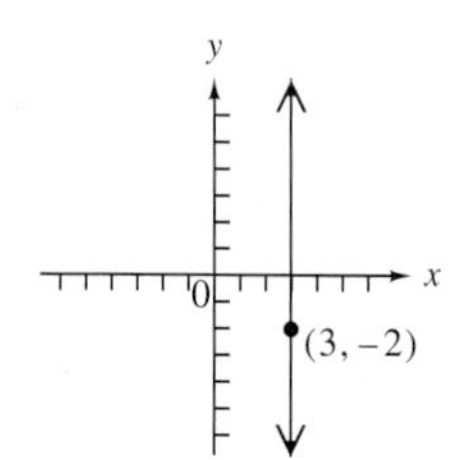

29. $y = \frac{2}{3}x$

31. $y = 2x - 7$ **33.** $y = -2x - 4$ **35.** $y = \frac{2}{3}x + \frac{19}{3}$

37. $y = x - 3$ **39.** $y = -\frac{5}{7}x - \frac{54}{7}$ **41.** $y = -\frac{2}{3}x - 2$
43. $x = 3$ (no slope-intercept form) **45.** $y = \frac{1}{3}x + \frac{4}{3}$
47. $y = \frac{3}{4}x - \frac{9}{2}$ **49.** $y = -2x - 3$ **51.** (0, 32); (100, 212)
52. $\frac{9}{5}$ **53.** $F - 32 = \frac{9}{5}(C - 0)$ **54.** $F = \frac{9}{5}C + 32$
55. $C = \frac{5}{9}(F - 32)$ **56.** 86° **57.** 10° **58.** $-40°$
59. (a) \$400 **(b)** \$.25 **(c)** $y = .25x + 400$ **(d)** \$425 **(e)** 1500
61. (a) (1, 1125), (3, 1239), (5, 1314), (7, 1338), (9, 1379)
(b) yes

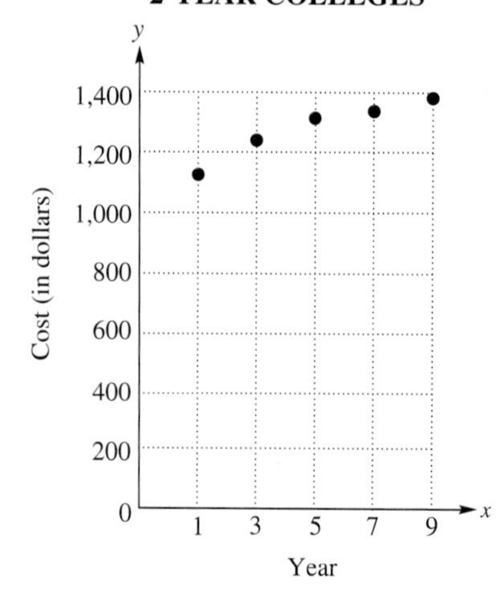

(c) $y = 23.33x + 1169$ **(d)** \$1426

Summary Exercises on Graphing Linear Equations (page 247)

1. $-3; (0, -6)$ **2.** $-2; (0, -4)$ **3.** $-4; (0, -3)$
4. $-5; (0, -8)$ **5.** $\frac{3}{2}; (0, 6)$ **6.** $\frac{5}{3}, (0, 5)$

7.

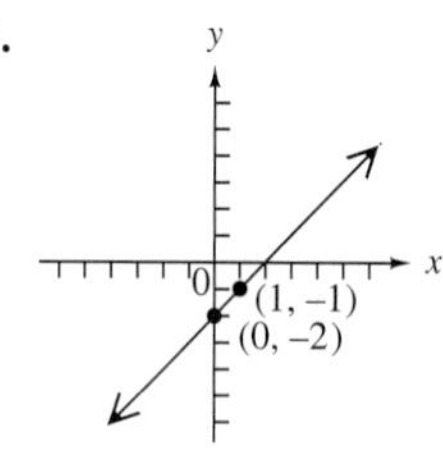

8.

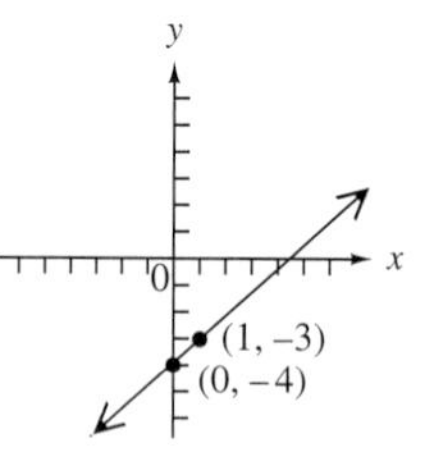

9.

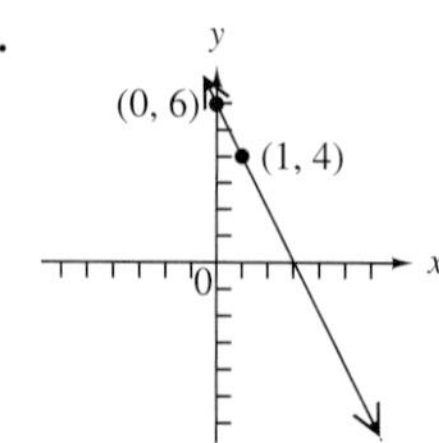

10.

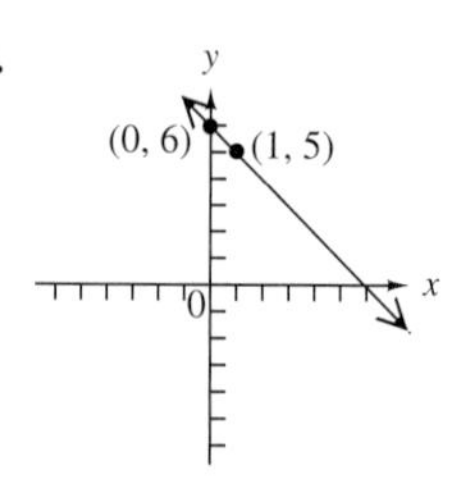

11.

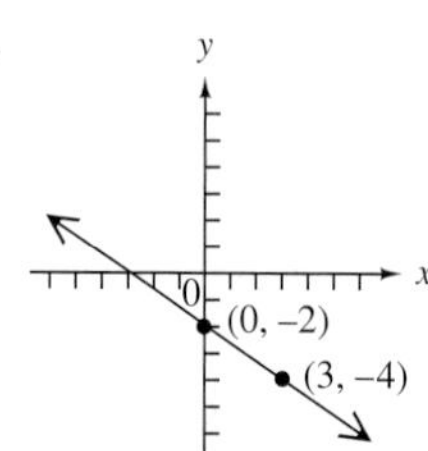

12.

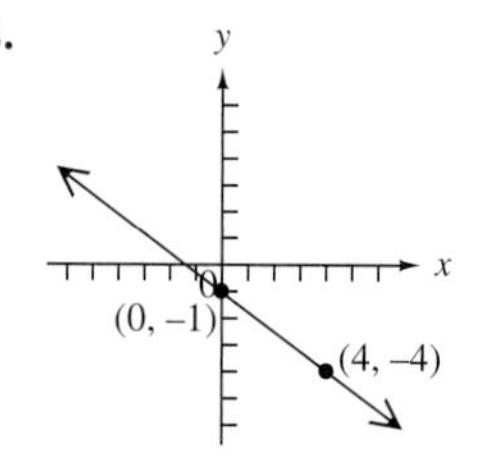

13. **(a)** 1 **(b)** $(0, -3)$ **(c)**

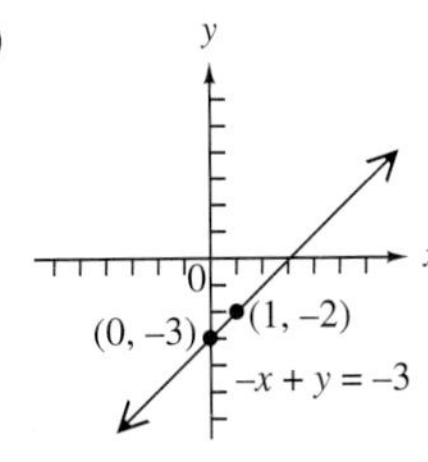

14. **(a)** 1 **(b)** $(0, -5)$ **(c)**

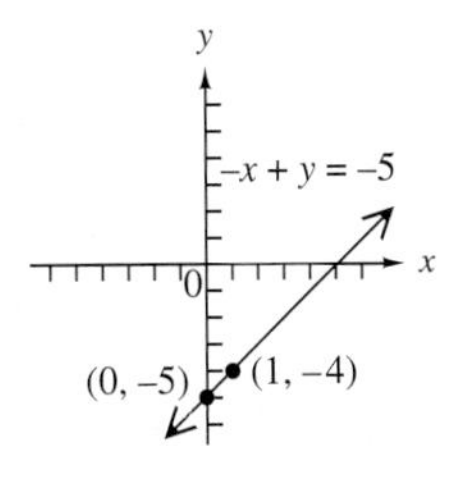

15. **(a)** $-\frac{1}{2}$ **(b)** $(0, 2)$ **(c)**

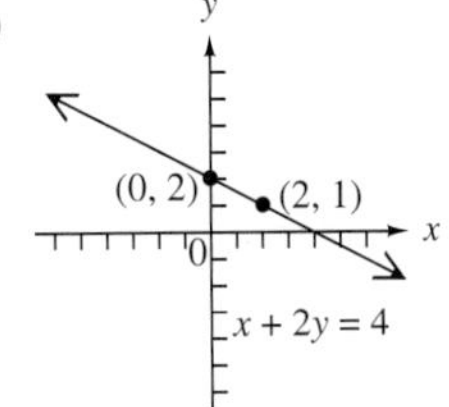

16. **(a)** $-\frac{1}{3}$ **(b)** $(0, -2)$ **(c)**

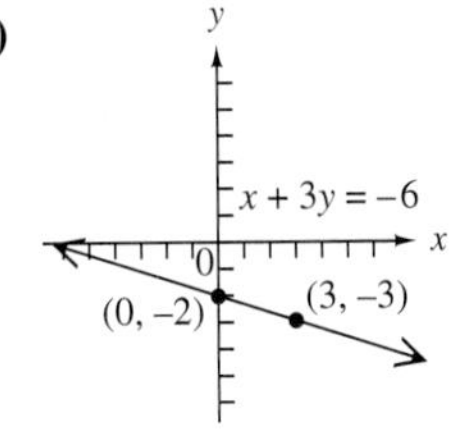

17. **(a)** $\frac{4}{5}$ **(b)** $(0, -4)$ **(c)**

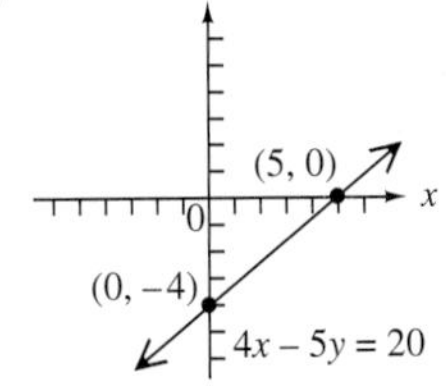

18. **(a)** $\frac{6}{5}$ **(b)** $(0, -6)$ **(c)**

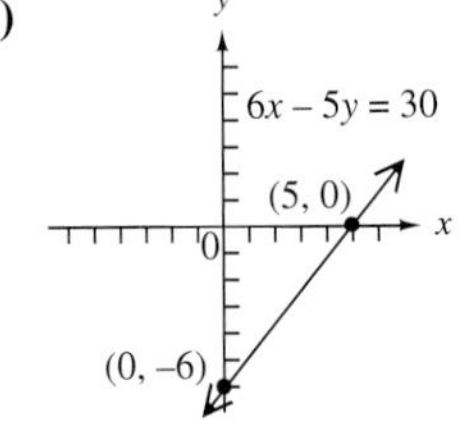

19. **(a)** $-\frac{2}{3}$ **(b)** $(0, 4)$ **(c)**

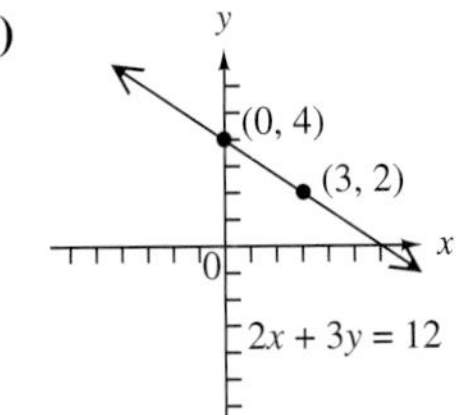

20. **(a)** $-\frac{5}{2}$ **(b)** $(0, 5)$ **(c)**

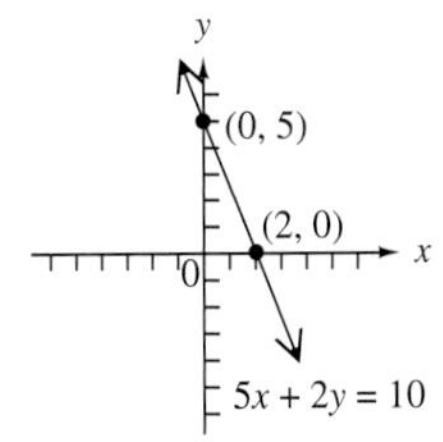

21. **(a)** $\frac{1}{3}$ **(b)** $(0, -2)$ **(c)**

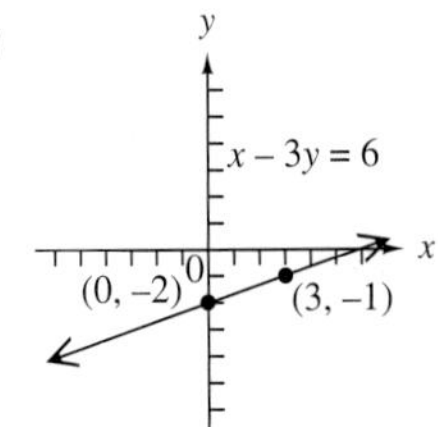

22. **(a)** $\frac{1}{2}$ **(b)** $(0, 2)$ **(c)**

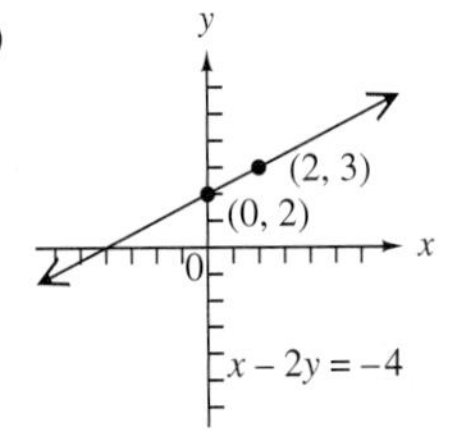

23. **(a)** $\frac{1}{4}$ **(b)** $(0, 0)$ **(c)**

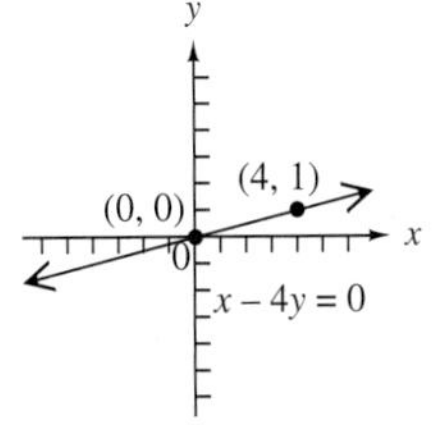

24. **(a)** $-\frac{1}{5}$ **(b)** $(0, 0)$ **(c)**

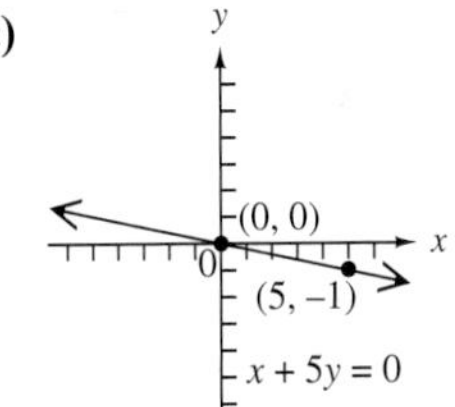

Section 3.5 (page 253)

1. false **3.** true **5.** $>$ **7.** $\leq$

9.

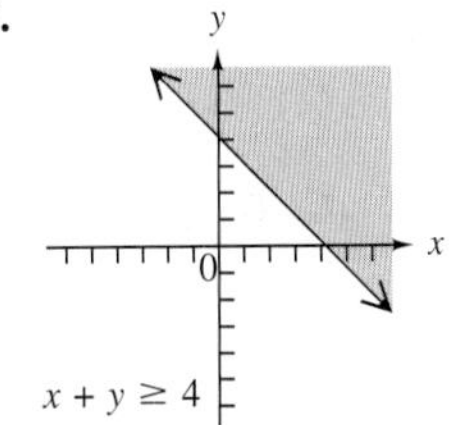

11.

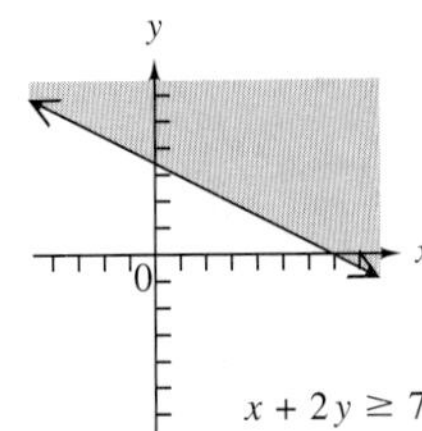

13.

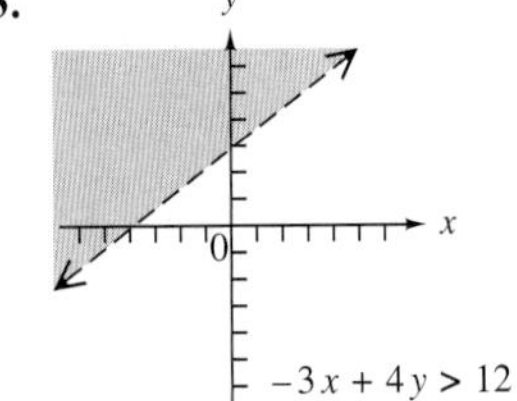

15.

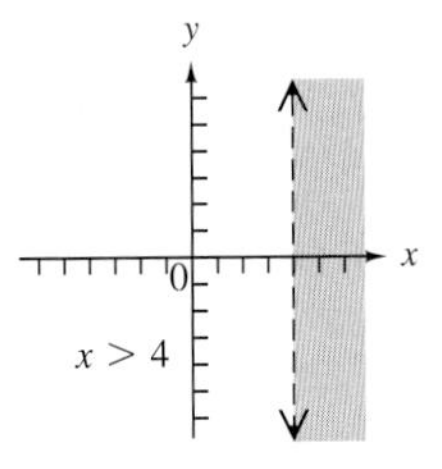

17. Use a dashed line if the symbol is $<$ or $>$. Use a solid line if the symbol is $\leq$ or $\geq$.

19.

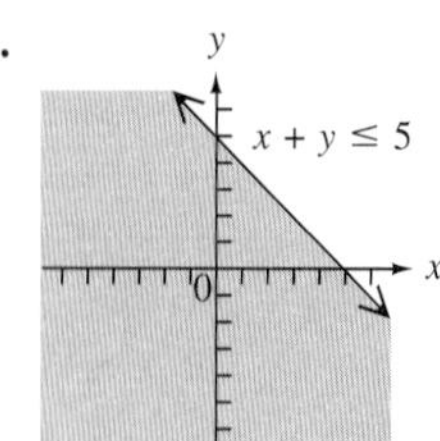

21.

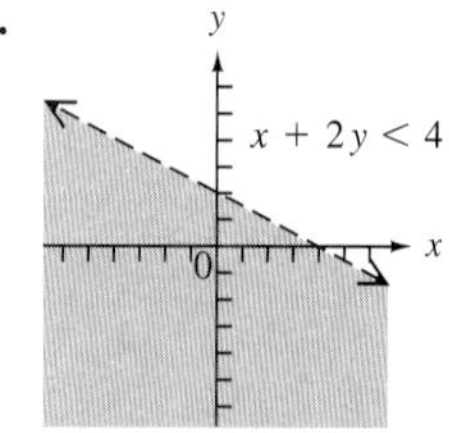

23.

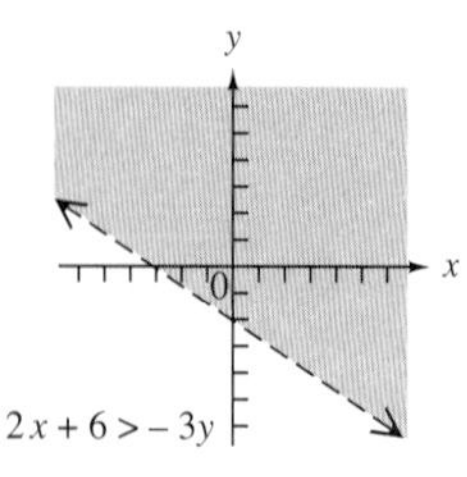

25.

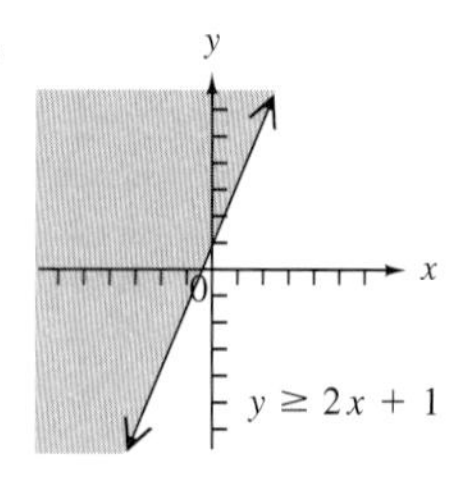

27.

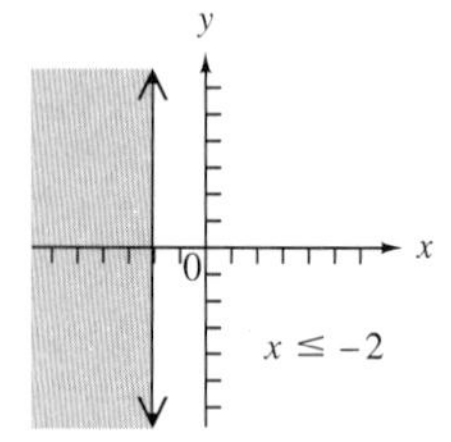

29.

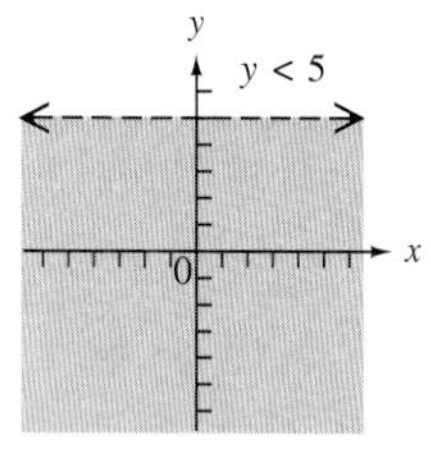

31.

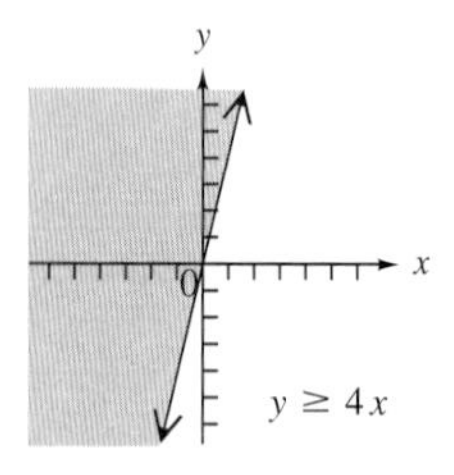

33. Every point in quadrant IV has a positive x-value and a negative y-value. Substituting into $y > x$ would imply that a negative number is greater than a positive number, which is always false. Thus, the graph of $y > x$ cannot lie in quadrant IV.

35. (a)

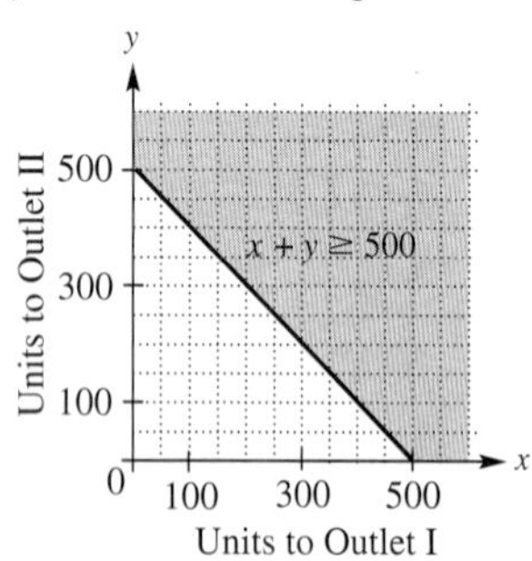

(b) (500, 0) and (200, 400); Other answers are possible.

Chapter 3 Review Exercises (page 261)

1. 2.0% **2.** (1996, 57.1), (1997, 56.6), (1998, 56.2), (1999, 55.8), (2000, 55.5), (2001, 55.1) **3.** 1997; .5% decrease **4.** There is a general trend of decreasing percents of private school students earning a degree within five years. **5.** -1; 2; 1 **6.** 2; $\frac{3}{2}$; $\frac{14}{3}$ **7.** 0; $\frac{8}{3}$; -9 **8.** 7; 7; 7 **9.** yes **10.** no **11.** yes

12.–15.

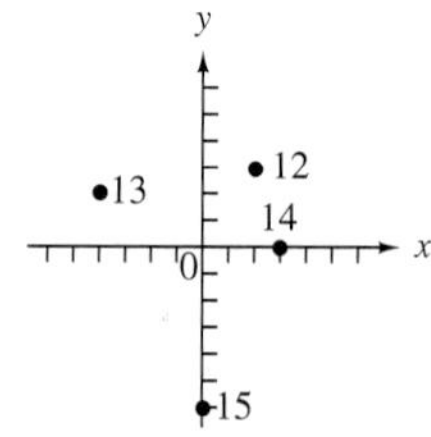

16. x is positive in quadrants I and IV; y is negative in quadrants III and IV. Thus, if x is positive and y is negative, (x, y) must lie in quadrant IV. **17.** In the ordered pair $(k, 0)$, the y-value is 0, so the point lies on the x-axis. In the ordered pair $(0, k)$, the x-value is 0, so the point lies on the y-axis. **18.** II **19.** III **20.** no quadrant

21. $\left(-\frac{5}{2}, 0\right)$; $(0, 5)$ **22.** $\left(-\frac{7}{2}, 0\right)$; $(0, -7)$ **23.** $\left(\frac{8}{3}, 0\right)$; $(0, 4)$

24.

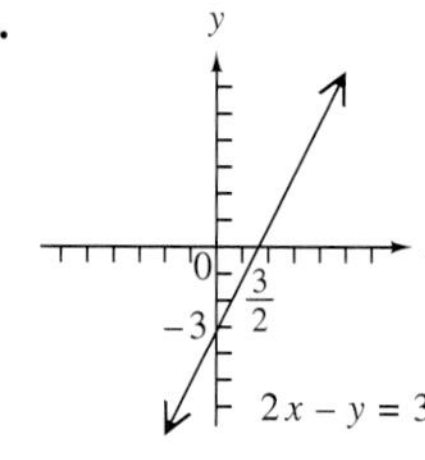

25.

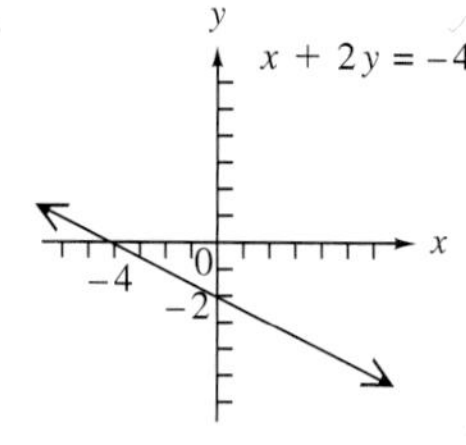

26.

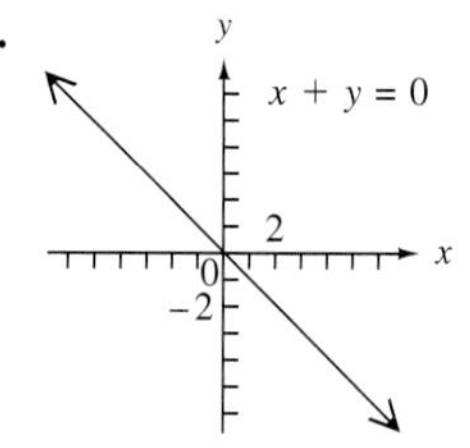

27. $-\frac{1}{2}$ **28.** $-\frac{2}{3}$ **29.** 0 **30.** undefined **31.** 3 **32.** $\frac{2}{3}$ **33.** $\frac{3}{2}$ **34.** $-\frac{1}{3}$ **35.** undefined **36.** 0 **37.** $\frac{3}{2}$ **38. (a)** 2 **(b)** $\frac{1}{3}$ **39.** parallel **40.** perpendicular **41.** neither **42.** 0

43. $y = -x + \frac{2}{3}$ **44.** $y = -\frac{1}{3}x + 1$ **45.** $y = x - 7$

46. $y = \frac{2}{3}x + \frac{14}{3}$ **47.** $y = -\frac{3}{4}x - \frac{1}{4}$ **48.** $y = -\frac{1}{4}x + \frac{3}{2}$

49. $y = 1$ **50.** $x = \frac{1}{3}$ **51. (a)** $y = -\frac{1}{3}x + 5$ **(b)** slope: $-\frac{1}{3}$; y-intercept: (0, 5) **(c)**

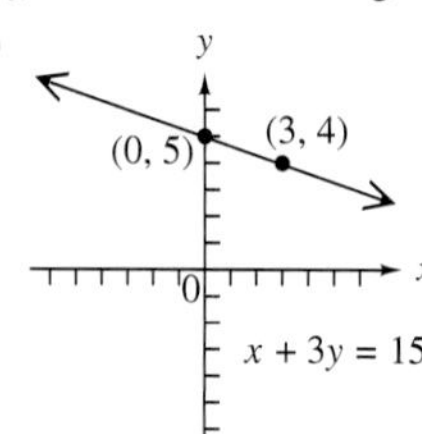

52.

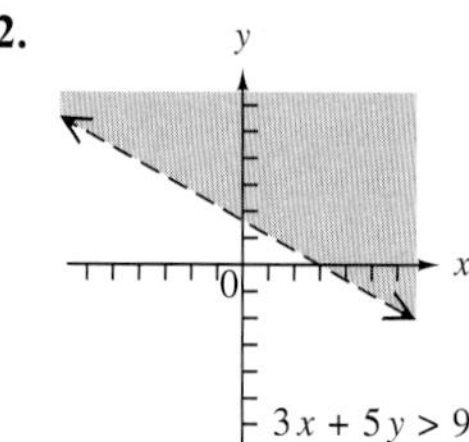

53.

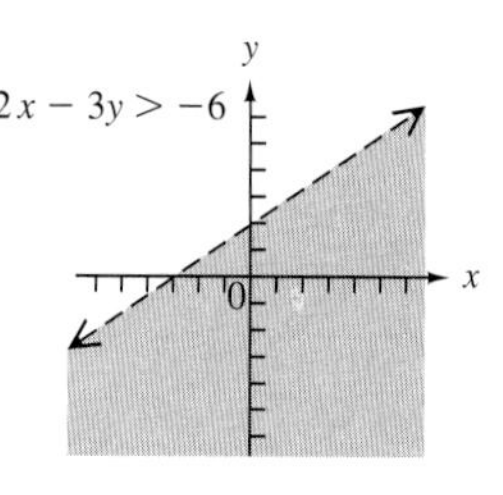

54.

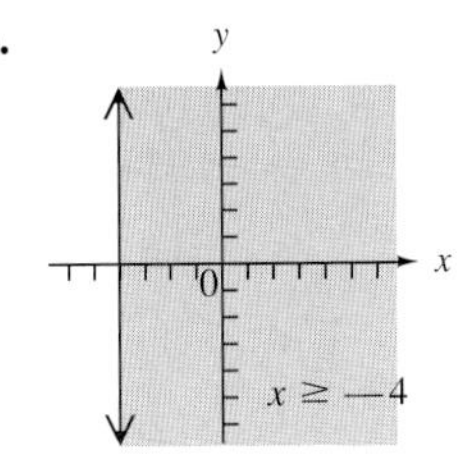

55. A **56.** C, D **57.** A, B, D **58.** D **59.** C **60.** B

61. 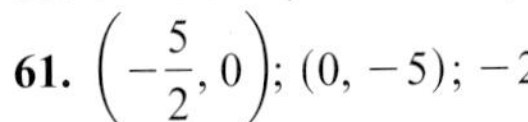 $\left(-\frac{5}{2}, 0\right)$; $(0, -5)$; -2

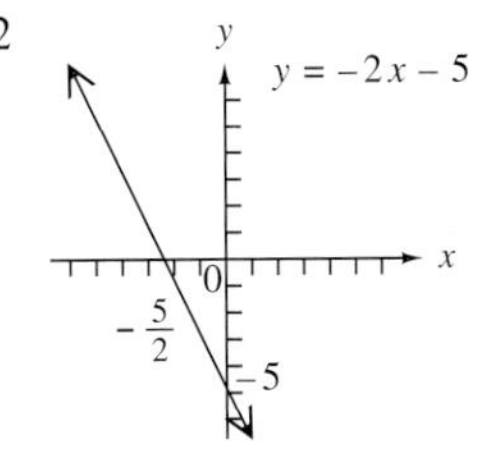

62. $(0, 0)$; $(0, 0)$; $-\frac{1}{3}$

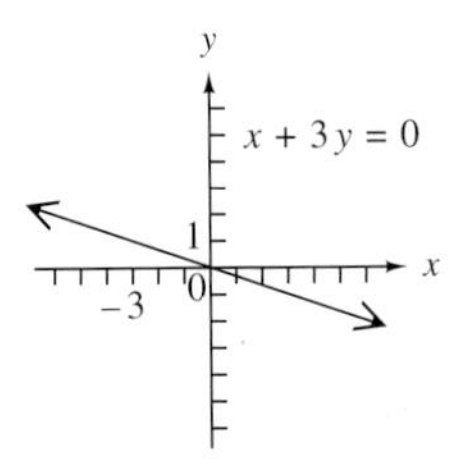

63. no x-intercept; $(0, 5)$; 0

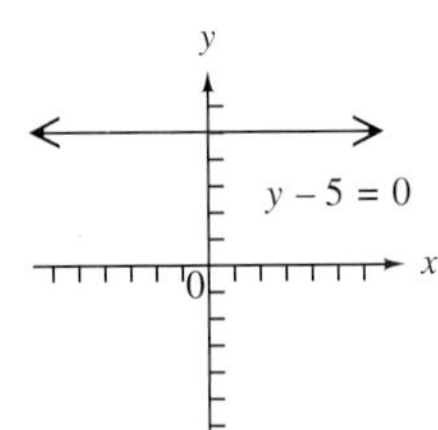

64. $y = -\frac{1}{4}x - \frac{5}{4}$ **65.** $y = -3x + 30$ **66.** $y = -\frac{4}{7}x - \frac{23}{7}$

67.

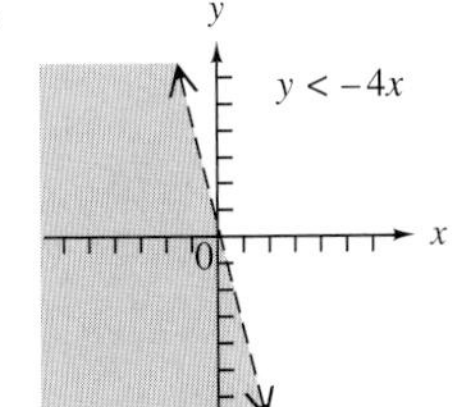

68.

y

x

0

x − 2y ≤ 6

69. 3.0% **70.** Since the graph falls from left to right, the slope is negative. **71.** (1997, 44.2), (2002, 41.2)
72. $y = -.6x + 1242.4$ **73.** $-.6$; yes **74.** 43.6, 43.0, 42.4, 41.8
75. 40.6%; No. The equation is based on data only for 1997 through 2002.

Chapter 3 Test (page 267)

1. \$1.06 **2.** \$.45 **3.** from 2000 to 2002
4. x-intercept: $(2, 0)$; y-intercept: $(0, 6)$

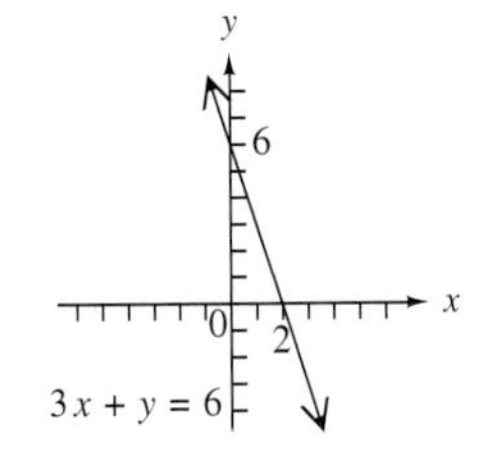

5. x-intercept: $(0, 0)$; y-intercept: $(0, 0)$

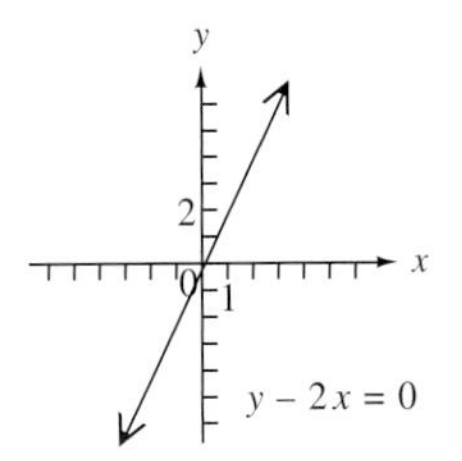

6. x-intercept: $(-3, 0)$; y-intercept: none

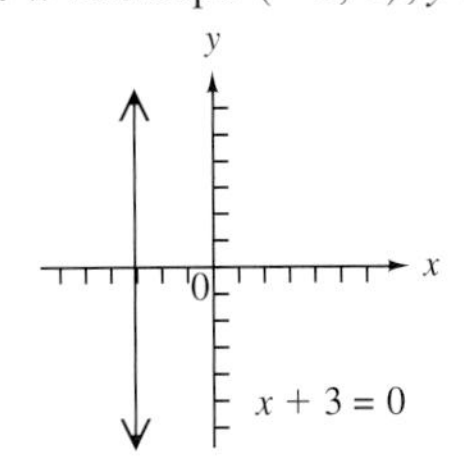

7. x-intercept: none; y-intercept: $(0, 1)$

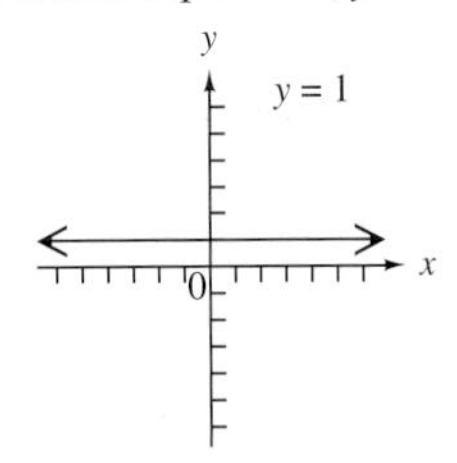

8. x-intercept: $(4, 0)$; y-intercept: $(0, -4)$

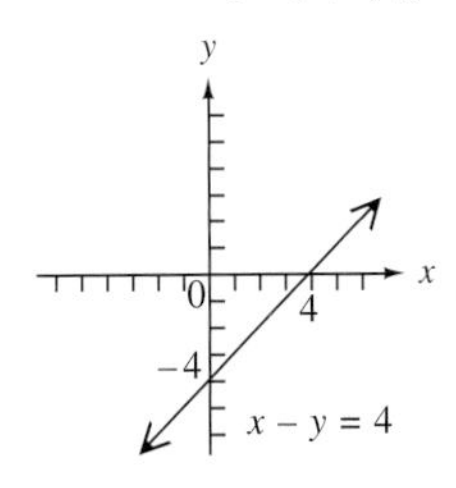

9. $-\frac{8}{3}$ **10.** -2 **11.** undefined **12.** $\frac{5}{2}$ **13.** 0 **14.** $y = 2x + 6$

15. $y = \frac{5}{2}x - 4$ **16.** $y = -9x + 12$ **17.** $y = -\frac{3}{2}x + \frac{9}{2}$

18.

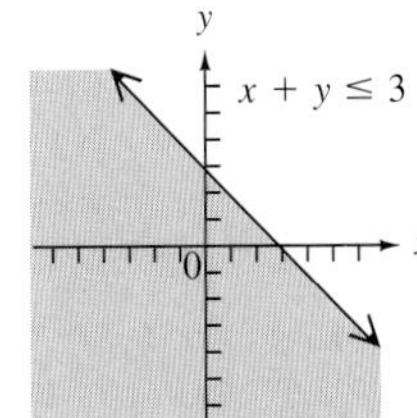

19.

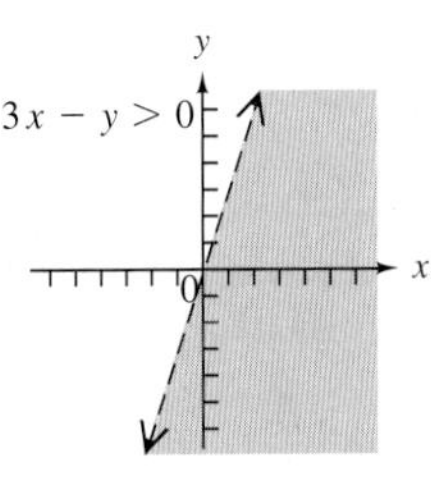

20. The slope is positive since food and drink sales are increasing.
21. (0, 43), (30, 376); 11.1 **22.** **(a)** 1990: \$265 billion; 1995: \$320.5 billion **(b)** In 2000, food and drink sales were \$376 billion.

Cumulative Review Exercises: Chapters R–3 (page 271)

1. $\frac{301}{40}$ or $7\frac{21}{40}$ **2.** 6 **3.** 7 **4.** $\frac{73}{18}$ or $4\frac{1}{18}$ **5.** true **6.** -43

7. distributive property **8.** $-p + 2$ **9.** $h = \frac{3V}{\pi r^2}$ **10.** -1

11. 2 **12.** -13 **13.** $x > -2.6$

−2.6

14. $x > 0$ 0

15. $t \leq -4$ -4

16. high school diploma: \$36,722; bachelor's degree: \$65,922
17. 13 mi **18.** **(a)** 89.33, 81.21, 78.16 **(b)** In 1980, the winning time was approximately 85.27 sec. **19.** **(a)** \$7000 **(b)** \$10,000 **(c)** about \$30,000 **20.** $(-4, 0)$; $(0, 3)$ **21.** $\frac{3}{4}$

22.

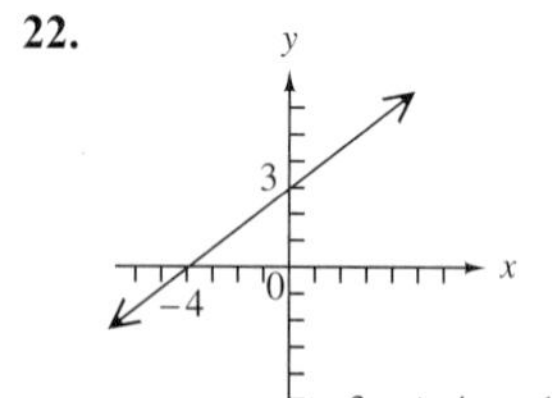

23. 6 **24.** perpendicular **25.** $y = 3x - 11$ **26.** $y = 4$

CHAPTER 4 SYSTEMS OF LINEAR EQUATIONS AND INEQUALITIES

Section 4.1 (page 279)

1. B, because the ordered pair must be in quadrant II. **3.** There is no way that the sum of two numbers can be both 2 and 4 at the same time. **5.** no **7.** yes **9.** yes **11.** no

We show the graphs here only for Exercises 13–17.

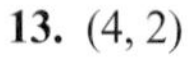

13. $(4, 2)$

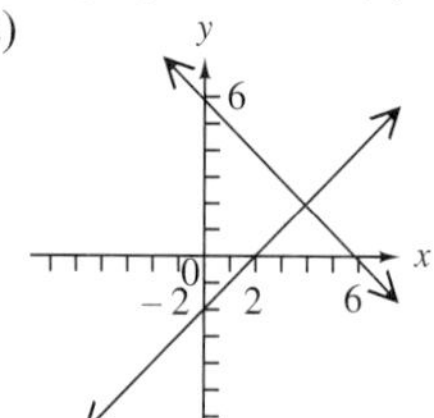

15. $(0, 4)$

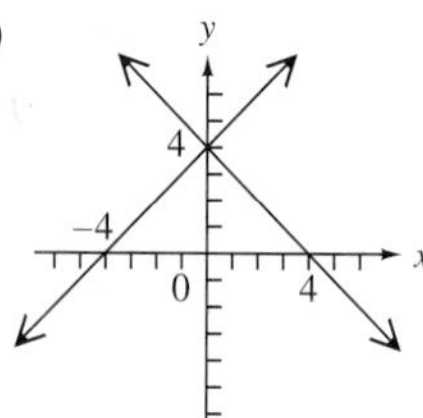

17. $(4, -1)$

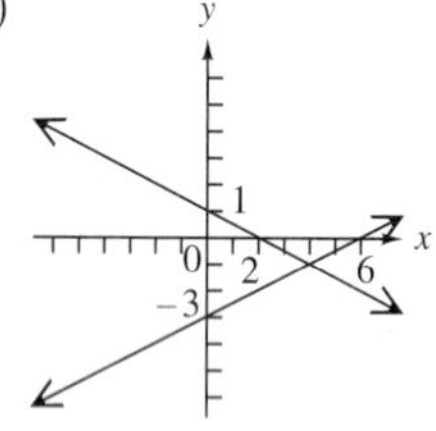

19. $(1, 3)$ **21.** $(0, 2)$ **23.** no solution **25.** infinite number of solutions **27.** $(4, -3)$ **29.** no solution

31. $y = -\frac{3}{2}x + 3$; $y = -\frac{3}{2}x + \frac{5}{2}$; The graphs are parallel lines.

32. $y = 2x - 4$; $y = 2x - 4$; The graphs are the same line.

33. $y = \frac{1}{3}x - \frac{5}{3}$; $y = -2x + 8$; The graphs are intersecting lines.

34. Exercise 31: no solution; Exercise 32: infinite number of solutions; Exercise 33: one solution **35.** 1989–1997 **37.** 1989: share 20%; 1998: share 16% **39.** If the coordinates of the point of intersection are not integers, the solution will be difficult to determine from a graph. **41.** Answers will vary, but the lines must intersect at $(-2, 3)$.

Section 4.2 (page 289)

1. No, it is not correct, because the solution is the ordered pair $(3, 0)$. The y-value must also be determined. **3.** $(3, 9)$ **5.** $(7, 3)$ **7.** $(-2, 4)$ **9.** $(-4, 8)$ **11.** $(3, -2)$ **13.** infinite number of solutions **15.** $\left(\frac{1}{3}, -\frac{1}{2}\right)$ **17.** no solution **19.** infinite number of solutions **21.** $(36, -35)$ **23.** no solution **25.** $(2, -3)$ **27.** $(10, -12)$ **29.** $(-4, 2)$ **30.** To find the total cost, multiply the number of bicycles (x) by the cost per bicycle (400 dollars) and add the fixed cost (5000 dollars). Thus, $y_1 = 400x + 5000$ gives this total cost (in dollars).
31. $y_2 = 600x$ **32.** $y_1 = 400x + 5000$, $y_2 = 600x$, solution: $(25, 15{,}000)$ **33.** 25; 15,000; 15,000

Section 4.3 (page 295)

1. true **3.** true **5.** $(-1, 3)$ **7.** $(-1, -3)$ **9.** $(-2, 3)$ **11.** $\left(\frac{1}{2}, 4\right)$ **13.** $(3, -6)$ **15.** $(7, 4)$ **17.** $(0, 4)$ **19.** $(-4, 0)$ **21.** $(0, 0)$ **23.** no solution **25.** infinite number of solutions **27.** $(2, 9)$ **29.** $(-6, 5)$ **31.** $\left(-\frac{6}{5}, \frac{4}{5}\right)$ **33.** $\left(\frac{1}{8}, -\frac{5}{6}\right)$ **35.** $(11, 15)$ **37.** no solution **39.** infinite number of solutions **41.** $1141 = 1991a + b$ **42.** $1465 = 1999a + b$ **43.** $1991a + b = 1141$, $1999a + b = 1465$; solution: $(40.5, -79{,}494.5)$ **44.** $y = 40.5x - 79{,}494.5$ **45.** 1424.5 (million); This is slightly less than the actual figure. **46.** Since the data do not lie in a perfectly straight line, the quantity obtained from an equation determined in this way will probably be "off" a bit. We cannot put too much faith in models such as this one, because not all sets of data points are linear in nature.

Summary Exercises on Solving Systems of Linear Equations (page 299)

1. **(a)** Use substitution since the second equation is solved for y. **(b)** Use elimination since the coefficients of the y-terms are opposites. **(c)** Use elimination since the equations are in standard form with no coefficients of 1 or -1. Solving by substitution would involve fractions. **2.** The system on the right is easier to solve by substitution because the second equation is already solved for y.
3. **(a)** $(1, 4)$ **(b)** $(1, 4)$ **(c)** Answers will vary. **4.** **(a)** $(-5, 2)$ **(b)** $(-5, 2)$ **(c)** Answers will vary. **5.** $(2, 6)$ **6.** $(-3, 2)$ **7.** $\left(\frac{1}{3}, \frac{1}{2}\right)$ **8.** no solution **9.** $(3, 0)$ **10.** $\left(\frac{3}{2}, -\frac{3}{2}\right)$ **11.** infinite number of solutions **12.** $(9, 4)$ **13.** $\left(-\frac{5}{7}, -\frac{2}{7}\right)$ **14.** $(4, -5)$ **15.** no solution **16.** $(-4, 6)$ **17.** $\left(\frac{19}{3}, -5\right)$ **18.** $\left(\frac{22}{13}, -\frac{23}{13}\right)$ **19.** $(-12, -60)$ **20.** $(2, -4)$ **21.** $(18, -12)$ **22.** $(-2, 1)$ **23.** $\left(13, -\frac{7}{5}\right)$ **24.** infinite number of solutions

Section 4.4 (page 307)

1. D **3.** B **5.** D **7.** C **9.** the second number; $x - y = 48$; The two numbers are 73 and 25. **11.** Cher: 84; Rolling Stones: 33 **13.** *The Lord of the Rings: The Return of the King:* \$361.1 million; *Finding Nemo:* \$339.7 million **15.** Terminal Tower: 708 ft; Key Tower: 950 ft **17.** **(a)** 45 units **(b)** Do not produce; the product will lead to a loss. **19.** 46 ones; 28 tens **21.** 2 DVDs of *Miracle*; 5 Linkin Park CDs **23.** \$2500 at 4%; \$5000 at 5% **25.** Japan: \$17.19; Switzerland: \$13.15 **27.** 80 L of 40% solution; 40 L of 70% solution **29.** 30 lb at \$6 per lb; 60 lb at \$3 per lb
31. 30 barrels at \$40 per barrel; 20 barrels at \$60 per barrel
33. bicycle: 15 mph; car: 55 mph **35.** car leaving Cincinnati: 55 mph; car leaving Toledo: 70 mph **37.** Roberto: 17.5 mph; Juana: 12.5 mph **39.** boat: 10 mph; current: 2 mph
41. plane: 470 mph; wind: 30 mph

Section 4.5 (page 317)

1. C **3.** B

5.

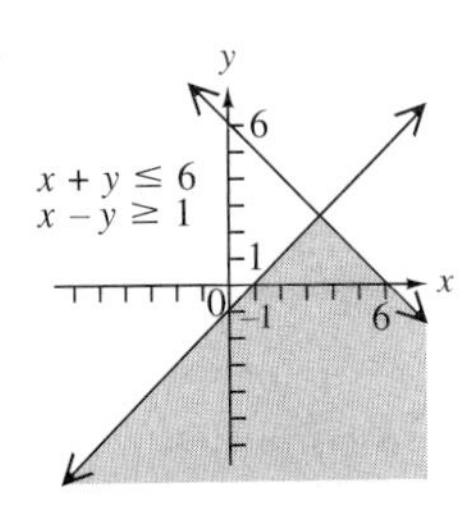

7.

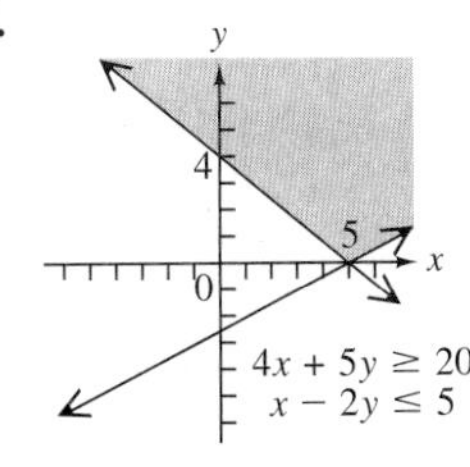

9.

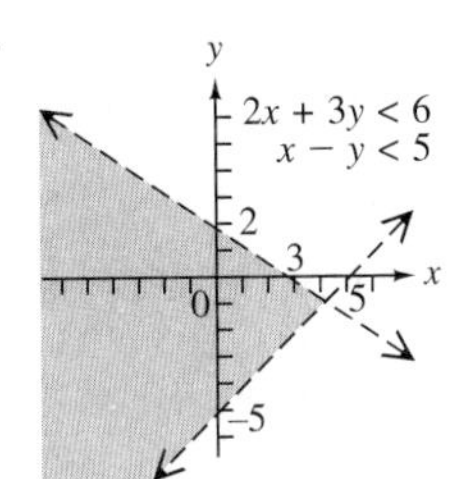

11.

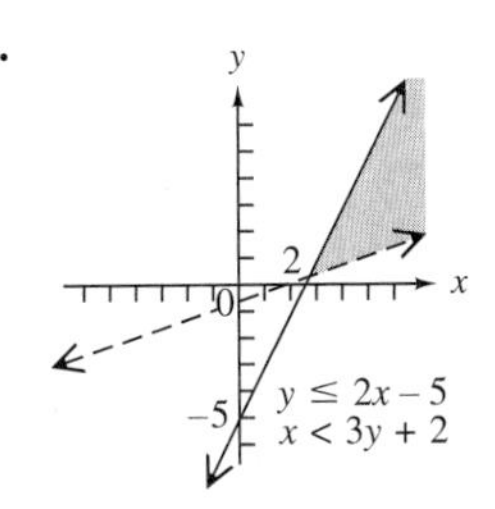

13.

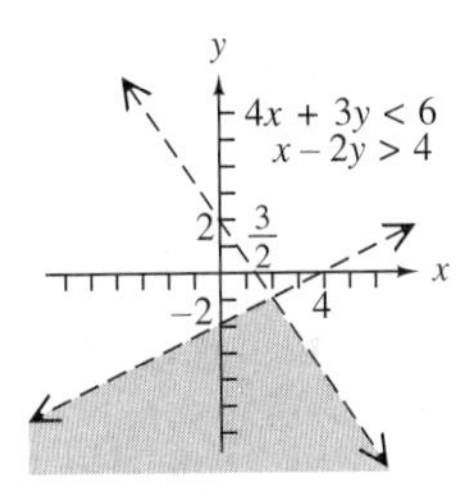

15.

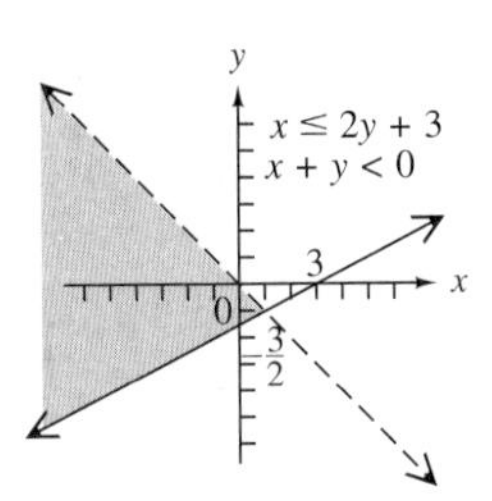

17.

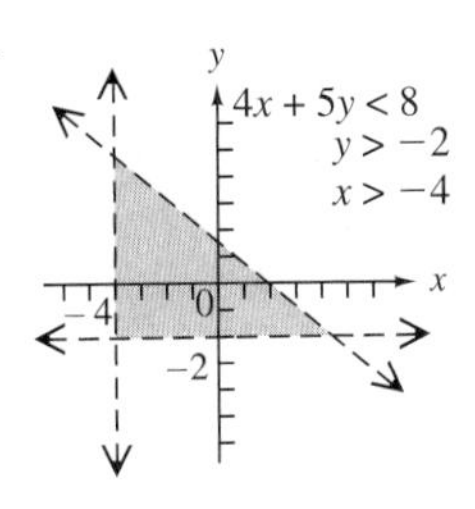

19.

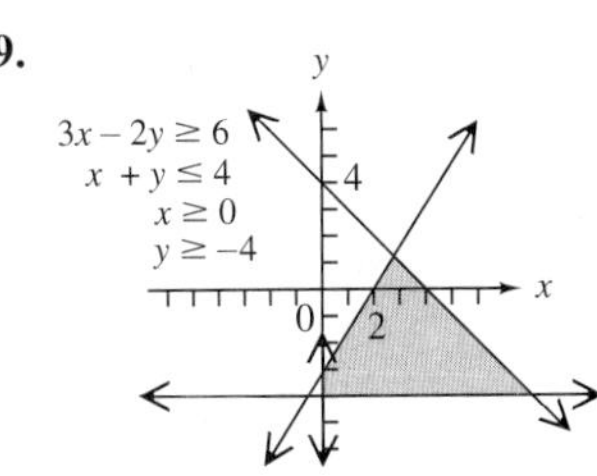

Chapter 4 Review Exercises (page 323)

1. yes **2.** no **3.** $(3, 1)$ **4.** $(0, -2)$ **5.** infinite number of solutions **6.** no solution **7.** It is not a solution of the system because it is not a solution of the second equation, $2x + y = 4$. **8.** $(2, 1)$ **9.** $(3, 5)$ **10.** $(6, 4)$ **11.** no solution **12.** $(7, 1)$ **13.** $(-5, -2)$ **14.** $(-4, 3)$ **15.** infinite number of solutions **16. (a)** 2 **(b)** 9 **17.** $(9, 2)$ **18.** $\left(\frac{10}{7}, -\frac{9}{7}\right)$ **19.** $(8, 9)$ **20.** $(2, 1)$ **21.** Subway: 13,247 restaurants; McDonald's: 13,099 restaurants **22.** *Modern Maturity:* 20.5 million; *Reader's Digest:* 15.1 million **23.** length: 27 m; width: 18 m **24.** 13 twenties; 7 tens **25.** 25 lb of $1.30 candy; 75 lb of $.90 candy **26.** plane: 250 mph; wind: 20 mph **27.** $7000 at 3%; $11,000 at 4% **28.** 60 L of 40% solution; 30 L of 70% solution

29.

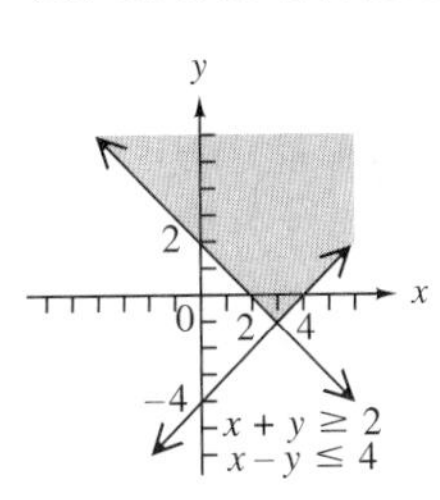

30.

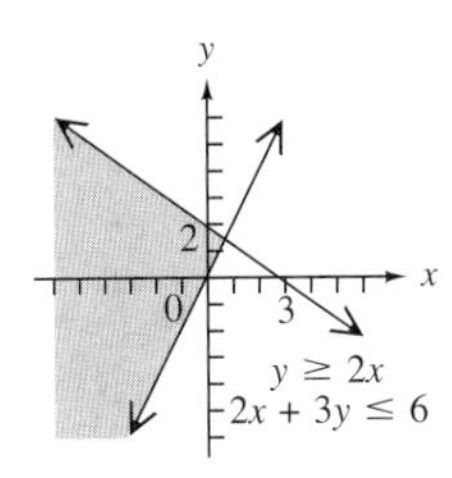

31.

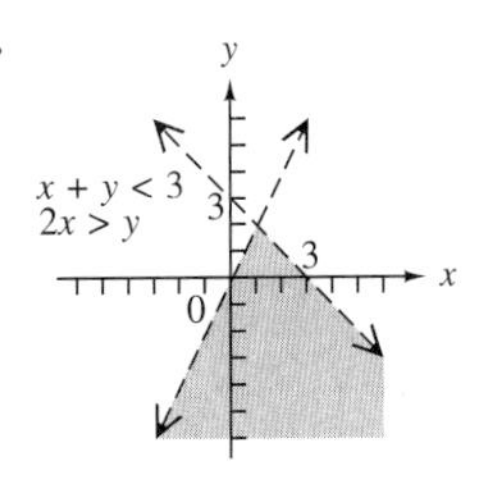

32. B **33.** B **34.** $(2, 0)$ **35.** $(-4, 15)$ **36.** no solution

37.

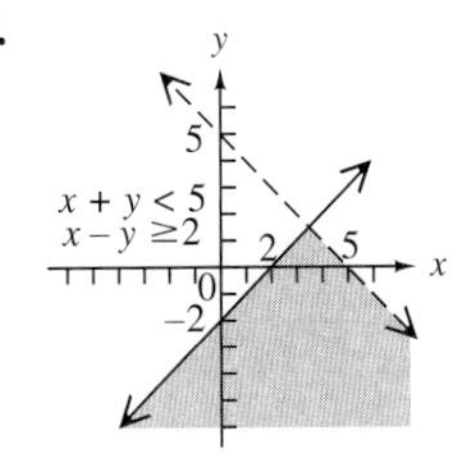

38.

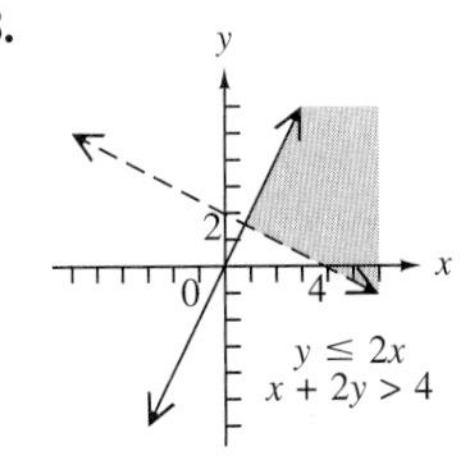

39.

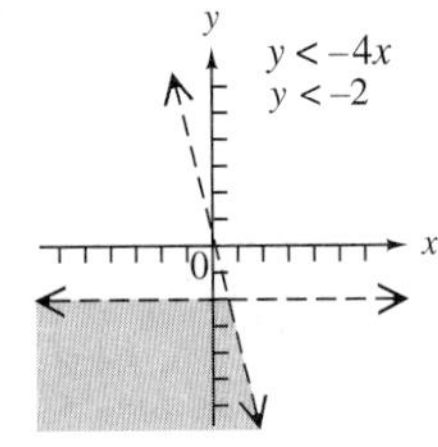

40. 8 in., 8 in., and 13 in. **41.** New England 24, Indianapolis 14 **42. (a)** years 0–6 **(b)** year 6; about $650

Chapter 4 Test (page 327)

1. $(2, -3)$ **2.** It has no solution. **3.** $(1, -6)$ **4.** $(-35, 35)$ **5.** $(5, 6)$ **6.** $(-1, 3)$ **7.** $(0, 0)$ **8.** no solution **9.** infinite number of solutions **10.** $(12, -4)$ **11.** Memphis and Atlanta: 394 mi; Minneapolis and Houston: 1176 mi **12.** Disneyland: 12.7 million; Magic Kingdom: 14.0 million **13.** $33\frac{1}{3}$ L of 25% solution; $16\frac{2}{3}$ L of 40% solution **14.** slower car: 45 mph; faster car: 60 mph

15.

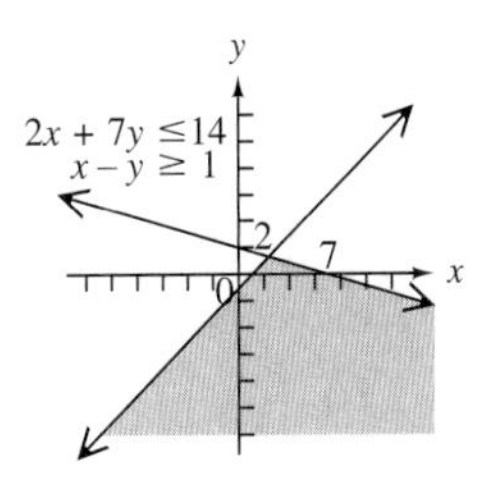

16.

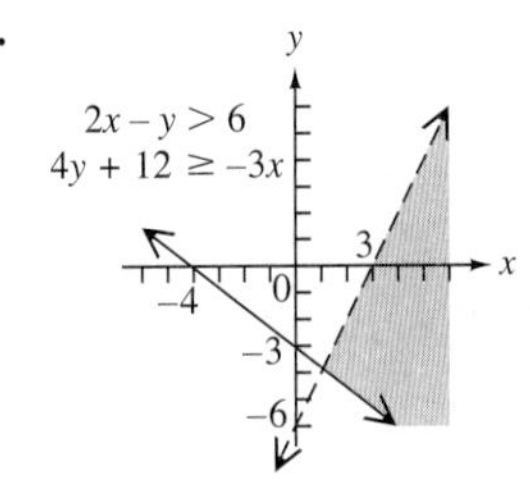

Cumulative Review Exercises: Chapters R–4 (page 329)

1. $-1, 1, -2, 2, -4, 4, -5, 5, -8, 8, -10, 10, -20, 20, -40, 40$ **2.** 1 **3.** commutative property **4.** distributive property **5.** inverse property **6.** 46 **7.** $-\frac{13}{11}$ **8.** $\frac{9}{11}$ **9.** $x > -18$ **10.** $x > -\frac{11}{2}$ **11.** width: 8.16 in.; length: 10.74 in.

12.

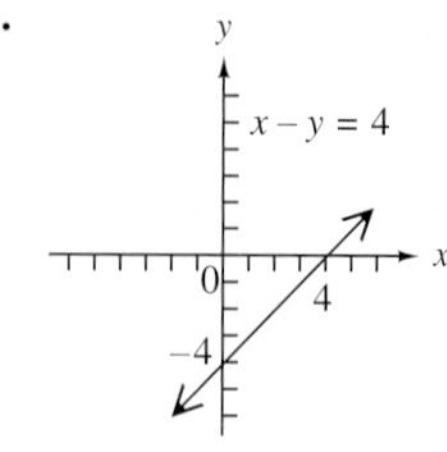

13.

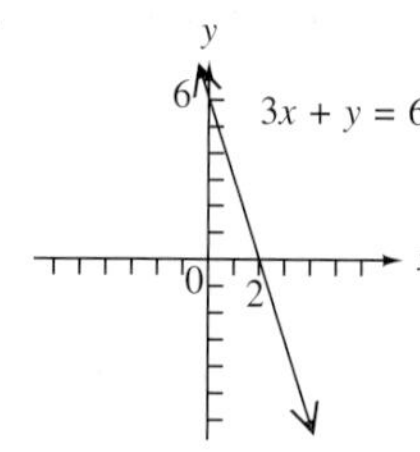

14. $-\frac{4}{3}$ **15.** $-\frac{1}{4}$ **16.** $y = \frac{1}{2}x + 3$ **17.** $y = 2x + 1$
18. **(a)** $x = 9$ **(b)** $y = -1$ **19.** $(-1, 6)$ **20.** $(3, -4)$
21. no solution **22.** 405 adults and 49 children
23. 19 in., 19 in., 15 in. **24.**

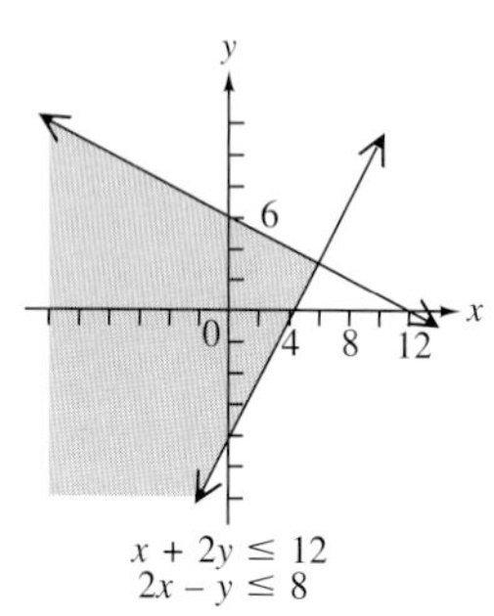

Chapter 5 Exponents and Polynomials

Section 5.1 (page 337)

1. 7; 5 **3.** 8 **5.** 26 **7.** 1; 6 **9.** 1; 1 **11.** 2; −19, −1
13. 2; 1, 8 **15.** $2m^5$ **17.** $-r^5$ **19.** cannot be simplified; $.2m^5 - .5m^2$ **21.** $-5x^5$ **23.** $5p^9 + 4p^7$ **25.** $-2y^2$
27. already simplified; 4; binomial **29.** already simplified; $6m^5 + 5m^4 - 7m^3 - 3m^2$; 5; none of these **31.** $x^4 + \frac{1}{3}x^2 - 4$; 4; trinomial **33.** 7; 0; monomial **35.** **(a)** −1 **(b)** 5 **37.** **(a)** 19 **(b)** −2 **39.** **(a)** 36 **(b)** −12 **41.** **(a)** −124 **(b)** 5
43. 175 ft **44.** 87 ft; (1, 87) **45.** \$7.20 **46.** \$27
47. $5m^2 + 3m$ **49.** $4x^4 - 4x^2$ **51.** $\frac{7}{6}x^2 - \frac{2}{15}x + \frac{5}{6}$
53. $12m^3 - 13m^2 + 6m + 11$ **55.** $8r^2 + 5r - 12$
57. $5m^2 - 14m + 6$ **59.** $4x^3 + 2x^2 + 5x$
61. $-18y^5 + 7y^4 + 5y^3 + 3y^2 + y$ **63.** $-2m^3 + 7m^2 + 8m - 9$
65. $8x^2 + 8x + 6$ **67.** $8t^2 + 8t + 13$ **69.** $-11x^2 - 3x - 3$
71. The degree of a term is determined by the exponents on the *variables*, but 3 is not a variable. The degree of $3^4 = 3^4x^0$ is 0.
73. $13a^2b - 7a^2 - b$ **75.** $c^4d - 5c^2d^2 + d^2$
77. $12m^3n - 11m^2n^2 - 4mn^2$

Section 5.2 (page 347)

1. 1 **3.** false **5.** false **7.** $(-2)^5$ **9.** $\left(\frac{1}{2}\right)^6$ **11.** $(-8p)^2$
13. The expression $(-3)^4$ means $(-3)(-3)(-3)(-3) = 81$, while -3^4 means $-(3 \cdot 3 \cdot 3 \cdot 3) = -81$. **15.** base: 3; exponent: 5; 243 **17.** base: −3; exponent: 5; −243 **19.** base: $-6x$; exponent: 4 **21.** base: x; exponent: 4 **23.** The product rule does not apply to $5^2 + 5^3$ because it is a *sum*, not a product. $5^2 + 5^3 = 25 + 125 = 150$ **25.** 5^8 **27.** 4^{12} **29.** $(-7)^9$
31. t^{24} **33.** $-56r^7$ **35.** $42p^{10}$ **37.** 4^6 **39.** t^{20} **41.** 7^3r^3
43. $5^5x^5y^5$ **45.** $8q^3r^3$ **47.** $\frac{1}{2^3}$ **49.** $\frac{a^3}{b^3}$ **51.** $\frac{9^8}{5^8}$ **53.** $(-2)^3x^6y^3$
55. $3^2a^6b^4$ **57.** $\frac{5^5}{2^5}$ **59.** $\frac{9^5}{8^3}$ **61.** $2^{12}x^{12}$ **63.** $(-6)^5p^5$
65. $6^5x^{10}y^{15}$ **67.** x^{21} **69.** $2^2w^4x^{26}y^7$ or $4w^4x^{26}y^7$ **71.** $-r^{18}s^{17}$
73. $\frac{5^3a^6b^{15}}{c^{18}}$ or $\frac{125a^6b^{15}}{c^{18}}$ **75.** $25m^6p^{14}q^5$ **77.** $16x^{10}y^{16}z^{10}$
79. $30x^7$

Section 5.3 (page 355)

1. $x^2 + 7x + 12$ **3.** $2x^3 + 7x^2 + 7x + 2$ **5.** distributive
7. $-6m^2 - 4m$ **9.** $6p - \frac{9}{2}p^2 + 9p^4$ **11.** $6y^5 + 4y^6 + 10y^9$
13. $12x^3 + 26x^2 + 10x + 1$ **15.** $20m^4 - m^3 - 8m^2 - 17m - 15$
17. $6x^6 - 3x^5 - 4x^4 + 4x^3 - 5x^2 + 8x - 3$
19. $5x^4 - 13x^3 + 20x^2 + 7x + 5$ **21.** $n^2 + n - 6$
23. $8r^2 - 10r - 3$ **25.** $9x^2 - 4$ **27.** $9q^2 + 6q + 1$
29. $6t^2 + 23st + 20s^2$ **31.** $-.3t^2 + .22t + .24$
33. $x^2 - \frac{5}{12}x - \frac{1}{6}$ **35.** $\frac{15}{16} - \frac{1}{4}r - 2r^2$ **37.** $6y^5 - 21y^4 - 45y^3$
39. $30x + 60$ yd² **40.** $30x + 60 = 600$; 18 **41.** 10 yd by 60 yd
42. \$2100 **43.** 140 yd **44.** \$1260 **45.** The answers are $x^2 - 16$, $y^2 - 4$, and $r^2 - 49$. Each product is the difference of the square of the first term and the square of the last term of the binomials.

Section 5.4 (page 361)

1. **(a)** $4x^2$ **(b)** $12x$ **(c)** 9 **(d)** $4x^2 + 12x + 9$ **3.** $a^2 - 2ac + c^2$
5. $p^2 + 4p + 4$ **7.** $16x^2 - 24x + 9$ **9.** $.64t^2 + 1.12ts + .49s^2$
11. $25x^2 + 4xy + \frac{4}{25}y^2$ **13.** $16a^2 - 12ab + \frac{9}{4}b^2$
15. $-16r^2 + 16r - 4$ **17.** **(a)** $49x^2$ **(b)** 0 **(c)** $-9y^2$ **(d)** $49x^2 - 9y^2$; Because 0 is the identity element for addition, it is not necessary to write "+ 0." **19.** $q^2 - 4$ **21.** $4w^2 - 25$
23. $100x^2 - 9y^2$ **25.** $4x^4 - 25$ **27.** $49x^2 - \frac{9}{49}$ **29.** $9p^3 - 49p$
31. $(a + b)^2$ **32.** a^2 **33.** $2ab$ **34.** b^2 **35.** $a^2 + 2ab + b^2$
36. They both represent the area of the entire large square.
37. 1225 **38.** $30^2 + 2(30)(5) + 5^2$ **39.** 1225 **40.** They are equal. **41.** $m^3 - 15m^2 + 75m - 125$ **43.** $8a^3 + 12a^2 + 6a + 1$
45. $81r^4 - 216r^3t + 216r^2t^2 - 96rt^3 + 16t^4$
47. $3x^5 - 27x^4 + 81x^3 - 81x^2$
49. $-8x^6y - 32x^5y^2 - 48x^4y^3 - 32x^3y^4 - 8x^2y^5$
51. 512 cu. units

Section 5.5 (page 371)

1. negative **3.** negative **5.** positive **7.** 0 **9.** 1 **11.** −1
13. 0 **15.** 0 **17.** 2 **19.** $\frac{1}{64}$ **21.** 16 **23.** $\frac{49}{36}$ **25.** $\frac{8}{15}$
27. $-\frac{7}{18}$ **29.** 1 **30.** $\frac{5^2}{5^2}$ **31.** 5^0 **32.** $5^0 = 1$; This supports the definition of a 0 exponent. **33.** $\frac{1}{9}$ **35.** $\frac{1}{6^5}$ **37.** 6^3 **39.** $2r^4$
41. $\frac{5^2}{4^3}$ **43.** $\frac{p^5}{q^8}$ **45.** r^9 **47.** $\frac{x^5}{6}$ **49.** $3y^2$ **51.** x^3 **53.** 7^3
55. $\frac{1}{x^2}$ **57.** $\frac{4^3x}{3^2}$ or $\frac{64x}{9}$ **59.** $\frac{x^2z^4}{y^2}$ **61.** $6x$ **63.** $\frac{1}{m^{10}n^5}$
65. $\frac{5}{16x^5}$ **67.** $\frac{36q^2}{m^4p^2}$

Summary Exercises on the Rules for Exponents (page 373)

1. $\frac{6^{12}x^{24}}{5^{12}}$ **2.** $\frac{r^6s^{12}}{729t^6}$ **3.** $10^5x^7y^{14}$ **4.** $-128a^{10}b^{15}c^4$ **5.** $\frac{729w^3x^9}{y^{12}}$
6. $\frac{x^4y^6}{16}$ **7.** c^{22} **8.** $\frac{1}{k^4t^{12}}$ **9.** $\frac{11}{30}$ **10.** $y^{12}z^3$ **11.** $\frac{x^6}{y^5}$ **12.** 0

13. $\frac{1}{z^2}$ **14.** $\frac{9}{r^2s^2t^{10}}$ **15.** $\frac{300x^3}{y^3}$ **16.** $\frac{3}{5x^6}$ **17.** x^8 **18.** $\frac{y^{11}}{x^{11}}$
19. $\frac{a^6}{b^4}$ **20.** $6ab$ **21.** $\frac{61}{900}$ **22.** 1 **23.** $\frac{343a^6b^9}{8}$ **24.** 1
25. -1 **26.** 0 **27.** $\frac{27y^{18}}{4x^8}$ **28.** $\frac{1}{a^8b^{12}c^{16}}$ **29.** $\frac{x^{15}}{216z^9}$ **30.** $\frac{q}{8p^6r^3}$
31. x^6y^6 **32.** 0 **33.** $\frac{343}{x^{15}}$ **34.** $\frac{9}{x^6}$ **35.** $5p^{10}q^9$ **36.** $\frac{7}{24}$
37. $\frac{r^{14}t}{2s^2}$ **38.** 1 **39.** $8p^{10}q$ **40.** $\frac{1}{mn^3p^3}$ **41.** -1 **42.** $\frac{3}{40}$

Section 5.6 (page 377)

1. $6x^2+8$; 2; $3x^2+4$ **3.** $3x^2+4$; 2 (These may be reversed.); $6x^2+8$ **5.** To use the method of this section, the divisor must be just one term. This is true of the first problem, but not the second. **7.** $30x^3-10x+5$ **9.** $-4m^3+2m^2-1$ **11.** $4t^4-2t^2+2t$ **13.** $a^4-a+\frac{2}{a}$ **15.** $-2x^3+\frac{2x^2}{3}-x$ **17.** $1+5x-9x^2$ **19.** $\frac{4x^2}{3}+x+\frac{2}{3x}$ **21.** $9r^3-12r^2-2r+1-\frac{2}{3r}$ **23.** $-m^2+3m-\frac{4}{m}$ **25.** $-4b^2+3ab-\frac{5}{a}$ **27.** $\frac{12}{x}-\frac{6}{x^2}+\frac{14}{x^3}-\frac{10}{x^4}$ **29.** No, $\frac{2}{3}x$ means $\frac{2x}{3}$, which is not the same as $\frac{2}{3x}$. In the first case we multiply by x; in the second case we divide by x. Yes, $\frac{4}{3}x^2=\frac{4x^2}{3}$. In both cases we are multiplying by x^2. **31.** $15x^5-35x^4+35x^3$ **33.** 1423 **34.** $(1\times10^3)+(4\times10^2)+(2\times10^1)+(3\times10^0)$ **35.** x^3+4x^2+2x+3 **36.** They are similar in that the coefficients of the powers of ten are equal to the coefficients of the powers of x. They are different in that one is a number while the other is a polynomial. They are equal if $x=10$.

Section 5.7 (page 383)

1. The divisor is $2x+5$; the quotient is $2x^3-4x^2+3x+2$. **3.** Divide $12m^2$ by $2m$ to get $6m$. **5.** $x+2$ **7.** $2y-5$ **9.** $p-4+\frac{44}{p+6}$ **11.** $r-5$ **13.** $2a-14+\frac{74}{2a+3}$ **15.** $4x^2-7x+3$ **17.** $3y^2-2y+2$ **19.** $3k-4+\frac{2}{k^2-2}$ **21.** x^2+1 **23.** $2p^2-5p+4+\frac{6}{3p^2+1}$ **25.** x^3+6x-7 **27.** x^2+1 **29.** $2x^2+\frac{3}{5}x+\frac{1}{5}$ **31.** x^2+x-3 units **33.** 33 **34.** 33 **35.** They are the same. **36.** The answers should agree.

Section 5.8 (page 389)

1. 6.1309×10^9; 5.8689×10^9 **3.** 6.9627×10^{10}; 1.8104×10^{11} **5.** in scientific notation **7.** not in scientific notation; 5.6×10^6 **9.** not in scientific notation; 4×10^{-3} **11.** not in scientific notation; 8×10^1 **13.** A number is written in scientific notation if it is the product of a number whose absolute value is between 1 and 10 (inclusive of 1) and a power of 10. **15.** 5.876×10^9 **17.** 8.235×10^4 **19.** 7×10^{-6} **21.** -2.03×10^{-3} **23.** 750,000 **25.** 5,677,000,000,000 **27.** -6.21 **29.** .00078 **31.** .000000005134 **33.** 6×10^{11}; 600,000,000,000 **35.** 1.5×10^7; 15,000,000 **37.** 6.426×10^4; 64,260 **39.** 3×10^{-4} **41.** 4×10^1 **43.** 2.6×10^{-3} **45.** about 3.3 **47.** about $63,000,000,000 **49.** about .276 lb

Chapter 5 Review Exercises (page 395)

1. $22m^2$; degree 2; monomial **2.** p^3-p^2+4p+2; degree 3; none of these **3.** already in descending powers; degree 5; none of these **4.** $-8y^5-7y^4+9y$; degree 5; trinomial **5.** $-5a^3+4a^2$ **6.** $2r^3-3r^2+9r$ **7.** $11y^2-10y+9$ **8.** $-13k^4-15k^2-4k-6$ **9.** $10m^3-6m^2-3$ **10.** $-y^2-4y+26$ **11.** $10p^2-3p-11$ **12.** $7r^4-4r^3-1$ **13.** 4^{11} **14.** $(-5)^{11}$ **15.** $-72x^7$ **16.** $10x^{14}$ **17.** 19^5x^5 **18.** $(-4)^7y^7$ **19.** $5p^4t^4$ **20.** $\frac{7^6}{5^6}$ **21.** $3^3x^6y^9$ **22.** t^{42} **23.** $6^2x^{16}y^4z^{16}$ **24.** The product rule for exponents does not apply here because we want the sum of 7^2 and 7^4, not their product. **25.** $10x^2+70x$ **26.** $-6p^5+15p^4$ **27.** $6r^3+8r^2-17r+6$ **28.** $8y^3+27$ **29.** $5p^5-2p^4-3p^3+25p^2+15p$ **30.** $6k^2-9k-6$ **31.** $12p^2-48pq+21q^2$ **32.** $2m^4+5m^3-16m^2-28m+9$ **33.** $a^2+8a+16$ **34.** $9p^2-12p+4$ **35.** $4r^2+20rs+25s^2$ **36.** $r^3+6r^2+12r+8$ **37.** $8x^3-12x^2+6x-1$ **38.** $36m^2-25$ **39.** $4z^2-49$ **40.** $25a^2-36b^2$ **41.** $4x^4-25$ **42.** $(a+b)^2=(a+b)(a+b)=a^2+2ab+b^2$. The term $2ab$ is not in a^2+b^2. **43.** 2 **44.** $\frac{1}{32}$ **45.** $\frac{5^2}{6^2}$ or $\frac{25}{36}$ **46.** $-\frac{3}{16}$ **47.** 6^2 **48.** x^2 **49.** $\frac{1}{p^{12}}$ **50.** r^4 **51.** 2^8 **52.** $\frac{1}{9^6}$ **53.** 5^8 **54.** $\frac{1}{8^{12}}$ **55.** $\frac{1}{m^2}$ **56.** y^7 **57.** r^{13} **58.** $(-5)^2m^6$ **59.** $\frac{y^{12}}{2^3}$ **60.** $\frac{1}{a^3b^5}$ **61.** $2\cdot6^2\cdot r^5$ **62.** $\frac{2^3n^{10}}{3m^{13}}$ **63.** $\frac{5y^2}{3}$ **64.** $-2x^2y$ **65.** $-y^3+2y-3$ **66.** $p-3+\frac{5}{2p}$ **67.** $-x^9+2x^8-4x^3+7x$ **68.** $-2m^2n+mn^2+\frac{6n^3}{5}$ **69.** $2r+7$ **70.** $4m+3+\frac{5}{3m-5}$ **71.** $2a+1+\frac{-8a+12}{5a^2-3}$ **72.** $k^2+2k+4+\frac{-2k-12}{2k^2+1}$ **73.** 4.8×10^7 **74.** 2.8988×10^{10} **75.** 6.5×10^{-5} **76.** 8.24×10^{-8} **77.** 24,000 **78.** 78,300,000 **79.** .000000897 **80.** .00000000000995 **81.** 800 **82.** 4,000,000 **83.** .025 **84.** .01 **85.** .0000000000016 **86.** **(a)** 1×10^3 **(b)** 2×10^3 **(c)** 5×10^4 **(d)** 1×10^5 **87.** 0 **88.** $\frac{3^5}{p^3}$ **89.** $\frac{1}{7^2}$ **90.** $49-28k+4k^2$ **91.** y^2+5y+1 **92.** $\frac{6^4r^8s^4}{5^4}$ **93.** $-8m^7-10m^6-6m^5$ **94.** 2^5 **95.** $5xy^3-\frac{8y^2}{5}+3x^2y$ **96.** $\frac{r^2}{6}$ **97.** $8x^3+12x^2y+6xy^2+y^3$ **98.** $\frac{3}{4}$ **99.** a^3-2a^2-7a+2 **100.** $8y^3-9y^2+5$ **101.** $10r^2+21r-10$ **102.** $144a^2-1$ **103.** $2x^2+x-6$ **104.** $25x^8+20x^6+4x^4$

Chapter 5 Test (page 399)

1. $4t^4+t^3-6t^2-t$ **2.** $-2y^2-9y+17$ **3.** $-12t^2+5t+8$ **4.** $(-2)^5$ or -2^5 **5.** $\frac{6^3}{m^6}$ **6.** $-27x^5+18x^4-6x^3+3x^2$ **7.** $2r^3+r^2-16r+15$ **8.** $t^2-5t-24$ **9.** $8x^2+2xy-3y^2$ **10.** $25x^2-20xy+4y^2$ **11.** $100v^2-9w^2$ **12.** x^3+3x^2+3x+1 **13.** $\frac{1}{625}$ **14.** 2 **15.** $\frac{7}{12}$ **16.** 8^5

17. x^2y^6 **18.** $4y^2 - 3y + 2 + \frac{5}{y}$ **19.** $-3xy^2 + 2x^3y^2 + 4y^2$
20. $2x + 9$ **21.** $3x^2 + 6x + 11 + \frac{26}{x - 2}$ **22. (a)** 3.44×10^{11}
(b) 5.57×10^{-6} **23. (a)** 29,600,000 **(b)** .0000000607
24. $9x^2 + 54x + 81$ **25.** Answers will vary. One example is $(-4x^4 + 3x^3 + 2x + 1) + (4x^4 - 8x^3 + 2x + 7) = -5x^3 + 4x + 8$.

Cumulative Review Exercises: Chapters R–5 (page 401)

1. $\frac{19}{24}$ **2.** $-\frac{1}{20}$ **3.** 3.72 **4.** 62.006 **5.** \$1836 **6.** -8 **7.** 24
8. $\frac{1}{2}$ **9.** -4 **10.** associative property **11.** inverse property
12. distributive property **13.** 10 **14.** $\frac{13}{4}$ **15.** no solution
16. $r = \frac{d}{t}$ **17.** -5 **18.** -12 **19.** 20 **20.** all real numbers
21. mouse: 160; elephant: 10 **22.** 4 **23.** $x \geq 10$
24. $x < -\frac{14}{5}$ **25.** $-4 \leq x < 2$ **26.** (0, 2) and $(-3, 0)$
27.

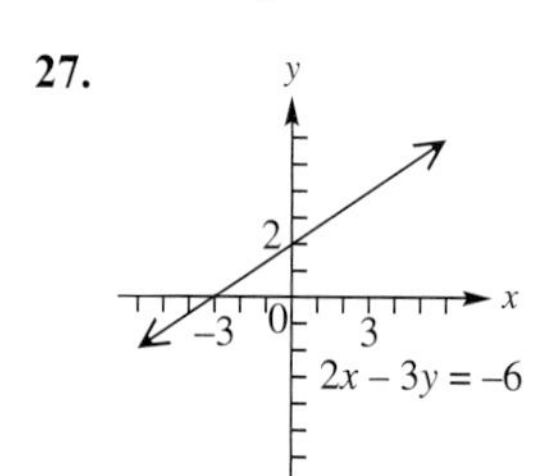

28. $\frac{2}{3}$ **29.** $(-3, -1)$ **30.** $(4, -5)$
31. $\frac{5}{4}$ or $1\frac{1}{4}$ **32.** 2 **33.** 1 **34.** $\frac{2b}{a^{10}}$ **35.** 3.45×10^4
36. $11x^3 - 14x^2 - x + 14$ **37.** $18x^7 - 54x^6 + 60x^5$
38. $63x^2 + 57x + 12$ **39.** $25x^2 + 80x + 64$ **40.** $y^2 - 2y + 6$

Chapter 6 Factoring and Applications

Section 6.1 (page 411)

1. 4 **3.** 4 **5.** 6 **7.** 1 **9.** 8 **11.** $10x^3$ **13.** xy^2 **15.** 6
17. $3m^2$ **19.** $2z^4$ **21.** $2mn^4$ **23.** $y + 2$ **25.** $a - 2$
27. $2 + 3xy$ **29.** $x(x - 4)$ **31.** $3t(2t + 5)$ **33.** $\frac{1}{4}d(d - 3)$
35. $6x^2(2x + 1)$ **37.** $5y^6(13y^4 + 7)$
39. no common factor (except 1) **41.** $8m^2n^2(n + 3)$
43. $2x(2x^2 - 5x + 3)$ **45.** $13y^2(y^6 + 2y^2 - 3)$
47. $9qp^3(5q^3p^2 + 4p^3 + 9q)$ **49.** $(x + 2)(c + d)$
51. $(2a + b)(a^2 - b)$ **53.** $(p + 4)(q - 1)$
55. $(5 + n)(m + 4)$ **57.** $(2y - 7)(3x + 4)$
59. $(y + 3)(3x + 1)$ **61.** $(z + 2)(7z - a)$
63. $(3r + 2y)(6r - x)$ **65.** $(w + 1)(w^2 + 9)$
67. $(a + 2)(3a^2 - 2)$ **69.** $(4m - p^2)(4m^2 - p)$
71. $(y + 3)(y + x)$ **73.** $(z - 2)(2z - 3w)$
75. commutative property **76.** $2x(y - 4) - 3(y - 4)$
77. No, because it is not a product. It is the difference between $2x(y - 4)$ and $3(y - 4)$. **78.** $(2x - 3)(y - 4)$; yes

Section 6.2 (page 417)

1. a and b must have different signs. **3.** A prime polynomial is one that cannot be factored using only integers in the factors.
5. 1 and 12, -1 and -12, 2 and 6, -2 and -6, 3 and 4, -3 and -4; the pair with a sum of 7 is 3 and 4. **7.** 1 and -24, -1 and 24, 2 and -12, -2 and 12, 3 and -8, -3 and 8, 4 and -6, -4 and 6; the pair with a sum of -5 is 3 and -8. **9.** C **11.** $x + 11$
13. $x - 8$ **15.** $y - 5$ **17.** $x + 11$ **19.** $y - 9$
21. $(y + 8)(y + 1)$ **23.** $(b + 3)(b + 5)$
25. $(m + 5)(m - 4)$ **27.** $(x + 8)(x - 5)$
29. $(y - 5)(y - 3)$ **31.** $(z - 8)(z - 7)$ **33.** $(r - 6)(r + 5)$
35. $(a - 12)(a + 4)$ **37.** prime **39.** $(r + 2a)(r + a)$
41. $(x + y)(x + 3y)$ **43.** $(t + 2z)(t - 3z)$
45. $(v - 5w)(v - 6w)$ **47.** $4(x + 5)(x - 2)$
49. $2t(t + 1)(t + 3)$ **51.** $2x^4(x - 3)(x + 7)$
53. $a^3(a + 4b)(a - b)$ **55.** $mn(m - 6n)(m - 4n)$
57. The factored form $(2x + 4)(x - 3)$ is incorrect because $2x + 4$ has a common factor of 2, which must be factored out for the trinomial to be *completely* factored.

Section 6.3 (page 421)

1. $(m + 6)(m + 2)$ **3.** $(a + 5)(a - 2)$ **5.** $(2t + 1)(5t + 2)$
7. $(3z - 2)(5z - 3)$ **9.** $(2s - t)(4s + 3t)$
11. $(3a + 2b)(5a + 4b)$ **13.** B **15. (a)** 2; 12; 24; 11
(b) 3; 8 (Order is irrelevant.) **(c)** $3m$; $8m$
(d) $2m^2 + 3m + 8m + 12$ **(e)** $(2m + 3)(m + 4)$
(f) $(2m + 3)(m + 4) = 2m^2 + 11m + 12$
17. $(2x + 1)(x + 3)$ **19.** $(4r - 3)(r + 1)$
21. $(4m + 1)(2m - 3)$ **23.** $(3m + 1)(7m + 2)$
25. $(2b + 1)(3b + 2)$ **27.** $(4y - 3)(3y - 1)$
29. $3(4x - 1)(2x - 3)$ **31.** $2m(m - 4)(m + 5)$
33. $4z^3(8z + 3)(z - 1)$ **35.** $(3p + 4q)(4p - 3q)$
37. $(3a - 5b)(2a + b)$ **39.** $(5 - x)(1 - x)$ **41.** The student stopped too soon. He needs to factor out the common factor $4x - 1$ to get $(4x - 1)(4x - 5)$ as the correct answer.

Section 6.4 (page 427)

1. B **3.** A **5.** A **7.** $2a + 5b$ **9.** $x^2 + 3x - 4$; $x + 4$, $x - 1$, or $x - 1$, $x + 4$ **11.** $2z^2 - 5z - 3$; $2z + 1$, $z - 3$, or $z - 3$, $2z + 1$ **13.** The binomial $2x - 6$ cannot be a factor because it has a common factor of 2, but the polynomial does not.
15. $(3a + 7)(a + 1)$ **17.** $(2y + 3)(y + 2)$
19. $(3m - 1)(5m + 2)$ **21.** $(3s - 1)(4s + 5)$
23. $(5m - 4)(2m - 3)$ **25.** $(4w - 1)(2w - 3)$
27. $(4y + 1)(5y - 11)$ **29.** prime **31.** $2(5x + 3)(2x + 1)$
33. $q(5m + 2)(8m - 3)$ **35.** $3n^2(5n - 3)(n - 2)$
37. $y^2(5x - 4)(3x + 1)$ **39.** $(5a + 3b)(a - 2b)$
41. $(4s + 5t)(3s - t)$ **43.** $m^4n(3m + 2n)(2m + n)$
45. $-1(x + 7)(x - 3)$ **47.** $-1(3x + 4)(x - 1)$
49. $-1(a + 2b)(2a + b)$ **51.** $5 \cdot 7$ **52.** $(-5)(-7)$
53. The product of $3x - 4$ and $2x - 1$ is $6x^2 - 11x + 4$.
54. The product of $4 - 3x$ and $1 - 2x$ is $6x^2 - 11x + 4$.
55. The factors in Exercise 53 are the opposites of the factors in Exercise 54. **56.** $(3 - 7t)(5 - 2t)$

Section 6.5 (page 433)

1. 1; 4; 9; 16; 25; 36; 49; 64; 81; 100; 121; 144; 169; 196; 225; 256; 289; 324; 361; 400 **3.** 2 **5.** $(y + 5)(y - 5)$
7. $\left(p + \frac{1}{3}\right)\left(p - \frac{1}{3}\right)$ **9.** prime **11.** $(3r + 2)(3r - 2)$
13. $\left(6m + \frac{4}{5}\right)\left(6m - \frac{4}{5}\right)$ **15.** $4(3x + 2)(3x - 2)$
17. $(14p + 15)(14p - 15)$ **19.** $(4r + 5a)(4r - 5a)$
21. prime **23.** $(p^2 + 7)(p^2 - 7)$ **25.** $(x^2 + 1)(x + 1)(x - 1)$
27. $(p^2 + 16)(p + 4)(p - 4)$ **29.** The teacher was justified, because it was not factored *completely*; $x^2 - 9$ can be factored as $(x + 3)(x - 3)$. The complete factored form is $(x^2 + 9)(x + 3)(x - 3)$. **31.** No, it is not a perfect square since the middle term would have to be $30y$. **33.** $(w + 1)^2$
35. $(x - 4)^2$ **37.** $\left(t + \frac{1}{2}\right)^2$ **39.** $(x - .5)^2$

41. $2(x + 6)^2$ **43.** $(4x - 5)^2$ **45.** $(7x - 2y)^2$ **47.** $(8x + 3y)^2$ **49.** $2h(5h - 2y)^2$ **51.** $(2x + 3)(5x - 2)$ **52.** $5x - 2$ **53.** Yes. We saw in Exercise 51 that $(2x + 3)(5x - 2) = 10x^2 + 11x - 6$. **54.** The quotient is $x^2 + x + 1$, so $x^3 - 1 = (x - 1)(x^2 + x + 1)$.

Summary Exercises on Factoring (page 435)

1. $8m^3(4m^6 + 2m^2 + 3)$ **2.** $2(m + 3)(m - 8)$ **3.** $7k(2k + 5)(k - 2)$ **4.** prime **5.** $(6z + 1)(z + 5)$ **6.** $(m + n)(m - 4n)$ **7.** $(7z + 4y)(7z - 4y)$ **8.** $10nr(10nr + 3r^2 - 5n)$ **9.** $4x(4x + 5)$ **10.** $(4 + m)(5 + 3n)$ **11.** $(5y - 6z)(2y + z)$ **12.** $(y^2 + 9)(y + 3)(y - 3)$ **13.** $(m - 3)(m + 5)$ **14.** $(2y + 1)(3y - 4)$ **15.** $8z(4z - 1)(z + 2)$ **16.** $5y(3y + 1)$ **17.** $(z - 6)^2$ **18.** $(3m + 8)(3m - 8)$ **19.** $(y - 6k)(y + 2k)$ **20.** $(4z - 1)^2$ **21.** $6(y - 2)(y + 1)$ **22.** $\left(x + \frac{1}{4}\right)^2$ **23.** $(p - 6)(p - 11)$ **24.** $(a + 8)(a + 9)$ **25.** prime **26.** $3(6m - 1)^2$ **27.** $(z + 2a)(z - 5a)$ **28.** $(2a + 1)(a^2 - 7)$ **29.** $(2k - 3)^2$ **30.** $(a - 7b)(a + 4b)$ **31.** $(4r + 3m)^2$ **32.** $(3k - 2)(k + 2)$ **33.** prime **34.** $(a^2 + 25)(a + 5)(a - 5)$ **35.** $4(2k - 3)^2$ **36.** $(4k + 1)(2k - 3)$ **37.** $6y^4(3y + 4)(2y - 5)$ **38.** $5z(z - 2)(z - 7)$ **39.** $(8p - 1)(p + 3)$ **40.** $(4k - 3h)(2k + h)$ **41.** $6(3m + 2z)(3m - 2z)$ **42.** $(2k - 5z)^2$ **43.** $2(3a - 1)(a + 2)$ **44.** $(3h - 2g)(5h + 7g)$ **45.** $(m + 9)(m - 9)$ **46.** $(5z - 6)(2z + 1)$ **47.** $5m^2(5m - 13n)(5m - 3n)$ **48.** $(3y - 1)(3y + 5)$ **49.** $(m - 2)^2$ **50.** prime **51.** $9p^8(3p + 7)(p - 4)$ **52.** $5(2m - 3)(m + 4)$ **53.** $(2 - q)(2 - 3p)$ **54.** $\left(k + \frac{8}{11}\right)\left(k - \frac{8}{11}\right)$ **55.** $4(4p + 5m)(4p - 5m)$ **56.** $(m + 4)(m^2 - 6)$ **57.** $(10a + 9y)(10a - 9y)$ **58.** $(8a - b)(a + 3b)$ **59.** $(a + 4)^2$ **60.** $(2y + 5)(2y - 5)$

Section 6.6 (page 443)

1. $-5, 2$ **3.** $3, \frac{7}{2}$ **5.** $-\frac{5}{6}, 0$ **7.** $0, \frac{4}{3}$ **9.** $-\frac{1}{2}, \frac{1}{6}$ **11.** $-.8, 2$ **13.** 9 **15.** Set each *variable* factor equal to 0, to get $2x = 0$ or $3x - 4 = 0$. The solutions are 0 and $\frac{4}{3}$. **17.** $-2, -1$ **19.** $1, 2$ **21.** $-8, 3$ **23.** $-1, 3$ **25.** $-2, -1$ **27.** -4 **29.** $-2, \frac{1}{3}$ **31.** $-\frac{4}{3}, \frac{1}{2}$ **33.** $-\frac{2}{3}$ **35.** $-3, 3$ **37.** $-\frac{7}{4}, \frac{7}{4}$ **39.** $-11, 11$ **41.** $0, 7$ **43.** $0, \frac{1}{2}$ **45.** $2, 5$ **47.** $-4, \frac{1}{2}$ **49.** $-12, \frac{11}{2}$ **51.** $-2, 0, 2$ **53.** $-\frac{7}{3}, 0, \frac{7}{3}$ **55.** $-\frac{5}{2}, \frac{1}{3}, 5$ **57.** $-\frac{7}{2}, -3, 1$ **59. (a)** 64; 144; 4; 6 **(b)** No time has elapsed, so the object hasn't fallen (been released) yet. **(c)** Time cannot be negative.

Section 6.7 (page 451)

1. Read; variable; equation; Solve; answer; Check; original **3.** *Step 3:* $45 = (2x + 1)(x + 1)$; *Step 4:* $x = 4$ or $x = -\frac{11}{2}$; *Step 5:* base: 9 units; height: 5 units; *Step 6:* $9 \cdot 5 = 45$ **5.** *Step 3:* $80 = (x + 8)(x - 8)$; *Step 4:* $x = 12$ or $x = -12$; *Step 5:* length: 20 units; width: 4 units; *Step 6:* $20 \cdot 4 = 80$ **7.** length: 7 in.; width: 4 in. **9.** length: 13 in.; width: 10 in. **11.** height: 13 in.; width: 10 in. **13.** mirror: 7 ft; painting: 9 ft **15.** 20, 21 **17.** $-3, -2$ or 4, 5 **19.** $-3, -1$ or 7, 9 **21.** $-2, 0, 2$ or 6, 8, 10 **23.** 12 cm **25.** 12 mi **27.** 8 ft **29. (a)** 1 sec **(b)** $\frac{1}{2}$ sec and $1\frac{1}{2}$ sec **(c)** 3 sec **(d)** The negative solution, -1, does not make sense since t represents time, which cannot be negative. **31. (a)** 42.6 million; The result using the model is a little less than 44 million, the actual number for 1996. **(b)** 12 **(c)** 142.2 million; The result is a little more than 140 million, the actual number for 2002. **(d)** 215.4 million **32.** 107.4 billion dollars; 40% **33.** 1997: 148.5 billion dollars; 1999: 230.1 billion dollars; 2000: 270.9 billion dollars. **34.** The answers using the linear equation are not at all close to the actual data. **35.** 1997: 111.2 billion dollars; 1999: 266.4 billion dollars; 2000: 399.5 billion dollars **36.** The answers in Exercise 35 are fairly close to the actual data. The quadratic equation models the data better. **37.** (0, 97.5), (1, 104.3), (2, 104.7), (3, 164.3), (4, 271.3), (5, 378.7) **38.** no

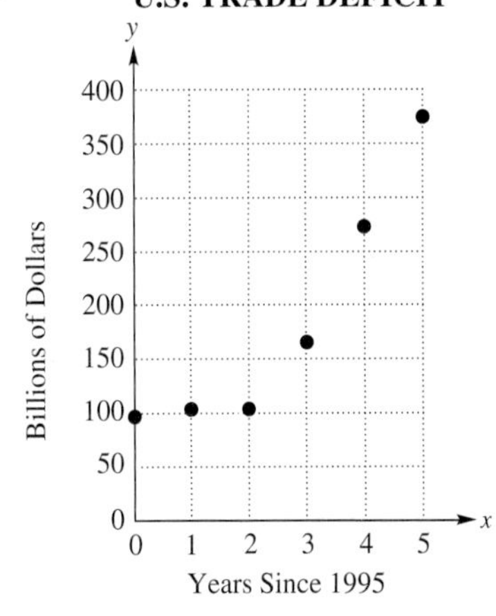

39. 776.7 billion dollars **40. (a)** The actual deficit is quite a bit less than the prediction. **(b)** No, data for later years might not follow the same pattern.

Chapter 6 Review Exercises (page 461)

1. $7(t + 2)$ **2.** $30z(2z^2 + 1)$ **3.** $35x^2(x + 2)$ **4.** $50m^2n^2(2n - mn^2 + 3)$ **5.** $(x - 4)(2y + 3)$ **6.** $(2y + 3)(3y + 2x)$ **7.** $(x + 3)(x + 2)$ **8.** $(y - 5)(y - 8)$ **9.** $(q + 9)(q - 3)$ **10.** $(r - 8)(r + 7)$ **11.** $(r + 8s)(r - 12s)$ **12.** $(p + 12q)(p - 10q)$ **13.** $8p(p + 2)(p - 5)$ **14.** $3x^2(x + 2)(x + 8)$ **15.** $(m + 3n)(m - 6n)$ **16.** $(y - 3z)(y - 5z)$ **17.** $p^5(p - 2q)(p + q)$ **18.** $3r^3(r + 3s)(r - 5s)$ **19.** prime **20.** $3(x^2 + 2x + 2)$ **21.** r and $6r$, $2r$ and $3r$ **22.** Factor out z. **23.** $(2k - 1)(k - 2)$ **24.** $(3r - 1)(r + 4)$ **25.** $(3r + 2)(2r - 3)$ **26.** $(5z + 1)(2z - 1)$ **27.** prime **28.** $4x^3(3x - 1)(2x - 1)$ **29.** $-3(x + 2)(2x - 5)$ **30.** $rs(5r + 6s)(2r + s)$ **31.** B **32.** D **33.** $(n + 8)(n - 8)$ **34.** $(5b + 11)(5b - 11)$ **35.** $(7y + 5w)(7y - 5w)$ **36.** $36(2p + q)(2p - q)$ **37.** prime **38.** $\left(x + \frac{7}{10}\right)\left(x - \frac{7}{10}\right)$ **39.** $(z + 5)^2$ **40.** $(r - 6)^2$ **41.** $(3t - 7)^2$ **42.** $(4m + 5n)^2$ **43.** $6x(3x - 2)^2$ **44.** $\left(x + \frac{1}{3}\right)^2$ **45.** $-\frac{3}{4}, 1$ **46.** $-7, -3, 4$ **47.** $0, \frac{5}{2}$ **48.** $-3, -1$ **49.** $1, 4$ **50.** $3, 5$ **51.** $-\frac{4}{3}, 5$ **52.** $-\frac{8}{9}, \frac{8}{9}$ **53.** $0, 8$ **54.** $-1, 6$ **55.** 7 **56.** 6 **57.** $-\frac{2}{5}, -2, -1$ **58.** $-3, 3$ **59.** length: 10 ft; width: 4 ft **60.** 5 ft **61.** length: 6 m; width: 2 m

62. length: 6 m; height: 5 m **63.** 6, 7 or $-5, -4$ **64.** 26 mi **65.** 112 ft **66.** 192 ft **67.** 256 ft **68.** after 8 sec **69. (a)** \$771.9 million; the answer using the model is a little high. **(b)** \$1727.6 million **(c)** No, the prediction seems low. If eBay revenues were \$1516.7 million through three quarters, they were approximately \$500 million per quarter, which would lead to annual revenue in 2003 of about \$2000 million. **70.** D **71.** The factor $2x + 8$ has a common factor of 2. The complete factored form is $2(x + 4)(3x - 4)$. **72.** $(z - x)(z - 10x)$ **73.** $(3k + 5)(k + 2)$ **74.** $(3m + 4p)(5m - 4)$ **75.** $(y^2 + 25)(y + 5)(y - 5)$ **76.** $3m(2m + 3)(m - 5)$ **77.** $8abc(3b^2c - 7ac^2 + 9ab)$ **78.** prime **79.** $6xyz(2xz^2 + 2y - 5x^2yz^3)$ **80.** $2a^3(a + 2)(a - 6)$ **81.** $(2r + 3q)(6r - 5q)$ **82.** $(10a + 3)(10a - 3)$ **83.** $(7t + 4)^2$ **84.** 0, 7 **85.** $-5, 2$ **86.** $-\frac{2}{5}$ **87.** 15 m, 36 m, 39 m **88.** length: 6 m; width: 4 m **89.** $-5, -4, -3$ or 5, 6, 7 **90. (a)** 256 ft **(b)** 1024 ft **91.** width: 10 m; length: 17 m **92.** 6 m **93. (a)** 481 thousand vehicles **(b)** The estimate may be unreliable because the conditions that prevailed in the years 1998–2001 may have changed, causing either a greater increase or a greater decrease in the numbers of alternative-fueled vehicles.

Chapter 6 Test (page 467)

1. D **2.** $6x(2x - 5)$ **3.** $m^2n(2mn + 3m - 5n)$ **4.** $(2x + y)(a - b)$ **5.** $(x - 7)(x - 2)$ **6.** $(2x + 3)(x - 1)$ **7.** $(3x + 1)(2x - 7)$ **8.** $3(x + 1)(x - 5)$ **9.** $(5z - 1)(2z - 3)$ **10.** prime **11.** prime **12.** $(y + 7)(y - 7)$ **13.** $(3y + 8)(3y - 8)$ **14.** $(x + 8)^2$ **15.** $(2x - 7y)^2$ **16.** $-2(x + 1)^2$ **17.** $3t^2(2t + 9)(t - 4)$ **18.** prime **19.** $4t(t + 4)^2$ **20.** $(x^2 + 9)(x + 3)(x - 3)$ **21.** $(p + 3)(p + 3) = p^2 + 6p + 9 \neq p^2 + 9$ **22.** $-3, 9$ **23.** $\frac{1}{2}, 6$ **24.** $-\frac{2}{5}, \frac{2}{5}$ **25.** 10 **26.** 0, 3 **27.** 6 ft by 9 ft **28.** $-2, -1$ **29.** 17 ft **30.** 243

Cumulative Review Exercises: Chapters R–6 (page 469)

1. 0 **2.** .05 **3.** 6 **4.** $P = \frac{A}{1 + rt}$ **5.** 345; 210; 38%; 15% **6.** gold: 12; silver: 16; bronze: 7 **7.** \$34,875 **8.** 110° and 70° **9. (a)** negative; positive **(b)** negative; negative **10.** $\left(-\frac{1}{4}, 0\right)$, $(0, 3)$ **11.** 12 **12.**

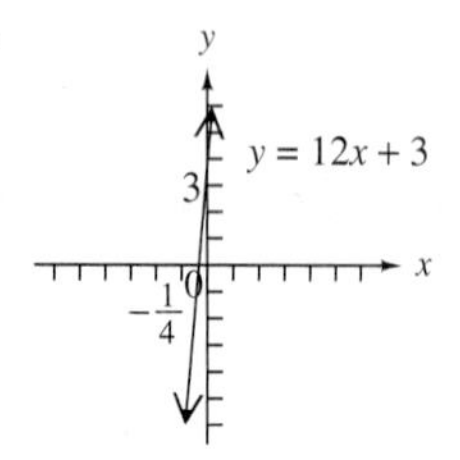

13. (a) 97; A slope of 97 means that the number of radio stations increased by about 97 stations per year. **(b)** (2000, 10,375) **14.** $(-1, 2)$ **15.** no solution **16.** 4 **17.** $\frac{16}{9}$ **18.** 1 **19.** 256 **20.** $\frac{1}{p^2}$ **21.** $\frac{1}{m^6}$ **22.** $-4k^2 - 4k + 8$ **23.** $45x^2 + 3x - 18$ **24.** $9p^2 + 12p + 4$ **25.** $4x^3 + 6x^2 - 3x + 10$ **26.** $(2a - 1)(a + 4)$ **27.** $(2m + 3)(5m + 2)$ **28.** $(4t + 3v)(2t + v)$ **29.** $(2p - 3)^2$ **30.** $(5r + 9t)(5r - 9t)$ **31.** $2pq(3p + 1)(p + 1)$ **32.** $-\frac{2}{3}, \frac{1}{2}$ **33.** 0, 8 **34.** 5 m, 12 m, 13 m

Chapter 7 Rational Expressions and Applications

Section 7.1 (page 479)

1. (a) 3; -5 **(b)** q; -1 **3.** A rational expression is a quotient of polynomials, such as $\frac{x + 3}{x^2 - 4}$. **5.** 0 **7.** $\frac{5}{3}$ **9.** $-3, 2$ **11.** never undefined **13. (a)** 1 **(b)** $\frac{17}{12}$ **15. (a)** 0 **(b)** $-\frac{10}{3}$ **17. (a)** $\frac{9}{5}$ **(b)** undefined **19. (a)** $\frac{2}{7}$ **(b)** $\frac{13}{3}$ **21.** No, not if the number is 0. Division by 0 is undefined. **23.** $3r^2$ **25.** $\frac{2}{5}$ **27.** $\frac{x - 1}{x + 1}$ **29.** $\frac{7}{5}$ **31.** $m - n$ **33.** $\frac{3(2m + 1)}{4}$ **35.** $\frac{3m}{5}$ **37.** $\frac{3r - 2s}{3}$ **39.** $\frac{x + 1}{x - 1}$ **41.** $\frac{z - 3}{z + 5}$ **43.** -1 **45.** $-(m + 1)$ **47.** -1

Answers may vary in Exercises 49–53.

49. $\frac{-(x + 4)}{x - 3}, \frac{-x - 4}{x - 3}, \frac{x + 4}{-(x - 3)}, \frac{x + 4}{-x + 3}$

51. $\frac{-(2x - 3)}{x + 3}, \frac{-2x + 3}{x + 3}, \frac{2x - 3}{-(x + 3)}, \frac{2x - 3}{-x - 3}$

53. $-\frac{3x - 1}{5x - 6}, \frac{-(3x - 1)}{5x - 6}, -\frac{-3x + 1}{-5x + 6}, \frac{3x - 1}{-5x + 6}$

55. $x^2 + 3$

Section 7.2 (page 487)

1. (a) B **(b)** D **(c)** C **(d)** A **3.** $\frac{4m}{3}$ **5.** $\frac{40y^2}{3}$ **7.** $\frac{2}{c + d}$ **9.** $\frac{16q}{3p^3}$ **11.** $\frac{7}{r^2 + rp}$ **13.** $\frac{z^2 - 9}{z^2 + 7z + 12}$ **15.** 5 **17.** $-\frac{3}{2t^4}$ **19.** $\frac{1}{4}$ **21.** To multiply two rational expressions, multiply the numerators and multiply the denominators. Write the answer in lowest terms. **23.** $\frac{10}{9}$ **25.** $-\frac{3}{4}$ **27.** -1 **29.** $\frac{-9(m - 2)}{m + 4}$ **31.** $\frac{p + 4}{p + 2}$ **33.** $\frac{(k - 1)^2}{(k + 1)(2k - 1)}$ **35.** $\frac{4k - 1}{3k - 2}$ **37.** $\frac{m + 4p}{m + p}$ **39.** $\frac{10}{x + 10}$ **41.** Division requires multiplying by the reciprocal of the second rational expression. In the reciprocal, $x + 7$ is in the denominator, so $x \neq -7$.

Section 7.3 (page 493)

1. C **3.** C **5.** 30 **7.** x^7 **9.** $72q$ **11.** $84r^5$ **13.** $2^3 \cdot 3 \cdot 5$ **15.** The least common denominator is their product. **17.** $28m^2(3m - 5)$ **19.** $30(b - 2)$ **21.** $c - d$ or $d - c$ **23.** $k(k + 5)(k - 2)$ **25.** $(p + 3)(p + 5)(p - 6)$ **27.** $\frac{20}{55}$ **29.** $\frac{-45}{9k}$ **31.** $\frac{26y^2}{80y^3}$ **33.** $\frac{35t^2r^3}{42r^4}$ **35.** $\frac{20}{8(m + 3)}$ **37.** $\frac{8t}{12 - 6t}$ **39.** $\frac{14(z - 2)}{z(z - 3)(z - 2)}$ **41.** $\frac{2(b - 1)(b + 2)}{b^3 + 3b^2 + 2b}$

Section 7.4 (page 501)

1. E **3.** C **5.** B **7.** G **9.** $\frac{11}{m}$ **11.** b **13.** x **15.** $y - 6$ **17.** To add or subtract rational expressions with the same denominator, combine the numerators and keep the same denominator. For example, $\frac{3x+2}{x-6} + \frac{-2x-8}{x-6} = \frac{x-6}{x-6}$. Then write in lowest terms: $\frac{x-6}{x-6} = 1$. **19.** $\frac{3z+5}{15}$ **21.** $\frac{10-7r}{14}$ **23.** $\frac{-3x-2}{4x}$ **25.** $\frac{x+1}{2}$ **27.** $\frac{5x+9}{6x}$ **29.** $\frac{3x+3}{x(x+3)}$ **31.** $\frac{x^2+6x-8}{(x-2)(x+2)}$ **33.** $\frac{3}{t}$ **35.** $m-2$ or $2-m$ **37.** $\frac{-2}{x-5}$ or $\frac{2}{5-x}$ **39.** -4 **41.** $\frac{-5}{x-y^2}$ or $\frac{5}{y^2-x}$ **43.** $\frac{x+y}{5x-3y}$ or $\frac{-x-y}{3y-5x}$ **45.** $\frac{-6}{4p-5}$ or $\frac{6}{5-4p}$ **47.** $\frac{-(m+n)}{2(m-n)}$ **49.** $\frac{-x^2+6x+11}{(x+3)(x-3)(x+1)}$ **51.** $\frac{-5q^2-13q+7}{(3q-2)(q+4)(2q-3)}$ **53.** $\frac{9r+2}{r(r+2)(r-1)}$ **55.** $\frac{2x^2+6xy+8y^2}{(x+y)(x+y)(x+3y)}$ or $\frac{2x^2+6xy+8y^2}{(x+y)^2(x+3y)}$ **57.** $\frac{15r^2+10ry-y^2}{(3r+2y)(6r-y)(6r+y)}$ **59. (a)** $\frac{9k^2+6k+26}{5(3k+1)}$ **(b)** $\frac{1}{4}$

Section 7.5 (page 511)

1. (a) 6; $\frac{1}{6}$ **(b)** 12; $\frac{3}{4}$ **(c)** $\frac{1}{6} \div \frac{3}{4}$ **(d)** $\frac{2}{9}$ **3.** -6 **5.** $\frac{1}{pq}$ **7.** $\frac{1}{xy}$ **9.** $\frac{2a^2b}{3}$ **11.** $\frac{m(m+2)}{3(m-4)}$ **13.** $\frac{2}{x}$ **15.** $\frac{8}{x}$ **17.** $\frac{a^2-5}{a^2+1}$ **19.** $\frac{3(p+2)}{2(2p+3)}$ **21.** $\frac{t(t-2)}{4}$ **23.** $\frac{-k}{2+k}$ **25.** $\frac{2x-7}{3x+1}$ **27.** $\frac{3m(m-3)}{(m-1)(m-8)}$ **29.** $\frac{6}{5}$

Section 7.6 (page 521)

1. expression; $\frac{43}{40}x$ **3.** equation; $\frac{40}{43}$ **5.** expression; $-\frac{1}{10}y$ **7.** When solving an equation, we multiply each side by the LCD, which eliminates all denominators. When adding or subtracting fractions, we multiply by 1 in the form $\frac{\text{LCD}}{\text{LCD}}$. The denominators are not eliminated. **9.** -6 **11.** 24 **13.** -15 **15.** 7 **17.** -15 **19.** -5 **21.** -6 **23.** 5 **25.** 12 **27.** 2 **29.** 0 and 4 **31.** -6 **33.** no solution **35.** 3 **37.** 3 **39.** -2, 12 **41.** no solution **43.** $-6, \frac{1}{2}$ **45.** $-\frac{1}{5}, 3$ **47.** $-\frac{3}{5}, 3$ **49.** Transform the equation so that the terms with k are on one side and the remaining term is on the other. **51.** $F = \frac{ma}{k}$ **53.** $a = \frac{kF}{m}$ **55.** $R = \frac{E-Ir}{I}$ **57.** $A = \frac{h(B+b)}{2}$ **59.** $a = \frac{2S-ndL}{nd}$ **61.** $t = \frac{rs}{rs-2s-3r}$ or $t = \frac{-rs}{-rs+2s+3r}$ **63.** $c = \frac{ab}{b-a-2ab}$ or $c = \frac{-ab}{-b+a+2ab}$ **65.** $z = \frac{3y}{5-9xy}$ or $z = \frac{-3y}{9xy-5}$

Summary Exercises on Rational Expressions and Equations (page 526)

1. expression; $\frac{10}{p}$ **2.** expression; $\frac{y^3}{x^3}$ **3.** expression; $\frac{1}{2x^2(x+2)}$ **4.** equation; 9 **5.** expression; $\frac{y+2}{y-1}$ **6.** expression; $\frac{5k+8}{k(k-4)(k+4)}$ **7.** equation; 39 **8.** expression; $\frac{t-5}{3(2t+1)}$ **9.** expression; $\frac{13}{3(p+2)}$ **10.** equation; $-1, \frac{12}{5}$ **11.** equation; $\frac{1}{7}, 2$ **12.** expression; $\frac{16}{3y}$ **13.** expression; $\frac{7}{12z}$ **14.** equation; 13 **15.** expression; $\frac{3m+5}{(m+2)(m+3)(m+1)}$ **16.** expression; $\frac{k+3}{5(k-1)}$ **17.** equation; no solution **18.** equation; -7

Section 7.7 (page 535)

1. (a) the amount **(b)** $5+x$ **(c)** $\frac{5+x}{6} = \frac{13}{3}$ **3.** $\frac{9}{5}$ **5.** $\frac{2}{6}$ **7.** -6 **9.** 36 **11.** 8.45 m per sec **13.** 3.735 hr **15.** 338.730 m per min **17.** $\frac{8}{4-x} = \frac{24}{4+x}$ **19.** into a headwind: $m-5$ mph; with a tailwind: $m+5$ mph **21.** 10 mph **23.** 18.5 mph **25.** $\frac{1}{2}t + \frac{1}{3}t = 1$ or $\frac{1}{2} + \frac{1}{3} = \frac{1}{t}$ **27.** $2\frac{2}{9}$ hr **29.** $4\frac{8}{19}$ hr **31.** 10 hr **33.** 36 hr **35.** $8\frac{1}{4}$ hr **37.** The equation would be erroneously written $\frac{8}{4+x} = \frac{24}{4-x}$. Solving for x gives $x = -2$. Because x represents the speed of the current, it cannot be negative. Therefore, the student should realize that there is a problem in the setup.

Section 7.8 (page 541)

1. (a) increases **(b)** decreases **3.** 15 **5.** 300 **7.** 4 **9.** 6 **11.** 15 in.^{2} **13.** $42\frac{2}{3}$ in. **15.** 15 ft **17.** 20 lb per ft^2 **19.** 25 kg per hr **21.** direct **23.** inverse **25.** inverse **27.** direct **29.** 8 **31.** 2 **33.** 80 ft

Chapter 7 Review Exercises (page 551)

1. 3 **2.** 0 **3.** $-1, 3$ **4.** $-5, -\frac{2}{3}$ **5. (a)** $-\frac{4}{7}$ **(b)** -16 **6. (a)** $\frac{11}{8}$ **(b)** $\frac{13}{22}$ **7. (a)** undefined **(b)** 1 **8. (a)** undefined **(b)** $\frac{1}{2}$ **9.** $\frac{b}{3a}$ **10.** -1 **11.** $\frac{-(2x+3)}{2}$ **12.** $\frac{2p+5q}{5p+q}$

Answers may vary in Exercises 13 and 14.

13. $\frac{-(4x-9)}{2x+3}, \frac{-4x+9}{2x+3}, \frac{4x-9}{-(2x+3)}, \frac{4x-9}{-2x-3}$ **14.** $\frac{-8+3x}{-3-6x}, \frac{-(-8+3x)}{3+6x}, \frac{8-3x}{-(-3-6x)}, -\frac{-8+3x}{3+6x}$ **15.** 2 **16.** $\frac{2}{3m^6}$ **17.** $\frac{5}{8}$ **18.** $\frac{r+4}{3}$ **19.** $\frac{3}{2}$ **20.** $\frac{y-2}{y-3}$ **21.** $\frac{p+5}{p+1}$ **22.** $\frac{3z+1}{z+3}$ **23.** 96 **24.** $108y^4$

25. $m(m+2)(m+5)$ **26.** $(x+3)(x+1)(x+4)$ **27.** $\frac{35}{56}$

28. $\frac{40}{4k}$ **29.** $\frac{15a}{10a^4}$ **30.** $\frac{-54}{18-6x}$ **31.** $\frac{15y}{50-10y}$

32. $\frac{4b(b+2)}{(b+3)(b-1)(b+2)}$ **33.** $\frac{15}{x}$ **34.** $-\frac{2}{p}$ **35.** $\frac{4k-45}{k(k-5)}$

36. $\frac{28+11y}{y(7+y)}$ **37.** $\frac{-2-3m}{6}$ **38.** $\frac{3(16-x)}{4x^2}$

39. $\frac{7a+6b}{(a-2b)(a+2b)}$ **40.** $\frac{-k^2-6k+3}{3(k+3)(k-3)}$

41. $\frac{5z-16}{z(z+6)(z-2)}$ **42.** $\frac{-13p+33}{p(p-2)(p-3)}$ **43.** $\frac{a}{b}$

44. $\frac{4(y-3)}{y+3}$ **45.** $\frac{6(3m+2)}{2m-5}$ **46.** $\frac{(q-p)^2}{pq}$ **47.** $\frac{xw+1}{xw-1}$

48. $\frac{1-r-t}{1+r+t}$ **49.** $\frac{35}{6}$ **50.** -16 **51.** -4 **52.** no solution

53. 3 **54.** $t=\frac{Ry}{m}$ **55.** $y=\frac{4x+5}{3}$ **56.** $t=\frac{rs}{s-r}$ **57.** 12

58. $\frac{20}{15}$ **59.** $\frac{3}{18}$ **60.** 1.527 hr **61.** $3\frac{1}{13}$ hr **62.** 10 hr **63.** 4 cm

64. $\frac{36}{5}$ **65.** $\frac{m+7}{(m-1)(m+1)}$ **66.** $8p^2$ **67.** $\frac{1}{6}$ **68.** 3

69. $\frac{z+7}{(z+1)(z-1)^2}$ **70.** $d=\frac{k+FD}{F}$ or $d=\frac{k}{F}+D$

71. $-2, 3$ **72.** 150 km per hr **73.** $1\frac{7}{8}$ hr **74.** 4 **75.** 24

Chapter 7 Test (page 555)

1. $-2, 4$ **2. (a)** $\frac{11}{6}$ **(b)** undefined **3.** (Answers may vary.) $\frac{-(6x-5)}{2x+3}, \frac{-6x+5}{2x+3}, \frac{6x-5}{-(2x+3)}, \frac{6x-5}{-2x-3}$ **4.** $-3x^2y^3$

5. $\frac{3a+2}{a-1}$ **6.** $\frac{25}{27}$ **7.** $\frac{3k-2}{3k+2}$ **8.** $\frac{a-1}{a+4}$ **9.** $150p^5$

10. $(2r+3)(r+2)(r-5)$ **11.** $\frac{240p^2}{64p^3}$ **12.** $\frac{21}{42m-84}$

13. 2 **14.** $\frac{-14}{5(y+2)}$ **15.** $\frac{x^2+x+1}{3-x}$ or $\frac{-x^2-x-1}{x-3}$

16. $\frac{-m^2+7m+2}{(2m+1)(m-5)(m-1)}$ **17.** $\frac{2k}{3p}$ **18.** $\frac{-2-x}{4+x}$ **19.** $-\frac{1}{2}$

20. $D=\frac{dF-k}{F}$ or $D=\frac{k-dF}{-F}$ **21.** -4 **22.** 3 mph

23. $2\frac{2}{9}$ hr **24.** 27 **25.** 27 days

Cumulative Review Exercises: Chapters R–7 (page 557)

1. 2 **2.** 17 **3.** $b=\frac{2A}{h}$ **4.** $-\frac{2}{7}$

5. $y \geq -8$

6. $m > 4$

7.

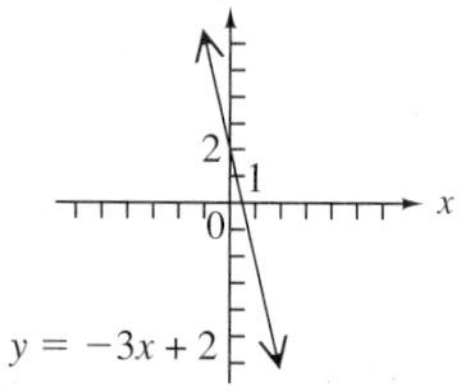

8.

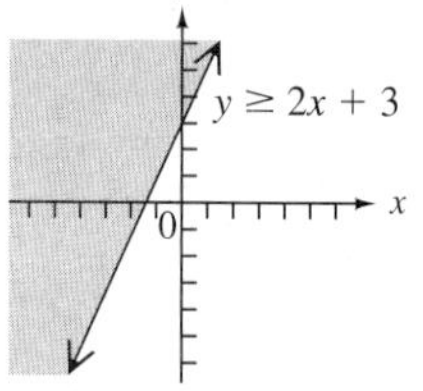

9. $(-1, 2)$ **10.** $(1, -2)$ **11.** $\frac{1}{2^4x^7}$ **12.** $\frac{1}{m^6}$ **13.** $\frac{q}{4p^2}$

14. k^2+2k+1 **15.** $72x^6y^7$ **16.** $4a^2-4ab+b^2$

17. $3y^3+8y^2+12y-5$ **18.** $6p^2+7p+1+\frac{3}{p-1}$

19. $(4t+3v)(2t+v)$ **20.** prime

21. $(4x^2+1)(2x+1)(2x-1)$ **22.** $-3, 5$ **23.** $5, -\frac{1}{2}, \frac{2}{3}$

24. -2 or -1 **25.** 6 m **26.** $-2, 2$ **27.** A **28.** D **29.** $\frac{4}{q}$

30. $\frac{3r+28}{7r}$ **31.** $\frac{7}{15(q-4)}$ **32.** $\frac{-k-5}{k(k+1)(k-1)}$

33. $\frac{7(2z+1)}{24}$ **34.** $\frac{195}{29}$ **35.** $\frac{21}{2}$ **36.** $-2, 1$ **37.** 150 mi

38. $1\frac{1}{5}$ hr

CHAPTER 8 ROOTS AND RADICALS

Section 8.1 (page 567)

1. true **3.** false; Zero has only one square root. **5.** true **7.** $-3, 3$ **9.** $-8, 8$ **11.** $-13, 13$ **13.** $-\frac{5}{14}, \frac{5}{14}$ **15.** $-30, 30$ **17.** 1 **19.** 7 **21.** -16 **23.** $-\frac{12}{11}$ **25.** .8 **27.** not a real number **29.** not a real number **31.** 100 **33.** 19 **35.** $\frac{2}{3}$ **37.** $3x^2+4$ **39.** a must be positive. **41.** a must be negative. **43.** rational; 5 **45.** irrational; 5.385 **47.** rational; -8 **49.** irrational; -17.321 **51.** not a real number **53.** irrational; 34.641 **55.** C **57.** $c=17$ **59.** $b=8$ **61.** $c=11.705$ **63.** 24 cm **65.** 80 ft **67.** 195 ft **69.** 9.434 **71.** 13.5 ft **73.** Answers will vary. For example, if $a=2$ and $b=7$, $\sqrt{a^2+b^2}=\sqrt{2^2+7^2}=\sqrt{53}$, while $a+b=2+7=9$. Therefore, $\sqrt{a^2+b^2} \neq a+b$ because $\sqrt{53} \neq 9$. **75.** 1 **77.** 5 **79.** -3 **81.** -6 **83.** -2 **85.** 4 **87.** 6 **89.** not a real number **91.** -3 **93.** -4

Section 8.2 (page 577)

1. false; $\sqrt{(-6)^2}=\sqrt{36}=6$ **3.** $\sqrt{15}$ **5.** $\sqrt{22}$ **7.** $\sqrt{42}$ **9.** $\sqrt{13r}$ **11.** A **13.** $3\sqrt{5}$ **15.** $2\sqrt{6}$ **17.** $3\sqrt{10}$ **19.** $5\sqrt{3}$ **21.** $5\sqrt{5}$ **23.** cannot be simplified **25.** $4\sqrt{10}$ **27.** $-10\sqrt{7}$ **29.** $3\sqrt{6}$ **31.** 24 **33.** $6\sqrt{10}$ **35.** $12\sqrt{5}$ **37.** $30\sqrt{5}$ **39.** $\sqrt{8}\cdot\sqrt{32}=\sqrt{8\cdot 32}=\sqrt{256}=16$. Also, $\sqrt{8}=2\sqrt{2}$ and $\sqrt{32}=4\sqrt{2}$, so $\sqrt{8}\cdot\sqrt{32}=2\sqrt{2}\cdot 4\sqrt{2}=8\cdot 2=16$. Both methods give the same answer, and the correct answer can always be obtained using either method. **41.** $\frac{4}{15}$ **43.** $\frac{\sqrt{7}}{4}$ **45.** 5 **47.** $\frac{25}{4}$ **49.** $6\sqrt{5}$ **51.** m **53.** y^2 **55.** $6z$ **57.** $20x^3$ **59.** $3x^4\sqrt{2}$ **61.** $3c^7\sqrt{5}$ **63.** $z^2\sqrt{z}$ **65.** $a^6\sqrt{a}$ **67.** $8x^3\sqrt{x}$ **69.** x^3y^6 **71.** $9m^2n$ **73.** $\frac{\sqrt{7}}{x^5}$ **75.** $\frac{y^2}{10}$ **77.** $\frac{x^3}{y^4}$ **79.** $2\sqrt[3]{5}$ **81.** $3\sqrt[3]{2}$ **83.** $4\sqrt[3]{2}$ **85.** $2\sqrt[4]{5}$ **87.** $\frac{2}{3}$ **89.** $-\frac{6}{5}$ **91.** p **93.** x^3 **95.** $4z^2$ **97.** $7a^3b$ **99.** $2t\sqrt[3]{2t^2}$ **101.** $\frac{m^4}{2}$ **103.** 6 cm **105.** 6 in. **107.** D

Section 8.3 (page 583)

1. distributive **3.** radicands **5.** $-5\sqrt{7}$ **7.** $5\sqrt{17}$ **9.** $5\sqrt{7}$ **11.** $11\sqrt{5}$ **13.** $15\sqrt{2}$ **15.** $-6\sqrt{2}$ **17.** $17\sqrt{7}$ **19.** $-16\sqrt{2}-8\sqrt{3}$ **21.** $20\sqrt{2}+6\sqrt{3}-15\sqrt{5}$ **23.** $4\sqrt{2}$ **25.** $22\sqrt{2}$ **27.** $11\sqrt{3}$ **29.** $5\sqrt{x}$ **31.** $3x\sqrt{6}$ **33.** 0 **35.** $-20\sqrt{2k}$ **37.** $42x\sqrt{5z}$ **39.** $-\sqrt[3]{2}$ **41.** $6\sqrt[3]{p^2}$ **43.** $21\sqrt[4]{m^3}$ **45.** $-6x^2y$ **46.** $-6(p-2q)^2(a+b)$ **47.** $-6a^2\sqrt{xy}$ **48.** The answers are alike because the numerical coefficient of the three answers is the same: -6. Also, the first variable factor is raised to the second power, and the second variable factor is raised to the first power. The answers are different because the variables are different: x and y, then $p-2q$ and $a+b$, and then a and $\sqrt{xy}$.

Section 8.4 (page 591)

1. $4\sqrt{2}$ **3.** $\frac{-\sqrt{33}}{3}$ **5.** $\frac{7\sqrt{15}}{5}$ **7.** $\frac{\sqrt{30}}{2}$ **9.** $\frac{16\sqrt{3}}{9}$ **11.** $\frac{-3\sqrt{2}}{10}$ **13.** $\frac{21\sqrt{5}}{5}$ **15.** $\sqrt{3}$ **17.** $\frac{\sqrt{2}}{2}$ **19.** $\frac{\sqrt{65}}{5}$ **21.** We are actually multiplying by 1. The identity property of multiplication justifies our result. **23.** $\frac{\sqrt{21}}{3}$ **25.** $\frac{3\sqrt{14}}{4}$ **27.** $\frac{1}{6}$ **29.** 1 **31.** $\frac{\sqrt{7x}}{x}$ **33.** $\frac{2x\sqrt{xy}}{y}$ **35.** $\frac{x\sqrt{30xz}}{6}$ **37.** $\frac{3ar^2\sqrt{7rt}}{7t}$ **39.** B **41.** $\frac{\sqrt[3]{12}}{2}$ **43.** $\frac{\sqrt[3]{196}}{7}$ **45.** $\frac{\sqrt[3]{6y}}{2y}$ **47.** $\frac{\sqrt[3]{42mn^2}}{6n}$ **49.** **(a)** $\frac{9\sqrt{2}}{4}$ sec **(b)** 3.182 sec

Section 8.5 (page 599)

1. 13 **3.** 4 **5.** $\sqrt{15}-\sqrt{35}$ **7.** $2\sqrt{10}+30$ **9.** $4\sqrt{7}$ **11.** $57+23\sqrt{6}$ **13.** $81+14\sqrt{21}$ **15.** $71-16\sqrt{7}$ **17.** $37+12\sqrt{7}$ **19.** $a+2\sqrt{a}+1$ **21.** 23 **23.** 1 **25.** $y-10$ **27.** $2\sqrt{3}-2+3\sqrt{2}-\sqrt{6}$ **29.** $15\sqrt{2}-15$ **31.** $\sqrt{30}+\sqrt{15}+6\sqrt{5}+3\sqrt{10}$ **33.** $\sqrt{5x}-\sqrt{10}-\sqrt{10x}+2\sqrt{5}$ **35.** Because multiplication must be performed before addition, it is incorrect to add -37 and -2. Only like radicals can be combined. **37.** $\frac{3-\sqrt{2}}{7}$ **39.** $-4-2\sqrt{11}$ **41.** $1+\sqrt{2}$ **43.** $-\sqrt{10}+\sqrt{15}$ **45.** $2\sqrt{5}+\sqrt{15}+4+2\sqrt{3}$ **47.** $\frac{12(\sqrt{x}-1)}{x-1}$ **49.** $\frac{3(7+\sqrt{x})}{49-x}$ **51.** $\sqrt{11}-2$ **53.** $\frac{\sqrt{3}+5}{8}$ **55.** $\frac{6-\sqrt{10}}{2}$ **57.** $30+18x$ **58.** They are not like terms. **59.** $30+18\sqrt{5}$ **60.** They are not like radicals. **61.** Make the first term $30x$, so that $30x+18x=48x$; make the first term $30\sqrt{5}$, so that $30\sqrt{5}+18\sqrt{5}=48\sqrt{5}$. **62.** Both like terms and like radicals are combined by adding their numerical coefficients. The variables in like terms are replaced by radicals in like radicals. **63.** 4 in.

Summary Exercises on Operations with Radicals (page 603)

1. $-3\sqrt{10}$ **2.** $5-\sqrt{15}$ **3.** $2-\sqrt{6}+2\sqrt{3}-3\sqrt{2}$ **4.** $6\sqrt{2}$ **5.** $73-12\sqrt{35}$ **6.** $\frac{\sqrt{6}}{2}$ **7.** $3\sqrt[3]{2t^2}$ **8.** $4\sqrt{7}+4\sqrt{5}$ **9.** $-3-2\sqrt{2}$ **10.** 4 **11.** -33 **12.** $\frac{\sqrt{t}-\sqrt{3}}{t-3}$ **13.** $2xyz^2\sqrt[3]{y^2}$ **14.** $4\sqrt[3]{3}$ **15.** $\sqrt{6}+1$ **16.** $\frac{\sqrt{6x}}{3x}$ **17.** $\frac{3}{5}$ **18.** $4\sqrt{2}$ **19.** $-2\sqrt[3]{2}$ **20.** $11-2\sqrt{30}$ **21.** $3\sqrt{3x}$ **22.** $52+30\sqrt{3}$ **23.** 1 **24.** $\frac{2\sqrt[3]{18}}{9}$ **25.** $-x^2\sqrt[4]{x}$ **26.** $2\sqrt{6}$ **27.** cannot be simplified further **28.** $12\sqrt{6}+6\sqrt{5}$ **29.** $\frac{\sqrt{15}}{10}$ **30.** $\sqrt{5}$ **31.** $20\sqrt[3]{3}$ **32.** $\frac{8(4+\sqrt{x})}{16-x}$ **33.** $2-3\sqrt[3]{4}$ **34.** $\sqrt{2x}$ **35.** $\frac{\sqrt{10}}{4}$ **36.** $49+14\sqrt{x}+x$ **37.** **(a)** 57 species **(b)** 858 species

Section 8.6 (page 611)

1. 49 **3.** 7 **5.** 85 **7.** -45 **9.** $-\frac{3}{2}$ **11.** no solution **13.** 121 **15.** 8 **17.** 1 **19.** 6 **21.** no solution **23.** 5 **25.** $x^2-14x+49$ **27.** 12 **29.** 5 **31.** 0, 3 **33.** $-1, 3$ **35.** 8 **37.** 4 **39.** 8 **41.** 9 **43.** 4, 20 **45.** -5 **47.** 21 **49.** 8 **51.** **(a)** 70.5 mph **(b)** 59.8 mph **(c)** 53.9 mph **53.** 158.6 ft **55.** $s=13$ units **56.** $6\sqrt{13}$ sq. units **57.** $h=\sqrt{13}$ units **58.** $3\sqrt{13}$ sq. units **59.** $6\sqrt{13}$ sq. units **60.** They are both $6\sqrt{13}$.

Chapter 8 Review Exercises (page 619)

1. $-7, 7$ **2.** $-9, 9$ **3.** $-14, 14$ **4.** $-11, 11$ **5.** $-15, 15$ **6.** $-27, 27$ **7.** 4 **8.** $-.6$ **9.** 10 **10.** 3 **11.** not a real number **12.** -65 **13.** $\frac{7}{6}$ **14.** $\frac{10}{9}$ **15.** B **16.** F **17.** D **18.** A **19.** C **20.** A **21.** 8 **22.** 40.6 cm **23.** irrational; 4.796 **24.** rational; 13 **25.** rational; -5 **26.** not a real number **27.** $\sqrt{14}$ **28.** $5\sqrt{3}$ **29.** $-3\sqrt{3}$ **30.** $4\sqrt{3}$ **31.** $4\sqrt{10}$ **32.** 18 **33.** $16\sqrt{6}$ **34.** $25\sqrt{10}$ **35.** $\frac{3}{2}$ **36.** $-\frac{11}{20}$ **37.** $\frac{\sqrt{7}}{13}$ **38.** $\frac{\sqrt{5}}{6}$ **39.** $\frac{2}{15}$ **40.** $3\sqrt{2}$ **41.** 8 **42.** $2\sqrt{2}$ **43.** p **44.** $\sqrt{km}$ **45.** r^9 **46.** x^5y^8 **47.** $x^4\sqrt{x}$ **48.** $\frac{6}{p}$ **49.** $a^7b^{10}\sqrt{ab}$ **50.** $11x^3y^5$ **51.** y^2 **52.** $6x^5$ **53.** Yes, because both approximations are .7071067812. **54.** $2\sqrt{11}$ **55.** $9\sqrt{2}$ **56.** $21\sqrt{3}$ **57.** $12\sqrt{3}$ **58.** 0 **59.** $3\sqrt{7}$ **60.** $2\sqrt{3}+3\sqrt{10}$ **61.** $2\sqrt{2}$ **62.** $6\sqrt{30}$ **63.** $5\sqrt{x}$ **64.** 0 **65.** $-m\sqrt{5}$ **66.** $11k^2\sqrt{2n}$ **67.** $\frac{10\sqrt{3}}{3}$ **68.** $\frac{8\sqrt{10}}{5}$ **69.** $\sqrt{6}$ **70.** $\frac{\sqrt{10}}{5}$ **71.** $\sqrt{10}$ **72.** $\frac{\sqrt{42}}{21}$ **73.** $\frac{r\sqrt{x}}{4x}$ **74.** $\frac{\sqrt[3]{9}}{3}$ **75.** $r=\frac{\sqrt{3V\pi h}}{\pi h}$ **76.** $r=\frac{\sqrt{S\pi}}{2\pi}$ **77.** $-\sqrt{15}-9$ **78.** $3\sqrt{6}+12$ **79.** $22-16\sqrt{3}$ **80.** $2\sqrt{21}-\sqrt{14}+12\sqrt{2}-4\sqrt{3}$ **81.** -13 **82.** $x+4\sqrt{x}+4$ **83.** $-2+\sqrt{5}$ **84.** $\frac{-2\sqrt{2}-6}{7}$ **85.** $\frac{3(1-\sqrt{x})}{1-x}$ **86.** $\frac{-2+6\sqrt{2}}{17}$

87. $\frac{-\sqrt{10}+3\sqrt{5}+\sqrt{2}-3}{7}$ **88.** $\frac{2\sqrt{3}+2+3\sqrt{2}+\sqrt{6}}{2}$ **89.** $\frac{3+2\sqrt{6}}{3}$ **90.** $\frac{1+3\sqrt{7}}{4}$ **91.** $3+4\sqrt{3}$ **92.** no solution **93.** 48 **94.** 1 **95.** 2 **96.** 6 **97.** $-3, -1$ **98.** -2 **99.** 4 **100.** 7 **101.** 9 **102.** $11\sqrt{3}$ **103.** $\frac{11}{t}$ **104.** $\frac{5-\sqrt{2}}{23}$ **105.** $\frac{2\sqrt{10}}{5}$ **106.** $5y\sqrt{2}$ **107.** -5 **108.** $-\sqrt{10}-5\sqrt{15}$ **109.** $\frac{4r\sqrt{3rs}}{3s}$ **110.** $\frac{2+\sqrt{13}}{2}$ **111.** $-7\sqrt{2}$ **112.** $7-2\sqrt{10}$ **113.** $166+2\sqrt{7}$ **114.** -11 **115.** $7\sqrt{2}$ **116.** 7 **117.** no solution **118.** 8 **119.** 39.2 mph **120.** $\frac{10\sqrt{7}-5\sqrt{7}}{-3\sqrt{14}-2\sqrt{14}}$ or $\frac{5\sqrt{7}-10\sqrt{7}}{2\sqrt{14}+3\sqrt{14}}$ **121.** $-\frac{5\sqrt{7}}{5\sqrt{14}}$ **122.** $-\sqrt{\frac{1}{2}}$ **123.** $-\frac{\sqrt{2}}{2}$ **124.** It falls from left to right.

Chapter 8 Test (page 625)

1. $-14, 14$ **2. (a)** irrational **(b)** 11.916 **3.** a must be negative. **4.** 6 **5.** $-3\sqrt{3}$ **6.** $\frac{8\sqrt{2}}{5}$ **7.** $2\sqrt[3]{4}$ **8.** $4\sqrt{6}$ **9.** $9\sqrt{7}$ **10.** $-5\sqrt{3x}$ **11.** $4xy\sqrt{2y}$ **12.** 31 **13.** $6\sqrt{2}+2-3\sqrt{14}-\sqrt{7}$ **14.** $11+2\sqrt{30}$ **15. (a)** $6\sqrt{2}$ in. **(b)** 8.485 in. **16.** 50 ohms **17.** $\frac{5\sqrt{14}}{7}$ **18.** $\frac{\sqrt{6x}}{3x}$ **19.** $-\sqrt[3]{2}$ **20.** $\frac{-3(4+\sqrt{3})}{13}$ **21.** $\frac{\sqrt{3}+12\sqrt{2}}{3}$ **22.** no solution **23.** 3 **24.** 1, 4 **25.** 12 is not a solution. A check shows that it does not satisfy the original equation.

Cumulative Review Exercises: Chapters R–8 (page 627)

1. 54 **2.** 6 **3.** 3 **4.** 3 **5.** $x \geq -16$ **6.** $z > 5$ **7.** Cody Ohl: \$296,419; Trevor Brazile: \$273,997

8.

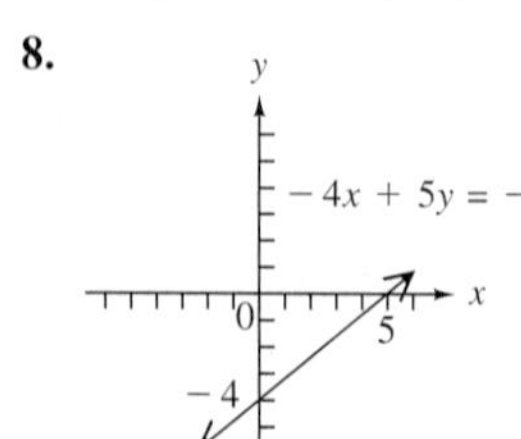

9.

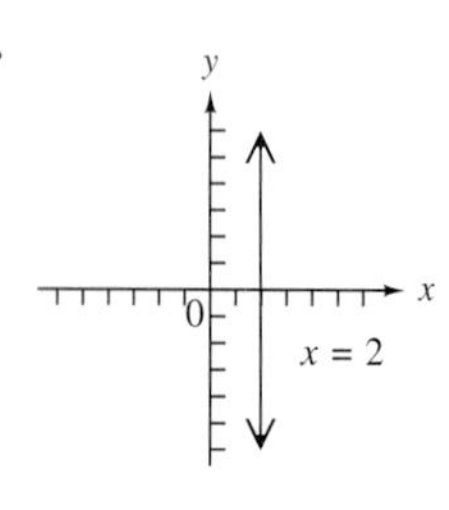

10.

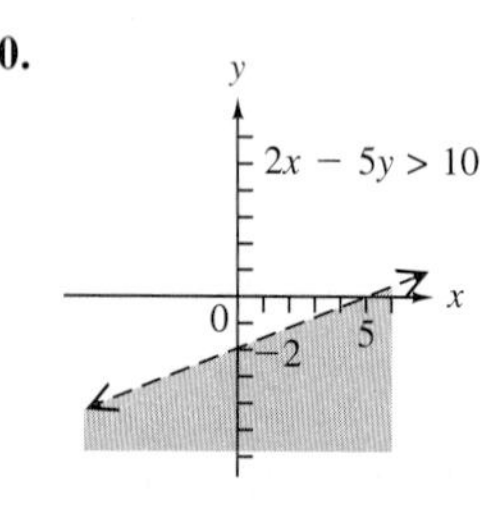

11. .47; Convention spending increased \$.47 million per year. **12.** $y = .47x - 926.5$ **13.** \$15.4 million **14.** $(3, -7)$ **15.** infinite number of solutions **16.** from Chicago: 61 mph; from Des Moines; 54 mph **17.** $12x^{10}y^2$ **18.** $\frac{y^{15}}{5832}$ **19.** $3x^3+11x^2-13$ **20.** $4t^2-8t+5$ **21.** \$6,900,000,000,000 **22.** $(m+8)(m+4)$ **23.** $(5t^2+6)(5t^2-6)$ **24.** $(6a+5b)(2a-b)$ **25.** $(9z+4)^2$ **26.** 3, 4 **27.** $-2, -1$ **28.** $\frac{x+1}{x}$ **29.** $(t+5)(t+3)$ **30.** $\frac{-2x-14}{(x+3)(x-1)}$ **31.** $29\sqrt{3}$ **32.** $-\sqrt{3}+\sqrt{5}$ **33.** $10xy^2\sqrt{2y}$ **34.** $21-5\sqrt{2}$ **35.** 16

CHAPTER 9 QUADRATIC EQUATIONS

Section 9.1 (page 633)

1. true **3.** false; If k is a positive integer that is not a perfect square, then the solutions will be irrational. **5.** false; For values of k that satisfy $0 \leq k < 10$, there are real solutions. **7.** $-9, 9$ **9.** $-\sqrt{14}, \sqrt{14}$ **11.** $-4\sqrt{3}, 4\sqrt{3}$ **13.** $-\frac{5}{2}, \frac{5}{2}$ **15.** no real number solution **17.** $-1.5, 1.5$ **19.** $-\sqrt{3}, \sqrt{3}$ **21.** $-\frac{2\sqrt{7}}{7}, \frac{2\sqrt{7}}{7}$ **23.** $-\frac{2\sqrt{5}}{5}, \frac{2\sqrt{5}}{5}$ **25.** $-2\sqrt{6}, 2\sqrt{6}$ **27.** $-2, 8$ **29.** no real number solution **31.** $8+3\sqrt{3}, 8-3\sqrt{3}$ **33.** $-3, \frac{5}{3}$ **35.** $0, \frac{3}{2}$ **37.** $\frac{5+\sqrt{30}}{2}, \frac{5-\sqrt{30}}{2}$ **39.** $\frac{-1+3\sqrt{2}}{3}, \frac{-1-3\sqrt{2}}{3}$ **41.** $-10+4\sqrt{3}, -10-4\sqrt{3}$ **43.** $\frac{1+4\sqrt{3}}{4}, \frac{1-4\sqrt{3}}{4}$ **45.** The answers are equivalent. If the answer of either student is multiplied by $\frac{-1}{-1}$, it will look like the answer of the other student. **47.** about $\frac{1}{2}$ sec **49.** 9 in. **51.** 5%

Section 9.2 (page 643)

1. $10x$ **3.** 49 **5.** D **7.** 100 **9.** $\frac{25}{4}$ **11.** $\frac{1}{16}$ **13.** 1, 3 **15.** $-3, -2$ **17.** $-1+\sqrt{6}, -1-\sqrt{6}$ **19.** -3 **21.** $\frac{-1+\sqrt{5}}{2}, \frac{-1-\sqrt{5}}{2}$ **23.** $-\frac{3}{2}, \frac{1}{2}$ **25.** $\frac{2+\sqrt{14}}{2}, \frac{2-\sqrt{14}}{2}$ **27.** no real number solution **29.** $\frac{-7+\sqrt{97}}{6}, \frac{-7-\sqrt{97}}{6}$ **31.** $-4, 2$ **33.** $1+\sqrt{6}, 1-\sqrt{6}$ **35.** 1 sec and 5 sec **37.** 3 sec and 5 sec **39.** 75 ft by 100 ft

Section 9.3 (page 651)

1. 4; 5; -9 **3.** 3; -4; -2 **5.** 3; 7; 0 **7.** $-13, 1$ **9.** 2 **11.** $\frac{-6+\sqrt{26}}{2}, \frac{-6-\sqrt{26}}{2}$ **13.** $-1, \frac{5}{2}$ **15.** $-1, 0$ **17.** no real number solution **19.** $\frac{-5+\sqrt{13}}{6}, \frac{-5-\sqrt{13}}{6}$ **21.** $0, \frac{12}{7}$ **23.** $-2\sqrt{6}, 2\sqrt{6}$ **25.** $-\frac{2}{5}, \frac{2}{5}$ **27.** $\frac{6+2\sqrt{6}}{3}, \frac{6-2\sqrt{6}}{3}$ **29.** no real number solution

31. There is no real number solution. **33.** $-\frac{2}{3}, \frac{4}{3}$

35. $\frac{-1+\sqrt{73}}{6}, \frac{-1-\sqrt{73}}{6}$ **37.** $1+\sqrt{2}, 1-\sqrt{2}$

39. no real number solution **41.** 3.5 ft

43. $-8, 16$; Only 16 board feet is a reasonable answer.

Summary Exercises on Quadratic Equations (page 653)

1. $-6, 6$ **2.** $\frac{-3+\sqrt{5}}{2}, \frac{-3-\sqrt{5}}{2}$ **3.** $-\frac{10}{9}, \frac{10}{9}$ **4.** $-\frac{7}{9}, \frac{7}{9}$

5. $1, 3$ **6.** $-2, -1$ **7.** $4, 5$ **8.** $\frac{-3+\sqrt{17}}{2}, \frac{-3-\sqrt{17}}{2}$

9. $-\frac{1}{3}, \frac{5}{3}$ **10.** $\frac{1+\sqrt{10}}{2}, \frac{1-\sqrt{10}}{2}$ **11.** $-17, 5$ **12.** $-\frac{7}{5}, 1$

13. $\frac{7+2\sqrt{6}}{3}, \frac{7-2\sqrt{6}}{3}$ **14.** $\frac{1+4\sqrt{2}}{7}, \frac{1-4\sqrt{2}}{7}$

15. no real number solution **16.** no real number solution

17. $-\frac{1}{2}, 2$ **18.** $-\frac{1}{2}, 1$ **19.** $-\frac{5}{4}, \frac{3}{2}$ **20.** $-3, \frac{1}{3}$

21. $1+\sqrt{2}, 1-\sqrt{2}$ **22.** $\frac{-5+\sqrt{13}}{6}, \frac{-5-\sqrt{13}}{6}$ **23.** $\frac{2}{5}, 4$

24. $-3+\sqrt{5}, -3-\sqrt{5}$ **25.** $\frac{-3+\sqrt{41}}{2}, \frac{-3-\sqrt{41}}{2}$

26. $-\frac{5}{4}$ **27.** $\frac{1}{4}, 1$ **28.** $\frac{1+\sqrt{3}}{2}, \frac{1-\sqrt{3}}{2}$

29. $\frac{-2+\sqrt{11}}{3}, \frac{-2-\sqrt{11}}{3}$ **30.** $\frac{-5+\sqrt{41}}{8}, \frac{-5-\sqrt{41}}{8}$

31. $\frac{-7+\sqrt{5}}{4}, \frac{-7-\sqrt{5}}{4}$ **32.** $-\frac{8}{3}, -\frac{6}{5}$

33. $\frac{8+8\sqrt{2}}{3}, \frac{8-8\sqrt{2}}{3}$ **34.** $\frac{-5+\sqrt{5}}{2}, \frac{-5-\sqrt{5}}{2}$

35. no real number solution **36.** no real number solution

37. $-\frac{2}{3}, 2$ **38.** $-\frac{1}{4}, \frac{2}{3}$ **39.** $-4, \frac{3}{5}$ **40.** $-3, 5$ **41.** $-\frac{2}{3}, \frac{2}{5}$

42. $-4, 6$

Section 9.4 (page 661)

1. The vertex of a parabola is the lowest or highest point on the graph.

3. $(0, 0)$

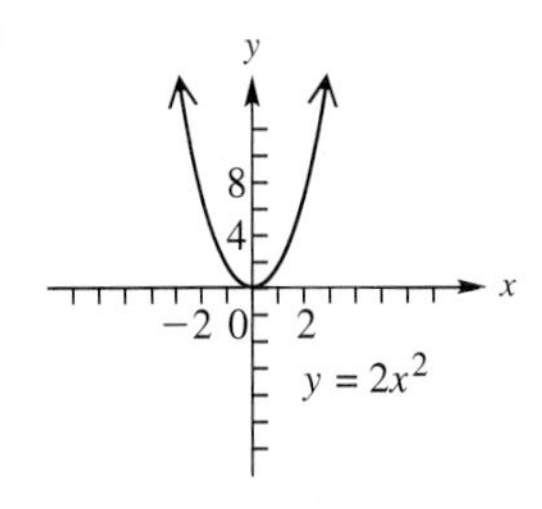

5. $(0, -4)$

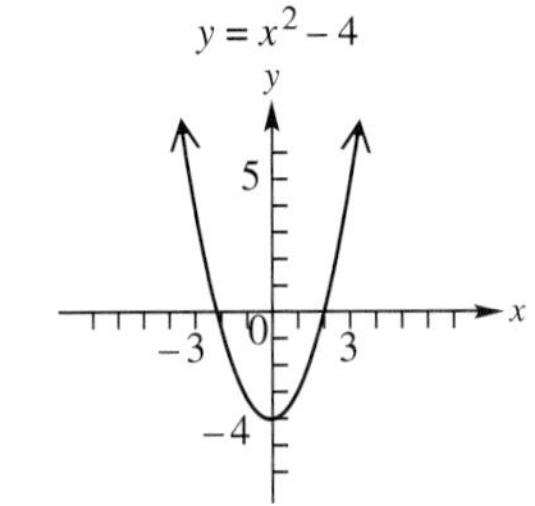

7. $(0, 2)$

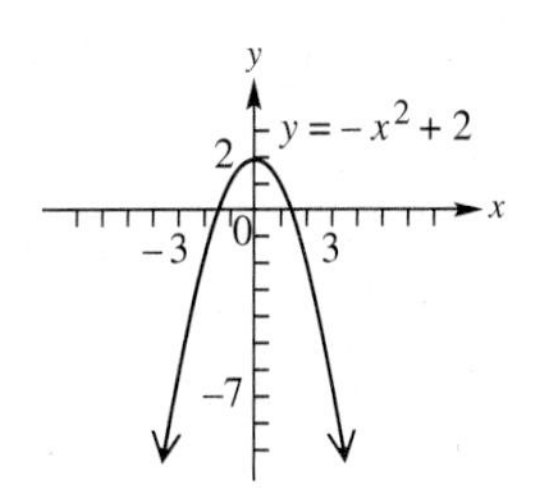

9. $(-3, 0)$

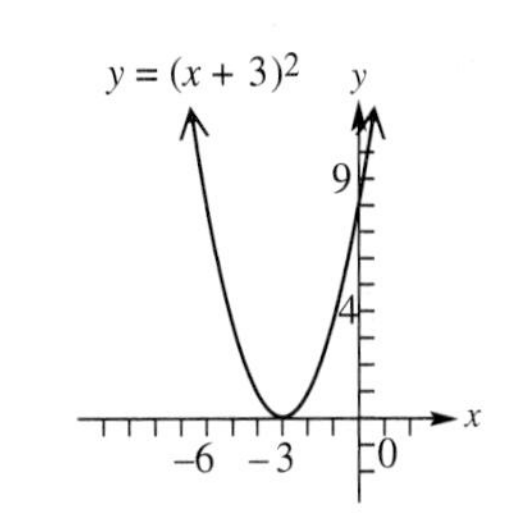

11. $(-1, 2)$

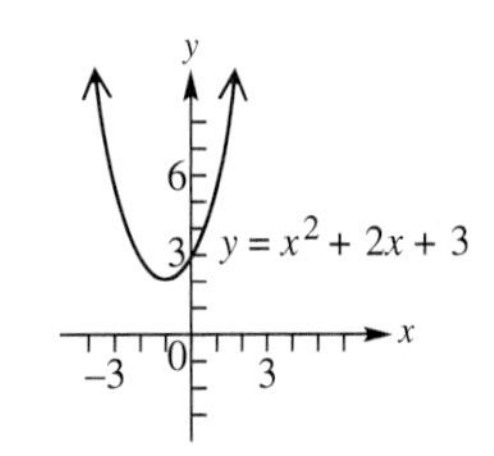

13. $(3, 4)$

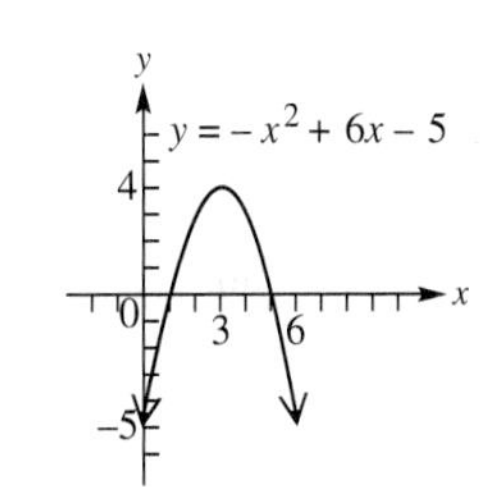

15. If $a > 0$, it opens upward, and if $a < 0$, it opens downward.

17. $y = \frac{11}{5625}x^2$

Section 9.5 (page 669)

1. 3; 3; $(1, 3)$ **3.** 5; 5; $(3, 5)$ **5.** The graph consists of the five points $(0, 2)$, $(1, 3)$, $(2, 4)$, $(3, 5)$, and $(4, 6)$. **7.** not a function; domain: $\{-4, -2, 0\}$; range: $\{3, 1, 5, -8\}$

9. function; domain: {A, B, C, D, E}; range: $\{2, 3, 6, 4\}$ **11.** not a function; domain: $\{-4, -2, 0, 2, 3\}$; range: $\{-2, 0, 1, 2, 3\}$

13. function **15.** not a function **17.** function

19. not a function **21.** $(2, 4)$ **22.** $(-1, -4)$ **23.** $\frac{8}{3}$

24. $f(x) = \frac{8}{3}x - \frac{4}{3}$ **25. (a)** 11 **(b)** 3 **(c)** -9 **27. (a)** 4 **(b)** 2 **(c)** 14 **29. (a)** 2 **(b)** 0 **(c)** 3

31. $\{(1970, 9.6), (1980, 14.1), (1990, 19.8), (2000, 28.4)\}$; yes

33. $g(1980) = 14.1$ (million); $g(1990) = 19.8$ (million)

35. For the year 2002, the function predicts 30.3 million foreign-born residents in the United States.

36. yes

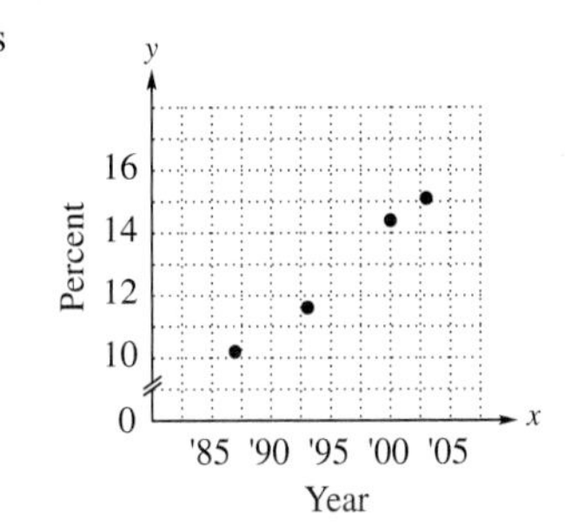

37. $y = .30625x - 598.32$ **38.** 1993: 12.0%; 2000: 14.2%
39. $y = .4x - 785.6$ **40.** 1987: 9.2%; 2003: 15.6%; The equation from Exercise 38 gives better approximations. The results in Exercise 38 vary by .4% and .2% from the data. The results here give answers that vary by 1.0% and .5%.

Chapter 9 Review Exercises (page 677)

1. $-12, 12$ **2.** $-\sqrt{37}, \sqrt{37}$ **3.** $-8\sqrt{2}, 8\sqrt{2}$ **4.** $-7, 3$
5. $3 + \sqrt{10}, 3 - \sqrt{10}$ **6.** $\frac{-1 + \sqrt{14}}{2}, \frac{-1 - \sqrt{14}}{2}$
7. no real number solution **8.** $-\frac{5}{3}$ **9.** $-5, -1$
10. $-2 + \sqrt{11}, -2 - \sqrt{11}$ **11.** $-1 + \sqrt{6}, -1 - \sqrt{6}$
12. $\frac{-4 + \sqrt{22}}{2}, \frac{-4 - \sqrt{22}}{2}$ **13.** $-\frac{7}{2}$ **14.** no real number solution **15.** 2.5 sec **16.** 6, 8, 10 **17.** $\left(\frac{k}{2}\right)^2$ or $\frac{k^2}{4}$
18. **(a)** $-3, 3$ **(b)** $-3, 3$ **(c)** $-3, 3$ **(d)** We will always get the same results, no matter which method of solution is used.
19. $\frac{-1 + \sqrt{29}}{4}, \frac{-1 - \sqrt{29}}{4}$ **20.** $\frac{2 + \sqrt{10}}{2}, \frac{2 - \sqrt{10}}{2}$
21. $\frac{1 + \sqrt{21}}{10}, \frac{1 - \sqrt{21}}{10}$ **22.** $\frac{-3 + \sqrt{41}}{2}, \frac{-3 - \sqrt{41}}{2}$
23. $-3 + \sqrt{19}, -3 - \sqrt{19}$ **24.** The $-b$ term should be above the fraction bar.
25. (0, 0)

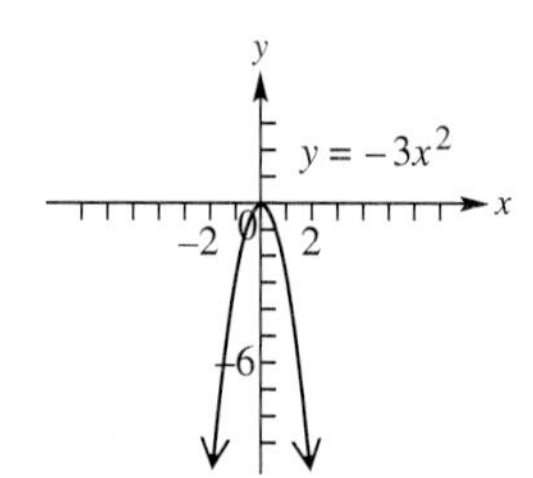

26. (0, 5)

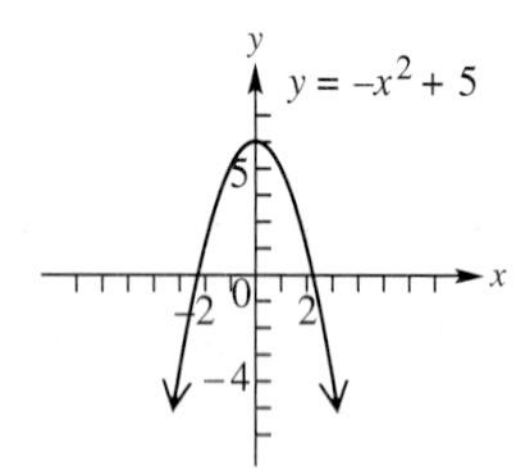

27. (1, 0)

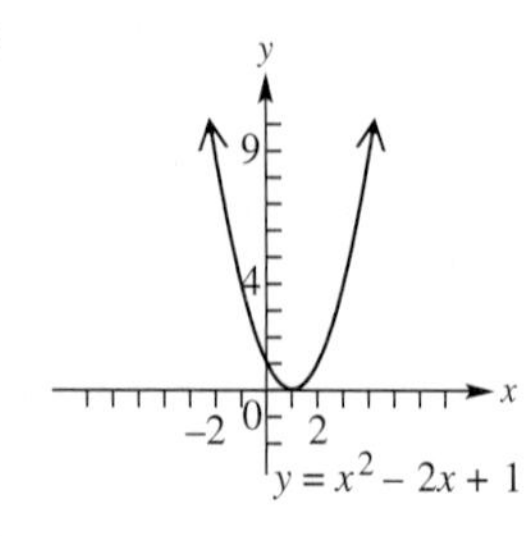

28. (1, 4)

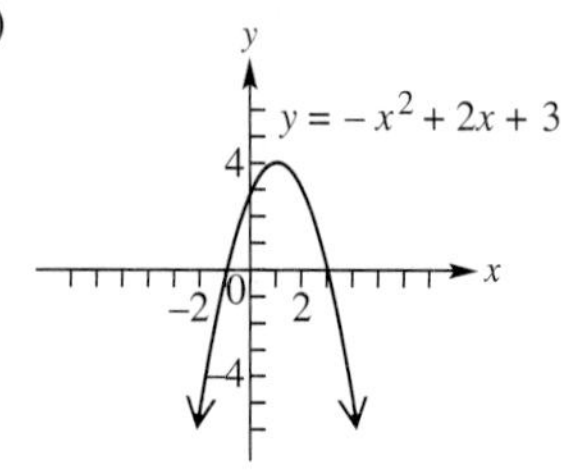

29. (−2, −2)

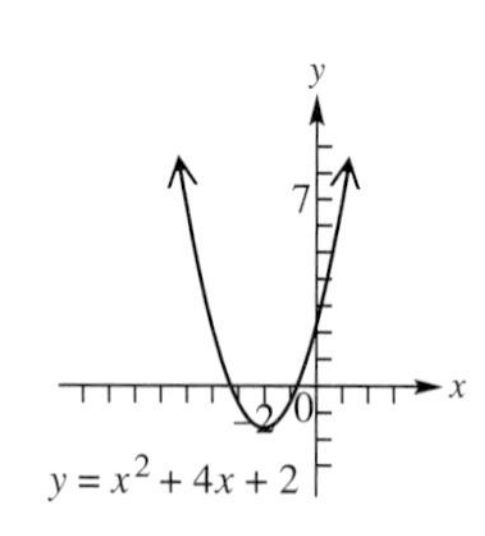

30. (−4, 0)

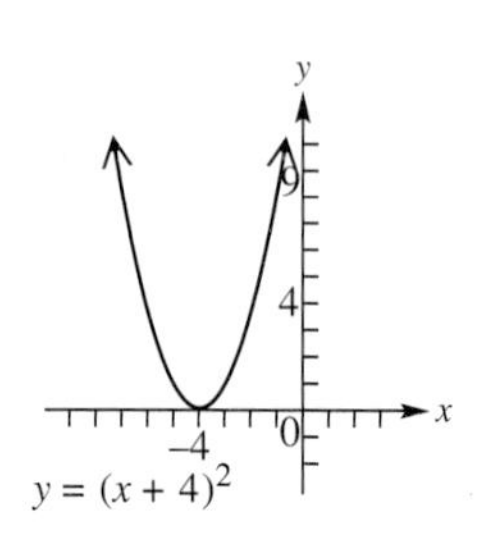

31. $y = \frac{3}{1000}x^2$ **32.** not a function; domain: $\{-2, 0, 2\}$; range: $\{4, 8, 5, 3\}$ **33.** function; domain: $\{8, 7, 6, 5, 4\}$; range: $\{3, 4, 5, 6, 7\}$ **34.** not a function **35.** function
36. function **37.** function **38.** not a function **39.** **(a)** 8 **(b)** -1 **40.** **(a)** 7 **(b)** 1 **41.** **(a)** 5 **(b)** 2 **42.** 400 or 800
43. (6, 10) **44.** demand: 600; price: \$10 **45.** $-\frac{11}{2}, 5$
46. $-\frac{11}{2}, \frac{9}{2}$ **47.** $\frac{-1 + \sqrt{21}}{2}, \frac{-1 - \sqrt{21}}{2}$ **48.** $-\frac{3}{2}, \frac{1}{3}$
49. $\frac{-5 + \sqrt{17}}{2}, \frac{-5 - \sqrt{17}}{2}$ **50.** $-1 + \sqrt{3}, -1 - \sqrt{3}$
51. no real number solution **52.** $\frac{9 + \sqrt{41}}{2}, \frac{9 - \sqrt{41}}{2}$ **53.** $-\frac{6}{5}$
54. $-1 + 2\sqrt{2}, -1 - 2\sqrt{2}$ **55.** $-2 + \sqrt{5}, -2 - \sqrt{5}$
56. $-2\sqrt{2}, 2\sqrt{2}$

Chapter 9 Test (page 681)

1. $-\sqrt{39}, \sqrt{39}$ **2.** $-11, 5$ **3.** $\frac{-3 + 2\sqrt{6}}{4}, \frac{-3 - 2\sqrt{6}}{4}$
4. $2 + \sqrt{10}, 2 - \sqrt{10}$ **5.** $\frac{-6 + \sqrt{42}}{2}, \frac{-6 - \sqrt{42}}{2}$ **6.** $-3, \frac{1}{2}$
7. $\frac{3 + \sqrt{3}}{3}, \frac{3 - \sqrt{3}}{3}$ **8.** no real number solution
9. $\frac{5 + \sqrt{13}}{6}, \frac{5 - \sqrt{13}}{6}$ **10.** $1 + \sqrt{2}, 1 - \sqrt{2}$
11. $\frac{-1 + 3\sqrt{2}}{2}, \frac{-1 - 3\sqrt{2}}{2}$ **12.** $\frac{11 + \sqrt{89}}{4}, \frac{11 - \sqrt{89}}{4}$
13. 5 **14.** 2 sec **15.** 12, 16, 20
16. vertex: (3, 0)

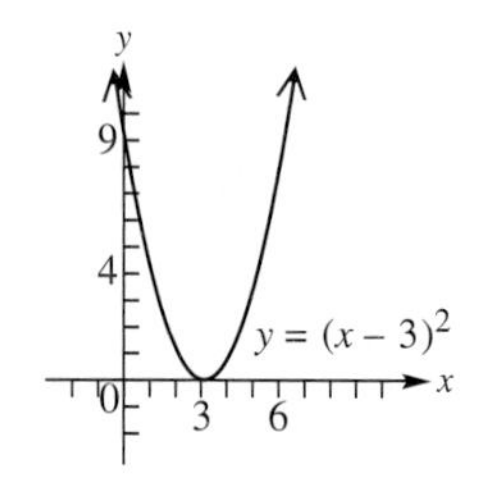

17. vertex: $(-1, -3)$

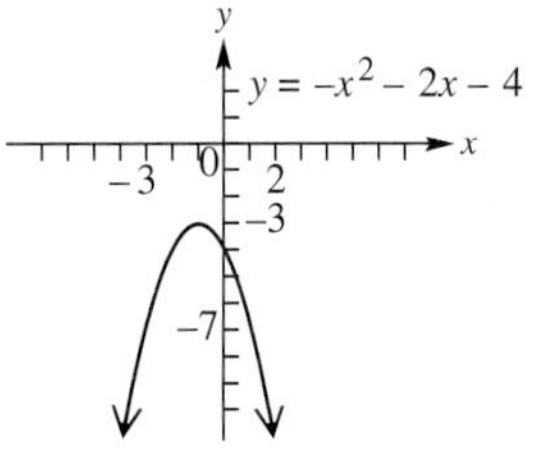

18. **(a)** not a function **(b)** function; domain: $\{0, 1, 2\}$; range: $\{2\}$
19. not a function **20.** 1

Cumulative Review Exercises: Chapters R–9 (page 683)

1. 15 **2.** 5 **3.** -2 **4.** $-r + 7$ **5.** $-2k$ **6.** $19m - 17$
7. 3 **8.** 5 **9.** 2 **10.** width: 50 ft; length: 94 ft **11.** 100°, 80°
12. $L = \frac{P - 2W}{2}$ or $L = \frac{P}{2} - W$
13. $m > -2$

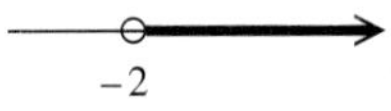

14. $p \leq 4$

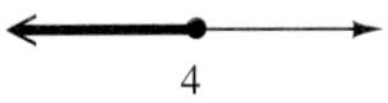

15.

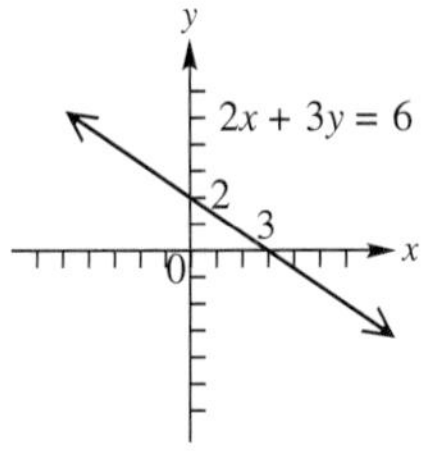

16.

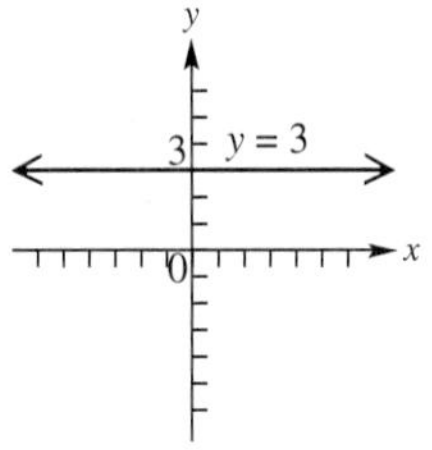

17.

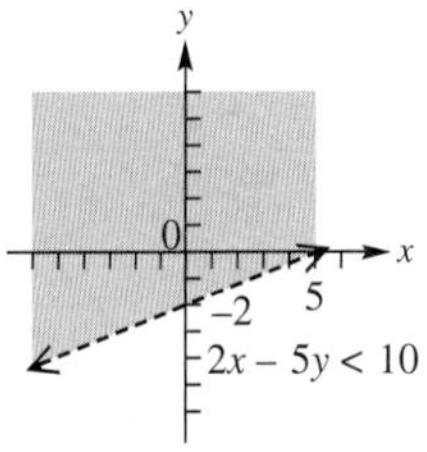

18. $-\frac{1}{3}$ **19.** $2x - y = -3$ **20.** $(-3, 2)$ **21.** no solution
22. Motorola: \$79.99; Kyocera: \$69.99
23.

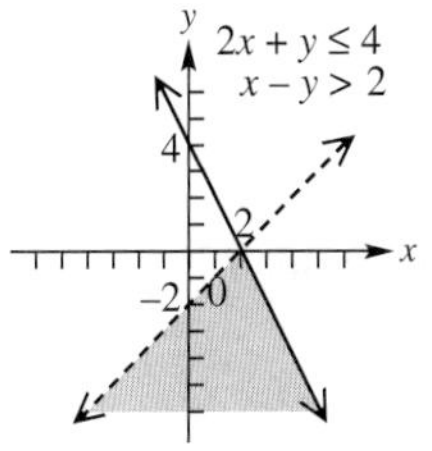

24. $\frac{x^4}{9}$ **25.** $\frac{b^{16}}{c^2}$ **26.** $\frac{27}{125}$ **27.** $8x^5 - 17x^4 - x^2$
28. $2x^4 + x^3 - 19x^2 + 2x + 20$ **29.** $25t^2 + 90t + 81$
30. $3x^2 - 2x + 1$ **31.** $16x^2(x - 3y)$
32. $(4x^2 + 1)(2x + 1)(2x - 1)$ **33.** $(2a + 1)(a - 3)$
34. $(5m - 2)^2$ **35.** $-9, 6$ **36.** $-1, \frac{4}{3}$ **37.** 50 m **38.** $\frac{4}{5}$
39. $\frac{-k - 1}{k(k - 1)}$ **40.** $\frac{5a + 2}{(a - 2)^2(a + 2)}$ **41.** $\frac{6x + 1}{3x - 1}$
42. $-\frac{15}{7}, 2$ **43.** 10 **44.** $\frac{6\sqrt{30}}{5}$ **45.** $4\sqrt{5}$ **46.** $-ab\sqrt[3]{2b}$
47. 7 **48.** $\frac{1 + \sqrt{3}}{2}, \frac{1 - \sqrt{3}}{2}$

49. $(0, -4)$; $(2, 0)$ and $(-2, 0)$

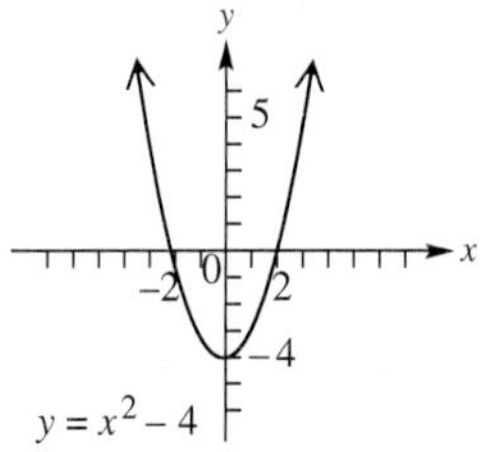

50. **(a)** yes **(b)** domain: $\{0, 1, 3\}$; range: $\{4, 2, 5\}$ **(c)** 11 **(d)** no

Appendix A Strategies for Problem Solving

(page 695)

1. You should choose a sock from the box labeled *red and green socks.* Since it is mislabeled, it contains only red socks or only green socks, determined by the sock you choose. If the sock is green, relabel this box *green socks.* Since the other two boxes were mislabeled, switch the remaining label to the other box and place the label that says *red and green socks* on the unlabeled box. No other choice guarantees a correct relabeling, since you can remove only one sock. **3.** 70 **5.** You must place the O in the bottom-left square. No other choice guarantees you a win. **7.** 18 **9.** 329,476 **11.** One possible sequence is shown here. The numbers represent the number of gallons in each bucket in each successive step.

Big	7	4	4	1	1	0	7	5	5
Small	0	3	0	3	0	1	1	3	0

13. 07

15. 17 days **17.** 16 **19.**

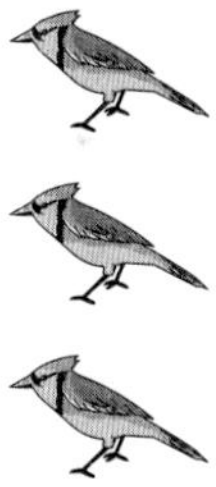

21.

6	1	8
7	5	3
2	9	4

23. 25 pitches (The visiting team's pitcher retires 24 consecutive batters through the first eight innings, using only one pitch per batter. His team does not score either. Going into the bottom of the ninth inning tied 0–0, the first batter for the home team hits his first pitch for a home run. The pitcher threw 25 pitches and loses the game by a score of 1–0.) **25.** For three weighings, first balance four against four. Of the lighter four, balance two against the other two. Finally, of the lighter two, balance them one against the other. To find the bad coin in two weighings, divide the eight coins into groups of 3, 3, 2. Weigh the groups of three against each other on the scale. If the groups weigh the same, the fake is in the two left out and can be found in one additional weighing. If the two groups of three do not weigh the same, pick the lighter group. Choose any two of the coins and weigh them. If one of these is lighter, it is the fake; if they weigh the same, then the third coin is the fake. **27.** Q **29.** 28
31.

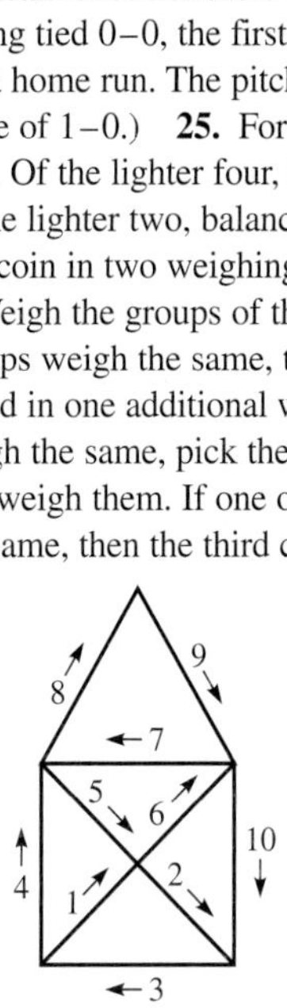

33. Dan (36) is married to Jessica (29); James (30) is married to Cathy (31). **35.** The CEO is a woman. **37.** None, since there is no dirt in a hole. **39.** One solution is $1 + 2 + 3 + 4 + 5 + 6 + 7 + 8 \times 9 = 100$. **41.** The correct problem follows.

$$\begin{array}{r} 402 \\ \times \quad 39 \\ \hline 15{,}678 \end{array}$$

43. 35 **45.** none **47.** Eve has \$5 and Adam has \$7.

Appendix B Sets

(page 705)

1. {1, 2, 3, 4, 5, 6, 7} **3.** {winter, spring, summer, fall} **5.** ∅ **7.** {L} **9.** {2, 4, 6, 8, 10, . . .} **11.** The sets in Exercises 9 and 10 are infinite sets. **13.** true **15.** false **17.** true **19.** true **21.** true **23.** true **25.** true **27.** false **29.** true **31.** true **33.** false **35.** true **37.** true **39.** false **41.** true **43.** false **45.** {g, h} **47.** {b, c, d, e, g, h} **49.** {a, c, e} = B **51.** {d} = D **53.** {a} **55.** {a, c, d, e} **57.** {a, c, e, f} **59.** ∅ **61.** B and D; C and D

Appendix C Mean, Median, and Mode

(page 711)

1. 7 **3.** 69.8 (rounded) **5.** \$39,622 **7.** \$58.24 **9.** \$35,500 **11.** 6.1 **13.** 17.2 **15.** 2.60 **17. (a)** 2.80 **(b)** 2.93 (rounded) **(c)** 3.13 (rounded) **19.** 15 **21.** 516 **23.** 48 **25.** 4142 mi **27. (a)** 4050 mi **(b)** The median is somewhat different from the mean; the mean is more affected by the high and low numbers. **29.** 8 **31.** 68 and 74; bimodal **33.** no mode

Appendix D Factoring Sums and Differences of Cubes

(page 717)

1. 1; 8; 27; 64; 125; 216; 343; 512; 729; 1000 **3.** C, D **5.** $(a + 1)(a^2 - a + 1)$ **7.** $(a - 1)(a^2 + a + 1)$ **9.** $(p + q)(p^2 - pq + q^2)$ **11.** $(y - 6)(y^2 + 6y + 36)$ **13.** $(k + 10)(k^2 - 10k + 100)$ **15.** $(3x - 1)(9x^2 + 3x + 1)$ **17.** $(5a + 2)(25a^2 - 10a + 4)$ **19.** $(y - 2x)(y^2 + 2xy + 4x^2)$ **21.** $(3a - 4b)(9a^2 + 12ab + 16b^2)$ **23.** $(2p + 9q)(4p^2 - 18pq + 81q^2)$ **25.** $2(2t - 1)(4t^2 + 2t + 1)$ **27.** $5(2w + 3)(4w^2 - 6w + 9)$ **29.** $(x + y^2)(x^2 - xy^2 + y^4)$ **31.** $(5k - 2m^3)(25k^2 + 10km^3 + 4m^6)$ **33.** $(x^3 - 1)(x^3 + 1)$ **34.** $(x - 1)(x^2 + x + 1)(x + 1)(x^2 - x + 1)$ **35.** $(x^2 - 1)(x^4 + x^2 + 1)$ **36.** $(x - 1)(x + 1)(x^4 + x^2 + 1)$ **37.** The result in Exercise 34 is completely factored. **38.** Show that $x^4 + x^2 + 1 = (x^2 + x + 1)(x^2 - x + 1)$. **39.** difference of squares **40.** $(x - 3)(x^2 + 3x + 9)(x + 3)(x^2 - 3x + 9)$

Index

O

P

Q

R

Triangles and Angles

Right Triangle

Triangle has one 90° (right) angle.

Pythagorean Formula (for right triangles)

$a^2 + b^2 = c^2$

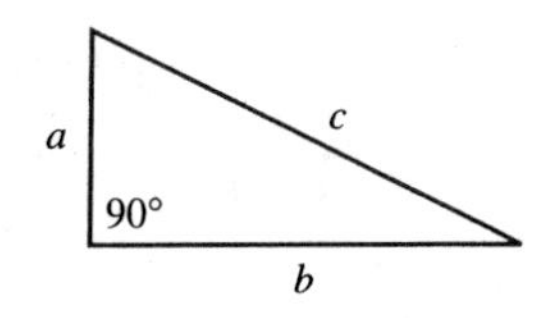

Right Angle

Measure is 90°.

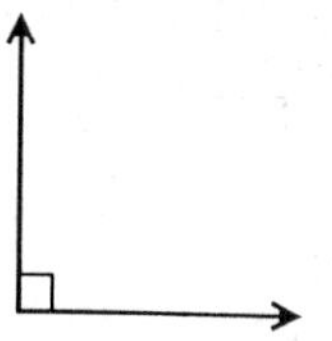

Isosceles Triangle

Two sides are equal.
$AB = BC$

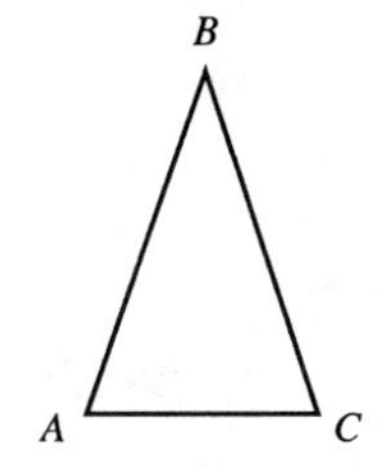

Straight Angle

Measure is 180°.

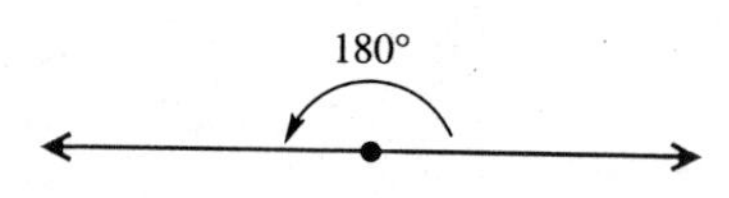

Equilateral Triangle

All sides are equal.
$AB = BC = CA$

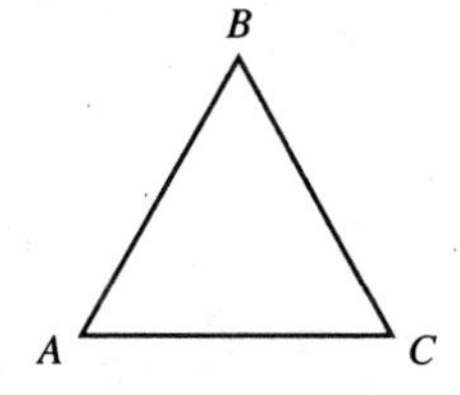

Complementary Angles

The sum of the measures of two complementary angles is 90°.

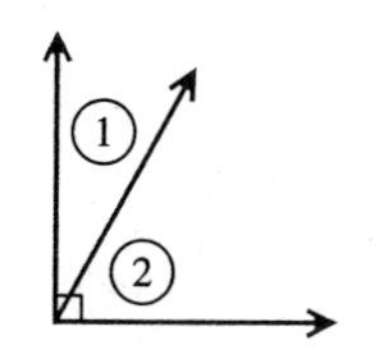

Angles ① and ② are complementary.

Sum of the Angles of Any Triangle

$A + B + C = 180°$

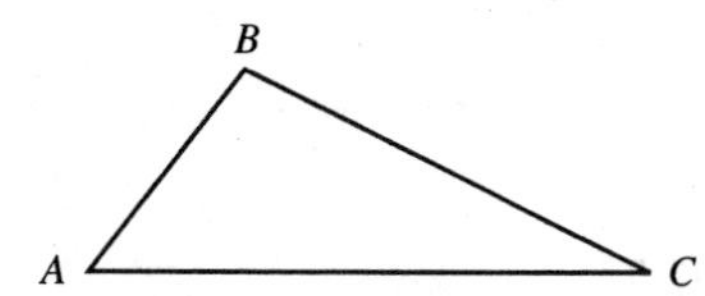

Supplementary Angles

The sum of the measures of two supplementary angles is 180°.

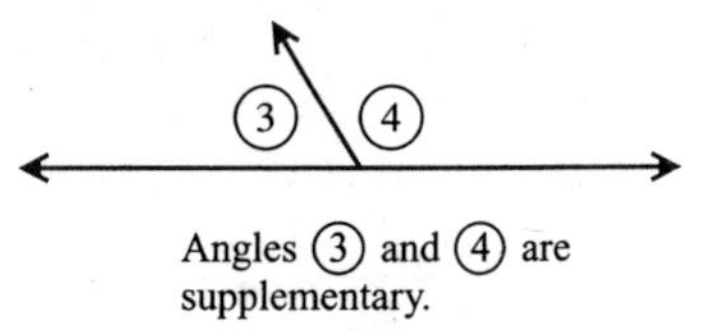

Angles ③ and ④ are supplementary.

Similar Triangles

Corresponding angles are equal; corresponding sides are proportional.

$A = D, B = E, C = F$

$$\frac{AB}{DE} = \frac{AC}{DF} = \frac{BC}{EF}$$

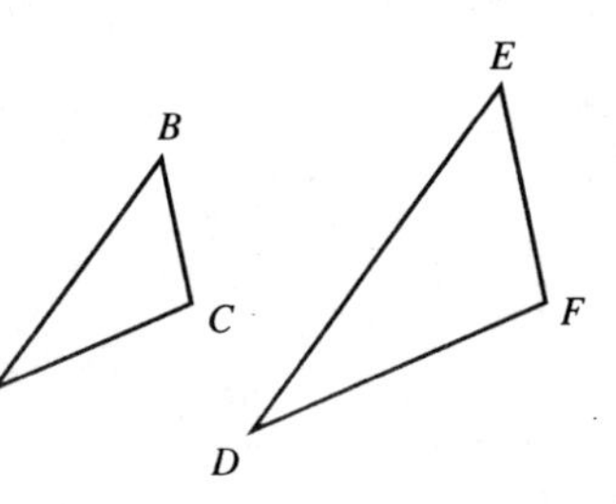

Vertical Angles

Vertical angles have equal measures.

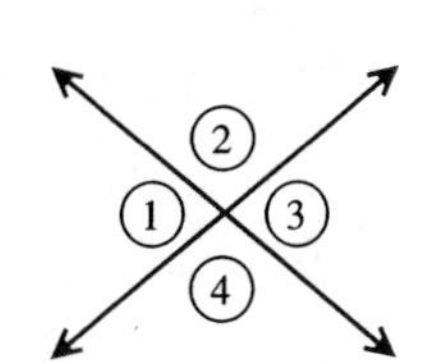

Angle ① = Angle ③
Angle ② = Angle ④